Special Functions
and
Fourier Series

With Boundary Value Problems and Curve Fitting

Special Functions and Fourier Series

With Boundary Value Problems and Curve Fitting

Salam Subhaschandra Singh

Associate Professor

Department of Physics, Imphal College, Imphal

Formerly Associate Professor, Dhanamanjuri College of Science, Imphal and

Lecturer (Selection Grade), Moirang College, Moirang

I.K. International Pvt. Ltd.

NEW DELHI

Published by

I.K. International Pvt. Ltd.
4435-36/7, Ansari Road, Daryaganj
New Delhi–110 002 (India)
E-mail: info@ikinternational.com
Website: www.techsarworld.com

ISBN: 978-93-90620-14-2

Published/Printed by Krishan Makhijani for I.K. International Pvt. Ltd., 4435-36/7, Ansari Road, Daryaganj, New Delhi–110 002 .

Dedicated to

My Late mother Yaimaleishabi Devi

Foreword

The contents of this book are really good and sumptuous as basic concepts of Mathematical Physics for graduate students. I congratulate the author for the remarkable piece of work he did. It will certainly serve the purpose for which he took up the task. May God bless him with all the happiness and success in his future life.

Laxmi Kant Singh
Retired Professor of Physics
Dr. Ram Manohar Lohia Avadh University, Ayodhya

Preface

This book is meant for B. Sc. Physics (Honours) and B. E. students. My experience of teaching Degree students for more than three decades has been utilized in writing the book. The topics included in this text broadly cover the syllabi of various universities.

While preparing this book, it is supposed that the students in need of this book are acquainted with fundamentals of calculus, co-ordinate geometry, trigonometry and the theory of complex variables. Utmost effort has been made to present the material in a systematic manner so that average students can easily follow it.

In preparing the manuscript of the book, I have consulted a number of books and, therefore, I acknowledge indebtedness to the authors and publishers of those books. Moreover, the graphs of Bessel functions of different kinds, shown in this book, are drawn with the help of either Wolfram Mathematica or Wolfram Alpha. Further, I acknowledge the constant enthusiastic support from my wife, K. Suniti Devi and son, S. Satischandra Singh.

I shall highly appreciate constructive criticisms, comments and suggestions for the improvement of the book.

S. Subhaschandra Singh

Contents

1
CHAPTER

Power Series and Their Convergence and Divergence

In this chapter, we define power series and discuss their convergence and divergence by suitable methods.

1.1 INFINITE SERIES

An infinite series means the sum of an infinite number of mathematical terms.

Partial Sum and Remainder of Infinite Series

In an infinite series, the sum of any finite number of terms is called a ***partial sum*** of the series. Thus, for an infinite series

$$\sum_{r=1}^{\infty} u_r = u_1 + u_2 + u_3 + \ldots + u_n + \ldots \infty,$$

the sum of the first n terms, viz.,

$$S_n = \sum_{r=1}^{n} u_r = u_1 + u_2 + u_3 + \ldots + u_n$$

is called the nth partial sum.

In such an infinite series, if we omit the terms of the nth partial sum (S_n), the remaining terms are collectively called the ***remainder*** (R_n) of the series after the nth term. Thus,

$$R_n = u_{n+1} + u_{n+2} + u_{n+3} + u_{n+4} + \ldots \infty.$$

Convergence and Divergence of Series

An infinite series

$$\sum_{r=1}^{\infty} u_r = u_1 + u_2 + u_3 + \ldots + u_n + \ldots \infty$$

is said to be ***convergent*** if its nth partial sum (S_n) tends to a finite limit (S) as n tends to infinity, i.e., if $\lim_{n\to\infty} S_n = S$. In such a case, the series is said to converge to a value, S.

On the other hand, a series $\sum_{r=1}^{\infty} u_r$ is said to be ***divergent*** if its nth partial sum (S_n) does not tend to any finite limit as n tends to infinity.

As an example, let us consider the geometric series

$$a + ax + ax^2 + ax^3 + \ldots ax^{n-1} + \ldots \infty.$$

Let the partial sum for n terms be S_n. Then

$$S_n = a + ax + ax^2 + ax^3 + \ldots + ax^{n-1}.$$

Multiplication by x on both sides results in

$$xS_n = ax + ax^2 + ax^3 + \ldots + ax^{n-1} + ax^n.$$

Subtraction of the last equation from the former one results in

$$(1 - x)S_n = a - ax^n.$$

Therefore,
$$S_n = \frac{a(1-x^n)}{1-x} = \frac{a}{1-x} - \frac{ax^n}{1-x} \tag{1.1}$$

If $|x| < 1$, then x^n decreases in absolute value as n increases so that

$$\lim_{n\to\infty} \frac{ax^n}{1-x} = 0.$$

Hence, $\lim_{n\to\infty} S_n = \frac{a}{1-x} = S$.

Thus, for $|x| < 1$, if the sum (S_n) of a geometric series approaches to a finite limit as the number of terms is increased indefinitely, then the series is said to be convergent.

If $|x| > 1$, x^n will become infinite as n tends to infinity and hence

$$\lim_{n\to\infty} S_n = \infty$$

and the series becomes divergent.

Oscillatory Series

In the geometrical series $a + ax + ax^2 + ax^3 + \ldots + ax^n + \ldots \infty$, if $x = -1$, the series will become

$$a - a + a - a + \ldots + (-1)^{n+1} a + \ldots \infty.$$

Then $S_n = 0$ if n is even and $S_n = a$ if n is odd. But, S_n does not approach to a finite limit when n tends to infinity. Such a series is called an ***oscillatory*** *series*. Such series are divergent in nature.

Alternating Series

A series whose terms are alternately positive and negative is called an ***alternating*** series. Such a series has the form

$$\sum_{n=1}^{\infty}(-1)^{n+1}u_n = u_1 - u_2 + u_3 - u_4 + u_5 - u_6 + \ldots + (-1)^{n+1}u_n + \ldots \infty.$$

Absolute (or Unconditional) and Conditional Convergence

A series $$\sum_{n=1}^{\infty}u_n = u_1 + u_2 + u_3 + u_4 + \ldots \infty$$

is said to be ***absolutely*** convergent if the corresponding series

$$\sum_{n=1}^{\infty}|u_n| = |u_1| + |u_2| + |u_3| + |u_4| + \ldots + |u_n| + \ldots \infty$$

(whose terms are real and non-negative) is convergent.

A series $$\sum_{n=1}^{\infty}u_n = u_1 + u_2 + u_3 + u_4 + \ldots + u_n + \ldots \infty$$

is said to be ***conditionally*** convergent if the series itself is convergent but the series

$$\sum_{n=1}^{\infty}|u_n| = |u_1| + |u_2| + |u_3| + |u_4| + \ldots + |u_n| + \ldots \infty$$

is divergent.

Tests for Convergence or Divergence of Series

There are various methods for testing the convergence or divergence of a given series. They are: (i) Comparison test, (ii) Cauchy's root test, (iii) D'Alembert's ratio test, (iv) Cauchy's integral test, (v) Kummer's test, (vi) Raabe's test, and (vii) Gauss' test. These are stated here without proof.

(i) Comparison Test

For a given series $\sum_{n=1}^{\infty}u_n = u_1 + u_2 + u_3 + u_4 + \ldots + u_n + \ldots \infty$, if we can find a convergent series $\sum_{n=1}^{\infty}a_n = a_1 + a_2 + a_3 + a_4 + \ldots + a_n + \ldots \infty$ with non-negative real terms such that $a_n \geq u_n$ for all $n(n = 1, 2, 3, \ldots)$, then the given series is convergent.

For a given series $\sum_{n=1}^{\infty} v_n = v_1 + v_2 + v_3 + v_4 + \ldots + v_n + \ldots \infty$, if we can find a divergent series $\sum_{n=1}^{\infty} b_n = b_1 + b_2 + b_3 + b_4 + \ldots + b_n + \ldots \infty$ such that $b_n \le v_n$ for all n (n = 1, 2, 3,...), then the given series is divergent.

(ii) Cauchy's Root Test

For an infinite series $\sum_{n=1}^{\infty} u_n = u_1 + u_2 + u_3 + u_4 + \ldots + u_n + \ldots \infty$,

(a) if $\lim_{n\to\infty} \sqrt[n]{|u_n|} < 1$, the series is convergent,

(b) if $\lim_{n\to\infty} \sqrt[n]{|u_n|} > 1$, the series is divergent;

(c) if $\lim_{n\to\infty} \sqrt[n]{|u_n|} = 1$, the test fails resulting in no conclusion.

(iii) D'Alembert's Ratio Test

For an infinite series $\sum_{n=1}^{\infty} u_n = u_1 + u_2 + u_3 + u_4 + \ldots + u_n + \ldots \infty$,

(a) if $\lim_{n\to\infty} \left|\frac{u_{n+1}}{u_n}\right| < 1$, the series is convergent,

(b) if $\lim_{n\to\infty} \left|\frac{u_{n+1}}{u_n}\right| > 1$, the series is divergent;

(c) if $\lim_{n\to\infty} \left|\frac{u_{n+1}}{u_n}\right| = 1$, the test fails resulting in no conclusion.

(iv) Cauchy's Integral Test

For a series of positive terms $\sum_{n=1}^{\infty} u_n = u_1 + u_2 + u_3 + u_4 + \ldots + u_n + \ldots \infty$ such that $u_{n+1} < u_n$, if there exists a positive function, $f(n) = u_n$, which decreases with increasing n, then the series is convergent if the integral, $I = \int_1^{\infty} f(n)\,dn$, exists and the series is divergent if this integral does not exist.

(v) Kummer's Test

For a given series $\sum_{n=1}^{\infty} u_n$ of positive terms u_n and a sequence of positive constants a_n, if $K = \lim_{n\to\infty}\left(a_n \frac{u_n}{u_{n+1}} - a_{n+1}\right)$ is

(a) greater than zero, the series is convergent
(b) less than zero, the series is divergent
(c) equal to zero, the test fails resulting in no conclusion.

(vi) Raabe's Test

For a given series $\sum_{n=1}^{\infty} u_n$ of positive terms u_n,

(a) if $R = \lim_{n\to\infty} n\left(\frac{u_n}{u_{n+1}} - 1\right) > 1$, the given series is convergent
(b) if $R < 1$, the given series is divergent
(c) if $R = 0$, the test fails resulting in no conclusion.

(vii) Gauss' Test

For a given series $\sum_{n=1}^{\infty} u_n$ of positive terms u_n, if we can find a relation

$$\frac{u_n}{u_{n+1}} = 1 + \frac{h}{n} + \frac{Q(n)}{n^2}$$

where $Q(n)$ is a bounded function of n for $n \to \infty$, then the given series is convergent if $h > 1$ and divergent if $h \le 1$.

EXAMPLES

1. Test the convergence or divergence of the series

$$\sum_{r=1}^{\infty} \frac{r^2}{2^r} = \frac{1}{2} + \frac{1}{1} + \frac{9}{8} + \frac{1}{1} + \frac{25}{32} + \ldots \infty.$$

Solution:

(i) Ratio test: Here, $u_n = \dfrac{n^2}{2^n}$, $u_{n+1} = \dfrac{(n+1)^2}{2^{n+1}}$.

Now, $\lim_{n\to\infty}\left|\frac{u_{n+1}}{u_n}\right| = \lim_{n\to\infty}\left|\frac{(n+1)^2}{2^{n+1}}\times\frac{2^n}{n^2}\right| = \lim_{n\to\infty}\left|\frac{(n+1)^2}{2n^2}\right|$

$$= \lim_{n\to\infty}\left|\frac{2(n+1)}{4n}\right| = \frac{2}{4} = \frac{1}{2} < 1.$$

Here, use has been made of L' Hospital's rule. The value of the limit shows that the given series is convergent.

(ii) Root test: In this case,

$$\lim_{n\to\infty}\sqrt[n]{|u_n|} = \lim_{n\to\infty}\sqrt[n]{\left|\frac{n^2}{2^n}\right|} = \lim_{n\to\infty}\frac{1}{2}n^{\frac{2}{n}}$$

$$= \lim_{n\to\infty}\frac{1}{2}e^{\ln\left(n^{\frac{2}{n}}\right)} = \lim_{n\to\infty}\frac{1}{2}e^{\frac{2}{n}\ln n} = \frac{1}{2}e^0 = \frac{1}{2} < 1.$$

Therefore, the given series is convergent.

2. Test the convergence or divergence of the series

$$\sum_{r=1}^{\infty}\frac{1}{\sqrt{r+1}} = \frac{1}{\sqrt{2}} + \frac{1}{\sqrt{3}} + \frac{1}{\sqrt{4}} + \frac{1}{\sqrt{5}} + \ldots \infty.$$

Solution:

(i) Ratio test: Here, $u_n = \frac{1}{\sqrt{n+1}}$ and $u_{n+1} = \frac{1}{\sqrt{n+2}}$.

Then, $\lim_{n\to\infty}\left|\frac{u_{n+1}}{u_n}\right| = \lim_{n\to\infty}\sqrt{\frac{n+2}{n+1}} = 1.$

Thus, ratio test fails in this problem.

(ii) Root test:

Here, $\lim_{n\to\infty}\sqrt[n]{|u_n|} = \lim_{n\to\infty}\sqrt[n]{\frac{1}{\sqrt{n+1}}} = \lim_{n\to\infty}\frac{1}{(n+1)^{\frac{1}{2n}}}$

$$= \lim_{n\to\infty}\frac{1}{e^{\ln\left\{(n+1)^{\frac{1}{2n}}\right\}}} = \lim_{n\to\infty}\frac{1}{e^{\frac{1}{2n}\ln(n+1)}} = \frac{1}{e^0} = 1.$$

Thus, the root test also fails in this problem and, therefore, other tests are necessary.

(iii) Cauchy's integral test: Let $f(n) = \frac{1}{\sqrt{n+1}} = (n+1)^{-\frac{1}{2}}$.

Then,

$$I = \int_1^{\infty} f(n)\,dn = \int_1^{\infty} (n+1)^{-\frac{1}{2}}\,dn$$

$$= \int_2^{\infty} v^{-\frac{1}{2}}\,dv \text{ where we put } v = n+1 \text{ and } dv = dn.$$

$$= 2\left[v^{\frac{1}{2}}\right]_2^{\infty} = \infty, \text{ i.e., the integral does not exist.}$$

Hence, the given series is divergent.

3. Determine the convergence or divergence of the series

$$\sum_{r=1}^{\infty} \frac{1}{r} = 1 + \frac{1}{2} + \frac{1}{3} + \frac{1}{4} + \ldots \infty.$$

Solution: As in the previous example, the ratio test and the root test fail in this example. Hence, we are to perform other tests.

(i) Cauchy's integral test: Let $f(n) = \frac{1}{n}$. Then, $I = \int_1^{\infty} \frac{1}{n}\,dn = [\ln n]_1^{\infty} = \infty$.

As the integral does not exist, the given series is divergent.

(ii) Gauss' test: We have, $u_n = \frac{1}{n}$, $u_{n+1} = \frac{1}{n+1}$.

Then, $\frac{u_n}{u_{n+1}} = \frac{n+1}{n} = 1 + \frac{1}{n} = 1 + \frac{h}{n}$ where $h = 1$.

According to the condition of Gauss' test, the given harmonic series is divergent.

1.2 POWER SERIES

A power series is an infinite series of the form

$$(x - x_0)^k [a_0 + a_1(x - x_0) + a_2(x - x_0)^2 + a_3(x - x_0)^3 + \ldots \infty] = \sum_{r=0}^{\infty} a_r (x - x_0)^{k+r},$$

where k is a number, x_0 is a fixed point on the abscissa of a Cartesian coordinate system and $a_r (r = 0, 1, 2, 3, \text{etc.})$ are constants. Some familiar examples of power series are

1. $y = 1 + x + x^2 + x^3 + \ldots \infty$ [Geometric series]

2. $y = e^x = 1 + \frac{x}{1!} + \frac{x^2}{2!} + \frac{x^3}{3!} + \frac{x^4}{4!} + \ldots \infty$ [Exponential series]

3. $y = \ln(1+x) = x - \frac{x^2}{2} + \frac{x^3}{3} - \frac{x^4}{4} + \ldots \infty$ [Logarithmic series]

4. $y = \cos x = 1 - \frac{x^2}{2!} + \frac{x^4}{4!} - \frac{x^6}{6!} + \ldots \infty$ [Cosine series]

5. $y = \sin x = x - \frac{x^3}{3!} + \frac{x^5}{5!} - \frac{x^7}{7!} + \ldots \infty$ [Sine series]

6. $y = \cosh x = 1 + \frac{x^2}{2!} + \frac{x^4}{4!} + \frac{x^6}{6!} + \ldots \infty$ [Hyperbolic cosine series]

7. $y = \sinh x = x + \frac{x^3}{3!} + \frac{x^5}{5!} + \frac{x^7}{7!} + \ldots \infty$ [Hyperbolic sine series]

etc.

Interval of Convergence

In a power series like $\sum_{r=0}^{\infty} a_r (x - x_0)^r$, if we set $x = x_0 = c$, the series reduces to a single term (a_0) showing that the series is convergent at $x = c$. In some cases, this may be the only value of x for which the above power series is convergent. Such a power series is of no physical interest. In many cases, there may be other values of x for which the power series converges. If a power series $\sum_{r=1}^{\infty} a_r (x - x_0)^r$ is convergent for certain values of x, the interval formed by these values of x is called the ***interval of convergence*** and the midpoint of the interval is at $x = x_0$. In the interval of convergence of a given power series, the midpoint is called the ***centre*** of the interval and the distance between the centre and any one of the two extreme points of the interval of convergence is called the ***radius of convergence, R***. The radius of convergence R can be determined from the coefficients of the series by means of each of the following formulae

(a) $\frac{1}{R} = \lim_{n \to \infty} \sqrt[n]{|a_n|}$ (b) $\frac{1}{R} = \lim_{n \to \infty} \left| \frac{a_{n+1}}{a_n} \right|$ (1.2)

provided the limits exist.

1.3 ORDINARY AND SINGULAR POINTS OF LINEAR, SECOND ORDER HOMOGENEOUS ODE

In this section, certain terms related to the second order linear and homogeneous ordinary differential equations are defined. These definitions help in determining the existence of power series solutions of such type of differential equations.

Ordinary and Singular Points

In a linear, second order, homogeneous ordinary differential equation in the form

$$\frac{d^2y}{dx^2}+P(x)\frac{dy}{dx}+Q(x)y=0,$$

if the coefficient functions $P(x)$ and $Q(x)$ are both finite at a point $x=x_0$, then the point $x=x_0$ is called an ***ordinary point*** of the given differential equation.

In general, an ***ordinary point*** of a given linear ordinary differential equation of any order in the standard (or normal) form is defined as a point at which all the coefficient functions remain finite.

Thus, for a linear ODE of the form

$$\frac{d^ny}{dx^n}+f_{n-1}(x)\frac{d^{n-1}y}{dx^{n-1}}+f_{n-2}(x)\frac{d^{n-2}y}{dx^{n-2}}+\ldots+f_2(x)\frac{d^2y}{dx^2}+f_1(x)\frac{dy}{dx}+f_0(x)y=0\,,$$

a point $x=x_0$ is said to be an ordinary point if all the coefficient functions $f_0(x), f_1(x), f_2(x),\ldots f_{n-1}(x)$ are all finite at that point.

On the other hand, in a linear, second order, homogeneous ordinary differential equation in the form

$$\frac{d^2y}{dx^2}+P(x)\frac{dy}{dx}+Q(x)y=0,$$

if either of the coefficient functions $P(x)$ or $Q(x)$ or both of them is/are infinite as $x\to x_0$, then the point $x=x_0$ is called a ***singular point (or a singularity)*** of the given differential equation.

In general, a ***singular point*** of a given linear ordinary differential equation of any order in the standard (or normal) form is defined as a point at which at least one, or more, or all of the coefficient functions is/are not finite.

Such a singularity is of three types: (1) ***regular*** (or ***removable*** or ***non-essential***) ***singular point***, (2) ***irregular*** (or ***non-removable*** or ***essential***) ***singular point*** and (3) ***apparent singularity***.

Regular singular point

In a linear, second order, homogeneous ordinary differential equation in the form

$$\frac{d^2y}{dx^2}+P(x)\frac{dy}{dx}+Q(x)y=0$$

if either $P(x)$ or $Q(x)$ or both go to infinity as $x\to x_0$, but $(x-x_0)P(x)$ and $(x-x_0)^2Q(x)$ remain both finite as $x\to x_0$, then the point $x=x_0$ is called a ***regular*** or ***removable*** or ***non-essential singular point***.

In general, a point $x = x_0$ is said to be a regular (or removable or non-essential) singular point of a linear ODE of the form

$$\frac{d^n y}{dx^n} + f_{n-1}(x)\frac{d^{n-1}y}{dx^{n-1}} + f_{n-2}(x)\frac{d^{n-2}y}{dx^{n-2}} + \ldots + f_2(x)\frac{d^2y}{dx^2} + f_1(x)\frac{dy}{dx} + f_0(x)y = 0$$

if anyone or more or all of the coefficient functions, $f_0(x), f_1(x), f_2(x), \ldots f_{n-1}(x)$, become infinite as $x \to x_0$, but all of the functions

$$(x - x_0)f_{n-1}(x), (x - x_0)^2 f_{n-2}(x), (x - x_0)^3 f_{n-3}(x), \ldots (x - x_0)^n f_0(x)$$

are finite as $x \to x_0$.

Irregular singular point

In a linear, second order, homogeneous ordinary differential equation in the form

$$\frac{d^2y}{dx^2} + P(x)\frac{dy}{dx} + Q(x)y = 0$$

if either or both of the coefficient functions $P(x)$ and $Q(x)$ become infinite as $x \to x_0$, and also either or both of the functions $(x - x_0)P(x)$ and $(x - x_0)^2 Q(x)$ become infinite as $x \to x_0$, then the point $x = x_0$ is said to be an ***irregular (or non-removable or essential) singular point.***

In general, a point $x = x_0$ is said to be an irregular (or non-removable or essential) singular point of a linear ODE of the form

$$\frac{d^n y}{dx^n} + f_{n-1}(x)\frac{d^{n-1}y}{dx^{n-1}} + f_{n-2}(x)\frac{d^{n-2}y}{dx^{n-2}} + \ldots + f_2(x)\frac{d^2y}{dx^2} + f_1(x)\frac{dy}{dx} + f_0(x)y = 0$$

if anyone or more or all of the coefficient functions $f_0(x), f_1(x), f_2(x) \ldots f_{n-1}(x)$ become infinite as $x \to x_0$, and also anyone or more or all of the functions

$$(x - x_0)f_{n-1}(x), (x - x_0)^2 f_{n-2}(x), (x - x_0)^3 f_{n-3}(x), \ldots (x - x_0)^n f_0(x)$$

become infinite as $x \to x_0$.

Apparent singularity

If a linear ordinary differential equation has a removable singularity at a certain point $x = x_0$, but its series solution is found to have no singularity at that point, then the point $x = x_0$ is said to be an ***apparent singularity*** of that differential equation.

1.4 POWER SERIES SOLUTIONS OF LINEAR AND HOMOGENEOUS SECOND ORDER ORDINARY DIFFERENTIAL EQUATIONS

This section studies the feasibility of finding a power series solution of a given linear, second order homogeneous ODE through the concept of ordinary point and singular point or singularity of the given differential equation.

In the case of linear and homogeneous second order ODEs, the existence of power series solution is determined by a theorem called ***Fuchs' theorem*** [introduced by the German mathematician, Lazarus Emmanuel Fuchs (May 5, 1833 – Apr. 26, 1902)] and such a power series solution is usually obtained by a method known as ***Frobenius method*** [introduced by the German mathematician, Ferdinand George Frobenius (Oct. 26, 1849 – Aug. 3, 1917)].

Fuchs' Theorem

Fuchs' theorem states that a given second order linear and homogeneous ordinary differential equation shall have a power series solution about a point $x = x_0$ if this point is either an ordinary point or a regular singular point of that differential equation.

If the point $x = x_0$ is an irregular singular point, then regular solutions of such a differential equation may not exist and no general method for solving such an equation is known.

Frobenius Method

In Mathematics, the ***Frobenius method*** is a way of finding an infinite series solution of a second order linear and homogeneous ordinary differential equation having the form

$$\frac{d^2y}{dx^2} + P(x)\frac{dy}{dx} + Q(x)y = 0$$

about an ordinary point or a regular singular point, $x = x_0$, of the differential equation, the form of the series solution being taken in the form

$$y(x) = \sum_{r=0}^{\infty} a_r (x - x_0)^{k+r}, a_0 \neq 0$$

where k is a constant to be determined [usually zero if x_0 is ordinary point] and the coefficients (a_r) are to be determined from two types of equations called indicial equation and recursion formula.

General Procedure of Frobenius Method

Let us consider a differential equation of the form

$$\frac{d^2y}{dx^2}+P(x)\frac{dy}{dx}+Q(x)y=0. \tag{1.3}$$

where $P(x)$ and $Q(x)$ are functions of x.

If $x=x_0$ is an ordinary point or a regular singular point of the above differential equation, then, by Fuchs' theorem, the equation has a power series solution about the point $x=x_0$.

Let us put $(x-x_0)P(x)=f(x)$ and $(x-x_0)^2Q(x)=g(x)$ where $f(x)$ and $g(x)$ are functions of x analytic (or finite) at $x=x_0$.

Then Eq. (1.3) becomes

$$\frac{d^2y}{dx^2}+\frac{f(x)}{(x-x_0)}\frac{dy}{dx}+\frac{g(x)}{(x-x_0)^2}y=0 \tag{1.4}$$

or

$$(x-x_0)^2\frac{d^2y}{dx^2}+(x-x_0)f(x)\frac{dy}{dx}+g(x)y=0. \tag{1.5}$$

Let us assume the power series solution to be

$$y=\sum_{r=0}^{\infty}a_r(x-x_0)^{k+r},\ a_0\neq 0 \tag{1.6}$$

where k is a constant to be determined latter.

We have

$$\frac{dy}{dx}=\sum_{r=0}^{\infty}(k+r)a_r(x-x_0)^{k+r-1} \tag{1.7}$$

and

$$\frac{d^2y}{dx^2}=\sum_{r=0}^{\infty}(k+r-1)(k+r)a_r(x-x_0)^{k+r-2} \tag{1.8}$$

Substitution of Eqs. (1.6) to (1.8) into Eq. (1.5) results in

$$\sum_{r=0}^{\infty}(k+r-1)(k+r)a_r(x-x_0)^{k+r}+\sum_{r=0}^{\infty}(k+r)a_r(x-x_0)^{k+r}f(x)+\sum_{r=0}^{\infty}a_r(x-x_0)^{k+r}g(x)=0. \tag{1.9}$$

Since $f(x)$ and $g(x)$ are analytic at $x=x_0$, they can be expanded in power series about $x=x_0$ in the forms

$$f(x)=\sum_{i=0}^{\infty}b_i(x-x_0)^i=b_0+b_1(x-x_0)+b_2(x-x_0)^2+\ldots+b_i(x-x_0)^i+\ldots\infty \tag{1.10}$$

and

$$g(x)=\sum_{j=1}^{\infty}c_j(x-x_0)^j=c_0+c_1(x-x_0)+c_2(x-x_0)^2+\ldots+c_j(x-x_0)^j+\ldots\infty \tag{1.11}$$

Substitution of Eqs. (1.10) and (1.11) into Eq. (1.9) results in

$$\sum_{r=0}^{\infty}(k+r-1)(k+r)a_r(x-x_0)^{k+r}$$

$$+\sum_{r=0}^{\infty}(k+r)\{b_0+b_1(x-x_0)+b_2(x-x_0)^2+\ldots\infty\}a_r(x-x_0)^{k+r}$$

$$+\sum_{r=0}^{\infty}\{c_0+c_1(x-x_0)+c_2(x-x_0)^2+\ldots\infty\}a_r(x-x_0)^{k+r}=0. \quad (1.12)$$

As the sum of terms with different powers of $(x-x_0)$ is zero, the coefficient of every power of $(x-x_0)$ should be zero. Starting from the lowest power, let us equate to zero the coefficients of various powers of $(x-x_0)$ as in the following.

$(x-x_0)^k$:

$$[(k-1)k+b_0k+c_0]a_0=0.$$

or

$$k^2+(b_0-1)k+c_0=0 \text{ [Indicial equation]} \quad (1.13)$$

[as a_0 is assumed to be non-zero]

Such a quadratic equation in k obtained by equating the coefficient of lowest power of $(x-x_0)$ to zero is called ***indicial equation*** of the given differential equation.

$(x-x_0)^{k+1}$:

$$k(k+1)a_1+(k+1)b_0a_1+kb_1a_0+c_0a_1+c_1a_0=0,$$

or

$$\{k(k+1)+(k+1)b_0+c_0\}a_1+(kb_1+c_1)a_0=0,$$

and so on.

From the various equations thus obtained, we can obtain a general relation between a coefficient (a_r) and a lower or higher coefficient. Such a relation is called ***recursion relation*** or ***formula***. From this formula, we can solve coefficients (a_r) in terms of lower coefficients and substitution of these coefficients into Eq. (1.6) will give a power series solution of the given differential equation.

Thus, one of the two solutions of the given second order linear and homogeneous ODE will be of the form given in Eq. (1.6) where k is a root of the indicial equation. The form of the other solution will depend on the nature of the remaining root of the indicial equation. There are three possible cases which are separately discussed as in the following.

CASE I: Distinct Roots of the Indicial Equation Differing by a Non-integer

This includes the case of complex conjugate roots also. In this case, let the roots of the indicial equation be $k=k_1$ and $k=k_2$, differing by a non-integer. Then, using Eq. (1.6), we obtain two independent solutions in the forms

$$y_1(x) = \sum_{r=0}^{\infty} a_r (x - x_0)^{k_1 + r}$$

$$= (x - x_0)^{k_1} [a_0 + a_1(x - x_0) + a_2(x - x_0)^2 + a_3(x - x_0)^3 + \ldots \infty] \tag{1.14}$$

and

$$y_2(x) = \sum_{r=0}^{\infty} \gamma_r (x - x_0)^{k_2 + r}$$

$$= (x - x_0)^{k_2} [\gamma_0 + \gamma_1(x - x_0) + \gamma_2(x - x_0)^2 + \gamma_3(x - x_0)^3 + \ldots \infty] \tag{1.15}$$

Here, in the second solution, the symbols γ_r $(r = 0, 1, 2, 3, \text{etc.})$ are used for the coefficients to make them different from the symbols used in the first solution as well as in the expressions for $f(x)$ and $g(x)$.

Then, the general solution of the given ODE is the linear combination

$$y(x) = Ay_1(x) + By_2(x) \tag{1.16}$$

where A and B are arbitrary constants. Thus, we summarize as

$$y(x) = A[y(k, x)]_{k=k_1} + B[y(k, x)]_{k=k_2} \tag{1.17}$$

CASE II: Double or Repeated Root (Equal Roots) of the Indicial Equation

As the indicial equation is a quadratic equation, it will have a double root (equal roots) if and only if its discriminant is zero, i.e., if

$$(b_0 - 1)^2 - 4c_0 = 0.$$

The roots of the indicial equation are

$$k = k_1 = \frac{-(b_0 - 1) + \sqrt{(b_0 - 1)^2 - 4c_0}}{2} \text{ and } k = k_2 = \frac{-(b_0 - 1) - \sqrt{(b_0 - 1)^2 - 4c_0}}{2},$$

i.e., $$k = k_1 = k_2 = \frac{1 - b_0}{2} = \lambda \text{ (say)} \tag{1.18}$$

In this case, the first series solution of the given ODE will be

$$y_1(x) = \sum_{r=0}^{\infty} a_r (x - x_0)^{\lambda + r}$$

$$= (x - x_0)^{\lambda} [a_0 + a_1(x - x_0) + a_2(x - x_0)^2 + a_3(x - x_0)^3 + \ldots \infty] \tag{1.19}$$

To find another solution $y_2(x)$, we have to apply the method of variation of parameter or the reduction of order taking

$$y_2(x) = u(x) y_1(x) \tag{1.20}$$

where $u(x)$ is to be determined latter.

Then, we have

$$\frac{dy_2}{dx} = \frac{du}{dx}y_1 + u\frac{dy_1}{dx}$$

and
$$\frac{d^2y_2}{dx^2} = \frac{d^2u}{dx^2}y_1 + 2\frac{du}{dx}\frac{dy_1}{dx} + u\frac{d^2y_1}{dx^2}.$$

As $y_2(x)$ is a solution of Eq. (1.5), substitution of $y_2(x)$ and its derivatives in place of $y(x)$ and its derivatives, will result in

$$(x-x_0)^2\left(\frac{d^2u}{dx^2}y_1 + 2\frac{du}{dx}\frac{dy_1}{dx} + u\frac{d^2y_1}{dx^2}\right) + (x-x_0)f(x)\left(\frac{du}{dx}y_1 + u\frac{dy_1}{dx}\right) + g(x)u(x)y_1(x) = 0,$$

or
$$(x-x_0)^2 y_1\frac{d^2u}{dx^2} + 2(x-x_0)^2\frac{du}{dx}\frac{dy_1}{dx} + (x-x_0)fy_1\frac{du}{dx}$$

$$+u\left\{(x-x_0)^2\frac{d^2y_1}{dx^2} + (x-x_0)f\frac{dy_1}{dx} + gy_1\right\} = 0 \qquad (1.21)$$

As $y_1(x)$ is a solution of Eq. (1.5), the expression within the curly brackets of Eq. (1.21) must vanish. Therefore, we have

$$(x-x_0)^2 y_1\frac{d^2u}{dx^2} + 2(x-x_0)^2\frac{du}{dx}\frac{dy_1}{dx} + (x-x_0)fy_1\frac{du}{dx} = 0.$$

Dividing both sides by $(x-x_0)^2 y_1$, we have

$$\frac{d^2u}{dx^2} + \left[\frac{2}{y_1}\frac{dy_1}{dx} + \frac{f}{x-x_0}\right]\frac{du}{dx} = 0 \qquad (1.22)$$

From Eq. (1.19), we see that

$$y_1(x) = (x-x_0)^\lambda[a_0 + a_1(x-x_0) + a_2(x-x_0)^2 + \ldots\infty]$$
$$= a_0(x-x_0)^\lambda + a_1(x-x_0)^{\lambda+1} + a_2(x-x_0)^{\lambda+2} + \ldots\infty.$$

Therefore, we have

$$\frac{dy_1}{dx} = \lambda a_0(x-x_0)^{\lambda-1} + (\lambda+1)a_1(x-x_0)^\lambda + (\lambda+2)a_2(x-x_0)^{\lambda+1} + \ldots$$
$$= (x-x_0)^{\lambda-1}[\lambda a_0 + (\lambda+1)a_1(x-x_0) + (\lambda+2)a_2(x-x_0)^2 + \ldots\infty].$$

and
$$\frac{1}{y_1}\frac{dy_1}{dx} = \frac{(x-x_0)^{\lambda-1}\left[\lambda a_0 + (\lambda+1)a_1(x-x_0) + (\lambda+2)a_2(x-x_0)^2 + \ldots\infty\right]}{(x-x_0)^\lambda[a_0 + a_1(x-x_0) + a_2(x-x_0)^2 + \ldots\infty]}$$

$$= \frac{1}{(x-x_0)}\left[\frac{\lambda a_0+(\lambda+1)a_1(x-x_0)+(\lambda+2)a_2(x-x_0)^2+\ldots\infty}{a_0+a_1(x-x_0)+a_2(x-x_0)^2+\ldots\infty}\right]$$

$$= \frac{\lambda}{x-x_0}+\ldots$$

where the dots denote a constant term plus terms containing positive powers of $(x-x_0)$.

Also, from Eq. (1.10), we see that

$$f(x)=b_0+b_1(x-x_0)+b_2(x-x_0)^2+b_3(x-x_0)^3+\ldots\infty.$$

Substituting the expressions for $\frac{1}{y_1}\frac{dy_1}{dx}$ and $f(x)$ into Eq. (1.22), we have

$$\frac{d^2u}{dx^2}+\left[2\left\{\frac{\lambda}{x-x_0}+\ldots\right\}+\frac{1}{(x-x_0)}\{b_0+b_1(x-x_0)+b_2(x-x_0)^2+\ldots\}\right]\frac{du}{dx}=0,$$

or

$$\frac{d^2u}{dx^2}+\left[\frac{2\lambda+b_0}{x-x_0}+\ldots\right]\frac{du}{dx}=0 \tag{1.23}$$

Since, from Eq. (1.18), $k=\lambda=\frac{1-b_0}{2}$, we have, $2\lambda+b_0=1$.

Therefore, we write

$$\frac{d^2u}{dx^2}+\left[\frac{1}{x-x_0}+\ldots\right]\frac{du}{dx}=0.$$

Putting $v=\frac{du}{dx}$, we write

$$\frac{dv}{dx}=-\left[\frac{1}{x-x_0}+\ldots\right]v,$$

or

$$\frac{dv}{v}=-\left[\frac{1}{x-x_0}+\ldots\right]dx$$

$$=-\frac{dx}{x-x_0}+\ldots=-\frac{dz}{z}+\ldots \text{ where } z=x-x_0.$$

Integrating both sides, we obtain

$$\int\frac{dv}{v}=-\int\frac{dz}{z}+\ldots$$

or $\ln v = -\ln z + \ldots$

or $\ln\dfrac{du}{dx} = -\ln(x-x_0) + \ldots = \ln(x-x_0)^{-1} + \ldots$

or $\dfrac{du}{dx} = \dfrac{1}{(x-x_0)} e^{(\ldots)}$, or, $du = \left[\dfrac{1}{x-x_0} e^{(\ldots)}\right] dx$

where $(\ldots)$ represents a constant term plus positive powers of $(x-x_0)$.

Expanding the exponential function in powers of $(x-x_0)$ and integrating again, we obtain

$$u = u(x) = \ln(x-x_0) + p_1(x-x_0) + p_2(x-x_0)^2 + p_3(x-x_0)^3 + \ldots$$

where $p_1, p_2, p_3, \ldots$ are constants.

Substituting this expression for $u(x)$ into Eq. (1.20), we obtain

$$y_2(x) = y_1 \ln(x-x_0) + [p_1(x-x_0) + p_2(x-x_0)^2 + \ldots][(x-x_0)^{\lambda}\{a_0 + a_1(x-x_0) + a_2(x-x_0)^2 + \ldots\}]$$

$$= y_1 \ln(x-x_0) + (x-x_0)^{\lambda} \sum_{m=1}^{\infty} A_m (x-x_0)^m \tag{1.24}$$

where $A_1 = a_0 p_1$,

$$A_2 = a_1 p_1 + a_0 p_2, \text{ etc.}$$

Thus, Eqs. (1:19) and (1:24) are the two independent solutions of the given ODE when the indicial equation has a double root and the general solution is the linear combination

$$y = y(x) = Ay_1(x) + By_2(x)$$

where A and B are arbitrary constants.

Here, the coefficients A_m can be determined by substituting back $y_2(x)$ for y into the given ODE.

[**Note:** Sometimes if $y_1(x)$ appears to be a simple elementary function, it may be much simpler to obtain the second solution $y_2(x)$ by starting right from choosing $y_2(x) = y_1(x)u(x)$ and perform the procedure of variation of parameters.

Alternative Approach

Alternatively, in a shorter way, the second independent solution $y_2(x)$ can also be obtained as

$$y_2 = y_2(x) = \left[\frac{\partial}{\partial k} y(k, x)\right]_{k=\lambda}$$

so that the general solution is

$$y = y(x) = Ay_1(x) + By_2(x)$$

$$= A[y(k, x)]_{k=\lambda} + B\left[\frac{\partial}{\partial k} y(k, x)\right]_{k=\lambda} \tag{1.25}$$

where $\lambda = k = k_1 = k_2$ is the repeated root of indicial equation.

CASE III: Distinct Roots of Indicial Equation Differing by an Integer

Let k_1 and $k_2 (k_1 > k_2)$ be the two distinct roots such that $k_1 - k_2 = q$ where q is an integer. Then, $k_2 = k_1 - q$.

Now, one solution of the given ODE will be

$$y_1(x) = \sum_{r=0}^{\infty} a_r (x - x_0)^{k_1 + r}$$

$$= (x - x_0)^{k_1} [a_0 + a_1(x - x_0) + a_2(x - x_0)^2 + \ldots \infty] \tag{1.26}$$

To obtain the second independent solution, $y_2(x)$, of the given ODE corresponding to the other root, $k_2 = k_1 - q$, we have to proceed as in Case II up to Eq. (1.23), viz.

$$\frac{d^2u}{dx^2} + \left[\frac{2k_1 + b_0}{x - x_0} + \ldots\right]\frac{du}{dx} = 0.$$

For the roots k_1 and k_2 of the indicial equation, which is a quadratic equation, we have

$$k_1 + k_2 = 1 - b_0 \text{ [using Eq. (1.13) and a property of quadratic equation]}$$

or
$$k_1 + k_1 - q = 1 - b_0,$$

or
$$2k_1 + b_0 = q + 1.$$

Therefore, we write

$$\frac{d^2u}{dx^2} + \left[\frac{q+1}{x - x_0} + \ldots\right]\frac{du}{dx} = 0.$$

Putting $v = \frac{du}{dx}$, we write

$$\frac{dv}{dx} = -\left[\frac{q+1}{x - x_0} + \ldots\right] v$$

or
$$\frac{dv}{v} = -(q+1)\frac{dx}{x - x_0} + \ldots = -(q+1)\frac{dz}{z} + \ldots$$

where $z = x - x_0$.

Integrating both sides, we obtain

$$\int\frac{dv}{v}=-(q+1)\int\frac{dz}{z}+\ldots$$

or
$$\ln v=-(q+1)\ln z+\ldots=\ln z^{-(q+1)}+\ldots$$

or
$$v=\frac{du}{dx}=z^{-(q+1)}e^{(\ldots)}=(x-x_0)^{-(q+1)}e^{(\ldots)}$$

where the dots represent terms with non-negative powers of $(x-x_0)$.

Expanding the exponential function in powers of $(x-x_0)$ as in Case II, we obtain a series

$$\frac{du}{dx}=\frac{1}{(x-x_0)^{q+1}}+\frac{s_1}{(x-x_0)^{q}}+\frac{s_2}{(x-x_0)^{q-1}}+\ldots+\frac{s_q}{(x-x_0)}$$
$$+s_{q+1}+s_{q+2}(x-x_0)+s_{q+3}(x-x_0)^2+\ldots$$

where $s_j (j=1,2,3,\ldots)$ are constants.

Integrating both sides once more, we get

$$u=u(x)=-\frac{1}{q(x-x_0)^{q}}-\frac{s_1}{(q-1)(x-x_0)^{q-1}}-\frac{s_2}{(q-2)(x-x_0)^{q-2}}+\ldots+s_q\ln(x-x_0)$$
$$+s_{q+1}(x-x_0)+\frac{s_{q+2}}{2}(x-x_0)^2+\frac{s_{q+3}}{3}(x-x_0)^3+\ldots$$

Substituting this expression for u into Eq. (1.20), we get

$$y_2(x)=u(x)y_1(x)=s_q y_1\ln(x-x_0)$$
$$+\left[\begin{array}{l}-\frac{1}{q}(x-x_0)^{-q}-\frac{s_1}{(q-1)}(x-x_0)^{-q+1}-\frac{s_2}{(q-2)}(x-x_0)^{-q+2}+\ldots\\ +s_{q+1}(x-x_0)+\frac{s_{q+2}}{2}(x-x_0)^2+\frac{s_{q+3}}{3}(x-x_0)^3+\ldots\end{array}\right]$$
$$\times(x-x_0)^{k_1}[a_0+a_1(x-x_0)+a_2(x-x_0)^2+a_3(x-x_0)^3+\ldots]$$
$$=s_q y_1\ln(x-x_0)$$
$$+(x-x_0)^{k_1-q}\left[\begin{array}{l}-\frac{1}{q}+\frac{s_1(\dot{x}-x_0)}{(1-q)}+\frac{s_2(x-x_0)^2}{(2-q)}+\ldots+s_{q+1}(x-x_0)^{q+1}\\ +\frac{s_{q+2}}{2}(x-x_0)^{q+2}+\frac{s_{q+3}}{3}(x-x_0)^{q+3}+\ldots\end{array}\right]$$
$$\times[a_0+a_1(x-x_0)+a_2(x-x_0)^2+a_3(x-x_0)^3+\ldots]$$

Thus, we write

$$y_2(x) = s_q y_1 \ln(x - x_0) + (x - x_0)^{k_2} \sum_{m=0}^{\infty} B_m (x - x_0)^m \tag{1.27}$$

where $B_0 = -\frac{a_0}{q}, B_1 = \frac{a_0 s_1}{1-q} - \frac{a_1}{q}, B_2 = \frac{a_0 s_2}{2-q} + \frac{a_1 s_1}{1-q} - \frac{a_2}{q}$, etc.

The general solution is then

$$y = y(x) = Dy_1(x) + Ey_2(x) \text{ where } D \text{ and } E \text{ are arbitrary constants.}$$

Alternative Approach

(i) When $k_1 > k_2$, if some coefficients become infinite at $k = k_2$, to overcome such a difficulty, we have to put

$$a_0 = d_0(k - k_2) \text{ (where } d_0 \text{ is a constant) in } y(x, k).$$

Then,

$$y_1(x) = [y(k, x)]_{k=k_1} \text{ and } y_2(x) = \left[\frac{\partial y(k, x)}{\partial k}\right]_{k=k_2} \quad \text{where } k_1 > k_2.$$

The general solution is their linear combination as indicated above.

(ii) If some coefficients become indeterminate when $k = k_2$ (smaller root), then the general solution of the given ODE is given by taking $k = k_2$ in $y = y(x, k)$. By putting $k = k_1$ (greater root), we will get a series which is merely a constant multiple of one of the series contained in the previous solution and generally, this solution is neglected. Thus, we have the general solution $y = [y(k, x)]_{k=k_2}$.

The solution $y_2(x) = [y(k, x)]_{k=k_1}$ where $k_1 > k_2$ will be a constant multiple of a part of the former solution and it is to be neglected.

1.5 SOLVED PROBLEMS

In most of the practical problems, series solutions of second order linear and homogeneous ODEs about the origin (i.e., $x_0 = 0$) are generally obtained.

Indicial Equation having Distinct Roots Differing by Non-integer

PROBLEM 1 Obtain the series solution of the differential equation

$$(x^3 + 2x)\frac{d^2y}{dx^2} - \frac{dy}{dx} - 6xy = 0. \tag{i}$$

Also find the region of convergence.

Solution: The given ODE is

$$(x^3+2x)\frac{d^2y}{dx^2}-\frac{dy}{dx}-6xy=0. \tag{i}$$

Writing in standard form, we have

$$\frac{d^2y}{dx^2}+\left(-\frac{1}{x^3+2x}\right)\frac{dy}{dx}+\left(\frac{-6}{x^2+2}\right)y=0. \tag{ii}$$

The function, $P(x)=-\dfrac{1}{x^3+2x}$, becomes infinite at $x=0$ and hence the point $x=0$ is a singular point of the given ODE.

But, $$\lim_{x\to 0} xP(x)=\lim_{x\to 0}\frac{-1}{(x^2+2)}=-\frac{1}{2}\text{ (finite).}$$

Also, $$\lim_{x\to 0} x^2Q(x)=\lim_{x\to 0} x^2\left(\frac{-6}{x^2+2}\right)=0\text{ (finite).}$$

Hence, the point $x=0$ is a regular singular point of the given ODE and, by Fuchs' theorem, the given ODE will have a power series solution.

We assume a power series solution of the form

$$y=\sum_{r=0}^{\infty}a_r x^{k+r}$$

where k is a constant to be determined latter and the coefficient a_0 is not zero.

Then, $$\frac{dy}{dx}=\sum_{r=0}^{\infty}(k+r)a_r x^{k+r-1}$$

and $$\frac{d^2y}{dx^2}=\sum_{r=0}^{\infty}(k+r)(k+r-1)a_r x^{k+r-2}.$$

Substituting these expressions for y and its derivatives into Eq. (i), we have

$$(x^3+2x)\sum_{r=0}^{\infty}(k+r)(k+r-1)a_r x^{k+r-2}-\sum_{r=0}^{\infty}(k+r)a_r x^{k+r-1}-6x\sum_{r=0}^{\infty}a_r x^{k+r}=0.$$

or $$\sum_{r=0}^{\infty}(k+r)(2k+2r-3)a_r x^{k+r-1}+\sum_{r=0}^{\infty}\{(k+r)(k+r-1)-6\}a_r x^{k+r+1}=0.$$

Here, the coefficients of various powers of x must vanish.

Thus, coefficient of lowest power of x, i.e., the coefficient of x^{k-1} is

$$k(2k-3)a_0=0. \qquad \text{[Indicial equation]}$$

The indicial equation has the roots $k = \frac{3}{2}, 0$ with $a_0 \neq 0$. We see that the roots are distinct and they differ by a non-integer.

Coefficient of next higher power of x, i.e., of x^k is

$$(k+1)(2k-1)a_1 = 0.$$

As $(k+1)(2k-1) \neq 0$ for both $k = 0$ and $k = \frac{3}{2}$, we must have $a_1 = 0$.

In general, the coefficient of x^{k+r+1} is

$$(k+r+2)(2k+2r+1)a_{r+2} + \{(k+r)(k+r-1)-6\}a_r = 0,$$

or

$$(k+r+2)(2k+2r+1)a_{r+2} + (k+r+2)(k+r-3)a_r = 0$$

The last equation gives

$$a_{r+2} = -\frac{(k+r-3)}{(2k+2r+1)}a_r.$$

Now, putting $r = 0, 1, 2, 3, 4, 5, \ldots$ respectively, we obtain

$$a_2 = -\frac{(k-3)}{(2k+1)}a_0, \qquad a_3 = -\frac{k-2}{(2k+3)}a_1 = 0 \text{ as } a_1 = 0.$$

$$a_4 = -\frac{(k-1)}{(2k+5)}a_2$$

$$= (-1)^2 \frac{(k-1)(k-3)}{(2k+5)(2k+1)}a_0, \qquad a_5 = 0$$

$$a_6 = -\frac{(k+1)}{(2k+9)}a_4$$

$$= (-1)^3 \frac{(k+1)(k-1)(k-3)}{(2k+9)(2k+5)(2k+1)}a_0, \qquad a_7 = 0$$

and so on.

Substituting these values of coefficients into the assumed solution, we obtain

$$y = y(x,k) = \sum_{r=0}^{\infty} a_r x^{k+r} = x^k[a_0 + a_1x + a_2x^2 + a_3x^3 + a_4x^4 + a_5x^5 + a_6x^6 + \ldots \infty]$$

$$= a_0x^k \left[1 - \frac{(k-3}{(2k+1)}x^2 + \frac{(k-3)(k-1)}{(2k+1)(2k+5)}x^4 - \frac{(k-3)(k-1)(k+1)}{(2k+1)(2k+5)(2k+9)}x^6 + \ldots \infty\right] \quad \text{(iii)}$$

First Solution

Substituting the first root $k = k_1 = \frac{3}{2}$ of the indicial equation into Eq. (iii), we get

$$y_1(x) = a_0 x^{\frac{3}{2}}\left[1 + \frac{3}{8}x^2 - \frac{1\times 3}{8\times 16}x^4 + \frac{1\times 3\times 5}{8\times 16\times 24}x^6 - \ldots\infty\right].$$

Second Solution

Substituting the second root $k = k_2 = 0$ of the indicial equation into Eq. (iii), we get

$$y_2(x) = a_0\left[1 + 3x^2 + \frac{3}{5}x^4 - \frac{1}{15}x^6 + \ldots\infty\right].$$

The general solution is the linear combination of $y_1(x)$ and $y_2(x)$, i.e.

$$y(x) = A\left[x^{\frac{3}{2}}\left[1 + \frac{3}{8}x^2 - \frac{1\times 3}{8\times 16}x^4 + \frac{1\times 3\times 5}{8\times 16\times 24}x^6 - \ldots\infty\right]\right]$$

$$+ B\left[1 + 3x^2 + \frac{3}{5}x^4 - \frac{1}{15}x^6 + \ldots\infty\right]$$

where the constant a_0 has been absorbed into the arbitrary constants A and B.

Region of convergence: To find the region of convergence, let us apply D'Alembert's ratio test for convergence to the series in Eq. (iii). If u_n denotes the nth term of the series, we have

$$|u_n| = a_0\,\frac{(k-3)(k-1)(k+1)\ldots(k+2n-7)}{(2k+1)(2k+5)(2k+9)\ldots(2k+4n-7)}x^{k+2n-2}$$

Similarly, if u_{n+1} denotes the $(n+1)$th term, we have

$$|u_{n+1}| = a_0\,\frac{(k-3)(k-1)(k+1)\ldots(k+2n-7)(k+2n-5)}{(2k+1)(2k+5)(2k+9)\ldots(2k+4n-7)(2k+4n-3)}x^{k+2n}$$

Then $\lim\limits_{n\to\infty}\frac{|u_{n+1}|}{|u_n|} = \lim\limits_{n\to\infty}\frac{k+2n-5}{2k+4n-3}x^2 = \frac{1}{2}x^2.$

Hence, the region of convergence is

$$\frac{1}{2}x^2 < 1 \text{ or } x^2 < 2, \text{i.e.,} -\sqrt{2} < x < \sqrt{2}.$$

PROBLEM 2 Obtain the series solution of the differential equation

$$2x^2\frac{d^2y}{dx^2} + x(2x-1)\frac{dy}{dx} + y = 0.$$

Solution: The point $x = 0$ is a regular singular point and hence by Fuchs' theorem, the given ODE will have a power series solution.

Let us assume the power series solution in the form

$$y = \sum_{r=0}^{\infty} a_r x^{k+r} \text{ where } k \text{ is a constant to be determined and } a_0 \neq 0.$$

Now, we have

$$\frac{dy}{dx} = \sum_{r=0}^{\infty}(k+r)a_r x^{k+r-1} \text{ and } \frac{d^2y}{dx^2} = \sum_{r=0}^{\infty}(k+r)(k+r-1)a_r x^{k+r-2}.$$

Substituting the expressions of y and its derivatives into the given ODE, we have

$$\sum_{r=0}^{\infty} 2(k+r)(k+r-1)a_r x^{k+r} + \sum_{r=0}^{\infty} 2(k+r)a_r x^{k+r+1} - \sum_{r=0}^{\infty}(k+r)a_r x^{k+r} + \sum_{r=0}^{\infty} a_r x^{k+r} = 0,$$

or

$$\sum_{r=0}^{\infty}[(k+r)(2k+2r-3)+1]a_r x^{k+r} + \sum_{r=0}^{\infty} 2(k+r)a_r x^{k+r+1} = 0. \quad \text{(i)}$$

In Eq. (i), the coefficient of each of the various powers of x must vanish. Coefficient of the lowest power of x, i.e., of x^k is

$$[k(2k-3)+1]a_0 = 0,\ a_0 \neq 0 \quad \text{[Indicial equation]}$$

or

$$2k^2 - 3k + 1 = 0 \text{ giving the roots } k = 1 \text{ and } k = \frac{1}{2}.$$

Coefficient of next higher power of x, i.e., of x^{k+1} is

$$[(k+1)(2k-1)+1]a_1 + 2ka_0 = 0, \text{ and so on.}$$

The coefficient of x^{k+r+1} is

$$[(k+r+1)(2k+2r-1)+1]a_{r+1} + 2(k+r)a_r = 0.$$

Thus, we get

$$a_{r+1} = -\frac{2(k+r)}{[(k+r+1)(2k+2r-1)+1]}a_r \quad \text{[Recursion relation]} \quad \text{(ii)}$$

Putting $r = 0, 1, 2, 3, 4, \ldots$, we get

$$a_1 = -\frac{2k}{(k+1)(2k-1)+1}a_0,$$

$$a_2 = -\frac{2(k+1)}{(k+2)(2k+1)+1}a_1 = (-1)^2\frac{2^2(k+1)k}{\{(k+2)(2k+1)+1\}\{(k+1)(2k-1)+1\}}a_0.$$

$$a_3 = -\frac{2(k+2)}{(k+3)(2k+3)+1}a_2$$

$$= (-1)^3 \frac{2^3(k+2)(k+1)k}{\{(k+3)(2k+3)+1\}\{(k+2)(2k+1)+1\}\{(k+1)(2k-1)+1\}}a_0 \text{ and so on.}$$

Substituting the values of these coefficients into the assumed power series solution, we have

$$y = a_0 x^k \left[1 - \frac{2k}{(k+1)(2k-1)+1}x + \frac{2^2(k+1)k}{\{(k+2)(2k+1)+1\}\{(k+1)(2k-1)+1\}}x^2 \right.$$

$$\left. - \frac{2^3(k+2)(k+1)k}{\{(k+3)(2k+3)+1\}\{(k+2)(2k+1)+1\}\{(k+1)(2k-1)+1\}}x^3 + \ldots\infty\right]. \text{(iii)}$$

First Solution

Substituting $k = k_1 = 1$ (larger root of indicial equation) into Eq. (iii), we obtain

$$y_1(x) = a_0 x\left[1 - \frac{2}{3}x + \frac{2^2}{3\times5}x^2 - \frac{2^3}{3\times5\times7}x^3 + \ldots\infty\right].$$

Second Solution

Substituting $k = k_2 = \frac{1}{2}$ (the smaller root of indicial equation) into Eq. (iii), we obtain

$$y_2(x) = a_0 x^{\frac{1}{2}}\left[1 - x + \frac{1}{2}x^2 - \frac{1}{6}x^3 + \ldots\infty\right].$$

The general solution is a linear combination of these two solutions, i.e.,

$$y(x) = Ax\left[1 - \frac{2}{3}x + \frac{2^2}{3\times5}x^2 - \frac{2^3}{3\times5\times7}x^3 + \ldots\infty\right] + B\sqrt{x}\left[1 - x + \frac{1}{2}x^2 - \frac{1}{6}x^3 + \ldots\infty\right]$$

where the constant a_0 has been absorbed into the arbitrary constants A and B.

Indicial Equation having a Double Root (Equal Roots)

PROBLEM 3 Discuss the nature of the point $x = 0$ in respect of the differential equation

$$x^2\frac{d^2y}{dx^2} - x\frac{dy}{dx} + (1-x)y = 0.$$

Obtain its series solutions and the general solution.

Solution: The given ODE is

$$x^2\frac{d^2y}{dx^2}-x\frac{dy}{dx}+(1-x)y=0 \tag{i}$$

which can be written as

$$\frac{d^2y}{dx^2}-\frac{1}{x}\frac{dy}{dx}+\frac{(1-x)}{x^2}y=0.$$

At $x=0, [P(x)]_{x=0}=\left[-\frac{1}{x}\right]_{x=0}=-\infty, [Q(x)]_{x=0}=\left[\frac{1-x}{x^2}\right]_{x=0}=\infty.$

Therefore, $x = 0$ is a singular point.

We have, $\quad [xP(x)]_{x\to 0}=-1\,(\text{finite})$ and $[x^2Q(x)]_{x\to 0}=1\,(\text{finite})$.

We see that the point $x=0$ is a regular singular point and, therefore, by Fuchs' theorem, the given ODE will have a power series solution about this singularity.

Let us assume the power series solution in the form

$$y=\sum_{r=0}^{\infty}a_r x^{k+r}\,(a_0\neq 0) \text{ where } k \text{ is a constant to be determined.}$$

Then $\quad \frac{dy}{dx}=\sum_{r=0}^{\infty}(k+r)a_r x^{k+r-1}$ and $\frac{d^2y}{dx^2}=\sum_{r=0}^{\infty}(k+r)(k+r-1)a_r x^{k+r-2}$.

Substituting the expressions for y and its derivatives into Eq. (i), we obtain

$$\sum_{r=0}^{\infty}[(k+r)(k+r-2)+1]a_r x^{k+r}-\sum_{r=0}^{\infty}a_r x^{k+r+1}=0. \tag{ii}$$

In this equation, each of the coefficients of various powers of x must vanish.

In Eq. (ii), the coefficient of lowest power of x, i.e., of x^k is

$$[k(k-2)+1]a_0=0. \quad \text{[Indicial equation]}$$

or $$k^2-2k+1=0 \quad \text{as } a_0\neq 0$$

giving the double root $k=\lambda=1,1$.

Coefficient of x^{k+1} is $[(k+1)(k-1)+1]a_1-a_0=0$

giving $$a_1=\frac{1}{k^2}a_0, \text{ etc.}$$

Coefficient of x^{k+r+1} is $[(k+r+1)(k+r-1)+1]a_{r+1}-a_r=0$

giving $$a_{r+1}=\frac{1}{(k+r)^2}a_r. \quad \text{[Recursion relation]}$$

Putting $r = 0, 1, 2, 3$, etc., we obtain

$$a_1 = \frac{1}{k^2} a_0, \qquad a_2 = \frac{1}{(k+1)^2} a_1 = \frac{1}{(k+1)^2 k^2} a_0,$$

$$a_3 = \frac{1}{(k+2)^2} a_2 = \frac{1}{(k+2)^2 (k+1)^2 k^2} a_0,$$

$$a_4 = \frac{1}{(k+3)^2} a_3 = \frac{1}{(k+3)^2 (k+2)^2 (k+1)^2 k^2} a_0, \text{ etc.}$$

Substituting the values of these coefficients into the assumed power series solution, we obtain

$$y = y(x, k) = x^k [a_0 + a_1 x + a_2 x^2 + a_3 x^3 + \dots \infty]$$

$$= a_0 x^k \left[1 + \frac{1}{k^2} x + \frac{1}{(k+1)^2 k^2} x^2 + \frac{1}{(k+2)^2 (k+1)^2 k^2} x^3 + \dots \infty\right]. \qquad \text{(iii)}$$

First Solution

Substituting $k = \lambda = 1$ (root of indicial equation) in Eq. (iii) , we obtain the first solution as

$$y_1(x) = a_0 \left[x + \frac{1}{(1!)^2} x^2 + \frac{1}{(2!)^2} x^3 + \frac{1}{(3!)^2} x^4 + \dots + \frac{1}{(m!)^2} x^{m+1} + \dots \infty\right] = a_0 \sum_{m=0}^{\infty} \frac{1}{(m!)^2} x^{m+1}.$$

If we take $a_0 = 1$, we obtain

$$y_1(x) = x + \frac{1}{(1!)^2} x^2 + \frac{1}{(2!)^2} x^3 + \frac{1}{(3!)^2} x^4 + \dots + \frac{1}{(m!)^2} x^{m+1} + \dots \infty = \sum_{m=0}^{\infty} \frac{1}{(m!)^2} x^{m+1} \qquad \text{(iv)}$$

Second Solution

To obtain the second solution, we can use different approaches as in the following.

First Approach

Let us assume the second solution as

$$y_2(x) = y_1 \ln x + x^{\lambda} \sum_{m=1}^{\infty} A_m x^m = y_1 \ln x + \sum_{m=1}^{\infty} A_m x^{m+1} \qquad \text{(v)}$$

where $k = \lambda = 1$ is the repeated root of the indicial equation and A_m are coefficients to be determined.

Then, we have

$$\frac{dy_2}{dx} = \frac{dy_1}{dx} \ln x + \frac{1}{x} y_1 + \sum_{m=1}^{\infty} (m+1) A_m x^m,$$

$$\frac{d^2y_2}{dx^2}=\frac{d^2y_1}{dx^2}\ln x+\frac{2}{x}\frac{dy_1}{dx}-\frac{1}{x^2}y_1+\sum_{m=1}^{\infty}(m+1)mA_m x^{m-1}.$$

Since $y_2(x)$ is a solution of the given ODE, we have

$$x^2\frac{d^2y_2}{dx^2}-x\frac{dy_2}{dx}+(1-x)y_2=0.$$

Substituting the expressions for $y_2(x)$ and its derivatives into this ODE, we obtain

$$\left[x^2\frac{d^2y_1}{dx^2}-x\frac{dy_1}{dx}+(1-x)y_1\right]\ln x+2x\frac{dy_1}{dx}-2y_1+\sum_{m=1}^{\infty}(m+1)mA_m x^{m+1}-\sum_{m=1}^{\infty}(m+1)A_m x^{m+1}$$

$$+\sum_{m=1}^{\infty}A_m x^{m+1}-\sum_{m=1}^{\infty}A_m x^{m+2}=0.$$

Since $y_1(x)$ is a solution of the given ODE, the quantities within the square bracket vanish and hence substituting the expression for $y_1(x)$ in the remaining expression, we write

$$2x\frac{d}{dx}\sum_{m=0}^{\infty}\frac{x^{m+1}}{(m!)^2}-2\sum_{m=0}^{\infty}\frac{x^{m+1}}{(m!)^2}+\sum_{m=1}^{\infty}m^2A_m x^{m+1}-\sum_{m=1}^{\infty}A_m x^{m+2}=0,$$

or

$$\sum_{m=0}^{\infty}\frac{2(m+1)x^{m+1}}{(m!)^2}-\sum_{m=0}^{\infty}\frac{2x^{m+1}}{(m!)^2}+\sum_{m=1}^{\infty}m^2A_m x^{m+1}-\sum_{m=1}^{\infty}A_m x^{m+2}=0,$$

or

$$\sum_{m=1}^{\infty}\frac{2m}{(m!)^2}x^{m+1}+\sum_{m=1}^{\infty}m^2A_m x^{m+1}-\sum_{m=1}^{\infty}A_m x^{m+2}=0.$$

Since the coefficients of various powers of x must separately vanish.

Coefficient of x^2 is $2+A_1=0$ giving $A_1=-2$,

Coefficient of x^3 is $1+4A_2-A_1=0$ giving $A_2=-\frac{3}{4}$,

Coefficient of x^4 is $\frac{1}{6}+9A_3-A_2=0$ giving $A_3=-\frac{11}{108}$, etc.

Substituting the expression for y_1 given in Eq. (iv) and also the values of these coefficients into Eq. (v), we obtain the second solution as

$$y_2(x)=\left[\sum_{m=0}^{\infty}\frac{x^{m+1}}{(m!)^2}\right]\ln x-\left[2x^2+\frac{3}{4}x^3+\frac{11}{108}x^4+\ldots\infty\right]$$

Second Approach

Using Eq. (iii), we write

$$y_2(x) = \left[\frac{\partial y(x,k)}{\partial k}\right]_{k=1}$$

$$= \left[a_0 x^k (\ln x)\left\{1 + \frac{1}{k^2}x + \frac{1}{(k+1)^2 k^2}x^2 + \frac{1}{(k+2)^2(k+1)^2 k^2}x^3 + \ldots\infty\right\}\right]_{k=1}$$

$$+ \left\{a_0 x^k \left[0 - \frac{2}{k^3}x - \frac{2(2k+1)}{(k+1)^3 k^3}x^2 - \frac{2\{(k+1)k + (k+2)k + (k+2)(k+1)\}}{(k+2)^3(k+1)^3 k^3}x^3 + \ldots\infty\right]\right\}_{k=1}$$

Choosing $a_0 = 1$, we obtain

$$y_2(x) = \left\{x + x^2 + \frac{1}{(2!)^2}x^3 + \frac{1}{(3!)^2}x^4 + \ldots\infty\right\}\ln x - \left\{2x^2 + \frac{3}{4}x^3 + \frac{11}{108}x^4 + \ldots\infty\right\}$$

$$= \left\{\sum_{m=0}^{\infty}\frac{x^{m+1}}{(m!)^2}\right\}\ln x - \left\{2x^2 + \frac{3}{4}x^3 + \frac{11}{108}x^4 + \ldots\infty\right\}.$$

The general solution is $y(x) = Ay_1(x) + By_2(x)$ where A and B are arbitrary constants.

PROBLEM 4 Discuss the nature of the point $x = 0$ in respect of the differential equation

$$x(x-1)\frac{d^2y}{dx^2} + (3x-1)\frac{dy}{dx} + y = 0.$$

If possible, obtain its independent solutions in power series and also obtain the general solution.

Solution: The given ODE is

$$x(x-1)\frac{d^2y}{dx^2} + (3x-1)\frac{dy}{dx} + y = 0 \quad \text{(i)}$$

which can be expressed in the form

$$\frac{d^2y}{dx^2} + P(x)\frac{dy}{dx} + Q(x)y = 0$$

where $P(x) = \dfrac{(3x-1)}{x(x-1)}$ and $Q(x) = \dfrac{1}{x(x-1)}$.

We see that $P(0)$ and $Q(0)$ are both infinite showing that the point $x = 0$ is a singular point of the given ODE.

But, $$\lim_{x\to 0} xP(x) = \lim_{x\to 0} \frac{3x-1}{(x-1)} = 1, \text{(finite)}$$

and $$\lim_{x\to 0} x^2 Q(x) = \lim_{x\to 0} \frac{x}{(x-1)} = 0, \text{(finite)}.$$

Thus, the point $x=0$ is a regular singular point and, hence, by Fuchs' theorem, the given ODE can have a power series solution in the form

$$y = \sum_{r=0}^{\infty} a_r x^{k+r} \quad (a_0 \neq 0) \tag{ii}$$

where k is a constant to be determined later.

Now, we have

$$\frac{dy}{dx} = \sum_{r=0}^{\infty} (k+r) a_r x^{k+r-1}$$

and $$\frac{d^2y}{dx^2} = \sum_{r=0}^{\infty} (k+r)(k+r-1) a_r x^{k+r-2}.$$

Substitution of the expressions for $y(x)$ and its derivatives into Eq. (i) results in

$$\sum_{r=0}^{\infty} (k+r+1)^2 a_r x^{k+r} - \sum_{r=0}^{\infty} (k+r)^2 a_r x^{k+r-1} = 0.$$

Each of the coefficients of various powers of x must separately vanish.

Coefficient of x^{k-1} is $-k^2 a_0 = 0\ (a_0 \neq 0)$ [Indicial equation]

giving the double root $k = \lambda = 0, 0$ as $a_0 \neq 0$.

Coefficient of x^k is $(k+1)^2 a_0 - (k+1)^2 a_1 = 0$

giving $a_1 = a_0$ ans so on.

Coefficient of x^{k+r} is $(k+r+1)^2 a_r - (k+r+1)^2 a_{r+1} = 0$

giving $a_{r+1} = a_r$ [Recursion relation].

Putting $r = 0, 1, 2, 3, \ldots$ etc., we get

$$a_1 = a_0,\ \ a_2 = a_1 = a_0,\ \ a_3 = a_0, \text{ etc.}$$

Substitution of the expressions for these coefficients into Eq. (ii) results in

$$y = a_0 x^k\, [1 + x + x^2 + x^3 + \ldots \infty] \tag{iii}$$

First Solution

Substituting $k=\lambda=0$ obtained from the indicial equation into Eq. (iii), we obtain

$$y_1(x)=a_0[1+x+x^2+x^3+\ldots\infty]$$

$$=\frac{a_0}{1-x} \tag{iv}$$

Choosing $a_0=1$, we have

$$y_1(x)=\frac{1}{1-x}=-\frac{1}{x-1} \tag{v}$$

Second Solution

In this problem, the second solution is being obtained by either of the following approaches.

First Approach

Let us assume the second solution as

$$y_2(x)=y_1\ln x+x^{\lambda}\sum_{m=1}^{\infty}A_m x^m=y_1\ln x+\sum_{m=1}^{\infty}A_m x^m \tag{vi}$$

where we use $k=\lambda=0$ (Root of indicial equation).

Now, we have

$$\frac{dy_2}{dx}=\frac{dy_1}{dx}\ln x+\frac{1}{x}y_1+\sum_{m=1}^{\infty}mA_m x^{m-1}$$

and
$$\frac{d^2y_2}{dx^2}=\frac{d^2y_1}{dx^2}\ln x+\frac{2}{x}\frac{dy_1}{dx}-\frac{1}{x^2}y_1+\sum_{m=2}^{\infty}m(m-1)A_m x^{m-2}.$$

As $y_2(x)$ is a solution of the given ODE, we have

$$x(x-1)\frac{d^2y_2}{dx^2}+(3x-1)\frac{dy_2}{dx}+y_2=0.$$

Substituting the expressions for $y_2(x)$ and its derivatives obtained above, we write

$$\left[x(x-1)\frac{d^2y_1}{dx^2}+(3x-1)\frac{dy_1}{dx}+y_1\right]\ln x+2(x-1)\frac{dy_1}{dx}$$

$$+\sum_{m=2}^{\infty}m(m-1)A_m x^m-\sum_{m=2}^{\infty}m(m-1)A_m x^{m-1}+2y_1+\sum_{m=1}^{\infty}3mA_m x^m$$

$$-\sum_{m=1}^{\infty}mA_m x^{m-1}+\sum_{m=1}^{\infty}A_m x^m=0.$$

As $y_1(x)$ is a solution of the given ODE, the quantities inside the square bracket vanish and we write

$$2(x-1)\frac{dy_1}{dx}+2y_1+4A_1x-A_1+\sum_{m=2}^{\infty}(m+1)^2A_mx^m-\sum_{m=2}^{\infty}m^2A_mx^{m-1}=0.$$

Using the expression $y_1(x)=\frac{a_0}{1-x}=-\frac{a_0}{x-1}$, we write

$$\frac{2(x-1)a_0}{(x-1)^2}-\frac{2a_0}{x-1}-A_1+4A_1x+\sum_{m=2}^{\infty}(m+1)^2A_mx^m-\sum_{m=2}^{\infty}m^2A_mx^{m-1}=0,$$

or
$$-A_1+4A_1x+\sum_{m=2}^{\infty}(m+1)^2A_mx^m-\sum_{m=2}^{\infty}m^2A_mx^{m-1}=0.$$

Here, the coefficients of various powers of x must vanish separately. Therefore,

Coefficient of x^0 : $A_1=0$,

Coefficient of x : $4A_1-4A_2=0$, giving $A_2=A_1=0$, etc.

In general, coefficient of x^m: $(m+1)^2A_m-(m+1)^2A_{m+1}=0$,

giving
$$A_{m+1}=A_m \tag{vii}$$

As $A_1=0$, putting $m=1,2,3,4,\dots$, we obtain

$$A_2=A_3=A_4=A_5=A_6=\dots=0.$$

Hence, Eqs. (iv) and (vi) yield

$$y_2(x)=y_1\ln x=\frac{a_0\ln x}{1-x}.$$

Second Approach

Let us assume the second solution in the form

$$y_2(x)=u(x)\,y_1(x) \tag{a}$$

where $u(x)$ is a function to be determined.

Then, we have

$$\frac{dy_2}{dx}=\frac{du}{dx}y_1+u\frac{dy_1}{dx}$$

and
$$\frac{d^2y_2}{dx^2}=\frac{d^2u}{dx^2}y_1+2\frac{du}{dx}\frac{dy_1}{dx}+u\frac{d^2y_1}{dx^2}.$$

Since $y_2(x)$ is a solution of the given ODE, we have

$$x(x-1)\frac{d^2y_2}{dx^2}+(3x-1)\frac{dy_2}{dx}+y_2=0.$$

Using Eq. (a) and its first and second derivatives, we write

$$x(x-1)\left(\frac{d^2u}{dx^2}y_1+2\frac{du}{dx}\frac{dy_1}{dx}+u\frac{d^2y_1}{dx^2}\right)+(3x-1)\left(\frac{du}{dx}y_1+u\frac{dy_1}{dx}\right)+uy_1=0,$$

or
$$u\left[x(x-1)\frac{d^2y_1}{dx^2}+(3x-1)\frac{dy_1}{dx}+y_1\right]+x(x-1)y_1\frac{d^2u}{dx^2}+2x(x-1)\frac{du}{dx}\frac{dy_1}{dx}$$
$$+(3x-1)y_1\frac{du}{dx}=0. \quad \text{(b)}$$

Since y_1 is a solution of the given ODE, the terms within the square brackets become zero.

Recalling that $y_1=-\frac{a_0}{x-1}$, we have, $\frac{dy_1}{dx}=\frac{a_0}{(x-1)^2}$.

Hence, substitution of these relations in equation (b) yields $-x\frac{d^2u}{dx^2}=\frac{du}{dx}$.

Now, putting $v=\frac{du}{dx}$, we write the above equation as

$$\frac{1}{v}\frac{dv}{dx}=-\frac{1}{x} \text{ or, } \frac{dv}{v}=-\frac{dx}{x}.$$

Integrating both sides and choosing the integration constant as zero, we obtain

$$\ln v=-\ln x=\ln x^{-1}$$

$$v=\frac{du}{dx}=\frac{1}{x} \text{ or, } du=\frac{dx}{x}.$$

Integrating again and choosing the integration constant as zero again, we get

$$u=\ln x.$$

Substituting the expressions for y_1 (first solution) and u into Eq. (a), we obtain the second solution of the given ODE as

$$y_2(x)=\frac{a_0\ln x}{1-x}=y_1\ln x.$$

Third Approach

Applying the formula $y_2(x)=\left[\frac{\partial y(k,x)}{\partial k}\right]_{k=0}$ to Eq. (iii), we write the second solution as

$$y_2(x)=[a_0x^k(\ln x)(1+x+x^2+x^3+x^4+\ldots\infty)]_{k=0}=\frac{a_0\ln x}{1-x}.$$

Indicial Equation having Distinct Roots Differing by an Integer

PROBLEM 5 Discuss the nature of the point $x=0$ in relation with the differential equation

$$x(x-1)\frac{d^2y}{dx^2}+3x\frac{dy}{dx}+y=0$$

and obtain its series solutions if it exists.

Solution: The given ODE can be written in standard form as

$$\frac{d^2y}{dx^2}+\frac{3}{x-1}\frac{dy}{dx}+\frac{1}{x(x-1)}y=0.$$

Here, $P(x)=\dfrac{3}{x-1}=-3$ (finite) at $x=0$ but

$$Q(x)=\frac{1}{x(x-1)} \text{ is infinite at } x=0.$$

Hence, the point $x=0$ is a singular point of the given ODE.

Further, we see that $[xP(x)]_{x\to 0}$ and $[x^2Q(x)]_{x\to 0}$ are both equal to zero and, hence, the point $x=0$ is a regular singular point.

Therefore, by Fuchs' theorem, the given ODE will have power series solutions. Let us assume a power series solution in the form

$$y=\sum_{r=0}^{\infty}a_r x^{k+r}, \text{(where } k \text{ is a constant to be determined and } a_0\neq 0). \quad \text{(i)}$$

Then, we have

$$\frac{dy}{dx}=\sum_{r=0}^{\infty}(k+r)a_r x^{k+r-1} \text{ and } \frac{d^2y}{dx^2}=\sum_{r=0}^{\infty}(k+r)(k+r-1)a_r x^{k+r-2}.$$

Substitution of the expressions for y and its derivatives into the given ODE results in

$$\sum_{r=0}^{\infty}(k+r+1)^2 a_r x^{k+r}-\sum_{r=0}^{\infty}(k+r)(k+r-1)a_r x^{k+r-1}=0.$$

In this equation, the coefficients of various power of x must separately vanish.

Coefficient of x^{k-1} : $k(k-1)a_0=0$, with $a_0\neq 0$ [Indicial equation]

giving $k=k_1=1$ and $k=k_2=0$ with $k_1-k_2=1$.

Coefficient of x^k : $(k+1)^2a_0-(k+1)ka_1=0$

giving $a_1=\dfrac{k+1}{k}a_0$.

In general, the coefficient of x^{k+r} is

$$(k+r+1)^2 a_r - (k+r+1)(k+r)a_{r+1} = 0$$

giving $a_{r+1} = \dfrac{(k+r+1)}{(k+r)} a_r$. [Recursion relation]

Putting $r = 0, 1, 2, 3, 4, \ldots$ etc., we have

$$a_1 = \frac{k+1}{k} a_0, \qquad a_2 = \frac{(k+2)}{(k+1)} \frac{(k+1)}{k} a_0 = \frac{k+2}{k} a_0$$

$$a_3 = \frac{k+3}{k} a_0, \qquad a_4 = \frac{k+4}{k} a_0, \text{ etc.}$$

Substituting the values of the various coefficients (with arbitrary a_0) into Eq. (i), we obtain

$$y = a_0 x^k \left[1 + \frac{k+1}{k} x + \frac{k+2}{k} x^2 + \frac{k+3}{k} x^3 + \frac{k+4}{k} x^4 + \ldots \infty \right]. \tag{ii}$$

First Solution

Substituting $k = k_1 = 1$ (larger root of indicial equation) into Eq. (ii), we obtain

$$y_1(x) = a_0 x[1 + 2x + 3x^2 + 4x^3 + 5x^4 + \ldots \infty] = \frac{a_0 x}{(1-x)^2}. \tag{iii}$$

Choosing the arbitrary constant a_0 as unity, we have

$$y_1(x) = x[1 + 2x + 3x^2 + 4x^3 + 5x^4 + \ldots \infty] = \frac{x}{(1-x)^2}. \tag{iv}$$

Second Solution

Let us assume the second solution in the form

$$y_2(x) = S y_1 \ln x + x^{k_2} \sum_{m=0}^{\infty} B_m x^m = S y_1 \ln x + \sum_{m=0}^{\infty} B_m x^m, \tag{v}$$

as $k_2 = 0$.

Then we have

$$\frac{dy_2}{dx} = S \frac{dy_1}{dx} \ln x + \frac{S}{x} y_1 + \sum_{m=0}^{\infty} m B_m x^{m-1}$$

and

$$\frac{d^2 y_2}{dx^2} = S \frac{d^2 y_1}{dx^2} \ln x + 2 \frac{S}{x} \frac{dy_1}{dx} - \frac{S}{x^2} y_1 + \sum_{m=0}^{\infty} m(m-1) B_m x^{m-2}.$$

As y_2 is a solution of the given ODE, we have

$$x(x-1)\frac{d^2y_2}{dx^2}+3x\frac{dy_2}{dx}+y_2=0 \quad \text{(vi)}$$

Substituting the expressions for y_2 and its derivatives into Eq. (vi), we obtain

$$S\ln x\left[x(x-1)\frac{d^2y_1}{dx^2}+3x\frac{dy_1}{dx}+y_1\right]+2Sx\frac{dy_1}{dx}-2S\frac{dy_1}{dx}+2Sy_1+\frac{S}{x}y_1$$
$$+\sum_{m=0}^{\infty}(m+1)^2B_m x^m-\sum_{m=0}^{\infty}m(m-1)B_m x^{m-1}=0.$$

As y_1 is a solution of the given ODE, the quantities within the square bracket becomes zero and then substituting the expression for y_1 obtained in Eq. (iv) and also its derivative into the above equation, we write

$$2S(x+4x^2+9x^3+16x^4+\ldots)-2S(1+4x+9x^2+16x^3+\ldots)$$
$$+2S(x+2x^2+3x^3+4x^4+\ldots)+S(1+2x+3x^2+4x^3+\ldots)$$
$$+\sum_{m=0}^{\infty}(m+1)^2B_m x^m-\sum_{m=0}^{\infty}m(m-1)B_m x^{m-1}=0.$$

In this equation, the coefficients of various powers of x must separately vanish. Therefore,

Coefficient of x^0 : $-2S+S+B_0+0B_1=0$,

giving $S=B_0$ with arbitrary B_1.

Coefficient of x^1 : $2S-8S+2S+2S+4B_1-2B_2=0$,

giving $B_2=2B_1-S=2B_1-B_0$.

Coefficient of x^2 : $8S-18S+4S+3S+9B_2-6B_3=0$,

giving $-S+3B_2=2B_3$

or $\quad B_3=\frac{1}{2}(3B_2-S)=3B_1-2B_0$; [using the values of S and B_2] and so on.

Substituting these coefficients into Eq. (v), we obtain

$$y_2(x)=B_0x(1+2x+3x^2+4x^3+\ldots\infty)\ln x$$
$$+[B_0+B_1x+(2B_1-B_0)x^2+(3B_1-2B_0)x^3+\ldots\infty].$$

Choosing the arbitrary constants B_0 and B_1 as unity, we write the second solution as

$$y_2(x) = x\ln x(1+2x+3x^2+4x^3+\ldots\infty)+(1+x+x^2+x^3+\ldots\infty)$$

$$= \frac{x\ln x}{(1-x)^2}+\frac{1}{1-x} \qquad \text{(vii)}$$

The general solution of the given ODE is the linear combination

$$y(x) = Ay_1 + By_2$$

where A and B are arbitrary constants.

Alternative Approach for Finding the Second Solution

As the first solution y_1 is a simple one, it should be easier to find the second solution via the method of reduction of order (variation of parameter) as in the following.

Let us assume the second solution in the form

$$y_2(x) = u(x)\, y_1(x) \qquad \text{(viii)}$$

As y_2 is assumed to be a solution of the given ODE, we must have

$$x(x-1)\frac{d^2y_2}{dx^2}+3x\frac{dy_2}{dx}+y_2 = 0,$$

or
$$x(x-1)\left(\frac{d^2u}{dx^2}y_1+2\frac{du}{dx}\frac{dy_1}{dx}+u\frac{d^2y_1}{dx^2}\right)+3x\left(\frac{du}{dx}y_1+u\frac{dy_1}{dx}\right)+uy_1 = 0.$$

or
$$u\left[x(x-1)\frac{d^2y_1}{dx^2}+3x\frac{dy_1}{dx}+y_1\right]+x(x-1)y_1\frac{d^2u}{dx^2}+\left[2x(x-1)\frac{dy_1}{dx}+3xy_1\right]\frac{du}{dx} = 0.$$

As y_1 is a solution of the given ODE, the terms within the square bracket will become zero. Then, substituting the value of y_1 from Eq. (iv) and also of its derivative into the above equation, we have

$$x(x-1)\frac{x}{(1-x)^2}\frac{d^2u}{dx^2}+\left[-2x(1-x)\frac{(1+x)}{(1-x)^3}+3x\frac{x}{(1-x)^2}\right]\frac{du}{dx} = 0$$

or
$$-\frac{x^2}{(1-x)}\frac{d^2u}{dx^2}+\frac{x(x-2)}{(1-x)^2}\frac{du}{dx} = 0.$$

Putting $v = \dfrac{du}{dx}$, we have

$$(x-2)v = x(1-x)\frac{dv}{dx},$$

or
$$\frac{(x-2)}{x(1-x)}dx = \frac{dv}{v},$$

or
$$\frac{dv}{v} = -\frac{2}{x} + \frac{1}{x-1}.$$

Integrating both sides and choosing the integration constant as zero, we have

$$\ln v = -2\ln x + \ln(x-1) = \ln[(x-1)x^{-2}]$$

giving
$$v = \frac{du}{dx} = (x-1)x^{-2}$$

or
$$du = (x-1)x^{-2}dx.$$

Integrating again and choosing the integration constant as zero again, we have

$$u = (x-1)\frac{x^{-1}}{-1} + \int \frac{dx}{x} = \frac{1-x}{x} + \ln x.$$

Substituting this expression for u into Eq. (viii), we write

$$y_2(x) = \frac{1}{1-x} + \frac{x\ln x}{(1-x)^2} \quad \text{which is Eq. (vii).}$$

Another Alternative Approach for Finding the Second Solution

Eq. (ii) reads

$$y = y(x,k) = a_0 x^k \left[1 + \frac{k+1}{k}x + \frac{k+2}{k}x^2 + \frac{k+3}{k}x^3 + \frac{k+4}{k}x^4 + \ldots\infty\right].$$

We see that at $k = k_2 = 0$, the coefficients become infinite starting from the second term and in order to eliminate the difficulty, let us put $a_0 = R_0 k$ where R_0 is a constant. Then, we have

$$y = R_0 x^k [k + (k+1)x + (k+2)x^2 + (k+3)x^3 + (k+4)x^4 + \ldots\infty].$$

The second solution is also given by

$$y_2(x) = \left[\frac{\partial y}{\partial k}\right]_{k=k_2} = [R_0 x^k (\ln x)\{k + (k+1)x + (k+2)x^2 + (k+3)x^3 + \ldots\infty\}$$
$$+ R_0 x^k \{1 + x + x^2 + x^3 + x^4 + \ldots\infty\}]_{k=0}$$

$$= xR_0(1 + 2x + 3x^2 + 4x^3 + \ldots\infty)\ln x$$
$$+ R_0(1 + x + x^2 + x^3 + x^4 + \ldots\infty) = R_0\frac{x\ln x}{(1-x)^2} + \frac{R_0}{1-x}.$$

If R_0 is taken as unity, we obtain

$$y_2(x) = \frac{x \ln x}{(1-x)^2} + \frac{1}{1-x}.$$

PROBLEM 6 Discuss the nature of the point $x = 0$ in relation with the differential equation

$$x^2(x^2-1)\frac{d^2y}{dx^2} - x(x^2+1)\frac{dy}{dx} + (x^2+1)y = 0$$

and, if possible, find its series solutions.

Solution: The given ODE can be written as

$$\frac{d^2y}{dx^2} + \frac{\{-(x^2+1)\}}{x(x^2-1)}\frac{dy}{dx} + \frac{(x^2+1)}{x^2(x^2-1)}y = 0.$$

We see that $P(x) = -\dfrac{(x^2+1)}{x(x^2-1)}$ and $Q(x) = \dfrac{(x^2+1)}{x^2(x^2-1)}$ are both infinite at $x = 0$.

Hence, the point $x = 0$ is a singular point of the given ODE. Further, we see that

$$[xP(x)]_{x\to 0} = 1 \,(\text{finite}) \text{ and } [x^2Q(x)]_{x\to 0} = -1 \,(\text{finite}).$$

Therefore, the point $x = 0$ is a regular singular point of the given ODE and, by Fuchs' theorem, it can have power series solutions. Let us assume a power series solution in the form

$$y = \sum_{r=0}^{\infty} a_r x^{k+r} \tag{i}$$

where k is a constant to be determined and $a_0 \neq 0$.

$$\frac{dy}{dx} = \sum_{r=0}^{\infty}(k+r)a_r x^{k+r-1}, \quad \frac{d^2y}{dx^2} = \sum_{k=0}^{\infty}(k+r)(k+r-1)a_r x^{k+r-2}.$$

Substitution of the expressions for y and its derivatives into the given ODE yields

$$\sum_{r=0}^{\infty}(k+r-1)^2 a_r x^{k+r+2} - \sum_{r=0}^{\infty}(k+r+1)(k+r-1)a_r x^{k+r} = 0.$$

Here, the coefficients of various powers of x must separately vanish.
Then,

Coefficient of x^k (lowest power):–

$$(k+1)(k-1)a_0 = 0, \; a_0 \neq 0 \quad [\text{Indicial equation}]$$

giving $k = k_1 = 1$ and $k = k_2 = -1$ such that $k_1 - k_2 = 2$ (integer).

Coefficient of x^{k+1} : $-(k+2)ka_1 = 0$,

giving $a_1 = 0$, since $(k+2)k$ is not zero at both $k = 1$ and $k = -1$;

and so on.

Coefficient of x^{k+r+2} : $(k+r-1)^2 a_r - (k+r+3)(k+r+1)a_{r+2} = 0$,

resulting in $a_{r+2} = \dfrac{(k+r-1)^2}{(k+r+3)(k+r+1)} a_r$ [Recursion relation]

Putting $r = 0, 1, 2, 3, 4, \ldots$ etc., the recursion relation yields

$$a_2 = \frac{(k-1)^2}{(k+3)(k+1)} a_0, \qquad a_3 = \frac{k^2}{(k+4)(k+2)} a_1 = 0 \text{ as } a_1 = 0;$$

$$a_4 = \frac{(k+1)^2(k-1)^2}{(k+5)(k+3)^2(k+1)} a_0, \qquad a_5 = 0;$$

$$a_6 = \frac{(k+3)^2(k+1)^2(k-1)^2}{(k+7)(k+5)^2(k+3)^2(k+1)} a_0, \qquad a_7 = 0; \text{ etc.}$$

Substituting these values of coefficients into Eq. (i), we obtain

$$y = y(x, k) = a_0 x^k \left[1 + \frac{(k-1)^2}{(k+3)(k+1)} x^2 + \frac{(k+1)^2(k-1)^2}{(k+5)(k+3)^2(k+1)} x^4 + \ldots \infty \right] \tag{ii}$$

as $a_1 = a_3 = a_5 = a_{2j+1} = \ldots = 0$ where $j = 0, 1, 2, 3, 4 \ldots$

First Solution

Substituting $k = k_1 = 1$ (greater root of indicial equation) into Eq. (ii), we obtain

$$y_1(x) = a_0 x[1 + 0 + 0 + 0 + \ldots] = a_0 x. \tag{iii}$$

If we take the arbitrary constant a_0 as unity, we have

$$y_1(x) = x \tag{iv}$$

Second Solution

Let us assume the second solution in the form

$$y_2(x) = Sy_1 \ln x + x^{k_2} \sum_{m=0}^{\infty} B_m x^m = Sx \ln x + \sum_{m=0}^{\infty} B_m x^{m-1}, \tag{v}$$

where S and B_m are constants to be determined.

We have

$$\frac{dy_2}{dx} = S \ln x + S + \sum_{m=0}^{\infty}(m-1)B_m x^{m-2}$$

and

$$\frac{d^2y_2}{dx^2} = \frac{S}{x} + \sum_{m=0}^{\infty}(m-1)(m-2)B_m x^{m-3}$$

As y_2 satisfies the given ODE, we have

$$(x^4 - x^2)\frac{d^2y_2}{dx^2} - (x^3 + x)\frac{dy_2}{dx} + (x^2 + 1)y_2 = 0. \quad \text{(vi)}$$

Substitution of the expressions for y_2 and its derivatives into Eq. (vi) results in

$$-2Sx - \sum_{m=0}^{\infty} m(m-2)B_m x^{m-1} + \sum_{m=0}^{\infty}(m-2)^2 B_m x^{m+1} = 0. \quad \text{(vii)}$$

In Eq. (vii), the coefficients of various powers of x must separately vanish. Hence,

Coefficient of x^0 : $-B_1 = 0$ giving $B_1 = 0$.

Coefficient of x^1 : $-2S - 0B_2 + 4B_0 = 0$, giving $S = 2B_0$ with arbitrary B_2.

Coefficient of x^2 : $-3B_3 + B_1 = 0$ giving $B_3 = \frac{1}{3}B_1 = 0$.

Coefficient of x^3 : $8B_4 + 0B_2 = 0$ giving $B_4 = 0$ with arbitrary B_2.

We see that in general,

Coefficient of x^m : $-(m+1)(m-1)B_{m+1} + (m-3)^2 B_{m-1} = 0$

$$\text{giving } B_{m+1} = \frac{(m-3)^2}{(m+1)(m-1)} B_{m-1}.$$

As B_3 and B_4 are equal to zero, putting $m = 4, 5, 6, 7,$ etc., we have

$$B_5 = 0 = B_6 = B_7 = B_8 = \ldots \text{etc.}$$

Substituting the values of S and other coefficients into Eq. (v), we obtain the second solution as

$$y_2(x) = 2B_0 x \ln x + B_0 x^{-1} + 0 + B_2 x + 0 = 2B_0 x \ln x + \frac{B_0}{x} + B_2 y_1 \text{ as } y_1 = x.$$

Since y_1 is an independent solution, it cannot be a part of y_2.
Hence, we have to choose $B_2 = 0$.
Therefore, we have

$$y_2(x) = 2B_0 x \ln x + \frac{B_0}{x}.$$

Choosing the arbitrary constant B_0 as $\frac{1}{2}$, we have

$$y_2(x) = x \ln x + \frac{1}{2x} \tag{viii}$$

General solution of the given ODE is the linear combination

$y = Ay_1 + By_2$ where A and B are arbitrary constants.

Alternative Approach for Finding the Second Solution

As the first solution ($y_1 = x$) is a simple one, it should be easier to find the second solution via the method of reduction of order (variation of parameter) as in the following.

Let us assume the second solution in the form

$$y_2(x) = u(x)y_1(x) = xu(x) \tag{ix}$$

where $u(x)$ is a function to be determined.

Now, we have

$$\frac{dy_2}{dx} = u + x\frac{du}{dx} \text{ and } \frac{d^2y_2}{dx^2} = 2\frac{du}{dx} + x\frac{d^2u}{dx^2}.$$

Since $y_2(x)$ is a solution of the given ODE, we must have

$$(x^4 - x^2)\frac{d^2y_2}{dx^2} - (x^3 + x)\frac{dy_2}{dx} + (x^2 + 1)y_2 = 0.$$

Substitution of the expressions for y_2 and its derivatives into this equation results in

$$(x^5 - x^3)\frac{d^2u}{dx^2} + (x^4 - 3x^2)\frac{du}{dx} = 0.$$

Putting $v = \frac{du}{dx}$, we write

$$(x^3 - x)\frac{dv}{dx} + (x^2 - 3)v = 0,$$

or

$$\frac{dv}{v} = \left(\frac{3 - x^2}{x^3 - x}\right)dx \tag{x}$$

Right-hand side of Eq. (x) can be written as

$$\left[\frac{3 - x^2}{x(x+1)(x-1)}\right]dx = \left[\frac{A}{x} + \frac{B}{x+1} + \frac{C}{x-1}\right]dx \text{ (say)},$$

where A, B, C are to be determined.

or
$$3-x^2=-A+(C-B)x+(A+B+C)x^2,$$

giving $3=-A, C-B=0, A+B+C=-1,$

i.e.,
$$A=-3, B=1, C=1.$$

Then, Eq. (x) becomes

$$\frac{dv}{v}=-\frac{3dx}{x}+\frac{dx}{x+1}+\frac{dx}{x-1}.$$

Integrating both sides and choosing the integration constant as zero, we obtain

$$\ln v=-3\ln x+\ln(x+1)+\ln(x-1)=\ln\left[\frac{(x+1)(x-1)}{x^3}\right],$$

giving $v=\dfrac{du}{dx}=\dfrac{x^2-1}{x^3}=\dfrac{1}{x}-\dfrac{1}{x^3};$

or
$$du=\frac{dx}{x}-x^{-3}dx.$$

Integrating both sides and choosing the integration constant as zero, we obtain

$$u=\ln x+\frac{1}{2x^2}$$

Substituting this into Eq. (ix), we obtain

$$y_2(x)=x\ln x+\frac{1}{2x},$$

which is Eq. (viii).

Another Alternative Approach for Finding the Second Solution

Second solution can also be obtained from Eq. (ii) as in the following.

With $k=k_2=-1$, coefficients become infinite starting from the second term.

To overcome the difficulty, let us put

$$a_0=R_0(k+1) \text{ where } R_0 \text{ is a constant.}$$

Then, Eq. (ii) becomes

$$y=y(x,k)=R_0x^k\left[(k+1)+\frac{(k-1)^2}{k+3}x^2+\frac{(k+1)^2(k-1)^2}{(k+5)(k+3)^2}x^4+\ldots\infty\right].$$

Then, the second solution is given by

$$y_2(x) = \left[\frac{\partial y}{\partial k}\right]_{k=k_2} = \left[\frac{\partial}{\partial k} y(x,k)\right]_{k=-1}$$

$$= \left[R_0 x^k \ln x \left\{ (k+1) + \frac{(k-1)^2}{k+3} x^2 + \frac{(k^2-1)^2}{(k+5)(k+3)^2} x^4 + \ldots \infty \right\} \right.$$

$$+ R_0 x^k \left\{ 1 + \frac{2(k-1)(k+3) - (k-1)^2}{(k+3)^2} x^2 \right.$$

$$\left.\left. + \frac{4k(k^2-1)(k+5)(k+3)^2 - (k^2-1)^2(k+3)^2 - 2(k^2-1)^2(k+5)(k+3)}{(k+5)^2(k+3)^4} x^4 + \ldots \infty \right\} \right]_{k=-1}$$

or

$$y_2(x) = \frac{R_0}{x} \ln x \{0 + 2x^2 + 0\} + \frac{R_0}{x} \{1 - 3x^2 + 0\}$$

$$= 2R_0 x \ln x + \frac{R_0}{x} - 3R_0 x = 2R_0 x \ln x + \frac{R_0}{x} - 3R_0 y_1$$

where the last term is a constant multiple of the first solution. As y_1 and y_2 are to be independent solutions, one solution should not contain a constant multiple of the other.

Hence, the constant multiple of y_1 contained in y_2 is neglected.

Hence, we write

$$y_2(x) = 2R_0 x \ln x + \frac{R_0}{x}.$$

Choosing the value of R_0 as $\frac{1}{2}$, we have

$$y_2(x) = x \ln x + \frac{1}{2x} \text{ which is Eq. (viii).}$$

PROBLEM 7 Discuss the nature of the point $x = 0$ in relation with the differential equation $x^2 \frac{d^2y}{dx^2} + x(x-3)\frac{dy}{dx} + 3y = 0$ and, if possible, solve it by Frobenius method.

Solution: The given ODE can be written in the standard form as

$$\frac{d^2y}{dx^2} + \frac{(x-3)}{x}\frac{dy}{dx} + \frac{3}{x^2} y = 0.$$

Here, $P(x) = \frac{x-3}{x}$ and $Q(x) = +\frac{3}{x^2}$ are not finite at $x = 0$ and hence the point $x = 0$ is a singular point of the given ODE.

We see that

$$[xP(x)]_{x\to 0} = -3 \text{ (finite) and } [x^2Q(x)]_{x\to 0} = +3 \text{ (finite)}.$$

Hence, the point $x = 0$ is a regular singular point of the given ODE.

Therefore, by Fuchs' theorem, the given ODE can have a power series solution and hence Frobenius method can be applied.

Let us assume a series solution in the form

$$y = \sum_{r=0}^{\infty} a_r x^{k+r}, \text{ [where } a_0 \neq 0 \text{ and } k \text{ is a constant to be determined]}. \quad \text{(i)}$$

Now, we have

$$\frac{dy}{dx} = \sum_{r=0}^{\infty}(k+r)a_r x^{k+r-1} \text{ and } \frac{d^2y}{dx^2} = \sum_{r=0}^{\infty}(k+r)(k+r-1)a_r x^{k+r-2}.$$

Substitution of the expressions for y and its derivatives into the given ODE yields

$$\sum_{r=0}^{\infty}(k+r-3)(k+r-1)a_r x^{k+r} + \sum_{r=0}^{\infty}(k+r)a_r x^{k+r+1} = 0.$$

In this equation, the coefficients of various powers of x must separately vanish.

Coefficient of x^k : $(k-3)(k-1)a_0 = 0$ with $a_0 \neq 0$ [Indicial equation]

giving $k = k_1 = 3$ and $k = k_2 = 1$.

Coefficient of x^{k+1} : $(k-2)ka_1 + ka_0 = 0$

giving $a_1 = -\frac{1}{k-2}a_0$.

In general, coefficient of

$$x^{k+r+1} : (k+r-2)(k+r)a_{r+1} + (k+r)a_r = 0$$

giving $a_{r+1} = -\frac{1}{(k+r-2)}a_r$. [Recursion relation]

Putting $r = 0, 1, 2, 3, 4, \ldots$; we obtain

$$a_1 = -\frac{1}{k-2}a_0, \qquad a_2 = -\frac{1}{k-1}a_1 = (-1)^2\frac{1}{(k-1)(k-2)}a_0,$$

$$a_3 = -\frac{1}{k}a_2 = (-1)^3\frac{1}{k(k-1)(k-2)}a_0, \qquad a_4 = (-1)^4\frac{1}{(k+1)k(k-1)(k-2)}a_0,$$

and so on.

Substituting these coefficients into Eq. (i), we have

$$y = y(x,k) = a_0 x^k\left[1 - \frac{1}{k-2}x + \frac{1}{(k-1)(k-2)}x^2 - \frac{1}{k(k-1)(k-2)}x^3 + \frac{1}{(k+1)k(k-1)(k-2)}x^4 - \ldots\infty\right]. \quad \text{(ii)}$$

First Solution

Substituting $k = k_1 = 3$ (larger root of indicial equation) into Eq. (ii), and choosing the arbitrary constant a_0 as unity, we obtain

$$y_1(x) = x^3\left[1 - x + \frac{1}{2!}x^2 - \frac{1}{3!}x^3 + \frac{1}{4!}x^4 - \frac{1}{5!}x^5 + \ldots\infty\right]$$

$$= x^3 e^{-x} \quad \text{(iii)}$$

Second Solution

Let us assume the second solution in the form

$$y_2(x) = u(x)\, y_1(x) \quad \text{(iv)}$$

where $u(x)$ is a function to be determined.

We have,

$$\frac{dy_2}{dx} = \frac{du}{dx}y_1 + u\frac{dy_1}{dx}, \quad \frac{d^2y_2}{dx^2} = \frac{d^2u}{dx^2}y_1 + 2\frac{du}{dx}\frac{dy_1}{dx} + u\frac{d^2y_1}{dx^2}$$

Substituting y_2 and its derivatives into the given ODE, we have

$$x^2\left(\frac{d^2u}{dx^2}y_1 + 2\frac{du}{dx}\frac{dy_1}{dx} + u\frac{d^2y_1}{dx^2}\right) + x(x-3)\left(\frac{du}{dx}y_1 + u\frac{dy_1}{dx}\right) + 3uy_1 = 0,$$

or

$$u\left[x^2\frac{d^2y_1}{dx^2} + x(x-3)\frac{dy_1}{dx} + 3y_1\right] + \left\{2x^2\frac{dy_1}{dx} + x(x-3)y_1\right\}\frac{du}{dx} + x^2 y_1\frac{d^2u}{dx^2} = 0.$$

As y_1 is a solution of the given ODE, the terms within the square bracket will vanish and substituting the expression for y_1 from Eq. (iii) in the above equation, we have

$$\left\{2x^2\frac{d}{dx}(x^3e^{-x}) + x(x-3)x^3e^{-x}\right\}\frac{du}{dx} + x^5e^{-x}\frac{d^2u}{dx^2} = 0.$$

Simplifying the equation and putting $v = \frac{du}{dx}$, we write

$$x\frac{dv}{dx} + (3 - x)v = 0,$$

or

$$\frac{dv}{v} = dx - 3\frac{dx}{x}.$$

Integrating both sides and choosing the integration constant as zero, we obtain

$$\ln v = x - 3\ln x,$$

or

$$v = \frac{du}{dx} = x^{-3}e^{x}.$$

Integrating further and choosing the integration constant as zero again, we obtain

$$u(x) = \int x^{-3}e^{x}dx = \int x^{-3}\left(1 + x + \frac{x^2}{2!} + \frac{x^3}{3!} + \frac{x^4}{4!} + \frac{x^5}{5!} + \ldots\infty\right)dx$$

$$= -\frac{x^{-2}}{2} - \frac{1}{x} + \frac{\ln x}{2!} + \frac{x}{3!} + \frac{x^2}{2(4!)} + \frac{x^3}{3(5!)} + \ldots\infty. \qquad \text{(v)}$$

Substituting Eqs. (iii) and (v) into Eq. (iv), we obtain the second solution as

$$y_2(x) = \left(-\frac{1}{2x^2} - \frac{1}{x} + \frac{1}{2}\ln x + \frac{x}{3!} + \frac{x^2}{2(4!)} + \frac{x^3}{3(5!)} + \ldots\infty\right)x^3e^{-x},$$

$$= \frac{1}{2}x^3e^{-x}\ln x + e^{-x}\left(-\frac{1}{2}x - x^2 + \frac{1}{3!}x^4 + \frac{1}{2(4!)}x^5 + \frac{1}{3(5!)}x^6 + \ldots\infty\right)$$

$$= \frac{1}{2}x^3e^{-x}\ln x + \left(1 - x + \frac{1}{2!}x^2 - \frac{1}{3!}x^3 + \frac{1}{4!}x^4 - \ldots\infty\right)$$

$$\times\left(-\frac{1}{2}x - x^2 + \frac{1}{6}x^4 + \frac{1}{48}x^5 + \frac{1}{360}x^6 + \ldots\infty\right),$$

$$= \frac{1}{2}x^3e^{-x}\ln x - \frac{1}{2}x + \left(-1 + \frac{1}{2}\right)x^2 + \left(1 - \frac{1}{4}\right)x^3 + \left(\frac{1}{6} - \frac{1}{2} + \frac{1}{12}\right)x^4 + \ldots\infty,$$

$$= \frac{1}{2}x^3e^{-x}\ln x - \frac{1}{2}x - \frac{1}{2}x^2 + \frac{3}{4}x^3 - \frac{1}{4}x^4 + \ldots\infty.$$

The general solution is the linear combination given by

$$y(x) = Ay_1(x) + By_2(x) \text{ where } A \text{ and } B \text{ are arbitrary constants.}$$

PROBLEM 8 Discuss the nature of the point $x = 1$ in respect of the following ordinary differential equation

$$x(1-x)\frac{d^2y}{dx^2} + 2(1-x)\frac{dy}{dx} + 2y = 0$$

and, if possible, obtain the series solutions of the differential equation about $x = 1$.

Solution: The given ODE can be written in the standard form as

$$\frac{d^2y}{dx^2} + \frac{2}{x}\frac{dy}{dx} + \frac{2}{x(1-x)}y = 0.$$

Here, at $x = 1$, although $P(x) = \frac{2}{x} = 2$ (finite), $Q(x) = \frac{2}{x(x-1)}$ is not finite.

Therefore, the point $x = 1$ is a singular point of the given ODE. But, we see that

$$[(x-1)P(x)]_{x\to 1} = 0 \text{ (finite) and } [(x-1)^2 Q(x)]_{x\to 1} = 0 \text{ (finite).}$$

Hence, the point $x = 1$ is a regular singular point of the given ODE and, therefore, by Fuchs' theorem, the given ODE has power series solutions.

Putting $t = x - 1$ or $x = t + 1$, the given ODE can be rewritten as

$$(1+t)(-t)\frac{d^2y}{dt^2} + 2(-t)\frac{dy}{dt} + 2y = 0,$$

or
$$t(1+t)\frac{d^2y}{dt^2} + 2t\frac{dy}{dt} - 2y = 0 \tag{i}$$

The solution of the given ODE about $x = 1$ is the same as the solution of Eq. (i) about $t = 0$. The point $(t = 0)$ is a regular singular point of the differential Eq. (i) and, therefore, by Fuchs' theorem, it will have power series solutions. Now, let us find the series solutions of Eq. (i) about $t = 0$.

Let us assume the series solution in the form

$$y = \sum_{r=0}^{\infty} a_r t^{k+r}, \text{ with } a_0 \neq 0, k \text{ being a constant to be determined.} \tag{ii}$$

Then,

$$\frac{dy}{dt} = \sum_{r=0}^{\infty}(k+r)a_r t^{k+r-1}, \text{ and } \frac{d^2y}{dt^2} = \sum_{r=0}^{\infty}(k+r)(k+r-1)a_r t^{k+r-2}.$$

Substituting these expressions for y and its derivatives into Eq. (i), we obtain

$$t(1+t)\sum_{r=0}^{\infty}(k+r)(k+r-1)a_r t^{k+r-2} + 2t\sum_{r=0}^{\infty}(k+r)a_r t^{k+r-1} - 2\sum_{r=0}^{\infty}a_r t^{k+r} = 0,$$

or
$$\sum_{r=0}^{\infty}(k+r)(k+r-1)a_r t^{k+r-1} + \sum_{r=0}^{\infty}(k+r-1)(k+r+2)a_r t^{k+r} = 0.$$

Here, the coefficients of various powers of t must separately vanish.

Coefficient of t^{k-1}: $k(k-1)a_0 = 0$ with $a_0 \neq 0$ [Indicial equation]

giving $k = k_1 = 1$ and $k = k_2 = 0$.

In general, coefficient of t^{k+r}:

$$(k+r+1)(k+r)a_{r+1} + (k+r-1)(k+r+2)a_r = 0,$$

giving
$$a_{r+1} = -\frac{(k+r-1)(k+r+2)}{(k+r)(k+r+1)}a_r \text{ [Recursion relation]}$$

Putting $r = 0, 1, 2, 3, 4,$ etc., we obtain

$$a_1 = -\frac{(k-1)(k+2)}{k(k+1)}a_0, \qquad a_2 = (-1)^2\frac{(k-1)(k+3)}{(k+1)^2}a_0,$$

$$a_3 = (-1)^3\frac{(k-1)(k+4)}{(k+1)(k+2)}a_0, \qquad a_4 = (-1)^4\frac{(k-1)(k+5)}{(k+1)(k+3)}a_0, \text{ etc.}$$

In general, we write,

$$a_n = (-1)^n\frac{(k-1)(k+n+1)}{(k+1)(k+n-1)}a_0, \; n = 0, 1, 2, 3, \ldots$$

Substitution of these coefficients into Eq. (ii) results in

$$y = a_0 t^k\left[1 - \frac{(k-1)(k+2)}{k(k+1)}t + \frac{(k-1)(k+3)}{(k+1)^2}t^2 - \frac{(k-1)(k+4)}{(k+1)(k+2)}t^3 + \ldots \right.$$
$$\left. + \ldots + (-1)^n\frac{(k-1)(k+n+1)}{(k+1)(k+n-1)}t^n + \ldots\infty\right]. \quad \text{(iii)}$$

First Solution

Substituting $k = k_1 = 1$ (greater root of indicial equation) in Eq. (iii), we obtain the first series solution of Eq. (i) as

$$y_1(t) = a_0 t[1 - 0] = a_0 t.$$

Starting from the second term, all terms vanish at $k = 1$.

Thus, the first power series solution of the given ODE about $x=1$ is given by

$$y_1(x)=a_0(x-1), \text{ as we put } t=x-1 \text{ earlier.}$$

Choosing the arbitrary constant a_0 as unity, we have

$$y_1(x) = x - 1 \tag{iv}$$

Second Solution

From Eq. (iii), we see that at $k=k_2=0$ (smaller root of indicial equation), the coefficients become infinite in the second term. To overcome the difficulty, let us put $a_0=R_0k$, where R_0 is a constant.

Then, Eq. (iii) can be written as

$$y=y(x,k)=R_0t^k\left[k-\frac{(k-1)(k+2)}{k+1}t+\frac{(k-1)k(k+3)}{(k+1)^2}t^2-\frac{(k-1)k(k+4)}{(k+1)(k+2)}t^3+\ldots\right.$$
$$\left.+(-1)^n\frac{(k+n+1)k(k-1)}{(k+n-1)(k+1)}t^n+\ldots\infty\right].$$

Now, the second series solution of Eq. (i) is given by

$$y_2(t)=\left[\frac{\partial y}{\partial k}\right]_{k=k_2}=\left[\frac{\partial}{\partial k}y(x,k)\right]_{k=0}$$

$$=\left[R_0t^k\ln t\left\{k-\frac{(k-1)(k+2)}{k+1}t+\sum_{n=2}^{\infty}(-1)^n\frac{(k-1)k(k+n+1)}{(k+n-1)(k+1)}t^n\right\}\right.$$

$$+R_0t^k\left\{1-\frac{(2k+1)(k+1)-(k-1)(k+2)}{(k+1)^2}t\right.$$

$$\left.\left.+\sum_{n=2}^{\infty}(-1)^n\frac{\{(k-1)k+(2k-1)(k+n+1)\}(k+n-1)(k+1)-(k-1)k(k+n+1)(2k+n)}{(k+n-1)^2(k+1)^2}t^n\right\}\right]_{k=0}$$

$$=R_0\ln t\,\{0+2t+0\}+R_0\left\{1-3t+\sum_{n=2}^{\infty}(-1)^n\left[\frac{-(n+1)}{(n-1)}\right]t^n\right\}$$

$$=R_0\left[2t\ln t+1-3t-\sum_{n=2}^{\infty}(-1)^n\frac{n+1}{n-1}t^n\right].$$

The second power series solution of the given ODE about $x=1$ is then given by

$$y_2(x)=R_0\left[2(x-1)\ln(x-1)+1-3(x-1)-\sum_{n=2}^{\infty}(-1)^n\frac{(n+1)}{n-1}(x-1)^n\right].$$

Choosing the arbitrary constant R_0 as unity, we write

$$y_2(x) = 2(x-1)\ln(x-1) + 1 - 3(x-1) - \sum_{n=2}^{\infty}(-1)^n \frac{(n+1)}{(n-1)}(x-1)^n.$$

The general solution of the given ODE is given by

$$y(x) = Ay_1 + By_2$$

where A and B are arbitrary constants.

[Try with another approach]

PROBLEM 9 In the case of the following ODE

$$\frac{d^2y}{dx^2} + 2x\frac{dy}{dx} + 4y = 0,$$

what is the nature of the point $x = 0$? If possible, obtain its power series solutions about $x = 0$.

Solution: At $x = 0$, both $P(x) = 2x$ and $Q(x) = 4$ are finite. Therefore, the point $x = 0$ is an ordinary point of the given ODE and hence the given ODE will have power series solutions about $x = 0$. Let us assume a solution in the form

$$y = \sum_{r=0}^{\infty} a_r x^{k+r} \tag{i}$$

where k is a constant to be determined and a_0 has non-zero value.

We have

$$\frac{dy}{dx} = \sum_{r=0}^{\infty}(k+r)a_r x^{k+r-1}, \quad \frac{d^2y}{dx^2} = \sum_{r=0}^{\infty}(k+r)(k+r-1)a_r x^{k+r-2}.$$

Substitution of the expressions for y and its derivatives into the given ODE results in

$$\sum_{r=0}^{\infty}(k+r)(k+r-1)a_r x^{k+r-2} + \sum_{r=0}^{\infty}2(k+r+2)a_r x^{k+r} = 0.$$

In this equation the coefficients of various powers of x must separately vanish.

Coefficient of x^{k-2}: $k(k-1)a_0 = 0$, with $a_0 \neq 0$ [Indicial equation]

giving $k = k_1 = 1$ and $k = k_2 = 0$.

Coefficient of x^{k-1}: $(k+1)ka_1 = 0$

giving $a_1 = 0$ for $k = 1$ and $a_1 =$ indeterminate for $k = 0$.

In general, coefficient of x^{k+r}: $(k+r+2)(k+r+1)a_{r+2}+2(k+r+2)a_r=0$

$$\text{giving } a_{r+2}=-\frac{2}{k+r+1}a_r. \text{ [Recursion relation]}$$

Putting $r=0,1,2,3,4,\ldots$ etc., we have

$$a_2=-\frac{2}{k+1}a_0, \qquad a_3=-\frac{2}{k+2}a_1,$$

$$a_4=-\frac{2}{k+3}a_2=(-1)^2\frac{2^2}{(k+3)(k+1)}a_0, \qquad a_5=(-1)^2\frac{2^2}{(k+4)(k+2)}a_1,$$

$$a_6=(-1)^3\frac{2^3}{(k+5)(k+3)(k+1)}a_0, \qquad a_7=(-1)^3\frac{2^3}{(k+6)(k+4)(k+2)}a_1, \text{ etc.}$$

Substituting these coefficients into Eq. (i) , we obtain

$$y=y(x,k)$$

$$=a_0x^k\left\{1-\frac{2}{k+1}x^2+\frac{2^2}{(k+3)(k+1)}x^4-\frac{2^3}{(k+5)(k+3)(k+1)}x^6+\ldots\infty\right\}$$

$$+a_1x^k\left\{x-\frac{2}{k+2}x^3+\frac{2^2}{(k+4)(k+2)}x^5-\frac{2^3}{(k+6)(k+4)(k+2)}x^7+\ldots\infty\right\}.$$

We have earlier seen that a_1 becomes indeterminate at $k=k_2=0$, the smaller root of indicial equation.

Thus, the general solution of the given ODE is given by

$$y=y(x)=[y(x,k)]_{k=0}=a_0\left\{1-2x^2+\frac{2^2}{1\cdot 3}x^4-\frac{2^3}{1\cdot 3\cdot 5}x^6+\ldots\infty\right\}$$

$$+a_1\left\{x-x^3+\frac{1}{2!}x^5-\frac{1}{3!}x^7+\ldots\infty\right\}. \qquad \text{(ii)}$$

We may write

$$y(x)=a_0y_1+a_1y_2$$

where a_0 and a_1 are arbitrary constants and

$$y_1=\left\{1-2x^2+\frac{2^2}{1\cdot 3}x^4-\frac{2^3}{1\cdot 3\cdot 5}x^6+\ldots\infty\right\} \text{ and } y_2=\left\{x-x^3+\frac{1}{2!}x^5-\frac{1}{3!}x^7+\ldots\infty\right\}.$$

Here, y_1 and y_2 are two independent solutions of the given ODE.

If one finds the solution with $k = k_1 = 1$ (the greater root of indicial equation), he will find a constant multiple of y_1.

PROBLEM 10 What is the nature of the point $x = 0$ in the case of the differential equation

$$x\frac{d^2y}{dx^2} + 2\frac{dy}{dx} + xy = 0?$$

Obtain power series solutions of the given ODE if it exists.

Solution: The given ODE can be written in standard form as

$$\frac{d^2y}{dx^2} + \frac{2}{x}\frac{dy}{dx} + y = 0.$$

At $x = 0$, $P(x) = \frac{2}{x}$ is not finite although $Q(x) = 1$ (finite).

Therefore, the point $x = 0$ is a singular point of the given ODE.

Here, we see that $[xP(x)]_{x\to 0} = 2$ (finite) and $[x^2Q(x)]_{x\to 0} = 0$ (finite). Hence, by Fuchs' theorem, the given ODE will have power series solutions. Let us assume a solution in the form

$$y = \sum_{r=0}^{\infty} a_r x^{k+r} \tag{i}$$

Here, $a_0 \neq 0$ and k is a constant to be determined.
Now, we have

$$\frac{dy}{dx} = \sum_{r=0}^{\infty}(k+r)a_r x^{k+r-1} \text{ and } \frac{d^2y}{dx^2} = \sum_{r=0}^{\infty}(k+r)(k+r-1)a_r x^{k+r-2}.$$

Substitution of the expressions for y and its derivatives into the given ODE results in

$$\sum_{r=0}^{\infty}(k+r)(k+r+1)a_r x^{k+r-1} + \sum_{r=0}^{\infty}a_r x^{k+r+1} = 0.$$

In this equation, the coefficients of various powers of x must separately vanish.

Coefficient of x^{k-1} : $k(k+1)a_0 = 0$, with $a_0 \neq 0$ [Indicial equation]
giving $k = k_1 = 0$ and $k = k_2 = -1$

Coefficient of x^k : $(k+1)(k+2)a_1 = 0$
showing that a_1 is indeteminate at $k = -1$ as $k+2$ is non-zero.

In general, coefficient of x^{k+r+1}: $(k+r+2)(k+r+3)a_{r+2}+a_r=0$,

$$\text{giving } a_{r+2}=-\frac{1}{(k+r+2)(k+r+3)}a_r. \text{ [Recursion relation]}$$

Putting $r=0,1,2,3,4,\dots$ etc., we obtain

$$a_2=-\frac{1}{(k+3)(k+2)}a_0,$$

$$a_3=-\frac{1}{(k+4)(k+3)}a_1,$$

$$a_4=-\frac{1}{(k+5)(k+4)}a_2=(-1)^2\frac{1}{(k+5)(k+4)(k+3)(k+2)}a_0,$$

$$a_5=(-1)^2\frac{1}{(k+6)(k+5)(k+4)(k+3)}a_1,$$

$$a_6=(-1)^3\frac{1}{(k+7)(k+6)(k+5)(k+4)(k+3)(k+2)}a_0,$$

$$a_7=(-1)^3\frac{1}{(k+8)(k+7)(k+6)(k+5)(k+4)(k+3)}a_1, \text{ etc.}$$

Substituting these coefficients into Eq. (i), we obtain

$$\begin{aligned} y &= y(x,k) \\ &= a_0x^k\left[1-\frac{1}{(k+3)(k+2)}x^2+\frac{1}{(k+5)(k+4)(k+3)(k+2)}x^4\right. \\ &\qquad \left.-\frac{1}{(k+7)(k+6)(k+5)(k+4)(k+3)(k+2)}x^6+\dots\infty\right] \\ &\quad +a_1x^k\left[x-\frac{1}{(k+4)(k+3)}x^3+\frac{1}{(k+6)(k+5)(k+4)(k+3)}x^5\right. \\ &\qquad \left.-\frac{1}{(k+8)(k+7)(k+6)(k+5)(k+4)(k+3)}x^7+\dots\infty\right]. \end{aligned}$$

Since a_1 is indeterminate at $k=k_2=-1$ (the smaller root of the indicial equation), the general solution of the given ODE is given by

$$y(x)=[y(x,k)]_{k=-1}$$

$$= a_0 x^{-1}\left\{1-\frac{x^2}{2!}+\frac{x^4}{4!}-\frac{x^6}{6!}+\ldots\infty\right\}+a_1 x^{-1}\left\{x-\frac{x^3}{3!}+\frac{x^5}{5!}-\frac{x^7}{7!}+\ldots\infty\right\}$$

$$= a_0 y_1 + a_1 y_2,$$

where a_0 and a_1 are arbitrary constants and

$$y_1 = x^{-1}\left\{1-\frac{x^2}{2!}+\frac{x^4}{4!}-\frac{x^6}{6!}+\ldots\infty\right\}=\frac{\cos x}{x}$$

and
$$y_2 = x^{-1}\left\{x-\frac{x^3}{3!}+\frac{x^5}{5!}-\frac{x^7}{7!}+\ldots\infty\right\}=\frac{\sin x}{x}$$

If one finds the solution with $k = k_1 = 0$ (the greater root of indicial equation), he will find a constant multiple of y_1.

PROBLEM 11 Obtain the series solution about $x = 0$ of the differential equation

$$\frac{d^2y}{dx^2} - y = x.$$

Solution: In the case of an ODE having a forcing function, finding its solution involves two steps: (i) finding the complementary function (C.F.), and (ii) finding the particular integral (P. I.).

To Find the Complementary Function

Here, the C.F. denoted by y_c is the solution of the homogeneous equation

$$\frac{d^2y}{dx^2} - y = 0 \tag{i}$$

Let us assume the series solution of Eq. (i) as

$$y_c = \sum_{r=0}^{\infty} a_r x^{k+r},\ a_0 \neq 0 \tag{ii}$$

Substituting this expression into Eq. (i) and executing the differentiation, we obtain

$$\sum_{r=0}^{\infty}(k+r)(k+r-1)a_r x^{k+r-2} - \sum_{r=0}^{\infty} a_r x^{k+r} = 0.$$

In this equation, the coefficients of various powers of x must separately vanish.

Coefficient of x^{k-2} : $k(k-1)a_0 = 0, a_0 \neq 0$ [Indicial equation]

giving $k = k_1 = 1$ and $k = k_2 = 0$.

Coefficient of x^{k-1} : $(k+1)ka_1 = 0$, indicating that $a_1 = 0$ for $k = k_1 = 1$,

and a_1 is indeterminate for $k = k_2 = 0$.

In general, the coefficient of x^{k+r} is

$$(k+r+2)(k+r+1)a_{r+2} - a_r = 0,$$

$$\text{giving } a_{r+2} = \frac{1}{(k+r+1)(k+r+2)}a_r.$$

Putting $r = 0, 1, 2, 3, 4,$ etc., we have

$$a_2 = \frac{1}{(k+1)(k+2)}a_0, \qquad a_3 = \frac{1}{(k+2)(k+3)}a_1,$$

$$a_4 = \frac{1}{(k+1)(k+2)(k+3)(k+4)}a_0, \qquad a_5 = \frac{1}{(k+2)(k+3)(k+4)(k+5)}a_1$$

$$a_6 = \frac{1}{(k+1)(k+2)(k+3)(k+4)(k+5)(k+6)}a_0, \quad a_7 = \frac{1}{(k+2)(k+3)(k+4)(k+5)(k+6)(k+7)}a_1,$$

etc.

Substitution of these coefficients into Eq. (ii) yields

$$y_c = y_c(x, k)$$

$$= a_0 x^k \left[\left[1 + \frac{x^2}{(k+1)(k+2)} + \frac{x^4}{(k+1)(k+2)(k+3)(k+4)} + \ldots \infty\right]\right.$$

$$\left. + a_1 x^k \left[x + \frac{x^3}{(k+2)(k+3)} + \frac{x^5}{(k+2)(k+3)(k+4)(k+5)} + \ldots \infty\right]\right]. \quad \text{(iii)}$$

Since a_1 is indeterminate at $k = k_2 = 0$, the general solution of Eq. (i), i.e., the C.F. of the given ODE is given by

$$y_c = [y_c(x, k)]_{k=k_2=0}$$

$$= a_0\left\{1 + \frac{x^2}{2!} + \frac{x^4}{4!} + \frac{x^6}{6!} + \ldots \infty\right\} + a_1\left\{x + \frac{x^3}{3!} + \frac{x^5}{5!} + \frac{x^7}{7!} + \ldots \infty\right\}. \quad \text{(iv)}$$

To Find the Particular Integral

Let us assume the particular integral of the given ODE as

$$y_p = \sum_{r=0}^{\infty} A_r x^{m+r} = x^m (A_0 + A_1 x + A_2 x^2 + A_3 x^3 + \ldots \infty), \tag{v}$$

where m is a constant to be determined and $A_0 \neq 0$.

Substitution of this expression in the given ODE and execution of the differentiation results in

$$\sum_{r=0}^{\infty} (m+r)(m+r-1) A_r x^{m+r-2} - \sum_{r=0}^{\infty} A_r x^{m+r} = x. \tag{vi}$$

In the left-hand side of Eq. (vi), the term with the lowest power of x must be x and the coefficients of other powers of x must vanish. This implies that

$$m - 2 = 1, \text{ or, } m = 3.$$

Also, $m(m-1)A_0 = 1$, giving $A_0 = \dfrac{1}{m(m-1)} = \dfrac{1}{3(2)} = \dfrac{1}{3!}$.

Now, Eq. (vi) becomes

$$\sum_{r=0}^{\infty} (3+r)(2+r) A_r x^{1+r} - \sum_{r=0}^{\infty} A_r x^{3+r} = x. \tag{vii}$$

From Eq. (vii), we have

Coefficient of x^2 : $(3+1)(2+1)A_1 = 0$ indicating that $A_1 = 0$.

Coefficient of x^3 : $(3+2)(2+2)A_2 - A_0 = 0$,

$$\text{giving } A_2 = \frac{1}{4 \times 5} A_0 = \frac{1}{5!}.$$

Coefficient of x^4 : $(6)(5)\, A_3 - A_1 = 0$, giving $A_3 = \dfrac{A_1}{(5)(6)} = 0$.

Coefficient of x^5 : $(7)(6)\, A_4 - A_2 = 0$, giving $A_4 = \dfrac{A_2}{6 \times 7} = \dfrac{1}{7!}$, etc.

Substituting the values of these coefficients and also the value of m in Eq. (v), we obtain

$$y_p = x^3 \left(\frac{1}{3!} + 0 + \frac{1}{5!} x^2 + 0 + \frac{1}{7!} x^4 + \ldots \infty \right)$$

$$= \frac{x^3}{3!} + \frac{x^5}{5!} + \frac{x^7}{7!} + \ldots \infty.$$

Thus, the general solution of the given ODE is given by

$$y = y_c + y_p$$

$$= a_0\left(1+\frac{x^2}{2!}+\frac{x^4}{4!}+\frac{x^6}{6!}+\ldots\infty\right)+a_1\left(x+\frac{x^3}{3!}+\frac{x^5}{5!}+\frac{x^7}{7!}+\ldots\infty\right)+\left(\frac{x^3}{3!}+\frac{x^5}{5!}+\frac{x^7}{7!}+\ldots\infty\right).$$

where a_0 and a_1 are arbitrary constants.

EXERCISE

1. Determine whether the following series are convergent or divergent.

 (i) $\frac{1}{1.2}+\frac{1}{2.3}+\frac{1}{3.4}+\frac{1}{4.5}+\ldots\infty.$

 (ii) $\frac{1}{\sqrt{2}}+\frac{1}{\sqrt{3}}+\frac{1}{\sqrt{4}}+\frac{1}{\sqrt{5}}+\ldots\infty.$

 (iii) $\frac{1}{\ln 2}+\frac{1}{\ln 3}+\frac{1}{\ln 4}+\frac{1}{\ln 5}+\ldots\infty.$

2. Determine the nature of the point $x=0$ in respect of the following ordinary differential equations and obtain their power series solutions if they exist.

 (i) $2x\frac{d^2y}{dx^2}+(x+1)\frac{dy}{dx}+3y=0.$

 Ans. $y_1 = x^{\frac{1}{2}}\left(1-\frac{7}{6}x+\frac{21}{40}x^2-\frac{11}{80}x^3+\ldots\infty\right)$, $y_2 = 1-3x+2x^2-\frac{2}{3}x^3+\ldots\infty.$

 (ii) $4x\frac{d^2y}{dx^2}+2\frac{dy}{dx}-y=0.$

 Ans. $y_1 = x^{\frac{1}{2}}\sum_{n=0}^{\infty}\frac{x^n}{(2n+1)!}$, $y_2 = \sum_{n=0}^{\infty}\frac{x^n}{(2n)!}.$

 (iii) $4x\frac{d^2y}{dx^2}+2\frac{dy}{dx}+y=0.$

 Ans. $y_1 = x^{\frac{1}{2}}\sum_{n=0}^{\infty}\frac{(-1)^n x^n}{(2n+1)!} = \sin\sqrt{x}$, $y_2 = \sum_{n=0}^{\infty}\frac{(-1)^n x^n}{(2n)!} = \cos\sqrt{x}.$

 (iv) $4x\frac{d^2y}{dx^2}+6\frac{dy}{dx}-y=0$

 Ans. $y_1 = \sum_{n=0}^{\infty}\frac{x^n}{(2n+1)!}$, $y_2 = x^{-\frac{1}{2}}\sum_{n=0}^{\infty}\frac{x^n}{(2n)!}.$

(v) $9x(1-x)\frac{d^2y}{dx^2}-12\frac{dy}{dx}+4y=0.$

$$\textbf{Ans. } y_1=1+\frac{1}{3}x+\frac{1\cdot4}{3\cdot6}x^2+\frac{1\cdot4\cdot7}{3\cdot6\cdot9}x^3+\ldots\infty,$$

$$y_2=x^{\frac{7}{3}}\left(1+\frac{8}{10}x+\frac{8\cdot11}{10\cdot13}x^2+\frac{8\cdot11\cdot14}{10\cdot13\cdot16}x^3+\ldots\infty\right).$$

(vi) $2x^2\frac{d^2y}{dx^2}+x\frac{dy}{dx}-(x+1)y=0.$

$$\textbf{Ans. } y_1=x\left(1+\frac{x}{5}+\frac{x^2}{5\cdot14}+\frac{x^3}{5\cdot14\cdot27}+\frac{x^4}{5.14\cdot27\cdot44}+\ldots\infty\right),$$

$$y_2=x^{-\frac{1}{2}}\left(1-x-\frac{x^2}{2}-\frac{x^3}{2\cdot9}-\frac{x^4}{2\cdot9\cdot20}+\ldots\infty\right).$$

(vii) $2x(1-x)\frac{d^2y}{dx^2}+(1-x)\frac{dy}{dx}+3y=0.$

$$\textbf{Ans. } y_1=x^{\frac{1}{2}}(1-x),\ y_2=1-3x+\frac{3x^2}{1.3}+\frac{3x^3}{3.5}+\frac{3x^4}{5.7}+\ldots\infty.$$

(viii) $3x\frac{d^2y}{dx^2}+2\frac{dy}{dx}+y=0.$

$$\textbf{Ans. } y_1=1-\frac{x}{4}+\frac{x^2}{4\cdot14}-\frac{x^3}{4\cdot14\cdot30}+\ldots\infty,\ y_2=1-\frac{x}{2}+\frac{x^2}{2!2\cdot5}-\frac{x^3}{3!2\cdot5\cdot8}\ldots\infty.$$

(ix) $x\frac{d^2y}{dx^2}+\frac{dy}{dx}-xy=0.$

$$\textbf{Ans. } y_1=\sum_{n=0}^{\infty}\frac{\left(\frac{x}{2}\right)^{2n}}{(n!)^2},\ y_2=y_1\ln x-\frac{1}{4}x^2-\frac{3}{8.16}x^4+\ldots$$

(x) $x(1-x)\frac{d^2y}{dx^2}+(1-5x)\frac{dy}{dx}-4y=0.$

$$\textbf{Ans. } y_1=\sum_{n=0}^{\infty}(n+1)^2x^n,\ y_2=y_1\ln x-2\sum_{n=1}^{\infty}n(n+1)x^n.$$

(xi) $x\frac{d^2y}{dx^2}+\frac{dy}{dx}-y=0.$

$$\textbf{Ans. } y_1=\sum_{n=0}^{\infty}\frac{x^n}{(n!)^2},$$

$$y_2 = y_1 \ln x - 2\left[x + \frac{1}{(2!)^2}\left(1+\frac{1}{2}\right)x^2 + \frac{1}{(3!)^2}\left(1+\frac{1}{2}+\frac{1}{3}\right)x^3 + \ldots \infty\right].$$

(xii) $x\dfrac{d^2y}{dx^2} + (1+2x)\dfrac{dy}{dx} + (1+x)y = 0.$

Ans. $y_1 = e^{-x},\ y_2 = e^{-x} \ln x.$

(xiii) $x(1-x)\dfrac{d^2y}{dx^2} + (1-x)\dfrac{dy}{dx} - y = 0.$

Ans. $y_1 = 1 + x + \dfrac{2}{2^2}x^2 + \dfrac{2\cdot 5}{2^2\cdot 3^2}x^3 + \ldots \infty,$

$$y_2 = y_1 \ln x - \left(2x + x^2 + \frac{14}{27}x^3 + \ldots \infty\right).$$

(xiv) $x\dfrac{d^2y}{dx^2} + \dfrac{dy}{dx} + x^2 y = 0.$

Ans. $y_1 = 1 - \dfrac{x^3}{3^2} + \dfrac{x^6}{3^2 6^2} - \dfrac{x^9}{3^2 6^2 9^2} + \ldots \infty,$

$$y_2 = y_1 \ln x + 2\left\{\frac{x^3}{3^3} - \left(\frac{1}{3}+\frac{1}{6}\right)\frac{x^6}{3^2 6^2} + \left(\frac{1}{3}+\frac{1}{6}+\frac{1}{9}\right)\frac{x^9}{3^2 6^2 9^2} + \ldots \infty\right\}.$$

(xv) $x\dfrac{d^2y}{dx^2} + 5\dfrac{dy}{dx} + xy = 0.$

Ans. $y_1 = 1 - \dfrac{x^2}{2\cdot 6} + \dfrac{x^4}{2\cdot 4\cdot 6\cdot 8} - \dfrac{x^6}{2\cdot 4\cdot 6^2\cdot 8\cdot 10} + \ldots \infty,$

$$y_2 = -\frac{1}{16}y_1 \ln x + x^{-4}\left(1 + \frac{x^2}{2^2} + \frac{x^4}{2^2 4^2} + \ldots \infty\right).$$

(xvi) $x^2\dfrac{d^2y}{dx^2} + x\dfrac{dy}{dx} + (x-1)y = 0.$

Ans. $y_1 = \displaystyle\sum_{n=1}^{\infty}(-1)^{n-1}\frac{2x^n}{(n-1)!(n+1)!},\ y_2 = \frac{1}{4}y_1 \ln x + \left(-\frac{1}{2x} - \frac{5}{6} + \frac{29}{144}x + \ldots \infty\right).$

(xvii) $x\dfrac{d^2y}{dx^2} + y = 0.$

Ans. $y_1 = \displaystyle\sum_{n=0}^{\infty}\frac{(-1)^n x^{n+1}}{(n!)^2(n+1)},\ y_2 = -y_1 \ln x + \left(1 + x - \frac{5}{4}x^2 + \ldots \infty\right).$

(xviii) $(x^2+2)\dfrac{d^2y}{dx^2}+x\dfrac{dy}{dx}+(1+x)y=0.$

$$\textbf{Ans. } y = A\left(1-\frac{x^2}{4}-\frac{x^3}{12}+\frac{5x^4}{96}+\ldots\infty\right)+B\left(x-\frac{x^3}{6}-\frac{x^4}{24}+\ldots\infty\right),$$

where A and B are arbitrary constants.

(xix) $x^2\dfrac{d^2y}{dx^2}+4x\dfrac{dy}{dx}+(x^2+2)y=0.$ Ans.: $y=x^{-2}(A\cos x+B\sin x).$

(xx) $\dfrac{d^2y}{dx^2}-2x\dfrac{dy}{dx}+4y=0.$ [Hermite Eqn. of 2nd order]

$$\textbf{Ans. } y=A(1-2x^2)+B\left(x-\frac{2}{3!}x^3-\frac{2^2}{5!}x^5-\frac{2^3\cdot 3}{7!}x^7-\ldots\infty\right).$$

(xxi) $2x^2\dfrac{d^2y}{dx^2}+3x\dfrac{dy}{dx}-(1+x)y=0.$

$$\textbf{Ans. } y=A\left(x^{\frac{1}{2}}+\frac{1}{5}x^{\frac{3}{2}}+\frac{1}{70}x^{\frac{5}{2}}+\frac{1}{1890}x^{\frac{7}{2}}+\ldots\right)+B\left(x^{-1}-1-\frac{1}{2}x-\frac{1}{18}x^2-\ldots\right).$$

(xxii) $x\dfrac{d^2y}{dx^2}+\dfrac{dy}{dx}+xy=0.$

$$\textbf{Ans. } y=A\sum_{m=0}^{\infty}\frac{(-1)^m}{2^{2m}(m!)^2}x^{2m}$$

$$+B\left[\left\{\sum_{m=0}^{\infty}\frac{(-1)^m}{2^{2m}(m!)^2}x^{2m}\right\}\ln x+\frac{x^2}{2^2}-\frac{\left(1+\frac{1}{2}\right)}{2^2\cdot 4^2}x^4+\frac{\left(1+\frac{1}{2}+\frac{1}{3}\right)}{2^2\cdot 4^2\cdot 6^2}x^6-\cdots\right]$$

3. Determine the nature of the point $x=1$ in the case of the ordinary differential equation

$$x(x-1)\frac{d^2y}{dx^2}+3\frac{dy}{dx}-2y=0$$

and hence obtain the series solution about that point , if it exists.

$$\textbf{Ans. } y=A[(x-1)^{-2}+4(x-1)^{-1}]+B\left[1+\frac{2}{3}(x-1)+\frac{1}{6}(x-1)^2\right].$$

4. Obtain the power series solution, if it exists, in the case of the differential equation

$$2x^2\frac{d^2y}{dx^2}-x\frac{dy}{dx}+(1-x^2)y=x^2 \text{ about the point } x=0.$$

Ans. $$y=Ax\left[1+\frac{x^2}{2\cdot 5}+\frac{x^4}{2\cdot 4\cdot 5\cdot 9}+\frac{x^6}{2\cdot 4\cdot 6\cdot 5\cdot 9\cdot 13}+\ldots\infty\right]$$

$$+Bx^{\frac{1}{2}}\left[1+\frac{x^2}{2\cdot 3}+\frac{x^4}{2\cdot 4\cdot 3\cdot 7}+\frac{x^6}{2\cdot 4\cdot 6\cdot 3\cdot 7\cdot 11}+\ldots\infty\right]$$

$$+\frac{1}{3}x^2\left[1+\frac{x^2}{3\cdot 7}+\frac{x^4}{3\cdot 7\cdot 5\cdot 11}+\ldots\infty\right].$$

2

CHAPTER

Legendre Polynomials

2.1 INNER OR SCALAR PRODUCT OF FUNCTIONS

We know that in vector analysis, the scalar or dot product of two vectors **A** and **B** in three dimensions is defined by

$$\mathbf{A}\cdot\mathbf{B}=\sum_{j=1}^{3}A_jB_j$$

where A_1, A_2, A_3 are the three components of the vector **A** along the three co-ordinate axes and similarly for **B** also.

With an extension for n-dimensional vectors. We write

$$\mathbf{A}\cdot\mathbf{B}=\sum_{j=1}^{n}A_jB_j.$$

As a generalization, we are to include cases in which the magnitude of a vector is a function of a complex variable z and we write

$$\mathbf{A}\cdot\mathbf{B}=\sum_{j=1}^{n}A_j^*(z)B_j(z)$$

in order to ensure that the scalar product of a vector with itself be non-negative. Here, the asterisk symbol denotes complex conjugation.

If $A_j(z)$ and $B_j(z)$ are continuous on a given interval, the summation can be replaced by integration on that interval.

Thus, in a similar fashion, the *scalar or inner product* of two functions $f(\xi)$ and $g(\xi)$ in a given interval $a\le\xi\le b$ is defined by

$$(f,g)=\int_a^b f^*(\xi)\,g(\xi)\,d\xi \tag{2.1}$$

In writing scalar products, the dot notation is generally used in the case of vectors whereas the parenthesis notation is used in the case of functions.

Norm of a Function

In vector analysis, we know that the norm of a vector **A** is defined as the non-negative square root of the scalar product of the vector with itself.

Analogously, the *norm of a function*, $f(\xi)$, is defined as the non-negative square root of the scalar product of the function with itself in a given interval say, $a \le \xi \le b$. It is generally denoted by $\|f\|$. Thus, we write

$$\|f\| = \sqrt{(f,f)} = \left[\int_a^b f^*(\xi)f(\xi)d\xi\right]^{\frac{1}{2}} \tag{2.2}$$

Orthogonal Functions

In the case of vectors, two vectors are said to be orthogonal to each other if their scalar product vanishes. Thus, two vectors **A** and **B** are orthogonal to each other if $\mathbf{A}\cdot\mathbf{B} = 0$

Analogously, two functions, $f(\xi)$ and $g(\xi)$, are said to be *orthogonal* in an interval $a \le \xi \le b$ if their scalar product vanishes, i.e., if

$$(f, g) = \int_a^b f^*(\xi)g(\xi)d\xi = 0 \tag{2.3}$$

Normalized Functions

A function, $f(\xi)$, is said to be *normalized* in an interval $a \le \xi \le b$ if its scalar product with itself is unity, i.e. if

$$(f, f) = \int_a^b f^*(\xi)f(\xi)d\xi = 1 \tag{2.4}$$

Normalization and Normalization Constant

If any function, $g(\xi)$, which is not normalized, can be made normalized by the multiplication of a suitable constant, the process is called *normalization* and the multiplying constant is called *normalization constant*. This multiplying constant is nothing but the reciprocal of the norm of the function and hence the process is termed *normalization*.

If a function $g(\xi)$ is such that

$$(g, g) = \int_a^b g^*(\xi)g(\xi)d\xi = |N| \quad \text{(a constant)}$$

Then, we can write

$$\int_a^b \left[\frac{1}{\sqrt{N^*}} g^*(\xi)\right]\left[\frac{1}{\sqrt{N}} g(\xi)\right] d\xi = 1,$$

or

$$\int_a^b f^*(\xi) f(\xi)\, d\xi = 1 \tag{2.5}$$

where

$$f(\xi) = \frac{1}{\sqrt{N}} g(\xi)$$

The multiplying constant $\frac{1}{\sqrt{N}}$ is called the normalization constant of the particular function, $g(\xi)$. Thus, the *normalization constant* of a function is defined as the reciprocal of the norm of that function.

Orthogonal Set of Functions

A set of functions, $g_1(\xi), g_2(\xi), g_3(\xi)\ldots$, is said to be an *orthogonal set* of functions in an interval $(a \le \xi \le b)$ if these functions are defined in that interval and if, in the given interval, all the scalar products of distinct pairs of these functions in that set are all zero; that is, if

$$(g_m, g_n) = \int_a^b g_m^*(\xi) g_n(\xi) d\xi = 0, \text{ with } m \ne n \tag{2.6}$$

$$m = 1, 2, 3 \ldots, n = 1, 2, 3, \ldots$$

Example

The functions, $g_m(x) = \sin mx$ with $m = 1, 2, 3, 4, \ldots,$ form an orthogonal set in the interval $-\pi \le x \le \pi$; since

$$(g_m, g_n) = \int_{-\pi}^{+\pi} \sin mx \sin nx\, dx$$

$$= \frac{1}{2}\int_{-\pi}^{+\pi} \cos(m-n)x dx - \frac{1}{2}\int_{-\pi}^{+\pi} \cos(m+n)x dx = 0 \tag{2.7}$$

Orthonormal Set of Functions

If every function of an orthogonal set of functions is normalized by dividing by the corresponding norm, then the resulting set of functions is called an *orthonormal set* of functions. The word *orthonormal* means orthogonal and also normalized. Thus, an *orthonormal set* of functions is defined as a set of functions in which its members are mutually orthogonal and each member is normalized.

Mathematically, a set of functions, $f_1(\xi), f_2(\xi), f_3(\xi), \ldots$, is said to be an orthonormal set of functions in the interval $a \le \xi \le b$ if

$$\text{(i)}\ (f_m, f_n) = \int_a^b f_m^*(\xi) f_n(\xi) d\xi = 0 \text{ for } m \neq n, \tag{2.8}$$

$$m = 1, 2, 3, \ldots, n = 1, 2, 3, \ldots$$

$$\text{and} \quad \text{(ii)}\ (f_m, f_m) = \int_a^b f_m^*(\xi) f_m(\xi) d\xi = (f_n, f_n) = \int_a^b f_n^*(\xi) f_n(\xi) d\xi = 1 \tag{2.9}$$

In a more general and shorter way, we say that a set of functions, $f_1(\xi), f_2(\xi), f_3(\xi), \ldots$ is said to be an orthonormal set of functions in an interval $a \le \xi \le b$ if

$$(f_m, f_n) = \int_a^b f_m^*(\xi) f_n(\xi) d\xi = \delta_{mn} = \begin{cases} 0, & \text{for } m \neq n \\ 1 & \text{for } m = n \end{cases} \tag{2.10}$$

The symbol (δ_{mn}) is known as the *Kronecker delta* symbol. Its value is zero if $m \neq n$ and unity if $m = n$. This symbol is taken out from *tensor analysis* which is beyond the scope of this book.

Thus, we can say that a set of orthonormal functions is a set of orthogonal functions which are normalized.

Example

In the above example of orthogonal set of functions, viz., the set of functions $g_m(x) = \sin mx, m = 1, 2, 3, \ldots,$ we see that the norm of every member function is $\sqrt{\pi}$. Then, the set of functions, $f_m(x) = \frac{\sin mx}{\sqrt{\pi}}, m = 1, 2, 3, \ldots,$ is an orthonormal set of functions.

We can say that the functions, $\sin x, \sin 2x, \sin 3x, \sin 4x, \ldots$, form an orthogonal set whereas the functions, $f_1(x) = \frac{\sin x}{\sqrt{\pi}}, f_2(x) = \frac{\sin 2x}{\sqrt{\pi}}, f_3(x) = \frac{\sin 3x}{\sqrt{\pi}}, f_4(x) = \frac{\sin 4x}{\sqrt{\pi}}, \ldots,$ form an orthonormal set.

Complete Set of Orthonormal Functions

A set of orthonormal functions is said to form a *complete set* in a given interval if there exists no other non-zero member (or element) orthonormal to all of the functions in that set and in the given interval. Thus, the complete set of orthonormal functions is analogous to a complete set of unit vectors $\hat{i}, \hat{j}, \hat{k}$ in three-dimensional *vector analysis*, as there cannot be any vector orthonormal to all these three unit vectors.

We know that any vector **A** can be expressed as $\mathbf{A} = A_1\hat{i} + A_2\hat{j} + A_3\hat{k}$ where A_1, A_2, A_3 are the components of the vector **A** along the axes of x, y, z, respectively and $\hat{i}, \hat{j}, \hat{k}$ are unit vectors along these axes, respectively. Likewise, for a complete set of orthonormal functions in a given interval, an arbitrary function can be expanded as a linear combination of the functions in that set. We can, thus, define a complete set in another way. An orthonormal set of functions $\{\Phi_n(\xi), n = 1, 2, 3, \ldots\}$ is said to be complete in a given interval $a \le \xi \le b$ if an arbitrary function, $\Psi(\xi)$, can be expanded in terms of $\phi_n(\xi)$ as

$$\Psi(\xi) = \sum_n A_n \Phi_n(\xi) \tag{2.11}$$

where the quantities A_n are called expansion coefficients and these are given by

$$A_n = (\Phi_n, \Psi) = \int_a^b \Phi_n^*(\xi)\, \Psi(\xi) d\xi \tag{2.12}$$

Example

The set of functions $\{f_n(x) = e^{2n\pi i x}, n = 0, \pm 1, \pm 2, \pm 3, \ldots\}$ form a complete orthonormal set in the interval $0 \le x \le 1$ since there can be no other non-zero function orthonormal to $f_n(x)$ for every n.

Orthogonality with Respect to a Weight Function

In the cases of some important sets of real functions, $g_1(x), g_2(x), g_3(x), \ldots$; occurring in *mathematics and physics*, if we have

$$\int_a^b g_m(x) g_n(x) dx \ne 0 \quad \text{for } m \ne n,$$

but

$$\int_a^b w(x) g_m(x) g_n(x) dx = \int_a^b \sqrt{w(x)}\, g_m(x) \sqrt{w(x)}\, g_n(x) dx = 0 \tag{2.13}$$

where $w(x)$ is a non-negative and non-zero function; then $w(x)$ is known as the weight function of the functions in the set in the interval $a \le x \le b$ and the functions $g_m(x)$ and $g_n(x)$ are said to be orthogonal with respect to the weight function $w(x)$. Then, the set is said to be orthogonal in the given interval with respect to the weight function $w(x)$.

If we have

$$\int_a^b w(x) g_n(x) g_n(x) dx = \int_a^b \left\{\sqrt{w(x)}\, g_n(x)\right\}\left\{\sqrt{w(x)}\, g_n(x)\right\} dx = 1 \tag{2.14}$$

then the set is said to be an orthonormal set in the given interval with respect to the weight function, $w(x)$. Some authors call $\sqrt{w(x)}$ the weight function.

Example

Let us consider two functions, $f_1(x) = 1 - x$ and $f_2(x) = 1 - 2x + \frac{x^2}{2}$ in an interval $0 \le x \le \infty$

We have

$$(f_1, f_2) = \int_0^\infty (1-x)\left(1 - 2x + \frac{x^2}{2}\right) dx = \infty,$$

and
$$\left(e^{-\frac{x}{2}} f_1, e^{-\frac{x}{2}} f_2\right) = \int_0^\infty e^{-x} f_1(x) f_2(x) dx$$

$$= \int_0^\infty e^{-x} (1-x)\left(1 - 2x + \frac{x^2}{2}\right) dx = 0$$

Thus, we see that the functions, $f_1(x) = 1 - x$ and $f_2(x) = 1 - 2x + \frac{x^2}{2}$, are orthogonal with respect to a weight function, $w(x) = e^{-x}$ in the interval, $0 \le x \le \infty$.

2.2 SPECIAL FUNCTIONS

Special functions are particular mathematical functions which have more or less established names and notations after their early investigators and having particular uses in *Mathematical Physics* or some other branches of mathematics and science including physics. Some examples are Beta function, Gamma function, Legendre polynomial, Hermite polynomial, Bessel function, Laguerre polynomial, Hypergeometric function, Mathieu function, Elliptic function, Theta function, Zeta function, Green function, etc.

2.3 LEGENDRE EQUATION

Legendre equation is a second order and homogeneous linear ordinary differential equation having the form

$$(1-x^2)\frac{d^2y}{dx^2} - 2x\frac{dy}{dx} + l(l+1)y = 0, \text{ with } l = 0, 1, 2, 3, \ldots \tag{2.15}$$

or
$$\frac{d}{dx}\left\{(1-x^2)\frac{dy}{dx}\right\} + l(l+1)y = 0 \tag{2.16}$$

Here, the symbol l is introduced to suit for the orbital angular momentum quantum number in the study of quantum mechanics. Some authors use the symbol n.

This equation was named after the French mathematician, Adrien Marie Legendre (Sep. 18, 1752 to Jan. 10, 1833). He began to use it in the year 1784.

2.4 POWER SERIES SOLUTION OF LEGENDRE EQUATION

Here, one can easily see that the point $x = 0$ is an ordinary point of the given ODE and hence, by Fuchs' theorem, the equation can have a power series solution. Let us assume a power series solution in the form

$$y = \sum_{r=0}^{\infty} a_r\, x^{k+r} \quad \text{(with } a_0 \neq 0\text{)} \tag{2.17}$$

where k is a constant to be determined.

Then, we have

$$\frac{dy}{dx} = \sum_{r=0}^{\infty} (k+r) a_r\, x^{k+r-1} \text{ and } \frac{d^2y}{dx^2} = \sum_{r=0}^{\infty} (k+r)(k+r-1) a_r\, x^{k+r-2}$$

Substituting the expressions for y and its derivatives into Eq. (2.15), we obtain

$$\sum_{r=0}^{\infty}(k+r)(k+r-1) a_r x^{k+r-2} - \sum_{r=0}^{\infty}\{(k+r)(k+r+1) - l(l+1)\}\, a_r x^{k+r} = 0$$

In this equation, the coefficients of various powers of x must separately vanish.

Coefficient of x^{k-2} (i.e., the lowest power):

$$k(k-1)a_0 = 0, \text{ with } a_0 \neq 0 \text{ [Indicial equation]}$$

giving $k = k_1 = 1$ and $k = k_2 = 0$.

Coefficient of x^{k-1}: $(k+1)ka_1 = 0$ giving $a_1 = 0$ when $k = k_1 = 1$

and $\quad a_1 =$ indeterminate when $k = k_2 = 0$

In general, Coefficient of x^{k+r}:

$$(k+r+2)(k+r+1)a_{r+2} - \{(k+r)^2 + k + r - l^2 - l\}\, a_r = 0$$

giving
$$a_{r+2} = -\frac{(l-k-r)(k+r+l+1)}{(k+r+1)(k+r+2)} a_r \text{ [Recursion relation]} \tag{2.18}$$

Putting $r = 0, 1, 2, 3, 4, \ldots,$ Eq. (2.18) yields

$$a_2 = -\frac{(l-k)(k+l+1)}{(k+1)(k+2)} a_0,$$

$$a_3 = -\frac{(l-k-1)(k+l+2)}{(k+2)(k+3)}a_1,$$

$$a_4 = (-1)^2\frac{(l-k-2)(l-k)(k+l+1)(k+l+3)}{(k+1)(k+2)(k+3)(k+4)}a_0,$$

$$a_5 = (-1)^2\frac{(l-k-3)(l-k-1)(k+l+2)(k+l+4)}{(k+2)(k+3)(k+4)(k+5)}a_1,$$

$$a_6 = (-1)^3\frac{(l-k-4)(l-k-2)(l-k)(k+l+1)(k+l+3)(k+l+5)}{(k+1)(k+2)(k+3)(k+4)(k+5)(k+6)}a_0$$

and $a_7 = (-1)^3\dfrac{(l-k-5)(l-k-3)(l-k-1)(l+k+2)(l+k+4)(l+k+6)}{(k+2)(k+3)(k+4)(k+5)(k+7)}a_1, \ldots$ etc. (2.19)

The assumed power series solution, i.e., Eq. (2.17) can be written as

$$\begin{aligned} y &= y(x,k) \\ &= x^k[\{a_0 + a_2x^2 + a_4x^4 + \ldots \infty\} + \{a_1x + a_3x^3 + a_5x^5 + \ldots \infty\}] \end{aligned} \tag{2.20}$$

Substitution of the various coefficients given in Eq. (2.19) into Eq. (2.20) results in

$$y = a_0x^k\left\{1 - \frac{(l-k)(k+l+1)}{(k+1)(k+2)}x^2 + \frac{(l-k-2)(l-k)(k+l+1)(k+l+3)}{(k+1)(k+2)(k+3)(k+4)}x^4 - \ldots \infty\right\}$$

$$+ a_1x^k\left\{x - \frac{(l-k-1)(k+l+2)}{(k+2)(k+3)}x^3 + \frac{(l-k-3)(l-k-1)(k+l+2)(k+l+4)}{(k+2)(k+3)(k+4)(k+5)}x^5 - \ldots \infty\right\}$$

As a_1 is indeterminate at $k = k_2 = 0$ (smaller root of indicial equation) the general solution of the Legendre equation is given by

$$\begin{aligned} y &= [y(x,k)]_{k=k_2} \\ &= a_0\left\{1 - \frac{l(l+1)}{2!}x^2 + \frac{(l-2)l(l+1)(l+3)}{4!}x^4 - \ldots \right. \\ &\quad \left. + \frac{(-1)^j(l-2j+2)\ldots(l-2)l(l+1)(l+3)\ldots(l+2j-1)}{(2j)!}x^{2j} + \ldots \infty\right\} \\ &\quad + a_1\left\{x - \frac{(l-1)(l+2)}{3!}x^3 + \frac{(l-3)(l-1)(l+2)(l+4)}{5!}x^5 - \ldots \right. \\ &\quad \left. + \frac{(-1)^j(l-2j+1)\ldots(l-3)(l-1)(l+2)(l+4)\ldots(l+2j)}{(2j+1)!}x^{2j+1} + \ldots \infty\right\} \end{aligned} \tag{2.21}$$

We write the general solution as

$$y = y(x) = a_0y_1(x) + a_1y_2(x) \tag{2.22}$$

where y_1 represents the series within the first curly bracket and y_2 represents the series within the second curly brackets in Eq. (2.21). If one finds a solution with the other value of k, i.e., with $k=1$, the solution will be constant multiple of y_1.

2.5 POLYNOMIAL SOLUTION OF LEGENDRE EQUATION

We see that for $k=k_2=0$, Eq. (2.18) [Recursion relation] becomes

$$a_{r+2}=-\frac{(l-r)(l+r+1)}{(r+1)(r+2)}a_r \tag{2.23}$$

This shows that the series terminates when $r=l$, because $a_{r+2}, a_{r+4}, a_{r+6}, \ldots$ become zero. The highest power of x in the resulting polynomial solution will be l. If l is a non-negative even integer, y_1 terminates and it becomes a polynomial solution and if l is an odd positive integer, y_2 terminates and it becomes a polynomial solution.

Thus, when $l=2j\,(j=0,1,2,3,\ldots)$, there will be a polynomial solution

$$y_{\text{pol, even}}(x)=a_0\left\{1+\sum_{j=1}^{\frac{l}{2}}\frac{(-1)^j(l-2j+2)\ldots(l-4)(l-2)l(l+1)(l+3)\ldots(l+2j-1)}{(2j)!}x^{2j}\right\}$$

$$=a_0\left\{1+\sum_{j=1}^{\frac{l}{2}}\frac{(-1)^j2^j\left(\frac{l}{2}-j+1\right)\ldots\left(\frac{l}{2}-2\right)\left(\frac{l}{2}-1\right)\frac{l}{2}(l+1)(l+2)(l+3)(l+4)\ldots(l+2j-1)(l+2j)}{(2j)!(l+2)(l+4)(l+6)\ldots(l+2j-2)(l+2j)}x^{2j}\right\}$$

$$=a_0\left\{1+\sum_{j=1}^{\frac{l}{2}}\frac{(-1)^j2^j\left(\frac{l}{2}\right)!\,(l+2j)!}{(2j)!\left(\frac{l}{2}-j\right)!\,l!\,2^j\left(\frac{l}{2}+1\right)\left(\frac{l}{2}+2\right)\left(\frac{l}{2}+3\right)\ldots\left(\frac{l}{2}+j\right)}x^{2j}\right\}$$

$$=a_0\left\{1+\sum_{j=1}^{\frac{l}{2}}\frac{(-1)^j\left[\left(\frac{l}{2}\right)!\right]^2(l+2j)!}{(2j)!\,l!\left(\frac{l}{2}-j\right)!\left(\frac{l}{2}+j\right)!}x^{2j}\right\}$$

$$=a_0\frac{\left[\left(\frac{l}{2}\right)!\right]^2}{l!}\sum_{j=0}^{\frac{l}{2}}\frac{(-1)^j(l+2j)!}{\left(\frac{l}{2}-j\right)!\left(\frac{l}{2}+j\right)!}\frac{x^{2j}}{(2j)!} \tag{2.24}$$

Introducing a new summation index, $r=\frac{l}{2}-j$, we can write Eq. (2.24) as

$$y_{\text{pol,even}}(x)=(-1)^{\frac{l}{2}}\frac{\left[\left(\frac{l}{2}\right)!\right]^2}{l!}a_0\sum_{r=0}^{\frac{l}{2}}(-1)^r\frac{(2l-2r)!}{r!(l-r)!}\frac{x^{l-2r}}{(l-2r)!} \tag{2.25}$$

Equation (2.25) is a polynomial solution of Legendre equation when l is a non-negative even integer.

Similarly, when $l=2j+1\,(j=0,1,2,3,\ldots)$, there will be a polynomial solution

$$y_{\text{pol,odd}}(x)=a_1\left\{x+\sum_{j=1}^{\frac{l-1}{2}}\frac{(-1)^j(l-2j+1)\ldots(l-3)(l-1)(l+2)(l+4)\ldots(l+2j)}{(2j+1)!}x^{2j+1}\right\}$$

$$=a_1\left\{x+\sum_{j=1}^{\frac{l-1}{2}}\frac{(-1)^j\,(l-1-2j+2)\ldots(l-1-2)(l-1)(l+2)(l+4)\ldots(l+2j)}{(2j+1)!}x^{2j+1}\right\}$$

$$=a_1\left\{x+\sum_{j=1}^{\frac{l-1}{2}}\frac{(-1)^j2^j\left(\frac{l-1}{2}-j+1\right)\ldots\left(\frac{l-1}{2}-1\right)\left(\frac{l-1}{2}\right)(l+2)(l+4)\ldots(l+2j)}{(2j+1)!}x^{2j+1}\right\}$$

$$=a_1\left\{x+\sum_{j=1}^{\frac{l-1}{2}}\frac{(-1)^j2^j\left(\frac{l-1}{2}\right)!(l+1)(l+2)(l+3)(l+4)\ldots(l+2j-1)(l+2j)}{(2j+1)!\left(\frac{l-1}{2}-j\right)!(l-1+2)(l-1+4)(l-1+6)\ldots(l-1+2j)}x^{2j+1}\right\}$$

$$=a_1\left\{x+\sum_{j=1}^{\frac{l-1}{2}}\frac{(-1)^j2^j\left(\frac{l-1}{2}\right)!\,(l+2j)!\,x^{2j+1}}{(2j+1)!\left(\frac{l-1}{2}-j\right)!\,l!\,2^j\left(\frac{l-1}{2}+1\right)\left(\frac{l-1}{2}+2\right)\left(\frac{l-1}{2}+3\right)\ldots\left(\frac{l-1}{2}+j\right)}\right\}$$

$$=a_1\left\{x+\sum_{j=1}^{\frac{l-1}{2}}\frac{(-1)^j\left[\left(\frac{l-1}{2}\right)!\right]^2(l+2j)!}{(2j+1)!\left(\frac{l-1}{2}-j\right)!\,l!\left(\frac{l-1}{2}+j\right)!}x^{2j+1}\right\}$$

$$=\frac{\left[\left(\frac{l-1}{2}\right)!\right]^2}{l!}a_1\sum_{j=0}^{\frac{l-1}{2}}\frac{(-1)^j(l+2j)!}{\left(\frac{l-1}{2}-j\right)!\left(\frac{l-1}{2}+j\right)!}\frac{x^{2j+1}}{(2j+1)!} \tag{2.26}$$

Equation (2.26) is a polynomial solution of Legendre equation when l is a positive odd integer.

Introducing a new summation index, $r = \frac{l-1}{2} - j$, we can write Eq. (2.26) as

$$y_{\text{pol,odd}}(x) = (-1)^{\frac{l-1}{2}} \frac{\left[\left(\frac{l-1}{2}\right)!\right]^2}{l!} a_1 \sum_{r=0}^{\frac{l-1}{2}} \frac{(-1)^r (2l-2r-1)!}{r!(l-r-1)!} \frac{x^{l-2r}}{(l-2r)!}$$

$$= (-1)^{\frac{l-1}{2}} \frac{\left[\left(\frac{l-1}{2}\right)!\right]^2}{l!} a_1 \sum_{r=0}^{\frac{l-1}{2}} \frac{(-1)^r (2l-2r)!}{r!\,2(l-r)(l-r-1)!} \frac{x^{l-2r}}{(l-2r)!}$$

$$= (-1)^{\frac{l-1}{2}} \frac{\left[\left(\frac{l-1}{2}\right)!\right]^2}{2\,(l!)} a_1 \sum_{r=0}^{\frac{l-1}{2}} \frac{(-1)^r (2l-2r)!}{r!(l-r)!} \frac{x^{l-2r}}{(l-2r)!} \tag{2.27}$$

Equations (2.25) and (2.27) can be represented in a general way by the single equation

$$y_{\text{pol}}(x) = k_l \sum_{r=0}^{\left[\frac{l}{2}\right]} \frac{(-1)^r (2l-2r)!}{r!(l-r)!} \frac{x^{l-2r}}{(l-2r)!} \tag{2.28}$$

where k_l represents the constant factors outside the summation in both the equations and the symbol $\left[\frac{l}{2}\right]$ equals $\frac{l}{2}$ if l is a non-zero even integer and $\frac{l-1}{2}$ if l is a positive odd integer, that is,

$$\left[\frac{l}{2}\right] = \begin{cases} \frac{l}{2}, & \text{for } l = \text{non-zero even integer} \\ \frac{l-1}{2}, & \text{for } l = \text{positive odd integer} \end{cases}$$

2.6 LEGENDRE POLYNOMIALS

A Legendre polynomial, denoted by $P_l(x)$, is a polynomial solution of the Legendre equation

$$(1-x^2)\frac{d^2y}{dx^2} - 2x\frac{dy}{dx} + l(l+1)y = 0, l = 0, 1, 2, 3, \ldots$$

and having the property that $P_l(1) = 1$ irrespective of the value of the degree l of the polynomial. Here, $P_l(x)$ is called Legendre polynomial of the first kind of degree l or Legendre function of the first kind. There exists another type of Legendre polynomial known as the Legendre polynomial of the second kind, denoted by $Q_l(x)$ which will not be considered in this context. A Legendre polynomial is given by (i) a sum formula, (ii) a differential formula known as Rodrigues' formula, and (iii) a generating function.

The sum formula and the differential formula of Legendre polynomial can be derived from Eq. (2.28) by assigning a certain value to the parameter k_l. Let us find such a particular value of k_l.

We know that

$$\frac{d^l}{dx^l}(x^{2l-2r}) = \frac{(2l-2r)!}{(l-2r)!}x^{l-2r}$$

Hence, Eq. (2.28) can be written as

$$y_{\text{pol}}(x) = k_l \sum_{r=0}^{\left[\frac{l}{2}\right]} \frac{(-1)^r}{r!(l-r)} \frac{d^l}{dx^l}(x^{2l-2r})$$

$$= \frac{k_l}{l!} \sum_{r=0}^{\left[\frac{l}{2}\right]} (-1)^r \frac{l!}{r!\,(l-r)!} \frac{d^l}{dx^l}(x^{2l-2r})$$

$$= \frac{k_l}{l!} \frac{d^l}{dx^l} \sum_{r=0}^{\left[\frac{l}{2}\right]} (-1)^r \frac{l!}{(l-r)!\,r!} x^{2l-2r}$$

$$= \frac{k_l}{l!} \frac{d^l}{dx^l}(x^2-1)^l \quad \text{[using binomial theorem]} \tag{2.29}$$

Further, we can write

$$y_{\text{pol}}(x) = \frac{k_l}{l!} \frac{d^l}{dx^l}\{(x+1)^l (x-1)^l\}$$

$$= \frac{k_l}{l!} \sum_{p=0}^{l} \frac{l!}{p!\,(l-p)!} \frac{d^p}{dx^p}(x+1)^l \frac{d^{l-p}}{dx^{l-p}}(x-1)^l$$

[By Leibnitz theorem]

$$= \frac{k_l}{l!}\Big[(x+1)^l\, l!$$

$$+ \text{ terms containing positive powers of } (x+1) \text{ and } (x-1)\Big].$$

If the value of k_l is so chosen that $y_{\text{pol}}\ (x)$ becomes unity at $x = 1$ for all possible values of l, then such a polynomial is called Legendre polynomial $P_l(x)$.

Therefore, we have

$$1 = P_l(1) = \frac{k_l}{l!}[2^l l! + 0]$$

giving $k_l = \frac{1}{2^l}$

Thus, if $k_l = \frac{1}{2^l}$, both of Eqs. (2.28) and (2.29) become the Legendre polynomial $P_l(x)$.

Substituting the above value of k_l into Eq. (2.28), we obtain the sum formula of Legendre polynomial as

$$P_l(x) = \sum_{r=0}^{\left[\frac{l}{2}\right]} \frac{(-1)^r (2l-2r)!}{2^l\, r!\,(l-r)!} \frac{x^{l-2r}}{(l-2r)!} \tag{2.30}$$

Substituting the above value of k_l into Eq. (2.29), we obtain the differential formula of Legendre polynomial as

$$P_l(x) = \frac{1}{2^l\, l!} \cdot \frac{d^l}{dx^l}(x^2-1)^l \tag{2.31}$$

Equation (2.31) is known as Rodrigues' formula for Legendre polynomials.

2.7 SCHLAEFLI'S INTEGRAL REPRESENTATION

Using the generalized Cauchy's integral formula for the derivative of a function of a complex variable, we have

$$\frac{d^l}{dz^l}(Z^2-1)^l = \frac{l!}{2\pi i} \oint_C \frac{(\xi^2-1)^l\, d\xi}{(\xi - z)^{l+1}} \tag{2.32}$$

where $i = \sqrt{-1}$, ξ is a complex variable and C is a closed contour enclosing the point z in the complex plane.

Using Eqs. (2.31) and (2.32), we write

$$P_l(z) = \frac{1}{2\pi i\, 2^l} \cdot \oint_C \frac{(\xi^2-1)^l\, d\xi}{(\xi - z)^{l+1}} \tag{2.33}$$

Equation (2.33) is known as Schlaefli's integral representation of Legendre polynomials. This integral representation is named after the Swiss mathematician, Ludwig Schlaefli (January 15, 1814 - March 20, 1895).

2.8 LAPLACE FIRST INTEGRAL REPRESENTATION

Let us take the contour C in Schlaefli's integral representation as a circle with centre at z and having a radius $|\sqrt{z^2-1}|$. Then, we have

$$|\xi - z| = |\sqrt{z^2-1}|, \text{ or } \xi - z = (\sqrt{z^2-1})e^{i\phi}, \tag{2.34}$$

where ϕ is the phase.

or
$$\xi = z + \left(\sqrt{z^2-1}\right)e^{i\phi} \tag{2.35}$$

We have

$$\begin{aligned}
\xi^2 - 1 &= (\xi+1)(\xi-1) \\
&= [z+1+(\sqrt{z^2-1})e^{i\phi}][z-1+(\sqrt{z^2-1})e^{i\phi}] \\
&= (z^2-1)+(z+1+z-1)(\sqrt{z^2-1})e^{i\phi}+(z^2-1)e^{2i\phi} \\
&= (z^2-1)(1+e^{2i\phi})+2z(\sqrt{z^2-1})e^{i\phi} \\
&= (z^2-1)e^{i\phi}(e^{-i\phi}+e^{i\phi})+2z(\sqrt{z^2-1})e^{i\phi} \\
&= 2(\sqrt{z^2-1})e^{i\phi}\{(\sqrt{z^2-1})\cos\phi+z\} \\
&= 2(\xi-z)\{z+(\sqrt{z^2-1})\cos\phi\}
\end{aligned} \tag{2.36}$$

Also, using Eq. (2.35), we have

$$d\xi = i(\sqrt{z^2-1})e^{i\phi}d\phi = i(\xi - z)d\phi \tag{2.37}$$

Substituting Eqs. (2:34) to (2.37) into Eq. (2.33), we obtain

$$\begin{aligned}
P_l(z) &= \frac{1}{2\pi}\int_0^{2\pi}(z+\sqrt{z^2-1}\cos\phi)^l\,d\phi \\
&= \frac{1}{\pi}\int_0^{\pi}(z+\sqrt{z^2-1}\cos\phi)^l\,d\phi
\end{aligned} \tag{2.38}$$

If we replace z by a real variable x, we have

$$P_l(x) = \frac{1}{\pi}\int_0^{\pi}(x+\sqrt{x^2-1}\cos\phi)^l\,d\phi \tag{2.39}$$

Equations (2.38) and (2.39) are known as Laplace first integral representation of Legendre polynomials.

If we put $x = \cos\theta$, then we write

$$P_l(\cos\theta) = \frac{1}{\pi}\int_0^{\pi}\left(\cos\theta + i\sin\theta\cos\phi\right)^l d\phi \tag{2.40}$$

2.9 GENERATING FUNCTION OF LEGENDRE POLYNOMIALS

Let us define the generating function of Legendre polynomials by a function

$$G(h,x) = \sum_{l=0}^{\infty} h^l P_l(x) \tag{2.41}$$

Using Laplace first integral representation, we can write

$$G(h,x) = \sum_{l=0}^{\infty} h^l \frac{1}{\pi}\int_0^{\pi}(x+\sqrt{x^2-1}\cos\phi)^l\, d\phi$$

$$= \frac{1}{\pi}\int_0^{\pi}\sum_{l=0}^{\infty} h^l (x+\sqrt{x^2-1}\cos\phi)^l\, d\phi$$

$$= \frac{1}{\pi}\int_0^{\pi}\{1-h(x+\sqrt{x^2-1}\cos\phi)\}^{-1}\, d\phi$$

$$= \frac{1}{\pi}\int_0^{\pi}\frac{d\phi}{(1-hx)-(h\sqrt{x^2-1})\cos\phi} = \frac{1}{\pi}\frac{\pi}{\sqrt{(1-hx)^2-h^2(x^2-1)}}$$

where use has been made of the formula

$$\int_0^{\pi}\frac{d\phi}{a-b\cos\phi} = \frac{\pi}{\sqrt{a^2-b^2}} \quad \text{for } a > b$$

Thus, we write

$$G(h,x) = (1-2hx+h^2)^{-\frac{1}{2}} = \sum_{l=0}^{\infty} h^l P_l(x) \tag{2.42}$$

This shows that $P_l(x)$ is the coefficient of h^l in the expansion of $(1-2hx+h^2)^{-\frac{1}{2}}$

The function $G(h,x) = (1-2hx+h^2)^{-\frac{1}{2}}$ is called the generating function of Legendre polynomials.

2.10 LAPLACE SECOND INTEGRAL REPRESENTATION

In the standard integral,

$$\int_0^{\pi} \frac{d\phi}{a \pm b\cos\phi} = \frac{\pi}{\sqrt{a^2 - b^2}}$$

if we put $a = hx - 1$ and $b = h\sqrt{x^2 - 1}$ we write

$$\frac{\pi}{\{(hx-1)^2 - (h\sqrt{x^2-1})^2\}^{\frac{1}{2}}} = \int_0^{\pi} \frac{d\phi}{(xh-1) \pm h(\sqrt{x^2-1})\cos\phi}$$

or
$$\frac{\pi}{\{h^2x^2 - 2hx + 1 - h^2x^2 + h^2\}^{\frac{1}{2}}} = \int_0^{\pi} \frac{d\phi}{h(x \pm \sqrt{x^2-1}\cos\phi) - 1}$$

or
$$\frac{\pi}{h}\left(1 - 2\frac{1}{h}x + \frac{1}{h^2}\right)^{-\frac{1}{2}} = \int_0^{\pi} \frac{\left\{1 - \dfrac{1}{h\left(x \pm \sqrt{x^2-1}\cos\right)}\right\}^{-1}}{h\left(x \pm \sqrt{x^2-1}\cos\phi\right)} d\phi$$

or
$$\frac{\pi}{h}\sum_{l=0}^{\infty}\left(\frac{1}{h}\right)^l P_l(x) = \int_0^{\pi} \frac{\sum_{l=0}^{\infty}\left\{\dfrac{1}{h\left(x \pm \sqrt{x^2-1}\cos\phi\right)}\right\}^1}{h\left(x \pm \sqrt{x^2-1}\cos\phi\right)} d\phi$$

or
$$\sum_{l=0}^{\infty} \pi\left(\frac{1}{h}\right)^{l+1} P_l(x) = \int_0^{\pi} \sum_{l=0}^{\infty} \frac{d\phi}{\{h(x \pm \sqrt{x^2-1}\cos\phi)\}^{l+1}}$$

$$= \sum_{l=0}^{\infty}\left(\frac{1}{h}\right)^{l+1} \int_0^{\pi} \frac{d\phi}{(x \pm \sqrt{x^2-1}\cos\phi)^{l+1}}$$

Equating coefficients of $\left(\frac{1}{h}\right)^{l+1}$ from both sides of the above equation, we obtain

$$\pi P_l(x) = \int_0^{\pi} \frac{d\phi}{(x \pm \sqrt{x^2-1}\cos\phi)^{l+1}}$$

or
$$P_l(x) = \frac{1}{\pi}\int_0^{\pi} \frac{d\phi}{\left(x \pm \sqrt{x^2-1}\cos\phi\right)^{l+1}} \tag{2.43}$$

2.11 RECURRENCE RELATIONS OF LEGENDRE POLYNOMIALS

I. We know that

$$\sum_{l=0}^{\infty} h^l P_l(x) = (1-2hx+h^2)^{-\frac{1}{2}}$$

Differentiating both sides once with respect to h, we have

$$\sum_{l=0}^{\infty} l h^{l-1} P_l(x) = (x-h)(1-2hx+h^2)^{-\frac{3}{2}}$$

Multiplying both sides by $(1-2hx+h^2)$ we have

$$\sum_{l=0}^{\infty} (1-2hx+h^2) l h^{l-1} P_l(x) = (x-h)(1-2hx+h^2)^{-\frac{1}{2}}$$

or
$$\sum_{l-0}^{\infty} \{l h^{l-1} P_l(x) - 2lx h^l P_l(x) + l h^{l+1} P_l(x)\} = (x-h)\sum_{l=0}^{\infty} h^l P_l(x)$$

$$= \sum_{l=0}^{\infty} \{h^l x P_l(x) - h^{l+1} P_l(x)\}$$

Equating the coefficients of h^l from both sides, we obtain

$$(l+1)P_{l+1}(x) - 2lxP_l(x) + (l-1)P_{l-1}(x) = xP_l(x) - P_{l-1}(x)$$

or
$$(l+1)P_{l+1}(x) + lP_{l-1}(x) = (2l+1)xP_l(x) \tag{2.44}$$

II. We know that

$$\sum_{l=0}^{\infty} h^l P_l(x) = (1-2hx+h^2)^{-\frac{1}{2}}$$

Differentiating both sides once with respect to x then dividing by h, we have

$$\sum_{l=0}^{\infty} h^{l-1} P_l'(x) = (1-2hx+h^2)^{-\frac{3}{2}} \quad \text{where } P_l'(x) = \frac{d}{dx}P_l(x)$$

Multiplying both sides by $(1-2hx+h^2)$, we have

$$\sum_{l=0}^{\infty} (1-2hx+h^2) h^{l-1} P_l'(x) = (1-2hx+h^2)^{-\frac{1}{2}} = \sum_{l=0}^{\infty} h^l P_l(x)$$

or
$$\sum_{l=0}^{\infty} [h^{l-1} P_l'(x) - 2xh^l P_l'(x) + h^{l+1} P_l'(x)] = \sum_{l=0}^{\infty} h^l P_l(x)$$

Equating the coefficients of h^l from both sides, we obtain

$$P_{l+1}'(x) - 2xP_l'(x) + P_{l-1}'(x) = P_l(x) \tag{2.45}$$

III. From the first recurrence relation, we know that

$$(2l+1)xP_l(x)-(l+1)P_{l+1}(x)-lP_{l-1}(x)=0$$

Differentiating both sides once with respect to x, we have

$$(2l+1)P_l(x)+(2l+1)xP_l'(x)-(l+1)P_{l+1}'(x)-lP_{l-1}'(x)=0 \tag{a}$$

Also, from the second recurrence relation, we have

$$P_{l+1}'(x)-2xP_l'(x)+P_{l-1}'(x)-P_l(x)=0 \tag{b}$$

Adding l times Eq. (b) to Eq. (a) , we obtain

$$(l+1)P_l(x)+xP_l'(x)-P_{l+1}'(x)=0$$

or
$$(l+1)P_l(x)=P_{l+1}'(x)-xP_l'(x) \tag{2.46}$$

IV. Proceed up to Eq. (b) of recurrence relation III and then add the following. Adding $(l+1)$ times Eq. (b) to Eq. (a), we obtain

$$lP_l(x)-xP_l'(x)+P_{l-1}'(x)=0$$

or
$$lP_l(x)=xP_l'(x)-P_{l-1}'(x) \tag{2.47}$$

V. Proceed up to Eq. (b) of recurrence relation III and then add the following. Adding $(2l+1)$ times Eq. (b) to twice of Eq. (a), we obtain

$$(2l+1)P_l(x)-P_{l+1}'(x)+P_{l-1}'(x)=0$$

or
$$(2l+1)P_l(x)=P_{l+1}'(x)-P_{l-1}'(x) \tag{2.48}$$

VI. From the third recurrence relation, we have

$$(l+1)P_l(x)=P_{l+1}'(x)-xP_l'(x)$$

Replacing l by $(l-1)$, we obtain

$$lP_{l-1}(x)=P_l'(x)-xP_{l-1}'(x) \tag{c}$$

Also, from the fourth recurrence relation, we have

$$lP_l(x)=xP_l'(x)-P_{l-1}'(x) \tag{d}$$

Subtracting x times Eq. (d) from Eq. (c), we obtain

$$l[P_{l-1}(x)-xP_l(x)]=(1-x^2)P_l'(x) \tag{2.49}$$

VII. From the first recurrence relation, we have

$$(2l+1)xP_l(x)=(l+1)P_{l+1}(x)+lP_{l-1}(x)$$

or
$$(l+1)xP_l(x)+lxP_l(x)=(l+1)P_{l+1}(x)+lP_{l-1}(x)$$

or $$(l+1)\{xP_l(x)-P_{l+1}(x)\} = l\{P_{l-1}(x)-xP_l(x)\}$$
$$= (1-x^2)P_l'(x) \tag{2.50}$$
[Using the sixth recurrence relation]

VIII. Laplace second integral representation of Legendre polynomials can be written as

$$P_l(x) = \frac{1}{\pi}\int_0^\pi \frac{d\phi}{\left(x+\sqrt{x^2-1}\cos\phi\right)^{l+1}}$$

If we replace l by $-(l+1)$, we write

$$P_{-(l+1)}(x) = \frac{1}{\pi}\int_0^\pi (x+\sqrt{x^2-1}\cos\phi)^l\, d\phi = P_l(x) \tag{2.51}$$
[Using Laplace first integral representation]

Alternatively, we know that the Legendre equation

$$(1-x^2)\frac{d^2y}{dx^2} - 2x\frac{dy}{dx} + l(l+1)y = 0$$

has the polynomial solution $P_l(x)$.
The same Legendre equation can also be written in the form

$$(1-x^2)\frac{d^2y}{dx^2} - 2x\frac{dy}{dx} + (-l)\{-(l+1)\}y = 0$$

or $$(1-x^2)\frac{d^2y}{dx^2} - 2x\frac{dy}{dx} + \{-(l+1)\}\{-(l+1)+1\}y = 0.$$

This equation is also a Legendre equation having a polynomial solution $P_{-(l+1)}(x)$
Hence, we write

$$P_{-(l+1)}(x) = P_l(x)$$

which is Eq. (2.51).

IX. Christoffel's Expansion
We know that

$$(2l+1)P_l(x) = P_{l+1}'(x) - P_{l-1}'(x)$$

or $$P_{l+1}'(x) = (2l+1)P_l(x) + P_{l-1}'(x)$$

Replacing l by $(l-1)$, we write

$$P_l'(x) = (2l-1)P_{l-1}(x) + P_{l-2}'(x).$$

Repeated use of the formula itself results in

$$\begin{aligned}
P_l'(x) &= (2l-1)P_{l-1}(x) + [\{2(l-2)-1\}P_{l-3}(x) + P_{l-4}'(x)] \\
&= (2l-1)P_{l-1}(x) + (2l-5)P_{l-3}(x) + P_{l-4}'(x) \\
&= (2l-1)P_{l-1}(x) + (2l-5)P_{l-3}(x) + [\{2(l-4)-1\}P_{l-5}(x) + P_{l-6}'(x)] \\
&= (2l-1)P_{l-1}(x) + (2l-5)P_{l-3}(x) + (2l-9)P_{l-5}(x) + P_{l-6}'(x) \\
&= (2l-1)P_{l-1}(x) + (2l-5)P_{l-3}(x) + (2l-9)P_{l-5}(x) \\
&\quad + (2l-13)P_{l-7}(x) + \ldots + \{2l-(2r-1)\}P_{l-r}(x) + \ldots + \begin{cases} 3P_1(x) \text{ if } l \text{ is even} \\ P_0(x) \text{ if } l \text{ is odd} \end{cases} \qquad (2.52)
\end{aligned}$$

When l is even, the last term in the right-hand side of Eq. (2.52) will be $3P_1(x)$ and the coefficients on the right-hand side form an arithmatical progression, $(2l-1), (2l-5), (2l-9), \ldots 7, 3$ with a common difference of (– 4).
If n is the number of terms, we have

$$3 = (2l-1) + (n-1)(-4) \text{ giving } n = \frac{l}{2}$$

Similarly, when l is odd, the last term is $P_0(x)$ and the coefficients form an arithmatic progression $(2l-1), (2l-5), (2l-9), \ldots 5, 1$ having a common difference (– 4).
If n is the number of terms, we have

$$1 = (2l-1) + (n-1)(-4) \text{ giving } n = \frac{l+1}{2}$$

X. To prove the Beltrami's relation,

$(2l+1)(x^2-1)P_l'(x) = l(l+1)\{P_{l+1}(x) - P_{l-1}(x)\}$ [Beltrami's relation]

where the prime symbol denotes differentiation with respect to x.

Proof: Let us recall the recurrence relations

$$(1-x^2)P_l'(x) = l\{P_{l-1}(x) - xP_l(x)\} \qquad \text{(i)}$$

and

$$(1-x^2)P_l'(x) = (l+1)\{xP_l(x) - P_{l+1}(x)\} \qquad \text{(ii)}$$

Adding $(l+1)$ times Eq. (i) to l times Eq. (ii) we obtain

$$(2l+1)(1-x^2)P_l'(x) = l(l+1)\{P_{l-1}(x) - P_{l+1}(x)\}$$

or

$$(2l+1)(x^2-1)P_l'(x) = l(l+1)\{P_{l+1}(x) - P_{l-1}(x)\}$$

2.12 ORTHOGONALITY AND ORTHONORMALITY OF LEGENDRE POLYNOMIALS

I. Orthogonality

Legendre polynomials, $P_l(x)$ and $P_m(x)$, are orthogonal in the interval $-1 \le x \le +1$,

i.e., $\int_{-1}^{+1} P_l(x) P_m(x)\, dx = 0$ when l and m are distinct integers

First Proof (Using Legendre Equation)

Legendre polynomial $P_l(x)$ satisfies the Legendre equation

$$\frac{d}{dx}\left\{(1-x^2)\frac{d}{dx}P_l(x)\right\} + l(l+1)P_l(x) = 0 \qquad \text{(a)}$$

Similarly, $P_m(x)$ satisfies

$$\frac{d}{dx}\left\{(1-x^2)\frac{d}{dx}P_m(x)\right\} + m(m+1)P_m(x) = 0 \qquad \text{(b)}$$

Subtracting $P_l(x)$ times equation (b) from $P_m(x)$ times equation (a) and integrating the result with respect to x within the limit from $x = -1$ to $x = +1$, we obtain

$$\int_{-1}^{+1} P_m(x)\frac{d}{dx}\left\{(1-x^2)\frac{d}{dx}P_l(x)\right\}dx - \int_{-1}^{+1} P_l(x)\frac{d}{dx}\left\{(1-x^2)\frac{d}{dx}P_m(x)\right\}dx$$
$$+\{l(l+1) - m(m+1)\}\int_{-1}^{+1} P_l(x)P_m(x)\,dx = 0$$

or
$$\left[P_m(x)(1-x^2)\frac{d}{dx}P_l(x)\right]_{-1}^{+1} - \int_{-1}^{+1}\left\{\frac{d}{dx}P_m(x)\right\}(1-x^2)\left\{\frac{d}{dx}P_l(x)\right\}dx$$
$$-\left[P_l(x)(1-x^2)\frac{d}{dx}P_m(x)\right]_{-1}^{+1} + \int_{-1}^{+1}\left\{\frac{d}{dx}P_l(x)\right\}(1-x^2)\left\{\frac{d}{dx}P_m(x)\right\}dx$$
$$+(l-m)(l+m+1)\int_{-1}^{+1} P_l(x)P_m(x)\,dx = 0,$$

or
$$0 + (l-m)(l+m+1)\int_{-1}^{+1} P_l(x)P_m(x)\,dx = 0.$$

If $l \neq m$, we have

$$\int_{-1}^{+1} P_l(x)P_m(x)\,dx = 0 \tag{2.53}$$

Equation (2.53) is known as the orthogonality relation of Legendre polynomials.

Second Proof (Using the Generating Function)

We know that

$$\sum_{l=0}^{\infty} h^l P_l(x) = (1-2xh+h^2)^{-\frac{1}{2}}$$

Similarly, we have

$$\sum_{m=0}^{\infty} h^m P_m(x) = (1-2hx+h^2)^{-\frac{1}{2}}$$

Hence, we write

$$\left[\sum_{l=0}^{\infty} h^l P_l(x)\right]\left[\sum_{m=0}^{\infty} h^m P_m(x)\right] = (1-2hx+h^2)^{-1},$$

or

$$\sum_{l=0}^{\infty}\sum_{m=0}^{\infty} h^{l+m} P_l(x)P_m(x) = (1-2hx+h^2)^{-1}$$

Integrating both sides with respect to x within the limit from $x=-1$ to $x=+1$, we have

$$\sum_{l=0}^{\infty}\sum_{m=0}^{\infty} h^{l+m}\int_{-1}^{+1} P_l(x)P_m(x)\,dx = \int_{-1}^{+1}\frac{dx}{1-2hx+h^2}$$

$$= -\frac{1}{2h}\int_{(1+h)^2}^{(1-h)^2}\frac{dy}{y} \quad \text{where we put } y = 1-2hx+h^2;$$

$$= \frac{1}{2h}\int_{(1-h)^2}^{(1+h)^2}\frac{dy}{y} = \frac{1}{2h}\left[\ln y\right]_{(1-h)^2}^{(1+h)^2} = \frac{1}{2h}\left\{\ln(1+h)^2 - \ln(1-h)^2\right\}$$

$$= \frac{1}{h}\left\{\ln(1+h) - \ln(1-h)\right\}$$

$$= \frac{1}{h}\left\{\left(h - \frac{h^2}{2} + \frac{h^3}{3} - \frac{h^4}{4} + \ldots\infty\right) + \left(h + \frac{h^2}{2} + \frac{h^3}{3} + \frac{h^4}{4} + \ldots\infty\right)\right\}$$

$$= \sum_{l=0}^{\infty} 2\frac{h^{2l}}{2l+1} \tag{2.54}$$

Equating the coefficients of h^{l+m} $(\text{with } l \neq m)$ from both sides of Eq. (2.54), we obtain

$$\int_{-1}^{+1} P_l(x) P_m(x) dx = 0, \text{ for } l \neq m$$

which is Eq. (2.53).

Third Method (Using Rodrigues' Formula)

By Rodrigues' formula, we have

$$P_l(x) = \frac{1}{2^l\, l!} \frac{d^l}{dx^l}\left(x^2-1\right)^l \quad \text{and} \quad P_m(x) = \frac{1}{2^m\, m!} \frac{d^m}{dx^m}\left(x^2-1\right)^m$$

Multiplying the two and then integrating both sides with respect to x within the limit from $x = -1$ to $x = +1$, we obtain

$$\int_{-1}^{+1} P_l(x) P_m(x) dx = \frac{1}{2^{l+m}\, l!m!} \int_{-1}^{+1} \frac{d^l}{dx^l}\left(x^2-1\right)^l \frac{d^m}{dx^m}\left(x^2-1\right)^m dx$$

Choosing $\frac{d^l}{dx^l}(x^2-1)^l$ as the first function and performing integration by parts once on the right-hand side, we have

$$\int_{-1}^{+1} P_l(x) P_m(x) dx = \frac{1}{2^{l+m}\, l!\, m!} \left\{ \left[\frac{d^l}{dx^l}\left(x^2-1\right)^l \frac{d^{m-1}}{dx^{m-1}}\left(x^2-1\right)^m \right]_{-1}^{+1} - \int_{-1}^{+1} \frac{d^{l+1}}{dx^{l+1}}\left(x^2-1\right)^l \frac{d^{m-1}}{dx^{m-1}}\left(x^2-1\right)^m dx \right\}$$

In the first term (the integrated out term), using Leibnitz theorem for successive differentiation of the product of two functions, it can be shown that all terms of $\frac{d^{m-1}}{dx^{m-1}}\left(x^2-1\right)^m$ will contain both of the factors $(x-1)$ and $(x+1)$ and these terms will vanish when we substitute the limits $x = +1$ and $x = -1$.

Therefore, we write,

$$\int_{-1}^{+1} P_l(x) P_m(x) dx = -\frac{1}{2^{l+m}\, l!\, m!} \int_{-1}^{+1} \frac{d^{l+1}}{dx^{l+1}}\left(x^2-1\right)^l \frac{d^{m-1}}{dx^{m-1}}\left(x^2-1\right)^m dx$$

If we repeat the integration by parts up to m times, we will get

$$\int_{-1}^{+1} P_l(x)P_m(x)dx = (-1)^m \frac{1}{2^{l+m}\, l!\, m!} \int_{-1}^{+1} \left\{ \frac{d^{l+m}}{dx^{l+m}} (x^2-1)^l \right\} (x^2-1)^m \, dx$$

$$= 0, \text{ if } m > l.$$

With an alternative choice of the first and the second functions in performing the integration by parts, it can also be shown that

$$\int_{-1}^{+1} P_l(x)P_m(x)dx = 0, \text{ if } l > m$$

Hence, we write

$$\int_{-1}^{+1} P_l(x)P_m(x)dx = 0 \text{ if } l \neq m,$$

which is also Eq. (2.53).

II. Orthonormality with a Normalization Constant of Legendre Polynomials

First Method (Using Generating Function)

Proceed as in the second proof of Eq. (2.53) in sub-section I up to Eq. (2.54) and then add the following.

Equating the coefficients of h^{2l} (when $l = m$) from both sides of Eq. (2.54), we obtain

$$\int_{-1}^{+1} P_l(x)P_l(x)dx = \frac{2}{2l+1} \tag{2.55}$$

or

$$\int_{-1}^{+1} \left\{ \sqrt{\frac{2l+1}{2}} P_l(x) \right\} \left\{ \sqrt{\frac{2l+1}{2}} P_l(x) \right\} dx = 1 \tag{2.56}$$

The function, $\sqrt{\frac{2l+1}{2}} P_l(x)$, is called normalized Legendre polynomial or Legendre function.

Second Method (Using Rodrigues' Formula)

Using Rodrigues' formula, we write

$$\int_{-1}^{+1} P_l(x)P_l(x)dx = \frac{1}{2^{2l}\, l!\, l!} \int_{-1}^{+1} \left\{ \frac{d^l}{dx^l} (x^2-1)^l \right\} \left\{ \frac{d^l}{dx^l} (x^2-1)^l \right\} dx$$

Performing integration by parts l times and remembering that the integrated out term in each step vanishes as in the case of the third proof in sub-section I, we arrive at

$$\int_{-1}^{+1} P_l(x)P_l(x)dx = \frac{(-1)^l}{2^{2l}\, l!\, l!} \int_{-1}^{+1} \left\{ \frac{d^{2l}}{dx^{2l}} \left(x^2-1\right)^l \right\} \left(x^2-1\right)^l dx$$

Since, $\frac{d^{2l}}{dx^{2l}}\left(x^2-1\right)^l = \frac{d^{2l}}{dx^{2l}}\left(x^{2l} + \text{terms having lower powers of } x\right) = (2l)!$, we write

$$\int_{-1}^{+1} P_l(x)P_l(x)dx = \frac{(-1)^l}{2^{2l}\, l!\, l!} \int_{-1}^{+1} (2l)!\left(x^2-1\right)^l dx$$

$$= \frac{(2l)!}{2^{2l}\,(l!)^2} \int_{-1}^{+1} (1+x)^l (1-x)^l dx$$

If we put $x = 2u - 1$ or $u = \frac{1}{2}(x+1)$, then, $dx = 2du$

Therefore, we write

$$\int_{-1}^{+1} P_l(x)P_l(x)dx = \frac{(2l)!}{2^{2l}\, l!\, l!} \int_{0}^{1} (2u)^l \{2(1-u)\}^l\, 2du$$

$$= \frac{2(2l)!}{l!\, l!} \int_0^1 u^l (1-u)^l du$$

$$= \frac{2(2l)!}{l!\, l!} B(l+1\; l+1)$$

$$= \frac{2(2l)!}{l!\, l!} \frac{\Gamma(l+1)\Gamma(l+1)}{\Gamma(2l+2)}$$

$$= \frac{2}{2l+1}$$

which is Eq. (2.55).

Here, B and Γ denote *beta* and *gamma* functions respectively.

III. Representation of Orthogonality and Orthonormality of Legendre Polynomials by a General equation

If Legendre polynomials are orthogonal and also normalized, they are said to be orthonormal.

Orthonormalization condition of Legendre polynomials is represented by the general formula

$$\int_{-1}^{+1} P_l(x) P_m(x) dx = \frac{2}{2l+1}\delta_{lm} = \frac{2}{2m+1}\delta_{lm} \tag{2.57}$$

where $\delta_{lm} = \begin{cases} 1 \text{ for } l = m, \\ 0 \text{ for } l \neq m \end{cases}$ is known as the Kronecker delta symbol.

2.13 TRIGONOMETRIC SERIES FOR LEGENDRE POLYNOMIALS

We know that

$$\sum_{l=0}^{\infty} h^l P_l(x) = \left(1 - 2hx + h^2\right)^{-\frac{1}{2}}$$

Putting $x = \cos\theta$, we have

$$\sum_{l=0}^{\infty} h^l P_l(\cos\theta) = \left(1 - 2h\cos\theta + h^2\right)^{-\frac{1}{2}}$$

$$= \left[1 - h\left(e^{i\theta} + e^{-i\theta}\right) + h^2\right]^{-\frac{1}{2}}$$

$$= \left[1 - he^{i\theta} - he^{-i\theta} + he^{i\theta}\, he^{-i\theta}\right]^{-\frac{1}{2}}$$

$$= \left[\left(1 - he^{i\theta}\right) - he^{-i\theta}\left(1 - he^{i\theta}\right)\right]^{-\frac{1}{2}}$$

$$= \left(1 - he^{i\theta}\right)^{-\frac{1}{2}}\left(1 - he^{-i\theta}\right)^{-\frac{1}{2}}$$

$$= \left[1 + \frac{1}{2}he^{i\theta} + \frac{1.3}{2.4}h^2 e^{2i\theta} + \frac{1.3.5}{2.4.6}h^3 e^{3i\theta} + \ldots + \frac{1.3.5\ldots(2l-1)}{2.4.6\ldots 2l}h^l e^{il\theta} + \ldots\right]$$

$$\times \left[1 + \frac{1}{2}he^{-2i\theta} + \frac{1.3}{2.4}h^2 e^{-2i\theta} + \frac{1.3.5}{2.4.6}h^3 e^{-3i\theta} + \ldots + \frac{1.3.5\ldots(2l-1)}{2.4.6\ldots 2l}h^l e^{-li\theta} + \ldots\right]$$

Equating coefficients of h^l from both sides, we obtain

$$P_l(\cos\theta) = \frac{1.3.5\ldots(2l-1)}{2.4.6\ldots 2l}\left\{e^{li\theta} + e^{-li\theta}\right\} + \frac{1.3.5\ldots(2l-3)}{2.4.6\ldots(2l-2)}\cdot\frac{1}{2}\left\{e^{(l-2)i\theta} + e^{-(l-2)i\theta}\right\}$$

$$+ \frac{1\cdot 3\cdot 5\ldots(2l-5)}{2\cdot 4\cdot 6\ldots(2l-4)}\,\frac{1\cdot 3}{2\cdot 4}\left\{e^{(l-4)i\theta} + e^{-(l-4)i\theta}\right\} + \ldots$$

$$= \frac{1.3.5\ldots(2l-1)}{2.4.6\ldots 2l} 2\cos l\theta + \frac{1.3.5\ldots(2l-3)}{2.4.6\ldots(2l-2)}\cos(l-2)\theta$$

$$+ \frac{1\cdot 3\cdot 5\ldots(2l-5)}{2\cdot 4\cdot 6\ldots(2l-4)}\frac{3}{4}\cos(l-4)\theta + \ldots$$

$$= \frac{1.3.5\ldots(2l-1)}{2.4.6\ldots 2l} 2\left[\cos l\theta + \frac{1}{2}\frac{2l}{2l-1}\cos(l-2)\theta + \frac{1}{2}\frac{3}{4}\frac{(2l-2)2l}{(2l-3)(2l-1)}\cos(l-4)\theta + \right.$$

$$\left. \ldots + \frac{\{1.3.5\ldots(2j-1)\}\{(2l-2j+2)\ldots(2l-4)(2l-2)(2l)\}}{\{2.4.6\ldots 2j\}\{(2l-2j+1)\ldots(2l-5)(2l-3)(2l-1)\}}\cos(l-2j)\theta + \ldots\right]$$

$$= \frac{(2l)!}{2^{2l-1}(l!)^2}\left[\begin{array}{l}\cos l\theta + \frac{1}{2}\frac{2l}{(2l-1)}\cos(l-2)\theta + \frac{1.3}{2.4}\frac{(2l-2)2l}{(2l-3)(2l-1)}\cos(l-4)\theta + \ldots \\ + \frac{\{1.3.5\ldots(2j-1)\}\{(2l-2j+2)\ldots(2l-4)(2l-2)2l\}}{\{2.4.6\ldots 2j\}\ \{(2l-2j+1)\ldots(2l-5)(2l-3)(2l-1)\}}\cos(l-2j)\theta + \ldots\end{array}\right] \tag{2.58}$$

2.14 EXPANSION OF AN ARBITRARY FUNCTION IN A SERIES OF LEGENDRE POLYNOMIALS

In addition to their orthogonality property, Legendre polynomials form a complete set. [*Complete Set*: If there exists some functions such that every one of them can be expanded as a linear combination of the members of a given set of functions $\{\phi_n\}$, then they are said to form a complete set. The set $\{\phi_n\}$ is called the basis.] Any arbitrary function, $f(x)$, which is sectionally continuous with at least a sectionally continuous derivative in the interval $(-1, 1)$ can be expressed as a linear combination of Legendre polynomials of various orders. [*Sectionally* (or piece-wise) continuous functions: A function $f(x)$ is said to be sectionally continuous in a closed interval (a, b), i.e., $a \le x \le b$, if the interval can be divided into a finite number of sub-intervals such that the function remains continuous in each of the sub- intervals and possesses finite limits everywhere on either side (i.e., finite left hand as well as right hand limits).]

Thus, the arbitrary function $f(x)$ can be expressed as

$$f(x) = \sum_{l=0}^{\infty} a_l P_l(x) \tag{2.59}$$

where a_l are coefficients to be determined.

Multiplying both sides of the above equation by $P_m(x)$ and integrating the results with respect to x within the limit from $x=-1$ to $x=+1$, we obtain

$$\int_{-1}^{+1} f(x)P_m(x)dx = \sum_{l=0}^{\infty}\left[a_l \int_{-1}^{+1} P_l(x)P_m(x)dx\right]$$

$$= a_m \frac{2}{2m+1} \quad \text{[Using the orthogonality of Legendre polynomials.]}$$

Therefore, we have

$$a_m = \frac{2m+1}{2}\int_{-1}^{+1} f(x)P_m(x)dx.$$

Changing the subscript, we write

$$a_l = \frac{2l+1}{2}\int_{-1}^{+1} f(x)P_l(x)dx \tag{2.60}$$

Substituting the values of $a_l\,(l=0,1,2,3,\ldots\text{etc.})$ thus obtained into Eq. (2.59), we can obtain the series of $f(x)$ in terms of Legendre polynomials.

2.15 SOLVED PROBLEMS (RELATED TO LEGENDRE EQUATION AND LEGENDRE POLYNOMIALS)

PROBLEM 1 Obtain the expressions for the first five Legendre polynomials.

Solution:

First method (Using the Sum Formula): We know that

$$P_l(x) = \sum_{r=0}^{\left[\frac{l}{2}\right]} (-1)^r \frac{(2l-2r)!}{2^l\, r!\,(l-r)!}\frac{x^{l-2r}}{(l-2r)!}$$

where $$\left[\frac{l}{2}\right] = \begin{cases} \dfrac{l}{2} & \text{if } l \text{ is non-negative even integer} \\ \dfrac{l-1}{2} & \text{if } l \text{ is positive odd integer} \end{cases}$$

Putting $l=0,1,2,3$ and 4 respectively, we obtain the first five Legendre polynomials as in the following.

$$P_0(x) = (-1)^0 \frac{0!}{2^0\,0!\,0!}\frac{x^0}{0!} = 1$$

$$P_1(x) = (-1)^0 \frac{(2-0)!}{2\cdot 0!(1-0)!} \frac{x^{1-0}}{(1-0)!} = x$$

$$P_2(x) = \sum_{r=0}^{1} (-1)^r \frac{(4-2r)!}{2^2\, r!(2-r)!} \frac{x^{2-2r}}{(2-2r)!}$$

$$= (-1)^0 \frac{4!}{4\cdot 0!\; 2!} \frac{x^2}{2!} + (-1) \frac{(4-2)!}{4\cdot 1!\; 1!} \frac{x^0}{0!}$$

$$= \frac{1}{2}\left(3x^2 - 1\right)$$

$$P_3(x) = \sum_{r=0}^{1} (-1)^r \frac{(6-2r)!}{2^3\, r!(3-r)!} \frac{x^{3-2r}}{(3-2r)!}$$

$$= (-1)^0 \frac{6!}{8\cdot 3!} \frac{x^3}{3!} + (-1) \frac{4!}{8\cdot 1!\; 2!} \frac{x}{1!}$$

$$= \frac{1}{2}\left(5x^3 - 3x\right)$$

$$P_4(x) = \sum_{r=0}^{2} (-1)^r \frac{(8-2r)!}{2^4\, r!(4-r)!} \frac{x^{4-2r}}{(4-2r)!}$$

$$= (-1)^0 \frac{8!}{16\cdot 0!\; 4!} \frac{x^4}{4!} + (-1) \frac{6!}{16\cdot 1!\, 3!} \frac{x^2}{2!} + (-1)^2 \frac{4!}{16\cdot 2!\; 2!} \frac{x^0}{0!}$$

$$= \frac{1}{8}\left(35x^4 - 30x^2 + 3\right)$$

Second Method (Using the Generating Function): We know that

$$\sum_{l=0}^{\infty} h^l P_l(x) = \left(1 - 2hx + h^2\right)^{-\frac{1}{2}} = \left\{1 - \left(2hx - h^2\right)\right\}^{-\frac{1}{2}}$$

or $P_0(x) + hP_1(x) + h^2 P_2(x) + h^3 P_3(x) + h^4 P_4(x) + \ldots\infty$

$$= 1 + \frac{1}{2}\left(2hx - h^2\right) + \frac{1}{2}\cdot\frac{3}{4}\left(2hx - h^2\right)^2 + \frac{1}{2}\cdot\frac{3}{4}\cdot\frac{5}{6}\left(2hx - h^2\right)^3$$

$$+ \frac{1}{2}\cdot\frac{3}{4}\cdot\frac{5}{6}\cdot\frac{7}{8}\left(2hx - h^2\right)^4 + \ldots\infty$$

$$= 1 + hx + \left(-\frac{1}{2} + \frac{3}{2}x^2\right)h^2 + \left(-\frac{3}{2}x + \frac{5}{2}x^3\right)h^3 + \left(\frac{3}{8} - \frac{15}{4}x^2 + \frac{35}{8}x^4\right)h^4 + \ldots\infty \qquad \text{(a)}$$

Equating the coefficients of h^0 from both sides of equation (a), we obtain

$$P_0(x) = 1.$$

Equating the coefficients of h from both sides of equation (a), we obtain

$$P_1(x) = x.$$

Equating the coefficients of h^2 from both sides of equation (a), we obtain

$$P_2(x) = \frac{1}{2}(3x^2 - 1).$$

Equating the coefficients of h^3 from both sides of equation (a), we obtain

$$P_3(x) = \frac{1}{2}(5x^3 - 3x).$$

Equating the coefficients of h^4 from both sides of equation (a), we obtain

$$P_4(x) = \frac{1}{8}(35x^4 - 30x^2 + 3).$$

Third Method (Using Rodrigues' Formula): Rodrigues' formula for Legendre polynomials is

$$P_l(x) = \frac{1}{2^l\, l!} \frac{d^l}{dx^l}(x^2 - 1)^l.$$

Putting $l = 0, 1, 2, 3$ and 4 respectively, we obtain the following.

$$P_0(x) = \frac{1}{2^0\, 0!}(x^2 - 1)^0 = 1$$

$$P_1(x) = \frac{1}{2 \cdot 1!} \frac{d}{dx}(x^2 - 1) = x$$

$$P_2(x) = \frac{1}{2^2 \cdot 2!} \frac{d^2}{dx^2}(x^2 - 1)^2$$

$$= \frac{1}{8} \frac{d^2}{dx^2}(x^4 - 2x^2 + 1) = \frac{1}{8}(12x^2 - 4) = \frac{1}{2}(3x^2 - 1)$$

$$P_3(x) = \frac{1}{2^3 \cdot 3!} \frac{d^3}{dx^3}(x^2 - 1)^3$$

$$= \frac{1}{48} \frac{d^3}{dx^3}(x^6 - 3x^4 + 3x^2 - 1)$$

$$= \frac{1}{48}\left(\frac{6!}{3!}x^3 - \frac{3 \times 4!}{1!}x\right) = \frac{1}{2}(5x^3 - 3x)$$

$$P_4(x) = \frac{1}{2^4 \cdot 4!} \frac{d^4}{dx^4} \left(x^2 - 1\right)^4$$

$$= \frac{1}{16 \cdot 4!} \frac{d^4}{dx^4} \left(x^8 - 4x^6 + \frac{4 \cdot 3}{2!} x^4 - \frac{4 \cdot 3 \cdot 2}{3!} x^2 + 1 \right)$$

$$= \frac{1}{16 \cdot 4!} \left(\frac{8!}{4!} x^4 - 4 \frac{6!}{2!} x^2 + \frac{4 \cdot 3 \cdot 4!}{2!} \right) = \frac{1}{8} \left(35x^4 - 30x^2 + 3\right)$$

PROBLEM 2 Prove that

(i) $P_l(1) = 1$

(ii) $P_l(-x) = (-1)^l P_l(x)$

(iii) $P_l(-1) = (-1)^l$

Solution:

(i) We know that

$$\sum_{l=0}^{\infty} h^l P_l(x) = \left(1 - 2hx + h^2\right)^{-\frac{1}{2}}$$

Putting $x = 1$ we obtain

$$\sum_{l=0}^{\infty} h^l P_l(1) = \left(1 - 2h + h^2\right)^{-\frac{1}{2}} = \left(1 - h\right)^{-1} = \sum_{l=0}^{\infty} h^l$$

Equating coefficients of h^l from both sides, we obtain

$$P_l(1) = 1$$

(ii) We have

$$\sum_{l=0}^{\infty} h^l P_l(x) = \left(1 - 2hx + h^2\right)^{-\frac{1}{2}} \tag{a}$$

and

$$\sum_{l=0}^{\infty} (-h)^l P_l(-x) = \left\{1 - 2(-h)(-x) + (-h)^2\right\}^{-\frac{1}{2}}$$

$$= \left(1 - 2hx + h^2\right)^{-\frac{1}{2}} \tag{b}$$

From equations (a) and (b), we obtain

$$\sum_{l=0}^{\infty} (-1)^l h^l P_l(-x) = \sum_{l=0}^{\infty} h^l P_l(x)$$

Equating coefficients of h^l from both sides, we obtain

$$(-1)^l P_l(-x) = P_l(x)$$

or
$$P_l(-x) = (-1)^l P_l(x)$$

(iii) In the result of (ii), if we put $x = 1$ we obtain

$$P_l(-1) = (-1)^l P_l(1) = (-1)^l \quad \text{[Using the result of (i)]}$$

PROBLEM 3 Prove that $\sum_{l=0}^{\infty} P_l(x) = \dfrac{1}{\sqrt{2-2x}}$

Solution: We know that $\sum_{l=0}^{\infty} h^l P_l(x) = \left(1 - 2hx + h^2\right)^{-\frac{1}{2}}$

Putting $h = 1$, we obtain

$$\sum_{l=0}^{\infty} P_l(x) = (1 - 2x + 1)^{-\frac{1}{2}} = \frac{1}{\sqrt{2-2x}}$$

PROBLEM 4 Prove that $\dfrac{1+h}{h\sqrt{1-2hx+h^2}} - \dfrac{1}{h} = \sum_{l=0}^{\infty} \{P_l(x) + P_{l+1}(x)\} h^l$

Solution: We have

$$\frac{1+h}{h\sqrt{1-2hx+h^2}} - \frac{1}{h} = \left(\frac{1}{h} + 1\right)\left(1 - 2hx + h^2\right)^{-\frac{1}{2}} - \frac{1}{h}$$

$$= \frac{1}{h}\left(1 - 2hx + h^2\right)^{-\frac{1}{2}} + \left(1 - 2hx + h^2\right)^{-\frac{1}{2}} - \frac{1}{h}$$

$$= \frac{1}{h}\sum_{l=0}^{\infty} h^l P_l(x) + \sum_{l=0}^{\infty} h^l P_l(x) - \frac{1}{h}$$

$$= \frac{1}{h}\left\{P_0(x) + \sum_{l=1}^{\infty} h^l P_l(x)\right\} + \sum_{l=0}^{\infty} h^l P_l(x) - \frac{1}{h}$$

$$= \frac{1}{h} + \sum_{l=1}^{\infty} h^{l-1} P_l(x) + \sum_{l=0}^{\infty} h^l P_l(x) - \frac{1}{h}, \text{ as } P_0(x) = 1;$$

$$= \sum_{l=1}^{\infty} h^{l-1} P_l(x) + \sum_{l=0}^{\infty} h^l P_l(x)$$

$$= \sum_{l=0}^{\infty} h^l P_{l+1}(x) + \sum_{l=0}^{\infty} h^l P_l(x)$$

$$= \sum_{l=0}^{\infty} \{P_l(x) + P_{l+1}(x)\} h^l$$

PROBLEM 5 Prove that $\dfrac{1-h^2}{\left(1-2hx+h^2\right)^{\frac{3}{2}}} = \sum_{l=0}^{\infty} (2l+1) P_l(x) h^l$

Solution: We know that

$$\left(1-2hx+h^2\right)^{-\frac{1}{2}} = \frac{1}{\left(1-2hx+h^2\right)^{\frac{1}{2}}} = \sum_{l=0}^{\infty} h^l P_l(x) \qquad \text{(i)}$$

Differentiating both sides once with respect to h, we obtain

$$\frac{x-h}{\left(1-2hx+h^2\right)^{\frac{3}{2}}} = \sum_{l=0}^{\infty} l h^{l-1} P_l(x)$$

Multiplying both sides by $2h$, we have

$$\frac{2hx-2h^2}{(1-2hx+h^2)^{\frac{3}{2}}} = \sum_{l=0}^{\infty} 2lh^l P_l(x) \qquad \text{(ii)}$$

Adding Eqs. (i) and (ii), we write

$$\frac{1-2hx+h^2+2hx-2h^2}{\left(1-2hx+h^2\right)^{\frac{3}{2}}} = \sum_{l=0}^{\infty} (2l+1) P_l(x) h^l$$

or

$$\frac{1-h^2}{\left(1-2hx+h^2\right)^{\frac{3}{2}}} = \sum_{l=0}^{\infty} (2l+1) P_l(x) h^l$$

PROBLEM 6 Prove that

$$P_l\left(-\frac{1}{2}\right) = P_0\left(-\frac{1}{2}\right)P_{2l}\left(\frac{1}{2}\right) + P_1\left(-\frac{1}{2}\right)P_{2l-1}\left(\frac{1}{2}\right) + P_2\left(-\frac{1}{2}\right)P_{2l-2}\left(\frac{1}{2}\right) + \dots$$
$$+ P_{2l-1}\left(-\frac{1}{2}\right)P_1\left(\frac{1}{2}\right) + P_{2l}\left(-\frac{1}{2}\right)P_0\left(\frac{1}{2}\right)$$

Solution: We know that

$$\left(1-2hx+h^2\right)^{-\frac{1}{2}} = \sum_{l=0}^{\infty} h^l P_l(x) \qquad \text{(i)}$$

Substituting x by $\frac{1}{2}$ into Eq. (i), we have

$$(1-h+h^2)^{-\frac{1}{2}} = \sum_{l=0}^{\infty} h^l P_l\left(\frac{1}{2}\right) \tag{ii}$$

Again, substituting x by $\left(-\frac{1}{2}\right)$ in Eq. (i), we obtain

$$\left(1+h+h^2\right)^{-\frac{1}{2}} = \sum_{l=0}^{\infty} h^l P_l\left(-\frac{1}{2}\right) \tag{iii}$$

Multiplying equation (ii) by Eq. (iii), we obtain

$$\left\{\left(1-h+h^2\right)^{-\frac{1}{2}}\right\}\left\{\left(1+h+h^2\right)^{-\frac{1}{2}}\right\} = \left[\sum_{l=0}^{\infty} h^l P_l\left(\frac{1}{2}\right)\right]\left[\sum_{m=0}^{\infty} h^m P_m\left(-\frac{1}{2}\right)\right]$$

or
$$\left\{\left(1+h^2\right)^2 - h^2\right\}^{-\frac{1}{2}} = \sum_{m=0}^{\infty}\sum_{l=0}^{\infty} h^{l+m} P_m\left(-\frac{1}{2}\right)P_l\left(\frac{1}{2}\right)$$

or
$$\left\{1+h^2+h^4\right\}^{-\frac{1}{2}} = \sum_{m=0}^{\infty}\sum_{l=0}^{\infty} h^{l+m} P_m\left(-\frac{1}{2}\right)P_l\left(\frac{1}{2}\right)$$

or
$$\left\{1-2\left(h^2\right)\left(-\frac{1}{2}\right)+\left(h^2\right)^2\right\}^{-\frac{1}{2}} = \sum_{m=0}^{\infty}\sum_{l=0}^{\infty} h^{l+m} P_m\left(-\frac{1}{2}\right)P_l\left(\frac{1}{2}\right)$$

or
$$\sum_{l=0}^{\infty}\left(h^2\right)^l P_l\left(-\frac{1}{2}\right) = \sum_{m=0}^{\infty}\sum_{l=0}^{\infty} h^{l+m} P_m\left(-\frac{1}{2}\right)P_l\left(\frac{1}{2}\right)$$

Equating the coefficients of h^{2l} from both sides, we obtain

$$P_l\left(-\frac{1}{2}\right) = P_0\left(-\frac{1}{2}\right)P_{2l}\left(\frac{1}{2}\right) + P_1\left(-\frac{1}{2}\right)P_{2l-1}\left(\frac{1}{2}\right) + P_2\left(-\frac{1}{2}\right)P_{2l-2}\left(\frac{1}{2}\right) + \ldots$$
$$+ P_{2l-1}\left(-\frac{1}{2}\right)P_1\left(\frac{1}{2}\right) + P_{2l}\left(-\frac{1}{2}\right)P_0\left(\frac{1}{2}\right)$$

PROBLEM 7 Prove that

$$P_l(0) = 0 \text{ for } l = \text{odd integer}$$

and $P_l(0) = \dfrac{(-1)^{\frac{l}{2}}\, l!}{2^l \left\{\left(\dfrac{l}{2}\right)!\right\}^2}$ for l = even integer

Alternatively, prove that

$$P_{2l+1}(0) = 0 \text{ and } P_{2l}(0) = \frac{(-1)^l (2l)!}{2^{2l}(l!)^2} \text{ for } l = 0, 1, 2, 3, 4, \text{etc.}$$

Solution: We know that

$$\sum_{l=0}^{\infty} h^l P_l(x) = \left(1 - 2hx + h^2\right)^{-\frac{1}{2}}$$

Putting $x = 0$, we obtain

$$\sum_{l=0}^{\infty} h^l P_l(0) = \left(1 + h^2\right)^{-\frac{1}{2}}$$

$$= 1 - \frac{1}{2}h^2 + \frac{1}{2!}\cdot\frac{1}{2}\cdot\frac{3}{2}h^4 - \frac{1}{3!}\cdot\frac{1}{2}\cdot\frac{3}{2}\cdot\frac{5}{2}h^6 + \ldots$$

$$+ \left[(-1)^l\right]\frac{1}{l!}\cdot\frac{1}{2}\cdot\frac{3}{2}\cdot\frac{5}{2}\cdot\frac{7}{2}\cdots\frac{(2l-1)}{2}h^{2l} + \ldots\infty$$

$$= \sum_{l=0}^{\infty}\left[(-1)^l\right]\frac{1\cdot 3\cdot 5\cdot 7\cdot(2l-1)}{2^l\, l!}h^{2l} \qquad \text{(i)}$$

Now, equating coefficients of h^l from both sides of Eq. (i) when l is a positive odd integer, we obtain

$$P_l(0) = 0 \quad \text{(where } l \text{ is a positive odd integer)}$$

Alternatively, equating coefficients of h^{2l+1} from both sides of Eq. (i) with l = 0,1,2,3,4, . . . etc., we write

$$P_{2l+1}(0) = 0 \quad \text{(where } l = 0, 1, 2, 3, 4, \ldots \text{ etc.)}$$

Again, equating coefficients of h^l from both sides of Eq. (i) when l is an even integer, we obtain

$$P_l(0) = \frac{(-1)^{\frac{l}{2}}(l-1)!!}{2^{\frac{l}{2}}\left(\dfrac{l}{2}\right)!} \quad \text{where } (l-1)!! = 1.3.5.7\ldots(l-1)$$

Thus, for even integer l, we write

$$P_l(0)=\frac{(-1)^{\frac{1}{2}}1.3.5.7\ldots(l-1)}{2^{\frac{l}{2}}\left(\frac{l}{2}\right)!}=\frac{(-1)^{\frac{1}{2}}\,1.2.3.4.5.6\ldots(l-1)l}{2.4.6.8.\ldots(l-2)l\,2^{\frac{l}{2}}\left(\frac{l}{2}\right)!}$$

$$=\frac{(-1)^{\frac{1}{2}}\,l!}{2^{\frac{l}{2}}\left[1.2.3.4\ldots\left(\frac{l}{2}-1\right)\frac{l}{2}\right]\cdot 2^{\frac{l}{2}}\left(\frac{l}{2}\right)!}=\frac{(-1)^{\frac{1}{2}}\,l!}{2^{l}\left\{\left(\frac{l}{2}\right)!\right\}^{2}}$$

Alternatively, equating coefficients of h^{2l} from both sides of Eq. (i) with $l = 0, 1, 2, 3, 4, \ldots$ etc., we obtain

$$P_{2l}(0)=\frac{(-1)^{l}(2l-1)!!}{2^{l}\,l!}\text{with}(2l-1)!!=1.3.5.7\ldots(2l-1),$$

That is,

$$P_{2l}(0)=\frac{(-1)^{l}\,1.2.3.4.5.6.7\ldots(2l-1)(2l)}{2.4.6.8\ldots(2l-2)(2l)\,2^{l}\,l!}=\frac{(-1)^{l}(2l)!}{2^{2l}(l!)^{2}}$$

where $l = 0, 1, 2, 3, 4, \ldots$ etc.

PROBLEM 8 Prove that

$$1+\frac{1}{2}P_1(\cos\theta)+\frac{1}{3}P_2(\cos\theta)+\frac{1}{4}P_3(\cos\theta)+\ldots\infty=\ln\left[\frac{1+\sin\frac{\theta}{2}}{\sin\frac{\theta}{2}}\right]$$

Solution: We know that

$$\sum_{l=0}^{\infty}h^{l}P_l(x)=\left(1-2hx+h^{2}\right)^{-\frac{1}{2}}=\left(h^{2}-2hx+x^{2}+1-x^{2}\right)^{-\frac{1}{2}}$$

$$=\left[(h-x)^{2}+\left(1-x^{2}\right)\right]^{-\frac{1}{2}}$$

Integrating both sides with respect to h within the limit from $h=0$ to $h=1$ we obtain

$$\sum_{l=0}^{\infty}P_l(x)\int_0^1 h^{l}\,dh=\int_0^1\frac{dh}{\sqrt{(h-x)^{2}+\left(\sqrt{1-x^{2}}\right)^{2}}},$$

or
$$\sum_{l=0}^{\infty}\frac{1}{l+1}P_l(x)=\ln\left[(h-x)+\sqrt{(h-x)^2+1-x^2}\right]_{h=0}^{1}$$
$$=\ln\left[(1-x)+\sqrt{(1-x)^2+1-x^2}\right]-\ln(1-x)$$
$$=\ln\left[(1-x)+\sqrt{2(1-x)}\right]-\ln(1-x)$$
$$=\ln\left[\frac{(1-x)+\sqrt{2(1-x)}}{1-x}\right] \quad \text{(i)}$$

If we put $x=\cos\theta$, then $1-x=1-\cos\theta=2\sin^2\frac{\theta}{2}$, and Eq. (i) becomes

$$\sum_{l=0}^{\infty}\frac{1}{l+1}P_l(\cos\theta)=\ln\left[\frac{2\sin^2\frac{\theta}{2}+2\sin\frac{\theta}{2}}{2\sin^2\frac{\theta}{2}}\right]=\ln\left[\frac{\sin\frac{\theta}{2}+1}{\sin\frac{\theta}{2}}\right],$$

or
$$\ln\left[\frac{1+\sin\frac{\theta}{2}}{\sin\frac{\theta}{2}}\right]=P_0(\cos\theta)+\frac{1}{2}P_1(\cos\theta)+\frac{1}{3}P_2(\cos\theta)+\frac{1}{4}P_3(\cos\theta)+\ldots\infty$$
$$=1+\frac{1}{2}P_1(\cos\theta)+\frac{1}{3}P_2(\cos\theta)+\frac{1}{4}P_3(\cos\theta)+\ldots\infty$$

PROBLEM 9 Prove that $P_n(x)=\sum_{r=0}^{\left[\frac{n}{2}\right]}\frac{n!(x^2-1)^r\,x^{n-2r}}{2^{2r}(r!)^2(n-2r)!}$

Solution: We know that

$$\sum_{n=0}^{\infty}P_n(x)h^n=(1-2xh+h^2)^{-\frac{1}{2}}$$
$$=\left[(1-hx)^2-h^2\left(x^2-1\right)\right]^{-\frac{1}{2}}$$
$$=(1-hx)^{-1}\left[1-\frac{h^2(x^2-1)}{(1-hx)^2}\right]^{-\frac{1}{2}}$$
$$=(1-hx)^{-1}\left[1+\sum_{r=1}^{\infty}\frac{\frac{1}{2}\cdot\frac{3}{2}\cdot\frac{5}{2}\cdots\left(\frac{2r-1}{2}\right)}{r!}\left\{\frac{h^2(x^2-1)}{(1-hx)^2}\right\}^r\right]$$

$$= (1-hx)^{-1}\left[1+\sum_{r=1}^{\infty}\frac{1.3.5\ldots(2r-1)}{2^r\, r!}h^{2r}(x^2-1)^r(1-hx)^{-2r}\right]$$

$$= (1-hx)^{-1}\left[1+\sum_{r=1}^{\infty}\frac{1.2.3.4.5\ldots(2r-1)2r}{2^r\, r!(2.4.6\ldots 2r)}h^{2r}(x^2-1)^r(1-hx)^{-2r}\right]$$

$$= (1-hx)^{-1}\left[1+\sum_{r=1}^{\infty}\frac{(2r)!}{2^{2r}\,(r!)^2}h^{2r}\left(x^2-1\right)^r(1-hx)^{-2r}\right]$$

$$= \sum_{r=0}^{\infty}\frac{(2r)!}{2^{2r}\,(r!)^2}h^{2r}\left(x^2-1\right)^r(1-hx)^{-(2r+1)}$$

$$= \sum_{r=0}^{\infty}\frac{(2r)!}{2^{2r}\,(r!)^2}h^{2r}\left(x^2-1\right)^r\left[1+\sum_{n=1}^{\infty}\frac{(2r+1)(2r+2)(2r+3)\ldots(2r+n)}{n!}h^n x^n\right]$$

$$= \sum_{r=0}^{\infty}\frac{(2r)!}{2^{2r}\,(r!)^2}h^{2r}\left(x^2-1\right)^r\left[1+\sum_{n=1}^{\infty}\frac{(n+2r)!}{n!(2r)!}h^n x^n\right]$$

$$= \sum_{r=0}^{\infty}\frac{(2r)!}{2^{2r}\,(r!)^2}h^{2r}\left(x^2-1\right)^r\sum_{n=0}^{\infty}\frac{(n+2r)!}{n!(2r)!}h^n x^n$$

$$= \sum_{n=0}^{\infty}\sum_{r=0}^{\infty}\frac{(n+2r)!\left(x^2-1\right)^r}{2^{2r}\,(r!)^2\, n!}h^{n+2r}x^n$$

$$= \sum_{n=0}^{\infty}\sum_{r=0}^{\left[\frac{n}{2}\right]}\frac{n!\left(x^2-1\right)^r}{2^{2r}\,(r!)^2\,(n-2r)!}h^n x^{n-2r}$$

which is obtained by replacing n by $n - 2r$

and writing $\left[\dfrac{n}{2}\right]=\begin{cases}\dfrac{n}{2}, & \text{if } n \text{ is even integer,}\\[2mm] \dfrac{n-1}{2}, & \text{if } n \text{ is odd integer}\end{cases}$

Equating coefficients of h^n from both sides of the above equation, we obtain

$$P_n(x)=\sum_{r=0}^{\left[\frac{n}{2}\right]}\frac{n!\left(x^2-1\right)^r x^{n-2r}}{2^{2r}\,(r!)^2\,(n-2r)!}.$$

PROBLEM 10 Using an appropriate recurrence relation among Legendre polynomials of different orders, obtain the expressions for $P_1(x)$ and $P_2(x)$ given that $P_0(x)=1$.

Solution: In the recurrence relation

$$(2l+1)x\,P_l(x)=(l+1)P_{l+1}(x)+l\,P_{l-1}(x),$$

if we put $l=0$ and 1 successively, we obtain

$$x\,P_0(x)=P_1(x)+0,\text{ giving } P_1(x)=x$$

and
$$3x\,P_1(x)=2P_2(x)+P_0(x)$$

or
$$3x^2=2P_2(x)+1 \text{ giving } P_2(x)=\frac{1}{2}\left(3x^2-1\right)$$

PROBLEM 11 Prove that

$$P'_{l+1}(x)+P'_l(x)=P_0(x)+3P_1(x)+5P_2(x)+\ldots+(2l+1)P_l(x)$$
$$=\sum_{r=0}^{l}(2r+1)P_r(x).$$

Solution: We know that

$$P'_{l+1}(x)-P'_{l-1}(x)=(2l+1)P_l(x).$$

Putting $l=1,2,3,4,\ldots(l-2),(l-1),l,$ we obtain successively

$$P'_2(x)-P'_0(x)=3P_1(x),$$
$$P'_3(x)-P'_1(x)=5P_2(x),$$
$$P'_4(x)-P'_2(x)=7P_3(x),$$
$$P'_5(x)-P'_3(x)=9P_4(x),$$
$$\ldots \quad \ldots \quad \ldots$$
$$\ldots \quad \ldots \quad \ldots$$
$$P'_{l-1}(x)-P'_{l-3}(x)=(2l-3)\,P'_{l-2}(x)$$
$$P'_l(x)-P'_{l-2}(x)=(2l-1)P_{l-1}(x),$$
$$P'_{l+1}(x)-P'_{l-1}(x)=(2l+1)P_l(x),$$

Adding all these together, we obtain

$$P'_{l+1}(x)+P'_l(x)-P'_0(x)-P'_1(x)=3P_1(x)+5P_2(x)+7P_3(x)+\cdots+(2l+1)P_l(x)$$

As $P_0(x)=1$ and $P_1(x)=x,$ we have, $P'_0(x)=0$ and $P'_1(x)=1=P_0(x).$

Hence, we write

$$P'_{l+1}(x)+P'_l(x)-0-P_0(x)=3P_1(x)+5P_2(x)+7P_3(x)+\cdots+(2l+1)P_l(x)$$

On transposition, we write

$$P'_{l+1}(x)+P'_l(x)=P_0(x)+3P_1(x)+5P_2(x)+7P_3(x)+\cdots+(2l+1)P_l(x)$$

$$=\sum_{r=0}^{l}(2r+1)P_r(x)$$

PROBLEM 12 Prove that $P_l(x)=\sum_{r=0}^{l}\left[\frac{l!}{(l-r)!r!}\right]^2\left(\frac{x-1}{2}\right)^r\left(\frac{x+1}{2}\right)^{l-r}$

Solution: We know that Rodrigues' formula for Legendre polynomials is

$$P_l(x)=\frac{1}{2^l\,l!}\frac{d^l}{dx^l}(x^2-1)^l=\frac{1}{2^l\,l!}\frac{d^l}{dx^l}\left\{(x-1)^l(x+1)^l\right\}$$

Using Leibnitz formula for successive differentiation, we write

$$P_l(x)=\frac{1}{2^l\,l!}\sum_{r=0}^{l}\frac{l!}{(l-r)!r!}\left\{\frac{d^{l-r}}{dx^{l-r}}(x-1)^l\right\}\left\{\frac{d^r}{dx^r}(x+1)^l\right\}$$

$$=\frac{1}{2^l\,l!}\sum_{r=0}^{l}\frac{l!}{(l-r)!r!}\frac{l!}{\{l-(l-r)\}!}(x-1)^{l-(l-r)}\frac{l!}{(l-r)!}(x+1)^{l-r}$$

$$=\frac{1}{2^l}\sum_{r=0}^{l}\frac{l!}{(l-r)!\,r!}\frac{l!}{r!(l-r)!}(x-1)^r(x+1)^{l-r}$$

$$=\sum_{r=0}^{l}\left[\frac{l!}{(l-r)!\,r!}\right]^2\left(\frac{x-1}{2}\right)^r\left(\frac{x+1}{2}\right)^{l-r}$$

Note: Writing $\frac{d^l}{dx^l}(x^2-1)^l=\frac{d^l}{dx^l}\{(x+1)^l(x-1)^l\}$, one can prove that

$$P_l(x)=\sum_{r=0}^{l}\left[\frac{l!}{(l-r)!\,r!}\right]^2\left(\frac{x-1}{2}\right)^{l-r}\left(\frac{x+1}{2}\right)^r$$

PROBLEM 13 If α is a zero of $P_l(x)$, then prove that $P_{l-1}(\alpha)$ and $P_{l+1}(\alpha)$ are of opposite signs.

Solution: Since α is a zero of $P_l(x)$, we have $P_l(\alpha)=0$

We know that $(2l+1)xP_l(x)=(l+1)P_{l+1}(x)+l\,P_{l-1}(x)$.

Putting $x=\alpha$, we write

$$(2l+1)\alpha\, P_l(\alpha)=(l+1)P_{l+1}(\alpha)+l\,P_{l-1}(\alpha),$$

or
$$0=(l+1)P_{l+1}(\alpha)+l\,P_{l-1}(\alpha), \qquad [\text{as } P_l(\alpha)=0];$$

or
$$-(l+1)P_{l+1}(\alpha)=l\,P_{l-1}(\alpha),$$

or
$$\frac{P_{l+1}(\alpha)}{P_{l-1}(\alpha)}=-\frac{l}{l+1}$$

As l is a non-negative integer, we conclude that $P_{l-1}(\alpha)$ and $P_{l+1}(\alpha)$ are of opposite signs.

PROBLEM 14 Prove that $\int_{-1}^{+1}\frac{P_l(x)}{\sqrt{1-2hx+h^2}}dx=\frac{2h^l}{2l+1}$ where l is a non-negative integer.

Solution: We write

$$\int_{-1}^{+1}\frac{P_l(x)}{\sqrt{1-2hx+h^2}}dx=\int_{-1}^{+1}P_l(x)\left\{\sum_{n=0}^{\infty}h^nP_n(x)\right\}dx$$

$$=\sum_{n=0}^{\infty}h^n\int_{-1}^{+1}P_l(x)P_n(x)dx$$

$$=h^l\frac{2}{2l+1}$$

since
$$\int_{-1}^{+1}P_l(x)P_n(x)dx=\begin{cases}0 \text{ for } l\neq n,\\ \dfrac{2}{2l+1}\text{ for } l=n\end{cases}$$

PROBLEM 15 Prove that

$$\text{(i) } P_l'(1)=\frac{1}{2}l(l+1) \quad\text{and}\quad \text{(ii) } P_l'(-1)=(-1)^{l+1}\frac{1}{2}l(l+1)$$

where the prime symbol denotes differentiation with respect to x.

Solution: We know that

$$(1-x^2)P_l''(x)-2x\,P_l'(x)+l(l+1)P_l(x)=0$$

(i) Putting $x=1$, we have

$$0-2P_l'(1)+l(l+1)\,P_l(1)=0$$

Since $P_l(1)=1$, we write

$$l(l+1)=2P_l'(1) \text{ and hence } P_l'(1)=\frac{1}{2}l(l+1)$$

(ii) Again, putting $x=-1$, we write

$$0-2(-1)P_l'(-1)+l(l+1)P_l(-1)=0$$

Since, $P_l(-1)=(-1)^l$, we write

$$2P_l'(-1)+(-1)^l l(l+1)=0,$$

or
$$2P_l'(-1)=-(-1)^l l(l+1),$$

or
$$P_l'(-1)=(-1)^{l+1}\frac{1}{2}l(l+1)$$

PROBLEM 16 Prove that all the zeroes of $P_l(x)$ are real and lie between -1 and $+1$.

Solution: Zeroes of $P_l(x)$ are the values of x at which $P_l(x)=0$, that is, the zeroes of $P_l(x)$ are the roots of $P_l(x)=0$.

The function $f(x)=\left(x^2-1\right)^l=(x-1)^l(x+1)^l$ vanishes at both the points $x=-1$ and $x=+1$.

By Rolle's theorem, $\dfrac{d^l f(x)}{dx^l}=\dfrac{d^l}{dx^l}\left(x^2-1\right)^l$ will vanish at l intermediate points between -1 and $+1$ but not at the points $x=-1$ and $x=+1$.

Thus, $P_l(x)=\dfrac{1}{2^l\, l!}\dfrac{d^l}{dx^l}\left(x^2-1\right)^l$ will vanish at l intermediate points between -1 and $+1$ but not at $x=-1$ and $x=+1$, since $P_l(1)=1$ and $P_l(-1)=(-1)^l$.

Hence, all the l zeroes $P_l(x)$ lie on the real axis between $x=-1$ and $x=+1$

That is, all the zeroes of $P_l(x)$ are real and lie between -1 and $+1$ indicating that the limiting values of x in the Legendre polynomial $P_l(x)$ are -1 and $+1$.

PROBLEM 17 Prove that the limiting values of x in the Legendre polynomial $P_l(x)$ are -1 and $+1$.

Solution: Same as in Problem 16.

PROBLEM 18 Prove that

$$(l+1)\left[P_l(x)P_{l+1}'(x)-P_l'(x)P_{l+1}(x)\right]=(l+1)^2\{P_l(x)\}^2+\left(1-x^2\right)\{P_l'(x)\}^2$$

Solution: Let us recall the recurrence relations

$$\left(1-x^2\right)P_l'(x)=(l+1)\left\{xP_l(x)-P_{l+1}(x)\right\} \tag{i}$$

and
$$(l+1)P_l(x)=P_{l+1}'(x)-xP_l'(x) \tag{ii}$$

Multiplying Eq. (i) by $P_l'(x)$ and adding the result to $(l+1)P_l(x)$ times Eq. (ii), we obtain

$$(l+1)^2\left\{P_l(x)\right\}^2+\left(1-x^2\right)\left\{P_l'(x)\right\}^2=(l+1)\left[P_l(x)P_{l+1}'(x)-P_l'(x)P_{l+1}(x)\right]$$

PROBLEM 19 Prove that all the zeroes of $P_l(x)$ are distinct.

Solution: Let us suppose that the zeroes of $P_l(x)$ are not all distinct such that at least two of them are equal. Let α be a repeated zero of $P_l(x)$, then α is a double root of

$$P_l(x)=0$$

Then, $P_l(\alpha)=0$ and $P_l'(\alpha)=0$ where the prime symbol denotes differentiation with respect to x.

We know that

$$\left(1-x^2\right)P_l''(x)-2x\,P_l'(x)+l(l+1)P_l(x)=0 \tag{i}$$

At $x=\alpha$, we have, $P_l(\alpha)=0$ and $P_l'(\alpha)=0$

Therefore, Eq. (i) yields

$$\left(1-\alpha^2\right)P_l''(\alpha)-2\alpha P_l'(\alpha)+l(l+1)P_l(\alpha)=0,$$

or
$$\left(1-\alpha^2\right)P_l''(\alpha)-0+0=0$$

or
$$P_l''(\alpha)=0$$

Differentiating Eq. (i) with respect to x, we obtain

$$\left(1-x^2\right)P_l'''(x)-4x\,P_l''(x)+\left(l^2+l-2\right)P_l'(x)=0$$

At $x=\alpha$, we have

$$\left(1-\alpha^2\right)P_l'''(\alpha)-4\alpha P_l''(\alpha)+\left(l^2+l-2\right)P_l'(\alpha)=0,$$

or
$$\left(1-\alpha^2\right)P_l'''(\alpha)-0+0=0,$$

or
$$P_l'''(\alpha)=0$$

Proceeding in this manner, we obtain

$$P_l^{(\text{iv})}(\alpha)=P_l^{(\text{v})}(\alpha)=\cdots=P_l^{(l)}(\alpha)=0$$

which is incorrect since the lth derivative of $P_l(x)$ at $x=\alpha$ is not zero.

It can be seen that

$$P_l^{(l)}(\alpha) = \frac{d^l}{dx^l} P_l(x)\bigg|_{x=\alpha} = \left[\frac{d^l}{dx^l} \sum_{r=0}^{\left[\frac{l}{2}\right]} \frac{(-1)^r (2l-2r)!}{2^l r!(l-r)!} \frac{x^{l-2r}}{(l-2r)!} \right]_{x=\alpha}$$

where
$$\left[\frac{l}{2}\right] = \begin{cases} \frac{l}{2} \text{ if } l \text{ is even,} \\ \frac{l-1}{2} \text{ if } l \text{ is odd} \end{cases}$$

or
$$P_l^{(l)}(\alpha) = \frac{(2l)!\, l!}{2^l\, l!\, l!} = \frac{(2l)!}{2^l\, l!} \neq 0$$

Therefore, our assumption that $P_l(x)$ has a repeated zero, that is the equation $P_l(x) = 0$ has a repeated root is incorrect.

Hence, all the zeroes of $P_l(x)$ are distinct.

PROBLEM 20 Prove that $\int_{-1}^{+1} P_l(x)dx = 0$ except for $l = 0$ in which case the value is 2.

Solution: For $l \neq 0$, using Rodrigues' formula, we write

$$\int_{-1}^{+1} P_l(x)dx = \int_{-1}^{+1} \frac{1}{2^l\, l!} \frac{d^l}{dx^l}(x^2-1)^l\, dx$$

$$= \frac{1}{2^l\, l!} \int_{-1}^{+1} \frac{d^l}{dx^l}(x^2-1)^l\, dx$$

$$= \frac{1}{2^l\, l!} \left[\frac{d^{l-1}}{dx^{l-1}}(x^2-1)^l \right]_{-1}^{+1}$$

$$= \frac{1}{2^l\, l!} \left[\frac{d^{l-1}}{dx^{l-1}} \left\{ (x+1)^l (x-1)^l \right\} \right]_{-1}^{+1}$$

$$= \frac{1}{2^l\, l!} \left[\left\{ \frac{d^{l-1}}{dx^{l-1}}(x+1)^l \right\} (x-1)^l + (l-1) \left\{ \frac{d^{l-2}}{dx^{l-2}}(x+1)^l \right\} \frac{d}{dx}(x-1)^l + \ldots \right.$$

$$\left. + (x+1)^l \frac{d^{l-1}}{dx^{l-1}}(x-1)^l \right]_{-1}^{+1}, \text{ by Leibnitz theorem}$$

$$=\frac{1}{2^l\, l!}\left[\frac{l!}{1!}(x+1)(x-1)^l+(l-1)\frac{l!}{2!}(x+1)^2\, l(x-1)^{l-1}+\cdots+(x+1)^l\frac{l!}{1!}(x-1)\right]_{-1}^{+1}$$

$=0$, as all terms contain $(x+1)$ and $(x-1)$

Alternative Method

For $l \neq 0$, we have

$$\int_{-1}^{+1} P_l(x)\,dx = \int_{-1}^{+1} 1\cdot P_l(x)\,dx$$

$$= \int_{-1}^{+1} P_0(x)P_l(x)\,dx = \delta_{0l} = 0 \text{ for } l \neq 0$$

For $l = 0$, we write

$$\int_{-1}^{+1} P_0(x)\,dx = \int_{-1}^{+1} 1\cdot dx = [x]_{-1}^{+1} = 2$$

PROBLEM 21 Evaluate $\int_{-1}^{+1} x^m P_l(x)\,dx$ for $m \leq l$

Solution: Using Rodrigues' formula, we have

$$I = \int_{-1}^{+1} x^m P_l(x)\,dx$$

$$= \frac{1}{2^l\, l!}\int_{-1}^{+1} x^m \frac{d^l}{dx^l}(x^2-1)^l\,dx$$

$$= \frac{1}{2^l\, l!}\left[\left\{x^m \frac{d^{l-1}}{dx^{l-1}}(x^2-1)^l\right\}_{-1}^{+1} - \int_{-1}^{+1} m x^{m-1}\frac{d^{l-1}}{dx^{l-1}}(x^2-1)^l\,dx\right],$$

Integrating by parts.

If we expand $\frac{d^{l-1}}{dx^{l-1}}(x^2-1)^l$ using Leibnitz theorem, all terms will contain the factors $(x+1)$ and $(x-1)$ and, therefore, on substitution of the limits $x=-1$ and $x=+1$, the first term within the square brackets will vanish.

Then, we have

$$I = 0 - \frac{m}{2^l\, l!}\int_{-1}^{+1} x^{m-1}\frac{d^{l-1}}{dx^{l-1}}\left(x^2-1\right)^l dx$$

$$= 0 + \frac{(-1)^2\, m(m-1)}{2^l\, l!}\int_{-1}^{+1} x^{m-2}\frac{d^{l-2}}{dx^{l-2}}\left(x^2-1\right)^l dx, \text{ integrating by parts once more.}$$

In this manner, if we continue the integration by parts in the right-hand side m times, we will get

$$I = \int_{-1}^{+1} x^m P_l(x)\,dx = \frac{(-1)^m\, m!}{2^l\, l!}\int_{-1}^{+1}\frac{d^{l-m}}{dx^{l-m}}\left(x^2-1\right)^l dx, \text{ if } m<l,$$

$$= \frac{(-1)^m m!}{2^l\, l!}\left[\frac{d^{l-m-1}}{dx^{l-m-1}}(x^2-1)^l\right]_{-1}^{+1} = 0 \text{ for } m<l \qquad \text{(i)}$$

For $m=l$, we have, as above

$$I = \int_{-1}^{+1} x^l P_l(x)\,dx = \frac{(-1)^l\, l!}{2^l\, l!}\int_{-1}^{+1}\frac{d^{l-l}}{dx^{l-l}}\left(x^2-1\right)^l dx$$

$$= \frac{1}{2^l}\int_{-1}^{+1}\left(1-x^2\right)^l dx$$

$$= \frac{1}{2^l}\int_{-1}^{+1}(1+x)^l(1-x)^l dx$$

If we put $x = 2u-1 \text{ or } u = \frac{1}{2}(x+1), \text{we have, } dx = 2du$

Then, we write

$$I = \int_{-1}^{+1} x^l P_l(x)\,dx = \frac{1}{2^l}\int_0^1 (2u)^l(2-2u)^l\, 2du$$

$$= 2^{l+1}\int_0^1 u^l(1-u)^l dx$$

$$= 2^{l+1}B(l+1, l+1)$$

$$= 2^{l+1}\frac{\Gamma(l+1)\Gamma(l+1)}{\Gamma(2l+2)} = \frac{2^{l+1}(l!)^2}{(2l+1)!} \text{ (with } m = l) \qquad \text{(ii)}$$

Here, $B(l+1, l+1)$ and $\Gamma(l+1)$ are beta and gamma functions.

PROBLEM 22 Prove that $\int_{-1}^{+1} [P_l'(x)]^2 dx = l(l+1)$ where the prime symbol denotes differentiation with respect to x.

Solution: By Christoffel's expansion, we have

$$P_l'(x) = (2l-1)P_{l-1}(x) + (2l-5)P_{l-3}(x) + (2l-9)P_{l-5}(x) + \ldots \quad \text{(i)}$$

The last term being $3P_1(x)$ or $P_0(x)$ according as l is even or odd.

For l even: If l is even, let m be the number of terms on the right-hand side of Eq. (i), then the last coefficient is

$$3 = (2l - 1) + (m - 1)(-4) \text{ as the coefficients are in A.P.}$$

giving $m = \dfrac{l}{2}$.

Now, we have

$$\int_{-1}^{+1} [P_l'(x)]^2 dx = (2l-1)^2 \int_{-1}^{+1} [P_{l-1}(x)]^2 dx + (2l-5)^2 \int_{-1}^{+1} [P_{l-3}(x)]^2 dx$$

$$+ (2l-9)^2 \int_{-1}^{+1} [P_{l-5}(x)]^2 dx + \ldots + 3^2 \int_{-1}^{+1} [P_1(x)]^2 dx, \left[\text{having } \frac{l}{2} \text{ terms}\right]$$

Here, other integrals vanish by orthogonality condition of Legendre polynomials. Now, we write

$$\int_{-1}^{+1} [P_l'(x)]^2 dx = (2l-1)^2 \frac{2}{\{2(l-1)+1\}} + (2l-5)^2 \frac{2}{\{2(l-3)+1\}} + (2l-9)^2 \frac{2}{\{2(l-5)+1\}} + \ldots$$

$$+ 3^2 \frac{2}{2(1)+1} \text{ having } l/2 \text{ terms.}$$

$$= 2[(2l-1) + (2l-5) + (2l-9) + \ldots + 3]$$

$$= 2\left[\frac{1}{2}\left(\frac{l}{2}\right)\{(2l-1)+3\}\right] = l(l+1), \text{ using sum formula of A.P. series}$$

For l odd: If l is odd, let n be the number of terms on the right-hand side of Eq. (i), then the last coefficient is

$$1 = (2l-1) + (n-1)(-4) \text{ as the coefficients are in A.P.}$$

giving $n = \dfrac{l+1}{2}$.

Then, we have

$$\int_{-1}^{+1}[P_l'(x)]^2\,dx=(2l-1)^2\int_{-1}^{+1}[P_{l-1}(x)]^2\,dx+(2l-5)^2\int_{-1}^{+1}[P_{l-3}(x)]^2\,dx$$

$$+(2l-9)^2\int_{-1}^{+1}[P_{l-5}(x)]^2\,dx+\ldots+1^2\int_{-1}^{+1}[P_0(x)]^2\,dx,\text{ having }\frac{l+1}{2}\text{ terms}$$

$$=(2l-1)^2\frac{2}{\{2(l-1)+1\}}+(2l-5)^2\frac{2}{\{2(l-3)+1\}}+(2l-9)^2\frac{2}{\{2(l-5)+1\}}+\ldots$$

$$+\ldots+2,\text{ having }\frac{l+1}{2}\text{ terms}$$

$$=2[(2l-1)+(2l-5)+(2l-9)+\ldots+1]$$

$$=2\left[\frac{1}{2}\frac{(l+1)}{2}\{(2l-1)+1\}\right]=l(l+1)$$

PROBLEM 23 Prove that $\int_{-1}^{+1}(x^2-1)P_{l+1}(x)P_l'(x)dx=\dfrac{2l(l+1)}{(2l+1)(2l+3)}$ where the prime symbol denotes differentiation with respect to x.

Solution: Let us recall the recurrence relation,

$$(1-x^2)P_l'(x)=l[P_{l-1}(x)-xP_l(x)]$$

or

$$(x^2-1)P_l'(x)=l[xP_l(x)-P_{l-1}(x)]$$

Multiplying both sides by $P_{l+1}(x)$ and then integrating the result with respect to x within the limit from $x=-1$ to $x=+1$, we write

$$\int_{-1}^{+1}(x^2-1)P_{l+1}(x)P_l'(x)dx=l\int_{-1}^{+1}xP_l(x)P_{l+1}(x)dx-l\int_{-1}^{+1}P_{l-1}(x)P_{l+1}(x)dx$$

$$=l\int_{-1}^{+1}xP_l(x)P_{l+1}(x)dx \qquad \text{(i)}$$

as the second integral vanishes by orthogonality condition.

Using the recurrence relation

$$(2l+1)x\,P_l(x)=(l+1)P_{l+1}(x)+l\,P_{l-1}(x)$$

we have

$$x\,P_l(x)=\frac{(l+1)}{(2l+1)}P_{l+1}(x)+\frac{l}{(2l+1)}P_{l-1}(x) \qquad \text{(ii)}$$

Substituting Eq. (ii) into Eq. (i), we obtain

$$\int_{-1}^{+1}(x^2-1)P_{l+1}(x)P_l'(x)dx = l\int_{-1}^{+1}\left[\frac{l+1}{2l+1}P_{l+1}(x)+\frac{l}{2l+1}P_{l-1}(x)\right]P_{l+1}(x)dx$$

$$=\frac{l(l+1)}{2l+1}\int_{-1}^{+1}\left[P_{l+1}(x)\right]^2 dx+\frac{l^2}{2l+1}\int_{-1}^{+1}P_{l-1}(x)P_{l+1}(x)dx$$

$$=\frac{l(l+1)}{(2l+1)}\frac{2}{2(l+1)+1}+0=\frac{2l(l+1)}{(2l+1)(2l+3)}$$

PROBLEM 24 Prove that $\int_{-1}^{+1} xP_l(x)P_{l-1}(x)dx=\frac{2l}{(2l-1)(2l+1)}$

Solution: From the recurrence relation

$$(2l+1)xP_l(x)=(l+1)P_{l+1}(x)+lP_{l-1}(x),$$

we have

$$xP_l(x)=\frac{1}{2l+1}\left[(l+1)P_{l+1}(x)+l\,P_{l-1}(x)\right]$$

Hence, we have

$$\int_{-1}^{+1} xP_l(x)P_{l-1}(x)dx=\int_{-1}^{+1}\frac{1}{2l+1}\left[(l+1)P_{l+1}(x)+l\,P_{l-1}(x)\right]P_{l-1}(x)dx$$

$$=\frac{l+1}{2l+1}\int_{-1}^{+1}P_{l+1}(x)P_{l-1}(x)dx+\frac{l}{2l+1}\int_{-1}^{+1}P_{l-1}(x)P_{l-1}(x)dx$$

$$=0+\frac{l}{(2l+1)}\frac{2}{\{2(l-1)+1\}}=\frac{2l}{(2l+1)(2l-1)}$$

PROBLEM 25 Prove that (a) $\int_{-1}^{+1}(1-x^2)P_l'(x)P_m'(x)dx=0$, for $l\neq m$,

(b) $\int_{-1}^{+1}(1-x^2)P_l'(x)P_l'(x)dx=\frac{2l(l+1)}{2l+1}$

where the prime denotes differentiation with respect to x .

Solution: We have

$$\int_{-1}^{+1}(1-x^2)P_l'(x)P_m'(x)dx=\left[(1-x^2)P_l'(x)P_m(x)\right]_{-1}^{+1}-\int_{-1}^{+1}\frac{d}{dx}\left\{(1-x^2)P_l'(x)\right\}P_m(x)dx,$$

$$= 0 - \int_{-1}^{+1} \frac{d}{dx}\left\{\left(1-x^2\right)P_l'(x)\right\}P_m(x)\,dx \qquad \text{(i)}$$

As the Legendre polynomial, $P_l(x)$, is the solution of the Legendre differential equation, we have

$$\frac{d}{dx}\left\{\left(1-x^2\right)P_l'(x)\right\} = -l(l+1)P_l(x) \qquad \text{(ii)}$$

Substituting Eq. (ii) into Eq. (i), we have

$$\int_{-1}^{+1}\left(1-x^2\right)P_l'(x)P_m'(x)\,dx = -\int_{-1}^{+1} -l(l+1)P_l(x)P_m(x)\,dx$$

$$= l(l+1)\int_{-1}^{+1} P_l(x)P_m(x)\,dx \qquad \text{(iii)}$$

(a) For $l \neq m$, we have

$$\int_{-1}^{+1}\left(1-x^2\right)P_l'(x)P_m'(x)\,dx = l(l+1)0 = 0$$

(b) For $l = m$, we have

$$\int_{-1}^{+1}\left(1-x^2\right)P_l'(x)P_l'(x)\,dx = l(l+1)\frac{2}{2l+1} = \frac{2l(l+1)}{2l+1}$$

Note: We can combine the results in (a) and (b) into the single equation

$\displaystyle\int_{-1}^{+1}\left(1-x^2\right)P_l'(x)P_m'(x)\,dx = \frac{2l(l+1)}{2l+1}\delta_{lm}$ where δ_{lm} is the *Kronecker delta* symbol.

PROBLEM 26 Prove that $\displaystyle\int_{-1}^{+1} x^2 P_{l+1}(x)P_{l-1}(x)\,dx = \frac{2l(l+1)}{(2l-1)(2l+1)(2l+3)}$ and deduce the value of $\displaystyle\int_0^1 x^2 P_{l+1}(x)P_{l-1}(x)\,dx$.

Solution: In the recurrence relation

$$(2l+1)\,x\,P_l(x) = (l+1)P_{l+1}(x) + l\,P_{l-1}(x),$$

Replacing l by $(l + 1)$ and $(l - 1)$ successively, we have

$$(2l+3)x\,P_{l+1}(x)=(l+2)P_{l+2}(x)+(l+1)P_l(x)$$

and
$$(2l-1)\,x\,P_{l-1}(x) = lP_l(x) + (l-1)\,P_{l-2}(x)$$

Multiplying these two equations together and integrating the result with respect to x within the limit from $x=-1$ to $x=+1$, we obtain

$$(2l+3)(2l-1)\int_{-1}^{+1} x^2 P_{l+1}(x)P_{l-1}(x)\,dx$$

$$=l(l+1)\int_{-1}^{+1} P_l(x)P_l(x)\,dx, \text{ as other integrals vanish by orthogonality condition}$$

$$=l(l+1)\frac{2}{2l+1}$$

Therefore, we write

$$\int_{-1}^{+1} x^2 P_{l+1}(x)P_{l-1}(x)\,dx=\frac{2l(l+1)}{(2l-1)(2l+1)(2l+3)}$$

Deduction: We know that the integrand is always an even function for any positive integer l.

Therefore, we have

$$\int_{0}^{1} x^2 P_{l+1}(x)P_{l-1}(x)\,dx=\frac{l(l+1)}{(2l-1)(2l+1)(2l+3)}$$

PROBLEM 27 Prove that $\int_{-1}^{+1} x^2\left[P_l(x)\right]^2 dx=\frac{1}{8(2l-1)}+\frac{3}{4(2l+1)}+\frac{1}{8(2l+3)}$

Solution: We know the recurrence relation

$$(2l+1)xP_l(x)=(l+1)P_{l+1}(x)+lP_{l-1}(x)$$

Squaring both sides and integrating the result with respect to x within the limit from $x = -1$ to $x = +1$, we have

$$(2l+1)^2\int_{-1}^{+1} x^2\left[P_l(x)\right]^2 dx=(l+1)^2\int_{-1}^{+1}\left[P_{l+1}(x)\right]^2 dx+l^2\int_{-1}^{+1}\left[P_{l-1}(x)\right]^2 dx,$$

since other integrals vanish by orthogonality condition.

$$=(l+1)^2\frac{2}{\{2(l+1)+1\}}+l^2\frac{2}{\{2(l-1)+1\}}=\frac{2(l+1)^2}{(2l+3)}+\frac{2l^2}{(2l-1)}$$

Therefore, we have

$$\int_{-1}^{+1} x^2[P_l(x)]^2\,dx = \frac{2(l+1)^2}{(2l+1)^2(2l+3)} + \frac{2l^2}{(2l+1)^2(2l-1)} \tag{i}$$

Using the theory of partial fractions, we obtain

$$\frac{2(l+1)^2}{(2l+1)^2(2l+3)} = \frac{3}{8(2l+1)} + \frac{1}{4(2l+1)^2} + \frac{1}{8(2l+3)} \tag{ii}$$

and

$$\frac{2l^2}{(2l+1)^2(2l-1)} = \frac{1}{8(2l-1)} + \frac{3}{8(2l+1)} - \frac{1}{4(2l+1)^2} \tag{iii}$$

Substituting Eqs. (ii) and (iii) into Eq. (i), we obtain

$$\int_{-1}^{+1} x^2\left[P_l(x)\right]^2 dx = \frac{1}{8(2l-1)} + \frac{3}{4(2l+1)} + \frac{1}{8(2l+3)}$$

PROBLEM 28 Prove that

$$xP_l'(x) = lP_l(x) + (2l-3)P_{l-2}(x) + (2l-7)P_{l-4}(x) + (2l-11)P_{l-6}(x) + \ldots$$

and hence deduce that

$$\int_{-1}^{+1} xP_l(x)P_m'(x)\,dx = 0,\ 2 \text{ or } \frac{2l}{2l+1}$$

where the prime denotes differentiation with respect to x.

Solution: We know the recurrence relation

$$(2l+1)P_l(x) = P_{l+1}'(x) - P_{l-1}'(x) \tag{i}$$

Replacing l by $(l-2), (l-4), (l-6), \ldots,$ in Eq. (i), we have

$$(2l-3)P_{l-2}(x) = P_{l-1}'(x) - P_{l-3}'(x) \tag{ii}$$

$$(2l-7)P_{l-4}(x) = P_{l-3}'(x) - P_{l-5}'(x) \tag{iii}$$

$$(2l-11)P_{l-6}(x) = P_{l-5}'(x) - P_{l-7}'(x), \tag{iv}$$

$$\ldots = \ldots \text{ etc.}$$

Also, we know the recurrence relation

$$lP_l(x) = xP_l'(x) - P_{l-1}'(x) \tag{v}$$

Adding all the Eqs. from (ii) to (v) together, we obtain

$$l\,P_l(x)+(2l-3)P_{l-2}(x)+(2l-7)P_{l-4}(x)+(2l-11)P_{l-6}(x)+\ldots$$

$$= xP_l'(x) \qquad \text{(vi)}$$

Deduction: In Eq. (vi), if we replace l by m, we obtain

$$mP_m(x)+(2m-3)P_{m-2}(x)+(2m-7)P_{m-4}(x)+(2m-11)P_{m-6}(x)+\ldots = xP_m'(x)$$

Multiplying both sides by $P_l(x)$ and integrating the result with respect to x within the limit from $x=-1$ to $x=+1$, we obtain

$$\int_{-1}^{+1} xP_l(x)P_m'(x)\,dx = m\int_{-1}^{+1} P_l(x)P_m(x)\,dx+(2m-3)\int_{-1}^{+1} P_l(x)P_{m-2}(x)\,dx$$

$$+(2m-7)\int_{-1}^{+1} P_l(x)P_{m-4}(x)\,dx+(2m-11)\int_{-1}^{+1} P_l(x)P_{m-6}(x)\,dx+\ldots \qquad \text{(vii)}$$

When $l=m-2,$ right-hand side of Eq. (vii) becomes

$$(2m-3)\frac{2}{\{2(m-2)+1\}}=2$$

When $l=m-4,$ right-hand side of Eq. (vii) becomes

$$(2m-7)\frac{2}{\{2(m-4)+1\}}=2$$

Similarly, when $l = m-6, m-8, m-10, \ldots$ etc. right-hand side of Eq. (vii) becomes 2.

When $l=m,$ right-hand side of Eq. (vii) becomes $\dfrac{2l}{2l+1}$

In other cases, the right-hand side of Eq. (vii) becomes zero.
Hence, we write

$$\int_{-1}^{+1} xP_l(x)P_m'(x)\,dx = 0,\ 2 \text{ or } \frac{2l}{2l+1}$$

PROBLEM 29 Obtain the value of

$$\int_{-1}^{+1}\left(1-x^2\right)P_l''(x)\,dx \text{ where } P_l''(x)=\frac{d^2}{dx^2}P_l(x) \text{ and } l\neq 0$$

Solution:

$$\int_{-1}^{+1}(1-x^2)P_l''(x)\,dx=\left[(1-x^2)P_l'(x)\right]_{-1}^{+1}-\int_{-1}^{+1}-2x\,P_l'(x)\,dx,$$

Integrating by parts.

$$=0+2\int_{-1}^{+1}x\,P_l'(x)\,dx$$

$$=2\left\{\left[x\,P_l(x)\right]_{-1}^{+1}-\int_{-1}^{+1}P_l(x)\,dx\right\},\ \text{integrating by parts again}$$

$$=2\{P_l(1)-(-1)P_l(-1)-0\},\ \text{as}\int_{-1}^{+1}P_l(x)\,dx=0\ \text{for}\ l\neq 0$$

$$=2\{1+(-1)^l\}=\begin{cases}4\ \text{if}\ l\ \text{is even,}\\ 0\ \text{if}\ l\ \text{is odd}\end{cases}$$

Here, use has been made of the relations $P_l(1)=1$ and $P_l(-1)=(-1)^l$

PROBLEM 30 Obtain the value of $\int_{-1}^{+1}xP_l(x)P_m(x)\,dx$

Solution: We know the recurrence relation

$$(2l+1)xP_l(x)=(l+1)P_{l+1}(x)+l\,P_{l-1}(x)$$

Multiplying both sides by $P_m(x)$ and integrating the result with respect to x within the limit from $x=-1$ to $x=+1$, we obtain

$$(2l+1)\int_{-1}^{+1}x\,P_l(x)P_m(x)\,dx=(l+1)\int_{-1}^{+1}P_{l+1}(x)P_m(x)\,dx+l\int_{-1}^{+1}P_{l-1}(x)P_m(x)\,dx$$

Therefore, we have

$$\int_{-1}^{+1}xP_l(x)P_m(x)\,dx=\frac{(l+1)}{(2l+1)}\int_{-1}^{+1}P_{l+1}(x)P_m(x)\,dx+\frac{l}{(2l+1)}\int_{-1}^{+1}P_{l-1}(x)P_m(x)\,dx$$

$$=\begin{cases}\dfrac{(l+1)}{(2l+1)}\dfrac{2}{\{2(l+1)+1\}}=\dfrac{2(l+1)}{(2l+1)(2l+3)} & \text{when } m=l+1;\\[2ex] \dfrac{l}{(2l+1)}\dfrac{2}{\{2(l-1)+1\}}=\dfrac{2l}{(2l+1)(2l-1)} & \text{when } m=l-1;\\[2ex] 0, & \text{in all other cases}\end{cases}$$

PROBLEM 31 Obtain the value of $\int_{-1}^{+1} x^2 P_l(x) P_m(x)\, dx$

Solution: We know the recurrence relation

$$(2l+1)xP_l(x) = (l+1)P_{l+1}(x) + lP_{l-1}(x) \quad \text{(i)}$$

Replacing l by m, we have

$$(2m+1)xP_m(x) = (m+1)P_{m+1}(x) + mP_{m-1}(x) \quad \text{(ii)}$$

Integrating the product of Eqs. (i) and (ii) with respect to x within the limit from $x=-1$ to $x=+1$, we obtain

$$(2l+1)(2m+1)\int_{-1}^{+1} x^2 P_l(x) P_m(x)\, dx$$

$$= (l+1)(m+1)\int_{-1}^{+1} P_{l+1}(x)P_{m+1}(x)\,dx + (l+1)m\int_{-1}^{+1} P_{l+1}(x)P_{m-1}(x)\,dx$$

$$+ l(m+1)\int_{-1}^{+1} P_{l-1}(x)P_{m+1}(x)\,dx + lm\int_{-1}^{+1} P_{l-1}(x)P_{m-1}(x)\,dx \quad \text{(iii)}$$

CASE I: When $m = l$, Eq. (iii) becomes

$$(2l+1)^2\int_{-1}^{+1} x^2\left[P_l(x)\right]^2 dx = (l+1)^2\int_{-1}^{+1}\left[P_{l+1}(x)\right]^2 dx + l^2\int_{-1}^{+1}\left[P_{l-1}(x)\right]^2 dx$$

[Other integrals vanish by orthogonality condition.]

$$= \frac{2(l+1)^2}{\{2(l+1)+1\}} + \frac{2l^2}{\{2(l-1)+1\}} = \frac{2(l+1)^2}{2l+3} + \frac{2l^2}{2l-1}$$

Therefore, we write

$$\int_{-1}^{+1} x^2\left[P_l(x)\right]^2 dx = \frac{2}{(2l+1)^2}\left[\frac{(l+1)^2}{2l+3} + \frac{l^2}{2l-1}\right]$$

CASE II: If $m-1 = l+1$ or, $m = l+2$, Eq. (iii) yields

$$(2l+1)(2l+5)\int_{-1}^{+1} x^2 P_l(x)P_{l+2}(x)\,dx = (l+1)(l+2)\int_{-1}^{+1} P_{l+1}(x)P_{l+1}(x)\,dx$$

$$= \frac{2(l+1)(l+2)}{2(l+1)+1}$$

or $$\int_{-1}^{+1} x^2 P_l(x) P_{l+2}(x)\,dx = \frac{2(l+1)(l+2)}{(2l+1)(2l+3)(2l+5)}$$

CASE III: If $m+1=l-1$ or, $m=l-2$, Eq. (iii) yields

$$(2l+1)(2l-3)\int_{-1}^{+1} x^2 P_l(x) P_{l-2}(x)\,dx = l(l-1)\int_{-1}^{+1} P_{l-1}(x) P_{l-1}(x)\,dx = \frac{2l(l-1)}{2(l-1)+1}$$

Therefore, $$\int_{-1}^{+1} x^2 P_l(x) P_{l-2}(x)\,dx = \frac{2l(l-1)}{(2l-3)(2l-1)(2l+1)}$$

CASE IV: We have

$$\int_{-1}^{+1} x^2 P_l(x) P_m(x)\,dx = 0 \quad \text{for all other values of } m.$$

PROBLEM 32 Obtain the value of $\int_0^1 P_l(x)\,dx$ when l is an integer.

Solution:

CASE I: When $l = 0$

In this case, we have

$$\int_0^1 P_0(x)\,dx = \int_0^1 1\,dx = 1$$

CASE II: When l is even integer greater than zero:

Let us recall the recurrence relation

$$(2l+1)P_l(x) = P'_{l+1}(x) - P'_{l-1}(x),$$

where the prime symbol denotes differentiation with respect to x.

or $$P_l(x) = \frac{1}{(2l+1)}\{P'_{l+1}(x) - P'_{l-1}(x)\}$$

Integrating both sides with respect to x within the limit from $x=0$ to $x=1$, we have

$$\int_0^1 P_l(x)\,dx = \frac{1}{(2l+1)}\int_0^1 \{P'_{l+1}(x) - P'_{l-1}(x)\}\,dx$$

$$= \frac{1}{(2l+1)}\left\{\left[P_{l+1}(x)\right]_0^1 - \left[P_{l-1}(x)\right]_0^1\right\}$$

$$= \frac{1}{(2l+1)}\{P_{l+1}(1) - P_{l+1}(0) - P_{l-1}(1) + P_{l-1}(0)\} \quad \text{(i)}$$

When l is an even integer, both of $l-1$ and $l+1$ are odd and we have

$$P_{l-1}(1)=1=P_{l+1}(1),\ P_{l-1}(0)=0=P_{l+1}(0)$$

Now, Eq. (i) becomes

$$\int_0^1 P_l(x)\,dx=0,\text{ for } l = \text{even integer greater than zero.}$$

CASE III: When l is odd integer

When l is odd, both $l-1$ and $l+1$ are even and we have

$$P_{l-1}(1)=1=P_{l+1}(1),\ P_{l-1}(0)=\frac{(-1)^{\frac{l-1}{2}}(l-1)!}{2^{l-1}\left[\left(\frac{l-1}{2}\right)!\right]^2}\text{ and } P_{l+1}(0)=\frac{(-1)^{\frac{l+1}{2}}(l+1)!}{2^{l+1}\left[\left(\frac{l+1}{2}\right)!\right]^2}$$

Then, Eq. (i) yields

$$\int_0^1 P_l(x)dx=\frac{1}{(2l+1)}\left\{1-\frac{(-1)^{\frac{l+1}{2}}(l+1)!}{2^{l+1}\left[\left(\frac{l+1}{2}\right)!\right]^2}-1+\frac{(-1)^{\frac{l-1}{2}}(l-1)!}{2^{l-1}\left[\left(\frac{l-1}{2}\right)!\right]^2}\right\}$$

$$=\frac{1}{(2l+1)}\frac{(-1)^{\frac{l-1}{2}}(l-1)!}{2^{l-1}\left[\left(\frac{l-1}{2}\right)!\right]^2}\left\{\frac{l(l+1)}{2^2\left(\frac{l+1}{2}\right)^2}+1\right\}$$

$$=\frac{(-1)^{\frac{l-1}{2}}(l-1)!}{2^{l-1}(2l+1)\left[\left(\frac{l-1}{2}\right)!\right]^2}\left\{\frac{l}{l+1}+1\right\}$$

$$=\frac{(-1)^{\frac{l-1}{2}}(l-1)!}{2^{l-1}\left[\left(\frac{l-1}{2}\right)!\right]^2}\frac{1}{(l+1)}=\frac{P_{l-1}(0)}{(l+1)}$$

$$=\frac{(-1)^{\frac{l-1}{2}}(l-1)!}{2^l\left(\frac{l-1}{2}\right)!\left(\frac{l-1}{2}\right)!\left(\frac{l+1}{2}\right)}=\frac{(-1)^{\frac{l-1}{2}}(l-1)!}{2^l\left(\frac{l-1}{2}\right)!\left(\frac{l+1}{2}\right)!}$$

Summarizing the results, we write

$$\int_0^1 P_l(x)dx = \begin{cases} 1, & \text{when } l=0, \\ 0, & \text{when } l \text{ is even integer,} \\ \dfrac{(-1)^{\frac{l-1}{2}}(l-1)!}{2^l\left(\frac{l-1}{2}\right)!\left(\frac{l+1}{2}\right)!}, & \text{when } l \text{ is odd integer} \end{cases}$$

Alternative Method

We know that

$$\sum_{l=0}^{\infty} h^l P_l(x) = \left(1-2hx+h^2\right)^{-\frac{1}{2}}$$

Integrating both sides with respect to x within the limit from $x = 0$ to $x = 1$, we obtain

$$\sum_{l=0}^{\infty} h^l \int_0^1 P_l(x)dx = \int_0^1 \frac{dx}{\sqrt{1-2hx+h^2}} = -\frac{1}{h}\int_{\sqrt{1+h^2}}^{1-h} du = \frac{1}{h}\int_{1-h}^{\sqrt{1+h^2}} du$$

$$= \frac{1}{h}\left[\sqrt{1+h^2}-1+h\right] \text{ [where we put } 1-2hx+h^2=u^2\text{]}$$

$$= \frac{1}{h}\left[1+\frac{1}{2}h^2+\frac{\frac{1}{2}\left(\frac{1}{2}-1\right)}{2!}h^4+\frac{\frac{1}{2}\left(\frac{1}{2}-1\right)\left(\frac{1}{2}-2\right)}{3!}h^6+\ldots\right.$$

$$\left.+\frac{\frac{1}{2}\left(\frac{1}{2}-1\right)\left(\frac{1}{2}-2\right)\left(\frac{1}{2}-3\right)\ldots\left(\frac{1}{2}-l+1\right)}{l!}h^{2l}+\ldots-1+h\right]$$

$$= 1+\frac{1}{2}h-\frac{1}{2^2\,2!}h^3+(-1)^2\frac{1.3}{2^3\,3!}h^5+(-1)^3\frac{1.3.5}{2^4 4!}h^7+\ldots$$

$$+(-1)^{l-1}\frac{1.3.5\ldots(2l-3)}{2^l\;l!}h^{2l-1}+\ldots \quad \text{(A)}$$

CASE I: Equating coefficients of h^0 from both sides of Eq. (A), we obtain

$$\int_0^1 P_0(x)dx = 1$$

CASE II: Equating coefficients of h^l (with l even) from both sides of Eq. (A), we obtain

$$\int_0^1 P_l(x)\,dx = 0, \text{ for } l \text{ even.}$$

CASE III: Equating coefficients of h^l (with l odd) from both sides of Eq. (A), we obtain

$$\int_0^1 P_l(x)dx = (-1)^{\frac{l-1}{2}} \frac{1.3.5\ldots(l-4)(l-2)}{2^{\frac{l+1}{2}}\left(\frac{l+1}{2}\right)!} \tag{B}$$

or $$\int_0^1 P_l(x)dx = (-1)^{\frac{l-1}{2}} \frac{1.2.3.4.5.6\ldots(l-4)(l-3)(l-2)(l-1)}{2^{\frac{l+1}{2}}\, 2.4.6.8\ldots\left\{2\left(\frac{l-3}{2}\right)\right\}\left\{2\left(\frac{l-1}{2}\right)\right\}\left(\frac{l+1}{2}\right)!}$$

$$= \frac{(-1)^{\frac{l-1}{2}}(l-1)!}{2^{\left(\frac{l+1}{2}\right)+\left(\frac{l-1}{2}\right)}\left(\frac{l-1}{2}\right)!\left(\frac{l+1}{2}\right)!} = \frac{(-1)^{\frac{l-1}{2}}(l-1)!}{2^l\left(\frac{l-1}{2}\right)!\left(\frac{l+1}{2}\right)!} \tag{C}$$

where use has been made of the formula

$$2.4.6.8\ldots 2n = 2^n n!$$

As in the first method, one can summarize the results of the three cases.

[**Note:** For Case I, Case II, if we put $l = 2s$, we write

$$\int_0^1 P_{2s}(x)dx = 1 \text{ for } s = 0$$

and $\int_0^1 P_{2s}(x)\,dx = 0$ for s = 1, 2, 3, 4, ... etc.

For Case III, if we put $l = 2s + 1$ or, $s = \frac{l-1}{2}$, with s = 0, 1, 2, 3, ... etc., we write

$$\int_0^1 P_{2s+1}(x)\,dx = \frac{(-1)^s(2s)!}{2^{2s+1}s!(s+1)!} = \frac{(-1)^s(2s)!}{2^s s!\,2^{s+1}(s+1)!}$$

$$= \frac{(-1)^s(2s)!}{\{2^s s!\}\{2.4.6\ldots(2s+2)\}}$$

$$= \frac{(-1)^s\,1.2.3.4\ldots(2s-4)(2s-3)(2s-2)(2s-1)(2s)}{\{2^s s!\}(2s+2)!!}$$

$$= \frac{(-1)^s \{1.3.5\ldots(2s-3)(2s-1)\}\{2.4.6\ldots(2s-2)(2s)\}}{2^s\, s!\,(2s+2)!!}$$

$$= \frac{(-1)^s\,(2s-1)!!\,2^s s!}{2^s s!\,(2s+2)!!} = \frac{(-1)^s\,(2s-1)!!}{(2s+2)!!}$$

Summarizing the results, we write

$$\int_0^1 P_{2s}(x)\,dx = \begin{cases} 1 & \text{for } s = 0; \\ 0 & \text{for } s = 1, 2, 3, \ldots \text{etc} \end{cases}$$

and
$$\int_0^1 P_{2s+1}(x)\,dx = \frac{(-1)^s\,(2s-1)!!}{(2s+2)!!},$$

where use has been made of the formulae

$$1.3.5\ldots(2s-1) = (2s-1)!!$$

and
$$2.4.6\ldots(2s+2) = (2s+2)!! = 2^{s+1}(s+1)!]$$

PROBLEM 33 Prove that $\int_{-1}^{+1} P_n(x)\frac{dx}{\sqrt{1-x}} = \frac{2\sqrt{2}}{2n+1}$

Solution: We know that

$$\left(1-2hx+h^2\right)^{-\frac{1}{2}} = \sum_{l=0}^{\infty} h^l P_l(x)$$

Putting $h=1$, we obtain

$$\{2(1-x)\}^{-\frac{1}{2}} = \sum_{l=0}^{\infty} P_l(x)$$

Multiplying both sides by $P_n(x)$ and integrating the results with respect to x within the limit from $x=-1$ to $x=+1$, we obtain

$$\frac{1}{\sqrt{2}}\int_{-1}^{+1} P_n(x)\frac{dx}{\sqrt{1-x}} = \sum_{l=0}^{\infty}\int_{-1}^{+1} P_l(x)P_n(x)\,dx$$

$$= \int_{-1}^{+1} P_n(x)P_n(x)\,dx, \text{ since other integrals vanish.}$$

$$= \frac{2}{2n+1}$$

Hence, we have

$$\int_{-1}^{+1} P_n(x)\frac{dx}{\sqrt{1-x}} = \frac{2\sqrt{2}}{2n+1}$$

PROBLEM 34 Prove that

$$\int_0^1 x^m P_l(x)dx = \frac{m(m-1)(m-2)(m-3)\dots(m-l+2)}{(m+l+1)(m+l-1)(m+l-3)\dots(m-l+3)}, \text{ for } m \geq l$$

where l and m are non-negative integers.

Solution: We know the Rodrigues' formula,

$$P_l(x) = \frac{1}{2^l\, l!}\frac{d^l}{dx^l}(x^2-1)^l$$

Therefore, we have

$$\int_0^1 x^m P_l(x)dx = \frac{1}{2^l\, l!}\int_0^1 x^m \frac{d^l}{dx^l}(x^2-1)^l\, dx$$

$$= \frac{1}{2^l\ l!}\left\{\left[x^m \frac{d^{l-1}}{dx^{l-1}}(x^2-1)^l\right]_0^1 - m\int_0^1 x^{m-1}\frac{d^{l-1}}{dx^{l-1}}(x^2-1)^l\, dx\right\},$$

[integrating by parts;]

$$= 0 - \frac{m}{2^l\, l!}\int_0^1 x^{m-1}\frac{d^{l-1}}{dx^{l-1}}(x^2-1)^l\, dx$$

Continuing the process of integration by parts up to l times, we obtain

$$\int_0^1 x^m P_l(x)dx = \frac{(-1)^l\, m(m-1)(m-2)(m-3)\dots(m-l+2)(m-l+1)}{2^l\, l!}\int_0^1 x^{m-l}(x^2-1)^l\, dx$$

$$= \frac{m(m-1)(m-2)(m-3)\dots(m-l+2)(m-l+1)}{2^l\, l!}\int_0^1 (x^2)^{\frac{m-l}{2}}(1-x^2)^l\, dx$$

Putting $x^2 = u$, we have $2xdx = du$ and $dx = \frac{1}{2}\frac{du}{\sqrt{u}}$

Now, we write

$$\int_0^1 x^m P_l(x)dx = \frac{m(m-1)(m-2)(m-3)\dots(m-l+2)(m-l+1)}{2^{l+1}\, l!}\int_0^1 u^{\frac{m-l-1}{2}}(1-u)^l\, du$$

$$= \frac{m(m-1)(m-2)(m-3)\ldots(m-l+2)(m-l+1)}{2^{l+1}l!} B\left(\frac{m-l+1}{2}, l+1\right)$$

$$= \frac{m(m-1)(m-2)(m-3)\ldots(m-l+2)(m-l+1)}{2^{l+1}l!} \frac{\Gamma\left(\frac{m-l+1}{2}\right)\Gamma(l+1)}{\Gamma\left(\frac{m+l+3}{2}\right)}$$

$$= \frac{m(m-1)(m-2)(m-3)\ldots(m-l+2)(m-l+1)\Gamma\left(\frac{m-l+1}{2}\right)l!}{2^{l+1}l!\left(\frac{m+l+1}{2}\right)\left(\frac{m+l-1}{2}\right)\left(\frac{m+l-3}{2}\right)\ldots\left(\frac{m-l+3}{2}\right)\left(\frac{m-l+1}{2}\right)\Gamma\left(\frac{m-l+1}{2}\right)}$$

$$= \frac{m(m-1)(m-2)(m-3)\ldots(m-l+2)(m-l+1)}{(m+l+1)(m+l-1)(m+l-3)\ldots(m-l+3)(m-l+1)}$$

$$= \frac{m(m-1)(m-2)(m-3)\ldots(m-l+2)}{(m+l+1)(m+l-1)(m+l-3)\ldots(m-l+3)}$$

PROBLEM 35 Prove that $\int_{-1}^{+1} x^l P_l(x)\,dx = \frac{2^{l+1}(l!)^2}{(2l+1)!}$

Solution: We know that the integrand is always an even function.

Hence, we write

$$\int_{-1}^{+1} x^l P_l(x)\,dx = 2\int_{0}^{1} x^l P_l(x)\,dx$$

$$= 2\frac{l(l-1)(l-2)(l-3)\ldots 3.2}{(2l+1)(2l-1)(2l-3)\ldots 5.3}$$

where use has been made of the result of Problem 34 with m replaced by l.

or $$\int_{-1}^{+1} x^l P_l(x)dx = 2\frac{l!\,(2l)(2l-2)(2l-4)(2l-6)\ldots 4.2}{(2l+1)\,(2l)\,(2l-1)\,(2l-2)\,(2l-3)\ldots 5.4.3.2} = \frac{2^{l+1}(l!)^2}{(2l+1)!}$$

PROBLEM 36 Prove that

$$\int_{-1}^{+1} x^{2r} P_{2n}(x)\,dx = \frac{2^{2n+1}(2r)!\,(n+r)!}{(2n+2r+1)!\,(r-n)!}, \text{ for } r \geq n$$

Solution: The integrand is always an even function and, therefore, we write

$$\int_{-1}^{+1} x^{2r} P_{2n}(x)\,dx = 2\int_{0}^{1} x^{2r} P_{2n}(x)\,dx$$

$$= 2\frac{(2r)(2r-1)(2r-2)(2r-3)\ldots(2r-2n+2)}{(2r+2n+1)(2r+2n-1)(2r+2n-3)\ldots(2r-2n+3)},$$

using the result of Problem 34 where m is replaced by $2r$ and l by $2n$

or $\displaystyle\int_{-1}^{+1} x^{2r} P_{2n}(x)\,dx$

$$= 2\frac{(2r)!}{(2r-2n+1)!}\left[\frac{(2r+2n)(2r+2n-2)(2r+2n-4)(2r+2n-6)\ldots(2r-2n+4)(2r-2n+2)}{(2r+2n+1)(2r+2n)(2r+2n-1)(2r+2n-2)\ldots(2r-2n+3)(2r-2n+2)}\right]$$

Here, for the factors in the numerator forming an arithmatical progression, the number of factors are calculated through the formula relating the smallest member, number of members, common difference and the greatest member.

Thus, the number of factors in the numerator forming an A.P. is equal to

$$\left[\frac{\text{greatest member} - \text{smallest member}}{\text{common difference}} + 1\right] = \frac{(2r+2n)-(2r-2n+2)}{2} + 1 = 2n$$

Thus, we write,

$$\int_{-1}^{+1} x^{2r} P_{2n}(x)\,dx$$

$$= 2\frac{(2r)!}{(2r-2n+1)!}\left[\frac{2^{2n}\left\{\left(\frac{2r+2n}{2}\right)\left(\frac{2r+2n-2}{2}\right)\left(\frac{2r+2n-4}{2}\right)\ldots\left(\frac{2r-2n+4}{2}\right)\left(\frac{2r-2n+2}{2}\right)\right\}}{(2r+2n+1)(2r+2n)(2r+2n-1)\ldots(2r-2n+3)(2r-2n+2)}\right]$$

$$= \frac{2^{2n+1}(2r)!\,(r+n)(r+n-1)(r+n-2)\ldots(r-n+2)(r-n+1)}{(2r+2n+1)!}$$

$$= \frac{2^{2n+1}(2r)!\,(r+n)!}{(2n+2r+1)!\,(r-n)!}$$

PROBLEM 37 Prove that $\displaystyle\int_{-1}^{+1} x^{2r+1} P_{2n+1}(x)\,dx = \frac{2^{2n+2}(2r+1)!\,(r+n+1)!}{(2r+2n+3)!\,(r-n)!}$

Solution: To the proof of Problem 36, add the following.

$$\int_{-1}^{+1} x^{2r} P_{2n}(x)\,dx = \frac{2^{2n+1}(2r)!\left(\frac{2r+2n}{2}\right)!}{(2n+2r+1)!\left(\frac{2r-2n}{2}\right)!}$$

Analogously, we write

$$\int_{-1}^{+1} x^{2r+1} P_{2n+1}(x)\,dx = \frac{2^{2n+2}(2r+1)!\left(\frac{2r+1+2n+1}{2}\right)!}{\{(2r+1)+(2n+1)+1\}!\left\{\frac{(2r+1)-(2n+1)}{2}\right\}!}$$

$$= \frac{2^{2n+2}(2r+1)!\,(r+n+1)!}{(2r+2n+3)!\,(r-n)!}$$

PROBLEM 38 Prove that $\int_0^{\pi} P_l(\cos\theta)P_n(\cos\theta)\sin\theta\,d\theta = \frac{2}{2l+1}\delta_{ln}$

Solution: First, see the proof of $\int_{-1}^{+1} P_l(x)P_n(x)\,dx = \frac{2}{2l+1}\delta_{ln}$

Put $x = \cos\theta$, then, we have, $dx = -\sin\theta\,d\theta$.
Now, we write

$$\int_{\pi}^{0} -P_l(\cos\theta)P_n(\cos\theta)\sin\theta\,d\theta = \frac{2}{2l+1}\delta_{ln},$$

or
$$\int_0^{\pi} P_l(\cos\theta)P_n(\cos\theta)\sin\theta\,d\theta = \frac{2}{2l+1}\delta_{ln}$$

PROBLEM 39 Prove that $\int_0^{\pi} P_n(\cos\theta)\sin 2\theta\,d\theta = 0,\ n>1$

Solution: We have

$$\int_0^{\pi} P_n(\cos\theta)\sin 2\theta\,d\theta = \int_0^{\pi} P_n(\cos\theta)\,2\sin\theta\cos\theta\,d\theta$$

$$= -2\int_0^{\pi} P_n(\cos\theta)\cos\theta\,d(\cos\theta)$$

$$= -2\int_{+1}^{-1} P_n(x)\,x\,dx, \text{ [where we put } x = \cos\theta\text{]}$$

$$= 2\int_{-1}^{+1} x P_n(x)dx = 2\int_{-1}^{+1} P_1(x)P_n(x)dx = 0, \text{ as } n > 1$$

PROBLEM 40 Prove that

$$x^{2r} = \sum_{n=0}^{r} \frac{2^{2n}(4n+1)(2r)!(r+n)!}{(2r+2n+1)!(r-n)!} P_{2n}(x),$$

$$x^{2r+1} = \sum_{n=0}^{r} \frac{2^{2n+1}(4n+3)(2r+1)!(r+n+1)!}{(2r+2n+3)!(r-n)!} P_{2n+1}(x);$$

where r is a non-negative integer.

Solution: For any non-negative integer s (even or odd), let us write

$$x^s = \ldots + a_{\lambda-2} P_{\lambda-2}(x) + a_\lambda P_\lambda(x) + a_{\lambda+2} P_{\lambda+2}(x) + \ldots + a_s P_s(x) \qquad \text{(i)}$$

where the first term of Eq. (i) above is

$a_0P_0(x)$ if s is even or $a_1P_1(x)$ if s is odd.

Multiplying both sides of Eq. (i) by $P_\lambda(x)$ and integrating the results with respect to x within the limit from $x = -1$ to $x = +1$, we obtain

$$\int_{-1}^{+1} x^s P_\lambda(x)dx = a_\lambda \frac{2}{2\lambda+1},$$

or

$$a_\lambda = \frac{(2\lambda+1)}{2}\int_{-1}^{+1} x^s P_\lambda(x)dx \qquad \text{(ii)}$$

(a) If s is an even integer, let us write $s = 2r$, where $r = 0, 1, 2, 3, \ldots$ etc.
Then, λ will also be an even integer and we write

$\lambda = 2n$ where $n = 0, 1, 2, 3, \ldots$ etc.

Now, Eq. (i) can be written as

$$x^{2r} = \sum_{n=0}^{r} a_{2n} P_{2n}(x) \qquad \text{(iii)}$$

where a_{2n} is obtained from Eq. (ii) as

$$a_{2n} = \frac{(4n+1)}{2}\int_{-1}^{+1} x^{2r} P_{2n}(x)dx \qquad \text{(iv)}$$

Using the result of Problem 36, we obtain

$$\int_{-1}^{+1} x^{2r} P_{2n}(x)dx = \frac{2^{2n+1}(2r)!(r+n)!}{(2r+2n+1)!(r-n)!}$$

Then, Eq. (iv) results in

$$a_{2n} = \frac{(4n+1)}{2} \frac{2^{2n+1}(2r)!\,(r+n)!}{(2r+2n+1)!\,(r-n)!} \tag{v}$$

Substituting Eq. (v) into Eq. (iii), we obtain

$$x^{2r} = \sum_{n=0}^{r} \frac{2^{2n}(4n+1)(2r)!\,(r+n)!}{(2r+2n+1)!\,(r-n)!} P_{2n}(x) \tag{vi}$$

(b) If s is an odd integer, we write, $s = 2r+1$ where $r = 0, 1, 2, 3, \ldots$ etc . Then, λ will also be an odd integer and we write $\lambda = 2n+1$, where n = 0, 1, 2, 3, ... etc. Now, Eq. (i) can be written as

$$x^{2r+1} = \sum_{n=0}^{r} a_{2n+1} P_{2n+1}(x) \tag{vii}$$

where a_{2n+1} is given by Eq. (ii) as

$$a_{2n+1} = \frac{\{2(2n+1)+1\}}{2} \int_{-1}^{+1} x^{2r+1} P_{2n+1}(x)\,dx$$

Using the result of Problem 37, we obtain

$$a_{2n+1} = \frac{(4n+3)}{2} \frac{2^{2n+2}(2r+1)!\,(r+n+1)!}{(2r+2n+3)!\,(r-n)!}$$

Substituting this value of a_{2n+1} into Eq. (vii), we obtain

$$x^{2r+1} = \sum_{n=0}^{r} \frac{2^{2n+1}(4n+3)(2r+1)!\,(r+n+1)!}{(2r+2n+3)!\,(r-n)!} P_{2n+1}(x)$$

PROBLEM 41 Expand $P_l(x)$ in powers of $\frac{1}{2}(1-x)$ [Murphy's formula].

Solution: Rodrigues' formula is written as

$$P_l(x) = \frac{1}{2^l\, l!} \frac{d^l}{dx^l}\left(x^2-1\right)^l = \frac{1}{l!} \frac{d^l}{dx^l}\left[(x-1)^l \left(\frac{x+1}{2}\right)^l\right]$$

$$= \frac{1}{l!} \frac{d^l}{dx^l}\left[(x-1)^l \left\{1 + \left(\frac{x-1}{2}\right)\right\}^l\right]$$

$$= \frac{1}{l!} \frac{d^l}{dx^l}\left[(x-1)^l \left\{1 + l\left(\frac{x-1}{2}\right) + \frac{l(l-1)}{2!}\left(\frac{x-1}{2}\right)^2 + \frac{l(l-1)(l-2)}{3!}\left(\frac{x-1}{2}\right)^3 + \ldots\right\}\right]$$

$$= \frac{1}{l!}\frac{d^l}{dx^l}\left[(x-1)^l + \frac{l(x-1)^{l+1}}{2} + \frac{l(l-1)}{2!}\frac{(x-1)^{l+2}}{2^2} + \frac{l(l-1)(l-2)}{3!}\frac{(x-1)^{l+3}}{2^3} + \ldots\right]$$

$$= \frac{1}{l!}\left[l! + l(l+1)!\left(\frac{x-1}{2}\right) + \frac{l(l-1)}{2!}\frac{(l+2)!}{2!}\frac{(x-1)^2}{2^2} + \frac{l(l-1)(l-2)}{3!}\frac{(l+3)!}{3!}\frac{(x-1)^3}{2^3} + \ldots\right]$$

$$= 1 - \frac{(l+1)!}{(l-1)!(1!)^2}\left\{\frac{1}{2}(1-x)\right\} + \frac{(l+2)!}{(l-2)!(2!)^2}\left\{\frac{1}{2}(1-x)\right\}^2 - \frac{(l+3)!}{(l-3)!\,(3!)^2}\left\{\frac{1}{2}(1-x)\right\}^3 + \ldots$$

This is known as Murphy's formula.

PROBLEM 42 Prove that $\int_0^{\pi} P_l(\cos\theta)\cos l\theta d\theta = \frac{(2l-1)!!}{(2l)!!}\pi = B\left(l+\frac{1}{2}, \frac{1}{2}\right)$

Solution:

First Part: From Section 2.13, we know that

$$P_l(\cos\theta) = \frac{1.3.5\ldots(2l-1)}{2.4.6\ldots 2l}\left[2\cos l\theta + \frac{1}{2}\frac{2l}{(2l-1)}2\cos(l-2)\theta + \frac{1.3}{2.4}\frac{(2l-2)2l}{(2l-3)(2l-1)}2\cos(l-4)\theta + \ldots\right]$$

Multiplying both sides by $\cos l\theta$ and integrating the results with respect to θ within the limit from $\theta = 0$ to $\theta = \pi$, we obtain

$$\int_0^{\pi} P_l(\cos\theta)\cos l\theta\, d\theta = \frac{1.3.5\ldots(2l-1)}{2.4.6\ldots 2l}\int_0^{\pi} 2\cos^2 l\theta\, d\theta,$$

[as other integrals vanish by orthogonality of cosine functions.]

$$= \frac{1.3.5\ldots(2l-1)}{2.4.6\ldots 2l}\pi = \frac{(2l-1)!!}{(2l)!!}\pi$$

Second Part: We proved that

$$\int_0^{\pi} P_l(\cos l\theta)\cos l\theta\, d\theta = \frac{1.3.5\ldots(2l-1)}{2.4.6\ldots 2l}\pi$$

$$= \frac{\frac{1}{2}\cdot\frac{3}{2}\cdot\frac{5}{2}\cdots\left(\frac{2l-3}{2}\right)\left(\frac{2l-1}{2}\right)\sqrt{\pi}\sqrt{\pi}}{1.2.3\ldots l}$$

$$= \frac{\left(\frac{2l-1}{2}\right)\left(\frac{2l-3}{2}\right)\cdots\frac{5}{2}\cdot\frac{3}{2}\frac{1}{2}\Gamma\left(\frac{1}{2}\right)\Gamma\left(\frac{1}{2}\right)}{\Gamma(l+1)}$$

$$= \frac{\Gamma\left(l+\frac{1}{2}\right)\Gamma\left(\frac{1}{2}\right)}{\Gamma(l+1)} = B\left(l+\frac{1}{2}, \frac{1}{2}\right)$$

PROBLEM 43 Prove that

$$\int_0^{\pi} P_l(\cos\theta)\,d\theta = \begin{cases} \dfrac{\pi}{2^{2l-1}} \dfrac{(l!)^2}{\left\{\left(\frac{l}{2}\right)!\right\}^4}, & \text{if } l \text{ is an even integer;} \\ 0, & \text{if } l \text{ is an odd integer} \end{cases}$$

Solution: We know that

$$P_l(\cos\,\theta) = \frac{(2l)!}{2^{2l-1}(l!)^2}\left[\cos l\theta + \frac{1}{2}\frac{2l}{(2l-1)}\cos(l-2)\theta + \frac{1.3}{2.4}\frac{(2l-2)(2l)}{(2l-3)(2l-1)}\cos(l-4)\theta + \ldots\right.$$

$$\left. \ldots + \frac{\{1.3.5\ldots(2j-1)\}}{\{2.4.6\ldots(2j)\}}\frac{\{(2l-2j+2)\ldots(2l-4)(2l-2)(2l)\}}{\{(2l-2j+1)\ldots(2l-5)(2l-3)(2l-1)\}}\cos(l-2j)\theta + \ldots\right] \quad \text{(i)}$$

CASE I: If l is an even integer

For l = even integer, in the expression within the square bracket of Eq. (i), the only term which will be independent of cosine function is obtained when $l = 2j$, or, $j = l/2$.

Hence, in Eq. (i), the term independent of cosine function is

$$\frac{(2l)!}{2^{2l-1}(l!)^2}\frac{1.3.5\ldots(l-1)}{2.4.6\ldots l}\frac{(2l-l+2)\ldots(2l-4)(2l-2)(2l)}{(2l-l+1)\ldots(2l-5)(2l-3)(2l-1)}$$

$$= \frac{(2l)!}{2^{2l-1}(l!)^2}\frac{1.2.3.4.5.6\ldots(l-1)\,l}{(2.4.6\ldots l)^2}\frac{(2l-l+2)\ldots(2l-4)(2l-2)(2l)}{(l+1)\ldots(2l-5)(2l-3)(2l-1)}$$

$$= \frac{(2l)!}{2^{2l-1}(l!)^2}\frac{l!}{\left\{2^{l/2}\left(\frac{l}{2}\right)!\right\}^2}\frac{\{(2l-l+2)\ldots(2l-4)(2l-2)(2l)\}^2}{(l+l)(l+2)\ldots(2l-5)(2l-4)(2l-3)(2l-2)(2l-1)(2l)}$$

$$= \frac{(2l)!}{2^{2l-1}\,2^l\left\{\left(\frac{l}{2}\right)!\right\}^2}\frac{\left\{2^{\frac{l}{2}}\left(\frac{l}{2}+1\right)\ldots(l-2)(l-1)l\right\}^2}{(2l)!}$$

$$= \frac{(l!)^2}{2^{2l-1}\left\{\left(\frac{l}{2}\right)!\right\}^4}$$

On integration of Eq. (i) with respect to θ within the limit from $\theta = 0$ to $\theta = \pi$, the integrals of cosine functions will all vanish leaving only the integral independent of cosine function.

Thus, we write

$$\int_0^{\pi} P_l(\cos\theta)\,d\theta = \frac{(l!)^2}{2^{2l-1}\left\{\left(\frac{l}{2}\right)!\right\}^4}\int_0^{\pi} d\theta = \frac{\pi}{2^{2l-1}}\frac{(l!)^2}{\left\{\left(\frac{l}{2}\right)!\right\}^4}, \text{ if } l \text{ is even integer.}$$

CASE II: If l is an odd integer

If l is an odd integer, all the terms within the square brackets of Eq. (i) contain cosine functions.

Therefore, on integration of Eq. (i) with respect to θ within the limit from $\theta = 0$ to $\theta = \pi$, the integrals of these cosine functions will all vanish.

Thus, we write

$$\int_0^{\pi} P_l(\cos\theta)\,d\theta = 0, \text{ if } l \text{ is an odd integer.}$$

In general, we can write

$$\int_0^{\pi} P_l(\cos\theta)d\theta = \begin{cases} \dfrac{\pi}{2^{2l-1}}\dfrac{(l!)^2}{\left\{\left(\frac{l}{2}\right)!\right\}^4}, & \text{if } l \text{ is an even integer;} \\ 0, & \text{if } l \text{ is an odd integer} \end{cases}$$

PROBLEM 44 Express $x^4 + x^2 - 7$ in a series of Legendre polynomials.

Solution: From the formula for Legendre polynomials, we have,

$$P_2(x) = \frac{1}{2}\left(3x^2 - 1\right),\ P_4(x) = \frac{1}{8}\left(35x^4 - 30x^2 + 3\right)$$

Therefore, we obtain

$$x^2 = \frac{1}{3}\left[2P_2(x) + 1\right] = \frac{2}{3}P_2(x) + \frac{1}{3}P_0(x),$$

$$x^4 = \frac{1}{35}\left[8P_4(x) + 30x^2 - 3\right] = \frac{8}{35}P_4(x) + \frac{6}{7}x^2 - \frac{3}{35}$$

$$= \frac{8}{35}P_4(x) + \frac{6}{7}\left\{\frac{2}{3}P_2(x) + \frac{1}{3}P_0(x)\right\} - \frac{3}{35}P_0(x)$$

$$= \frac{8}{35}P_4(x) + \frac{4}{7}P_2(x) + \frac{2}{7}P_0(x) - \frac{3}{35}P_0(x)$$

Now, we write

$$x^4+x^2-7=\frac{8}{35}P_4(x)+\frac{4}{7}P_2(x)+\frac{2}{7}P_0(x)-\frac{3}{35}P_0(x)+\frac{2}{3}P_2(x)+\frac{1}{3}P_0(x)-7P_0(x)$$

$$=\frac{8}{35}P_4(x)+\frac{26}{21}P_2(x)-\frac{97}{15}P_0(x)$$

PROBLEM 45 Express x^8 as a series in Legendre polynomials.

Solution: We have,

$$x^{2r}=\sum_{n=0}^{r}\frac{2^{2n}(4n+1)(2r)!(r+n)!}{(2r+2n+1)!(r-n)!}P_{2n}(x), \text{ [From Problem 40]}$$

Thus, $$x^8=\sum_{n=0}^{4}\frac{2^{2n}(4n+1)\times 8!\times(4+n)!}{(8+2n+1)!(4-n)!}P_{2n}(x)$$

$$=\frac{8!\,4!}{9!\,4!}P_0(x)+\frac{2^2\times5\times8!\times5!}{11!\times3!}P_2(x)+\frac{2^4\times9\times8!\times6!}{13!\times2!}P_4(x)$$

$$+\frac{2^6\times13\times8!\times7!}{15!\times1!}P_6(x)+\frac{2^8\times17\times8!\times8!}{17!\times0!}P_8(x)$$

$$=\frac{1}{9}P_0(x)+\frac{40}{99}P_2(x)+\frac{48}{143}P_4(x)+\frac{64}{495}P_6(x)+\frac{128}{6435}P_8(x)$$

PROBLEM 46 Express x^7 as a series in Legendre polynomials.

Solution: We have

$$x^{2r+1}=\sum_{n=0}^{r}\frac{2^{2n+1}(4n+3)(2r+1)!(r+n+1)!}{(2r+2n+3)!(r-n)!}P_{2n+1}(x) \text{ [Using Problem 40]}$$

Therefore, we have

$$x^7=\sum_{n=0}^{3}\frac{2^{2n+1}(4n+3)7!(3+n+1)!}{(6+2n+3)!(3-n)!}P_{2n+1}(x)$$

$$=\frac{2\times3\times7!\times4!}{9!\times3!}P_1(x)+\frac{2^3\times7\times7!\times5!}{11!\times2!}P_3(x)+\frac{2^5\times11\times7!\times6!}{13!\times1!}P_5(x)$$

$$+\frac{2^7\times15\times7!\times7!}{15!\times0!}P_7(x)$$

$$=\frac{1}{3}P_1(x)+\frac{14}{33}P_3(x)+\frac{8}{39}P_5(x)+\frac{16}{429}P_7(x)$$

PROBLEM 47 Expand the following functions in series of Legendre polynomials,

(i) $f(x)=\begin{cases}-1, & -1<x<0,\\ 1, & 0<x<1\end{cases}$

(ii) $f(x)=\begin{cases}0, & -1<x<0,\\ x, & 0<x<1\end{cases}$

(iii) $f(x)=\begin{cases}0, & -1<x<\alpha\\ 1, & \alpha<x<1\end{cases}$

where $-1 < \alpha < 1$.

Solution: Let us write

$$f(x)=\sum_{l=0}^{\infty} a_l P_l(x)$$

$$=a_0P_0(x)+a_1P_1(x)+a_2P_2(x)+a_3P_3(x)+a_4P_4(x)+a_5P_5(x)+\dots \quad \text{(i)}$$

where $$a_l=\frac{2l+1}{2}\int_{-1}^{+1} f(x)\,P_l(x)\,dx \quad \text{(ii)}$$

(i) Here, the function $f(x)$ is an odd function and, therefore,

$$P_0(x), P_2(x), P_4(x)\dots P_{2r}(x),\dots$$

are to be absent. Therefore, we must have

$$0=a_0=a_2=a_4=a_6=\dots=a_{2r}=\dots \quad \text{(iii)}$$

Now, putting $l=1, 3, 5, 7, \dots, 2r+1, \dots$ in Eq. (ii), we have

$$a_1=\frac{3}{2}\int_{-1}^{+1} f(x)P_1(x)\,dx=\frac{3}{2}\left\{\int_{-1}^{0}(-1)x\,dx+\int_{0}^{+1}1\cdot x dx\right\}$$

$$=\frac{3}{2}\left\{\int_{0}^{-1}x\,dx+\int_{0}^{1}x\,dx\right\}=\frac{3}{2}, \quad \text{(iv)}$$

$$a_3=\frac{7}{2}\int_{-1}^{+1} f(x)P_3(x)\,dx=\frac{7}{2}\left\{\int_{-1}^{0}-P_3(x)\,dx+\int_{0}^{1}P_3(x)\,dx\right\}$$

$$=\frac{7}{2}\left\{\int_{0}^{-1}P_3(x)\,dx+\int_{0}^{1}P_3(x)\,dx\right\}$$

$$=\frac{7}{2}\left\{\int_{0}^{-1}\frac{1}{2}\left(5x^3-3x\right)dx+\int_{0}^{1}\frac{1}{2}\left(5x^3-3x\right)dx\right\}=-\frac{7}{8}, \quad \text{(v)}$$

$$a_5 = \frac{11}{2}\int_{-1}^{+1} f(x)P_5(x)dx = \frac{11}{2}\left\{\int_0^{-1} P_5(x)dx + \int_0^1 P_5(x)dx\right\}$$

$$= \frac{11}{2}\left\{\int_0^{-1}\frac{1}{8}\left(63x^5 - 70x^3 + 15x\right)dx + \int_0^1 \frac{1}{8}\left(63x^5 - 70x^3 + 15x\right)dx\right\}$$

$$\frac{11}{16}, \qquad \text{(vi)}$$

and so on.

Substituting the various coefficients into Eq. (i), we obtain

$$f(x) = \frac{3}{2}P_1(x) - \frac{7}{8}P_3(x) + \frac{11}{16}P_5(x) + \ldots, -1 < x < 1$$

(ii) Putting $l = 0, 1, 2, 3, 4, \ldots$ etc. in Eq. (ii), we obtain

$$a_0 = \frac{1}{2}\int_{-1}^{+1} f(x)P_0(x)dx = \frac{1}{2}\left\{\int_{-1}^{0} 0\cdot dx + \int_0^1 x\,dx\right\} = \frac{1}{4},$$

$$a_1 = \frac{3}{2}\int_{-1}^{+1} f(x)P_1(x)dx = \frac{3}{2}\left\{\int_{-1}^{0} 0\cdot x\,dx + \int_0^1 x^2\,dx\right\} = \frac{1}{2},$$

$$a_2 = \frac{5}{2}\int_{-1}^{+1} f(x)P_2(x)dx = \frac{5}{2}\left\{0 + \int_0^1 x\frac{1}{2}\left(3x^2 - 1\right)dx\right\} = \frac{5}{16},$$

$$a_3 = \frac{7}{2}\int_{-1}^{+1} f(x)P_3(x)dx = \frac{7}{2}\left\{0 + \int_0^1 x\frac{1}{2}\left(5x^3 - 3x\right)dx\right\} = 0,$$

$$a_4 = \frac{9}{2}\int_{-1}^{+1} f(x)P_4(x)dx = \frac{9}{2}\left\{0 + \int_0^1 x\frac{1}{8}\left(35x^4 - 30x^2 + 3\right)dx\right\} = -\frac{3}{32},$$

and so on.

Substituting the various coefficients into Eq. (ii), we obtain

$$f(x) = \frac{1}{4}P_0(x) + \frac{1}{2}P_1(x) + \frac{5}{16}P_2(x) - \frac{3}{32}P_4(x) + \ldots$$

(iii) Let us write

$$f(x) = \sum_{l=0}^{\infty} a_l P_l(x) = a_0 P_0(x) + \sum_{l=1}^{\infty} a_l P_l(x)$$

where $$a_l = \frac{2l+1}{2}\int_{-1}^{+1} f(x)P_l(x)\,dx$$

Putting $l=0$, we have

$$a_0 = \frac{1}{2}\int_{-1}^{+1} f(x)P_0(x)\,dx = \frac{1}{2}\left[\int_{-1}^{\alpha}(0)P_0(x)\,dx + \int_{\alpha}^{1}(1)P_0(x)\,dx\right]$$

$$= \frac{1}{2}\left[0 + \int_{\alpha}^{1} dx\right]$$

$$= \frac{1}{2}(1-\alpha), \text{ as } P_0(x) = 1;$$

In general, we have

$$a_l = \frac{2l+1}{2}\left[\int_{-1}^{\alpha}(0)P_l(x)\,dx + \int_{\alpha}^{1}(1)P_l(x)\,dx\right]$$

$$= \frac{2l+1}{2}\left[0 + \int_{\alpha}^{1} P_l(x)\,dx\right]$$

We know that

$$(2l+1)P_l(x) = P'_{l+1}(x) - P'_{l-1}(x)$$

Therefore, we write for $l > 0$,

$$a_l = \frac{2l+1}{2}\int_{\alpha}^{1}\frac{1}{(2l+1)}\left[P'_{l+1}(x) - P'_{l-1}(x)\right]dx$$

$$= \frac{1}{2}\left\{\left[P_{l+1}(x)\right]_{\alpha}^{1} - \left[P_{l-1}(x)\right]_{\alpha}^{1}\right\}$$

$$= \frac{1}{2}\left\{P_{l+1}(1) - P_{l+1}(\alpha) - P_{l-1}(1) + P_{l-1}(\alpha)\right\}$$

$$= \frac{1}{2}\left\{P_{l-1}(\alpha) - P_{l+1}(\alpha)\right\}, \text{ as } P_{l+1}(1) = 1 = P_{l-1}(1)$$

Using the values of the coefficients a_0 and $a_l\,(l>0)$, the function $f(x)$ can be written as

$$f(x) = \frac{1}{2}(1-\alpha)P_0(x) + \frac{1}{2}\sum_{l=1}^{\infty}\left\{P_{l-1}(\alpha) - P_{l+1}(\alpha)\right\}P_l(x)$$

PROBLEM 48 Determine the coefficients a_0, a_1, a_2, a_3 in the expansion

$$f(x) = \sin\frac{\pi x}{2} = \sum_{n=0}^{\infty} a_n P_n(x), -1 < x < 1$$

Solution: Here, $f(x)$ is an odd function. Further, $P_n(x)$ is an (i) even function if n is an even integer and (ii) odd function if n is an odd integer.

Therefore,

$$a_0, a_2, a_4, \ldots a_{2r} \ldots = 0 \tag{i}$$

The value of a_n is given by

$$a_n = \frac{2n+1}{2}\int_{-1}^{+1} f(x) P_n(x)\,dx = \frac{2n+1}{2}\int_{-1}^{+1} \sin\frac{\pi x}{2} P_n(x)\,dx \tag{ii}$$

Putting $n = 1$, Eq. (ii) yields

$$a_1 = \frac{3}{2}\int_{-1}^{+1} \sin\frac{\pi x}{2} P_1(x)\,dx = \frac{3}{2}\int_{-1}^{+1} x\sin\frac{\pi x}{2}\,dx$$

$$= 3\int_0^1 x\sin\frac{\pi x}{2}\,dx = 3\left\{\left[-\frac{2}{\pi}x\cos\frac{\pi x}{2}\right]_0^1 + \int_0^1 \frac{2}{\pi}\cos\frac{\pi x}{2}\,dx\right\}$$

$$= 3\left\{0 + \left[\frac{4}{\pi^2}\sin\frac{\pi x}{2}\right]_0^1\right\} = \frac{12}{\pi^2} \tag{iii}$$

Putting $n = 3$, Eq. (ii) yields

$$a_3 = \frac{7}{2}\int_{-1}^{+1} \sin\frac{\pi x}{2} P_3(x)\,dx$$

$$= \frac{7}{2}\int_{-1}^{+1} \left(\sin\frac{\pi x}{2}\right)\frac{1}{2}\left(5x^3 - 3x\right)dx$$

$$= \frac{7}{2}\int_0^1 \left(5x^3 - 3x\right)\sin\frac{\pi x}{2}\,dx, \text{ as the integrand is even function;}$$

$$= \frac{7}{2}\left\{5\int_0^1 x^3\sin\frac{\pi x}{2}\,dx - 3\int_0^1 x\sin\frac{\pi x}{2}\,dx\right\}$$

$$= \frac{7}{2}\left\{\left[-\frac{10}{\pi}x^3\cos\frac{\pi x}{2}\right]_0^1 + \frac{30}{\pi}\int_0^1 x^2\cos\frac{\pi x}{2}\,dx - 3\int_0^1 x\sin\frac{\pi x}{2}\,dx\right\}$$

$$= \frac{7}{2}\left\{0 + \frac{60}{\pi^2}\left[x^2 \sin\frac{\pi x}{2}\right]_0^1 - \frac{120}{\pi^2}\int_0^1 x \sin\frac{\pi x}{2}dx - 3\int_0^1 x\sin\frac{\pi x}{2}dx\right\}$$

$$= \frac{7}{2}\left\{\frac{60}{\pi^2} - \left(3 + \frac{120}{\pi^2}\right)\int_0^1 x\sin\frac{\pi x}{2}dx\right\}$$

$$= \frac{7}{2}\left\{\frac{60}{\pi^2} - \left(3 + \frac{120}{\pi^2}\right)\frac{4}{\pi^2}\right\} \quad \text{[Using the result in (iii)]}$$

$$= \frac{168}{\pi^2}\left(1 - \frac{10}{\pi^2}\right)$$

PROBLEM 49 Express the reciprocal of the distance between two points having position vectors $\mathbf{r}_1$ and $\mathbf{r}_2$ in three-dimensional space in terms of Legendre polynomials.

Solution: Let $\mathbf{r}_1$ and $\mathbf{r}_2$ be the position vectors of two points in three-dimensional space and let θ be the angle included between these two position vectors. Then, the reciprocal of the distance between these two points can be expressed as

$$\frac{1}{|\mathbf{r}_1 - \mathbf{r}_2|} = \frac{1}{\left(r_1^2 - 2r_1 r_2\cos\theta + r_2^2\right)^{\frac{1}{2}}}$$

where r_1 and r_2 represent the magnitudes of the position vectors $\mathbf{r}_1$ and $\mathbf{r}_2$ respectively.

CASE I: $r_1 > r_2$

In this case, we have

$$\frac{1}{|\mathbf{r}_1 - \mathbf{r}_2|} = \frac{1}{r_1}\left[1 - \frac{2r_2}{r_1}\cos\theta + \left(\frac{r_2}{r_1}\right)^2\right]^{-\frac{1}{2}}$$

$$= \frac{1}{r_1}\sum_{l=0}^{\infty}\left(\frac{r_2}{r_1}\right)^l P_l(\cos\theta),$$

[Using the formula for generating function of Legendre polynomials]

$$= \sum_{l=0}^{\infty}\frac{(r_2)^l}{(r_1)^{l+1}} P_l(\cos\theta)$$

CASE II: $r_1 < r_2$

Following a procedure similar to that in Case I, we can obtain

$$\frac{1}{|\mathbf{r}_1-\mathbf{r}_2|}=\sum_{l=0}^{\infty}\frac{(r_1)^l}{(r_2)^{l+1}}P_l(\cos\theta)$$

For both the cases, we can write a general form as in the following.

$$\frac{1}{|\mathbf{r}_1-\mathbf{r}_2|}=\sum_{l=0}^{\infty}\frac{r_<^l}{r_>^{l+1}}P_l(\cos\theta)$$

where $r_<$ and $r_>$, respectively denote the smaller and the greater of r_1 and r_2.

PROBLEM 50 An electric dipole consists of two charges $+Q$ and $-Q$ separated by a distance $2a$. Find the electrostatic potential at an arbitrary point P distant r from the midpoint of the dipole.

Solution: Let $\mathbf{r}_1$ be the position vector from the charge $+Q$ to the point P and let $\mathbf{r}_2$ be the position vector from the charge $-Q$ to the point P. Let $\mathbf{r}$ be the position vector from the midpoint O of the dipole to the point P. Let θ be the angle included between the positive side of the dipole and the position vector $\mathbf{r}$ Let us take O (mid-point of dipole) as the origin of co-ordinates.

The potential at the point P will be the algebraic sum of the potentials due to the two charges.

Thus, the potential in SI units at the point P is given by

$$V(r,\theta)=\frac{Q}{4\pi\varepsilon_0 r_1}-\frac{Q}{4\pi\varepsilon_0 r_2},$$

where ε_0 is the pemittivity free space, $r_1 = |\mathbf{r}_1|$ and $r_2 = |\mathbf{r}_2|$.

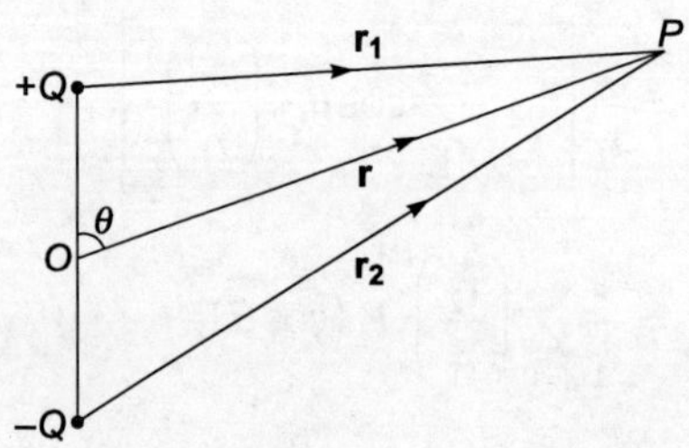

Fig. 2.1

We know that each of the charges $+Q$ and $-Q$ are situated at points distant a from the midpoint O.

Further, we know that

$$r_1=\left(r^2-2ar\cos\theta+a^2\right)^{\frac{1}{2}}=r\left\{1-2\frac{a}{r}\cos\theta+\left(\frac{a}{r}\right)^2\right\}^{\frac{1}{2}},$$

and $$r_2 = r\left\{1+2\frac{a}{r}\cos\theta+\left(\frac{a}{r}\right)^2\right\}^{\frac{1}{2}}$$

$$= r\left\{1-2\left(-\frac{a}{r}\right)\cos\theta+\left(-\frac{a}{r}\right)^2\right\}^{\frac{1}{2}}$$

Substituting the values of r_1 and r_2 into the expression for $V(r, \theta)$, we have

$$V(r,\theta)=\frac{Q}{4\pi\varepsilon_0 r}\left[\left\{1-2\frac{a}{r}\cos\theta+\left(\frac{a}{r}\right)^2\right\}^{-\frac{1}{2}}-\left\{1-2\left(-\frac{a}{r}\right)\cos\theta+\left(-\frac{a}{r}\right)^2\right\}^{-\frac{1}{2}}\right]$$

$$=\frac{Q}{4\pi\varepsilon_0 r}\left[\sum_{l=0}^{\infty}\left(\frac{a}{r}\right)^l P_l(\cos\theta)-\sum_{l=0}^{\infty}\left(-\frac{a}{r}\right)^l P_l(\cos\theta)\right]$$

$$=\frac{Q}{4\pi\varepsilon_0}\sum_{l=0}^{\infty}\left[1-(-1)^l\right]\frac{a^l}{r^{l+1}}P_l(\cos\theta)$$

$$=\frac{Q}{2\pi\varepsilon_0}\left[\frac{a}{r^2}P_1(\cos\theta)+\frac{a^3}{r^4}P_3(\cos\theta)+\frac{a^5}{r^6}P_5(\cos\theta)+\ldots\right]$$

PROBLEM 51 If $u_n=\int_{-1}^{+1}x^{-1}P_n(x)P_{n-1}(x)dx$, show that $(n + 1)u_{n+1} + nu_n = 2$ and hence find u_n.

Solution: We have

$$(n+1)u_{n+1}=(n+1)\int_{-1}^{+1}x^{-1}P_{n+1}(x)P_n(x)dx$$

$$=\int_{-1}^{+1}x^{-1}(n+1)P_{n+1}(x)P_n(x)dx \tag{i}$$

From the recurrence relation

$$(2n+1)xP_n(x)=(n+1)P_{n+1}(x)+nP_{n-1}(x)$$

we have

$$(n+1)P_{n+1}(x)=(2n+1)xP_n(x)-nP_{n-1}(x) \tag{ii}$$

Substituting Eq. (ii) into Eq. (i), we obtain

$$(n+1)u_{n+1} = \int_{-1}^{+1} x^{-1}\left[(2n+1)xP_n(x) - nP_{n-1}(x)\right]P_n(x)dx$$

$$= (2n+1)\int_{-1}^{+1} P_n(x)P_n(x)dx - n\int_{-1}^{+1} x^{-1}P_n(x)P_{n-1}(x)dx$$

$$= (2n+1)\frac{2}{(2n+1)} - nu_n$$

where use have been made of the orthogonality of Legendre polynomials and the definition of u_n given in the problem.

On transposition, we obtain

$$(n+1)u_{n+1} + nu_n = 2 \tag{iii}$$

Now, putting $n = 0, 1, 2, 3, 4, \ldots$ etc. successively, we obtain

$$u_1 = 2,$$

$$2u_2 + u_1 = 2, \text{ giving } u_2 = 0;$$

$$3u_3 + 2u_2 = 2, \text{ giving } u_3 = \frac{2}{3};$$

$$4u_4 + 3u_3 = 2, \text{ giving } u_4 = 0;$$

$$5u_5 + 4u_4 = 2, \text{ giving } u_5 = \frac{2}{5}; \text{ etc.}$$

In general, we write

$$u_n = \begin{cases} 0 & \text{if } n \text{ is even integer} \\ \dfrac{2}{n} & \text{if } n \text{ is odd integer} \end{cases}$$

PROBLEM 52 Prove that $P_{-(l+1)}(x) = P_l(x)$

Solution: See Recurrence Relation VIII.

EXERCISE

1. Show that $f(x) = \dfrac{d^l}{dx^l}(x^2-1)^l$ satisfies the Legendre differential equation

$$\left(1-x^2\right)\frac{d^2 f}{dx^2} - 2x\frac{df}{dx} + l(l+1)f = 0$$

2. Use the change of variable $u(x)=2x-1$ to prove that the differential equation

$$x(1-x)\frac{d^2y}{dx^2}+(1-2x)\frac{dy}{dx}+l(l+1)y=0$$

leads to the usual form of Legendre equation.

3. Using (a) sum formula, and (b) Rodrigues' formula for Legendre polynomials, prove that

$$P_l(-x)=(-1)^l P_l(x)$$

4. Obtain the values of the integrals

$$\text{(i)} \int_{-1}^{+1} xP_0(x)dx, \quad \text{(ii)} \int_{-1}^{+1} xP_1(x)dx, \quad \text{(iii)} \int_{-1}^{+1} xP_2(x)dx$$

$$\text{(iv)} \int_{-1}^{+1} x^2P_3(x)dx, \quad \text{(v)} \int_{-1}^{+1} x^2P_2(x)dx \quad \text{(vi)} \int_{-1}^{+1} (1+x)P_l(x)dx \text{ for } l>1$$

5. Prove that $P_l(\cos\theta)=\dfrac{1}{\pi}\displaystyle\int_0^{\pi}(\cos\theta+i\sin\theta\cos\phi)^l\,d\phi$

[*Hint:* Use Laplace's first integral representation.]

6. Prove that

$$(2l+1)\left[xP_l'(x)+P_l(x)\right]=(l+1)P_{l+1}'(x)+lP_{l-1}'(x),$$

where the prime symbol denotes differentiation with respect to x.

7. Prove that $P_l(x)=\displaystyle\sum_{r=0}^{l}\left[\frac{l!}{r!(l-r)!}\right]^2\left(\frac{x-1}{2}\right)^{l-r}\left(\frac{x+1}{2}\right)^r$

8. Prove that $\displaystyle\int_0^1 P_l(x)dx=\frac{P_{l-1}(0)}{l+1}$ if l is an odd integer.

[*Hint:* Use the recurrence relation $(2l+1)P_l(x)=P_{l+1}'(x)-P_{l-1}'(x)$ or the generating function.]

9. Prove that

$$\int_{-1}^{+1}\left[(2l-1)P_{l-1}(x)+(2l-5)P_{l-3}(x)+(2l-9)P_{l-5}(x)+\ldots\right]^2 dx=l(l+1)$$

[*Hint:* Use Christoffel's expansion.]

10. Prove that $\int_{-1}^{+1} x P_l(x) P_{l+1}(x) dx = \dfrac{2(l+1)}{(2l+1)(2l+3)}$

11. Prove that $\int_{0}^{\pi} P_k(\cos\theta) P_l(\cos\theta) \sin\theta \, d\theta = \dfrac{2}{(2l+1)} \delta_{kl}$

12. Prove that $\int_{0}^{\pi} P_l(\cos\theta) \sin 2\theta \, d\theta = 0$ for $l > 1$

13. Express (i) x^4 (ii) x^5 and (iii) x^6 in series of Legendre polynomials.

Ans. (i) $x^4 = \dfrac{1}{5} P_0(x) + \dfrac{4}{7} P_2(x) + \dfrac{8}{35} P_4(x)$

(ii) $x^5 = \dfrac{3}{7} P_1(x) + \dfrac{2}{3} P_3(x) + \dfrac{8}{63} P_5(x)$

(iii) $x^6 = \dfrac{1}{7} P_0(x) + \dfrac{10}{21} P_2(x) + \dfrac{24}{77} P_4(x) + \dfrac{16}{231} P_6(x)$

14. Expand the following functions in series of Legendre polynomials

(i) $f(x) = \begin{cases} 0, -1 < x < 0 \\ x^2, 0 < x < 1 \end{cases}$

Ans. $f(x) = \dfrac{1}{6} P_0(x) + \dfrac{3}{8} P_1(x) + \dfrac{1}{3} P_2(x) + \ldots$

(ii) $f(x) = \begin{cases} 0, -1 < x < 0 \\ 1, \quad 0 < x < 1 \end{cases}$

Ans. $f(x) = \dfrac{1}{2} P_0(x) + \dfrac{3}{4} P_1(x) - \dfrac{7}{16} P_3(x) + \ldots$

(iii) $f(x) = \begin{cases} 0, & -1 < x < 0 \\ 2x+1, & 0 < x < 1 \end{cases}$

Ans. $f(x) = P_0(x) + \dfrac{7}{4} P_1(x) + \dfrac{5}{8} P_2(x) - \dfrac{7}{16} P_3(x) + \ldots$

15. If $u_n = \int_{-1}^{+1} x^{-1} P_n(x) P_{n-1}(x) dx$, show that $n u_n + (n-1) u_{n-1} = 2$ and hence find u_n.

16. Prove that $\int_{0}^{\pi} P_l(\cos\theta) \cos l\theta \, d\theta = \dfrac{(2l)!}{2^{2l} (l!)^2}$.

17. Using binomial expansion, prove that $P_l(x)$ is the coefficient of h^l in the expansion of $(1 - 2xh + h^2)^{-\frac{1}{2}}$.

3 CHAPTER

Associated Legendre Polynomials

3.1 ASSOCIATED LEGENDRE EQUATION

Associated Legendre equation is a homogeneous linear second order ordinary differential equation of the form

$$\left(1-x^2\right)\frac{d^2y}{dx^2}-2x\frac{dy}{dx}+\left\{l(l+1)-\frac{m^2}{1-x^2}\right\}y=0, \tag{3.1}$$

with $l=0,1,2,3\ldots$ and m has any integral value within the range from $-l$ to $+l$.

This equation can be derived from the ordinary Legendre differential equation as in the following.

Since the Legendre polynomial, $P_l(x)$, is the solution of the Legendre equation, we write

$$(1-x^2)\frac{d^2}{dx^2}P_l(x)-2x\frac{d}{dx}P_l(x)+l(l+1)P_l(x)=0$$

Differentiating this equation m times with respect to x using Leibnitz theorem, we obtain

$$(1-x^2)\frac{d^{m+2}}{dx^{m+2}}P_l(x)-2(m+1)x\frac{d^{m+1}}{dx^{m+1}}P_l(x)$$
$$+\{l(l+1)-m(m+1)\}\frac{d^m}{dx^m}P_l(x)=0 \quad \ldots(3.2)$$

Let us put

$$y=(1-x^2)^{\frac{m}{2}}\frac{d^m}{dx^m}P_l(x)=P_l^m(x)$$

Then, we have

$$\frac{d^m}{dx^m}P_l(x)=y(1-x^2)^{-\frac{m}{2}} \tag{3.3}$$

$$\frac{d^{m+1}}{dx^{m+1}}P_l(x)=\frac{dy}{dx}(1-x^2)^{-\frac{m}{2}}+mxy(1-x^2)^{-\frac{m}{2}-1} \tag{3.4}$$

and
$$\frac{d^{m+2}}{dx^{m+2}}P_l(x)=(1-x^2)^{-\frac{m}{2}}\frac{d^2y}{dx^2}+2mx\,(1-x^2)^{-\frac{m}{2}-1}\frac{dy}{dx}$$
$$+m(m+2)x^2y(1-x^2)^{-\frac{m}{2}-2}+my(1-x^2)^{-\frac{m}{2}-1} \tag{3.5}$$

Substituting Eqs. (3.3), (3.4) and (3.5) into Eq. (3.2), we obtain

$$(1-x^2)\left[(1-x^2)^{-\frac{m}{2}}\frac{d^2y}{dx^2}+2mx(1-x^2)^{-\frac{m}{2}-1}\frac{dy}{dx}\right.$$
$$\left.+m(m+2)x^2y(1-x^2)^{-\frac{m}{2}-2}+my(1-x^2)^{-\frac{m}{2}-1}\right]$$
$$-2(m+1)x\left[(1-x^2)^{-\frac{m}{2}}\frac{dy}{dx}+mxy(1-x^2)^{-\frac{m}{2}-1}\right]$$
$$+\left[l(l+1)-m(m+1)\right](1-x^2)^{-\frac{m}{2}}y=0$$

or
$$(1-x^2)^{-\frac{m}{2}}\left[\left(1-x^2\right)\frac{d^2y}{dx^2}-2x\frac{dy}{dx}+\left\{l(l+1)-\left(m^2+\frac{m^2x^2}{1-x^2}\right)\right\}y\right]=0$$

or
$$(1-x^2)\frac{d^2y}{dx^2}-2x\frac{dy}{dx}+\left\{l(l+1)-\frac{m^2}{1-x^2}\right\}y=0$$

which is known as associated Legendre equation.

3.2 ASSOCIATED LEGENDRE POLYNOMIAL

The associated Legendre polynomial, $P_l^m(x)$, [also known as associated Legendre function] is defined as

$$P_l^m(x)=(1-x^2)^{\frac{m}{2}}\frac{d^m}{dx^m}P_l(x) \tag{3.6}$$

It satisfies the associated Legendre equation. Some authors used to define the associated Legendre polynomial as

$$P_l^m(x)=(-1)^m\left(1-x^2\right)^{\frac{m}{2}}\frac{d^m}{dx^m}P_l(x) \tag{3.7}$$

The associated Legendre polynomial defined here is known as associated Legendre polynomial of the first kind. There exists another type of associated Legendre polynomial known as associated Legendre polynomial of the second kind which is denoted by $Q_l^m(x)$.

3.3 PROPERTIES OF ASSOCIATED LEGENDRE POLYNOMIALS

$P_l^m(x)$ has the properties

(i) $P_l^0(x) = P_l(x)$ and (ii) $P_l^m(x) = 0$ for $m > l$.

3.4 EXPRESSIONS FOR SOME ASSOCIATED LEGENDRE POLYNOMIALS

(i) $P_1^0(x) = x,\ P_1^0(\cos\theta) = \cos\theta$

(ii) $P_1^1(x) = (1-x^2)^{\frac{1}{2}},\ P_1^1(\cos\theta) = \sin\theta$

(iii) $P_2^1(x) = 3x(1-x^2)^{\frac{1}{2}},\ P_2^1(\cos\theta) = \frac{3}{2}\sin 2\theta$

(iv) $P_2^2(x) = 3(1-x^2),\ P_2^2(\cos\theta) = 3\sin^2\theta$

3.5 RODRIGUES' FORMULA FOR ASSOCIATED LEGENDRE POLYNOMIALS

Rodrigues' formula for ordinary Legendre polynomials is

$$P_l(x) = \frac{1}{2^l\, l!}\frac{d^l}{dx^l}(x^2-1)^l$$

Now, Rodrigues' formula for associated Legendre polynomials can be written as

$$P_l^m(x) = (1-x^2)^{\frac{m}{2}}\frac{d^m}{dx^m}P_l(x)$$

$$= (1-x^2)^{\frac{m}{2}}\frac{d^m}{dx^m}\left[\frac{1}{2^l\, l!}\frac{d^l}{dx^l}(x^2-1)^l\right]$$

$$= \frac{1}{2^l\, l!}(1-x^2)^{\frac{m}{2}}\frac{d^{l+m}}{dx^{l+m}}(x^2-1)^l \tag{3.8}$$

3.6 ASSOCIATED LEGENDRE POLYNOMIALS WITH NEGATIVE SUPERSCRIPT

The subscript l and the superscript m of an associated Legendre polynomial $P_l^m(x)$ have special physical significances in Quantum Mechanics. In quantum mechanics, l stands for orbital angular momentum quantum number and m stands for magnetic quantum number of a quantum mechanical system. The quantum number m has integral values from $-l$ to $+l$. Therefore, it is necessary to find an expression for $P_l^{-m}(x)$.

The associated Legendre polynomial with negative upper index is defined from Eq. (3.8) as

$$P_l^{-m}(x) = \frac{1}{2^l\, l!}(1-x^2)^{-\frac{m}{2}} \frac{d^{l-m}}{dx^{l-m}}(x^2-1)^l \tag{3.9}$$

3.7 RELATION BETWEEN ASSOCIATED LEGENDRE POLYNOMIALS WITH POSITIVE AND NEGATIVE SUPERSCRIPTS ($P_l^m(x)$ and $P_l^{-m}(x)$)

Such a relation can be obtained via two methods as detailed below.

First Method

We have

$$P_l^m(x) = \frac{1}{2^l\, l!}(1-x^2)^{\frac{m}{2}} \frac{d^{l+m}}{dx^{l+m}}(x^2-1)^l$$

$$= \frac{1}{2^l\, l!}(1-x^2)^{-\frac{m}{2}}(1-x^2)^m \frac{d^{l+m}}{dx^{l+m}}(x^2-1)^l$$

$$= \frac{(-1)^m}{2^l\, l!}(1-x^2)^{-\frac{m}{2}}(x^2-1)^m \frac{d^{l+m}}{dx^{l+m}}\left\{(x+1)^l(x-1)^l\right\}$$

Performing the differentiation in the right-hand side via Leibnitz theorem, we obtain

$$P_l^m(x) = \frac{(-1)^m}{2^l\, l!}(1-x^2)^{-\frac{m}{2}}(x^2-1)^m \sum_{r=0}^{l+m} \frac{(l+m)!}{r!\,(l+m-r)!} \frac{d^r}{dx^r}(x+1)^l \frac{d^{l+m-r}}{dx^{l+m-r}}(x-1)^l$$

$$= \frac{(-1)^m}{2^l\, l!}\left(1-x^2\right)^{-\frac{m}{2}}\left(x^2-1\right)^m \sum_{r=0}^{l+m} \frac{(l+m)!}{r!(l+m-r)!} \frac{l!}{(l-r)!}(x+1)^{l-r} \frac{l!}{(r-m)!}(x-1)^{r-m}$$

$$= \frac{(-1)^m}{2^l\, l!}(1-x^2)^{-\frac{m}{2}} \sum_{r=0}^{l+m} \frac{(l+m)!\,l!\,l!}{r!(l+m-r)!(l-r)!(r-m)!}(x+1)^{l+m-r}(x-1)^r$$

Putting $j = r - m$ in the summation, we write,

$$P_l^m(x) = \frac{(-1)^m}{2^l\, l!}(1-x^2)^{-\frac{m}{2}} \sum_{j=0}^{l-m} \frac{(l+m)!\, l!\, l!}{(j+m)!(l-j)!(l-m-j)!\, j!}(x+1)^{l-j}(x-1)^{j+m}$$

[as the denominator becomes infinite if it contains the factorial of a negative integer].

After rearrangement, we write,

$$P_l^m(x)$$

$$= \frac{(-1)^m\left(1-x^2\right)^{-\frac{m}{2}}}{2^l\, l!} \frac{(l+m)!}{(l-m)!} \sum_{j=0}^{l-m} \frac{(l-m)!}{j!(l-m-j)!} \frac{l!}{(l-j)!}(x+1)^{l-j} \frac{l!}{\{l-(l-m-j)\}!}(x-1)^{l-(l-m-j)}$$

$$= \frac{(-1)^m\left(1-x^2\right)^{-\frac{m}{2}}}{2^l\, l!} \frac{(l+m)!}{(l-m)!} \sum_{j=0}^{l-m} \frac{(l-m)!}{j!(l-m-j)!} \frac{d^j}{dx^j}(x+1)^l \frac{d^{l-m-j}}{dx^{l-m-j}}(x-1)^l$$

$$= (-1)^m \frac{(l+m)!}{(l-m)!} \frac{1}{2^l\, l!}(1-x^2)^{-\frac{m}{2}} \frac{d^{l-m}}{dx^{l-m}}[(x+1)^l(x-1)^l]$$

$$= (-1)^m \frac{(l+m)!}{(l-m)!} \frac{1}{2^l\, l!}(1-x^2)^{-\frac{m}{2}} \frac{d^{l-m}}{dx^{l-m}}(x^2-1)^l$$

$$= (-1)^m \frac{(l+m)!}{(l-m)!} P_l^{-m}(x) \tag{3.10}$$

Also, we have

$$P_l^{-m}(x) = (-1)^m \frac{(l-m)!}{(l+m)!} P_l^m(x) \tag{3.11}$$

Second method

We have

$$P_l^m(x) = \frac{1}{2^l l!}(1-x^2)^{\frac{m}{2}} \frac{d^{l+m}}{dx^{l+m}}(x^2-1)^l$$

$$= \frac{1}{2^l\, l!}(1-x^2)^{\frac{m}{2}} \frac{d^{l+m}}{dx^{l+m}}\{(x+1)^l(x-1)^l\}$$

$$= \frac{(-1)^{\frac{m}{2}}}{2^l\, l!}(x^2-1)^{\frac{m}{2}} \sum_{r=0}^{l+m} \frac{(l+m)!}{r!(l+m-r)!} \frac{d^r}{dx^r}(x+1)^l \frac{d^{l+m-r}}{dx^{l+m-r}}(x-1)^l.$$

Regarding the two derivatives, the first one will be non-zero if $r \le l$ and the second one will be non-zero if $l+m-r \le l$ or $m \le r$. Combining these two conditions, we have,

$$m \le r \le l$$

Therefore, we have

$$P_l^m(x) = \frac{(-1)^{\frac{m}{2}}}{2^l\, l!}(x^2-1)^{\frac{m}{2}} \sum_{r=m}^{l} \frac{(l+m)!}{r!(l+m-r)!} \frac{l!}{(l-r)!}(x+1)^{l-r} \frac{l!}{(r-m)!}(x-1)^{r-m}$$

$$= (-1)^{\frac{m}{2}} \frac{l!(l+m)!}{2^l} \sum_{r=m}^{l} \frac{(x+1)^{l+\frac{m}{2}-r}(x-1)^{r-\frac{m}{2}}}{r!(l+m-r)(l-r)!(r-m)!} \tag{3.12}$$

Also, we have

$$P_l^{-m}(x) = \frac{1}{2^l\, l!}(1-x^2)^{-\frac{m}{2}} \frac{d^{l-m}}{dx^{l-m}}(x^2-1)^l$$

$$= \frac{(-1)^{-\frac{m}{2}}}{2^l\, l!}(x^2-1)^{-\frac{m}{2}} \sum_{s=0}^{l-m} \frac{(l-m)!}{s!(l-m-s)!} \frac{d^s}{dx^s}(x+1)^l \frac{d^{l-m-s}}{dx^{l-m-s}}(x-1)^l$$

For non-zero derivatives, we must have the conditions,

$$s \le l \text{ and } l-m-s \le l \text{ or}, -m \le s$$

Combining these two conditions, we have

$$-m \le s \le l$$

Then, we write

$$P_l^{-m}(x) = \frac{(-1)^{-\frac{m}{2}}}{2^l\, l!}(x^2-1)^{-\frac{m}{2}} \sum_{s=-m}^{l} \frac{(l-m)!}{s!(l-m-s)!} \frac{l!}{(l-s)!}(x+1)^{l-s} \frac{l!}{(s+m)!}(x-1)^{s+m}$$

$$= (-1)^{-\frac{m}{2}} \frac{l!\,(l-m)!}{2^l} \sum_{s=-m}^{l} \frac{(x+1)^{l-s-\frac{m}{2}}(x-1)^{\frac{m}{2}+s}}{s!\,(l-m-s)!\,(l-s)!\,(m+s)!}$$

If we put $r = s+m$, we write

$$P_l^{-m}(x) = (-1)^{-\frac{m}{2}} \frac{l!\,(l-m)!}{2^l} \sum_{r=0}^{l} \frac{(x+1)^{l+\frac{m}{2}-r}(x-1)^{r-\frac{m}{2}}}{(r-m)!\,(l-r)!\,(l+m-r)!\,r!}$$

$$= (-1)^{-\frac{m}{2}} \frac{l!\,(l-m)!}{2^l} \sum_{r=m}^{l} \frac{(x+1)^{l+\frac{m}{2}-r}(x-1)^{r-\frac{m}{2}}}{r!\,(l+m-r)!\,(l-r)!(r-m)!} \tag{3.13}$$

Dividing Eq. (3.12) by Eq. (3.13), we obtain

$$\frac{P_l^m(x)}{P_l^{-m}(x)} = (-1)^m \frac{(l+m)!}{(l-m)!}$$

or

$$P_l^m(x) = (-1)^m \frac{(l+m)!}{(l-m)!} P_l^{-m}(x)$$

which is Eq. (3.10).

3.8 RECURRENCE RELATIONS AMONG ASSOCIATED LEGENDRE POLYNOMIALS

I. $P_l^{m+1}(x) - \frac{2mx}{\sqrt{1-x^2}} P_l^m(x) + \{l(l+1) - m(m-1)\} P_l^{m-1}(x) = 0$

Proof: Since the ordinary Legendre polynomial, $P_l(x)$, satisfies the Legendre equation, we write,

$$(1-x^2)\frac{d^2}{dx^2}P_l(x) - 2x\frac{d}{dx}P_l(x) + l(l+1)P_l(x) = 0$$

Differentiating both sides of this equation $(m-1)$ times with respect to x using Leibnitz theorem, we write

$$(1-x^2)\frac{d^{m+1}}{dx^{m+1}}P_l(x) - 2(m-1)x\frac{d^m}{dx^m}P_l(x) - \frac{(m-1)(m-2)}{2}\cdot 2\cdot\frac{d^{m-1}}{dx^{m-1}}P_l(x)$$

$$-\left\{2x\frac{d^m}{dx^m}P_l(x) + 2(m-1)\cdot 1\cdot\frac{d^{m-1}}{dx^{m-1}}P_l(x)\right\} + l(l+1)\frac{d^{m-1}}{dx^{m-1}}P_l(x) = 0$$

or

$$(1-x^2)\frac{d^{m+1}}{dx^{m+1}}P_l(x) - 2mx\frac{d^m}{dx^m}P_l(x) + \{l(l+1) - m(m-1)\}\frac{d^{m-1}}{dx^{m-1}}P_l(x) = 0$$

Multiplying both sides by $(1-x^2)^{\frac{m-1}{2}}$ we obtain

$$(1-x^2)^{\frac{m+1}{2}}\frac{d^{m+1}}{dx^{m+1}}P_l(x) - \frac{2mx}{\sqrt{1-x^2}}(1-x^2)^{\frac{m}{2}}\frac{d^m}{dx^m}P_l(x)$$

$$+\{l(l+1) - m(m-1)\}(1-x^2)^{\frac{m-1}{2}}\frac{d^{m-1}}{dx^{m-1}}P_l(x) = 0$$

or

$$P_l^{m+1}(x) - \frac{2mx}{\sqrt{1-x^2}}P_l^m(x) + \{l(l+1) - m(m-1)\}P_l^{m-1}(x) = 0 \tag{3.14}$$

II. $(2l+1)(1-x^2)^{\frac{1}{2}} P_l^m(x) = P_{l+1}^{m+1}(x) - P_{l-1}^{m+1}(x)$

Proof: In the case of ordinary Legendre polynomials, we know that

$$(2l+1)P_l(x) = P'_{l+1}(x) - P'_{l-1}(x)$$

where the primes denote differentiation with respect to x.

Differentiating both sides of this equation m times with respect to x, we obtain

$$(2l+1)\frac{d^m}{dx^m}P_l(x) = \frac{d^{m+1}}{dx^{m+1}}P_{l+1}(x) - \frac{d^{m+1}}{dx^{m+1}}P_{l-1}(x)$$

Multiplying both sides of this equation by $(1-x^2)^{\frac{m+1}{2}}$, we obtain

$$(2l+1)(1-x^2)^{\frac{1}{2}} P_l^m(x) = P_{l+1}^{m+1}(x) - P_{l-1}^{m+1}(x) \tag{3.15}$$

III. $(2l+1)x P_l^m(x) = (l-m+1)P_{l+1}^m(x) + (l+m)P_{l-1}^m(x)$

Proof: We know, in the case of ordinary Legendre polynomials, that

$$(2l+1)xP_l(x) = (l+1)P_{l+1}(x) + lP_{l-1}(x) \tag{i}$$

and

$$(2l+1)P_l(x) = P'_{l+1}(x) - P'_{l-1}(x) \tag{ii}$$

where the primes denote differentiation with respect to x.

Differentiating both sides of Eq. (i) m times with respect to x using Leibnitz theorem, we have

$$(2l+1)\left\{x\frac{d^m}{dx^m}P_l(x) + m\frac{d^{m-1}}{dx^{m-1}}P_l(x)\right\} = (l+1)\frac{d^m}{dx^m}P_{l+1}(x) + l\frac{d^m}{dx^m}P_{l-1}(x) \tag{iii}$$

Again, differentiating both sides of Eq. (ii) $(m-1)$ times with respect to x, we obtain

$$(2l+1)\frac{d^{m-1}}{dx^{m-1}}P_l(x) = \frac{d^m}{dx^m}P_{l+1}(x) - \frac{d^m}{dx^m}P_{l-1}(x) \tag{iv}$$

Subtracting m times Eq. (iv) from Eq. (iii), we obtain

$$(2l+1)x\frac{d^m}{dx^m}P_l(x) = (l-m+1)\frac{d^m}{dx^m}P_{l+1}(x) + (l+m)\frac{d^m}{dx^m}P_{l-1}(x)$$

Multiplying both sides by $(1-x^2)^{\frac{m}{2}}$ we obtain

$$(2l+1)xP_l^m(x) = (l-m+1)P_{l+1}^m(x) + (l+m)P_{l-1}^m(x) \tag{3.16}$$

IV. $(2l+1)(1-x^2)^{\frac{1}{2}} P_l^m(x) = (l+m)(l+m-1) P_{l-1}^{m-1}(x) - (l-m+1)(l-m+2) P_{l+1}^{m-1}(x)$

Proof: From Eq. (3.14), we have

$$P_l^{m+1}(x) - \frac{2m}{\sqrt{1-x^2}} x P_l^m(x) + \{l(l+1) - m(m-1)\} P_l^{m-1}(x) = 0 \tag{i}$$

Also, from Eq. (3.16), we have

$$x P_l^m(x) = \frac{1}{(2l+1)}\left[(l+m) P_{l-1}^m(x) + (l-m+1) P_{l+1}^m(x)\right] \tag{ii}$$

Substitution of Eq. (ii) into Eq. (i) results in

$$P_l^{m+1}(x) - \frac{2m}{\sqrt{1-x^2}} \frac{1}{(2l+1)}\left[(l+m) P_{l-1}^m(x) + (l-m+1) P_{l+1}^m(x)\right] + \{l(l+1) - m(m-1)\} P_l^{m-1}(x) = 0 \tag{iii}$$

Further, from Eq. (3.15), we have

$$(2l+1)(1-x^2)^{\frac{1}{2}} P_l^m(x) = P_{l+1}^{m+1}(x) - P_{l-1}^{m+1}(x)$$

Replacing m by $(m-1)$, we obtain

$$(2l+1)(1-x^2)^{\frac{1}{2}} P_l^{m-1}(x) = P_{l+1}^m(x) - P_{l-1}^m(x)$$

or

$$P_l^{m-1}(x) = \frac{1}{(2l+1)(1-x^2)^{\frac{1}{2}}}\left[P_{l+1}^m(x) - P_{l-1}^m(x)\right] \tag{iv}$$

Substituting Eq. (iv) into Eq. (iii), we obtain

$$P_l^{m+1}(x) - \frac{2m}{(2l+1)(1-x^2)^{\frac{1}{2}}}\left[(l+m) P_{l-1}^m(x) + (l-m+1) P_{l+1}^m(x)\right] + \frac{\{l(l+1) - m(m-1)\}}{(2l+1)(1-x^2)^{\frac{1}{2}}}\left[P_{l+1}^m(x) - P_{l-1}^m(x)\right] = 0$$

or

$$(2l+1)(1-x^2)^{\frac{1}{2}} P_l^{m+1}(x) - \left[\{2m(l+m) + l(l+1) - m(m-1)\} P_{l-1}^m(x) + \{l(l+1) - m(m-1) - 2m(l-m+1)\} P_{l+1}^m(x)\right] = 0$$

or

$$(2l+1)(1-x^2)^{\frac{1}{2}} P_l^{m+1}(x) - (l+m)(l+m+1) P_{l-1}^m(x) + (l-m)(l-m+1) P_{l+1}^m(x) = 0$$

Replacing m by $(m-1)$ we obtain

$$(2l+1)(1-x^2)^{\frac{1}{2}} P_l^m(x) = (l+m)(l+m-1)P_{l-1}^{m-1}(x) - (l-m+1)(l-m+2)P_{l+1}^{m-1}(x) \quad (3.17)$$

3.9 GENERATING FUNCTION FOR ASSOCIATED LEGENDRE POLYNOMIALS

We know that the generating function for ordinary Legendre polynomials, $P_l(x)$, is

$$(1-2hx+h^2)^{-\frac{1}{2}} = \sum_{l=0}^{\infty} P_l(x)h^l$$

Differentiating both sides m times with respect to x, we obtain

$$1.3.5.7\ldots(2m-1)h^m(1-2hx+h^2)^{-\frac{1}{2}-m} = \sum_{l=m}^{\infty}\left[\frac{d^m}{dx^m}P_l(x)\right]h^l$$

Here, we use the fact that the derivative in the expression on the right-hand side vanishes for $l<m$

or $$\frac{1.2.3.4.5.6.7.8\ldots(2m-1)(2m)}{2.4.6.8\ldots(2m-2)(2m)}\frac{h^m}{(1-2hx+h^2)^{m+\frac{1}{2}}} = \sum_{l=m}^{\infty}\left[\frac{d^m}{dx^m}P_l(x)\right]h^l$$

or $$\frac{(2m)!}{2^m m!\left(1-2hx+h^2\right)^{m+\frac{1}{2}}} = \sum_{l=m}^{\infty}\left[\frac{d^m}{dx^m}P_l(x)\right]h^{l-m}$$

Multiplying both sides by $(1-x^2)^{\frac{m}{2}}$ we obtain

$$\frac{(2m)!(1-x^2)^{\frac{m}{2}}}{2^m m!\left(1-2hx+h^2\right)^{m+\frac{1}{2}}} = \sum_{l=m}^{\infty}P_l^m(x)h^{l-m} = \sum_{r=0}^{\infty}P_{m+r}^m(x)h^r \quad (3.18)$$

where we put $r=l-m$.

3.10 ORTHOGONAL PROPERTIES OF ASSOCIATED LEGENDRE POLYNOMIALS

The associated Legendre polynomials have the following orthogonal properties:

(a) $$I_{kl}^m = \int_{-1}^{+1} P_k^m(x)P_l^m(x)dx = \frac{2}{2l+1}\frac{(l+m)!}{(l-m)!}\delta_{kl} \quad (3.19a)$$

where $\delta_{kl} = \begin{cases} 0 \text{ for } k \neq l, \\ 1 \text{ for } k = l \end{cases}$

and (b) $I_l^{mn} = \int_{-1}^{+1} \frac{P_l^m(x) P_l^n(x)}{(1-x^2)} dx = \frac{(l+m)!}{m(l+m)!} \delta_{mn}, m \neq 0$ (3.19 b)

The second orthogonality condition with the weight function $(1-x^2)^{-1}$, i.e., Eq. (3.19b) is of no physical interest.

(a) Proof of the First Orthogonality Condition

Using Rodrigues' formula, we have

$$I_{kl}^m = \int_{-1}^{+1} P_k^m(x) P_l^m(x) dx$$

$$= \frac{1}{2^{k+l} k!\, l!} \int_{-1}^{+1} (1-x^2)^m \frac{d^{k+m}}{dx^{k+m}} (x^2-1)^k \frac{d^{l+m}}{dx^{l+m}} (x^2-1)^l dx$$

$$= \frac{(-1)^m}{2^{k+l} k!\, l!} \int_{-1}^{+1} \left\{ (x^2-1)^m \frac{d^{k+m}}{dx^{k+m}} (x^2-1)^k \right\} \frac{d^{l+m}}{dx^{l+m}} (x^2-1)^l dx$$

CASE I: $k \neq l$

First, let us assume that $k < l$.

If we repeatedly integrate by parts, the integrated parts will vanish as long as there is the factor, (x^2-1). Thus, if we integrate $(l+m)$ times, we will get

$$\int_{-1}^{+1} P_k^m(x) P_l^m(x) dx = \frac{(-1)^m (-1)^{l+m}}{2^{k+l} k!\, l!} \int_{-1}^{+1} \left[\frac{d^{l+m}}{dx^{l+m}} \left\{ (x^2-1)^m \frac{d^{k+m}}{dx^{k+m}} (x^2-1)^k \right\} \right] (x^2-1)^l dx \quad (3.20)$$

If we evaluate the derivative in the integrand on the right-hand side of the above equation using Leibnitz theorem, we obtain

$$\frac{d^{l+m}}{dx^{l+m}} \left\{ (x^2-1)^m \frac{d^{k+m}}{dx^{k+m}} (x^2-1)^k \right\}$$

$$= \sum_{r=0}^{l+m} \frac{(l+m)!}{r!\,(l+m-r)!} \frac{d^{l+m-r}}{dx^{l+m-r}} (x^2-1)^m \frac{d^{k+m+r}}{dx^{k+m+r}} (x^2-1)^k \quad (3.21)$$

For the preceding derivative to have non-zero value, we must have the condition $l+m-r \leq 2m$, or, $l-m \leq r$, otherwise, the derivative will vanish.

For the latter derivative to have non-zero value, we must have the condition

$k+m+r \le 2k$, or, $r \le k-m$ otherwise, the derivative will vanish.

Adding these two conditions, we obtain $l \le k$ which contradicts our earlier assumption that $k < l$.

Hence, there is no solution for r and we conclude that the integral vanishes. Similar conclusion will be reached if we start with the assumption that $k > l$.

Thus, we can write

$$I_{kl}^m = \int_{-1}^{+1} P_k^m(x) P_l^m(x)\, dx = 0 \quad \text{for } k \neq l \tag{3.22}$$

[**Note:** An alternative method using associated Legendre equation is presented in Solved Problems.]

CASE II: $k = l$

In this case, the above two conditions yield $r = (l-m)$ and the right-hand side of Eq. (3.21) will have the single term corresponding to $r = l-m$.

Thus, Eq. (3.21) will become

$$\frac{d^{l+m}}{dx^{l+m}}\left\{(x^2-1)^m \frac{d^{l+m}}{dx^{l+m}}(x^2-1)^l\right\} = \frac{(l+m)!}{(l-m)!\,(2m)!}\frac{d^{2m}}{dx^{2m}}(x^2-1)^m \frac{d^{2l}}{dx^{2l}}(x^2-1)^l$$

$$= \frac{(l+m)!}{(l-m)!\,(2m)!}(2m)!\,(2l)! \tag{3.23}$$

Substituting Eq. (3.23) into Eq. (3.20) with $k = l$, we obtain

$$I_{ll}^m = \int_{-1}^{+1} P_l^m(x) P_l^m(x)\, dx \quad [\text{with } k = l]$$

$$= \frac{(-1)^l}{2^{2l}\, l!\, l!} \frac{(l+m)!\,(2l)!}{(l-m)!} \int_{-1}^{+1} (x^2-1)^l\, dx$$

$$= \frac{(l+m)!\,(2l)!}{(l-m)!\, 2^{2l}\, l!\, l!} \int_{-1}^{+1} (1+x)^l (1-x)^l\, dx$$

If we put $x = 2u - 1$, or, $u = \dfrac{x+1}{2}$, then, $dx = 2du$ and hence, we write

$$I_{ll}^m = \int_{-1}^{+1} P_l^m(x) P_l^m(x)\, dx = \frac{(l+m)!\,(2l)!}{(l-m)!\, 2^{2l}\, l!\, l!} \int_0^1 (2u)^l\, 2^l (1-u)^l\, 2du$$

$$= 2\frac{(l+m)!\,(2l)!}{(l-m)!\, l!\, l!} \int_0^1 u^l (1-u)^l\, du$$

$$= 2\frac{(l+m)!\,(2l)!}{(l-m)\,l!\,l!}B(l+1,l+1)$$

$$= 2\frac{(l+m)!\,(2l)!}{(l-m)!\,l!\,l!}\frac{\Gamma(l+1)\,\Gamma(l+1)}{\Gamma(2l+2)} = \frac{2}{(2l+1)}\frac{(l+m)!}{(l-m)!}. \tag{3.24}$$

Equations (3.22) and (3.24) can be expressed in a general equation

$$I_{kl}^{m} = \int_{-1}^{+1} P_k^m(x)\,P_l^m(x)\,dx = \frac{2}{2l+1}\frac{(l+m)!}{(l-m)!}\delta_{kl}$$

which is Eq. (3.19a).

(b) Proof of the Second Orthogonality Condition

CASE I: $m \neq n$

We know that $P_l^m(x)$ satisfies the associated Legendre equation

$$\frac{d}{dx}\left\{(1-x^2)\frac{d}{dx}P_l^m(x)\right\} + \left\{l(l+1) - \frac{m^2}{1-x^2}\right\}P_l^m(x) = 0. \tag{i}$$

Similarly, $P_l^n(x)$ satisfies the equation

$$\frac{d}{dx}\left\{(1-x^2)\frac{d}{dx}P_l^n(x)\right\} + \left\{l(l+1) - \frac{n^2}{1-x^2}\right\}P_l^n(x) = 0. \tag{ii}$$

Subtracting $P_l^m(x)$ times equation (ii) from $P_l^n(x)$ times equation (i) and integrating the results with respect to x within the interval from $x = -1$ to $x = +1$, we obtain

$$\int_{-1}^{+1} P_l^n(x)\frac{d}{dx}\left\{(1-x^2)\frac{d}{dx}P_l^m(x)\right\}dx - \int_{-1}^{+1} P_l^m(x)\frac{d}{dx}\left\{(1-x^2)\frac{d}{dx}P_l^n(x)\right\}dx$$

$$-\left(m^2 - n^2\right)\int_{-1}^{+1} P_l^m(x)\,P_l^n(x)\frac{dx}{(1-x^2)} = 0,$$

or

$$0 = \left(m^2 - n^2\right)\int_{-1}^{+1} P_l^m(x)\,P_l^n(x)\frac{dx}{(1-x^2)}$$

For $m \neq n$, we write

$$I_l^{mn} = \int_{-1}^{+1} P_l^m(x)\,P_l^n(x)\frac{dx}{(1-x^2)} = 0\,(m \neq n). \tag{3.25}$$

CASE II: $m = n$

Using Rodrigues' formula, we write

$$I_l^{mm} = \int_{-1}^{+1} \frac{P_l^m(x) P_l^m(x)}{(1-x^2)} dx$$

$$= \frac{1}{2^{2l}(l!)^2} \int_{-1}^{+1} \left\{ (1-x^2)^{m-1} \frac{d^{l+m}}{dx^{l+m}} (x^2-1)^l \right\} \left\{ \frac{d^{l+m}}{dx^{l+m}} (x^2-1)^l \right\} dx \tag{3.26}$$

If we integrate by parts $(m-1)$ times in the integral on the right-hand side of Eq. (3.26), the integrated out terms will vanish at the given limits because of the factor $(1-x^2)$.

Thus, Eq. (3.26) becomes

$$I_l^{mm} = \int_{-1}^{+1} \frac{P_l^m(x) P_l^m(x)}{(1-x^2)} dx$$

$$= \frac{1}{2^{2l}(l!)^2} \left[(-1)^{m-1} \int_{-1}^{+1} \frac{d^{m-1}}{dx^{m-1}} \left\{ (1-x^2)^{m-1} \frac{d^{l+m}}{dx^{l+m}} (x^2-1)^l \right\} \left\{ \frac{d^{l+1}}{dx^{l+1}} (x^2-1)^l \right\} dx \right]$$

$$= \frac{1}{2^{2l}(l!)^2} \left[\int_{-1}^{+1} \frac{d^{m-1}}{dx^{m-1}} \left\{ (x^2-1)^{m-1} \frac{d^{l+m}}{dx^{l+m}} (x^2-1)^l \right\} \left\{ \frac{d^{l+1}}{dx^{l+1}} (x^2-1)^l \right\} dx \right]$$

Further integration by parts results in

$$I_l^{mm} = \int_{-1}^{+1} \frac{P_l^m(x) P_l^m(x)}{(1-x^2)} dx$$

$$= \frac{1}{2^{2l}(l!)^2} \left[\frac{d^{m-1}}{dx^{m-1}} \left\{ (x^2-1)^{m-1} \frac{d^{l+m}}{dx^{l+m}} (x^2-1)^l \right\} \left\{ \frac{d^l}{dx^l} (x^2-1)^l \right\} \right]_{-1}^{+1}$$

$$- \frac{1}{2^{2l}(l!)^2} \int_{-1}^{+1} \frac{d^m}{dx^m} \left\{ (x^2-1)^{m-1} \frac{d^{l+m}}{dx^{l+m}} (x^2-1)^l \right\} \left\{ \frac{d^l}{dx^l} (x^2-1)^l \right\} dx$$

$$= S_l^m - \frac{1}{2^{2l}(l!)^2} \int_{-1}^{+1} \frac{d^m}{dx^m} \left\{ (x^2-1)^{m-1} \frac{d^{l+m}}{dx^{l+m}} (x^2-1)^l \right\} \left\{ \frac{d^l}{dx^l} (x^2-1)^l \right\} dx \tag{3.27}$$

where $S_l^m = \frac{1}{2^{2l}(l!)^2} \left[\frac{d^{m-1}}{dx^{m-1}} \left\{ (x^2-1)^{m-1} \frac{d^{l+m}}{dx^{l+m}} (x^2-1)^l \right\} \left\{ \frac{d^l}{dx^l} (x^2-1)^l \right\} \right]_{-1}^{+1}$ (3.28)

In performing further integration by parts l times in Eq. (3.27), the integrated out terms vanish and we write Eq. (3.27) as

$$I_l^{mm} = \int_{-1}^{+1} \frac{P_l^m(x) P_l^m(x)}{(1-x^2)} dx$$

$$= S_l^m - \frac{(-1)^l}{2^{2l}(l!)^2} \int_{-1}^{+1} \frac{d^{l+m}}{dx^{l+m}} \left\{ (x^2-1)^{m-1} \frac{d^{l+m}}{dx^{l+m}} (x^2-1)^l \right\} (x^2-1)^l dx \qquad (3.29)$$

In the above integral, the highest power of x in the expression within the curly braces is

$$(2m-2) + 2l - (l+m) = l + m - 2$$

Therefore, $(l+m)$ times differentiation of the expression within the curly braces has zero value and we write Eq. (3.29) as

$$I_l^{mm} = \int_{-1}^{+1} \frac{P_l^m(x) P_l^m(x)}{1-x^2} dx = S_l^m$$

$$= \frac{1}{2^{2l}(l!)^2} \left[\frac{d^{m-1}}{dx^{m-1}} \left\{ (x^2-1)^{m-1} \frac{d^{l+m}}{dx^{l+m}} (x^2-1)^l \right\} \frac{d^l}{dx^l} (x^2-1)^l \right]_{-1}^{+1}$$

$$= \frac{2}{2^{2l}(l!)^2} \left[\frac{d^{m-1}}{dx^{m-1}} \left\{ (x^2-1)^{m-1} \frac{d^{l+m}}{dx^{l+m}} (x^2-1)^l \right\} \left\{ \frac{d^l}{dx^l} (x^2-1)^l \right\} \right]_{x=1} \qquad (3.30)$$

We have

$$\frac{d^l}{dx^l}(x^2-1)^l = \frac{d^l}{dx^l}\left\{ (x+1)^l (x-1)^l \right\}$$

$$= \sum_{r=0}^{l} \frac{l!}{r!(l-r)!} \frac{d^r}{dx^r}(x+1)^l \frac{d^{l-r}}{dx^{l-r}}(x-1)^l$$

$$= (x+1)^l \frac{d^l}{dx^l}(x-1)^l + l \frac{d}{dx}(x+1)^l \frac{d^{l-1}}{dx^{l-1}}(x-1)^l + \ldots$$

$$= l!(x+1)^l + \text{terms having positive powers of } (x-1). \qquad (3.31)$$

Also, we have

$$\frac{d^{l+m}}{dx^{l+m}}(x^2-1)^l = \sum_{r=0}^{l+m} \frac{(l+m)!}{r!(l+m-r)!} \frac{d^r}{dx^r}(x+1)^l \frac{d^{l+m-r}}{dx^{l+m-r}}(x-1)^l$$

$$= \frac{(l+m)!}{m!\,l!} \frac{d^m}{dx^m}(x+1)^l \frac{d^l}{dx^l}(x-1)^l + \ldots$$

$$= \frac{(l+m)!}{l!\,m!} \cdot l! \cdot \frac{l!}{(l-m)!}(x+1)^{l-m} + \text{terms having positive powers of } (x-1) \quad (3.32)$$

Further, using Eq. (3.32), we have

$$\frac{d^{m-1}}{dx^{m-1}}\left\{(x^2-1)^{m-1}\frac{d^{l+m}}{dx^{l+m}}(x^2-1)^l\right\}$$

$$= \frac{d^{m-1}}{dx^{m-1}}\left\{\frac{(l+m)!\,l!}{m!\,(l-m)!}(x+1)^{l-1}(x-1)^{m-1} + \text{terms with} + \text{ve powers of}\,(x-1)\right\}$$

$$= \frac{(l+m)!\,l!}{m!\,(l-m)!}\sum_{r=0}^{m-1}\frac{(m-1)!}{r!\,(m-1-r)!}\frac{d^r}{dx^r}(x+1)^{l-1}\frac{d^{m-1-r}}{dx^{m-1-r}}(x-1)^{m-1} + \ldots$$

$$= \frac{(l+m)!\,l!}{m!\,(l-m)!}(x+1)^{l-1}\frac{d^{m-1}}{dx^{m-1}}(x-1)^{m-1} + \ldots \text{terms with +ve powers of } (x-1)$$

$$= \frac{(l+m)!\,l!}{m!\,(l-m)!}\cdot(m-1)!\,(x+1)^{l-1} + \text{terms with positive powers of } (x-1) \quad (3.33)$$

Substituting Eqs. (3.31) and (3.33) into Eq. (3.30), we obtain

$$I_l^{mm} = \int_{-1}^{+1}\frac{P_l^m(x)P_l^m(x)}{1-x^2}dx$$

$$= \frac{2}{2^{2l}(l!)^2}\frac{(l+m)!\,l!}{m!\,(l-m)!}(m-1)!\,2^{l-1}\,l!\,2^l$$

$$= \frac{(l+m)!}{m(l-m)!} \quad (3.34)$$

In general, Eqs. (3.25) and (3.34) can be generalized as

$$I_l^{mn} = \int_{-1}^{+1}\frac{P_l^m(x)P_l^n(x)}{1-x^2}dx = \frac{(l+m)!}{m(l-m)!}\delta_{mn} \quad (3.35)$$

which is Eq. (3.19b).

3.11 EXPANSION OF ARBITRARY FUNCTION IN SERIES OF ASSOCIATED LEGENDRE POLYNOMIALS

The associated Legendre polynomials with fixed m form a complete set in the interval $(-1,+1)$ and, therefore, any arbitrary function $f(x)$ can be expanded in a series of associated Legendre polynomials in the form

$$f(x)=\sum_{l=0}^{\infty} a_l P_l^m(x) \tag{3.36}$$

where the coefficients a_l are to be determined as in the following.

Multiplying both sides of Eq. (3.36) by $P_k^m(x)$ and then integrating the results with respect to x within the limit from $x=-1$ to $x=+1,$ we obtain

$$\int_{-1}^{+1} f(x)P_k^m(x)dx=\sum_{l=0}^{\infty} a_l \int_{-1}^{+1} P_k^m(x)P_l^m(x)dx$$

$$=a_k \frac{2}{(2k+1)} \frac{(k+m)!}{(k-m)!}$$

where use has been made of the orthogonality condition of associated Legendre polynomials.

Thus, we obtain

$$a_k=\frac{(2k+1)}{2} \frac{(k-m)!}{(k+m)!} \int_{-1}^{+1} f(x)P_k^m(x)dx$$

Changing the subscript, we write

$$a_l=\frac{(2l+1)}{2} \frac{(l-m)!}{(l+m)!} \int_{-1}^{+1} f(x)\, P_l^m(x)\, dx \tag{3.37}$$

3.12 SOLVED PROBLEMS

PROBLEM 1 Prove that $P_l^m(-x)=(-1)^{l-m} P_l^m(x)$ where l and m are integers.

Solution:

First Method: We know that

$$P_l^m(x)=(1-x^2)^{\frac{m}{2}} \frac{d^m}{dx^m} P_l(x)$$

Replacing x by $(-x)$, we write

$$P_l^m(-x)=\left\{1-(-x)^2\right\}^{\frac{m}{2}} \frac{d^m P_l(-x)}{(-1)^m dx^m}$$

$$=(1-x^2)^{\frac{m}{2}} (-1)^{-m} \frac{d^m}{dx^m}\left\{(-1)^l P_l(x)\right\}$$

$$=(-1)^{l-m} (1-x^2)^{\frac{m}{2}} \frac{d^m}{dx^m} P_l(x)=(-1)^{l-m} P_l^m(x)$$

Second method: Rodrigues' formula for associated Legendre polynomials is

$$P_l^m(x) = \frac{1}{2^l\, l!}(1-x^2)^{\frac{m}{2}} \frac{d^{l+m}}{dx^{l+m}}(x^2-1)^l$$

Replacing x by $-x$, we obtain

$$P_l^m(-x) = \frac{1}{2^l\, l!}\left\{1-(-x)^2\right\}^{\frac{m}{2}} \frac{1}{(-1)^{l+m}} \frac{d^{l+m}}{dx^{l+m}}(x^2-1)^l$$

$$= (-1)^{-l-m} \frac{1}{2^l\, l!}(1-x^2)^{\frac{m}{2}} \frac{d^{l+m}}{dx^{l+m}}(x^2-1)^l$$

$$= (-1)^{-l-m} P_l^m(x) = \frac{(-1)^{2l}(-1)^{-l-m}}{(-1)^{2l}} P_l^m(x) = (-1)^{l-m} P_l^m(x)$$

PROBLEM 2 Prove that

$$P_l^m(0) = \begin{cases} \dfrac{(-1)^{\frac{l-m}{2}}(l+m)!}{2^l\left(\dfrac{l-m}{2}\right)!\left(\dfrac{l+m}{2}\right)!} & \text{for } l+m = \text{even integer} \\ 0 & \text{for } l+m = \text{odd integer} \end{cases}$$

Solution: Rodrigues' formula for associated Legendre polynomials is given by

$$P_l^m(x) = \frac{1}{2^l\, l!}(1-x^2)^{\frac{m}{2}} \frac{d^{l+m}}{dx^{l+m}}(x^2-1)^l \tag{i}$$

Also, using Binomial theorem, we know that

$$(x^2-1)^l = \sum_{r=0}^{l} \frac{l!}{r!\,(l-r)!}(x^2)^r(-1)^{l-r}$$

Now, Eq. (i) becomes

$$P_l^m(x) = \frac{1}{2^l\, l!}\sum_{r=0}^{l} \frac{(-1)^{l-r}\, l!}{r!\,(l-r)!}(1-x^2)^{\frac{m}{2}} \frac{d^{l+m}}{dx^{l+m}} x^{2r}$$

$$= \frac{1}{2^l\, l!}\sum_{r=0}^{l} \frac{(-1)^{l-r}\, l!}{r!\,(l-r)!}(1-x^2)^{\frac{m}{2}} \frac{(2r)!}{(2r-l-m)} x^{2r-l-m}$$

Putting $x=0$, we obtain

$$P_l^m(0) = \frac{1}{2^l\, l!}\left[\sum_{r=0}^{l} \frac{(-1)^{l-r}\, l!}{r!\,(l-r)!}(1-x^2)^{\frac{m}{2}} \frac{(2r)!}{(2r-l-m)!} x^{2r-l-m}\right]_{x=0}$$

For $l+m>2r$ all the derivatives will be zero. Again, for $l+m<2r$ all the derivatives will contain x and they will become zero at $x=0$.

Therefore, the only non-vanishing term will be that for $2r=l+m$, i.e, $r=\frac{l+m}{2}$. Thus, we write

$$P_l^m(0)=\frac{1}{2^l}\frac{(-1)^{\frac{l-m}{2}}(l+m)!}{\left(\frac{l+m}{2}\right)!\left(\frac{l-m}{2}\right)!}$$

As $r=0,1,2,3,4\ldots$ etc., we see that $l+m=2r$ is always an even integer. Hence, we conclude that

$$P_l^m(0)=\begin{cases}\dfrac{(-1)^{\frac{l-m}{2}}(l+m)!}{2^l\left(\frac{l+m}{2}\right)!\left(\frac{l-m}{2}\right)!} & \text{for } l+m=\text{even integer}\\[2ex] 0 & \text{for } l+m=\text{odd integer}\end{cases}$$

PROBLEM 3 Using the associated Legendre equation, prove that

(a) $\int_{-1}^{+1}P_k^m(x)P_l^m(x)\,dx=0,\ k\neq l$

and (b) $\int_{-1}^{+1}P_l^m(x)P_l^n(x)\frac{dx}{(1-x^2)}=0,\ m\neq n$

Solution:

(a) We know that $P_l^m(x)$ satisfies the associated Legendre equation

$$(1-x^2)\frac{d^2}{dx^2}P_l^m(x)-2x\frac{d}{dx}P_l^m(x)+\left\{l(l+1)-\frac{m^2}{1-x^2}\right\}P_l^m(x)=0$$

or

$$\frac{d}{dx}\left\{(1-x^2)\frac{d}{dx}P_l^m(x)\right\}+\left\{l(l+1)-\frac{m^2}{1-x^2}\right\}P_l^m(x)=0 \qquad \text{(i)}$$

Similarly, $P_k^m(x)$ satisfies the equation

$$\frac{d}{dx}\left\{(1-x^2)\frac{d}{dx}P_k^m(x)\right\}+\left\{k(k+1)-\frac{m^2}{1-x^2}\right\}P_k^m(x)=0 \qquad \text{(ii)}$$

Subtracting $P_l^m(x)$ times equation (ii) from $P_k^m(x)$ times equation (i), we obtain

$$P_k^m(x)\frac{d}{dx}\left\{(1-x^2)\frac{d}{dx}P_l^m(x)\right\}-P_l^m(x)\frac{d}{dx}\left\{(1-x^2)\frac{d}{dx}P_k^m(x)\right\}$$
$$+\{l(l+1)-k(k+1)\}P_k^m(x)P_l^m(x)=0$$

Integrating both sides with respect to x within the interval from $x=-1$ to $x=+1$, we obtain

$$\int_{-1}^{+1}P_k^m(x)\frac{d}{dx}\left\{\left(1-x^2\right)\frac{d}{dx}P_l^m(x)\right\}dx-\int_{-1}^{+1}P_l^m(x)\frac{d}{dx}\left\{\left(1-x^2\right)\frac{d}{dx}P_k^m(x)\right\}dx$$
$$+(l-k)(k+l+1)\int_{-1}^{+1}P_k^m(x)P_l^m(x)\,dx=0$$

or
$$\left[P_k^m(x)\left(1-x^2\right)\frac{d}{dx}P_l^m(x)\right]_{-1}^{+1}-\int_{-1}^{+1}\left\{\frac{d}{dx}P_k^m(x)\right\}\left(1-x^2\right)\left\{\frac{d}{dx}P_l^m(x)\right\}dx$$
$$-\left[P_l^m(x)\left(1-x^2\right)\frac{d}{dx}P_k^m(x)\right]_{-1}^{+1}+\int_{-1}^{+1}\left\{\frac{d}{dx}P_l^m(x)\right\}(1-x^2)\left\{\frac{d}{dx}P_k^m(x)\right\}dx$$
$$+(l-k)(k+l+1)\int_{-1}^{+1}P_k^m(x)P_l^m(x)\,dx=0$$

or
$$(l-k)(k+l+1)\int_{-1}^{+1}P_k^m(x)P_l^m(x)\,dx=0$$

For $k\neq l$, we have

$$\int_{-1}^{+1}P_k^m(x)P_l^m(x)\,dx=0$$

(b) See Section 3.10: (b) Proof of second orthogonality condition: Case I: $m\neq n$

PROBLEM 4 Using the relation between $P_l^m(x)$ and $P_l^{-m}(x)$ prove that

$$\int_{-1}^{+1}P_l^m(x)P_l^m(x)\,dx=\frac{2(l+m)!}{(2l+1)(l-m)!}$$

Solution: We know that $P_l^m(x)=\dfrac{(-1)^m(l+m)!}{(l-m)!}P_l^{-m}(x)$

Multiplying both sides by $P_l^m(x)$ and integrating the results with respect to x within the interval from $x=-1$ to $x=+1$, we obtain

$$\int_{-1}^{+1} P_l^m(x)P_l^m(x)dx = \frac{(-1)^m(l+m)!}{(l-m)!}\int_{-1}^{+1} P_l^m(x)P_l^{-m}(x)dx \qquad \text{(i)}$$

Rodrigues' formula gives

$$P_l^m(x) = \frac{1}{2^l\, l!}(1-x^2)^{\frac{m}{2}} \frac{d^{l+m}}{dx^{l+m}}(x^2-1)^l$$

and

$$P_l^{-m}(x) = \frac{1}{2^l\, l!}(1-x^2)^{\frac{-m}{2}} \frac{d^{l-m}}{dx^{l-m}}(x^2-1)^l$$

Substituting these into the integral on the right-hand side of Eq. (i), we obtain

$$\int_{-1}^{+1} P_l^m(x)P_l^m(x)\,dx = \frac{(-1)^m(l+m)!}{2^{2l}(l!)^2(l-m)!}\int_{-1}^{+1}\left\{\frac{d^{l+m}}{dx^{l+m}}(x^2-1)^l\right\}\left\{\frac{d^{l-m}}{dx^{l-m}}(x^2-1)^l\right\}dx$$

If the integral on the right-hand side of the above equation is integrated by parts $(l-m)$ times, the integrated part at each step will vanish at the given limits.

Therefore, we write

$$\int_{-1}^{+1} P_l^m(x)P_l^m(x)dx = \frac{(-1)^m(l+m)!}{2^{2l}(l!)^2(l-m)!}(-1)^{l-m}\int_{-1}^{+1}\left\{\frac{d^{2l}}{dx^{2l}}(x^2-1)^l\right\}(x^2-1)^l\,dx$$

$$= \frac{(2l)!(l+m)!}{2^{2l}(l!)^2(l-m)!}\int_{-1}^{+1}(1-x^2)^l\,dx$$

$$= \frac{(2l)!(l+m)!}{2^{2l}(l!)^2(l-m)!}\int_{-1}^{+1}(1+x)^l(1-x)^l\,dx \qquad \text{(ii)}$$

In the integral on the right-hand side of Eq. (ii), if we put $x=2u-1$, i.e., $u=\frac{x+1}{2}$, we have, $dx=2du$

Therefore, we have

$$\int_{-1}^{+1}(1+x)^l(1-x)^l\,dx = \int_0^1 (2u)^l\{2(1-u)\}^l\,2du$$

$$= 2^{2l+1}\int_0^1 u^l(1-u)^l\,du$$

$$= 2^{2l+1} B(l+1, l+1)$$

$$= 2^{2l+1} \frac{\Gamma(l+1)\Gamma(l+1)}{\Gamma(2l+2)} = 2^{2l+1} \frac{(l!)^2}{(2l+1)!} \qquad \text{(iii)}$$

Substituting Eq. (iii) into Eq. (ii), we obtain

$$\int_{-1}^{+1} P_l^m(x) P_l^m(x) dx = \frac{2(l+m)!}{(2l+1)(l-m)!}$$

PROBLEM 5 Prove that $\int_{-1}^{+1} x\left[P_l^m(x)\right]^2 dx = 0$

Solution: We know that

$$(2l+1)xP_l^m(x) = (l-m+1)P_{l+1}^m(x) + (l+m)P_{l-1}^m(x)$$

Multiplying both sides by $P_l^m(x)$ and then integrating the results with respect to x in the limit from $x=-1$ to $x=+1$, we obtain

$$(2l+1)\int_{-1}^{+1} x\left[P_l^m(x)\right]^2 dx = (l-m+1)\int_{-1}^{+1} P_l^m(x) P_{l+1}^m(x) dx + (l+m)\int_{-1}^{+1} P_{l-1}^m(x) P_l^m(x) dx = 0$$

(Using orthogonality property of associated Legendre polynomials.)

Therefore, $\int_{-1}^{+1} x\left[P_l^m(x)\right]^2 dx = 0$.

EXERCISE

1. Prove that

   (a) $P_{-l-1}^m(x) = P_l^m(x)$,

   (b) $P_l^m(-x) = (-1)^{l+m} P_l^m(x)$ when l and m are positive integers.

2. Prove that $P_l^l(\cos\theta) = (2l-1)!!\sin^l\theta$ where l is a positive integer.

   [**Note:** $(2l-1)!! = 1.3.5.7.9\ldots(2l-1)$.

3. Prove that $P_l^{m+1}(x) - \frac{2mx}{\sqrt{1-x^2}} P_l^m(x) + (l+m)(l-m+1)P_l^{m-1}(x) = 0$

4. Prove that $\int_{-1}^{+1} P_l^{-m}(x) P_l^{-m}(x)\, dx = \frac{2(l-m)!}{(2l+1)(l+m)!}$ where m is a positive integer.

5. Prove that

(i) $\int_{-1}^{+1} x P_l^m(x) P_{l+1}^m(x)\, dx = \frac{2(l+m+1)!}{(2l+1)(2l+3)(l-m)!}$

(ii) $\int_{-1}^{+1} x P_l^m(x) P_{l-1}^m(x)\, dx = \frac{2(l+m)!}{(2l-1)(2l+1)(l-m-1)!}$

6. Prove that $\int_{-1}^{+1} x(1-x^2)^{-3/2} \left[P_l^m(x)\right]^2 dx = 0$

Hint: Use the first recurrence relation and the second orthogonality condition.

7. Prove that $\int_{0}^{\pi} P_k^m(\cos\theta) P_l^m(\cos\theta) \sin\theta\, d\theta = \frac{2}{(2l+1)} \frac{(l+m)!}{(l-m)!} \delta_{kl}$

8. Prove that $\int_{-1}^{+1} P_l^m(x) P_k^{-m}(x)\, dx = (-1)^m \frac{2}{(2l+1)} \frac{(k-m)!\,(l+m)!}{(k+m)!\;(l-m)!} \delta_{kl}$

4

CHAPTER

Spherical Harmonics

4.1 DEFINITION OF SPHERICAL HARMONICS

A spherical harmonic is an orthonormal solution of the angular part of three-dimensional Laplace's differential equation in spherical polar coordinates. It is denoted by $Y_l^m(\theta, \phi)$ or $Y_{lm}(\theta, \phi)$.

The spherical harmonic $Y_{l,m}(\theta, \phi)$ is defined by

$$Y_{l,m}(\theta, \phi) = (-1)^m \sqrt{\frac{(2l+1)}{4\pi} \frac{(l-m)!}{(l+m)!}} P_l^m(\cos\theta) e^{im\phi} \tag{4.1}$$

with $-l \le m \le l$ where l is a non-negative integer. In quantum physics, l is the orbital angular momentum quantum number and m is the magnetic quantum number.

Since this function is defined over the surface of a sphere with polar angle θ and azimuth angle ϕ, and it is a solution of Laplace's equation; this function is given the name 'spherical harmonic'. The factor $(-1)^m$ is introduced for convenience in the study of Quantum Theory of Angular Momentum.

Special Cases

(i) All spherical harmonics with zero value of azimuth angle (ϕ) are termed *zonal harmonics*. In other words, A *zonal harmonic* is a spherical harmonic which reduces to a Legendre polynomial of the cosine of the polar angle (θ). Zonal harmonics are spherical harmonics which assume constant values on parallel unit circles centred on an axis of symmetry.

(ii) *Sectoral harmonics* are those spherical harmonics having the form $Y_{l,l}(\theta, \phi)$.

The first a few spherical harmonics are given below.

$$Y_{0,0}(\theta, \phi) = \sqrt{\frac{1}{4\pi}}, \qquad Y_{1,0}(\theta, \phi) = \sqrt{\frac{3}{4\pi}} \cos\theta,$$

$$Y_{1,1}(\theta, \phi) = -\sqrt{\frac{3}{8\pi}} (\sin\theta) e^{i\phi} \qquad Y_{1,-1}(\theta, \phi) = \sqrt{\frac{3}{8\pi}} (\sin\theta) e^{-i\phi},$$

$$Y_{2,0}(\theta,\phi)=\sqrt{\frac{5}{16\pi}}\left(3\cos^2\theta-1\right) \qquad Y_{2,1}(\theta,\phi)=-\sqrt{\frac{15}{8\pi}}(\sin\theta\cos\theta)e^{i\phi},$$

$$Y_{2,-1}(\theta,\phi)=\sqrt{\frac{15}{8\pi}}(\sin\theta\cos\theta)e^{-i\phi} \qquad Y_{2,2}(\theta,\phi)=\sqrt{\frac{15}{32\pi}}\left(\sin^2\theta\right)e^{2i\phi},$$

$$Y_{2,-2}(\theta,\phi)=\sqrt{\frac{15}{32\pi}}\left(\sin^2\theta\right)e^{-2i\phi}$$

4.2 TO PROVE THAT $Y_{l,-m}(\theta,\phi)=(-1)^m Y_{l,m}^*(\theta,\phi)$ WHERE THE ASTERISK DENOTES COMPLEX CONJUGATION

We know that

$$Y_{l,m}(\theta,\phi)=(-1)^m\sqrt{\frac{(2l+1)}{4\pi}\frac{(l-m)!}{(l+m)!}}e^{im\phi}P_l^m(\cos\theta)$$

Replacing m by $-m$, we have

$$Y_{l,-m}(\theta,\phi)=(-1)^{-m}\sqrt{\frac{(2l+1)}{4\pi}\frac{(l+m)!}{(l-m)!}}e^{-im\phi}P_l^{-m}(\cos\theta)$$

$$=(-1)^{-m}\sqrt{\frac{(2l+1)}{4\pi}\frac{(l+m)!}{(l-m)!}}e^{-im\phi}(-1)^m\frac{(l-m)!}{(l+m)!}P_l^m(\cos\theta)$$

$$=\sqrt{\frac{(2l+1)}{4\pi}\frac{(l-m)!}{(l+m)!}}e^{-im\phi}P_l^m(\cos\theta)$$

$$=(-1)^m\left[(-1)^m\sqrt{\frac{(2l+1)}{4\pi}\frac{(l-m)!}{(l+m)!}}e^{-im\phi}P_l^m(\cos\theta)\right]$$

$$=(-1)^m Y_{l,m}^*(\theta,\phi) \tag{4.2}$$

4.3 TO PROVE THAT SPHERICAL HARMONICS ARE SOLUTIONS OF THE ANGULAR PART OF THREE-DIMENSIONAL LAPLACE'S EQUATION IN SPHERICAL POLAR COORDINATES

In spherical polar coordinates (r,θ,ϕ), Laplace's equation is written as

$$\frac{1}{r^2}\frac{\partial}{\partial r}\left(r^2\frac{\partial\Psi}{\partial r}\right)+\frac{1}{r^2\sin\theta}\frac{\partial}{\partial\theta}\left(\sin\theta\frac{\partial\Psi}{\partial\theta}\right)+\frac{1}{r^2\sin^2\theta}\frac{\partial^2\Psi}{\partial\phi^2}=0$$

or
$$\frac{\partial}{\partial r}\left(r^2\frac{\partial\Psi}{\partial r}\right)+\frac{1}{\sin\theta}\frac{\partial}{\partial\theta}\left(\sin\theta\frac{\partial\Psi}{\partial\theta}\right)+\frac{1}{\sin^2\theta}\frac{\partial^2\Psi}{\partial\phi^2}=0 \tag{4.3}$$

where $\Psi=\Psi(r,\theta,\phi)$.

To separate the variables, let us put

$$\Psi=R(r)\,S(\theta,\phi). \tag{4.4}$$

Substituting Eq. (4.4) into Eq. (4.3), we obtain

$$S\frac{d}{dr}\left(r^2\frac{dR}{dr}\right)+\frac{R}{\sin\theta}\frac{\partial}{\partial\theta}\left(\sin\theta\frac{\partial S}{\partial\theta}\right)+\frac{R}{\sin^2\theta}\frac{\partial^2 S}{\partial\phi^2}=0.$$

Dividing both sides by *RS*, we obtain

$$\frac{1}{R}\frac{d}{dr}\left(r^2\frac{dR}{dr}\right)+\frac{1}{S}\left[\frac{1}{\sin\theta}\frac{\partial}{\partial\theta}\left(\sin\theta\frac{\partial S}{\partial\theta}\right)+\frac{1}{\sin^2\theta}\frac{\partial^2 S}{\partial\phi^2}\right]=0$$

or
$$\frac{1}{R}\frac{d}{dr}\left(r^2\frac{dR}{dr}\right)=-\frac{1}{S}\left[\frac{1}{\sin\theta}\frac{\partial}{\partial\theta}\left(\sin\theta\frac{\partial S}{\partial\theta}\right)+\frac{1}{\sin^2\theta}\frac{\partial^2 S}{\partial\phi^2}\right]=\beta \text{ (a constant).} \tag{4.5}$$

From Eq. (4.5), we obtain two equations

$$\frac{d}{dr}\left(r^2\frac{dR}{dr}\right)-\beta R=0$$

or
$$r^2\frac{d^2R}{dr^2}+2r\frac{dR}{dr}-\beta R=0, \text{ (Radial part)} \tag{4.6}$$

and
$$\frac{1}{\sin\theta}\frac{\partial}{\partial\theta}\left(\sin\theta\frac{\partial S}{\partial\theta}\right)+\frac{1}{\sin^2\theta}\frac{\partial^2 S}{\partial\phi^2}+\beta S=0 \text{ (Angular part).} \tag{4.7}$$

To solve the angular part, let us put

$$S=S(\theta,\phi)=\Theta(\theta)\Phi(\phi). \tag{4.8}$$

Now, Eq. (4.7) can be written as

$$\frac{\Phi}{\sin\theta}\frac{d}{d\theta}\left(\sin\theta\frac{d\Theta}{d\theta}\right)+\frac{\Theta}{\sin^2\theta}\frac{d^2\Phi}{d\phi^2}+\beta\Theta\Phi=0.$$

Multiplying both sides by $\frac{\sin^2\theta}{\Theta\Phi}$, we obtain

$$\frac{\sin\theta}{\Theta}\frac{d}{d\theta}\left(\sin\theta\frac{d\Theta}{d\theta}\right)+\frac{1}{\Phi}\frac{d^2\Phi}{d\phi^2}+\beta\sin^2\theta=0.$$

Hence, we write

$$\frac{\sin\theta}{\Theta}\frac{d}{d\theta}\left(\sin\theta\frac{d\Theta}{d\theta}\right)+\beta\sin^2\theta=-\frac{1}{\Phi}\frac{d^2\Phi}{d\phi^2}=m^2 \text{ (a constant).} \tag{4.9}$$

From Eq. (4.9), we obtain two equations, namely

$$\sin\theta\frac{d}{d\theta}\left(\sin\theta\frac{d\Theta}{d\theta}\right)+\left(\beta\sin^2\theta-m^2\right)\Theta=0 \tag{4.10}$$

and

$$\frac{d^2\Phi}{d\phi^2}+m^2\Phi=0. \tag{4.11}$$

Dividing both sides of Eq. (4.10) by $\sin^2\theta$, we write

$$\frac{1}{\sin\theta}\frac{d}{d\theta}\left(\sin\theta\frac{d\Theta}{d\theta}\right)+\left(\beta-\frac{m^2}{\sin^2\theta}\right)\Theta=0,$$

or

$$\frac{d^2\Theta}{d\theta^2}+\frac{\cos\theta}{\sqrt{1-\cos^2\theta}}\frac{d\Theta}{d\theta}+\left(\beta-\frac{m^2}{1-\cos^2\theta}\right)\Theta=0. \tag{4.12}$$

Further, let us put

$$\cos\theta=q \text{ and } \Theta=\Theta(\theta)=f(q). \tag{4.13}$$

Then,

$$\frac{d\Theta}{d\theta}=\frac{d\Theta}{dq}\frac{dq}{d\theta}=-\sin\theta\frac{d\Theta}{dq}=-\sqrt{1-q^2}\frac{df}{dq} \tag{4.14}$$

and similarly,

$$\frac{d^2\Theta}{d\theta^2}=\left(1-q^2\right)\frac{d^2f}{dq^2}-q\frac{df}{dq}.$$

Using the above relations, we write Eq. (4.12) as

$$(1-q^2)\frac{d^2f}{dq^2}-2q\frac{df}{dq}+\left(\beta-\frac{m^2}{1-q^2}\right)f=0. \tag{4.15}$$

Considering a special case in which the constant m^2 is zero, we have

$$(1-q^2)\frac{d^2f}{dq^2}-2q\frac{df}{dq}+\beta f=0. \tag{4.16}$$

Let us assume the series solution of Eq. (4.16) in the form

$$f(q)=\sum_{l=0}^{\infty}a_l\,q^l.$$

Then, $\frac{df}{dq} = \sum_{l=0}^{\infty} l a_l q^{l-1}$ and $\frac{d^2 f}{dq^2} = \sum_{l=0}^{\infty} (l-1) l a_l q^{l-2}$.

Substituting the expressions for $q, \frac{df}{dq}$ and $\frac{d^2 f}{dq^2}$ into Eq. (4.16), we obtain

$$\sum_{l=0}^{\infty} (l-1) l a_l q^{l-2} + \sum_{l=0}^{\infty} \{\beta - l(l+1)\} a_l q^l = 0 .$$

Coefficients of various powers of q should vanish and, therefore, equating coefficient of q^l to zero, we write

$$(l+1)(l+2) a_{l+2} + \{\beta - l(l+1)\} a_l = 0 ,$$

giving $\frac{a_{l+2}}{a_l} = \frac{l(l+1) - \beta}{(l+1)(l+2)}$.

If the series terminates at the term containing the coefficient a_l then it becomes a polynomial and we must have

$$\frac{a_{l+2}}{a_l} = 0, \text{ i.e., } \beta = l(l+1) .$$

Substituting this value of β into Eq. (4.15), we obtain

$$(1-q^2)\frac{d^2 f}{dq^2} - 2q\frac{df}{dq} + \left\{ l(l+1) - \frac{m^2}{1-q^2} \right\} f = 0 . \tag{4.17}$$

This equation is associated Legendre equation and hence its normalized solution is

$$f(q) = \sqrt{\frac{(2l+1)}{2} \frac{(l-m)!}{(l+m)!}} P_l^m(q) .$$

Using Eq. (4.13), we have

$$\Theta = \Theta_{lm}(\theta) = \sqrt{\frac{(2l+1)}{2} \frac{(l-m)!}{(l+m)!}} P_l^m(\cos\theta) . \tag{4.18}$$

Also we know that the general solution of Eq. (4.11) is

$$\Phi = \Phi_m(\phi) = A e^{-im\phi} + B e^{im\phi}$$

where A and B are arbitrary constants.

Without loss of generality, the functions $Ae^{-im\phi}$ and $Be^{im\phi}$ can be represented by a single function $Ce^{im\phi}$ where C is a constant and m has negative and positive integral values.

Therefore, we write

$$\Phi = \Phi_m(\phi) = Ce^{im\phi}.$$

Normalization condition demands that

$\int_0^{2\pi} \Phi^* \Phi \, d\phi = 1$, where the asterisk denotes complex conjugation.

Therefore, we obtain

$$C = \frac{1}{\sqrt{2\pi}}$$

and
$$\Phi = \Phi_m = \frac{1}{\sqrt{2\pi}} e^{im\phi}. \tag{4.19}$$

Substituting Eqs. (4.18) and (4.19) into Eq. (4.8), we obtain

$$S = S_{lm}(\theta, \phi) = \sqrt{\frac{(2l+1)}{4\pi} \frac{(l-m)!}{(l+m)!}} e^{im\phi} P_l^m(\cos\theta) \tag{4.20}$$

which is a solution of the angular part of the three-dimensional Laplace's equation in spherical polar co-ordinates.

In order to agree with the literature used in the quantum theory of angular momentum, we define the spherical harmonic by multiplying Eq. (4.20) by a factor $(-1)^m$ without affecting the normalization condition and without loss of generality.

Thus, we define the spherical harmonics by the function

$$Y_{l,m}(\theta, \phi) = (-1)^m \sqrt{\frac{(2l+1)}{4\pi} \frac{(l-m)!}{(l+m)!}} e^{im\phi} P_l^m(\cos\theta) \tag{4.21}$$

where the factor $(-1)^m$ is known as the phase factor.

Thus, we see that $Y_{l,m}(\theta, \phi)$ is solution of the angular part of the three-dimensional Laplace's equation in spherical polar co-ordinates.

4.4 RECURRENCE RELATIONS FOR SPHERICAL HARMONICS

I. $\cos\theta \, Y_{l,m}(\theta, \phi)$

$$= \left[\frac{(l-m)(l+m)}{(2l-1)(2l+1)}\right]^{\frac{1}{2}} Y_{l-1,m}(\theta, \phi) + \left[\frac{(l-m+1)(l+m+1)}{(2l+1)(2l+3)}\right]^{\frac{1}{2}} Y_{l+1,m}(\theta, \phi) \tag{4.22}$$

Proof: In the case of the associated Legendre polynomials, we know the recurrence relation

$$(2l+1)xP_l^m(x)=(l+m)P_{l-1}^m(x)+(l-m+1)P_{l+1}^m(x).$$

Replacing x by $\cos\theta$ we write

$$(2l+1)\cos\theta P_l^m(\cos\theta)=(l+m)P_{l-1}^m(\cos\theta)+(l-m+1)P_{l+1}^m(\cos\theta). \tag{4.23}$$

Also, we know that

$$Y_{l,m}(\theta,\phi)=(-1)^m\left[\frac{(2l+1)(l-m)!}{4\pi\,(l+m)!}\right]^{\frac{1}{2}}e^{im\phi}P_l^m(\cos\theta).$$

Therefore, we have

$$P_l^m(\cos\theta)=(-1)^m\left[\frac{4\pi(l+m)!}{(2l+1)\,(l-m)!}\right]^{\frac{1}{2}}e^{-im\phi}\,Y_{l,m}(\theta,\phi).$$

Replacing l successively by $(l-1)$ and $(l+1)$, we obtain

$$P_{l-1}^m(\cos\theta)=(-1)^m\left[\frac{4\pi(l+m-1)!}{(2l-1)(l-m-1)!}\right]^{\frac{1}{2}}e^{-im\phi}\,Y_{l-1,m}(\theta,\phi)$$

and
$$P_{l+1}^m(\theta,\phi)=(-1)^m\left[\frac{4\pi(l+m+1)!}{(2l+3)(l-m+1)!}\right]^{\frac{1}{2}}e^{-im\phi}\,Y_{l+1,m}(\theta,\phi).$$

Substituting these relations into Eq. (4.23), we obtain

$$(-1)^m\sqrt{\frac{4\pi(2l+1)(l+m)!}{(l-m)!}}\cos\theta\, e^{-im\phi}\,Y_{l,m}(\theta,\phi)$$
$$=(-1)^m\sqrt{\frac{4\pi(l+m)(l+m)!}{(2l-1)(l-m-1)!}}\,e^{-im\phi}\,Y_{l-1,m}(\theta,\phi)$$
$$+(-1)^m\sqrt{\frac{4\pi\,(l-m+1)\,(l+m+1)!}{(2l+3)\,(l-m)!}}\,e^{-im\phi}\,Y_{l+1,m}(\theta,\phi).$$

Dividing both sides by $(-1)^m\sqrt{\dfrac{4\pi(2l+1)(l+m)!}{(l-m)!}}\,e^{-im\phi}$, we obtain

$$\cos\theta\,Y_{l,m}(\theta,\phi)=\sqrt{\frac{(l-m)(l+m)}{(2l-1)(2l+1)}}\,Y_{l-1,m}(\theta,\phi)+\sqrt{\frac{(l-m+1)(l+m+1)}{(2l+1)(2l+3)}}\,Y_{l+1,m}(\theta,\phi)$$

which is Eq. (4.22).

II. $\sin\theta\, Y_{l,m}(\theta,\phi)$

$$=\left[\frac{(l-m-1)(l-m)}{(2l-1)(2l+1)}\right]^{\frac{1}{2}} e^{-i\phi}\, Y_{l-1,m+1}(\theta,\phi)-\left[\frac{(l+m+1)(l+m+2)}{(2l+1)(2l+3)}\right]^{\frac{1}{2}} e^{-i\phi}\, Y_{l+1,m+1}(\theta,\phi) \tag{4.24}$$

Proof: In the case of associated Legendre polynomials, we know that

$$(2l+1)\left(1-x^2\right)^{1/2} P_l^m(x)=P_{l+1}^{m+1}(x)-P_{l-1}^{m+1}(x).$$

Replacing x by $\cos\theta$, we write

$$(2l+1)\sin\theta\, P_l^m(\cos\theta)=P_{l+1}^{m+1}(\cos\theta)-P_{l-1}^{m+1}(\cos\theta). \tag{4.25}$$

Also, we know that

$$P_l^m(\cos\theta)=(-1)^m\left[\frac{4\pi(l+m)!}{(2l+1)(l-m)!}\right]^{\frac{1}{2}} e^{-im\phi}\, Y_{l,m}(\theta,\phi),$$

$$P_{l-1}^{m+1}(\cos\theta)=(-1)^{m+1}\left[\frac{4\pi\,(l+m)!}{(2l-1)(l-m-2)!}\right]^{\frac{1}{2}} e^{-i(m+1)\phi}\, Y_{l-1,m+1}(\theta,\phi),$$

$$P_{l+1}^{m+1}(\cos\theta)=(-1)^{m+1}\left[\frac{4\pi\,(l+m+2)!}{(2l+3)(l-m)!}\right]^{\frac{1}{2}} e^{-i(m+1)\phi}\, Y_{l+1,m+1}(\theta,\phi).$$

Substituting these relations into Eq. (4.25), we obtain

$$(-1)^m\sin\theta\left[\frac{4\pi\,(2l+1)(l+m)!}{(l-m)!}\right]^{\frac{1}{2}} e^{-im\phi}\, Y_{l,m}(\theta,\phi)$$

$$=(-1)^{m+1}\left[\frac{4\pi\,(l+m+2)!}{(2l+3)(l-m)!}\right]^{\frac{1}{2}} e^{-i(m+1)\phi}\, Y_{l+1,m+1}(\theta,\phi)$$

$$-(-1)^{m+1}\left[\frac{4\pi(l+m)!}{(2l-1)(l-m-2)!}\right]^{\frac{1}{2}} e^{-i(m+1)\phi}\, Y_{l-1,m+1}(\theta,\phi).$$

Dividing both sides by $(-1)^m\left[\frac{4\pi(2l+1)(l+m)!}{(l-m)!}\right]^{\frac{1}{2}} e^{-im\phi}$, we obtain

$$\sin\theta\, Y_{l,m}(\theta,\phi) = \left[\frac{(l-m-1)(l-m)}{(2l-1)(2l+1)}\right]^{\frac{1}{2}} e^{-i\phi}\, Y_{l-1,m+1}(\theta,\phi)$$

$$-\left[\frac{(l+m+1)(l+m+2)}{(2l+1)(2l+3)}\right]^{\frac{1}{2}} e^{-i\phi}\, Y_{l+1,m+1}(\theta,\phi) \text{ which is Eq. (4.24).}$$

III. $$\sin\theta\, Y_{l,m}(\theta,\phi) = \left[\frac{(l-m+1)(l-m+2)}{(2l+1)(2l+3)}\right]^{\frac{1}{2}} e^{i\phi}\, Y_{l+1,m-1}(\theta,\phi)$$

$$-\left[\frac{(l+m-1)(l+m)}{(2l-1)(2l+1)}\right]^{\frac{1}{2}} e^{i\phi}\, Y_{l-1,m-1}(\theta,\phi) \quad (4.26)$$

Proof: In the case of associated Legendre polynomials, we know the recurrence relation,

$$(2l+1)(1-x^2)^{\frac{1}{2}} P_l^m(x) = (l+m)(l+m-1)P_{l-1}^{m-1}(x) - (l-m+1)(l-m+2)P_{l+1}^{m-1}(x).$$

Replacing x by $\cos\theta$ we write,

$$(2l+1)\sin\theta P_l^m(\cos\theta) = (l+m)(l+m-1)P_{l-1}^{m-1}(\cos\theta)$$

$$-(l-m+1)(l-m+2)P_{l+1}^{m-1}(\cos\theta). \quad (4.27)$$

Further, we know that

$$P_l^m(\cos\theta) = (-1)^m \left[\frac{4\pi(l+m)!}{(2l+1)(l-m)!}\right]^{\frac{1}{2}} e^{-im\phi}\, Y_{l,m}(\theta,\phi),$$

$$P_{l-1}^{m-1}(\cos\theta) = (-1)^{m-1} \left[\frac{4\pi\,(l+m-2)!}{(2l-1)(l-m)!}\right]^{\frac{1}{2}} e^{-i(m-1)\phi}\, Y_{l-1,m-1}(\theta,\phi),$$

$$P_{l+1}^{m-1}(\cos\theta) = (-1)^{m-1} \left[\frac{4\pi\,(l+m)!}{(2l+3)(l-m+2)!}\right]^{\frac{1}{2}} e^{-i(m-1)\phi}\, Y_{l+1,m-1}(\theta,\phi).$$

Substituting these relations into Eq. (4.27), we obtain

$$(-1)^m \sin\theta \left[\frac{4\pi\,(2l+1)(l+m)!}{(l-m)!}\right]^{\frac{1}{2}} e^{-im\phi}\, Y_{l,m}(\theta,\phi)$$

$$= (-1)^{m-1}\left[\frac{4\pi(l+m-1)(l+m)(l+m)!}{(2l-1)(l-m)!}\right]^{\frac{1}{2}} e^{-i(m-1)\phi}\, Y_{l-1,m-1}(\theta,\phi)$$

$$-(-1)^{m-1}\left[\frac{4\pi(l-m+1)(l-m+2)(l+m)!}{(2l+3)(l-m)!}\right]^{\frac{1}{2}} e^{-i(m-1)\phi}\, Y_{l+1,m-1}(\theta,\phi).$$

Dividing both sides by $(-1)^m\left[\dfrac{4\pi\,(2l+1)(l+m)!}{(l-m)!}\right]^{\frac{1}{2}} e^{-im\phi}$,

we obtain, $\sin\theta\, Y_{l,m}(\theta,\phi) = \left[\dfrac{(l-m+1)(l-m+2)}{(2l+1)(2l+3)}\right]^{\frac{1}{2}} e^{i\phi}\, Y_{l+1,m-1}(\theta,\phi)$

$$-\left[\frac{(l+m-1)(l+m)}{(2l-1)(2l+1)}\right]^{\frac{1}{2}} e^{i\phi}\, Y_{l-1,m-1}(\theta,\phi)$$ which is Eq. (4.26).

4.5 ORTHOGONALITY OF SPHERICAL HARMONICS

The spherical harmonics $Y_{l,m}(\theta,\phi)$ form an orthonormal set, i. e.,

$$\int_{\phi=0}^{2\pi}\int_{-1}^{+1} Y^*_{l',m'}(\theta,\phi)\,Y_{l,m}(\theta,\phi)\,d(\cos\theta)\,d\phi = \delta_{ll'}\,\delta_{mm'} \tag{4.28a}$$

or

$$\int_{\phi=0}^{2\pi}\int_{\theta=0}^{\pi} Y^*_{l',m'}(\theta,\phi)\,Y_{l,m}(\theta,\phi)\sin\theta\, d\theta\, d\phi = \delta_{ll'}\,\delta_{mm'} \tag{4.28b}$$

where the asterisk denotes complex conjugation.

Proof: We know that

$$Y_{lm}(\theta,\phi) = (-1)^m\left[\frac{(2l+1)(l-m)!}{4\pi(l+m)!}\right]^{\frac{1}{2}} e^{im\phi}\, P_l^m(\cos\theta)$$

and

$$Y^*_{l'm'}(\theta,\phi) = (-1)^{m'}\left[\frac{(2l'+1)(l'-m')!}{4\pi(l'+m')!}\right]^{\frac{1}{2}} e^{-im'\phi}\, P_{l'}^{m'}(\cos\theta).$$

Therefore, we have

$$\int_{\phi=0}^{2\pi}\int_{\theta=0}^{\pi} Y^*_{l'm'}(\theta,\phi)Y_{lm}(\theta,\phi)\sin\theta\, d\theta\, d\phi = \frac{(-1)^{m+m'}}{4\pi}\left[\frac{(2l'+1)(l'-m')!(2l+1)(l-m)!}{(l'+m')!(l+m)!}\right]^{\frac{1}{2}}$$

$$\times \int_{\phi=0}^{2\pi} e^{i(m-m')\phi}\, d\phi \int_{\theta=0}^{\pi} P_{l'}^{m'}(\cos\theta)P_l^m(\cos\theta)\sin\theta\, d\theta\,. \quad (4.29)$$

When $m \neq m'$, we have

$$\int_{\phi=0}^{2\pi} e^{i(m-m')\phi} d\phi = \int_0^{2\pi}\left[\cos(m-m')\phi + i\sin(m-m')\phi\right]d\phi = 0\,.$$

Hence, Eq. (4.29) becomes

$$\int_{\phi=0}^{2\pi}\int_{\theta=0}^{\pi} Y^*_{l'm'}(\theta,\phi)Y_{lm}(\theta,\phi)\sin\theta\, d\theta\, d\phi = 0, \text{ for } m \neq m'. \quad (4.30a)$$

When $m = m'$, we have

$$\int_{\phi=0}^{2\pi} e^{i(m-m')\phi} d\phi = \int_0^{2\pi} d\phi = 2\pi\,.$$

Also, when $m = m'$, we have

$$\int_{\theta=0}^{\pi} P_{l'}^{m'}(\cos\theta)P_l^m(\cos\theta)\sin\theta\, d\theta = \frac{2(l+m)!}{(2l+1)(l-m)!}\delta_{ll'}\,.$$

Therefore, when $m = m'$ Eq. (4.29) becomes

$$\int_{\phi=0}^{2\pi}\int_{\theta=0}^{\pi} Y^*_{l'm'}(\theta,\phi)Y_{lm}(\theta,\phi)\sin\theta\, d\theta\, d\phi = \delta_{ll'}\,. \quad (4.30b)$$

Combining Eqs. (4.30 a) and (4.30 b), we write a generalized equation as

$$\int_{\phi=0}^{2\pi}\int_{\theta=0}^{\pi} Y^*_{l'm'}(\theta,\phi)Y_{lm}(\theta,\phi)\sin\theta\, d\theta\, d\phi = \delta_{ll'}\,\delta_{mm'}$$

which is Eq. (4.28).

We can also write in the form

$$\int_{\Omega} Y^*_{l'm'}(\theta,\phi)Y_{lm}(\theta,\phi)\, d\Omega = \delta_{ll'}\,\delta_{mm'} \quad (4.31)$$

where $d\Omega = \sin\theta\, d\theta\, d\phi$ is the differential solid angle and Ω is the total solid angle within the angular limits $0 \le \theta \le \pi$ and $0 \le \phi \le 2\pi$.

4.6 EXPANSION OF ARBITRARY FUNCTION IN SERIES OF SPHERICAL HARMONICS (LAPLACE SERIES)

It is well known that the spherical harmonics possess the completeness property, i.e., these functions are linearly independent and there does not exist any function of θ and ϕ that is orthogonal to all $Y_{lm}(\theta,\phi)$, where l ranges over all possible values $l = 0, 1, 2, 3 \ldots \infty$ and m ranges over all possible values $m = -l \ldots -3, -2, -1, 0, 1, 2, 3 \ldots +l$.

The completeness property of spherical harmonics implies that any well-behaved function of θ and ϕ can be expressed as

$$f(\theta,\phi) = \sum_{l=0}^{\infty} \sum_{m=-l}^{l} a_{lm} Y_{lm}(\theta,\phi) \tag{4.32}$$

where the coefficients (a_{lm}) are to be determined through the orthogonality condition (4.28). This series is known as Laplace series.

Multiplying both sides of Eq. (4.32) by $Y^*_{l'm'}(\theta,\phi)$ and integrating over the whole solid angle of the interval, we obtain

$$\int_{\phi=0}^{2\pi} \int_{\theta=0}^{\pi} f(\theta,\phi) Y^*_{l'm'}(\theta,\phi) \sin\theta \, d\theta \, d\phi$$

$$= \sum_{l=0}^{\infty} \sum_{m=-l}^{l} a_{lm} \int_{\phi=0}^{2\pi} \int_{\theta=0}^{\pi} Y^*_{l'm'}(\theta,\phi) Y_{lm}(\theta,\phi) \sin\theta \, d\theta \, d\phi$$

$$= \sum_{l=0}^{\infty} \sum_{m=-l}^{l} a_{lm} \, \delta_{ll'} \, \delta_{mm'} = a_{l'm'} .$$

Therefore, dropping the primes, we write

$$a_{lm} = \int_{\phi=0}^{2\pi} \int_{\theta=0}^{\pi} f(\theta,\phi) Y^{"}_{lm}(\theta,\phi) \sin\theta d\theta d\phi . \tag{4.33}$$

Substituting Eq. (4.33) into Eq. (4.32), we will get the (Laplace) series.

4.7 ADDITION THEOREM OF SPHERICAL HARMONICS

The addition theorem of spherical harmonics is a relation between two spherical harmonics referred to two differently oriented coordinate systems having a common origin. It states that if two different directions specified by two pairs of angular coordinates (θ,ϕ) and (θ',ϕ') are separated by an angle γ, then

$$P_l(\cos\gamma) = \frac{4\pi}{(2l+1)} \sum_{m=-l}^{l} Y_{lm}(\theta,\phi) Y^*_{lm}(\theta',\phi') \tag{4.34}$$

where the asterisk denotes complex conjugation.

Proof: Let us consider two vectors

$$\mathbf{r} = \mathbf{r}(r, \theta, \phi) \text{ and } \mathbf{r}' = \mathbf{r}'(r', \theta', \phi')$$

where $$r = |\mathbf{r}| \text{ and } r' = |\mathbf{r}'|.$$

Then, one can write the unit vectors along **r** and **r′** respectively as

$$\hat{r} = \hat{i}\sin\theta\cos\phi + \hat{j}\sin\theta\sin\phi + \hat{k}\cos\theta$$

and $$\hat{r}' = \hat{i}\sin\theta'\cos\phi' + \hat{j}\sin\theta'\sin\phi' + \hat{k}\cos\theta'$$

Thus, if γ is the angle included between these two vectors, one finds that

$$\begin{aligned}\cos\gamma = r \cdot r' &= \sin\theta\cos\phi\sin\theta'\cos\phi' + \sin\theta\sin\phi\sin\theta'\sin\phi' + \cos\theta\cos\theta' \\ &= \sin\theta\sin\theta'[\cos\phi\cos\phi' + \sin\phi\sin\phi'] + \cos\theta\cos\theta' \\ &= \cos\theta\cos\theta' + \sin\theta\sin\theta'\cos(\phi - \phi'). \qquad (4.35)\end{aligned}$$

The orientations of these two vectors are illustrated in Figure 4.1. Here, γ is the angle included between them. By virtue of Eq. (4.35), the Legendre function, $P_l(\cos\gamma)$, can be viewed as a function of θ and ϕ while θ' and ϕ' are kept fixed. Thus, we can expand this function in a Laplace series of the form,

$$P_l(\cos\gamma) = \sum_{m=-l}^{l} b_m(\theta', \phi') Y_{lm}(\theta, \phi) \qquad (4.36)$$

where the coefficients (b_m) depend on the fixed parameters θ' and ϕ'.

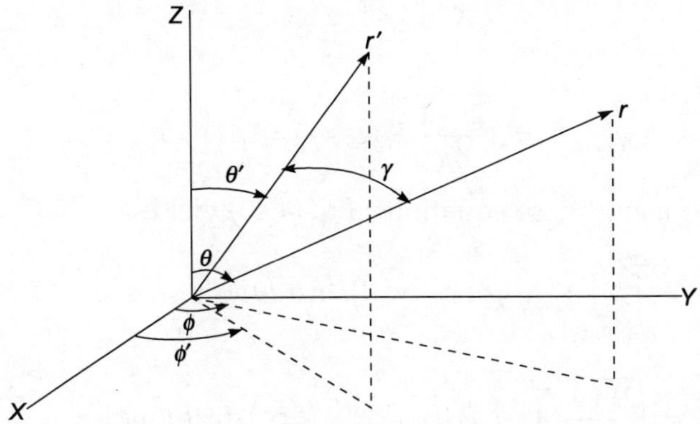

Fig. 4.1 Orientations of the vectors **r** and **r′** with an angular separation γ

Multiplying both sides of Eq. (4.36) by $Y^*_{lm}(\theta, \phi)$ and integrating over the whole solid angle, we obtain

$$b_m(\theta', \phi') = \int_{\phi=0}^{2\pi} \int_{\theta=0}^{\pi} P_l(\cos\gamma) Y^*_{lm}(\theta, \phi) \sin\theta \, d\theta \, d\phi. \qquad (4.37)$$

Using Eq. (4.35), we can also write

$$b_m(\theta',\phi') = \int_{\phi=0}^{2\pi}\int_{\theta=0}^{\pi} P_l\left(\cos\theta\cos\theta' + \sin\theta\sin\theta'\cos(\phi-\phi')\right)\sin\theta\, d\theta\, d\phi . \tag{4.38}$$

Since direct evaluation of the integral in Eq. (4.38) is difficult, we shall use an indirect approach as in the following.

In the coordinate system in which the z–axis lies along the vector $\mathbf{r}'$, the vector $\mathbf{r}$ will have a polar angle γ with a corresponding azimuthal angle, β. Then the angles θ and ϕ will be functions of the angles γ and β. Therefore, we can write

$$Y_{lm}^*(\theta,\phi) = \sum_{m'=-l}^{l} B_{m'}\, Y_{lm'}(\gamma,\beta) \tag{4.39}$$

where we use the coefficients $(B_{m'})$ and dummy index (m') in order to avoid confusion with those used earlier. As usual, the coefficients $B_{m'}$ are found to be

$$B_{m'} = \int_{\phi=0}^{2\pi}\int_{\theta=0}^{\pi} Y_{lm}^*(\theta,\phi)\, Y_{lm'}^*(\gamma,\beta)\sin\theta\, d\theta\, d\phi \tag{4.40}$$

where use has been made of the fact that a rigid rotation of a coordinate system leaves the infinitesimal solid angle element invariant.

In particular, if we set $m'=0$ and also if we remember that

$$Y_{l0}(\theta,\phi) = \sqrt{\frac{2l+1}{4\pi}}\, P_l(\cos\theta),$$

we can write,

$$Y_{l0}(\gamma,\beta) = \sqrt{\frac{2l+1}{4\pi}}\, P_l(\cos\gamma) = Y_{l0}^*(\gamma,\beta).$$

Thus, for $m'=0$, using these equations, Eq. (4.40) yields

$$\begin{aligned} B_0 &= \int_{\phi=0}^{2\pi}\int_{\theta=0}^{\pi} Y_{lm}^*(\theta,\phi)\, Y_{l0}^*(\gamma,\beta)\sin\theta\, d\theta\, d\phi \\ &= \sqrt{\frac{2l+1}{4\pi}} \int_{\phi=0}^{2\pi}\int_{\theta=0}^{\pi} P_l(\cos\gamma)\, Y_{lm}^*(\theta,\phi)\sin\theta\, d\theta\, d\phi \\ &= \sqrt{\frac{2l+1}{4\pi}}\, b_m(\theta',\phi') \end{aligned} \tag{4.41}$$

where use has been made of Eq. (4.37).

Therefore, we write

$$b_m(\theta',\phi') = \sqrt{\frac{4\pi}{2l+1}}\, B_0 . \tag{4.42}$$

We also see that B_0 can be directly calculated from Eq. (4.39) by letting $m' = 0$ and $\gamma = 0$ as in the following.

$$\left[Y_{lm}^*(\theta,\phi)\right]_{\gamma=0} = B_0 Y_{l0}(0,\beta) = B_0\sqrt{\frac{2l+1}{4\pi}}\, e^0 P_l^0(\cos 0)$$

$$= \sqrt{\frac{2l+1}{4\pi}}\, P_l(1) B_0 = \sqrt{\frac{2l+1}{4\pi}}\, B_0 .$$

It is also seen that when $\gamma = 0$, the vectors **r** and **r**′ coincide and then $\theta = \theta'$ $\phi = \phi'$. Then, we write

$$Y_{lm}^*(\theta',\phi') = \sqrt{\frac{2l+1}{4\pi}}\, B_0$$

or

$$B_0 = \sqrt{\frac{4\pi}{2l+1}}\, Y_{lm}^*(\theta',\phi').$$

Substituting this value of B_0 into Eq. (4.42), we obtain

$$b_m(\theta',\phi') = \frac{4\pi}{(2l+1)} Y_{lm}^*(\theta',\phi').$$

Substituting this result into Eq. (4.36), we obtain

$$P_l(\cos\gamma) = \frac{4\pi}{(2l+1)} \sum_{m=-l}^{l} Y_{lm}(\theta,\phi) Y_{lm}^*(\theta',\phi') \text{ which is Eq. (4.34).}$$

Thus, the addition theorem is proved.

4.8 SOLVED PROBLEMS

PROBLEM 1 Prove that $\int_{\phi=0}^{2\pi}\int_{\theta=0}^{\pi} Y_{lm}(\theta,\phi)\sin\theta\, d\theta\, d\phi = \sqrt{4\pi}\,\delta_{l0}\,\delta_{m0}$

Solution: We have

$$\int_{\phi=0}^{2\pi}\int_{\theta=0}^{\pi} Y_{lm}(\theta,\phi)\sin\theta\, d\theta\, d\phi = \sqrt{4\pi}\int_{\phi=0}^{2\pi}\int_{\theta=0}^{\pi}\sqrt{\frac{1}{4\pi}}\, Y_{lm}(\theta,\phi)\sin\theta\, d\theta\, d\phi$$

$$= \sqrt{4\pi}\int_{\phi=0}^{2\pi}\int_{\theta=0}^{\pi} Y_{00}^*(\theta,\phi) Y_{lm}(\theta,\phi)\sin\theta\, d\theta\, d\phi$$

where use has been made of $Y_{00}(\theta,\phi) = \sqrt{\frac{1}{4\pi}} = Y_{00}^*(\theta,\phi)$.

Now, using the orthogonality property of spherical harmonics, we write

$$\int_{\phi=0}^{2\pi}\int_{\theta=0}^{\pi} Y_{lm}(\theta,\phi)\sin\theta\, d\theta\, d\phi = \sqrt{4\pi}\,\delta_{l0}\,\delta_{m0}.$$

PROBLEM 2 Prove that $Y_{lm}(0,\phi)=\left(\frac{2l+1}{4\pi}\right)^{\frac{1}{2}}\delta_{m0}$

Solution: We know that

$$Y_{lm}(\theta,\phi)=(-1)^m\sqrt{\frac{(2l+1)(l-m)!}{4\pi(l+m)!}}\,e^{im\phi}P_l^m(\cos\theta).$$

Taking $\theta=0$, we have

$$Y_{lm}(0,\phi)=(-1)^m\sqrt{\frac{(2l+1)(l-m)!}{4\pi(l+m)!}}\,e^{im\phi}P_l^m(1).$$

(i) When $m\neq 0$ we have, $P_l^m(1)=0$ and hence

$$Y_{lm}(0,\phi)=0, \text{ for } m\neq 0.$$

(ii) When $m = 0$, we have, $P_l^m(1)=P_l^0(1)=P_l(1)=1$ and hence

$$Y_{lm}(0,\phi)=(-1)^0\sqrt{\frac{(2l+1)\,l!}{4\pi\, l!}}\,e^0\cdot 1=\sqrt{\frac{2l+1}{4\pi}} \text{ for } m=0.$$

The above two results can be expressed by the general expression,

$$Y_{lm}(0,\phi)=\left(\frac{2l+1}{4\pi}\right)^{\frac{1}{2}}\delta_{m0}.$$

PROBLEM 3 Prove that the parity of $Y_{lm}(\theta,\phi)$ is $(-1)^l$.

Solution: We know that

$$Y_{lm}(\theta,\phi)=(-1)^m\sqrt{\frac{(2l+1)(l-m)!}{4\pi(l+m)!}}\,e^{im\phi}P_l^m(\cos\theta). \qquad \text{(i)}$$

Parity operation in spherical polar coordinates is achieved by the following changes,

$$\theta\to\pi-\theta \text{ and } \phi\to\phi\pm\pi.$$

Therefore, we have

$$Y_{lm}(\pi-\theta,\phi\pm\pi)=(-1)^m\sqrt{\frac{(2l+1)(l-m)!}{4\pi(l+m)!}}\,e^{im(\phi\pm\pi)}P_l^m(\cos(\pi-\theta))$$

$$= (-1)^m \sqrt{\frac{(2l+1)(l-m)!}{4\pi\,(l+m)!}}\, e^{im\phi}\, e^{\pm im\pi}\, P_l^m(-\cos\theta)\,. \qquad \text{(ii)}$$

Further, we know that

$$e^{\pm im\pi} = \cos m\pi \pm i \sin m\pi = (-1)^m \text{ and } P_l^m(-\cos\theta) = (-1)^{l-m} P_l^m(\cos\theta)\,.$$

Thus, Eq. (ii) becomes,

$$Y_{lm}(\pi-\theta, \phi\pm\pi) = (-1)^l \left[(-1)^m \sqrt{\frac{(2l+1)(l-m)!}{4\pi(l+m)!}}\, e^{im\phi} P_l^m(\cos\theta) \right]$$

$$= (-1)^l\, Y_{lm}(\theta,\phi), \text{ using Eq. (i).}$$

Hence, we conclude that the parity of $Y_{lm}(\theta,\phi)$ is $(-1)^l$.

PROBLEM 4 Prove that $Y_{ll}(\theta,\phi) = \dfrac{(-1)^l}{2^l\, l!}\sqrt{\dfrac{(2l+1)!}{4\pi}}\, e^{il\phi} \sin^l\theta$.

Solution: We know that

$$Y_{lm}(\theta,\phi) = (-1)^m \sqrt{\frac{(2l+1)(l-m)!}{4\pi\,(l+m)!}}\, e^{im\phi}\, P_l^m(\cos\theta).$$

Taking $m = l$, we have

$$Y_{ll}(\theta,\phi) = (-1)^l \sqrt{\frac{(2l+1)}{4\pi(2l)!}}\, e^{il\phi} P_l^l(\cos\theta)\,. \qquad \text{(i)}$$

Let us recall that

$$P_l^l(\cos\theta) = (2l-1)!!\sin^l\theta = 1.3.5.7 \ldots (2l-1)\sin^l\theta$$

$$= \frac{1.2.3.4.5.6\ldots(2l-1)(2l)}{2.4.6.8\ldots 2l}\sin^l\theta = \frac{(2l)!}{2^l l!}\sin^l\theta\,. \qquad \text{(ii)}$$

Substitution of Eq. (ii) into Eq. (i) yields

$$Y_{ll}(\theta,\phi) = \frac{(-1)^l}{2^l\, l!}\sqrt{\frac{(2l+1)!}{4\pi}}\, e^{il\phi}\sin^l\theta\,.$$

PROBLEM 5 Prove that $\displaystyle\sum_{m=-l}^{l} \left|Y_{lm}(\theta,\phi)\right|^2 = \frac{2l+1}{4\pi}$

Solution: Addition formula of spherical harmonics is

$$P_l(\cos\gamma) = \frac{4\pi}{2l+1}\sum_{m=-l}^{l} Y_{lm}(\theta,\phi)\, Y_{lm}^*(\theta',\phi')$$

where γ is the angle between two directions (θ,ϕ) and (θ',ϕ').

When $\gamma = 0$, we have, $\theta' = \theta$ and $\phi' = \phi$, then we write

$$P_l(1) = \frac{4\pi}{2l+1}\sum_{m=-l}^{l} Y_{lm}(\theta,\phi)Y_{lm}^{*}(\theta,\phi)$$

or

$$1 = \frac{4\pi}{2l+1}\sum_{m=-l}^{l} \left|Y_{lm}(\theta,\phi)\right|^2$$

or

$$\sum_{m=-l}^{l} \left|Y_{lm}(\theta,\phi)\right|^2 = \frac{2l+1}{4\pi}.$$

PROBLEM 6 Show that the reciprocal of the distance between any two points having position vectors **r** and **r′** can be expressed as

$$\frac{1}{|\mathbf{r}-\mathbf{r}'|} = \sum_{l=0}^{\infty}\frac{4\pi}{2l+1}\frac{r_<^l}{r_>^{l+1}}\sum_{m=-l}^{l} Y_{lm}(\theta,\phi)Y_{lm}^{*}(\theta,\phi)$$

where $r_<$ and $r_>$ respectively denote the smaller and the greater of the magnitudes of the two position vectors.

Solution: The reciprocal of the distance between two points with position vectors **r** and **r′** is given by

$$\frac{1}{|\mathbf{r}-\mathbf{r}'|} = \frac{1}{\sqrt{r^2 + r'^2 - 2rr'\cos\gamma}}$$

where $r = |\mathbf{r}|$, $r' = |\mathbf{r}'|$ and γ is the angle between the two vectors.

If $r > r'$, we have

$$\frac{1}{|\mathbf{r}-\mathbf{r}'|} = \frac{1}{r}\left[1 - 2\frac{r'}{r}\cos\gamma + \left(\frac{r'}{r}\right)^2\right]^{-\frac{1}{2}} = \frac{1}{r}\sum_{l=0}^{\infty}\left(\frac{r'}{r}\right)^l P_l(\cos\gamma) = \sum_{l=0}^{\infty}\frac{(r')^l}{r^{l+1}}P_l(\cos\gamma) \qquad \text{(i)}$$

Similarly, if $r < r'$, we will obtain

$$\frac{1}{|\mathbf{r}-\mathbf{r}'|} = \sum_{l=0}^{\infty}\frac{r^l}{(r')^{l+1}}P_l(\cos\gamma) \qquad \text{(ii)}$$

In general, we write

$$\frac{1}{|\mathbf{r}-\mathbf{r}'|} = \sum_{l=0}^{\infty}\frac{r_<^l}{r_>^{l+1}}P_l(\cos\gamma) \text{ for all } r \neq r' \qquad \text{(iii)}$$

where $r_<$ and $r_>$ are respectively the smaller and the greater of the two quantities r and r'.

Using the addition theorem of spherical harmonics, we write

$$\frac{1}{|\mathbf{r}-\mathbf{r}'|}=\sum_{l=0}^{\infty}\frac{4\pi}{2l+1}\frac{r_<^l}{r_>^{l+1}}\sum_{m=-l}^{l}Y_{lm}(\theta,\phi)\,Y_{lm}^*(\theta',\phi') \qquad \text{(iv)}$$

where (θ,ϕ) and (θ',ϕ') gives the orientations of the directions of the vectors **r** and **r**′, respectively.

EXERCISE

1. Prove that $Y_{l,-m}(\theta,\phi)=(-1)^m Y_{lm}^*(\theta,\phi)$ and deduce that

$$Y_{l,-l}(\theta,\phi)=\frac{1}{2^l\, l!}\sqrt{\frac{(2l+1)!}{4\pi}}\;e^{-il\phi}\sin^l\phi\,.$$

2. From a suitable recurrence relation of spherical harmonics deduce that

(a) $\cos\theta\, Y_{1,0}(\theta,\phi)=\sqrt{\frac{1}{3}}\,Y_{0,0}(\theta,\phi)+\sqrt{\frac{4}{15}}\,Y_{2,0}(\theta,\phi)$,

(b) $\cos\theta\, Y_{1,1}(\theta,\phi)=\sqrt{\frac{1}{5}}\,Y_{2,1}(\theta,\phi)$.

3. In spherical polar coordinates, if γ is the angle between two different directions respectively specified by (θ,ϕ) and (θ',ϕ'), use the addition theorem of spherical harmonics to prove that

$$\cos\gamma=\cos\theta\cos\theta'+\sin\theta\sin\theta'\cos(\phi-\phi').$$

4. Starting from the addition theorem of spherical harmonics, prove that

$$P_l(\cos\gamma)=P_l(\cos\theta)P_l(\cos\theta')+2\sum_{m=1}^{l}\frac{(l-m)!}{(l+m)!}P_l^m(\cos\theta)P_l^m(\cos\theta')\cos m(\phi-\phi'),$$

where γ is the angle between two different directions (θ,ϕ) and (θ',ϕ').

5. For the cases of $l=0$ and $l=1$, verify the addition formula of spherical harmonics

$$P_l(\cos\gamma)=\frac{4\pi}{2l+1}\sum_{m=-l}^{l}Y_{lm}(\theta,\phi)Y_{lm}^*(\theta',\phi')$$

where γ is the angle between the two directions (θ,ϕ) and (θ',ϕ').

5
CHAPTER

Hermite Polynomials

5.1 HERMITE EQUATION

Hermite equation is a homogeneous second order linear ordinary differential equation of the form

$$\frac{d^2y}{dx^2}-2x\frac{dy}{dx}+2n\,y=0 \tag{5.1}$$

where $y=y(x)$ is a function of x and n is a non-negative integer.

This equation was named after the French mathematician, Charles Hermite Circa (Dec. 24, 1822 - Jan. 14, 1901).

5.2 EXISTENCE OF POWER SERIES SOLUTION

In the above Hermite equation, the coefficient functions $P(x)=-2x$ and $Q(x)=2n$ are both finite at $x = 0$, therefore $x = 0$ is an ordinary point of the equation and hence, by Fuchs' theorem, the equation can have a power series solution about $x = 0$. Let us assume the power series solution as

$$y=y(x)=\sum_{r=0}^{\infty} a_r\, x^{k+r} \qquad (a_0 \neq 0) \tag{5.2}$$

where the coefficients a_r and the constant k are to be determined.

We have

$$\frac{dy}{dx}=\sum_{r=0}^{\infty}(k+r)a_r\, x^{k+r-1} \text{ and } \frac{d^2y}{dx^2}=\sum_{r=0}^{\infty}(k+r)(k+r-1)a_r\, x^{k+r-2}.$$

Substituting the expressions for y, $\frac{dy}{dx}$ and $\frac{d^2y}{dx^2}$ respectively into Eq. (5.1), we obtain

$$\sum_{r=0}^{\infty}(k+r)(k+r-1)a_r x^{k+r-2}+\sum_{r=0}^{\infty}2(n-k-r)a_r\, x^{k+r}=0.$$

To satisfy this identity, all the coefficients of various powers of x must vanish. Therefore, coefficient of x^{k-2}(i.e., coefficient of lowest power) is

$$k(k-1)a_0 = 0, a_0 \neq 0 \quad \text{[Indicial equation]}$$

giving $k = 0, 1$.

As the point $x = 0$ is an ordinary point of the Hermite equation, we can consider only the case with $k = 0$.

Now, the coefficient of x^{k+r} (i.e., coefficient of general term) is

$$(k+r+2)(k+r+1)a_{r+2} + 2(n-k-r)a_r = 0.$$

Therefore, with $k = 0$, we have

$$a_{r+2} = -\frac{2(n-r)}{(r+1)(r+2)} a_r \text{ (Recursion relation).} \tag{5.3}$$

Putting $r = 0, 1, 2, 3, 4, 5$, etc., Eq. (5.3) yields

$$a_2 = -\frac{2n}{2!} a_0 \qquad a_3 = -\frac{2(n-1)}{3!} a_1$$

$$a_4 = \frac{(-1)^2 2^2 n(n-2)}{4!} a_0 \qquad a_5 = \frac{(-1)^2 2^2 (n-1)(n-3)}{5!} a_1$$

$$a_6 = \frac{(-1)^3 2^3 n(n-2)(n-4)}{6!} a_0 \qquad a_7 = \frac{(-1)^3 2^3 (n-1)(n-3)(n-5)}{7!} a_1$$

... etc. ... etc.

In general, we write, for $j = 0, 1, 2, 3 \ldots,$

$$a_{2j} = \frac{(-1)^j 2^j n(n-2)(n-4)(n-6)\ldots(n-2j+2)}{(2j)!} a_0 \tag{5.4}$$

and
$$a_{2j+1} = \frac{(-1)^j 2^j (n-1)(n-3)(n-5)\ldots(n-2j+1)}{(2j+1)!} a_1. \tag{5.5}$$

With $k = 0$, Eq. (5.2) can be written as

$$y = \left(a_0 + a_2 x^2 + a_4 x^4 + a_6 x^6 + \ldots \infty\right) + \left(a_1 x + a_3 x^3 + a_5 x^5 + a_7 x^7 + \ldots \infty\right). \tag{5.6}$$

Substituting the values of various coefficients obtained above, we obtain the solution of Hermite equation as

$$y = a_0\left[1 - \frac{2n}{2!}x^2 + \frac{2^2 n(n-2)}{4!}x^4 - \ldots + \frac{(-1)^j 2^j n(n-2)(n-4)(n-6)\ldots(n-2j+2)}{(2j)!}x^{2j} + \ldots\right]$$

$$+a_1\left[x - \frac{2(n-1)}{3!}x^3 + \frac{2^2(n-1)(n-3)}{5!}x^5 - \ldots + \frac{(-1)^j 2^j (n-1)(n-3)(n-5)\ldots(n-2j+1)}{(2j+1)!}x^{2j+1} + \ldots\right] \tag{5.7}$$

We know that the Hermite equation is a second order linear ordinary differential equation and its general solution should contain two arbitrary constants. Since the above solution contains two arbitrary constants, a_0 and a_1, it represents the general solution of Hermite equation and hence further consideration of the case with $k=1$ is not necessary. If one considers the case of $k=1$ he will get one part of the above general solution, which is not an independent solution.

5.3 POLYNOMIAL SOLUTION: THE HERMITE POLYNOMIAL

If the above series solution of Hermite equation terminates at a term containing the nth power of x, the solution is said to be an nth order *polynomial solution.*

In the polynomial solution of Hermite equation, if the value of the arbitrary constant appearing in that solution is so chosen that the coefficient of x^n becomes 2^n, then the resulting polynomial solution is known as a *Hermite polynomial*. It is denoted by $H_n(x)$.

CASE I: When n is an even integer

In such a case, from Eq. (5.7), we see that the polynomial solution of Hermite equation will be in the form,

$$y_{n(\text{even})} = a_0\left[1 + \sum_{j=1}^{\frac{n}{2}} \frac{(-1)^j 2^j n(n-2)(n-4)(n-6)\ldots(n-2j+2)}{(2j)!}x^{2j}\right]$$

$$= a_0\left[1 + \sum_{j=1}^{\frac{n}{2}} \frac{(-1)^j 2^{2j}\left(\frac{n}{2}\right)\left(\frac{n}{2}-1\right)\left(\frac{n}{2}-2\right)\left(\frac{n}{2}-3\right)\ldots\left(\frac{n}{2}-j+1\right)}{(2j)!}x^{2j}\right]$$

$$= a_0\left[1 + \sum_{j=1}^{\frac{n}{2}} \frac{(-1)^j 2^{2j}\left(\frac{n}{2}\right)!}{\left(\frac{n}{2}-j\right)!(2j)!}x^{2j}\right]$$

$$= \sum_{j=0}^{\frac{n}{2}} a_0 \frac{(-1)^j \left(\frac{n}{2}\right)!}{\left(\frac{n}{2}-j\right)!\,(2j)!}(2x)^{2j}. \tag{5.8}$$

From Eq. (5.8) we will get the Hermite polynomial $H_n(x)$ if we choose the coefficient x^n to be 2^n; that is, if we choose

$$a_0 \frac{(-1)^{\frac{n}{2}} \left(\frac{n}{2}\right)!}{0!\,n!} = 1$$

giving $a_0 = \dfrac{n!}{(-1)^{\frac{n}{2}}\left(\frac{n}{2}\right)!}$.

Substituting this value of a_0 into Eq. (5.8), we obtain

$$H_n(x) = \sum_{j=0}^{\frac{n}{2}} \frac{(-1)^{j-\frac{n}{2}}\, n!}{\left(\frac{n}{2}-j\right)!\,(2j)!}(2x)^{2j}.$$

Putting $r = \frac{n}{2} - j$ or, $j = \frac{n}{2} - r$, we write

$$H_n(x) = \sum_{r=0}^{\frac{n}{2}} \frac{(-1)^{-r}\, n!}{r!\,(n-2r)!}(2x)^{n-2r}$$

$$= \sum_{r=0}^{\frac{n}{2}} \frac{(-1)^{r}\, n!}{r!\,(n-2r)!}(2x)^{n-2r} \tag{5.9}$$

when n is an even integer.

Equation (5.9) is the formula for Hermite polynomial, $H_n(x)$, when n is an even integer.

CASE II: When n is an odd integer

For such a case, from Eq. (5.7), we see that the polynomial solution of Hermite equation will be in the form,

$$y_{n(\text{odd})} = a_1 \left[x + \sum_{j=1}^{\frac{n-1}{2}} \frac{(-1)^j 2^j (n-1)(n-3)(n-5)\ldots(n-2j+1)}{(2j+1)!} x^{2j+1} \right]$$

$$= a_1 \left[x + \sum_{j=1}^{\frac{n-1}{2}} \frac{(-1)^j 2^j (n+1-2)(n+1-4)(n+1-6)\ldots(n+1-2j)}{(2j+1)!} x^{2j+1} \right]$$

$$= a_1 \left[x + \sum_{j=1}^{\frac{n-1}{2}} \frac{(-1)^j \, 2^{2j+1} \left(\frac{n+1}{2}\right)\left(\frac{n+1}{2}-1\right)\left(\frac{n+1}{2}-2\right)\left(\frac{n+1}{2}-3\right)\cdots\left(\frac{n+1}{2}-j\right)}{(n+1)\,(2j+1)!} x^{2j+1} \right]$$

$$= a_1 \left[x + \sum_{j=1}^{\frac{n-1}{2}} \frac{(-1)^j \, 2^{2j+1} \left(\frac{n+1}{2}\right)!}{(n+1)\left(\frac{n-1}{2}-j\right)!(2j+1)!} x^{2j+1} \right]$$

$$= \sum_{j=0}^{\frac{n-1}{2}} a_1 \frac{(-1)^j \left(\frac{n+1}{2}\right)!}{(n+1)\left(\frac{n-1}{2}-j\right)!\,(2j+1)!} (2x)^{2j+1} . \tag{5.10}$$

From Eq. (5.10), we will get the Hermite polynomial, $H_n(x)$, if we choose the coefficient of x^n to be 2^n; that is, if we choose

$$a_1 = \frac{(n+1)!}{(-1)^{\frac{n-1}{2}} \left(\frac{n+1}{2}\right)!} .$$

Substituting this value of a_1 into Eq. (5.10), we obtain

$$H_n(x) = \sum_{j=0}^{\frac{n-1}{2}} \frac{(-1)^{j-\frac{n-1}{2}} \, n!}{\left(\frac{n-1}{2}-j\right)!\,(2j+1)!} (2x)^{2j+1} .$$

Putting $r = \frac{n-1}{2} - j$, or, $j = \frac{n-1}{2} - r$, we write

$$H_n(x) = \sum_{r=0}^{\frac{n-1}{2}} \frac{(-1)^{-r} \, n!}{r!\,(n-2r)!} (2x)^{n-2r}$$

$$= \sum_{r=0}^{\frac{n-1}{2}} \frac{(-1)^{r} \, n!}{r!\,(n-2r)!} (2x)^{n-2r} \tag{5.11}$$

when n is an odd integer.

Equation (5.9) gives Hermite polynomial, $H_n(x)$, when n is an even integer whereas Eq. (5.11) gives the Hermite polynomial, $H_n(x)$, when n is an odd integer.

In a generalized form, we write the expression for a Hermite polynomial as

$$H_n(x)=\sum_{r=0}^{\left[\frac{n}{2}\right]}\frac{(-1)^r\,n!}{r!(n-2r)!}(2x)^{n-2r} \tag{5.12}$$

where
$$\left[\frac{n}{2}\right]=\begin{cases}\dfrac{n}{2}, & \text{if } n \text{ is an even integer,}\\[2mm] \dfrac{n-1}{2}, & \text{if } n \text{ is an odd integer.}\end{cases}$$

5.4 GENERATING FUNCTION OF HERMITE POLYNOMIALS

Let us define the generating function $G(x, h)$ of Hermite polynomials as

$$G(x,h)=\sum_{n=0}^{\infty}H_n(x)\frac{h^n}{n!}. \tag{5.13}$$

Using Eq. (5.12), we write

$$G(x,h)=\sum_{n=0}^{\infty}\frac{h^n}{n!}\sum_{r=0}^{\left[\frac{n}{2}\right]}\frac{(-1)^r\,n!}{r!(n-2r)!}(2x)^{n-2r}, \text{ where } \left[\frac{n}{2}\right] \text{ is as defined above;}$$

$$=\sum_{n=0}^{\infty}\sum_{r=0}^{\left[\frac{n}{2}\right]}\frac{(-1)^r\,h^n}{r!(n-2r)!}(2x)^{n-2r}$$

If we put $n-r=s$, or, $n=r+s$, we write

$$G(x,h)=\sum_{s=-r}^{\infty}\sum_{r=0}^{s}\frac{(-1)^r\,h^{r+s}}{r!(s-r)!}(2x)^{s-r}$$

$$=\sum_{s=0}^{\infty}\sum_{r=0}^{s}\frac{h^s}{s!}\frac{s!}{r!(s-r)!}(-h)^r(2x)^{s-r}.$$

Using the combination formula, $\binom{s}{r}=\dfrac{s!}{r!(s-r)!}$, we write

$$G(x,h)=\sum_{s=0}^{\infty}\frac{h^s}{s!}\sum_{r=0}^{s}\binom{s}{r}(2x)^{s-r}(-h)^r$$

$$=\sum_{s=0}^{\infty}\frac{h^s}{s!}(2x-h)^s, \text{ using Binomial expansion formula;}$$

$$=\sum_{s=0}^{\infty}\frac{\left(2xh-h^2\right)^s}{s!}=e^{2xh-h^2}. \tag{5.14}$$

The function $G(x,h)=e^{2xh-h^2}$ is called the generating function of Hermite polynomials in the sense that it can generate Hermite polynomials of various orders through Eq. (5.13) as

$$e^{2xh-h^2}=\sum_{n=0}^{\infty}H_n(x)\frac{h^n}{n!}. \tag{5.15}$$

Thus, $H_n(x)$ is the coefficient of $\frac{h^n}{n!}$ in the expansion of the exponential function e^{2xh-h^2}.

5.5 RODRIGUES' FORMULA AND YET ANOTHER DIFFERENTIAL FORMULA FOR HERMITE POLYNOMIALS

(a) Rodrigues' Formula

We know that

$$e^{2xh-h^2}=\sum_{n=0}^{\infty}H_n(x)\frac{h^n}{n!}$$

or

$$e^{x^2}\cdot e^{-(x-h)^2}=H_0(x)+H_1(x)h+H_2(x)\frac{h^2}{2!}+\ldots$$

$$+H_n(x)\frac{h^n}{n!}+H_{n+1}(x)\frac{h^{n+1}}{(n+1)!}+H_{n+2}(x)\frac{h^{n+2}}{(n+2)!}+\ldots$$

Performing n times partial differentiation with respect to h on both sides of the above equation, we obtain

$$e^{x^2}\frac{\partial^n}{\partial h^n}e^{-(x-h)^2}=0+H_n(x)+H_{n+1}(x)h+H_{n+2}(x)\frac{h^2}{2!}+\ldots$$

In the limit $h\to 0$ the right-hand side becomes $H_n(x)$.
Thus, we write

$$H_n(x)=e^{x^2}\left[\frac{\partial^n}{\partial h^n}e^{-(x-h)^2}\right]_{h=0}. \tag{5.16}$$

Let us put $u=x-h$. Then, for any arbitrary function $f(x,h)$, we have

$$\frac{\partial f}{\partial x}=\frac{\partial f}{\partial u}\frac{\partial u}{\partial x}=\frac{df}{du}, \text{ as } \frac{\partial u}{\partial x}=1$$

and

$$\frac{\partial f}{\partial h}=\frac{\partial f}{\partial u}\frac{\partial u}{\partial h}=-\frac{df}{du} \text{ as } \frac{\partial u}{\partial h}=-1.$$

Therefore, we obtain

$$\frac{\partial f}{\partial h} = -\frac{\partial f}{\partial x}$$

or, in operator form, $\dfrac{\partial}{\partial h} \equiv -\dfrac{\partial}{\partial x}$

and, therefore, $\dfrac{\partial^n}{\partial h^n} \equiv (-1)^n \dfrac{\partial^n}{\partial x^n}$.

Substituting this operator into Eq. (5.16), we obtain

$$H_n(x) = (-1)^n e^{x^2} \left[\frac{\partial^n}{\partial x^n} e^{-(x-h)^2} \right]_{h=0} = (-1)^n e^{x^2} \frac{d^n}{dx^n} e^{-x^2}. \tag{5.17}$$

Equation (5.17) is called Rodrigues' formula for Hermite polynomials.

(b) Another Differential Formula for Hermite Polynomials

We know that

$$\sum_{n=0}^{\infty} H_n(x) \frac{h^n}{n!} = e^{2xh-h^2} = e^{2xh} \cdot e^{-h^2} = e^{2xh} \sum_{n=0}^{\infty} \frac{\left(-h^2\right)^n}{n!}$$

$$= \sum_{n=0}^{\infty} \frac{(-1)^n h^{2n}}{n!} e^{2xh} = \sum_{n=0}^{\infty} \frac{1}{2^{2n}} \frac{(-1)^n}{n!} (2h)^{2n} e^{2xh}$$

$$= \sum_{n=0}^{\infty} \left[\frac{(-1)^n}{4^n \, n!} \frac{d^{2n}}{dx^{2n}} \right] e^{2xh} = \sum_{n=0}^{\infty} \frac{1}{n!} \left(-\frac{1}{4} \frac{d^2}{dx^2} \right)^n [e^{2xh}]$$

$$= \left[\exp.\left(-\frac{1}{4} \frac{d^2}{dx^2} \right) \right] e^{2xh} = \left[\exp.\left(-\frac{1}{4} \frac{d^2}{dx^2} \right) \right] \left[\sum_{n=0}^{\infty} \frac{(2xh)^n}{n!} \right]$$

$$= \sum_{n=0}^{\infty} \frac{2^n h^n}{n!} \left[\exp.\left(-\frac{1}{4} \frac{d^2}{dx^2} \right) \right] x^n.$$

Equating coefficients of $\dfrac{h^n}{n!}$ from both sides, we have

$$H_n(x) = 2^n \left[\exp.\left(-\frac{1}{4} \frac{d^2}{dx^2} \right) \right] x^n. \tag{5.18}$$

5.6 INTEGRAL REPRESENTATIONS OF HERMITE POLYNOMIALS

(a) From the Generating Function

In a complex plane, we have

$$e^{2xz-z^2}=\sum_{n=0}^{\infty}H_n(x)\frac{z^n}{n!}.$$

Multiplying both sides by z^{-m-1} and then performing complex integration with respect to z about a contour of unit circle (C) around the origin, we have

$$\oint_C z^{-m-1}\,e^{2xz-z^2}\,dz=\sum_{n=0}^{\infty}\oint_C\frac{1}{n!}H_n(x)z^{n-m-1}dz.$$

Let us put $z=e^{i\theta}$, then $dz=ie^{i\theta}d\theta$.
Therefore, we write

$$\oint_C z^{-m-1}\,e^{2xz-z^2}\,dz=\sum_{n=0}^{\infty}\frac{i}{n!}H_n(x)\int_0^{2\pi}e^{i(n-m)\theta}d\theta.$$

The integral on the right-hand side has a non-vanishing value of 2π only when $n=m$.
Thus, we obtain

$$\oint_C z^{-m-1}\,e^{2xz-z^2}\,dz=\frac{2\pi i}{m!}H_m(x).$$

Hence, we obtain

$$H_m(x)=\frac{m!}{2\pi i}\oint_C z^{-m-1}e^{2xz-z^2}\,dz. \tag{5.19}$$

(b) From Rodrigues' Formula

Rodrigues' formula for Hermite polynomials is

$$H_n(x)=(-1)^n e^{x^2}\frac{\partial^n}{\partial x^n}e^{-x^2}=e^{x^2}\frac{\partial^n}{(-1)^n\,\partial x^n}(e^{-x^2}).$$

Using Cauchy's generalized formula for the derivative of a complex function, we write

$$H_n(x)=e^{x^2}\cdot\frac{n!}{2\pi i}\oint_C\frac{e^{-z^2}\,dz}{\{z-(-x)\}^{n+1}}$$

where C is a closed contour enclosing the point $z=-x$.

Thus, we write

$$H_n(x) = \frac{n!}{2\pi i} \oint_C \frac{e^{-(z^2-x^2)}}{(z+x)^{n+1}} dz. \tag{5.20}$$

5.7 RECURRENCE RELATIONS OF HERMITE POLYNOMIALS

I. $\frac{d}{dx} H_n(x) = 2n H_{n-1}(x)$

Proof:

First method: Using generating function

We know that $\sum_{n=0}^{\infty} H_n(x) \frac{h^n}{n!} = e^{2xh-h^2}$.

Differentiating both sides once with respect to x, we obtain

$$\sum_{n=0}^{\infty} \frac{d}{dx} H_n(x) \frac{h^n}{n!} = 2h e^{2xh-h^2} = 2h \sum_{n=0}^{\infty} H_n(x) \frac{h^n}{n!} = \sum_{n=0}^{\infty} 2H_n(x) \frac{h^{n+1}}{n!}.$$

Equating coefficients of h^n from both sides, we have

$$\frac{1}{n!} \frac{d}{dx} H_n(x) = \frac{2}{(n-1)!} H_{n-1}(x)$$

or

$$\frac{d}{dx} H_n(x) = 2n H_{n-1}(x). \tag{5.21}$$

Second method: Using the sum formula

We know that the Hermite polynomial, $H_n(x)$, satisfies the Hermite equation,

$$\frac{d^2}{dx^2} H_n(x) - 2x \frac{d}{dx} H_n(x) + 2nH_n(x) = 0. \tag{i}$$

Differentiating both sides once with respect to x, we have

$$\frac{d^2}{dx^2}\left\{\frac{d}{dx} H_n(x)\right\} - 2x \frac{d}{dx}\left\{\frac{d}{dx} H_n(x)\right\} + 2(n-1)\left\{\frac{d}{dx} H_n(x)\right\} = 0 \tag{ii}$$

which also a Hermite equation in $\frac{d}{dx} H_n(x)$

In Eq. (i), if we replace n by $n-1$, we have

$$\frac{d^2}{dx^2} H_{n-1}(x) - 2x \frac{d}{dx} H_{n-1}(x) + 2(n-1) H_{n-1}(x) = 0. \tag{iii}$$

As $\frac{d}{dx}H_n(x)$ and $H_{n-1}(x)$ satisfy the same Hermite equation, one of them can be a constant multiple of the other. Hence, we can write

$$\frac{d}{dx}H_n(x) = C\,H_{n-1}(x) \tag{iv}$$

where C is a constant to be determined.

Using the sum formula for Hermite polynomials, we have

$$\frac{d}{dx}\sum_{r=0}^{\left[\frac{n}{2}\right]}\frac{(-1)^r\,n!}{r!(n-2r)!}(2x)^{n-2r} = C\sum_{r=0}^{\left[\frac{n-1}{2}\right]}\frac{(-1)^r\,(n-1)!}{r!(n-1-2r)!}(2x)^{n-1-2r},$$

where

$$\left[\frac{n}{2}\right] = \begin{cases}\dfrac{n}{2} & \text{if } n \text{ is even integer}\\[2mm] \dfrac{n-1}{2} & \text{if } n \text{ is odd integer}\end{cases}$$

or

$$\sum_{r=0}^{\left[\frac{n-1}{2}\right]}\frac{2(-1)^r\,n!}{r!(n-2r-1)!}(2x)^{n-2r-1} = C\sum_{r=0}^{\left[\frac{n-1}{2}\right]}\frac{(-1)^r\,(n-1)!}{r!(n-1-2r)!}(2x)^{n-1-2r}.$$

Equating coefficients of x^{n-1} from both sides, we have

$$C = 2n.$$

Hence, Eq. (iv) becomes

$$\frac{d}{dx}H_n(x) = 2n\,H_{n-1}(x).$$

II. $\frac{d^2}{dx^2}H_n(x) = 4n(n-1)H_{n-2}(x)$

Proof: Using recurrence relation I, we write

$$\frac{d^2}{dx^2}H_n(x) = \frac{d}{dx}\left\{\frac{d}{dx}H_n(x)\right\} = \frac{d}{dx}\{2n\,H_{n-1}(x)\}$$

$$= 2n\left\{\frac{d}{dx}H_{n-1}(x)\right\} = 2n\{2(n-1)H_{n-2}(x)\}$$

$$= 4n(n-1)H_{n-2}(x). \tag{5.22}$$

III. $H_{n+1}(x)-2xH_n(x)+2n\,H_{n-1}(x)=0$

Proof:

First method: Using generating function

We know that

$$e^{2xh-h^2}=\sum_{n=0}^{\infty}H_n(x)\frac{h^n}{n!}.$$

Differentiating both sides once with respect to *h*, we have

$$2(x-h)e^{2xh-h^2}=\sum_{n=1}^{\infty}H_n(x)\frac{h^{n-1}}{(n-1)!}$$

or
$$2xe^{2xh-h^2}-2he^{2xh-h^2}=\sum_{n=1}^{\infty}H_n(x)\frac{h^{n-1}}{(n-1)!}$$

or
$$2x\sum_{n=0}^{\infty}H_n(x)\frac{h^n}{n!}-2h\sum_{n=0}^{\infty}H_n(x)\frac{h^n}{n!}=\sum_{n=0}^{\infty}H_{n+1}(x)\frac{h^n}{n!}$$

or
$$\sum_{n=0}^{\infty}2xH_n(x)\frac{h^n}{n!}-\sum_{n=0}^{\infty}2H_n(x)\frac{h^{n+1}}{n!}=\sum_{n=0}^{\infty}H_{n+1}(x)\frac{h^n}{n!}$$

Equating coefficients of h^n from both sides, we have

$$\frac{1}{n!}2xH_n(x)-\frac{1}{(n-1)!}2H_{n-1}(x)=\frac{1}{n!}H_{n+1}(x)$$

or
$$H_{n+1}(x)-2xH_n(x)+2nH_{n-1}(x)=0. \tag{5.23}$$

Second method: Using the previous recurrence relations

As the Hermite polynomial, $H_n(x)$, is to satisfy the Hermite equation, we have

$$\frac{d^2}{dx^2}H_n(x)-2x\frac{d}{dx}H_n(x)+2nH_n(x)=0$$

Using recurrence relations I and II, we have

$$4n(n-1)H_{n-2}(x)-2x\cdot 2nH_{n-1}(x)+2nH_n(x)=0$$

Dividing both sides by *2n*, we obtain

$$2(n-1)H_{n-2}(x)-2xH_{n-1}(x)+H_n(x)=0$$

Replacing n by $(n+1)$, we obtain

$$2nH_{n-1}(x)-2xH_n(x)+H_{n+1}(x)=0. \tag{5.24}$$

5.8 ORTHOGONALITY OF HERMITE POLYNOMIALS

With a weight function, e^{-x^2}, Hermite polynomials are orthogonal in the interval $(-\infty, +\infty)$. Hermite polynomials are also orthonormal in that interval with a certain normalization constant.

Using the generating function for Hermite polynomials, we write

$$\sum_{m=0}^{\infty} H_m(x)\frac{h_1^m}{m!} = e^{2xh_1-h_1^2} \text{ and } \sum_{n=0}^{\infty} H_n(x)\frac{h_2^n}{n!} = e^{2xh_2-h_2^2}$$

from which we obtain

$$\sum_{m=0}^{\infty} H_m(x)\frac{h_1^m}{m!}\sum_{n=0}^{\infty} H_n(x)\frac{h_2^n}{n!} = e^{2xh_1+2xh_2-\left(h_1^2+h_2^2\right)}.$$

Multiplying both sides by e^{-x^2}, we have

$$\sum_{m=0}^{\infty}\sum_{n=0}^{\infty} e^{-x^2} H_m(x)H_n(x)\frac{h_1^m\, h_2^n}{m!\,n!} = e^{-(x^2+h_1^2+h_2^2-2xh_1-2xh_2)}$$

$$= e^{-(x-h_1-h_2)^2+2h_1h_2}.$$

Integrating both sides with respect to x within the limit from $x = -\infty$ to $x = +\infty$, we have

$$\sum_{m=0}^{\infty}\sum_{n=0}^{\infty}\frac{h_1^m\, h_2^n}{m!\,n!}\int_{-\infty}^{+\infty} e^{-x^2} H_m(x)H_n(x)\,dx = e^{2h_1h_2}\int_{-\infty}^{+\infty} e^{-(x-h_1-h_2)^2}\,dx$$

$$= e^{2h_1h_2}\int_{-\infty}^{+\infty} e^{-u^2}\,du \text{ where } u = x - h_1 - h_2$$

or

$$\sum_{m=0}^{\infty}\sum_{n=0}^{\infty}\frac{h_1^m\, h_2^n}{m!\,n!}\int_{-\infty}^{+\infty} e^{-x^2} H_m(x)H_n(x)\,dx = e^{2h_1h_2}\Gamma\left(\frac{1}{2}\right)$$

$$= \sum_{n=0}^{\infty}\frac{2^n\, h_1^n\, h_2^n}{n!}\sqrt{\pi}. \tag{5.25}$$

Equating coefficients of $h_1^m\, h_2^n\,(m \neq n)$ from both sides of Eq. (5.25), we have

$$\int_{-\infty}^{+\infty} e^{-x^2} H_m(x)H_n(x)\,dx = 0,\ (m \neq n). \tag{5.26}$$

This equation shows the orthogonality of Hermite polynomials $H_m(x)$ and $H_n(x)$ (if $m \neq n$) with a weight function, e^{-x^2}, in an interval $(-\infty, +\infty)$.

Again, equating coefficients of $h_1^n\, h_2^n\,(m=n)$ from both sides of Eq. (5.25), we have

$$\int_{-\infty}^{+\infty} e^{-x^2} H_n(x) H_n(x)\,dx = 2^n n!\sqrt{\pi}\,. \tag{5.27}$$

Thus, we see that the functions, $\sqrt{\dfrac{1}{2^n n!\sqrt{\pi}}}H_n(x)$, are orthonormal functions with a weight function, e^{-x^2}, in an interval $(-\infty,+\infty)$.

In other words, we say that the functions, $\Phi(x)=\sqrt{\dfrac{1}{2^n\, n!\sqrt{\pi}}}\, e^{-\frac{x^2}{2}} H_n(x)$, are orthonormal functions in an interval $(-\infty,+\infty)$. Such a function is known as *Hermite-Gauss function* or simply *Hermite function.*

Equations (5.26) and (5.27) can be represented by the general equation

$$\int_{-\infty}^{+\infty} e^{-x^2} H_m(x) H_n(x)\,dx = 2^n n!\sqrt{\pi}\,\delta_{mn} \tag{5.28}$$

where the Kronecker delta symbol, $\delta_{mn}=\begin{cases}1 \text{ for } m=n,\\ 0 \text{ for } m\neq n.\end{cases}$

5.9 EXPANSION OF ARBITRARY FUNCTION IN A SERIES OF HERMITE POLYNOMIALS

An arbitrary function, $f(x)$, can be expanded in a series of Hermite polynomials as

$$f(x)=\sum_{n=0}^{\infty} c_n H_n(x) \tag{5.29}$$

where the coefficients (c_n) are to be determined from the orthonormality condition.

Multiplying both sides by $e^{-x^2} H_m(x)$ and then integrating with respect to x within the limit from $x=-\infty$ to $x=+\infty$, we have

$$\int_{-\infty}^{+\infty} f(x)e^{-x^2} H_m(x)\,dx = \sum_{n=0}^{\infty} c_n \int_{-\infty}^{+\infty} e^{-x^2} H_m(x) H_n(x)\,dx$$

$$= c_m 2^m m!\sqrt{\pi}\,.$$

Thus, we have

$$c_m = \frac{1}{2^m\, m!\sqrt{\pi}} \int_{-\infty}^{+\infty} f(x)e^{-x^2} H_m(x)\,dx.$$

Changing the subscript, we write

$$c_n = \frac{1}{2^n n! \sqrt{\pi}} \int_{-\infty}^{+\infty} f(x) e^{-x^2} H_n(x) dx. \tag{5.30}$$

Substituting the values of these coefficients obtained from Eq. (5.30) into Eq. (5.29), we will get the required series.

Thus, a polynomial of degree (k) given by $f(x) = a_0 + a_1 x + a_2 x^2 + \ldots + a_k x^k$ can be expressed as

$$f(x) = a_0 + a_1 x + a_2 x^2 + \ldots + a_k x^k = \sum_{n=0}^{k} c_n H_n(x)$$

where the coefficients (c_n) are to be calculated as in Eq. (5.30).

5.10 SOLVED PROBLEMS

PROBLEM 1 Obtain the expressions for the first five Hermite polynomials.

Solution:

First method: Using the sum formula

We know that

$$H_n(x) = \sum_{r=0}^{\left[\frac{n}{2}\right]} \frac{(-1)^r n!}{r!(n-2r!)} (2x)^{n-2r}, \text{where } \left[\frac{n}{2}\right] = \begin{cases} \frac{n}{2} \text{ if } n \text{ is even integer} \\ \frac{n-1}{2} \text{ if } n \text{ is odd integer} \end{cases}$$

Therefore, $H_0(x) = (2x)^0 = 1.\quad H_1(x) = 2x.$

$$H_2(x) = \sum_{r=0}^{1} \frac{(-1)^r 2!}{r!(2-2r)!} (2x)^{2-2r} = (2x)^2 - 2.$$

$$H_3(x) = \sum_{r=0}^{1} \frac{(-1)^r 3!}{r!(3-2r)!} (2x)^{3-2r} = (2x)^3 - 12x.$$

$$H_4(x) = \sum_{r=0}^{2} \frac{(-1)^r 4!}{r!(4-2r)!} (2x)^{4-2r} = (2x)^4 - 12(2x)^2 + 12.$$

Second method: Using Rodrigues' formula

Rodrigues' formula for Hermite polynomials is

$$H_n(x) = (-1)^n e^{x^2} \frac{d^n}{dx^n} e^{-x^2}.$$

Thus, putting $n=0,1,2,3,4$, we have

$$H_0(x)=e^{x^2}\cdot e^{-x^2}=1.\ H_1(x)=-e^{x^2}\frac{d}{dx}e^{-x^2}=-e^{x^2}(-2xe^{-x^2})=2x.$$

$$H_2(x)=(-1)^2 e^{x^2}\frac{d^2}{dx^2}e^{-x^2}=e^{x^2}\frac{d}{dx}(-2xe^{-x^2})=e^{x^2}(4x^2-2)e^{-x^2}=4x^2-2.$$

$$H_3(x)=(-1)^3 e^{x^2}\frac{d^3}{dx^3}e^{-x^2}=-e^{x^2}\frac{d}{dx}\left[\left(4x^2-2\right)e^{-x^2}\right]$$

$$=-e^{x^2}\left[8x-2x\left(4x^2-2\right)\right]e^{-x^2}=8x^3-12x.$$

$$H_4(x)=(-1)^4 e^{x^2}\frac{d^4}{dx^4}e^{-x^2}=e^{x^2}\frac{d}{dx}\left[\left(12x-8x^3\right)e^{-x^2}\right]$$

$$=e^{x^2}\left[\left(12-24x^2\right)-2x\left(12x-8x^3\right)\right]e^{-x^2}=16x^4-48x^2+12.$$

Third method: Using the formula

$$H_n(x)=2^n\left[\exp.\left(-\frac{1}{4}\frac{d^2}{dx^2}\right)\right]x^n,$$

we have,

$$H_0(x)=2^0\left[\exp.\left(-\frac{1}{4}\frac{d^2}{dx^2}\right)\right]x^0=1\left[1-\frac{1}{4}\frac{d^2}{dx^2}+\ldots\right]1=1,$$

$$H_1(x)=2\left[\exp.\left(-\frac{1}{4}\frac{d^2}{dx^2}\right)\right]x=2\left[1-\frac{1}{4}\frac{d^2}{dx^2}+\frac{1}{2!}\frac{1}{4^2}\frac{d^4}{dx^4}-\ldots\right]x=2x,$$

$$H_2(x)=2^2\left[\exp.\left(-\frac{1}{4}\frac{d^2}{dx^2}\right)\right]x^2=4\left[1-\frac{1}{4}\frac{d^2}{dx^2}+\frac{1}{2!}\frac{1}{4^2}\frac{d^4}{dx^4}-\ldots\right]x^2$$

$$=4\left(x^2-\frac{1}{2}\right)=4x^2-2,$$

$$H_3(x)=2^3\left[\exp.\left(-\frac{1}{4}\frac{d^2}{dx^2}\right)\right]x^3=8\left[1-\frac{1}{4}\frac{d^2}{dx^2}+\frac{1}{2!}\frac{1}{4^2}\frac{d^4}{dx^4}+\ldots\right]x^3$$

$$=8\left(x^3-\frac{1}{4}\cdot 6x\right)=8x^3-12x,$$

$$H_4(x)=2^4\left[\exp.\left(-\frac{1}{4}\frac{d^2}{dx^2}\right)\right]x^4$$

$$=16\left[1-\frac{1}{4}\frac{d^2}{dx^2}+\frac{1}{2!}\frac{1}{4^2}\frac{d^4}{dx^4}-\frac{1}{3!}\frac{1}{4^3}\frac{d^6}{dx^6}+\ldots\right]x^4$$

$$=16\left(x^4-\frac{1}{4}\cdot 12x^2+\frac{1}{32}\cdot 24\right)=16x^4-48x^2+12.$$

PROBLEM 2 Prove that $H_n(x)=(-1)^n H_n(x)$ where n is a non-negative integer.

Solution:

First Method: Using generating function

We have

$$\sum_{n=0}^{\infty} H_n(x)\frac{h^n}{n!}=e^{2xh-h^2}. \tag{i}$$

Replacing x by $(-x)$ and h by $(-h)$, we have

$$\sum_{n=0}^{\infty} H_n(-x)\frac{(-h)^n}{n!}=e^{2(-x)(-h)-(-h)^2}=e^{2xh-h^2}. \tag{ii}$$

From Eqs. (i) and (ii), we have

$$\sum_{n=0}^{\infty} H_n(-x)\frac{(-1)^n h^n}{n!}=\sum_{n=0}^{\infty} H_n(x)\frac{h^n}{n!}.$$

Equating coefficients of h^n from both sides, we obtain

$$\frac{(-1)^n}{n!}H_n(-x)=\frac{1}{n!}H_n(x).$$

Therefore, $H_n(-x)=(-1)^n H_n(x)$.

Second Method: Using the sum formula

We have

$$H_n(x)=\sum_{r=0}^{\left[\frac{n}{2}\right]}\frac{(-1)^r n!}{r!(n-2r)!}(2x)^{n-2r}, \text{where} \left[\frac{n}{2}\right]=\begin{cases}\frac{n}{2} \text{ if } n \text{ is even integer,}\\ \frac{n-1}{2} \text{ if } n \text{ is odd integer.}\end{cases}$$

Replacing x by $(-x)$, we obtain

$$H_n(-x) = \sum_{r=0}^{\left[\frac{n}{2}\right]} \frac{(-1)^r \, n!}{r!\,(n-2r)!}(-2x)^{n-2r}$$

$$= (-1)^n \sum_{r=0}^{\left[\frac{n}{2}\right]} \frac{(-1)^r (-1)^{-2r} \, n!}{r!\,(n-2r)!}(2x)^{n-2r}$$

$$= (-1)^n \sum_{r=0}^{\left[\frac{n}{2}\right]} \frac{(-1)^r \, n!}{r!\,(n-2r)!}(2x)^{n-2r} = (-1)^n H_n(x).$$

PROBLEM 3 Prove that

(a) $H_{2n}(0) = \dfrac{(-1)^n (2n)!}{n!}$,

(b) $H_{2n+1}(0) = 0$,

(c) $H'_{2n}(0) = 0$,

(d) $H'_{2n+1}(0) = \dfrac{(-1)^n \, 2 \cdot (2n+1)!}{n!}$,

where n is a non-negative integer and the primes denote differentiation with respect to x.

Solution: (a) and (b)

We know that

$$\sum_{n=0}^{\infty} H_n(x) \frac{h^n}{n!} = e^{2xh-h^2}.$$

Putting $x = 0$, we have

$$\sum_{n=0}^{\infty} H_n(0) \frac{h^n}{n!} = e^{-h^2} = \sum_{n=0}^{\infty} \frac{1}{n!}(-h^2)^n = \sum_{n=0}^{\infty} \frac{(-1)^n}{n!} h^{2n}. \tag{i}$$

(a) Equating coefficients of h^{2n} from both sides of Eq. (i), we have

$$\frac{1}{(2n)!} H_{2n}(0) = \frac{(-1)^n}{n!}.$$

Therefore,

$$H_{2n}(0) = \frac{(-1)^n (2n)!}{n!}. \tag{ii}$$

(b) Equating coefficients of h^{2n+1} from both sides of Eq. (i), we have

$$\frac{1}{(2n+1)!}H_{2n+1}(0)=0$$

or
$$H_{2n+1}(0)=0. \tag{iii}$$

(c) We know that

$$H_n'(x)=2nH_{n-1}(x).$$

Replacing n by $2n$ and also putting $x=0$, we have

$$H_{2n}'(0)=2\cdot 2nH_{2n-1}(0). \tag{iv}$$

From (b), we see that

$$H_{2n+1}(0)=0.$$

Replacing n by $(n-1)$, we have

$$H_{2n-1}(0)=0.$$

Hence, Eq. (iv) yields,

$$H_{2n}'(0)=0. \tag{v}$$

(d) In the recurrence relation

$$H_n'(x)=2n\,H_{n-1}(x)$$

replacing n by $(2n+1)$ and putting x = 0, we have

$$H_{2n+1}'(0)=2(2n+1)H_{2n}(0)=2(2n+1)\frac{(-1)^n(2n)!}{n!}, \text{ using the result in (a);}$$

$$=(-1)^n\frac{2\cdot(2n+1)!}{n!}.$$

PROBLEM 4 Prove that

(a) $e^{-1}\cosh 2x=\sum_{n=0}^{\infty}\frac{1}{(2n)!}H_{2n}(x)$,

(b) $e^{-1}\sinh 2x=\sum_{n=0}^{\infty}\frac{1}{(2n+1)!}H_{2n+1}(x)$,

(c) $e\cos 2x=\sum_{n=0}^{\infty}\frac{(-1)^n}{(2n)!}H_{2n}(x)$,

(d) $e\sin 2x=\sum_{n=0}^{\infty}\frac{(-1)^n}{(2n+1)!}H_{2n+1}(x)$.

Solution: The generating function for Hermite polynomials is

$$e^{2xh-h^2} = \sum_{n=0}^{\infty} H_n(x)\frac{h^n}{n!}. \tag{i}$$

(a) Putting $h=1$ in Eq. (i), we have

$$e^{-1}e^{2x} = \sum_{n=0}^{\infty} \frac{1}{n!} H_n(x)$$

or $$e^{-1}\left[1+2x+\frac{(2x)^2}{2!}+\frac{(2x)^3}{3!}+\frac{(2x)^4}{4!}+\frac{(2x)^5}{5!}+\ldots\infty\right]$$

$$= H_0(x)+H_1(x)+\frac{1}{2!}H_2(x)+\frac{1}{3!}H_3(x)+\frac{1}{4!}H_4(x)+\ldots\infty. \tag{ii}$$

Putting $h=-1$ in Eq. (i), we have

$$e^{-1}\left[1-2x+\frac{(2x)^2}{2!}-\frac{(2x)^3}{3!}+\frac{(2x)^4}{4!}-\frac{(2x)^5}{5!}+\ldots\infty\right]$$

$$= H_0(x)-H_1(x)+\frac{1}{2!}H_2(x)-\frac{1}{3!}H_3(x)+\frac{1}{4}H_4(x)-\ldots\infty. \tag{iii}$$

Adding Eqs. (ii) and (iii), we obtain

$$2e^{-1}\left[1+\frac{(2x)^2}{2!}+\frac{(2x)^4}{4!}+\ldots\infty\right] = 2\left[H_0(x)+\frac{1}{2!}H_2(x)+\frac{1}{4!}H_4(x)+\ldots\infty\right]$$

or $$e^{-1}\cosh 2x = \sum_{n=0}^{\infty}\frac{1}{(2n)!}H_{2n}(x).$$

(b) Subtracting Eq. (iii) from Eq. (ii), we will get

$$e^{-1}\sinh 2x = \sum_{n=0}^{\infty}\frac{1}{(2n+1)!}H_{2n+1}(x).$$

(c) Putting $h=i$ in Eq. (i), we have

$$e\cdot e^{2xi} = \sum_{n=0}^{\infty} H_n(x)\frac{i^n}{n!}$$

or $e\cdot(\cos 2x + i\sin 2x)$

$$= H_0(x)+iH_1(x)+\frac{i^2}{2!}H_2(x)+\frac{i^3}{3!}H_3(x)+\frac{i^4}{4!}H_4(x)+\frac{i^5}{5!}H_5(x)+\ldots\infty$$

$$=\left[H_0(x)-\frac{1}{2!}H_2(x)+\frac{1}{4!}H_4(x)-\frac{1}{6!}H_6(x)+\ldots\infty\right]$$

$$+i\left[H_1(x)-\frac{1}{3!}H_3(x)+\frac{1}{5!}H_5(x)-\frac{1}{7!}H_7(x)+\ldots\infty\right]$$

$$=\sum_{n=0}^{\infty}\frac{(-1)^n}{(2n)!}H_{2n}(x)+i\sum_{n=0}^{\infty}\frac{(-1)^n}{(2n+1)!}H_{2n+1}(x). \quad \text{(iv)}$$

Equating the real parts on both sides of Eq. (iv), we get

$$e\cdot\cos 2x=\sum_{n=0}^{\infty}\frac{(-1)^n}{(2n)!}H_{2n}(x).$$

(d) Equating the imaginary parts on both sides of Eq. (iv), we get

$$e\cdot\sin 2x=\sum_{n=0}^{\infty}\frac{(-1)^n}{(2n+1)!}H_{2n+1}(x).$$

PROBLEM 5 Prove that

(i) $x^n=\dfrac{n!}{2^n}\displaystyle\sum_{k=0}^{\left[\frac{n}{2}\right]}\frac{H_{n-2k}(x)}{k!(n-2k)!}$, where $\left[\dfrac{n}{2}\right]=\begin{cases}\dfrac{n}{2} & \text{if } n \text{ is even integer,}\\ \dfrac{n-1}{2} & \text{if } n \text{ is odd integer.}\end{cases}$

(ii) $x^{2m}=\dfrac{(2m)!}{2^{2m}}\displaystyle\sum_{p=0}^{m}\frac{H_{2p}(x)}{(2p)!(m-p)!}$.

Solution:

(i) The generating function for Hermite polynomials is

$$e^{2xh-h^2}=\sum_{n=0}^{\infty}H_n(x)\frac{h^n}{n!}.$$

Multiplying both sides by e^{h^2}, we have

$$e^{2xh}=e^{h^2}\sum_{n=0}^{\infty}H_n(x)\frac{h^n}{n!}=\sum_{k=0}^{\infty}\frac{\left(h^2\right)^k}{k!}\sum_{n=0}^{\infty}H_n(x)\frac{h^n}{n!},$$

or

$$\sum_{n=0}^{\infty}\frac{2^n x^n h^n}{n!}=\sum_{k=0}^{\infty}\sum_{n=0}^{\infty}H_n(x)\frac{h^{n+2k}}{k!\,n!}.$$

Equating coefficients of h^n from both sides of the above equation, we have

$$\frac{2^n x^n}{n!} = \sum_{k=0}^{\left[\frac{n}{2}\right]} \frac{H_{n-2k}(x)}{k!(n-2k)!}$$

or
$$x^n = \frac{n!}{2^n} \sum_{k=0}^{\left[\frac{n}{2}\right]} \frac{H_{n-2k}(x)}{k!(n-2k)!}.$$

(ii) In the above result, if we put $n = 2m$, we have

$$x^{2m} = \frac{(2m)!}{2^{2m}} \sum_{k=0}^{m} \frac{H_{2m-2k}(x)}{k!(2m-2k)!}.$$

Putting $p = m - k$, we write

$$x^{2m} = \frac{(2m)!}{2^{2m}} \sum_{p=0}^{m} \frac{H_{2p}(x)}{(m-p)!(2p)!}.$$

PROBLEM 6 Prove that

$$H_n(x+y) = 2^{-\frac{n}{2}} \sum_{p=0}^{n} \frac{n!}{p!(n-p)!} H_{n-p}\left(\sqrt{2}\,x\right) H_p\left(\sqrt{2}\,y\right)$$ and deduce that

$$H_n\left(\sqrt{2}\,t\right) = 2^{-\frac{n}{2}} \sum_{p=0}^{n} \frac{n!}{p!(n-p)!} H_{n-p}(t) H_p(t)$$

Solution: Using the generating function for Hermite polynomials, we write

$$\sum_{n=0}^{\infty} H_n(x+y) \frac{h^n}{n!} = e^{2(x+y)h-h^2} = e^{2xh-\frac{h^2}{2}} e^{2yh-\frac{h^2}{2}}$$

$$= e^{2\left(\sqrt{2}\,x\right)\left(\frac{h}{\sqrt{2}}\right)-\left(\frac{h}{\sqrt{2}}\right)^2} e^{2\left(\sqrt{2}\,y\right)\left(\frac{h}{\sqrt{2}}\right)-\left(\frac{h}{\sqrt{2}}\right)^2}$$

$$= \sum_{n=0}^{\infty} H_n\left(\sqrt{2}\,x\right) \frac{\left(\frac{h}{\sqrt{2}}\right)^n}{n!} \sum_{p=0}^{\infty} H_p\left(\sqrt{2}\,y\right) \frac{\left(\frac{h}{\sqrt{2}}\right)^p}{p!}$$

$$= \sum_{n=0}^{\infty} \sum_{p=0}^{\infty} \frac{1}{n!\,p!} H_n\left(\sqrt{2}\,x\right) H_p\left(\sqrt{2}\,y\right) \frac{h^{n+p}}{\left(\sqrt{2}\right)^{n+p}}.$$

Equating coefficients of h^n from both sides of the equation, we have

$$\frac{1}{n!}H_n(x+y)=\sum_{p=0}^{n}\frac{1}{(n-p)!\,p!}H_{n-p}\left(\sqrt{2}\,x\right)H_p\left(\sqrt{2}\,y\right)\frac{1}{\left(\sqrt{2}\right)^n}$$

or
$$H_n(x+y)=2^{-\frac{n}{2}}\sum_{p=0}^{n}\frac{n!}{p!(n-p)!}H_{n-p}\left(\sqrt{2}\,x\right)H_p\left(\sqrt{2}\,y\right).$$

[**Note:** For $p>n$, the denominator becomes infinite.]

Deduction: In the above result, if we put $x=y=\frac{t}{\sqrt{2}}$, we obtain

$$H_n\left(\sqrt{2}\,t\right)=2^{-\frac{n}{2}}\sum_{p=0}^{n}\frac{n!}{p!(n-p)!}H_{n-p}(t)H_p(t).$$

PROBLEM 7 Prove that

(i) $$H_{n_1}(x)H_{n_2}(x)=\sum_{p=0}^{n_<}2^p p!\binom{n_1}{p}\binom{n_2}{p}H_{n_1+n_2-2p}(x),$$

where $n_<$ denotes the smaller of n_1 and n_2 and $\binom{n}{p}=\frac{n!}{p!(n-p)!}$,

(ii) $$\left[H_n(x)\right]^2=\sum_{k=0}^{n}\frac{2^{n-k}(n!)^2}{(n-k)!(k!)^2}H_{2k}(x).$$

Solution:

(i) Using the generating function of Hermite polynomials, we have

$$\sum_{n_1=0}^{\infty}H_{n_1}(x)\frac{h_1^{n_1}}{n_1!}\sum_{n_2=0}^{\infty}H_{n_2}(x)\frac{h_2^{n_2}}{n_2!}=e^{2xh_1-h_1^2}e^{2xh_2-h_2^2}$$

$$=e^{2x(h_1+h_2)-(h_1+h_2)^2}e^{2h_1h_2}$$

or
$$\sum_{n_1=0}^{\infty}\sum_{n_2=0}^{\infty}H_{n_1}(x)H_{n_2}(x)\frac{h_1^{n_1}h_2^{n_2}}{n_1!\,n_2!}=\sum_{N=0}^{\infty}H_N(x)\frac{(h_1+h_2)^N}{N!}\sum_{p=0}^{\infty}\frac{2^p h_1^p h_2^p}{p!}$$

$$=\sum_{N=0}^{\infty}\sum_{p=0}^{\infty}\frac{2^p h_1^p h_2^p}{N!\,p!}H_N(x)\sum_{r=0}^{N}\frac{N!}{r!(N-r)!}h_1^{N-r}h_2^r$$

$$=\sum_{N=0}^{\infty}\sum_{p=0}^{\infty}\sum_{r=0}^{N}H_N(x)\frac{2^p\,h_1^{N+p-r}\,h_2^{p+r}}{p!\,r!(N-r)!}. \quad \text{(i)}$$

When $N+p-r=n_1$ and $p+r=n_2$ we have, by addition,

$$N+2p=n_1+n_2, \text{ and then } N=n_1+n_2-2p.$$

Also, we have, $r=n_2-p, N-r=n_1-p$.

Now, equating coefficients of $h_1^{n_1} h_2^{n_2}$ from both sides of Eq. (i), we obtain,

$$\frac{1}{n_1!\,n_2!}H_{n_1}(x)H_{n_2}(x)=\sum_{p=0}^{n_<}H_{n_1+n_2-2p}(x)\frac{2^p}{p!(n_2-p)!(n_1-p)!}$$

where $n_<$ denotes the smaller of n_1 and n_2.
Thus, we have

$$H_{n_1}(x)H_{n_2}(x)=\sum_{p=0}^{n_<}2^p p!\frac{n_1!}{p!(n_1-p)!}\frac{n_2!}{p!(n_2-p)!}H_{n_1+n_2-2p}(x)$$

$$=\sum_{p=0}^{n_<}2^p p!\binom{n_1}{p}\binom{n_2}{p}H_{n_1+n_2-2p}(x).$$

(ii) In the above result, if we put $n_1=n_2=n$ and $(n-p)=k$, we obtain

$$\left[H_n(x)\right]^2=\sum_{k=0}^{n}\frac{2^{n-k}(n!)^2}{(n-k)!(k!)^2}H_{2k}(x).$$

PROBLEM 8 Prove that for $k<n$,

$$\frac{d^k}{dx^k}H_n(x)=\frac{2^k\, n!}{(n-k)!}H_{n-k}(x).$$

Solution: We know that

$$\sum_{n=0}^{\infty}H_n(x)\frac{h^n}{n!}=e^{2xh-h^2}.$$

Differentiating both sides k times with respect to x, we have

$$\sum_{n=0}^{\infty}\frac{h^n}{n!}\frac{d^k}{dx^k}H_n(x)=e^{2xh-h^2}(2h)^k=2^k h^k\sum_{n=0}^{\infty}H_n(x)\frac{h^n}{n!}$$

$$=2^k\sum_{n=0}^{\infty}H_n(x)\frac{h^{n+k}}{n!}.$$

Equating coefficients of h^n from both sides of the above equation, we obtain

$$\frac{1}{n!}\frac{d^k}{dx^k}H_n(x)=\frac{2^k}{(n-k)!}H_{n-k}(x).$$

Hence, we have,

$$\frac{d^k}{dx^k}H_n(x)=\frac{2^k\, n!}{(n-k)!}H_{n-k}(x).$$

PROBLEM 9 Show that the function $\Phi(x)=e^{-\frac{x^2}{2}}H_n(x)$ satisfies the differential equation

$$\frac{d^2\Phi}{dx^2}+\left(2n+1-x^2\right)\Phi=0.$$

Solution: We have, $\Phi(x)=e^{-\frac{x^2}{2}}H_n(x)$

and $$\frac{d^2\Phi}{dx^2}=e^{-\frac{x^2}{2}}\left[\frac{d^2}{dx^2}H_n(x)-2x\frac{d}{dx}H_n(x)+\left(x^2-1\right)H_n(x)\right]$$

$$=e^{-\frac{x^2}{2}}\left[\left\{\frac{d^2}{dx^2}H_n(x)-2x\frac{d}{dx}H_n(x)+2n\,H_n(x)\right\}+\left(x^2-1-2n\right)H_n(x)\right]$$

$$=e^{-\frac{x^2}{2}}\left[0+\left(x^2-1-2n\right)\right]H_n(x)$$

$$=\left(x^2-1-2n\right)\Phi(x).$$

Therefore, we have

$$\frac{d^2\Phi}{dx^2}+\left(2n+1-x^2\right)\Phi=0.$$

PROBLEM 10 Prove that $\int\limits_{-\infty}^{+\infty}e^{-x^2}H_{2n}(\alpha x)\,dx=\frac{(2n)!}{n!}\sqrt{\pi}\left(\alpha^2-1\right)^n,\ \alpha\neq 1.$

Solution: We have

$$\sum_{n=0}^{\infty}H_n(\alpha x)\frac{h^n}{n!}=e^{2\alpha xh-h^2}.$$

Multiplying both sides by e^{-x^2} and then integrating both sides with respect to x within the limits from $x=-\infty$ to $x=+\infty$, we have

$$\sum_{n=0}^{\infty}\frac{h^n}{n!}\int\limits_{-\infty}^{+\infty}e^{-x^2}H_n(\alpha x)\,dx=e^{-h^2}\int\limits_{-\infty}^{+\infty}e^{-x^2+2\alpha xh}\,dx$$

$$=e^{-h^2}\sqrt{\pi}\,e^{\alpha^2h^2}$$ where we use the formula

$$\int_{-\infty}^{+\infty} e^{-ax^2+bx} dx = \sqrt{\frac{\pi}{a}}\, e^{\frac{b^2}{4a}}.$$

Now, we write

$$\sum_{n=0}^{\infty} \frac{h^n}{n!} \int_{-\infty}^{+\infty} e^{-x^2} H_n(\alpha x) dx = \sqrt{\pi}\, e^{(\alpha^2-1)h^2} = \sqrt{\pi} \sum_{n=0}^{\infty} \frac{(\alpha^2-1)^n h^{2n}}{n!}.$$

Equating coefficients of h^{2n} from both sides of the above equation, we have

$$\frac{1}{(2n)!} \int_{-\infty}^{+\infty} e^{-x^2} H_{2n}(\alpha x) dx = \sqrt{\pi}\, \frac{(\alpha^2-1)^n}{n!}$$

or

$$\int_{-\infty}^{+\infty} e^{-x^2} H_{2n}(\alpha x) dx = \frac{(2n)!}{n!} \sqrt{\pi}\, (\alpha^2-1)^n.$$

PROBLEM 11 Show that

(i) $\displaystyle\int_{-\infty}^{+\infty} e^{-\frac{x^2}{2}} H_{2n}(x) dx = \frac{(2n)!}{n!} \sqrt{2\pi},$

(ii) $\displaystyle\int_{-\infty}^{+\infty} e^{-\frac{x^2}{2}} H_{2n+1}(x) dx = 0.$

Solution: We know that

$$\sum_{n=0}^{\infty} H_n(x) \frac{h^n}{n!} = e^{2xh-h^2}.$$

Multiplying both sides by $e^{-\frac{x^2}{2}}$ and the integrating with respect to x in the limit from $x=-\infty$ to $x=+\infty$, we obtain

$$\begin{aligned}\sum_{n=0}^{\infty} \frac{h^n}{n!} \int_{-\infty}^{+\infty} e^{-\frac{x^2}{2}} H_n(x) dx &= e^{-h^2} \int_{-\infty}^{+\infty} e^{-\frac{x^2}{2}+2hx} dx \\ &= e^{-h^2} \sqrt{2\pi}\, e^{2h^2} \\ &= \sqrt{2\pi}\, e^{h^2} = \sqrt{2\pi} \sum_{n=0}^{\infty} \frac{h^{2n}}{n!}, \qquad \text{(i)}\end{aligned}$$

where we use the formula $\displaystyle\int_{-\infty}^{+\infty} e^{-ax^2+bx} dx = \sqrt{\frac{\pi}{a}}\, e^{\frac{b^2}{4a}}.$

(i) Equating coefficients of h^{2n} from both sides of Eq. (i), we obtain

$$\frac{1}{(2n)!}\int_{-\infty}^{+\infty} e^{-\frac{x^2}{2}} H_{2n}(x)\,dx = \frac{\sqrt{2\pi}}{n!}$$

or
$$\int_{-\infty}^{+\infty} e^{-\frac{x^2}{2}} H_{2n}(x)\,dx = \frac{(2n)!}{n!}\sqrt{2\pi}\,.$$

(ii) Equating coefficients of h^{2n+1} from both sides of Eq. (i), we obtain

$$\int_{-\infty}^{+\infty} e^{-\frac{x^2}{2}} H_{2n+1}(x)\,dx = 0\,.$$

PROBLEM 12 Evaluate $\int_{-\infty}^{+\infty} x^2 e^{-x^2} H_n(x) H_m(x)\,dx$

Solution: From the recurrence relation,

$$H_{n+1}(x) - 2xH_n(x) + 2nH_{n-1}(x) = 0\,,$$

we have,

$$xH_n(x) = nH_{n-1}(x) + \frac{1}{2}H_{n+1}(x)\,. \qquad \text{(i)}$$

Replacing n by m, we have

$$xH_m(x) = mH_{m-1}(x) + \frac{1}{2}H_{m+1}(x)\,. \qquad \text{(ii)}$$

The product of Eqs. (i) and (ii) is,

$$x^2 H_m(x) H_n(x) = mnH_{m-1}(x)H_{n-1}(x) + \frac{1}{4}H_{m+1}(x)H_{n+1}(x)$$
$$+\frac{m}{2}H_{m-1}(x)H_{n+1}(x) + \frac{n}{2}H_{m+1}(x)H_{n-1}(x)\,.$$

Multiplying both sides by e^{-x^2} and then integrating with respect to x within the limit from $x=-\infty$ to $x=+\infty$, we obtain

$$\int_{-\infty}^{+\infty} x^2 e^{-x^2} H_m(x)H_n(x)\,dx = mn\int_{-\infty}^{+\infty} e^{-x^2} H_{m-1}(x)H_{n-1}(x)\,dx + \frac{1}{4}\int_{-\infty}^{+\infty} e^{-x^2} H_{m+1}(x)H_{n+1}(x)\,dx$$

$$+\frac{m}{2}\int_{-\infty}^{+\infty} e^{-x^2} H_{m-1}(x)H_{n+1}(x)\,dx + \frac{n}{2}\int_{-\infty}^{+\infty} e^{-x^2} H_{m+1}(x)H_{n-1}(x)\,dx\,. \qquad \text{(i)}$$

CASE I: When $m = n$:

In this case, Eq. (i) yields,

$$\int_{-\infty}^{+\infty} x^2 e^{-x^2} H_m(x) H_n(x) dx = n^2 2^{n-1}(n-1)!\sqrt{\pi} + \frac{1}{4} 2^{n+1}(n+1)!\sqrt{\pi} + 0$$

$$= 2^n n! \left(n + \frac{1}{2}\right)\sqrt{\pi}.$$

CASE II: When $n - 1 = m + 1$ or, $m = n - 2$:

In this case, Eq. (i) yields,

$$\int_{-\infty}^{+\infty} x^2 e^{-x^2} H_m(x) H_n(x) dx = 0 + \frac{n}{2} \cdot 2^{n-1}(n-1)!\sqrt{\pi} + 0 = 2^{n-2} n!\sqrt{\pi}.$$

CASE III: When $m - 1 = n + 1$, or, $m = n + 2$:

In this case, Eq. (i) yields,

$$\int_{-\infty}^{+\infty} x^2 e^{-x^2} H_n(x) H_m(x) dx = 0 + \frac{(n+2)}{2} 2^{n+1}(n+1)!\sqrt{\pi} = 2^n (n+2)!\sqrt{\pi}.$$

In all other cases, the integral in the right-hand side of Eq. (i) vanishes. We summarize the results as,

$$\int_{-\infty}^{+\infty} x^2 e^{-x^2} H_n(x) H_m(x) dx = \begin{cases} 2^n n!\left(n + \frac{1}{2}\right)\sqrt{\pi} & \text{when } m = n, \\ 2^{n-2} n!\sqrt{\pi}, & \text{when } m = n-2, \\ 2^n (n+2)!\sqrt{\pi}, & \text{when } m = n+2, \\ 0 & \text{for all other cases.} \end{cases}$$

PROBLEM 13 Prove that

$$\int_{-\infty}^{+\infty} e^{-x^2} x H_n(x)\left[H_{n+1}(x) - H_{n-1}(x)\right] dx = 2^n n!\left(n + \frac{1}{2}\right)\sqrt{\pi}.$$

Solution: From the recurrence relation

$$H_{n+1}(x) - 2xH_n(x) + 2nH_{n-1}(x) = 0,$$

we have

$$xH_n(x) = \frac{1}{2} H_{n+1}(x) + nH_{n-1}(x).$$

Substituting this equation into the given integral, we obtain

$$\int_{-\infty}^{+\infty} e^{-x^2} xH_n(x)\left[H_{n+1}(x)-H_{n-1}(x)\right]dx$$

$$= \int_{-\infty}^{+\infty} e^{-x^2}\left[\frac{1}{2}H_{n+1}(x)+nH_{n-1}(x)\right]\left[H_{n+1}(x)-H_{n-1}(x)\right]dx$$

$$= \frac{1}{2}\int_{-\infty}^{+\infty} e^{-x^2}\left[H_{n+1}(x)\right]^2 dx - n\int_{-\infty}^{+\infty} e^{-x^2}\left[H_{n-1}(x)\right]^2 dx$$

$$-\frac{1}{2}\int_{-\infty}^{+\infty} e^{-x^2} H_{n+1}(x)H_{n-1}(x)dx + n\int_{-\infty}^{+\infty} e^{-x^2} H_{n-1}(x)H_{n+1}(x)dx$$

$$= \frac{1}{2}\cdot 2^{n+1}(n+1)!\sqrt{\pi} - n2^{n-1}(n-1)!\sqrt{\pi} - 0 - 0$$

$$= 2^n n!\left(n+\frac{1}{2}\right)\sqrt{\pi}.$$

PROBLEM 14 Prove that

$$\int_{-\infty}^{+\infty} xe^{-x^2} H_n(x)H_m(x)dx = 2^{n-1}n!\sqrt{\pi}\,\delta_{m,\,n-1} + 2^n(n+1)!\sqrt{\pi}\,\delta_{m,\,n+1}.$$

Solution: From the recurrence relation,

$$H_{n+1}(x) - 2xH_n(x) + 2nH_{n-1}(x) = 0,$$

we have

$$xH_n(x) = nH_{n-1}(x) + \frac{1}{2}H_{n+1}(x).$$

Substituting this relation into the given integral, we have

$$\int_{-\infty}^{+\infty} e^{-x^2}\left[nH_{n-1}(x)+\frac{1}{2}H_{n+1}(x)\right]H_m(x)dx$$

$$= n\int_{-\infty}^{+\infty} e^{-x^2} H_m(x)H_{n-1}(x)dx + \frac{1}{2}\int_{-\infty}^{+\infty} e^{-x^2} H_m(x)H_{n+1}(x)dx$$

$$= n2^{n-1}(n-1)!\sqrt{\pi}\,\delta_{m,n-1} + \frac{1}{2}\cdot 2^{n+1}(n+1)!\sqrt{\pi}\,\delta_{m,n+1}$$

$$= 2^{n-1}n!\sqrt{\pi}\,\delta_{m,n-1} + 2^n(n+1)!\sqrt{\pi}\,\delta_{m,n+1}.$$

PROBLEM 15 Prove that

$$e^{-b^2x^2} = \sum_{n=0}^{\infty} \frac{(-1)^n b^{2n}}{2^{2n} n! \left(1+b^2\right)^{n+\frac{1}{2}}} H_{2n}(x).$$

Solution: Let us write,

$$e^{-b^2x^2} = \sum_{n=0}^{\infty} c_n H_{2n}(x) \tag{i}$$

where c_n are to be determined from the orthogonality condition.

Multiplying both sides of Eq. (i) by $e^{-x^2} H_k(x)$ and then integrating both sides with respect to x within the limit from $x = -\infty$ to $x = +\infty$, we have

$$\int_{-\infty}^{+\infty} e^{-\left(1+b^2\right)x^2} H_k(x)dx = \sum_{n=0}^{\infty} c_n \int_{-\infty}^{+\infty} e^{-x^2} H_k(x) H_{2n}(x) dx$$

$$= \sum_{n=0}^{\infty} c_n 2^k k! \sqrt{\pi}\, \delta_{k,2n} = c_{k/2} 2^k k! \sqrt{\pi}\,.$$

With $k = 2n$, we write

$$\int_{-\infty}^{+\infty} e^{-\left(1+b^2\right)x^2} H_{2n}(x) dx = c_n 2^{2n} (2n)! \sqrt{\pi}$$

or
$$c_n = \frac{1}{2^{2n}(2n)!\sqrt{\pi}} \int_{-\infty}^{+\infty} e^{-(1+b^2)x^2} H_{2n}(x) dx\,. \tag{ii}$$

To evaluate the integral in Eq. (ii), let us put

$$\sqrt{1+b^2}\, x = u, \text{ then } x = \frac{u}{\sqrt{1+b^2}} \text{ and } dx = \frac{du}{\sqrt{1+b^2}}\,.$$

Then the integral becomes

$$\int_{-\infty}^{+\infty} e^{-\left(1+b^2\right)x^2} H_{2n}(x) dx = \frac{1}{\sqrt{1+b^2}} \int_{-\infty}^{+\infty} e^{-u^2} H_{2n}(\alpha u) du \tag{iii}$$

where we put $\alpha = \dfrac{1}{\sqrt{1+b^2}}$.

Recalling the result of Problem 10, viz.

$$\int_{-\infty}^{+\infty} e^{-x^2} H_{2n}(\alpha x)\,dx = \frac{(2n)!}{n!}\sqrt{\pi}\,(\alpha^2-1)^n, \text{ we write Eq. (iii) as}$$

$$\int_{-\infty}^{+\infty} e^{-(1+b^2)x^2} H_{2n}(x)\,dx = \frac{\sqrt{\pi}}{\sqrt{1+b^2}}\frac{(2n)!}{n!}\left[\frac{1}{1+b^2}-1\right]^n$$

$$= \sqrt{\pi}\,\frac{(2n)!}{n!}\frac{(-1)^n b^{2n}}{(1+b^2)^{n+\frac{1}{2}}}. \qquad \text{(iv)}$$

Substitution of Eq. (iv) into Eq. (ii) yields

$$c_n = \frac{(-1)^n b^{2n}}{2^{2n} n!(1+b^2)^{n+\frac{1}{2}}}.$$

Substituting this value of c_n into Eq. (i), we obtain

$$e^{-b^2x^2} = \sum_{n=0}^{\infty} \frac{(-1)^n b^{2n}}{2^{2n} n!(1+b^2)^{n+\frac{1}{2}}} H_{2n}(x).$$

PROBLEM 16 Evaluate $\int_{-\infty}^{+\infty} e^{-x^2} f(x) H_n(x)\,dx$

where $f(x)$ is a polynomial of degree p having unit positive coefficients such that $0 < p \le n$.

Solution: In this problem, we have $f(x) = 1 + x + x^2 + x^3 + \ldots + x^p$

Using Rodrigues' formula, we write

$$\int_{-\infty}^{+\infty} e^{-x^2} f(x) H_n(x)\,dx = \int_{-\infty}^{+\infty} e^{-x^2} f(x)(-1)^n e^{x^2} \frac{d^n}{dx^n}(e^{-x^2})\,dx$$

$$= (-1)^n \int_{-\infty}^{+\infty} f(x) \frac{d^n}{dx^n}(e^{-x^2})\,dx.$$

Performing integration by parts n times in the integral on the right-hand side, we get

$$\int_{-\infty}^{+\infty} e^{-x^2} f(x) H_n(x)\,dx = (-1)^n \left\{0 + (-1)^n \int_{-\infty}^{+\infty} \frac{d^n f}{dx^n} e^{-x^2}\,dx\right\}.$$

CASE I: $p<n$:

If the degree (p) of the polynomial function $f(x)$ is less than n, we have

$$\frac{d^n f(x)}{dx^n}=0, \text{ and hence } \int_{-\infty}^{+\infty} e^{-x^2} f(x)H_n(x)dx=0, \text{ for } p<n.$$

CASE II: $p=n$:

If the degree (p) of the polynomial function $f(x)$ is equal to n, we have

$$\frac{d^n f(x)}{dx^n}=n!.$$

Hence, we have

$$\int_{-\infty}^{+\infty} e^{-x^2} f(x)H_n(x)dx=(-1)^{2n}\, n!\int_{-\infty}^{+\infty} e^{-x^2} dx=n!\sqrt{\pi}.$$

The results of cases I and II can be expressed as

$$\int_{-\infty}^{+\infty} e^{-x^2} f(x)H_n(x)dx=\begin{cases} 0 \text{ if } p=0,1,2,3\ldots(n-1); \\ n!\sqrt{\pi} \text{ if } p=n. \end{cases}$$

PROBLEM 17 Show that

$$\int_{-\infty}^{+\infty} x^r e^{-x^2} H_n(x)H_{n+s}(x)dx=\begin{cases} 0, & \text{if } r<s; \\ 2^n(n+s)!\sqrt{\pi}, & \text{if } r=s; \end{cases}$$

where n, r and s are non-negative integers.

Solution: From Problem 7, we know that,

$$H_{n_1}(x)H_{n_2}(x)=\sum_{p=0}^{n_<} 2^p\, p!\,\frac{n_1!}{p!(n_1-p)!}\frac{n_2!}{p!(n_2-p)!}H_{n_1+n_2-2p}(x)$$

where $n_<$ denotes the smaller of n_1 and n_2.

Replacing n_1 by n and n_2 by $(n+s)$, we have

$$H_n(x)H_{n+s}(x)=\sum_{p=0}^{n} 2^p\, p!\frac{n!}{p!(n-p)!}\frac{(n+s)!}{p!(n+s-p)!}H_{2n+s-2p}(x).$$

Substituting this relation into the given integral, we have

$$\int_{-\infty}^{+\infty} x^r e^{-x^2} H_n(x) H_{n+s}(x)\,dx$$

$$= \int_{-\infty}^{+\infty} x^r e^{-x^2} \sum_{p=0}^{n} 2^p p! \frac{n!}{p!(n-p)!} \frac{(n+s)!}{p!(n+s-p)!} H_{2n+s-2p}(x)\,dx$$

$$= \sum_{p=0}^{n} 2^p p! \frac{n!}{p!(n-p)!} \frac{(n+s)!}{p!(2n+s-2p)!} \int_{-\infty}^{+\infty} x^r e^{-x^2} H_{2n+s-2p}(x)\,dx. \quad \text{(i)}$$

CASE I: $r < s$

If $r < s$, then $r < 2n+s-2p$ for $p = 0, 1, 2, 3 \ldots n$.

In this case, by the result of Problem 16, the integral on the right-hand side of Eq. (i) becomes zero and, hence, we write

$$\int_{-\infty}^{+\infty} x^r e^{-x^2} H_n(x) H_{n+s}(x)\,dx = 0 \text{ for } r < s.$$

CASE II: $r = s$

In this case, by the result of Problem 16, the integral on the right-hand side of Eq. (i) does not vanish when

$$r = 2n+s-2p$$

that is, $s = 2n+s-2p$, or, $p = n$ then, Eq. (i) become,

$$\int_{-\infty}^{+\infty} x^r e^{-x^2} H_n(x) H_{n+s}(x)\,dx = 2^n n! \frac{n!}{n!\,0!} \frac{(n+s)!}{n!\,s!} \int_{-\infty}^{+\infty} x^s e^{-x^2} H_s(x)\,dx$$

$$= 2^n \frac{(n+s)!}{s!} s! \sqrt{\pi} \quad \text{[Using Problem 16]}$$

$$= 2^n (n+s)! \sqrt{\pi}.$$

The results in Case I and Case II can be generally expressed as

$$\int_{-\infty}^{+\infty} x^r e^{-x^2} H_n(x) H_{n+s}(x)\,dx = \begin{cases} 0, & \text{if } r < s; \\ 2^n (n+s)! \sqrt{\pi} & \text{if } r = s. \end{cases}$$

PROBLEM 18 Prove that $P_n(x) = \dfrac{2}{\sqrt{\pi}\, n!} \displaystyle\int_0^{\infty} t^n e^{-t^2} H_n(xt)\,dt$.

Solution: We have

$$\frac{2}{\sqrt{\pi}\,n!}\int_0^\infty t^n e^{-t^2} H_n(xt)\,dt = \frac{2}{\sqrt{\pi}\,n!}\int_0^\infty t^n e^{-t^2} \sum_{r=0}^{\left[\frac{n}{2}\right]} \frac{(-1)^r\, n!}{r!(n-2r)!}(2xt)^{n-2r}\,dt$$

$$= \sum_{r=0}^{\left[\frac{n}{2}\right]} \frac{(-1)^r\, 2^{n-2r} x^{n-2r}}{\sqrt{\pi}\, r!(n-2r)!}\left[2\int_0^\infty t^{2\left(n-r+\frac{1}{2}\right)-1} e^{-t^2}\,dt\right]$$

$$= \sum_{r=0}^{\left[\frac{n}{2}\right]} \frac{(-1)^r\, 2^{n-2r} x^{n-2r}}{\sqrt{\pi}\, r!(n-2r)!}\Gamma\left(n-r+\frac{1}{2}\right) \qquad \text{(i)}$$

where $\Gamma\left(n-r+\frac{1}{2}\right)$ is a gamma function.

The duplication formula for gamma functions gives

$$\Gamma\left(m+\frac{1}{2}\right) = \frac{\sqrt{\pi}\,(2m)!}{2^{2m}\; m!}.$$

Therefore, we have

$$\Gamma\left(n-r+\frac{1}{2}\right) = \frac{\sqrt{\pi}\,(2n-2r)!}{2^{2n-2r}\,(n-r)!}.$$

Substituting this value of gamma function into Eq. (i), we obtain

$$\frac{2}{\sqrt{\pi}\,n!}\int_0^\infty t^n e^{-t^2} H_n(xt)\,dt = \sum_{r=0}^{\left[\frac{n}{2}\right]} \frac{(-1)^r\,(2n-2r)!}{2^n\, r!(n-r)!}\,\frac{x^{n-2r}}{(n-2r)!} = P_n(x).$$

PROBLEM 19 Show that $\frac{d^n}{dx^n}H_n(x) = 2^n n!\, H_0(x)$.

Solution: We know that

$$\sum_{n=0}^\infty H_n(x)\frac{h^n}{n!} = e^{2xh-h^2}.$$

Differentiating both sides n times with respect to x, we have

$$\sum_{n=0}^\infty \frac{d^n}{dx^n}H_n(x)\frac{h^n}{n!} = (2h)^n e^{2xh-h^2} = 2^n h^n \sum_{k=0}^\infty H_k(x)\frac{h^k}{k!}$$

$$= \sum_{k=0}^\infty 2^n H_k(x)\frac{h^{n+k}}{k!}.$$

Equating coefficients of h^n from both sides, we have

$$\frac{1}{n!}\frac{d^n}{dx^n}H_n(x) = 2^n H_0(x)$$

or

$$\frac{d^n}{dx^n}H_n(x) = 2^n n!\, H_0(x).$$

PROBLEM 20 Express the polynomial $x^4 + 2x^3 + 2x^2 - x - 3$ in a series of Hermite polynomials.

Solution: We have

$$H_0(x) = 1,\ H_1(x) = 2x, \qquad H_2(x) = 4x^2 - 2,$$

$$H_3(x) = 8x^3 - 12x, \qquad H_4(x) = 16x^4 - 48x^2 + 12.$$

Therefore, the given polynomial

$$= \frac{1}{16}H_4(x) + 3x^2 - \frac{3}{4} + 2\left\{\frac{1}{8}H_3(x) + \frac{3}{2}x\right\} + 2x^2 - x - 3$$

$$= \frac{1}{16}H_4(x) + \frac{1}{4}H_3(x) + 5x^2 + 2x - \frac{15}{4}$$

$$= \frac{1}{16}H_4(x) + \frac{1}{4}H_3(x) + 5\left\{\frac{1}{4}H_2(x) + \frac{1}{2}\right\} + H_1(x) - \frac{15}{4}$$

$$= \frac{1}{16}H_4(x) + \frac{1}{4}H_3(x) + \frac{5}{4}H_2(x) + H_1(x) - \frac{5}{4}H_0(x).$$

PROBLEM 21 Express x^6 in a series of Hermite polynomials.

Solution: In the formula,

$$x^{2m} = \frac{(2m)!}{2^{2m}}\sum_{p=0}^{m}\frac{H_{2p}(x)}{(2p)!\,(m-p)!}.$$

If we put $m = 3$, we have

$$x^6 = \frac{6!}{2^6}\sum_{p=0}^{3}\frac{H_{2p}(x)}{(2p)!\,(3-p)!}$$

$$= \frac{6!}{2^6}\left[\frac{H_0(x)}{0!\,3!} + \frac{H_2(x)}{2!\,2!} + \frac{H_4(x)}{4!\,1!} + \frac{H_6(x)}{6!\,0!}\right]$$

$$= \frac{15}{8}H_0(x) + \frac{45}{16}H_2(x) + \frac{15}{32}H_4(x) + \frac{1}{64}H_6(x).$$

PROBLEM 22 Express x^5 in a series of Hermite polynomials.

Solution: In the formula,

$$x^{2m+1} = \frac{(2m+1)!}{2^{2m+1}} \sum_{p=0}^{m} \frac{H_{2p+1}(x)}{(2p+1)!(m-p)!}.$$

If we put $m = 2$, we have

$$x^5 = \frac{5!}{2^5} \sum_{p=0}^{2} \frac{H_{2p+1}(x)}{(2p+1)!(2-p)!}$$

$$= \frac{5!}{2^5}\left[\frac{H_1(x)}{1!2!} + \frac{H_3(x)}{3!1!} + \frac{H_5(x)}{5!0!}\right] = \frac{15}{8}H_1(x) + \frac{5}{8}H_3(x) + \frac{1}{32}H_5(x).$$

PROBLEM 23 Prove that

$$\int_{-\infty}^{+\infty} H_n\left(x + \frac{1}{2}x_0\right) e^{-x^2}\,dx = \sqrt{\pi}\, x_0^n.$$

Solution: We have

$$\sum_{n=0}^{\infty} H_n(u)\frac{h^n}{n!} = e^{2uh-h^2}.$$

Multiplying both sides by $e^{-\left(u-\frac{1}{2}x_0\right)^2}$ and then integrating with respect to u within the limit from $u = -\infty$ to $u = +\infty$, we have

$$\sum_{n=0}^{\infty} \frac{h^n}{n!} \int_{-\infty}^{+\infty} H_n(u) e^{-\left(u-\frac{1}{2}x_0\right)^2} du = \int_{-\infty}^{+\infty} e^{2uh-h^2-\left(u-\frac{1}{2}x_0\right)^2} du$$

$$= e^{-h^2} \int_{-\infty}^{+\infty} e^{-\left(u-\frac{1}{2}x_0\right)^2 + 2uh} du.$$

Putting $x = u - \frac{1}{2}x_0$, we have, $u = x + \frac{1}{2}x_0$ and $du = dx$.

Now, we write

$$\sum_{n=0}^{\infty} \frac{h^n}{n!} \int_{-\infty}^{+\infty} H_n\left(x + \frac{1}{2}x_0\right) e^{-x^2} dx = e^{-h^2} \int_{-\infty}^{+\infty} e^{-x^2+2h\left(x+\frac{1}{2}x_0\right)} dx$$

$$= e^{hx_0-h^2} \int_{-\infty}^{+\infty} e^{-x^2+2hx} dx$$

$$= e^{hx_0-h^2}\sqrt{\pi}\, e^{\frac{4h^2}{4}} = \sqrt{\pi}\, e^{hx_0}$$

$$= \sqrt{\pi}\sum_{n=0}^{\infty}\frac{h^n x_0^n}{n!}.$$

Equating coefficients of h^n from both sides, we obtain

$$\int_{-\infty}^{+\infty} H_n\left(x+\frac{1}{2}x_0\right)e^{-x^2}\,dx = \sqrt{\pi}\, x_0^n .$$

PROBLEM 24 Show that

$$H_n(x) = \frac{(-i)^n}{\sqrt{2\pi}} e^{\frac{x^2}{2}} \int_{-\infty}^{+\infty} e^{-\frac{1}{2}k^2+ikx} H_n(k)\,dk .$$

Hence, deduce

(a) $H_{2n}(x) = (-1)^n \sqrt{\frac{2}{\pi}}\, e^{\frac{x^2}{2}} \int_0^{\infty} e^{-\frac{1}{2}k^2} H_{2n}(k)\cos kx\, dk$,

(b) $H_{2n+1}(x) = (-1)^n e^{\frac{x^2}{2}} \sqrt{\frac{2}{\pi}} \int_0^{\infty} e^{-\frac{1}{2}k^2} H_{2n+1}(x)\sin kx\, dk$.

Solution: We know that,

$$\sum_{n=0}^{\infty} H_n(k)\frac{t^n}{n!} = e^{2kt-t^2}.$$

Multiplying both sides by $e^{ikx-\frac{1}{2}k^2}$ and integrating with respect to k in the interval from $k=-\infty$ to $k=\infty$, we have

$$\sum_{n=0}^{\infty}\frac{t^n}{n!}\int_{-\infty}^{+\infty} e^{-\frac{1}{2}k^2+ikx} H_n(k)\,dk = e^{-t^2}\int_{-\infty}^{+\infty} e^{-\frac{1}{2}k^2+(2t+ix)k}\,dk$$

$$= e^{-t^2}\sqrt{2\pi}\, e^{\frac{(2t+ix)^2}{2}}$$

$$= \sqrt{2\pi}\, e^{-\frac{x^2}{2}}\, e^{t^2+2ixt}$$

$$= \sqrt{2\pi}\, e^{-\frac{x^2}{2}}\, e^{2x(it)-(it)^2}$$

$$= \sqrt{2\pi}\, e^{-\frac{x^2}{2}} \sum_{n=0}^{\infty} H_n(x) \frac{(it)^n}{n!}.$$

Equating coefficients of $\frac{t^n}{n!}$ from both sides, we obtain

$$\int_{-\infty}^{+\infty} e^{-\frac{1}{2}k^2 + ikx} H_n(k)\, dk = \sqrt{2\pi}\, i^n e^{-\frac{x^2}{2}} H_n(x)$$

or
$$H_n(x) = \frac{(-i)^n}{\sqrt{2\pi}} e^{\frac{x^2}{2}} \int_{-\infty}^{+\infty} e^{-\frac{1}{2}k^2 + ikx} H_n(k)\, dk. \qquad \text{(i)}$$

Deductions:

(a) In Eq. (i), replacing n by $2n$, we write

$$H_{2n}(x) = \frac{(-i)^{2n}}{\sqrt{2\pi}} e^{\frac{x^2}{2}} \int_{-\infty}^{+\infty} e^{-\frac{1}{2}k^2} H_{2n}(k)\, e^{ikx} dk$$

$$= \frac{(-1)^n}{\sqrt{2\pi}} e^{\frac{x^2}{2}} \int_{-\infty}^{+\infty} e^{-\frac{1}{2}k^2} H_{2n}(k) [\cos kx + i \sin kx]\, dk$$

Equating real parts from both sides, we have

$$H_{2n}(x) = \frac{(-1)^n}{\sqrt{2\pi}} e^{\frac{x^2}{2}} \int_{-\infty}^{+\infty} e^{-\frac{1}{2}k^2} H_{2n}(k) \cos kx\, dk$$

$$= (-1)^n \sqrt{\frac{2}{\pi}}\, e^{\frac{x^2}{2}} \int_{0}^{\infty} e^{-\frac{1}{2}k^2} H_{2n}(k) \cos kx\, dk$$

as the integrand is even function of k.

(b) In Eq. (i), replacing n by $2n+1$, we obtain

$$H_{2n+1}(x) = \frac{(-i)^{2n+1}}{\sqrt{2\pi}} e^{\frac{x^2}{2}} \int_{-\infty}^{+\infty} e^{-\frac{1}{2}k^2} H_{2n+1}(k) [\cos kx + i \sin kx]\, dk$$

$$= \frac{(-1)^n}{\sqrt{2\pi}} e^{\frac{x^2}{2}} \int_{-\infty}^{+\infty} e^{-\frac{1}{2}k^2} H_{2n+1}(k) [-i \cos kx + \sin kx]\, dk.$$

Equating real parts from both sides, we have

$$H_{2n+1}(x)=\frac{(-1)^n}{\sqrt{2\pi}}e^{\frac{x^2}{2}}\int_{-\infty}^{+\infty}e^{-\frac{1}{2}k^2}H_{2n+1}(k)\sin kx\,dk$$

$$=(-1)^n\sqrt{\frac{2}{\pi}}e^{\frac{x^2}{2}}\int_{0}^{\infty}e^{-\frac{1}{2}k^2}H_{2n+1}(k)\sin kx\,dk$$

as the integrand is even function of k.

PROBLEM 25 Prove that $\left(2x-\frac{d}{dx}\right)^n 1=H_n(x)$.

Solution: When $n = 0$, left-hand side $=\left(2x-\frac{d}{dx}\right)^0 1=1\times1=1=H_0(x)$.

Thus, the given relation is true for $n=0$.

When $n = 1$, left-hand side $=\left(2x-\frac{d}{dx}\right)1=2x-\frac{d}{dx}(1)=2x-0=H_1(x)$.

This shows that the given relation is true for $n=1$.
When $n=2$ left-hand side of the given equation is

$$\left(2x-\frac{d}{dx}\right)^2 1=\left(2x-\frac{d}{dx}\right)\left(2x-\frac{d}{dx}\right)1$$

$$=\left(2x-\frac{d}{dx}\right)(2x-0)=4x^2-\frac{d}{dx}(2x)=4x^2-2=H_2(x).$$

This shows that the given relation is true for $n=2$.
Now, assuming that the given relation is true for $n=k$ let us prove that it is true for $n=k+1$.
For $n=k$, the given relation becomes

$$\left(2x-\frac{d}{dx}\right)^k 1=H_k(x). \tag{i}$$

We are to prove that

$$\left(2x-\frac{d}{dx}\right)^{k+1} 1=H_{k+1}(x). \tag{ii}$$

We have

$$\left(2x-\frac{d}{dx}\right)^{k+1} 1=\left(2x-\frac{d}{dx}\right)\left[\left(2x-\frac{d}{dx}\right)^k 1\right]$$

$$= \left(2x - \frac{d}{dx}\right) H_k(x), \text{ using Eq. (i);}$$

$$= 2xH_k(x) - \frac{d}{dx} H_k(x) = 2xH_k(x) - 2kH_{k-1}(x)$$

$$= H_{k+1}(x)$$

using recurrence relations of Hermite polynomials.

Thus, the given relation is proved for $n = k + 1$ assuming that the relation is true for $n = k$.

According to the method of mathematical induction, the given relation is true for all non-negative integers, n.

PROBLEM 26 Prove that

(a) $$H_{2n}(x) = \frac{(-1)^n 2^{2n+1} e^{x^2}}{\sqrt{\pi}} \int_0^{\infty} e^{-t^2} t^{2n} \cos 2xt \, dt,$$

(b) $$H_{2n+1}(x) = \frac{(-1)^n 2^{2n+2} e^{x^2}}{\sqrt{\pi}} \int_0^{\infty} e^{-t^2} t^{2n+1} \sin 2xt \, dt.$$

Solution: Recalling the integral $\int_{-\infty}^{+\infty} e^{-au^2 + bu} du = \sqrt{\frac{\pi}{a}} e^{\frac{b^2}{4a}}$,

let us consider the integral

$$\int_{-\infty}^{+\infty} e^{-t^2 + 2ixt} dt = \sqrt{\pi} e^{-x^2}.$$

Therefore, we write

$$e^{-x^2} = \frac{1}{\sqrt{\pi}} \int_{-\infty}^{+\infty} e^{-t^2 + 2ixt} dt. \quad \text{(i)}$$

(i) Differentiating both sides of Eq. (i) $2n$ times with respect to x, we have

$$\frac{d^{2n}}{dx^{2n}}(e^{-x^2}) = \frac{1}{\sqrt{\pi}} \int_{-\infty}^{+\infty} (2it)^{2n} e^{-t^2 + 2ixt} dt$$

$$= \frac{(-1)^n 2^{2n}}{\sqrt{\pi}} \int_{-\infty}^{+\infty} t^{2n} e^{-t^2} e^{2ixt} dt$$

$$= \frac{(-1)^n 2^{2n}}{\sqrt{\pi}} \int_{-\infty}^{+\infty} e^{-t^2} t^{2n} (\cos 2xt + i \sin 2xt) dt.$$

Multiplying both sides of the equation by $(-1)^{2n} e^{x^2}$ we obtain

$$(-1)^{2n} e^{x^2} \frac{d^{2n}}{dx^{2n}}(e^{-x^2}) = \frac{(-1)^{2n} (-1)^n 2^{2n} e^{x^2}}{\sqrt{\pi}} \int_{-\infty}^{+\infty} e^{-t^2} t^{2n} (\cos 2xt + i \sin 2xt) dt$$

or
$$H_{2n}(x) = \frac{(-1)^n 2^{2n} e^{x^2}}{\sqrt{\pi}} \int_{-\infty}^{+\infty} e^{-t^2} t^{2n} (\cos 2xt + i \sin 2xt) dt,$$

where use has been made of Rodrigues' formula.

Equating real parts from both sides, we have

$$H_{2n}(x) = \frac{(-1)^n 2^{2n} e^{x^2}}{\sqrt{\pi}} \int_{-\infty}^{+\infty} e^{-t^2} t^{2n} \cos 2xt \, dt$$

$$= \frac{(-1)^n 2^{2n+1} e^{x^2}}{\sqrt{\pi}} \int_{0}^{\infty} e^{-t^2} t^{2n} \cdot \cos 2xt \, dt$$

as the integrand is even function of t.

(ii) In a similar manner, differentiating both sides of Eq. (i) $2n + 1$ times with respect to x, we have

$$\frac{d^{2n+1}}{dx^{2n+1}}(e^{-x^2}) = \frac{i(-1)^n 2^{2n+1}}{\sqrt{\pi}} \int_{-\infty}^{+\infty} e^{-t^2} t^{2n+1} (\cos 2xt + i \sin 2xt) dt$$

$$= \frac{(-1)^n 2^{2n+1}}{\sqrt{\pi}} \int_{-\infty}^{+\infty} e^{-t^2} t^{2n+1} (-\sin 2xt + i \cos 2xt) dt.$$

Multiplying both sides by $(-1)^{2n+1} e^{x^2}$, we obtain

$$(-1)^{2n+1} e^{x^2} \frac{d^{2n+1}}{dx^{2n+1}}(e^{-x^2}) = \frac{(-1)^n 2^{2n+1} e^{x^2}}{\sqrt{\pi}} \int_{-\infty}^{+\infty} e^{-t^2} t^{2n+1} (\sin 2xt - i \cos 2xt) dt$$

or
$$H_{2n+1}(x) = \frac{(-1)^n 2^{2n+1} e^{x^2}}{\sqrt{\pi}} \int_{-\infty}^{+\infty} e^{-t^2} t^{2n+1} (\sin 2xt - i \cos 2xt) dt.$$

Equating real parts from both sides, we have

$$H_{2n+1}(x) = \frac{(-1)^n 2^{2n+1} e^{x^2}}{\sqrt{\pi}} \int_{-\infty}^{+\infty} e^{-t^2} t^{2n+1} \sin 2xt \, dt$$

$$= \frac{(-1)^n 2^{2n+2} e^{x^2}}{\sqrt{\pi}} \int_{0}^{\infty} e^{-t^2} t^{2n+1} \sin 2xt \, dt,$$

as the integrand is even function of t.

PROBLEM 27 A function is defined by

$$f(x) = \begin{cases} -1 \text{ for } x < 0, \\ +1 \text{ for } x > 0. \end{cases}$$

Expand this function in a series of Hermite polynomials.

Solution: Here, the function $f(x)$ is an odd function and, therefore, we write

$$f(x) = \sum_{n=0}^{\infty} c_{2n+1} H_{2n+1}(x) \tag{i}$$

where $c_{2n+1} = \dfrac{1}{2^{2n+1}(2n+1)!\sqrt{\pi}} \displaystyle\int_{-\infty}^{+\infty} e^{-x^2} f(x) H_{2n+1}(x) dx$

$$= \frac{1}{2^{2n+1}(2n+1)!\sqrt{\pi}} \left\{ \int_{-\infty}^{0} -e^{-x^2} H_{2n+1}(x) dx + \int_{0}^{\infty} e^{-x^2} H_{2n+1}(x) dx \right\}$$

$$= \frac{1}{2^{2n+1}(2n+1)!\sqrt{\pi}} \left\{ \int_{0}^{-\infty} e^{-x^2} H_{2n+1}(x) dx + \int_{0}^{\infty} e^{-x^2} H_{2n+1}(x) dx \right\}$$

$$= \frac{1}{2^{2n+1}(2n+1)!\sqrt{\pi}} \left\{ \int_{0}^{\infty} e^{-x^2} H_{2n+1}(-x)(-dx) + \int_{0}^{\infty} e^{-x^2} H_{2n+1}(x) dx \right\}$$

$$= \frac{1}{2^{2n+1}(2n+1)!\sqrt{\pi}} \left\{ \int_{0}^{\infty} e^{-x^2} (-1)^{2n+1} H_{2n+1}(x)(-dx) + \int_{0}^{\infty} e^{-x^2} H_{2n+1}(x) dx \right\}$$

$$= \frac{1}{2^{2n}(2n+1)!\sqrt{\pi}} \int_{0}^{\infty} e^{-x^2} H_{2n+1}(x) dx. \tag{ii}$$

We know that

$$H_{2n+1}(x) = 2xH_{2n}(x) - 2(2n)H_{2n-1}(x)$$

$$= 2x\,H_{2n}(x) - \frac{d}{dx}H_{2n}(x).$$

Multiplying both sides by e^{-x^2}, we have

$$e^{-x^2}H_{2n+1}(x) = 2x\,e^{-x^2}H_{2n}(x) - e^{-x^2}\frac{d}{dx}H_{2n}(x)$$

$$= -\frac{d}{dx}\left[e^{-x^2}H_{2n}(x)\right].$$

Using this relation, Eq. (ii) can be written as,

$$c_{2n+1} = \frac{1}{2^{2n}(2n+1)!\sqrt{\pi}}\left[-e^{-x^2}H_{2n}(x)\right]_0^{\infty}$$

$$= \frac{H_{2n}(0)}{2^{2n}(2n+1)!\sqrt{\pi}} = \frac{1}{2^{2n}\,(2n+1)!\sqrt{\pi}}\frac{(-1)^n(2n)!}{n!} = \frac{(-1)^n}{2^{2n}(2n+1)n!\sqrt{\pi}}.$$

Substituting this value of c_{2n+1} into Eq. (i), we get

$$f(x) = \frac{1}{\sqrt{\pi}}\sum_{n=0}^{\infty}\frac{(-1)^n}{2^{2n}(2n+1)n!}H_{2n+1}(x) \text{ with } -\infty < x < +\infty.$$

PROBLEM 28 Using the generating function for Hermite polynomials, prove that

$$H_n(x) = \sum_{r=0}^{\left[\frac{n}{2}\right]}\frac{n!\,H_{n-2r}(1)\left(x^2-1\right)^r x^{n-2r}}{r!(n-2r)!}.$$

Solution: We have

$$\sum_{n=0}^{\infty}H_n(x)\frac{h^n}{n!} = e^{2hx-h^2} = e^{2hx-h^2x^2}e^{h^2(x^2-1)}$$

$$= \sum_{n=0}^{\infty}H_n(1)\frac{(hx)^n}{n!}\sum_{r=0}^{\infty}\frac{\left[h^2(x^2-1)\right]^r}{r!}$$

$$= \sum_{n=0}^{\infty}\sum_{r=0}^{\infty}\frac{H_n(1)h^{n+2r}(x^2-1)^r x^n}{n!\,r!}$$

$$= \sum_{n=0}^{\infty} \sum_{r=0}^{\left[\frac{n}{2}\right]} \frac{H_{n-2r}(1)\, h^n (x^2-1)^r x^{n-2r}}{r!(n-2r)!}.$$

Equating coefficients of h^n from both sides of the above equation, we have

$$H_n(x) = \sum_{r=0}^{\left[\frac{n}{2}\right]} \frac{n! H_{n-2r}(1)(x^2-1)^r x^{n-2r}}{r!(n-2r)!}.$$

PROBLEM 29 Prove that $H_n(x) = 2^{n+1} e^{x^2} \int_x^{\infty} e^{-t^2} t^{n+1} P_n\left(\frac{x}{t}\right) dt$.

Solution: From Problem 9 of chapter 2, we know that

$$P_n(x) = \sum_{r=0}^{\left[\frac{n}{2}\right]} \frac{n!(x^2-1)^r x^{n-2r}}{2^{2r}(r!)^2(n-2r)!}.$$

Therefore, $P_n\left(\frac{x}{t}\right) = \sum_{r=0}^{\left[\frac{n}{2}\right]} \frac{n!\left(\frac{x^2}{t^2}-1\right)^r \left(\frac{x}{t}\right)^{n-2r}}{2^{2r}(r!)^2(n-2r)!}$.

Thus, we write

$$2^{n+1} e^{x^2} \int_x^{\infty} e^{-t^2} t^{n+1} P_n\left(\frac{x}{t}\right) dt = \int_x^{\infty} 2^{n+1} e^{-(t^2-x^2)}\, t^{n+1} \sum_{r=0}^{\left[\frac{n}{2}\right]} \frac{n!(x^2-t^2)^r x^{n-2r}}{2^{2r}(r!)^2(n-2r)!\, t^n} dt$$

$$= \sum_{r=0}^{\left[\frac{n}{2}\right]} \frac{(-1)^r n!}{(r!)^2(n-2r)!} \int_x^{\infty} e^{-(t^2-x^2)} (t^2-x^2)^r (2x)^{n-2r} (2tdt).$$

Let us put $t^2 - x^2 = u$. Then, $2tdt = du$.
Now, we write

$$2^{n+1} e^{x^2} \int_x^{\infty} e^{-t^2} t^{n+1} P_n\left(\frac{x}{t}\right) dt = \sum_{r=0}^{\left[\frac{n}{2}\right]} \frac{(-1)^r n!(2x)^{n-2r}}{(r!)^2(n-2r)!} \int_0^{\infty} e^{-u} u^r du$$

$$= \sum_{r=0}^{\left[\frac{n}{2}\right]} \frac{(-1)^r n!(2x)^{n-2r}\, \Gamma(r+1)}{(r!)^2(n-2r)!}$$

$$= \sum_{r=0}^{\left[\frac{n}{2}\right]} \frac{(-1)^r\, n!}{r!\,(n-2r)!}(2x)^{n-2r} = H_n(x).$$

EXERCISE

1. For what value(s) of n, the Hermite polynomial $H_n(0)$ is equal to zero?
2. Prove that $H_0'(x) = 0$ and $H_0(0) = 1$ where the prime denotes differentiation with respect to x.
3. Prove that

$$H_n(0) = \begin{cases} \dfrac{(-1)^{\frac{n}{2}}\, n!}{\left(\dfrac{n}{2}\right)!} \text{ if } n \text{ is even integer,} \\ 0 \quad \text{if } n \text{ is odd integer.} \end{cases}$$

4. Prove that

(i) $e^{t^2} \cos 2tx = \sum_{n=0}^{\infty} (-1)^n H_{2n}(x) \dfrac{t^{2n}}{(2n)!}$,

(ii) $e^{t^2} \sin 2tx = \sum_{n=0}^{\infty} (-1)^n H_{2n+1}(x) \dfrac{t^{2n+1}}{(2n+1)!}$.

[*Hint:* In the generating function given earlier, take $h = it$]

5. What are the values of $\int_{-\infty}^{+\infty} e^{-x^2} H_5(x) H_{10}(x)\, dx$ and $\int_{-\infty}^{+\infty} e^{-x^2} [H_5(x)]^2\, dx$?
6. Evaluate

$$\int_{-\infty}^{+\infty} e^{-x^2} x^2 H_n(x) H_m(x)\, dx$$

in terms of appropriate Kronecker delta functions and prove that

$$\int_{-\infty}^{+\infty} e^{-x^2} x^2 [H_n(x)]^2\, dx = \sqrt{\pi}\, 2^n n! \left(n + \frac{1}{2}\right).$$

7. Prove that

$$x^{2m+1} = \frac{(2m+1)!}{2^{2m+1}} \sum_{p=0}^{m} \frac{H_{2p+1}(x)}{(2p+1)!\,(m-p)!}$$

where m is a non-negative integer.

8. Prove that $\int_0^\infty e^{-x^2} H_{2n+1}(x)\,dx = \frac{(-1)^n (2n)!}{n!}$.

9. Prove that $H_n(x) = \frac{(-i)^n 2^n e^{x^2}}{\sqrt{\pi}} \int_{-\infty}^{+\infty} e^{-t^2+2itx} t^n\,dt$.

   [*Hint:* Proceed as in Problem 26 performing the differentiation n times in place of $2n$ or $2n+1$.]

10. Express x^7 and x^8 in a series of Hermite polynomials.

11. Use the expansion of x^n in a series of Hermite polynomials to prove that

$$\int_{-\infty}^{+\infty} e^{-x^2} x^n H_{n-2p}(x)\,dx = \frac{n!\sqrt{\pi}}{2^{2p}\,p!}.$$

12. Prove that an integral representation of Hermite polynomials can be written as

$$H_n(x) = \frac{(-1)^n n!}{2\pi i} \oint_C \frac{e^{-z^2+x^2}}{(z-x)^{n+1}}\,dz$$

   where C is a closed contour enclosing the point $z = x$.

13. Express the polynomial $8x^3 + 8x^2 - 6x + 2$ in terms of Hermite polynomials.

14. Prove that

$$\int_{-\infty}^{+\infty} e^{-\frac{x^2}{2}} H_n(x)\,dx = \begin{cases} \dfrac{n!\sqrt{2\pi}}{\left(\dfrac{n}{2}\right)!} & \text{if } n \text{ is an even integer,} \\ 0 & \text{if } n \text{ is an odd integer.} \end{cases}$$

15. Expand the function

$$f(x) = e^{ax}$$

   where a is a real or complex number into a series of Hermite polynomials.

   $\left[\textbf{Ans. } e^{a^2/4} \sum_{n=0}^{\infty} \frac{a^n}{2^n n!} H_n(x), \text{with} -\infty < x < +\infty\right]$

16. Prove that $\int_{-\infty}^{+\infty} e^{-x^2} x^n H_n(x)\,dx = n!\sqrt{\pi}$.

6
CHAPTER

Laguerre Polynomials

6.1 LAGUERRE EQUATION AND ITS SERIES SOLUTION

Laguerre equation is a homogeneous and linear second order ordinary differential equation of the form

$$x\frac{d^2y}{dx^2}+(1-x)\frac{dy}{dx}+ny=0 \tag{6.1}$$

where n is a non-negative integer.

This equation was named after the French mathematician, Edmond Nicolas Laguerre (April 9, 1834 - August 14, 1886).

Equation (6.1) can be written as

$$\frac{d^2y}{dx^2}+P(x)\frac{dy}{dx}+Qy=0 \text{ where } P(x)=\frac{1-x}{x} \text{ and } Q(x)=\frac{n}{x}.$$

Both $P(x)$ and $Q(x)$ are not finite at $x = 0$ and, therefore, the point $x=0$ is a singularity of Laguerre equation.

Here, we have

$$\lim_{x\to 0} xP(x)=1(\text{finite}) \text{ and also } \lim_{x\to 0} x^2Q(x)=0(\text{finite}).$$

Therefore, the point $x=0$ is a regular singular point of the above Laguerre equation and, hence, by Fuchs' theorem, this equation can have a power series solution of the form,

$$y=y(x)=\sum_{r=0}^{\infty} a_r\, x^{k+r} \quad (a_0 \neq 0) \tag{6.2}$$

where k is a number to be determined.

Now, we have

$$\frac{dy}{dx}=\sum_{r=0}^{\infty}(k+r)a_r\, x^{k+r-1} \text{ and } \frac{d^2y}{dx^2}=\sum_{r=0}^{\infty}(k+r)(k+r-1)a_r\, x^{k+r-2}.$$

Substituting the values of y and its derivatives into Eq. (6.1), we obtain

$$\sum_{r=0}^{\infty}(k+r)^2 a_r x^{k+r-1} - \sum_{r=0}^{\infty}(k+r-n)a_r x^{k+r} = 0 .$$

In this identity, the coefficients of various powers of x must vanish.

Equating the coefficient of lowest power of x, i.e., of x^{k-1} to zero, we have

$$a_0 k^2 = 0 \; (a_0 \neq 0) \quad \text{[Indicial equation]}$$

giving $k = 0, 0$ (double root).

Continuing the process, if we equate the coefficient of x^{k+r} to zero, we will have

$$(k+r+1)^2 a_{r+1} - (k+r-n)a_r = 0$$

or
$$a_{r+1} = \frac{(k+r-n)}{(k+r+1)^2} a_r \quad \text{[Recursion relation]}.$$

Substituting $k = 0$ into the recursion relation, we have

$$a_{r+1} = \frac{-(n-r)}{(r+1)^2} a_r . \tag{6.3}$$

Putting $r = 0, 1, 2, 3 \ldots$ Eq. (6.3) yields

$$a_1 = \frac{-n}{(1!)^2} a_0, \qquad a_2 = \frac{(-1)^2 n(n-1)}{(2!)^2} a_0$$

$$a_3 = \frac{(-1)^3 n(n-1)(n-2)}{(3!)^2} a_0, \quad a_4 = \frac{(-1)^4 n(n-1)(n-2)(n-3)}{(4!)^2} a_0$$

and so on.

In general, we have

$$a_r = \frac{(-1)^r n(n-1)(n-2)(n-3)\ldots(n-r+1)}{(r!)^2} a_0 .$$

Substituting the values of these various coefficients into Eq. (6.2), we obtain

$$y = \sum_{r=0}^{\infty} a_r x^r \quad [\text{since } k = 0]$$

$$= a_0 + a_1 x + a_2 x^2 + a_3 x^3 + \ldots + a_r x^r + \ldots \infty$$

$$= a_0 \left[1 - \frac{n}{(1!)^2} x + \frac{n(n-1)}{(2!)^2} x^2 - \frac{n(n-1)(n-2)}{(3!)^2} x^3 + \ldots \right.$$

$$\left. \ldots + (-1)^r \frac{n(n-1)(n-2)\ldots(n-r+1)}{(r!)^2} x^r + \ldots \infty \right] \tag{6.4}$$

which is the series solution of Laguerre equation.

6.2 POLYNOMIAL SOLUTION OF LAGUERRE EQUATION AND LAGUERRE POLYNOMIAL

The above series solution can be written as

$$y = a_0 \sum_{r=0}^{\infty} \frac{(-1)^r \, n!}{(n-r)!\,(r!)^2} x^r . \tag{6.5}$$

If the series terminates at a term containing x^n as a factor, the series becomes a polynomial, i.e., the solution becomes a polynomial solution.

From such a polynomial solution of Laguerre equation, we define Laguerre polynomial by assigning a certain fixed value to the arbitrary constant a_0. Various authors use to define Laguerre polynomial in two different ways.

1. In the first definition, Laguerre polynomial is defined from the above polynomial solution of Laguerre equation by assigning the value of the arbitrary constant a_0 as unity such that the coefficient of x^n in that polynomial is $\dfrac{(-1)^n}{n!}$.

   Thus, from Eq. (6.5), taking $a_0 = 1$ and denoting the Laguerre polynomial of degree n by $L_n(x)$, we write

$$L_n(x) = \sum_{r=0}^{n} \frac{(-1)^r \, n!}{(n-r)!\,(r!)^2} x^r . \tag{6.6}$$

2. In the second definition, Laguerre polynomial is defined from the above polynomial solution of Laguerre equation by assigning the value of the arbitrary constant a_0 as $n!$ such that the coefficient of x^n in that polynomial is $(-1)^n$.

   Thus, from Eq. (6.5), taking $a_0 = n!$ and denoting the Laguerre polynomial of degree n by $\mathcal{L}_n(x)$, we write

$$\mathcal{L}_n(x) = \sum_{r=0}^{n} \frac{(-1)^r \, (n!)^2}{(n-r)!\,(r!)^2} x^r . \tag{6.7}$$

   Thus, we write

$$\mathcal{L}_n(x) = n! \; L_n(x).$$

In physical uses, the first definition is more common and, therefore, in theorems and solved examples in this text, the first definition is being employed. However, hints are given elsewhere for interested persons to solve the same by employing the second definition as well.

6.3 RODRIGUES' FORMULA FOR LAGUERRE POLYNOMIALS

I. Using the First Definition of Laguerre Polynomials

We have,

$$
\begin{aligned}
L_n(x) &= \sum_{r=0}^{n} \frac{(-1)^r\, n!}{(n-r)!\,(r!)^2} x^r \\
&= \frac{e^x}{n!} \sum_{r=0}^{n} \frac{n!}{(n-r)!\,r!} \cdot \frac{n!}{r!} x^r (-1)^r e^{-x} \\
&= \frac{e^x}{n!} \sum_{r=0}^{n} \frac{n!}{(n-r)!\,r!} \frac{d^{n-r}}{dx^{n-r}} (x^n) \frac{d^r}{dx^r} (e^{-x}) \\
&= \frac{e^x}{n!} \frac{d^n}{dx^n} (x^n e^{-x}).
\end{aligned}
\tag{6.8}
$$

II. Using the Second Definition of Laguerre Polynomials

In a similar procedure, using the second definition of Laguerre polynomials, we can arrive at

$$
\mathcal{L}_n(x) = e^x \frac{d^n}{dx^n} (x^n\, e^{-x}). \tag{6.8a}
$$

Equations (6.8) and (6.8a) are called Rodrigues' formulae for Laguerre polynomials for the two definitions of Laguerre polynomial.

6.4 INTEGRAL REPRESENTATIONS OF LAGUERRE POLYNOMIALS

I. Using Rodrigues' Formula for the First Definition of Laguerre Polynomials

Rodrigues' formula for the first definition of Laguerre polynomials gives

$$
L_n(x) = \frac{e^x}{n!} \frac{d^n}{dx^n} (x^n\, e^{-x}).
$$

Using the generalized Cauchy's integral formula for the derivative of a function of a complex variable (z), we write

$$
L_n(x) = \frac{e^x}{n!} \frac{n!}{2\pi i} \oint_C \frac{z^n e^{-z} dz}{(z-x)^{n+1}} = \frac{1}{2\pi i} \oint_C \frac{z^n e^{-(z-x)} dz}{(z-x)^{n+1}} \tag{6.9}
$$

where C is a closed contour enclosing the point $z = x$.

Equation (6.9) is an integral representation of Laguerre polynomial of the first definition. Introducing a new complex variable $\xi = \frac{z-x}{z}$, we have

$$z = \frac{x}{1-\xi}, z - x = z\xi = \frac{x\xi}{1-\xi} \text{ and } dz = \frac{xd\xi}{(1-\xi)^2}.$$

Substitution of these quantities into Eq. (6.9) results in

$$L_n(x) = \frac{1}{2\pi i}\oint_C \frac{\left(\frac{x}{1-\xi}\right)^n e^{-\frac{x\xi}{1-\xi}}}{\left(\frac{x\xi}{1-\xi}\right)^{n+1}} \frac{xd\xi}{(1-\xi)^2} = \frac{1}{2\pi i}\oint_C \frac{e^{-\frac{x\xi}{1-\xi}}}{(1-\xi)\,\xi^{n+1}}\,d\xi \tag{6.10}$$

where ξ is within or on a circle C with centre at the origin.

Equation (6.10) is another integral representation of Laguerre polynomial of the first definition.

II. Using Rodrigues' Formula for the Second Definition of Laguerre Polynomials

Rodrigues' formula for the second definition of Laguerre polynomials gives

$$\mathcal{L}_n(x) = e^x \frac{d^n}{dx^n}(x^n e^{-x}).$$

Following a procedure similar to that in 6.4 I, we can obtain

$$\mathcal{L}_n(x) = \frac{n!}{2\pi i}\oint_C \frac{z^n e^{-(z-x)}}{(z-x)^{n+1}}\,dz \tag{6.9a}$$

where C is a closed contour enclosing the point $z = x$.

And, introducing a new variable ξ as in *I*, we write

$$\mathcal{L}_n(x) = \frac{n!}{2\pi i}\oint_C \frac{e^{-\frac{x\xi}{1-\xi}}}{(1-\xi)\,\xi^{n+1}}\,d\xi \tag{6.10a}$$

where ξ is within or on a circle C with centre at the origin.

6.5 GENERATING FUNCTION OF LAGUERRE POLYNOMIALS

I. Using Rodrigues' Formula for the First Definition of Laguerre Polynomials

Let us put

$$f(\xi) = \frac{e^{-\frac{x\xi}{1-\xi}}}{1-\xi} \text{ and } f^{(n)}(0) = \left[\frac{d^n}{d\xi^n} f(\xi)\right]_{\xi=0}.$$

Equation (6.10) can be written as,

$$L_n(x)=\frac{1}{n!}\left[\frac{n!}{2\pi i}\oint_C \frac{f(\xi)}{\xi^{n+1}}\,d\xi\right]=\frac{f^{(n)}(0)}{n!}. \tag{6.11}$$

Multiplying both sides of Eq. (6.11) by ξ^n and the taking sum over n from $n=0$ to $n=\infty$, we have

$$\sum_{n=0}^{\infty}\xi^n L_n(x)=\sum_{n=0}^{\infty}\frac{f^{(n)}(0)}{n!}\xi^n=f(\xi)=\frac{e^{-\frac{x\xi}{1-\xi}}}{1-\xi}$$

where use has been made of Maclaurin series.

Now, replacing ξ by a parameter h, we write

$$\sum_{n=0}^{\infty}L_n(x)h^n=\frac{1}{1-h}e^{-\frac{hx}{1-h}}. \tag{6.12}$$

The function on the right-hand side of Eq. (6.12) is called the generating function of Laguerre polynomials with the first definition.

II. Using Rodrigues' Formula for the Second Definition of Laguerre Polynomials

From Eq. (6.10a), following a procedure similar to 6.5I above, one can obtain

$$\sum_{n=0}^{\infty}\mathcal{L}_n(x)\frac{h^n}{n!}=\frac{1}{1-h}e^{-\frac{hx}{1-h}}. \tag{6.12a}$$

Equation (6.12a) gives the generating function of Laguerre polynomials with the second definition.

6.6 RECURRENCE RELATIONS FOR LAGUERRE POLYNOMIALS

I: A

$$(n+1)L_{n+1}(x)=(2n+1-x)L_n(x)-nL_{n-1}(x) \tag{6.13}$$

Solution: The generating function for the first definition of Laguerre polynomials gives

$$\sum_{n=0}^{\infty}L_n(x)h^n=\frac{1}{1-h}e^{-\frac{hx}{1-h}}.$$

Differentiating both sides once with respect to h, we have

$$\sum_{n=0}^{\infty} n L_n(x) h^{n-1} = \frac{1}{(1-h)^2} e^{-\frac{hx}{1-h}} + \frac{1}{1-h} e^{-\frac{hx}{1-h}} \left\{ \frac{-x(1-h)-(-hx)(-1)}{(1-h)^2} \right\}$$

$$= \frac{1}{1-h}\left[\frac{1}{1-h} e^{-\frac{hx}{1-h}}\right] - \frac{x}{(1-h)^2}\left[\frac{1}{1-h} e^{-\frac{hx}{1-h}}\right]$$

$$= \frac{1}{1-h}\sum_{n=0}^{\infty} L_n(x) h^n - \frac{x}{(1-h)^2}\sum_{n=0}^{\infty} L_n(x) h^n.$$

Multiplying both sides by $(1-h)^2$, we have

$$\sum_{n=0}^{\infty}(1-2h+h^2)\, nL_n(x) h^{n-1} = \sum_{n=0}^{\infty}(1-h)L_n(x)h^n - \sum_{n=0}^{\infty} xL_n(x)h^n$$

or
$$\sum_{n=0}^{\infty} nL_n(x)h^{n-1} - \sum_{n=0}^{\infty} 2nL_n(x)h^n + \sum_{n=0}^{\infty} nL_n(x)h^{n+1}$$

$$= \sum_{n=0}^{\infty} L_n(x)h^n - \sum_{n=0}^{\infty} L_n(x)h^{n+1} - \sum_{n=0}^{\infty} xL_n(x)h^n$$

or
$$\sum_{n=0}^{\infty} nL_n(x)h^{n-1} = \sum_{n=0}^{\infty}(2n+1-x)L_n(x)h^n - \sum_{n=0}^{\infty}(n+1)L_n(x)h^{n+1}.$$

Equating coefficients of h^n from both sides of the above equation, we have

$$(n+1)L_{n+1}(x) = (2n+1-x)L_n(x) - nL_{n-1}(x).$$

I: B

$$\mathcal{L}_{n+1}(x) = (2n+1-x)\,\mathcal{L}_n(x) - n^2\,\mathcal{L}_{n-1}(x). \tag{6.13a}$$

Solution: The generating function for the second definition of Laguerre polynomials gives

$$\sum_{n=0}^{\infty} \mathcal{L}_n(x)\frac{h^n}{n!} = \frac{1}{1-h} e^{-\frac{hx}{1-h}}.$$

Following a procedure similar to that used in I: A above, we can arrive at Eq. (6.13a).

II: A

$$L_n'(x) = L_{n-1}'(x) - L_{n-1}(x) \tag{6.14}$$

where the prime denotes differentiation with respect to x.

Solution: The generating function for the first definition of Laguerre polynomials gives

$$\sum_{n=0}^{\infty} L_n(x)h^n = \frac{1}{1-h}e^{-\frac{hx}{1-h}}$$

or

$$(1-h)\sum_{n=0}^{\infty} L_n(x)h^n = e^{-\frac{hx}{1-h}}.$$

Differentiating both sides once with respect to x, we have

$$\sum_{n=0}^{\infty}(1-h)L_n'(x)h^n = -\frac{h}{1-h}e^{-\frac{hx}{1-h}} = -h\sum_{n=0}^{\infty} L_n(x)h^n$$

or

$$\sum_{n=0}^{\infty} L_n'(x)h^n - \sum_{n=0}^{\infty} L_n'(x)h^{n+1} = -\sum_{n=0}^{\infty} L_n(x)h^{n+1}.$$

Equating coefficients of h^n from both sides, we have

$$L_n'(x) = L_{n-1}'(x) - L_{n-1}(x)$$

which is Eq. (6.14).

II: B

$$\mathcal{L}_n'(x) = n\mathcal{L}_{n-1}'(x) - n\mathcal{L}_{n-1}(x) \qquad (6.14a)$$

Solution: The generating function for the second definition of Laguerre polynomials gives

$$\sum_{n=0}^{\infty} \mathcal{L}_n(x)\frac{h^n}{n!} = \frac{1}{1-h}e^{-\frac{hx}{1-h}}$$

Following a procedure similar to that used in II: A above, we can arrive at Eq. (6.14a).

III: A

$$xL_n'(x) = nL_n(x) - nL_{n-1}(x). \qquad (6.15)$$

Solution: From Eq. (6.13), we have

$$(n+1)L_{n+1}(x) = (2n+1-x)L_n(x) - nL_{n-1}(x).$$

Differentiating both sides once with respect to x, we have

$$(n+1)L_{n+1}'(x) = (2n+1-x)L_n'(x) - L_n(x) - nL_{n-1}'(x). \qquad (6.16)$$

From Eq. (6.14), we have

$$L_n'(x) = L_{n-1}'(x) - L_{n-1}(x) \tag{6.17}$$

or

$$L_{n-1}'(x) = L_n'(x) + L_{n-1}(x).$$

From Eq. (6.17), replacing n by $n+1$, we have

$$L_{n+1}'(x) = L_n'(x) - L_n(x).$$

Substituting these values of $L_{n-1}'(x)$ and $L_{n+1}'(x)$ into Eq. (6.16), we obtain

$$(n+1)\{L_n'(x) - L_n(x)\} = (2n+1-x)L_n'(x) - L_n(x) - n\{L_n'(x) + L_{n-1}(x)\}$$

or

$$xL_n'(x) = nL_n(x) - nL_{n-1}(x).$$

III: B

Through a similar procedure, using Eqs. (6.13a) and (6.14a), we can arrive at

$$x\mathcal{L}_n'(x) = n\mathcal{L}_n(x) - n^2\mathcal{L}_{n-1}(x). \tag{6.15a}$$

IV: A

$$L_n'(x) = -\sum_{r=0}^{n-1} L_r(x). \tag{6.18}$$

Solution: The generating function for the first definition of Laguerre polynomials gives

$$\sum_{n=0}^{\infty} L_n(x)h^n = \frac{1}{1-h}e^{-\frac{hx}{1-h}}.$$

Differentiating both sides once with respect to x, we have

$$\sum_{n=0}^{\infty} L_n'(x)h^n = -\frac{h}{1-h}\left[\frac{1}{1-h}e^{-\frac{hx}{1-h}}\right] = -h(1-h)^{-1}\sum_{r=0}^{\infty} L_r(x)h^r$$

$$= -h\sum_{s=0}^{\infty} h^s \sum_{r=0}^{\infty} L_r(x)h^r$$

$$= -\sum_{s=0}^{\infty}\sum_{r=0}^{\infty} L_r(x)h^{s+r+1}.$$

On the right-hand side, if we put $n = s+r+1$, then $s = n - r - 1$. Since, $s \geq 0$ and $r \geq 0$, we have, $n - r - 1 \geq 0$, i.e., $0 \leq r \leq n-1$.

Therefore, the coefficient of h^n on the right-hand side is

$$-\sum_{r=0}^{n-1} L_r(x).$$

Now, equating coefficients of h^n from both sides of the above equation, we have

$$L_n'(x) = -\sum_{r=0}^{n-1} L_r(x).$$

IV: B

$$\mathcal{L}_n'(x) = -\sum_{r=0}^{n-1} \frac{n!}{r!} \mathcal{L}_r(x). \tag{6.18a}$$

Solution: The generating function for the second definition of Laguerre polynomials gives

$$\sum_{n=0}^{\infty} \mathcal{L}_n(x)\frac{h^n}{n!} = \frac{1}{1-h} e^{-\frac{hx}{1-h}}$$

Following a procedure similar to that in IVA above, we can arrive at Eq. (6.18a).

6.7 ORTHOGONALITY (AND ORTHONORMALITY) OF LAGUERRE POLYNOMIALS

Laguerre polynomials are orthogonal over the interval from $x = 0$ to $x = \infty$ with respect to a weight function, e^{-x}

Thus, the orthogonality relation is

$$\int_0^{\infty} e^{-x} L_m(x) L_n(x) dx = 0 \text{ if } m \neq n \tag{6.19}$$

which is valid for both definitions of Laguerre polynomials.

Proof:

I. Using the Generating Function for the First Definition of Laguerre Polynomials

Using the generating function for the first definition of Laguerre polynomials, we write,

$$\sum_{m=0}^{\infty} L_m(x) h^m = \frac{1}{1-h} e^{-\frac{hx}{1-h}}$$

and

$$\sum_{n=0}^{\infty} L_n(x) t^n = \frac{1}{1-t} e^{-\frac{tx}{1-t}}.$$

The product of these two equations is written as

$$\sum_{m=0}^{\infty}\sum_{n=0}^{\infty} L_m(x)L_n(x)h^m t^n = \frac{1}{(1-h)(1-t)} e^{-x\left(\frac{h}{1-h}+\frac{t}{1-t}\right)}.$$

Multiplying both sides by e^{-x} and then integrating with respect to x within the interval $x=0$ to $x=\infty$, we have

$$\sum_{m=0}^{\infty}\sum_{n=0}^{\infty} h^m t^n \int_0^{\infty} e^{-x} L_m(x)L_n(x)\,dx = \frac{1}{(1-h)(1-t)}\int_0^{\infty} e^{-x\left(1+\frac{h}{1-h}+\frac{t}{1-t}\right)}dx$$

$$= \frac{1}{(1-h)(1-t)}\int_0^{\infty} e^{-\frac{x(1-ht)}{(1-h)(1-t)}}dx$$

$$= \frac{1}{(1-h)(1-t)}\left[-\frac{(1-h)(1-t)}{1-ht}\right]\left[e^{-\frac{x(1-ht)}{(1-h)(1-t)}}\right]_{x=0}^{\infty}$$

$$= (1-ht)^{-1} = \sum_{n=0}^{\infty} h^n t^n. \tag{6.20}$$

Equating coefficients of $h^m t^n$ from both sides with $m \neq n$, we obtain

$\int_0^{\infty} e^{-x} L_m(x)L_n(x)\,dx = 0$, for $m \neq n$, which is Eq. (6.19).

II. Using the Generating Function for the Second Definition of Laguerre Polynomials

Similarly, starting from the generating function for the second definition of Laguerre polynomials, viz.

$$\sum_{n=0}^{\infty} \mathcal{L}_n(x)\frac{h^n}{n!} = \frac{1}{1-h} e^{-\frac{hx}{1-h}},$$

we can arrive at the result, $\int_0^{\infty} e^{-x}\,\mathcal{L}_m(x)\,\mathcal{L}_n(x)\,dx = 0.$

6.8A ORTHONORMALITY CONDITION

I. For Laguerre Polynomials of the First Definition

The Laguerre polynomials of the first definition are orthonormal in the interval $x=0$ to $x=\infty$, with respect to a weight function e^{-x}, that is,

$$\int_0^\infty e^{-x} L_m(x) L_n(x) dx = 1. \tag{6.21}$$

Proof: From Eq. (6.20), equating coefficients of $h^n t^n$ (i.e., with $m = n$) from both sides, we obtain

$$\int_0^\infty e^{-x} L_n(x) L_n(x) dx = 1 \text{ which is Eq. (6.21).}$$

II. For Laguerre Polynomials of the Second Definition

In this case, Eq. (6.21) can be written as

$$\int_0^\infty e^{-x} \left[\frac{1}{n!}\mathcal{L}_n(x)\right]\left[\frac{1}{n!}\mathcal{L}_n(x)\right] dx = 1$$

Hence, we write

$$\int_0^\infty e^{-x} \mathcal{L}_n(x) \mathcal{L}_n(x) dx = (n!)^2. \tag{6.21a}$$

One can also arrive at this result starting from the generating function.

6.8B GENERALIZED EXPRESSIONS FOR ORTHOGONALITY (AND ORTHONORMALITY) OF LAGUERRE POLYNOMIALS

I. For Laguerre Polynomials of the First Definition

The orthogonality (and orthonormality) condition of such Laguerre polynomials are generally expressed as

$$\int_0^\infty e^{-x} L_m(x) L_n(x) dx = \delta_{mn} \tag{6.22}$$

where the Kronecker delta symbol $(\delta_{mn}) = \begin{cases} 0 \text{ if } m \neq n, \\ 1 \text{ if } m = 1. \end{cases}$

II. For Laguerre Polynomials of the Second Definition

The orthogonality (and orthonormality) condition of such Laguerre polynomials are generally expressed as

$$\int_0^\infty e^{-x} \mathcal{L}_m(x) \mathcal{L}_n(x) dx = m!\, n!\, \delta_{mn}. \tag{6.22a}$$

6.9 EXPANSION OF AN ARBITRARY FUNCTION IN A SERIES OF LAGUERRE POLYNOMIALS

If an arbitrary function $f(x)$ defined in an infinite interval $(0,\infty)$ is piecewise smooth in every finite sub-interval (x_1,x_2) where $0<x_1<x_2<\infty$ and if the integral $\int_0^\infty e^{-x}f^2(x)dx$ is finite, then the function $f(x)$ can be expanded in a series of Laguerre polynomials in the form

$$f(x)=\sum_{n=0}^{\infty} c_n L_n(x) \tag{6.23}$$

where c_n are to be determined.

Multiplying both sides by $e^{-x}L_k(x)$ and then integrating with respect to x within the interval from $x=0$ to $x=\infty$, we have

$$\int_0^\infty e^{-x}f(x)L_k(x)dx=\sum_{n=0}^{\infty} c_n\int_0^\infty e^{-x}L_k(x)L_n(x)dx$$

$$=c_k$$

if the *first* definition of Laguerre polynomial is used.

Thus, changing the subscript, we have

$$c_n=\int_0^\infty e^{-x}f(x)L_n(x)dx. \tag{6.24}$$

Substitution of Eq. (6.24) into Eq. (6.23) will give the required series expansion of $f(x)$. On the other hand, if we use the *second* definition of Laguerre polynomial, we will get

$$f(x)=\sum_{n=0}^{\infty} c_n\mathcal{L}_n(x) \tag{6.23a}$$

where $$c_n=\frac{1}{(n!)^2}\int_0^\infty e^{-x}f(x)\mathcal{L}_n(x)dx \tag{6.24a}$$

Substitution of Eq. (6.24a) into Eq. (6.23a) will give the required series expansion of $f(x)$.

6.10 SOLVED PROBLEMS

PROBLEM 1 Obtain the first five Laguerre polynomials.

Solution:

First method: Using the sum formula:

(a) For the first definition: We have

$$L_n(x)=\sum_{r=0}^{n}\frac{(-1)^r\,n!}{(n-r)!\,(r!)^2}x^r.$$

Therefore, we have

$$L_0(x)=\frac{(-1)^0\,0!}{(0-0)!\,(0!)^2}x^0=1,$$

$$L_1(x)=\sum_{r=0}^{1}\frac{(-1)^r\,1!}{((1-r)!)!\,(r!)^2}x^r=1-x,$$

$$L_2(x)=\sum_{r=0}^{2}\frac{(-1)^r\,2!}{(2-r)!\,(r!)^2}x^r=\frac{(-1)^0\,2!}{2!\,(0!)^2}-\frac{2!}{1!\,(1!)^2}x+\frac{2!}{0!\,(2!)^2}x^2$$

$$=1-2x+\frac{1}{2}x^2,$$

$$L_3(x)=\sum_{r=0}^{3}\frac{(-1)^r\,3!}{(3-r)!\,(r!)^2}x^r=\frac{(-1)^0\,3!}{3!\,(0!)^2}-\frac{3!}{2!\,(1!)^2}x+\frac{3!}{1!\,(2!)^2}x^2-\frac{3!}{0!\,(3!)^2}x^3$$

$$=1-3x+\frac{3}{2}x^2-\frac{1}{6}x^3,$$

$$L_4(x)=\sum_{r=0}^{4}\frac{(-1)^r\,4!}{(4-r)!\,(r!)^2}x^r$$

$$=\frac{4!}{4!\,(0!)^2}-\frac{4!}{3!\,(1!)^2}x+\frac{4!}{2!\,(2!)^2}x^2-\frac{4!}{1!\,(3!)^2}x^3+\frac{4!}{0!\,(4!)^2}x^4$$

$$=1-4x+3x^2-\frac{2}{3}x^3+\frac{1}{24}x^4.$$

(b) For the second definition: We have

$$\mathcal{L}_n(x)=\sum_{r=0}^{n}\frac{(-1)^r\,(n!)^2}{(n-r)!\,(r!)^2}x^r.$$

Therefore, we obtain,

$\mathcal{L}_0(x)=1,$ $\mathcal{L}_1(x)=1-x,$

$\mathcal{L}_2(x)=2-4x+x^2,$ $\mathcal{L}_3(x)=6-18x+9x^2-x^3,$

$\mathcal{L}_4(x)=24-96x+72x^2-16x^3+x^4.$

Second method: Using Rodrigues' formula

(a) For the first definition: We have

$$L_n(x)=\frac{e^x}{n!}\frac{d^n}{dx^n}(x^n e^{-x}).$$

For $n=0$, we have

$$L_0(x)=\frac{e^x}{0!}(x^0 e^{-x})=1.$$

For $n=1$, we have

$$L_1(x)=\frac{e^x}{1!}\frac{d}{dx}(xe^{-x})=e^x(e^{-x}-xe^{-x})=1-x.$$

For $n=2$, we have

$$L_2(x)=\frac{e^x}{2!}\frac{d^2}{dx^2}(x^2e^{-x})=\frac{e^x}{2}(2e^{-x}-4xe^{-x}+x^2e^{-x})$$

$$=1-2x+\frac{1}{2}x^2.$$

For $n=3$, we have

$$L_3(x)=\frac{e^x}{3!}\frac{d^3}{dx^3}(x^3e^{-x})=\frac{e^x}{3!}\left\{3!\,e^{-x}-3(6x)e^{-x}+3(3x^2)e^{-x}-x^3e^{-x}\right\}$$

$$=1-3x+\frac{3}{2}x^2-\frac{1}{6}x^3.$$

For $n=4$, we have

$$L_4(x)=\frac{e^x}{4!}\frac{d^4}{dx^4}(x^4e^{-x})$$

$$=\frac{e^x}{4!}\left\{4!e^{-x}-4(4!)xe^{-x}+(3!)(12x^2)e^{-x}-4(4x^3)e^{-x}+x^4e^{-x}\right\}$$

$$=1-4x+3x^2-\frac{2}{3}x^3+\frac{1}{4!}x^4.$$

(b) For the second definition: Using the Rodrigues' formula,

$$\mathcal{L}_n(x)=e^x\frac{d^n}{dx^n}(x^n e^{-x}),$$ one can obtain

$\mathcal{L}_0(x)=1$ $\qquad$ $\mathcal{L}_1(x)=1-x$

$\mathcal{L}_2(x)=2-4x+x^2$ $\qquad$ $\mathcal{L}_3(x)=6-18x+9x^2-x^3$

$\mathcal{L}_4(x)=24-96x+72x^2-16x^3+x^4$

PROBLEM 2 Prove that

$L_n(0) = 1$ for the first definition of Lguerre polynomials

whereas, $\mathcal{L}_n(0) = n!$ for the second definition.

Solution:

First method: From sum formula

(a) For the first definition: In this case, we have

$$L_n(x) = \sum_{r=0}^{n} \frac{(-1)^r\, n!}{(n-r)!\,(r!)^2} x^r$$

$$= 1 - \frac{n!}{(n-1)!} x + \frac{n!}{(n-2)!\,(2!)^2} x^2 - \ldots$$

When $x = 0$, we have

$$L_n(0) = 1 + 0 = 1.$$

(b) For the second definition: In this case, we have

$$\mathcal{L}_n(x) = \sum_{r=0}^{n} \frac{(-1)^r\,(n!)^2}{(n-r)!\,(r!)^2} x^r$$

$$= (n!)^2 \left[\frac{1}{n!} - \frac{1}{(n-1)!} x + \frac{1}{(n-2)!\,(2!)^2} x^2 - \ldots \right].$$

When $x = 0$, we have

$$\mathcal{L}_n(0) = n!.$$

Second method: From Rodrigues' formula

(a) For the first definition: In this case, Rodrigues' formula is

$$L_n(x) = \frac{e^x}{n!} \frac{d^n}{dx^n} (x^n e^{-x})$$

$$= \frac{e^x}{n!} \left[(n!) e^{-x} - n(n!)\, x e^{-x} + \frac{n(n-1)}{2!} \frac{n!}{2!} x^2 e^{-x} - \ldots + (-1)^n x^n e^{-x} \right]$$

When $x = 0$, we have

$$L_n(0) = 1.$$

(b) For the second definition: In this case, Rodrigues' formula is

$$\mathcal{L}_n(x) = e^x \frac{d^n}{dx^n}(x^n e^{-x})$$

$$= e^x \left[(n!)e^{-x} - n(n!)xe^{-x} + \frac{n(n-1)}{2!}\frac{n!}{2!}x^2 e^{-x} - \ldots + (-1)^n x^n e^{-x} \right]$$

When $x = 0$, we have

$$\mathcal{L}_n(0) = n!.$$

Third method: From the generating function

(a) For the first definition: In this case, we have

$$\sum_{n=0}^{\infty} L_n(x)h^n = \frac{1}{1-h} e^{-\frac{hx}{1-h}}.$$

When $x = 0$, we have

$$\sum_{n=0}^{\infty} L_n(0)h^n = \frac{1}{1-h} = (1-h)^{-1} = \sum_{n=0}^{\infty} h^n.$$

Equating coefficients of h^n from both sides of the equation, we have

$$L_n(0) = 1.$$

(b) For the second definition: In this case, we have

$$\sum_{n=0}^{\infty} \mathcal{L}_n(x)\frac{h^n}{n!} = \frac{1}{1-h} e^{-\frac{hx}{1-h}}.$$

When $x = 0$, we have

$$\sum_{n=0}^{\infty} \mathcal{L}_n(0)\frac{h^n}{n!} = (1-h)^{-1} = \sum_{n=0}^{\infty} h^n.$$

Equating coefficients of h^n from both sides, we get

$$\mathcal{L}_n(0) = n!$$

PROBLEM 3 Prove that

(i) $L_n'(0) = -n$

(ii) $L_n''(0) = \dfrac{n(n-1)}{2}.$

Solution:

(i) Since $L_n(x)$ satisfies the Laguerre equation, we write

$$xL_n''(x)+(1-x)L_n'(x)+nL_n(x)=0.$$

Putting $x = 0$, we write

$$L_n'(0)=-nL_n(0)=-n, \quad [\text{as } L_n(0) = 1].$$

Alternatively, we have, for the first definition,

$$\sum_{n=0}^{\infty} L_n(x)h^n = \frac{1}{1-h}e^{-\frac{hx}{1-h}}.$$

Differentiating both sides once with respect to x, we have

$$\sum_{n=0}^{\infty} L_n'(x)h^n = -\frac{h}{(1-h)^2}e^{-\frac{hx}{1-h}}.$$

Putting $x=0$, we have

$$\sum_{n=0}^{\infty} L_n'(0)h^n = -h(1-h)^{-2} = -h(1+2h+3h^2+\dots\infty)$$

$$= -\sum_{n=0}^{\infty} nh^n.$$

Equating coefficients of h^n from both sides, we get

$$L_n'(0)=-n$$

(ii) We have, for the first definition

$$\sum_{n=0}^{\infty} L_n(x)h^n = \frac{1}{1-h}e^{-\frac{hx}{1-h}}.$$

Differentiating both sides twice with respect to x, we have

$$\sum_{n=0}^{\infty} L_n''(x)h^n = h^2(1-h)^{-3}e^{-\frac{hx}{1-h}}.$$

Putting $x=0$, we have

$$\sum_{n=0}^{\infty} L_n''(0)h^n = h^2(1-h)^{-3}$$

$$= h^2\left[1+3h+\frac{3.4}{2!}h^2+\frac{3.4.5}{3!}h^3+\dots+\frac{3.4.5\dots(3+n-1)}{n!}h^n+\dots\infty\right]$$

$$= \sum_{n=0}^{\infty} \frac{(n+2)!}{2!\,n!}h^{n+2}.$$

Equating coefficients of h^n from both sides, we have

$$L_n''(0)=\frac{n!}{2!(n-2)!}=\frac{(n-1)n}{2}.$$

PROBLEM 4 Prove that

(i) $L_n^{(n)}(0)=(-1)^n$ for the first definition of Laguerre polynomials,

(ii) $\mathcal{L}_n^{(n)}(0)=(-1)^n n!$ for the second definition of Laguerre polynomials,

Here, $L_n^{(n)}(0)$ denotes the nth order derivative of $L_n(x)$ at $x=0$.

Solution:

(i) We have

$$\sum_{n=0}^{\infty} L_n(x)h^n=\frac{1}{1-h}e^{-\frac{hx}{1-h}}.$$

Differentiating both sides n times with respect to x, we obtain

$$\sum_{n=0}^{\infty} L_n^{(n)}(x)h^n=\frac{1}{(1-h)}\left(\frac{-h}{1-h}\right)^n e^{-\frac{hx}{1-h}}.$$

Putting $x=0$, we have

$$\sum_{n=0}^{\infty} L_n^{(n)}(0)h^n=(-1)^n h^n (1-h)^{-(n+1)}$$

$$=(-1)^n h^n\left[1+(n+1)h+\frac{(n+1)(n+2)}{2!}h^2+\ldots+\frac{(n+1)(n+2)(n+3)\ldots(n+r)}{r!}h^r+\ldots\infty\right].$$

Equating coefficients of h^n from both sides, we obtain

$$L_n^{(n)}(0)=(-1)^n.$$

(ii) We have

$$\sum_{n=0}^{\infty} \mathcal{L}_n(x)\frac{h^n}{n!}=\frac{1}{1-h}e^{-\frac{hx}{1-h}}.$$

Proceeding as in (i), we can obtain the result.

PROBLEM 5 Directly prove that $L_n(x)$ is the coefficient of h^n in the expansion of $\frac{1}{1-h}e^{-\frac{hx}{1-h}}$.

Solution: We have

$$\frac{1}{1-h}e^{-\frac{hx}{1-h}} = \frac{1}{1-h}\sum_{n=0}^{\infty}\frac{1}{n!}\left(-\frac{hx}{1-h}\right)^n = \sum_{n=0}^{\infty}\frac{(-1)^n h^n x^n}{n!}(1-h)^{-(n+1)}$$

$$= \sum_{n=0}^{\infty}\frac{(-1)^n h^n x^n}{n!}\left[1+(n+1)h+\frac{(n+1)(n+2)}{2!}h^2+\ldots+\ldots\right.$$

$$\left.+\frac{(n+1)(n+2)(n+3)\ldots(n+p)}{p!}h^p+\ldots\infty\right]$$

$$= \sum_{n=0}^{\infty}\frac{(-1)^n h^n x^n}{n!}\sum_{p=0}^{\infty}\frac{(n+p)!}{n!\,p!}h^p$$

$$= \sum_{n=0}^{\infty}\sum_{p=0}^{\infty}\frac{(n+p)!(-1)^n x^n}{(n!)^2\,p!}h^{n+p}.$$

The coefficient of h^n is

$$\sum_{p=0}^{\infty}\frac{n!\,(-1)^{n-p}\,x^{n-p}}{\left[(n-p)!\right]^2 p!} \quad \text{[replacing } n \text{ by } (n-p)]$$

$$= \sum_{p=0}^{n}\frac{(-1)^{n-p}\,n!}{p!\left[(n-p)!\right]^2}x^{n-p}$$

$$= \sum_{r=0}^{n}\frac{(-1)^r\,n!}{(n-r)!\,(r!)^2}x^r \quad \text{where we put } r = n-p.$$

The above sum represents the Laguerre polynomial $L_n(x)$ [the first definition]. Hence, we see that the Laguerre polynomial, $L_n(x)$, is the coefficient of h^n in the expansion of $\frac{1}{1-h}e^{-\frac{hx}{1-h}}$ and we can write,

$$\sum_{n=0}^{\infty}L_n(x)h^n = \frac{1}{1-h}e^{-\frac{hx}{1-h}}.$$

Note: [Here, we use the first definition of Laguerre polynomials.] Using the second definition of Laguerre polynomial $\mathcal{L}_n(x)$ we can prove that $\mathcal{L}_n(x)$ is $n!$ times the coefficient of h^n in the expansion of

$$\frac{1}{1-h}e^{-\frac{hx}{1-h}}$$

and we can write,

$$\sum_{n=0}^{\infty} \mathcal{L}_n(x)\frac{h^n}{n!} = \frac{1}{1-h} e^{-\frac{hx}{1-h}}.$$

PROBLEM 6 Show that $\int_0^{\infty} e^{-st} L_n(t)\, dt = \frac{1}{s}\left(1-\frac{1}{s}\right)^n$.

Solution: Using the sum formula for Laguerre polynomials $L_n(x)$ of the first definition, we have

$$\begin{aligned}
\int_0^{\infty} e^{-st} L_n(t) dt &= \int_0^{\infty} e^{-st} \sum_{r=0}^{n} \frac{(-1)^r\, n!\, t^r}{(n-r)!(r!)^2} dt \\
&= \sum_{r=0}^{n} \frac{(-1)^r\, n!}{(n-r)!\,(r!)^2} \int_0^{\infty} e^{-u}\left(\frac{u}{s}\right)^r \frac{du}{s} \quad \text{[we put } u = st\text{]} \\
&= \sum_{r=0}^{n} \frac{(-1)^r\, n!}{(n-r)!\,(r!)^2\, s^{r+1}} \int_0^{\infty} e^{-u} u^r du \\
&= \sum_{r=0}^{n} \frac{(-1)^r\, n!}{(n-r)!\,(r!)^2\, s^{r+1}} \Gamma(r+1) \\
&= \frac{1}{s}\sum_{r=0}^{n} \frac{n!}{(n-r)!\, r!} 1^{n-r}\left(-\frac{1}{s}\right)^r = \frac{1}{s}\left(1-\frac{1}{s}\right)^n.
\end{aligned}$$

PROBLEM 7 Prove that $\int_x^{\infty} e^{-t} L_n(t) dt = e^{-x}\left[L_n(x) - L_{n-1}(x)\right]$.

Solution: Performing integration by parts once [with $L_n(t)$ as the first function] in the integral on the left-hand side, we write

$$\begin{aligned}
\int_x^{\infty} e^{-t} L_n(t) dt &= \left[-L_n(t) e^{-t}\right]_x^{\infty} + \int_x^{\infty} L_n'(t) e^{-t} dt \\
&= 0 + e^{-x} L_n(x) + \int_x^{\infty} -\left\{\sum_{r=0}^{n-1} L_r(t)\right\} e^{-t} dt \\
&= e^{-x} L_n(x) - \sum_{r=0}^{n-1} \int_x^{\infty} e^{-t} L_r(t) dt. \qquad \text{(i)}
\end{aligned}$$

From Eq. (i), we write

$$\int_x^{\infty} e^{-t} L_n(t)dt + \sum_{r=0}^{n-1} \int_x^{\infty} e^{-t} L_r(t)dt = e^{-x} L_n(x)$$

or
$$\sum_{r=0}^{n} \int_x^{\infty} e^{-t} L_r(t)dt = e^{-x} L_n(x). \quad \text{(ii)}$$

In Eq. (ii), replacing n by $(n-1)$, we write

$$\sum_{r=0}^{n-1} \int_x^{\infty} e^{-t} L_r(t)dt = e^{-x} L_{n-1}(t). \quad \text{(iii)}$$

Substitution of Eq. (iii) into Eq. (i) yields

$$\int_x^{\infty} e^{-t} L_n(t)dt = e^{-x}\left[L_n(x) - L_{n-1}(x)\right].$$

PROBLEM 8 If $f(x)$ is a polynomial of degree m, show that $f(x)$ may be expressed in the form

$$f(x) = \sum_{n=0}^{m} c_n L_n(x)$$

where
$$c_n = \int_0^{\infty} f(x) e^{-x} L_n(x)dx.$$

Deduce that $\int_0^{\infty} x^p e^{-x} L_n(x)dx = \begin{cases} 0, & \text{if } p < n, \\ (-1)^n n!, & \text{if } p = n. \end{cases}$

Solution: The function, $f(x)$, can be expressed in a series of Laguerre polynomials $L_n(x)$. Since $f(x)$ is a polynomial of degree m, n will have values from 0 to m such that we write,

$f(x) = \sum_{n=0}^{m} c_n L_n(x)$, where the coefficients (c_n) are to be determined.

Multiplying both sides by $e^{-x} L_k(x)$ and then integrating both sides with respect to x within the interval from $x = 0$ to $x = \infty$, we have

$$\int_0^{\infty} f(x) e^{-x} L_k(x)dx = \sum_{n=0}^{m} c_n \int_0^{\infty} e^{-x} L_k(x) L_n(x)dx = c_k.$$

Changing the subscript, we write

$$c_n = \int_0^\infty f(x)e^{-x}L_n(x)dx. \quad \text{(i)}$$

Deduction: Using Rodrigues' formula for the first definition of Laguerre polynomials $L_n(x)$, we have

$$\int_0^\infty x^p e^{-x} L_n(x)dx = \int_0^\infty x^p e^{-x} \frac{e^x}{n!}\frac{d^n}{dx^n}(e^{-x}x^n)dx$$

$$= \frac{1}{n!}\int_0^\infty x^p \frac{d^n}{dx^n}(e^{-x}x^n)dx.$$

In the integral on the right-hand side, performing integration by parts n times, the integrated out term vanishes at each stage and we arrive at

$$\int_0^\infty x^p e^{-x} L_n(x)dx = \frac{(-1)^n}{n!}\int_0^\infty \left\{\frac{d^n}{dx^n}x^p\right\} e^{-x}x^n dx. \quad \text{(ii)}$$

If $p < n$, we have $\frac{d^n}{dx^n}x^p = 0$, then Eq. (ii) becomes

$$\int_0^\infty x^p e^{-x} L_n(x)dx = 0. \quad \text{(iii)}$$

If $p = n$, we have $\frac{d^n}{dx^n}x^p = \frac{d^n}{dx^n}x^n = n!$. Then, Eq. (ii) becomes

$$\int_0^\infty x^p e^{-x} L_n(x)dx = \frac{(-1)^n}{n!}\int_0^\infty n!\, e^{-x}x^n dx$$

$$= (-1)^n \Gamma(n+1) = (-1)^n n!. \quad \text{(iv)}$$

Equations (iii) and (iv) can be generally expressed in the form,

$$\int_0^\infty x^p e^{-x} L_n(x)dx = \begin{cases} 0, & \text{if } p < n, \\ (-1)^n n!, & \text{if } p = n. \end{cases}$$

Note: If we use the Rodrigues' formula for the second definition of Laguerre polynomials, we will obtain

$$\int_0^\infty x^p e^{-x} \mathcal{L}_n(x)dx = \begin{cases} 0, & \text{if } p < n, \\ (-1)^n (n!)^2, & \text{if } p = n. \end{cases}$$

PROBLEM 9 Show that $e^{-ax} = \sum_{n=0}^{\infty} \frac{a^n}{(1+a)^{n+1}} L_n(x)$.

Solution: Let us write

$$e^{-ax} = \sum_{n=0}^{\infty} c_n L_n(x) \tag{i}$$

where c_n are coefficients to be determined.

Multiplying both sides of Eq. (i) by $e^{-x}L_k(x)$ and the integrating the results with respect to x within the interval $x = 0$ to $x = \infty$, we have

$$\int_0^{\infty} e^{-(1+a)x} L_k(x) dx = \sum_{n=0}^{\infty} c_n \int_0^{\infty} e^{-x} L_k(x) L_n(x) dx = c_k .$$

or

$$c_k = \int_0^{\infty} e^{-(1+a)x} \frac{e^x}{k!} \frac{d^k}{dx^k}(e^{-x}x^k) dx$$

$$= \frac{1}{k!} \int_0^{\infty} e^{-ax} \frac{d^k}{dx^k}(e^{-x}x^k) dx .$$

Performing integration by parts k times in the right-hand side, the integrated out term vanishes at each stage and, hence, we write

$$c_k = \frac{(-1)^k}{k!} \int_0^{\infty} \frac{d^k}{dx^k}(e^{-ax}) e^{-x} x^k dx$$

$$= \frac{(-1)^k}{k!} \int_0^{\infty} (-a)^k e^{-ax} e^{-x} x^k dx$$

$$= \frac{(-1)^{2k} a^k}{k!} \int_0^{\infty} e^{-(1+a)x} x^k dx$$

$$= \frac{a^k}{k!} \int_0^{\infty} e^{-u} \left(\frac{u}{1+a}\right)^k \frac{du}{1+a}, \text{ where we put } u = (1 + a)x;$$

$$= \frac{a^k}{(1+a)^{k+1} k!} \int_0^{\infty} e^{-u} u^k du = \frac{a^k}{(1+a)^{k+1} k!} \Gamma(k+1)$$

$$= \frac{a^k}{(1+a)^{k+1}}, \text{ since } \Gamma(k + 1) = k! \text{ if } k \text{ is a non-negative integer.}$$

Thus, we write

$$c_n = \frac{a^n}{(1+a)^{n+1}} \text{ and, hence, Eq. (i) can be written as}$$

$$e^{-ax} = \sum_{n=0}^{\infty} \frac{a^n}{(1+a)^{n+1}} L_n(x).$$

PROBLEM 10 Prove that $L_p(x^2+y^2) = \frac{(-1)^p}{2^{2p}} \sum_{n=0}^{p} \frac{H_{2n}(x) H_{2p-2n}(y)}{n!(p-n)!}$.

Solution: In terms of generating function of Laguerre polynomials (of first definition), we have

$$\sum_{p=0}^{\infty} L_p(x^2+y^2)h^p = \frac{1}{1-h} e^{-\frac{(x^2+y^2)h}{1-h}} = \frac{1}{1-h} e^{-\frac{hx^2}{1-h}} \cdot e^{-\frac{hy^2}{1-h}}. \tag{i}$$

In the case of Hermite polynomials discussed in chapter 5, we have seen that

$$e^{-b^2x^2} = \sum_{n=0}^{\infty} \frac{(-1)^n b^{2n} H_{2n}(x)}{2^{2n} n! (1+b^2)^{n+\frac{1}{2}}}. \tag{ii}$$

Therefore, we write

$$e^{-\frac{hx^2}{1-h}} = \sum_{p=0}^{\infty} \frac{(-1)^p \left(\frac{h}{1-h}\right)^p H_{2p}(x)}{2^{2p} p! \left(1+\frac{h}{1-h}\right)^{p+\frac{1}{2}}} \tag{iii}$$

and

$$e^{-\frac{hy^2}{1-h}} = \sum_{r=0}^{\infty} \frac{(-1)^r \left(\frac{h}{1-h}\right)^r H_{2r}(y)}{2^{2r} r! \left(1+\frac{h}{1-h}\right)^{r+\frac{1}{2}}}. \tag{iv}$$

Substituting Eqs. (iii) and (iv) into Eq. (i), we obtain

$$\sum_{p=0}^{\infty} L_p(x^2+y^2)h^p = \frac{1}{1-h} \sum_{p=0}^{\infty} \frac{(-1)^p \left(\frac{h}{1-h}\right)^p H_{2p}(x)}{2^{2p} p! \left(\frac{1}{1-h}\right)^{p+\frac{1}{2}}} \sum_{r=0}^{\infty} \frac{(-1)^r \left(\frac{h}{1-h}\right)^r H_{2r}(y)}{2^{2r} r! \left(\frac{1}{1-h}\right)^{r+\frac{1}{2}}}$$

$$= \sum_{p=0}^{\infty} \sum_{r=0}^{\infty} \frac{(-1)^{p+r} h^{p+r} H_{2p}(x) H_{2r}(y)}{2^{2p+2r} p! r!}.$$

Equating coefficients of h^p from both sides of the equation, we obtain

$$L_p(x^2+y^2)=\sum_{r=0}^{p}\frac{(-1)^p H_{2p-2r}(x)H_{2r}(y)}{2^{2p}(p-r)!\,r!}.$$

If we put $n=p-r$, we can write

$$L_p(x^2+y^2)=\frac{(-1)^p}{2^{2p}}\sum_{n=0}^{p}\frac{H_{2n}(x)H_{2p-2n}(y)}{n!\,(p-n)!}.$$

PROBLEM 11 Prove that $L_n(2x)=n!\sum_{p=0}^{n}\frac{(-1)^p 2^{n-p}}{p!\,(n-p)!}L_{n-p}(x)$.

Solution: Using the generating function, we write

$$\sum_{n=0}^{\infty}L_n(2x)h^n=\frac{1}{1-h}e^{-\frac{2xh}{1-h}}. \tag{i}$$

Using the result of Problem 7, we have

$$e^{-ax}=\sum_{n=0}^{\infty}\frac{a^n}{(1+a)^{n+1}}L_n(x). \tag{ii}$$

Therefore, we write

$$e^{-\frac{2xh}{1-h}}=\sum_{n=0}^{\infty}\frac{\left(\frac{2h}{1-h}\right)^n}{\left(1+\frac{2h}{1-h}\right)^{n+1}}L_n(x). \tag{iii}$$

Substituting Eq. (iii) into Eq. (i), we obtain

$$\sum_{n=0}^{\infty}L_n(2x)h^n=\frac{1}{1-h}\sum_{n=0}^{\infty}\frac{\left(\frac{2h}{1-h}\right)^n}{\left(\frac{1+h}{1-h}\right)^{n+1}}L_n(x)$$

$$=\sum_{n=0}^{\infty}2^n h^n L_n(x)(1+h)^{-(n+1)}$$

$$=\sum_{n=0}^{\infty}2^n h^n L_n(x)\left[1-(n+1)h+\frac{(n+1)(n+2)}{2!}h^2-\ldots+\ldots\right.$$

$$\left.+(-1)^p\frac{(n+1)(n+2)(n+3)\ldots(n+p)}{p!}h^p+\ldots\infty\right]$$

$$= \sum_{n=0}^{\infty} 2^n h^n L_n(x) \sum_{p=0}^{\infty} \frac{(-1)^p (n+p)!}{n!\,p!} h^p$$

$$= \sum_{n=0}^{\infty} \sum_{p=0}^{\infty} \frac{(-1)^p 2^n (n+p)!}{n!\,p!} L_n(x) h^{n+p}.$$

Equating coefficients of h^n from both sides of the above equation, we obtain

$$L_n(2x) = \sum_{p=0}^{\infty} \frac{(-1)^p 2^{n-p} n!}{(n-p)!\,p!} L_{n-p}(x)$$

$$= n! \sum_{p=0}^{n} \frac{(-1)^p 2^{n-p}}{p!\,(n-p)!} L_{n-p}(x).$$

PROBLEM 12 Express the polynomial $x^3 + x^2 - 3x + 2$ in a series of Laguerre polynomials.

Solution: We know that

$$L_0(x) = 1, \qquad L_1(x) = 1 - x,$$

$$L_2(x) = \frac{1}{2!}(2 - 4x + x^2), \qquad L_3(x) = \frac{1}{3!}(6 - 18x + 9x^2 - x^3).$$

Thus, we have

$$x = L_0(x) - L_1(x),$$

$$x^2 = 2L_2(x) + 4\{L_0(x) - L_1(x)\} - 2L_0(x) = 2L_0(x) - 4L_1(x) + 2L_2(x),$$

$$x^3 = 6L_0(x) - 18\{L_0(x) - L_1(x)\} + 9\{2L_2(x) - 4L_1(x) + 2L_0(x)\} - 6L_3(x)$$

$$= 6L_0(x) - 18L_1(x) + 18L_2(x) - 6L_3(x).$$

Now, we have

$$x^3 + x^2 - 3x + 2 = 6L_0(x) - 18L_1(x) + 18L_2(x) - 6L_3(x) + 2L_0(x)$$

$$-4L_1(x) + 2L_2(x) - 3L_0(x) + 3L_1(x) + 2L_0(x)$$

$$= 7L_0(x) - 19L_1(x) + 20L_2(x) - 6L_3(x).$$

Alternative Method

Let $x^3 + x^2 - 3x + 2 = c_0 L_0(x) + c_1 L_1(x) + c_2 L_2(x) + c_3 L_3(x)$ where the coefficients c_0, c_1, c_2 and c_3 are to be determined.

We have

$$x^3+x^2-3x+2=c_0+c_1(1-x)+\frac{1}{2}c_2(2-4x+x^2)+\frac{1}{6}c_3(6-18x+9x^2-x^3)$$

$$=c_0+c_1+c_2+c_3-(c_1+2c_2+3c_3)x+\left(\frac{1}{2}c_2+\frac{3}{2}c_3\right)x^2-\frac{1}{6}c_3x^3.$$

Thus, we obtain the set of equations,

$$c_0+c_1+c_2+c_3=2,$$
$$c_1+2c_2+3c_3=3,$$
$$\frac{1}{2}c_2+\frac{3}{2}c_3=1,$$
$$-\frac{1}{6}c_3=1.$$

Solving these equations, we obtain

$$c_3=-6, c_2=20, c_1=-19, c_0=7.$$

Now, the given polynomial can be expressed as

$$x^3+x^2-3x+2=7L_0(x)-19L_1(x)+20L_2(x)-6L_3(x).$$

EXERCISE

[Remember $\mathcal{L}_n(x)=n!\,L_n(x)$]

1. Prove that (i) $\mathcal{L}_n'(0) = -n\,(n!)$ and (ii) $\mathcal{L}_n''(0)=\dfrac{n(n-1)}{2}n!$

2. Prove that $\sum\limits_{n=0}^{\infty}\mathcal{L}_n(x)\dfrac{h^n}{n!}=\dfrac{1}{1-h}e^{-\frac{hx}{1-h}}$

3. Prove that $\int\limits_0^{\infty}e^{-st}\mathcal{L}_n(t)dt=\dfrac{n!}{s}\left(1-\dfrac{1}{s}\right)^n$

4. Prove that $\int\limits_x^{\infty}e^{-t}\mathcal{L}_n(t)dt=e^{-x}\left[\mathcal{L}_n(x)-n\mathcal{L}_{n-1}(x)\right]$

5. Prove that (i) $e^{-x}L_n(x)=\sum\limits_{r=0}^{n}\int\limits_x^{\infty}e^{-t}L_r(t)dt$

   (ii) $e^{-x}\mathcal{L}_n(x)=n!\sum\limits_{r=0}^{n}\dfrac{1}{r!}\int\limits_x^{\infty}e^{-t}\mathcal{L}_r(t)dt$

6. Prove that (i) $x^p = (p!)^2 \sum_{n=0}^{p} \frac{(-1)^n}{n!(p-n)!} L_n(x)$,

(ii) $x^p = (p!)^2 \sum_{n=0}^{p} \frac{(-1)^n}{(n!)^2 (p-n)!} \mathcal{L}_n(x)$.

7. Prove that (i) $\int_0^{\frac{\pi}{2}} e^{-\tan\theta} L_n(\tan\theta) L_m(\tan\theta) \sec^2\theta \, d\theta = \delta_{mn}$,

(ii) $\int_0^{\frac{\pi}{2}} e^{-\tan\theta} \mathcal{L}_n(\tan\theta) \mathcal{L}_m(\tan\theta) \sec^2\theta d\theta = m!\, n!\, \delta_{mn}$.

8. Prove that $\int_0^{\infty} e^{-x} x^p \mathcal{L}_n(x) dx = \begin{cases} 0, & \text{if } p < n, \\ (-1)^n (n!)^2, & \text{if } p = n. \end{cases}$

9. Prove that (i) $e^{-x} = \sum_{n=0}^{\infty} \frac{1}{2^{n+1}} L_n(x)$,

(ii) $e^{-x} = \sum_{n=0}^{\infty} \frac{1}{2^{n+1} n!} \mathcal{L}_n(x)$

(iii) $e^{-ax} = \sum_{n=0}^{\infty} \frac{a^n}{n!(1+a)^{n+1}} \mathcal{L}_n(x)$.

10. Express the polynomial $18 - 41x + 20x^2 - 2x^3$ in a series of Laguerre polynomials $L_n(x)$.

[**Ans.** $5L_0(x) - 3L_1(x) + 4L_2(x) + 12L_3(x)$]

11. What are the values of

(i) $\int_0^{\infty} e^{-x} x^3 L_5(x) dx$ (ii) $\int_0^{\infty} e^{-x} x^4 L_4(x) dx$

(iii) $\int_0^{\infty} e^{-x} L_2(x) L_3(x) dx$ (iv) $\int_0^{\infty} e^{-x} [L_5(x)]^2 dx$?

7

CHAPTER

Associated Laguerre Polynomials

7.1 ASSOCIATED LAGUERRE EQUATION AND ITS POWER SERIES SOLUTION

The associated (or generalized) Laguerre equation is a homogeneous linear second order ordinary differential equation of the form

$$x\frac{d^2y}{dx^2}+(\alpha+1-x)\frac{dy}{dx}+ny=0 \tag{7.1}$$

where α is a number greater than (-1) and n is a non-negative integer.

The restriction that $\alpha>-1$ is justified later.

Power Series Solution

For Eq. (7.1), the point $x = 0$ is a removable singularity and hence, by Fuchs' theorem, this equation will have a power series solution. Let us assume the power series solution in the form

$$y=\sum_{r=0}^{\infty}c_r x^{k+r}\ , (c_0\neq 0) \tag{7.2}$$

where the coefficients c_r and the constant k are to be determined.

Then, we have

$$\frac{dy}{dx}=\sum_{r=0}^{\infty}(k+r)c_r x^{k+r-1}$$

and

$$\frac{d^2y}{dx^2}=\sum_{r=0}^{\infty}(k+r)(k+r-1)c_r x^{k+r-2}.$$

Substituting the expressions for y and its derivatives into Eq. (7.1), we obtain

$$\sum_{r=0}^{\infty}(k+r)(k+\alpha+r)c_r\, x^{k+r-1}+\sum_{r=0}^{\infty}(n-k-r)c_r x^{k+r}=0\,.$$

In this identity, the coefficients of various powers of x must separately vanish. Thus, equating the coefficient of x^{k-1} [lowest power] to zero, we have

$$k(k+\alpha)c_0 = 0, \ (c_0 \neq 0) \text{ [Indicial equation]}$$

giving $k = 0, -\alpha$.

Next, equating the coefficient of x^k to zero, we have

$$(k+1)(k+\alpha+1)c_1 + (n-k)c_0 = 0$$

giving $c_1 = \dfrac{k-n}{(k+1)(k+\alpha+1)} c_0$.

In general, equating the coefficient of x^{k+r} to zero, we obtain

$$(k+r+1)(k+\alpha+r+1)c_{r+1} + (n-k-r)c_r = 0$$

giving $c_{r+1} = \dfrac{-(n-k-r)}{(k+r+1)(k+\alpha+r+1)} c_r$ [Recursion relation].

We are interested only in the solution with $k = 0$ and, therefore, we have

$$c_{r+1} = \frac{-(n-r)}{(r+1)(\alpha+r+1)} c_r. \tag{7.3}$$

Now, putting $r = 0, 1, 2, 3\ldots$ etc., we obtain

$$c_1 = \frac{-n}{1!(\alpha+1)} c_0,$$

$$c_2 = \frac{(-1)^2 n(n-1)}{2!(\alpha+1)(\alpha+2)} c_0,$$

$$c_3 = \frac{(-1)^3 n(n-1)(n-2)}{3!(\alpha+1)(\alpha+2)(\alpha+3)} c_0$$

and so on.

In general, we write

$$c_r = \frac{(-1)^r n(n-1)(n-2)(n-3)\ldots(n-r+1)}{r!(\alpha+1)(\alpha+2)(\alpha+3)\ldots(\alpha+r)} c_0$$

$$= \frac{(-1)^r n!\,\Gamma(\alpha+1)}{r!(n-r)!\,\Gamma(\alpha+r+1)} c_0. \tag{7.4}$$

As α may or may not be an integer, gamma function is used in place of factorial.

In the factor $\Gamma(\alpha+r+1)$ appearing in the denominator, the quantity $(\alpha+r+1)$ must always be greater than zero for any value of r from zero to infinity. Therefore, for $r=0$, we must have

$$\alpha+1>0\text{, i.e., } \alpha>-1.$$

Substituting Eq. (7.4) into Eq. (7.2) together with $k = 0$, we obtain

$$y=\sum_{r=0}^{\infty}\frac{(-1)^r c_0\, n!\,\Gamma(\alpha+1)}{r!\,(n-r)!\,\Gamma(\alpha+r+1)}x^r.$$

This equation is a power series solution of the associated (or generalized) Laguerre equation.

Polynomial Solution

As the series solution terminates at some term with $r = n$, it becomes the polynomial solution

$$y=\sum_{r=0}^{n}\frac{(-1)^r c_0\, n!\,\Gamma(\alpha+1)}{r!\,(n-r)!\,\Gamma(\alpha+r+1)}x^r. \tag{7.5}$$

This power series solution is a polynomial of order n.

7.2 ASSOCIATED (OR GENERALIZED) LAGUERRE POLYNOMIAL

Various authors define the associated (or generalized) Laguerre polynomial in two different ways.

A. First Convention

In one convention, the associated Laguerre polynomial $L_n^\alpha(x)$ is defined as the polynomial solution of the associated Laguerre equation of the type given in Eq. (7.1) when the value of c_0 in Eq. (7.5) is so chosen that the coefficient of x^n in the resulting polynomial is $\frac{(-1)^n}{n!}$.

Thus, using Eq. (7.5), we write

$$\frac{(-1)^n c_0\, n!\,\Gamma(\alpha+1)}{n!\,(n-n)!\,\Gamma(\alpha+n+1)}=\frac{(-1)^n}{n!}$$

or

$$c_0=\frac{\Gamma(\alpha+n+1)}{n!\,\Gamma(\alpha+1)}. \tag{7.6}$$

Substitution of Eq. (7.6) into Eq. (7.5) yields a polynomial known as the associated Laguerre polynomial, $L_n^\alpha(x)$. Thus, we define the associated Laguerre polynomial, $L_n^\alpha(x)$, as the polynomial given by

$$L_n^\alpha(x) = \sum_{r=0}^{n} \frac{(-1)^r \,\Gamma(\alpha+n+1)}{r!\,(n-r)!\,\Gamma(\alpha+r+1)} x^r . \tag{7.7}$$

It can be shown that if k is a non-negative integer, then

$$L_n^k(x) = (-1)^k \frac{d^k}{dx^k} L_{n+k}(x). \tag{7.8}$$

It is to be remembered that $L_n^\alpha(x)$ is a polynomial of degree, n.

B. Second Convention

In another convention, the associated Laguerre polynomial, $\mathcal{L}_n^\alpha(x)$, is defined in such a way that the coefficient of x^n in Eq. (7.5) becomes $(-1)^n$.

From Eq. (7.5), the coefficient of x^n is given by

$$\frac{(-1)^n \, n!\,\Gamma(\alpha+1)}{n!\,\Gamma(n+\alpha+1)} c_0 = (-1)^n$$

$$\text{yielding } c_0 = \frac{\Gamma(n+\alpha+1)}{\Gamma(\alpha+1)} . \tag{7.9}$$

Substitution of this value of c_0 into Eq. (7.5) yields the associated Laguerre polynomial in the second convention in the form

$$\mathcal{L}_n^\alpha(x) = \sum_{r=0}^{n} \frac{(-1)^r \, n!\,\Gamma(n+\alpha+1)}{r!\,(n-r)!\,\Gamma(\alpha+r+1)} x^r . \tag{7.10}$$

It is readily seen that

$$\mathcal{L}_n^\alpha(x) = n!\, L_n^\alpha(x). \tag{7.11}$$

If k is a non-negative integer, we have

$$\mathcal{L}_n^k(x) = (-1)^k \frac{n!}{(n+k)!} \frac{d^k}{dx^k} L_{n+k}(x). \tag{7.12}$$

7.3 RODRIGUES' FORMULA FOR ASSOCIATED LAGUERRE POLYNOMIALS

A. For the First Convention

In the first convention, the sum formula for associated Laguerre polynomials is given by

$$L_n^\alpha(x)=\sum_{r=0}^{n}\frac{(-1)^r\,\Gamma(n+\alpha+1)}{r!\,(n-r)!\,\Gamma(r+\alpha+1)}x^r$$

$$=\frac{\Gamma(n+\alpha+1)}{n!\,\Gamma(\alpha+1)}+\frac{\Gamma(n+\alpha+1)}{(n-1)!\,\Gamma(\alpha+2)}\frac{(-x)}{1!}+\ldots+\frac{\Gamma(n+\alpha+1)}{2!\,\Gamma(n+\alpha-1)}\frac{(-x)^{n-2}}{(n-2)!}$$

$$+\frac{\Gamma(n+\alpha+1)}{1!\,\Gamma(n+\alpha)}\frac{(-x)^{n-1}}{(n-1)!}+\frac{(-x)^n}{n!}.$$

Rewriting the above equation in reverse order, we have

$$L_n^\alpha(x)=\frac{(-x)^n}{n!}+\frac{(n+\alpha)}{1!\,(n-1)!}(-x)^{n-1}+\frac{(n+\alpha-1)(n+\alpha)}{2!\,(n-2)!}(-x)^{n-2}+\ldots$$

$$+\frac{(\alpha+1)(\alpha+2)(\alpha+3)\ldots(n+\alpha)}{n!}$$

$$=\frac{1}{n!}\Big[(-x)^n+n(n+\alpha)(-x)^{n-1}+\frac{n(n-1)}{2!}(n+\alpha)(n+\alpha-1)(-x)^{n-2}+\ldots$$

$$+\ldots+(\alpha+n)(\alpha+n-1)\ldots(\alpha+3)(\alpha+2)(\alpha+1)\Big]$$

$$=\frac{x^{-\alpha}e^x}{n!}\Big[x^{n+\alpha}(-1)^n e^{-x}+n(n+\alpha)x^{n+\alpha-1}(-1)^{n-1}e^{-x}$$

$$+\frac{n(n-1)}{2!}(n+\alpha)(n+\alpha-1)x^{n+\alpha-2}(-1)^{n-2}e^{-x}+\ldots$$

$$\ldots+(n+\alpha)(n+\alpha-1)(n+\alpha-2)\ldots(\alpha+1)x^\alpha e^{-x}\Big]$$

$$=\frac{x^{-\alpha}e^x}{n!}\left[\left\{\frac{d^n}{dx^n}(e^{-x})\right\}x^{n+\alpha}+n\left\{\frac{d^{n-1}}{dx^{n-1}}(e^{-x})\right\}\left\{\frac{d}{dx}x^{n+\alpha}\right\}\right.$$

$$\left.+\frac{n(n-1)}{2!}\left\{\frac{d^{n-2}}{dx^{n-2}}(e^{-x})\right\}\left\{\frac{d^2}{dx^2}x^{n+\alpha}\right\}+\ldots+e^{-x}\frac{d^n}{dx^n}x^{n+\alpha}\right]$$

$$=\frac{x^{-\alpha}e^x}{n!}\frac{d^n}{dx^n}(e^{-x}x^{n+\alpha})\ \text{[using Leibnitz theorem].}\qquad(7.13)$$

B. For the Second Convention

Following a procedure similar to that in A above, we can obtain

$$\mathcal{L}_n^{\alpha}(x) = x^{-\alpha} e^{x} \frac{d^n}{dx^n}\left(e^{-x} x^{n+\alpha}\right). \tag{7.14}$$

7.4 INTEGRAL REPRESENTATION AND GENERATING FUNCTION OF ASSOCIATED LAGUERRE POLYNOMIALS

A. Using the First Convention

Rodrigues' formula for associated Laguerre polynomials with the first convention is given by

$$L_n^{\alpha}(x) = \frac{x^{-\alpha} e^{x}}{n!} \frac{d^n}{dx^n}\left(e^{-x} x^{n+\alpha}\right).$$

Using generalized Cauchy's integral formula for the derivative of a function of a complex variable z, we write

$$L_n^{\alpha}(x) = \frac{x^{-\alpha} e^{x}}{n!}\left[\frac{n!}{2\pi i}\oint_C \frac{e^{-z} z^{n+\alpha}}{(z-x)^{n+1}} dz\right]$$

where C is a closed contour enclosing the point $z = x$ and i is the imaginary number, $\sqrt{-1}$.

Now, we write

$$L_n^{\alpha}(x) = \frac{1}{2\pi i}\oint_C \frac{e^{-(z-x)} x^{-\alpha} z^{n+\alpha}}{(z-x)^{n+1}} dz.$$

Introducing a new variable

$$\xi = \frac{z-x}{z} \text{ so that } z = \frac{x}{1-\xi},\ dz = \frac{x\,d\xi}{(1-\xi)^2} \text{ and } (z-x) = \frac{x\xi}{1-\xi};$$

we write

$$L_n^{\alpha}(x) = \frac{1}{2\pi i}\oint_C \frac{e^{-\frac{x\xi}{1-\xi}} x^{-\alpha}\left(\frac{x}{1-\xi}\right)^{n+\alpha}}{\left(\frac{x\xi}{1-\xi}\right)^{n+1}} \frac{x\,d\xi}{(1-\xi)^2}$$

$$= \frac{1}{2\pi i}\oint_C \frac{e^{-\frac{x\xi}{1-\xi}}}{(1-\xi)^{\alpha+1}} \frac{d\xi}{\xi^{n+1}}. \tag{7.15}$$

Equation (7.15) is the integral representation of associated Laguerre polynomial of the first convention. Now, we can write

$$L_n^\alpha(x) = \frac{1}{n!}\left[\frac{n!}{2\pi i}\oint_C \frac{f(\xi)d\xi}{(\xi-0)^{n+1}}\right] = \frac{f^{(n)}(0)}{n!} \tag{7.16}$$

where
$$f(\xi) = \frac{e^{-\frac{x\xi}{1-\xi}}}{(1-\xi)^{\alpha+1}} \tag{7.17}$$

and
$$f^{(n)}(0) = \left[\frac{d^n}{d\xi^n} f(\xi)\right]_{\xi=0}. \tag{7.18}$$

Multiplying both sides of Eq. (7.16) by ξ^n and then taking sum over n from zero to infinity, we write

$$\sum_{n=0}^{\infty} \xi^n L_n^\alpha(x) = \sum_{n=0}^{\infty} \frac{f^{(n)}(0)}{n!}\xi^n = f(\xi) = \frac{1}{(1-\xi)^{\alpha+1}} e^{-\frac{x\xi}{1-\xi}} \tag{7.19}$$

where use has been made of Maclaurin series.

The function on the right-hand side of Eq. (7.19), viz.

$$f(\xi) = g(x,\xi) = \frac{1}{(1-\xi)^{\alpha+1}} e^{-\frac{x\xi}{1-\xi}} \tag{7.20}$$

is called the generating function of associated Laguerre polynomials.

If we replace the complex variable ξ by a real parameter h, Eqs. (7.19) and (7.20) become respectively

$$\sum_{n=0}^{\infty} h^n L_n^\alpha(x) = \frac{1}{(1-h)^{\alpha+1}} e^{-\frac{xh}{1-h}} \tag{7.21}$$

and
$$g(x,h) = \frac{1}{(1-h)^{\alpha+1}} e^{-\frac{xh}{1-h}}.$$

Here, the function $g(x,h)$ is known as the generating function of associated Laguerre polynomials.

B. Using the Second Convention

Rodrigues' formula for associated Laguerre polynomials with the second convention is given by

$$\mathcal{L}_n^\alpha(x) = x^{-\alpha} e^x \frac{d^n}{dx^n}(e^{-x} x^{n+\alpha})$$

Following a procedure similar to that in A above, we can obtain the integral representation and the generating function respectively as in the following.

$$\mathcal{L}_n^{\alpha}(x) = \frac{n!}{2\pi i} \oint_C \frac{e^{-\frac{x\xi}{1-\xi}}}{(1-\xi)^{\alpha+1}} \frac{d\xi}{\xi^{n+1}}. \tag{7.15a}$$

Equation (7.15a) is the integral representation of associated Laguerre polynomial of the second convention. Following a procedure similar to that in A, we can arrive at

$$\sum_{n=0}^{\infty} \mathcal{L}_n^{\alpha}(x) \frac{h^n}{n!} = \frac{1}{(1-h)^{\alpha+1}} e^{-\frac{xh}{1-h}}. \tag{7.21a}$$

The function on the right-hand side of Eq. (7.21a) is called the generating function.

7.5 RECURRENCE RELATIONS OF ASSOCIATED LAGUERRE POLYNOMIALS

(i) A. For the First Convention

We know that

$$\sum_{n=0}^{\infty} L_n^{\alpha}(x) h^n = \frac{1}{(1-h)^{\alpha+1}} e^{-\frac{xh}{1-h}}.$$

Multiplying both sides by $(1-h)$, we obtain

$$\sum_{n=0}^{\infty} (1-h) L_n^{\alpha}(x) h^n = \frac{1}{(1-h)^{\alpha-1+1}} e^{-\frac{xh}{1-h}}$$

or

$$\sum_{n=0}^{\infty} L_n^{\alpha}(x) h^n - \sum_{n=0}^{\infty} L_n^{\alpha}(x) h^{n+1} = \sum_{n=0}^{\infty} L_n^{\alpha-1}(x) h^n.$$

Equating coefficients of h^n from both sides, we obtain

$$L_n^{\alpha}(x) = L_n^{\alpha-1}(x) + L_{n-1}^{\alpha}(x). \tag{7.22}$$

B. For the Second Convention

With the second convention, we know that

$$\sum_{n=0}^{\infty} \mathcal{L}_n^{\alpha}(x) \frac{h^n}{n!} = \frac{1}{(1-h)^{\alpha+1}} e^{-\frac{xh}{1-h}}.$$

Following a procedure similar to that in A, we will obtain

$$\mathcal{L}_n^{\alpha}(x) = \mathcal{L}_n^{\alpha-1}(x) + n\mathcal{L}_{n-1}^{\alpha}(x). \tag{7.22a}$$

(ii) A. For the First Convention

We know that

$$\sum_{n=0}^{\infty} L_n^{\alpha}(x) h^n = \frac{1}{(1-h)^{\alpha+1}} e^{-\frac{xh}{1-h}}.$$

Multiplying both sides by $(1-h)$, we obtain

$$\sum_{n=0}^{\infty} L_n^{\alpha}(x) h^n - \sum_{n=0}^{\infty} L_n^{\alpha}(x) h^{n+1} = \frac{1}{(1-h)^{\alpha}} e^{-\frac{xh}{1-h}}.$$

Differentiating both sides once with respect to h, we obtain

$$\sum_{n=0}^{\infty} n L_n^{\alpha}(x) h^{n-1} - \sum_{n=0}^{\infty} (n+1) L_n^{\alpha}(x) h^n = \frac{\alpha}{(1-h)^{\alpha+1}} e^{-\frac{xh}{1-h}} - \frac{x}{(1-h)^{\alpha+2}} e^{-\frac{xh}{1-h}}$$

$$= \sum_{n=0}^{\infty} \alpha L_n^{\alpha}(x) h^n - \sum_{n=0}^{\infty} x L_n^{\alpha+1}(x) h^n.$$

Now, equating coefficients of h^n from both sides, we obtain

$$(n+1) L_{n+1}^{\alpha}(x) - (n+1) L_n^{\alpha}(x) = \alpha L_n^{\alpha}(x) - x L_n^{\alpha+1}(x)$$

or

$$x L_n^{\alpha+1}(x) + (n+1) L_{n+1}^{\alpha}(x) = (n+\alpha+1) L_n^{\alpha}(x). \tag{7.23}$$

Replacing α by $\alpha-1$, we obtain

$$x L_n^{\alpha}(x) + (n+1) L_{n+1}^{\alpha-1}(x) = (n+\alpha) L_n^{\alpha-1}(x). \tag{7.24}$$

B. For the Second Convention

In this case, we have

$$\sum_{n=0}^{\infty} \mathcal{L}_n^{\alpha}(x) \frac{h^n}{n!} = \frac{1}{(1-h)^{\alpha+1}} e^{-\frac{xh}{1-h}}.$$

Following a procedure similar to that in A, we can obtain the relation

$$x \mathcal{L}_n^{\alpha+1}(x) + \mathcal{L}_{n+1}^{\alpha}(x) = (n+\alpha+1) \mathcal{L}_n^{\alpha}(x). \tag{7.23a}$$

If we replace α by $\alpha-1$, we will get

$$x \mathcal{L}_n^{\alpha}(x) + \mathcal{L}_{n+1}^{\alpha-1}(x) = (n+\alpha) \mathcal{L}_n^{\alpha-1}(x). \tag{7.24a}$$

(iii) A. For the First Convention

We know that

$$\sum_{n=0}^{\infty} L_n^{\alpha}(x)h^n = \frac{1}{(1-h)^{\alpha+1}} e^{-\frac{xh}{1-h}}.$$

Differentiating both sides once with respect to x, we obtain

$$\sum_{n=0}^{\infty} h^n \frac{d}{dx} L_n^{\alpha}(x) = -\frac{h}{(1-h)^{\alpha+2}} e^{-\frac{hx}{1-h}} = -\sum_{n=0}^{\infty} h^{n+1} L_n^{\alpha+1}(x).$$

Equating coefficients of h^n from both sides, we obtain

$$\frac{d}{dx} L_n^{\alpha}(x) = -L_{n-1}^{\alpha+1}(x). \tag{7.25}$$

B. For the Second Convention

We know that

$$\sum_{n=0}^{\infty} \mathcal{L}_n^{\alpha}(x) \frac{h^n}{n!} = \frac{1}{(1-h)^{\alpha+1}} e^{-\frac{xh}{1-h}}.$$

Differentiating both sides with respect to x once and then equating coefficients of h^n from both sides, we obtain

$$\frac{d}{dx} \mathcal{L}_n^{\alpha}(x) = -n\, \mathcal{L}_{n-1}^{\alpha+1}(x). \tag{7.25a}$$

(iv) A. For the First Convention

We know that

$$L_n^{\alpha}(x) = L_n^{\alpha-1}(x) + L_{n-1}^{\alpha}(x)$$

or

$$L_n^{\alpha}(x) - L_{n-1}^{\alpha}(x) = L_n^{\alpha-1}(x). \tag{i}$$

In Eq. (i), replacing n by $n+1$, we obtain

$$L_{n+1}^{\alpha}(x) - L_n^{\alpha}(x) = L_{n+1}^{\alpha-1}(x). \tag{ii}$$

Also, we know that

$$xL_n^{\alpha}(x) + (n+1) L_{n+1}^{\alpha-1}(x) = (n+\alpha) L_n^{\alpha-1}(x). \tag{iii}$$

Substituting Eqs. (i) and (ii) into Eq. (iii), we obtain

$$xL_n^{\alpha}(x) + (n+1)\left[L_{n+1}^{\alpha}(x) - L_n^{\alpha}(x)\right] = (n+\alpha)\left[L_n^{\alpha}(x) - L_{n-1}^{\alpha}(x)\right]$$

or $$(2n+\alpha+1-x)L_n^{\alpha}(x)=(n+1)L_{n+1}^{\alpha}(x)+(n+\alpha)L_{n-1}^{\alpha}(x). \tag{7.26}$$

B. For the Second Convention

Through a procedure similar to that in A, one can obtain

$$(2n+\alpha+1-x)\mathcal{L}_n^{\alpha}(x)=\mathcal{L}_{n+1}^{\alpha}(x)+n(n+\alpha)\mathcal{L}_{n-1}^{\alpha}(x). \tag{7.26a}$$

(v) A. For the First Convention

Equation (7.24) gives

$$xL_n^{\alpha}(x)+(n+1)L_{n+1}^{\alpha-1}(x)=(n+\alpha)L_n^{\alpha-1}(x).$$

Replacing n by $n-1$ and α by $\alpha+1$ and then transposing, we have

$$xL_{n-1}^{\alpha+1}(x)=(n+\alpha)L_{n-1}^{\alpha}(x)-nL_n^{\alpha}(x). \tag{i}$$

Since $$L_n^{\alpha}(x)=L_n^{\alpha-1}(x)+L_{n-1}^{\alpha}(x),$$

we have $$L_{n-1}^{\alpha}(x)=L_n^{\alpha}(x)-L_n^{\alpha-1}(x) \tag{ii}$$

Substitution of Eq. (ii) into Eq (i) yields

$$xL_{n-1}^{\alpha+1}(x)=\alpha L_n^{\alpha}(x)-(n+\alpha)L_n^{\alpha-1}(x). \tag{iii}$$

Also, from Eq. (7.25), we have

$$\frac{d}{dx}L_n^{\alpha}(x)=-L_{n-1}^{\alpha+1}(x).$$

Multiplying both sides by $-x$, we obtain

$$-x\frac{d}{dx}L_n^{\alpha}(x)=xL_{n-1}^{\alpha+1}(x).$$

Substituting this Eq. into Eq. (iii) above, we write

$$x\frac{d}{dx}L_n^{\alpha}(x)+\alpha L_n^{\alpha}(x)=(n+\alpha)L_n^{\alpha-1}(x). \tag{7.27}$$

B. For the Second Convention

By a similar procedure, we can obtain for the second convention, the following relation

$$x\frac{d}{dx}\mathcal{L}_n^{\alpha}(x)+\alpha\mathcal{L}_n^{\alpha}(x)=(n+\alpha)\mathcal{L}_n^{\alpha-1}(x). \tag{7.27a}$$

(vi) A. For the First Convention

The second recurrence relation, i.e., Eq. (7.24) is written as

$$xL_n^{\alpha}(x)+(n+1)L_{n+1}^{\alpha-1}(x)=(n+\alpha)L_n^{\alpha-1}(x).$$

The fifth recurrence relation, i.e., Eq. (7.27) is written as

$$x\frac{d}{dx}L_n^{\alpha}(x)+\alpha L_n^{\alpha}(x)=(n+\alpha)L_n^{\alpha-1}(x).$$

Subtracting the first equation from the later equation here, we obtain

$$x\frac{d}{dx}L_n^{\alpha}(x)+(\alpha-x)L_n^{\alpha}(x)=(n+1)L_{n+1}^{\alpha-1}(x). \tag{7.28}$$

B. For the Second Convention

In a similar manner, subtracting Eq. (7.24a) from Eq. (7.27a), we obtain

$$x\frac{d}{dx}\mathcal{L}_n^{\alpha}(x)+(\alpha-x)\mathcal{L}_n^{\alpha}(x)=\mathcal{L}_{n+1}^{\alpha-1}(x). \tag{7.28a}$$

(vii) A. For the First Convention

From the fifth recurrence relation, i.e., Eq. (7.27), we have

$$x\frac{d}{dx}L_n^{\alpha}(x)+\alpha L_n^{\alpha}(x)=(n+\alpha)L_n^{\alpha-1}(x).$$

Also, from the first recurrence relation, i.e., Eq. (7.22), we have

$$L_n^{\alpha}(x)=L_n^{\alpha-1}(x)+L_{n-1}^{\alpha}(x)$$

or

$$L_n^{\alpha-1}(x)=L_n^{\alpha}(x)-L_{n-1}^{\alpha}(x).$$

Substituting this equation into the former equation, we obtain

$$x\frac{d}{dx}L_n^{\alpha}(x)+\alpha L_n^{\alpha}(x)=(n+\alpha)\left[L_n^{\alpha}(x)-L_{n-1}^{\alpha}(x)\right]$$

or

$$x\frac{d}{dx}L_n^{\alpha}(x)-nL_n^{\alpha}(x)=-(n+\alpha)L_{n-1}^{\alpha}(x). \tag{7.29}$$

B. For the Second Convention

By a similar procedure, using Eqs. (7.22a) and (7.27a), we will obtain

$$x\frac{d}{dx}\mathcal{L}_n^{\alpha}(x)-n\mathcal{L}_n^{\alpha}(x)=-n(n+\alpha)\mathcal{L}_{n-1}^{\alpha}(x). \tag{7.29a}$$

(viii) A. For the First Convention

From the sixth recurrence relation, i.e., Eq. (7.28), we have

$$x\frac{d}{dx}L_n^\alpha(x)+(\alpha-x)L_n^\alpha(x)=(n+1)L_{n+1}^{\alpha-1}(x). \tag{a}$$

Also, from Eq. (7.22), we have

$$L_n^\alpha(x)=L_n^{\alpha-1}(x)+L_{n-1}^\alpha(x)$$

or

$$L_n^{\alpha-1}(x)=L_n^\alpha(x)-L_{n-1}^\alpha(x).$$

Replacing n by $n+1$, we have

$$L_{n+1}^{\alpha-1}(x)=L_{n+1}^\alpha(x)-L_n^\alpha(x). \tag{b}$$

Substituting Eq. (b) into Eq. (a), we obtain

$$x\frac{d}{dx}L_n^\alpha(x)+(\alpha-x)L_n^\alpha(x)=(n+1)\left[L_{n+1}^\alpha(x)-L_n^\alpha(x)\right]$$

or

$$x\frac{d}{dx}L_n^\alpha(x)+(\alpha+n+1-x)L_n^\alpha(x)=(n+1)L_{n+1}^\alpha(x). \tag{7.30}$$

B. For the Second Convention

In a similar manner, using Eqs. (7.22a) and (7.28a), we can obtain the relation,

$$x\frac{d}{dx}\mathcal{L}_n^\alpha(x)+(\alpha+n+1-x)\mathcal{L}_n^\alpha(x)=\mathcal{L}_{n+1}^\alpha(x). \tag{7.30a}$$

(ix) A. For the First Convention

From the first recurrence relation, viz., Eq. (7.22), we have

$$L_n^\alpha(x)=L_n^{\alpha-1}(x)+L_{n-1}^\alpha(x).$$

Also, from the third recurrence relation, viz., Eq. (7.25), we have

$$\frac{d}{dx}L_n^\alpha(x)=-L_{n-1}^{\alpha+1}(x).$$

Therefore, we have

$$\frac{d}{dx}L_n^\alpha(x)=\frac{d}{dx}L_n^{\alpha-1}(x)+\frac{d}{dx}L_{n-1}^\alpha(x)$$

or

$$\frac{d}{dx}L_n^\alpha(x)=-L_{n-1}^\alpha(x)+\frac{d}{dx}L_{n-1}^\alpha(x)$$

or

$$\frac{d}{dx}\left[L_n^\alpha(x)-L_{n-1}^\alpha(x)\right]=-L_{n-1}^\alpha(x). \tag{7.31}$$

B. For the Second Convention

Similarly, using Eqs. (7.22a) and (7.25a), we can obtain the relation,

$$\frac{d}{dx}\left[\mathcal{L}_n^{\alpha}(x) - n\mathcal{L}_{n-1}^{\alpha}(x)\right] = -n\mathcal{L}_{n-1}^{\alpha}(x). \tag{7.31a}$$

(x) A. For the First Convention

Using the generating function for the first convention, we write

$$\sum_{n=0}^{\infty} L_n^{\alpha}(x)h^n = \frac{1}{(1-h)^{\alpha+1}} e^{-\frac{hx}{1-h}}.$$

Differentiating both sides once with respect to x, we obtain

$$\begin{aligned}\sum_{n=0}^{\infty}\left[\frac{d}{dx}L_n^{\alpha}(x)\right]h^n &= -\frac{h}{1-h}\left[\frac{1}{(1-h)^{\alpha+1}} e^{-\frac{hx}{1-h}}\right] \\ &= -h(1-h)^{-1}\sum_{r=0}^{\infty} L_r^{\alpha}(x)h^r \\ &= -h\sum_{s=0}^{\infty} h^s \sum_{r=0}^{\infty} L_r^{\alpha}(x)h^r \\ &= -\sum_{s=0}^{\infty}\sum_{r=0}^{\infty} L_r^{\alpha}(x)h^{r+s+1}.\end{aligned} \tag{7.32}$$

On the right-hand side, if we put $n = r + s + 1$ then $s = n - r - 1$.
Since, $s \geq 0$ and $r \geq 0$, we have, $n - r - 1 \geq 0$, i.e., $0 \leq r \leq n-1$
Therefore, the coefficient of h^n on the right-hand side of Eq. (7.32) is

$$-\sum_{r=0}^{n-1} L_r^{\alpha}(x).$$

Then, equating coefficients of h^n from both sides of Eq. (7.32), we obtain

$$\frac{d}{dx}L_n^{\alpha}(x) = -\sum_{r=0}^{n-1} L_r^{\alpha}(x). \tag{7.33}$$

B. For the Second Convention

Similarly, if we start from the generating function for the second convention, we will obtain

$$\frac{d}{dx}\mathcal{L}_n^{\alpha}(x) = -\sum_{r=0}^{n-1}\frac{n!}{r!}\mathcal{L}_r^{\alpha}(x). \tag{7.33a}$$

7.6 ORTHOGONALITY PROPERTY OF ASSOCIATED LAGUERRE POLYNOMIALS

Associated Laguerre polynomials $L_n^\alpha(x)$ can be made orthogonal in an interval, $0 \le x \le \infty$, by introducing a weight function, $e^{-x}x^\alpha$. They can also be made orthonormal with a certain normalization constant.

A. For the First Convention

Using the generating function of associated Laguerre polynomials, we write

$$\sum_{n=0}^{\infty} L_n^\alpha(x)h^n = \frac{1}{(1-h)^{\alpha+1}} e^{-\frac{hx}{1-h}}$$

and

$$\sum_{m=0}^{\infty} L_m^\alpha(x)k^m = \frac{1}{(1-k)^{\alpha+1}} e^{-\frac{kx}{1-k}}.$$

Multiplying the product of the above two equations by $e^{-x}x^\alpha$ and integrating the result with respect to x within the limit from $x = 0$ to $x = \infty$, we obtain

$$\sum_{m=0}^{\infty}\sum_{n=0}^{\infty} h^n k^m \int_0^\infty e^{-x}x^\alpha L_m^\alpha(x)L_n^\alpha(x)dx$$

$$= \frac{1}{\{(1-h)(1-k)\}^{\alpha+1}} \int_0^\infty x^\alpha e^{-\left(1+\frac{h}{1-h}+\frac{k}{1-k}\right)x} dx$$

$$= \frac{1}{\{(1-h)(1-k)\}^{\alpha+1}} \int_0^\infty x^\alpha e^{-\frac{(1-hk)x}{(1-h)(1-k)}} dx.$$

Putting $u = \dfrac{(1-hk)x}{(1-h)(1-k)}$, we have, $x = \dfrac{(1-h)(1-k)u}{1-hk}$ and $dx = \dfrac{(1-h)(1-k)du}{1-hk}$.

Therefore, we write

$$\sum_{m=0}^{\infty}\sum_{n=0}^{\infty} h^n k^m \int_0^\infty e^{-x}x^\alpha L_m^\alpha(x)L_n^\alpha(x)dx$$

$$= \frac{1}{(1-hk)^{\alpha+1}} \int_0^\infty e^{-u}u^\alpha du$$

$$= [1-hk]^{-(\alpha+1)}\Gamma(\alpha+1)$$

$$= \Gamma(\alpha+1)\left[1+(\alpha+1)hk+\frac{(\alpha+1)(\alpha+2)}{2!}h^2k^2+\frac{(\alpha+1)(\alpha+2)(\alpha+3)}{3!}h^3k^3+\ldots\infty\right]$$

$$= \Gamma(\alpha+1)\left[1+\sum_{r=1}^{\infty}\frac{(\alpha+1)(\alpha+2)(\alpha+3)\ldots(\alpha+r)}{r!}h^r k^r\right]$$

$$= \left[\Gamma(\alpha+1)+\sum_{r=1}^{\infty}\frac{\Gamma(\alpha+r+1)}{r!}h^r k^r\right]$$

$$= \sum_{r=0}^{\infty}\frac{\Gamma(\alpha+r+1)}{r!}h^r k^r. \tag{7.34}$$

Equating coefficients of $h^n k^m\,(m \neq n)$ from both sides of Eq. (7.34), we obtain

$$\int_0^{\infty} e^{-x}x^{\alpha}L_m^{\alpha}(x)L_n^{\alpha}(x)dx = 0. \tag{7.35}$$

Equation (7.35) is known as the orthogonality relation for associated Laguerre polynomials in the interval $\{0, \infty\}$.

If we equate the coefficients of $h^n k^n\,(m = n)$ from both sides of Eq. (7.34), we will obtain

$$\int_0^{\infty} e^{-x}x^{\alpha}L_n^{\alpha}(x)L_n^{\alpha}(x)dx = \frac{\Gamma(\alpha+n+1)}{n!}. \tag{7.36}$$

From Eq. (7.36), we say that associated Laguerre polynomials are orthonormal in the interval {0, ∞} with a normalization constant $\sqrt{\frac{n!}{\Gamma(\alpha+n+1)}}$.

Thus, the functions $\Lambda_n^{\alpha}(x) = \sqrt{\frac{n!}{\Gamma(\alpha+n+1)}}\, e^{-\frac{x}{2}} x^{\frac{\alpha}{2}} L_n^{\alpha}(x)$

and $\Lambda_m^{\alpha}(x) = \sqrt{\frac{m!}{\Gamma(\alpha+m+1)}}\, e^{-\frac{x}{2}} x^{\frac{\alpha}{2}} L_m^{\alpha}(x)$ are orthonormal in the interval {0, ∞}.

Thus, we can write

$$\int_0^{\infty} e^{-x}x^{\alpha}L_m^{\alpha}(x)L_n^{\alpha}(x)dx = \frac{\Gamma(\alpha+n+1)}{n!}\delta_{mn} \tag{7.37}$$

i.e.,

$$\int_0^{\infty} \Lambda_m^{\alpha}(x)\,\Lambda_n^{\alpha}(x)\,dx = \delta_{mn}. \tag{7.38}$$

B. For the Second Convention

Using the generating function for the second convention of associated Laguerre polynomials and following a similar procedure, we can obtain

$$\int_0^\infty e^{-x}x^\alpha \mathcal{L}_m^\alpha(x)\mathcal{L}_n^\alpha(x)dx = n!\,\Gamma(n+\alpha+1)\delta_{mn} \tag{7.37a}$$

or

$$\int_0^\infty \mathcal{Q}_m^\alpha(x)\,\mathcal{Q}_n^\alpha(x)\,dx = \delta_{mn} \tag{7.38a}$$

where

$$\mathcal{Q}_m^\alpha(x) = \frac{1}{\sqrt{m!\,\Gamma(m+\alpha+1)}} e^{-\frac{x}{2}} x^{\frac{\alpha}{2}} \mathcal{L}_m^\alpha(x) \text{ etc.}$$

7.7 EVALUATION OF INTEGRALS INVOLVING ASSOCIATED LAGUERRE POLYNOMIALS

A. For the First Convention

Using the first convention of associated Laguerre polynomials, let us evaluate

$$I = \int_0^\infty e^{-x}x^\alpha L_n^\alpha(x)f(x)dx\,, \text{ [where } f(x) \text{ is a function of } x]$$

$$= \int_0^\infty e^{-x}x^\alpha \left[\frac{e^x x^{-\alpha}}{n!}\frac{d^n}{dx^n}(e^{-x}x^{n+\alpha})\right]f(x)dx$$

$$= \frac{1}{n!}\int_0^\infty f(x)\frac{d^n}{dx^n}(e^{-x}x^{n+\alpha})dx$$

$$= \frac{(-1)^n}{n!}\int_0^\infty \left[\frac{d^n}{dx^n}f(x)\right]e^{-x}x^{n+\alpha}dx\,. \tag{7.39}$$

Here, Eq. (7.39) is the result of integration by parts performed n times.

Some Different Choices of $f(x)$:

I. $f(x) = e^{(1-k)x}$ where k is a parameter.

In this case, we have, $\frac{d^n}{dx^n}f(x) = (1-k)^n\, e^{(1-k)x}$.

Using this, Eq. (7.39) becomes

$$I = \int_0^\infty x^\alpha e^{-kx}L_n^\alpha(x)dx$$

$$=\frac{(k-1)^n}{n!}\int_0^\infty e^{-kx}x^{n+\alpha}dx$$

$$=\frac{(k-1)^n}{n!}\int_0^\infty e^{-u}\left(\frac{u}{k}\right)^{n+\alpha}\frac{du}{k}\quad \text{where we put } u = kx;$$

$$=\frac{(k-1)^n\Gamma(n+\alpha+1)}{n!\,k^{n+\alpha+1}}. \tag{7.40}$$

A more general type of integral,

$$\int_0^\infty x^{a+r}e^{-kx}L_n^\alpha(x)\,dx \tag{7.41}$$

can be evaluated if r is an integer.

II. $f(x)=x^r$ where r is an integer.

(a) For $r\geq n$, we have

$$\frac{d^n}{dx^n}f(x)=\frac{r!}{(r-n)!}x^{r-n}.$$

Using this, Eq. (7.39) becomes

$$I=\int_0^\infty x^{\alpha+r}e^{-x}L_n^\alpha(x)dx=\frac{(-1)^n}{n!}\frac{r!}{(r-n)!}\int_0^\infty x^{\alpha+r}e^{-x}dx$$

$$=\frac{(-1)^n\,r!}{n!\,(r-n)!}\Gamma(\alpha+r+1). \tag{7.42}$$

(b) If $r = 0$, we have, $f(x)=x^0=1$ and $\frac{d^n}{dx^n}f(x)=0$.

Therefore, using Eq. (7.39), we obtain

$$\int_0^\infty x^\alpha e^{-x}L_n^\alpha(x)\,dx=0. \tag{7.43}$$

(c) If $r < n$, we have, $\frac{d^n}{dx^n}f(x)=0$ and, therefore, Eq. (7.39) gives

$$\int_0^\infty x^{\alpha+r}e^{-x}L_n^\alpha(x)dx=0. \tag{7.44}$$

III. $f(x) = L_n^\alpha(x)$

We have $$f(x) = L_n^\alpha(x) = \sum_{r=0}^{n} \frac{(-1)^r \Gamma(n+\alpha+1)}{r!(n-r)!\Gamma(\alpha+r+1)} x^r$$

$$= \frac{(-1)^n}{n!} x^n + \frac{(-1)^{n-1}(n+\alpha)}{(n-1)!} x^{n-1} + \ldots + \frac{\Gamma(n+\alpha+1)}{n!\Gamma(\alpha+1)}.$$

Therefore, $$\frac{d^n}{dx^n} f(x) = (-1)^n.$$

Substituting this into Eq. (7.39), we obtain

$$\int_0^\infty x^\alpha e^{-x} L_n^\alpha(x) L_n^\alpha(x) dx = \frac{(-1)^n}{n!} \int_0^\infty (-1)^n e^{-x} x^{n+\alpha} dx$$

$$= \frac{1}{n!} \Gamma(n+\alpha+1). \tag{7.45}$$

The cases for $f(x) = x^2 L_n^\alpha(x), xL_n^\alpha(x), x^{-1} L_n^\alpha(x), x^{-2} L_n^\alpha(x)$, etc. are to be discussed in the section for Solved Problems.

B. For the Second Convention

Using the second convention of associated Laguerre polynomials, let us evaluate

$$I = \int_0^\infty e^{-x} x^\alpha \mathcal{L}_n^\alpha(x) f(x) dx = (-1)^n \int_0^\infty \left[\frac{d^n}{dx^\alpha} f(x) \right] e^{-x} x^{n+\alpha} dx. \tag{7.39a}$$

Some different choices of $f(x)$:

I. $f(x) = e^{(1-k)x}$

In this case, Eq. (7.39a) yields

$$\int_0^\infty x^\alpha e^{-kx} \mathcal{L}_n^\alpha(x) dx = \frac{(k-1)^n \Gamma(n+\alpha+1)}{k^{n+\alpha+1}}. \tag{7.40a}$$

As in the first convention, a more general type of integral,

$$\int_0^\infty x^{\alpha+r} e^{-kx} \mathcal{L}_n^\alpha(x) dx \tag{7.41a}$$

can be evaluated if r is an integer.

II. $f(x)=x^r$ where r is an integer.

(a) For $r \geq n$ Eq. (7.39a) yields

$$\int_0^\infty x^{\alpha+r} e^{-x} \mathcal{L}_n^\alpha(x)dx = (-1)^n \frac{r!}{(r-n)!}\Gamma(\alpha+r+1). \tag{7.42a}$$

(b) For $r = 0$, $f(x)=x^0=1$ and hence Eq. (7.39a) yields

$$\int_0^\infty x^\alpha e^{-x} \mathcal{L}_n^\alpha(x)dx = 0. \tag{7.43a}$$

(c) If $r<n$, we have $\frac{d^n}{dx^n}f(x)=0$ and hence Eq. (7.39a) yields

$$\int_0^\infty x^{\alpha+r} e^{-x} \mathcal{L}_n^\alpha(x)dx = 0. \tag{7.44a}$$

III. If we choose $f(x)=\mathcal{L}_n^\alpha(x)$ Eq. (7.39a) yields

$$\int_0^\infty e^{-x} x^\alpha \mathcal{L}_n^\alpha(x)\mathcal{L}_n^\alpha(x)dx = n!\,\Gamma(n+\alpha+1). \tag{7.45a}$$

7.8 EXPANSION OF AN ARBITRARY FUNCTION IN A SERIES OF ASSOCIATED LAGUERRE POLYNOMIALS

If an arbitrary function $f(x)$ is piecewise smooth in every finite interval $[x_1, x_2]$ and suitably well behaved near the points $x=0$ and $x=\infty$, then it can be expanded as

$$f(x)=\sum_{n=0}^{\infty} c_n L_n^\alpha(x) \tag{7.46}$$

where c_n are to be determined.

Multiplying both sides of Eq. (7.46) by $x^\alpha e^{-x} L_m^\alpha(x)$ and integrating with respect to x from $x=0$ to $x=\infty$, we have

$$\int_0^\infty f(x)x^\alpha e^{-x} L_m^\alpha(x)dx = \sum_{n=0}^{\infty} c_n \int_0^\infty x^\alpha e^{-x} L_m^\alpha(x) L_n^\alpha(x)dx = \frac{\Gamma(m+\alpha+1)}{m!}c_m.$$

Therefore, we have $c_m = \dfrac{m!}{\Gamma(m+\alpha+1)}\displaystyle\int_0^\infty e^{-x} x^\alpha f(x) L_m^\alpha(x)dx.$

Changing the subscript m to n, we have

$$c_n = \frac{n!}{\Gamma(n+\alpha+1)}\int_0^\infty e^{-x} x^\alpha f(x) L_n^\alpha(x)dx. \tag{7.47}$$

In the above case, we use the first convention of associated Laguerre polynomials. If we use the second convention of associated Laguerre polynomials, we will have

$$f(x)=\sum_{n=0}^{\infty} b_n \mathcal{L}_n^{\alpha}(x) \tag{7.46a}$$

where
$$b_n=\frac{1}{n!\,\Gamma(n+\alpha+1)}\int_0^{\infty} e^{-x}x^{\alpha} f(x)\mathcal{L}_n^{\alpha}(x)dx. \tag{7.47a}$$

7.9 SOLVED PROBLEMS

PROBLEM 1 Prove that $L_n^{k+\beta+1}(x+y)=\sum_{r=0}^{n} L_r^k(x)L_{n-r}^{\beta}(y)$ where n is a non-negative integer.

Solution: Using the generating function for associated Laguerre polynomials, we write

$$\sum_{n=0}^{\infty} L_n^{k+\beta+1}(x+y)h^n=\frac{1}{(1-h)^{k+\beta+2}}e^{-\frac{h(x+y)}{1-h}}$$

$$=\frac{1}{(1-h)^{k+1}}e^{-\frac{hx}{1-h}}\frac{1}{(1-h)^{\beta+1}}e^{-\frac{hy}{1-h}}$$

$$=\sum_{r=0}^{\infty} L_r^k(x)h^r\sum_{n=0}^{\infty} L_n^{\beta}(y)h^n=\sum_{r=0}^{\infty}\sum_{n=0}^{\infty} L_r^k(x)L_n^{\beta}(y)h^{r+n}. \qquad \text{...(i)}$$

Now, equating coefficients of h^n from both sides of Eq. (i), we obtain

$$L_n^{k+\beta+1}(x+y)=\sum_{r=0}^{n} L_r^k(x)L_{n-r}^{\beta}(y).$$

PROBLEM 2 Prove that $L_n^{\alpha}(x)=\sum_{k=0}^{n}\frac{(\alpha-\beta)_k}{k!}L_{n-k}^{\beta}(x)$ noting that the symbol

$$(c)_n=c(c+1)(c+2)(c+3)\dots(c+n-1)=\frac{\Gamma(c+n)}{\Gamma(c)}$$

is the so called Pochhammer symbol.

Solution: We have

$$\sum_{n=0}^{\infty} L_n^{\alpha}(x)h^n=\frac{1}{(1-h)^{\alpha+1}}e^{-\frac{hx}{1-h}}$$

$$= \left[(1-h)^{-(\alpha-\beta)}\right] \frac{1}{(1-h)^{\beta+1}} e^{-\frac{hx}{1-h}}$$

$$= \left[1 + (\alpha-\beta)h + \frac{(\alpha-\beta)(\alpha-\beta+1)}{2!}h^2 + \frac{(\alpha-\beta)(\alpha-\beta+1)(\alpha-\beta+2)}{3!}h^3 + \ldots\right.$$

$$\left. + \frac{(\alpha-\beta)(\alpha-\beta+1)(\alpha-\beta+2)\ldots(\alpha-\beta+k-1)}{k!}h^k + \ldots\infty\right] \sum_{n=0}^{\infty} L_n^{\beta}(x)h^n$$

$$= \left[1 + \sum_{k=1}^{\infty} \frac{\Gamma(\alpha-\beta+k)}{\Gamma(\alpha-\beta)\,k!}h^k\right] \sum_{n=0}^{\infty} L_n^{\beta}(x)h^n$$

$$= \sum_{k=0}^{\infty}\sum_{n=0}^{\infty} \frac{(\alpha-\beta)_k}{k!} L_n^{\beta}(x)h^{k+n}.$$

Equating coefficients of h^n from both sides of the above equation, we obtain

$$L_n^{\alpha}(x) = \sum_{k=0}^{n} \frac{(\alpha-\beta)_k}{k!} L_{n-k}^{\beta}(x).$$

PROBLEM 3 Prove that $n!\dfrac{d^k}{dx^k}\left[e^{-x}x^{\alpha}L_n^{\alpha}(x)\right] = (n+k)!\,e^{-x}x^{\alpha-k}L_{n+k}^{\alpha-k}(x).$

Solution: By Rodrigues' formula, we write

$$L_n^{\alpha}(x) = \frac{x^{-\alpha}e^{x}}{n!}\frac{d^n}{dx^n}(x^{n+\alpha}e^{-x}).$$

Thus, we have

$$n!\,e^{-x}x^{\alpha}L_n^{\alpha}(x) = \frac{d^n}{dx^n}(x^{n+\alpha}e^{-x}).$$

Differentiating both sides k times with respect to x, we obtain

$$n!\frac{d^k}{dx^k}\left[e^{-x}x^{\alpha}L_n^{\alpha}(x)\right] = \frac{d^{n+k}}{dx^{n+k}}(x^{n+\alpha}e^{-x})$$

$$= \sum_{r=0}^{n+k} \frac{(n+k)!}{r!(n+k-r)!}\left(\frac{d^{n+k-r}}{dx^{n+k-r}}x^{n+\alpha}\right)\left(\frac{d^r}{dx^r}e^{-x}\right) \text{ (by Leibnitz theorem);}$$

$$=\sum_{r=0}^{n+k}\frac{(n+k)!}{r!(n+k-r)!}(n+\alpha)(n+\alpha-1)\ldots(\alpha-k+r+1)x^{\alpha-k+r}(-1)^r e^{-x}$$

$$=\sum_{r=0}^{n+k}\frac{(n+k)!}{r!(n+k-r)!}\frac{\Gamma(n+\alpha+1)}{\Gamma(\alpha-k+r+1)}x^{\alpha-k+r}(-1)^r e^{-x}$$

$$=(n+k)!\,e^{-x}x^{\alpha-k}\sum_{r=0}^{n+k}\frac{(-1)^r\,\Gamma(n+k+\alpha-k+1)}{r!(n+k-r)!\,\Gamma(\alpha-k+r+1)}x^r$$

$$=(n+k)!\,e^{-x}x^{\alpha-k}L_{n+k}^{\alpha-k}(x).$$

Here, use has been made of the formula

$$L_n^{\alpha}(x)=\sum_{r=0}^{n}\frac{(-1)^r\,\Gamma(n+\alpha+1)}{r!(n-r)!\,\Gamma(r+\alpha+1)}x^r.$$

PROBLEM 4 If $J_k(x)$ is the Bessel function defined by

$$J_k(x)=\sum_{r=0}^{\infty}\frac{(-1)^r}{r!\,\Gamma(k+r+1)}\left(\frac{x}{2}\right)^{k+2r},$$

then prove that $e^t(xt)^{-\frac{k}{2}}J_k(2\sqrt{xt})=\sum_{n=0}^{\infty}\frac{L_n^k(x)}{\Gamma(n+k+1)}t^n$.

Solution: We have

$$e^t(xt)^{-\frac{k}{2}}J_k(2\sqrt{xt})=e^t(xt)^{-\frac{k}{2}}\sum_{r=0}^{\infty}\frac{(-1)^r(\sqrt{xt})^{k+2r}}{r!\,\Gamma(k+r+1)}$$

$$=\sum_{r=0}^{\infty}\frac{(-1)^r e^t(xt)^r}{r!\,\Gamma(k+r+1)}$$

$$=\sum_{r=0}^{\infty}\frac{(-1)^r(xt)^r}{r!\,\Gamma(k+r+1)}\sum_{n=0}^{\infty}\frac{t^n}{n!}$$

$$=\sum_{n=0}^{\infty}\sum_{r=0}^{\infty}\frac{(-1)^r x^r t^{n+r}}{r!\,n!\,\Gamma(k+r+1)}.$$

In the summation on the right-hand side, if we replace n by $(n-r)$, we write

$$e^t(xt)^{-\frac{k}{2}}J_k(2\sqrt{xt})=\sum_{n=0}^{\infty}\sum_{r=0}^{\infty}\frac{(-1)^r x^r t^n}{r!(n-r)!\,\Gamma(k+r+1)}$$

$$= \sum_{n=0}^{\infty} \frac{1}{\Gamma(n+k+1)} \left[\sum_{r=o}^{n} \frac{(-1)^r \Gamma(n+k+1)}{r!(n-r)!\Gamma(k+r+1)} x^r \right] t^n$$

$$= \sum_{n=0}^{\infty} \frac{L_n^k(x)}{\Gamma(n+k+1)} t^n .$$

PROBLEM 5 Prove that $L_n^\alpha(x) = \dfrac{e^x x^{-\frac{\alpha}{2}}}{n!} \int_0^\infty t^{n+\frac{\alpha}{2}} J_\alpha(2\sqrt{xt}) e^{-t} dt$.

Solution: We know that

$$e^{-x} x^{n+\alpha} = \int_0^\infty (\sqrt{xt})^{n+\alpha} e^{-t} J_{n+\alpha}(2\sqrt{xt}) dt$$

which is to be proved in the chapter on Bessel functions.

Differentiating both sides n times with respect to x, we obtain

$$\frac{d^n}{dx^n}(e^{-x} x^{n+\alpha}) = \frac{d^n}{dx^n} \int_0^\infty (\sqrt{xt})^{n+\alpha} e^{-t} J_{n+\alpha}(2\sqrt{xt}) dt .$$

Putting $2\sqrt{xt} = u$ in the expression on right-hand side of the above equation, we write

$$\frac{d^n}{dx^n}(e^{-x} x^{n+\alpha}) = \int_0^\infty \left(\frac{du}{dx} \right)^n \frac{d^n}{du^n} \left[\left(\frac{u}{2} \right)^{n+\alpha} J_{n+\alpha}(u) \right] e^{-t} dt$$

$$= \frac{1}{2^{n+\alpha}} \int_0^\infty (\sqrt{xt})^{-n} t^n \frac{d^n}{du^n} \left[u^{n+\alpha} J_{n+\alpha}(u) \right] e^{-t} dt .$$

n times use of the recurrence relation of Bessel functions, namely

$\dfrac{d}{du}\left[u^k J_k(u) \right] = u^k J_{k-1}(u)$ results in

$$\frac{d^n}{dx^n}(e^{-x} x^{n+\alpha}) = \frac{1}{2^{n+\alpha}} \int_0^\infty (\sqrt{xt})^{-n} t^n u^{n+\alpha} J_\alpha(u) e^{-t} dt$$

$$= \frac{1}{2^{n+\alpha}} \int_0^\infty (\sqrt{xt})^{-n} t^n (2\sqrt{xt})^{n+\alpha} J_\alpha(2\sqrt{xt}) e^{-t} dt$$

$$= x^{\frac{\alpha}{2}} \int_0^\infty t^{n+\frac{\alpha}{2}} J_\alpha(2\sqrt{xt}) e^{-t} dt .$$

Multiplying both sides by $\dfrac{e^x x^{-\alpha}}{n!}$, we obtain

$$\frac{e^x x^{-\alpha}}{n!}\frac{d^n}{dx^n}(e^{-x}x^{n+\alpha}) = \frac{e^x x^{-\frac{\alpha}{2}}}{n!}\int_0^\infty t^{n+\frac{\alpha}{2}} J_\alpha(2\sqrt{xt})e^{-t}dt$$

or

$$L_n^\alpha(x) = \frac{e^x x^{-\frac{\alpha}{2}}}{n!}\int_0^\infty t^{n+\frac{\alpha}{2}} J_\alpha(2\sqrt{xt})e^{-t}dt.$$

PROBLEM 6 Prove that

$$\int_{-1}^{+1}(1-t^2)^{\alpha-\frac{1}{2}} H_{2n}(\sqrt{x}\,t)dt = (-1)^n\sqrt{\pi}\,\frac{(2n)!\,\Gamma\left(\alpha+\frac{1}{2}\right)}{\Gamma(n+\alpha+1)}L_n^\alpha(x).$$

Solution:

$$\begin{aligned}
\text{L.H.S.} &= \int_{-1}^{+1}(1-t^2)^{\alpha-\frac{1}{2}} H_{2n}(\sqrt{x}\,t)dt \\
&= \int_{-1}^{+1}(1-t^2)^{\alpha-\frac{1}{2}}\sum_{r=0}^{n}\frac{(-1)^r(2n)!}{r!(2n-2r)!}(2\sqrt{x}\,t)^{2n-2r}dt \\
&= \int_{-1}^{+1}\sum_{r=0}^{n}\frac{(-1)^r(2n)!\,2^{2n-2r}x^{n-r}}{r!(2n-2r)!}(1-t^2)^{\alpha-\frac{1}{2}}t^{2n-2r}dt \\
&= \sum_{r=0}^{n}\frac{(-1)^r(2n)!\,2^{2n-2r}x^{n-r}}{r!(2n-2r)!}2\int_0^1(1-t^2)^{\alpha-\frac{1}{2}}t^{2n-2r}dt.
\end{aligned}$$

Putting $s = n-r$ or $r = n-s$ in the expression on the right-hand side of the above equation, we write

$$\begin{aligned}
\int_{-1}^{+1}(1-t^2)^{\alpha-\frac{1}{2}} H_{2n}(\sqrt{x}\,t)dt &= \sum_{s=0}^{n}\frac{(-1)^{n-s}(2n)!\,2^{2s}x^s}{(n-s)!(2s)!}2\int_0^1(1-t^2)^{\alpha-\frac{1}{2}}t^{2s}dt \\
&= (-1)^n(2n)!\sum_{s=0}^{n}\frac{(-1)^{-s}2^{2s}x^s}{(n-s)!(2s)!}2\int_0^1(1-t^2)^{\alpha-\frac{1}{2}}t^{2s}dt \\
&= (-1)^n(2n)!\sum_{s=0}^{n}\frac{(-1)^s 2^{2s}x^s}{(n-s)!(2s)!}2\int_0^1(1-t^2)^{\alpha-\frac{1}{2}}t^{2s}dt. \qquad \text{(i)}
\end{aligned}$$

Putting $t = \sin\theta$ the integral on the right-hand side of the above equation can be written as

$$I = 2\int_0^1 (1-t^2)^{\alpha-\frac{1}{2}} t^{2s}\,dt = 2\int_0^{\frac{\pi}{2}} \cos^{2\alpha}\theta \sin^{2s}\theta\,d\theta$$

$$= B\left(\frac{2\alpha+1}{2}, \frac{2s+1}{2}\right) = \frac{\Gamma\left(\alpha+\frac{1}{2}\right)\Gamma\left(s+\frac{1}{2}\right)}{\Gamma(\alpha+s+1)}.$$

Substituting the value of this integral into Eq. (i), we write

$$\int_{-1}^{+1} (1-t^2)^{\alpha-\frac{1}{2}} H_{2n}(\sqrt{x}\,t)dt = (-1)^n (2n)! \sum_{s=0}^{n} \frac{(-1)^s 2^{2s} x^s}{(n-s)!\,(2s)!} \frac{\Gamma\left(\alpha+\frac{1}{2}\right)\Gamma\left(s+\frac{1}{2}\right)}{\Gamma(\alpha+s+1)}. \quad \text{(ii)}$$

We know that

$$\Gamma\left(s+\frac{1}{2}\right) = \Gamma\left(\frac{2s+1}{2}\right) = \left(\frac{2s-1}{2}\right)\left(\frac{2s-3}{2}\right)\left(\frac{2s-5}{2}\right)\ldots\frac{5}{2}\cdot\frac{3}{2}\cdot\frac{1}{2}\Gamma\left(\frac{1}{2}\right)$$

$$= \frac{(2s-1)(2s-3)(2s-5)\ldots 5\cdot 3\cdot 1}{2^s}\Gamma\left(\frac{1}{2}\right)$$

$$= \frac{(2s)(2s-1)(2s-2)(2s-3)(2s-4)(2s-5)\ldots 5.4.3.2.1}{2^s(2s)(2s-2)(2s-4)\ldots 6.4.2}\sqrt{\pi}$$

$$= \frac{(2s)!\sqrt{\pi}}{2^{2s}\,s!}. \quad \text{(iii)}$$

Substituting Eq (iii) into Eq. (ii), we obtain

$$\int_{-1}^{+1} (1-t^2)^{\alpha-\frac{1}{2}} H_{2n}(\sqrt{x}\,t)dt = (-1)^n (2n)! \sum_{s=0}^{n} \frac{(-1)^s x^s}{(n-s)!\,s!} \frac{\sqrt{\pi}\Gamma\left(\alpha+\frac{1}{2}\right)}{\Gamma(\alpha+s+1)}$$

$$= \frac{(-1)^n (2n)!\sqrt{\pi}\,\Gamma\left(\alpha+\frac{1}{2}\right)}{\Gamma(n+\alpha+1)} \sum_{s=0}^{n} \frac{(-1)^s \Gamma(n+\alpha+1)}{s!\,(n-s)!\,\Gamma(\alpha+s+1)} x^s$$

$$= (-1)^n \sqrt{\pi} \frac{(2n)!\,\Gamma\left(\alpha+\frac{1}{2}\right)}{\Gamma(n+\alpha+1)} L_n^{\alpha}(x).$$

PROBLEM 7 Prove that $\int_0^1 t^{\alpha}(1-t)^{\beta-1} L_n^{\alpha}(xt)\,dt = \dfrac{\Gamma(\beta)\Gamma(\alpha+n+1)}{\Gamma(\alpha+\beta+n+1)} L_n^{\alpha+\beta}(x)$.

Solution:

$$\int_0^1 t^{\alpha}(1-t)^{\beta-1} L_n^{\alpha}(xt)\,dt = \int_0^1 t^{\alpha}(1-t)^{\beta-1} \sum_{r=0}^{n} \frac{(-1)^r \Gamma(n+\alpha+1)}{r!(n-r)!\Gamma(\alpha+r+1)}(xt)^r\,dt$$

$$= \sum_{r=0}^{n} \frac{(-1)^r \Gamma(n+\alpha+1) x^r}{r!(n-r)!\Gamma(\alpha+r+1)} \int_0^1 t^{\alpha+r}(1-t)^{\beta-1}\,dt. \tag{i}$$

The integral in the expression on the right-hand side can be evaluated through beta and gamma functions such that

$$\int_0^1 t^{\alpha+r}(1-t)^{\beta-1}\,dt = B(\alpha+r+1,\beta) = \frac{\Gamma(\alpha+r+1)\Gamma(\beta)}{\Gamma(\alpha+\beta+r+1)}. \tag{ii}$$

Substitution of Eq. (ii) into Eq. (i) yields

$$\int_0^1 t^{\alpha}(1-t)^{\beta-1} L_n^{\alpha}(xt)\,dt = \sum_{r=0}^{n} \frac{(-1)^r \Gamma(n+\alpha+1)\Gamma(\beta)}{r!(n-r)!\Gamma(\alpha+\beta+r+1)} x^r$$

$$= \frac{\Gamma(\beta)\Gamma(n+\alpha+1)}{\Gamma(\alpha+\beta+n+1)} \sum_{r=0}^{n} \frac{(-1)^r \Gamma(\alpha+\beta+n+1)}{r!(n-r)!\Gamma(\alpha+\beta+r+1)} x^r$$

$$= \frac{\Gamma(\beta)\Gamma(n+\alpha+1)}{\Gamma(\alpha+\beta+n+1)} L_n^{\alpha+\beta}(x).$$

PROBLEM 8 Prove that the Hermite and associated Laguerre polynomials are related by the following relations

$$H_{2m}(x) = (-1)^m 2^{2m} m!\, L_m^{-\frac{1}{2}}(x^2)$$

and

$$H_{2m+1}(x) = (-1)^m 2^{2m+1} m!\, x L_m^{\frac{1}{2}}(x^2).$$

Solution:

First part: A Hermite polynomial is written as

$$H_n(x) = \sum_{r=0}^{\left[\frac{n}{2}\right]} \frac{(-1)^r n!}{r!(n-2r)!}(2x)^{n-2r}$$

where
$$\left[\frac{n}{2}\right]=\begin{cases}\dfrac{n}{2} \text{ if } n \text{ is even integer}\\ \dfrac{n-1}{2} \text{ if } n \text{ is odd integer}.\end{cases}$$

Therefore, we write

$$H_{2m}(x)=\sum_{r=0}^{m}\frac{(-1)^r(2m)!}{r!(2m-2r)!}(2x)^{2m-2r}$$

$$=\sum_{r=0}^{m}\frac{(-1)^r 1.2.3.4.5.6\ldots(2m-1)(2m)}{r!1.2.3.4\ldots(2m-2r-1)(2m-2r)}(2x)^{2m-2r}$$

$$=\sum_{r=0}^{m}\frac{(-1)^r\left[1.3.5\ldots(2m-1)\right][2.4.6\ldots 2m]}{r!\left[1.3.5\ldots(2m-2r-1)\right]\left[2.4.6\ldots(2m-2r)\right]}(2x)^{2m-2r}$$

$$=\sum_{r=0}^{m}\frac{(-1)^r 2^m\left[\frac{1}{2}\cdot\frac{3}{2}\cdot\frac{5}{2}\ldots\frac{(2m-1)}{2}\right]2^m(1.2.3\ldots m)}{r!\,2^{m-r}\left[\frac{1}{2}\cdot\frac{3}{2}\cdot\frac{5}{2}\ldots\frac{(2m-2r-1)}{2}\right]2^{m-r}\left[(1.2.3\ldots m-r)\right]}(2x)^{2m-2r}$$

$$=\sum_{r=0}^{m}\frac{(-1)^r 2^{2m}m!\,\Gamma\left(\frac{1}{2}\right)\left[\frac{1}{2}\cdot\frac{3}{2}\cdot\frac{5}{2}\ldots\frac{(2m-1)}{2}\right]}{r!\,2^{2m-2r}(m-r)!\,\Gamma\left(\frac{1}{2}\right)\left[\frac{1}{2}\cdot\frac{3}{2}\cdot\frac{5}{2}\ldots\frac{(2m-2r-1)}{2}\right]}(2x)^{2m-2r}$$

$$=\sum_{r=0}^{m}\frac{(-1)^r 2^{2m}m!\,\Gamma\left(m+\frac{1}{2}\right)}{r!(m-r)!\,\Gamma\left(m-r+\frac{1}{2}\right)}(x^2)^{m-r}.$$

Putting $s=m-r$ or $r=m-s$ in the expression on the right-hand side, we write

$$H_{2m}(x)=\sum_{s=0}^{m}\frac{(-1)^{m-s}2^{2m}m!\,\Gamma\left(m+\frac{1}{2}\right)}{(m-s)!\,s!\,\Gamma\left(s+\frac{1}{2}\right)}(x^2)^s$$

$$=(-1)^m 2^{2m}m!\sum_{s=0}^{m}\frac{(-1)^s\,\Gamma\left(m-\frac{1}{2}+1\right)}{s!\,(m-s)!\,\Gamma\left(s-\frac{1}{2}+1\right)}(x^2)^s$$

$$=(-1)^m 2^{2m}m!\,L_m^{-\frac{1}{2}}(x^2).$$

Second part: We have

$$H_{2m+1}(x)=\sum_{r=0}^{m}\frac{(-1)^r(2m+1)!}{r!(2m+1-2r)!}(2x)^{2m+1-2r}$$

$$=\sum_{r=0}^{m}\frac{(-1)^r 1.2.3.4\ldots(2m)(2m+1)}{r!1.2.3.4\ldots(2m-2r)(2m-2r+1)}(2x)^{2m-2r+1}$$

$$=\sum_{r=0}^{m}\frac{(-1)^r[1.3.5\ldots(2m+1)][2.4.6\ldots(2m)]}{r![1.3.5\ldots(2m-2r+1)][2.4.6\ldots(2m-2r)]}(2x)^{2m-2r+1}$$

$$=\sum_{r=0}^{m}\frac{(-1)^r 2^m m!\,2^{m+1}\Gamma\left(\frac{1}{2}\right)\left[\frac{1}{2}\cdot\frac{3}{2}\cdot\frac{5}{2}\cdots\frac{(2m+1)}{2}\right]}{r!\,2^{m-r}(m-r)!2^{m-r+1}\Gamma\left(\frac{1}{2}\right)\left[\frac{1}{2}\cdot\frac{3}{2}\cdot\frac{5}{2}\cdots\frac{(2m-2r+1)}{2}\right]}(2x)^{2m-2r+1}$$

$$=\sum_{r=0}^{m}\frac{(-1)^r 2^{2m+1}m!\,\Gamma\left(m+\frac{3}{2}\right)}{r!(m-r)!\,\Gamma\left(m-r+\frac{3}{2}\right)}(x)^{2m-2r+1}.$$

Putting $s=m-r$ or $r=m-s$ in the expression on the right-hand side, we write

$$H_{2m+1}(x)=\sum_{s=0}^{m}\frac{(-1)^{m-s}2^{2m+1}m!\,\Gamma\left(m+\frac{3}{2}\right)}{(m-s)!\,s!\,\Gamma\left(s+\frac{3}{2}\right)}x^{2s+1}$$

$$=(-1)^m 2^{2m+1}m!\,x\sum_{s=0}^{m}\frac{(-1)^s\Gamma\left(m+\frac{1}{2}+1\right)}{s!(m-s)!\,\Gamma\left(\frac{1}{2}+s+1\right)}(x^2)^s$$

$$=(-1)^m 2^{2m+1}m!\,x\,L_m^{\frac{1}{2}}(x^2).$$

PROBLEM 9 Prove that $L_n^{1/2}(x)=\dfrac{2e^x}{n!\sqrt{\pi x}}\displaystyle\int_0^\infty t^{2n+1}e^{-t^2}\sin(2t\sqrt{x})\,dt.$

Solution: From Problem 5, we know that

$$L_n^\alpha(x)=\frac{e^x x^{-\alpha/2}}{n!}\int_0^\infty t^{n+\frac{\alpha}{2}}J_\alpha(2\sqrt{xt})e^{-t}dt.$$

Putting $\alpha = \frac{1}{2}$, we write

$$L_n^{1/2}(x) = \frac{e^x x^{-\frac{1}{4}}}{n!} \int_0^\infty t^{n+\frac{1}{4}} J_{\frac{1}{2}}(2\sqrt{xt}) e^{-t} dt.$$

Since, $J_{\frac{1}{2}}(2\sqrt{xt}) = \sqrt{\frac{2}{\pi(2\sqrt{xt})}} \sin(2\sqrt{xt})$, we write

$$L_n^{1/2}(x) = \frac{e^x}{n!\sqrt{\pi x}} \int_0^\infty t^n e^{-t} \sin(2\sqrt{xt}) dt.$$

Let us put $t = u^2$, then $dt = 2u\, du$.
Thus, we write

$$L_n^{1/2}(x) = \frac{2e^x}{n!\sqrt{\pi x}} \int_0^\infty u^{2n+1} e^{-u^2} \sin(2u\sqrt{x}) du.$$

Renaming the variable, we write

$$L_n^{1/2}(x) = \frac{2e^x}{n!\sqrt{\pi x}} \int_0^\infty t^{2n+1} e^{-t^2} \sin(2t\sqrt{x}) dt.$$

PROBLEM 10 Evaluate the integrals

(a) $\int_0^\infty e^{-x} x^{\alpha+1} \left[L_n^\alpha(x)\right]^2 dx$, (b) $\int_0^\infty e^{-x} x^{\alpha+2} \left[L_n^\alpha(x)\right]^2 dx$,

(b) $\int_0^\infty e^{-x}\, x^{\alpha-1} \left[L_n^\alpha(x)\right]^2 dx$, (d) $\int_0^\infty e^{-x} x^{\alpha-2} \left[L_n^\alpha(x)\right]^2 dx$.

Solution:

(a) Using Rodrigues' formula, we write the given integral as

$$\int_0^\infty e^{-x} x^{\alpha+1} \left[L_n^\alpha(x)\right]^2 dx = \int_0^\infty e^{-x} x^{\alpha+1} L_n^\alpha(x) \left[\frac{x^{-\alpha} e^x}{n!} \frac{d^n}{dx^n}(e^{-x} x^{n+\alpha})\right] dx$$

$$= \frac{1}{n!} \int_0^\infty x L_n^\alpha(x) \frac{d^n}{dx^n}(e^{-x} x^{n+\alpha}) dx$$

$$= \frac{(-1)^n}{n!} \int_0^\infty \frac{d^n}{dx^n}\left[x L_n^\alpha(x)\right] e^{-x} x^{n+\alpha} dx. \tag{i}$$

Equation (i) results from integration by parts n times.
We know that

$$x L_n^\alpha(x) = x\sum_{r=0}^{n}\frac{(-1)^r\,\Gamma(\alpha+n+1)}{r!\,(n-r)!\,\Gamma(\alpha+r+1)}x^r$$

$$= x\left[\frac{(-1)^n x^n}{n!}+\frac{(-1)^{n-1}(\alpha+n)x^{n-1}}{(n-1)!\,1!}+\frac{(-1)^{n-2}(\alpha+n)(\alpha+n-1)x^{n-2}}{(n-2)!\,2!}+\ldots\right]$$

$$= \frac{(-1)^n}{n!}\left[x^{n+1}-n(\alpha+n)x^n+\frac{n(n-1)(\alpha+n)(\alpha+n-1)}{2}x^{n-1}-\ldots\right].$$

Therefore,
$$\frac{d^n}{dx^n}\left[xL_n^\alpha(x)\right]=(-1)^n\left[(n+1)x-n(\alpha+n)+0\right]. \qquad \text{(ii)}$$

Substituting Eq. (ii) into Eq. (i), we obtain

$$\int_0^\infty e^{-x}x^{\alpha+1}\left[L_n^\alpha(x)\right]^2 dx = \frac{1}{n!}\left[(n+1)\int_0^\infty e^{-x}x^{n+\alpha+1}dx - n(n+\alpha)\int_0^\infty e^{-x}x^{n+\alpha}dx\right]$$

$$= \frac{1}{n!}\left[(n+1)\Gamma(n+\alpha+2)-n(n+\alpha)\Gamma(n+\alpha+1)\right]$$

$$= \frac{(\alpha+2n+1)\Gamma(\alpha+n+1)}{n!}.$$

(b) Proceeding as in (a), we obtain

$$\int_0^\infty e^{-x}x^{\alpha+2}\left[L_n^\alpha(x)\right]^2 dx = \frac{1}{n!}\int_0^\infty x^2 L_n^\alpha(x)\frac{d^n}{dx^n}(e^{-x}x^{n+\alpha})dx$$

$$= \frac{(-1)^n}{n!}\int_0^\infty \frac{d^n}{dx^n}\left[x^2 L_n^\alpha(x)\right]e^{-x}x^{n+\alpha}dx. \qquad \text{(iii)}$$

We have

$$x^2 L_n^\alpha(x) = x^2\sum_{r=0}^{n}\frac{(-1)^r\,\Gamma(\alpha+n+1)}{r!\,(n-r)!\,\Gamma(\alpha+r+1)}x^r$$

$$= \frac{(-1)^n}{n!}\left[x^{n+2}-n(n+\alpha)x^{n+1}+\frac{n(n-1)(n+\alpha)(n+\alpha-1)}{2}x^n-\ldots\right]$$

where the omitted terms represent a polynomial of degree lower than n.

Therefore,

$$\frac{d^n}{dx^n}\left[x^2 L_n^\alpha(x)\right]$$

$$=\frac{(-1)^n}{n!}\left[\frac{(n+2)!}{2!}x^2-n(n+\alpha)(n+1)!x+\frac{n(n-1)(n+\alpha)(n+\alpha-1)n!}{2}\right]$$

$$=(-1)^n\left[\frac{(n+1)(n+2)}{2}x^2-n(n+1)(n+\alpha)x+\frac{n(n-1)(n+\alpha-1)(n+\alpha)}{2}\right]. \quad \text{(iv)}$$

Substituting Eq. (iv) into Eq. (iii), we obtain

$$\int_0^\infty e^{-x}x^{\alpha+2}\left[L_n^\alpha(x)\right]^2 dx$$

$$=\frac{1}{n!}\left[\frac{(n+1)(n+2)}{2}\int_0^\infty e^{-x}x^{n+\alpha+2}\,dx-n(n+1)(n+\alpha)\int_0^\infty e^{-x}x^{n+\alpha+1}dx\right.$$

$$\left.+\frac{n(n-1)(n+\alpha-1)(n+\alpha)}{2}\int_0^\infty e^{-x}x^{n+\alpha}dx\right]$$

$$=\frac{1}{n!}\left[\frac{(n+1)(n+2)}{2}\Gamma(n+\alpha+3)-n(n+1)(n+\alpha)\Gamma(n+\alpha+2)\right.$$

$$\left.+\frac{n(n-1)(n+\alpha-1)(n+\alpha)}{2}\Gamma(n+\alpha+1)\right]$$

$$=\frac{\Gamma(n+\alpha+1)}{n!}\left[6n^2+6n(\alpha+1)+(\alpha+1)(\alpha+2)\right].$$

(c) Proceeding as in (a), we obtain

$$\int_0^\infty e^{-x}x^{\alpha-1}\left[L_n^\alpha(x)\right]^2 dx=\frac{1}{n!}\int_0^\infty x^{-1}L_n^\alpha(x)\frac{d^n}{dx^n}(e^{-x}x^{n+\alpha})dx$$

$$=\frac{(-1)^n}{n!}\int_0^\infty\frac{d^n}{dx^n}\left[x^{-1}L_n^\alpha(x)\right]e^{-x}x^{n+\alpha}dx. \quad \text{(v)}$$

We know that

$$x^{-1}L_n^\alpha(x)=x^{-1}\sum_{r=0}^{n}\frac{(-1)^r\,\Gamma(\alpha+n+1)}{r!(n-r)!\Gamma(\alpha+r+1)}x^r$$

$$=\frac{\Gamma(\alpha+n+1)}{n!\,\Gamma(\alpha+1)}x^{-1}-\frac{\Gamma(\alpha+n+1)}{(n-1)!\,\Gamma(\alpha+2)}+\dots$$

where the omitted terms contain positive powers of x from 2 to $n-1$.

Therefore, we have

$$\frac{d^n}{dx^n}\left[x^{-1}L_n^\alpha(x)\right]=\frac{\Gamma(\alpha+n+1)}{n!\,\Gamma(\alpha+1)}(-1)^n\, n!\, x^{-(n+1)}+0$$

$$=\frac{(-1)^n\,\Gamma(\alpha+n+1)}{\Gamma(\alpha+1)}x^{-(n+1)}. \qquad \text{(vi)}$$

Substituting Eq. (vi) into Eq. (v), we obtain

$$\int_0^\infty e^{-x}x^{\alpha-1}\left[L_n^\alpha(x)\right]^2 dx=\frac{\Gamma(\alpha+n+1)}{n!\,\Gamma(\alpha+1)}\int_0^\infty e^{-x}x^{\alpha-1}dx$$

$$=\frac{\Gamma(\alpha+n+1)\Gamma(\alpha)}{n!\,\Gamma(\alpha+1)}=\frac{\Gamma(\alpha+n+1)}{\alpha\, n!}.$$

(d) Following the same procedure, we write

$$\int_0^\infty e^{-x}x^{\alpha-2}\left[L_n^\alpha(x)\right]^2 dx=\int_0^\infty e^{-x}x^{\alpha-2}L_n^\alpha(x)\frac{e^x x^{-\alpha}}{n!}\frac{d^n}{dx^n}(e^{-x}x^{n+\alpha})dx$$

$$=\frac{1}{n!}\int_0^\infty x^{-2}L_n^\alpha(x)\frac{d^n}{dx^n}(e^{-x}x^{n+\alpha})dx$$

$$=\frac{(-1)^n}{n!}\int_0^\infty \frac{d^n}{dx^n}\left[x^{-2}L_n^\alpha(x)\right]e^{-x}x^{n+\alpha}dx. \qquad \text{(vii)}$$

We know that,

$$x^{-2}L_n^\alpha(x)=x^{-2}\sum_{r=0}^{n}\frac{(-1)^r\,\Gamma(\alpha+n+1)}{r!(n-r)!\,\Gamma(\alpha+r+1)}x^r$$

$$=\frac{\Gamma(\alpha+n+1)}{n!\,\Gamma(\alpha+1)}x^{-2}-\frac{\Gamma(\alpha+n+1)}{(n-1)!\,\Gamma(\alpha+2)}x^{-1}+\ldots$$

$$=\frac{\Gamma(\alpha+n+1)}{n!\,\Gamma(\alpha+1)}\left[x^{-2}-\frac{n}{\alpha+1}x^{-1}+\ldots\right]$$

where the omitted terms represent terms with powers of x ranging from 0 to $n-2$. Therefore, we write

$$\frac{d^n}{dx^n}\left[x^{-2}L_n^\alpha(x)\right]=\frac{\Gamma(\alpha+n+1)}{n!\,\Gamma(\alpha+1)}\left[(-1)^n(n+1)!\,x^{-(n+2)}-\frac{n}{\alpha+1}(-1)^n\, n!\,x^{-(n+1)}\right]+0$$

$$=\frac{(-1)^n\,\Gamma(\alpha+n+1)}{\Gamma(\alpha+1)}\left[(n+1)x^{-(n+2)}-\frac{n}{\alpha+1}x^{-(n+1)}\right]. \qquad \text{(viii)}$$

Substituting Eq. (viii) into Eq. (vii), we obtain

$$\int_0^\infty e^{-x}x^{\alpha-2}\left[L_n^\alpha(x)\right]^2 dx$$

$$=\frac{\Gamma(\alpha+n+1)}{n!\,\Gamma(\alpha+1)}\left[(n+1)\int_0^\infty e^{-x}x^{\alpha-2}dx-\frac{n}{\alpha+1}\int_0^\infty e^{-x}x^{\alpha-1}dx\right]$$

$$=\frac{\Gamma(\alpha+n+1)}{n!\,\Gamma(\alpha+1)}\left[(n+1)\Gamma(\alpha-1)-\frac{n}{\alpha+1}\Gamma(\alpha)\right]=\frac{(\alpha+2n+1)\Gamma(\alpha+n+1)}{(\alpha-1)\alpha(\alpha+1)n!}.$$

PROBLEM 11 Expand the functions (a) $f(x) = e^{-kx}$ and (b) $f(x) = x^p$ (where p is a positive integer) in series of associated Laguerre polynomials.

Solution: We write

$$f(x)=\sum_{n=0}^{\infty} c_n L_n^\alpha(x) \tag{i}$$

where

$$c_n=\frac{n!}{\Gamma(\alpha+n+1)}\int_0^\infty e^{-x}x^\alpha f(x)L_n^\alpha(x)dx. \tag{ii}$$

(a) We write,

$$f(x)=e^{-kx}=\sum_{n=0}^{\infty} c_n L_n^\alpha(x) \tag{iii}$$

where

$$c_n=\frac{n!}{\Gamma(\alpha+n+1)}\int_0^\infty e^{-x}x^\alpha f(x)L_n^\alpha(x)dx$$

$$=\frac{n!}{\Gamma(\alpha+n+1)}\int_0^\infty e^{-(k+1)x}x^\alpha L_n^\alpha(x)dx$$

$$=\frac{n!}{\Gamma(\alpha+n+1)}\int_0^\infty e^{-(k+1)x}x^\alpha\frac{e^x x^{-\alpha}}{n!}\frac{d^n}{dx^n}(e^{-x}x^{n+\alpha})dx$$

$$=\frac{1}{\Gamma(\alpha+n+1)}\int_0^\infty e^{-kx}\frac{d^n}{dx^n}(e^{-x}x^{n+\alpha})dx$$

$$=\frac{(-1)^n}{\Gamma(\alpha+n+1)}\int_0^\infty \frac{d^n}{dx^n}(e^{-kx})e^{-x}x^{n+\alpha}dx$$

$$=\frac{k^n}{\Gamma(\alpha+n+1)}\int_0^\infty e^{-(k+1)x}x^{n+\alpha}dx$$

$$= \frac{k^n}{\Gamma(\alpha+n+1)(k+1)^{n+\alpha+1}} \int_0^\infty e^{-u} u^{n+\alpha} du \text{, where } u = (k+1)x;$$

$$= \frac{k^n}{(k+1)^{n+\alpha+1}} \text{ as } \int_0^\infty e^{-u} u^{n+\alpha} du = \Gamma(n+\alpha+1).$$

Substituting this value of c_n into Eq. (iii), we obtain

$$e^{-kx} = \frac{1}{(k+1)^{\alpha+1}} \sum_{n=0}^{\infty} \left(\frac{k}{k+1}\right)^n L_n^\alpha(x) \text{ where } \alpha > -1.$$

(b) We write,

$$f(x) = x^p = \sum_{n=0}^{p} c_n L_n^\alpha(x) \tag{iv}$$

where
$$c_n = \frac{n!}{\Gamma(\alpha+n+1)} \int_0^\infty e^{-x} x^\alpha f(x) L_n^\alpha(x) dx$$

$$= \frac{n!}{\Gamma(\alpha+n+1)} \int_0^\infty e^{-x} x^{\alpha+p} \frac{e^x x^{-\alpha}}{n!} \frac{d^n}{dx^n}(e^{-x} x^{n+\alpha}) dx$$

$$= \frac{1}{\Gamma(\alpha+n+1)} \int_0^\infty x^p \frac{d^n}{dx^n}(e^{-x} x^{n+\alpha}) dx$$

$$= \frac{(-1)^n}{\Gamma(\alpha+n+1)} \int_0^\infty \left[\frac{d^n}{dx^n} x^p\right] e^{-x} x^{n+\alpha} dx$$

$$= \frac{(-1)^n}{\Gamma(\alpha+n+1)} \int_0^\infty \frac{p!}{(p-n)!} x^{p-n} e^{-x} x^{n+\alpha} dx$$

$$= \frac{(-1)^n p!}{(p-n)!\,\Gamma(\alpha+n+1)} \int_0^\infty e^{-x} x^{p+\alpha} dx = \frac{(-1)^n p!\,\Gamma(p+\alpha+1)}{(p-n)!\,\Gamma(\alpha+n+1)}.$$

Substituting this value of c_n into Eq. (iv), we obtain

$$x^p = p!\,\Gamma(p+\alpha+1) \sum_{n=0}^{p} \frac{(-1)^n L_n^\alpha(x)}{(p-n)!\,\Gamma(\alpha+n+1)}.$$

EXERCISE

1. Show that $L_n^{\alpha}(0)=\dfrac{\Gamma(n+\alpha+1)}{n!\,\Gamma(\alpha+1)}$.

2. Obtain the expressions for

$$L_0^{1/2}(x),\, L_0^{1}(x),\, L_0^{3/2}(x),\, L_1^{1/2}(x),\, L_1^{1}(x),\, L_1^{3/2}(x),\, L_n^{n}(0).$$

3. Prove that

$$L_n^{k}(x)=\sum_{r=0}^{n}\frac{\Gamma(k+n-\alpha-r)}{(n-r)!\,\Gamma(k-\alpha)}L_r^{\alpha}(x).$$

4. Prove that

(i) $L_n^{1/2}\left(\dfrac{t^2}{4}\right)=\dfrac{(-1)^n H_{2n+1}\left(\dfrac{t}{2}\right)}{2^{2n}\,t\,n!}$,

(ii) $L_n^{-1/2}(x)=\dfrac{2e^x}{n!\sqrt{\pi}}\displaystyle\int_0^{\infty}e^{-t^2}t^{2n}\cos(2t\sqrt{x})dt$.

5. Prove that $\displaystyle\int_0^{\infty}e^{-x}x^3\left[L_n^2(x)\right]^2dx=(n+1)(n+2)(2n+3)$.

6. Prove that

(i) $L_n^{1/2}(x)=\dfrac{(-1)^n H_{2n+1}(\sqrt{x})}{2^{2n+1}\,n!\sqrt{x}}$,

(ii) $L_n^{-1/2}(x)=\dfrac{(-1)^n H_{2n}(\sqrt{x})}{2^{2n}\,n!}$.

7. Evaluate

$\displaystyle\int_0^{\infty}e^{-x}x^{\alpha+r}L_n^{\alpha}(x)dx$ if $\alpha>-1$, n and r are non-negative integers.

[**Ans.** 0 for $0\le r<n$,
$(-1)^n\Gamma(n+\alpha+1)$ for $r=n$,
$\dfrac{(-1)^n\,r!\,\Gamma(r+\alpha+1)}{n!\,(r-n)!}$ for $r>n$]

8. Expand the functions e^{-x}, x^7 and x^8 in series of associated Laguerre polynomials.

9. Prove that $\int_x^\infty e^{-y} L_n^\alpha(y)dy = e^{-x}\left[L_n^\alpha(x) - L_{n-1}^\alpha(x)\right]$.

10. Prove that $\int_0^\infty e^{-kx} x^\alpha L_n^\alpha(x)dx = \frac{1}{k^{\alpha+1}}\left(\frac{k-1}{k}\right)^n \frac{\Gamma(n+\alpha+1)}{n!}$.

CHAPTER 8

Bessel Functions

8.1 BESSEL EQUATION

Bessel equation is a linear and homogeneous second order ordinary differential equation having the form

$$x^2 \frac{d^2y}{dx^2} + x\frac{dy}{dx} + (x^2 - p^2)y = 0 \tag{8.1}$$

where p is a constant.

This differential equation is called Bessel equation and its solutions are called Bessel functions in honour of the famous German astronomer, Friedrich Wilhelm Bessel (July 22, 1784 – March 17, 1846) who studied this equation and its solutions extensively.

8.2 POWER SERIES SOLUTION OF BESSEL EQUATION AND THE BESSEL FUNCTIONS OF THE FIRST KIND

We see that the point $x = 0$ is a removable singularity of the Bessel equation and, therefore, by Fuchs' theorem, the equation can have a power series solution in the form

$$y = \sum_{r=0}^{\infty} a_r x^{k+r} \quad (a_0 \neq 0) \tag{8.2}$$

where k is a constant and a_r are coefficients, all of which are to be determined.

Then, we have

$$\frac{dy}{dx} = \sum_{r=0}^{\infty} (k+r)a_r x^{k+r-1}$$

and

$$\frac{d^2y}{dx^2} = \sum_{r=0}^{\infty} (k+r)(k+r-1)a_r x^{k+r-2}.$$

Substituting the expressions for $y, \frac{dy}{dx}$ and $\frac{d^2y}{dx^2}$ into Eq. (8.1), we obtain

$$\sum_{r=0}^{\infty}(k+r+p)(k+r-p)a_r x^{k+r}+\sum_{r=0}^{\infty}a_r x^{k+r+2}=0\,. \tag{8.3}$$

Here, all the coefficients of various powers of x must separately vanish.

Equating the coefficient of x^k to zero, we have

$$(k+p)(k-p)a_0=0 \quad \text{[Indicial equation]}$$

giving $k=+p,-p$ for arbitrary a_0.

Equating the coefficient of x^{k+1} to zero, we obtain

$$(k+1+p)(k+1-p)a_1=0 \text{ giving } a_1=0.$$

Equating the coefficient of x^{k+r+2} to zero, we obtain

$$(k+r+2+p)(k+r+2-p)a_{r+2}+a_r=0$$

or

$$a_{r+2}=-\frac{1}{(k+r+2+p)(k+r+2-p)}a_r \quad \text{[Recursion relation].}$$

CASE I: $k=p, a_0\neq 0, a_1=0$: With the choice $k=p$, the recursion relation becomes

$$a_{r+2}=-\frac{1}{(2p+r+2)(r+2)}a_r\,. \tag{8.4}$$

Putting $r=0,1,2,3,4$, etc., Eq. (8.4) yields,

$$a_2=-\frac{1}{2^2\,1!(p+1)}a_0, \qquad a_3=0,$$

$$a_4=(-1)^2\frac{1}{2^4\,2!(p+2)(p+1)}a_0, \qquad a_5=0,$$

$$a_6=(-1)^3\frac{1}{2^6\,3!(p+3)(p+2)(p+1)}a_0, \qquad a_7=0, \text{ etc.}$$

In general, we write

$$a_{2\nu}=(-1)^{\nu}\frac{1}{2^{2\nu}\nu!(p+\nu)\ldots(p+3)(p+2)(p+1)}a_0$$

and $a_{2\nu+1}=0$; with $\nu=0, 1, 2, 3 \ldots$ etc.

Substituting the values of these various coefficients and also $k=p$ into Eq. (8.2), we obtain

$$y=x^p\left[a_0+a_2x^2+a_4x^4+a_6x^6+\ldots+a_{2\nu}x^{2\nu}+\ldots\infty\right]$$

$$=\sum_{\nu=0}^{\infty}\frac{(-1)^{\nu}a_0}{2^{2\nu}\nu!(p+\nu)\ldots(p+3)(p+2)(p+1)}x^{p+2\nu}\,. \tag{8.5}$$

Conventionally, a_0 is chosen as $\frac{1}{2^p\Gamma(p+1)}$. With this choice of a_0, the resulting solution given in Eq. (8.5) is known as a **Bessel function of the first kind of order *p*** and it is denoted by $J_p(x)$. We generally write

$$J_p(x)=\sum_{\nu=0}^{\infty}\frac{(-1)^\nu}{\nu!\Gamma(p+\nu+1)}\left(\frac{x}{2}\right)^{p+2\nu}.$$

Changing the summation index, we can also write

$$J_p(x)=\sum_{r=0}^{\infty}\frac{(-1)^r}{r!\Gamma(p+r+1)}\left(\frac{x}{2}\right)^{p+2r}. \tag{8.6}$$

CASE II: $k=-p, a_0\neq 0, a_1=0$: In this case, if p is not an integer, the solution is obtained by replacing p by $-p$ (non-integer) in the solution of Case I. Such solution is known as Bessel function of the first kind with negative order $(-p)$. It is denoted by $J_{-p}(x)$. Thus, we write

$$J_{-p}(x)=\sum_{r=0}^{\infty}\frac{(-1)^r}{r!\Gamma(r-p+1)}\left(\frac{x}{2}\right)^{2r-p}. \tag{8.7}$$

Thus, for non-integral values of p, the general solution of the above Bessel equation is given by

$$y=AJ_p(x)+BJ_{-p}(x), \text{ where } A \text{ and } B \text{ are arbitrary constants.}$$

If p is an integer, the solutions $J_p(x)$ and $J_{-p}(x)$ are not independent solutions. In such a case, to obtain a general solution of the Bessel equation, we need another independent solution known as Bessel function of the second kind or Neumann function. This function will be studied latter.

8.3 BESSEL FUNCTION OF THE SECOND KIND

When $p \neq$ Integer

We see that as $x\to 0^+$, the Bessel functions $J_p(x)$ and $J_{-p}(x)$ become

$$J_p(x)\sim\frac{(x/2)^p}{\Gamma(p+1)},\ x\to 0^+ \tag{8.8}$$

and

$$J_{-p}(x)\sim\frac{(x/2)^{-p}}{\Gamma(1-p)},\ x\to 0^+. \tag{8.9}$$

For $p=$ non-integer, it follows that the function $J_{-p}(x)$ becomes infinite as $x\to 0^+$ whereas the function $J_p(x)$ approaches zero. Hence, $J_{-p}(x)$ cannot be a multiple of $J_p(x)$ or vice versa. Hence, $J_p(x)$ and $J_{-p}(x)$ are linearly independent solutions when p is not an integer. Cases with half integral values of p are of special interest because the Bessel functions corresponding to these half integral values of p are actually elementary functions.

Thus, for non-integral values of p, the general solution of Eq. (8.1) can be written as

$$y = A J_p(x) + B J_{-p}(x) \tag{8.10}$$

where A and B are arbitrary constants.

When $p = n$ = Positive Integer

In this case, as n and r are positive integers, we have

$$J_{-n}(x) = \sum_{r=0}^{\infty} \frac{(-1)^r}{r!\,\Gamma(-n+r+1)} \left(\frac{x}{2}\right)^{2r-n} = \sum_{r=0}^{\infty} \frac{(-1)^r}{r!\,(r-n)!} \left(\frac{x}{2}\right)^{2r-n}$$

$$= \sum_{r=n}^{\infty} \frac{(-1)^r}{r!\,(r-n)!} \left(\frac{x}{2}\right)^{2r-n}.$$

If we put $(r-n) = s$ or $r = n+s$ we write

$$J_{-n}(x) = \sum_{s=0}^{\infty} \frac{(-1)^{n+s}}{(n+s)!\,s!} \left(\frac{x}{2}\right)^{n+2s}$$

$$= (-1)^n \sum_{s=0}^{\infty} \frac{(-1)^s}{s!\,\Gamma(n+s+1)} \left(\frac{x}{2}\right)^{n+2s} = (-1)^n J_n(x). \tag{8.11}$$

From Eq. (8.11), it is seen that for n equal to an integer, the functions $J_n(x)$ and $J_{-n}(x)$ are not linearly independent solutions. Therefore, we are to find a second linearly independent solution of the Bessel equation. Such a solution is defined by a function

$$Y_p(x) = N_p(x) = \frac{(\cos p\pi) J_p(x) - J_{-p}(x)}{\sin p\pi}$$

$$= (\cot p\pi) J_p(x) - (\operatorname{cosec} p\pi) J_{-p}(x). \tag{8.12}$$

It is seen that this function is a linear combination of $J_p(x)$ and $J_{-p}(x)$ and hence it satisfies Eq. (8.1). But, there is a difficulty that this function has the indeterminate form $0/0$ when $p = n$ (an integer). In such cases, we define

$$Y_n(x) = \lim_{p \to n} Y_p(x)$$

$$= N_n(x) = \lim_{p \to n} N_p(x) = \lim_{p \to n} \frac{(\cos p\pi) J_p(x) - J_{-p}(x)}{\sin p\pi}. \tag{8.13}$$

Applying L' Hospital's rule, we obtain

$$Y_n(x) = N_n(x) = \lim_{p \to n} \frac{(-\pi \sin p\pi) J_p(x) + (\cos p\pi) \frac{\partial}{\partial p} J_p(x) - \frac{\partial}{\partial p} J_{-p}(x)}{\pi \cos p\pi}$$

$$= \lim_{p \to n} \frac{1}{\pi} \left[\frac{\partial}{\partial p} J_p(x) - (-1)^p \frac{\partial}{\partial p} J_{-p}(x) \right]. \tag{8.14}$$

Thus, the general solution of Bessel equation (8.1) for all values of p can be written as

$$y = C_1 J_p(x) + C_2 Y_p(x) \tag{8.15}$$

where C_1 and C_2 are arbitrary constants. The function $Y_p(x)$ is known as **Bessel function of the second kind of order *p***. It is also known as Neumann function, $N_p(x)$, in honour of the German mathematician, Carl Gottfried Neumann (May 7, 1832 - March 27, 1925) for his extensive studies in Bessel functions. This function is also known as Weber function in honour of another German mathematician, Heinrich Martin Weber (March 5, 1842 – May 17, 1913).

8.4 SERIES EXPANSION FOR BESSEL FUNCTION OF THE SECOND KIND $Y_n(x)$ or $N_n(x)$

Let us find an expression for Neumann function (Bessel function of the second kind) when p takes an integral value n.

For $x > 0$, the derivative of Bessel function of the first kind with respect to its order yields

$$\frac{\partial}{\partial p} J_p(x) = \frac{\partial}{\partial p} \sum_{r=0}^{\infty} \frac{(-1)^r}{r!\,\Gamma(r+p+1)} \left(\frac{x}{2}\right)^{p+2r}$$

$$= \sum_{r=0}^{\infty} \frac{(-1)^r \ln(x/2)}{r!\,\Gamma(r+p+1)} \left(\frac{x}{2}\right)^{p+2r} - \sum_{r=0}^{\infty} \frac{(-1)^r}{r!} \left(\frac{x}{2}\right)^{p+2r} \frac{\Gamma'(r+p+1)}{\left[\Gamma(r+p+1)\right]^2}$$

where we are using the notation $\Gamma'(t) = \frac{d}{dt}\Gamma(t)$.

We write

$$\frac{\partial}{\partial p} J_p(x) = J_p(x) \ln\left(\frac{x}{2}\right) - \sum_{r=0}^{\infty} \frac{(-1)^r (x/2)^{p+2r}}{r!\,\Gamma(r+p+1)} \psi(r+p+1) \tag{8.16}$$

where the notation $\psi(t)$ denotes the digamma function, $\frac{\Gamma'(t)}{\Gamma(t)}$.

By a similar process, we can obtain

$$\frac{\partial}{\partial p}J_{-p}(x) = -J_{-p}(x)\ln\left(\frac{x}{2}\right) + \sum_{r=0}^{\infty}\frac{(-1)^r (x/2)^{2r-p}}{r!\,\Gamma(r-p+1)}\psi(r-p+1). \tag{8.17}$$

By this time, let us consider the special case $p \to 0$ so that Eq. (8.14) yields

$$N_0(x) = \lim_{p\to 0}\frac{1}{\pi}\left[\frac{\partial}{\partial p}J_p(x) - \frac{\partial}{\partial p}J_{-p}(x)\right].$$

Using Eqs. (8.16) and (8.17), we write

$$N_0(x) = \frac{2}{\pi}J_0(x)\ln\left(\frac{x}{2}\right) - \frac{2}{\pi}\sum_{r=0}^{\infty}\frac{(-1)^r (x/2)^{2r}}{(r!)^2}\psi(r+1). \tag{8.18}$$

Recalling the study of digamma functions, we can write

$$\sum_{r=0}^{\infty}\frac{(-1)^r (x/2)^{2r}}{(r!)^2}\psi(r+1) = -\gamma + \sum_{r=1}^{\infty}\frac{(-1)^r (x/2)^{2r}}{(r!)^2}\left(-\gamma + 1 + \frac{1}{2} + \frac{1}{3} + \ldots + \frac{1}{r}\right)$$

$$= -\gamma\sum_{r=0}^{\infty}\frac{(-1)^r (x/2)^{2r}}{(r!)^2} + \sum_{r=1}^{\infty}\frac{(-1)^r (x/2)^{2r}}{(r!)^2}\left(1 + \frac{1}{2} + \frac{1}{3} + \ldots + \frac{1}{r}\right)$$

$$= -\gamma J_0(x) + \sum_{r=1}^{\infty}\frac{(-1)^r (x/2)^{2r}}{(r!)^2}\left(1 + \frac{1}{2} + \frac{1}{3} + \ldots + \frac{1}{r}\right)$$

where γ is a constant known as Euler constant.

Substituting this last relation into Eq. (8.18), we obtain

$$N_0(x) = \frac{2}{\pi}J_0(x)\left[\ln\left(\frac{x}{2}\right) + \gamma\right] - \frac{2}{\pi}\sum_{r=1}^{\infty}\frac{(-1)^r (x/2)^{2r}}{(r!)^2}\left(1 + \frac{1}{2} + \frac{1}{3} + \ldots + \frac{1}{r}\right). \tag{8.19}$$

The derivation of the general series for $N_n(x)$ $(n = 1, 2, 3 \ldots)$ is a more difficult task. Substitution of Eqs. (8.16) and (8.17) in Eq. (8.14) yields

$$N_n(x) = \frac{2}{\pi}J_n(x)\ln\left(\frac{x}{2}\right) - \frac{1}{\pi}\lim_{p\to n}\sum_{r=0}^{\infty}\frac{(-1)^r (x/2)^{2r+p}}{r!\,\Gamma(r+p+1)}\psi(r+p+1)$$

$$-\frac{1}{\pi}\lim_{p\to n}(-1)^p\sum_{r=0}^{\infty}\frac{(-1)^r (x/2)^{2r-p}}{r!\,\Gamma(r-p+1)}\psi(r-p+1). \tag{8.20}$$

Here, we use the fact that if n is a positive integer, $J_n(x)=(-1)^n J_{-n}(x)$.

In the study of gamma and digamma functions, we know that

$$\Gamma(t)\Gamma(1-t)=\frac{\pi}{\sin \pi t}$$

and $\psi(1-t)-\psi(t)=\pi \cot \pi t$.

Therefore, we write

$$\psi(r-p+1)=\psi\left[1-(p-r)\right]=\psi(p-r)+\pi \cot(p-r)\pi.$$

Also, we have

$$\frac{1}{\Gamma(r-p+1)}=\frac{1}{\Gamma\left[1-(p-r)\right]}=\frac{\Gamma(p-r)\sin(p-r)\pi}{\pi}.$$

Now, we have

$$\lim_{p\to n}\frac{\psi(r-p+1)}{\Gamma(r-p+1)}=\lim_{p\to n}\frac{\Gamma(p-r)\sin(p-r)\pi}{\pi}\cdot\left[\psi(p-r)+\pi\cot(p-r)\pi\right]$$

$$=(-1)^{n-r}\Gamma(n-r)=(-1)^{n-r}(n-r-1)! \tag{8.21}$$

with $r=0,1,2,3\ldots n-1$.

Using Eq. (8.21), we write the last series part of Eq. (8.20) as

$$-\frac{1}{\pi}\lim_{p\to n}(-1)^p\sum_{r=0}^{\infty}\frac{(-1)^r(x/2)^{2r-p}}{r!\,\Gamma(r-p+1)}\psi(r-p+1)$$

$$=-\frac{1}{\pi}\left[\sum_{r=0}^{n-1}\frac{(n-r-1)!}{r!}\left(\frac{x}{2}\right)^{2r-n}+(-1)^n\sum_{r=n}^{\infty}\frac{(-1)^r(x/2)^{2r-n}}{r!\,\Gamma(r-n+1)}\psi(r-n+1)\right]$$

$$=-\frac{1}{\pi}\left[\sum_{r=0}^{n-1}\frac{(n-r-1)!}{r!}\left(\frac{x}{2}\right)^{2r-n}+\sum_{m=0}^{\infty}\frac{(-1)^m(x/2)^{2m+n}}{(m+n)!\,m!}\psi(m+1)\right] \tag{8.22}$$

where we put $m=r-n$.

Substitution of Eq. (8.22) into Eq. (8.20) results in

$$N_n(x)=\frac{2}{\pi}J_n(x)\ln\left(\frac{x}{2}\right)-\frac{1}{\pi}\sum_{r=0}^{\infty}\frac{(-1)^r(x/2)^{2r+n}}{r!\,(r+n)!}\psi(r+n+1)$$

$$-\frac{1}{\pi}\sum_{r=0}^{n-1}\frac{(n-r-1)!}{r!}\left(\frac{x}{2}\right)^{2r-n}-\frac{1}{\pi}\sum_{m=0}^{\infty}\frac{(-1)^m(x/2)^{2m+n}}{m!\,(m+n)!}\psi(m+1)$$

$$= \frac{2}{\pi} J_n(x) \ln\left(\frac{x}{2}\right) - \frac{1}{\pi} \sum_{r=0}^{n-1} \frac{(n-r-1)!}{r!} \left(\frac{x}{2}\right)^{2r-n}$$

$$- \frac{1}{\pi} \sum_{m=0}^{\infty} \frac{(-1)^m (x/2)^{2m+n}}{m!(m+n)!} \left[\psi(m+n+1) + \psi(m+1)\right] \tag{8.23}$$

where the summation index (r) in the second series expression is changed to m.

8.5 BESSEL FUNCTIONS OF THE THIRD KIND OR HANKEL FUNCTIONS

Using Bessel functions of the first and the second kinds, we define two types of functions known as the Hankel functions of the first and the second kinds, also known as Bessel functions of the third kind.

We define Hankel functions of the first kind of order p as

$$H_p^{(1)}(x) = J_p(x) + iN_p(x) = \frac{i}{\sin p\pi} \left[e^{-p\pi i} J_p(x) - J_{-p}(x)\right] \tag{8.24}$$

and Hankel function of the second kind of order p as

$$H_p^{(2)}(x) = J_p(x) - iN_p(x) = -\frac{i}{\sin p\pi} \left[e^{p\pi i} J_p(x) - J_{-p}(x)\right]. \tag{8.25}$$

Hankel functions are named after the German mathematician, Hermann Hankel (February 14, 1839 - August 29, 1873). These functions are infinite at $x = 0$.

8.6 MODIFIED (OR HYPERBOLIC) BESSEL FUNCTIONS

In Bessel equation (8.1), if we replace x by ix, we will get the equation

$$x^2 \frac{d^2y}{dx^2} + x\frac{dy}{dx} - \left(x^2 + p^2\right) y = 0, p \geq 0. \tag{8.26}$$

This equation is known as *modified Bessel equation*.

Thus, we see that $J_p(ix)$ is a solution of the above modified Bessel equation.

We have

$$J_p(ix) = \sum_{r=0}^{\infty} \frac{(-1)^r}{r!\Gamma(r+p+1)} \left(\frac{ix}{2}\right)^{p+2r}$$

$$= e^{p\pi i/2} \sum_{r=0}^{\infty} \frac{1}{r!\Gamma(r+p+1)} \left(\frac{x}{2}\right)^{p+2r}, p \geq 0. \tag{8.27}$$

It is seen that

$$I_p(x) = e^{-\frac{p\pi i}{2}} J_p(ix) = \sum_{r=0}^{\infty} \frac{1}{r!\Gamma(r+p+1)} \left(\frac{x}{2}\right)^{p+2r} \tag{8.28}$$

is also a solution of Eq. (8.26).

Here, the function $I_p(x)$ defined in Eq. (8.28) is known as *modified Bessel function* of the first kind of order p. Modified Bessel functions are also known as *hyperbolic Bessel functions*. When p is equal to an integer n, $e^{-\frac{p\pi i}{2}} = i^{-p}$.

The modified Bessel function of the second kind of order p (also called Macdonald function) is defined by

$$K_p(x) = \frac{\pi}{2}\left[\frac{I_{-p}(x) - I_p(x)}{\sin p\pi}\right], \ (p \neq \text{integer}). \tag{8.29}$$

If p = an integer (n), then

$$K_n(x) = \lim_{p\to n} K_p(x) = \frac{(-1)^n}{2}\left[\frac{\partial}{\partial p} I_{-p}(x) - \frac{\partial}{\partial p} I_p(x)\right]_{p=n}. \tag{8.29a}$$

In calculating the limit in the above equation, use has been made of L' Hospital's rule.

It can be shown that $K_{-p}(x) = K_p(x)$ for all values of p as

$$K_{-p}(x) = \frac{\pi}{2}\left[\frac{I_p(x) - I_{-p}(x)}{\sin(-p\pi)}\right] = \frac{\pi}{2}\left[\frac{I_p(x) - I_{-p}(x)}{-\sin p\pi}\right] = \frac{\pi}{2}\left[\frac{I_{-p}(x) - I_p(x)}{\sin p\pi}\right] = K_p(x).$$

In the definitions of $I_p(x)$ and $K_p(x)$ i is generally replaced by $e^{\pi i/2}$.

8.7 SPHERICAL BESSEL FUNCTIONS

Spherical Bessel functions are generally associated with solving the Helmholtz equation in spherical polar coordinates. When one is to solve the **Helmholtz equation** $(\nabla^2 u + k^2 u = 0)$ in spherical polar coordinates using the method of separation of variables, he will find for solutions which are finite on the polar axis (r) that the radial part $R(r)$ of the solution must satisfy the equation

$$r^2\frac{d^2R}{dr^2} + 2r\frac{dR}{dr} + \left[k^2r^2 - l(l+1)\right]R = 0 \tag{8.30}$$

where l is a non-negative integer.

If we put

$$R(r) = r^{-\frac{1}{2}} f(r), \tag{8.31}$$

then $f(r)$ will satisfy the equation

$$r^2\frac{d^2 f}{dr^2} + r\frac{df}{dr} + \left[k^2r^2 - \left(l + \frac{1}{2}\right)^2\right] f = 0. \tag{8.32}$$

If we change variables such as $s = kr$ and $f(r) = u(s)$, Eq. (8.32) can be written as

$$s^2 \frac{d^2u}{ds^2} + s\frac{du}{ds} + \left[s^2 - \left(l + \frac{1}{2}\right)^2\right]u = 0 \tag{8.33}$$

Solution of Eq. (8.33) is given by

$$u(s) = C_1 J_{l+\frac{1}{2}}(s) + C_2 N_{l+\frac{1}{2}}(s)$$

where C_1 and C_2 are arbitrary constants.

or
$$f(r) = C_1 J_{l+\frac{1}{2}}(kr) + C_2 N_{l+\frac{1}{2}}(kr). \tag{8.34}$$

Substitution of Eq. (8.34) into Eq. (8.31) yields the solution of Eq. (8.30) as

$$R(r) = r^{-\frac{1}{2}}\left[C_1 J_{l+\frac{1}{2}}(kr) + C_2 N_{l+\frac{1}{2}}(kr)\right]. \tag{8.35}$$

Since l takes integral values, the Bessel functions in the above equations are of half integral orders. We have earlier found that the particular Bessel function of half integral order $J_{\frac{1}{2}}(kr)$ is an elementary function given by

$$J_{\frac{1}{2}}(kr) = \sqrt{\frac{2}{\pi kr}}\ \sin(kr).$$

Moreover, it is also known that all half-integral order Bessel functions can be reduced to elementary functions. Therefore, to combine the multiplicative factor $r^{-\frac{1}{2}}$ appearing in Eq. (8.35) with the half-integral order Bessel functions, it is customary to introduce functions defined by

$$j_l(kr) = \sqrt{\frac{\pi}{2kr}} J_{l+\frac{1}{2}}(kr) \tag{8.36}$$

and
$$n_l(kr) = \sqrt{\frac{\pi}{2kr}} N_{l+\frac{1}{2}}(kr). \tag{8.37}$$

The function $j_l(kr)$ defined here is called the spherical Bessel function of the first kind of order l and the function $n_l(kr)$ is called the spherical Bessel function of the second kind or spherical Neumann function of order l. Thus, a general solution of Eq. (8.30) can be expressed as

$$R(r) = Aj_l(kr) + Bn_l(kr)$$

where A and B are arbitrary constants.

If we have $k=1$, the function

$$R(r)=Aj_l(r)+Bn_l(r)$$

is a general solution of the spherical Bessel equation

$$r^2\frac{d^2R}{dr^2}+2r\frac{dR}{dr}+\left\{r^2-l(l+1)\right\}R=0.$$

8.7.1 Spherical Hankel Functions

The spherical Hankel function of the first kind of order l is defined by

$$h_l^{(1)}(r)=j_l(r)+in_l(r) \tag{8.38}$$

and the spherical Hankel function of the second kind of order l is defined by

$$h_l^{(2)}(r)=j_l(r)-in_l(r). \tag{8.39}$$

8.7.2 Modified Spherical Bessel Functions

The modified spherical Bessel functions of the first and second kinds are respectively defined by

$$i_l(r)=\sqrt{\frac{\pi}{2r}}\,I_{l+\frac{1}{2}}(r) \tag{8.40}$$

and

$$k_l(r)=\sqrt{\frac{2}{\pi r}}\,K_{l+\frac{1}{2}}(r). \tag{8.41}$$

8.8 KELVIN FUNCTIONS

In the study of *skin effect* of alternating current flowing in a wire, it is sometimes necessary to solve differential equations of the form

$$x\frac{d^2y}{dx^2}+\frac{dy}{dx}-i\lambda^2xy=0 \quad (\lambda \text{ being a real constant}) \tag{8.42}$$

which is a special case of the differential equation

$$x^2\frac{d^2y}{dx^2}+x\frac{dy}{dx}-\left(i\lambda^2x^2+p^2\right)y=0 \tag{8.43}$$

with $p=0$.

Equation (8.43) has the form of the modified Bessel's equation

$$x^2\frac{d^2y}{dx^2}+x\frac{dy}{dx}-\left(\alpha^2x^2+p^2\right)y=0 \tag{8.44}$$

with $\alpha=i^{1/2}\lambda$.

Since a general solution of Eq. (8.44) is

$$y=A_1I_p(\alpha x)+A_2K_p(\alpha x)$$

where A_1 and A_2 are arbitrary constants, the general solution of Eq. (8.43) must be

$$y=A_1I_p(i^{1/2}\lambda x)+A_2K_p(i^{1/2}\lambda x). \tag{8.45}$$

Since $I_p(x)=i^{-p}J_p(ix)$, we can also take the independent solutions of Eq. (8.43) as

$$J_p(i^{3/2}\lambda x) \text{ and } K_p(i^{1/2}\lambda x).$$

Of the two values $e^{\pi i/4}$ and $e^{5\pi i/4}$ of $i^{1/2}$, we select the value $e^{\pi i/4}$.

When x is real, $J_p(i^{3/2}\lambda x)$ and $K_p(i^{1/2}\lambda x)$ may necessarily not be real.

Thus, we can obtain four real functions respectively defined by

$$\begin{aligned}
Ber_p(x)&=\text{Re}\,J_p(xe^{3\pi i/4})\\
&=\text{Re}\sum_{r=0}^{\infty}\frac{(-1)^r}{r!\,\Gamma(p+r+1)}\left(\frac{xe^{3\pi i/4}}{2}\right)^{p+2r}\\
&=\text{Re}\sum_{r=0}^{\infty}\frac{(-1)^r\left(\frac{x}{2}\right)^{p+2r}}{r!\,\Gamma(p+r+1)}\left(\cos\frac{3\pi}{4}+i\sin\frac{3\pi}{4}\right)^{p+2r}\\
&=\text{Re}\sum_{r=0}^{\infty}\frac{(-1)^r\left(\frac{x}{2}\right)^{p+2r}}{r!\,\Gamma(p+r+1)}\left[\cos\left\{(p+2r)\frac{3\pi}{4}\right\}+i\sin\left\{(p+2r)\frac{3\pi}{4}\right\}\right]\\
&=\sum_{r=0}^{\infty}\frac{(-1)^r\left(\frac{x}{2}\right)^{p+2r}}{r!\,\Gamma(p+r+1)}\cos\left(\frac{3\pi p}{4}+\frac{3\pi r}{2}\right)\\
&=\sum_{r=0}^{\infty}\frac{\cos\left(\frac{3p\pi}{4}+\frac{\pi r}{2}\right)}{r!\,\Gamma(p+r+1)}\left(\frac{x}{2}\right)^{p+2r}
\end{aligned} \tag{8.46}$$

and $\quad Bei_p(x) = Im\, J_p(xe^{3\pi i/4})$

$$= \sum_{r=0}^{\infty} \frac{(-1)^r \left(\frac{x}{2}\right)^{p+2r}}{r!\,\Gamma(p+r+1)} \sin\left(\frac{3\pi p}{4} + \frac{3\pi r}{2}\right)$$

$$= \sum_{r=0}^{\infty} \frac{\sin\left(\frac{3p\pi}{4} + \frac{\pi r}{2}\right)}{r!\,\Gamma(p+r+1)} \left(\frac{x}{2}\right)^{p+2r}. \tag{8.47}$$

Thus, we have

$$J_p(xe^{3\pi i/4}) = Ber_p(x) + i\, Bei_p(x)$$

$$= e^{p\pi i} J_p(xe^{-\pi i/4}) = e^{p\pi i/2} I_p(xe^{\pi i/4}). \tag{8.48}$$

Also, we have

$$Ker_p(x) = \text{Re}\, e^{-p\pi i/2} K_p(xe^{\pi i/4}), \tag{8.49}$$

$$Kei_p(x) = Im\, e^{-p\pi i/2} K_p(xe^{\pi i/4}) \tag{8.50}$$

such that

$$Ker_p(x) + i\, Kei_p(x) = e^{-p\pi i/2} K_p(x e^{\pi i/4}). \tag{8.51}$$

When $p = 0$ the series for $I_0\left(xi^{\frac{1}{2}}\right)$ can be found to be

$$I_0\left(xi^{\frac{1}{2}}\right) = I_0(xe^{\pi i/4}) = \sum_{m=0}^{\infty} \frac{i^m \left(\frac{x}{2}\right)^{2m}}{(m!)^2}. \tag{8.51a}$$

If we split this series into real and imaginary parts, the real part is known as *Ber* function and it is given by

$$Ber(x) = \sum_{r=0}^{\infty} \frac{(-1)^r \left(\frac{x}{2}\right)^{4r}}{[(2r)!]^2} \tag{8.51b}$$

whereas the imaginary part of the series is known as *Bei* function. It is given by

$$Bei(x) = \sum_{r=0}^{\infty} \frac{(-1)^r \left(\frac{x}{2}\right)^{4r+2}}{[(2r+1)!]^2}. \tag{8.51c}$$

Ber and *Bei* denote Ber_0 and Bei_0, respectively. However, the subscript zero is generally omitted by many authors without loss of generality.

Ber, *Bei*, *Ker* and *Kei* functions are named the *Kelvin functions* or *Thomson functions* in honour of the British physicist, *Lord William Thomson Kelvin* (June 26, 1824 - December17, 1907). The notation *Ber* denotes 'Bessel function real' and *Bei* denotes Bessel function imaginary'. *Ker* denotes Kelvin real and *Kei* denotes 'Kelvin imaginary'. Kelvin functions are also called modified Hankel functions.

Omitting the cumbersome task of calculations, we write the power series for $Ker_p(x)$, $Kei_p(x)$, $Ker(x)$ and $Kei(x)$ as in the following.

$$Ker_p(x)=\frac{1}{2}\left(\frac{x}{2}\right)^{-p}\sum_{r=0}^{p-1}\frac{(p-r-1)!}{r!}\cos\left(\frac{3p\pi}{4}+\frac{r\pi}{2}\right)\left(\frac{x}{2}\right)^{2r}-\ln\left(\frac{x}{2}\right)Ber_p(x)+\frac{\pi}{4}Bei_p(x)$$

$$+\frac{1}{2}\left(\frac{x}{2}\right)^{p}\sum_{r=0}^{\infty}\frac{\psi(r+1)+\psi(p+r+1)}{r!\,\Gamma(p+r+1)}\cos\left(\frac{3p\pi}{4}+\frac{r\pi}{2}\right)\left(\frac{x}{2}\right)^{2r}, \quad (8.51d)$$

$$Kei_p(x)=-\frac{1}{2}\left(\frac{x}{2}\right)^{-p}\sum_{r=0}^{p-1}\frac{(p-r-1)!}{r!}\sin\left(\frac{3p\pi}{4}+\frac{r\pi}{2}\right)\left(\frac{x}{2}\right)^{2r}-\ln\left(\frac{x}{2}\right)Bei_p(x)$$

$$-\frac{\pi}{4}Ber_p(x)+\frac{1}{2}\left(\frac{x}{2}\right)^{p}\sum_{r=0}^{\infty}\frac{\psi(r+1)+\psi(p+r+1)}{r!\,\Gamma(p+r+1)}\sin\left(\frac{3p\pi}{4}+\frac{r\pi}{2}\right)\left(\frac{x}{2}\right)^{2r}, \quad (8.51e)$$

$$Ker(x)=-\ln\left(\frac{x}{2}\right)Ber(x)+\frac{\pi}{4}Bei(x)+\sum_{r=0}^{\infty}(-1)^r\frac{\psi(2r+1)}{\{(2r)!\}^2}\left(\frac{x^2}{4}\right)^{2r} \quad (8.51f)$$

and $$Kei(x)=-\ln\left(\frac{x}{2}\right)Bei(x)-\frac{\pi}{4}Ber(x)+\sum_{r=0}^{\infty}(-1)^r\frac{\psi(2r+2)}{\{(2r+1)!\}^2}\left(\frac{x^2}{4}\right)^{2r+1} \quad (8.51g)$$

where $\psi(x)=\frac{\Gamma'(x)}{\Gamma(x)}$, the prime symbol denoting first order derivative.

8.9 STRUVE FUNCTIONS

Struve functions are found to occur in certain problems in optics and in loudspeaker design. These functions are independent solutions of the inhomogeneous Bessel equation

$$x^2\frac{d^2y}{dx^2}+x\frac{dy}{dx}+(x^2-p^2)y=\frac{4}{\sqrt{\pi}\,\Gamma\left(p+\frac{1}{2}\right)}\left(\frac{x}{2}\right)^{p+1}. \quad (8.52)$$

Struve function is defined by the series

$$\mathcal{H}_p(x) = \sum_{r=0}^{\infty} \frac{(-1)^r}{\Gamma\left(r+\frac{3}{2}\right)\Gamma\left(p+r+\frac{3}{2}\right)}\left(\frac{x}{2}\right)^{p+2r+1} \tag{8.53}$$

and the modified Struve function is defined by the series

$$\mathbb{L}_p(x) = i^{-(p+1)}\mathcal{H}_p(ix) = \sum_{r=0}^{\infty} \frac{1}{\Gamma\left(r+\frac{3}{2}\right)\Gamma\left(p+r+\frac{3}{2}\right)}\left(\frac{x}{2}\right)^{p+2r+1}. \tag{8.54}$$

The notations of these functions should not be confused with those of Hermite and Laguerre polynomials. Struve functions are named after the Russian astronomer, Karl Hermann Struve (October 3, 1854 - August 12, 1920).

8.10 AIRY FUNCTIONS

Airy functions are solutions of the homogeneous second order linear ODE of the form

$$\frac{d^2u(z)}{dz^2} - zu(z) = 0. \tag{8.55}$$

Such an equation is known as Airy equation which is named after the English mathematician-cum-astronomer and physicist, George Biddell Airy (July 27, 1801 – January 2, 1892).

These functions are encountered in a variety of applications in mathematical physics including the diffraction of radio-waves around the earth's surface.

Equation (8.55) is a special case of modified Lommel's equation, viz.

$$z^2\frac{d^2u(z)}{dz^2} + (1-2a)z\frac{du(z)}{dz} - \left[b^2c^2z^{2c} - (a^2 - p^2c^2)\right]u(z) = 0 \tag{8.55a}$$

when $a = \frac{1}{2}, b = \frac{2}{3}, c = \frac{3}{2}$ and $p = \pm\frac{1}{3}$.

Since modified Lommel's equation, i.e., Eq. (8.55a) has a solution in the form $u(z) = z^a v(bz^c)$, (which is to be proved in the section for solved problems), where v is a solution of modified Bessel equation, we write the solution of Eq. (8.55) in the form $u(z) = z^{\frac{1}{2}} v\left(\frac{2}{3}z^{\frac{3}{2}}\right)$.

With standard choices, Airy functions are defined as the following functions:

$$Ai(z) = \frac{\sqrt{z}}{3}\left[I_{-\frac{1}{3}}\left(\frac{2z^{3/2}}{3}\right) - I_{\frac{1}{3}}\left(\frac{2z^{3/2}}{3}\right)\right]$$

$$= \frac{1}{\pi}\sqrt{\frac{z}{3}}K_{\frac{1}{3}}\left(\frac{2z^{3/2}}{3}\right) \tag{8.56}$$

and $$Bi(z) = \frac{\sqrt{z}}{3}\left[I_{-\frac{1}{3}}\left(\frac{2z^{3/2}}{3}\right) + I_{\frac{1}{3}}\left(\frac{2z^{3/2}}{3}\right)\right]. \tag{8.57}$$

The functions $Ai(z)$ and $Bi(z)$ are called Airy functions of the first and second kinds respectively. These functions are named after the British astronomer, George Biddell Airy.

Using the series expansions of $I_{\pm\frac{1}{3}}(z)$, we write

$$Ai(z) = \sum_{r=0}^{\infty}\frac{z^{3r}}{3^{2r+\frac{2}{3}}r!\,\Gamma\left(r+\frac{2}{3}\right)} - \sum_{r=0}^{\infty}\frac{z^{3r+1}}{3^{2r+\frac{4}{3}}r!\,\Gamma\left(r+\frac{4}{3}\right)}, |z| < \infty \tag{8.58}$$

and $$Bi(z) = \sqrt{3}\left[\sum_{r=0}^{\infty}\frac{z^{3r}}{3^{2r+\frac{2}{3}}r!\,\Gamma\left(r+\frac{2}{3}\right)} + \sum_{r=0}^{\infty}\frac{z^{3r+1}}{3^{2r+\frac{4}{3}}r!\,\Gamma\left(r+\frac{4}{3}\right)}\right], |z| < \infty. \tag{8.59}$$

The general solution of Eq. (8.55) can be expressed in the form

$y = C_1\,Ai(z) + C_2\,Bi(z)$ where C_1 and C_2 are arbitrary constants.

8.11 CYLINDER (OR CYLINDRICAL) FUNCTIONS AS ALTERNATIVE NAMES OF BESSEL FUNCTIONS

Bessel functions are closely associated with physical problems possessing circular or cylindrical symmetry. For example, these functions arise in the studies of free transverse vibrations of a circular membrane, temperature distribution in a circular cylinder, reactor physics, electromagnetic theory, etc. Hence, these functions are also known as *cylinder* (or *cylindrical*) functions. Bessel functions are undoubtedly the most important functions used in many areas of physics and engineering.

8.12 GENERATING FUNCTION FOR BESSEL FUNCTION OF THE FIRST KIND

The *generating function* of Bessel function of the first kind is a particular function which, when expanded in a series, can generate Bessel functions of the first kind. Let us assume such a generating function as

$$G(x,h) = \sum_{n=-\infty}^{+\infty} J_n(x)h^n, \ (h \neq 0). \tag{8.60}$$

We remember that

$$J_n(x) = \sum_{r=0}^{\infty} \frac{(-1)^r}{r!\,\Gamma(n+r+1)} \left(\frac{x}{2}\right)^{n+2r}.$$

Taking $(n+r)$ as non-negative integer, we have $\Gamma\,(n + r + 1) = (n + r)!$.
Now, Eq. (8.60) can be written as

$$G(x,h) = \sum_{n=-\infty}^{+\infty} J_n(x)h^n = \sum_{n=-\infty}^{+\infty} \left[\sum_{r=0}^{\infty} \frac{(-1)^r}{r!\,(n+r)!} \left(\frac{x}{2}\right)^{n+2r} \right] h^n$$

$$= \sum_{s=-\infty}^{+\infty} \sum_{r=0}^{\infty} \frac{(-1)^r}{r!\,s!} \left(\frac{x}{2}\right)^{r+s} h^{s-r} \text{ where we put } n + r = s,$$

$$= \sum_{s=-\infty}^{+\infty} \sum_{r=0}^{\infty} \frac{(-1)^r \left(\frac{x}{2h}\right)^r}{r!} \frac{\left(\frac{x}{2}\right)^s h^s}{s!}$$

$$= \sum_{r=0}^{\infty} \frac{\left(-\frac{x}{2h}\right)^r}{r!} \sum_{s=0}^{+\infty} \frac{\left(\frac{xh}{2}\right)^s}{s!} \text{ as terms with negative } s \text{ vanish,}$$

$$= e^{-\frac{x}{2h}} e^{\frac{xh}{2}} = e^{\frac{x}{2}\left(h-\frac{1}{h}\right)}.$$

Hence, Eq. (8.60) can be written as

$$G(x,h) = e^{\frac{x}{2}\left(h-\frac{1}{h}\right)} = \sum_{n=-\infty}^{+\infty} J_n(x)h^n. \tag{8.61}$$

We say that $J_n(x)$ is the coefficient of h^n in the series expansion of $e^{\frac{x}{2}\left(h-\frac{1}{h}\right)}$.

8.13 RECURRENCE RELATIONS FOR BESSEL FUNCTIONS OF THE FIRST KIND

For simplicity, we are deriving relations for $p = n =$ integer, even though they are satisfied for non-integral values as well.

I. We know that

$$e^{\frac{x}{2}\left(h-\frac{1}{h}\right)} = \sum_{n=-\infty}^{+\infty} J_n(x)h^n.$$

Differentiating both sides once with respect to h, we obtain

$$\frac{x}{2}\left(1+\frac{1}{h^2}\right)e^{\frac{x}{2}\left(h-\frac{1}{h}\right)} = \sum_{n=-\infty}^{+\infty} nJ_n(x)h^{n-1}$$

or
$$\frac{x}{2}\left(1+\frac{1}{h^2}\right)\sum_{n=-\infty}^{+\infty} J_n(x)h^n = \sum_{n=-\infty}^{+\infty} nJ_n(x)h^{n-1}$$

or
$$\sum_{n=-\infty}^{+\infty} \frac{x}{2}J_n(x)h^n + \sum_{n=-\infty}^{+\infty} \frac{x}{2}J_n(x)h^{n-2} = \sum_{n=-\infty}^{+\infty} nJ_n(x)h^{n-1}.$$

Equating the coefficients of h^{n-1} from both sides, we obtain

$$\frac{x}{2}J_{n-1}(x)+\frac{x}{2}J_{n+1}(x) = nJ_n(x)$$

or
$$J_{n-1}(x)+J_{n+1}(x) = \frac{2n}{x}J_n(x). \tag{8.62}$$

II. We know that

$$e^{\frac{x}{2}\left(h-\frac{1}{h}\right)} = \sum_{n=-\infty}^{+\infty} J_n(x)h^n.$$

Differentiating both sides once with respect to x, we obtain

$$\frac{1}{2}\left(h-\frac{1}{h}\right)e^{\frac{x}{2}\left(h-\frac{1}{h}\right)} = \sum_{n=-\infty}^{+\infty} J_n'(x)h^n \text{ where } J_n'(x) = \frac{d}{dx}J_n(x)$$

or
$$\frac{1}{2}\left(h-\frac{1}{h}\right)\sum_{n=-\infty}^{+\infty} J_n(x)h^n = \sum_{n=-\infty}^{+\infty} J_n'(x)h^n$$

or
$$\sum_{n=-\infty}^{+\infty} \frac{1}{2}J_n(x)h^{n+1} - \sum_{n=-\infty}^{+\infty} \frac{1}{2}J_n(x)h^{n-1} = \sum_{n=-\infty}^{+\infty} J_n'(x)h^n.$$

Equating the coefficients of h^n from both sides, we obtain

$$\frac{1}{2}J_{n-1}(x)-\frac{1}{2}J_{n+1}(x)=J_n'(x)$$

or

$$J_{n-1}(x)-J_{n+1}(x)=2J_n'(x). \tag{8.63}$$

III. Adding Eq. (8.62) and Eq. (8.63) together, we obtain

$$2J_{n-1}(x)=2\left[J_n'(x)+\frac{n}{x}J_n(x)\right]$$

or

$$J_{n-1}(x)=\frac{n}{x}J_n(x)+J_n'(x). \tag{8.64}$$

IV. Subtracting Eq. (8.63) from Eq. (8.62), we obtain

$$J_{n+1}(x)=\frac{n}{x}J_n(x)-J_n'(x). \tag{8.65}$$

V. We have

$$\frac{d}{dx}\left[x^nJ_n(x)\right]=nx^{n-1}J_n(x)+x^nJ_n'(x)$$

$$=x^n\left[\frac{n}{x}J_n(x)+J_n'(x)\right]=x^nJ_{n-1}(x). \tag{8.66}$$

VI. From Eq. (8.62), we have

$$J_{n-1}(x)=\frac{2n}{x}J_n(x)-J_{n+1}(x).$$

Substituting this relation into Eq. (8.66), we obtain

$$\frac{d}{dx}\left[x^nJ_n(x)\right]=x^n\left[\frac{2n}{x}J_n(x)-J_{n+1}(x)\right]. \tag{8.67}$$

VII. We have

$$\frac{d}{dx}\left[x^{-n}J_n(x)\right]=-nx^{-n-1}J_n(x)+x^{-n}J_n'(x)$$

$$=x^{-n}\left[J_n'(x)-\frac{n}{x}J_n(x)\right]=-x^{-n}J_{n+1}(x) \tag{8.68}$$

where use has been made of Eq. (8.65).

VIII. While discussing the need of Bessel function of the second kind, it was proved earlier that if n is an integer, then

$$J_{-n}(x) = (-1)^n J_n(x).$$

IX. If we integrate both sides of Eq. (8.63) once with respect to x, we will obtain

$$J_n(x) = \frac{1}{2}\left[\int J_{n-1}(x)\,dx - \int J_{n+1}(x)\,dx\right] + C \tag{8.69}$$

where C is an integration constant.

X. If we integrate both sides of Eq. (8.66) once with respect to x, we will obtain

$$x^n J_n(x) + C = \int x^n J_{n-1}(x)\,dx. \tag{8.70}$$

XI. Similarly, if we integrate both sides of Eq. (8.68) once with respect to x, we will obtain

$$-x^{-n} J_n(x) + C = \int x^{-n} J_{n+1}(x)\,dx. \tag{8.71}$$

8.14 INTEGRAL REPRESENTATIONS OF BESSEL FUNCTION OF THE FIRST KIND

I. Schlaefli's Integral Representation

We know that the Bessel function of the first kind is given by

$$J_p(x) = \sum_{r=0}^{\infty} \frac{(-1)^r}{r!\,\Gamma(p+r+1)}\left(\frac{x}{2}\right)^{p+2r}.$$

From the study of gamma functions, it is known that

$$\frac{1}{\Gamma(z)} = \frac{1}{2\pi i}\oint_{\gamma} \zeta^{-z} e^{\zeta}\,d\zeta$$

where γ is some contour in the complex ζ-plane (Figure 8.1) running below the real axis from $-\infty$ towards the origin, turning around the origin along a circle C in the positive sense and then going above the real axis back to $-\infty$. The contour used here is known as a Hankel contour.

Therefore, we have

$$\frac{1}{\Gamma(p+r+1)} = \frac{1}{2\pi i}\oint_{\gamma} \zeta^{-p-r-1} e^{\zeta}\,d\zeta.$$

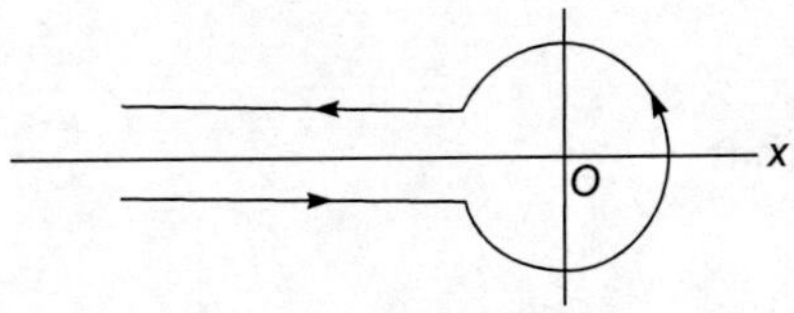

Fig. 8.1 Hankel contour

Substituting the expression for the reciprocal of gamma function into the series expression of Bessel function of the first kind, we write

$$J_p(x)=\sum_{r=0}^{\infty}\frac{(-1)^r\left(\frac{x}{2}\right)^{p+2r}}{r!}\frac{1}{2\pi i}\oint_{\gamma}\zeta^{-p-r-1}e^{\zeta}d\zeta$$

$$=\left(\frac{x}{2}\right)^p\frac{1}{2\pi i}\oint_{\gamma}\sum_{r=0}^{\infty}\frac{\left(-\frac{x^2}{4\zeta}\right)^r}{r!}\zeta^{-p-1}e^{\zeta}d\zeta$$

$$=\left(\frac{x}{2}\right)^p\frac{1}{2\pi i}\oint_{\gamma}\zeta^{-p-1}e^{\zeta-\frac{x^2}{4\zeta}}d\zeta$$

$$=\left(\frac{x}{2}\right)^p\frac{1}{2\pi i}\oint_{\gamma}\zeta^{-p-1}e^{\frac{x}{2}\left(\frac{2\zeta}{x}-\frac{x}{2\zeta}\right)}d\zeta.$$

In the integrand, putting $\frac{2\zeta}{x}=t$, i.e., $\zeta=\frac{x}{2}t$, we have $d\zeta=\frac{x}{2}dt$.

Thus, we write

$$J_p(x)=\left(\frac{x}{2}\right)^p\frac{1}{2\pi i}\oint_{\gamma'}\left(\frac{x}{2}t\right)^{-p-1}e^{\frac{x}{2}\left(t-\frac{1}{t}\right)}\frac{x}{2}dt$$

where γ' is a corresponding contour in the t-plane.

Now, we write

$$J_p(x)=\frac{1}{2\pi i}\oint_{\gamma'}e^{\frac{x}{2}\left(t-\frac{1}{t}\right)}t^{-p-1}dt. \tag{8.72}$$

Equation (8.72) represents the general form of Schlaefli's integral representation of Bessel function of the first kind.

II. Bessel's Integral

If n is an integer, we know that

$$e^{\frac{x}{2}\left(h-\frac{1}{h}\right)} = \sum_{n=-\infty}^{+\infty} J_n(x)h^n$$

$$= -\infty \ldots + J_{-3}(x)h^{-3} + J_{-2}(x)h^{-2} + J_{-1}(x)h^{-1} + J_0(x) + J_1(x)h + J_2(x)h^2 + J_3(x)h^3 + \ldots \infty$$

$$= J_0(x) + \sum_{n=1}^{\infty}\left[h^n + (-1)^n h^{-n}\right]J_n(x)$$

where use has been made of the formula $J_{-n}(x) = (-1)^n J_n(x)$ if n is an integer.

If we put $h = e^{i\phi}$ we can write

$$e^{\frac{x}{2}(e^{i\phi}-e^{-i\phi})} = J_0(x) + \sum_{n=1}^{\infty}\left[e^{in\phi} + (-1)^n e^{-in\phi}\right]J_n(x). \quad (8.73)$$

When n is an even integer, we have

$$e^{in\phi} + (-1)^n e^{-in\phi} = e^{in\phi} + e^{-in\phi} = 2\cos n\phi. \quad (8.74)$$

When n is an odd integer, we have

$$e^{in\phi} + (-1)^n e^{-in\phi} = e^{in\phi} - e^{-in\phi} = 2i\sin n\phi. \quad (8.75)$$

Using Eqs. (8.74) and (8.75), we can rewrite Eq. (8.73) as

$$e^{ix\sin\phi} = J_0(x) + \sum_{n=2,4,6\ldots}^{\infty} 2J_n(x)\cos n\phi + i\sum_{n=1,3,5\ldots}^{\infty} 2J_n(x)\sin n\phi$$

or $\cos(x\sin\phi) + i\sin(x\sin\phi)$

$$= J_0(x) + \sum_{k=1}^{\infty} 2J_{2k}(x)\cos 2k\phi + i\sum_{k=1}^{\infty} 2J_{2k-1}(x)\sin(2k-1)\phi. \quad (8.76)$$

Equating real parts from both sides of Eq. (8.76), we obtain

$$\cos(x\sin\phi) = J_0(x) + \sum_{k=1}^{\infty} 2J_{2k}(x)\cos 2k\phi. \quad (8.77)$$

Equating imaginary parts from both sides of Eq. (8.76), we obtain

$$\sin(x\sin\phi) = \sum_{k=1}^{\infty} 2J_{2k-1}(x)\sin(2k-1)\phi. \quad (8.78)$$

Equations (8.77) and (8.78) are known as Jacobi series.

If we multiply both sides of Eq. (8.77) by $\cos n\phi\ (n>0)$, integrate with respect to ϕ from $\phi=0$ to $\phi=\pi$ and use the identity

$$\int_0^{\pi}\cos p\phi\cos q\phi\,d\phi=\begin{cases}0, \text{ if } p\neq q,\\ \dfrac{\pi}{2}, \text{ if } p=q\neq 0,\\ \pi, \text{ if } p=q=0;\end{cases}$$

we will obtain

$$\int_0^{\pi}\cos n\phi\cos(x\sin\phi)d\phi=\begin{cases}\pi J_n(x) & \text{for } n=2k=\text{even integer},\\ 0 & \text{for } n=\text{odd integer}.\end{cases}\tag{8.79}$$

Similarly, if we multiply both sides of Eq. (8.78) by $\sin n\phi(n\geq 1)$, integrate with respect to ϕ from $\phi=0$ to $\phi=\pi$ and use the identity

$$\int_0^{\pi}\sin p\phi\sin q\phi\,d\phi=\begin{cases}0 & \text{if } p\neq q,\\ \dfrac{\pi}{2} & \text{if } p=q\neq 0;\end{cases}$$

we will obtain

$$\int_0^{\pi}\sin n\phi\sin(x\sin\phi)d\phi=\begin{cases}0 & \text{for } n=\text{even integer},\\ \pi J_n(x) & \text{for } n=2k-1=\text{odd integer}.\end{cases}\tag{8.80}$$

Adding Eqs. (8.79) and (8.80), we obtain

$$\int_0^{\pi}\left[\cos n\phi\cos(x\sin\phi)+\sin n\phi\sin(x\sin\phi)\right]d\phi=\pi J_n(x) \quad \text{for } n \text{ even or odd.}$$

or
$$\int_0^{\pi}\cos(n\phi-x\sin\phi)d\phi=\pi J_n(x)$$

giving
$$J_n(x)=\frac{1}{\pi}\int_0^{\pi}\cos(n\phi-x\sin\phi)d\phi.\tag{8.81}$$

This integral is called Bessel integral and it is valid when n is an integer.

IIa. Extension of this Integral for Bessel function of the First Kind of Non-integral Order

In the general form of Schlaefli's integral representation of Bessel function of the first kind (Eq. (8.72)), we now deform the contour in the complex t-plane so that it runs below the real axis from $-\infty$ to -1 encircles the origin O in the positive sense through the unit circle, returns above the real axis to -1 and then to $-\infty$

Thus, Eq. (8.72) can be written as

$$J_p(x) = \frac{1}{2\pi i}\oint_{\gamma'} t^{-p-1} e^{\frac{x}{2}\left(t-\frac{1}{t}\right)} dt$$

$$= \frac{1}{2\pi i}\int_{-\infty}^{-1} t^{-p-1} e^{\frac{x}{2}\left(t-\frac{1}{t}\right)} dt + \frac{1}{2\pi i}\int_{\text{unit circle}} t^{-p-1} e^{\frac{x}{2}\left(t-\frac{1}{t}\right)} dt + \frac{1}{2\pi i}\int_{-1}^{-\infty} e^{-t-1} e^{\frac{x}{2}\left(t-\frac{1}{t}\right)} dt.$$

If we take $t = \sigma e^{i\phi}$ then in the first integral on the right-hand side of the above equation, we have $t = \sigma e^{-\pi i}$, $dt = e^{-\pi i}\, d\sigma, 1 < \sigma < +\infty$; in the second integral, we have $t = e^{i\phi}$, $dt = ie^{i\phi} d\phi, -\pi < \phi < +\pi$; and in the third integral, we have $t = \sigma e^{\pi i}$, $dt = e^{\pi i} d\sigma, 1 < \sigma < +\infty$. Thus, we write

$$J_p(x) = \frac{1}{2\pi i}\int_{\infty}^{1} (\sigma e^{\pi i})^{-p-1} e^{\frac{x}{2}\left(\sigma e^{-\pi i} - \frac{1}{\sigma} e^{\pi i}\right)} e^{-\pi i} d\sigma + \frac{1}{2\pi i}\int_{-\pi}^{+\pi} (e^{i\phi})^{-p-1} e^{\frac{x}{2}(e^{i\phi} - e^{-i\phi})} ie^{i\phi} d\phi$$

$$+ \frac{1}{2\pi i}\int_{1}^{\infty} (\sigma e^{\pi i})^{-p-1} e^{\frac{x}{2}\left(\sigma e^{\pi i} - \frac{1}{\sigma} e^{-\pi i}\right)} e^{\pi i} d\sigma$$

$$= \frac{1}{2\pi i}\int_{\infty}^{1} \sigma^{-p-1} e^{p\pi i} e^{\frac{x}{2}\left(-\sigma + \frac{1}{\sigma}\right)} d\sigma + \frac{1}{2\pi}\int_{-\pi}^{+\pi} e^{-p\phi i} e^{ix\sin\phi} d\phi + \frac{1}{2\pi i}\int_{1}^{\infty} \sigma^{-p-1} e^{-p\pi i} e^{\frac{x}{2}\left(-\sigma + \frac{1}{\sigma}\right)} d\sigma$$

$$= \frac{1}{2\pi}\int_{-\pi}^{+\pi} e^{i(x\sin\phi - p\phi)} d\phi - \frac{1}{2\pi i} e^{p\pi i}\int_{1}^{\infty} \sigma^{-p-1} e^{\frac{x}{2}\left(-\sigma + \frac{1}{\sigma}\right)} d\sigma + \frac{1}{2\pi i} e^{-p\pi i}\int_{1}^{\infty} \sigma^{-p-1} e^{\frac{x}{2}\left(-\sigma + \frac{1}{\sigma}\right)} d\sigma$$

$$= \frac{1}{2\pi}\int_{-\pi}^{+\pi} \cos(x\sin\phi - p\phi) d\phi + \frac{i}{2\pi}\int_{-\pi}^{+\pi} \sin(x\sin\phi - p\phi) d\phi$$

$$- \frac{1}{2\pi i}\left\{e^{p\pi i} - e^{-p\pi i}\right\}\int_{1}^{\infty} \sigma^{-p-1} e^{\frac{x}{2}\left(-\sigma + \frac{1}{\sigma}\right)} d\sigma.$$

Here, the second integral vanishes as the integrand is an odd function. Therefore, we write

$$J_p(x) = \frac{1}{\pi}\int_{0}^{\pi} \cos(x\sin\phi - p\phi) d\phi - \frac{\sin p\pi}{\pi}\int_{1}^{\infty} \sigma^{-p-1} e^{\frac{x}{2}\left(-\sigma + \frac{1}{\sigma}\right)} d\sigma.$$

Let $\sigma = e^{\lambda}$, then $d\sigma = e^{\lambda} d\lambda$. Hence, we write

$$J_p(x) = \frac{1}{\pi}\int_0^{\pi} \cos\left(x\sin\phi - p\phi\right) d\phi - \frac{\sin p\pi}{\pi}\int_0^{\infty} e^{-p\lambda} e^{\frac{-x}{2}\left(e^{\lambda}-e^{-\lambda}\right)} d\lambda$$

$$= \frac{1}{\pi}\int_0^{\pi} \cos\left(x\sin\phi - p\phi\right) d\phi - \frac{\sin p\pi}{\pi}\int_0^{\infty} e^{-p\lambda - x\sinh\lambda} d\lambda. \tag{8.81a}$$

Equation (8.81a) is the general form Bessel integral for a Bessel function of the first kind of any order (whether integral or non-integral).

When p is an integer, the last term in Eq. (8.81a) will vanish and the equation will reduce to Eq. (8.81). Hence, Eq. (8.81a) can be considered a general form of integral representation of $J_p(x)$ whether p is an integer or not.

III. Poisson's Integral Representations of Bessel Function of the First Kind

We know that

$$J_p(x) = \sum_{r=0}^{\infty} \frac{(-1)^r \left(\dfrac{x}{2}\right)^{p+2r}}{r!\,\Gamma(p+r+1)}. \tag{8.82}$$

We know that beta and gamma functions are related through the relation

$$B(m, n) = \frac{\Gamma(m)\,\Gamma(n)}{\Gamma(m+n)}$$

or

$$\frac{1}{\Gamma(m+n)} = \frac{B(m, n)}{\Gamma(m)\,\Gamma(n)}.$$

Therefore, we have

$$\frac{1}{\Gamma(p+r+1)} = \frac{1}{\Gamma\left(p+\frac{1}{2}+r+\frac{1}{2}\right)} = \frac{B\left(p+\frac{1}{2}, r+\frac{1}{2}\right)}{\Gamma\left(p+\frac{1}{2}\right)\Gamma\left(r+\frac{1}{2}\right)}$$

$$= \frac{1}{\Gamma\left(p+\frac{1}{2}\right)\Gamma\left(r+\frac{1}{2}\right)}\int_0^1 u^{r-\frac{1}{2}}(1-u)^{p-\frac{1}{2}}\,du.$$

If we put $\sqrt{u} = t$, or $u = t^2$, we have $du = 2t\,dt$.

Therefore, we write

$$\frac{1}{\Gamma(p+r+1)}=\frac{1}{\Gamma\left(p+\frac{1}{2}\right)\Gamma\left(r+\frac{1}{2}\right)}\int_0^1 (t^2)^{r-\frac{1}{2}}(1-t^2)^{p-\frac{1}{2}}2t\,dt$$

$$=\frac{1}{\Gamma\left(p+\frac{1}{2}\right)\Gamma\left(r+\frac{1}{2}\right)}\int_{-1}^{+1} t^{2r}(1-t^2)^{p-\frac{1}{2}}dt \tag{8.83}$$

Substituting Eq. (8.83) into Eq. (8.82), we obtain

$$J_p(x)=\sum_{r=0}^{\infty}\frac{(-1)^r\left(\frac{x}{2}\right)^{p+2r}}{r!\,\Gamma\left(p+\frac{1}{2}\right)\Gamma\left(r+\frac{1}{2}\right)}\int_{-1}^{+1} t^{2r}(1-t^2)^{p-\frac{1}{2}}dt$$

$$=\frac{\left(\frac{x}{2}\right)^p}{\Gamma\left(p+\frac{1}{2}\right)}\int_{-1}^{+1}\sum_{r=0}^{\infty}\frac{(-1)^r(xt)^{2r}}{2^{2r}\,\Gamma(r+1)\Gamma\left(r+\frac{1}{2}\right)}(1-t^2)^{p-\frac{1}{2}}dt\,.$$

Duplication formula of gamma functions gives

$$2^{2r}\Gamma(r+1)\Gamma\left(r+\frac{1}{2}\right)=\Gamma\left(\frac{1}{2}\right)\Gamma(2r+1)=\sqrt{\pi}\,(2r)!.$$

Using this formula, we write

$$J_p(x)=\frac{\left(\frac{x}{2}\right)^p}{\sqrt{\pi}\,\Gamma\left(p+\frac{1}{2}\right)}\int_{-1}^{+1}\sum_{r=0}^{\infty}\frac{(i^2)^r(xt)^{2r}}{(2r)!}(1-t^2)^{p-\frac{1}{2}}dt$$

$$=\frac{\left(\frac{x}{2}\right)^p}{\sqrt{\pi}\,\Gamma\left(p+\frac{1}{2}\right)}\int_{-1}^{+1}\sum_{r=0}^{\infty}\left\{\frac{(ixt)^{2r}}{(2r)!}\right\}(1-t^2)^{p-\frac{1}{2}}dt \tag{8.84}$$

or

$$J_p(x)=\frac{\left(\frac{x}{2}\right)^p}{\sqrt{\pi}\,\Gamma\left(p+\frac{1}{2}\right)}\int_{-1}^{+1}(1-t^2)^{p-\frac{1}{2}}\cos(xt)\,dt \text{ with } p>-\frac{1}{2},\, t>0. \tag{8.85}$$

Equation (8.85) is another integral representation of Bessel function of the first kind $J_p(x)$.

It is known as *Poisson's integral formula* of Bessel function of the first kind $J_p(x)$.

If we put $t=\cos\phi$, Eq. (8.85) becomes

$$J_p(x)=\frac{\left(\frac{x}{2}\right)^p}{\sqrt{\pi}\,\Gamma\left(p+\frac{1}{2}\right)}\int_0^{\pi}\cos(x\cos\phi)\sin^{2p}\phi\,d\phi. \tag{8.86}$$

Further, Eq. (8.85) can be written as

$$J_p(x)=\frac{\left(\frac{x}{2}\right)^p}{\sqrt{\pi}\,\Gamma\left(p+\frac{1}{2}\right)}2\int_0^{1}(1-t^2)^{p-\frac{1}{2}}\cos(xt)\,dt. \tag{8.87}$$

Putting $t=\sin\phi$, Eq. (8.87) can be written as

$$J_p(x)=\frac{\left(\frac{x}{2}\right)^p}{\sqrt{\pi}\,\Gamma\left(p+\frac{1}{2}\right)}2\int_0^{\frac{\pi}{2}}\cos(x\sin\phi)\cos^{2p}\phi\,d\phi$$

$$=\frac{\left(\frac{x}{2}\right)^p}{\sqrt{\pi}\,\Gamma\left(p+\frac{1}{2}\right)}\int_0^{\pi}\cos(x\sin\phi)\cos^{2p}\phi\,d\phi. \tag{8.88}$$

Another Form of Poisson's Integral Representation of Bessel Function of the First Kind

Putting $s=2r$, Eq. (8.84) can be written as

$$J_p(x)=\frac{\left(\frac{x}{2}\right)^p}{\sqrt{\pi}\,\Gamma\left(p+\frac{1}{2}\right)}\int_{-1}^{+1}\sum_{s=0}^{\infty}\frac{(ixt)^s}{s!}(1-t^2)^{p-\frac{1}{2}}dt.$$

As the integral on the right-hand side of the above equation vanishes for s equal to an odd integer, we can write the above equation in the form

$$J_p(x)=\frac{\left(\frac{x}{2}\right)^p}{\sqrt{\pi}\,\Gamma\left(p+\frac{1}{2}\right)}\int_{-1}^{+1}e^{ixt}(1-t^2)^{p-\frac{1}{2}}dt,\ p>-\frac{1}{2},\ x>0. \tag{8.89}$$

Equation (8.89) is yet another form of Poisson's integral representation of Bessel function of the first kind.

8.15 ZEROES OF BESSEL FUNCTIONS OF THE FIRST KIND

The zeroes of a Bessel function of the first kind $J_p(x)$ are the values of x at which $J_p(x)$ is equal to zero. Thus, for the zeroes α_i of the Bessel function $J_p(x)$, we have $J_p(\alpha_i)=0$. The first ten zeroes of $J_0(x)$, $J_1(x)$, $J_2(x)$ and $J_3(x)$ are shown below.

Sl. No.	Zeroes of			
	$J_0(x)$	$J_1(x)$	$J_2(x)$	$J_3(x)$
1	2.40483	0	0	0
2	5.52008	3.83171	5.13562	6.38016
3	8.65373	7.01559	8.41724	9.76102
4	11.79153	10.17347	11.61984	13.01520
5	14.93092	13.32369	14.79595	16.22347
6	18.0710	16.47063	17.95982	19.40942
7	21.21164	19.61586	21.11700	22.58273
8	24.35247	22.76008	24.27011	25.74817
9	27.49348	25.90367	27.42057	28.90835
10	30.63461	29.04683	30.56920	32.06485

8.16 ORTHOGONALITY OF BESSEL FUNCTIONS OF THE FIRST KIND

If k_m and k_n are two distinct roots of $J_p(ka)=0$, then $J_p(k_m x)$ and $J_p(k_n x)$ are orthogonal with a weight function $w(x)=x$ in an interval $[0,a]$, that is,

$$\int_0^a J_p(k_m x) J_p(k_n x)\, x\, dx = 0 \text{ if } k_m \neq k_n. \tag{8.90}$$

We know that $J_p(k_m x)$ and $J_p(k_n x)$ respectively satisfy the equations

$$x^2 \frac{d^2}{dx^2} J_p(k_m x) + x\frac{d}{dx} J_p(k_m x) + \left(k_m^2 x^2 - p^2\right) J_p(k_m x) = 0$$

and

$$x^2 \frac{d^2}{dx^2} J_p(k_n x) + x\frac{d}{dx} J_p(k_n x) + \left(k_n^2 x^2 - p^2\right) J_p(k_n x) = 0.$$

If we put $f(x) = J_p(k_m x)$ and $g(x) = J_p(k_n x)$, the above two equations can be written in the forms

$$x\frac{d^2 f}{dx^2} + \frac{df}{dx} + x\left(k_m^2 - \frac{p^2}{x^2}\right)f = 0 \tag{8.91}$$

and

$$x\frac{d^2 g}{dx^2} + \frac{dg}{dx} + x\left(k_n^2 - \frac{p^2}{x^2}\right)g = 0. \tag{8.92}$$

Subtracting g times Eq. (8.91) from f times Eq. (8.92), we obtain

$$xf\frac{d^2 g}{dx^2} - xg\frac{d^2 f}{dx^2} + f\frac{dg}{dx} - g\frac{df}{dx} + \left(k_n^2 - k_m^2\right)xfg = 0$$

or

$$\frac{d}{dx}\left(xf\frac{dg}{dx} - xg\frac{df}{dx}\right) = \left(k_m^2 - k_n^2\right)xfg.$$

Integrating both sides once with respect to x in the limit from $x = 0$ to $x = a$, we obtain

$$\left[xf\frac{dg}{dx} - xg\frac{df}{dx}\right]_0^a = \left(k_m^2 - k_n^2\right)\int_0^a xfg\,dx$$

or

$$\int_0^a fg\,x dx = \frac{1}{\left(k_m^2 - k_n^2\right)}\left[xf\frac{dg}{dx} - xg\frac{df}{dx}\right]_0^a$$

or

$$\int_0^a J_p(k_m x)J_p(k_n x)\,x dx$$

$$= \frac{1}{\left(k_m^2 - k_n^2\right)}\left[xJ_p(k_m x)\frac{d}{dx}J_p(k_n x) - xJ_p(k_n x)\frac{d}{dx}J_p(k_m x)\right]_0^a$$

If $k_m \neq k_n$, we write

$$\int_0^a J_p(k_m x)J_p(k_n x)x dx = 0 \tag{8.93}$$

because the expression within the square brackets in the above equation vanishes at both the limits as $J_p(k_m x)$ and $J_p(k_n x)$ vanish at $x = a$.

If we take $k = k_m$ and $l = k_n$ then

$$\int_0^a J_p(kx)J_p(lx)x dx = 0. \tag{8.94}$$

Thus, we say that $J_p(kx)$ and $J_p(lx)$ are orthogonal with a weight function x if $J_p(kx)$ and $J_p(lx)$ vanish at $x = a$.

Orthonormality with a Normalization Constant

In the case of $k_m = k_n = k(\text{say})$, we have, $f(x) = g(x) = J_p(kx)$

Then, $f(x)$ satisfies

$$x^2\frac{d^2f}{dx^2} + x\frac{df}{dx} + \left(k^2x^2 - p^2\right)f = 0.$$

Multiplying both sides by $2\dfrac{df}{dx}$, we obtain

$$2x^2\frac{df}{dx}\frac{d^2f}{dx^2} + 2x\left(\frac{df}{dx}\right)^2 + 2\left(k^2x^2 - p^2\right)f\frac{df}{dx} = 0$$

or

$$\frac{d}{dx}\left[x^2\left(\frac{df}{dx}\right)^2 + k^2x^2f^2 - p^2f^2\right] - 2k^2xf^2 = 0.$$

Integrating both sides with respect to x within the interval from $x = 0$ to $x = a$, we obtain

$$\left[x^2\left(\frac{df}{dx}\right)^2 + k^2x^2f^2 - p^2f^2\right]_0^a = 2k^2\int_0^a xf^2dx$$

or

$$\left[x^2\left\{\frac{d}{dx}J_p(kx)\right\}^2 + k^2x^2\left\{J_p(kx)\right\}^2 - p^2\left\{J_p(kx)\right\}^2\right]_0^a = 2k^2\int_0^a\left\{J_p(kx)\right\}^2 xdx.$$

The second and the third terms within the square brackets on the left-hand side vanish at both the limits since $J_p(ka) = 0$ and $pJ_p(0) = 0$ for all values of p.

Therefore, we write

$$\left[\left\{x\frac{d}{dx}J_p(kx)\right\}^2\right]_0^a = 2k^2\int_0^a\left\{J_p(kx)\right\}^2 xdx. \qquad (8.95)$$

Since

$$x\frac{d}{dx}J_p(x) = pJ_p(x) - xJ_{p+1}(x),$$

we have

$$kx\frac{d}{d(kx)}J_p(kx) = pJ_p(kx) - kxJ_{p+1}(kx)$$

or

$$x\frac{d}{dx}J_p(kx) = pJ_p(kx) - kxJ_{p+1}(kx).$$

Hence, Eq. (8.95) can be written as

$$\left[\left\{pJ_p(kx) - kx\,J_{p+1}(kx)\right\}^2\right]_0^a = 2k^2\int_0^a\left\{J_p(kx)\right\}^2 xdx$$

or
$$k^2a^2\{J_{p+1}(ka)\}^2 = 2k^2\int_0^a\{J_p(kx)\}^2\, xdx$$

where use has been made of the fact that $J_p(ka)=0$.

Now, we obtain
$$\int_0^a\{J_p(kx)\}^2\, xdx = \frac{a^2}{2}\{J_{p+1}(ka)\}^2 \tag{8.96}$$

or
$$\int_0^a\left\{\frac{\sqrt{2}\,J_p(kx)}{a\,J_{p+1}(ka)}\right\}^2 xdx = 1. \tag{8.97}$$

8.17 EXPANSION OF AN ARBITRARY FUNCTION IN A SERIES OF BESSEL FUNCTIONS

An arbitrary function, $f(x)$ defined in an interval $0<x<a$, can be expanded in a series of Bessel functions as
$$f(x)=\sum_{n=1}^{\infty}c_nJ_p(k_nx),\, 0<x<a, \quad p>-\frac{1}{2}, \tag{8.98}$$

with $J_p(k_na)=0, (n=1,2,3\ldots)$. Such a series is known as Fourier-Bessel series or simply Bessel series.

Multiplying both sides of Eq. (8.98) by $xJ_p(k_mx)$ where $J_p(k_ma)=0$ and integrating with respect to x from $x=0$ to $x=a$, we obtain
$$\int_0^a f(x)xJ_p(k_mx)dx = \sum_{n=1}^{\infty}c_n\int_0^a xJ_p(k_mx)J_p(k_nx)dx$$
$$= c_m\int_0^a x\{J_p(k_mx)\}^2\, dx = \frac{a^2}{2}\{J_{p+1}(k_ma)\}^2\, c_m.$$

Therefore, we have
$$c_m = \frac{2}{a^2\{J_{p+1}(k_ma)\}^2}\int_0^a f(x)J_p(k_mx)xdx$$

Changing the subscript, we write
$$c_n = \frac{2}{a^2\{J_{p+1}(k_na)\}^2}\int_0^a f(x)J_p(k_nx)x\,dx,\ n=1,2,3\ldots \tag{8.99}$$

Substituting the values of the coefficients thus calculated into the assumed series, we can get the required series.

8.18 RECURRENCE RELATIONS FOR NEUMANN FUNCTIONS (BESSEL FUNCTION OF THE SECOND KIND)

As Neumann function (or Bessel function of the second kind) is a linear combination of two Bessel functions of the first kind, it will satisfy recurrence relations similar to those of the Bessel functions of the first kind. Relevant recurrence relations are derived as in the following.

I. From Eq. (8.14), we know that if p is equal to an integer n, then Neumann function of order n is given by

$$N_n(x) = \lim_{p\to n}\frac{1}{\pi}\left[\frac{\partial}{\partial p}J_p(x) - (-1)^n\frac{\partial}{\partial p}J_{-p}(x)\right].$$

Therefore, we have

$$\begin{aligned}
N_{-n}(x) &= \lim_{p\to -n}\frac{1}{\pi}\left[\frac{\partial}{\partial p}J_p(x) - (-1)^n\frac{\partial}{\partial p}J_{-p}(x)\right]\\
&= \lim_{p\to n}\frac{1}{\pi}\left[\frac{\partial}{-\partial p}J_{-p}(x) - (-1)^n\frac{\partial}{-\partial p}J_p(x)\right]\\
&= \lim_{p\to n}\frac{1}{\pi}\left[-\frac{\partial}{\partial p}J_{-p}(x) + (-1)^n\frac{\partial}{\partial p}J_p(x)\right]\\
&= \lim_{p\to n}\frac{1}{\pi}(-1)^n\left[-(-1)^n\frac{\partial}{\partial p}J_{-p}(x) + \frac{\partial}{\partial p}J_p(x)\right]\\
&= (-1)^n\lim_{p\to n}\frac{1}{\pi}\left[\frac{\partial}{\partial p}J_p(x) - (-1)^n\frac{\partial}{\partial p}J_{-p}(x)\right]\\
&= (-1)^n N_n(x). \qquad (8.100)
\end{aligned}$$

II. In this case also, let us consider Neumann functions of integral orders.

If n is an integer, we have

$$N_n(x) = \lim_{p\to n}\frac{1}{\pi}\left[\frac{\partial}{\partial p}J_p(x) - (-1)^n\frac{\partial}{\partial p}J_{-p}(x)\right].$$

Therefore, we have

$$N_{n-1}(x)=\lim_{p\to n-1}\frac{1}{\pi}\left[\frac{\partial}{\partial p}J_p(x)-(-1)^{n-1}\frac{\partial}{\partial p}J_{-p}(x)\right]$$

$$=\lim_{p\to n}\frac{1}{\pi}\left[\frac{\partial}{\partial p}J_{p-1}(x)+(-1)^{n}\frac{\partial}{\partial p}J_{-(p-1)}(x)\right]. \tag{8.101}$$

Similarly, we have

$$N_{n+1}(x)=\lim_{p\to n}\frac{1}{\pi}\left[\frac{\partial}{\partial p}J_{p+1}(x)+(-1)^{n}\frac{\partial}{\partial p}J_{-(p+1)}(x)\right]. \tag{8.102}$$

Adding Eq. (8.101) and Eq. (8.102), we obtain

$$N_{n-1}(x)+N_{n+1}(x)$$

$$=\lim_{p\to n}\frac{1}{\pi}\left[\frac{\partial}{\partial p}\left\{J_{p-1}(x)+J_{p+1}(x)\right\}+(-1)^{n}\frac{\partial}{\partial p}\left\{J_{-p+1}(x)+J_{-p-1}(x)\right\}\right]$$

$$=\lim_{p\to n}\frac{1}{\pi}\left[\frac{\partial}{\partial p}\left\{\frac{2p}{x}J_p(x)\right\}+(-1)^{n}\frac{\partial}{\partial p}\left\{\frac{-2p}{x}J_{-p}(x)\right\}\right]$$

$$=\lim_{p\to n}\frac{1}{\pi}\left[\left\{\frac{2}{x}J_p(x)+\frac{2p}{x}\frac{\partial}{\partial p}J_p(x)\right\}+(-1)^{n}\left\{-\frac{2}{x}J_{-p}(x)-\frac{2p}{x}\frac{\partial}{\partial p}J_{-p}(x)\right\}\right]$$

$$=\frac{2}{\pi x}\left[J_n(x)-(-1)^{n}J_{-n}(x)\right]+\lim_{p\to n}\frac{1}{\pi}\frac{2p}{x}\left[\frac{\partial}{\partial p}J_p(x)-(-1)^{n}\frac{\partial}{\partial p}J_{-p}(x)\right]$$

$$=\frac{2}{\pi x}\left[J_n(x)-(-1)^{2n}J_n(x)\right]+\frac{2n}{x}\lim_{p\to n}\frac{1}{\pi}\left[\frac{\partial}{\partial p}J_p(x)-(-1)^{n}\frac{\partial}{\partial p}J_{-p}(x)\right].$$

Finally, we write

$$N_{n-1}(x)+N_{n+1}(x)=\frac{2n}{x}N_n(x). \tag{8.103}$$

III. Subtracting Eq. (8.102) from Eq. (8.101), we obtain

$$N_{n-1}(x)-N_{n+1}(x)$$

$$=\lim_{p\to n}\frac{1}{\pi}\left[\frac{\partial}{\partial p}\left\{J_{p-1}(x)-J_{p+1}(x)\right\}+(-1)^{n}\frac{\partial}{\partial p}\left\{J_{-p+1}(x)-J_{-p-1}(x)\right\}\right]$$

$$= \lim_{p \to n} \frac{1}{\pi} \left[\frac{\partial}{\partial p} \left\{ 2 \frac{\partial}{\partial x} J_p(x) \right\} + (-1)^n \frac{\partial}{\partial p} \left\{ -2 \frac{\partial}{\partial x} J_{-p}(x) \right\} \right]$$

$$= 2 \frac{\partial}{\partial x} \lim_{p \to n} \frac{1}{\pi} \left[\frac{\partial}{\partial p} J_p(x) - (-1)^n \frac{\partial}{\partial p} J_{-p}(x) \right]$$

$$= 2 \frac{d}{dx} N_n(x) = 2N_n'(x). \tag{8.104}$$

IV. Adding Eqs. (8.103) and (8.104), we obtain

$$N_{n-1}(x) = \frac{n}{x} N_n(x) + N_n'(x). \tag{8.105}$$

V. Subtracting Eq. (8.104) from Eq. (8.103), we obtain

$$N_{n+1}(x) = \frac{n}{x} N_n(x) - N_n'(x). \tag{8.106}$$

VI. (a) For p = non-integer

We have

$$\frac{d}{dx}\left\{x^p N_p(x)\right\} = \frac{d}{dx}\left[x^p \frac{\left\{\cos p\pi J_p(x) - J_{-p}(x)\right\}}{\sin p\pi} \right]$$

$$= \frac{1}{\sin p\pi} \frac{d}{dx}\left[\cos p\pi \, x^p J_p(x) - x^p J_{-p}(x) \right]$$

$$= \frac{1}{\sin p\pi}\left[\cos p\pi \frac{d}{dx}\left\{x^p J_p(x)\right\} - \frac{d}{dx}\left\{x^p J_{-p}(x)\right\} \right]$$

$$= \frac{1}{\sin p\pi}\left[\left\{\cos p\pi\right\} x^p J_{p-1}(x) - \left\{-x^p J_{-p+1}(x)\right\} \right]$$

$$= x^p \left[\frac{\cos p\pi J_{p-1}(x) + J_{-(p-1)}(x)}{\sin p\pi} \right]$$

$$= x^p \left[\frac{\cos\left\{\pi + (p-1)\pi\right\} J_{p-1}(x) + J_{-(p-1)}(x)}{\sin\left\{\pi + (p-1)\pi\right\}} \right]$$

$$= x^p \left[\frac{-\cos(p-1)\pi J_{p-1}(x) + J_{-(p-1)}(x)}{-\sin(p-1)\pi} \right]$$

$$= x^p \left[\frac{\cos(p-1)\pi J_{p-1}(x) - J_{-(p-1)}(x)}{\sin(p-1)\pi} \right]$$

$$= x^p N_{p-1}(x). \tag{8.107}$$

(b) For p = integer:

We know that for $p = n$ (an integer),

$$N_n(x) = \lim_{p \to n} N_p(x)$$

Therefore, we write

$$\lim_{p \to n} \frac{d}{dx}\left\{x^p N_p(x)\right\} = \lim_{p \to n}\left[x^p N_{p-1}(x)\right]$$

or

$$\frac{d}{dx}\{x^n N_n(x)\} = x^n N_{n-1}(x). \tag{8.107a}$$

VII. (a) For p = non-integer:

$$\frac{d}{dx}\left\{x^{-p} N_p(x)\right\} = \frac{d}{dx}\left[x^{-p}\left\{\frac{\cos p\pi J_p(x) - J_{-p}(x)}{\sin p\pi}\right\}\right]$$

$$= \frac{1}{\sin p\pi}\left[\cos p\pi \frac{d}{dx}\left\{x^{-p} J_p(x)\right\} - \frac{d}{dx}\left\{x^{-p} J_{-p}(x)\right\}\right]$$

$$= \frac{1}{\sin p\pi}\left[(\cos p\pi)\left\{-x^{-p} J_{p+1}(x)\right\} - x^{-p} J_{-p-1}(x)\right]$$

$$= \frac{-x^{-p}}{\sin p\pi}\left[\cos p\pi\, J_{p+1}(x) + J_{-(p+1)}(x)\right]$$

$$= \frac{-x^{-p}}{\sin\left\{(p+1)\pi - \pi\right\}}\left[\cos\left\{(p+1)\pi - \pi\right\} J_{p+1}(x) + J_{-(p+1)}(x)\right]$$

$$= -x^{-p}\left[\frac{-\cos(p+1)\pi\, J_{p+1}(x) + J_{-(p+1)}(x)}{-\sin(p+1)\pi}\right]$$

$$= -x^{-p}\left[\frac{\cos(p+1)\pi J_{p+1}(x) - J_{-(p+1)}(x)}{\sin(p+1)\pi}\right] = -x^{-p} N_{p+1}(x). \tag{8.108}$$

(b) For p = integer

We know that for $p = n$ (an integer)

$$N_n(x) = \lim_{p \to n} N_p(x).$$

Therefore, we write

$$\lim_{p \to n} \frac{d}{dx}\left[x^{-p} N_p(x)\right] = \lim_{p \to n}\left[-x^{-p} N_{p+1}(x)\right]$$

or
$$\frac{d}{dx}\left[x^{-n} N_n(x)\right] = -x^{-n} N_{n+1}(x). \tag{8.108a}$$

8.19 RECURRENCE RELATIONS FOR HANKEL FUNCTIONS (OR BESSEL FUNCTIONS OF THE THIRD KIND)

We have known earlier in Section 8.5 that Hankel functions of the first and second kinds of order p are defined by

$$H_p^{(1)}(x) = J_p(x) + iN_p(x) \text{ and } H_p^{(2)}(x) = J_p(x) - iN_p(x)$$

Since Hankel functions are linear combinations of Bessel function of the first kind and Neumann function (Bessel function of the second kind), these functions will satisfy recurrence relations similar to those of Bessel functions of the first kind and of Neumann functions.

I. Considering Hankel functions of integral order, we have

$$\begin{aligned} H_{-n}^{(1)}(x) &= J_{-n}(x) + iN_{-n}(x) \\ &= (-1)^n J_n(x) + i(-1)^n N_n(x) \\ &= (-1)^n \left[J_n(x) + iN_n(x)\right] \\ &= (-1)^n H_n^{(1)}(x) = e^{i\pi n} H_n^{(1)}(x). \end{aligned} \tag{8.109}$$

Similarly, we have

$$H_{-n}^{(2)}(x) = (-1)^n H_n^{(2)}(x) = e^{-i\pi n} H_n^{(2)}(x). \tag{8.109a}$$

II. We have

$$\begin{aligned} H_{p-1}^{(1)}(x) + H_{p+1}^{(1)}(x) &= \left\{J_{p-1}(x) + iN_{p-1}(x)\right\} + \left\{J_{p+1}(x) + iN_{p+1}(x)\right\} \\ &= \left\{J_{p-1}(x) + J_{p+1}(x)\right\} + i\left\{N_{p-1}(x) + N_{p+1}(x)\right\} \end{aligned}$$

$$= \frac{2p}{x} J_p(x) + i \frac{2p}{x} N_p(x)$$

$$= \frac{2p}{x} \left\{ J_p(x) + iN_p(x) \right\} = \frac{2p}{x} H_p^{(1)}(x). \tag{8.110}$$

Similarly, we have

$$H_{p-1}^{(2)}(x) + H_{p+1}^{(2)}(x) = \frac{2p}{x} H_p^{(2)}(x). \tag{8.110a}$$

III. We have

$$H_{p-1}^{(1)}(x) - H_{p+1}^{(1)}(x) = \left\{ J_{p-1}(x) + iN_{p-1}(x) \right\} - \left\{ J_{p+1}(x) + iN_{p+1}(x) \right\}$$

$$= \left\{ J_{p-1}(x) - J_{p+1}(x) \right\} + i \left\{ N_{p-1}(x) - N_{p+1}(x) \right\}$$

$$= 2 \frac{d}{dx} J_p(x) + 2i \frac{d}{dx} N_p(x)$$

$$= 2 \frac{d}{dx} \left\{ J_p(x) + iN_p(x) \right\} = 2 \frac{d}{dx} H_p^{(1)}(x). \tag{8.111}$$

Similarly, we have

$$H_{p-1}^{(2)}(x) - H_{p+1}^{(2)}(x) = 2 \frac{d}{dx} H_p^{(2)}(x). \tag{8.111a}$$

IV. Addition of Eqs. (8.110) and (8.111) will result in

$$H_{p-1}^{(\alpha)}(x) = \frac{p}{x} H_p^{(\alpha)}(x) + \frac{d}{dx} H_p^{(\alpha)}(x), \ \alpha = 1, 2. \tag{8.112}$$

V. Subtraction of Eq. (8.111) from Eq. (8.110) will result in

$$H_{p+1}^{(\alpha)}(x) = \frac{p}{x} H_p^{(\alpha)}(x) - \frac{d}{dx} H_p^{(\alpha)}(x), \ \alpha = 1, 2. \tag{8.113}$$

VI. We have

$$\frac{d}{dx} \left[x^p H_p^{(1)}(x) \right] = \frac{d}{dx} \left[x^p \left\{ J_p(x) + iN_p(x) \right\} \right]$$

$$= \frac{d}{dx} \left[x^p J_p(x) \right] + i \frac{d}{dx} \left[x^p N_p(x) \right]$$

$$= x^p J_{p-1}(x) + i x^p N_{p-1}(x)$$

$$= x^p \left[J_{p-1}(x) + iN_{p-1}(x) \right] = x^p H_{p-1}^{(1)}(x). \tag{8.114}$$

Similarly, we have

$$\frac{d}{dx}\left[x^p H_p^{(2)}(x)\right] = x^p H_{p-1}^{(2)}(x). \tag{8.114a}$$

VII. We have

$$\begin{aligned}\frac{d}{dx}\left[x^{-p} H_p^{(1)}(x)\right] &= \frac{d}{dx}\left[x^{-p}\left\{J_p(x) + iN_p(x)\right\}\right]\\ &= \frac{d}{dx}\left[x^{-p} J_p(x)\right] + i\frac{d}{dx}\left[x^{-p} N_p(x)\right]\\ &= -x^{-p} J_{p+1}(x) + i\left[-x^{-p} N_{p+1}(x)\right]\\ &= -x^{-p}\left[J_{p+1}(x) + iN_{p+1}(x)\right]\\ &= -x^{-p} H_{p+1}^{(1)}(x).\end{aligned} \tag{8.115}$$

Similarly, we have

$$\frac{d}{dx}\left[x^{-p} H_p^{(2)}(x)\right] = -x^{-p} H_{p+1}^{(2)}(x). \tag{8.115a}$$

8.20 RECURRENCE RELATIONS FOR MODIFIED OR HYPERBOLIC BESSEL FUNCTIONS

(A) For Modified Bessel Functions of the First Kind

I. If n is an integer, we have

$$I_n(x) = i^{\,n} J_n(ix)$$

Therefore, we have

$$\begin{aligned}I_{-n}(x) &= i^n J_{-n}(ix) = i^n (-1)^n J_n(ix)\\ &= i^n (-1)^n\, i^n I_n(x) = (-1)^{2n} I_n(x) = I_n(x).\end{aligned} \tag{8.116}$$

II. As $I_p(x) = e^{-p\pi i/2} J_p(ix)$ we have

$$I_{p-1}(x) + I_{p+1}(x) = e^{-\frac{(p-1)\pi i}{2}} J_{p-1}(ix) + e^{-\frac{(p+1)\pi i}{2}} J_{p+1}(ix)$$

$$= e^{-\frac{(p-1)\pi i}{2}}\left[J_{p-1}(ix) + e^{-\pi i} J_{p+1}(ix)\right]$$

$$= e^{-\frac{(p-1)\pi i}{2}} \left[J_{p-1}(ix) - J_{p+1}(ix) \right]$$

$$= e^{-\frac{(p-1)\pi i}{2}} \left[2 \frac{d}{d(ix)} J_p(ix) \right]$$

$$= e^{-\frac{p\pi i}{2}} e^{\frac{\pi i}{2}} \left[\frac{2}{e^{\frac{\pi i}{2}}} \frac{d}{dx} J_p(ix) \right]$$

$$= 2\frac{d}{dx}\left[e^{-\frac{p\pi i}{2}} J_p(ix) \right] = 2\frac{d}{dx} I_p(x) = 2I_p'(x) . \tag{8.117}$$

III. Similarly, we have

$$I_{p-1}(x) - I_{p+1}(x) = e^{-\frac{(p-1)\pi i}{2}} J_{p-1}(ix) - e^{-\frac{(p+1)\pi i}{2}} J_{p+1}(ix)$$

$$= e^{-\frac{(p-1)\pi i}{2}} \left[J_{p-1}(ix) - e^{-\pi i} J_{p+1}(ix) \right]$$

$$= e^{-\frac{(p-1)\pi i}{2}} \left[J_{p-1}(ix) + J_{p+1}(ix) \right]$$

$$= e^{-\frac{(p-1)\pi i}{2}} \left[\frac{2p}{ix} J_p(ix) \right] = e^{-\frac{p\pi i}{2}} e^{\frac{\pi i}{2}} \left[\frac{2p}{x e^{\frac{\pi i}{2}}} J_p(ix) \right]$$

$$= \frac{2p}{x}\left[e^{-\frac{p\pi i}{2}} J_p(ix) \right] = \frac{2p}{x} I_p(x) . \tag{8.118}$$

IV. Addition of Eqs. (8.117) and (8.118) will result in

$$I_{p-1}(x) = I_p'(x) + \frac{p}{x} I_p(x) . \tag{8.119}$$

V. Subtraction of Eq. (8.118) from Eq. (8.117) will result in

$$I_{p+1}(x) = I_p'(x) - \frac{p}{x} I_p(x) . \tag{8.120}$$

VI. We have

$$\frac{d}{dx}\left[x^p I_p(x) \right] = \frac{d}{dx}\left[x^p e^{-p\pi i/2} J_p(ix) \right]$$

$$= e^{-\frac{p\pi i}{2}} \frac{d}{dx}\left[x^p J_p(ix)\right] = e^{-\frac{p\pi i}{2}} i^{1-p} \frac{d}{d(ix)}\left[(ix)^p J_p(ix)\right]$$

$$= e^{-\frac{p\pi i}{2}} i^{1-p}\left[(ix)^p J_{p-1}(ix)\right]$$

$$= i x^p e^{-\frac{p\pi i}{2}}\left[J_{p-1}(ix)\right] = x^p e^{\frac{\pi i}{2}} e^{-\frac{p\pi i}{2}}\left[J_{p-1}(ix)\right]$$

$$= x^p\left[e^{-\frac{(p-1)\pi i}{2}} J_{p-1}(ix)\right] = x^p I_{p-1}(x). \tag{8.121}$$

VII. We have

$$\frac{d}{dx}\left[x^{-p} I_p(x)\right] = \frac{d}{dx}\left[x^{-p} e^{-\frac{p\pi i}{2}} J_p(ix)\right] = \frac{d}{dx}\left[(ix)^{-p} J_p(ix)\right]$$

$$= i\frac{d}{d(ix)}\left[(ix)^{-p} J_p(ix)\right] = i\left[-(ix)^{-p} J_{p+1}(ix)\right]$$

$$= -i x^{-p}\left[i^{-p} J_{p+1}(ix)\right] = e^{-\frac{\pi i}{2}} x^{-p}\left[e^{-\frac{p\pi i}{2}} J_{p+1}(ix)\right]$$

$$= x^{-p}\left[e^{-\frac{(p+1)\pi i}{2}} J_{p+1}(ix)\right] = x^{-p} I_{p+1}(x). \tag{8.122}$$

(B) For Modified Bessel Functions of the Second Kind

I. In Section 8.6, it was proved that

$$K_{-p}(x) = K_p(x). \tag{8.123}$$

II. We have

$$K_p(x) = \frac{\pi}{2}\left[\frac{I_{-p}(x) - I_p(x)}{\sin p\pi}\right]$$

Therefore, we have

$$K_{p-1}(x) = \frac{\pi}{2}\left[\frac{I_{-(p-1)}(x) - I_{p-1}(x)}{\sin(p-1)\pi}\right]$$

$$= \frac{\pi}{2}\left[\frac{I_{-(p-1)}(x) - I_{p-1}(x)}{-\sin p\pi}\right] = \frac{\pi}{2}\left[\frac{-I_{-p+1}(x) + I_{p-1}(x)}{\sin p\pi}\right]. \tag{8.124}$$

Also, we have

$$K_{p+1}(x)=\frac{\pi}{2}\left[\frac{I_{-(p+1)}(x)-I_{p+1}(x)}{\sin(p+1)\pi}\right]$$

$$=\frac{\pi}{2}\left[\frac{I_{-p-1}(x)-I_{p+1}(x)}{-\sin p\pi}\right]=\frac{\pi}{2}\left[\frac{-I_{-p-1}(x)+I_{p+1}(x)}{\sin p\pi}\right]. \tag{8.125}$$

Adding Eqs. (8.124) and (8.125), we obtain

$$K_{p-1}(x)+K_{p+1}(x)$$

$$=\frac{\pi}{2\sin p\pi}\left[-\left\{I_{-p-1}(x)+I_{-p+1}(x)\right\}+\left\{I_{p-1}(x)+I_{p+1}(x)\right\}\right]$$

$$=\frac{\pi}{2\sin p\pi}\left[-2I'_{-p}(x)+2I'_{p}(x)\right]$$

$$=-2\left[\frac{\pi}{2}\frac{I'_{-p}(x)-I'_{p}(x)}{\sin p\pi}\right]=-2\frac{d}{dx}\left[\frac{\pi}{2}\frac{I_{-p}(x)-I_{p}(x)}{\sin p\pi}\right]=-2K'_{p}(x). \tag{8.126}$$

III. Subtracting Eq. (8.125) from Eq. (8.124), we obtain

$$K_{p-1}(x)-K_{p+1}(x)$$

$$=\frac{\pi}{2\sin p\pi}\left[\left\{I_{-p-1}(x)-I_{-p+1}(x)\right\}+\left\{I_{p-1}(x)-I_{p+1}(x)\right\}\right],$$

$$=\frac{\pi}{2\sin p\pi}\left[\frac{-2p}{x}I_{-p}(x)+\frac{2p}{x}I_{p}(x)\right],\ \text{(using Eq. (8.118)}$$

$$=-\frac{2p}{x}\frac{\pi}{2}\left[\frac{I_{-p}(x)-I_{p}(x)}{\sin p\pi}\right]=-\frac{2p}{x}K_{p}(x). \tag{8.127}$$

IV. Adding Eqs. (8.126) and (8.127), we obtain

$$K_{p-1}(x)=-\frac{p}{x}K_{p}(x)-K'_{p}(x). \tag{8.128}$$

V. Subtracting Eq. (8.127) from Eq. (8.126), we obtain

$$K_{p+1}(x)=\frac{p}{x}K_{p}(x)-K'_{p}(x). \tag{8.129}$$

VI. We have

$$\frac{d}{dx}\left[x^p K_p(x)\right] = px^{p-1}K_p(x) + x^p K_p'(x)$$

$$= x^p\left[\frac{p}{x}K_p(x) + K_p'(x)\right]$$

$$= -x^p K_{p-1}(x) \text{ [using Eq. (8.128)].} \qquad (8.130)$$

VII. We have

$$\frac{d}{dx}\left[x^{-p} K_p(x)\right] = -px^{-p-1}K_p(x) + x^{-p} K_p'(x)$$

$$= -x^{-p}\left[\frac{p}{x}K_p(x) - K_p'(x)\right]$$

$$= -x^{-p} K_{p+1}(x) \text{ [using Eq. (8.129)].} \qquad (8.131)$$

8.21 RECURRENCE RELATIONS FOR SPHERICAL BESSEL FUNCTIONS $j_n(x)$ WHERE n IS A NON-NEGATIVE INTEGER

I. For Bessel functions of the first kind $J_n(x)$, we know that

$$\frac{d}{dx}\left[x^n J_n(x)\right] = x^n J_{n-1}(x)$$

Replacing n by $n+\frac{1}{2}$, we write

$$\frac{d}{dx}\left[x^{n+\frac{1}{2}} J_{n+\frac{1}{2}}(x)\right] = x^{n+\frac{1}{2}} J_{n+\frac{1}{2}-1}(x). \qquad (8.132)$$

Since we define spherical Bessel functions by

$$j_n(x) = \sqrt{\frac{\pi}{2x}}\, J_{n+\frac{1}{2}}(x)$$

We have

$$J_{n+\frac{1}{2}}(x) = \sqrt{\frac{2}{\pi}}\, x^{\frac{1}{2}}\, j_n(x) \text{ and } J_{n-1+\frac{1}{2}}(x) = \sqrt{\frac{2}{\pi}}\, x^{\frac{1}{2}}\, j_{n-1}(x). \qquad (8.133)$$

Substituting Eq. (8.133) into Eq. (8.132), we obtain

$$\frac{d}{dx}\left[x^{n+\frac{1}{2}}\sqrt{\frac{2}{\pi}}\,x^{\frac{1}{2}}\,j_n(x)\right]=x^{n+\frac{1}{2}}\sqrt{\frac{2}{\pi}}\,x^{\frac{1}{2}}\,j_{n-1}(x)$$

or

$$\frac{d}{dx}\left[x^{n+1}j_n(x)\right]=x^{n+1}j_{n-1}(x). \tag{8.134}$$

II. We have

$$\frac{d}{dx}\left[x^{-n}J_n(x)\right]=-x^{-n}J_{n+1}(x).$$

Replacing n by $n+\frac{1}{2}$, we have

$$\frac{d}{dx}\left[x^{-n-\frac{1}{2}}J_{n+\frac{1}{2}}(x)\right]=-x^{-n-\frac{1}{2}}J_{n+1+\frac{1}{2}}(x). \tag{8.135}$$

As $j_n(x)=\sqrt{\frac{\pi}{2x}}\,J_{n+\frac{1}{2}}(x)$ we have

$$J_{n+\frac{1}{2}}(x)=\sqrt{\frac{2}{\pi}}\,x^{\frac{1}{2}}j_n(x) \text{ and } J_{n+1+\frac{1}{2}}(x)=\sqrt{\frac{2}{\pi}}\,x^{\frac{1}{2}}j_{n+1}(x).$$

Substituting these two relations into Eq. (8.135), we obtain

$$\frac{d}{dx}\left[x^{-n}j_n(x)\right]=-x^{-n}j_{n+1}(x). \tag{8.136}$$

III. We have

$$\frac{d}{dx}\left[x^{n+1}j_n(x)\right]=x^{n+1}j_{n-1}(x)$$

or

$$x^{n+1}j_n'(x)+(n+1)x^n j_n(x)=x^{n+1}j_{n-1}(x)$$

Dividing both sides by x^{n+1}, we obtain

$$j_n'(x)+\frac{n+1}{x}j_n(x)=j_{n-1}(x). \tag{8.137}$$

IV. We have

$$\frac{d}{dx}\left[x^{-n}j_n(x)\right]=-x^{-n}j_{n+1}(x)$$

or $$x^{-n} j_n'(x) - n x^{-n-1} j_n(x) = -x^{-n} j_{n+1}(x).$$

Dividing both sides by x^{-n}, we obtain

$$j_n'(x) - \frac{n}{x} j_n(x) = -j_{n+1}(x). \tag{8.138}$$

V. Adding n times Eq. (8.137) to $n + 1$ times Eq. (8.138), we obtain

$$(2n+1) j_n'(x) = n j_{n-1}(x) - (n+1) j_{n+1}(x). \tag{8.139}$$

VI. Subtracting Eq. (8.138) from Eq. (8.137), we obtain

$$\frac{2n+1}{x} j_n(x) = j_{n-1}(x) + j_{n+1}(x). \tag{8.140}$$

8.22 RECURRENCE RELATIONS FOR SPHERICAL NEUMANN FUNCTIONS $n_n(x)$ WHERE n IS A NON-NEGATIVE INTEGER

We know that spherical Neumann functions are defined by

$n_n(x) = \sqrt{\frac{\pi}{2x}} N_{n+\frac{1}{2}}(x)$, where $N_{n+\frac{1}{2}}(x)$ are Neumann functions (or Bessel functions of the second kind) which satisfy recurrence relations similar to those satisfied by Bessel functions of the first kind. Thus, spherical Neumann functions $n_n(x)$ will satisfy recurrence relations similar to those satisfied by spherical Bessel functions $j_n(x)$.

Therefore, we can derive recurrence relations of spherical Neumann functions $n_n(x)$ by processes similar to those in Section 8.21.

8.23 RECURRENCE RELATIONS FOR SPHERICAL HANKEL FUNCTIONS OF THE FIRST AND THE SECOND KINDS

As spherical Hankel functions of the first and second kinds are respectively defined by

$$h_n^{(1)}(x) = j_n(x) + i n_n(x) \text{ and } h_n^{(2)}(x) = j_n(x) - i n_n(x),$$

they will satisfy recurrence relations similar to those satisfied by spherical Bessel functions $j_n(x)$ and spherical Neumann functions $n_n(x)$.

Therefore, we can derive recurrence relations of spherical Hankel functions $h_n^{(\alpha)}(x)$ ($\alpha = 1, 2$) by processes similar to those in Section 8.21.

As a generalized case, if $f_n(x)$ represents any one of the functions

$$j_n(x), n_n(x), h_n^{(1)}(x) \text{ and } h_n^{(2)}(x),$$

we write

(i) $$\frac{d}{dx}\left[x^{n+1} f_n(x)\right] = x^{n+1} f_{n-1}(x), \tag{8.141}$$

(ii) $$\frac{d}{dx}\left[x^{-n} f_n(x)\right] = -x^{-n} f_{n+1}(x), \tag{8.142}$$

(iii) $$f_n'(x) + \frac{n+1}{x} f_n(x) = f_{n-1}(x), \tag{8.143}$$

(iv) $$f_n'(x) - \frac{n}{x} f_n(x) = -f_{n+1}(x), \tag{8.144}$$

(v) $$(2n+1) f_n'(x) = n f_{n-1}(x) - (n+1) f_{n+1}(x), \tag{8.145}$$

(vi) $$\frac{2n+1}{x} f_n(x) = f_{n-1}(x) + f_{n+1}(x). \tag{8.146}$$

8.24 MORE INTEGRAL REPRESENTATIONS OF BESSEL FUNCTIONS OF THE FIRST KIND (CONTOUR INTEGRAL FORM) AND OF OTHER KINDS

(A) Integral Representation of Neumann Function (Bessel Function of the Second Kind)

We can derive integral representations of Bessel functions of the second and third kinds from the corresponding integral representations of Bessel function of the first kind.

Since for non-integral values of p, we define the Bessel function of the second kind (or Neumann function) by

$$N_p(x) = \frac{J_p(x)\cos p\pi - J_{-p}(x)}{\sin p\pi} = J_p(x)\cot p\pi - J_{-p}(x)\operatorname{cosec} p\pi,$$

we can derive an integral representation of Bessel function of the second kind or Neumann function of non-integral order p using Eq. (8.81a).

Thus, we write

$$N_p(x) = \frac{\cot p\pi}{\pi}\int_0^{\pi} \cos(x\sin\phi - p\phi)\,d\phi - \frac{\cos p\pi}{\pi}\int_0^{\infty} e^{-x\sinh\lambda - p\lambda}\,d\lambda$$

$$-\frac{\operatorname{cosec} p\pi}{\pi}\int_0^{\pi} \cos(x\sin\phi + p\phi)\,d\phi - \frac{1}{\pi}\int_0^{\infty} e^{-x\sinh\lambda + p\lambda}\,d\lambda. \tag{8.147}$$

Putting $\phi = \pi - \theta$ in the third integral, we write

$$-\frac{\operatorname{cosec} p\pi}{\pi}\int_0^{\pi}\cos(x\sin\phi + p\phi)\,d\phi$$

$$= -\frac{\operatorname{cosec} p\pi}{\pi}\int_{\pi}^{0}\cos(x\sin\theta + p\pi - p\theta)(-d\theta)$$

$$= -\frac{\operatorname{cosec} p\pi}{\pi}\int_0^{\pi}\{\cos(x\sin\theta - p\theta)\cos p\pi - \sin(x\sin\theta - p\theta)\sin p\pi\}\,d\theta$$

$$= -\frac{\cot p\pi}{\pi}\int_0^{\pi}\cos(x\sin\phi - p\phi)\,d\phi + \frac{1}{\pi}\int_0^{\pi}\sin(x\sin\phi - p\phi)\,d\phi.$$

Here, the variable is again renamed.

Substituting this result into Eq. (8.147), we obtain

$$N_p(x) = \frac{1}{\pi}\int_0^{\pi}\sin(x\sin\phi - p\phi)\,d\phi - \frac{1}{\pi}\int_0^{\infty}(e^{-p\lambda}\cos p\pi + e^{p\lambda})e^{-x\sinh\lambda}\,d\lambda. \qquad (8.148)$$

Equation (8.148) is known as integral representation of Neumann function or Bessel function of the second kind of order p. Although it was derived for non-integral values of p, the formula holds for arbitrary p by the principle of analytic continuation.

(B) Contour Integral Forms of Bessel Functions of the First Kind and of the Third Kind (Hankel Functions) of Order *p*

Bessel differential equation can be written in the form

$$\frac{d^2w}{dz^2} + \frac{1}{z}\frac{dw}{dz} + \left(1 - \frac{P^2}{z^2}\right)w = 0. \qquad (8.149)$$

We seek a solution of Eq. (8:149) in the form

$$w(z) = z^p u(z) \qquad (8.150)$$

where u is a function to be determined later.

Substitution of Eq. (8.150) into Eq. (8.149) results in

$$z\frac{d^2u}{dz^2} + (2p+1)\frac{du}{dz} + zu = 0. \qquad (8.151)$$

Let us put

$$u = \int_C e^{iz\zeta} f(\zeta)\, d\zeta \tag{8.152}$$

where the contour C and the function *f* are to be determined latter.

Substituting Eq. (8.152) into Eq. (8.151), we obtain

$$0 = \int_C f(\zeta)\left\{-z\zeta^2 + i(2p+1)\zeta + z\right\}e^{iz\zeta}d\zeta$$

$$= -i\int_C f(\zeta)(1-\zeta^2)ize^{iz\zeta}\,d\zeta + i\int_C (2p+1)\zeta f(\zeta)e^{iz\zeta}d\zeta$$

$$= i\int_C \left(\zeta^2 - 1\right)f(\zeta)\frac{\partial}{\partial\zeta}e^{iz\zeta}\,d\zeta + i\int_C (2p+1)\zeta f(\zeta)e^{iz\zeta}d\zeta$$

$$= i\left[\left(\zeta^2-1\right)f(\zeta)e^{iz\zeta}\right]_C - i\int_C \frac{d}{d\zeta}\left\{\left(\zeta^2-1\right)f(\zeta)\right\}e^{iz\zeta}d\zeta + i\int_C (2p+1)\zeta f(\zeta)e^{iz\zeta}d\zeta.$$

Thus, we must have the conditions

$$\left[\left(\zeta^2-1\right)f(\zeta)e^{iz\zeta}\right]_C = 0 \tag{8.153}$$

and

$$\frac{d}{d\zeta}\left\{\left(\zeta^2-1\right)f(\zeta)\right\} = (2p+1)\zeta f(\zeta). \tag{8.154}$$

Condition (8.154) holds when

$$f(\zeta) = (\zeta^2-1)^{p-\frac{1}{2}}. \tag{8.155}$$

Condition (8.153) can be satisfied by choosing the contour C in such a way that

$$\psi(\zeta, z) = (\zeta^2-1)^{p+\frac{1}{2}}e^{iz\zeta} \tag{8.156}$$

returns to its initial value after one complete circuit.

Using Eqs. (8.150), (8.152) and (8.155), we obtain

$$w = z^p \int_C e^{iz\zeta}(\zeta^2-1)^{p-\frac{1}{2}}d\zeta. \tag{8.157}$$

Thus, Eq. (8.157) is a solution of Bessel equation (8.149).

The choice of the contour C can be made as in any one of the following two cases.

(i) In one choice, the amplitude of $\psi(\zeta, z)$ returns to its initial value on completion of the circuit, and

(ii) In another choice, $\psi(\zeta, z)$ vanishes at the initial and final points of the contour.

In a simple example, a contour satisfying condition (i) is shown in Figure 8.2(a) provided $\text{Re}\left(p+\frac{1}{2}\right)>0$.

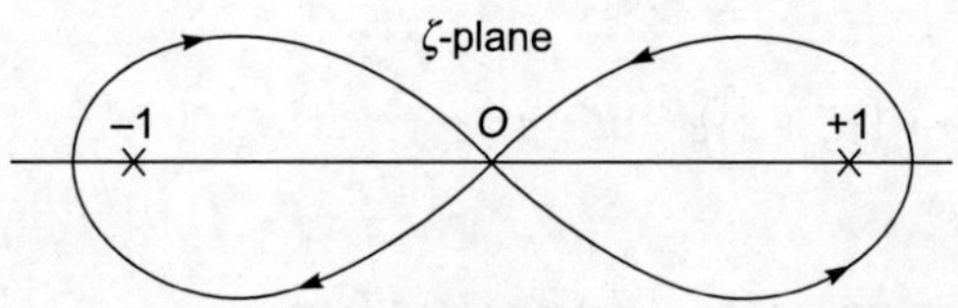

Fig. 8.2(a)

Here, the contour C consists of a loop around $(\zeta=-1)$ in the negative sense and another loop around $(\zeta=+1)$ in the positive sense. The loops are supposed to meet at the origin O which is taken as the initial point. The positive loop increases $\text{amp}(\zeta-1)$ by 2π whereas the negative loop changes $\text{amp}(\zeta+1)$ by -2π and hence $\text{amp}\,\psi(\zeta, z)$ remains unchanged after a complete circuit. Eventually, the contour shown in Figure 8.2(a) can be replaced by the contour shown in Figure 8.2(b).

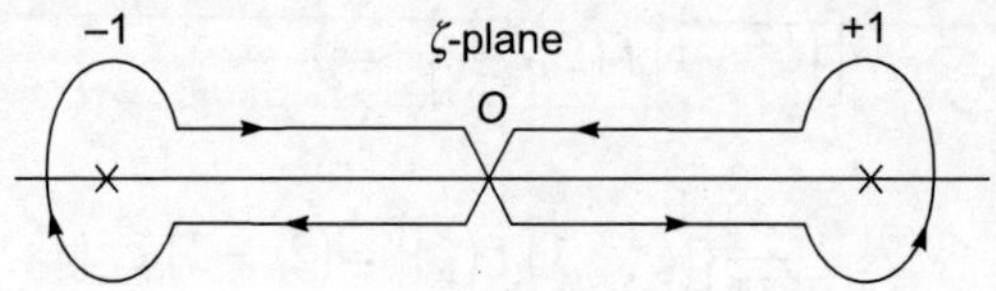

Fig. 8.2(b)

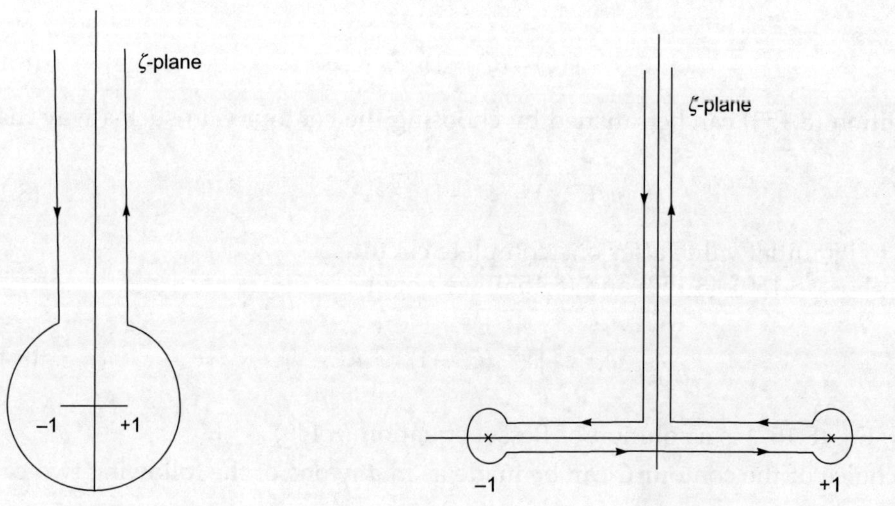

Fig. 8.3

Another example satisfying condition (ii) which is suitable for Re $z>0$ is shown in Figure 8.3(a). In this case, the contour C consists of a loop around both the points $\zeta=\pm 1$ with initial and final points at $z=i\infty$. The value of $\psi(\zeta, z)$ at these end points is zero. This contour can also be replaced without loss of generality by the one shown in Figure 8.3(b).

(a) Expression of $J_p(z)$ as a Contour Integral

It is seen that condition (8.153) holds when the contour C is taken as a straight line from -1 to $+1$ and back.

Now, Eq. (8.157) can be written as

$$w = z^p\left[\int_{-1}^{+1} e^{iz\zeta}\left\{\left(1-\zeta^2\right)e^{-\pi i}\right\}^{p-\frac{1}{2}} d\zeta + \int_{+1}^{-1} e^{iz\zeta}\left\{\left(1-\zeta^2\right)e^{\pi i}\right\}^{p-\frac{1}{2}} d\zeta\right]$$

$$= z^p\left[\int_{-1}^{+1} e^{iz\zeta}\left(1-\zeta^2\right)^{p-\frac{1}{2}} e^{\left(\frac{1}{2}-p\right)\pi i} d\zeta - \int_{-1}^{+1} e^{iz\zeta}\left(1-\zeta^2\right)^{p-\frac{1}{2}} e^{-\left(\frac{1}{2}-p\right)\pi i} d\zeta\right]$$

$$= \left\{e^{\left(\frac{1}{2}-p\right)\pi i} - e^{-\left(\frac{1}{2}-p\right)\pi i}\right\} z^p \int_{-1}^{+1} e^{iz\zeta}\left(1-\zeta^2\right)^{p-\frac{1}{2}} d\zeta$$

$$= 2i\sin\left\{\left(\frac{1}{2}-p\right)\pi\right\}\int_{-1}^{+1} z^p e^{iz\zeta}\left(1-\zeta^2\right)^{p-\frac{1}{2}} d\zeta .$$

Expanding the exponential factor inside the integral, we write

$$w = 2i\sin\left\{\left(\frac{1}{2}-p\right)\pi\right\}\sum_{s=0}^{\infty}\frac{i^s z^{p+s}}{s!}\int_{-1}^{+1}\zeta^s\left(1-\zeta^2\right)^{p-\frac{1}{2}} d\zeta . \tag{8.158}$$

For odd values of s say, $s=2k+1$ $(k=0,1,2,3\ldots)$ we know that $\int_{-1}^{+1}\zeta^{2k+1}\left(1-\zeta^2\right)^{p-\frac{1}{2}} d\zeta = 0$ as the integrand is an odd function.

Therefore, Eq. (8.158) can be written as

$$w = 2i\sin\left\{\left(\frac{1}{2}-p\right)\pi\right\}\sum_{k=0}^{\infty}\frac{i^{2k} z^{p+2k}}{(2k)!}\int_{-1}^{+1}\zeta^{2k}\left(1-\zeta^2\right)^{p-\frac{1}{2}} d\zeta$$

$$= 2i\sin\left\{\left(\frac{1}{2}-p\right)\pi\right\}\sum_{k=0}^{\infty}\frac{(-1)^k z^{p+2k}}{(2k)!} 2\int_{0}^{1}\zeta^{2k}\left(1-\zeta^2\right)^{p-\frac{1}{2}} d\zeta$$

as the integrand is an even function.

If we put $\zeta^2 = v$, then $d\zeta = \frac{1}{2} v^{-\frac{1}{2}} dv$ and, therefore, we write

$$w = 2i \sin\left\{\left(\frac{1}{2} - p\right)\pi\right\} \sum_{k=0}^{\infty} \frac{(-1)^k z^{p+2k}}{(2k)!} \int_0^1 v^{k-\frac{1}{2}} (1-v)^{p-\frac{1}{2}} dv$$

$$= 2i \sin\left\{\left(\frac{1}{2} - p\right)\pi\right\} \sum_{k=0}^{\infty} \frac{(-1)^k z^{p+2k}}{(2k)!} \frac{\Gamma\left(k+\frac{1}{2}\right)\Gamma\left(p+\frac{1}{2}\right)}{\Gamma(k+p+1)}.$$

Legendre's duplication formula for gamma functions gives

$$(2k)! = \frac{2^{2k} k!\, \Gamma\left(k+\frac{1}{2}\right)}{\sqrt{\pi}}.$$

Using this value of $(2k)!$ we write

$$w = 2i \sin\left\{\left(\frac{1}{2} - p\right)\pi\right\} 2^p \sqrt{\pi}\, \Gamma\left(p+\frac{1}{2}\right) J_p(z) \tag{8.159}$$

where $J_p(z) = \sum_{k=0}^{\infty} \frac{(-1)^k \left(\frac{z}{2}\right)^{p+2k}}{k!\, \Gamma(k+p+1)}$ is the Bessel function of the first kind of order p.

By reflection formula of gamma functions, we know that

$$\Gamma\left(\frac{1}{2} - p\right)\Gamma\left\{1-\left(\frac{1}{2} - p\right)\right\} = \frac{\pi}{\sin\left\{\left(\frac{1}{2} - p\right)\pi\right\}}$$

or

$$\Gamma\left(p+\frac{1}{2}\right)\sin\left\{\left(\frac{1}{2} - p\right)\pi\right\} = \frac{\pi}{\Gamma\left(\frac{1}{2} - p\right)}.$$

Using this value, Eq. (8.159) can be written as

$$w = \frac{2\pi i \sqrt{\pi}\, 2^p}{\Gamma\left(\frac{1}{2} - p\right)} J_p(z)$$

or

$$J_p(z) = \frac{\Gamma\left(\frac{1}{2} - p\right) w}{2\pi i\, \Gamma\left(\frac{1}{2}\right) 2^p}.$$

Using Eq. (8.157), we write

$$J_p(z)=\frac{\Gamma\left(\frac{1}{2}-p\right)\left(\frac{z}{2}\right)^p}{2\pi i\,\Gamma\left(\frac{1}{2}\right)}\int_C e^{iz\zeta}\left(\zeta^2-1\right)^{p-\frac{1}{2}}d\zeta,\ \text{Re}\,p>-\frac{1}{2} \tag{8.160}$$

where the contour C is that shown in Figures 8.2(a) or 8.2(b).

As the contour C encloses the point $\zeta=+1$ in the positive sense and the point $\zeta=-1$ in the negative sense, we may write Eq. (8.160) in an alternative form as

$$J_p(z)=\frac{\Gamma\left(\frac{1}{2}-p\right)\left(\frac{z}{2}\right)^p}{2\pi i\,\Gamma\left(\frac{1}{2}\right)}\int_A^{(1+,\,-1-)} e^{iz\zeta}\left(\zeta^2-1\right)^{p-\frac{1}{2}}d\zeta,\ \text{Re}\,p>-\frac{1}{2} \tag{8.160a}$$

where A is a point on the right of $\zeta=+1$ at which the contour crosses the real axis and the phases of the functions $\zeta-1$ and $\zeta+1$ become zero.

Earlier, we have put the restriction that $\text{Re}\left(p+\frac{1}{2}\right)>0$. But this restriction can be dropped through analytic continuation and Eq. (8.160) holds for all values of p except for those for which $\left(p+\frac{1}{2}\right)$ is a positive integer.

An Alternative form of $J_p(z)$

In Eq. (8.160), if we replace z by $-z$, we obtain

$$J_p(-z)=\frac{(-1)^p\,\Gamma\left(\frac{1}{2}-p\right)}{2\pi i\,\Gamma\left(\frac{1}{2}\right)}\left(\frac{z}{2}\right)^p\int_C e^{-iz\zeta}\left(\zeta^2-1\right)^{p-\frac{1}{2}}d\zeta=(-1)^p J_p(z)$$

where p is considered to be an integer.

Therefore, we write

$$J_p(z)=\frac{\Gamma\left(\frac{1}{2}-p\right)}{2\pi i\,\Gamma\left(\frac{1}{2}\right)}\left(\frac{z}{2}\right)^p\int_C e^{-iz\zeta}(\zeta^2-1)^{p-\frac{1}{2}}d\zeta. \tag{8.161}$$

Addition of Eqs. (8.160) and (8.161) yields

$$J_p(z)=\frac{\Gamma\left(\frac{1}{2}-p\right)}{2\pi i\,\Gamma\left(\frac{1}{2}\right)}\left(\frac{z}{2}\right)^p\int_C(\zeta^2-1)^{p-\frac{1}{2}}\cos(z\zeta)\,d\zeta\,. \tag{8.162}$$

Let us recall that the contour C encloses $\zeta=+1$ in the positive sense and $\zeta=-1$ in the negative sense. If we assume that the phase of ζ^2-1 vanishes when ζ is on the real axis on the right of $\zeta=1$,

$$J_p(z)=\frac{\Gamma\left(\frac{1}{2}-p\right)\left(\frac{z}{2}\right)^p}{2\pi i\,\Gamma\left(\frac{1}{2}\right)}\int_0^{(1+,-1-)}\left(\zeta^2-1\right)^{p-\frac{1}{2}}\cos(z\zeta)\,d\zeta\,. \tag{8.162a}$$

(b) Contour Integral Form of $J_{-p}(z)$

In the equation

$$w=z^p\int_C e^{iz\zeta}\left(\zeta^2-1\right)^{p-\frac{1}{2}}d\zeta,$$

let the contour C be in the form shown in Figure 8.3(b). This contour is assumed to start from $i\infty$ and enclose the circle $|\zeta|=1$.

Now, we write

$$w=z^p\int_C e^{iz\zeta}\left\{\zeta^2\left(1-\frac{1}{\zeta^2}\right)\right\}^{p-\frac{1}{2}}d\zeta$$

$$=z^p\int_C e^{iz\zeta}\,\zeta^{2p-1}\left(1-\frac{1}{\zeta^2}\right)^{-\left(\frac{1}{2}-p\right)}d\zeta$$

$$=z^p\int_C e^{iz\zeta}\zeta^{2p-1}\sum_{s=0}^{\infty}\frac{\Gamma\left(\frac{1}{2}-p+s\right)}{\Gamma\left(\frac{1}{2}-p\right)s!}\zeta^{-2s} \quad \text{[using Binomial expansion]}$$

$$=\sum_{s=0}^{\infty}\frac{\Gamma\left(\frac{1}{2}-p+s\right)}{s!\,\Gamma\left(\frac{1}{2}-p\right)}z^p\int_C e^{iz\zeta}\,\zeta^{2p-2s-1}d\zeta \tag{8.163}$$

with $-\frac{3\pi}{2}<\operatorname{amp}\zeta<\frac{\pi}{2}$.

If we set $\tau = iz\zeta = e^{\frac{\pi i}{2}} z\zeta$, then $\zeta = \tau\, z^{-1} e^{-\frac{\pi i}{2}}$ and $d\zeta = e^{-\frac{\pi i}{2}} \frac{d\tau}{z}$ and we write

$$w = \sum_{s=0}^{\infty} \frac{\Gamma\left(\frac{1}{2} - p + s\right)}{s!\,\Gamma\left(\frac{1}{2} - p\right)} z^p \int_C e^\tau \left(z^{-1} e^{-\frac{\pi i}{2}} \tau \right)^{2p-2s-1} z^{-1} e^{-\frac{\pi i}{2}} d\tau$$

$$= \sum_{s=0}^{\infty} \frac{\Gamma\left(\frac{1}{2} - p + s\right)}{s!\,\Gamma\left(\frac{1}{2} - p\right)} z^{2s-p} \int_C e^{(s-p)\pi i} e^\tau \tau^{-(2s-2p+1)} d\tau$$

$$= \sum_{s=0}^{\infty} \frac{(-1)^s\, \Gamma\left(\frac{1}{2} - p + s\right) e^{-p\pi i}}{s!\,\Gamma\left(\frac{1}{2} - p\right)} z^{2s-p} \int_C e^\tau \tau^{-(2s-2p+1)} d\tau, \text{ since } e^{s\pi i} = (-1)^s .$$

Since, $\dfrac{1}{\Gamma(2s - 2p + 1)} = \dfrac{1}{2\pi i} \displaystyle\int_C e^\tau \tau^{-(2s-2p+1)} d\tau,$

we write

$$w = \sum_{s=0}^{\infty} \frac{(-1)^s\, 2\pi i\, e^{-p\pi i}\, \Gamma\left(\frac{1}{2} - p + s\right)}{s!\,\Gamma\left(\frac{1}{2} - p\right) \Gamma(2s - 2p + 1)} z^{2s-p} . \tag{8.164}$$

By Legendre's duplication formula for gamma functions, we have

$$\Gamma(2s - 2p + 1) = \frac{\Gamma(s - p + 1)\, \Gamma\left(s - p + \frac{1}{2}\right)}{\sqrt{\pi}} 2^{2s-2p} .$$

Substitution of this relation into Eq. (8.164) results in

$$w = \frac{2\pi i\, \sqrt{\pi} e^{-p\pi i}\, 2^p}{\Gamma\left(\frac{1}{2} - p\right)} \sum_{s=0}^{\infty} \frac{(-1)^s \left(\frac{z}{2}\right)^{2s-p}}{s!\,\Gamma(s - p + 1)}$$

$$= \frac{2\pi i\, \sqrt{\pi}\, e^{-p\pi i}\, 2^p}{\Gamma\left(\frac{1}{2} - p\right)} J_{-p}(z) .$$

Therefore, we have

$$J_{-p}(z)=\frac{\Gamma\left(\frac{1}{2}-p\right)e^{p\pi i}}{2\pi i\sqrt{\pi}\,2^p}w.$$

Using w from Eq. (8.157), we write

$$J_{-p}(z)=\frac{\Gamma\left(\frac{1}{2}-p\right)e^{p\pi i}}{2\pi i\sqrt{\pi}}\left(\frac{z}{2}\right)^p\int_C e^{iz\zeta}(\zeta^2-1)^{p-\frac{1}{2}}d\zeta \tag{8.165}$$

where the contour C is shown in Figures 8.3(a) or 8.3(b).

(c) Contour Integral for Hankel Functions

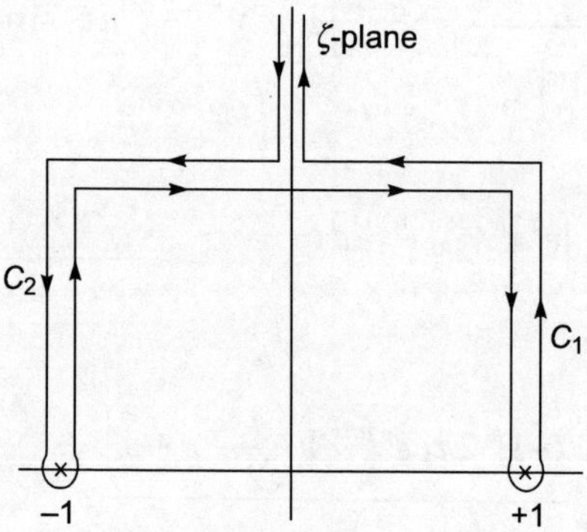

Fig. 8.4

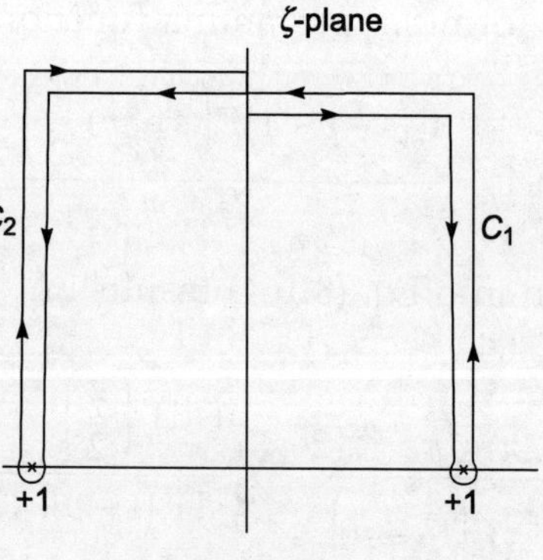

Fig. 8.5

Replacing the contours in Figures 8.2(b) and 8.3(b) by those shown in Figures 8.4 and 8.5 respectively, we can write

$$J_p(z)=\frac{\Gamma\left(\frac{1}{2}-p\right)}{2\pi i\sqrt{\pi}}\left(\frac{z}{2}\right)^p\left[\int_{C_1} e^{iz\zeta}\left(\zeta^2-1\right)^{p-\frac{1}{2}}d\zeta+\int_{C_2} e^{iz\zeta}\left(\zeta^2-1\right)^{p-\frac{1}{2}}d\zeta\right] \quad (8.166)$$

where the contours C_1 and C_2 are shown in Figure 8.4
and

$$J_{-p}(z)=\frac{\Gamma\left(\frac{1}{2}-p\right)}{2\pi i\sqrt{\pi}}\left(\frac{z}{2}\right)^p\left[e^{p\pi i}\int_{C_1} e^{iz\zeta}\left(\zeta^2-1\right)^{p-\frac{1}{2}}d\zeta+e^{-p\pi i}\int_{C_2} e^{iz\zeta}\left(\zeta^2-1\right)^{p-\frac{1}{2}}d\zeta\right]$$

where the contours C_1 and C_2 are shown in Figure 8.5. (8.167)

As Hankel functions of the first and the second kinds of order p are respectively defined by

$$H_p^{(1)}(z)=\frac{i}{\sin p\pi}\left[e^{-p\pi i}J_p(z)-J_{-p}(z)\right]$$

and

$$H_p^{(2)}(z)=-\frac{i}{\sin p\pi}\left[e^{p\pi i}J_p(z)-J_{-p}(z)\right],$$

using these definitions and Eqs. (8.166) and (8.167), we obtain

$$H_p^{(1)}(z)=\frac{\Gamma\left(\frac{1}{2}-p\right)}{\pi i\sqrt{\pi}}\left(\frac{z}{2}\right)^p\int_{C_1} e^{iz\zeta}\left(\zeta^2-1\right)^{p-\frac{1}{2}}d\zeta \quad (8.168)$$

and

$$H_p^{(2)}(z)=\frac{\Gamma\left(\frac{1}{2}-p\right)}{\pi i\sqrt{\pi}}\left(\frac{z}{2}\right)^p\int_{C_2} e^{iz\zeta}\left(\zeta^2-1\right)^{p-\frac{1}{2}}d\zeta \quad (8.169)$$

with Re $z > 0$, $-\frac{\pi}{2}<\text{amp}\, z<\frac{\pi}{2}$.

Alternative Contour Integral Form of Hankel Functions

From Eqs. (8.81a) and (8.148), we write

$$J_p(z)=\frac{1}{\pi}\int_0^{\pi}\cos(z\sin\phi-p\phi)\,d\phi-\frac{\sin p\pi}{\pi}\int_0^{\infty}e^{-z\sinh\lambda-p\lambda}d\lambda,\ \text{Re}\, z>0$$

and

$$N_p(z)=\frac{1}{\pi}\int_0^{\pi}\sin\left(z\sin\phi-p\phi\right)d\phi-\frac{1}{\pi}\int_0^{\infty}\left(e^{p\lambda}+e^{-p\lambda}\cos p\pi\right)e^{-z\sinh\lambda}d\lambda.$$

Now, Hankel function of the first kind of order p is obtained as

$$H_p^{(1)}(z)=J_p(z)+i\,N_p(z)$$

$$=\frac{1}{\pi}\int_0^{\pi}e^{i(z\sin\phi-p\phi)}d\phi+\frac{1}{\pi i}\int_0^{\infty}\left\{e^{p\lambda}+(\cos p\pi-i\sin p\pi)e^{-p\lambda}\right\}e^{-z\sinh\lambda}d\lambda$$

$$=\frac{1}{\pi}\int_0^{\pi}e^{iz\sin\phi-pi\phi}d\phi+\frac{1}{\pi i}\int_0^{\infty}\left\{e^{p\lambda}+e^{-p(\lambda+\pi i)}\right\}e^{-z\sinh\lambda}d\lambda$$

$$=\frac{1}{\pi i}\int_0^{\infty}e^{-z\sinh\lambda+p\lambda}d\lambda+\frac{1}{\pi}\int_0^{\pi}e^{z\sinh(i\phi)-pi\phi}d\phi+\frac{1}{\pi i}\int_0^{\infty}e^{-z\sinh\lambda-p(\lambda+\pi i)}d\lambda$$

$$=I_1+I_2+I_3\text{ (say)}\tag{8.170}$$

where I_1, I_2 and I_3 represent the respective terms on the right-hand side of the above equation.

In the expression for I_1 if we replace λ by $-\lambda$, we obtain

$$I_1=\frac{1}{\pi i}\int_0^{-\infty}e^{z\sinh\lambda-p\lambda}d(-\lambda)$$

$$=\frac{1}{\pi i}\int_{-\infty}^{0}e^{z\sinh\lambda-p\lambda}d\lambda.$$

The expression for I_2 can be written as

$$I_2=\frac{1}{\pi i}\int_{\phi=0}^{\pi}e^{z\sinh(i\phi)-p(i\phi)}d(i\phi).$$

Also, we have

$$I_3=\frac{1}{\pi i}\int_0^{\infty}e^{z\sinh(\lambda+\pi i)-p(\lambda+\pi i)}d\lambda.$$

Substituting these expressions for I_1, I_2 and I_3 into Eq. (8.170), we obtain

$$H_p^{(1)}(z)=\frac{1}{\pi i}\int_{-\infty}^{0}e^{z\sinh\lambda-p\lambda}d\lambda+\frac{1}{\pi i}\int_{\phi=0}^{\pi}e^{z\sinh(i\phi)-p(i\phi)}d(i\phi)+\frac{1}{\pi i}\int_0^{\infty}e^{z\sinh(\lambda+\pi i)-p(\lambda+\pi i)}d\lambda.$$

Introducing a variable $\tau = \lambda + i\phi$ and considering a contour C_1 in the τ–plane as shown in Figure 8.6 comprising of three sections namely

(i) from $(-\infty, 0)$ to $(0, 0)$,
(ii) from $(0, 0)$ to $(0, \pi i)$ and
(iii) from $(0, \pi i)$ to $(\infty, \pi i)$, we can write

$$H_p^{(1)}(z) = \frac{1}{\pi i}\int_{C_1} e^{z\sinh\tau - p\tau}\,d\tau \tag{8.171}$$

where the contour C_1 is shown in Figure 8.6.

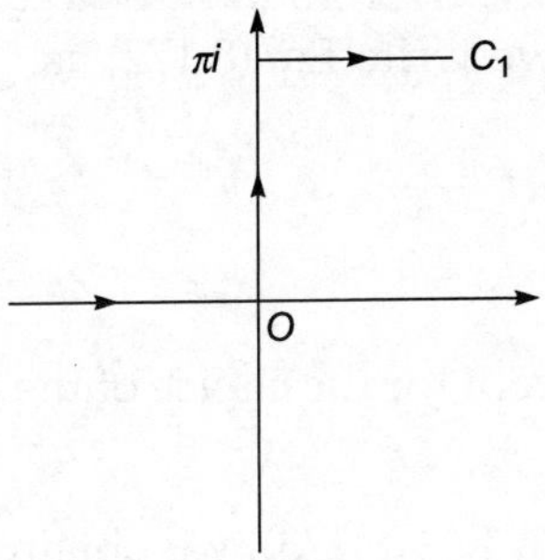

Fig. 8.6

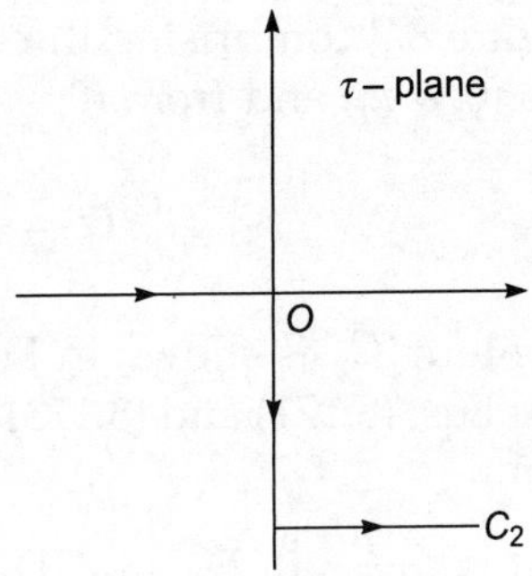

Fig. 8.7

Also, we can arrive at

$$H_p^{(2)}(z) = J_p(z) - i\,N_p(z)$$

$$= \frac{1}{\pi}\int_0^{\pi} e^{-i(z\sin\phi - p\phi)}d\phi - \frac{1}{\pi i}\int_0^{\infty}\left\{e^{p\lambda} + (\cos p\pi + i\sin p\pi)e^{-p\lambda}\right\}e^{-z\sinh\lambda}d\lambda$$

$$= -\frac{1}{\pi i}\int_0^{\infty} e^{-z\sinh\lambda + p\lambda}d\lambda + \frac{1}{\pi}\int_0^{\pi} e^{-iz\sin\phi + ip\phi}d\phi - \frac{1}{\pi i}\int_0^{\infty} e^{-z\sinh\lambda - p\lambda + ip\pi}$$

$$= -\frac{1}{\pi i}\int_0^{\infty} e^{z\sinh(-\lambda) + p\lambda}d\lambda + \frac{1}{\pi}\int_0^{\pi} e^{z\sinh(-i\phi) + pi\phi}d\phi - \frac{1}{\pi i}\int_0^{\infty} e^{z\sinh(\lambda - \pi i) - p(\lambda - \pi i)}d\lambda$$

In the first integral, we replace λ by $-\lambda$ and in the second integral, we replace ϕ by $-\phi$ to yield

$$H_p^{(2)}(z) = -\frac{1}{\pi i}\int_0^{-\infty} e^{z\sinh\lambda - p\lambda}d(-\lambda) + \frac{1}{\pi i}\int_{\phi=0}^{-\pi} e^{z\sinh(i\phi) - pi\phi}d(-i\phi) - \frac{1}{\pi i}\int_0^{\infty} e^{z\sinh(\lambda - \pi i) - p(\lambda - \pi i)}d\lambda$$

$$= I_4 + I_5 + I_6 \text{ (say)} \tag{8.172}$$

where $$I_4 = -\frac{1}{\pi i}\int_0^{-\infty} e^{z\sinh\lambda - p\lambda}\, d(-\lambda) = -\frac{1}{\pi i}\int_{-\infty}^{0} e^{z\sinh\lambda - p\lambda}\, d\lambda,$$

$$I_5 = \frac{1}{\pi i}\int_{\phi=0}^{-\pi} e^{z\sinh(i\phi) - pi\phi}\, d(-i\phi) = -\frac{1}{\pi i}\int_{\phi=0}^{-\pi} e^{z\sinh(i\phi) - pi\phi}\, d(i\phi)$$

and $$I_6 = -\frac{1}{\pi i}\int_0^{\infty} e^{z\sinh(\lambda-\pi i) - p(\lambda-\pi i)}\, d(\lambda - \pi i).$$

Introducing a variable $\tau = \lambda + i\phi$ and considering a contour C_2 in the τ-plane as shown in Figure 8.7 comprising three sections, namely: (i) from $(-\infty, 0)$ to $(0, 0)$, (ii) from $(0, 0)$ to $(0, -\pi i)$ and from $(0, -\pi i)$ to $(\infty, -\pi i)$, we write Eq. (8.172) as

$$H_p^{(2)}(z) = -\frac{1}{\pi i}\int_{C_2} e^{z\sinh\tau - p\tau}\, d\tau \tag{8.173}$$

where the contour C_2 is shown in Figure 8.7.

We see that Eqs. (8.171) and (8.173) are the same except for the choice of the contour of integration.

If we put $\tau = u + \dfrac{\pi i}{2}$ in Eq. (8.171) and $\tau = u - \dfrac{\pi i}{2}$ in Eq. (8.173), we obtain

$$H_p^{(1)}(z) = \frac{e^{-p\pi i/2}}{\pi i}\int_{C_3} e^{iz\cosh u - pu}\, du, \ \text{Re } z > 0 \tag{8.174}$$

and

$$H_p^{(2)}(z) = -\frac{e^{p\pi i/2}}{\pi i}\int_{C_4} e^{-iz\cosh u - pu}\, du \ \ \text{Re } z > 0 \tag{8.175}$$

where the contours C_3 and C_4 are shown in Figures 8.8 and 8.9.

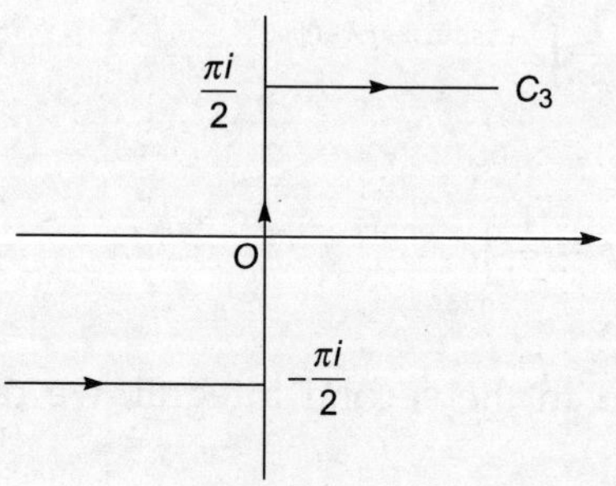

Fig. 8.8

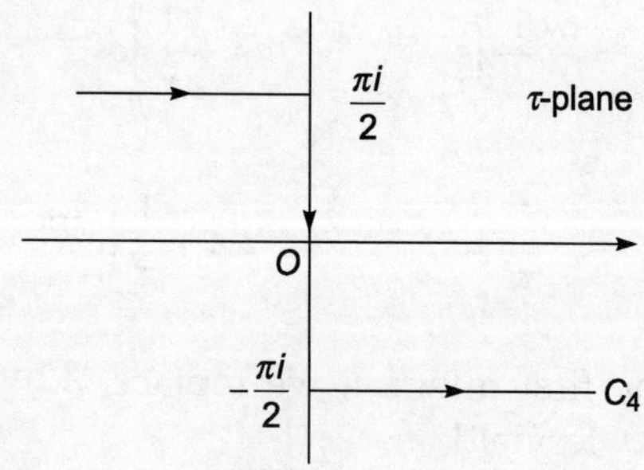

Fig. 8.9

For further transformation of the above integrals, let us temporarily assume that z is a positive real number and the parameter p is confined to the strip $-1 < \text{Re } z < +1$.

According to Cauchy's integral theorem, the integral along the left-hand part of the contour C_3 or C_4 up to the point $u=0$ can be replaced by an integral along the negative real axis and the integral along the right-hand part can be replaced by an integral along the positive real axis, the integral along the vertical segment needed to complete each contour approaching zero as the segment moves indefinitely far to the left or right of the imaginary axis.

Now, Eqs. (8.174) and (8.175) can be written as

$$H_p^{(1)}(z)=\frac{e^{-p\pi i/2}}{\pi i}\int_{-\infty}^{+\infty} e^{iz\cosh u-pu}\,du \tag{8.176}$$

and

$$H_p^{(2)}(z)=-\frac{e^{p\pi i/2}}{\pi i}\int_{-\infty}^{+\infty} e^{-iz\cosh u-pu}\,du \tag{8.177}$$

where $z>0$ and $-1<\text{Re}\,z<+1$

By analytic continuation, it is seen that Eq. (8.176) is valid for $0\le \text{amp}\,z<\pi$ whereas Eq. (8.177) is valid for $-\pi<\text{amp}\,z\le 0$ since in each case, both sides of the equations are analytic functions of z in the indicated regions. Also the condition $-1<\text{Re}\,p<1$ can be dropped if $Im\,z>0$ in Eq. (8.176) or if $Im\,z<0$ in Eq. (8.177). Therefore, we have the integrals

$$H_p^{(1)}(z)=\frac{e^{-p\pi i/2}}{\pi i}\int_{-\infty}^{+\infty} e^{iz\cosh u-pu}\,du,\ Im\,z>0 \tag{8.178}$$

and

$$H_p^{(2)}(z)=-\frac{e^{p\pi i/2}}{\pi i}\int_{-\infty}^{+\infty} e^{-iz\cosh u-pu}\,du,\ Im\,z<0 \tag{8.179}$$

for arbitrary p.

These two equations can be considered as basic integral representations of Hankel functions.

(C) Integral Representations of Modified Bessel Functions

(a) Contour Integral for $I_p(z)$

Let us recall Eq. (8.160), viz.

$$J_p(z)=\frac{1}{2\pi i}\frac{\Gamma\left(\frac{1}{2}-p\right)}{\sqrt{\pi}}\left(\frac{z}{2}\right)^p\int_C e^{iz\zeta}\left(\zeta^2-1\right)^{p-\frac{1}{2}}d\zeta$$

with the contour C stated there.

Replacing z by iz, we write

$$J_p(iz)=\frac{1}{2\pi i}\frac{\Gamma\left(\frac{1}{2}-p\right)}{\sqrt{\pi}}\left(\frac{iz}{2}\right)^p\int_C e^{-z\zeta}\left(\zeta^2-1\right)^{p-\frac{1}{2}}d\zeta \tag{8.180}$$

Now, we obtain

$$I_p(z)=i^{-p}J_p(iz)=\frac{1}{2\pi i}\frac{\Gamma\left(\frac{1}{2}-p\right)}{\sqrt{\pi}}\left(\frac{z}{2}\right)^p\int_C e^{-z\zeta}\left(\zeta^2-1\right)^{p-\frac{1}{2}}d\zeta. \tag{8.181}$$

Equation (8.181) represents $I_p(z)$ as a contour integral.
Replacing z by $-z$ and supposing p as integer, Eq. (8.181) gives

$$I_p(-z)=(-1)^p I_p(z)=\frac{1}{2\pi i}\frac{\Gamma\left(\frac{1}{2}-p\right)}{\sqrt{\pi}}\left(\frac{-z}{2}\right)^p\int_C e^{z\zeta}\left(\zeta^2-1\right)^{p-\frac{1}{2}}d\zeta$$

or

$$I_p(z)=\frac{1}{2\pi i}\frac{\Gamma\left(\frac{1}{2}-p\right)}{\sqrt{\pi}}\left(\frac{z}{2}\right)^p\int_C e^{z\zeta}(\zeta^2-1)^{p-\frac{1}{2}}d\zeta. \tag{8.182}$$

Adding Eqs. (8.181) and (8.182), we write

$$2I_p(z)=\frac{1}{2\pi i}\frac{\Gamma\left(\frac{1}{2}-p\right)}{\sqrt{\pi}}\left(\frac{z}{2}\right)^p\int_C\left(e^{z\zeta}+e^{-z\zeta}\right)\left(\zeta^2-1\right)^{p-\frac{1}{2}}d\zeta$$

or

$$I_p(z)=\frac{1}{2\pi i}\frac{\Gamma\left(\frac{1}{2}-p\right)}{\sqrt{\pi}}\left(\frac{z}{2}\right)^p\int_C\left(\zeta^2-1\right)^{p-\frac{1}{2}}\cosh(z\zeta)\,d\zeta. \tag{8.183}$$

Equation (8.183) represents another contour integral form of $I_p(z)$

(b) Definite Integrals for $I_p(z)$

(i) Recalling Eq. (8.89), we write

$$J_p(z)=\frac{1}{\sqrt{\pi}\,\Gamma\left(p+\frac{1}{2}\right)}\left(\frac{z}{2}\right)^p\int_{-1}^{+1}\left(1-t^2\right)^{p-\frac{1}{2}}e^{izt}dt.$$

We have

$$I_p(z)=i^{-p}J_p(iz)=i^{-p}\frac{1}{\sqrt{\pi}\,\Gamma\left(p+\frac{1}{2}\right)}\left(\frac{iz}{2}\right)^p\int_{-1}^{+1}\left(1-t^2\right)^{p-\frac{1}{2}}e^{-zt}dt$$

$$= \frac{1}{\sqrt{\pi}\,\Gamma\left(p+\frac{1}{2}\right)}\left(\frac{z}{2}\right)^{p}\int_{-1}^{+1}\left(1-t^2\right)^{p-\frac{1}{2}} e^{-zt}\,dt. \qquad (8.184)$$

(ii) Replacing z by $-z$ and assuming p as integer, Eq. (8.184) yields,

$$(-1)^p I_p(z) = \frac{1}{\sqrt{\pi}\,\Gamma\left(p+\frac{1}{2}\right)}\left(\frac{-z}{2}\right)^{p}\int_{-1}^{+1}\left(1-t^2\right)^{p-\frac{1}{2}} e^{zt}\,dt$$

or

$$I_p(z) = \frac{1}{\sqrt{\pi}\,\Gamma\left(p+\frac{1}{2}\right)}\left(\frac{z}{2}\right)^{p}\int_{-1}^{+1}\left(1-t^2\right)^{p-\frac{1}{2}} e^{zt}\,dt. \qquad (8.185)$$

Addition of Eqs. (8.184) and (8.185) yields

$$2I_p(z) = \frac{1}{\sqrt{\pi}\,\Gamma\left(p+\frac{1}{2}\right)}\left(\frac{z}{2}\right)^{p}\int_{-1}^{+1}(e^{zt}+e^{-zt})(1-t^2)^{p-\frac{1}{2}}\,dt$$

or

$$I_p(z) = \frac{1}{\sqrt{\pi}\,\Gamma\left(p+\frac{1}{2}\right)}\left(\frac{z}{2}\right)^{p}\int_{-1}^{+1}(1-t^2)^{p-\frac{1}{2}}\cosh(zt)\,dt. \qquad (8.186)$$

The above formulae are valid for $\operatorname{Re}\left(p+\frac{1}{2}\right)>0, |\operatorname{amp} z|<\frac{\pi}{2}$.

(iii) Generating function and integral representation of modified Bessel functions

The generating function for Bessel functions of the first kind is expressed as

$$e^{\frac{z}{2}\left(t-\frac{1}{t}\right)} = \sum_{n=-\infty}^{+\infty} J_n(z)t^n.$$

Replacing z by (iz) and t by $(-it)$, we write

$$e^{\frac{iz}{2}\left(-it-\frac{1}{-it}\right)} = \sum_{n=-\infty}^{+\infty} J_n(iz)(-it)^n$$

or

$$e^{\frac{z}{2}\left(t+\frac{1}{t}\right)} = \sum_{n=-\infty}^{+\infty} J_n(iz)(-1)^n i^n t^n$$

$$= \sum_{n=-\infty}^{+\infty} J_n(iz)i^{-n}(-1)^n i^{2n} t^n = \sum_{n=-\infty}^{+\infty} I_n(z)t^n. \qquad (8.187)$$

Here, the function on the left-hand side of Eq. (8.187) is called the generating function for modified Bessel functions.

Again replacing t by $e^{-i\phi}$, Eq. (8.187) yields

$$e^{z\cos\phi} = \sum_{n=-\infty}^{+\infty} I_n(z) e^{-in\phi}.$$

Multiplying both sides by $e^{im\phi}$ (where m is an integer) and integrating the results on both sides with respect to ϕ within the limit from $\phi = 0$ to $\phi = \pi$, we obtain

$$\int_0^{\pi} e^{z\cos\phi} e^{im\phi}\, d\phi = \sum_{n=-\infty}^{+\infty} I_n(z) \int_0^{\pi} e^{i(m-n)\phi}\, d\phi$$

or

$$\int_0^{\pi} e^{z\cos\phi} (\cos m\phi + i \sin m\phi)\, d\phi$$

$$= \sum_{n=-\infty}^{+\infty} I_n(z) \int_0^{\pi} \{\cos(m-n)\phi + i \sin(m-n)\phi\}\, d\phi.$$

Equating real parts from both sides of this equation, we get

$$\int_0^{\pi} e^{z\cos\phi} \cos(m\phi)\, d\phi = I_m(z) \int_0^{\pi} d\phi = \pi I_m(z)$$

or

$$I_m(z) = \frac{1}{\pi} \int_0^{\pi} e^{z\cos\phi} \cos(m\phi)\, d\phi.$$

Changing the subscript m to n, we write

$$I_n(z) = \frac{1}{\pi} \int_0^{\pi} e^{z\cos\phi} \cos(n\phi)\, d\phi. \tag{8.188}$$

(c) Contour Integral for $I_{-p}(z)$

Let us rewrite Eq. (8.165), viz.

$$J_{-p}(z) = \frac{\Gamma\left(\frac{1}{2} - p\right) e^{p\pi i}}{2\pi i \sqrt{\pi}} \left(\frac{z}{2}\right)^p \int_C e^{iz\zeta} \left(\zeta^2 - 1\right)^{p - \frac{1}{2}} d\zeta$$

where the contour was indicated therein.

Replacing z by iz, we write

$$J_{-p}(iz)=\frac{\Gamma\left(\frac{1}{2}-p\right)e^{p\pi i}}{2\pi i\sqrt{\pi}}i^{p}\left(\frac{z}{2}\right)^{p}\int\limits_{C}e^{-z\zeta}\left(\zeta^{2}-1\right)^{p-\frac{1}{2}}d\zeta.$$

Therefore, we have

$$I_{-p}(z)=i^{p}J_{-p}(iz)$$

$$=\frac{\Gamma\left(\frac{1}{2}-p\right)e^{2p\pi i}}{2\pi i\sqrt{\pi}}\left(\frac{z}{2}\right)^{p}\int\limits_{C}e^{-z\zeta}(\zeta^{2}-1)^{p-\frac{1}{2}}d\zeta \tag{8.189}$$

where the contour C is as indicated in Figure 8.10.

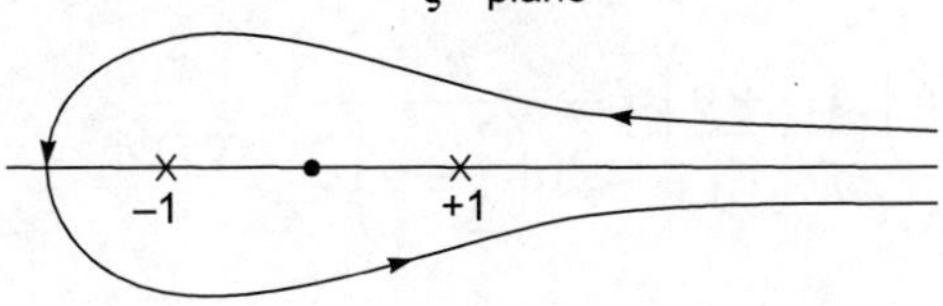

Fig. 8.10

(d) Definite Integral for $I_{-p}(z)$ and $K_p(z)$

Let us take $\text{Re}\left(p+\frac{1}{2}\right)>0$ and the path of integration as the contour of the type shown in Figure 8.11 in which the radii of the circles shown may be made to tend to zero. We know that the phase of (ζ^2-1) at the point where ζ crosses the negative real axis is -2π.

We have

$$I_{-p}(z)=\frac{\Gamma\left(\frac{1}{2}-p\right)e^{2p\pi i}}{2\pi i\sqrt{\pi}}\left(\frac{z}{2}\right)^{p}\left[(1-e^{-4p\pi i})\int\limits_{1}^{\infty}e^{-z\zeta}(\zeta^{2}-1)^{p-\frac{1}{2}}d\zeta\right.$$

$$\left.+i(e^{-p\pi i}+e^{-3p\pi i})\int\limits_{-1}^{+1}e^{-z\zeta}\left\{(1-\zeta^{2})\right\}^{p-\frac{1}{2}}d\zeta\right]$$

$$=\frac{\Gamma\left(\frac{1}{2}-p\right)}{\pi\sqrt{\pi}}\left(\frac{z}{2}\right)^{p}\left[\frac{(e^{2p\pi i}-e^{-2p\pi i})}{2i}\int\limits_{1}^{\infty}e^{-z\zeta}(\zeta^{2}-1)^{p-\frac{1}{2}}d\zeta\right.$$

$$\left.+\frac{(e^{p\pi i}+e^{-p\pi i})}{2}\int\limits_{-1}^{+1}e^{-z\zeta}(1-\zeta^{2})^{p-\frac{1}{2}}d\zeta\right]$$

$$= \frac{\Gamma\left(\frac{1}{2}-p\right)\sin 2p\pi}{\pi\sqrt{\pi}}\left(\frac{z}{2}\right)^p \int_1^\infty e^{-z\zeta}(\zeta^2-1)^{p-\frac{1}{2}}d\zeta$$

$$+\frac{\Gamma\left(\frac{1}{2}-p\right)\cos p\pi}{\pi\sqrt{\pi}}\left(\frac{z}{2}\right)^p \int_{-1}^{+1} e^{-z\zeta}(1-\zeta^2)^{p-\frac{1}{2}}d\zeta. \quad (8.190)$$

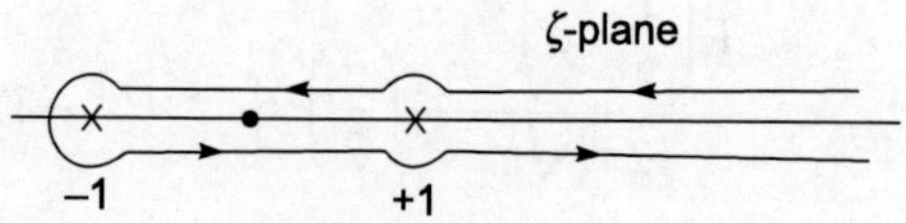

Fig. 8.11

Since $$\Gamma\left(\frac{1}{2}-p\right)\Gamma\left(\frac{1}{2}+p\right)=\frac{\pi}{\sin\left(\frac{1}{2}+p\right)\pi}=\frac{\pi}{\cos p\pi},$$

we have $$\Gamma\left(\frac{1}{2}-p\right)=\frac{\pi}{\Gamma\left(\frac{1}{2}+p\right)\cos p\pi}. \quad (8.191)$$

Using Eq. (8.191), the last term of Eq. (8.190) can be written as

$$\frac{1}{\sqrt{\pi}\,\Gamma\left(p+\frac{1}{2}\right)}\left(\frac{z}{2}\right)^p \int_{-1}^{+1} e^{-z\zeta}(1-\zeta^2)^{p-\frac{1}{2}}d\zeta.$$

Substituting this value into Eq. (8.190), we obtain

$$I_{-p}(z)=\frac{\Gamma\left(\frac{1}{2}-p\right)\sin 2p\pi}{\pi\sqrt{\pi}}\left(\frac{z}{2}\right)^p \int_1^\infty e^{-z\zeta}(\zeta^2-1)^{p-\frac{1}{2}}d\zeta$$

$$+\frac{1}{\sqrt{\pi}\,\Gamma\left(p+\frac{1}{2}\right)}\left(\frac{z}{2}\right)^p \int_{-1}^{+1} e^{-z\zeta}(1-\zeta^2)^{p-\frac{1}{2}}d\zeta. \quad (8.192)$$

Let us rewrite Eq. (8.184), viz.

$$I_p(z)=\frac{1}{\sqrt{\pi}\,\Gamma\left(p+\frac{1}{2}\right)}\left(\frac{z}{2}\right)^p \int_{-1}^{+1} e^{-z\zeta}(1-\zeta^2)^{p-\frac{1}{2}}d\zeta.$$

Subtracting this equation from Eq. (8.192), we obtain

$$I_{-p}(z)-I_p(z)=\frac{\Gamma\left(\frac{1}{2}-p\right)\sin 2p\pi}{\pi\sqrt{\pi}}\left(\frac{z}{2}\right)^p\int_1^\infty e^{-z\zeta}(\zeta^2-1)^{p-\frac{1}{2}}d\zeta$$

$$=\frac{2\Gamma\left(\frac{1}{2}-p\right)\sin p\pi\cos p\pi}{\pi\sqrt{\pi}}\left(\frac{z}{2}\right)^p\int_1^\infty e^{-z\zeta}(\zeta^2-1)^{p-\frac{1}{2}}d\zeta. \tag{8.193}$$

Using Eq. (8.191), we write Eq. (8.193) as

$$I_{-p}(z)-I_p(z)=\frac{2\sin p\pi}{\sqrt{\pi}\,\Gamma\left(p+\frac{1}{2}\right)}\left(\frac{z}{2}\right)^p\int_1^\infty e^{-z\zeta}(\zeta^2-1)^{p-\frac{1}{2}}d\zeta.$$

Now, we write

$$K_p(z)=\frac{\pi}{2}\left[\frac{I_{-p}(z)-I_p(z)}{\sin p\pi}\right]$$

$$=\frac{\sqrt{\pi}}{\Gamma\left(p+\frac{1}{2}\right)}\left(\frac{z}{2}\right)^p\int_1^\infty e^{-z\zeta}(\zeta^2-1)^{p-\frac{1}{2}}d\zeta. \tag{8.194}$$

(e) Other Integral Forms of $K_p(z)$ (Macdonald Function)

In Eq. (8.194), if we put $\zeta=\cosh\phi$, we have

$$K_p(z)=\frac{\sqrt{\pi}}{\Gamma\left(p+\frac{1}{2}\right)}\left(\frac{z}{2}\right)^p\int_0^\infty e^{-z\cosh\phi}(\sinh^2\phi)^{p-\frac{1}{2}}\sinh\phi\,d\phi$$

$$=\frac{\sqrt{\pi}}{\Gamma\left(p+\frac{1}{2}\right)}\left(\frac{z}{2}\right)^p\int_0^\infty e^{-z\cosh\phi}\sinh^{2p}\phi\,d\phi. \tag{8.195}$$

The above formulae are valid for $\mathrm{Re}\left(p+\frac{1}{2}\right)>0, |\mathrm{amp}\ z|<\frac{\pi}{2}$.

Again, let us put $\zeta=1+\frac{u}{z}$, then $d\zeta=\frac{du}{z}$. Then, at $\zeta=1$, $u=0$ and at $\zeta=\infty$, $u=\infty$. Now, Eq. (8.194) can be written as

$$K_p(z)=\frac{\sqrt{\pi}}{\Gamma\left(p+\frac{1}{2}\right)}\left(\frac{z}{2}\right)^p\int_0^{\infty}e^{-z\left(1+\frac{u}{z}\right)}\left\{\left(1+\frac{u}{z}\right)^2-1\right\}^{p-\frac{1}{2}}\frac{du}{z}$$

$$=\sqrt{\frac{\pi}{2z}}\frac{e^{-z}}{\Gamma\left(p+\frac{1}{2}\right)}\int_0^{\infty}e^{-u}u^{p-\frac{1}{2}}\left(1+\frac{u}{2z}\right)^{p-\frac{1}{2}}du. \qquad (8.196)$$

Also, we have

$$K_p(z)=\frac{\pi i}{2}e^{p\pi i/2}H_p^{(1)}(iz) \text{ [to be proved in solved problem.].}$$

Using Eq. (8.176), we write

$$K_p(z)=\frac{1}{2}\int_{-\infty}^{+\infty}e^{-z\cosh u-pu}du. \qquad (8.197)$$

Further, we write

$$K_p(z)=\frac{1}{2}\int_{-\infty}^{+\infty}e^{-z\cosh u}\cosh(pu)du-\frac{1}{2}\int_{-\infty}^{+\infty}e^{-z\cosh u}\sinh(pu)\,du$$

$$=\int_0^{\infty}e^{-z\cosh u}\cosh(pu)du \quad (\text{Re } z>0) \qquad (8.198)$$

since the last integral is equal to zero as the integrand is an odd function and the integrand of the first integral is an even function. Next, Eq. (8.197) can also be written as

$$K_p(z)=\frac{1}{2}\int_{-\infty}^{+\infty}e^{-\frac{z}{2}(e^u+e^{-u})}(e^u)^{-p}\,du.$$

Putting $t=e^u$ we write $u=\ln t,\ du=\frac{dt}{t}$ Then, we write

$$K_p(z)=\frac{1}{2}\int_0^{\infty}e^{-\frac{z}{2}\left(t+\frac{1}{t}\right)}t^{-p-1}dt. \qquad (8.199)$$

Again, let us put $v=\frac{zt}{2}$, or, $t=\frac{2v}{z}$, then

$$K_p(z)=\frac{1}{2}\left(\frac{z}{2}\right)^p\int_0^{\infty}e^{-v-\frac{z^2}{4v}}v^{-p-1}dv,\quad\left(\text{amp } z<\frac{\pi}{4}\right). \qquad (8.200)$$

(f) To obtain yet Another Integral Representation of Macdonald Function

If we put $e^u = v$, then $u = \ln v$ and $du = \dfrac{dv}{v}$.

Now, for $-\pi < \omega < \pi$ and $-\dfrac{\pi}{2} + \omega < \operatorname{amp} z < \dfrac{\pi}{2} + \omega$, we write Eq. (8.197) as the contour integral

$$K_p(z) = \frac{1}{2} \int_{0\exp(i\omega)}^{\infty \exp(-i\omega)} e^{-\frac{z}{2}\left(v + \frac{1}{v}\right)} v^{-p-1} dv. \tag{8.201}$$

The path of integration is shown in Figure 8.12.

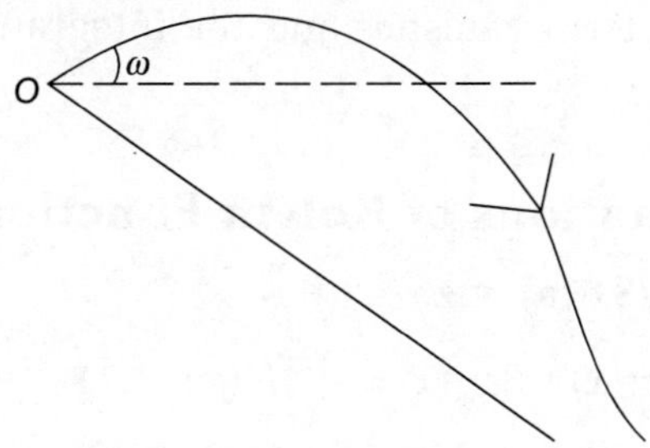

Fig. 8.12

When z is positive $(= x)$ and $-1 < \operatorname{Re} p < 1$, the path of integration in Eq. (8.201) may be swung around until it becomes the positive half of the imaginary axis. Thus, putting $v = ie^t$, we write

$$K_p(x) = \frac{1}{2} e^{-\frac{p\pi i}{2}} \int_{-\infty}^{\infty} e^{-\frac{ix}{2}(e^t - e^{-t})} e^{-pt} dt$$

$$= \frac{1}{2} e^{-\frac{p\pi i}{2}} \int_{-\infty}^{+\infty} e^{-ix \sinh t - pt} dt \tag{8.202}$$

or

$$2e^{\frac{p\pi i}{2}} K_p(x) = \int_{-\infty}^{+\infty} e^{-ix \sinh t - pt} dt. \tag{8.203}$$

The change of the sign of p results in

$$K_{-p}(x) = K_p(x) = \frac{1}{2} e^{\frac{p\pi i}{2}} \int_{-\infty}^{+\infty} e^{-ix \sinh t + pt} dt \tag{8.204}$$

or

$$2e^{-\frac{p\pi i}{2}} K_p(x) = \int_{-\infty}^{+\infty} e^{-ix \sinh t + pt} dt. \tag{8.205}$$

Addition of Eqs. (8.203) and (8.205) results in

$$K_p(x) = \frac{1}{\cos\left(\frac{p\pi}{2}\right)} \frac{1}{2} \int_{-\infty}^{+\infty} e^{-ix\sinh t} \cosh(pt)\,dt$$

$$= \sec\left(\frac{p\pi}{2}\right) \frac{1}{2} \int_{-\infty}^{+\infty} \left[\cos(x\sinh t) - i\sin(x\sinh t)\right] \cosh(pt)\,dt$$

$$= \sec\left(\frac{p\pi}{2}\right) \int_{0}^{\infty} \cos(x\sinh t)\cosh(pt)\,dt \tag{8.206}$$

since the integral of the last term vanishes and the integrand of the first term is an even function.

(D) Integral Representations of Kelvin Functions

Let us recall Eqs. (8.48) and (8.81a), viz.

$$Ber_p(x) + Bei_p(x) = e^{p\pi i} J_p(xe^{-i\pi/4}) \tag{8.48}$$

and

$$J_p(x) = \frac{1}{\pi}\int_0^{\pi} \cos(x\sin\phi - p\phi)\,d\phi - \frac{\sin p\pi}{\pi}\int_0^{\infty} e^{-p\lambda - x\sinh\lambda}\,d\lambda \tag{8.81a}$$

From Eqs. (8.48) and (8.81a), after long but rather elementary transformations, we can obtain ***integral representations*** of ***Kelvin*** functions as

$$Ber_p(x\sqrt{2})$$

$$= \frac{1}{\pi}\int_0^{\pi}\left[\cos(p\pi)\cos(x\sin t - pt)\cosh(x\sin t) - \sin(p\pi)\sin(x\sin t - pt)\sinh(x\sin t)\right]dt$$

$$- \frac{\sin(p\pi)}{\pi}\int_0^{\infty} e^{-pt - x\sinh t}\cos(x\sinh t + p\pi)\,dt \tag{8.207}$$

and

$$Bei_p(x\sqrt{2})$$

$$= \frac{1}{\pi}\int_0^{\pi} [\cos(p\pi)\sin(x\sin t - pt)\sinh(x\sin t)$$

$$+ \sin(p\pi)\cos(x\sin t - pt)\cosh(x\sin t)]\,dt$$

$$- \frac{\sin(p\pi)}{\pi}\int_0^{\infty} e^{-pt - x\sinh t}\sin(x\sinh t + p\pi)\,dt. \tag{8.208}$$

When $p = 0$ or a positive integer n, Eqs. (8.207) and (8.208) are respectively reduced to

$$Ber_n(x\sqrt{2}) = \frac{(-1)^n}{\pi}\int_0^{\pi} \cos(x\sin t - nt)\cosh(x\sin t)\,dt \tag{8.209}$$

and

$$Bei_n(x\sqrt{2}) = \frac{(-1)^n}{\pi}\int_0^{\pi} \sin(x\sin t - nt)\sinh(x\sin t)dt. \tag{8.210}$$

(E) Integral Representations of Struve Functions

Recalling Eq. (8.53), we write

$$\mathcal{H}_p(z) = \sum_{r=0}^{\infty} \frac{(-1)^r}{\Gamma\left(r+\frac{3}{2}\right)\Gamma\left(p+r+\frac{3}{2}\right)}\left(\frac{z}{2}\right)^{p+2r+1}, \quad \text{Re}\, p > -\frac{1}{2}.$$

By duplication formula of gamma functions, we have

$$\sqrt{\pi}\,\Gamma(2m+1) = 2^{2m}\Gamma(m+1)\Gamma\left(m+\frac{1}{2}\right).$$

Taking $m = r + \frac{1}{2}$, we write

$$\Gamma\left(r+\frac{3}{2}\right) = \frac{\sqrt{\pi}\,\Gamma(2r+2)}{2^{2r+1}\Gamma(r+1)}.$$

Substituting this value into the above equation for $\mathcal{H}_p(z)$, we have

$$\mathcal{H}_p(z) = \sum_{r=0}^{\infty} \frac{(-1)^r\left(\frac{z}{2}\right)^{p+2r+1} 2^{2r+1}}{\sqrt{\pi}\,\Gamma(2r+2)\Gamma\left(p+\frac{1}{2}\right)} \frac{\Gamma(r+1)\Gamma\left(p+\frac{1}{2}\right)}{\Gamma\left(r+1+p+\frac{1}{2}\right)}$$

$$= \frac{\left(\frac{z}{2}\right)^p}{\sqrt{\pi}\,\Gamma\left(p+\frac{1}{2}\right)} \sum_{r=0}^{\infty} \frac{(-1)^r z^{2r+1}}{(2r+1)!} B\left(r+1, p+\frac{1}{2}\right)$$

$$= \frac{\left(\frac{z}{2}\right)^p}{\sqrt{\pi}\,\Gamma\left(p+\frac{1}{2}\right)} \sum_{r=0}^{\infty} \frac{(-1)^r z^{2r+1}}{(2r+1)!} 2\int_0^{\frac{\pi}{2}} \cos^{2r+1}\theta \sin^{2p}\theta\, d\theta$$

$$= \frac{2\left(\frac{z}{2}\right)^p}{\sqrt{\pi}\,\Gamma\left(p+\frac{1}{2}\right)} \int_0^{\frac{\pi}{2}} \sum_{r=0}^{\infty} \frac{(-1)^r (z\cos\theta)^{2r+1}}{(2r+1)!} \sin^{2p}\theta\, d\theta$$

$$= \frac{2\left(\frac{z}{2}\right)^p}{\Gamma\left(\frac{1}{2}\right)\Gamma\left(p+\frac{1}{2}\right)} \int_0^{\frac{\pi}{2}} \sin(z\cos\theta)\sin^{2p}\theta\, d\theta,\ \text{Re}\, p > -\frac{1}{2}. \tag{8.211}$$

If we put $\cos\theta = \zeta$, we obtain another integral representation

$$\mathcal{H}_p(z) = \frac{2\left(\frac{z}{2}\right)^p}{\Gamma\left(\frac{1}{2}\right)\Gamma\left(p+\frac{1}{2}\right)} \int_0^1 \left(1-\zeta^2\right)^{p-\frac{1}{2}} \sin(z\zeta)\, d\zeta,\ \text{Re}\, p > -\frac{1}{2}. \tag{8.212}$$

Contour Integral form of Struve Function

The above definition of Struve function as definite integral fails when $p \le -\frac{1}{2}$. We can avoid the difficulty by considering a contour integral in place of the definite integral.

Now, let us consider the integral

$$\int_0^{(1+)} (\zeta^2-1)^{p-\frac{1}{2}} \sin(z\zeta)\, d\zeta$$

where the phase of ζ^2-1 vanishes at the point on the right of $\zeta=1$ at which the contour crosses the real axis. Assuming that $\text{Re}\, p > -\frac{1}{2}$, we can deform the contour into the segment of real axis from 0 to 1 and back to zero such that

$$\int_0^{(1+)} (\zeta^2-1)^{p-\frac{1}{2}} \sin(z\zeta)\, d\zeta$$

$$= \int_0^1 \left\{(1-\zeta^2)e^{-\pi i}\right\}^{p-\frac{1}{2}} \sin(z\zeta)\, d\zeta + \int_1^0 \left\{(1-\zeta^2)e^{\pi i}\right\}^{p-\frac{1}{2}} \sin(z\zeta)\, d\zeta$$

$$= \int_0^1 (1-\zeta^2)^{p-\frac{1}{2}} e^{-p\pi i} e^{\frac{\pi i}{2}} \sin(z\zeta)\, d\zeta - \int_0^1 (1-\zeta^2)^{p-\frac{1}{2}} e^{p\pi i} e^{-\frac{\pi i}{2}} \sin(z\zeta)\, d\zeta$$

$$= i\,(e^{-p\pi i} + e^{p\pi i}) \int_0^1 (1-\zeta^2)^{p-\frac{1}{2}} \sin(z\zeta)\, d\zeta$$

$$= 2i\cos(p\pi)\int_0^1 (1-\zeta^2)^{p-\frac{1}{2}} \sin(z\zeta)\, d\zeta$$

$$= 2i\frac{\pi}{\Gamma\left(p+\frac{1}{2}\right)\Gamma\left(\frac{1}{2}-p\right)}\int_0^1 (1-\zeta^2)^{p-\frac{1}{2}} \sin(z\zeta)\, d\zeta$$

where use has been made of the formulae

$$\cos p\pi = \sin\left(p+\frac{1}{2}\right)\pi \text{ and } \Gamma\left(\frac{1}{2}+p\right)\Gamma\left(\frac{1}{2}-p\right) = \frac{\pi}{\sin\left\{\left(\frac{1}{2}+p\right)\pi\right\}}.$$

Therefore, we write

$$\int_0^1 (1-\zeta^2)^{p-\frac{1}{2}} \sin(z\zeta)\, d\zeta = \frac{\Gamma\left(p+\frac{1}{2}\right)\Gamma\left(\frac{1}{2}-p\right)}{2\pi i}\int_0^{(1+)} (\zeta^2-1)^{p-\frac{1}{2}} \sin(z\zeta)\, d\zeta.$$

Substituting the above integral into Eq. (8.212), we obtain

$$\mathcal{H}_p(z) = \frac{\Gamma\left(\frac{1}{2}-p\right)\left(\frac{z}{2}\right)^p}{\pi i\,\Gamma\left(\frac{1}{2}\right)}\int_0^{(1+)} (\zeta^2-1)^{p-\frac{1}{2}} \sin(z\zeta)\, d\zeta. \tag{8.212a}$$

In Eq. (8.162a), if we bisect the path of integration, we obtain

$$J_p(z) = \frac{\Gamma\left(\frac{1}{2}-p\right)\left(\frac{z}{2}\right)^p}{\pi i\,\Gamma\left(\frac{1}{2}\right)}\int_0^{(1+)} (\zeta^2-1)^{p-\frac{1}{2}} \cos(z\zeta)\, d\zeta$$

where it is supposed that the phase of ζ^2-1 vanishes when ζ is on the real axis on the right of $\zeta=1$.

Adding the above equation to i times Eq. (8.212a), we obtain

$$J_p(z) + i\,\mathcal{H}_p(z) = \frac{\Gamma\left(\frac{1}{2}-p\right)\left(\frac{z}{2}\right)^p}{\pi i\,\Gamma\left(\frac{1}{2}\right)}\int_0^{(1+)} (\zeta^2-1)^{p-\frac{1}{2}} e^{iz\zeta}\, d\zeta. \tag{8.213}$$

If ω is an acute angle (positive or negative) and the phase of z lies between $-\frac{\pi}{2}+\omega$ and $\frac{\pi}{2}+\omega$, we can deform the contour into the form as shown in Figure 8.13, in which the four parallel lines make an angle of $-\omega$ with the imaginary axis. In that contour, as the lines parallel to the real axis move off to infinity, the integrals along these lines tend to zero. Recalling Eq. (8.168), we know that the integral along the path that starts from and returns to $1+\infty ie^{-i\omega}$ is equal to the Hankel function of the first kind, $H_p^{(1)}(z)$ On the lines through the origin, let us put $\zeta = iu$ so that on these lines

$$(\zeta^2-1)^{p-\frac{1}{2}} = e^{\mp\left(p-\frac{1}{2}\right)\pi i}(1+u^2)^{p-\frac{1}{2}}.$$

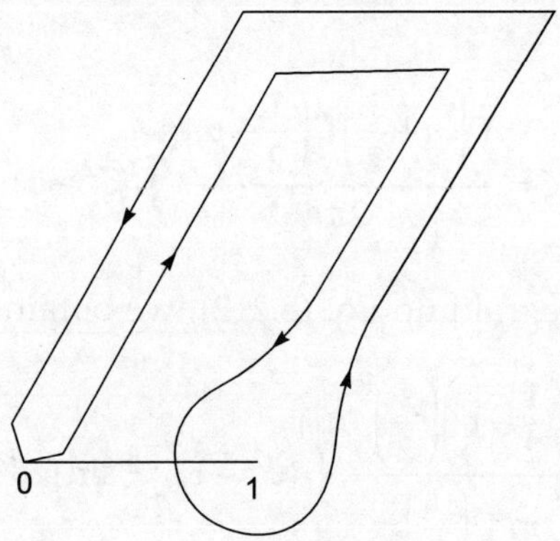

Fig. 8.13

Substituting this relation into Eq. (8:213) and performing calculations with the use of the above cited formula of gamma function, we write

$$J_p(z)+i\mathcal{H}_p(z) = H_p^{(1)}(z) + \frac{2i\left(\frac{z}{2}\right)^p}{\Gamma\left(\frac{1}{2}\right)\Gamma\left(p+\frac{1}{2}\right)}\int_0^{\infty\exp(-i\omega)} e^{-zu}(1+u^2)^{p-\frac{1}{2}}\,du$$

where the phase of $1+u^2$ has its principal value.

Recalling that $H_p^{(1)}(z) = J_p(z)+iN_p(z)$ where $N_p(z)$ is the Neumann function or the Bessel function of the second kind (also denoted by $Y_p(z)$), we write

$$\mathcal{H}_p(z) = N_p(z) + \frac{2\left(\frac{z}{2}\right)^p}{\Gamma\left(\frac{1}{2}\right)\Gamma\left(p+\frac{1}{2}\right)}\int_0^{\infty\exp(-i\omega)} e^{-zu}(1+u^2)^{p-\frac{1}{2}}\,du. \qquad (8.214)$$

If we replace ω by arg $z-\beta$ and also put $zu=q$, we can write Eq. (8.214) as

$$\mathcal{H}_p(z)=N_p(z)+\frac{\left(\frac{z}{2}\right)^{p-1}}{\Gamma\left(\frac{1}{2}\right)\Gamma\left(p+\frac{1}{2}\right)}\int_0^{\infty\exp i\beta} e^{-q}\left(1+\frac{q^2}{z^2}\right)^{p-\frac{1}{2}} dq$$

$$=N_p(z)+\frac{\left(\frac{z}{2}\right)^{p-1}}{\Gamma\left(\frac{1}{2}\right)\Gamma\left(p+\frac{1}{2}\right)}\int_0^{\infty\exp i\beta} e^{-u}\left(1+\frac{u^2}{z^2}\right)^{p-\frac{1}{2}} du \qquad (8.215)$$

where $-\frac{\pi}{2}<\beta<\frac{\pi}{2}$ and $-\frac{\pi}{2}+\beta<\arg z<\frac{\pi}{2}+\beta$.

Integral Representation of Modified Struve Function

Recalling Eq. (8.54), we write

$$\mathbb{L}_p(z)=i^{-p-1}\mathcal{H}_p(iz)$$

Using Eq. (8.211) in this equation, we obtain an integral representation of the modified Struve function in the form

$$\mathbb{L}_p(z)=i^{-p-1}\frac{2\left(\frac{iz}{2}\right)^p}{\sqrt{\pi}\,\Gamma\left(p+\frac{1}{2}\right)}\int_0^{\frac{\pi}{2}}\sin\left(iz\cos\theta\right)\sin^{2p}\theta\,d\theta$$

$$=\frac{2\left(\frac{z}{2}\right)^p}{\Gamma\left(\frac{1}{2}\right)\Gamma\left(p+\frac{1}{2}\right)}\int_0^{\frac{\pi}{2}}\sinh(z\cos\theta)\sin^{2p}\theta\,d\theta,\ \mathrm{Re}\,p>-\frac{1}{2} \qquad (8.216)$$

where use has been made of $\sin(i\phi)=i\sinh\phi$.

Further, using Eq. (8.212), we obtain another integral representation of modified Struve function in the form

$$\mathbb{L}_p(z)=\frac{2\left(\frac{z}{2}\right)^p}{\Gamma\left(\frac{1}{2}\right)\Gamma\left(p+\frac{1}{2}\right)}\int_0^1(1-\tau^2)^{p-\frac{1}{2}}\sinh(z\tau)\,d\tau,\ \mathrm{Re}\,p>-\frac{1}{2}. \qquad (8.217)$$

(F) Integral Representations of Airy Functions

The Airy function of the first kind is given by

$$Ai(x)=\frac{1}{\pi}\sqrt{\frac{x}{3}}\,K_{\frac{1}{3}}\left(\frac{2}{3}x^{\frac{3}{2}}\right),\ x>0$$

$$=\frac{1}{\pi}\sqrt{\frac{x}{3}}\sec\frac{\pi}{6}\int_0^\infty \cos\left(\frac{2x^{\frac{3}{2}}}{3}\sinh t\right)\cosh\frac{t}{3}\,dt$$

$$=\frac{2\sqrt{x}}{3\pi}\int_0^\infty \cos\left\{\frac{2x^{\frac{3}{2}}}{3}\left(4\sinh^3\frac{t}{3}+3\sinh\frac{t}{3}\right)\right\}\cosh\frac{t}{3}\,dt.$$

If we put $\sinh\frac{t}{3}=\frac{\xi}{2}x^{-\frac{1}{2}}$ we have $\cosh\frac{t}{3}dt=\frac{3}{2}x^{-\frac{1}{2}}d\xi$.

Thus, we write

$$Ai(x)=\frac{1}{\pi}\int_0^\infty \cos\left(\frac{\xi^3}{3}+x\xi\right)d\xi,\ x\geq 0. \tag{8.218}$$

A somewhat more complicated calculation leads to the integral representation of the Airy function of the second kind in the form

$$Bi(x)=\frac{1}{\pi}\int_0^\infty \left[e^{-\frac{t^3}{3}+xt}+\sin\left(\frac{t^3}{3}+xt\right)\right]dt,\ x\geq 0. \tag{8.219}$$

For derivation of this integral representation, one is suggested to see Jeffreys' *Methods of Mathematical Physics*, p. 510.

8.25 BEHAVIOUR OF BESSEL FUNCTIONS WITH SMALL AND LARGE VALUES OF ARGUMENTS (ASYMPTOTIC FORMS)

(A) Small Argument: $x \to 0$

(i) For Bessel functions of the first kind $J_p(x)$

Let us recall the series expression for $J_p(x)$, viz.

$$J_p(x)=\sum_{r=0}^{\infty}\frac{(-1)^r}{r!\,\Gamma(p+r+1)}\left(\frac{x}{2}\right)^{p+2r}.$$

Therefore, when x tends to zero, we have

$$\lim_{x\to 0} J_p(x) \sim \frac{\left(\frac{x}{2}\right)^p}{\Gamma(1+p)}. \tag{8.220}$$

Similarly, $$\lim_{x\to 0} J_{-p}(x) \sim \frac{\left(\frac{x}{2}\right)^{-p}}{\Gamma(1-p)}. \tag{8.221}$$

(ii) For Neumann functions (Bessel functions of the second kind) $N_p(x)$

We know that when p is not an integer,

$$N_p(x) = \frac{\cos p\pi\, J_p(x) - J_{-p}(x)}{\sin p\pi}$$

For $x \to 0$, we are to pick up the lowest power of x in the above formula yielding

$$\lim_{x\to 0} N_p(x) \sim -\frac{\left(\frac{x}{2}\right)^{-p}}{\Gamma(-p+1)\sin p\pi}.$$

Recalling $\Gamma(p)\Gamma(1-p) = \dfrac{\pi}{\sin p\pi}$ we write

$$\lim_{x\to 0} N_p(x) \sim -\frac{\Gamma(p)}{\pi}\left(\frac{2}{x}\right)^p, \quad p \neq \text{integer}. \tag{8.222}$$

When $p = n$ is an integer, taking the dominant term from Eq. (8.20) and remembering that the power of x dominates in the logarithmic term, we write

$$\lim_{x\to 0} N_n(x) \sim \frac{2}{\pi}\ln x \text{ when } n = 0 \tag{8.223}$$

and $$\lim_{x\to 0} N_n(x) \sim -\frac{\Gamma(n)}{\pi}\left(\frac{2}{x}\right)^n \text{ when } n \neq 0. \tag{8.224}$$

(iii) For Hankel functions $H_p^{(1)}(x)$ **and** $H_p^{(2)}(x)$

Since $H_p^{(1)}(x) = J_p(x) + iN_p(x)$ and $H_p^{(2)}(x) = J_p(x) - iN_p(x)$

We have

(a) $$\lim_{x\to 0} H_p^{(1)}(x) = \lim_{x\to 0} J_p(x) + i\lim_{x\to 0} N_p(x) \sim i\lim_{x\to 0} N_p(x) + \lim_{x\to 0} O(x^p) \tag{8.225}$$

and (b) $\lim_{x\to 0} H_p^{(2)}(x) \sim -i \lim_{x\to 0} N_p(x) + \lim_{x\to 0} O(x^p)$ (8.226)

where $O(x^p)$ is the Bachmann-Landao symbol which denotes a function of the order of the magnitude of x^p.

(iv) For modified Bessel functions $I_p(x)$ and $K_p(x)$

(a) Since, $I_p(x) = i^{-p} J_p(ix)$, we have

$$I_p(x) = \sum_{r=0}^{\infty} \frac{1}{r!\,\Gamma(p+r+1)} \left(\frac{x}{2}\right)^{p+2r}$$

Now, $\lim_{x\to 0} I_p(x) = \lim_{x\to 0}\left[i^{-p} J_p(ix)\right] \sim \frac{1}{\Gamma(p+1)}\left(\frac{x}{2}\right)^p$, ($p \neq$ integer). (8.227)

(b) We know that

$$K_p(x) = \frac{\pi}{2} \frac{I_{-p}(x) - I_p(x)}{\sin p\pi}.$$

In the limit x tends to zero, retaining only the leading term, we have

$$\lim_{x\to 0} K_p(x) \sim \frac{\pi}{2\sin p\pi} \frac{1}{\Gamma(1-p)} \left(\frac{x}{2}\right)^{-p} \quad (p > 0).$$

Using the formula $\Gamma(p)\Gamma(1-p) = \frac{\pi}{\sin p\pi}$, we write

$$\lim_{x\to 0} K_p(x) = \frac{\Gamma(p)}{2}\left(\frac{2}{x}\right)^p. \tag{8.228}$$

For $p = 0$, this approach fails but one can prove by another approach,

$$\lim_{x\to 0} K_0(x) \sim -\ln x + O(1). \tag{8.229}$$

(v) For spherical Bessel functions $j_l(x)$, $n_l(x)$, $i_l(x)$ *and* $k_l(x)$

(a) As $j_l(x) = \sqrt{\frac{\pi}{2x}} J_{l+\frac{1}{2}}(x)$, we have

$$\lim_{x\to 0} j_l(x) \sim \sqrt{\frac{\pi}{2x}} \frac{1}{\Gamma\left(l+\frac{3}{2}\right)} \left(\frac{x}{2}\right)^{l+\frac{1}{2}}$$

$$= \frac{\sqrt{\pi}\, x^l}{2^{l+1}\left(l+\frac{1}{2}\right)\left(l-\frac{1}{2}\right)\left(l-\frac{3}{2}\right)\ldots\frac{5}{2}\cdot\frac{3}{2}\cdot\frac{1}{2}\sqrt{\pi}}$$

$$= \frac{x^l}{(2l+1)(2l-1)(2l-3)\ldots 5.3.1}$$

$$= \frac{l!\,(2x)^l}{(2l+1)!}, \; (l = \text{non-negative integer}). \tag{8.230}$$

(b) As $n_l(x) = \sqrt{\frac{\pi}{2x}}\, N_{l+\frac{1}{2}}(x)$, we have

$$\lim_{x\to 0} n_l(x) \sim \sqrt{\frac{\pi}{2x}}\left[-\frac{\Gamma\left(l+\frac{1}{2}\right)}{\pi}\left(\frac{2}{x}\right)^{l+\frac{1}{2}}\right]$$

$$= -\frac{2^l\left(l-\frac{1}{2}\right)\left(l-\frac{3}{2}\right)\left(l-\frac{5}{2}\right)\ldots\frac{5}{2}\cdot\frac{3}{2}\cdot\frac{1}{2}\cdot\Gamma\left(\frac{1}{2}\right)}{\sqrt{\pi}\, x^{l+1}}$$

$$= -\frac{(2l-1)(2l-3)(2l-5)\ldots 5.3.1}{x^{l+1}}$$

$$= -\frac{2(2l)!}{l!\,(2x)^{l+1}}, \; (l = \text{non-negative integer}). \tag{8.231}$$

(c) As $i_l(x) = \sqrt{\frac{\pi}{2x}}\, I_{l+\frac{1}{2}}(x)$, we have

$$\lim_{x\to 0} i_l(x) \sim \sqrt{\frac{\pi}{2x}}\, \frac{\left(\frac{x}{2}\right)^{l+\frac{1}{2}}}{\Gamma\left(l+\frac{3}{2}\right)}$$

$$= \frac{l!\,(2x)^l}{(2l+1)!}, \; (l = \text{non-negative integer}). \tag{8.232}$$

(d) As $k_l(x)=\sqrt{\frac{2}{\pi x}}\,K_{l+\frac{1}{2}}(x)$, we have

$$\lim_{x\to 0} k_l(x) \sim \sqrt{\frac{2}{\pi x}}\,\frac{\Gamma\left(l+\frac{1}{2}\right)}{2}\left(\frac{2}{x}\right)^{l+\frac{1}{2}}$$

$$=\frac{2^l\left(l-\frac{1}{2}\right)\left(l-\frac{3}{2}\right)\left(l-\frac{5}{2}\right)\cdots\frac{5}{2}\cdot\frac{3}{2}\cdot\frac{1}{2}\Gamma\left(\frac{1}{2}\right)}{\sqrt{\pi}\,x^{l+1}}$$

$$=\frac{2(2l)!}{l!\,(2x)^{l+1}},\ (l = \text{non-negative integer}). \tag{8.233}$$

(vi) For Kelvin functions

We know that

$$\text{Ber}_p(x)=\sum_{r=0}^{\infty}\frac{\cos\left(\frac{3}{4}p\pi+\frac{1}{2}r\pi\right)}{r!\,\Gamma(p+r+1)}\left(\frac{x}{2}\right)^{p+2r}$$

$$\text{Bei}_p(x)=\sum_{r=0}^{\infty}\frac{\sin\left(\frac{3}{4}p\pi+\frac{1}{2}r\pi\right)}{r!\,\Gamma(p+r+1)}\left(\frac{x}{2}\right)^{p+2r}$$

$$\text{Ber}(x)=\sum_{r=0}^{\infty}\frac{(-1)^r\left(\frac{x}{2}\right)^{4r}}{[(2r)!]^2}$$

and

$$\text{Bei}(x)=\sum_{r=0}^{\infty}\frac{(-1)^r\left(\frac{x}{2}\right)^{4r+2}}{[(2r+1)!]^2}.$$

Therefore, we have

$$\lim_{x\to 0}\text{Ber}_p(x)=\lim_{x\to 0}\sum_{r=0}^{\infty}\frac{\cos\left(\frac{3}{4}p\pi+\frac{1}{2}r\pi\right)}{r!\,\Gamma(p+r+1)}\left(\frac{x}{2}\right)^{p+2r}$$

$$=\frac{\cos\left(\frac{3}{4}p\pi\right)}{\Gamma(p+1)}\left(\frac{x}{2}\right)^{p} \tag{8.234}$$

$$\lim_{x\to 0} \text{Bei}_p(x) = \frac{\sin\left(\frac{3}{4}p\pi\right)}{\Gamma(p+1)}\left(\frac{x}{2}\right)^p, \tag{8.235}$$

$$\lim_{x\to 0} \text{Ber}(x) \sim 1 \tag{8.236}$$

and
$$\lim_{x\to 0} \text{Bei}(x) \sim \frac{x^2}{4}. \tag{8.237}$$

Expressions for $Ker_p(x)$, $Kei_p(x)$, $Ker(x)$ and $Kei(x)$ with small arguments can be obtained using Eq. (8.51).

(vii) For Struve functions

We know that the Struve function and the modified Struve functions are given by

$$\mathcal{H}_p(x) = \sum_{r=0}^{\infty} \frac{(-1)^r \left(\frac{x}{2}\right)^{p+2r+1}}{\Gamma\left(r+\frac{3}{2}\right)\Gamma\left(p+r+\frac{3}{2}\right)}$$

and
$$\mathbb{L}_p(x) = \sum_{r=0}^{\infty} \frac{\left(\frac{x}{2}\right)^{p+2r+1}}{\Gamma\left(r+\frac{3}{2}\right)\Gamma\left(p+r+\frac{3}{2}\right)}.$$

Therefore, we have

$$\lim_{x\to 0} \mathcal{H}_p(x) \sim \frac{2\left(\frac{x}{2}\right)^{p+1}}{\sqrt{\pi}\,\Gamma\left(p+\frac{3}{2}\right)}, \ (p>0) \tag{8.238}$$

and
$$\lim_{x\to 0} \mathcal{H}_0(x) \sim \frac{2}{\pi}x. \tag{8.239}$$

Also, we have

$$\lim_{x\to 0} \mathbb{L}_p(x) \sim \frac{2\left(\frac{x}{2}\right)^{p+1}}{\sqrt{\pi}\,\Gamma\left(p+\frac{3}{2}\right)}. \tag{8.240}$$

(viii) For Airy functions

From Eq. (8.58), we write

$$Ai(x) = 3^{-2/3}\left[\sum_{n=0}^{\infty}\frac{3^{-2n}}{n!\,\Gamma\left(n+\frac{2}{3}\right)}x^{3n} - \sum_{n=0}^{\infty}\frac{3^{-2n-\frac{2}{3}}}{n!\,\Gamma\left(n+\frac{4}{3}\right)}x^{3n+1}\right]$$

$$= 3^{-2/3}\left[\left\{\frac{1}{\Gamma\left(\frac{2}{3}\right)} + \frac{3^{-2}}{1!\,\Gamma\left(1+\frac{2}{3}\right)}x^3 + \frac{3^{-4}}{2!\,\Gamma\left(2+\frac{2}{3}\right)}x^6 + \ldots\infty\right\}\right.$$

$$\left. - \left\{\frac{3^{-2/3}}{\Gamma\left(\frac{4}{3}\right)}x + \frac{3^{-8/3}}{1!\,\Gamma\left(1+\frac{4}{3}\right)}x^4 + \frac{3^{-14/3}}{2!\,\Gamma\left(2+\frac{4}{3}\right)}x^7 + \ldots\infty\right\}\right]$$

$$= 3^{-2/3}\left[\frac{1}{\Gamma\left(\frac{2}{3}\right)} - \frac{3^{-2/3}}{\frac{1}{3}\Gamma\left(\frac{1}{3}\right)}x + \frac{3^{-2}}{\frac{2}{3}\Gamma\left(\frac{2}{3}\right)}x^3 - \frac{3^{-8/3}}{\frac{4}{3}\cdot\frac{1}{3}\Gamma\left(\frac{1}{3}\right)}x^4 + \frac{3^{-4}}{2!\frac{5}{3}\cdot\frac{2}{3}\Gamma\left(\frac{2}{3}\right)}x^6 - \ldots\right]$$

Using the relation $\Gamma\left(\frac{2}{3}\right)\Gamma\left(\frac{1}{3}\right) = \dfrac{\pi}{\sin\left(\frac{2\pi}{3}\right)}$, we write

$$\text{Ai}(x) = 3^{-2/3}\left[\frac{\Gamma\left(\frac{1}{3}\right)\sin\left(\frac{2\pi}{3}\right)}{\pi} - \frac{(3^{1/3})\Gamma\left(\frac{2}{3}\right)\sin\left(\frac{2\pi}{3}\right)}{\pi}x + \frac{3^{-1}\Gamma\left(\frac{1}{3}\right)\sin\left(\frac{2\pi}{3}\right)}{2\pi}x^3\right.$$

$$\left. - \frac{3^{-2/3}\Gamma\left(\frac{2}{3}\right)\sin\left(\frac{2\pi}{3}\right)}{1.4\,\pi}x^4 + \ldots\infty\right]$$

$$= \frac{1}{(3^{2/3})\pi}\left[\frac{\Gamma\left(\frac{1}{3}\right)\sin\left(\frac{2\pi}{3}\right)}{0!} + \frac{\Gamma\left(\frac{2}{3}\right)\sin\left(\frac{4\pi}{3}\right)}{1!}(3^{1/3})x + \frac{\Gamma\left(\frac{3}{3}\right)\sin\left(\frac{6\pi}{3}\right)}{2!}(3^{2/3})x^2\right.$$

$$\left. + \frac{\Gamma\left(\frac{4}{3}\right)\sin\left(\frac{8\pi}{3}\right)}{3!}(3x^3) + \frac{\Gamma\left(\frac{5}{3}\right)\sin\left(\frac{10\pi}{3}\right)}{4!}3^{4/3}x^4 + \ldots\infty\right]$$

$$= \frac{1}{(3^{2/3})\pi} \sum_{n=0}^{\infty} \frac{1}{n!} \Gamma\left(\frac{n+1}{3}\right) \sin\left\{\frac{2\pi}{3}(n+1)\right\} (3^{1/3} x)^n . \tag{8.241}$$

Now, we have

$$\lim_{x\to 0} \text{Ai}(x) \sim \frac{\Gamma\left(\frac{1}{3}\right)}{(3^{2/3})\pi} \sin\left(\frac{2\pi}{3}\right) + \frac{\Gamma\left(\frac{2}{3}\right)}{(3^{1/3})\pi} x \sin\left(\frac{4\pi}{3}\right). \tag{8.241a}$$

By a similar procedure, using Eq. (8.59), we can obtain

$$Bi(x) = \frac{1}{(3^{1/6})\pi} \sum_{n=0}^{\infty} \frac{1}{n!} \Gamma\left(\frac{n+1}{3}\right) \sin\left\{\frac{2\pi}{3}(n+1)\right\} (3^{1/3} x)^n . \tag{8.242}$$

Now, we have

$$\lim_{x\to 0} Bi(x) \sim \frac{\Gamma\left(\frac{1}{3}\right)}{(3^{1/6})\pi} \sin\left(\frac{2\pi}{3}\right) + \frac{(3^{1/6})\Gamma\left(\frac{2}{3}\right)}{\pi} x \sin\left(\frac{4\pi}{3}\right). \tag{8.242a}$$

(B) Large Argument: $x \to \infty$

(i) For Hankel functions

(a) For Hankel function of the first kind

We know that Hankel function of the first kind with imaginary argument, $H_p^{(1)}(ix)$ is related to the Macdonald function $K_p(x)$ by the relation

$$K_p(x) = \frac{\pi i}{2} e^{p\pi i/2} H_p^{(1)}(ix)$$

[See Problem 7 of Art. 8.28: Solved problems on Bessel functions of other kinds and use $i^p = e^{p\pi i/2}$]

or
$$H_p^{(1)}(ix) = \frac{2}{\pi i} e^{-p\pi i/2} K_p(x).$$

Replacing x by $-ix$, we write

$$H_p^{(1)}(x) = \frac{2}{\pi i} e^{-p\pi i/2} K_p(-ix).$$

When $x \to \infty$ using $K_p(x) = \sqrt{\frac{\pi}{2x}}\, e^{-x}$ (to be proved in iv (b)), we have

$$\lim_{x\to\infty} H_p^{(1)}(x) \sim \frac{2}{\pi i} e^{-p\pi i/2} \left(\frac{\pi}{-2ix}\right)^{\frac{1}{2}} e^{ix}$$

$$= \left(\frac{2}{\pi x}\right)^{\frac{1}{2}} (-i)^{\frac{1}{2}} e^{i\left\{x - \frac{p\pi}{2}\right\}}$$

$$= \left(\frac{2}{\pi x}\right)^{\frac{1}{2}} e^{i\left\{x - \left(p + \frac{1}{2}\right)\frac{\pi}{2}\right\}}. \tag{8.243}$$

(b) For Hankel functions of the second kind:

We know that $H_p^{(2)}(x)$ is the complex conjugate of $H_p^{(1)}(x)$.

Hence, we write

$$H_p^{(2)}(x) = \left[H_p^{(1)}(x)\right]^*$$

and
$$\lim_{x\to\infty} H_p^{(2)}(x) \sim \left(\frac{2}{\pi x}\right)^{\frac{1}{2}} e^{-i\left\{x - \left(p + \frac{1}{2}\right)\frac{\pi}{2}\right\}}. \tag{8.244}$$

(ii) For Bessel functions of the first kind

We know that $J_p(x)$ is the real part of either $H_p^{(1)}(x)$ or $H_p^{(2)}(x)$.

Therefore, from Eq. (8.243) or (8.244), we have

$$\lim_{x\to\infty} J_p(x) = \left(\frac{2}{\pi x}\right)^{\frac{1}{2}} \cos\left\{x - \left(p + \frac{1}{2}\right)\frac{\pi}{2}\right\}. \tag{8.245}$$

(iii) For Neumann functions (Bessel functions of the second kind)

We know that the Neumann function $N_p(x)$ is the imaginary part of the Hankel function of the first kind $H_p^{(1)}(x)$

Therefore, from Eq. (8.243), we have

$$\lim_{x\to\infty} N_p(x) \sim \left(\frac{2}{\pi x}\right)^{\frac{1}{2}} \sin\left\{x - \left(p + \frac{1}{2}\right)\frac{\pi}{2}\right\}. \tag{8.246}$$

(iv) For modified Bessel functions

(a) For modified Bessel function of the first kind $I_p(x)$

Since $I_p(x) = i^{-p} J_p(ix)$ using Eq. (8.245), we write

$$\lim_{x\to\infty} I_p(x) \sim i^{-p} \left(\frac{2}{\pi i x}\right)^{\frac{1}{2}} \cos\left\{ix - \left(p + \frac{1}{2}\right)\frac{\pi}{2}\right\}$$

$$= i^{-p-\frac{1}{2}} \left(\frac{2}{\pi x}\right)^{\frac{1}{2}} \frac{1}{2}\left[e^{i\left\{ix - \left(p + \frac{1}{2}\right)\frac{\pi}{2}\right\}} + e^{-i\left\{ix - \left(p + \frac{1}{2}\right)\frac{\pi}{2}\right\}}\right]$$

$$= i^{-\left(p+\frac{1}{2}\right)}\left(\frac{1}{2\pi x}\right)^{\frac{1}{2}}\left[e^{-x-\left(p+\frac{1}{2}\right)\frac{\pi i}{2}} + e^{x+\left(p+\frac{1}{2}\right)\frac{\pi i}{2}}\right].$$

When $x \to \infty$, retaining only the dominant term, we write

$$\lim_{x\to\infty} I_p(x) \sim \left(\frac{1}{2\pi x}\right)^{\frac{1}{2}} i^{-\left(p+\frac{1}{2}\right)} e^{x+\left(p+\frac{1}{2}\right)\frac{\pi i}{2}}$$

$$= \left(\frac{1}{2\pi x}\right)^{\frac{1}{2}}\left(e^{\frac{\pi i}{2}}\right)^{-\left(p+\frac{1}{2}\right)} e^{x+\left(p+\frac{1}{2}\right)\frac{\pi i}{2}}$$

$$= \left(\frac{1}{2\pi x}\right)^{\frac{1}{2}} e^{x}. \tag{8.247}$$

(b) For modified Bessel function of the second kind (Macdonald function):

From Eq. (8.196), we have

$$K_p(x) = \sqrt{\frac{\pi}{2x}}\,\frac{e^{-x}}{\Gamma\left(p+\frac{1}{2}\right)}\int_0^\infty e^{-u} u^{p-\frac{1}{2}}\left(1+\frac{u}{2x}\right)^{p-\frac{1}{2}} du.$$

At very large x, we have $\left(1+\frac{u}{2x}\right)^{p-\frac{1}{2}} \sim 1$ and we write

$$\lim_{x\to\infty} K_p(x) = \lim_{x\to\infty}\sqrt{\frac{\pi}{2x}}\,\frac{e^{-x}}{\Gamma\left(p+\frac{1}{2}\right)}\int_0^\infty e^{-u} u^{p-\frac{1}{2}} du$$

$$= \lim_{x\to\infty}\sqrt{\frac{\pi}{2x}}\,\frac{e^{-x}}{\Gamma\left(p+\frac{1}{2}\right)}\Gamma\left(p+\frac{1}{2}\right)$$

$$= \sqrt{\frac{\pi}{2x}}\, e^{-x}. \tag{8.248}$$

(v) For spherical Bessel functions

(a) For spherical Bessel function of the first kind $j_l(x)$ (l = integer)

By definition, we have $j_l(x) = \left(\frac{\pi}{2x}\right)^{\frac{1}{2}} J_{l+\frac{1}{2}}(x)$.

Therefore, using Eq. (8.245), we have

$$\lim_{x\to\infty} j_l(x) \sim \left(\frac{\pi}{2x}\right)^{\frac{1}{2}}\left(\frac{2}{\pi x}\right)^{\frac{1}{2}}\cos\left\{x-(l+1)\frac{\pi}{2}\right\}$$

$$=\frac{1}{x}\cos\left\{x-\frac{l\pi}{2}-\frac{\pi}{2}\right\}$$

$$=\frac{1}{x}\cos\left\{\frac{\pi}{2}-\left(x-\frac{l\pi}{2}\right)\right\}$$

$$=\frac{1}{x}\sin\left(x-\frac{l\pi}{2}\right). \tag{8.249}$$

(b) For spherical Bessel function of the second kind (or spherical Neumann function) $n_l(x)$

By definition, we have

$$n_l(x)=\left(\frac{\pi}{2x}\right)^{\frac{1}{2}} N_{l+\frac{1}{2}}(x).$$

Therefore, using Eq. (8.246), we have

$$\lim_{x\to\infty} n_l(x) \sim \left(\frac{\pi}{2x}\right)^{\frac{1}{2}}\left(\frac{2}{\pi x}\right)^{\frac{1}{2}}\sin\left\{x-(l+1)\frac{\pi}{2}\right\}$$

$$=\frac{1}{x}\sin\left[x-\frac{l\pi}{2}-\frac{\pi}{2}\right]$$

$$=-\frac{1}{x}\cos\left(x-\frac{l\pi}{2}\right). \tag{8.250}$$

(c) For spherical Hankel function of the first kind $h_l^{(1)}(x)$

We know that

$$h_l^{(1)}(x)=j_l(x)+in_l(x)$$

Using Eqs. (8.249) and (8.250), we write

$$\lim_{x\to\infty} h_l^{(1)}(x) \sim \frac{1}{x}\sin\left(x-\frac{l\pi}{2}\right)-\frac{i}{x}\cos\left(x-\frac{l\pi}{2}\right)$$

$$=-\frac{i}{x}\left[\cos\left(x-\frac{l\pi}{2}\right)+i\sin\left(x-\frac{l\pi}{2}\right)\right]$$

$$= -\frac{i}{x} e^{i\left(x - \frac{l\pi}{2}\right)}. \tag{8.251}$$

(d) For spherical Hankel function of the second kind $h_l^{(2)}(x)$

We have

$$h_l^{(2)}(x) = \left[h_l^{(1)}(x)\right]^* \sim \frac{i}{x} e^{-i\left(x - \frac{l\pi}{2}\right)}. \tag{8.252}$$

(vi) For Kelvin functions

For large arguments, Kelvin functions can be approximated by the first terms in their asymptotic expansions. On account of their complexity, the cumbersome tasks of calculations are omitted and the relevant formulas are simply reproduced here (Copson, Dingle, Olver, etc). Thus, for $x \to \infty$, we have

$$\text{Ber}_p(x) \sim \frac{e^{\frac{x}{\sqrt{2}}}}{\sqrt{2\pi x}} \sum_{s=0}^{\infty} \frac{a_s(p)}{x^s} \cos\left\{\frac{x}{\sqrt{2}} + \left(\frac{p}{2} + \frac{3s}{4} - \frac{1}{8}\right)\pi\right\}, \tag{8.253}$$

$$\text{Bei}_p(x) \sim \frac{e^{\frac{x}{\sqrt{2}}}}{\sqrt{2\pi x}} \sum_{s=0}^{\infty} \frac{a_s(p)}{x^s} \sin\left\{\frac{x}{\sqrt{2}} + \left(\frac{p}{2} + \frac{3s}{4} - \frac{1}{8}\right)\pi\right\}, \tag{8.254}$$

$$\text{Ker}_p(x) \sim \sqrt{\frac{\pi}{2x}} e^{-\frac{x}{\sqrt{2}}} \sum_{s=0}^{\infty} \frac{a_s(p)}{x^s} \cos\left\{\frac{x}{\sqrt{2}} + \left(\frac{p}{2} + \frac{s}{4} + \frac{1}{8}\right)\pi\right\} \tag{8.255}$$

and $$\text{Kei}_p(x) \sim -\sqrt{\frac{\pi}{2x}}\, e^{-\frac{x}{\sqrt{2}}} \sum_{s=0}^{\infty} \frac{a_s(p)}{x^s} \sin\left\{\frac{x}{\sqrt{2}} + \left(\frac{p}{2} + \frac{s}{4} + \frac{1}{8}\right)\pi\right\} \tag{8.256}$$

where $$a_s(p) = \frac{(4p^2 - 1^2)(4p^2 - 3^2)(4p^2 - 5^2)\ldots\{4p^2 - (2s-1)^2\}}{s!\,8^s}, \; s \geq 1 \tag{8.257}$$

and $a_0(p) = 1.$ (8.258)

When $p = 0$, we have

$$\text{Ber}(x) \sim \frac{e^{\frac{x}{\sqrt{2}}}}{\sqrt{2\pi x}} \cos\left(\frac{x}{\sqrt{2}} - \frac{\pi}{8}\right), \tag{8.259}$$

$$\text{Bei}(x) \sim \frac{e^{\frac{x}{\sqrt{2}}}}{\sqrt{2\pi x}} \sin\left(\frac{x}{\sqrt{2}} - \frac{\pi}{8}\right), \tag{8.260}$$

$$\operatorname{Ker}(x) \sim \sqrt{\frac{\pi}{2x}}\, e^{-\frac{x}{\sqrt{2}}} \cos\left(\frac{x}{\sqrt{2}} + \frac{\pi}{8}\right) \tag{8.261}$$

and
$$\operatorname{Kei}(x) \sim -\sqrt{\frac{\pi}{2x}}\, e^{-\frac{x}{\sqrt{2}}} \sin\left(\frac{x}{\sqrt{2}} + \frac{\pi}{8}\right). \tag{8.262}$$

(vii) For Struve functions

On the real axis, Eq. (8:215) yields

$$\mathcal{H}_p(x) = N_p(x) + \frac{\left(\frac{x}{2}\right)^{p-1}}{\Gamma\left(\frac{1}{2}\right)\Gamma\left(p+\frac{1}{2}\right)} \int_0^{\infty} e^{-u}\left(1+\frac{u^2}{x^2}\right)^{p-\frac{1}{2}} du. \tag{8.263}$$

We have $\left(1+\dfrac{u^2}{x^2}\right)^{p-\frac{1}{2}}$

$$= 1 + \frac{\left(p-\frac{1}{2}\right)\Gamma\left(p-\frac{1}{2}\right)}{\Gamma\left(p-\frac{1}{2}\right)}\frac{u^2}{x^2} + \frac{\left(p-\frac{1}{2}\right)\left(p-\frac{3}{2}\right)\Gamma\left(p-\frac{3}{2}\right)}{2!\,\Gamma\left(p-\frac{3}{2}\right)}\frac{u^4}{x^4}$$

$$+ \frac{\left(p-\frac{1}{2}\right)\left(p-\frac{3}{2}\right)\left(p-\frac{5}{2}\right)\Gamma\left(p-\frac{5}{2}\right)}{3!\,\Gamma\left(p-\frac{5}{2}\right)}\frac{u^6}{x^6} + \ldots$$

$$+ \frac{\left(p-\frac{1}{2}\right)\left(p-\frac{3}{2}\right)\left(p-\frac{5}{2}\right)\ldots\left(p-\frac{2r-1}{2}\right)\Gamma\left(p-\frac{2r-1}{2}\right)}{r!\,\Gamma\left(p-\frac{2r-1}{2}\right)}\frac{u^{2r}}{x^{2r}} + \ldots$$

$$= 1 + \frac{\Gamma\left(p+\frac{1}{2}\right)}{\Gamma\left(p-\frac{1}{2}\right)}\frac{u^2}{x^2} + \frac{\Gamma\left(p+\frac{1}{2}\right)}{2!\,\Gamma\left(p-\frac{3}{2}\right)}\frac{u^4}{x^4} + \frac{\Gamma\left(p+\frac{1}{2}\right)}{3!\,\Gamma\left(p-\frac{5}{2}\right)}\frac{u^6}{x^6} + \ldots + \frac{\Gamma\left(p+\frac{1}{2}\right)}{r!\,\Gamma\left(p-r+\frac{1}{2}\right)}\frac{u^{2r}}{x^{2r}} + \ldots$$

Substituting this expansion into Eq. (8.263) and recalling the definition of gamma functions, we write

$$\mathcal{H}_p(x) = N_p(x)$$

$$+\frac{\left(\frac{x}{2}\right)^{p-1}}{\Gamma\left(\frac{1}{2}\right)\Gamma\left(p+\frac{1}{2}\right)}\left[1+\frac{2!\,\Gamma\left(p+\frac{1}{2}\right)}{\Gamma\left(p-\frac{1}{2}\right)x^2}+\frac{4!\,\Gamma\left(p+\frac{1}{2}\right)}{2!\,\Gamma\left(p-\frac{3}{2}\right)x^4}+\frac{6!\,\Gamma\left(p+\frac{1}{2}\right)}{3!\,\Gamma\left(p-\frac{5}{2}\right)x^6}+\ldots+\frac{(2r)!\,\Gamma\left(p+\frac{1}{2}\right)}{r!\,\Gamma\left(p-r+\frac{1}{2}\right)x^{2r}}+\ldots\right]$$

$$=N_p(x)+\frac{\left(\frac{x}{2}\right)^{p-1}}{\Gamma\left(\frac{1}{2}\right)\Gamma\left(p+\frac{1}{2}\right)}\sum_{r=0}^{m-1}\frac{(2r)!\,\Gamma\left(p+\frac{1}{2}\right)}{r!\,\Gamma\left(p-r+\frac{1}{2}\right)x^{2r}}+O(x^{-2m})$$

$$=N_p(x)+\frac{\left(\frac{x}{2}\right)^{p-1}}{\Gamma\left(\frac{1}{2}\right)}\sum_{r=0}^{m-1}\frac{2^{2r}r!\,\Gamma\left(r+\frac{1}{2}\right)}{r!\,\Gamma\left(\frac{1}{2}\right)\Gamma\left(p-r+\frac{1}{2}\right)x^{2r}}+O(x^{-2m})$$

$$=N_p(x)+\frac{\left(\frac{x}{2}\right)^{p-1}}{\pi}\sum_{r=0}^{m-1}\frac{\Gamma\left(r+\frac{1}{2}\right)}{\Gamma\left(p-r+\frac{1}{2}\right)\left(\frac{x}{2}\right)^{2r}}+O(x^{-2m}). \tag{8.264}$$

Here, we have used the formulae

(a) $\Gamma\left(\frac{1}{2}\right)=\sqrt{\pi}$ and

(b) $(2r)!=\dfrac{2^{2r}r!\,\Gamma\left(r+\frac{1}{2}\right)}{\Gamma\left(\frac{1}{2}\right)}$.

Equation (8.264) can also be written as

$$\mathcal{H}_p(x)=N_p(x)+\frac{1}{\pi}\sum_{r=0}^{m-1}\frac{\Gamma\left(r+\frac{1}{2}\right)\left(\frac{x}{2}\right)^{p-1-2r}}{\Gamma\left(p+\frac{1}{2}-r\right)}+O(x^{p-2m-1}). \tag{8.265}$$

In the limit $x \to \infty$, using Eq. (8.246) and retaining only the leading term, we write

$$\lim_{x\to\infty} \mathcal{H}_p(x) \sim \left(\frac{2}{\pi x}\right)^{\frac{1}{2}} \sin\left\{x - \left(p+\frac{1}{2}\right)\frac{\pi}{2}\right\} + \frac{\left(\frac{x}{2}\right)^{p-1}}{\sqrt{\pi}\,\Gamma\left(p+\frac{1}{2}\right)}. \tag{8.266}$$

(viii) For Airy functions

Using Eq. (8.218), we write, for a complex z,

$$\begin{aligned} Ai(z) &= \frac{1}{\pi}\int_0^{\infty} \cos\left(\frac{t^3}{3} + zt\right) dt \\ &= \frac{1}{2\pi}\int_{-\infty}^{\infty} \cos\left(\frac{t^3}{3} + zt\right) dt \\ &= \frac{1}{2\pi}\int_{-\infty}^{\infty} e^{i\left(\frac{t^3}{3}+zt\right)} dt. \end{aligned}$$

Putting $it = s$ or $t = -is$, $dt = -\,ids$, then, we obtain

$$\begin{aligned} Ai(z) &= \frac{1}{2\pi i}\int_{-i\infty}^{+i\infty} e^{zs-\frac{s^3}{3}} ds \\ &= \frac{1}{2\pi i}\int_I e^{zs-\frac{s^3}{3}} ds \end{aligned} \tag{8.267}$$

where I is the path of integration along the imaginary axis from $-\,i\infty$ to $+\,i\infty$. Putting $z = v^2$, we write

$$Ai(v^2) = \frac{1}{2\pi i}\int_I e^{v^2 s - \frac{1}{3}s^3} ds, \; v > 0. \tag{8.268}$$

In the complex s-plane, the point $s = -v$ corresponds to a saddle point and, by Cauchy's integral theorem, the path of integration I (the imaginary axis) can be deformed into a parallel line through $s = -v$ if we are able to show that the line integral along the straight line from $s = -v + it$ to $s = it$ tends to zero when $t \to \pm\infty$. To show this, let us write $s = \sigma + it$ and consider the integral

$$\int_{-v}^{0} e^{v^2(\sigma+it) - \frac{1}{3}(\sigma+it)^3} d\sigma.$$

The absolute value of this integral will not exceed

$$\int_{-v}^{0} e^{v^2\sigma-\frac{1}{3}\sigma^3+\sigma t^2}\,d\sigma \le e^{\frac{1}{3}v^3}\int_{-v}^{0} e^{\sigma t^2}\,d\sigma < \frac{e^{\frac{1}{3}v^3}}{t^2}$$

which tends to zero as $t\to\pm\infty$.

For $v>0$ putting $s=-v+it$, we write Eq. (8.268) as

$$Ai(v^2)=\frac{e^{-\frac{2}{3}v^3}}{2\pi}\int_{-\infty}^{+\infty} e^{-vt^2+\frac{1}{3}it^3}\,dt$$

$$=\frac{e^{-\frac{2}{3}v^3}}{2\pi}\int_{-\infty}^{+\infty} e^{-vt^2}\cos\left(\frac{t^3}{3}\right)dt+i\frac{e^{-\frac{2}{3}v^3}}{2\pi}\int_{-\infty}^{+\infty} e^{-vt^2}\sin\left(\frac{t^3}{3}\right)dt$$

$$=\frac{e^{-\frac{2}{3}v^3}}{\pi}\int_{0}^{\infty} e^{-vt^2}\cos\left(\frac{t^3}{3}\right)dt. \qquad (8.269)$$

We see that when v is complex, the above integral uniformly converges on any compact set in $|v|>0, |\arg v|<\frac{\pi}{2}$. It follows that Eq. (8.269) do hold in the upper half plane $(\operatorname{Re} v>0)$.

Equation (8.269) can be written as

$$Ai(v^2)=\frac{e^{-\frac{2}{3}v^3}}{\pi}\int_{0}^{\infty} e^{-vt^2}\left[\sum_{n=0}^{\infty}\frac{(-1)^n\,t^{6n}}{(2n)!\,3^{2n}}\right]dt.$$

Putting $\sqrt{v}\,t=u$ we write

$$Ai(v^2)=\frac{e^{-\frac{2}{3}v^3}}{2\pi}\sum_{n=0}^{\infty}\left[\frac{(-1)^n}{(2n)!\,3^{2n}\,v^{3n+\frac{1}{2}}}2\int_{0}^{\infty} e^{-u^2}u^{6n}\,du\right]$$

$$=\frac{e^{-\frac{2}{3}v^3}}{2\pi}\sum_{n=0}^{\infty}\frac{(-1)^n\Gamma\left(3n+\frac{1}{2}\right)}{(2n)!3^{2n}\,v^{3n+\frac{1}{2}}}. \qquad (8.270)$$

Restoring the original variable z, we write

$$Ai(z) = \frac{e^{-\frac{2}{3}z^{3/2}}}{2\pi\, z^{1/4}} \sum_{n=0}^{\infty} \frac{(-1)^n \Gamma\left(3n+\frac{1}{2}\right)}{(2n)!3^{2n} z^{3n/2}} \text{ as } |z| \to \infty \text{ with } \arg z < \pi. \tag{8.271}$$

When z is real such as $z = x$, then in the limit $x \to \infty$, retaining the leading term with $n = 0$, we write the asymptotic form as

$$\lim_{x\to\infty} Ai(x) \sim \frac{e^{-\frac{2}{3}x^{3/2}}}{2\sqrt{\pi}\, x^{1/4}}. \tag{8.272}$$

8.26 GRAPHS OF BESSEL FUNCTIONS OF VARIOUS KINDS

(i) Graphs of Bessel Functions of the First Kind

The graphs of Bessel functions of the first kind roughly look like those of damply oscillating sine or cosine functions although their roots are not generally periodic, except asymptotically for large x. Graphs of the first four members of the Bessel function of the first kind having non-negative integral orders are shown in Figure 8.14(a). Next, graphs of the first four members of the Bessel function of the first kind having negative integral orders are shown in Figure 8.14(b). Graphs of four members of the Bessel function of the first kind having half integral orders are shown in Figure 8.14(c).

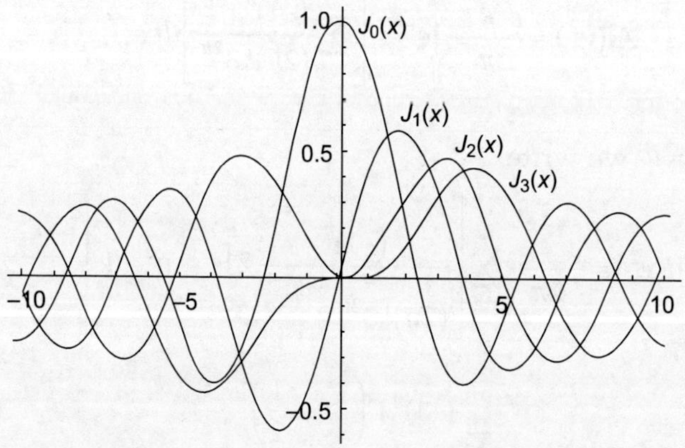

Fig. 8.14(a) Graphs of Bessel functions of the first kind $J_n(x)$ for $n = 0, 1, 2, 3$ in the interval from $x = -10$ to $x = 10$.

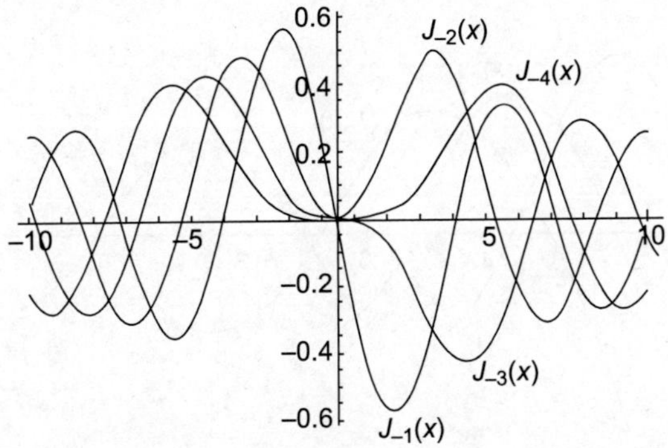

Fig. 814(b) Graphs of Bessel functions of the first kind $J_n(x)$ for $n = -1, -2, -3, -4$. Note that $J_{-n}(x)$ is equal to $J_n(x)$ if n is an even integer and $-J_n(x)$ if n is an odd integer.

Graphs of Bessel functions of the first kind $J_p(x)$ for $p = \pm\frac{1}{2}, \pm\frac{3}{2}$ are shown in Figure 8.14(c). Before looking at Figure 8.14(c), recall the formulae, viz.

$$J_{-\frac{1}{2}}(x) = \sqrt{\frac{2}{\pi x}} \cos x, \quad J_{\frac{1}{2}}(x) = \sqrt{\frac{2}{\pi x}} \sin x$$

$$J_{-\frac{3}{2}}(x) = \sqrt{\frac{2}{\pi x}} \left[-\sin x - \frac{\cos x}{x} \right], \quad J_{\frac{3}{2}}(x) = \sqrt{\frac{2}{\pi x}} \left[\frac{\sin x}{x} - \cos x \right].$$

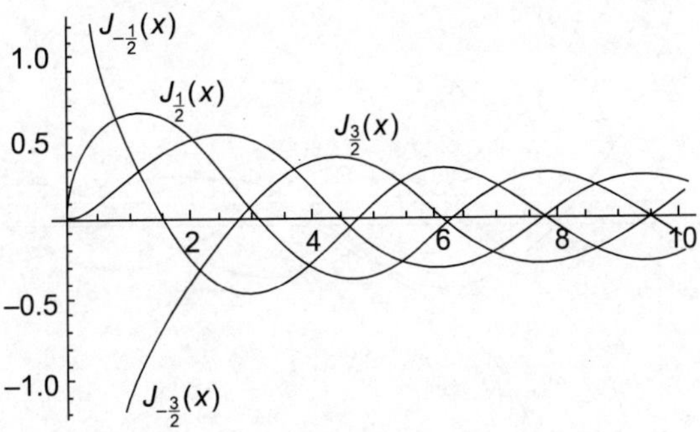

Fig. 8.14(c) Graphs of Bessel functions of the first kind $J_p(x)$ for $p = \pm\frac{1}{2}, \pm\frac{3}{2}$.

(ii) Graphs of Bessel Functions of the Second Kind or Neumann Functions

In Figure 8.15(a), graphs of Bessel functions of the second kind or Neumann functions $N_n(x)$ for $n = 0, 1, 2$, are shown.

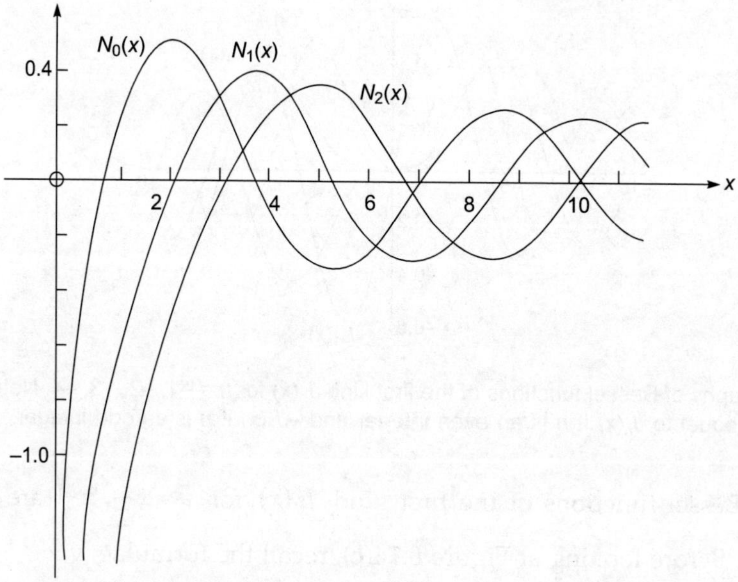

Fig. 8.15(a) Graphs of Bessel functions of the second kind or Neumann function $N_n(x)$ for $n = 0, 1, 2$.

In Figure 8.15(b), graphs of Bessel functions of the second kind or Neumann functions $N_n(x)$ for $n = -1, -2, -3, -4$ are shown. Note that $N_{-n}(x)$ is equal to $N_n(x)$ if n is an even integer and $-N_n(x)$ if n is an odd integer.

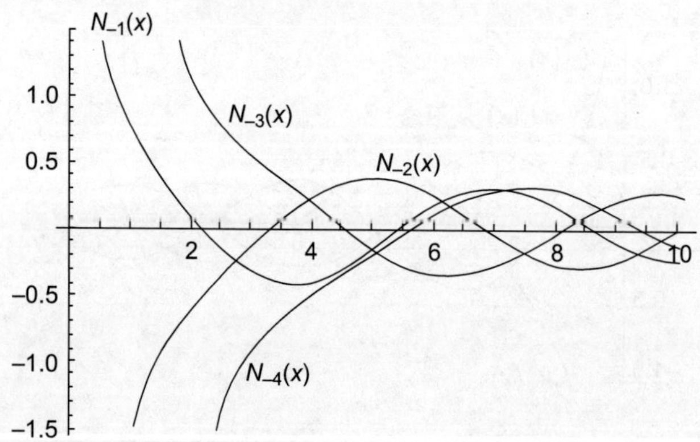

Fig. 8.15(b) Graphs of Bessel functions of the second kind or Neumann functions $N_n(x)$ for $n = -1, -2, -3, -4$.

In Figure 8.15(c), graphs of Bessel functions of the second kind or Neumann functions $N_n(x)$ for $n = \pm\frac{1}{2}, \pm\frac{3}{2}$ are shown. In looking at these graphs, one should note that

$$N_{-\frac{1}{2}}(x) = J_{\frac{1}{2}}(x) = \sqrt{\frac{2}{\pi x}}\sin x,$$

$$N_{\frac{1}{2}}(x) = -J_{-\frac{1}{2}}(x) = -\sqrt{\frac{2}{\pi x}}\cos x,$$

$$N_{\frac{3}{2}}(x) = J_{-\frac{3}{2}}(x) = \sqrt{\frac{2}{\pi x}}\left(-\sin x - \frac{\cos x}{x}\right)$$

and

$$N_{-\frac{3}{2}}(x) = -J_{\frac{3}{2}}(x) = \sqrt{\frac{2}{\pi x}}\left(\cos x - \frac{\sin x}{x}\right).$$

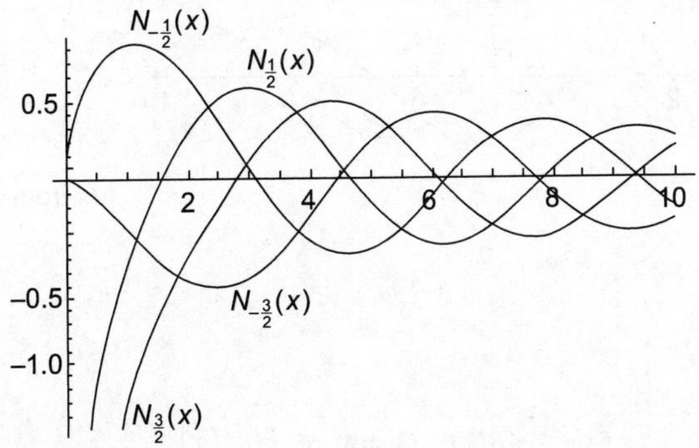

Fig. 8.15(c) Graphs of Bessel functions of the second kind or Neumann functions $N_n(x)$ for $n = \pm\frac{1}{2}, \pm\frac{3}{2}$ are shown.

(iii) Graphs of Bessel Functions of the Third Kind or Hankel Functions

(A) Graphs of Hankel functions of the first kind, symbolized by

$$H_0^{(1)}(x), H_1^{(1)}(x), H_2^{(1)}(x) \text{ and } H_3^{(1)}(x)$$

are shown in Figure 8.16. For each of the functions, the real and the imaginary parts are separately plotted.

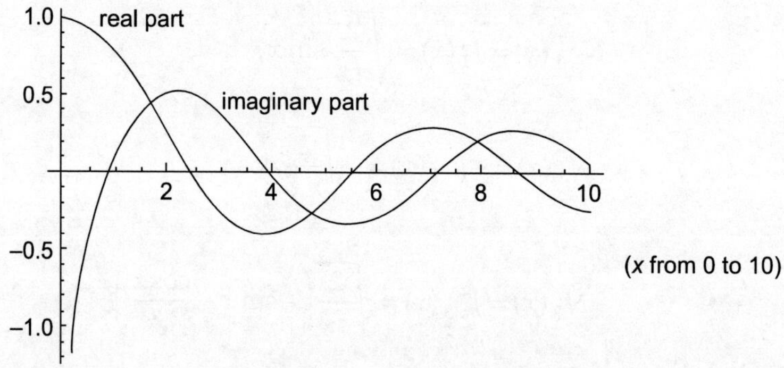

Fig. 8.16(a) Graph of $H_0^{(1)}(x)$.

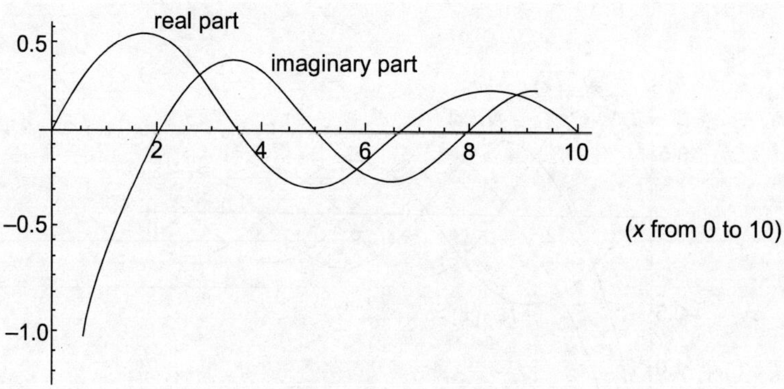

Fig. 8.16(b) Graph of $H_1^{(1)}(x)$.

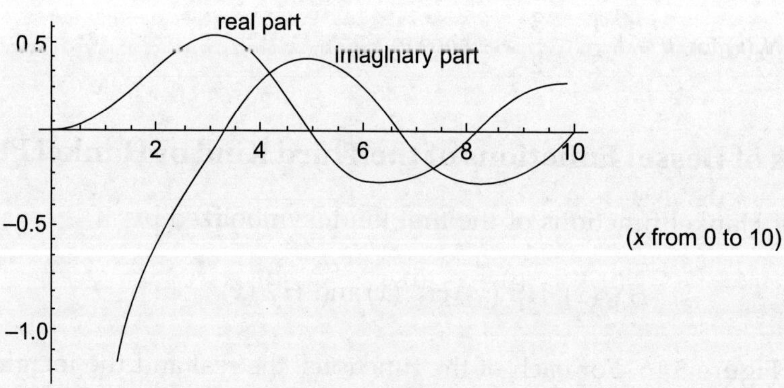

Fig. 8.16(c) Graph of $H_2^{(1)}(x)$.

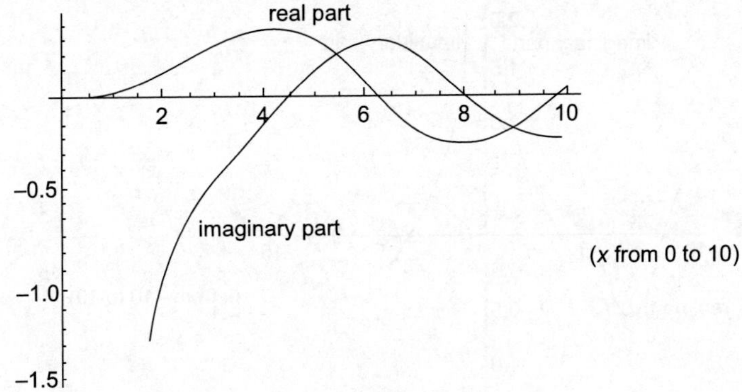

Fig. 8.16(d) Graph of $H_3^{(1)}(x)$.

(B) Graphs of Hankel functions of the second kind, symbolized by $H_0^{(2)}(x), H_1^{(2)}(x)$, $H_2^{(2)}(x)$ and $H_3^{(2)}(x)$ are shown in Figure 8.17.

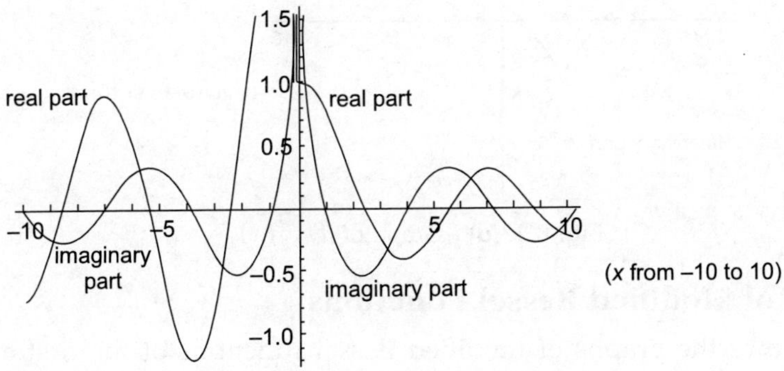

Fig. 8.17(a) Graph of $H_0^{(2)}(x)$.

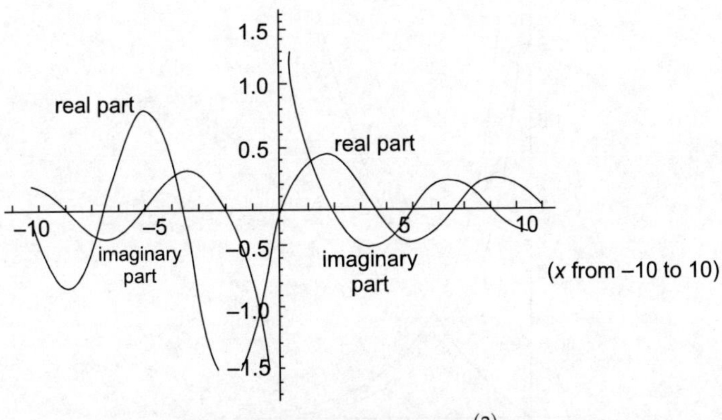

Fig. 8.17(b) Graph of $H_1^{(2)}(x)$.

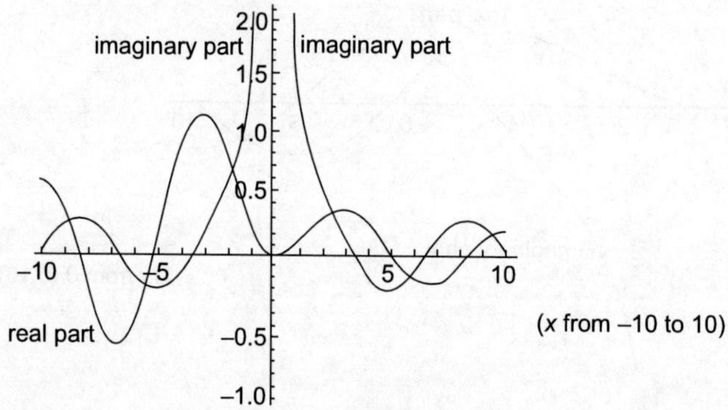

Fig. 8.17(c) Graph of $H_2^{(2)}(x)$.

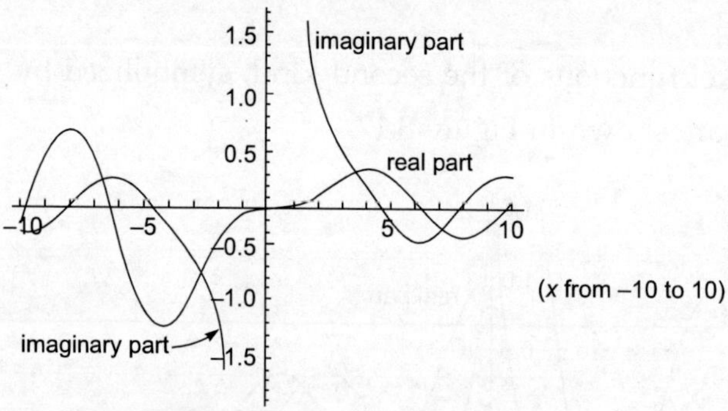

Fig. 8.17(d) Graph of $H_3^{(2)}(x)$.

(iv) Graphs of Modified Bessel Functions

Figure 8.18 shows the graphs of modified Bessel functions of the first and second kinds.

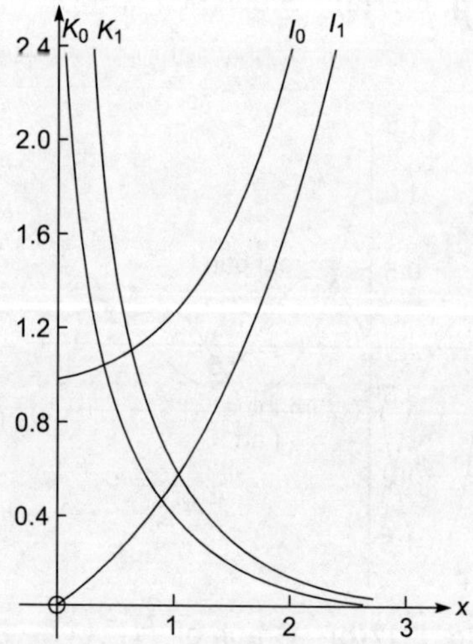

Fig. 8.18 Graphs of modified Bessel functions of the first kind $I_p(x)$ and the second kind $K_p(x)$ for p = 0, 1.

(v) Graphs of Spherical Bessel Functions

(a) Graphs of spherical Bessel functions of the first kind $j_n(x)$ for $n = 0, 1, 2$ are shown in Figure 8.19(a).

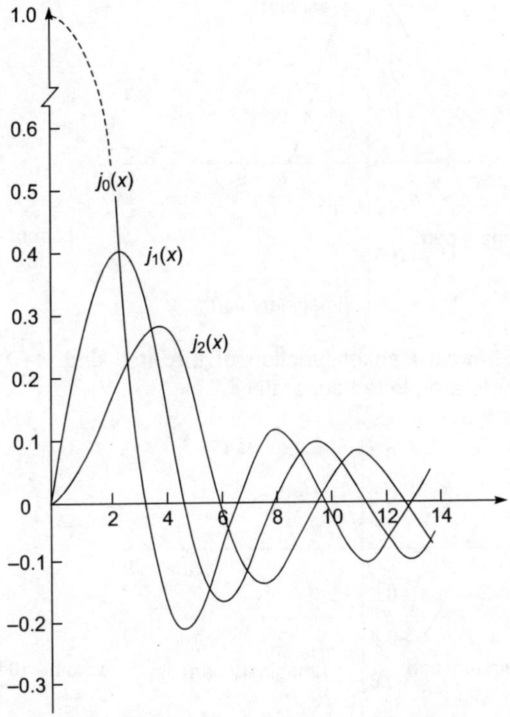

Fig. 8.19(a) Graphs of spherical Bessel functions of the first kind $j_n(x)$ for $n = 0, 1, 2$.

(b) Graphs of spherical Bessel functions of the second kind $y_n(x)$ (also known as spherical Neumann functions $n_n(x)$) are shown in Figure 8.19(b).

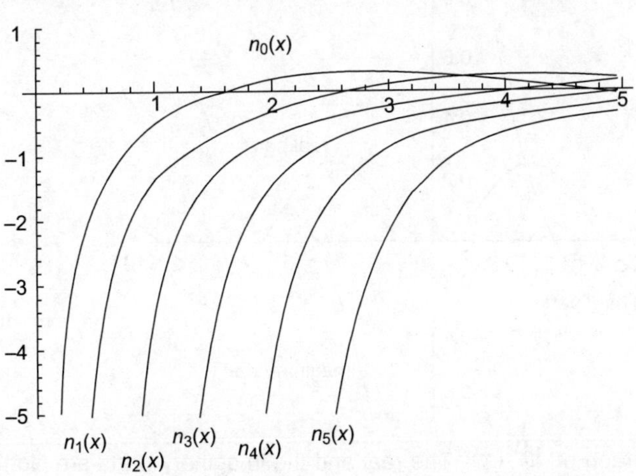

Fig. 8.19(b) Graphs of spherical Neumann functions $n_n(x)$ for $n = 0, 1, 2, 3, 4, 5$.

(c) Graphs of spherical Hankel functions of the first kind $h_0^{(1)}(x), h_1^{(1)}(x), h_2^{(1)}(x), h_3^{(1)}(x)$ are shown in Figure 8.19(c)I, II, III, IV.

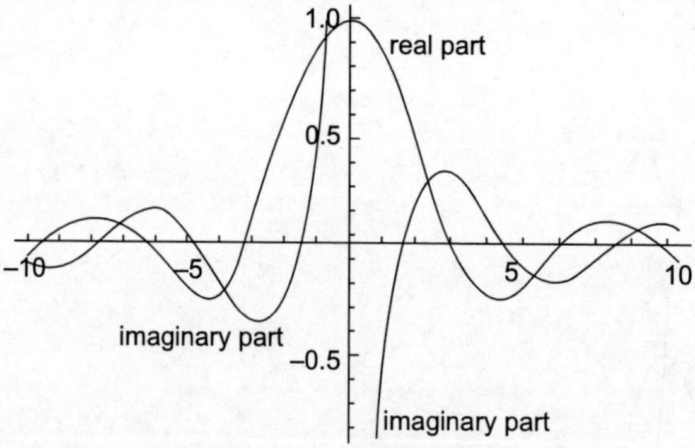

Fig. 8.19(c)I Graph of spherical Hankel function of the first kind $h_0^{(1)}(x)$. The real and the imaginary parts are plotted separately.

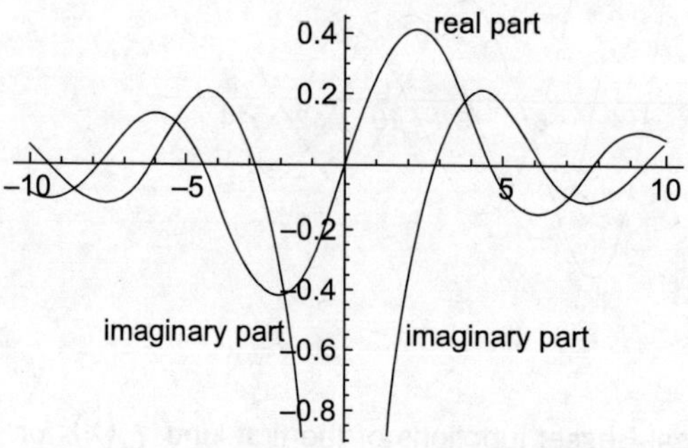

Fig. 8.19(c) II Graph of $h_1^{(1)}(x)$. The real and the imaginary parts are plotted separately.

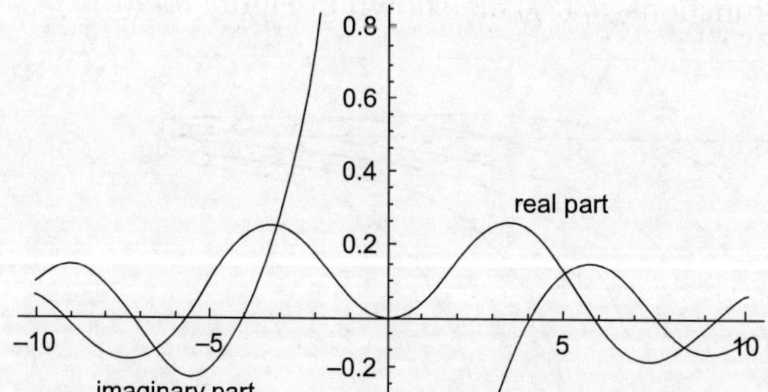

Fig. 8.19(c) III Graph of $h_2^{(1)}(x)$. The real and the imaginary parts are plotted separately.

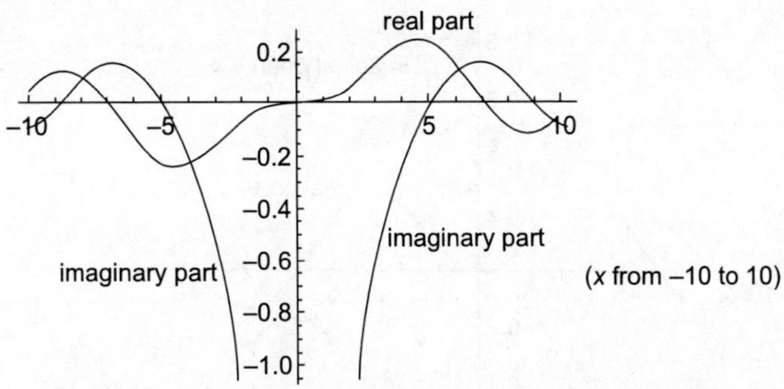

Fig. 8.19(c) IV Graph of $h_3^{(1)}(x)$. The real and the imaginary parts are plotted separately.

(d) Graphs of spherical Hankel functions of the second kind $h_n^{(2)}(x)$ for $n = 0, 1, 2, 3$ are shown in Figure 8.19(d).

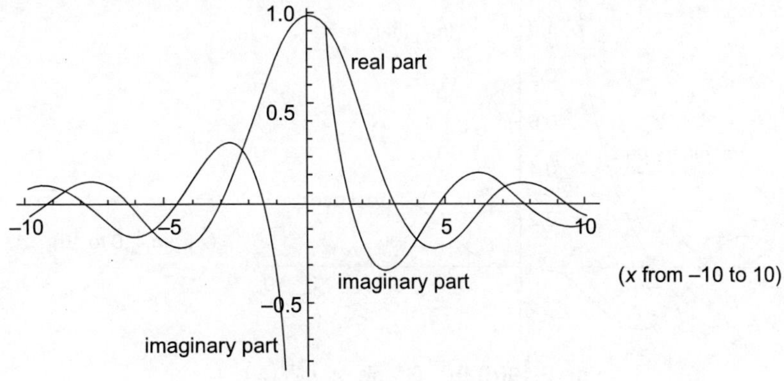

Fig. 8.19(d) I Graph of $h_0^{(2)}(x)$.

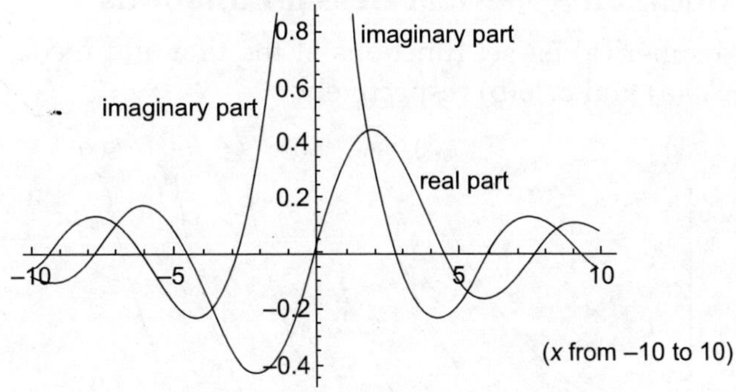

Fig. 8.19(d) II Graph of $h_1^{(2)}(x)$.

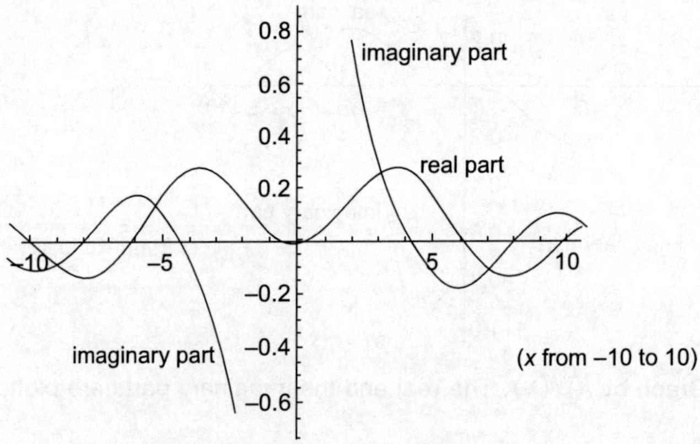

Fig. 8.19(d) III Graph of $h_2^{(2)}(x)$.

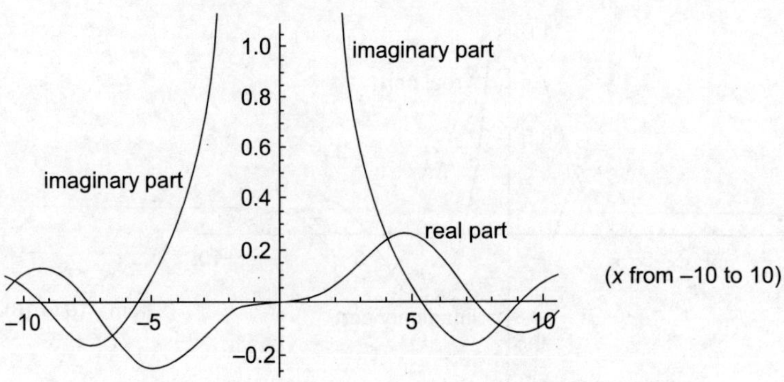

Fig. 8.19(d) IV Graph of $h_3^{(2)}(x)$.

(vi) Graphs of Modified Spherical Bessel Functions

Graphs of modified spherical Bessel functions of the first and the second kinds are shown in Figures 8.20(a) and 8.20(b) respectively.

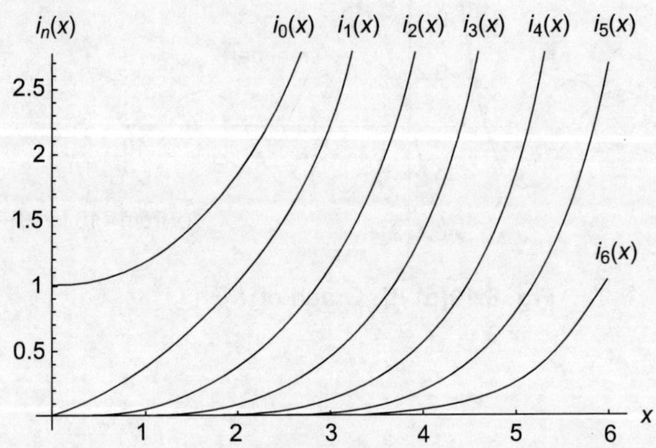

Fig. 8.20(a) Graphs of modified spherical Bessel functions of the first kind $i_l(x)$ for $l = 0, 1, 2, \ldots, 6$.

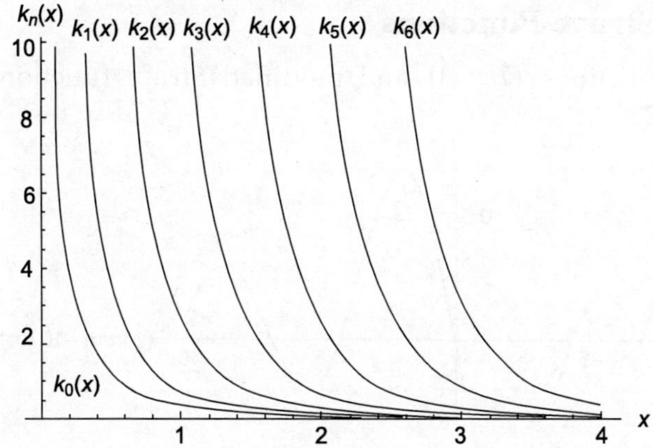

Fig. 8.20(b) Graphs of modified spherical Bessel functions of the second kind $k_l(x)$ for l = 0, 1, 2, ... 6.

(vii) Graphs of Kelvin Functions

Graphs of $Ber_n(x)$ and $Bei_n(x)$ for $n = 0,1,2,3$ are shown in Figures 8.21(a) and 8.21(b) respectively.

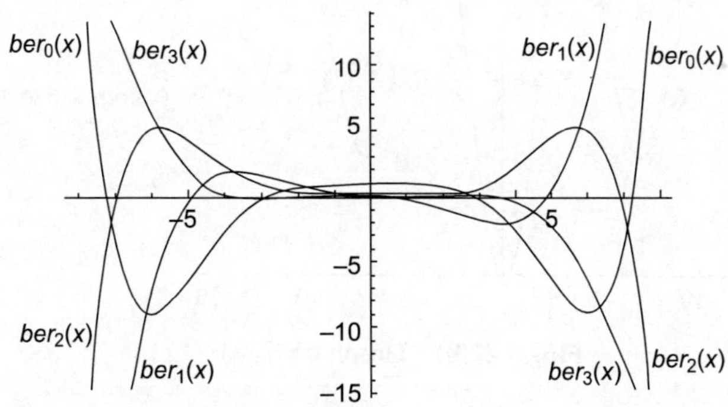

Fig. 8.21(a) Graphs of $Ber_n(x)$ for $n = 0, 1, 2, 3$.

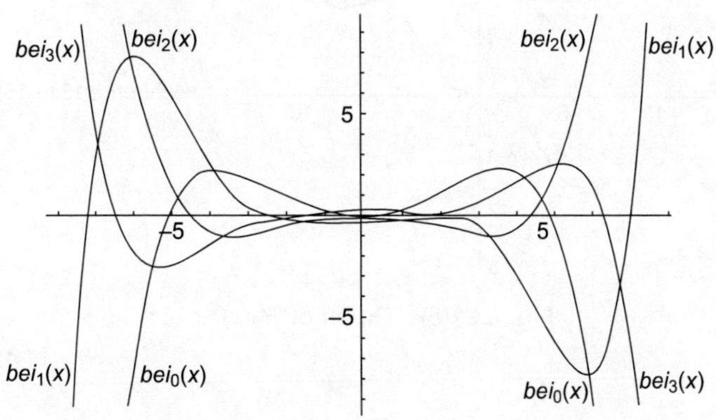

Fig. 8.21(b) Graphs of $Bei_n(x)$ for $n = 0, 1, 2, 3$.

(viii) Graphs of Struve Functions

Graphs of Struve functions $\left(\mathcal{H}_n(x)\right)$ and modified *Struve* functions $\left(\mathbb{L}_{n\mathbb{P}}(x)\right)$ are shown in Figures 8.22.

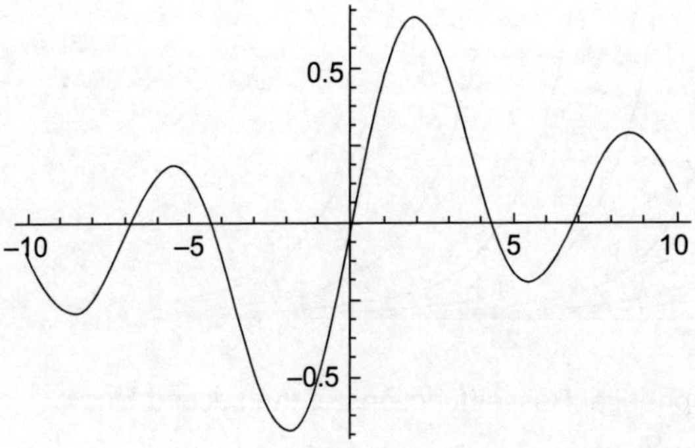

(x from −10 to 10)

Fig. 8.22(a) Graph of $\mathcal{H}_0(x)$.

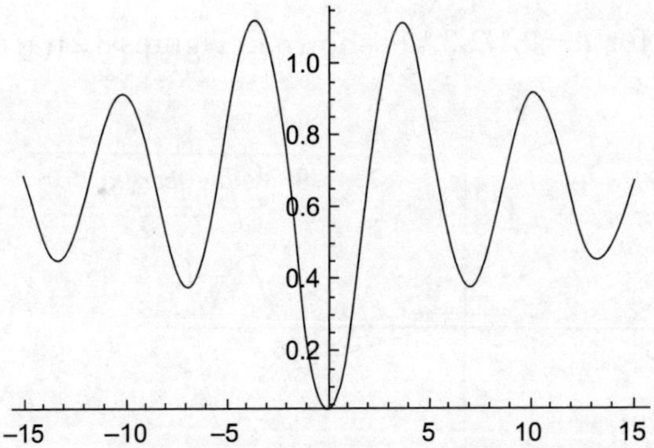

(*x* from −15 to 15)

Fig. 8.22(b) Graph of $\mathcal{H}_1(x)$.

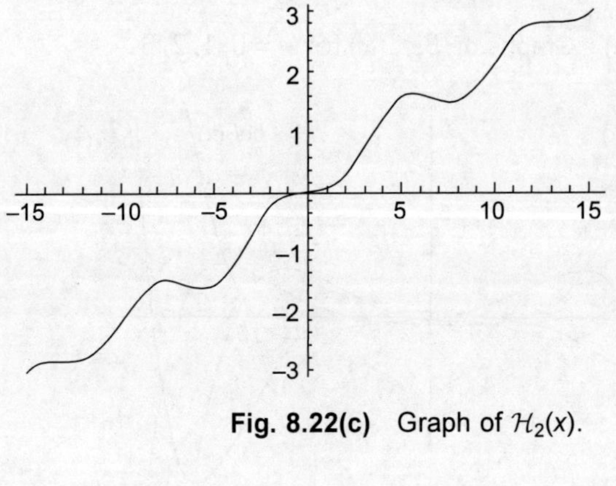

(*x* from −15 to 15)

Fig. 8.22(c) Graph of $\mathcal{H}_2(x)$.

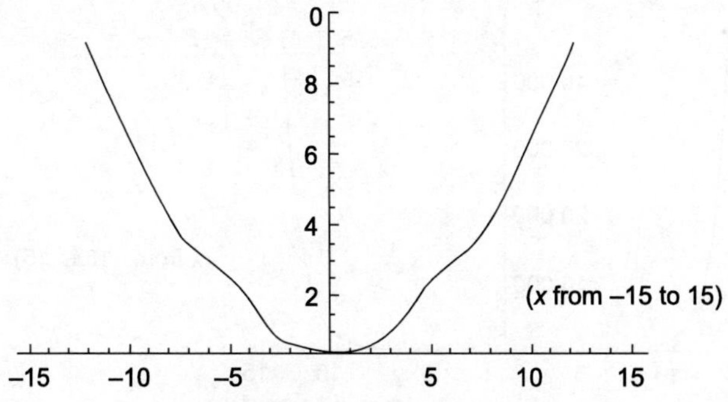

Fig. 8.22(d) Graph of $\mathcal{H}_3(x)$.

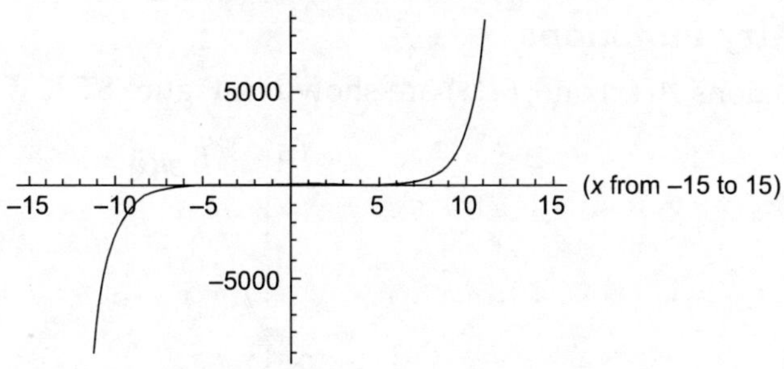

Fig. 8.22(e) Graph of $\mathbb{L}_0(x)$.

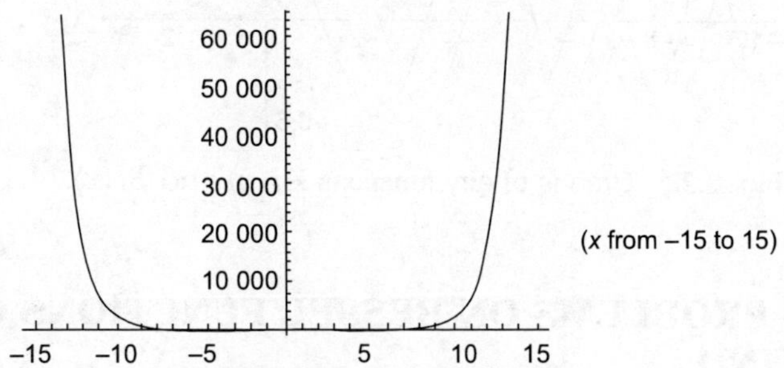

Fig. 8.22(f) Graph of $\mathbb{L}_1(x)$.

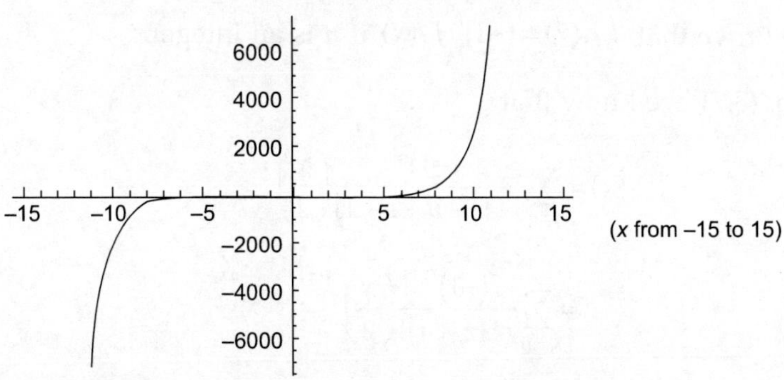

Fig. 8.22(g) Graph of $\mathbb{L}_2(x)$.

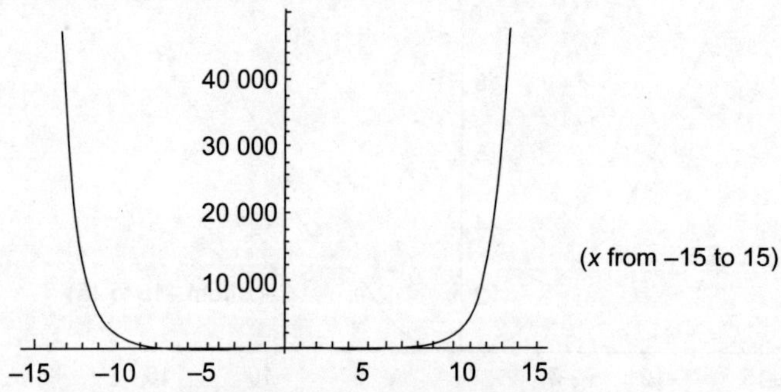

Fig. 8.22(h) Graph of $\mathbb{L}_3(x)$.

(ix) Graphs of Airy Functions

Graphs of Airy functions $Ai(x)$ and $Bi(x)$ are shown in Figure 8.23.

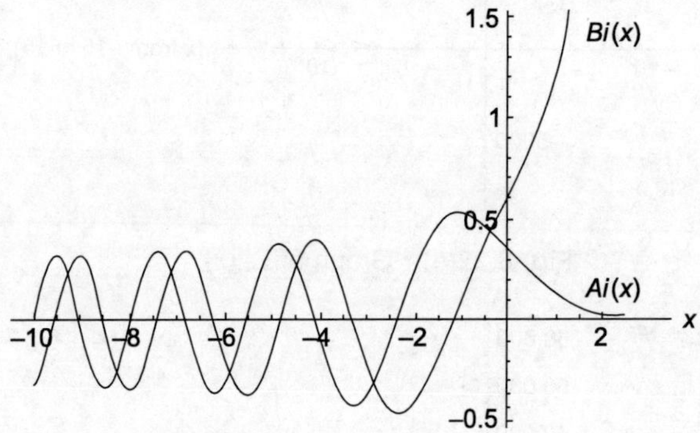

Fig. 8.23 Graphs of Airy functions $Ai(x)$ and $Bi(x)$.

8.27 SOLVED PROBLEMS ON BESSEL FUNCTIONS OF THE FIRST KIND

PROBLEM 1 Prove that $J_{-n}(x) = (-1)^n J_n(x)$ if n is an integer.

Proof: From Eq. (8.7), we know that

$$J_{-n}(x) = \sum_{r=0}^{\infty} \frac{(-1)^r}{r!\,\Gamma(-n+r+1)} \left(\frac{x}{2}\right)^{-n+2r}$$

$$= \sum_{r=0}^{\infty} \frac{(-1)^r}{r!\,(r-n)!} \left(\frac{x}{2}\right)^{-n+2r}$$

$$= \sum_{s=-n}^{\infty} \frac{(-1)^{n+s}}{(n+s)!\,s!}\left(\frac{x}{2}\right)^{n+2s}$$

where we put $s = r - n$,

$$= (-1)^n \sum_{s=0}^{\infty} \frac{(-1)^s}{s!\,(n+s)!}\left(\frac{x}{2}\right)^{n+2s} = (-1)^n J_n(x).$$

PROBLEM 2 Show that $J_n(-x) = (-1)^n J_n(x)$ *for* $n = 0,1,2,3\ldots$

Solution:

First Method: From infinite series formula

We know that

$$J_n(-x) = \sum_{r=0}^{\infty} \frac{(-1)^r}{r!\,\Gamma(n+r+1)}\left(\frac{-x}{2}\right)^{n+2r}$$

$$= (-1)^n \sum_{r=0}^{\infty} \frac{(-1)^r (-1)^{2r}}{r!\,\Gamma(n+r+1)}\left(\frac{x}{2}\right)^{n+2r}$$

$$= (-1)^n J_n(x).$$

Second Method: From generating function

We know that

$$\sum_{n=-\infty}^{\infty} J_n(x)h^n = e^{\frac{x}{2}\left(h-\frac{1}{h}\right)}$$

Replacing x by $-x$, we write

$$\sum_{n=-\infty}^{\infty} J_n(-x)h^n = e^{-\frac{x}{2}\left(h-\frac{1}{h}\right)}$$

$$= e^{\frac{x}{2}\left\{(-h)-\frac{1}{(-h)}\right\}}$$

$$= \sum_{n=-\infty}^{\infty} J_n(x)(-h)^n$$

$$= \sum_{n=-\infty}^{\infty} (-1)^n J_n(x)h^n.$$

Equating coefficients of h^n on both sides of the above equation, we obtain

$$J_n(-x) = (-1)^n J_n(x).$$

PROBLEM 3 Prove that

(i) $J_n(x+y)=\sum_{r=-\infty}^{\infty} J_r(x)J_{n-r}(y)$

(ii) $J_0(2x)=J_0^2(x)+2\sum_{r=1}^{\infty}(-1)^r J_r^2(x)$

(iii) $J_0(0)=1$ and $J_0(\infty)=0$.

Proof:

(i) Using the generating function, we can write

$$\sum_{n=-\infty}^{\infty} J_n(x+y)h^n = e^{\frac{(x+y)}{2}\left(h-\frac{1}{h}\right)}$$

$$= e^{\frac{x}{2}\left(h-\frac{1}{h}\right)} \cdot e^{\frac{y}{2}\left(h-\frac{1}{h}\right)}$$

$$= \sum_{r=-\infty}^{\infty} J_r(x)h^r \sum_{n=-\infty}^{\infty} J_n(y)h^n$$

$$= \sum_{r=-\infty}^{\infty}\sum_{n=-\infty}^{\infty} h^{n+r} J_r(x)J_n(y).$$

Equating coefficients of h^n from both sides of the above equation, we obtain

$$J_n(x+y)=\sum_{r=-\infty}^{\infty} J_r(x)J_{n-r}(y).$$

(ii) Putting $n=0$ and $x=y$ in the above result, we have

$$J_0(2x)=\sum_{r=-\infty}^{\infty} J_r(x)J_{-r}(x)$$

$$= \sum_{r=-\infty}^{-1} J_r(x)J_{-r}(x)+J_0^2(x)+\sum_{r=1}^{\infty} J_r(x)J_{-r}(x)$$

$$= J_0^2(x)+2\sum_{r=1}^{\infty} J_r(x)J_{-r}(x)$$

$$= J_0^2(x)+2\sum_{r=1}^{\infty}(-1)^r J_r^2(x).$$

(iii) We know that

$$J_0(x)=\sum_{r=0}^{\infty}\frac{(-1)^r}{(r!)^2}\left(\frac{x}{2}\right)^{2r}=1-\frac{x^2}{(1!)^2 2^2}+\frac{x^4}{(2!)^2 2^4}-\ldots\infty .$$

Therefore, $J_0(0)=1$.

Also, for large x, the asymptotic formula of Bessel functions is written as

$$J_n(x)\sim\sqrt{\frac{2}{\pi x}}\cos\left\{x-\left(n+\frac{1}{2}\right)\frac{\pi}{2}\right\}.$$

Therefore $J_n(\infty)=0$ for $n>0$.

PROBLEM 4 Show that for $k\neq 0, t\neq 0$

$$e^{-\frac{x}{2t}\left(k-\frac{1}{k}\right)}\sum_{n=-\infty}^{\infty}J_n(x)k^n t^n=\sum_{n=-\infty}^{\infty}J_n(kx)t^n$$

Solution: We know that

$$e^{-\frac{x}{2t}\left(k-\frac{1}{k}\right)}=e^{-\frac{x}{2}\left(\frac{k}{t}-\frac{1}{kt}\right)}$$

$$=e^{-\frac{x}{2}\left(kt-\frac{1}{kt}-kt+\frac{k}{t}\right)}$$

$$=e^{-\frac{x}{2}\left(kt-\frac{1}{kt}\right)}e^{\frac{kx}{2}\left(t-\frac{1}{t}\right)}.$$

Therefore, we write

$$e^{-\frac{x}{2t}\left(k-\frac{1}{k}\right)}e^{\frac{x}{2}\left(kt-\frac{1}{kt}\right)}=e^{\frac{kx}{2}\left(t-\frac{1}{t}\right)}$$

or

$$e^{-\frac{x}{2t}\left(k-\frac{1}{k}\right)}\sum_{n=-\infty}^{\infty}J_n(x)k^n t^n=\sum_{n=-\infty}^{\infty}J_n(kx)t^n .$$

PROBLEM 5 Obtain expressions for $J_{-\frac{1}{2}}(x), J_{\frac{1}{2}}(x), J_{-\frac{3}{2}}(x)$ and $J_{\frac{3}{2}}(x)$.

Solution: We know that $J_n(x)=\sum_{r=0}^{\infty}\frac{(-1)^r}{r!\,\Gamma(n+r+1)}\left(\frac{x}{2}\right)^{n+2r}$.

(i) Putting $n = -\frac{1}{2}$, we write

$$J_{-\frac{1}{2}}(x) = \sum_{r=0}^{\infty} \frac{(-1)^r}{r!\,\Gamma\left(r+\frac{1}{2}\right)}\left(\frac{x}{2}\right)^{-\frac{1}{2}+2r}$$

$$= \frac{\left(\frac{x}{2}\right)^{-\frac{1}{2}}}{\Gamma\left(\frac{1}{2}\right)} - \frac{\left(\frac{x}{2}\right)^{\frac{3}{2}}}{1!\,\Gamma\left(\frac{3}{2}\right)} + \frac{\left(\frac{x}{2}\right)^{\frac{7}{2}}}{2!\,\Gamma\left(\frac{5}{2}\right)} - \frac{\left(\frac{x}{2}\right)^{\frac{11}{2}}}{3!\,\Gamma\left(\frac{7}{2}\right)} + \ldots \infty$$

$$= \frac{\left(\frac{x}{2}\right)^{-\frac{1}{2}}}{\Gamma\left(\frac{1}{2}\right)}\left[1 - \frac{\left(\frac{x}{2}\right)^{2}}{1!\,\frac{1}{2}} + \frac{\left(\frac{x}{2}\right)^{4}}{2!\,\frac{1}{2}\cdot\frac{3}{2}} - \frac{\left(\frac{x}{2}\right)^{6}}{3!\,\frac{1}{2}\cdot\frac{3}{2}\cdot\frac{5}{2}} + \ldots \infty\right]$$

$$= \sqrt{\frac{2}{\pi x}}\left[1 - \frac{x^2}{2!} + \frac{x^4}{4!} - \frac{x^6}{6!} + \ldots \infty\right]$$

$$= \sqrt{\frac{2}{\pi x}}\cos x.$$

(ii) Similarly, putting $n = \frac{1}{2}$, we write

$$J_{\frac{1}{2}}(x) = \sum_{r=0}^{\infty} \frac{(-1)^r}{r!\,\Gamma\left(r+\frac{3}{2}\right)}\left(\frac{x}{2}\right)^{\frac{1}{2}+2r}$$

$$= \frac{\left(\frac{x}{2}\right)^{\frac{1}{2}}}{\Gamma\left(\frac{3}{2}\right)} - \frac{\left(\frac{x}{2}\right)^{\frac{5}{2}}}{1!\,\Gamma\left(\frac{5}{2}\right)} + \frac{\left(\frac{x}{2}\right)^{\frac{9}{2}}}{2!\,\Gamma\left(\frac{7}{2}\right)} - \frac{\left(\frac{x}{2}\right)^{\frac{13}{2}}}{3!\,\Gamma\left(\frac{9}{2}\right)} + \ldots \infty$$

$$= \frac{\left(\frac{x}{2}\right)^{-\frac{1}{2}}}{\Gamma\left(\frac{1}{2}\right)}\left[\frac{\frac{x}{2}}{\frac{1}{2}} - \frac{\left(\frac{x}{2}\right)^{3}}{1!\,\frac{1}{2}\cdot\frac{3}{2}} + \frac{\left(\frac{x}{2}\right)^{5}}{2!\,\frac{1}{2}\cdot\frac{3}{2}\cdot\frac{5}{2}} - \frac{\left(\frac{x}{2}\right)^{7}}{3!\,\frac{1}{2}\cdot\frac{3}{2}\cdot\frac{5}{2}\cdot\frac{7}{2}} + \ldots \infty\right]$$

$$= \sqrt{\frac{2}{\pi x}}\left[x - \frac{x^3}{3!} + \frac{x^5}{5!} - \frac{x^7}{7!} + \ldots \infty\right] = \sqrt{\frac{2}{\pi x}} \sin x.$$

(iii) In the recurrence relation

$$J_{n-1}(x) + J_{n+1}(x) = \frac{2n}{x} J_n(x)$$

if n is replaced by $-\frac{1}{2}$, we write

$$J_{-\frac{3}{2}}(x) + J_{\frac{1}{2}}(x) = -\frac{1}{x} J_{-\frac{1}{2}}(x)$$

or
$$J_{-\frac{3}{2}}(x) = -J_{\frac{1}{2}}(x) - \frac{1}{x} J_{-\frac{1}{2}}(x)$$

$$= -\sqrt{\frac{2}{\pi x}} \sin x - \frac{1}{x}\sqrt{\frac{2}{\pi x}} \cos x$$

$$= \sqrt{\frac{2}{\pi x}}\left(-\frac{\cos x}{x} - \sin x\right).$$

(iv) Similarly, replacing n by $\frac{1}{2}$, we have

$$J_{-\frac{1}{2}}(x) + J_{\frac{3}{2}}(x) = \frac{1}{x} J_{\frac{1}{2}}(x)$$

or
$$J_{\frac{3}{2}}(x) = \frac{1}{x} J_{\frac{1}{2}}(x) - J_{-\frac{1}{2}}(x)$$

$$= \sqrt{\frac{2}{\pi x}}\left(\frac{\sin x}{x} - \cos x\right).$$

PROBLEM 6 Prove that $J_0'(x) = -J_1(x)$ where the prime denotes differentiation with respect to x.

Proof:

First method: (Using the infinite series formula)

we have $J_n(x) = \sum_{r=0}^{\infty} \frac{(-1)^r}{r!\,\Gamma(n+r+1)}\left(\frac{x}{2}\right)^{n+2r}.$

Therefore, we write

$$J_0(x)=\sum_{r=0}^{\infty}\frac{(-1)^r}{(r!)^2}\left(\frac{x}{2}\right)^{2r}=1-\frac{\left(\frac{x}{2}\right)^2}{(1!)^2}+\frac{\left(\frac{x}{2}\right)^4}{(2!)^2}\quad\frac{\left(\frac{x}{2}\right)^6}{(3!)^2}+\ldots\infty$$

and $\quad J_1(x)=\sum_{r=0}^{\infty}\frac{(-1)^r}{r!\,(r+1)!}\left(\frac{x}{2}\right)^{1+2r}$

$$=\frac{x}{2}-\frac{\left(\frac{x}{2}\right)^3}{1!\,2!}+\frac{\left(\frac{x}{2}\right)^5}{2!\,3!}-\frac{\left(\frac{x}{2}\right)^7}{3!\,4!}+\ldots\infty$$

Differentiating $J_0(x)$ with respect to x, we obtain

$$J_0'(x)=0-\frac{x}{2}+\frac{\left(\frac{x}{2}\right)^3}{1!\,2!}-\frac{\left(\frac{x}{2}\right)^5}{2!\,3!}+\ldots\infty=-J_1(x).$$

Second method: (Using recurrence relations)

We have $\frac{d}{dx}\left[x^n J_n(x)\right]=x^n J_{n-1}(x)$

Taking $n = 0$, we write

$$\frac{d}{dx}J_0(x)=J_{-1}(x)=(-1)^1 J_1(x)=-J_1(x).$$

Alternatively, we have

$$\frac{d}{dx}\left[x^{-n}J_n(x)\right]=-x^n J_{n+1}(x)$$

Putting $n=0$, we write

$$\frac{d}{dx}J_0(x)=-J_1(x).$$

In another alternative method, we have

$$2J_n'(x)=J_{n-1}(x)-J_{n+1}(x)$$

Putting n = 0, we have

$$2J_0'(x)=J_{-1}(x)-J_1(x)=-J_1(x)-J_1(x)$$

Therefore, we have

$$J_0'(x)=-J_1(x).$$

PROBLEM 7 Using appropriate recurrence relation for Bessel functions, evaluate

$$\int x^4 J_1(x)dx.$$

Solution: Let us consider the recurrence relation

$$\frac{d}{dx}\left[x^n J_n(x)\right] = x^n J_{n-1}(x).$$

Integrating both sides with respect to x, we obtain

$$\int x^n J_{n-1}(x)dx = x^n J_n(x) + C \text{ (constant)}.$$

Therefore, the given integral can be written as

$$\int x^4 J_1(x)dx = \int x^2\left\{x^2 J_1(x)\right\}dx.$$

Integration by parts yields

$$\int x^4 J_1(x)dx = x^2\int x^2 J_1(x)dx - \int 2x\left\{\int x^2 J_1(x)dx\right\}dx + C$$

$$= x^2\left[x^2 J_2(x)\right] - 2\int x\left\{x^2 J_2(x)\right\}dx + C$$

$$= x^4 J_2(x) - 2\int x^3 J_2(x)dx + C$$

$$= x^4 J_2(x) - 2x^3 J_3(x) + C.$$

PROBLEM 8 Prove that $4J_n''(x) = J_{n-2}(x) - 2J_n(x) + J_{n+2}(x)$ where the primes denote differentiation with respect to x.

Proof: Let us consider the recurrence relation

$$2J_n'(x) = J_{n-1}(x) - J_{n+1}(x). \quad \text{...(i)}$$

Differentiating both sides with respect to x, we obtain

$$2J_n''(x) = J_{n-1}'(x) - J_{n+1}'(x)$$

or

$$4J_n''(x) = 2J_{n-1}'(x) - 2J_{n+1}'(x). \quad \text{...(ii)}$$

Replacing n by $n-1$ and $n+1$ respectively, Eq. (i) yields

$$2J_{n-1}'(x) = J_{n-2}(x) - J_n(x) \quad \text{...(iii)}$$

and

$$2J_{n+1}'(x) = J_n(x) - J_{n+2}(x). \quad \text{...(iv)}$$

Substituting Eqs. (iii) and (iv) into Eq. (ii), we obtain

$$4J_n''(x) = J_{n-2}(x) - 2J_n(x) + J_{n+2}(x).$$

PROBLEM 9 Prove that

(a) $\dfrac{d}{dx}\left[J_n^2(x)+J_{n+1}^2(x)\right]=\dfrac{2}{x}\left[nJ_n^2(x)-(n+1)J_{n+1}^2(x)\right],$

(b) $\dfrac{d}{dx}\left[xJ_n(x)J_{n+1}(x)\right]=x\left[J_n^2(x)-J_{n+1}^2(x)\right].$

Proof: Let us consider the recurrence relations

$$J_{n-1}(x)+J_{n+1}(x)=\frac{2n}{x}J_n(x)$$

and

$$J_{n-1}(x)-J_{n+1}(x)=2J_n'(x).$$

Subtraction and addition of the two relations respectively yield

$$J_n'(x)=\frac{n}{x}J_n(x)-J_{n+1}(x) \qquad \text{...(i)}$$

and

$$J_n'(x)=J_{n-1}(x)-\frac{n}{x}J_n(x). \qquad \text{...(ii)}$$

From Eq. (ii), replacing n by $n+1$, we obtain

$$J_{n+1}'(x)=J_n(x)-\frac{(n+1)}{x}J_{n+1}(x) \qquad \text{...(iii)}$$

(a) L.H.S. $=\dfrac{d}{dx}\left[J_n^2(x)+J_{n+1}^2(x)\right]$

$$=2J_n(x)J_n'(x)+2J_{n+1}(x)J_{n+1}'(x). \qquad \text{...(iv)}$$

Substituting Eqs. (i) and (iii) into Eq. (iv), we obtain

$$\frac{d}{dx}\left[J_n^2(x)+J_{n+1}^2(x)\right]=2J_n(x)\left[\frac{n}{x}J_n(x)-J_{n+1}(x)\right]$$

$$+2J_{n+1}(x)\left[J_n(x)-\frac{(n+1)}{x}J_{n+1}(x)\right]$$

$$=\frac{2}{x}\left[nJ_n^2(x)-(n+1)J_{n+1}^2(x)\right].$$

(b) L.H.S. $=\dfrac{d}{dx}\left[xJ_n(x)J_{n+1}(x)\right]$

$$=J_n(x)J_{n+1}(x)+xJ_n'(x)J_{n+1}(x)+xJ_n(x)J_{n+1}'(x). \qquad \text{...(v)}$$

Substituting Eqs. (i) and (iii) into Eq. (v), we obtain

$$\frac{d}{dx}\left[xJ_n(x)J_{n+1}(x)\right]=J_n(x)J_{n+1}(x)+x\left[\frac{n}{x}J_n(x)-J_{n+1}(x)\right]J_{n+1}(x)$$

$$+xJ_n(x)\left[J_n(x)-\frac{(n+1)}{x}J_{n+1}(x)\right]$$

$$=x\left[J_n^2(x)-J_{n+1}^2(x)\right].$$

PROBLEM 10 Using $J_0(2)=0.22389$ and $J_1(2)=0.57672$, calculate $J_2(2)$ and $J_3(3)$

Solution: In the recurrence relation,

$$J_{n-1}(x)+J_{n+1}(x)=\frac{2n}{x}J_n(x),$$

let us put $n=1$ and $x=2$ to yield

$$J_0(2)+J_2(2)=J_1(2)$$

or

$$J_2(2)=J_1(2)-J_0(2)$$

$$= 0.57672 - 0.22389 = 0.35283.$$

Again putting $n=x=2$, we have

$$J_1(2)+J_3(2)=2J_2(2)$$

or

$$J_3(2)=2J_2(2)-J_1(2)$$

$$= 0.70566 - 0.57672 = 0.12894.$$

PROBLEM 11 Prove that

(a) $J_0(\sqrt{x^2-2xt})=\sum_{s=0}^{\infty}J_s(x)\frac{t^s}{s!}$,

(b) $\left(\frac{x-2t}{x}\right)^{-\frac{p}{2}}J_p\left(\sqrt{x^2-2xt}\right)=\sum_{s=0}^{\infty}J_{p+s}(x)\frac{t^s}{s!}$.

Proof:

(a) We know that $J_n(x)=\sum_{r=0}^{\infty}\frac{(-1)^r}{r!\,\Gamma(n+r+1)}\left(\frac{x}{2}\right)^{n+2r}$.

Therefore, we write

$$J_0\left(\sqrt{x^2-2xt}\right)=\sum_{r=0}^{\infty}\frac{(-1)^r}{(r!)^2}\left(\frac{\sqrt{x^2-2xt}}{2}\right)^{2r}$$

$$=\sum_{r=0}^{\infty}\frac{(-1)^r}{(r!)^2\,2^{2r}}(x^2-2xt)^r$$

$$=\sum_{r=0}^{\infty}\frac{(-1)^r}{(r!)^2\,2^{2r}}\sum_{s=0}^{r}\frac{r!}{s!\,(r-s)!}(x^2)^{r-s}(-2xt)^s$$

$$=\sum_{r=0}^{\infty}\sum_{s=0}^{\infty}\frac{(-1)^{r+s}}{r!\,s!\,(r-s)!}\left(\frac{x}{2}\right)^{2r-s}t^s$$

$$=\sum_{s=0}^{\infty}\sum_{r=0}^{\infty}\frac{(-1)^{r+s}}{r!\,s!\,(r-s)!}\left(\frac{x}{2}\right)^{2r-s}t^s.$$

Putting $r - s = k$, we write

$$J_0\left(\sqrt{x^2-2xt}\right)=\sum_{s=0}^{\infty}\sum_{k=-s}^{\infty}\frac{(-1)^{k+2s}}{(k+s)!\,s!\,k!}\left(\frac{x}{2}\right)^{s+2k}t^s$$

$$=\sum_{s=0}^{\infty}\sum_{k=0}^{\infty}\frac{(-1)^{k}}{k!\,(k+s)!\,s!}\left(\frac{x}{2}\right)^{s+2k}t^s$$

$$=\sum_{s=0}^{\infty}\frac{t^s}{s!}\sum_{k=0}^{\infty}\frac{(-1)^{k}}{k!\,\Gamma(s+k+1)}\left(\frac{x}{2}\right)^{s+2k}$$

$$=\sum_{s=0}^{\infty}J_s(x)\frac{t^s}{s!}.$$

(b) We have

$$\left(\frac{x-2t}{x}\right)^{-\frac{p}{2}}J_p\left(\sqrt{x^2-2xt}\right)$$

$$=(x-2t)^{-\frac{p}{2}}x^{\frac{p}{2}}\sum_{r=0}^{\infty}\frac{(-1)^r}{r!\,\Gamma(p+r+1)}\left(\frac{\sqrt{x^2-2xt}}{2}\right)^{p+2r}$$

$$=\sum_{r=0}^{\infty}\frac{(-1)^r\,x^{p+r}\,(x-2t)^r}{r!\,\Gamma(p+r+1)\,2^{p+2r}}$$

$$=\sum_{r=0}^{\infty}\frac{(-1)^r x^{p+r}}{r!\,\Gamma(p+r+1)2^{p+2r}}\sum_{s=0}^{r}\frac{r!}{s!\,(r-s)!}x^{r-s}(-2t)^s$$

$$=\sum_{s=0}^{r}\sum_{r=0}^{\infty}\frac{(-1)^{r+s}\,t^s}{\Gamma(p+r+1)s!\,(r-s)!}\left(\frac{x}{2}\right)^{p+2r-s}.$$

As in (a), let us put $r-s=k$ to yield

$$\left(\frac{x-2t}{x}\right)^{-\frac{p}{2}}J_p\left(\sqrt{x^2-2xt}\right)=\sum_{s=0}^{\infty}\sum_{k=-s}^{\infty}\frac{(-1)^{k+2s}}{\Gamma(p+s+k+1)k!}\left(\frac{x}{2}\right)^{p+s+2k}\frac{t^s}{s!}$$

$$=\sum_{s=0}^{\infty}\frac{t^s}{s!}\sum_{k=0}^{\infty}\frac{(-1)^k}{k!\,\Gamma(p+s+k+1)}\left(\frac{x}{2}\right)^{p+s+2k}$$

$$=\sum_{s=0}^{\infty}J_{p+s}(x)\frac{t^s}{s!}.$$

PROBLEM 12 Prove that $\frac{d}{dx}J_p(kx)=\frac{p}{x}J_p(kx)-kJ_{p+1}(kx),\ k>1.$

Proof: Putting $y=kx$ in the expression on the left-hand side, we write

$$\frac{d}{dx}J_p(kx)=k\frac{d}{dy}J_p(y)=kJ_p'(y). \qquad \text{...(i)}$$

We know that

$$J_{p-1}(y)+J_{p+1}(y)=\frac{2p}{y}J_p(y)$$

and

$$J_{p-1}(y)-J_{p+1}(y)=2J_p'(y)$$

Subtraction of the second equation from the first one yields

$$J_p'(y)=\frac{p}{y}J_p(y)-J_{p+1}(y). \qquad \text{...(ii)}$$

Substituting Eq. (ii) into Eq. (i), we obtain

$$\frac{d}{dx}J_p(kx)=\frac{kp}{y}J_p(y)-k\,J_{p+1}(y)$$

$$=\frac{p}{x}J_p(kx)-k\,J_{p+1}(kx)\ \text{(as } y=kx\text{)}.$$

PROBLEM 13 Prove that

(a) $$e^{ix\sin\theta} = \sum_{n=-\infty}^{\infty} J_n(x)e^{in\theta}$$

and (b) $$e^{ix\cos\theta} = J_0(x) + 2\sum_{n=1}^{\infty} i^n J_n(x)\cos n\theta.$$

Proof: The generating function for Bessel functions $J_n(x)$ can be written as

$$e^{\frac{x}{2}\left(h-\frac{1}{h}\right)} = \sum_{n=-\infty}^{\infty} J_n(x)h^n$$

(a) In the above generating function, let us put $h = e^{i\theta}$ so as to yield

$$e^{ix\sin\theta} = \sum_{n=-\infty}^{\infty} J_n(x)e^{in\theta}.$$

(b) In this case, let us put $h = ie^{i\theta}$ Then, we write the above generating function as

$$\begin{aligned}
e^{ix\cos\theta} &= \sum_{n=-\infty}^{\infty} J_n(x)i^n e^{in\theta} \\
&= \sum_{n=-\infty}^{-1} J_n(x)i^n e^{in\theta} + J_0(x) + \sum_{n=1}^{\infty} J_n(x)i^n e^{in\theta} \\
&= \sum_{n=1}^{\infty} J_{-n}(x)i^{-n}e^{-in\theta} + J_0(x) + \sum_{n=1}^{\infty} J_n(x)i^n e^{in\theta} \\
&= \sum_{n=1}^{\infty} (-1)^n J_n(x)(-1)^n i^n e^{-in\theta} + J_0(x) + \sum_{n=1}^{\infty} J_n(x)\, i^n e^{in\theta} \\
&= J_0(x) + \sum_{n=1}^{\infty} J_n(x)i^n\left(e^{-in\theta} + e^{in\theta}\right) \\
&= J_0(x) + 2\sum_{n=1}^{\infty} i^n J_n(x)\cos n\theta.
\end{aligned}$$

PROBLEM 14 Prove that

(a) $$\cos(x\sin\theta) = J_0(x) + 2\sum_{n=1}^{\infty} J_{2n}(x)\cos(2n\theta),$$

(b) $$\sin(x\sin\theta) = 2\sum_{n=1}^{\infty} J_{2n-1}(x)\sin\{(2n-1)\theta\},$$

(c) $\cos(x\cos\theta) = J_0(x) + 2\sum_{n=1}^{\infty}(-1)^n J_{2n}(x)\cos(2n\theta)$,

(d) $\sin(x\cos\theta) = 2\sum_{n=0}^{\infty}(-1)^n J_{2n+1}(x)\cos\{(2n+1)\theta\}$,

(e) $\cos x = J_0(x) + 2\sum_{n=1}^{\infty}(-1)^n J_{2n}(x)$,

(f) $\sin x = 2\sum_{n=1}^{\infty}(-1)^{n-1} J_{2n-1}(x) = 2\sum_{n=0}^{\infty}(-1)^n J_{2n+1}(x)$,

(g) $x = 2\sum_{n=1}^{\infty}(2n-1)J_{2n-1}(x)$,

(h) $x\sin x = 2\sum_{n=1}^{\infty}(-1)^{n+1}(2n)^2 J_{2n}(x)$.

Proof: For (a) and (b), see **Article 8.14 II** (Eqs. 8.77 and 8.78) where replace ϕ by θ.

(c) In (a), replacing θ by $\left(\frac{\pi}{2}-\theta\right)$, we obtain

$$\cos(x\cos\theta) = J_0(x) + 2\sum_{n=1}^{\infty}(-1)^n J_{2n}(x)\cos(2n\theta).$$

(d) In (b), replacing θ by $\left(\frac{\pi}{2}-\theta\right)$, we obtain

$$\sin(x\cos\theta) = 2\sum_{n=0}^{\infty}(-1)^n J_{2n+1}(x)\cos\{(2n+1)\theta\}.$$

(e) Putting $\theta = \frac{\pi}{2}$ in (a) $\theta = 0$ in (c), we obtain

$$\cos x = J_0(x) + 2\sum_{n=1}^{\infty}(-1)^n J_{2n}(x).$$

(f) Putting $\theta = \frac{\pi}{2}$ in (b) or $\theta = 0$ in (d), we obtain

$$\sin x = 2\sum_{n=1}^{\infty}(-1)^{n-1} J_{2n-1}(x) = 2\sum_{n=0}^{\infty}(-1)^n J_{2n+1}(x).$$

(g) Differentiating both sides of relation (b) with respect to θ and putting $\theta = 0$, we obtain

$$x = 2\sum_{n=1}^{\infty}(2n-1)J_{2n-1}(x).$$

Alternatively, differentiating both sides of relation (d) with respect to θ and putting $\theta = \frac{\pi}{2}$, we can obtain the same result.

(h) Differentiating relation (a) twice with respect to θ and putting $\theta = \frac{\pi}{2}$, we obtain

$$x\sin x = 2\sum_{n=1}^{\infty}(-1)^{n+1}(2n)^2 J_{2n}(x).$$

Alternatively, the same relation can be derived by differentiating relation (c) twice with respect to θ and putting $\theta = 0$.

PROBLEM 15 Prove that

$$J_0^2(x) + 2\left[J_1^2(x) + J_2^2(x) + J_3^2(x) + \ldots\infty\right] = 1$$

and deduce that $|J_0(x)| \le 1$ and $|J_n(x)| \le 2^{-\frac{1}{2}}$, $(n \ge 1)$.

Proof:

First method: We know the Jacobi series, viz.

$$\cos(x\sin\theta) = J_0(x) + 2\sum_{n=1}^{\infty}J_{2n}(x)\cos(2n\theta)$$

$$= J_0(x) + 2J_2(x)\cos(2\theta) + 2J_4(x)\cos(4\theta) + 2J_6(x)\cos(6\theta) + \ldots\infty \qquad \ldots\text{(i)}$$

and $\sin(x\sin\theta) = 2\sum_{n=1}^{\infty}J_{2n-1}(x)\sin\{(2n-1)\theta\}$

$$= 2J_1(x)\sin\theta + 2J_3(x)\sin(3\theta) + 2J_5(x)\sin(5\theta) + \ldots\infty \qquad \ldots\text{(ii)}$$

Squaring Eq. (i) and integrating the result with respect to θ within the limit from $\theta = 0$ to $\theta = 2\pi$, we obtain

$$\int_0^{2\pi}\cos^2(x\sin\theta)\,d\theta$$

$$= J_0^2(x)\int_0^{2\pi}d\theta + 4J_2^2(x)\int_0^{2\pi}\cos^2(2\theta)\,d\theta + 4J_4^2(x)\int_0^{2\pi}\cos^2(4\theta)\,d\theta + \ldots\infty$$

$$= 2\pi[J_0^2(x) + 2\,J_2^2(x) + 2\,J_4^2\,(x) + \dots\,\infty] \qquad \dots\text{(iii)}$$

since the integrals of cross terms vanish.

Similarly, squaring Eq. (ii) and integrating the result with respect to θ within the limit from $\theta = 0$ to $\theta = 2\pi$, we obtain

$$\int_0^{2\pi} \sin^2(x\sin\theta)\,d\theta$$

$$= 4J_1^2(x)\int_0^{2\pi}\sin^2\theta\,d\theta + 4J_3^2(x)\int_0^{2\pi}\sin^2(3\theta)\,d\theta + 4J_5^2(x)\int_0^{2\pi}\sin^2(5\theta)\,d\theta + \dots\infty$$

$$= 2\pi\,[2\,J_1^2\,(x) + 2\,J_3^2\,(x) + 2\,J_5^2\,(x) + \dots\,\infty]. \qquad \dots\text{(iv)}$$

Adding Eqs.(iii) and (iv), we obtain

$$\int_0^{2\pi}\left[\cos^2(x\sin\theta) + \sin^2(x\sin\theta)\right]d\theta = \int_0^{2\pi} d\theta$$

$$= 2\pi = 2\pi\left[J_0^2(x) + 2\left\{J_1^2(x) + J_2^2(x) + J_3^2(x) + \dots\infty\right\}\right].$$

Thus, we have

$$J_0^2(x) + 2\left[J_1^2(x) + J_2^2(x) + J_3^2(x) + \dots\infty\right] = 1.$$

Second method: From Problem (9.a), we know that

$$\frac{d}{dx}\left[J_n^2(x) + J_{n+1}^2(x)\right] = \frac{2}{x}\left[nJ_n^2(x) - (n+1)J_{n+1}^2(x)\right].$$

Putting $n = 0, 1, 2, 3\dots(n-1)\dots\infty$, we write

$$\frac{d}{dx}\left[J_0^2(x) + J_1^2(x)\right] = \frac{2}{x}\left[0 - J_1^2(x)\right]$$

$$\frac{d}{dx}\left[J_1^2(x) + J_2^2(x)\right] = \frac{2}{x}\left[J_1^2(x) - 2J_2^2(x)\right]$$

$$\frac{d}{dx}\left[J_2^2(x) + J_3^2(x)\right] = \frac{2}{x}\left[2J_2^2(x) - 3J_3^2(x)\right]$$

$$\dots \quad \dots \quad \dots \quad \dots \quad \dots \quad \dots$$

$$\frac{d}{dx}\left[J_{n-1}^2(x) + J_n^2(x)\right] = \frac{2}{x}\left[(n-1)J_{n-1}^2(x) - nJ_n^2(x)\right].$$

Adding together, we obtain

$$\frac{d}{dx}\left[J_0^2(x)+2\left\{J_1^2(x)+J_2^2(x)+J_3^2(x)+\ldots+J_n^2(x)\right\}\right]=\frac{2}{x}\left[0-nJ_n^2(x)\right].$$

Since $J_n(x)\to 0$ as $n\to\infty$ in the limit $n\to\infty$, we write

$$\frac{d}{dx}\left[J_0^2(x)+2\left\{J_1^2(x)+J_2^2(x)+J_3^2(x)+\ldots\infty\right\}\right]=0.$$

Integrating both sides with respect to x, we have

$$J_0^2(x)+2\left\{J_1^2(x)+J_2^2(x)+J_3^2(x)+\ldots\infty\right\}=C \text{ (a constant)}.$$

We know that $J_0(0) = 1$ and $J_n(0) = 0$ for $n > 0$.
Therefore, when $x = 0$, we have

$$1 + 0 = C.$$

Hence, we write

$$J_0^2(x)+2\left[J_1^2(x)+J_2^2(x)+J_3^2(x)+\ldots\infty\right]=1.$$

Deduction: Since all terms in the above sum are positive, we see that

$$J_0^2(x)<1 \text{ and } 2J_n^2(x)<1 \text{ for } n\ge 1.$$

Hence, we have $|J_0(x)|\le 1$ and $|J_n(x)|\le 2^{-\frac{1}{2}}$ for $n\ge 1$.

PROBLEM 16 Prove that

$$J_n(x)=\frac{2}{x}\left[(n+1)J_{n+1}(x)-(n+3)J_{n+3}(x)+(n+5)J_{n+5}(x)-\ldots\infty\right].$$

Proof: We know that

$$J_{n-1}(x)+J_{n+1}(x)=\frac{2n}{x}J_n(x). \qquad \ldots\text{(i)}$$

In Eq. (i), replacing n by $(n+1),(n+3),(n+5)$ etc., we write

$$J_n(x)+J_{n+2}(x)=\frac{2(n+1)}{x}J_{n+1}(x),$$

$$-J_{n+2}(x)-J_{n+4}(x)=-\frac{2(n+3)}{x}J_{n+3}(x),$$

$$J_{n+4}(x)+J_{n+6}(x)=\frac{2(n+5)}{x}J_{n+5}(x),$$

$$\ldots \quad + \quad \ldots \quad = \quad \ldots$$

Adding all these equations together excluding Eq. (i) and noting that

$J_n(x) \to 0$ as $n \to \infty$, we obtain

$$J_n(x) = \frac{2}{x}\left[(n+1)J_{n+1}(x) - (n+3)J_{n+3}(x) + (n+5)J_{n+5}(x) - \ldots \infty\right]$$

$$= \frac{2}{x}\sum_{r=0}^{\infty}(-1)^r (n+2r+1)J_{n+2r+1}(x).$$

PROBLEM 17 Prove that

$$x = 2J_0(x)J_1(x) + 6J_1(x)J_2(x) + 10J_2(x)J_3(x) + \ldots + 2(2n+1)J_n(x)J_{n+1}(x) + \ldots \infty$$

Proof: From Problem 9(b), we have

$$\frac{d}{dx}\left[xJ_n(x)J_{n+1}(x)\right] = x\left[J_n^2(x) - J_{n+1}^2(x)\right]$$

Putting $n = 0,1,2,3\ \ldots$, we have

$$\frac{d}{dx}\left[xJ_0(x)J_1(x)\right] = x\left[J_0^2(x) - J_1^2(x)\right] \quad \ldots\text{(i)}$$

$$\frac{d}{dx}\left[xJ_1(x)J_2(x)\right] = x\left[J_1^2(x) - J_2^2(x)\right] \quad \ldots\text{(ii)}$$

$$\frac{d}{dx}\left[xJ_2(x)J_3(x)\right] = x\left[J_2^2(x) - J_3^2(x)\right] \quad \ldots\text{(iii)}$$

$$\frac{d}{dx}\left[J_3(x)J_4(x)\right] = x\left[J_3^2(x) - J_4^2(x)\right] \quad \ldots\text{(iv)}$$

$$\ldots \quad = \quad \ldots \quad \ldots$$

Multiplying Eqs. (i), (ii), (iii), (iv), ... by 1, 3, 5, 7, ... respectively, and then adding together, we obtain

$$\frac{d}{dx}\left[x\{J_0(x)\,J_1(x) + 3\,J_1(x)\,J_2\,(x) + 5\,J_2\,(x)\,J_3\,(x) + 7\,J_3\,(x)\,J_4\,(x) + \ldots \infty\}\right]$$

$$= x\left[J_0^2(x) + 2\left\{J_1^2(x) + J_2^2(x) + J_3^2(x) + J_4^2(x) + \ldots \infty\right\}\right] = x \cdot 1 = x.$$

Here, use has been made of the result of Problem 15.
Integrating both sides with respect to x, we obtain

$$x[J_0\,J_1 + 3\,J_1\,J_2 + 5\,J_2\,J_3 + 7\,J_3\,J_4 + \ldots \infty] = \frac{x^2}{2} + C \text{ (a constant)}.$$

When $x = 0$, we get $C = 0$.

Therefore, we write

$$x = 2J_0(x)J_1(x) + 6J_1(x)J_2(x) + 10J_2(x)J_3(x) + 14J_3(x)J_4(x)$$
$$+ \ldots + 2(2n+1)J_n(x)J_{n+1}(x) + \ldots \infty$$

or
$$x = \sum_{n=0}^{\infty} 2(2n+1)J_n J_{n+1}.$$

PROBLEM 18 Prove that

$$2^r \frac{d^r}{dx^r} J_n(x) = \sum_{s=0}^{r} \frac{(-1)^s r!}{(r-s)!\, s!} J_{n-r+2s}(x).$$

Proof: We know that

$$2\frac{d}{dx} J_n(x) = J_{n-1}(x) - J_{n+1}(x). \qquad \ldots \text{(i)}$$

Differentiating Eq. (i) with respect to x and then multiplying both sides by 2, we have

$$2^2 \frac{d^2}{dx^2} J_n(x) = 2\frac{d}{dx} J_{n-1}(x) - 2\frac{d}{dx} J_{n+1}(x). \qquad \ldots \text{(ii)}$$

Replacing n by $(n-1)$ and $(n+1)$ successively in Eq. (i), we obtain

$$2\frac{d}{dx} J_{n-1}(x) = J_{n-2}(x) - J_n(x) \qquad \ldots \text{(iii)}$$

and
$$2\frac{d}{dx} J_{n+1}(x) = J_n(x) - J_{n+2}(x). \qquad \ldots \text{(iv)}$$

Substituting Eqs. (iii) and (iv) into Eq. (ii), we have

$$2^2 \frac{d^2}{dx^2} J_n(x) = J_{n-2}(x) - 2J_n(x) + J_{n+2}(x). \qquad \ldots \text{(v)}$$

Differentiating Eq. (v) with respect to x and then multiplying both sides by 2, we have

$$2^3 \frac{d^3}{dx^3} J_n(x) = 2\frac{d}{dx} J_{n-2}(x) - 4\frac{d}{dx} J_n(x) + 2\frac{d}{dx} J_{n+2}(x). \qquad \ldots \text{(vi)}$$

Replacing n by $(n-2)$ and $(n+2)$ successively in Eq. (i), we obtain

$$2\frac{d}{dx} J_{n-2}(x) = J_{n-3}(x) - J_{n-1}(x) \qquad \ldots \text{(vii)}$$

and $$2\frac{d}{dx}J_{n+2}(x)=J_{n+1}(x)-J_{n+3}(x). \qquad \ldots\text{(viii)}$$

Substituting Eqs. (i), (vii) and (viii) into Eq. (vi), we have

$$\begin{aligned}2^3\frac{d^3}{dx^3}J_n(x)&=J_{n-3}(x)-J_{n-1}(x)-2\{J_{n-1}(x)-J_{n+1}(x)\}+J_{n+1}(x)-J_{n+3}(x)\\&=J_{n-3}(x)-3J_{n-1}(x)+3J_{n+1}(x)-J_{n+3}(x)\\&=(-1)^0J_{n-3}(x)+\frac{(-1)^1\,3!}{1!\,2!}J_{n-1}(x)+\frac{(-1)^2\,3!}{2!\,1!}J_{n+1}(x)+(-1)^3J_{n+3}(x).\end{aligned}$$

In general, continuing the differentiation and the multiplication by 2 at each step upto r times, we write

$$\begin{aligned}2^r\frac{d^r}{dx^r}J_n(x)&=(-1)^0J_{n-r}(x)+(-1)^1\frac{r!}{1!(r-1)!}J_{n-r+2}(x)+(-1)^2\frac{r!}{2!(r-2)!}J_{n-r+4}(x)\\&\quad+(-1)^3\frac{r!}{3!(r-3)!}J_{n-r+6}(x)+\ldots+(-1)^rJ_{n+r}(x)\\&=\sum_{s=0}^{r}(-1)^s\frac{r!}{s!(r-s)!}J_{n-r+2s}(x).\end{aligned}$$

PROBLEM 19 Prove the relation

$$\left(\frac{d}{xdx}\right)^r\left[x^{-n}J_n(x)\right]=(-1)^r x^{-n-r}J_{n+r}(x)$$

and deduce that

$$J_{l+\frac{1}{2}}(x)=\frac{(-1)^l(2x)^{l+\frac{1}{2}}}{\sqrt{\pi}}\left[\frac{d}{d(x^2)}\right]^l\left(\frac{\sin x}{x}\right).$$

Proof: Let us consider the recurrence relation

$$\frac{d}{dx}\{x^{-n}J_n(x)\}=-x^{-n}J_{n+1}(x).$$

Dividing both sides by x, we have

$$\frac{d}{xdx}\{x^{-n}J_n(x)\}=-x^{-n-1}J_{n+1}(x)=-x^{-(n+1)}J_{n+1}(x).$$

Thus, the given relation is true for $r=1$.

Differentiating both sides with respect to x, we have

$$\frac{d}{dx}\left[\frac{d}{xdx}\left\{x^{-n}J_n(x)\right\}\right]=-\frac{d}{dx}\left[x^{-(n+1)}J_{n+1}(x)\right]$$

$$=-\left[-x^{-(n+1)}J_{n+2}(x)\right]=(-1)^2\,x^{-n-1}J_{n+2}(x).$$

Dividing both sides by x, we have

$$\left(\frac{d}{xdx}\right)^2\left[x^{-n}J_n(x)\right]=(-1)^2\,x^{-n-2}J_{n+2}(x).$$

Thus, the given relation is true for $r=2$.

Assuming that the given relation is true for $r=s$, we have

$$\left(\frac{d}{xdx}\right)^s\left\{x^{-n}J_n(x)\right\}=(-1)^s\,x^{-n-s}J_{n+s}(x).$$

Differentiating both sides with respect to x, we have

$$\frac{d}{dx}\left[\left(\frac{d}{xdx}\right)^s\left\{x^{-n}J_n(x)\right\}\right]=(-1)^s\,\frac{d}{dx}\left[x^{-n-s}J_{n+s}(x)\right]$$

$$=(-1)^{s+1}\,x^{-n-s}J_{n+s+1}(x).$$

Dividing both sides by x, we have

$$\left(\frac{d}{xdx}\right)^{s+1}\left[x^{-n}J_n(x)\right]=(-1)^{s+1}\,x^{-n-(s+1)}J_{n+s+1}(x).$$

This shows that the given relation is true for $r=s+1$ if it is true for $r=s$.

Hence, by mathematical induction, we conclude that the given relation is true for all positive integral values of r. Thus, we complete the proof of the relation

$$\left(\frac{d}{xdx}\right)^r\left[x^{-n}J_n(x)\right]=(-1)^r\,x^{-n-r}J_{n+r}(x).$$

Deduction: Putting $n=\dfrac{1}{2}, r=l$, the given relation yields

$$\left(\frac{d}{xdx}\right)^l\left[x^{-½}J_{\frac{1}{2}}(x)\right]=(-1)^l\,x^{-\frac{1}{2}-l}J_{\frac{1}{2}+l}(x).$$

Therefore, we have

$$J_{l+\frac{1}{2}}(x)=(-1)^l\,x^{l+\frac{1}{2}}\left(\frac{d}{xdx}\right)^l\left[x^{-\frac{1}{2}}J_{\frac{1}{2}}(x)\right]$$

$$= (-1)^l x^{l+\frac{1}{2}} \left(\frac{d}{xdx}\right)^l \left[\frac{1}{\sqrt{x}}\sqrt{\frac{2}{\pi x}} \sin x\right]$$

$$= (-1)^l \frac{(2x)^{l+\frac{1}{2}}}{\sqrt{\pi}} \left(\frac{d}{2xdx}\right)^l \left[\frac{\sin x}{x}\right]$$

$$= (-1)^l \frac{(2x)^{l+\frac{1}{2}}}{\sqrt{\pi}} \left[\frac{d}{d(x^2)}\right]^l \left(\frac{\sin x}{x}\right).$$

PROBLEM 20 Prove that

$$\lim_{x\to 0} \frac{J_n(x)}{x^n} = \frac{1}{2^n \Gamma(n+1)}.$$

Proof: We know that

$$J_n(x) = \sum_{r=0}^{\infty} \frac{(-1)^r}{r!\Gamma(n+r+1)} \left(\frac{x}{2}\right)^{n+2r}.$$

Therefore, we write

$$\lim_{x\to 0} \frac{J_n(x)}{x^n} = \lim_{x\to 0} \frac{1}{x^n} \sum_{r=0}^{\infty} \frac{(-1)^r}{r!\Gamma(n+r+1)} \left(\frac{x}{2}\right)^{n+2r}$$

$$= \lim_{x\to 0} \frac{1}{2^n \left(\frac{x}{2}\right)^n} \sum_{r=0}^{\infty} \frac{(-1)^r}{r!\Gamma(n+r+1)} \left(\frac{x}{2}\right)^{n+2r}$$

$$= \frac{1}{2^n} \lim_{x\to 0} \sum_{r=0}^{\infty} \frac{(-1)^r}{r!\Gamma(n+r+1)} \left(\frac{x}{2}\right)^{2r}$$

$$= \frac{1}{2^n} \lim_{x\to 0} \left[\frac{1}{\Gamma(n+1)} - \frac{1}{\Gamma(n+2)}\left(\frac{x}{2}\right)^2 + \frac{1}{2!\Gamma(n+3)}\left(\frac{x}{2}\right)^4 - \ldots \infty\right]$$

$$= \frac{1}{2^n \Gamma(n+1)}.$$

PROBLEM 21 Prove that $J_r(x) = (-1)^r (2x)^r \left[\frac{d}{d(x^2)}\right]^r J_0(x)$.

Proof: In Problem 19, it was proved that

$$\left(\frac{d}{xdx}\right)^r \left[x^{-n} J_n(x)\right] = (-1)^r x^{-n-r} J_{n+r}(x).$$

Putting $n = 0$, we have

$$\left(\frac{d}{xdx}\right)^r J_0(x) = (-1)^r x^{-r} J_r(x)$$

or

$$J_r(x) = (-1)^r x^r \left(\frac{d}{xdx}\right)^r J_0(x)$$

$$= (-1)^r (2x)^r \left(\frac{d}{2xdx}\right)^r J_0(x)$$

$$= (-1)^r (2x)^r \left[\frac{d}{d(x^2)}\right]^r J_0(x).$$

PROBLEM 22 Prove that $\int_0^t J_0\left(\sqrt{x(t-x)}\right) dx = 2\sin\frac{t}{2}$.

Proof: We know that

$$J_0(x) = \sum_{r=0}^{\infty} \frac{(-1)^r}{(r!)^2} \left(\frac{x}{2}\right)^{2r}.$$

Replacing x by $\sqrt{x(t-x)}$, we write

$$J_0\left(\sqrt{x(t-x)}\right) = \sum_{r=0}^{\infty} \frac{(-1)^r}{(r!)^2} \left[\frac{\sqrt{x(t-x)}}{2}\right]^{2r}$$

$$= \sum_{r=0}^{\infty} \frac{(-1)^r}{2^{2r}(r!)^2} x^r (t-x)^r.$$

Integrating both sides with respect to x within the limit from $x = 0$ to $x = t$, we have

$$\int_0^t J_0\left(\sqrt{x(t-x)}\right) dx = \sum_{r=0}^{\infty} \frac{(-1)^r}{2^{2r}(r!)^2} \int_0^t x^r (t-x)^r \, dx.$$

Let us put $x = tu$ on the right-hand side and write

$$\int_0^t J_0\left(\sqrt{x(t-x)}\right) dx = \sum_{r=0}^{\infty} \frac{(-1)^r t^{2r+1}}{2^{2r}(r!)^2} \int_0^1 u^r (1-u)^r \, du$$

$$= \sum_{r=0}^{\infty} \frac{(-1)^r t^{2r+1}}{2^{2r}(r!)^2} B(r+1, r+1)$$

$$= \sum_{r=0}^{\infty} \frac{(-1)^r t^{2r+1}}{2^{2r}(r!)^2} \frac{\Gamma(r+1)\Gamma(r+1)}{\Gamma(2r+2)}$$

$$= 2\sum_{r=0}^{\infty} \frac{(-1)^r \left(\frac{t}{2}\right)^{2r+1}}{(2r+1)!} = 2\sin\frac{t}{2}.$$

PROBLEM 23 Prove that $\int_0^z J_1(ax)\,dx = \frac{1}{a}\left[1 - J_0(az)\right]$.

Proof: Let us consider the recurrence relation

$$\frac{d}{dx}\left[x^n J_n(x)\right] = x^n J_{n-1}(x).$$

Putting $n = 0$, we write

$$\frac{d}{dx} J_0(x) = J_{-1}(x) = -J_1(x).$$

Replacing x by ax, we write

$$J_1(ax) = -\frac{d}{a\,dx} J_0(ax).$$

Integrating both sides with respect to x within the limit from $x = 0$ to $x = z$, we have

$$\int_0^z J_1(ax)\,dx = -\frac{1}{a}\left[J_0(ax)\right]_0^z = \frac{1}{a}\left[1 - J_0(az)\right].$$

PROBLEM 24 Prove that

$$J_p(x) = \frac{\left(\frac{x}{2}\right)^p}{\Gamma\left(\frac{1}{2}\right)\Gamma\left(p+\frac{1}{2}\right)} 2\int_0^1 (1-t^2)^{p-1/2} \cos(xt)\,dt$$

and deduce that $J_0(x) = \frac{2}{\pi}\int_0^1 \frac{\cos(xt)\,dt}{\sqrt{1-t^2}}$.

Proof: See the derivation of Eq. 8.87 and put $p = 0$.

PROBLEM 25 Prove that

$$J_{m+n+1}(x)=\frac{2}{\Gamma(m+1)}\left(\frac{x}{2}\right)^{m+1}\int_0^1(1-t^2)^m\, t^{n+1}J_n(xt)\,dt \text{ for } m>-1,\ n>-1$$

and deduce that $J_0(x)=\dfrac{2}{\pi}\displaystyle\int_0^1\frac{\cos(xt)\,dt}{\sqrt{1-t^2}}$.

Proof: We know that

$$J_n(xt)=\sum_{r=0}^{\infty}\frac{(-1)^r}{r!\,\Gamma(n+r+1)}\left(\frac{xt}{2}\right)^{n+2r}.$$

Therefore, we write

$$\int_0^1(1-t^2)^m\, t^{n+1}J_n(xt)\,dt$$

$$=\int_0^1(1-t^2)^m\, t^{n+1}\sum_{r=0}^{\infty}\frac{(-1)^r}{r!\,\Gamma(n+r+1)}\left(\frac{xt}{2}\right)^{n+2r}dt$$

$$=\sum_{r=0}^{\infty}\frac{(-1)^r\left(\frac{x}{2}\right)^{n+2r}}{r!\,\Gamma(n+r+1)}\int_0^1(1-t^2)^m(t^2)^{n+r}\,t\,dt$$

$$=\sum_{r=0}^{\infty}\frac{(-1)^r\left(\frac{x}{2}\right)^{n+2r}}{r!\,\Gamma(n+r+1)}\frac{1}{2}\int_0^1(1-u)^m\, u^{n+r}\,du \quad \text{[Here, we put } u=t^2\text{]}$$

$$=\sum_{r=0}^{\infty}\frac{(-1)^r\left(\frac{x}{2}\right)^{n+2r}}{r!\,\Gamma(n+r+1)}\frac{1}{2}B(m+1,n+r+1)$$

$$=\frac{1}{2}\sum_{r=0}^{\infty}\frac{(-1)^r\left(\frac{x}{2}\right)^{n+2r}}{r!\,\Gamma(n+r+1)}\,\frac{\Gamma(m+1)\Gamma(n+r+1)}{\Gamma(m+n+r+2)}$$

$$=\frac{\Gamma(m+1)}{2}\left(\frac{x}{2}\right)^{-m-1}\sum_{r=0}^{\infty}\frac{(-1)^r}{r!\,\Gamma(m+n+r+2)}\left(\frac{x}{2}\right)^{m+n+1+2r}$$

$$=\frac{\Gamma(m+1)}{2}\left(\frac{x}{2}\right)^{-m-1}J_{m+n+1}(x).$$

Therefore, we write

$$J_{m+n+1}(x)=\frac{2}{\Gamma(m+1)}\left(\frac{x}{2}\right)^{m+1}\int_0^1(1-t^2)^m t^{n+1}J_n(xt)\,dt.$$

Deduction: Putting $m=n=-\frac{1}{2}$, we obtain

$$J_0(x)=\frac{2}{\Gamma\left(\frac{1}{2}\right)}\left(\frac{x}{2}\right)^{\frac{1}{2}}\int_0^1(1-t^2)^{-\frac{1}{2}}t^{\frac{1}{2}}J_{-\frac{1}{2}}(xt)\,dt$$

$$=\frac{2}{\sqrt{\pi}}\left(\frac{x}{2}\right)^{\frac{1}{2}}\int_0^1(1-t^2)^{-\frac{1}{2}}t^{\frac{1}{2}}\sqrt{\frac{2}{\pi xt}}\cos(xt)\,dt$$

$$=\frac{2}{\pi}\int_0^1\frac{\cos(xt)\,dt}{\sqrt{1-t^2}}.$$

PROBLEM 26 Prove that $\int_0^\infty e^{-ax^2}x^{n+1}J_n(x)dx=(2a)^{-n-1}e^{-\frac{1}{4a}}$.

Proof: The given integral can be written as

$$\int_0^\infty e^{-ax^2}x^{n+1}J_n(x)dx$$

$$=\int_0^\infty e^{-ax^2}x^{n+1}\left[\sum_{r=0}^\infty\frac{(-1)^r}{r!\,\Gamma(n+r+1)}\left(\frac{x}{2}\right)^{n+2r}\right]dx$$

$$=\sum_{r=0}^\infty\frac{(-1)^r}{2^{n+2r}\,r!\,\Gamma(n+r+1)}\int_0^\infty e^{-ax^2}x^{2n+2r+1}\,dx$$

$$=\sum_{r=0}^\infty\frac{(-1)^r}{2^{n+2r}\,r!\,\Gamma(n+r+1)}\int_0^\infty e^{-u^2}\left(\frac{u}{\sqrt{a}}\right)^{2n+2r+1}\frac{du}{\sqrt{a}}\quad\text{[Here, we put } ax^2=u^2\text{]}$$

$$=\sum_{r=0}^\infty\frac{(-1)^r}{2^{n+2r}\,r!\,\Gamma(n+r+1)a^{n+r+1}}\int_0^\infty e^{-u^2}u^{2(n+r+1)-1}\,du$$

$$=\sum_{r=0}^\infty\frac{(-1)^r}{2^{n+2r}\,r!\,\Gamma(n+r+1)a^{n+r+1}}\cdot\frac{1}{2}\Gamma(n+r+1)$$

$$= \frac{1}{(2a)^{n+1}} \sum_{r=0}^{\infty} \frac{\left(-\frac{1}{4a}\right)^r}{r!} = (2a)^{-n-1} e^{-\frac{1}{4a}}.$$

PROBLEM 27 Prove that $\int_0^\infty e^{-ax} x^{n+1} J_n(bx)\,dx = \dfrac{2^{n+1} a b^n \Gamma\left(n+\frac{3}{2}\right)}{\sqrt{\pi}\,(a^2+b^2)^{n+\frac{3}{2}}}$ and deduce that

$$\int_0^\infty e^{-ax} x J_0(bx)\,dx = \frac{a}{(a^2+b^2)^{\frac{3}{2}}}.$$

Proof: We know that

$$J_n(bx) = \sum_{r=0}^{\infty} \frac{(-1)^r}{r!\,\Gamma(n+r+1)} \left(\frac{bx}{2}\right)^{n+2r}.$$

Therefore, we write

$$\int_0^\infty e^{-ax} x^{n+1} J_n(bx)\,dx$$

$$= \int_0^\infty e^{-ax} x^{n+1} \left[\sum_{r=0}^{\infty} \frac{(-1)^r}{r!\,\Gamma(n+r+1)} \left(\frac{bx}{2}\right)^{n+2r}\right] dx$$

$$= \sum_{r=0}^{\infty} \frac{(-1)^r\, b^{n+2r}}{2^{n+2r}\, r!\,\Gamma(n+r+1)} \int_0^\infty e^{-ax} x^{2n+2r+1}\,dx$$

$$= \sum_{r=0}^{\infty} \frac{(-1)^r\, b^{n+2r}}{2^{n+2r}\, r!\,\Gamma(n+r+1)\, a^{2n+2r+2}} \int_0^\infty e^{-u} u^{2n+2r+1}\,du \quad \text{(where } u = ax\text{)}$$

$$= \sum_{r=0}^{\infty} \frac{(-1)^r\, b^{n+2r}\, \Gamma(2n+2r+2)}{2^{n+2r}\, a^{2n+2r+2}\, r!\,\Gamma(n+r+1)}.$$

By Legendre's duplication formula of gamma functions, we have

$$\frac{\Gamma(2n)}{\Gamma(n)} = \frac{2^{2n-1}}{\sqrt{\pi}} \Gamma\left(n+\frac{1}{2}\right).$$

Thus, we write the above integral as

$$\int_0^\infty e^{-ax}x^{n+1}J_n(bx)\,dx$$

$$=\sum_{r=0}^{\infty}\frac{(-1)^r\, b^{n+2r}\, 2^{2n+2r+1}\,\Gamma\left(n+r+\frac{3}{2}\right)}{2^{n+2r}\,a^{2n+2r+2}\,r!\,\sqrt{\pi}}$$

$$=\frac{2^{n+1}b^n}{\sqrt{\pi}\,a^{2n+2}}\sum_{r=0}^{\infty}\frac{(-1)^r\,\Gamma\left(n+r+\frac{3}{2}\right)\left(\frac{b^2}{a^2}\right)^r}{\Gamma(r+1)}$$

$$=\frac{2^{n+1}b^n\Gamma\left(n+\frac{3}{2}\right)}{\sqrt{\pi}\,a^{2n+2}}\sum_{r=0}^{\infty}\frac{(-1)^r\,\Gamma\left(n+r+\frac{3}{2}\right)}{\Gamma(r+1)\Gamma\left(n+\frac{3}{2}\right)}\left(\frac{b^2}{a^2}\right)^r$$

$$=\frac{2^{n+1}ab^n\Gamma\left(n+\frac{3}{2}\right)}{\sqrt{\pi}(a^2)^{n+\frac{3}{2}}}\left(1+\frac{b^2}{a^2}\right)^{-\left(n+\frac{3}{2}\right)}$$

$$=\frac{2^{n+1}ab^n\,\Gamma\left(n+\frac{3}{2}\right)}{\sqrt{\pi}\,(a^2+b^2)^{n+\frac{3}{2}}}.$$

PROBLEM 28 Prove that $\int_0^\infty e^{-ax}J_0(bx)\,dx=\frac{1}{\sqrt{a^2+b^2}}$ $(a>0)$.

Proof: We have

$$\int_0^\infty e^{-ax}J_0(bx)\,dx$$

$$=\int_0^\infty e^{-ax}\sum_{r=0}^{\infty}\frac{(-1)^r}{(r!)^2}\left(\frac{bx}{2}\right)^{2r}$$

$$=\sum_{r=0}^{\infty}\frac{(-1)^r\,b^{2r}}{2^{2r}\,(r!)^2}\int_0^\infty e^{-ax}x^{2r}\,dx$$

$$= \sum_{r=0}^{\infty} \frac{(-1)^r b^{2r}}{2^{2r} (r!)^2 a^{2r+1}} \int_0^{\infty} e^{-u} u^{2r} du \quad (\text{where } u = ax)$$

$$= \sum_{r=0}^{\infty} \frac{(-1)^r b^{2r}}{2^{2r} (r!)^2 a^{2r+1}} \Gamma(2r+1)$$

$$= \frac{1}{a} \sum_{r=0}^{\infty} \frac{\left(-\frac{b^2}{a^2}\right)^r (2r)!}{2^{2r} (r!)^2}$$

$$= \frac{1}{a} \sum_{r=0}^{\infty} \frac{\left(-\frac{b^2}{a^2}\right)^r 2^{2r} r!\,\Gamma\left(r+\frac{1}{2}\right)}{2^{2r} (r!)^2 \Gamma\left(\frac{1}{2}\right)} \quad \left[\text{using } (2r)! = \frac{r!\,2^{2r}}{\Gamma\left(\frac{1}{2}\right)} \Gamma\left(r+\frac{1}{2}\right)\right]$$

$$= \frac{1}{a} \sum_{r=0}^{\infty} \frac{\Gamma\left(r+\frac{1}{2}\right)}{\Gamma(r+1)\Gamma\left(\frac{1}{2}\right)} \left(-\frac{b^2}{a^2}\right)^r = \frac{1}{a}\left(1+\frac{b^2}{a^2}\right)^{-\frac{1}{2}} = \frac{1}{\sqrt{a^2+b^2}}.$$

PROBLEM 29 Prove that $\int_0^{\infty} \frac{J_n(x)}{x} dx = \frac{1}{n}$, $(n > 0)$.

Proof: We have

$$\frac{2n}{x} J_n(x) = J_{n-1}(x) + J_{n+1}(x)$$

Integrating both sides with respect to x within the limit from $x = 0$ to $x = \infty$, we obtain

$$2n \int_0^{\infty} \frac{J_n(x)}{x} dx = \int_0^{\infty} J_{n-1}(x) dx + \int_0^{\infty} J_{n+1}(x) dx$$

$$= 1 + 1 = 2.$$

Here, use has been made of the formula $\int_0^{\infty} J_n(x) dx = 1$ for n = non-negative integer, which is to be proved in the next problem.

Therefore, we obtain

$$\int_0^{\infty} \frac{J_n(x)}{x} dx = \frac{1}{n}.$$

PROBLEM 30 Prove that $\int_0^\infty J_n(bx)dx = \frac{1}{b}$ (n being a non-negative integer) and deduce that $\int_0^\infty J_n(x)dx = 1$.

Proof: Let us use the method of mathematical induction.

From the result of Problem 28, i.e.,

$$\int_0^\infty e^{-ax} J_0(bx)dx = \frac{1}{\sqrt{a^2+b^2}}.$$

We see that for $a=0$ $\int_0^\infty J_0(bx)dx = \frac{1}{b}$.

This shows that the given relation is true for $n = 0$.

For $n = 1$, let us consider the relation

$$\frac{d}{dx}\left[x^{-n}J_n(x)\right] = -x^{-n}J_{n+1}(x).$$

Putting $n = 0$, we obtain

$$\frac{d}{dx}J_0(x) = -J_1(x).$$

Replacing x by bx, we write

$$\frac{d}{d(bx)}J_0(bx) = \frac{1}{b}\frac{d}{dx}J_0(bx) = -J_1(bx).$$

Integrating both sides with respect to x within the limit from $x = 0$ to $x = \infty$, we obtain

$$\int_0^\infty J_1(bx)dx = -\frac{1}{b}\left[J_0(bx)\right]_0^\infty = \frac{1}{b}$$

as $J_0(\infty) = 0$ and $J_0(0) = 1$ [known from the asymptotic expressions].

This shows that the given relation is true for $n = 1$.

We are to show that if the result is true for $n = N$, it is also true for $n = N + 2$, thus completing the proof of the given relation for all non-negative integral n.

Let us consider the recurrence relation for Bessel functions, viz.

$$2J_n'(x) = J_{n-1}(x) - J_{n+1}(x).$$

Integrating both sides with respect to x within the limit from $x = 0$ to $x = \infty$, we obtain

$$2[J_n(x)]_0^\infty = \int_0^\infty J_{n-1}(x)dx - \int_0^\infty J_{n+1}(x)dx \quad (n \neq 0)$$

or
$$0 = \int_0^\infty J_{n-1}(x)dx - \int_0^\infty J_{n+1}(x)dx$$

or
$$\int_0^\infty J_{n+1}(x)dx = \int_0^\infty J_{n-1}(x)dx.$$

Replacing $n - 1$ by N and x by bx, we write

$$\int_0^\infty J_{N+2}(bx)\,dx = \int_0^\infty J_N(bx)\,dx.$$

If the given relation is true for $n = N$, we have

$$\int_0^\infty J_{N+2}(bx)\,dx = \int_0^\infty J_N(bx)\,dx = \frac{1}{b}.$$

Thus, the given relation is true for $n = N + 2$.

Hence, we write, for arbitrary n, $\int_0^\infty J_n(bx)\,dx = \frac{1}{b}$.

This completes the proof of the given relation.

Deduction: Putting $b = 1$, we write $\int_0^\infty J_n(x)dx = 1$.

PROBLEM 31 Prove that

(i) $$\int_0^\infty \sin(ax)J_0(bx)\,dx = \begin{cases} 0 & \text{for } a < b, \\ \dfrac{1}{\sqrt{a^2 - b^2}} & \text{for } a > b, \end{cases}$$

(ii) $$\int_0^\infty \cos(ax)J_0(bx)\,dx = \begin{cases} \dfrac{1}{\sqrt{b^2 - a^2}}, & \text{for } a < b, \\ 0 & \text{for } a > b. \end{cases}$$

Proof: In the result of Problem 28, replace *a* by *ia* to yield

$$\int_0^\infty e^{-iax} J_0(bx)\,dx = \frac{1}{\sqrt{(ia)^2 + b^2}}$$

or $$\int_0^\infty \cos(ax) J_0(bx)\,dx - i\int_0^\infty \sin(ax) J_0(bx)\,dx = \frac{1}{\sqrt{b^2 - a^2}}. \qquad \ldots\text{(i)}$$

CASE I: $a < b$

In this case, equating real parts from both sides of Eq. (i), we obtain

$$\int_0^\infty \cos(ax) J_0(bx)\,dx = \frac{1}{\sqrt{b^2 - a^2}}.$$

Equating imaginary parts from both sides of Eq. (i), we obtain

$$\int_0^\infty \sin(ax) J_0(kx)\,dx = 0.$$

CASE II: $a > b$

In this case, Eq. (i) becomes

$$\int_0^\infty \cos(ax) J_0(bx)\,dx - i\int_0^\infty \sin(ax) J_0(bx)\,dx = -\frac{i}{\sqrt{a^2 - b^2}}. \qquad \ldots\text{(ii)}$$

Equating real parts from both sides of Eq. (ii), we obtain

$$\int_0^\infty \cos(ax) J_0(bx)\,dx = 0.$$

Further, equating imaginary parts from both sides of Eq. (ii), we obtain

$$\int_0^\infty \sin(ax) J_0(bx)\,dx = \frac{1}{\sqrt{a^2 - b^2}}.$$

We summarize as

(i) $$\int_0^\infty \sin(ax) J_0(bx)\,dx = \begin{cases} 0 & \text{for } a < b, \\ \dfrac{1}{\sqrt{a^2 - b^2}} & \text{for } a > b, \end{cases}$$

(ii) $$\int_0^\infty \cos(ax) J_0(bx)\,dx = \begin{cases} \dfrac{1}{\sqrt{b^2 - a^2}} & \text{for } a < b, \\ 0 & \text{for } a > b. \end{cases}$$

PROBLEM 32 Prove that

$$J_\alpha(2\sqrt{xt}) = e^{-t}(xt)^{\frac{\alpha}{2}} \sum_{n=0}^{\infty} \frac{L_n^\alpha(x)}{\Gamma(n+\alpha+1)} t^n.$$

Proof: It was proved in the Solved Problem No. 4 of chapter 7.

PROBLEM 33 If an arbitrary function $f(x)$ is defined in a region $0 \le x \le b$ and if the Bessel function $J_p(\alpha_n b) = 0$ ($n = 1, 2, 3 \ldots$), prove that the function $f(x)$ can be expanded in a series of Bessel functions in the form

$$f(x) = \sum_{n=1}^{\infty} c_n J_p(\alpha_n x)$$

where $c_n = \dfrac{2}{b^2\left[J_{p+1}(\alpha_n b)\right]^2} \displaystyle\int_0^b x f(x) J_p(\alpha_n x)\,dx$ with $n = 1, 2, 3 \ldots$

Proof: Let us write

$$f(x) = \sum_{n=1}^{\infty} c_n J_p(\alpha_n x).$$

Multiplying both sides by $x J_p(\alpha_m x)$ and integrating the result with respect to x within the limit from $x = 0$ to $x = b$, we have

$$\int_0^b x f(x) J_p(\alpha_m x)\,dx = \sum_{n=1}^{\infty} c_n \int_0^b J_p(\alpha_m x) J_p(\alpha_n x)\, x\,dx$$

$$= c_m \frac{b^2}{2}\left[J_{p+1}(\alpha_m b)\right]^2$$

as integrals with $m \ne n$ vanish.

Therefore, we have

$$c_m = \frac{2}{b^2\left[J_{p+1}(\alpha_m b)\right]^2} \int_0^b x f(x) J_p(\alpha_m x)\,dx.$$

Replacing the subscript m by n, we write

$$c_n = \frac{2}{b^2\left[J_{p+1}(\alpha_n b)\right]^2} \int_0^b f(x) J_p(\alpha_n x)\, x\,dx.$$

PROBLEM 34 Prove that $x^r = \sum_{n=1}^{\infty} \frac{2}{\alpha_n J_{r+1}(\alpha_n)} J_r(\alpha_n x)$ if $0 \le x \le 1$, α_n are the zeroes of $J_r(x)$ and r is a non-negative integer.

Proof: We write

$$x^r = \sum_{n=1}^{\infty} c_n J_r(\alpha_n x). \qquad \text{...(i)}$$

Multiplying both sides by $x J_r(\alpha_m x)$ and then integrating both sides with respect to x within the limit from $x = 0$ to $x = 1$, we have

$$\int_0^1 x^{r+1} J_r(\alpha_m x)\,dx = \sum_{n=1}^{\infty} c_n \int_0^1 J_r(\alpha_m x) J_r(\alpha_n x)\, x\, dx$$

$$= c_m \frac{1^2}{2}\left[J_{r+1}(\alpha_m)\right]^2.$$

Therefore, we have

$$c_m = \frac{2}{\left[J_{r+1}(\alpha_m)\right]^2} \int_0^1 x^{r+1} J_r(\alpha_m x)\,dx.$$

Changing the subscript, we write

$$c_n = \frac{2}{\left[J_{r+1}(\alpha_n)\right]^2} \int_0^1 x^{r+1} J_r(\alpha_n x)\,dx. \qquad \text{...(ii)}$$

Putting $u = \alpha_n x$, we write

$$\int_0^1 x^{r+1} J_r(\alpha_n x)\,dx = \frac{1}{\alpha_n^{r+2}} \int_0^{\alpha_n} u^{r+1} J_r(u)\,du$$

$$= \frac{1}{\alpha_n^{r+2}} \int_0^{\alpha_n} \frac{d}{du}\left[u^{r+1} J_{r+1}(u)\right] du$$

$$= \frac{1}{\alpha_n^{r+2}} \left[u^{r+1} J_{r+1}(u)\right]_0^{\alpha_n}$$

$$= \frac{1}{\alpha_n^{r+2}} \cdot \alpha_n^{r+1} J_{r+1}(\alpha_n)$$

$$= \frac{J_{r+1}(\alpha_n)}{\alpha_n}.$$

In evaluating the above integral, use has been made of the relation

$$\frac{d}{dx}\left[x^n J_n(x)\right]=x^n J_{n-1}(x).$$

Substituting the value of the above integral into Eq. (ii), we obtain

$$c_n=\frac{2}{\alpha_n J_{r+1}(\alpha_n)}.$$

Substituting the value of c_n into Eq. (i), we write

$$x^r=\sum_{n=1}^{\infty}\frac{2}{\alpha_n J_{r+1}(\alpha_n)}J_r(\alpha_n x).$$

PROBLEM 35 If α_n are the zeroes of $J_0(x)$, prove that

$$\ln x=-2\sum_{n=1}^{\infty}\frac{J_0(\alpha_n x)}{\left[\alpha_n J_1(\alpha_n)\right]^2}\quad(0\le x\le 1).$$

Proof: Let us write

$$\ln x=\sum_{n=1}^{\infty}c_n J_0(\alpha_n x) \qquad \text{...(i)}$$

where the coefficients c_n are to be determined.

It is known that the coefficients c_n are given by the formula

$$c_n=\frac{2}{a^2\left[J_{p+1}(\alpha_n a)\right]^2}\int_0^a f(x)J_p(\alpha_n x)\,x\,dx.$$

Here, $a=1$, $f(x)=\ln x$, $p=0$

Therefore, we have

$$c_n=\frac{2}{\left[J_1(\alpha_n)\right]^2}\int_0^1 \ln x\, J_0(\alpha_n x)\,x\,dx$$

$$=\frac{2}{\left[J_1(\alpha_n)\right]^2}\int_0^1 \ln x\sum_{r=0}^{\infty}\frac{(-1)^r}{(r!)^2}\left(\frac{\alpha_n x}{2}\right)^{2r}x\,dx$$

$$=\frac{2}{\left[J_1(\alpha_n)\right]^2}\sum_{r=0}^{\infty}\frac{(-1)^r}{(r!)^2}\left(\frac{\alpha_n}{2}\right)^{2r}\int_0^1(\ln x)\,x^{2r+1}dx$$

$$=\frac{2}{\left[J_1(\alpha_n)\right]^2}\sum_{r=0}^{\infty}\frac{(-1)^r}{(r!)^2}\left(\frac{\alpha_n}{2}\right)^{2r}\left\{\left[(\ln x)\frac{x^{2r+2}}{(2r+2)}\right]_0^1-\int_0^1\frac{1}{x}\frac{x^{2r+2}}{(2r+2)}dx\right\}$$

$$=\frac{2}{\left[J_1(\alpha_n)\right]^2}\sum_{r=0}^{\infty}\frac{(-1)^r}{(r!)^2}\left(\frac{\alpha_n}{2}\right)^{2r}\left\{0-\frac{1}{(2r+2)}\int_0^1 x^{2r+1}dx\right\}$$

$$=\frac{2}{\left[J_1(\alpha_n)\right]^2}\sum_{r=0}^{\infty}\frac{(-1)^r}{(r!)^2}\left(\frac{\alpha_n}{2}\right)^{2r}\left\{-\frac{1}{(2r+2)^2}\right\}$$

$$=\frac{2}{\left[\alpha_n J_1(\alpha_n)\right]^2}\sum_{r=0}^{\infty}\frac{(-1)^{r+1}}{\{(r+1)!\}^2}\left(\frac{\alpha_n}{2}\right)^{2r+2}$$

$$=\frac{2}{\left[\alpha_n J_1(\alpha_n)\right]^2}\sum_{r=1}^{\infty}\frac{(-1)^r}{(r!)^2}\left(\frac{\alpha_n}{2}\right)^{2r}$$

$$=\frac{2}{\left[\alpha_n J_1(\alpha_n)\right]^2}\left\{\sum_{r=0}^{\infty}\frac{(-1)^r}{(r!)^2}\left(\frac{\alpha_n}{2}\right)^{2r}-1\right\}$$

$$=\frac{2}{\left[\alpha_n J_1(\alpha_n)\right]^2}\{J_0(\alpha_n)-1\}$$

$$=-\frac{2}{\left[\alpha_n J_1(\alpha_n)\right]^2}\quad \text{as } J_0(\alpha_n)=0.$$

Substituting this value of c_n into Eq. (i), we obtain

$$\ln x=-2\sum_{n=1}^{\infty}\frac{1}{\left[\alpha_n J_1(\alpha_n)\right]^2}J_0(\alpha_n x).$$

PROBLEM 36 Expand the function $f(x)=x^2$ $(0\le x\le a)$ in a series of Bessel functions if α_n are the positive roots of $J_2(x)=0$.

Solution: We write

$$f(x)=x^2=\sum_{n=1}^{\infty}c_n J_2(\alpha_n x)$$

where
$$c_n=\frac{2}{a^2\left[J_3(\alpha_n a)\right]^2}\int_0^a x^3 J_2(\alpha_n x)dx$$

$$= \frac{2}{a^2\left[J_3(\alpha_n a)\right]^2 \alpha_n^4} \int_0^{a\alpha_n} u^3 J_2(u)\,du \quad (\text{where } u = \alpha_n x)$$

$$= \frac{2}{a^2\left[J_3(\alpha_n a)\right]^2 \alpha_n^4} \left[u^3 J_3(u)\right]_0^{a\alpha_n}$$

$$= \frac{2}{a^2 \alpha_n^4 \left[J_3(\alpha_n a)\right]^2} a^3 \alpha_n^3 J_3(a\alpha_n) = \frac{2a}{\alpha_n J_3(a\alpha_n)}.$$

Hence, we have

$$x^2 = \sum_{n=1}^{\infty} \frac{2a}{\alpha_n J_3(a\alpha_n)} J_2(\alpha_n x) \; (0 \le x \le a).$$

PROBLEM 37 In a cylindrical nuclear reactor of height h along the z-axis and radius a, the diffusion of thermal neutrons has a radial symmetry and obeys thc differential equation

$$\nabla^2 \phi(r, z) + B^2 \phi(r, z) = 0$$

where ϕ (r, z) is the flux at the point (r, z) and B^2 is a positive constant known as buckling constant, a property of the material, its shape and size. Using the boundary condition that the flux vanishes at the surface of the cylinder, prove that the above equation has a solution

$$\phi(r, z) = A \cos\left(\frac{\pi z}{h}\right) J_0\left(\frac{2.405 r}{a}\right)$$

where A is the maximum flux that exists at the centre and r and z are cylindrical coordinates having radial symmetry.

Obtain the relations

I. $B^2 = \left(\frac{\pi}{h}\right)^2 + \left(\frac{2.405}{a}\right)^2,$

II. $B^2 h^2 = 3\pi^2,$

III. $2a^2 B^2 = 3(2.405)^2.$

Prove that the minimum volume of a cylindrical core of fixed buckling is obtained when the diameter is approximately equal to the height and also find the minimum volume in terms of buckling constant.

Solution: In cylindrical coordinates with radial symmetry, the given reactor equation can be written as

$$\frac{\partial^2 \phi}{\partial r^2}+\frac{1}{r}\frac{\partial \phi}{\partial r}+\frac{\partial^2 \phi}{\partial z^2}+B^2\phi=0. \qquad \text{...(i)}$$

Let us put
$$\phi=\phi(r,z)=F(r)\,G(z). \qquad \text{...(ii)}$$

Substitution of Eq. (ii) into Eq. (i) yields

$$G\frac{d^2F}{dr^2}+\frac{G}{r}\frac{dF}{dr}+F\frac{d^2G}{dz^2}+B^2FG=0.$$

Dividing both sides by $F(r)\,G(z)$, we obtain

$$\frac{1}{F}\frac{d^2F}{dr^2}+\frac{1}{rF}\frac{dF}{dr}+\frac{1}{G}\frac{d^2G}{dz^2}+B^2=0$$

or
$$\frac{1}{F}\frac{d^2F}{dr^2}+\frac{1}{rF}\frac{dF}{dr}=-\frac{1}{G}\frac{d^2G}{dz^2}-B^2=-\alpha^2 \qquad \text{...(iii)}$$

where α^2 is a positive constant.

From Eq. (iii), we obtain

$$\frac{d^2F}{dr^2}+\frac{1}{r}\frac{dF}{dr}+\alpha^2F=0.$$

Multiplying both sides by r^2, we write

$$\alpha^2r^2\frac{d^2F}{\alpha^2dr^2}+\alpha r\frac{dF}{\alpha dr}+\alpha^2r^2F=0$$

or
$$\rho^2\frac{d^2F}{d\rho^2}+\rho\frac{dF}{d\rho}+\rho^2F=0 \text{ (where } \rho=\alpha r). \qquad \text{...(iv)}$$

Equation (iv) is Bessel's differential equation and its general solution can be written as

$$F(\rho)=A_1J_0(\rho)+A_2N_0(\rho)$$

where A_1 and A_2 are arbitrary constants and $J_0(\rho)$ and $N_0(\rho)$ are respectively Bessel functions of the first and second kinds of order zero. [N_0 is also called Neumann function of order zero.] Substituting ρ by αr, we obtain

$$F(r)=A_1J_0(\alpha r)+A_2N_0(\alpha r).$$

We know that $N_0(\alpha r)\to\infty$ as $\alpha r\to 0$ and, therefore, to give a physically acceptable solution, we must have $A_2=0$.

Therefore, we have

$$F(r) = A_1 J_0(\alpha r). \qquad \text{...(v)}$$

Again, from Eq. (iii), we have

$$\frac{d^2G}{dz^2} + \beta^2 G = 0 \text{ (where } \beta^2 = B^2 - \alpha^2).$$

This equation has the general solution

$$G(z) = C_1 \cos(\beta z) + C_2 \sin(\beta z)$$

where C_1 and C_2 are arbitrary constants.

As the solution must be symmetric above and below the central plane normal to the cylinder axis, we must have $C_2 = 0$.

Therefore, we have

$$G(z) = C_1 \cos(\beta z). \qquad \text{...(vi)}$$

Substituting Eqs. (v) and (vi) into Eq. (ii), we obtain

$$\phi(r, z) = A J_0(\alpha r) \cos(\beta z) \text{ (where } A = A_1 C_1). \qquad \text{...(vii)}$$

The boundary conditions are

(i) $\phi(r, z) = 0$ at $r = a$

and (ii) $\phi(r, z) = 0$ at $z = \pm\frac{h}{2}$.

Therefore, from condition (i), we have $\phi(a, z) = 0$ for all z inside the cylinder.

or

$$J_0(\alpha a) = 0.$$

Considering only the first zero of J_0, we have

$$\alpha a = 2.405$$

or

$$\alpha = \frac{2.405}{a}. \qquad \text{(viii)}$$

From condition (ii), we have $\phi\left(r, \pm\frac{h}{2}\right) = 0$ for all r inside the cylinder.

Therefore, we have

$$\cos\left(\pm\frac{\beta h}{2}\right) = \cos\left(\frac{\beta h}{2}\right) = 0$$

or

$$\frac{\beta h}{2} = \frac{n\pi}{2} \text{ with } n = 1, 3, 5 \ldots$$

Of the various solutions, only the fundamental solution with $n = 1$ is of interest. Thus, we have

$$\beta = \frac{n\pi}{h} = \frac{\pi}{h}. \qquad \text{...(ix)}$$

Substituting the values of α and β in Eq. (vii), we obtain the solution of the given reactor equation as

$$\phi(r, z) = A\cos\left(\frac{\pi z}{h}\right) J_0\left(\frac{2.405r}{a}\right). \qquad \text{...(x)}$$

Significance of A: Maximum flux ϕ_{max} occurs at $r = z = 0$.

Therefore, $\phi_{\text{max}} = A\cos 0\, J_0(0) = A$.

I. Since $\beta^2 = B^2 - \alpha^2$, substituting the values of α and β in this equation, we obtain

$$B^2 = \beta^2 + \alpha^2 = \left(\frac{\pi}{h}\right)^2 + \left(\frac{2.405}{a}\right)^2. \qquad \text{(xi)}$$

II. From Eq. (xi), we have

$$a^2 = \frac{(2.405)^2}{B^2 - \dfrac{\pi^2}{h^2}}. \qquad \text{(xii)}$$

The volume of the cylindrical core is given by

$$V = \pi a^2 h = \frac{\pi h \times (2.405)^2}{B^2 - \dfrac{\pi^2}{h^2}} = \frac{(2.405)^2 \pi h^3}{B^2 h^2 - \pi^2}. \qquad \text{...(xiii)}$$

It is an essential condition that this volume is to be minimized.

For such a minimum volume, we have

$$\frac{dV}{dh} = \frac{(2.405)^2\left\{3\pi h^2\left(B^2h^2 - \pi^2\right) - 2\pi B^2 h^4\right\}}{\left(B^2h^2 - \pi^2\right)^2} = 0$$

or

$$B^2h^2 = 3\pi^2. \qquad \text{(xiv)}$$

III. From Eq. (xiv), we have

$$h = \frac{\sqrt{3}\,\pi}{B}. \qquad \text{...(xv)}$$

Substituting this value of h into Eq. (xii), we have

$$a^2 = \frac{(2.405)^2}{B^2 - \dfrac{B^2}{3}} = \frac{3}{2}\frac{(2.405)^2}{B^2}$$

or

$$2a^2B^2 = 3(2.405)^2. \qquad \text{...(xvi)}$$

Dividing Eq. (xvi) by Eq. (xiv), we obtain

$$\frac{2a^2}{h^2} = \frac{(2.405)^2}{\pi^2}$$

yielding

$$\frac{2a}{h} = 1.08.$$

Thus, the minimum volume of a cylindrical reactor core of fixed buckling constant is obtained when the height is approximately equal to the diameter.

Now, using the values of a^2 and h obtained in the minimum volume condition, we obtain the minimum volume as

$$V_{\text{min}} = \pi \times \frac{3}{2}\frac{(2.405)^2}{B^2} \times \frac{\sqrt{3}\,\pi}{B} = \frac{148.31}{B^3}.$$

PROBLEM 38 Hanging chain or rope undergoing small oscillations.

A flexible uniform chain or a heavy rope of length (L) and a constant linear density (mass per unit length) ρ, is fixed at the upper end and free at the lower end. Choosing the equilibrium position of the free end of the chain as the origin ($x = 0$), we let the x-axis be vertical such that the upper fixed end is represented by ($x = L$). If at a point distant x from the lower end, the chain is slightly disturbed from its position of equilibrium in a vertical plane, it undergoes "small" oscillations. Derive the equation of motion and find its solution.

Derivation of the Equation

Let $u(x, t)$ be the horizontal deflection of the chain at a point with vertical coordinate x at time t. As u is much smaller than L, we do not need to consider the difference between distances measured along the chain and distances measured along the x axis, i.e., we can neglect the difference $\sqrt{x^2+u^2} - x = x\left(1+\dfrac{u^2}{x^2}\right)^{\frac{1}{2}} - x \sim \dfrac{u^2}{2x}$. For the same reason we can neglect the vertical displacement due to oscillations. We also assume that the angle $\alpha(x)$ between the local direction of the chain and x-axis is small, thus we write

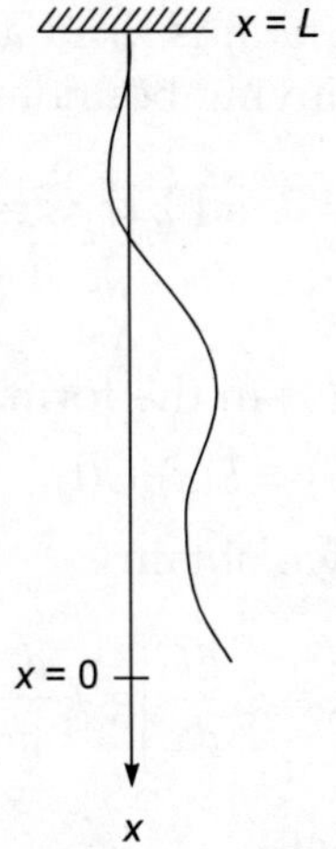

Fig. 8.24 Oscillating chain

$$\sin\alpha = \tan\alpha = \frac{\partial u}{\partial x}. \quad \text{...(i)}$$

The horizontal component of the net force acting on a segment of length Δx of the chain at time t due to the internal tension $T(x)$ is

$$T(x+\Delta x)\frac{\partial u(x+\Delta x,t)}{\partial x} - T(x)\frac{\partial u(x,t)}{\partial x} \cong \frac{\partial}{\partial x}\left\{T(x)\frac{\partial u}{\partial x}\right\}\Delta x. \quad \text{...(ii)}$$

By Newton's second law of motion, we have

$$\frac{\partial}{\partial x}\left\{T(x)\frac{\partial u}{\partial x}\right\}\Delta x = \rho\Delta x\frac{\partial^2 u}{\partial t^2} \quad \text{(where } t \text{ is the time elapsed)}$$

or

$$\frac{\partial}{\partial x}\left\{T(x)\frac{\partial u}{\partial x}\right\} = \rho\frac{\partial^2 u}{\partial t^2}. \quad \text{...(iii)}$$

The tension at the point x is the weight of the part of the chain or rope below x. Therefore, we have

$$T = \rho g x \quad \text{...(iv)}$$

where g is the acceleration due to gravity.

Substitution of Eq. (iv) into Eq. (iii) yields

$$g\left[x\frac{\partial^2 u}{\partial x^2} + \frac{\partial u}{\partial x}\right] = \frac{\partial^2 u}{\partial t^2}. \quad \text{...(v)}$$

Equation (v) is the equation of motion of the oscillating vertical chain.

Solution of the equation: Since the chain is fixed at the top where $x = L$ and free at the bottom where $x = 0$, the accompanying boundary conditions are given by

$$u(L,t)=0,\ |u(0,t)| << \infty \qquad \text{...(vi)}$$

for all values of t.

Let us assume the solution of Eq. (v) in the form

$$u(x, t) = F(x)\, G(t). \qquad \text{...(vii)}$$

Substituting Eq. (vii) into Eq. (v), we obtain

$$g\left[xG\frac{d^2F}{dx^2}+G\frac{dF}{dx}\right]=F\frac{d^2G}{dt^2}.$$

Dividing both sides by FG, we obtain

$$\frac{g}{F}\left[x\frac{d^2F}{dx^2}+\frac{dF}{dx}\right]=\frac{1}{G}\frac{d^2G}{dt^2}=-\omega^2 \qquad \text{...(viii)}$$

where ω is a constant usually the angular frequency of the oscillations.

From Eq. (viii), we obtain

$$\frac{d^2G}{dt^2}+\omega^2G=0 \qquad \text{...(ix)}$$

and

$$x\frac{d^2F}{dx^2}+\frac{dF}{dx}+\frac{\omega^2}{g}F=0. \qquad \text{...(x)}$$

A general solution of Eq. (ix) has the form

$$G(t)=C_1\cos(\omega t)+C_2\sin(\omega t) \qquad \text{...(xi)}$$

where C_1 and C_2 are arbitrary constants.

To solve Eq. (x), let us introduce a variable z such that

$$z^2=\frac{4x}{g} \text{ and let } F=F(x)=P(z).$$

With the new independent variable z, we rewrite Eq. (x) as

$$\frac{gz^2}{4}\cdot\frac{4}{g^2z^2}\left[\frac{d^2P(z)}{dz^2}-\frac{1}{z}\frac{dP(z)}{dz}\right]+\frac{2}{gz}\frac{dP(z)}{dz}+\frac{\omega^2}{g}P(z)=0$$

or

$$\frac{d^2P(z)}{dz^2}+\frac{1}{z}\frac{dP(z)}{dz}+\omega^2P(z)=0$$

or $$z^2\frac{d^2P(z)}{dz^2}+z\frac{dP(z)}{dz}+\omega^2z^2P(z)=0\,.$$

This equation is a Bessel equation having a solution

$$P(z)=C_3J_0(\omega z)+C_4N_0(\omega z)$$

where C_3 and C_4 are arbitrary constants and J_0 and N_0 are respectively the Bessel functions of the first and second kinds.

In terms of the old variable x, we write

$$P(z)=F(x)=C_3J_0\left(2\omega\sqrt{\frac{x}{g}}\right)+C_4N_0\left(2\omega\sqrt{\frac{x}{g}}\right).$$

Putting $k=\dfrac{\omega}{\sqrt{g}}$, we write

$$F(x)=C_3J_0(2k\sqrt{x})+C_4N_0(2k\sqrt{x})\,. \qquad \text{...(xii)}$$

Substituting Eqs. (xi) and (xii) in Eq. (vii), we obtain the solution of the equation of motion of the oscillating chain or rope as

$$u(x,t)=\left\{C_3J_0(2k\sqrt{x})+C_4N_0(2k\sqrt{x})\right\}\left\{C_1\cos(\omega t)+C_2\sin(\omega t)\right\}. \qquad \text{...(xiii)}$$

Since the Bessel function of the second kind or Neumann function $N_0(r)$ is not finite at $r = 0$, the second boundary condition given in (vi) requires that $C_4=0$.

If $\alpha_n\,(n=1,2,3\ldots)$ are the zeroes of J_0, the first boundary condition given in (vi) yields

$$J_0(2k_n\sqrt{L})=J_0(\alpha_n)=0$$

where $k_n=\dfrac{\omega_n}{\sqrt{g}}$, ($\omega_n$ being the angular frequency of nth mode of oscillation).

Therefore, we have

$$2k_n\sqrt{L}=\alpha_n \text{ or, } k_n=\frac{\alpha_n}{2\sqrt{L}} \text{ which yields } \omega_n=\frac{\alpha_n}{2}\sqrt{\frac{g}{L}}\,. \qquad \text{...(xiv)}$$

From Eq. (xiii), using the values of C_4 and ω_n we write for a particular n mode of normal oscillation,

$$u_n(x,t)=J_0\left(2\omega_n\sqrt{\frac{x}{g}}\right)\left[A_n\cos(\omega_nt)+B_n\sin(\omega_nt)\right] \qquad \text{...(xv)}$$

where $$A_n=C_1C_3,\;B_n=C_2C_3\,.$$

A superposition of the solutions for all possible oscillation modes gives the complete solution of the problem as

$$u(x,t)=\sum_{n=1}^{\infty}u_n(x,t)$$

$$=\sum_{n=1}^{\infty}J_0\left(2\omega_n\sqrt{\frac{x}{g}}\right)\left[A_n\cos(\omega_n t)+B_n\sin(\omega_n t)\right]. \qquad \text{...(xvi)}$$

Here, A_n and B_n are to be determined from the initial conditions of the problem.

PROBLEM 39 Show that for nth mode of normal oscillation in the previous problem, the horizontal deflection at a point x at time t is also given by

$$u_n(x,t)=C_nJ_0\left(\alpha_n\sqrt{\frac{x}{L}}\right)\cos\left\{\left(\frac{\alpha_n}{2}\sqrt{\frac{g}{L}}\right)t+\phi_n\right\}$$

where C_n and ϕ_n are arbitrary constants and hence write down the complete solution of the oscillating-chain problem.

Solution: In Eq. (xv) of previous problem, let us put $A_n=C_n\cos\phi_n$ and $B_n=-C_n\sin\phi_n$ such that $C_n=\sqrt{A_n^2+B_n^2}$ and $\phi_n=-\tan^{-1}\dfrac{B_n}{A_n}$.

Further, if α_n are the positive roots of J_0 then, we have

$$2\omega_n\sqrt{\frac{L}{g}}=\alpha_n \text{ or, } \omega_n=\frac{\alpha_n}{2}\sqrt{\frac{g}{L}} \quad \text{(by boundary condition).}$$

Then, we write Eq. (xv) as

$$u_n(x,t)=C_n\,J_0\left(\alpha_n\sqrt{\frac{x}{L}}\right)\cos\left\{\left(\frac{\alpha_n}{2}\sqrt{\frac{g}{L}}\right)t+\phi_n\right\}.$$

The complete solution is then written as

$$u(x,t)=\sum_{n=1}^{\infty}C_nJ_0\left(\alpha_n\sqrt{\frac{x}{L}}\right)\cos\left\{\left(\frac{\alpha_n}{2}\sqrt{\frac{g}{L}}\right)t+\phi_n\right\}.$$

PROBLEM 40 A perfectly flexible chain of length (L) and constant linear density (ρ) is fastened at one end to a fixed point from which it hangs vertically at rest along the x-axis taking the origin ($x = 0$) at the lower end of the chain. It is then set to small oscillation in a fixed vertical plane by imparting to each point of it at time $t = 0$ an initial horizontal velocity of magnitude $v = f(x)$ where $f(x)$ is continuous and differentiable for $0 \le x \le L$ and $f(L) = 0$. Determine the horizontal deflection $u(x, t)$

of a point at distance x from the lower end at time t. Assume that initially, all points have zero deflection.

Solution: See the solution of Problem 38 to obtain the solution

$$u(x,t)=\sum_{n=1}^{\infty}J_0\left(2\omega_n\sqrt{\frac{x}{g}}\right)\left[A_n\cos(\omega_n t)+B_n\sin(\omega_n t)\right].$$

The initial condition that $u(x, 0) = 0$ requires that $A_n = 0$.
Therefore, we have

$$u(x,t)=\sum_{n=1}^{\infty}B_nJ_0\left(2\omega_n\sqrt{\frac{x}{g}}\right)\sin(\omega_n t). \qquad \ldots\text{(i)}$$

Another initial condition that $v=\left(\frac{\partial u}{\partial t}\right)_{t=0}=f(x)$ yields

$$f(x)=\sum_{n=1}^{\infty}B_n\omega_nJ_0\left(2\omega_n\sqrt{\frac{x}{g}}\right). \qquad \ldots\text{(ii)}$$

Multiplying both sides of Eq. (ii) by $xJ_0\left(2\omega_m\sqrt{\frac{x}{g}}\right)$ and integrating the result with respect to x within the limit from $x = 0$ to $x = L$, we obtain

$$\int_0^L f(x)J_0\left(2\omega_m\sqrt{\frac{x}{g}}\right)x\,dx=\sum_{n=1}^{\infty}B_n\omega_n\int_0^L J_0\left(2\omega_m\sqrt{\frac{x}{g}}\right)J_0\left(2\omega_n\sqrt{\frac{x}{g}}\right)x\,dx$$

$$=B_m\omega_m\frac{L^2}{2}\left[J_1\left(2\omega_m\sqrt{\frac{L}{g}}\right)\right]^2.$$

Therefore, we have

$$B_m=\frac{2}{\omega_mL^2\left[J_1\left(2\omega_m\sqrt{\frac{L}{g}}\right)\right]^2}\int_0^L f(x)J_0\left(2\omega_m\sqrt{\frac{x}{g}}\right)x\,dx.$$

Changing the subscript, we write

$$B_n=\frac{2}{\omega_nL^2\left[J_1\left(2\omega_n\sqrt{\frac{L}{g}}\right)\right]^2}\int_0^L f(x)J_0\left(2\omega_n\sqrt{\frac{x}{g}}\right)x\,dx. \qquad \ldots\text{(iii)}$$

The horizontal displacement $u(x, t)$ of a point at distance x from the lower end at time t is given by Eq. (i) where B_n is given by Eq. (iii).

PROBLEM 41 For the second mode ($n = 2$) of the oscillating chain problem, there is a node (stationary point) located on the interval $0 < x < L$. Show that this node always occurs at the point $x \cong 0.19\,L$.

Solution: For the $n = 2$ mode of oscillation, the deflection at the point x at time t is given by

$$u_2(x,t) = C_2 J_0\left(\alpha_2\sqrt{\frac{x}{L}}\right)\cos\left\{\left(\frac{\alpha_2}{2}\sqrt{\frac{g}{L}}\right)t + \phi_2\right\} \quad \text{(see Problem 39)}$$

where C_2, ϕ_2 are arbitrary constants and α_2 is the second positive zero of the Bessel function of zero order.

By boundary condition, we know that at $x = L, u_2(L,t) = 0$, i.e., $J_0(\alpha_2) = 0$.

A node shall occur at a lower zero (α_1) of the same Bessel function.

$$\text{This will occur when } \alpha_2\sqrt{\frac{x}{L}} = \alpha_1 \text{ or, } \frac{x}{L} = \left(\frac{\alpha_1}{\alpha_2}\right)^2 = \left(\frac{2.4048}{5.5201}\right)^2 \cong 0.19$$

i.e., when $x \cong 0.19\,L$.

PROBLEM 42 The complete solution of the oscillating chain problem consists of the superposition of all vibration modes, i.e.,

$$u(x,t) = \sum_{n=1}^{\infty} C_n J_0\left(\alpha_n\sqrt{\frac{x}{L}}\right)\cos\left\{\left(\frac{\alpha_n}{2}\sqrt{\frac{g}{L}}\right)t + \phi_n\right\}$$

where α_n are zeroes of Bessel function of the first kind of zero order, g is the acceleration due to gravity, L is the length of the chain, t is the time elapsed, x is the distance from the free end and ϕ_n is a phase constant.

(a) If the initial speed is $\left[\frac{\partial u}{\partial t}\right]_{t=0} = 0$, show that a sufficient condition is $\phi_n = 0$ ($n = 1, 2, 3 \ldots$),

(b) Further, if the initial deflection is $u(x,0) = \varepsilon(L - x)$), where ε is a small constant, show that

$$u(x,t) = 8\varepsilon L \sum_{n=1}^{\infty} \frac{J_0\left(\alpha_n\sqrt{\frac{x}{L}}\right)\cos\left\{\left(\frac{\alpha_n}{2}\sqrt{\frac{g}{L}}\right)t\right\}}{\alpha_n^3 J_1(\alpha_n)}.$$

Solution:

(a) We have

$$u(x,t)=\sum_{n=1}^{\infty}C_n J_0\left(\alpha_n\sqrt{\frac{x}{L}}\right)\cos\left\{\left(\frac{\alpha_n}{2}\sqrt{\frac{g}{L}}\right)t+\phi_n\right\} \qquad \text{...(i)}$$

The initial speed is

$$\left[\frac{\partial u}{\partial t}\right]_{t=0}=-\left[\sum_{n=1}^{\infty}\frac{\alpha_n}{2}\sqrt{\frac{g}{L}}C_nJ_0\left(\alpha_n\sqrt{\frac{x}{L}}\right)\sin\left\{\left(\frac{\alpha_n}{2}\sqrt{\frac{g}{L}}\right)t+\phi_n\right\}\right]_{t=0}=0$$

or $\sin\phi_n=0$.

Therefore, a sufficient condition is $\phi_n = 0$. ...(ii)

(b) Using the result of (a), we have

$$\varepsilon(L-x)=u(x,0)=\sum_{n=1}^{\infty}C_nJ_0\left(\alpha_n\sqrt{\frac{x}{L}}\right).$$

Putting $\sqrt{\frac{x}{L}}=q$ or, $x=Lq^2$, we write

$$\varepsilon L\left(1-q^2\right)=\sum_{n=1}^{\infty}C_nJ_0\left(\alpha_n q\right).$$

Multiplying both sides by $qJ_0\left(\alpha_m q\right)$ and integrating the result with respect to q within the limit from $q = 0$ to $q = 1$, we obtain

$$\varepsilon L\int_0^1\left(1-q^2\right)qJ_0\left(\alpha_m q\right)dq=\sum_{n=1}^{\infty}C_n\int_0^1 J_0\left(\alpha_m q\right)J_0\left(\alpha_n q\right)q\,dq$$

or $$\varepsilon L\left\{\int_0^1 qJ_0\left(\alpha_m q\right)dq-\int_0^1 q^3J_0\left(\alpha_m q\right)dq\right\}=C_m\frac{1}{2}J_1^2\left(\alpha_m\right). \qquad \text{...(iii)}$$

Equation (iii) can be written as

$$\varepsilon L\{I_1-I_2\}=\frac{1}{2}C_mJ_1^2(\alpha_m) \qquad \text{...(iv)}$$

where $$I_1=\int_0^1 qJ_0(\alpha_m q)dq \qquad \text{...(v)}$$

and $$I_2=\int_0^1 q^3J_0(\alpha_m q)dq. \qquad \text{...(vi)}$$

Putting $\alpha_m q = z$, or, $q = \dfrac{z}{\alpha_m}$, Eq. (v) can be written as

$$I_1 = \frac{1}{\alpha_m^2}\int_0^{\alpha_m} zJ_0(z)\,dz$$

$$= \frac{1}{\alpha_m^2}\left[zJ_1(z)\right]_0^{\alpha_m}$$

$$= \frac{1}{\alpha_m}J_1(\alpha_m).$$

Here, for evaluating the integral, use has been made of the relation

$$\frac{d}{dz}\{z^n J_n(z)\} = z^n J_{n-1}(z).$$

Similarly, Eq. (vi) can be written as

$$I_2 = \frac{1}{\alpha_m^4}\int_0^{\alpha_m} z^3 J_0(z)\,dz$$

$$= \frac{1}{\alpha_m^4}\int_0^{\alpha_m} z^2\{zJ_0(z)\}\,dz.$$

Integrating by parts on the right-hand side, we write

$$I_2 = \frac{1}{\alpha_m^4}\left\{\left[z^3 J_1(z)\right]_0^{\alpha_m} - 2\int_0^{\alpha_m} z^2 J_1(z)\,dz\right\}$$

$$= \frac{1}{\alpha_m}J_1(\alpha_m) - \frac{2}{\alpha_m^4}\left[z^2 J_2(z)\right]_0^{\alpha_m}$$

$$= \frac{1}{\alpha_m}J_1(\alpha_m) - \frac{2}{\alpha_m^2}J_2(\alpha_m). \qquad \text{...(vii)}$$

In the relation

$$J_{n-1}(z) + J_{n+1}(z) = \frac{2n}{z}J_n(z),$$

putting $n = 1$, we obtain

$$J_0(z) + J_2(z) = \frac{2}{z}J_1(z)$$

or

$$J_2(z) = \frac{2}{z}J_1(z) - J_0(z).$$

Therefore, we have

$$J_2(\alpha_m) = \frac{2}{\alpha_m} J_1(\alpha_m) - J_0(\alpha_m). \qquad \text{...(viii)}$$

Substituting Eq. (viii) into Eq. (vii), we write

$$I_2 = \frac{1}{\alpha_m} J_1(\alpha_m) - \frac{2}{\alpha_m^2}\left[\frac{2}{\alpha_m} J_1(\alpha_m) - J_0(\alpha_m)\right].$$

Recalling the value of I_1, we write

$$I_1 - I_2 = \frac{4}{\alpha_m^3} J_1(\alpha_m) - \frac{2}{\alpha_m^2} J_0(\alpha_m). \qquad \text{...(ix)}$$

Substituting Eq. (ix) into Eq. (iv), we obtain

$$C_m = \frac{2\varepsilon L}{\{J_1(\alpha_m)\}^2}\left[\frac{4}{\alpha_m^3} J_1(\alpha_m) - \frac{2}{\alpha_m^2} J_0(\alpha_m)\right].$$

Changing the subscript m to n, we write

$$C_n = \frac{2\varepsilon L}{\{J_1(\alpha_n)\}^2}\left[\frac{4}{\alpha_n^3} J_1(\alpha_n) - \frac{2}{\alpha_n^2} J_0(\alpha_n)\right] = \frac{8\varepsilon L}{\alpha_n^3 J_1(\alpha_n)} \qquad \text{...(x)}$$

as $$J_0(\alpha_n) = 0.$$

Substituting Eqs. (ii) and (x) into Eq. (i), we obtain

$$u(x,t) = 8\varepsilon L \sum_{n=1}^{\infty} \frac{J_0\left(\alpha_n \sqrt{\frac{x}{L}}\right) \cos\left\{\left(\frac{\alpha_n}{2}\sqrt{\frac{g}{L}}\right)t\right\}}{\alpha_n^3 J_1(\alpha_n)} \quad [\text{as } \phi_n = 0].$$

PROBLEM 43 Prove that $J_n(x) = 0$ has no repeated roots except at $x = 0$.

Solution: If possible, let α be a repeated root of $J_n(x) = 0$.

Therefore, $J_n(\alpha) = 0$ and $J_n'(\alpha) = 0$.

From the recurrence relations for Bessel functions, we know that

$$J_{n+1}(\alpha) = \frac{n}{\alpha} J_n(\alpha) - J_n'(\alpha) = 0$$

and

$$J_{n-1}(\alpha) = \frac{n}{\alpha} J_n(\alpha) + J_n'(\alpha) = 0.$$

Thus, we see that $J_{n-1}(x)$, $J_n(x)$ and $_{n+1}(x)$ are all zero for the same value of x which is absurd except when $x = 0$ for $n > 0$, since two or more different power series cannot represent the same function. Hence, we conclude that $J_n(x) = 0$ cannot have a repeated root except at $x = 0$.

PROBLEM 44 Prove that the Wronskian of $J_p(x)$ and $J_{-p}(x)$ is equal to $-\dfrac{2\sin(p\pi)}{\pi x}$.

Solution: We know that

$$J_p(x) = \sum_{r=0}^{\infty} \frac{(-1)^r}{r!\,\Gamma(p+r+1)} \left(\frac{x}{2}\right)^{p+2r}$$

and
$$J_{-p}(x) = \sum_{r=0}^{\infty} \frac{(-1)^r}{r!\,\Gamma(r-p+1)} \left(\frac{x}{2}\right)^{2r-p}.$$

For simplicity, let us consider only the first terms of the above series for deriving the Wronskian.

Thus, we write

$$J_p(x) \cong \frac{x^p}{2^p\,\Gamma(p+1)}, \quad J_{-p}(x) \cong \frac{2^p x^{-p}}{\Gamma(1-p)}.$$

Therefore, we have

$$J_p'(x) \cong \frac{d}{dx} J_p(x) = \frac{p x^{p-1}}{2^p\,\Gamma(1+p)}, \quad J_{-p}'(x) \cong \frac{-p\,2^p x^{-p-1}}{\Gamma(1-p)}.$$

Now, the Wronskian is given by

$$W(J_p, J_{-p}) = J_p(x) J_{-p}'(x) - J_p'(x) J_{-p}(x) = -\frac{p x^{-1}}{\Gamma(1-p)\Gamma(1+p)} - \frac{p x^{-1}}{\Gamma(1-p)\Gamma(1+p)}$$

$$= -\frac{2p}{x\,\Gamma(1-p)\Gamma(1+p)} = -\frac{2\sin(p\pi)}{x\pi} \quad \text{[assuming that } p \neq \text{integer]}$$

where use has been made of the relation

$$\Gamma(1-p)\Gamma(1+p) = \frac{p\pi}{\sin(p\pi)}.$$

PROBLEM 45 If Laplace first integral representation of Legendre polynomials is given by

$$P_n(x) = \frac{1}{\pi} \int_0^{\pi} \left\{ x \pm \left(\sqrt{x^2 - 1}\right) \cos\theta \right\}^n d\theta,$$

prove that $$\sum_{n=0}^{\infty} P_n(x)\frac{t^n}{n!} = e^{tx} J_0\left(t\sqrt{1-x^2}\right).$$

Solution: We have

$$\sum_{n=0}^{\infty} P_n(x)\frac{t^n}{n!} = \sum_{n=0}^{\infty} \frac{1}{\pi}\int_0^{\pi}\left\{x \pm \left(\sqrt{x^2-1}\right)\cos\theta\right\}^n \frac{t^n}{n!}\,d\theta$$

$$= \frac{1}{\pi}\int_0^{\pi} e^{t\left\{x\pm\left(\sqrt{x^2-1}\right)\cos\theta\right\}}\,d\theta$$

$$= e^{tx}\frac{1}{\pi}\int_0^{\pi} e^{\pm it\left(\sqrt{1-x^2}\right)\cos\theta}\,d\theta$$

$$= e^{tx}\left[\frac{1}{\pi}\int_0^{\pi}\cos\left\{t\left(\sqrt{1-x^2}\right)\cos\theta\right\}d\theta \pm i\frac{1}{\pi}\int_0^{\pi}\sin\left\{t\left(\sqrt{1-x^2}\right)\cos\theta\right\}d\theta\right]$$

$$= e^{tx}\left[\frac{1}{\pi}\int_0^{\pi}\left\{J_0\left(t\sqrt{1-x^2}\right) + \sum_{n=1}^{\infty}(-1)^n\, 2J_{2n}\left(t\sqrt{1-x^2}\right)\cos(2n\theta)\right\}d\theta\right.$$

$$\left.\pm\frac{1}{\pi}i\int_0^{\pi}\sum_{n=0}^{\infty}(-1)^n\, 2J_{2n+1}\left(t\sqrt{1-x^2}\right)\cos(2n+1)\theta\,d\theta\right]$$

$$= e^{tx} J_0\left(t\sqrt{1-x^2}\right).$$

Here, results of **Problem 14 (c)** and **(d)** have been used and the cosine integrals vanish. The right-hand side can be considered a generating function of Legendre polynomials.

PROBLEM 46 Prove that $\sum_{n=1}^{\infty}\frac{2J_0(\alpha_n x)}{\alpha_n J_1(\alpha_n)} = 1$ where α_n are zeroes of $J_0(x)$.

Solution: For an arbitrary function $f(x)$, we have

$$f(x) = \sum_{n=1}^{\infty} c_n J_p(\alpha_n x)$$

where $$c_n = \frac{2}{\left[J_{p+1}(\alpha_n)\right]^2}\int_0^1 x f(x) J_p(\alpha_n x)\,dx.$$

If we put $f(x)=1$ and $p=0$, we have

$$1=\sum_{n=1}^{\infty} c_n J_0(\alpha_n x) \tag{i}$$

with
$$c_n=\frac{2}{[J_1(\alpha_n)]^2}\int_0^1 xJ_0(\alpha_n x)dx. \tag{ii}$$

Putting $u=\alpha_n x$, we have $x=\dfrac{u}{\alpha_n}$ and $dx=\dfrac{du}{\alpha_n}$.

Then, Eq. (ii) yields

$$c_n=\frac{2}{[\alpha_n J_1(\alpha_n)]^2}\int_0^{\alpha_n} uJ_0(u)du. \tag{iii}$$

In the recurrence relation

$$\frac{d}{du}\left[u^n J_n(u)\right]=u^n J_{n-1}(u),$$

putting n = 1, we have

$$\frac{d}{du}\left[uJ_1(u)\right]=uJ_0(u).$$

Substituting this into Eq. (iii), we have

$$c_n=\frac{2}{\left[\alpha_n J_1(\alpha_n)\right]^2}\int_0^{\alpha_n}\frac{d}{du}\left[uJ_1(u)\right]du$$

$$=\frac{2}{[\alpha_n J_1(\alpha_n)]^2}[uJ_1(u)]_0^{\alpha_n}=\frac{2}{\alpha_n J_1(\alpha_n)}. \tag{iv}$$

Substituting Eq. (iv) into Eq. (i), we obtain

$$1=\sum_{n=1}^{\infty}\frac{2J_0(\alpha_n x)}{\alpha_n J_1(\alpha_n)}.$$

PROBLEM 47 Prove that $J_{l+\frac{1}{2}}(x)=\dfrac{1}{i^l}\sqrt{\dfrac{x}{2\pi}}\displaystyle\int_{-1}^{+1} e^{ixt}P_l(t)dt$.

Solution: To be proved in **Problem 21: Section 8.28: Solved Problems on Bessel functions of other kinds.**

EXERCISE

1. Prove that $J_1'(0)=\frac{1}{2}$ and $J_n'(0) = 0$ for $n > 1$ where the prime denotes differentiation with respect to x.
2. Prove that

(a) $x\sin x = 2\sum_{n=1}^{\infty}(-1)^{n-1}(2n)^2 J_{2n}(x)$,

(b) $x\cos x = 2\sum_{n=1}^{\infty}(-1)^{n-1}(2n-1)^2 J_{2n-1}(x)$,

(c) $\cosh(x\sinh\theta) = J_0(x) + 2\sum_{n=1}^{\infty} J_{2n}(x)\cosh(2n\theta)$,

(d) $\sinh(x\sinh\theta) = 2\sum_{n=1}^{\infty} J_{2n-1}(x)\sinh\{(2n-1)\theta\}$.

3. Using the series formula of Bessel functions of the first kind, prove that

(i) $J_{p-1}(x) + J_{p+1}(x) = \frac{2p}{x}J_p(x)$,

(ii) $J_{p-1}(x) - J_{p+1}(x) = 2\frac{d}{dx}J_p(x)$.

4. Prove that

$$xJ_{n-1}(x) = 2\left[nJ_n(x) - (n+2)J_{n+2}(x) + (n+4)J_{n+4}(x) - \ldots\infty\right].$$

5. Prove that

(i) $2J_0''(x) = J_2(x) - J_0(x)$,

(ii) $J_0'(x) = -J_1(x)$,

(iii) $J_2(x) = J_0''(x) - x^{-1}J_0'(x)$,

(iv) $J_3(x) + 3J_0'(x) + 4J_0'''(x) = 0$,

(v) $\int J_3(x)dx + J_2(x) + 2x^{-1}J_1(x) = 0$.

6. Prove that

$$\frac{d}{dx}J_n(x) = \frac{2}{x}\left[\frac{n}{2}J_n(x) - (n+2)J_{n+2}(x) + (n+4)J_{n+4}(x) - \ldots\infty\right].$$

7. Prove that $\left(\frac{d}{xdx}\right)^r\left[x^nJ_n(x)\right] = x^{n-r}J_{n-r}(x)$ and deduce that

$$J_{-r-\frac{1}{2}}(x)=\frac{(2x)^{r+\frac{1}{2}}}{\sqrt{\pi}}\left[\frac{d}{d(x^2)}\right]^r\left(\frac{\cos x}{x}\right).$$

8. Using the infinite series formula for Bessel functions, prove that $\frac{1}{x}\frac{d}{dx}\left[\frac{J_n(x)}{x^n}\right]=-\frac{J_{n+1}(x)}{x^{n+1}}$ and deduce the relations

$$\left(\frac{1}{x}\frac{d}{dx}\right)^m\left[\frac{J_n(x)}{x^n}\right]=(-1)^m\frac{J_{n+m}(x)}{x^{n+m}} \text{ and } J_m(x)=(-1)^m x^m\left(\frac{1}{x}\frac{d}{dx}\right)^m J_0(x).$$

9. Obtain the expressions for $J_{-\frac{1}{2}}(x)$ and $J_{\frac{1}{2}}(x)$. Deduce that

(i) $J^2_{-\frac{1}{2}}(x)-J^2_{\frac{1}{2}}(x)=\frac{2}{\pi x}\cos 2x$,

(ii) $J^2_{-\frac{1}{2}}(x)+J^2_{\frac{1}{2}}(x)=\frac{2}{\pi x}$,

(iii) $J_{-\frac{1}{2}}(x)J_{\frac{1}{2}}(x)=\frac{\sin 2x}{\pi x}$.

10. Prove that

(i) $J_{-\frac{5}{2}}(x)=\sqrt{\frac{2}{\pi x}}\left[\frac{(3-x^2)\cos x}{x^2}-\frac{3\sin x}{x}\right]$,

(ii) $J_{5/2}(x)=\sqrt{\frac{2}{\pi x}}\left[\frac{(3-x^2)\sin x}{x^2}-\frac{3\cos x}{x}\right]$.

11. Prove that $J_n(-x)=J_{-n}(x)$ if n is a positive integer.

12. Prove that $\int J_1(x)dx=-J_0(x)$ + Constant.

13. Prove that

(i) $\int x^n J_{n-1}(x)dx=x^n J_n(x)+$ Constant,

(ii) $\int x^{-n}J_{n+1}(x)dx=-x^{-n}J_n(x)$ + Constant.

14. Prove that $\int_0^{\frac{\pi}{2}}\sqrt{\pi x}\,J_{\frac{1}{2}}(2x)dx=1$.

15. Prove that

(i) $\int_0^z x^k J_{k-1}(ax)\,dx = \frac{z^k}{a} J_k(az)$ where k is a positive integer,

(ii) $\int_0^z x J_0(ax)\,dx = \frac{z}{a} J_1(az)$.

16. Prove that $\int_0^\infty e^{-ax} x^n J_n(bx)\,dx = \frac{(2b)^n \Gamma\left(n+\frac{1}{2}\right)}{\sqrt{\pi}(a^2+b^2)^{n+\frac{1}{2}}}$

and deduce that

(i) $\int_0^\infty e^{-ax} J_0(bx)\,dx = \frac{1}{\sqrt{a^2+b^2}}$ $(a > 0)$,

(ii) $\int_0^\infty e^{-ax} x J_1(bx) = \frac{b}{(a^2+b^2)^{\frac{3}{2}}}$,

(iii) $\int_0^\infty e^{-ax} J_0(x)dx = \frac{1}{\sqrt{1+a^2}}$.

17. Prove that $\int_0^\infty e^{-ax^2} x^{n+1} J_n(bx)\,dx = \frac{b^n}{(2a)^{n+1}} e^{-\frac{b^2}{4a}}$.

18. Prove that if α_n are the zeroes of $J_0(x)$ then

$$1 = \sum_{n=1}^{\infty} \frac{2J_0(\alpha_n x)}{\alpha_n J_1(\alpha_n)},\ 0 < x < 1$$

hence deduce that

$$\sum_{n=1}^{\infty} \frac{1}{\alpha_n J_1(\alpha_n)} = \frac{1}{2}.$$

Is the expansion valid for $x = 1$?

19. What orders of Bessel function shall you choose to expand (i) x^2, (ii) x^3, (iii) x^4?

20. Obtain the value of $\int_0^a x^{n+1} J_n(x)dx$.

21. Prove that $x^2 = \sum_{n=1}^{\infty} \frac{4J_2(\alpha_n x)}{\alpha_n J_3(2\alpha_n)}$ $(0 \le x \le 2)$ where α_n are the positive roots of $J_2(x) = 0$.

22. Prove that the Wronskian of $J_p(x)$ and $J_{-p}(x)$ is equal to $-\dfrac{2\sin(p\pi)}{\pi x}$. Hence, deduce that $\dfrac{d}{dx}\left\{\dfrac{J_{-p}(x)}{J_p(x)}\right\}=-\dfrac{2\sin(p\pi)}{\pi x\left[J_p(x)\right]^2}$.

23. Using the result of No. 21, prove that

(i) $J_p(x)J_{-p-1}(x)+J_{-p}(x)J_{p+1}(x)=-\dfrac{2\sin(p\pi)}{\pi x}$,

(ii) $J_p(x)J_{-p+1}(x)+J_{-p}(x)J_{p-1}(x)=\dfrac{2\sin(p\pi)}{\pi x}$.

24. Expand the following functions in series of Bessel functions within the interval $0 \le x \le 1$ if α_n $(n = 1, 2, 3 \ldots)$ are zeroes of the Bessel function $J_0(x)$;

(i) $f(x)=1-x^2$,

(ii) $f(x)=x^2$.

Ans. (i) $1-x^2=8\sum_{n=1}^{\infty}\dfrac{J_0(\alpha_n x)}{\alpha_n^2 J_1(\alpha_n)}$,

(ii) $x^2=2\sum_{n=1}^{\infty}\dfrac{\alpha_n^2-4}{\alpha_n^3 J_1(\alpha_n)}J_0(\alpha_n x)$.

25. If α_n is a typical positive root of $J_0(x)=0$ and $-a \le r \le a$, show that

$$a^2-r^2=8a^2\sum_{n=1}^{\infty}\frac{1}{\alpha_n^3 J_1(\alpha_n)}J_0\left(\frac{\alpha_n r}{a}\right).$$

26. Using Laplace first integral representation of $P_n(x)$ prove that

$$\sum_{n=0}^{\infty}\frac{t^n}{n!}P_n(\cos\phi)=e^{t\cos\phi}J_0(t\sin\phi).$$

27. Prove that $(1-2xt+t^2)^{-\frac{1}{2}}=e^{tx}J_0(t\sqrt{1-x^2})$.

28. If Laplace first integral representation of Legendre polynomials is given by

$$P_n(x)=\frac{1}{\pi}\int_0^{\pi}\left\{x\pm(\sqrt{x^2-1})\cos\theta\right\}^n d\theta,$$

prove that $(1-2xt+t^2)^{-\frac{1}{2}}=e^{tx}J_0\left(t\sqrt{1-x^2}\right)$.

[*Hint:* Same as Solved Problem 45.]

8.28 SOLVED PROBLEMS ON BESSEL FUNCTIONS OF OTHER KINDS

PROBLEM 1 Obtain the expressions for (i) $N_{\frac{1}{2}}(x)$ and (ii) $N_{-\frac{1}{2}}(x)$

and (iii) show that $\left\{N_{\frac{1}{2}}(x)\right\}^2+\left\{N_{-\frac{1}{2}}(x)\right\}^2=\dfrac{2}{\pi x}$.

Solution:

(i) We know that

$$N_p(x)=\frac{\cos(p\pi)J_p(x)-J_{-p}(x)}{\sin(p\pi)}.$$

Putting $p=\frac{1}{2}$, we obtain

$$N_{\frac{1}{2}}(x)=-J_{-\frac{1}{2}}(x)=-\sqrt{\frac{2}{\pi x}}\cos x.$$

(ii) Putting $p=-\frac{1}{2}$, we obtain

$$N_{-\frac{1}{2}}(x)=J_{\frac{1}{2}}(x)=\sqrt{\frac{2}{\pi x}}\sin x.$$

(iii) Squaring the results of (i) and (ii) and adding together, we obtain

$$\left\{N_{\frac{1}{2}}(x)\right\}^2+\left\{N_{-\frac{1}{2}}(x)\right\}^2=\frac{2}{\pi x}.$$

PROBLEM 2 Prove that the Wronskian of $J_p(x)$ and $N_p(x)$ is equal to $\dfrac{2}{\pi x}$.

Solution: For simplicity, let us assume that p is not an integer.

$$N_p(x)=\frac{1}{\sin(p\pi)}\left[J_p(x)\cos(p\pi)-J_{-p}(x)\right]$$

$$=\cot(p\pi)J_p(x)-\operatorname{cosec}(p\pi)J_{-p}(x).$$

Differentiating both sides with respect to x, we have

$$N_p'(x)=\cot(p\pi)J_p'(x)-\operatorname{cosec}(p\pi)J_{-p}'(x).$$

Now, the Wronskian is given by

$$W = J_p(x)N_p'(x) - J_p'(x)N_p(x)$$
$$= J_p(x)\cot(p\pi)J_p'(x) - J_p(x)\,\text{cosec}(p\pi)J_{-p}'(x)$$
$$-J_p'(x)\cot(p\pi)J_p(x) + J_p'(x)\,\text{cosec}(p\pi)J_{-p}(x)$$
$$= -\text{cosec}(p\pi)\left[J_p(x)J_{-p}'(x) - J_p'(x)J_{-p}(x)\right]$$
$$= -\text{cosec}(p\pi)\left[-\frac{2\sin(p\pi)}{\pi x}\right]$$
$$= \frac{2}{\pi x}.$$

PROBLEM 3

Lengthening simple pendulum: If a simple pendulum has its length increased at a steady rate, find its equation of motion for small oscillations and find a solution in terms of Bessel functions.

Solution: Let a simple pendulum, oscillating with small amplitude, increase its length at the steady rate v. If L is the initial length, the length after time t is $r = L + vt$.

If θ is the instantaneous angular displacement, the transverse acceleration of the mass m is $r\frac{d^2\theta}{dt^2} + 2\frac{dr}{dt}\frac{d\theta}{dt}$, perpendicular to the string in the direction of increasing θ. This acceleration is also retarded by the component of the weight, thereby giving the equation of motion as $m\left(r\frac{d^2\theta}{dt^2} + 2\frac{dr}{dt}\frac{d\theta}{dt}\right) = -mg\sin\theta \approx -mg\theta$ (for small θ) where g is the acceleration due to gravity.

As $r = L + vt$ we have, $\frac{dr}{dt} = v, \frac{d\theta}{dt} = \frac{d\theta}{dr}\frac{dr}{dt} = v\frac{d\theta}{dr}, \frac{d^2\theta}{dt^2} = v^2\frac{d^2\theta}{dr^2}.$

Therefore, we write the equation of motion as

$$v^2 r\frac{d^2\theta}{dr^2} + 2v^2\frac{d\theta}{dr} + g\theta = 0$$

or

$$r\frac{d^2\theta}{dr^2} + 2\frac{d\theta}{dr} + \lambda^2\theta = 0 \tag{i}$$

where $\lambda = \frac{\sqrt{g}}{v}$.

Let us put $x = 2\lambda\sqrt{r}$. Then, $\frac{d\theta}{dr} = \frac{d\theta}{dx}\frac{dx}{dr} = \frac{2\lambda^2}{x}\frac{d\theta}{dx}, \frac{d^2\theta}{dr^2} = \frac{4\lambda^4}{x^2}\left(\frac{d^2\theta}{dx^2} - \frac{1}{x}\frac{d\theta}{dx}\right)$.

Using these relations, Eq. (i) can be written as

$$x\frac{d^2\theta}{dx^2} + 3\frac{d\theta}{dx} + x\theta = 0. \tag{ii}$$

Further, let us put $y = \theta x$ or $\theta = \frac{y}{x}$. Then, we have

$$\frac{d\theta}{dx} = \frac{1}{x}\frac{dy}{dx} - \frac{y}{x^2}, \frac{d^2\theta}{dx^2} = \frac{1}{x}\frac{d^2y}{dx^2} - \frac{2}{x^2}\frac{dy}{dx} + \frac{2y}{x^3}.$$

Then, we write Eq. (ii) as

$$x^2\frac{d^2y}{dx^2} + x\frac{dy}{dx} + (x^2 - 1)y = 0. \tag{iii}$$

Equation (iii) is a Bessel equation whose solution can be expressed as

$$y(x) = C_1 J_1(x) + C_2 N_1(x)$$

where C_1 and C_2 are arbitrary constants and $J_1(x)$ and $N_1(x)$ are Bessel functions of the first and second kinds of order one.

Therefore, solution of Eq. (ii) is

$$\theta = \frac{y}{x} = \frac{C_1}{2\lambda\sqrt{r}} J_1\left(2\lambda\sqrt{r}\right) + \frac{C_2}{2\lambda\sqrt{r}} N_1\left(2\lambda\sqrt{r}\right)$$

$$= \frac{A}{\sqrt{r}} J_1\left(2\lambda\sqrt{r}\right) + \frac{B}{\sqrt{r}} N_1\left(2\lambda\sqrt{r}\right) \tag{iv}$$

where $A = \frac{C_1}{2\lambda}, B = \frac{C_2}{2\lambda}$.

If we take the origin of time when the angular velocity is temporarily zero when the corresponding angular displacement is β, we have

at $$t = 0, r = L, \theta = \beta, \left[\frac{d\theta}{dt}\right]_{t=0} = 0, \lambda = \frac{\sqrt{g}}{v}.$$

Substituting β for θ when $r = L$, Eq. (iv) yields

$$\beta = \frac{A}{\sqrt{L}} J_1(\mu) + \frac{B}{\sqrt{L}} N_1(\mu), \left(\mu = 2\lambda\sqrt{L} = \frac{2\sqrt{gL}}{v}\right)$$

or $$AJ_1(\mu) + B\,N_1(\mu) = \beta\sqrt{L}. \tag{v}$$

As $x = 2\lambda\sqrt{r}$ or, $\sqrt{r} = \dfrac{x}{2\lambda}$,

we write Eq. (iv) as

$$\theta = \frac{2\lambda A}{x} J_1(x) + \frac{2\lambda B}{x} N_1(x).$$

We know that

$$\frac{d\theta}{dt} = \frac{d\theta}{dr}\frac{dr}{dt} = v\frac{d\theta}{dr} = v\frac{d\theta}{dx}\frac{dx}{dr}$$

$$= \frac{2v\lambda^2}{x}\left[2\lambda A\frac{d}{dx}\{x^{-1}J_1(x)\} + 2\lambda B\frac{d}{dx}\{x^{-1}N_1(x)\}\right]$$

$$= -\frac{4v\lambda^3}{x}\left[Ax^{-1}J_2(x) + Bx^{-1}N_2(x)\right].$$

Applying the condition, $\left[\dfrac{d\theta}{dt}\right]_{t=0} = 0$ and recalling that $r = L$ at $t = 0$, we have

$$AJ_2(\mu) + B\,N_2(\mu) = 0, \quad \left(\mu = 2\lambda\sqrt{L} = \frac{2\sqrt{gL}}{v}\right). \qquad \text{(vi)}$$

Solving Eqs. (v) and (vi), we obtain

$$A = -\frac{\pi\beta L\sqrt{g}}{v} N_2\left(\frac{2\sqrt{gL}}{v}\right), B = \frac{\pi\beta L\sqrt{g}}{v} J_2\left(\frac{2\sqrt{gL}}{v}\right).$$

Here, use has been made of the formula

$$J_p(x)N_{p+1}(x) - J_{p+1}(x)N_p(x) = -\frac{2}{\pi x}.$$

Substituting these values of A and B into Eq, (iv), we write

$$\theta = -\frac{\pi\beta L}{v}\sqrt{\frac{g}{r}}\, N_2\left(\frac{2\sqrt{gL}}{v}\right) J_1\left(\frac{2\sqrt{gr}}{v}\right) + \frac{\pi\beta L}{v}\sqrt{\frac{g}{r}}\, J_2\left(\frac{2\sqrt{gL}}{v}\right) N_2\left(\frac{2\sqrt{gr}}{v}\right).$$

PROBLEM 4

An equation reducible to Bessel equation: A differentiable equation has the form

$$x^2\frac{d^2y}{dx^2} + (1-2a)x\frac{dy}{dx} + \left[b^2c^2x^{2c} + \left(a^2 - c^2p^2\right)\right]y = 0, b > 0, p \ge 0,$$

find its solution in terms of Bessel functions.

Solution: Let us put

$$y = x^a Z. \tag{i}$$

Then, $x\dfrac{dy}{dx} = x^{a+1}\dfrac{dZ}{dx} + ax^a Z$,

$$x^2\frac{d^2y}{dx^2} = x^{a+2}\frac{d^2Z}{dx^2} + 2ax^{a+1}\frac{dZ}{dx} + a(a-1)x^a Z.$$

Substituting these expressions into the given equation, we obtain

$$x^2\frac{d^2Z}{dx^2} + x\frac{dZ}{dx} + (b^2c^2x^{2c} - c^2p^2)Z = 0. \tag{ii}$$

Next, let us change the independent variable to

$$t = x^c \tag{iii}$$

from which we obtain

$$\frac{dZ}{dx} = \frac{dZ}{dt}\frac{dt}{dx} = ct^{1-\frac{1}{c}}\frac{dZ}{dt},$$

$$\frac{d^2Z}{dx^2} = c^2t^{2-\frac{2}{c}}\frac{d^2Z}{dt^2} + c(c-1)t^{1-\frac{2}{c}}\frac{dZ}{dt}.$$

Substituting these relations into Eq. (ii), we obtain

$$t^2\frac{d^2Z}{dt^2} + t\frac{dZ}{dt} + \left(b^2t^2 - p^2\right)Z = 0 \tag{iv}$$

which is a Bessel equation having a general solution

$$Z(t) = C_1J_p(bt) + C_2\,N_p(bt) \tag{v}$$

where C_1, C_2 are arbitrary constants and $J_p(bt)$ and $N_p(bt)$ are Bessel functions of the first and the second kinds of order p.

Now, substituting Eqs. (iii) and (v) into Eq. (i), we obtain the solution of the given equation as

$$y = x^a\left[C_1J_p(bx^c) + C_2N_p(bx^c)\right].$$

PROBLEM 5 Prove that the Neumann function $N_p(x)$ satisfies Bessel equation.

Solution: We know that

$$N_p(x) = \frac{\cos(p\pi)J_p(x) - J_{-p}(x)}{\sin(p\pi)} = \cot(p\pi)J_p(x) - \operatorname{cosec}(p\pi)J_{-p}(x).$$

Then, $x\dfrac{dN_p(x)}{dx} = x\cot(p\pi)J_p'(x) - x\operatorname{cosec}(p\pi)J_{-p}'(x)$,

$$x^2\frac{d^2N_p(x)}{dx^2} = x^2\cot(p\pi)J_p''(x) - x^2\operatorname{cosec}(p\pi)J_{-p}''(x)$$

and $$(x^2 - p^2)N_p(x) = (x^2 - p^2)\cot(p\pi)J_p(x) - (x^2 - p^2)\operatorname{cosec}(p\pi)J_{-p}(x).$$

Adding all these, we obtain

$$x^2\frac{d^2N_p(x)}{dx^2} + x\frac{dN_p(x)}{dx} + (x^2 - p^2)N_p(x)$$

$$= \cot(p\pi)\left[x^2J_p''(x) + xJ_p'(x) + (x^2 - p^2)J_p(x)\right]$$

$$-\operatorname{cosec}(p\pi)\left[x^2J_{-p}''(x) + xJ_{-p}'(x) + \left\{x^2 - (-p)^2\right\}J_{-p}(x)\right]$$

$$= 0 + 0 = 0$$

Thus, we see that $N_p(x)$ satisfies Bessel equation.

PROBLEM 6 Obtain expressions for

(i) $N_{\frac{1}{2}}(x)$, (ii) $N_{-\frac{1}{2}}(x)$, (iii) $H_{\frac{1}{2}}^{(1)}(x)$, (iv) $H_{-\frac{1}{2}}^{(1)}(x)$, (v) $H_{\frac{1}{2}}^{(2)}(x)$ and (vi) $H_{-\frac{1}{2}}^{(2)}(x)$.

Solution: We know that

$$N_p(x) = \cot(p\pi)J_p(x) - \operatorname{cosec}(p\pi)J_{-p}(x)$$

(i) Putting $p = \dfrac{1}{2}$, we have

$$N_{\frac{1}{2}}(x) = 0 - J_{-\frac{1}{2}}(x) = -\sqrt{\frac{2}{\pi x}}\cos x.$$

(ii) Putting $p = -\dfrac{1}{2}$, we have

$$N_{-\frac{1}{2}}(x) = 0 + J_{\frac{1}{2}}(x) = \sqrt{\frac{2}{\pi x}}\sin x.$$

(iii) Also, we know that

$$H_p^{(1)}(x) = J_p(x) + i\,N_p(x).$$

Therefore, putting $p=\frac{1}{2}$, we have

$$H^{(1)}_{\frac{1}{2}}(x)=J_{\frac{1}{2}}(x)+i\,N_{\frac{1}{2}}(x)$$

$$=\sqrt{\frac{2}{\pi x}}\,(\sin x-i\cos x)$$

$$=-i\sqrt{\frac{2}{\pi x}}\,(i\sin x+\cos x)=-i\sqrt{\frac{2}{\pi x}}\,e^{ix}.$$

(iv) Similarly, putting $p=-\frac{1}{2}$, we obtain

$$H^{(1)}_{-\frac{1}{2}}(x)=J_{-\frac{1}{2}}(x)+iN_{-\frac{1}{2}}(x)=\sqrt{\frac{2}{\pi x}}\,e^{ix}.$$

(v) We have

$$H^{(2)}_p(x)=J_p(x)-iN_p(x)$$

Putting $p=\frac{1}{2}$, we have

$$H^{(2)}_{\frac{1}{2}}(x)=J_{\frac{1}{2}}(x)-i\,N_{\frac{1}{2}}(x)=i\sqrt{\frac{2}{\pi x}}\,e^{-ix}.$$

(vi) Putting $p=-\frac{1}{2}$, we obtain

$$H^{(2)}_{-\frac{1}{2}}(x)=J_{-\frac{1}{2}}(x)-i\,N_{-\frac{1}{2}}(x)=\sqrt{\frac{2}{\pi x}}\,e^{-ix}.$$

PROBLEM 7 Prove that

$$K_p(x)=\frac{\pi}{2}i^{p+1}H^{(1)}_p(ix).$$

Solution: We know that

$$K_p(x)=\frac{\pi}{2}\left[\frac{I_{-p}(x)-I_p(x)}{\sin(p\pi)}\right]=\frac{\pi}{2}\left[\frac{i^pJ_{-p}(ix)-i^{-p}J_p(ix)}{\sin(p\pi)}\right]. \qquad \text{(i)}$$

Also, we know that

$$N_p(x) = \frac{\cos(p\pi) J_p(x) - J_{-p}(x)}{\sin(p\pi)}$$

or

$$J_{-p}(x) = \cos(p\pi) J_p(x) - \sin(p\pi) N_p(x).$$

Replacing x by ix, we write

$$J_{-p}(ix) = \cos(p\pi) J_p(ix) - \sin(p\pi) N_p(ix). \tag{ii}$$

Substituting Eq. (ii) into Eq. (i), we write

$$K_p(x) = \frac{\pi}{2}\left[\frac{i^p \cos(p\pi) J_p(ix) + i^{p+2} \sin(p\pi) N_p(ix) - i^{-p} J_p(ix)}{\sin(p\pi)}\right] \tag{iii}$$

$$= \frac{\pi\, i^{p+1}}{2}\left[i\, N_p(ix) + \left\{\frac{-i\cos(p\pi) - i^{-2p-1}}{\sin(p\pi)}\right\} J_p(ix)\right]$$

$$= \frac{\pi\, i^{p+1}}{2}\left[iN_p(ix) + \left\{\frac{-i\cos(p\pi) + i.i^{-2p}}{\sin(p\pi)}\right\} J_p(ix)\right]$$

$$= \frac{\pi\, i^{p+1}}{2}\left[iN_p(ix) + \left\{\frac{-i\cos(p\pi) + i\cdot\left(e^{\frac{\pi i}{2}}\right)^{-2p}}{\sin(p\pi)}\right\} J_p(ix)\right]$$

$$= \frac{\pi\, i^{p+1}}{2}\left[i\, N_p(ix) + \left\{\frac{-i\cos(p\pi) + i\, e^{-p\pi i}}{\sin(p\pi)}\right\} J_p(ix)\right]$$

$$= \frac{\pi\, i^{p+1}}{2}\left[i\, N_p(ix) + \left\{\frac{-i\cos(p\pi) + i\,(\cos p\pi - i\sin p\pi)}{\sin(p\pi)}\right\} J_p(ix)\right]$$

$$= \frac{\pi\, i^{p+1}}{2}\left[J_p(ix) + i\, N_p(ix)\right]$$

$$= \frac{\pi}{2} i^{p+1} H_p^{(1)}(ix).$$

Alternatively,

We have $H_p^{(1)}(ix) = J_p(ix) + i\,N_p(ix)$

$$= J_p(ix) + i\left[\frac{\cos(p\pi)J_p(ix) - J_{-p}(ix)}{\sin(p\pi)}\right]$$

$$= \frac{\{\sin(p\pi) + i\cos(p\pi)\}J_p(ix) - i\,J_{-p}(ix)}{\sin(p\pi)}$$

$$= \frac{i\{\cos(p\pi) - i\sin(p\pi)\}J_p(ix) - i\,J_{-p}(ix)}{\sin(p\pi)}$$

$$= \frac{i\,e^{-p\pi i}J_p(ix) - i\,J_{-p}(ix)}{\sin(p\pi)}$$

$$= \frac{J_{-p}(ix) - e^{-p\pi i}J_p(ix)}{i\sin(p\pi)}$$

$$= \frac{i^{-p}I_{-p}(x) - e^{-p\pi i}i^{p}I_p(x)}{i\sin(p\pi)}$$

$$= \frac{\left(e^{\frac{\pi i}{2}}\right)^{-p} I_{-p}(x) - e^{-p\pi i}\left(e^{\frac{\pi i}{2}}\right)^{p} I_p(x)}{i\sin(p\pi)}$$

$$= \frac{2}{\pi}\frac{e^{-\frac{p\pi i}{2}}}{i}\frac{\pi}{2}\left[\frac{I_{-p}(x) - I_p(x)}{\sin(p\pi)}\right]$$

$$= \frac{2}{\pi i}e^{-\frac{p\pi i}{2}}K_p(x).$$

Therefore, we write

$$K_p(x) = \frac{\pi i}{2}e^{\frac{p\pi i}{2}}H_p^{(1)}(ix) = \frac{\pi}{2}i^{p+1}H_p^{(1)}(ix).$$

PROBLEM 8 Prove that $I_p(x) = e^{\frac{p\pi i}{2}}J_p(-ix)$

and deduce that $K_p(x) = -\frac{\pi i}{2}e^{-\frac{p\pi i}{2}}H_p^{(2)}(-ix)$.

Solution: We know that

$$J_p(-ix)=\sum_{r=0}^{\infty}\frac{(-1)^r}{r!\,\Gamma(p+r+1)}\left(\frac{-ix}{2}\right)^{p+2r}$$

$$=(-i)^p\sum_{r=0}^{\infty}\frac{1}{r!\,\Gamma(p+r+1)}\left(\frac{x}{2}\right)^{p+2r}=e^{-\frac{p\pi i}{2}}I_p(x).$$

Therefore, we write

$$I_p(x)=e^{\frac{p\pi i}{2}}J_p(-ix). \tag{i}$$

Deduction: In Eq. (i), replacing p by $-p$ we write

$$I_{-p}(x)=e^{-\frac{p\pi i}{2}}J_{-p}(-ix). \tag{ii}$$

Further, we know that

$$H_p^{(2)}(-ix)=J_p(-ix)-iN_p(-ix)$$

$$=J_p(-ix)-i\left[\frac{\cos(p\pi)J_p(-ix)-J_{-p}(-ix)}{\sin(p\pi)}\right]$$

$$=\frac{-i\left[\cos(p\pi)+i\sin(p\pi)\right]J_p(-ix)+i\,J_{-p}(-ix)}{\sin(p\pi)}$$

$$=i\left[\frac{J_{-p}(-ix)-e^{p\pi i}J_p(-ix)}{\sin(p\pi)}\right].$$

Using Eqs. (i) and (ii), we write

$$H_p^{(2)}(-ix)=i\left[\frac{e^{\frac{p\pi i}{2}}I_{-p}(x)-e^{p\pi i}e^{-\frac{p\pi i}{2}}I_p(x)}{\sin(p\pi)}\right]$$

or
$$\frac{\pi}{2}H_p^{(2)}(-ix)=i\,e^{\frac{p\pi i}{2}}\frac{\pi}{2}\left[\frac{I_{-p}(x)-I_p(x)}{\sin(p\pi)}\right]$$

$$=i\,e^{\frac{p\pi i}{2}}K_p(x)$$

Therefore, we have

$$K_p(x) = -\frac{\pi i}{2} e^{-\frac{p\pi i}{2}} H_p^{(2)}(-ix).$$

PROBLEM 9 Prove that the function $g(x,t) = e^{\frac{x}{2}\left(t+\frac{1}{t}\right)}$ is the generating function for modified Bessel functions.

Solution: We know that the generating function for Bessel function of the first kind is

$$e^{\frac{x}{2}\left(t-\frac{1}{t}\right)} = \sum_{n=-\infty}^{\infty} J_n(x) t^n \quad (t \neq 0).$$

Replacing x by ix and t by $(-it)$, we obtain

$$\begin{aligned} e^{\frac{x}{2}\left(t+\frac{1}{t}\right)} &= \sum_{n=-\infty}^{\infty} J_n(ix)(-1)^n i^n t^n \\ &= \sum_{n=-\infty}^{\infty} i^n I_n(x)(-1)^n i^n t^n \\ &= \sum_{n=-\infty}^{\infty} I_n(x) t^n \quad (t \neq 0). \end{aligned}$$

PROBLEM 10 Prove that

(a) $I_{\frac{1}{2}}(x) = \sqrt{\frac{2}{\pi x}} \sinh x,$

(b) $I_{-\frac{1}{2}}(x) = \sqrt{\frac{2}{\pi x}} \cosh x,$

(c) $\left[I_{-\frac{1}{2}}(x)\right]^2 - \left[I_{\frac{1}{2}}(x)\right]^2 = \frac{2}{\pi x},$

(d) $K_{\frac{1}{2}}(x) = K_{-\frac{1}{2}}(x) = \sqrt{\frac{\pi}{2x}}\, e^{-x}.$

Solution: Since $I_p(x) = i^{-p} J_p(ix) = e^{-\frac{p\pi i}{2}} J_p(ix)$, we have

(a) $I_{\frac{1}{2}}(x) = i^{-\frac{1}{2}} J_{\frac{1}{2}}(ix)$

$$= i^{-\frac{1}{2}}\sqrt{\frac{2}{\pi ix}}\sin(ix) = \frac{1}{i}\sqrt{\frac{2}{\pi x}}\, i\sinh x$$

$$= \sqrt{\frac{2}{\pi x}}\sinh x.$$

(b) $I_{-\frac{1}{2}}(x) = i^{\frac{1}{2}} J_{-\frac{1}{2}}(ix) = i^{\frac{1}{2}}\sqrt{\frac{2}{\pi ix}}\cos(ix)$

$$= \sqrt{\frac{2}{\pi x}}\cosh x.$$

(c) $\left[I_{-\frac{1}{2}}(x)\right]^2 - \left[I_{\frac{1}{2}}(x)\right]^2 = \frac{2}{\pi x}\left[\cosh^2 x - \sinh^2 x\right] = \frac{2}{\pi x}.$

(d) $K_p(x) = \frac{\pi}{2}\left[\frac{I_{-p}(x) - I_p(x)}{\sin(p\pi)}\right]$

Therefore, we have

$$K_{\frac{1}{2}}(x) = \frac{\pi}{2}\left[\frac{I_{-\frac{1}{2}}(x) - I_{\frac{1}{2}}(x)}{\sin\left(\frac{\pi}{2}\right)}\right]$$

$$= \frac{\pi}{2}\left[\sqrt{\frac{2}{\pi x}}(\cosh x - \sinh x)\right] = \sqrt{\frac{\pi}{2x}}\, e^{-x}.$$

Also, we have

$$K_{-\frac{1}{2}}(x) = \frac{\pi}{2}\left[\frac{I_{\frac{1}{2}}(x) - I_{-\frac{1}{2}}(x)}{\sin\left(-\frac{\pi}{2}\right)}\right] = \frac{\pi}{2}\left[I_{-\frac{1}{2}}(x) - I_{\frac{1}{2}}(x)\right] = \sqrt{\frac{\pi}{2x}}\, e^{-x}.$$

PROBLEM 11 Prove that the Wronskian of the two modified Bessel functions $I_p(x)$ and $K_p(x)$ is equal to minus the reciprocal of x.

Solution: The modified Bessel functions of the first and the second kind are respectively given by

$$I_p(x)=e^{-\frac{p\pi i}{2}}J_p(ix) \text{ and } K_p(x)=\frac{\pi}{2}\frac{I_{-p}(x)-I_p(x)}{\sin(p\pi)}.$$

Therefore, we have

$$K_p'(x)=\frac{d}{dx}K_p(x)=\frac{\pi}{2\sin(p\pi)}\left\{I_{-p}'(x)-I_p'(x)\right\}.$$

Then, we have

$$\begin{aligned}
W\left[I_p(x),K_p(x)\right]&=I_p(x)K_p'(x)-I_p'(x)K_p(x)\\
&=\frac{\pi}{2\sin(p\pi)}\left\{I_p(x)I_{-p}'(x)-I_p(x)I_p'(x)-I_p'(x)I_{-p}(x)+I_p'(x)I_p(x)\right\}\\
&=\frac{\pi}{2\sin(p\pi)}\left\{I_p(x)I_{-p}'(x)-I_p'(x)I_{-p}(x)\right\}\\
&=\frac{\pi}{2\sin(p\pi)}W\left[I_p(x),I_{-p}(x)\right]\\
&=\frac{\pi}{2\sin(p\pi)}W\left[e^{-\frac{p\pi i}{2}}J_p(ix),e^{\frac{p\pi i}{2}}J_{-p}(ix)\right]\\
&=\frac{\pi}{2\sin(p\pi)}\left[J_p(ix)\frac{d}{dx}J_{-p}(ix)-\left\{\frac{d}{dx}J_p(ix)\right\}J_{-p}(ix)\right]\\
&=\frac{\pi}{2\sin(p\pi)}\left\{J_p(ix)i\,J_{-p}'(ix)-i\,J_p'(ix)J_{-p}(ix)\right\}\\
&=\frac{\pi i}{2\sin(p\pi)}W\left[J_p(ix),J_{-p}(ix)\right]\\
&=\frac{\pi i}{2\sin(p\pi)}\left\{-\frac{2\sin(p\pi)}{\pi ix}\right\}=-\frac{1}{x}.
\end{aligned}$$

PROBLEM 12 Prove that $I_p(x)I_{-p}'(x)-I_p'(x)I_{-p}(x)=-\dfrac{2\sin(p\pi)}{\pi x}$.

Solution: We know that

$$I_p(x)=e^{-\frac{p\pi i}{2}}J_p(ix) \quad \text{and} \quad I_{-p}(x)=e^{\frac{p\pi i}{2}}J_{-p}(ix).$$

Then we have

$$I_p'(x)=i\,e^{-\frac{p\pi i}{2}}J_p'(ix),\quad I_{-p}'(x)=i\,e^{\frac{p\pi i}{2}}J_{-p}'(ix)$$

where the prime denotes differentiation with respect to x in the cases of $I_p(x)$ and $I_{-p}(x)$ whereas it denotes differentiation with respect to ix in the cases of $J_p(ix)$ and $J_{-p}(ix)$.

Now, we have

$$W = I_p(x)I'_{-p}(x) - I'_p(x)I_{-p}(x)$$
$$= i\{J_p(ix)J'_{-p}(ix) - J'_p(ix)J_{-p}(ix)\}$$
$$= i\left\{-\frac{2\sin(p\pi)}{\pi ix}\right\} = -\frac{2\sin(p\pi)}{\pi x}.$$

PROBLEM 13 Prove that $\int \frac{dx}{x\{I_p(x)\}^2} = -\frac{\pi}{2\sin(p\pi)}\frac{I_{-p}(x)}{I_p(x)}$.

Solution: From Problem 12, we have

$$\frac{I_p(x)I'_{-p}(x) - I'_p(x)I_{-p}(x)}{\{I_p(x)\}^2} = -\frac{2\sin(p\pi)}{\pi}\frac{1}{x\{I_p(x)\}^2}$$

or
$$\frac{1}{x\{I_p(x)\}^2} = -\frac{\pi}{2\sin(p\pi)}\frac{d}{dx}\left[\frac{I_{-p}(x)}{I_p(x)}\right]$$

Integrating both sides with respect to x and choosing the integration constant as zero, we obtain

$$\int \frac{dx}{x\{I_p(x)\}^2} = -\frac{\pi}{2\sin(p\pi)}\frac{I_{-p}(x)}{I_p(x)}.$$

PROBLEM 14 Prove that the spherical Bessel function of the first kind satisfies the differential equation

$$x^2\frac{d^2y}{dx^2} + 2x\frac{dy}{dx} + \{x^2 - l(l+1)\}y = 0$$

Solution: In Problem 4, we have seen that for various kinds of Bessel functions denoted by $Z(x)$ the function $x^a Z(x)$ satisfies a general differential equation which is expressed as

$$x^2\frac{d^2y}{dx^2}+(1-2a)x\frac{dy}{dx}+\left[b^2c^2x^{2c}+(a^2-c^2p^2)\right]y=0,\ (b>0,\ p>0)$$

where a, b, c, p are constants.

If we put $a=-\dfrac{1}{2}, b=c=1, p=\left\{\dfrac{1}{4}+l(l+1)\right\}^{\frac{1}{2}}, Z(x)=J_{l+\frac{1}{2}}(x)$

where $l = 0, 1, 2, 3$ … etc., we see that $y=x^{-\frac{1}{2}}J_{l+\frac{1}{2}}(x)$ satisfies the differential equation

$$x^2\frac{d^2y}{dx^2}+2x\frac{dy}{dx}+\left\{x^2-l(l+1)\right\}y=0.$$

Hence, the spherical Bessel function of the first kind $j_l(x)=\sqrt{\dfrac{\pi}{2x}}J_{l+\frac{1}{2}}(x)$ also satisfies the same equation.

PROBLEM 15 Obtain the expressions for the spherical Bessel functions of the first kind (i) $j_0(x)$ and (ii) $j_1(x)$.

Solution: We know that $j_l(x)=\sqrt{\dfrac{\pi}{2x}}J_{l+\frac{1}{2}}(x)$

(i) For $l = 0$, we have

$$j_0(x)=\sqrt{\frac{\pi}{2x}}J_{\frac{1}{2}}(x)=\sqrt{\frac{\pi}{2x}}\sqrt{\frac{2}{\pi x}}\sin x=\frac{\sin x}{x}.$$

(ii) For $l = 1$, we have

$$j_1(x)=\sqrt{\frac{\pi}{2x}}J_{\frac{3}{2}}(x)=\sqrt{\frac{\pi}{2x}}\sqrt{\frac{2}{\pi x}}\left(\frac{\sin x}{x}-\cos x\right)$$

$$=\frac{\sin x}{x^2}-\frac{\cos x}{x}.$$

PROBLEM 16 Obtain the expressions for the spherical Neumann functions [spherical Bessel functions of second kind] (i) $n_0(x)$ and (ii) $n_1(x)$.

Solution: We know that

$$n_l(x)=\sqrt{\frac{\pi}{2x}}N_{l+\frac{1}{2}}(x)$$

(i) Putting $l = 0$, we have

$$n_0(x) = \sqrt{\frac{\pi}{2x}}\, N_{\frac{1}{2}}(x)$$

Therefore, we have

$$n_0(x) = \sqrt{\frac{\pi}{2x}}\left[-\sqrt{\frac{2}{\pi x}}\cos x\right] = -\frac{\cos x}{x}.$$

(ii) Putting $l = 1$, we have

$$n_1(x) = \sqrt{\frac{\pi}{2x}}\, N_{\frac{3}{2}}(x).$$

We know that

$$N_{p-1}(x) + N_{p+1}(x) = \frac{2p}{x} N_p(x).$$

Putting $p = \frac{1}{2}$, we obtain

$$N_{\frac{3}{2}}(x) = \frac{1}{x} N_{\frac{1}{2}}(x) - N_{-\frac{1}{2}}(x) = \frac{1}{x}\left\{-\sqrt{\frac{2}{\pi x}}\cos x\right\} - \sqrt{\frac{2}{\pi x}}\sin x$$

$$= \sqrt{\frac{2}{\pi x}}\left\{-\frac{\cos x}{x} - \sin x\right\}.$$

Therefore, we have

$$n_1(x) = \sqrt{\frac{\pi}{2x}}\sqrt{\frac{2}{\pi x}}\left\{-\frac{\cos x}{x} - \sin x\right\} = -\frac{\cos x}{x^2} - \frac{\sin x}{x}.$$

PROBLEM 17 Obtain the expressions for the spherical Hankel functions of the first and the second kinds $h_0^{(1)}(x)$ and $h_0^{(2)}(x)$.

Solution: We have

$$h_0^{(1)}(x) = j_0(x) + i n_0(x) = \frac{\sin x}{x} + i\left(-\frac{\cos x}{x}\right) = \frac{e^{ix}}{ix}.$$

Also, we have

$$h_0^{(2)}(x) = j_0(x) - i n_0(x) = \frac{\sin x}{x} - i\left(-\frac{\cos x}{x}\right) = -\frac{e^{-ix}}{ix}.$$

PROBLEM 18 Obtain the expressions for the modified spherical Bessel functions of the first and the second kinds $i_0(x)$ and $k_0(s)$.

Solution: By definition, we have

$$i_l(x) = \sqrt{\frac{\pi}{2x}}\, I_{l+\frac{1}{2}}(x) \text{ and } k_l(x) = \sqrt{\frac{2}{\pi x}}\, K_{l+\frac{1}{2}}(x).$$

Taking $l = 0$, we have

$$i_0(x) = \sqrt{\frac{\pi}{2x}}\, I_{\frac{1}{2}}(x) \text{ and } k_0(x) = \sqrt{\frac{2}{\pi x}}\, K_{\frac{1}{2}}(x).$$

Using the results obtained in Problem 10, we write

$$i_0(x) = \sqrt{\frac{\pi}{2x}}\sqrt{\frac{2}{\pi x}}\sinh x = \frac{\sinh x}{x}$$

and

$$k_0(x) = \sqrt{\frac{2}{\pi x}}\sqrt{\frac{\pi}{2x}}\, e^{-x} = \frac{e^{-x}}{x}.$$

PROBLEM 19 Prove that

$$j_l(x) = (-1)^l x^l \left(\frac{1}{x}\frac{d}{dx}\right)^l \left(\frac{\sin x}{x}\right).$$

Solution: See the solution of Problem 19 of Solved Problems on **Bessel functions of the first kind** and add the following.

Now we have

$$j_l(x) = \sqrt{\frac{\pi}{2x}} J_{l+\frac{1}{2}}(x)$$

$$= \sqrt{\frac{\pi}{2x}}(-1)^l \frac{(2x)^{l+\frac{1}{2}}}{\sqrt{\pi}}\left(\frac{1}{2x}\frac{d}{dx}\right)^l \left(\frac{\sin x}{x}\right)$$

$$= (-1)^l x^l \left(\frac{1}{x}\frac{d}{dx}\right)^l \left(\frac{\sin x}{x}\right).$$

PROBLEM 20 Starting from the series formula of Bessel function of the first kind, prove that

$$J_l(x) = \frac{x^l}{2^l \sqrt{\pi}\,\Gamma\left(l+\frac{1}{2}\right)} \int_{-1}^{+1} e^{ixt}(1-t^2)^{l-\frac{1}{2}}\, dt \text{ for } l > -\frac{1}{2}$$

and deduce that $j_l(x) = \dfrac{x^l}{2^{l+1}\Gamma(l+1)} \displaystyle\int_{-1}^{+1} e^{ixt}(1-t^2)^l\, dt$.

Solution: Series formula of Bessel function of the first kind is

$$J_l(x) = \sum_{r=0}^{\infty} \frac{(-1)^r}{r!\,\Gamma(l+r+1)}\left(\frac{x}{2}\right)^{l+2r}$$

$$= \left(\frac{x}{2}\right)^l \sum_{r=0}^{\infty} \frac{(ix)^{2r}}{2^{2r} r!\,\Gamma\left(l+\frac{1}{2}+r+\frac{1}{2}\right)}$$

$$= \left(\frac{x}{2}\right)^l \sum_{r=0}^{\infty} \frac{(ix)^{2r}}{2^{2r} r!} \frac{B\left(l+\frac{1}{2}, r+\frac{1}{2}\right)}{\Gamma\left(l+\frac{1}{2}\right)\Gamma\left(r+\frac{1}{2}\right)} \text{ [using } B(m,n) = \frac{\Gamma(m)\Gamma(n)}{\Gamma(m+n)}\text{]}.$$

Duplication formula of gamma functions yields

$$\Gamma\left(r+\frac{1}{2}\right) = \frac{(2r)!\sqrt{\pi}}{r!\,2^{2r}}.$$

Using this relation, we write

$$J_l(x) = \frac{x^l}{2^l\,\Gamma\left(l+\frac{1}{2}\right)} \sum_{r=0}^{\infty} \frac{(ix)^{2r}}{(2r)!\sqrt{\pi}} B\left(l+\frac{1}{2}, r+\frac{1}{2}\right)$$

$$= \frac{x^l}{2^l\sqrt{\pi}\,\Gamma\left(l+\frac{1}{2}\right)} \sum_{r=0}^{\infty} \frac{(ix)^{2r}}{(2r)!} \int_0^1 u^{r-\frac{1}{2}} (1-u)^{l-\frac{1}{2}}\, du$$

$$= \frac{x^l}{2^l\sqrt{\pi}\,\Gamma\left(l+\frac{1}{2}\right)} \int_0^1 \sum_{r=0}^{\infty} \frac{(ix)^{2r}}{(2r)!} u^{r-\frac{1}{2}} (1-u)^{l-\frac{1}{2}}\, du.$$

Putting $u = t^2$, we have $du = 2t\,dt$.
Using these relations, we write

$$J_l(x) = \frac{x^l}{2^l\sqrt{\pi}\,\Gamma\left(l+\frac{1}{2}\right)} \cdot 2\int_0^1 \sum_{r=0}^{\infty} \frac{(ix)^{2r}}{(2r)!} t^{2r} (1-t^2)^{l-\frac{1}{2}}\, dt$$

$$= \frac{x^l}{2^l\sqrt{\pi}\,\Gamma\left(l+\frac{1}{2}\right)} \int_{-1}^{+1} \left[\sum_{r=0}^{\infty} \frac{(ixt)^{2r}}{(2r)!}\right] (1-t^2)^{l-\frac{1}{2}}\, dt$$

$$= \frac{x^l}{2^l \sqrt{\pi}\, \Gamma\left(l+\frac{1}{2}\right)} \int_{-1}^{+1} e^{ixt} (1-t^2)^{l-\frac{1}{2}} dt.$$

Here, use has been made of the fact that $\int_{-1}^{+1} \left[\sum_{s=0}^{\infty} \frac{(ixt)^s}{s!} \right] (1-t^2)^{l-\frac{1}{2}} dt$ vanishes for s = odd integer.

Deduction: Replacing l by $l+\frac{1}{2}$, we write

$$J_{l+\frac{1}{2}}(x) = \frac{x^{l+\frac{1}{2}}}{2^{l+\frac{1}{2}} \sqrt{\pi}\, \Gamma(l+1)} \int_{-1}^{+1} e^{ixt} (1-t^2)^l dt.$$

Hence, we have

$$j_l(x) = \sqrt{\frac{\pi}{2x}}\, J_{l+\frac{1}{2}}(x) = \frac{x^l}{2^{l+1} \Gamma(l+1)} \int_{-1}^{+1} e^{ixt} (1-t^2)^l dt.$$

PROBLEM 21 Prove that

$$J_{l+\frac{1}{2}}(x) = \frac{1}{i^l} \sqrt{\frac{x}{2\pi}} \int_{-1}^{+1} e^{ixt} P_l(t)\, dt$$

and deduce that $j_l(x) = \frac{1}{2\, i^l} \int_{-1}^{+1} e^{ixt} P_l(t)\, dt$.

Solution: From Problem 20, we know that

$$J_{l+\frac{1}{2}}(x) = \frac{x^{l+\frac{1}{2}}}{2^{l+\frac{1}{2}} \sqrt{\pi}\, \Gamma(l+1)} \int_{-1}^{+1} e^{ixt} (1-t^2)^l dt.$$

Performing integration by parts l times (l = integer) with $(1-t^2)^l$ as the first function and noting that the integrated out terms vanish at both the limits, we write

$$J_{l+\frac{1}{2}}(x) = \frac{(-1)^l\, x^{l+\frac{1}{2}}}{2^{l+\frac{1}{2}} \sqrt{\pi}\, l!} \int_{-1}^{+1} \frac{d^l}{dt^l} (1-t^2)^l \frac{e^{ixt}}{(ix)^l} dt$$

$$= \frac{1}{i^l} \sqrt{\frac{x}{2\pi}} \int_{-1}^{+1} \left[\frac{1}{2^l\, l!} \frac{d^l}{dx^l} (t^2 - 1)^l \right] e^{ixt} dt$$

$$= \frac{1}{i^l} \sqrt{\frac{x}{2\pi}} \int_{-1}^{+1} e^{ixt} P_l(t) dt.$$

Deduction:

$$j_l(x) = \sqrt{\frac{\pi}{2x}}\, J_{l+\frac{1}{2}}(x)$$

$$= \frac{1}{2\, i^l} \int_{-1}^{+1} e^{ixt} P_l(t) dt.$$

Note: This relation can also be derived by l times integration by parts of the integral

$$j_l(x) = \frac{x^l}{2^{l+1}\, l!} \int_{-1}^{+1} e^{ixt} (1 - t^2)^l dt.$$

PROBLEM 22 Prove Rayleigh's plane wave expansion formula

$$e^{ikr\cos\theta} = \sum_{l=0}^{\infty} (2l+1) i^l j_l(kr) P_l(\cos\theta)$$

where $j_l(kr)$ are spherical Bessel functions and $P_l(\cos\theta)$ are Legendre polynomials.
Generalize the formula for arbitrary direction of a wave vector **k.**

Solution: Let us write

$$e^{ikr\cos\theta} = \sum_{l=0}^{\infty} a_l j_l(kr) P_l(\cos\theta) \tag{i}$$

where the coefficient a_l are to be determined.

Multiplying both sides of Eq. (i) by $P_m(\cos\theta)\sin\theta$ and integrating both sides with respect to θ within the limit from $\theta = 0$ to $\theta = \pi$, we have

$$\int_0^{\pi} e^{ikr\cos\theta} P_m(\cos\theta)\sin\theta\, d\theta = \sum_{l=0}^{\infty} a_l j_l(kr) \int_0^{\pi} P_l(\cos\theta) P_m(\cos\theta)\sin\theta\, d\theta$$

$$= \sum_{l=0}^{\infty} a_l j_l(kr) \frac{2}{(2l+1)} \delta_{lm}$$

$$= a_m j_m(kr) \frac{2}{(2m+1)}. \tag{ii}$$

If we are more interested in the limit $kr \to 0$ by Eq. (8.230), we have

$$j_m(kr) \approx \frac{m!(2kr)^m}{(2m+1)!}.$$

Therefore, in the limit $kr \to 0$, we write

$$\lim_{(kr)\to 0} \int_0^{\pi} e^{ikr\cos\theta} P_m(\cos\theta)\sin\theta\, d\theta = \frac{m!\,2^m (kr)^m}{(2m+1)!} \cdot \frac{2}{(2m+1)} a_m.$$

Differentiating both sides of this equation m times with respect to (kr), we have

$$\int_0^{\pi} \lim_{(kr)\to 0} (i\cos\theta)^m e^{ikr\cos\theta} P_m(\cos\theta)\sin\theta\, d\theta = \frac{m!m!\,2^{m+1}}{(2m+1)(2m+1)!} a_m$$

or $$i^m \int_{-1}^{+1} x^m P_m(x) dx = \frac{m!\,m!2^{m+1}}{(2m+1)(2m+1)!} a_m \tag{iii}$$

where we put $x = \cos\theta$.

In the chapter of Legendre polynomials, we have known that

$$\int_{-1}^{+1} x^m P_m(x) dx = \frac{2^{m+1} m!\,m!}{(2m+1)!}.$$

Hence, Eq. (iii) can be written as

$$\frac{2^{m+1} m!\,m!}{(2m+1)!} i^m = \frac{m!\,m!\,2^{m+1}}{(2m+1)(2m+1)!} a_m$$

or $$a_m = i^m (2m+1).$$

Changing the subscript, we write

$$a_l = i^l (2l+1).$$

Substituting this value of a_l in Eq. (i), we write

$$e^{ikr\cos\theta} = \sum_{l=0}^{\infty} (2l+1) i^l j_l(kr) P_l(\cos\theta). \tag{iv}$$

Generalization: We know that the addition formula of spherical harmonics gives

$$P_l(\cos\theta) = \frac{4\pi}{2l+1} \sum_{m=-l}^{+l} Y_{lm}^*(\theta_2, \phi_2) Y_{lm}(\theta_1, \phi_1) \tag{v}$$

where (r, θ_1, ϕ_1) are the spherical co-ordinates of the head of the vector **r**, (k, θ_2, ϕ_2) are the spherical co-ordinates of the head of the vector **k** and θ is the angle between the two directions.

Using relation (v), we can generalize Eq. (iv) in the form

$$e^{i\mathbf{k}\cdot\mathbf{r}} = 4\pi \sum_{l=0}^{\infty} \sum_{m=-l}^{+l} i^l j_l(kr) Y^*_{lm}(\theta_2, \phi_2) Y_{lm}(\theta_1, \phi_1) \quad \text{(vi)}$$

where the arguments of Y^*_{lm} and Y_{lm} are the angular coordinates of **k** and **r** respectively.

Defining the directions indicated by the angular coordinates (θ_1, ϕ_1) and (θ_2, ϕ_2) by the unit vectors $\hat{r}$ and $\hat{k}$ respectively, we write

$$e^{i\mathbf{k}\cdot\mathbf{r}} = 4\pi \sum_{l=0}^{\infty} \sum_{m=-l}^{+l} i^l j_l(kr) Y^*_{lm}(\hat{k}) Y_{lm}(\hat{r}) \, . \quad \text{(vii)}$$

If the directions of the unit vectors $\hat{r}$ and $\hat{k}$ are associated with the solid angles Ω_r and Ω_k respectively, we write in a yet another form

$$e^{i\mathbf{k}\cdot\mathbf{r}} = 4\pi \sum_{l=0}^{\infty} \sum_{m=-l}^{+l} i^l j_l(kr) Y^*_{lm}(\Omega_k) Y_{lm}(\Omega_r) \, . \quad \text{(viii)}$$

PROBLEM 23 Prove that the spherical Bessel function $j_l(x)$ is the Fourier transform of the function

$$f_l(\mu) = \begin{cases} \dfrac{\sqrt{2\pi}}{2i^l} P_l(\mu), & |\mu| \le 1 \\ 0, & |\mu| > 1 \end{cases}$$

where $P_l(\mu)$ is a Legendre polynomial.

Hence, using Parseval's theorem, show that

$$\int_{-1}^{+1} j_l(x) j_m(x) dx = \frac{\pi}{2l+1} \delta_{lm} \, .$$

Solution:

First Part: From Problem 20, we have

$$e^{ikr\cos\theta} = \sum_{l=0}^{\infty} (2l+1) i^l j_l(kr) P_l(\cos\theta) \, .$$

Putting $kr = x, \cos\theta = \mu$, we write

$$e^{ix\mu} = \sum_{l=0}^{\infty} (2l+1) i^l j_l(x) P_l(\mu) \, .$$

Multiplying both sides by $P_m(\mu)$ and then integrating the resulting products with respect to μ within the limit from $\mu=-\infty$ to $\mu=+\infty$, we obtain

$$\int_{-\infty}^{+\infty} P_m(\mu)e^{ix\mu}d\mu = \sum_{l=0}^{\infty}(2l+1)i^l j_l(x)\int_{-\infty}^{+\infty} P_l(\mu)P_m(\mu)d\mu$$

$$= \sum_{l=0}^{\infty}(2l+1)i^l j_l(x)\int_{-1}^{+1} P_l(\mu)P_m(\mu)d\mu$$

$$= \sum_{l=0}^{\infty}(2l+1)i^l j_l(x)\frac{2}{2l+1}\delta_{lm}$$

$$= (2m+1)i^m j_m(x)\frac{2}{2m+1}$$

$$= 2i^m j_m(x).$$

Therefore, we write

$$j_m(x) = \frac{1}{2i^m}\int_{-\infty}^{+\infty} P_m(\mu)e^{ix\mu}d\mu. \qquad \text{(i)}$$

Changing the subscript, we write

$$j_l(x) = \frac{1}{\sqrt{2\pi}}\int_{-\infty}^{+\infty} f_l(\mu)e^{ix\mu}d\mu \qquad \text{(ii)}$$

where

$$f_l(\mu) = \begin{cases} \dfrac{\sqrt{2\pi}}{2i^l}P_l(\mu), & |\mu| \le 1 \\ 0, & |\mu| > 1. \end{cases} \qquad \text{(iii)}$$

Here, Eq. (ii) gives the Fourier transform of the function $f_l(\mu)$ that is defined in Eq. (iii).

Second Part: We write

$$\int_{-\infty}^{+\infty} j_l(x)j_m(x)dx = \int_{-\infty}^{+\infty}\left[j_l(x)\right]^* j_m(x)dx$$

$$= \int_{-\infty}^{+\infty}\left[f_l(\mu)\right]^* f_m(\mu)d\mu \quad \text{(By Parseval's theorem)}$$

$$= \int_{-1}^{+1}\left[\frac{\sqrt{2\pi}}{2(-i)^l}P_l(\mu)\right]\left[\frac{\sqrt{2\pi}}{2i^m}P_m(\mu)\right]d\mu$$

$$= \frac{\pi}{2} i^{l-m} \int_{-1}^{+1} P_l(\mu) P_m(\mu) d\mu$$

$$= \frac{\pi}{2} i^{l-m} \frac{2}{2l+1} \delta_{lm} = \frac{\pi}{2l+1} i^{l-m} \delta_{lm}.$$

When $l \neq m$, we have $\int_{-\infty}^{+\infty} j_l(x) j_m(x) dx = 0$

and when $l = m$, we obtain

$$\int_{-\infty}^{+\infty} j_l(x) j_l(x) dx = \frac{\pi}{2l+1}.$$

Hence, we write the general form

$$\int_{-\infty}^{+\infty} j_l(x) j_m(x) dx = \frac{\pi}{2l+1} \delta_{lm}.$$

PROBLEM 24 Prove by an alternative method the orthogonality condition for spherical Bessel functions,

$$\int_{-\infty}^{+\infty} j_l(x) j_m(x) dx = \frac{\pi}{2l+1} \delta_{lm}.$$

Solution: The spherical Bessel functions $j_l(x)$ and $j_m(x)$ satisfy the respective Bessel equations

$$x^2 \frac{d^2}{dx^2} j_l(x) + 2x \frac{d}{dx} j_l(x) + \left\{x^2 - l(l+1)\right\} j_l(x) = 0$$

and
$$x^2 \frac{d^2}{dx^2} j_m(x) + 2x \frac{d}{dx} j_m(x) + \left\{x^2 - m(m+1)\right\} j_m(x) = 0$$

Subtracting $j_l(x)$ times the second equation from $j_m(x)$ times the first equation, we obtain

$$x^2 \left[j_m(x) \frac{d^2}{dx^2} j_l(x) - j_l(x) \frac{d^2}{dx^2} j_m(x) \right] + 2x \left[j_m(x) \frac{d}{dx} j_l(x) - j_l(x) \frac{d}{dx} j_m(x) \right]$$
$$+ \left[m(m+1) - l(l+1) \right] j_l(x) j_m(x) = 0$$

or
$$\frac{d}{dx}\left[x^2 \left\{ j_m(x) \frac{d}{dx} j_l(x) - j_l(x) \frac{d}{dx} j_m(x) \right\} \right] = (l-m)(l+m+1) j_l(x) j_m(x)$$

Integrating both sides with respect to x within the limit from $x=0$ to $x=\infty$, we obtain

$$\left[x^2\left\{j_m(x)\frac{d}{dx}j_l(x)-j_l(x)\frac{d}{dx}j_m(x)\right\}\right]_0^\infty=(l-m)(l+m+1)\int_0^\infty j_l(x)j_m(x)dx. \quad \text{(i)}$$

To find the value of the left-hand side at $x=\infty$, we have to use the asymptotic forms of $j_l(x)$ and $j_m(x)$.

The asymptotic forms are

$$\lim_{x\to\infty} j_l(x)\approx\frac{1}{x}\sin\left(x-\frac{l\pi}{2}\right),$$

$$\lim_{x\to\infty}\frac{d}{dx}j_l(x)\approx\frac{1}{x}\cos\left(x-\frac{l\pi}{2}\right),$$

$$\lim_{x\to\infty} j_m(x)\approx\frac{1}{x}\sin\left(x-\frac{m\pi}{2}\right),$$

$$\lim_{x\to\infty}\frac{d}{dx}j_m(x)\approx\frac{1}{x}\cos\left(x-\frac{m\pi}{2}\right).$$

Using these expressions, the left-hand side of Eq. (i) with the upper limit becomes

$$\sin\left(x-\frac{m\pi}{2}\right)\cos\left(x-\frac{l\pi}{2}\right)-\cos\left(x-\frac{m\pi}{2}\right)\sin\left(x-\frac{l\pi}{2}\right)$$

$$=\sin\left\{\left(x-\frac{m\pi}{2}\right)-\left(x-\frac{l\pi}{2}\right)\right\}=\sin\left\{(l-m)\frac{\pi}{2}\right\}.$$

The left-hand side of Eq. (i) with the lower limit is equal to zero.

Hence, using the above values of the upper and the lower limits of the integrated out terms on the left-hand side of Eq. (i), we have

$$\int_0^\infty j_l(x)j_m(x)dx=\frac{\sin\left\{(l-m)\frac{\pi}{2}\right\}}{(l-m)(l+m+1)}. \quad \text{(ii)}$$

If $(l-m)$ is even, the right-hand side of Eq. (ii) is equal to zero.

Therefore, we have $\int_0^\infty j_l(x)j_m(x)dx=0, (l-m)=\text{even integer.}$ (iii)

Thus, we write

$$2\int_0^\infty j_l(x)j_m(x)dx=\int_{-\infty}^\infty j_l(x)j_m(x)dx=0, \text{ if } (l-m)=\text{even integer.} \quad \text{(iv)}$$

If $(l-m)$ is an odd integer, one of the spherical Bessel functions in the integrand of the left-hand side of Eq. (ii) will be an odd function while the other will be an even function so that the product will be an odd function.

Hence, we have

$$\int_{-\infty}^{+\infty} j_l(x) j_m(x) dx = 0, \text{ if } (l - m) = \text{odd integer.} \tag{v}$$

Combining Eqs. (iv) and (v), we have

$$\int_{-\infty}^{+\infty} j_l(x) j_m(x) dx = 0, \ (l \neq m, \ l, \ m \geq 0). \tag{vi}$$

PROBLEM 25 Obtain the integral form of the spherical Bessel function

$$j_l(x) = \frac{x^l}{2^{l+1}\, l!} \int_{-1}^{+1} e^{ix\mu} (1-\mu^2)^l \, d\mu .$$

Solution: From deduction part of Problem 21, we know that

$$j_l(x) = \frac{1}{2i^l} \int_{-\infty}^{+\infty} e^{ix\mu} P_l(\mu) d\mu .$$

Since $P_l(\mu)$ is defined for $-1 \leq \mu \leq 1$, we write

$$j_l(x) = \frac{1}{2i^l} \int_{-1}^{+1} e^{ix\mu} P_l(\mu) d\mu .$$

Using Rodrigues' formula for $P_l(\mu)$, we write

$$j_l(x) = \frac{1}{2i^l} \int_{-1}^{+1} e^{ix\mu} \frac{1}{2^l\, l!} \frac{d^l}{d\mu^l} (\mu^2 - 1)^l \, d\mu .$$

Performing integration by parts on the right-hand side l times and noting that the integrated out terms vanish at both the limits, we write

$$j_l(x) = \frac{(-1)^l}{2^{l+1}\, i^l\, l!} \int_{-1}^{+1} \left[\frac{d^l}{d\mu^l} (e^{ix\mu}) \right] (\mu^2 - 1)^l \, d\mu$$

$$= \frac{x^l}{2^{l+1}\, l!} \int_{-1}^{+1} e^{ix\mu} (1-\mu^2)^l \, d\mu .$$

PROBLEM 26 Prove the closure relation for spherical Bessel functions, viz.

$$\int_0^\infty j_l(kr)\, j_l(k'r) r^2 dr = \frac{\pi}{2k^2}\delta(k-k').$$

Solution: From Eq. (vii) of Problem 22, we have

$$e^{i\mathbf{k}\cdot\mathbf{r}} = 4\pi \sum_{l=0}^{\infty}\sum_{m=-l}^{+l} i^l j_l(kr) Y_{lm}^*(\Omega_\mathbf{k}) Y_{lm}(\Omega_\mathbf{r}). \tag{i}$$

From the study of Dirac-Delta functions, we know that

$$\int e^{i\mathbf{k}\cdot\mathbf{r}} e^{-i\mathbf{k}'\cdot\mathbf{r}} d^3r = \int e^{i(\mathbf{k}-\mathbf{k}')\cdot\mathbf{r}} d^3r = (2\pi)^3 \delta^3(\mathbf{k}-\mathbf{k}') \tag{ii}$$

where the integral is meant for all space.

Substituting Eq. (i) into the left-hand side of Eq. (ii), we have

$$(2\pi)^3 \delta^3(\mathbf{k}-\mathbf{k}') = \sum_{l=0}^{\infty}\sum_{m=-l}^{+l}\sum_{l'=0}^{\infty}\sum_{m'=-l'}^{+l'} (4\pi)^2 i^l (-i)^{l'} \int_{r=0}^{\infty} j_l(kr)\, j_{l'}(k'r) r^2 dr$$

$$\times Y_{lm}^*(\Omega_\mathbf{k}) Y_{l'm'}(\Omega_{\mathbf{k}'}) \int_{\Omega_r} Y_{lm}(\Omega_\mathbf{r}) Y_{l'm'}^*(\Omega_\mathbf{r}) d\Omega_\mathbf{r}.$$

Let us recall the orthonormality of spherical harmonics, viz.

$$\int Y_{l'm'}^*(\Omega_\mathbf{r}) Y_{lm}(\Omega_\mathbf{r}) d\Omega_\mathbf{r} = \delta_{ll'}\delta_{mm'}. \tag{iii}$$

Then, the above equation can be written as

$$(2\pi)^3 \delta^3(\mathbf{k}-\mathbf{k}') = (4\pi)^2 \sum_{l=0}^{\infty}\sum_{m=-l}^{+l} Y_{lm}^*(\Omega_\mathbf{k}) Y_{lm}(\Omega_{\mathbf{k}'}) \int_{r=0}^{\infty} j_l(kr)\, j_l(k'r) r^2 dr.$$

Recalling that

$$\sum_{l,m} Y_{lm}^*(\Omega_\mathbf{k}) Y_{lm}(\Omega_{\mathbf{k}'}) = \delta(\Omega_\mathbf{k} - \Omega_{\mathbf{k}'}), \tag{iv}$$

we write

$$(2\pi)^3 \delta^3(\mathbf{k}-\mathbf{k}') = (4\pi)^2 \delta(\Omega_\mathbf{k} - \Omega_{\mathbf{k}'}) \int_0^\infty j_l(kr)\, j_l(k'r) r^2 dr$$

or

$$(2\pi)^3 \frac{\delta(k-k')}{k^2} \delta(\Omega_k - \Omega_{k'}) = (4\pi)^2 \delta(\Omega_k - \Omega_{k'}) \int_0^\infty j_l(kr)\, j_l(k'r) r^2 dr.$$

Therefore, we obtain

$$\int_0^\infty j_l(kr)\, j_l(k'r) r^2 dr = \frac{\pi}{2k^2}\delta(k-k'). \tag{v}$$

PROBLEM 27 Derive the formula

$$j_l(r) = 2^l r^l \sum_{s=0}^{\infty} \frac{(-1)^s (l+s)!}{s!(2l+2s+1)!} r^{2s}.$$

Solution: We know that Bessel function of the first kind is given by

$$J_l(r) = \sum_{s=0}^{\infty} \frac{(-1)^s}{s!\,\Gamma(l+s+1)} \left(\frac{r}{2}\right)^{l+2s}.$$

By the definition of spherical Bessel function, we have

$$j_l(r) = \sqrt{\frac{\pi}{2r}}\, J_{l+\frac{1}{2}}(r) = \sqrt{\frac{\pi}{2r}} \sum_{s=0}^{\infty} \frac{(-1)^s}{s!\,\Gamma\left(l+s+\frac{3}{2}\right)} \left(\frac{r}{2}\right)^{l+\frac{1}{2}+2s}. \qquad \text{(i)}$$

Let us recall the duplication formula of gamma functions, viz.

$$\Gamma(n)\Gamma\left(n+\frac{1}{2}\right) = \frac{\sqrt{\pi}}{2^{2n-1}} \Gamma(2n) \qquad \text{(ii)}$$

where n is a positive integer.

Using Eq. (ii), we write

$$\Gamma\left(l+s+\frac{3}{2}\right) = \frac{\sqrt{\pi}\,\Gamma(2l+2s+2)}{2^{2l+2s+1}\,\Gamma(l+s+1)}. \qquad \text{(iii)}$$

Substitution of Eq. (iii) into Eq. (i) yields

$$j_l(r) = 2^l r^l \sum_{s=0}^{\infty} \frac{(-1)^s (l+s)!}{s!(2l+2s+1)!} r^{2s}. \qquad \text{(iv)}$$

PROBLEM 28 Obtain the expression for $Ber'(x) = \frac{d}{dx} Ber(x)$ and show that

$$\frac{d}{dx}\left[x\,Ber'(x)\right] = -x\,Bei(x).$$

Solution: We know that

$$Ber(x) = \sum_{r=0}^{\infty} \frac{(-1)^r \left(\frac{x}{2}\right)^{4r}}{\left[(2r)!\right]^2}$$

$$= 1 - \frac{x^4}{2^2 \cdot 4^2} + \frac{x^8}{2^2 \cdot 4^2 \cdot 6^2 \cdot 8^2} - \frac{x^{12}}{2^2 \cdot 4^2 \cdot 6^2 \cdot 8^2 \cdot 10^2 \cdot 12^2} + \ldots \infty.$$

Differentiating both sides with respect to x, we obtain

$$Ber'(x) = -\frac{x^3}{2^2 \cdot 4} + \frac{x^7}{2^2 \cdot 4^2 \cdot 6^2 \cdot 8} - \frac{x^{11}}{2^2 \cdot 4^2 \cdot 6^2 \cdot 8^2 \cdot 10^2 \cdot 12} + \ldots \infty.$$

Therefore, we have

$$xBer'(x) = -\frac{x^4}{2^2 \cdot 4} + \frac{x^8}{2^2 \cdot 4^2 \cdot 6^2 \cdot 8} - \frac{x^{12}}{2^2 \cdot 4^2 \cdot 6^2 \cdot 8^2 \cdot 10^2 \cdot 12} + \ldots \infty.$$

Now, we have

$$\frac{d}{dx}\left[xBer'(x)\right] = -\frac{x^3}{2^2} + \frac{x^7}{2^2 \cdot 4^2 \cdot 6^2} - \frac{x^{11}}{2^2 \cdot 4^2 \cdot 6^2 \cdot 8^2 \cdot 10^2} + \ldots \infty$$

$$= -x\left[\frac{x^2}{2^2} - \frac{x^6}{2^2 \cdot 4^2 \cdot 6^2} + \frac{x^{10}}{2^2 \cdot 4^2 \cdot 6^2 \cdot 8^2 \cdot 10^2} - \ldots \infty\right]$$

$$= -x\,Bei(x).$$

PROBLEM 29 Prove that $\frac{d}{dx}\left[x\,Bei'(x)\right] = xBer(x)$

Solution: We know that

$$Bei(x) = \frac{x^2}{2^2} - \frac{x^6}{2^2 \cdot 4^2 \cdot 6^2} + \frac{x^{10}}{2^2 \cdot 4^2 \cdot 6^2 \cdot 8^2 \cdot 10^2} - \ldots \infty.$$

Then, $$\frac{d}{dx}Bei(x) = Bei'(x) = \frac{x}{2} - \frac{x^5}{2^2 \cdot 4^2 \cdot 6} + \frac{x^9}{2^2 \cdot 4^2 \cdot 6^2 \cdot 8^2 \cdot 10} - \ldots \infty.$$

Therefore, we have

$$x\,Bei'(x) = \frac{x^2}{2} - \frac{x^6}{2^2 \cdot 4^2 \cdot 6} + \frac{x^{10}}{2^2 \cdot 4^2 \cdot 6^2 \cdot 8^2 \cdot 10} - \ldots \infty.$$

Finally, we have

$$\frac{d}{dx}\left[x\,Bei'(x)\right] = x - \frac{x^5}{2^2 \cdot 4^2} + \frac{x^9}{2^2 \cdot 4^2 \cdot 6^2 \cdot 8^2} - \ldots \infty$$

$$= x\left[1 - \frac{x^4}{2^2 \cdot 4^2} + \frac{x^8}{2^2 \cdot 4^2 \cdot 6^2 \cdot 8^2} - \ldots \infty\right]$$

$$= x\,Ber(x).$$

PROBLEM 30 Show that

(a) $\mathcal{H}_{-\frac{1}{2}}(x)=\sqrt{\dfrac{2}{\pi x}}\sin x$,

(b) $\mathcal{H}_{\frac{1}{2}}(x)=\sqrt{\dfrac{2}{\pi x}}(1-\cos x)$,

(c) $\mathbb{L}_{-\frac{1}{2}}(x)=\sqrt{\dfrac{2}{\pi x}}\sinh x$.

Solution:

(a) We know that

$$\mathcal{H}_p(x)=\sum_{r=0}^{\infty}\frac{(-1)^r\left(\frac{x}{2}\right)^{p+2r+1}}{\Gamma\left(r+\frac{3}{2}\right)\Gamma\left(p+r+\frac{3}{2}\right)}. \tag{i}$$

Putting $p=-\dfrac{1}{2}$, we write

$$\mathcal{H}_{-\frac{1}{2}}(x)=\sum_{r=0}^{\infty}\frac{(-1)^r\left(\frac{x}{2}\right)^{2r+\frac{1}{2}}}{\Gamma\left(r+\frac{3}{2}\right)r!}$$

$$=\frac{\left(\frac{x}{2}\right)^{\frac{1}{2}}}{\Gamma\left(\frac{3}{2}\right)0!}-\frac{\left(\frac{x}{2}\right)^{2+\frac{1}{2}}}{\Gamma\left(\frac{5}{2}\right)1!}+\frac{\left(\frac{x}{2}\right)^{4+\frac{1}{2}}}{\Gamma\left(\frac{7}{2}\right)2!}-\frac{\left(\frac{x}{2}\right)^{6+\frac{1}{2}}}{\Gamma\left(\frac{9}{2}\right)3!}+\ldots\infty$$

$$=\frac{\sqrt{\frac{x}{2}}}{\Gamma\left(\frac{1}{2}\right)}\left[2-\frac{x^2}{3}+\frac{x^4}{3\cdot4\cdot5}-\frac{x^6}{3\cdot4\cdot5\cdot6\cdot7}+\ldots\infty\right]$$

$$=\frac{2}{x}\sqrt{\frac{x}{2\pi}}\left[x-\frac{x^3}{3!}+\frac{x^5}{5!}-\frac{x^7}{7!}+\ldots\infty\right]$$

$$=\sqrt{\frac{2}{\pi x}}\sin x.$$

(b) Putting $p=\frac{1}{2}$ in Eq. (i) , we write

$$\mathcal{H}_{\frac{1}{2}}(x)=\sum_{r=0}^{\infty}\frac{(-1)^r\left(\frac{x}{2}\right)^{2r+\frac{3}{2}}}{\Gamma\left(r+\frac{3}{2}\right)(r+1)!}$$

$$=\frac{\left(\frac{x}{2}\right)^{\frac{3}{2}}}{\Gamma\left(\frac{3}{2}\right)1!}-\frac{\left(\frac{x}{2}\right)^{2+\frac{3}{2}}}{\Gamma\left(\frac{5}{2}\right)2!}+\frac{\left(\frac{x}{2}\right)^{4+\frac{3}{2}}}{\Gamma\left(\frac{7}{2}\right)3!}-\ldots\infty$$

$$=\frac{\sqrt{\frac{x}{2}}}{\Gamma\left(\frac{1}{2}\right)}\left[\frac{\frac{x}{2}}{\frac{1}{2}}-\frac{\left(\frac{x}{2}\right)^3}{\frac{3}{2}\cdot\frac{1}{2}\cdot 2!}+\frac{\left(\frac{x}{2}\right)^5}{\frac{5}{2}\cdot\frac{3}{2}\cdot\frac{1}{2}\cdot 3!}-\ldots\infty\right]$$

$$=\sqrt{\frac{x}{2\pi}}\left[x-\frac{x^3}{3\cdot 4}+\frac{x^5}{3\cdot 4\cdot 5\cdot 6}-\ldots\infty\right]$$

$$=\frac{2}{x}\cdot\sqrt{\frac{x}{2\pi}}\left[\frac{x^2}{2!}-\frac{x^4}{4!}+\frac{x^6}{6!}-\ldots\infty\right]$$

$$=\sqrt{\frac{2}{\pi x}}\left[1-\left\{1-\frac{x^2}{2!}+\frac{x^4}{4!}-\frac{x^6}{6!}+\ldots\infty\right\}\right]$$

$$=\sqrt{\frac{2}{\pi x}}(1-\cos x).$$

(c) We know that

$$\mathbb{L}_p(x)=\sum_{r=0}^{\infty}\frac{\left(\frac{x}{2}\right)^{2r+p+1}}{\Gamma\left(r+\frac{3}{2}\right)\Gamma\left(p+r+\frac{3}{2}\right)}$$

Putting $p=-\frac{1}{2}$, we obtain

$$\mathbb{L}_{-\frac{1}{2}}(x)=\sum_{r=0}^{\infty}\frac{\left(\frac{x}{2}\right)^{2r+\frac{1}{2}}}{\Gamma\left(r+\frac{3}{2}\right)r!}$$

$$=\sqrt{\frac{x}{2}}\left[\frac{1}{\Gamma\left(\frac{3}{2}\right)0!}+\frac{\left(\frac{x}{2}\right)^2}{\Gamma\left(\frac{5}{2}\right)1!}+\frac{\left(\frac{x}{2}\right)^4}{\Gamma\left(\frac{7}{2}\right)2!}+\frac{\left(\frac{x}{2}\right)^6}{\Gamma\left(\frac{9}{2}\right)3!}+\ldots\infty\right]$$

$$=\frac{1}{\Gamma\left(\frac{1}{2}\right)}\sqrt{\frac{x}{2}}\left[\frac{1}{\frac{1}{2}}+\frac{x^2}{2^2\cdot\frac{3}{2}\cdot\frac{1}{2}}+\frac{x^4}{2^4\cdot\frac{5}{2}\cdot\frac{3}{2}\cdot\frac{1}{2}\cdot 2}+\frac{x^6}{2^6\cdot\frac{7}{2}\cdot\frac{5}{2}\cdot\frac{3}{2}\cdot\frac{1}{2}\cdot 6}+\ldots\infty\right]$$

$$=\sqrt{\frac{x}{2\pi}}\left[2+\frac{x^2}{3}+\frac{x^4}{3\cdot 4\cdot 5}+\frac{x^6}{3\cdot 4\cdot 5\cdot 6\cdot 7}+\ldots\infty\right]$$

$$=\frac{2}{x}\sqrt{\frac{x}{2\pi}}\left[x+\frac{x^3}{3!}+\frac{x^5}{5!}+\frac{x^7}{7!}+\ldots\infty\right]$$

$$=\sqrt{\frac{2}{\pi x}}\sinh x.$$

Alternatively, we have

$$\mathcal{H}_{-\frac{1}{2}}(x)=\sqrt{\frac{2}{\pi x}}\sin x.$$

Therefore, putting $p=-\frac{1}{2}$ in the relation

$$\mathbb{L}_p(x)=i^{-(p+1)}\mathcal{H}_p(ix),$$

we obtain

$$\mathbb{L}_{-\frac{1}{2}}(x)=i^{-\frac{1}{2}}\sqrt{\frac{2}{\pi ix}}\sin(ix)=\sqrt{\frac{2}{\pi x}}\sinh x.$$

PROBLEM 31 Prove that

(i) $\mathcal{H}_{p-1}(x)+\mathcal{H}_{p+1}(x)=\frac{2p}{x}\mathcal{H}_p(x)+\frac{\left(\frac{x}{2}\right)^p}{\sqrt{\pi}\,\Gamma\left(p+\frac{3}{2}\right)}$,

(ii) $\mathcal{H}_{p-1}(x)-\mathcal{H}_{p+1}(x)=2\mathcal{H}_p'(x)-\frac{\left(\frac{x}{2}\right)^p}{\sqrt{\pi}\,\Gamma\left(p+\frac{3}{2}\right)}$,

(iii) $\mathcal{H}_{p-1}(x)=\frac{p}{x}\mathcal{H}_p(x)+\mathcal{H}_p'(x)$,

(iv) $\mathcal{H}_{p+1}(x)=\frac{p}{x}\mathcal{H}_p(x)-\mathcal{H}_p'(x)+\frac{\left(\frac{x}{2}\right)^p}{\sqrt{\pi}\,\Gamma\left(p+\frac{3}{2}\right)}$,

(v) $\mathcal{H}_0'(x)=\frac{2}{\pi}-\mathcal{H}_1(x)$.

Solution:

(i) We know that

$$\mathcal{H}_p(x)=\sum_{r=0}^{\infty}\frac{(-1)^r\left(\frac{r}{2}\right)^{2r+p+1}}{\Gamma\left(r+\frac{3}{2}\right)\Gamma\left(p+r+\frac{3}{2}\right)}$$

Therefore, we have

$\mathcal{H}_{p-1}(x)+\mathcal{H}_{p+1}(x)$

$$=\sum_{r=0}^{\infty}\frac{(-1)^r\left(\frac{x}{2}\right)^{2r+p}}{\Gamma\left(r+\frac{3}{2}\right)\Gamma\left(p+r+\frac{1}{2}\right)}+\sum_{r=0}^{\infty}\frac{(-1)^r\left(\frac{x}{2}\right)^{2r+p+2}}{\Gamma\left(r+\frac{3}{2}\right)\Gamma\left(p+r+\frac{5}{2}\right)}$$

$$=\left(\frac{x}{2}\right)^p\left[\sum_{r=0}^{\infty}\frac{(-1)^r\left(\frac{x}{2}\right)^{2r}}{\Gamma\left(r+\frac{3}{2}\right)\Gamma\left(p+r+\frac{1}{2}\right)}+\sum_{r=0}^{\infty}\frac{(-1)^r\left(\frac{x}{2}\right)^{2r+2}}{\Gamma\left(r+\frac{3}{2}\right)\Gamma\left(p+r+\frac{5}{2}\right)}\right]$$

$$=\left(\frac{x}{2}\right)^p\left[\frac{1}{\Gamma\left(\frac{3}{2}\right)\Gamma\left(p+\frac{1}{2}\right)}+\sum_{r=1}^{\infty}\frac{(-1)^r\left(\frac{x}{2}\right)^{2r}}{\Gamma\left(r+\frac{3}{2}\right)\Gamma\left(p+r+\frac{1}{2}\right)}+\sum_{r=1}^{\infty}\frac{(-1)^{r-1}\left(\frac{x}{2}\right)^{2r}}{\Gamma\left(r+\frac{1}{2}\right)\Gamma\left(p+r+\frac{3}{2}\right)}\right]$$

$$=\frac{\left(\frac{x}{2}\right)^p}{\Gamma\left(\frac{3}{2}\right)\Gamma\left(p+\frac{1}{2}\right)}+\sum_{r=1}^{\infty}(-1)^r\left(\frac{x}{2}\right)^{p+2r}\left[\frac{1}{\Gamma\left(r+\frac{3}{2}\right)\Gamma\left(p+r+\frac{1}{2}\right)}-\frac{1}{\Gamma\left(r+\frac{1}{2}\right)\Gamma\left(p+r+\frac{3}{2}\right)}\right]$$

$$=\frac{\left(\frac{x}{2}\right)^p}{\Gamma\left(\frac{3}{2}\right)\Gamma\left(p+\frac{1}{2}\right)}+\sum_{r=1}^{\infty}(-1)^r\left(\frac{x}{2}\right)^{p+2r}\left[\frac{p}{\Gamma\left(r+\frac{3}{2}\right)\Gamma\left(p+r+\frac{3}{2}\right)}\right]$$

$$=\frac{\left(p+\frac{1}{2}\right)\left(\frac{x}{2}\right)^p}{\Gamma\left(\frac{3}{2}\right)\Gamma\left(p+\frac{3}{2}\right)}+p\sum_{r=1}^{\infty}\frac{(-1)^r\left(\frac{x}{2}\right)^{p+2r}}{\Gamma\left(r+\frac{3}{2}\right)\Gamma\left(p+r+\frac{3}{2}\right)}$$

$$=\frac{\frac{1}{2}\left(\frac{x}{2}\right)^p}{\Gamma\left(\frac{3}{2}\right)\Gamma\left(p+\frac{3}{2}\right)}+p\left[\frac{\left(\frac{x}{2}\right)^p}{\Gamma\left(\frac{3}{2}\right)\Gamma\left(p+\frac{3}{2}\right)}+\sum_{r=1}^{\infty}\frac{(-1)^r\left(\frac{x}{2}\right)^{p+2r}}{\Gamma\left(r+\frac{3}{2}\right)\Gamma\left(p+r+\frac{3}{2}\right)}\right]$$

$$=\frac{\left(\frac{x}{2}\right)^p}{\sqrt{\pi}\,\Gamma\left(p+\frac{3}{2}\right)}+p\sum_{r=0}^{\infty}\frac{(-1)^r\left(\frac{x}{2}\right)^{p+2r}}{\Gamma\left(r+\frac{3}{2}\right)\Gamma\left(p+r+\frac{3}{2}\right)}$$

$$=\frac{2p}{x}\sum_{r=0}^{\infty}\frac{(-1)^r\left(\frac{x}{2}\right)^{p+2r+1}}{\Gamma\left(r+\frac{3}{2}\right)\Gamma\left(p+r+\frac{3}{2}\right)}+\frac{\left(\frac{x}{2}\right)^p}{\sqrt{\pi}\,\Gamma\left(p+\frac{3}{2}\right)}$$

$$=\frac{2p}{x}\mathcal{H}_p(x)+\frac{\left(\frac{x}{2}\right)^p}{\sqrt{\pi}\,\Gamma\left(p+\frac{3}{2}\right)}.$$

(ii) $\mathcal{H}_{p-1}(x)-\mathcal{H}_{p+1}(x)$

$$=\sum_{r=0}^{\infty}\frac{(-1)^r\left(\frac{x}{2}\right)^{p+2r}}{\Gamma\left(r+\frac{3}{2}\right)\Gamma\left(p+r+\frac{1}{2}\right)}-\sum_{r=0}^{\infty}\frac{(-1)^r\left(\frac{x}{2}\right)^{p+2r+2}}{\Gamma\left(r+\frac{3}{2}\right)\Gamma\left(p+r+\frac{5}{2}\right)}$$

$$=\sum_{r=0}^{\infty}\frac{(-1)^r\left(\frac{x}{2}\right)^{p+2r}}{\Gamma\left(r+\frac{3}{2}\right)\Gamma\left(p+r+\frac{1}{2}\right)}-\sum_{r=1}^{\infty}\frac{(-1)^{r-1}\left(\frac{x}{2}\right)^{p+2r}}{\Gamma\left(r+\frac{1}{2}\right)\Gamma\left(p+r+\frac{3}{2}\right)}$$

$$=\left(\frac{x}{2}\right)^p\left[\frac{1}{\Gamma\left(\frac{3}{2}\right)\Gamma\left(p+\frac{1}{2}\right)}+\sum_{r=1}^{\infty}\left\{\frac{(-1)^r\left(\frac{x}{2}\right)^{2r}}{\Gamma\left(r+\frac{3}{2}\right)\Gamma\left(p+r+\frac{1}{2}\right)}+\frac{(-1)^r\left(\frac{x}{2}\right)^{2r}}{\Gamma\left(r+\frac{1}{2}\right)\Gamma\left(p+r+\frac{3}{2}\right)}\right\}\right]$$

$$=\left(\frac{x}{2}\right)^p\left[\frac{p+\frac{1}{2}}{\Gamma\left(\frac{3}{2}\right)\Gamma\left(p+\frac{3}{2}\right)}+\sum_{r=1}^{\infty}\frac{(p+2r+1)(-1)^r\left(\frac{x}{2}\right)^{2r}}{\Gamma\left(r+\frac{3}{2}\right)\Gamma\left(p+r+\frac{3}{2}\right)}\right]$$

$$=\left(\frac{x}{2}\right)^p\left[\frac{(p+1)-\frac{1}{2}}{\Gamma\left(\frac{3}{2}\right)\Gamma\left(p+\frac{3}{2}\right)}+\sum_{r=1}^{\infty}\frac{(p+2r+1)(-1)^r\left(\frac{x}{2}\right)^{2r}}{\Gamma\left(r+\frac{3}{2}\right)\Gamma\left(p+r+\frac{3}{2}\right)}\right]$$

$$=\sum_{r=0}^{\infty}\frac{(p+2r+1)(-1)^r\left(\frac{x}{2}\right)^{2r+p}}{\Gamma\left(r+\frac{3}{2}\right)\Gamma\left(p+r+\frac{3}{2}\right)}-\frac{\frac{1}{2}\left(\frac{x}{2}\right)^p}{\Gamma\left(\frac{3}{2}\right)\Gamma\left(p+\frac{3}{2}\right)}$$

$$=2\frac{d}{dx}\sum_{r=0}^{\infty}\frac{(-1)^r\left(\frac{x}{2}\right)^{2r+p+1}}{\Gamma\left(r+\frac{3}{2}\right)\Gamma\left(p+r+\frac{3}{2}\right)}-\frac{\left(\frac{x}{2}\right)^p}{\sqrt{\pi}\,\Gamma\left(p+\frac{3}{2}\right)}$$

$$=2\mathcal{H}_p'(x)-\frac{\left(\frac{x}{2}\right)^p}{\sqrt{\pi}\,\Gamma\left(p+\frac{3}{2}\right)}.$$

(iii) Adding relations (i) and (ii) together, we obtain

$$\mathcal{H}_{p-1}(x)=\frac{p}{x}\mathcal{H}_p(x)+\mathcal{H}_p'(x).$$

(iv) Subtracting relation (ii) from relation (i), we obtain

$$\mathcal{H}_{p+1}(x)=\frac{p}{x}\mathcal{H}_p(x)-\mathcal{H}_p'(x)+\frac{\left(\frac{x}{2}\right)^p}{\sqrt{\pi}\,\Gamma\left(p+\frac{3}{2}\right)}.$$

(v) Putting $p=0$ in relation (iv), we obtain

$$\mathcal{H}_1(x)=0-\mathcal{H}_0'(x)+\frac{1}{\sqrt{\pi}\,\Gamma\left(\frac{3}{2}\right)}$$

or $$\mathcal{H}_0'(x)=\frac{2}{\pi}-\mathcal{H}_1(x).$$

PROBLEM 32 Find the values of $Ai(0)$, $Ai'(0)$, $Bi(0)$ and $Bi'(0)$.

Solution: From Eq. (8.58), (8.59), we know that Airy function of the first kind is given by

$$Ai(x)=\sum_{r=0}^{\infty}\frac{1}{3^{2r+\frac{2}{3}}\,r!\,\Gamma\left(r+\frac{2}{3}\right)}x^{3r}-\sum_{r=0}^{\infty}\frac{1}{3^{2r+\frac{4}{3}}\,r!\,\Gamma\left(r+\frac{4}{3}\right)}x^{3r+1}. \tag{i}$$

When $x = 0$ only the non-vanishing term is the term with $r = 0$ in the first sum. Thus, we have

$$Ai(0)=\frac{1}{3^{\frac{2}{3}}\Gamma\left(\frac{2}{3}\right)}. \tag{ii}$$

The first order derivative of $Ai(x)$ is given by

$$Ai'(x)=\sum_{r=1}^{\infty}\frac{3}{3^{2r+\frac{2}{3}}(r-1)!\,\Gamma\left(r+\frac{2}{3}\right)}x^{3r-1}-\sum_{r=0}^{\infty}\frac{3\left(r+\frac{1}{3}\right)}{3^{2r+\frac{4}{3}}\,r!\,\Gamma\left(r+\frac{4}{3}\right)}x^{3r}$$

$$=\sum_{r=0}^{\infty}\frac{1}{3^{2r+\frac{5}{3}}\,r!\,\Gamma\left(r+\frac{5}{3}\right)}x^{3r+2}-\sum_{r=0}^{\infty}\frac{1}{3^{2r+\frac{1}{3}}\,r!\,\Gamma\left(r+\frac{1}{3}\right)}x^{3r}. \tag{iii}$$

In this case also, when $x = 0$ only the non-vanishing term is the term with $r = 0$ in the second sum.

Thus, we have

$$Ai'(0) = -\frac{1}{3^{\frac{1}{3}}\Gamma\left(\frac{1}{3}\right)}. \tag{iv}$$

Further, from Eq. (8.59), we have

$$Bi(x) = \sqrt{3}\left[\sum_{r=0}^{\infty}\frac{1}{3^{2r+\frac{2}{3}}r!\,\Gamma\left(r+\frac{2}{3}\right)}x^{3r} + \sum_{r=0}^{\infty}\frac{1}{3^{2r+\frac{4}{3}}r!\,\Gamma\left(r+\frac{4}{3}\right)}x^{3r+1}\right] \tag{v}$$

and $$Bi'(x) = \sqrt{3}\left[\sum_{r=1}^{\infty}\frac{3}{3^{2r+\frac{2}{3}}(r-1)!\,\Gamma\left(r+\frac{2}{3}\right)}x^{3r-1} + \sum_{r=0}^{\infty}\frac{3\left(r+\frac{1}{3}\right)}{3^{2r+\frac{4}{3}}r!\,\Gamma\left(r+\frac{4}{3}\right)}x^{3r}\right]$$

$$= \sum_{r=0}^{\infty}\frac{1}{3^{2r+\frac{7}{6}}r!\,\Gamma\left(r+\frac{5}{3}\right)}x^{3r+2} + \sum_{r=0}^{\infty}\frac{1}{3^{2r-\frac{1}{6}}r!\,\Gamma\left(r+\frac{1}{3}\right)}x^{3r}. \tag{vi}$$

From Eqs. (v) and (vi), we obtain

$$Bi(0) = \frac{1}{3^{\frac{1}{6}}\Gamma\left(\frac{2}{3}\right)} \text{ and } Bi'(0) = \frac{1}{3^{-\frac{1}{6}}\Gamma\left(\frac{1}{3}\right)}. \tag{vii}$$

PROBLEM 33 Prove that the Wronskian of $Ai(x)$ and $Bi(x)$ is equal to $\frac{1}{\pi}$.

Solution: We know that $Ai(x)$ and $Bi(x)$ satisfy the Airy equation.

Thus, we write

$$\frac{d^2}{dx^2}Ai(x) - xAi(x) = 0 \tag{i}$$

and $$\frac{d^2}{dx^2}Bi(x) - xBi(x) = 0. \tag{ii}$$

Subtracting $Bi(x)$ times Eq. (i) from $Ai(x)$ times Eq. (ii) and integrating the result with respect to x, we obtain

$$\int Ai(x)\frac{d^2}{dx^2}Bi(x)\,dx - \int Bi(x)\frac{d^2}{dx^2}Ai(x)\,dx = C \text{ (integration constant).}$$

Performing integration by parts in both the integrals, we obtain

$$Ai(x)\frac{d}{dx}Bi(x) - \int\left\{\frac{d}{dx}Ai(x)\right\}\left\{\frac{d}{dx}Bi(x)\right\}dx$$

$$-Bi(x)\frac{d}{dx}Ai(x) + \int\left\{\frac{d}{dx}Bi(x)\right\}\left\{\frac{d}{dx}Ai(x)\right\}dx = C$$

or
$$Ai(x)\frac{d}{dx}Bi(x) - Bi(x)\frac{d}{dx}Ai(x) = C$$

or
$$W(Ai(x), Bi(x)) = C \cdot \qquad \text{(iii)}$$

Here, the constant C is given by $W\left(Ai(0), Bi(0)\right)$.
Using the results of Problem 32, we have

$$\begin{aligned} C &= W\left(Ai(0), Bi(0)\right) \\ &= Ai(0)Bi'(0) - Bi(0)Ai'(0) \\ &= \frac{1}{3^{\frac{2}{3}}\Gamma\left(\frac{2}{3}\right)} \cdot \frac{1}{3^{-\frac{1}{6}}\Gamma\left(\frac{1}{3}\right)} - \frac{1}{3^{\frac{1}{6}}\Gamma\left(\frac{2}{3}\right)} \cdot \left\{-\frac{1}{3^{\frac{1}{3}}\Gamma\left(\frac{1}{3}\right)}\right\} \\ &= \frac{2}{3^{\frac{1}{2}}\Gamma\left(\frac{1}{3}\right)\Gamma\left(\frac{2}{3}\right)}. \qquad \text{(iv)} \end{aligned}$$

Using the reflection formula of gamma functions, viz.

$$\Gamma(n)\Gamma(1-n) = \frac{\pi}{\sin n\pi} \quad (n \neq \text{integer}),$$

we have $\Gamma\left(\frac{1}{3}\right)\Gamma\left(\frac{2}{3}\right) = \Gamma\left(\frac{1}{3}\right)\Gamma\left(1-\frac{1}{3}\right) = \dfrac{\pi}{\sin\frac{\pi}{3}} = \dfrac{2\pi}{\sqrt{3}}$.

Using this result in Eq. (iv), we have

$$C = \frac{1}{\pi}.$$

Substituting this value of C into Eq. (iii), we write

$$W\left(Ai(x), Bi(x)\right) = Ai(x)Bi'(x) - Bi(x)Ai'(x) = \frac{1}{\pi}.$$

PROBLEM 34 Prove that $Ai'(x) = -\frac{x}{\pi\sqrt{3}} K_{\frac{2}{3}}(\xi)\left(\xi = \frac{2}{3}x^{\frac{3}{2}}\right)$.

Solution: We know that

$$Ai(x) = \frac{\sqrt{x}}{3}\left[I_{-1/3}(\xi) - I_{1/3}(\xi)\right].$$

Therefore, its first order derivative is given by

$$Ai'(x) = \frac{1}{3}\cdot\frac{1}{2}x^{-1/2}\left[I_{-1/3}(\xi) - I_{1/3}(\xi)\right] + \frac{\sqrt{x}}{3}\frac{d}{dx}\left[I_{-1/3}(\xi) - I_{1/3}(\xi)\right].$$

Since $\frac{d}{dx} \equiv \frac{d\xi}{dx}\frac{d}{d\xi} \equiv \sqrt{x}\frac{d}{d\xi}$, we write

$$Ai'(x) = \frac{x^{-½}}{6}\left[I_{-\frac{1}{3}}(\xi) - I_{\frac{1}{3}}(\xi)\right] + \frac{x}{3}\left[I'_{-\frac{1}{3}}(\xi) - I'_{\frac{1}{3}}(\xi)\right]$$

$$= \frac{x}{3}\left[\frac{x^{-\frac{3}{2}}}{2}I_{-\frac{1}{3}}(\xi) - \frac{x^{-\frac{3}{2}}}{2}I_{\frac{1}{3}}(\xi) + I'_{-\frac{1}{3}}(\xi) - I'_{\frac{1}{3}}(\xi)\right]$$

$$= \frac{x}{3}\left[\frac{1}{3\xi}I_{-\frac{1}{3}}(\xi) - \frac{1}{3\xi}I_{\frac{1}{3}}(\xi) + I'_{-\frac{1}{3}}(\xi) - I'_{\frac{1}{3}}(\xi)\right]. \quad \text{(i)}$$

In the recurrence relation

$$I'_p(\xi) = I_{p+1}(\xi) + \frac{p}{\xi}I_p(\xi),$$

if we put $p = -\frac{1}{3}$, we obtain

$$I'_{-1/3}(\xi) = I_{2/3}(\xi) - \frac{1}{3\xi}I_{-\frac{1}{3}}(\xi). \quad \text{(ii)}$$

Similarly, in the recurrence relation

$$I'_p(\xi) = I_{p-1}(\xi) - \frac{p}{\xi}I_p(\xi),$$

if we put $p=\frac{1}{3}$, we obtain

$$I'_{1/3}(\xi)=I_{-2/3}(\xi)-\frac{1}{3\xi}I_{\frac{1}{3}}(\xi). \quad \text{(iii)}$$

Substituting Eqs. (ii) and (iii) into Eq. (i), we obtain

$$Ai'(x)=\frac{x}{3}\left[\frac{1}{3\xi}I_{-\frac{1}{3}}(\xi)-\frac{1}{3\xi}I_{\frac{1}{3}}(\xi)+I_{\frac{2}{3}}(\xi)-\frac{1}{3\xi}I_{-\frac{1}{3}}(\xi)-I_{-\frac{2}{3}}(\xi)+\frac{1}{3\xi}I_{\frac{1}{3}}(\xi)\right]$$

$$\left(\text{since}\frac{x^{-3/2}}{2}=\frac{1}{3\xi}\right)$$

$$=-\frac{x}{3}\left[I_{-\frac{2}{3}}(\xi)-I_{\frac{2}{3}}(\xi)\right]$$

$$=-\frac{x}{3}\cdot\frac{2}{\pi}\cdot\sin\left(\frac{2\pi}{3}\right)\left[\frac{\pi}{2}\frac{I_{-\frac{2}{3}}(\xi)-I_{\frac{2}{3}}(\xi)}{\sin\left(\frac{2\pi}{3}\right)}\right]$$

$$=-\frac{x}{\pi\sqrt{3}}K_{\frac{2}{3}}(\xi),\left(\xi=\frac{2}{3}x^{3/2}\right).$$

EXERCISE

1. Using the Wronskian of $J_p(x)$ and $N_p(x)$ prove that

$$J_p(x)N_{p+1}(x)-J_{p+1}(x)N_p(x)=-\frac{2}{\pi x}.$$

2. Prove that $J_{p+1}(x)N_p(x)-J_p(x)N_{p+1}(x)=\frac{1}{x}$.

3. Prove that $W\left(H_p^{(1)}(x),H_p^{(2)}(x)\right)=-\frac{4}{\pi i x}$ where W denotes the Wronskian.

4. Prove that (i) $H_{-n}^{(1)}(x)=e^{i\pi n}H_n^{(1)}(x)$ and (ii) $H_{-n}^{(2)}(x)=e^{-i\pi n}H_n^{(2)}(x)$

5. Prove that Hankel functions satisfy Bessel equation.

6. Prove that (i) $J_p(x)=\frac{1}{2}\left[H_p^{(1)}(x)+H_p^{(2)}(x)\right]$ and (ii) $N_p(x)=-\frac{i}{2}\left[H_p^{(1)}(x)-H_p^{(2)}(x)\right]$.

7. Prove that $I_p(x)K_{p+1}(x)+I_{p+1}(x)K_p(x)=\dfrac{1}{x}$.

8. Prove that $I_n(x+y)=\sum\limits_{s=-\infty}^{\infty} I_s(x)I_{n-s}(y)$.

9. Prove that (a) $I_p(x)=e^{-\frac{p\pi i}{2}}J_p(ix)$, (b) $K_p(x)=\dfrac{\pi i}{2}e^{\frac{p\pi i}{2}}H_p^{(1)}(ix)$.

10. Prove that (a) $I_p(x)=i^p J_p(-ix)$, (b) $K_p(x)=-\dfrac{\pi}{2}i^{1-p}H_p^{(2)}(-ix)$.

11. Prove that

$$I_p(x)I_{-p-1}(x)-I_{-p}(x)I_{p+1}(x)=I_p(x)I_{-p+1}(x)-I_{-p}(x)I_{p-1}(x)=-\frac{2\sin(p\pi)}{\pi x}.$$

12. Prove that (i) $\dfrac{d}{dx}\left\{\dfrac{K_p(x)}{I_p(x)}\right\}=-\dfrac{1}{x\{I_p(x)\}^2}$, (ii) $K_{\frac{3}{2}}(x)=\left(1+\dfrac{1}{x}\right)K_{\frac{1}{2}}(x)$.

13. Prove that modified Bessel functions $I_p(x)$ and $K_p(x)$ satisfy the differential equation

$$x^2\frac{d^2y}{dx^2}+x\frac{dy}{dx}-(x^2+p^2)y=0 \quad (p\geq 0).$$

14. Obtain the expressions for

(i) $J_{\frac{3}{2}}(x)$, (ii) $J_{-\frac{3}{2}}(x)$, (iii) $J_{\frac{5}{2}}(x)$, (iv) $J_{-\frac{5}{2}}(x)$,

(v) $N_{\frac{3}{2}}(x)$, (vi) $N_{-\frac{3}{2}}(x)$, (vii) $N_{\frac{5}{2}}(x)$, (viii) $N_{-\frac{5}{2}}(x)$,

(ix) $j_2(x)$, (x) $j_3(x)$, (xi) $n_2(x)$, (xii) $n_3(x)$.

Ans. (ix) $j_2(x)=\left(\dfrac{3}{x^3}-\dfrac{1}{x}\right)\sin x-\dfrac{3\cos x}{x^2}$,

(x) $j_3(x)=\left(\dfrac{15}{x^4}-\dfrac{6}{x^2}\right)\sin x-\left(\dfrac{15}{x^3}-\dfrac{1}{x}\right)\cos x$,

(xi) $n_2(x)=-\left(\dfrac{3}{x^3}-\dfrac{1}{x}\right)\cos x-\dfrac{3\sin x}{x^2}$,

(xii) $n_3(x)=-\left(\dfrac{15}{x^4}-\dfrac{6}{x^2}\right)\cos x-\left(\dfrac{15}{x^3}-\dfrac{1}{x}\right)\sin x$.

15. Given that $j_0(x) = \dfrac{\sin x}{x}$, obtain the expressions for $j_1(x)$ and $j_2(x)$

Ans. $j_1(x) = \dfrac{\sin x}{x^2} - \dfrac{\cos x}{x}$,

$$j_2(x) = \left(\frac{3}{x^3} - \frac{1}{x}\right)\sin x - \frac{3}{x^2}\cos x.$$

16. Obtain the expressions for $h_1^{(1)}(x)$ and $h_2^{(1)}(x)$

Ans. $h_1^{(1)}(x) = -\left(\dfrac{1}{x} + \dfrac{i}{x^2}\right)e^{ix}$, $h_2^{(1)}(x) = \left(\dfrac{i}{x} - \dfrac{3}{x^2} - \dfrac{3i}{x^3}\right)e^{ix}$.

17. Prove that

$$j_l(x) = 2^l x^l \sum_{m=0}^{\infty} \frac{(-1)^m (l+m)!\, x^{2m}}{m!\,(2l+2m+1)!}, \quad (l = 0, 1, 2 \ldots).$$

18. By direct differentiations of the spherical Bessel function $j_l(x) = \sqrt{\dfrac{\pi}{2x}}\, J_{l+\frac{1}{2}}(x)$, prove that $x^2 j_l''(x) + 2x\, j_l'(x) + \{x^2 - l(l+1)\}\, j_l(x) = 0$.

19. Prove that

(a) $i_1(x) = -\dfrac{\sinh x}{x^2} + \dfrac{\cosh x}{x}$,

(b) $i_2(x) = \left(\dfrac{3}{x^3} + \dfrac{1}{x}\right)\sinh x - \dfrac{3\cosh x}{x^2}$,

(c) $k_1(x) = \left(\dfrac{1}{x^2} + \dfrac{1}{x}\right)e^{-x}$.

20. Prove that

(i) $n_l(x) = -(-1)^l x^l \left(\dfrac{1}{x}\dfrac{d}{dx}\right)^l \left(\dfrac{\cos x}{x}\right)$,

(ii) $h_l^{(1)}(x) = -i(-1)^l x^l \left(\dfrac{1}{x}\dfrac{d}{dx}\right)^l \left(\dfrac{e^{ix}}{x}\right)$,

(iii) $h_2^{(2)}(x) = i(-1)^l x^l \left(\dfrac{1}{x}\dfrac{d}{dx}\right)^l \left(\dfrac{e^{-ix}}{x}\right)$.

21. Prove that

(i) $\displaystyle\int_0^a j_l\left(\alpha_{lm}\frac{r}{a}\right) j_l\left(\alpha_{ln}\frac{r}{a}\right) r^2\,dr = \frac{a^3}{2}\left[j_{l+1}(\alpha_{lm})\right]^2 \delta_{mn}$,

(ii) $\int_0^1 j_l(\alpha_{lm} r) j_l(\alpha_{ln} r) r^2 dr = \frac{1}{2}[j_{l+1}(\alpha_{lm})]^2 \delta_{mn}$

where α_{lm} is the mth zero of $j_l(x)$.

22. Prove that

$$\int_0^a x\left[Ber^2(x) + Bei^2(x)\right]dx = a\left[Ber(a)Bei'(a) - Ber'(a)Bei(a)\right].$$

23. Prove that $\int_0^a x\,Ber(x)dx = a\,Bei'(a)$.

24. Show that

(i) $Ber_1(x) = -\dfrac{x}{2\sqrt{2}}\left[1 + \dfrac{\left(\frac{x}{2}\right)^2}{1!2!} - \dfrac{\left(\frac{x}{2}\right)^4}{2!3!} - \dfrac{\left(\frac{x}{2}\right)^6}{3!4!} + \infty\right]$,

(ii) $Bei_1(x) = \dfrac{x}{2\sqrt{2}}\left[1 - \dfrac{\left(\frac{x}{2}\right)^2}{1!2!} - \dfrac{\left(\frac{x}{2}\right)^4}{2!3!} + \dfrac{\left(\frac{x}{2}\right)^6}{3!4!} + \dots\infty\right]$.

25. Prove that

(i) $\mathcal{H}_0(x) = \dfrac{2}{\pi} - \mathcal{H}_1(x)$, (ii) $\mathbb{L}_0'(x) = \dfrac{2}{\pi} + \mathbb{L}_1(x)$

26. Obtain the expressions for (a) $\mathcal{H}_0(x)$ and (b) $\mathbb{L}_{\frac{1}{2}}(x)$

Ans. (a) $\dfrac{2}{\pi}\left(\dfrac{x}{1^2} - \dfrac{x^3}{1^2\cdot3^2} + \dfrac{x^5}{1^2\cdot3^2\cdot5^2} - \dfrac{x^7}{1^2\cdot3^2\cdot5^2\cdot7^2} + \dots\infty\right)$,

(b) $\sqrt{\dfrac{2}{\pi x}}(\cosh x - 1)$.

27. Prove that

(i) $\mathbb{L}_{p-1}(x) - \mathbb{L}_{p+1}(x) = \dfrac{2p}{x}\mathbb{L}_p(x) + \dfrac{\left(\frac{x}{2}\right)^p}{\sqrt{\pi}\,\Gamma\left(p + \frac{3}{2}\right)}$

(ii) $\mathbb{L}_{p-1}(x)+\mathbb{L}_{p+1}(x)=2\mathbb{L}_p'(x)-\dfrac{\left(\dfrac{x}{2}\right)^p}{\sqrt{\pi}\,\Gamma\left(p+\dfrac{3}{2}\right)}$

(iii) $\mathbb{L}_{p-1}(x)=\dfrac{p}{x}\mathbb{L}_p(x)+\mathbb{L}_p'(x)$

(iv) $\mathbb{L}_{p+1}(x)=\mathbb{L}_p'(x)-\dfrac{p}{x}\mathbb{L}_p(x)-\dfrac{\left(\dfrac{x}{2}\right)^p}{\sqrt{\pi}\,\Gamma\left(p+\dfrac{3}{2}\right)}$

(v) $\mathbb{L}_0'(x)=\dfrac{2}{\pi}+\mathbb{L}_1(x)$

28. Find the power series solution of the Airy equation, $\dfrac{d^2y}{dx^2}-xy=0$.

Ans. $y=a_0\left[1+\dfrac{1}{3!}x^3+\dfrac{1.4}{6!}x^6+\dfrac{1.4.7}{9!}x^9+\ldots\infty\right]$

$+a_1\left[x+\dfrac{2}{4!}x^4+\dfrac{2.5}{7!}x^7+\dfrac{2.5.8}{10!}x^{10}+\ldots\infty\right].$

29. Starting from $Bi(x)=\sqrt{\dfrac{x}{3}}\left[I_{-1/3}(\xi)+I_{1/3}(\xi)\right]$ with $\xi=\dfrac{2}{3}x^{3/2}$, prove that

$$Bi'(x)=\frac{x}{\sqrt{3}}\left[I_{-2/3}(\xi)+I_{2/3}(\xi)\right].$$

9

CHAPTER

Hypergeometric Functions

9.1 HYPERGEOMETRIC EQUATION, ITS POWER SERIES SOLUTION AND HYPERGEOMETRIC FUNCTION

A hypergeometric equation is a homogeneous second order linear ordinary differential equation having the form

$$x(x-1)\frac{d^2y}{dx^2}+[(\alpha+\beta+1)x-\lambda]\frac{dy}{dx}+\alpha\beta y=0 \tag{9.1}$$

where α, β, λ are suitable constants. This equation was extensively studied by the German mathematician-cum-physicist, Johann Carl Friedrich Gauss (April 30, 1777-February 23, 1855) and, therefore, this equation is known as Gauss' hypergeometric equation.

One can see that the above differential equation has removable singularities at $x=0,1,\infty$.

To verify the removable singularities at $x=0,1$, we see that the above equation can be recast in the form

$$\frac{d^2y}{dx^2}+P(x)\frac{dy}{dx}+Q(x)y=0$$

where $$P(x)=\frac{(\alpha+\beta+1)x-\lambda}{x(x-1)} \text{ and } Q(x)=\frac{\alpha\beta}{x(x-1)}.$$

Here, $P(x) = \infty$, $Q(x) = \infty$ at $x = 0, 1$.

Hence, the points $x = 0, 1$ are singular points of the hypergeometric equation.

Further, we see that

$$\lim_{x\to 0} xP(x)=\lambda \text{ (finite)}, \quad \lim_{x\to 0} x^2Q(x)=0 \text{ (finite)}$$

$$\lim_{x\to 1}(x-1)P(x)=\alpha+\beta-\lambda+1 \text{(finite)}, \quad \lim_{x\to 1}(x-1)^2 Q(x)=0 \text{ (finite)}.$$

Hence, we confirm that the points $x = 0, 1$ are removable singular points of hypergeometric equation.

To study the singularity at $x = \infty$, we put $x = \frac{1}{z}$ such that

$$\frac{dy}{dx} = -z^2\frac{dy}{dz} \quad \text{and} \quad \frac{d^2y}{dx^2} = z^4\frac{d^2y}{dz^2} + 2z^3\frac{dy}{dz}.$$

Then, the hypergeometric Eq. (9.1) transforms into

$$z^2(z-1)\frac{d^2y}{dz^2} + \left[(2-\lambda)z^2 + (\alpha+\beta-1)z\right]\frac{dy}{dz} - \alpha\beta y = 0 \tag{9.2}$$

which can be recast into the form

$$\frac{d^2y}{dz^2} + P(z)\frac{dy}{dz} + Q(z)y = 0$$

where
$$P(z) = \frac{(2-\lambda)z + \alpha + \beta - 1}{z(z-1)}, \; Q(z) = -\frac{\alpha\beta}{z^2(z-1)}.$$

Here, we see that

$$P(0) = \infty, \; Q(0) = -\infty$$

implying that the point $z = 0$ is a singular point of the transformed Eq. (9.2).
Further, we see that

$$\lim_{z\to 0} zP(z) = 1 - \alpha - \beta (\text{finite}), \lim_{z\to 0} z^2Q(z) = \alpha\beta\,(\text{finite}).$$

Hence, the point $z = 0$ is a removable singular point of the transformed Eq. (9.2) and correspondingly, the point $x = \infty$ is a removable singular point of the hypergeometric Eq. (9.1).

Thus, we see that the points $x = 0, 1, \infty$ are removable singular points of hypergeometric Eq. (9.1) and, therefore, by Fuchs' theorem, Eq. (9.1) can have power series solutions about these points. The power series solutions of hypergeometric equation about these three points are successively found out as in the following.

(a) Series Solution about $x = 0$

Let the series solution of Eq. (9.1) be

$$y = \sum_{r=0}^{\infty} a_r x^{k+r} \; (a_0 \neq 0) \tag{9.3}$$

where the constant k and the coefficients a_r are to be determined.

Then, we have

$$\frac{dy}{dx}=\sum_{r=0}^{\infty}(k+r)a_r x^{k+r-1},\ \frac{d^2y}{dx^2}=\sum_{r=0}^{\infty}(k+r)(k+r-1)a_r x^{k+r-2}.$$

Substituting the expressions for $y, \frac{dy}{dx}, \frac{d^2y}{dx^2}$ into Eq. (9.1), we have

$$\sum_{r=0}^{\infty}(k+r+\alpha)(k+r+\beta)a_r x^{k+r}-\sum_{r=0}^{\infty}(k+r)(k+r+\lambda-1)a_r x^{k+r-1}=0.$$

The above equation, being an identity, will be satisfied when all the coefficients of various powers of x separately vanish.

Thus, equating the coefficient of the lowest power of x, i.e., the coefficient of x^{k-1} to zero, we obtain the indicial equation

$$k(k+\lambda-1)a_0=0 \text{ yielding } k=0,\ 1-\lambda \text{ with } a_0\neq 0.$$

Equating the coefficient of next higher power of x, i.e., of x^k to zero, we have

$$(k+\alpha)(k+\beta)a_0-(k+1)(k+\lambda)a_1=0$$

yielding $a_1=\frac{(k+\alpha)(k+\beta)}{(k+1)(k+\lambda)}a_0$ and so on.

In general, equating the coefficient of x^{k+r} to zero, we obtain

$$(k+r+\alpha)(k+r+\beta)a_r-(k+r+1)(k+r+\lambda)a_{r+1}=0$$

yielding

$$a_{r+1}=\frac{(k+r+\alpha)(k+r+\beta)}{(k+r+1)(k+r+\lambda)}a_r. \tag{9.4}$$

Choice I: When $k = 0$, Eq. (9.4) yields

$$a_{r+1}=\frac{(r+\alpha)(r+\beta)}{(r+1)(r+\lambda)}a_r. \tag{9.5}$$

Putting $r = 0, 1, 2, 3$ … etc., in Eq. (9.5), we obtain

$$a_1=\frac{\alpha\beta}{1\cdot\lambda}a_0$$

$$a_2=\frac{\alpha(\alpha+1)\beta(\beta+1)}{1\cdot 2\lambda(\lambda+1)}a_0=\frac{(\alpha)_2(\beta)_2}{2!(\lambda)_2}a_0,$$

$$a_3 = \frac{\alpha(\alpha+1)(\alpha+2)\beta(\beta+1)(\beta+2)}{1\cdot 2\cdot 3\,\lambda(\lambda+1)(\lambda+2)}a_0 = \frac{(\alpha)_3(\beta)_3}{3!(\lambda)_3}a_0 \text{ and so on.}$$

In general, we write

$$a_n = \frac{\alpha(\alpha+1)(\alpha+2)(\alpha+3)\ldots(\alpha+n-1)\beta(\beta+1)(\beta+2)(\beta+3)\ldots(\beta+n-1)}{n!\,\lambda(\lambda+1)(\lambda+2)(\lambda+3)\ldots(\lambda+n-1)}a_0$$

$$= \frac{(\alpha)_n(\beta)_n}{n!(\lambda)_n}a_0 \tag{9.6}$$

where $n = 1, 2, 3 \ldots$

Here, the symbol, viz.

$$(\alpha)_n = \frac{\Gamma(\alpha+n)}{\Gamma(\alpha)} = \alpha(\alpha+1)(\alpha+2)(\alpha+3)\ldots(\alpha+n-1)$$

is known as Pochhammer symbol .

Substituting the values of these coefficients into the assumed solution (Eq. (9.3)), we obtain the power series solution of Eq. (9.1) as

$$y_1 = a_0 + a_1x + a_2x^2 + a_3x^3 + \ldots + a_nx^n + \ldots \infty \text{ (as } k = 0)$$

$$= a_0 \sum_{n=0}^{\infty} \frac{(\alpha)_n(\beta)_n}{(\lambda)_n}\frac{x^n}{n!}. \tag{9.7}$$

The series represented by $\sum_{n=0}^{\infty} \frac{(\alpha)_n(\beta)_n}{(\lambda)_n}\frac{x^n}{n!}$ is known as hypergeometric function or hypergeometric series.

It is denoted by

$${}_2F_1(\alpha, \beta; \lambda; x) = \sum_{n=0}^{\infty} \frac{(\alpha)_n(\beta)_n}{(\lambda)_n}\frac{x^n}{n!}$$

$$= 1 + \frac{\alpha\cdot\beta}{\lambda}\frac{x}{1!} + \frac{\alpha(\alpha+1)\beta(\beta+1)}{\lambda(\lambda+1)}\frac{x^2}{2!} + \ldots \tag{9.8}$$

with $\lambda \neq 0, -1, -2, -3 \ldots$

Thus, Eq. (9.7) can be written as

$$y_1 = a_0\, {}_2F_1(\alpha, \beta; \lambda; x) \tag{9.9}$$

Here, the leading subscript 2 indicates that there are two Pochhammer symbols in the numerator and the rear subscript 1 indicates that there is one Pochhammer symbol in the denominator. The semicolons separate the parameters in the numerator, that in the denominator and the variable.

The hypergeometric function is so named because of the fact that the series reduces to the elementary geometric series when $\alpha = 1$ and $\beta = \lambda$.

In such a case, we have

$$ {}_2F_1(1, \beta; \beta; x) = \sum_{n=0}^{\infty} x^n = 1 + x + x^2 + x^3 + \ldots + x^n + \ldots \tag{9.10} $$

Convergence of the Hypergeometric Series

The nth term of the hypergeometric series is

$$ T_n = \frac{(\alpha)_n (\beta)_n}{n!(\lambda)_n} x^n . $$

The $(n + 1)th$ term is

$$ T_{n+1} = \frac{(\alpha)_{n+1} (\beta)_{n+1}}{(n+1)!(\lambda)_{n+1}} x^{n+1} . $$

Therefore, we have

$$ \lim_{n\to\infty} \left| \frac{T_{n+1}}{T_n} \right| = \lim_{n\to\infty} \frac{(\alpha+n)(\beta+n)}{(n+1)\ (\lambda+n)} |x| = |x| . $$

The series is convergent for $|x| < 1$.

Choice II: When $k = 1 - \lambda$, Eq. (9.4) yields

$$ a_{r+1} = \frac{(1-\lambda+r+\alpha)(1-\lambda+r+\beta)}{(2-\lambda+r)(1+r)} a_r . \tag{9.11} $$

Putting $r = 0,1,2,3,4 \ldots$ etc., we write

$$ a_1 = \frac{(\alpha-\lambda+1)(\beta-\lambda+1)}{1!(2-\lambda)} a_0 , $$

$$ a_2 = \frac{(\alpha-\lambda+1)(\alpha-\lambda+2)(\beta-\lambda+1)(\beta-\lambda+2)}{2!(2-\lambda)(3-\lambda)} a_0 = \frac{(\alpha-\lambda+1)_2 (\beta-\lambda+1)_2}{2!(2-\lambda)_2} a_0 , $$

$$ a_3 = \frac{(\alpha-\lambda+1)(\alpha-\lambda+2)(\alpha-\lambda+3)(\beta-\lambda+1)(\beta-\lambda+2)(\beta-\lambda+3)}{3!(2-\lambda)(3-\lambda)(4-\lambda)} a_0 $$

$$ = \frac{(\alpha-\lambda+1)_3 (\beta-\lambda+1)_3}{3!(2-\lambda)_3} a_0 \text{ and so on.} $$

In general, we have

$$ a_n = \frac{(\alpha-\lambda+1)_n (\beta-\lambda+1)_n}{n!(2-\lambda)_n} a_0 . $$

Substituting the values these coefficients into Eq. (9.3) (assumed solution), we obtain

$$y_2 = x^{1-\lambda}\left[a_0 + a_1 x + a_2 x^2 + a_3 x^3 + \ldots + a_n x^n + \ldots \infty\right]$$

$$= a_0 x^{1-\lambda} \sum_{n=0}^{\infty} \frac{(\alpha+1-\lambda)_n (\beta+1-\lambda)_n}{n!(2-\lambda)_n} x^n$$

$$= a_0 x^{1-\lambda} {}_2F_1(\alpha+1-\lambda, \beta+1-\lambda; 2-\lambda; x) \tag{9.12}$$

with $\lambda \neq 1$ or negative integer.

Now, the general solution of Eq. (9.1) about the regular singular point $x = 0$ is

$$y = C_1 {}_2F_1(\alpha, \beta; \lambda; x) + C_2 x^{1-\lambda} {}_2F_1(\alpha+1-\lambda, \beta+1-\lambda; 2-\lambda; x) \tag{9.13}$$

where C_1 and C_2 are arbitrary constants in which the constant a_0 has been absorbed.

When $\lambda = 1$ Eqs. (9.9) and (9.12) become identical and they cannot be independent solutions.

Further, when λ is equal to a negative integer, the hypergeometric function (${}_2F_1$) is not defined. Hence, these values of λ are discarded.

(b) Series Solution about $x = 1$

In this case, let us put $t = 1 - x$. Then, we have

$$\frac{dy}{dx} = \frac{dy}{dt}\frac{dt}{dx} = -\frac{dy}{dt} \text{ and } \frac{d^2y}{dx^2} = \frac{d^2y}{dt^2}.$$

Using these expressions, Eq. (9.1) transforms into the equation

$$t(t-1)\frac{d^2y}{dt^2} + [(\alpha+\beta+1)t - (\alpha+\beta+1-\lambda)]\frac{dy}{dt} + \alpha\beta y = 0. \tag{9.14}$$

Equation (9.14) is similar to Eq. (9.1). Here, the quantities x and λ are replaced by t and $\lambda' = (\alpha + \beta + 1 - \lambda)$, respectively.

Let us assume the series solution of Eq. (9.14) as

$$y = \sum_{r=0}^{\infty} a_r x^{k+r}.$$

Following the procedure of Frobenius method, the roots of the indicial equation are found to be

$$k = 0, 1 - \lambda' = \lambda - \alpha - \beta.$$

Thus, as in (a), the two independent series solutions of Eq. (9.14) and hence of Eq. (9.1) are found to be

$$y_3 = a_0\,{}_2F_1(\alpha, \beta; \lambda'; t) = a_0\,{}_2F_1(\alpha, \beta; \alpha+\beta+1-\lambda; 1-x) \tag{9.15}$$

and

$$\begin{aligned} y_4 &= a_0 t^{1-\lambda'}\,{}_2F_1(\alpha+1-\lambda', \beta+1-\lambda'; 2-\lambda'; t) \\ &= a_0 (1-x)^{\lambda-\alpha-\beta}\,{}_2F_1(\lambda-\beta, \lambda-\alpha; 1-\alpha-\beta+\lambda; 1-x) \\ &= a_0 (1-x)^{\lambda-\alpha-\beta}\,{}_2F_1(\lambda-\alpha, \lambda-\beta; 1-\alpha-\beta+\lambda; 1-x). \end{aligned} \tag{9.16}$$

As in (a), the cases with $\lambda' = 1$, i.e., $\lambda = \alpha + \beta$ and $\lambda' =$ negative integer are discarded. Thus, the general solution of Eq. (9.1) about $x = 1$ is written as

$$\begin{aligned} y &= C_3\,{}_2F_1(\alpha, \beta; \alpha+\beta+1-\lambda; 1-x) \\ &\quad + C_4 (1-x)^{\lambda-\alpha-\beta}\,{}_2F_1(\lambda-\alpha, \lambda-\beta; 1-\alpha-\beta+\lambda; 1-x) \end{aligned} \tag{9.17}$$

where C_3 and C_4 are arbitrary constants in which the constant (a_0) has been absorbed.

(c) Series Solution about $x = \infty$

We see that the series solution of Eq. (9.1) about $x = \infty$ is the same as the series solution of the transformed equation, viz., Eq. (9.2) about $z = 0$. Hence, instead of solving Eq. (9.1), we are to solve Eq. (9.2) as it is more convenient.

Let us assume the series solution of Eq. (9.2) about $z = 0$ as

$$y = \sum_{r=0}^{\infty} a_r z^{k+r} \quad (a_0 \neq 0) \tag{9.18}$$

where the coefficients a_r and the constant k are to be determined later.

Now, we have

$$\frac{dy}{dz} = \sum_{r=0}^{\infty} (k+r) a_r z^{k+r-1}$$

and

$$\frac{d^2y}{dz^2} = \sum_{r=0}^{\infty} (k+r)(k+r-1) a_r z^{k+r-2}.$$

Substituting the expressions for $y, \dfrac{dy}{dz}, \dfrac{d^2y}{dz^2}$ into Eq. (9.2), we obtain

$$\sum_{r=0}^{\infty} (k+r)(k+r+1-\lambda) a_r z^{k+r+1} - \sum_{r=0}^{\infty} (k+r-\alpha)(k+r-\beta) a_r z^{k+r} = 0.$$

In this equation, the coefficients of various powers of z must separately vanish.

Equating the coefficient of lowest power of z, i.e., of z^k to zero, we obtain the indicial equation

$$(k-\alpha)(k-\beta) a_0 = 0 \quad (a_0 \neq 0)$$

yielding $k = \alpha, \beta$ for arbitrary a_0.

In this way, equating the coefficient of z^{k+r+1} to zero, we obtain

$$(k+r)(k+r+1-\lambda)a_r-(k+r+1-\alpha)(k+r+1-\beta)a_{r+1}=0$$

or
$$a_{r+1}=\frac{(k+r)(k-\lambda+1+r)}{(k-\alpha+1+r)(k-\beta+1+r)}a_r. \tag{9.19}$$

Choice I: We choose $k=\alpha$

With this choice, we have

$$a_{r+1}=\frac{(\alpha+r)(\alpha-\lambda+1+r)}{(1+r)(\alpha-\beta+1+r)}a_r. \tag{9.20}$$

Putting $r=0,1,2,3,4\ldots$, we obtain

$$a_1=\frac{\alpha(\alpha-\lambda+1)}{1!(\alpha-\beta+1)}a_0, \qquad a_2=\frac{\alpha(\alpha+1)(\alpha-\lambda+1)(\alpha-\lambda+2)}{2!(\alpha-\beta+1)(\alpha-\beta+2)}a_0$$

$$a_3=\frac{\alpha(\alpha+1)(\alpha+2)(\alpha-\lambda+1)(\alpha-\lambda+2)(\alpha-\lambda+3)}{3!(\alpha-\beta+1)(\alpha-\beta+2)(\alpha-\beta+3)}a_0=\frac{(\alpha)_3(\alpha-\lambda+1)_3}{3!(\alpha-\beta+1)_3}a_0 \text{ and so on.}$$

In general, we write

$$a_n=\frac{(\alpha)_n(\alpha-\lambda+1)_n}{n!(\alpha-\beta+1)_n}a_0.$$

Substituting the value of k and those of the above various coefficients into Eq. (9.18), we obtain one series solution of Eq. (9.2) about $z=0$ as

$$y_5(z)=z^{\alpha}\left[a_0+a_1z+a_2z^2+a_3z^3+\ldots+a_nz^n+\ldots\infty\right]$$

$$=a_0z^{\alpha}\sum_{n=0}^{\infty}\frac{(\alpha)_n(\alpha-\lambda+1)_n}{n!(\alpha-\beta+1)_n}z^n.$$

Therefore, one series solution of Eq. (9.1) about $x=\infty$ is obtained as

$$y_5(x)=a_0x^{-\alpha}\sum_{n=0}^{\infty}\frac{(\alpha)_n(\alpha-\lambda+1)_n}{n!(\alpha-\beta+1)_n}x^{-n},\left(\text{as } z=x^{-1}\right)$$

$$=a_0x^{-\alpha}\,{}_2F_1(\alpha,\alpha-\lambda+1;\alpha-\beta+1;x^{-1}). \tag{9.21}$$

Choice II: Choosing $k=\beta$, another series solution of Eq. (9.1) about $x=\infty$ is obtained as

$$y_6(x)=a_0x^{-\beta}\,{}_2F_1(\beta,\beta-\lambda+1;\beta-\alpha+1;x^{-1}). \tag{9.22}$$

Thus, about the singularity at $x = \infty$, the general solution of hypergeometric Eq. (9.1) is:

$$y = C_5 x^{-\alpha}\, {}_2F_1(\alpha, \alpha-\lambda+1; \alpha-\beta+1; x^{-1}) + C_6 x^{-\beta}\, {}_2F_1(\beta, \beta-\lambda+1; \beta-\alpha+1; x^{-1}) \quad (9.23)$$

where C_5 and C_6 are arbitrary constants in which the constant a_0 has been absorbed.

9.2 SYMMETRY PROPERTY OF HYPERGEOMETRIC FUNCTION

From the expression for hypergeometric function itself, it is obvious that the function remains unchanged if the parameters, α and β, are interchanged keeping λ fixed, i.e.,

$${}_2F_1(\alpha, \beta; \lambda; x) = {}_2F_1(\beta, \alpha; \lambda; x). \quad (9.24)$$

This property of hypergeometric function is known as the symmetry property.

9.3 INTEGRAL REPRESENTATION OF HYPERGEOMETRIC FUNCTION

We know that

$${}_2F_1(\alpha, \beta; \lambda; x) = \sum_{n=0}^{\infty} \frac{(\alpha)_n (\beta)_n}{n!(\lambda)_n} x^n$$

$$= \sum_{n=0}^{\infty} \frac{(\beta)_n}{(\lambda)_n} \left\{ \frac{(\alpha)_n x^n}{n!} \right\}$$

$$= \sum_{n=0}^{\infty} \frac{\Gamma(\beta+n)}{\Gamma(\beta)} \cdot \frac{\Gamma(\lambda)}{\Gamma(\lambda+n)} \left\{ \frac{(\alpha)_n x^n}{n!} \right\}$$

$$= \sum_{n=0}^{\infty} \frac{\Gamma(\lambda)}{\Gamma(\beta)} \cdot \frac{\Gamma(\beta+n)\Gamma(\lambda-\beta)}{\Gamma(\lambda+n)\Gamma(\lambda-\beta)} \left\{ \frac{(\alpha)_n x^n}{n!} \right\}$$

$$= \sum_{n=0}^{\infty} \frac{\Gamma(\lambda)}{\Gamma(\beta)\Gamma(\lambda-\beta)} \frac{\Gamma(\beta+n)\Gamma(\lambda-\beta)}{\Gamma(\beta+n+\lambda-\beta)} \left\{ \frac{\Gamma(\alpha+n)}{n!\Gamma(\alpha)} x^n \right\}$$

$$= \sum_{n=0}^{\infty} \frac{\Gamma(\lambda)}{\Gamma(\beta)\Gamma(\lambda-\beta)} B(\beta+n, \lambda-\beta) \left\{ \frac{\Gamma(\alpha+n)}{n!\,\Gamma(\alpha)} x^n \right\}$$

$$= \sum_{n=0}^{\infty} \frac{\Gamma(\lambda)}{\Gamma(\beta)\Gamma(\lambda-\beta)} \int_0^1 t^{\beta+n-1} (1-t)^{\lambda-\beta-1}\, dt \left\{ \frac{\Gamma(\alpha+n)}{n!\Gamma(\alpha)} x^n \right\}$$

$$=\frac{\Gamma(\lambda)}{\Gamma(\beta)\Gamma(\lambda-\beta)}\int_0^1 t^{\beta-1}(1-t)^{\lambda-\beta-1}\sum_{n=0}^{\infty}\left\{\frac{\Gamma(\alpha+n)}{n!\Gamma(\alpha)}(tx)^n\right\}dt$$

$$=\frac{\Gamma(\lambda)}{\Gamma(\beta)\Gamma(\lambda-\beta)}\int_0^1 t^{\beta-1}(1-t)^{\lambda-\beta-1}\left[1+\frac{\alpha}{1!}tx+\frac{\alpha(\alpha+1)}{2!}(tx)^2+\ldots\infty\right]dt$$

$$=\frac{\Gamma(\lambda)}{\Gamma(\beta)\Gamma(\lambda-\beta)}\int_0^1 t^{\beta-1}(1-t)^{\lambda-\beta-1}(1-tx)^{-\alpha}\,dt. \qquad (9.25)$$

9.4 GAUSS' FORMULA

Putting $x = 1$ in Eq. (9.25), we write

$${}_2F_1(\alpha,\beta;\lambda;1)=\frac{\Gamma(\lambda)}{\Gamma(\beta)\Gamma(\lambda-\beta)}\int_0^1 t^{\beta-1}(1-t)^{\lambda-\alpha-\beta-1}\,dt$$

$$=\frac{\Gamma(\lambda)}{\Gamma(\beta)\Gamma(\lambda-\beta)}B(\beta,\lambda-\alpha-\beta)$$

$$=\frac{\Gamma(\lambda)}{\Gamma(\beta)\Gamma(\lambda-\beta)}\frac{\Gamma(\beta)\Gamma(\lambda-\alpha-\beta)}{\Gamma(\lambda-\alpha)}=\frac{\Gamma(\lambda)\Gamma(\lambda-\alpha-\beta)}{\Gamma(\lambda-\alpha)\,\Gamma(\lambda-\beta)}. \qquad (9.26)$$

Equation (9.26) is known as Gauss' formula .

9.5 VANDERMONDE'S FORMULA

In Gauss' formula, i.e., in Eq. (9.26), if we put $\alpha=-n$ (where n is a positive integer), we write

$${}_2F_1(-n,\beta;\lambda;1)=\frac{\Gamma(\lambda)\Gamma(\lambda-\beta+n)}{\Gamma(\lambda+n)\,\Gamma(\lambda-\beta)}=\frac{(\lambda-\beta)_n}{(\lambda)_n}. \qquad (9.27)$$

This formula is known as Vandermonde's formula.

9.6 KUMMER'S FORMULA

Putting $x=-1,\lambda=\beta-\alpha+1$ in the integral representation of hypergeometric function, i.e., in Eq. (9.25), we have

$${}_2F_1(\alpha,\beta;\beta-\alpha+1;-1)=\frac{\Gamma(\beta-\alpha+1)}{\Gamma(\beta)\,\Gamma(1-\alpha)}\int_0^1 t^{\beta-1}(1-t)^{-\alpha}(1+t)^{-\alpha}\,dt$$

$$= \frac{\Gamma(\beta-\alpha+1)}{\Gamma(\beta)\Gamma(1-\alpha)} \int_0^1 t^{\beta-1}\left(1-t^2\right)^{-\alpha} dt$$

$$= \frac{\Gamma(\beta-\alpha+1)}{\Gamma(\beta)\Gamma(1-\alpha)} \int_0^1 p^{\frac{\beta-1}{2}} (1-p)^{-\alpha} \frac{dp}{2\sqrt{p}} \text{ (where we put } p=t^2 \text{)}$$

$$= \frac{1}{2}\frac{\Gamma(\beta-\alpha+1)}{\Gamma(\beta)\Gamma(1-\alpha)} \int_0^1 p^{\frac{\beta}{2}-1} (1-p)^{-\alpha} dp$$

$$= \frac{1}{2}\frac{\Gamma(\beta-\alpha+1)}{\Gamma(\beta)\Gamma(1-\alpha)} B\left(\frac{\beta}{2}, 1-\alpha\right)$$

$$= \frac{1}{2}\frac{\Gamma(\beta-\alpha+1)}{\Gamma(\beta)\Gamma(1-\alpha)} \frac{\Gamma\left(\frac{\beta}{2}\right)\Gamma(1-\alpha)}{\Gamma\left(1-\alpha+\frac{\beta}{2}\right)}$$

$$= \frac{\Gamma(\beta-\alpha+1)\frac{\beta}{2}\Gamma\left(\frac{\beta}{2}\right)}{\beta\Gamma(\beta)\Gamma\left(1-\alpha+\frac{\beta}{2}\right)} = \frac{\Gamma(\beta-\alpha+1)\Gamma\left(\frac{\beta}{2}+1\right)}{\Gamma(\beta+1)\Gamma\left(\frac{\beta}{2}-\alpha+1\right)}. \tag{9.28}$$

9.7 DIFFERENTIATION OF HYPERGEOMETRIC FUNCTIONS

We know that

$${}_2F_1(\alpha, \beta; \lambda; x) = \sum_{n=0}^{\infty} \frac{(\alpha)_n (\beta)_n}{(\lambda)_n} \frac{x^n}{n!}.$$

Differentiating both sides m times with respect to x, we obtain

$$\frac{d^m}{dx^m}\, {}_2F_1(\alpha, \beta; \lambda; x) = \sum_{n=0}^{\infty} \frac{(\alpha)_n (\beta)_n}{(\lambda)_n} \frac{n!}{(n-m)!} \frac{x^{n-m}}{n!}$$

$$= \sum_{n=m}^{\infty} \frac{(\alpha)_n (\beta)_n}{(\lambda)_n} \frac{x^{n-m}}{(n-m)!}$$

$$= \sum_{k=0}^{\infty} \frac{(\alpha)_{m+k} (\beta)_{m+k}}{(\lambda)_{m+k}} \frac{x^k}{k!} \quad \text{where we put } k=n-m.$$

It is known that

$$(\alpha)_{m+k} = \left[\alpha(\alpha+1)(\alpha+2)(\alpha+3)\dots(\alpha+m-1)\right]\left[(\alpha+m)(\alpha+m+1)(\alpha+m+2)\right.$$
$$\left.\times(\alpha+m+3)(\alpha+m+4)\dots(\alpha+m+k-1)\right]$$
$$= (\alpha)_m (\alpha+m)_k.$$

Now, we write

$$\frac{d^m}{dx^m}\,{}_2F_1(\alpha,\beta;\lambda;x) = \sum_{k=0}^{\infty} \frac{(\alpha)_m (\alpha+m)_k (\beta)_m (\beta+m)_k}{(\lambda)_m (\lambda+m)_k} \frac{x^k}{k!}$$
$$= \frac{(\alpha)_m (\beta)_m}{(\lambda)_m} \sum_{k=0}^{\infty} \frac{(\alpha+m)_k (\beta+m)_k}{(\lambda+m)_k} \frac{x^k}{k!}$$
$$= \frac{(\alpha)_m (\beta)_m}{(\lambda)_m}\,{}_2F_1(\alpha+m,\beta+m;\lambda+m;x). \tag{9.29}$$

9.8 LINEAR TRANSFORMATION OF HYPERGEOMETRIC FUNCTION

The integral formula for hypergeometric function is

$${}_2F_1(\alpha,\beta;\lambda;x) = \frac{\Gamma(\lambda)}{\Gamma(\beta)\Gamma(\lambda-\beta)} \int_0^1 t^{\beta-1}(1-t)^{\lambda-\beta-1}(1-xt)^{-\alpha}\,dt$$

Putting $\tau = 1-t$, we write

$${}_2F_1(\alpha,\beta;\lambda;x) = \frac{\Gamma(\lambda)}{\Gamma(\beta)\Gamma(\lambda-\beta)} \int_1^0 (1-\tau)^{\beta-1}\,\tau^{\lambda-\beta-1}(1-x+x\tau)^{-\alpha}(-d\tau)$$
$$= (1-x)^{-\alpha} \frac{\Gamma(\lambda)}{\Gamma(\beta)\Gamma(\lambda-\beta)} \int_0^1 \tau^{\lambda-\beta-1}(1-\tau)^{\beta-1}\left(1+\frac{x\tau}{1-x}\right)^{-\alpha} d\tau$$

Putting $\beta' = \lambda - \beta$ or, $\beta = \lambda - \beta'$ in the above integral, we write

$${}_2F_1(\alpha,\beta;\lambda;x) = (1-x)^{-\alpha} \frac{\Gamma(\lambda)}{\Gamma(\beta)\Gamma(\lambda-\beta)} \int_0^1 \tau^{\beta'-1}(1-\tau)^{\lambda-\beta'-1}\left(1-\frac{x}{x-1}\tau\right)^{-\alpha} d\tau$$
$$= (1-x)^{-\alpha}\,{}_2F_1\left(\alpha,\beta';\lambda;\frac{x}{x-1}\right)$$
$$= (1-x)^{-\alpha}\,{}_2F_1\left(\alpha,\lambda-\beta;\lambda;\frac{x}{x-1}\right). \tag{9.30}$$

9.9 CONFLUENT HYPERGEOMETRIC EQUATION AND CONFLUENT HYPERGEOMETRIC FUNCTION

In the ordinary hypergeometric equation, viz.,

$$x(1-x)\frac{d^2y}{dx^2}+\left[\lambda-(\alpha+\beta+1)x\right]\frac{dy}{dx}-\alpha\beta y=0.$$

If we replace x by $\frac{x}{\beta}$, we will obtain

$$\frac{x}{\beta}\left(1-\frac{x}{\beta}\right)\beta^2\frac{d^2y}{dx^2}+\left[\lambda-(\alpha+\beta+1)\frac{x}{\beta}\right]\beta\frac{dy}{dx}-\alpha\beta y=0.$$

Dividing both sides by β, we obtain

$$x\left(1-\frac{x}{\beta}\right)\frac{d^2y}{dx^2}+\left[\lambda-\left(1+\frac{\alpha+1}{\beta}\right)x\right]\frac{dy}{dx}-\alpha y=0.$$

Making the confluence $\beta\to\infty$, we obtain

$$x\frac{d^2y}{dx^2}+(\lambda-x)\frac{dy}{dx}-\alpha y=0. \tag{9.31}$$

Equation (9.31) is known as confluent hypergeometric equation. This equation has a removable singularity at $x=0$.

As the confluent hypergeometric equation is obtained from the ordinary hypergeometric equation by replacing x by $\frac{x}{\beta}$ and making $\beta\to\infty$, the solution(s) of the confluent hypergeometric equation can be obtained from the solution(s) of the ordinary hypergeometric equation by making the necessary changes.

Thus, one solution of Eq. (9.31) is

$$y_1=\lim_{\beta\to\infty}{}_2F_1\left(\alpha,\beta;\lambda;\frac{x}{\beta}\right)=\lim_{\beta\to\infty}\sum_{n=0}^{\infty}\frac{(\alpha)_n(\beta)_n}{(\lambda)_n}\frac{1}{n!}\left(\frac{x}{\beta}\right)^n$$

$$=\sum_{n=0}^{\infty}\frac{(\alpha)_n}{(\lambda)_n}\frac{x^n}{n!} \tag{9.32}$$

where we use the relation $\lim_{\beta\to\infty}\frac{(\beta)_n}{\beta^n}=\lim_{\beta\to\infty}\frac{\beta(\beta+1)(\beta+2)(\beta+3)\ldots(\beta+n-1)}{\beta^n}=1.$

The above solution in Eq. (9.32) is known as confluent hypergeometric function and it is denoted by ${}_1F_1(\alpha;\lambda;x)$.

Thus, we write

$$y_1 = {}_1F_1(\alpha; \lambda; x) = \sum_{n=0}^{\infty} \frac{(\alpha)_n}{(\lambda)_n} \frac{x^n}{n!}. \tag{9.33}$$

The series solution of confluent hypergeometric equation can also be directly obtained by the Frobenius method. It can also be easily proved that the confluent hypergeometric series is convergent.

As in the case of ordinary hypergeometric function, a second independent solution is found as

$$y_2 = x^{1-\lambda}\, {}_1F_1(\alpha + 1 - \lambda; 2 - \lambda; x). \tag{9.34}$$

The general solution of confluent hypergeometric equation about the removable singularity $x = 0$, is a linear combination of y_1 and y_2

9.10 DIFFERENTIATION OF CONFLUENT HYPERGEOMETRIC FUNCTION

We know that

$${}_1F_1(\alpha; \lambda; x) = \sum_{n=0}^{\infty} \frac{(\alpha)_n}{(\lambda)_n} \frac{x^n}{n!}.$$

Differentiating both sides m times with respect to x, we obtain

$$\frac{d^m}{dx^m}\left[{}_1F_1(\alpha; \lambda; x)\right] = \sum_{n=0}^{\infty} \frac{(\alpha)_n}{(\lambda)_n} \frac{n!}{(n-m)!} \frac{x^{n-m}}{n!}$$

$$= \sum_{n=m}^{\infty} \frac{(\alpha)_n}{(\lambda)_n} \frac{x^{n-m}}{(n-m)!}$$

$$= \sum_{k=0}^{\infty} \frac{(\alpha)_{m+k}}{(\lambda)_{m+k}} \frac{x^k}{k!} \quad \text{where we put } k = n - m;$$

$$= \frac{(\alpha)_m}{(\lambda)_m} \sum_{k=0}^{\infty} \frac{(\alpha + m)_k}{((\lambda + m))_k} \frac{x^k}{k!} \quad \text{since } (\alpha)_{m+k} = (\alpha)_m (\alpha + m)_k.$$

Hence, we write

$$\frac{d^m}{dx^m}\left[{}_1F_1(\alpha;\lambda;x)\right] = \frac{(\alpha)_m}{(\lambda)_m}\, {}_1F_1(\alpha + m; \lambda + m; x). \tag{9.35}$$

9.11 INTEGRAL REPRESENTATION OF CONFLUENT HYPERGEOMETRIC FUNCTION

We know that

$$
\begin{aligned}
{}_1F_1(\alpha;\lambda;x) &= \sum_{n=0}^{\infty}\frac{(\alpha)_n}{(\lambda)_n}\frac{x^n}{n!} = \sum_{n=0}^{\infty}\frac{\Gamma(\alpha+n)}{\Gamma(\alpha)}\frac{\Gamma(\lambda)}{\Gamma(\lambda+n)}\frac{x^n}{n!} \\
&= \sum_{n=0}^{\infty}\frac{\Gamma(\lambda)}{\Gamma(\alpha)\Gamma(\lambda-\alpha)}\frac{\Gamma(\alpha+n)\Gamma(\lambda-\alpha)}{\Gamma(\lambda+n)}\frac{x^n}{n!} \\
&= \sum_{n=0}^{\infty}\frac{\Gamma(\lambda)}{\Gamma(\alpha)\Gamma(\lambda-\alpha)}B(\alpha+n,\lambda-\alpha)\frac{x^n}{n!} \\
&= \sum_{n=0}^{\infty}\frac{\Gamma(\lambda)}{\Gamma(\alpha)\Gamma(\lambda-\alpha)}\frac{x^n}{n!}\int_0^1 t^{\alpha+n-1}(1-t)^{\lambda-\alpha-1}\,dt \\
&= \frac{\Gamma(\lambda)}{\Gamma(\alpha)\Gamma(\lambda-\alpha)}\int_0^1 t^{\alpha-1}(1-t)^{\lambda-\alpha-1}\left[\sum_{n=0}^{\infty}\frac{(xt)^n}{n!}\right]dt \\
&= \frac{\Gamma(\lambda)}{\Gamma(\alpha)\Gamma(\lambda-\alpha)}\int_0^1 t^{\alpha-1}(1-t)^{\lambda-\alpha-1}e^{xt}\,dt. \qquad (9.36)
\end{aligned}
$$

Equation (9.36) is the integral representation of confluent hypergeometric function.

9.12 SOLVED PROBLEMS

PROBLEM 1 Prove that $(1-x)^n = {}_2F_1(-n, 1; 1; x)$.

Solution: We know that

$$
\begin{aligned}
(1-x)^n &= 1 - nx + \frac{n(n-1)}{2!}x^2 - \frac{n(n-1)(n-2)}{3!}x^3 + \ldots + (-1)^n x^n \\
&= 1 + \frac{-n\cdot 1}{1}\frac{x}{1!} + \frac{-n(-n+1)1\cdot 2}{1\cdot 2}\frac{x^2}{2!} + \frac{(-n)(-n+1)(-n+2)1\cdot 2\cdot 3}{1\cdot 2\cdot 3}\frac{x^3}{3!} + \ldots \\
&\quad + \frac{(-n)(-n+1)(-n+2)(-n+3)\ldots(-n+n-1)1\cdot 2\cdot 3\ldots(1+n-1)}{1\cdot 2\cdot 3\ldots(1+n-1)}\frac{x^n}{n!}
\end{aligned}
$$

Comparing this with the equation

$$
{}_2F_1(\alpha,\beta;\lambda;x) = 1 + \frac{\alpha\cdot\beta}{\lambda}\frac{x}{1!} + \frac{\alpha(\alpha+1)\beta(\beta+1)}{\lambda(\lambda+1)}\frac{x^2}{2!} + \frac{\alpha(\alpha+1)(\alpha+2)\beta(\beta+1)(\beta+2)}{\lambda(\lambda+1)(\lambda+2)}\frac{x^3}{3!} + \ldots
$$

We write

$$ {}_2F_1(-n, 1; 1; x) = (1-x)^n. $$

PROBLEM 2 Prove that

(i) $\ln(1-x) = -x\,{}_2F_1(1, 1; 2; x)$,

(ii) $\ln\left(\dfrac{1+x}{1-x}\right) = 2x\,{}_2F_1\left(\dfrac{1}{2}, 1; \dfrac{3}{2}; x^2\right)$.

Solution:

First part: We know that

$$ \begin{aligned} \ln(1-x) &= -x - \frac{x^2}{2} - \frac{x^3}{3} - \frac{x^4}{4} - \ldots \infty \\ &= -x\left[1 + \frac{1\cdot 1}{2}\frac{x}{1!} + \frac{1(2)1(2)}{(2)(3)}\frac{x^2}{2!} + \frac{1(2)(3)1(2)(3)}{(2)(3)(4)}\frac{x^3}{3!} + \ldots \infty\right] \\ &= -x\,{}_2F_1(1, 1; 2; x). \end{aligned} $$

Second part:

$$ \begin{aligned} \ln\left(\frac{1+x}{1-x}\right) &= \ln(1+x) - \ln(1-x) \\ &= \left(x - \frac{x^2}{2} + \frac{x^3}{3} - \frac{x^4}{4} + \ldots \infty\right) - \left(-x - \frac{x^2}{2} - \frac{x^3}{3} - \frac{x^4}{4} - \ldots \infty\right) \\ &= 2\left[x + \frac{x^3}{3} + \frac{x^5}{5} + \ldots \infty\right] \\ &= 2x\left[1 + \frac{\left(\frac{1}{2}\right)(1)}{\frac{3}{2}}\frac{x^2}{1!} + \frac{\left(\frac{1}{2}\right)\left(\frac{3}{2}\right)\cdot(1)(2)}{\left(\frac{3}{2}\right)\left(\frac{5}{2}\right)}\frac{(x^2)^2}{2!} + \ldots \infty\right]. \end{aligned} $$

Comparing this equation with the equation

$$ {}_2F_1(\alpha, \beta; \lambda; x) = 1 + \frac{\alpha\beta}{\lambda}\frac{x}{1!} + \frac{\alpha(\alpha+1)\beta(\beta+1)}{\lambda(\lambda+1)}\frac{x^2}{2!} + \ldots \infty. $$

We write

$$ \ln\left(\frac{1+x}{1-x}\right) = 2x_2F_1\left(\frac{1}{2}, 1; \frac{3}{2}; x^2\right). $$

PROBLEM 3 Show that $\sin^{-1} x = x\,{}_2F_1\left(\frac{1}{2},\frac{1}{2};\frac{3}{2};x^2\right)$.

Solution: We know that

$$\sin^{-1} x = x + \frac{x^3}{3!} + 3^2\frac{x^5}{5!} + 15^2\frac{x^7}{7!} + \ldots \infty$$

$$= x\left[1 + \frac{x^2}{3!} + \frac{9x^4}{5!} + \frac{225x^6}{7!} + \ldots \infty\right]$$

$$= x\left[1 + \frac{\frac{1}{2}\cdot\frac{1}{2}}{\frac{3}{2}}\frac{x^2}{1!} + \frac{\left(\frac{1}{2}\right)\left(\frac{3}{2}\right)\cdot\left(\frac{1}{2}\right)\left(\frac{3}{2}\right)}{\left(\frac{3}{2}\right)\left(\frac{5}{2}\right)}\frac{(x^2)^2}{2!} + \ldots \infty\right]$$

$$= x\,{}_2F_1\left(\frac{1}{2},\frac{1}{2};\frac{3}{2};x^2\right).$$

PROBLEM 4 Prove that $P_n(x) = {}_2F_1\left(-n, n+1; 1; \frac{1-x}{2}\right)$.

Solution: By Rodrigues's formula for Legendre polynomials, we know that

$$P_n(x) = \frac{1}{2^n n!}\frac{d^n}{dx^n}(x^2-1)^n = \frac{1}{n!}\frac{d^n}{dx^n}\left[(x-1)^n\left(\frac{x+1}{2}\right)^n\right]$$

$$= \frac{1}{n!}\frac{d^n}{dx^n}\left[(x-1)^n\left(1+\frac{x-1}{2}\right)^n\right]$$

$$= \frac{1}{n!}\frac{d^n}{dx^n}\left[(x-1)^n\left\{1 + n\left(\frac{x-1}{2}\right) + \frac{n(n-1)}{2!}\left(\frac{x-1}{2}\right)^2 + \frac{n(n-1)(n-2)}{3!}\left(\frac{x-1}{2}\right)^3 + \ldots + \left(\frac{x-1}{2}\right)^n\right\}\right]$$

$$= \frac{1}{n!}\frac{d^n}{dx^n}\left[(x-1)^n + n\frac{(x-1)^{n+1}}{2} + \frac{n(n-1)}{2!}\frac{(x-1)^{n+2}}{2^2} + \frac{n(n-1)(n-2)}{3!}\frac{(x-1)^{n+3}}{2^3} + \ldots + \frac{(x-1)^{2n}}{2^n}\right]$$

$$= \frac{1}{n!}\left[n! + \frac{n(n+1)!}{1!}\left(\frac{x-1}{2}\right) + \frac{n(n-1)}{2!}\frac{(n+2)!}{2!}\left(\frac{x-1}{2}\right)^2 + \right.$$

$$\left. + \frac{n(n-1)(n-2)}{3!}\frac{(n+3)!}{3!}\left(\frac{x-1}{2}\right)^3 + \ldots\right]$$

$$= 1 + \frac{(-n)(n+1)}{1}\frac{\frac{1-x}{2}}{1!} + \frac{(-n)(-n+1)\cdot(n+1)(n+2)}{1\cdot 2}\frac{\left(\frac{1-x}{2}\right)^2}{2!}$$

$$+ \frac{(-n)(-n+1)(-n+2)\cdot(n+1)(n+2)(n+3)}{1\cdot 2\cdot 3}\frac{\left(\frac{1-x}{2}\right)^3}{3!} + \ldots$$

$$= {}_2F_1\left(-n, n+1; 1; \frac{1-x}{2}\right).$$

PROBLEM 5 Prove that

(i) $P_l(\cos\theta) = {}_2F_1\left(-l, l+1; 1; \sin^2\left(\frac{\theta}{2}\right)\right),$

(ii) $P_l(\cos\theta) = (-1)^l\, {}_2F_1\left(-l, l+1; 1; \cos^2\left(\frac{\theta}{2}\right)\right)$

if l is a positive integer.

Solution: From the previous problem, we know that

$$P_n(x) = {}_2F_1\left(-n, n+1; 1; \frac{1-x}{2}\right)$$

(i) Putting $n = l, x = \cos\theta$, we write

$$P_l(\cos\theta) = {}_2F_1\left(-l, l+1; 1; \frac{1-\cos\theta}{2}\right)$$

$$= {}_2F_1\left(-l, l+1; 1; \sin^2\left(\frac{\theta}{2}\right)\right).$$

(ii) Similarly, putting $n = l$, $x = -\cos\theta$, we have

$$P_l(-\cos\theta) = {}_2F_1\left(-l, l+1; 1; \frac{1+\cos\theta}{2}\right)$$

or $$(-1)^l P_l(\cos\theta) = {}_2F_1\left(-l, l+1; 1; \cos^2\left(\frac{\theta}{2}\right)\right)$$

or $$P_l(\cos\theta) = (-1)^l \, {}_2F_1\left(-l, l+1; 1; \cos^2\left(\frac{\theta}{2}\right)\right).$$

PROBLEM 6 Prove that

$$P_l(\cos\theta) = \cos^{2l}\left(\frac{\theta}{2}\right) {}_2F_1\left(-l, -l; 1; -\tan^2\left(\frac{\theta}{2}\right)\right) \text{ if } l \text{ is a positive integer.}$$

Solution: In Section 9.8, we have seen that

$${}_2F_1(\alpha, \beta; \lambda; x) = (1-x)^{-\alpha} \, {}_2F_1\left(\alpha, \lambda-\beta; \lambda; \frac{x}{x-1}\right). \qquad \text{(i)}$$

From the previous problem, we know that

$$P_l(\cos\theta) = {}_2F_1\left(-l, l+1; 1; \sin^2\left(\frac{\theta}{2}\right)\right). \qquad \text{(ii)}$$

Using Eq. (i) in Eq. (ii), we obtain

$$P_l(\cos\theta) = \left\{1 - \sin^2\left(\frac{\theta}{2}\right)\right\}^l \, {}_2F_1\left(-l, 1-l-1; 1; \frac{\sin^2\left(\frac{\theta}{2}\right)}{\sin^2\left(\frac{\theta}{2}\right) - 1}\right)$$

$$= \cos^{2l}\left(\frac{\theta}{2}\right) {}_2F_1\left(-l, -l; 1; -\tan^2\left(\frac{\theta}{2}\right)\right).$$

PROBLEM 7 Prove that

$$x\frac{d}{dx}\, {}_2F_1(\alpha, \beta; \lambda; x) + \alpha \, {}_2F_1(\alpha, \beta; \lambda; x) = \alpha \, {}_2F_1(\alpha+1, \beta; \lambda; x).$$

Solution: We know that

$$\frac{d^m}{dx^m} \, {}_2F_1(\alpha, \beta; \lambda; x) = \frac{(\alpha)_m (\beta)_m}{(\lambda)_m} \, {}_2F_1(\alpha+m, \beta+m; \lambda+m; x)$$

Therefore, we have

$$\frac{d}{dx} \, {}_2F_1(\alpha, \beta; \lambda; x) = \frac{\alpha\beta}{\lambda} \, {}_2F_1(\alpha+1, \beta+1; \lambda+1; x)$$

$$= \frac{\alpha\beta}{\lambda}\left[1 + \frac{(\alpha+1)(\beta+1)}{(\lambda+1)}\frac{x}{1!} + \frac{(\alpha+1)(\alpha+2)(\beta+1)(\beta+2)}{(\lambda+1)(\lambda+2)}\frac{x^2}{2!} + \ldots\right].$$

Now, we write

$$x\frac{d}{dx}\,{}_2F_1(\alpha, \beta; \lambda; x) + \alpha\,{}_2F_1(\alpha, \beta; \lambda; x)$$

$$=\alpha\left[\frac{\beta}{\lambda}x+\frac{(\alpha+1)\beta(\beta+1)}{\lambda(\lambda+1)}\frac{x^2}{1!}+\frac{(\alpha+1)(\alpha+2)\beta(\beta+1)(\beta+2)}{\lambda(\lambda+1)(\lambda+2)}\frac{x^3}{2!}+\dots\right]$$

$$+\alpha\left[1+\frac{\alpha\beta}{\lambda}\frac{x}{1!}+\frac{\alpha(\alpha+1)\beta(\beta+1)}{\lambda(\lambda+1)}\frac{x^2}{2!}+\dots\right]$$

$$=\alpha\left[1+\frac{(\alpha+1)\beta}{\lambda}\frac{x}{1!}+\frac{(\alpha+1)(\alpha+2)\beta(\beta+1)}{\lambda(\lambda+1)}\frac{x^2}{2!}+\dots\right]$$

$$=\alpha\,{}_2F_1(\alpha+1, \beta; \lambda; x).$$

PROBLEM 8 Express the associated Legendre polynomial $P_l^m(x)$ in terms of hyperbolic function.

Solution: As in Problem 4, we obtain

$$P_l(x)={}_2F_1\left(-l, l+1; 1; \frac{1-x}{2}\right).$$

Then, we have

$$P_l^m(x)=(1-x^2)^{\frac{m}{2}}\frac{d^m}{dx^m}P_l(x)$$

$$=(1-x^2)^{\frac{m}{2}}\frac{d^m}{dx^m}\,{}_2F_1\left(-l, l+1; 1; \frac{1-x}{2}\right). \tag{i}$$

Putting $u=\dfrac{1-x}{2}$ or, $x = 1 - 2u$, we write

$$\frac{d^m}{dx^m}\,{}_2F_1\left(-l, l+1; 1; \frac{1-x}{2}\right)$$

$$=\frac{1}{(-2)^m}\frac{d^m}{du^m}\,{}_2F_1(-l, l+1; 1; u)$$

$$=\frac{1}{(-2)^m}\frac{(-l)_m(l+1)_m}{(1)_m}\,{}_2F_1(-l+m, l+1+m; 1+m; u)$$

$$= \frac{1}{(-2)^m} \frac{\left[(-l)(-l+1)(-l+2)\ldots(-l+m-1)\right]\left[(l+1)(l+2)(l+3)\ldots(l+1+m-1)\right]}{1\cdot 2\cdot 3\ldots(1+m-1)}$$

$$\times\, {}_2F_1(-l+m, l+m+1; m+1; u)$$

$$= \frac{1}{2^m} \frac{\left[l(l-1)(l-2)\ldots(l-m+1)\right]\left[(l+1)(l+2)(l+3)\ldots(l+m)\right]}{m!}$$

$$\times\, {}_2F_1(-l+m, l+m+1; m+1; u)$$

$$= \frac{1}{2^m} \frac{(l+m)!}{(l-m)!\, m!}\, {}_2F_1(-l+m, l+m+1; m+1; u). \qquad \text{(ii)}$$

Substituting Eq. (ii) into Eq. (i), we obtain

$$P_l^m(x) = \frac{(l+m)!}{2^m m!(l-m)!}(1-x^2)^{\frac{m}{2}}\, {}_2F_1\left(-l+m, l+m+1; m+1; \frac{1-x}{2}\right)$$

where we restore $u = \dfrac{1-x}{2}$.

PROBLEM 9 Prove that (a) $e^x = {}_1F_1(\alpha; \alpha; x)$, (b) ${}_1F_1(1; 2; x) = \dfrac{1}{x}(e^x - 1)$.

Solution:

(a) We know that

$$ {}_1F_1(\alpha; \lambda; x) = 1 + \frac{\alpha}{\lambda}\frac{x}{1!} + \frac{\alpha(\alpha+1)}{\lambda(\lambda+1)}\frac{x^2}{2!} + \frac{\alpha(\alpha+1)(\alpha+2)}{\lambda(\lambda+1)(\lambda+2)}\frac{x^3}{3!} + \ldots\infty.$$

Putting $\alpha = \lambda$ in the above Eq., we write

$$ {}_1F_1(\alpha; \alpha; x) = 1 + \frac{x}{1!} + \frac{x^2}{2!} + \frac{x^3}{3!} + \ldots\infty.$$

$$= e^x.$$

(b) Putting $\alpha = 1, \lambda = 2$, we have

$$ {}_1F_1(1; 2; x) = 1 + \frac{1}{2}\frac{x}{1!} + \frac{1\cdot 2}{2\cdot 3}\frac{x^2}{2!} + \frac{1\cdot 2\cdot 3}{2\cdot 3\cdot 4}\frac{x^3}{3!} + \ldots\infty$$

$$= \frac{1}{x}\left(x + \frac{x^2}{2!} + \frac{x^3}{3!} + \frac{x^4}{4!} + \ldots\infty\right)$$

$$= \frac{1}{x}(e^x - 1).$$

PROBLEM 10 Prove that

(a) $H_{2n}(x) = (-1)^n \dfrac{(2n)!}{n!}\, {}_1F_1\left(-n; \dfrac{1}{2}; x^2\right),$

(b) $H_{2n+1}(x) = (-1)^n \dfrac{(2n+1)!}{n!} 2x\, {}_1F_1\left(-n; \dfrac{3}{2}; x^2\right).$

Solution:

(a) For even order Hermite polynomial, we have

$$H_{2n}(x) = \sum_{r=0}^{n} \frac{(-1)^r (2n)!}{r!(2n-2r)!}(2x)^{2n-2r}.$$

Putting $k = n - r$ we write

$$H_{2n}(x) = (2n)! \sum_{k=0}^{n} \frac{(-1)^{n-k} (2x)^{2k}}{(n-k)!(2k)!}$$

$$= \frac{(-1)^n (2n)!}{n!} \sum_{k=0}^{\infty} \frac{(-1)^{-k} n!(2x)^{2k}}{(n-k)!(2k)!}$$

$$= \frac{(-1)^n (2n)!}{n!} \sum_{k=0}^{\infty} \frac{(-1)^k (n-k+1)(n-k+2)\ldots(n-1)n}{(2k)!}(2x)^{2k}$$

$$= \frac{(-1)^n (2n)!}{n!} \sum_{k=0}^{\infty} \frac{(-n+k-1)(-n+k-2)(-n+k-2)\ldots(-n+1)(-n)}{(2k)!}(2x)^{2k}$$

$$= \frac{(-1)^n (2n)!}{n!} \sum_{k=0}^{\infty} \frac{(-n)_k}{(2k)!}(2x)^{2k}. \quad \text{(i)}$$

Also, we know that

$$(2k)! = 2^{2k}\,(1.2.3\ \ldots\ k)\left\{\frac{1}{2}\cdot\frac{3}{2}\cdot\frac{5}{2}\ldots\left(k+\frac{1}{2}-1\right)\right\}$$

$$= 2^{2k} k!\left(\frac{1}{2}\right)_k. \quad \text{(ii)}$$

Substituting Eq, (ii) into Eq. (i), we obtain

$$H_{2n}(x) = \frac{(-1)^n (2n)!}{n!} \sum_{k=0}^{\infty} \frac{(-n)_k}{\left(\frac{1}{2}\right)_k} \frac{(x^2)^k}{k!}$$

$$= \frac{(-1)^n (2n)!}{n!} \, {}_1F_1\left(-n; \frac{1}{2}; x^2\right).$$

(b) For odd order Hermite polynomial, we have

$$H_{2n+1}(x) = \sum_{r=0}^{n} \frac{(-1)^r (2n+1)!}{r!(2n+1-2r)!} (2x)^{2n+1-2r}$$

$$= \frac{(2n+1)!}{n!} \sum_{r=0}^{\infty} \frac{(-1)^r \, n!(2x)^{2n-2r+1}}{r!(2n-2r+1)!}.$$

Putting $n-r=k$, we write

$$H_{2n+1}(x) = \frac{(2n+1)!}{n!} \sum_{k=0}^{\infty} \frac{(-1)^{n-k} (n-k+1)(n-k+2)(n-k+3)\dots(n-1)(n)}{(2k+1)!} (2x)^{2k+1}$$

$$= \frac{(-1)^n (2n+1)!}{n!} \sum_{k=0}^{\infty} \frac{(-1)^k (n-k+1)(n-k+2)(n-k+3)\dots(n-1)(n)}{(2k+1)!} (2x)^{2k+1}$$

$$= \frac{(-1)^n (2n+1)!}{n!} \sum_{k=0}^{\infty} \frac{(-n+k-1)(-n+k-2)(-n+k-3)\dots(-n+1)(-n)}{(2k+1)!} (2x)^{2k+1}$$

$$= \frac{(-1)^n (2n+1)!}{n!} \sum_{k=0}^{\infty} \frac{(-n)_k}{(2k+1)!} (2x)^{2k+1}. \qquad \text{(iii)}$$

We know that

$$(2k+1)! = (2 \cdot 4 \cdot 6 \dots 2k)\{1 \cdot 3 \cdot 5 \; \dots \; (2k+1\}$$

$$= 2^{2k} k! \frac{3}{2} \cdot \frac{5}{2} \cdot \frac{7}{2} \dots \left(k + \frac{3}{2} - 1\right) = 2^{2k} k! \left(\frac{3}{2}\right)_k. \qquad \text{(iv)}$$

Substituting Eq. (iv) into Eq. (iii), we obtain

$$H_{2n+1}(x) = \frac{(-1)^n (2n+1)!}{n!} 2x \sum_{k=0}^{\infty} \frac{(-n)_k}{\left(\frac{3}{2}\right)_k} \frac{\left(x^2\right)^k}{k!}$$

$$= \frac{(-1)^n (2n+1)!}{n!} 2x \, {}_1F_1\left(-n; \frac{3}{2}; x^2\right).$$

PROBLEM 11 Prove that

$$L_n^k(x) = \frac{\Gamma(n+k+1)}{n!\Gamma(k+1)} \, {}_1F_1(-n; k+1; x).$$

Solution: We know that

$$L_n^k(x)=\sum_{r=0}^{n}\frac{(-1)^r\,\Gamma(n+k+1)}{(n-r)!\,r!\,\Gamma(k+r+1)}x^r$$

$$=\frac{\Gamma(n+k+1)}{n!}\sum_{r=0}^{\infty}\frac{(-1)^r\,n!}{(n-r)!\,r!\,\Gamma(k+r+1)}x^r$$

$$=\frac{\Gamma(n+k+1)}{n!}\sum_{r=0}^{\infty}\frac{(-1)^r\,(n-r+1)(n-r+2)(n-r+3)\ldots(n-1)n}{\Gamma(k+1)\{(k+1)(k+2)(k+3)\ldots(k+r-1)(k+r)\}}\frac{x^r}{r!}$$

$$=\frac{\Gamma(n+k+1)}{n!\,\Gamma(k+1)}\sum_{r=0}^{\infty}\frac{(-n+r-1)(-n+r-2)(-n+r-3)\ldots(-n+1)(-n)}{(k+1)_r}\frac{x^r}{r!}$$

$$=\frac{\Gamma(n+k+1)}{n!\,\Gamma(k+1)}\sum_{r=0}^{\infty}\frac{(-n)_r}{(k+1)_r}\frac{x^r}{r!}$$

$$=\frac{\Gamma(n+k+1)}{n!\,\Gamma(k+1)}\;{}_1F_1(-n;\,k+1;\,x).$$

PROBLEM 12 Prove that $J_n(x)=\dfrac{e^{-ix}}{\Gamma(n+1)}\left(\dfrac{x}{2}\right)^n {}_1F_1\left(n+\dfrac{1}{2};\,2n+1;\,2ix\right).$

Solution: We know that

$$J_n(x)=\frac{1}{\Gamma\left(\frac{1}{2}\right)\Gamma\left(n+\frac{1}{2}\right)}\left(\frac{x}{2}\right)^n\int_{-1}^{+1}(1-t^2)^{n-\frac{1}{2}}e^{ixt}\,dt.$$

Let us put $t=2v-1$.

Then we write

$$J_n(x)=\frac{e^{-ix}}{\Gamma\left(\frac{1}{2}\right)\Gamma\left(n+\frac{1}{2}\right)}\left(\frac{x}{2}\right)^n\int_0^1\{2^2v(1-v)\}^{n-\frac{1}{2}}e^{2ivx}\,2dv$$

$$=\frac{2^{2n}e^{-ix}\Gamma\left(n+\frac{1}{2}\right)\Gamma(2n+1)}{\Gamma\left(\frac{1}{2}\right)\Gamma(2n+1)\Gamma\left(n+\frac{1}{2}\right)\Gamma\left(n+\frac{1}{2}\right)}\left(\frac{x}{2}\right)^n\int_0^1 e^{2ivx}v^{\left(n+\frac{1}{2}\right)-1}(1-v)^{(2n+1)-\left(n+\frac{1}{2}\right)-1}\,dv$$

$$= \frac{2^{2n} e^{-ix} \Gamma\left(n+\frac{1}{2}\right)}{\Gamma\left(\frac{1}{2}\right)\Gamma(2n+1)} \left(\frac{x}{2}\right)^n {}_1F_1\left(n+\frac{1}{2}; 2n+1; 2ix\right) \qquad \text{(i)}$$

where we have recalled that

$${}_1F_1(\alpha;\lambda;x) = \frac{\Gamma(\lambda)}{\Gamma(\alpha)\Gamma(\lambda-\alpha)} \int_0^1 e^{xt} t^{\alpha-1} (1-t)^{\lambda-\alpha-1}\, dt. \qquad \text{(ii)}$$

Further, recalling that

$$\Gamma(2n+1) = \frac{2^{2n}\Gamma(n+1)\Gamma\left(n+\frac{1}{2}\right)}{\Gamma\left(\frac{1}{2}\right)},$$

we write Eq. (i) as

$$J_n(x) = \frac{e^{-ix}}{\Gamma(n+1)} \left(\frac{x}{2}\right)^n {}_1F_1\left(n+\frac{1}{2}; 2n+1; 2ix\right).$$

PROBLEM 13 Prove that

$$erf(x) = \frac{2x}{\sqrt{\pi}}\, {}_1F_1\left(\frac{1}{2}; \frac{3}{2}; -x^2\right)$$

where $erf(x) = \frac{2}{\sqrt{\pi}} \int_0^x e^{-t^2}\, dt$ is known as error function.

Solution: We have

$$erf(x) = \frac{2}{\sqrt{\pi}} \int_0^x e^{-t^2}\, dt$$

$$= \frac{2}{\sqrt{\pi}} \sum_{n=0}^{\infty} \frac{(-1)^n}{n!} \int_0^x t^{2n}\, dt = \frac{2}{\sqrt{\pi}} \sum_{n=0}^{\infty} \frac{(-1)^n}{n!} \frac{x^{2n+1}}{2n+1}$$

$$= \frac{2x}{\sqrt{\pi}} \sum_{n=0}^{\infty} \frac{(-x^2)^n}{(2n+1)n!}. \qquad \text{(i)}$$

We know that

$$(2n+1) = \frac{(2n+1)\{(2n-1)(2n-3)(2n-5)\ldots 7\cdot 5\cdot 3\cdot 1\}}{(2n-1)(2n-3)(2n-5)\ldots 7\cdot 5\cdot 3\cdot 1}$$

$$= \frac{(2n+1)\left\{\left(n-\frac{1}{2}\right)\left(n-\frac{3}{2}\right)\left(n-\frac{5}{2}\right)\cdots\frac{7}{2}\cdot\frac{5}{2}\cdot\frac{3}{2}\cdot\frac{1}{2}\right\}}{\left(n-\frac{1}{2}\right)\left(n-\frac{3}{2}\right)\left(n-\frac{5}{2}\right)\cdots\frac{7}{2}\cdot\frac{5}{2}\cdot\frac{3}{2}\cdot\frac{1}{2}}$$

$$= \frac{\left(n+\frac{1}{2}\right)\left(n-\frac{1}{2}\right)\left(n-\frac{3}{2}\right)\cdots\frac{7}{2}\cdot\frac{5}{2}\cdot\frac{3}{2}}{\left(n-\frac{1}{2}\right)\left(n-\frac{3}{2}\right)\left(n-\frac{5}{2}\right)\cdots\frac{7}{2}\cdot\frac{5}{2}\cdot\frac{3}{2}\cdot\frac{1}{2}}$$

$$= \frac{\frac{3}{2}\left(\frac{3}{2}+1\right)\left(\frac{3}{2}+2\right)\cdots\left(\frac{3}{2}+n-2\right)\left(\frac{3}{2}+n-1\right)}{\frac{1}{2}\left(\frac{1}{2}+1\right)\left(\frac{1}{2}+2\right)\cdots\left(\frac{1}{2}+n-2\right)\left(\frac{1}{2}+n-1\right)}$$

$$= \frac{\left(\frac{3}{2}\right)_n}{\left(\frac{1}{2}\right)_n}. \tag{ii}$$

Substituting Eq. (ii) into Eq. (i), we write

$$erf(x) = \frac{2x}{\sqrt{\pi}}\sum_{n=0}^{\infty}\frac{\left(\frac{1}{2}\right)_n}{\left(\frac{3}{2}\right)_n}\frac{\left(-x^2\right)^n}{n!} = \frac{2x}{\sqrt{\pi}}\,{}_1F_1\left(\frac{1}{2};\frac{3}{2};-x^2\right).$$

PROBLEM 14 Show that

$${}_1F_1(\alpha;\lambda;x) = e^x\,{}_1F_1(\lambda-\alpha;\lambda;-x).$$

Solution: We know that

$${}_1F_1(\alpha;\lambda;x) = \frac{\Gamma(\lambda)}{\Gamma(\alpha)\Gamma(\lambda-\alpha)}\int_0^1 e^{xt}t^{\alpha-1}(1-t)^{\lambda-\alpha-1}\,dt.$$

Replacing α by $\lambda-\alpha$ and x by $(-x)$, we have

$${}_1F_1(\lambda-\alpha;\lambda;-x) = \frac{\Gamma(\lambda)}{\Gamma(\lambda-\alpha)\Gamma(\alpha)}\int_0^1 e^{-xt}t^{\lambda-\alpha-1}(1-t)^{\alpha-1}\,dt$$

$$= e^{-x}\frac{\Gamma(\lambda)}{\Gamma(\lambda-\alpha)\Gamma(\alpha)}\int_0^1 e^{x(1-t)}t^{\lambda-\alpha-1}(1-t)^{\alpha-1}\,dt.$$

Putting $u=1-t$ so that $dt=-du$, we write

$$ {}_1F_1(\lambda-\alpha;\lambda;-x)=e^{-x}\frac{\Gamma(\lambda)}{\Gamma(\alpha)\Gamma(\lambda-\alpha)}\int_0^1 e^{xu}u^{\alpha-1}(1-u)^{\lambda-\alpha-1}\,du $$

$$ =e^{-x}\,{}_1F_1(\alpha;\lambda;x). $$

Therefore, we obtain

$$ e^{x}\,{}_1F_1(\lambda-\alpha;\lambda;-x)={}_1F_1(\alpha;\lambda;x). $$

PROBLEM 15 Show that the confluent hypergeometric function ${}_1F_1(\alpha;\lambda;x)$ is convergent for all values of $|x|$. Obtain its asymptotic form as $x\to\infty$.

Solution: We know that

$$ {}_1F_1(\alpha;\lambda;x)=1+\frac{\alpha}{\lambda}x+\frac{\alpha(\alpha+1)}{\lambda(\lambda+1)}\frac{x^2}{2!}+\frac{\alpha(\alpha+1)(\alpha+2)}{\lambda(\lambda+1)(\lambda+2)}\frac{x^3}{3!}+\ldots\infty $$

Here, the nth term is

$$ T_n=\frac{\alpha(\alpha+1)(\alpha+2)(\alpha+3)\ldots(\alpha+n-2)}{\lambda(\lambda+1)(\lambda+2)(\lambda+3)\ldots(\lambda+n-2)}\frac{x^{n-1}}{(n-1)!} $$

$$ =\frac{(\lambda-1)!(\alpha+n-2)!}{(\alpha-1)!(\lambda+n-2)!}\frac{x^{n-1}}{(n-1)!} \tag{i} $$

and the $(n+1)$th term is

$$ T_{n+1}=\frac{(\lambda-1)!\,(\alpha+n-1)!}{(\alpha-1)!\,(\lambda+n-1)!}\frac{x^n}{n!}. \tag{ii} $$

Now, we have

$$ \lim_{n\to\infty}\frac{T_{n+1}}{T_n}=\lim_{n\to\infty}\frac{(\alpha+n-1)}{n(\lambda+n-1)}x<1 $$

for all values of $|x|$ provided that λ is not a negative integer.

For $n\gg\alpha,\lambda$; we have

$$ \lim_{n\to\infty}\frac{(\alpha+n-1)!}{(\lambda+n-1)!}=\lim_{n\to\infty}\frac{(n-1)!(\alpha+n-\alpha)(\alpha+n-\alpha+1)\ldots(\alpha+n-2)(\alpha+n-1)}{(n-1)!(\lambda+n-\lambda)(\lambda+n-\lambda+1)\ldots(\lambda+n-2)(\lambda+n-1)} $$

$$ =\frac{n^\alpha}{n^\lambda}=n^{\alpha-\lambda}. \tag{iii} $$

In the limit $n \to \infty$, substituting this result into Eq. (ii), we write

$$\lim_{n\to\infty} T_{n+1} = \lim_{n\to\infty} \frac{(\lambda-1)!}{(\alpha-1)!}\frac{n^{\alpha-\lambda}}{n!}x^n$$

$$= \lim_{n\to\infty} \frac{(\lambda-1)!}{(\alpha-1)!}\frac{x^n}{(n-\alpha+\lambda)!}.$$

Therefore, for $x \to \infty$, we have

$${}_1F_1(\alpha;\lambda;x) \approx \frac{(\lambda-1)!}{(\alpha-1)!}\sum_{n=\alpha-\lambda}^{\infty}\frac{x^n}{(n-\alpha+\lambda)!}$$

$$= \frac{\Gamma(\lambda)}{\Gamma(\alpha)}x^{\alpha-\lambda}\left[1+x+\frac{x^2}{2!}+\ldots\infty\right].$$

Thus, the asymptotic form of confluent hypergeometric function as x tends to infinity is

$$\lim_{x\to\infty} {}_1F_1(\alpha;\lambda;x) = \frac{\Gamma(\lambda)}{\Gamma(\alpha)}x^{\alpha-\lambda}e^x. \qquad \text{(iv)}$$

Another asymptotic formula can be obtained using the result of Problem 14. Thus, we have

$$\lim_{x\to-\infty} {}_1F_1(\alpha;\lambda;x) = \lim_{x\to\infty} {}_1F_1(\alpha;\lambda;-x)$$

$$= e^{-x}\,{}_1F_1(\lambda-\alpha;\lambda;x)$$

$$= e^{-x}\frac{\Gamma(\lambda)}{\Gamma(\lambda-\alpha)}x^{\lambda-\alpha-\lambda}e^x = \frac{\Gamma(\lambda)}{\Gamma(\lambda-\alpha)}x^{-\alpha}.$$

EXERCISE

1. Prove that $\dfrac{d^m}{dx^m}\left[{}_2F_1(\alpha,\beta;\lambda;0)\right] = \dfrac{(\alpha)_m(\beta)_m}{(\lambda)_m}$.
2. Prove that
   (i) $(1-x)^{-n} = {}_2F_1(n,1;1;x)$,
   (ii) $(1-x)^{-1/2} = {}_2F_1\left(\frac{1}{2},1;1;x\right)$,
   (iii) $t^n = {}_2F_1(-n,1;1;1-t)$.

3. Using integral representation of confluent hypergeometric function, prove that

   (i) ${}_1F_1(\alpha;\lambda;x) = e^x\,{}_1F_1(\lambda-\alpha;\lambda;-x)$,

   (ii) ${}_1F_1(\alpha;\lambda;0) = 1$.

4. Prove that

   (i) $(1+x)e^x = {}_1F_1(2;1;x)$,

   (ii) $1-2x+\dfrac{x^2}{2} = {}_1F_1(-2;1;x)$.

5. Prove that

$$x\frac{d}{dx}\,{}_1F_1(\alpha;\lambda;x)+\alpha\,{}_1F_1(\alpha;\lambda;x)=\alpha\,{}_1F_1(\alpha+1;\lambda;x).$$

6. Prove that

   (i) $P_{2l}(x)=\dfrac{(-1)^l(2l)!}{2^{2l}(l!)^2}\,{}_2F_1\left(-l,l+\dfrac{1}{2};\dfrac{1}{2};x^2\right)$,

   (ii) $P_{2l+1}(x)=\dfrac{(-1)^l(2l+1)!}{2^{2l}(l!)^2}x\,{}_2F_1\left(-l,l+\dfrac{3}{2};\dfrac{3}{2};x^2\right)$

   where $l=0,1,2,3\ldots$

7. Prove that $L_n(x)={}_1F_1(-n;1;x)$ where the symbols have usual meanings.
8. Using Frobenius method, obtain a power series solution of confluent hypergeometric equation and express the solution in terms of confluent hypergeometric function.

10
CHAPTER

Dirac-Delta Function and Heaviside Unit Step Function

10.1 INTRODUCTION

In mechanics and physics, we often deal with the idea of an action occurring at a point. The concept of an impulsive force, i.e., a very large force acting for a very short interval of time, or of a large load acting at a point of a beam generating a very large pressure over a very small area of the beam, plays a very important role in mechanics. In order to deal with such basic ideas mathematically, the English physicist, Paul Adrien Maurice Dirac (August 8, 1902 – October 20, 1984) introduced a certain type of function which is known after him as **Dirac-Delta** function or simply delta function, in the first edition of his famous book, 'The Princples of Quantum Mechanics' published in 1930. The Dirac-Delta function or simply the delta function is defined by

$$\delta(x) = \begin{cases} \infty \text{ for } x = 0 \\ 0 \text{ for } x \neq 0 \end{cases} \tag{10.1}$$

together with the condition that

$$\int_{-\infty}^{+\infty} \delta(x)dx = 1. \tag{10.2}$$

Thus, delta function is a function which is very large in a small neighbourhood of a point but is zero at all other points. The simplest way to visualize this is a rectangular pulse from $-\varepsilon/2$ to $+\varepsilon/2$ with a height of $1/\varepsilon$ when $\varepsilon \to 0$. A simple example of such a pulse is the hammering of a thin iron bar. As we take the limit of this setup as ε approaches 0, we see that the width tends to zero and the height tends to infinity as the total area remains constant at one. In such a limiting case, the rectangular pulse becomes a delta function.

Mathematically, we define the above rectangular pulse by a function.

$$\delta_\varepsilon(x)=\begin{cases}\dfrac{1}{\varepsilon}\ \text{for}\ -\dfrac{\varepsilon}{2}<x<+\dfrac{\varepsilon}{2}, \\ 0, \qquad \text{for}\ |x|>\dfrac{\varepsilon}{2}.\end{cases} \tag{10.3}$$

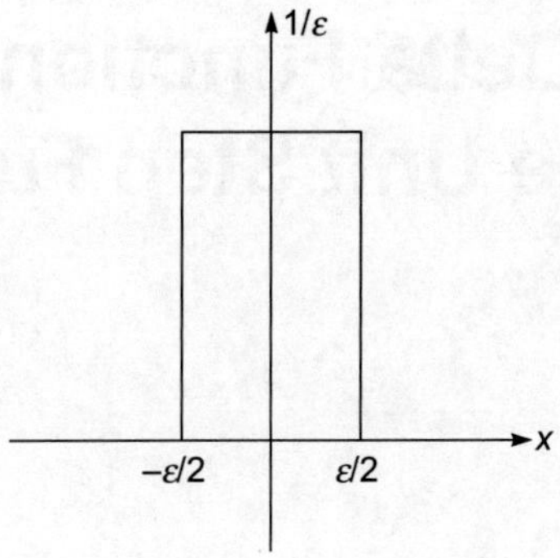

Fig. 10.1 Graphical representation of rectangular pulse

We know that

$$\int_{-\infty}^{+\infty}\delta_\varepsilon(x)dx=\int_{-\infty}^{-\frac{\varepsilon}{2}}\delta_\varepsilon(x)dx+\int_{-\frac{\varepsilon}{2}}^{+\frac{\varepsilon}{2}}\delta_\varepsilon(x)dx+\int_{+\frac{\varepsilon}{2}}^{+\infty}\delta_\varepsilon(x)dx$$

$$=0+\int_{-\varepsilon/2}^{+\varepsilon/2}\delta_\varepsilon(x)dx+0=\int_{-\varepsilon/2}^{\varepsilon/2}\frac{1}{\varepsilon}dx=1. \tag{10.4}$$

It follows that the delta function $\delta(x)$ is the limiting value of $\delta_\varepsilon(x)$ given in Eq. (10.3) when ε tends to zero.

Thus, we have

$$\delta(x)=\lim_{\varepsilon\to 0}\delta_\varepsilon(x)=\begin{cases}\infty\ \text{for}\ x=0 \\ 0\ \text{for}\ x\neq 0\end{cases} \tag{10.5}$$

together with $\int_{-\infty}^{+\infty}\delta(x)dx=1.$

Replacing x by $(x-a)$, we write

$$\delta(x-a)=\begin{cases}\infty\ \text{for}\ x=a \\ 0\ \text{for}\ x\neq a\end{cases} \tag{10.6}$$

together with

$$\int_{-\infty}^{+\infty} \delta(x-a)dx = 1. \tag{10.7}$$

Since it is quite difficult to draw something that is infinitely tall, we represent the Dirac-Delta function by an arrow (or spike) with tail at the origin.

Strictly speaking, the Dirac-Delta function is not an ordinary mathematical function, but a distribution. However, in Engineering and Physics, it can be treated as an ordinary mathematical function. Its properties are discussed in the next section.

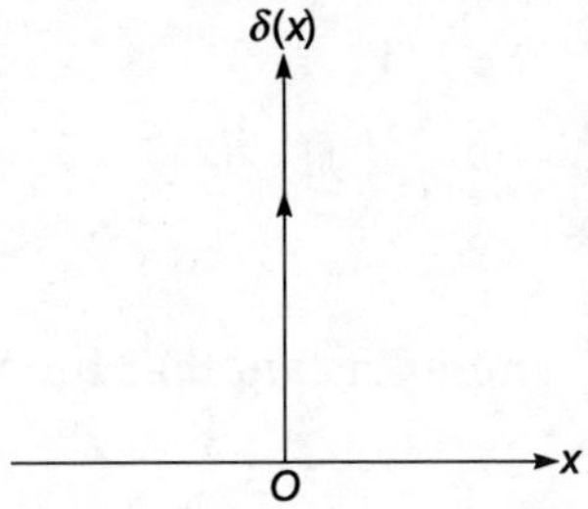

Fig. 10.2 Graphical representation of Dirac-Delta function.

10.2 PROPERTIES OF DELTA FUNCTIONS

I. Delta function integrates to unity, i.e.,

$$\int_{-\infty}^{+\infty} \delta(x)dx = 1.$$

Proof: Using Eqs. (10.4) and (10.5), we write

$$\int_{-\infty}^{+\infty} \delta(x)dx = \int_{-\infty}^{+\infty} \lim_{\varepsilon\to 0} \delta_\varepsilon(x)dx$$

$$= \lim_{\varepsilon\to 0} \int_{-\frac{\varepsilon}{2}}^{+\frac{\varepsilon}{2}} \frac{1}{\varepsilon} dx = \lim_{\varepsilon\to 0} \frac{1}{\varepsilon} \int_{-\frac{\varepsilon}{2}}^{+\frac{\varepsilon}{2}} dx = 1. \tag{10.8}$$

Similarly, one can prove that

$$\int_{-\infty}^{+\infty} \delta(x-a)dx = 1. \tag{10.9}$$

II. The integral of any continuous function multiplied by a delta function located about the origin is just the value of the function at the origin.

We are to prove that

$$\int_{-\infty}^{+\infty} f(x)\delta(x)dx = f(0).$$

Proof: As in the case of I, we have

$$\int_{-\infty}^{+\infty} f(x)\delta(x)dx = \lim_{\varepsilon\to 0}\int_{-\frac{\varepsilon}{2}}^{+\frac{\varepsilon}{2}} \frac{1}{\varepsilon} f(x)dx$$

$$= \lim_{\varepsilon\to 0}\frac{1}{\varepsilon}\int_{-\frac{\varepsilon}{2}}^{+\frac{\varepsilon}{2}} f(x)dx.$$

Putting $f(x) = \frac{d}{dx}F(x) = F'(x)$ where F is another function of x, we write

$$\int_{-\infty}^{+\infty} f(x)\delta(x)dx = \lim_{\varepsilon\to 0}\frac{1}{\varepsilon}\int_{-\frac{\varepsilon}{2}}^{+\frac{\varepsilon}{2}} F'(x)dx$$

$$= \lim_{\varepsilon\to 0}\frac{1}{\varepsilon}\left[F(x)\right]_{-\frac{\varepsilon}{2}}^{+\frac{\varepsilon}{2}}$$

$$= \lim_{\varepsilon\to 0}\left[\frac{F\left(\frac{\varepsilon}{2}\right) - F\left(-\frac{\varepsilon}{2}\right)}{\varepsilon}\right]$$

$$= \lim_{\varepsilon\to 0}\left[\frac{F\left(-\frac{\varepsilon}{2}+\varepsilon\right) - F\left(-\frac{\varepsilon}{2}\right)}{\varepsilon}\right]$$

$$= F'(0) = f(0). \tag{10.10}$$

III. Delta function possesses the shifting property.

Proof: From II, we know that

$$\int_{-\infty}^{+\infty} f(x)\delta(x)dx = f(0). \tag{10.11}$$

Now, we can write

$$\int_{-\infty}^{+\infty} f(x)\delta(x-a)\,dx = \int_{-\infty}^{+\infty} f(u+a)\delta(u)\,du \text{ where we put } u = x-a.$$

Putting $f(u+a) = F(u)$, we write

$$\int_{-\infty}^{+\infty} f(x)\delta(x-a)\,dx = \int_{-\infty}^{+\infty} F(u)\delta(u)\,du = F(0) = f(a). \quad (10.12)$$

IV. Delta function is an even function, i.e., $\delta(-x) = \delta(x)$. In other words, it has a symmetry property.

Proof: Putting $-x = t$, we see that

$$\int_{-\infty}^{+\infty} \delta(-x)\,dx = \int_{+\infty}^{-\infty} \delta(t)(-dt) \text{ [since } dx = -dt \text{]}$$

$$= \int_{-\infty}^{+\infty} \delta(t)\,dt = \int_{-\infty}^{+\infty} \delta(x)\,dx = 1.$$

Hence, we write

$$\delta(-x) = \delta(x). \quad (10.13)$$

Further, we see that

$$\delta(a-x) = \delta\{-(x-a)\} = \delta(x-a). \quad (10.14)$$

V. Delta function has the property that $\delta(ax) = \frac{1}{|a|}\delta(x), (a \neq 0)$.

Proof: By property I, we know that

$$\int_{-\infty}^{+\infty} a\delta(ax)\,dx = \int_{-\infty}^{+\infty} \delta(ax)\,d(ax)$$

$$= \int_{-\infty}^{+\infty} \delta(t)\,dt = 1 \text{ where we put } t = ax$$

$$= \int_{-\infty}^{+\infty} \delta(x)\,dx.$$

Thus, we write

$$a\delta(ax) = \delta(x)$$

or
$$\delta(ax) = \frac{1}{a}\delta(x).$$

Similarly, for negative a, we can obtain

$$\delta(-ax) = \delta(ax) = \frac{1}{a}\delta(x).$$

In general, we write

$$\delta(ax) = \frac{1}{|a|}\delta(x). \tag{10.15}$$

VI. Delta function has the property that

$$\int_{-\infty}^{+\infty} f(x)\delta'(x-a)\,dx = -f'(a)$$

where the prime denotes derivative with respect to x.

Proof: Performing integration by parts, we have

$$\int_{-\infty}^{+\infty} f(x)\delta'(x-a)\,dx = \left[f(x)\delta(x-a)\right]_{-\infty}^{+\infty} - \int_{-\infty}^{+\infty} f'(x)\delta(x-a)\,dx$$

$$= 0 - f'(a) = -f'(a). \tag{10.16}$$

With the nth order derivative of delta function, we have

$$\int_{-\infty}^{+\infty} f(x)\delta^{(n)}(x-a)\,dx = (-1)^n f^{(n)}(a). \tag{10.17}$$

VII. Delta function has the property that $x\delta(x) - 0$ for all values of x.

Proof: We see that for $x = 0,\ x\delta(x) = 0$.

Also, for $x \neq 0,\ \delta(x) = 0$ and, therefore, $x\delta(x) = 0$.

Hence, $x\delta(x) = 0$ for all values of x. (10.18)

In general, we have

$$(x-a)\delta(x-a) = 0. \tag{10.19}$$

VIII. Delta function has the property that $\frac{d}{dx}\delta(x) = \delta'(x) = -\frac{1}{x}\delta(x)$.

Proof: Let us consider the integral

$$\int_{-\infty}^{+\infty}\left\{f(x)x\frac{d}{dx}\delta(x)\right\}dx = \left[f(x)x\delta(x)\right]_{-\infty}^{+\infty} - \int_{-\infty}^{+\infty}\left[\frac{d}{dx}\{xf(x)\}\right]\delta(x)\,dx$$

$$= 0 - \int_{-\infty}^{+\infty} \left\{ f(x) + x \frac{d}{dx} f(x) \right\} \delta(x) dx$$

$$= - \int_{-\infty}^{+\infty} f(x)\delta(x)dx - \int_{-\infty}^{+\infty} \left\{ \frac{d}{dx} f(x) \right\} x\delta(x)dx$$

$$= \int_{-\infty}^{+\infty} - f(x)\delta(x)dx .$$

Here, use has been made of the fact that $x\delta(x)$ is always zero.

Thus, we obtain

$$f(x)x\frac{d}{dx}\delta(x) = -f(x)\delta(x)$$

or
$$\frac{d}{dx}\delta(x) = -\frac{1}{x}\delta(x). \tag{10.20}$$

IX. Delta function has the property that $\delta\left(f(x)\right) = \sum_i \frac{1}{\left|f'(x_i)\right|}\delta\left(x - x_i\right)$

where $f(x)$ is a continuous and differentiable function having zeroes at x_i and the quantity

$$f'(x_i) = \left[\frac{d}{dx} f(x)\right]_{x=x_i} \quad \text{is not zero.}$$

Proof: Let $f(x)$ be a continuous and differentiable function having zeroes at $x = x_i$.

The Taylor series of the function $f(x)$ about the first zero (x_0) of the function $f(x)$ is given by

$$f(x) = f\left(x_0\right) + f'\left(x_0\right)\left(x - x_0\right) \tag{10.21}$$

where we neglect the square and higher powers of $(x - x_0)$ assuming that they are very small.

Since, x_0 is a zero of $f(x)$ we have, $f(x_0) = 0$ and we write

$$f(x) = f'(x_0)(x - x_0).$$

Therefore, we have

$$\delta\left(f(x)\right) = \delta\left(f'(x_0)(x - x_0)\right) = \frac{1}{\left|f'(x_0)\right|}\delta(x - x_0), \text{ [using Eq. (10.15)].}$$

Thus, considering all the zeroes $x_i\,(i=0,1,2,3\ldots)$, we write

$$\delta(f(x))=\sum_i \frac{1}{|f'(x_i)|}\delta(x-x_i). \tag{10.22}$$

X. Delta function has the property that

$$\delta((x-a)(x-b))=\frac{1}{|a-b|}\left[\delta(x-a)+\delta(x-b)\right].$$

Proof: Let us put $f(x)=(x-a)(x-b)$.

Then a and b are the zeroes of $f(x)$.

Now, we have

$$f'(a)=\left[\frac{d}{dx}f(x)\right]_{x=a}=\left[1\cdot(x-b)+(x-a)\cdot 1\right]_{x=a}=a-b$$

Similarly, we have

$$f'(b)=b-a.$$

Substituting $f(x)$ and its derivatives into Eq. (10.22), we obtain

$$\delta((x-a)(x-b))=\frac{1}{|a-b|}\delta(x-a)+\frac{1}{|b-a|}\delta(x-b)$$

$$=\frac{1}{|a-b|}\left[\delta(x-a)+\delta(x-b)\right]. \tag{10.23}$$

XI. Delta function has the property that $f(x)\delta(x-a)=f(a)\delta(x-a)$.

Proof: We have

$$\int_{-\infty}^{+\infty} f(x)\delta(x-a)dx=f(a) \qquad \text{[by Eq. (10.12)]}$$

$$=f(a)\int_{-\infty}^{+\infty}\delta(x-a)dx=\int_{-\infty}^{+\infty} f(a)\delta(x-a)dx.$$

Hence, we have

$$f(x)\delta(x-a)=f(a)\delta(x-a). \tag{10.24}$$

XII. Delta function has the property that $\int_{-\infty}^{+\infty}\delta(x-a)\delta(x-b)dx=\delta(a-b)$.

Proof: Using Eq. (10.12), we have

$$\int_{-\infty}^{+\infty} \delta(x-a)\delta(x-b)dx = \int_{-\infty}^{+\infty} \{\delta(x-b)\}\delta(x-a)dx$$

$$= \delta(a-b). \qquad (10.25)$$

10.3 DELTA FUNCTION IN DIFFERENT FORMS

(i) Integral Form

The *Fourier transform* of a delta function $\delta(x-x_0)$ is given by

$$FT\{\delta(x-x_0)\} = \frac{1}{\sqrt{2\pi}} \int_{-\infty}^{+\infty} \delta(x-x_0)e^{ikx}dx$$

$$= \frac{1}{\sqrt{2\pi}} e^{ikx_0}. \qquad \text{[by Eq. (10.12)].}$$

The reverse Fourier transform (RFT) of the above gives the delta function

$$\delta(x-x_0).$$

Thus, we have

$$\delta(x-x_0) = RFT\left\{\frac{e^{ikx_0}}{\sqrt{2\pi}}\right\} = \frac{1}{\sqrt{2\pi}} \int_{-\infty}^{+\infty} \left(\frac{e^{ikx_0}}{\sqrt{2\pi}}\right) e^{-ikx} dk$$

$$= \frac{1}{2\pi} \int_{-\infty}^{+\infty} e^{ik(x_0-x)} dk. \qquad (10.26a)$$

By the symmetry property of delta functions, we write

$$\delta(x-x_0) = \delta(x_0-x) = \frac{1}{2\pi} \int_{-\infty}^{+\infty} e^{ik(x-x_0)} dk. \qquad (10.26b)$$

Also, the Fourier transform of $\delta(x)$ is $\frac{1}{\sqrt{2\pi}}$.

The reverse Fourier transform is then $\frac{1}{2\pi} \int_{-\infty}^{+\infty} e^{ikx} dk$.

Thus, we have

$$\delta(x)=\frac{1}{2\pi}\int_{-\infty}^{+\infty} e^{ikx}\,dk. \tag{10.26c}$$

(ii) A Limiting Form

Equation (10.26b) can be written as

$$\begin{aligned}\delta(x-x_0)&=\frac{1}{2\pi}\lim_{p\to\infty}\int_{-p}^{+p} e^{ik(x-x_0)}dk\\ &=\frac{1}{2\pi}\lim_{p\to\infty}\frac{e^{ip(x-x_0)}-e^{-ip(x-x_0)}}{i(x-x_0)}\\ &=\frac{1}{\pi}\lim_{p\to\infty}\frac{1}{(x-x_0)}\left[\frac{e^{ip(x-x_0)}-e^{-ip(x-x_0)}}{2i}\right]\\ &=\lim_{p\to\infty}\frac{\sin\{p(x-x_0)\}}{\pi\,(x-x_0)}.\end{aligned} \tag{10.27a}$$

Correspondingly, we write

$$\delta(x)=\lim_{p\to\infty}\frac{\sin(px)}{\pi x}. \tag{10.27b}$$

(iii) Another Limiting Form

We know that

$$\begin{aligned}\delta(x)&=\frac{1}{2\pi}\int_{-\infty}^{+\infty} e^{ikx}dk\\ &=\frac{1}{2\pi}\lim_{\varepsilon\to 0^+}\int_{-\infty}^{+\infty} e^{ikx-\varepsilon|k|}dk\\ &=\frac{1}{2\pi}\lim_{\varepsilon\to 0^+}\left[\int_{-\infty}^{0} e^{\varepsilon k}e^{ikx}dk+\int_{0}^{\infty} e^{-\varepsilon k}e^{ikx}dk\right]\\ &=\frac{1}{2\pi}\lim_{\varepsilon\to 0^+}\left[\int_{-\infty}^{0} e^{k(\varepsilon+ix)}dk+\int_{0}^{\infty} e^{-k(\varepsilon-ix)}dk\right]\end{aligned}$$

$$=\frac{1}{2\pi}\lim_{\varepsilon\to 0^+}\left[\frac{1}{\varepsilon+ix}+\frac{1}{\varepsilon-ix}\right]$$

$$=\frac{1}{\pi}\lim_{\varepsilon\to 0^+}\left[\frac{\varepsilon}{(x^2+\varepsilon^2)}\right]. \tag{10.28a}$$

Correspondingly, by shifting property, we write

$$\delta(x-x_0)=\frac{1}{\pi}\lim_{\varepsilon\to 0^+}\left[\frac{\varepsilon}{(x-x_0)^2+\varepsilon^2}\right]. \tag{10.28b}$$

(iv) Yet Another Limiting Form (Gaussian Form)

We can write

$$\delta(x)=\frac{1}{2\pi}\int_{-\infty}^{+\infty}e^{ikx}dk$$

$$=\frac{1}{2\pi}\lim_{\varepsilon\to 0}\int_{-\infty}^{+\infty}e^{-\left(\varepsilon^2k^2-ikx\right)}dk$$

$$=\frac{1}{2\pi}\lim_{\varepsilon\to 0}e^{-\frac{x^2}{4\varepsilon^2}}\int_{-\infty}^{+\infty}e^{-\left(\varepsilon k-\frac{ix}{2\varepsilon}\right)^2}dk.$$

Putting $u=\varepsilon k-\dfrac{ix}{2\varepsilon}$ we have $dk=\dfrac{du}{\varepsilon}$.

Hence, we write

$$\delta(x)=\frac{1}{2\pi}\lim_{\varepsilon\to 0}\frac{1}{\varepsilon}e^{-\frac{x^2}{4\varepsilon^2}}\int_{-\infty}^{+\infty}e^{-u^2}du$$

$$=\frac{1}{2\pi}\lim_{\varepsilon\to 0}\frac{\sqrt{\pi}}{\varepsilon}e^{-\frac{x^2}{4\varepsilon^2}}=\lim_{\varepsilon\to 0}\frac{1}{2\varepsilon\sqrt{\pi}}e^{-\frac{x^2}{4\varepsilon^2}}.$$

If we put $\sigma=4\varepsilon^2$, we can write

$$\delta(x)=\lim_{\sigma\to 0}\frac{1}{\sqrt{\pi\sigma}}e^{-\frac{x^2}{\sigma}}. \tag{10.29a}$$

If we put $\lambda=\sqrt{\sigma}=2\varepsilon$, we can write

$$\delta(x)=\lim_{\lambda\to 0}\frac{1}{\lambda\sqrt{\pi}}e^{-\frac{x^2}{\lambda^2}}. \tag{10.29b}$$

By the shifting property, we write

$$\delta(x-x_0)=\lim_{\sigma\to 0}\frac{1}{\sqrt{\pi\sigma}}e^{-\frac{(x-x_0)^2}{\sigma}} \tag{10.29c}$$

and
$$\delta(x-x_0)=\lim_{\lambda\to 0}\frac{1}{\lambda\sqrt{\pi}}e^{-\frac{(x-x_0)^2}{\lambda^2}}. \tag{10.29d}$$

10.4 HEAVISIDE (UNIT) STEP FUNCTION

The Heaviside (Unit) step function is defined by the function

$$H(x-a)=\begin{cases}0 \text{ for } x<a,\\ 1 \text{ for } x>a.\end{cases} \tag{10.30}$$

If we take $a=0$, we have

$$H(x)=\begin{cases}0 \text{ for } x<0,\\ 1 \text{ for } x>0.\end{cases} \tag{10.31}$$

This function was first introduced by the English physicist-cum-engineer, Oliver Heaviside (18 May, 1850 - 3 February, 1925) and hence it was named after him.

The derivative of a Heaviside step function gives a Dirac-Delta function and the proof will be presented in the section of Solved Problems.

10.5 GENERALIZED DERIVATIVE OF A PIECEWISE CONTINUOUS FUNCTION

In this section, we will define a piecewise continuous function and will introduce a generalized function and a generalized derivative.

A function $f(x)$ (on an interval) is said to be **piecewise continuous (or sectionally smooth)** if (i) it is continuous everywhere in its interval of definition ecxept at a sparse collection of points $x=x_i$ and (ii) at every such point, both the left-hand and right-hand limits do exist. Such a function is said to have a **jump discontinuity** at a point $x=x_0$ if its graph is continuous on both the left and the right of this point and there is a jump at that point. This means that in the case of a sectionally continuous function, a **jump discontinuity** is said to exist at a point $x=x_0$ if both the left- and the right-hand limits do exist there but are different. In the case of cuves, a **corner** is a point where two converging lines meet at some angle.

A function $f(x)$ is piecewise differentiable if (1) it is piecewise continuous, (2) it is differentiable everywhere except at a sparse collection of points, and its derivative is piecewise continuous.

The generalized derivative of a piecewise continuous function that is smooth except for some jump discontinuties and corners is the regular derivative of the function defined in the regions away from the jumps and corners plus a linear combination of delta functions at the jumps and corners, the coefficient of each delta function being the size or amount of jump at the associated discontinuity or corner.

10.6 THREE-DIMENSIONAL DELTA FUNCTION IN CARTESIAN COORDINATES

The three-dimensional Dirac-Delta function in Cartesian coordinates is defined by

$$\delta(\mathbf{r}-\mathbf{r}')=\delta(x-x')\delta(y-y')\delta(z-z') \tag{10.32}$$

where $\mathbf{r}$ and $\mathbf{r}'$ are position vectors having Cartesian coordinates (x,y,z) and (x',y',z'), respectively.

Just like the various representations and formulae of one-dimensional delta-function discussed earlier, we have many representations and formulae of the three-dimensional delta function as well. A few examples are given below.

$$\int_{-\infty}^{+\infty}\int_{-\infty}^{+\infty}\int_{-\infty}^{+\infty} f(\mathbf{r})\delta(\mathbf{r}-\mathbf{r}')dx\,dy\,dz = f(\mathbf{r}') \tag{10.33}$$

$$\delta(\mathbf{r}-\mathbf{r}')=\frac{1}{(2\pi)^3}\int_{-\infty}^{+\infty}\int_{-\infty}^{+\infty}\int_{-\infty}^{+\infty} e^{i\mathbf{k}\cdot(\mathbf{r}-\mathbf{r}')}dk_x\,dk_y\,dk_z \tag{10.34}$$

where $\mathbf{k}\cdot(\mathbf{r}-\mathbf{r}')=k_x(x-x')+k_y(y-y')+k_z(z-z')$.

Sometimes, the volume element, $dk_x\,dk_y\,dk_z$, is represented by d^3k and the tripple integral symbol is also represented by a single integral symbol even though the integral is a triple integral and we write

$$\delta(\mathbf{r}-\mathbf{r}')=\frac{1}{(2\pi)^3}\int e^{ik\cdot(\mathbf{r}-\mathbf{r}')}d^3k. \tag{10.35}$$

Another representation is

$$\delta(\mathbf{r}-\mathbf{r}')=\lim_{\sigma\to 0}\frac{1}{(\pi\sigma)^{\frac{3}{2}}}\exp\left[-\frac{(x-x')^2+(y-y')^2+(z-z')^2}{\sigma}\right]. \tag{10.36}$$

If we denote the volume element $dx\,dy\,dz$ by d^3r we have the important property of delta function

$$\int\delta(\mathbf{r}-\mathbf{r}')d^3r=1. \tag{10.37}$$

Thus, there are many representations and formulae analogous to those in one dimension.

10.7 THREE-DIMENSIONAL DIRAC -DELTA FUNCTION IN OTHER COORDINATE SYSTEMS

Using the theory of coordinate transformation, the three-dimensional delta function in a coordinate system described by (u, v, w) is given by

$$\delta(\mathbf{r}-\mathbf{r}')=\delta(x-x')\delta(y-y')\delta(z-z')$$
$$=\frac{\delta(u-u')\delta(v-v')\delta(w-w')}{|J|} \tag{10.38}$$

where J is the Jacobian of the transformation.

It is noteworthy to know that the Jacobian J of a transformation of a coordinate system (x, y, z) to a new coordinate system (u,v,w) is given by

$$j=\begin{vmatrix} \frac{\partial x}{\partial u} & \frac{\partial x}{\partial v} & \frac{\partial x}{\partial w} \\ \frac{\partial y}{\partial u} & \frac{\partial y}{\partial v} & \frac{\partial y}{\partial w} \\ \frac{\partial z}{\partial u} & \frac{\partial z}{\partial v} & \frac{\partial z}{\partial w} \end{vmatrix}. \tag{10.39}$$

(A) Dirac-Delta Function in Circular Cylindrical Coordinatc System

Thc coordinates (x, y, z) of three-dimensional Cartesian coordinate system are related to the coordinates (ρ, ϕ, z) of circular cylindrical coordinate system through the transformations

$$x=\rho\cos\phi, \qquad y=\rho\sin\phi, \qquad z=z.$$

The Jacobian is given by

$$J=\begin{vmatrix} \frac{\partial x}{\partial \rho} & \frac{\partial x}{\partial \phi} & \frac{\partial x}{\partial z} \\ \frac{\partial y}{\partial \rho} & \frac{\partial y}{\partial \phi} & \frac{\partial y}{\partial z} \\ \frac{\partial z}{\partial \rho} & \frac{\partial z}{\partial \phi} & \frac{\partial z}{\partial z} \end{vmatrix}=\begin{vmatrix} \cos\phi & -\rho\sin\phi & 0 \\ \sin\phi & \rho\cos\phi & 0 \\ 0 & 0 & 1 \end{vmatrix}=\rho.$$

Then, Eq. (10.38) yields

$$\delta(\mathbf{r}-\mathbf{r}')=\delta(x-x')\delta(y-y')\delta(z-z')$$
$$=\frac{\delta(\rho-\rho')\delta(\phi-\phi')\delta(z-z')}{\rho}. \tag{10.40}$$

Also, we have

$$\delta(\mathbf{r})=\frac{\delta(\rho)\delta(\phi)\delta(z)}{\rho}. \tag{10.41}$$

Corollary: From the Dirac-Delta function in circular cylindrical coordinates (ρ,ϕ,z) if we omit $\delta(z-z)$ then we will get delta function in plane polar coordinates.

(B) Dirac-Delta Function in Spherical Polar Coordinate System

The coordinates (x,y,z) of three-dimensional Cartesian coordinate system are related to the coordinates (r,θ,ϕ) of spherical polar coordinate system through the transformations

$$x=r\sin\theta\cos\phi,\qquad y=r\sin\theta\sin\phi\,,\qquad z=r\cos\theta.$$

The Jacobian is given by

$$J=\begin{vmatrix}\dfrac{\partial x}{\partial r} & \dfrac{\partial x}{\partial\theta} & \dfrac{\partial x}{\partial\phi}\\ \dfrac{\partial y}{\partial r} & \dfrac{\partial y}{\partial\theta} & \dfrac{\partial y}{\partial\phi}\\ \dfrac{\partial z}{\partial r} & \dfrac{\partial z}{\partial\theta} & \dfrac{\partial z}{\partial\phi}\end{vmatrix}=\begin{vmatrix}\sin\theta\cos\phi & r\cos\theta\cos\phi & -r\sin\theta\sin\phi\\ \sin\theta\sin\phi & r\cos\theta\sin\phi & r\sin\theta\cos\phi\\ \cos\theta & -r\sin\theta & 0\end{vmatrix}$$

$$=\cos\theta\left(r^2\cos\theta\sin\theta\cos^2\phi+r^2\cos\theta\sin\theta\sin^2\phi\right)$$
$$+r\sin\theta\left(r\sin^2\theta\cos^2\phi+r\sin^2\theta\sin^2\phi\right)$$

$$=r^2\cos^2\theta\sin\theta\left(\cos^2\phi+\sin^2\phi\right)+r^2\sin^3\theta\left(\cos^2\phi+\sin^2\phi\right)$$

$$=r^2\sin\theta\left(\cos^2\theta+\sin^2\theta\right)=r^2\sin\theta$$

Using Eq. (10.38), we write

$$\delta(\mathbf{r}-\mathbf{r}')=\delta(x-x')\delta(y-y')\delta(z-z')$$
$$=\frac{\delta(r-r')\delta(\theta-\theta')\delta(\phi-\phi')}{r^2\sin\theta}. \tag{10.42}$$

10.8 COMPLETENESS AND CLOSURE OF FUNCTIONS

A set of functions $\{\Psi_n(x)\}$ is said to be a *complete* set if an arbitrary function, $f(x)$, can be expanded as

$$f(x)=\sum_n c_n \Psi_n(x) \tag{10.43}$$

where c_n are known as expansion coefficients. These coefficients are to be determined via the orthonomality condition. Such a property that a certain function can be expanded in a series of orthonormal functions is known as completeness.

A relation that expresses the completeness of a set of functions $\{\Psi_n(x)\}$ is known as a closure relation. A closure relation can be obtained as in the following.

Multiplying both sides of Eq. (10.43) by $\Psi^*_m(x)$ and integrating the result with respect to x in the interval from $x=-\infty$ to $x=+\infty$, we obtain

$$\int_{-\infty}^{+\infty}\Psi^*_m(x)f(x)dx=\sum_n c_n\int_{-\infty}^{+\infty}\Psi^*_m(x)\Psi_n(x)dx$$

$$=\sum_n c_n\delta_{mn}=c_m.$$

Here, $\Psi^*_m(x)$ is the complex conjugate of $\Psi_m(x)$.

Renaming the subscript, we write

$$c_n=\int_{-\infty}^{+\infty}\Psi^*_n(x)f(x)dx=\int_{-\infty}^{+\infty}\Psi^*_n(x')f(x')dx'. \tag{10.44}$$

Substituting Eq. (10.44) into Eq. (10.43), we obtain

$$f(x)=\sum_n\left[\int_{-\infty}^{+\infty}\Psi^*_n(x')f(x')dx'\right]\Psi_n(x).$$

Here, we put a prime on the integration variable within the square bracket in order that it will not be confused with the variable of $\Psi_n(x)$

An interchange of the summation and the integration results in

$$f(x)=\int_{-\infty}^{+\infty}\left[\sum_n\Psi^*_n(x')\Psi_n(x)\right]f(x')dx'. \tag{10.45}$$

Comparing Eq. (10.45) with the formula

$$f(x)=\int_{-\infty}^{+\infty}\delta(x-x')f(x')dx',$$

we obtain

$$\sum_n \Psi_n^*(x')\Psi_n(x) = \delta(x - x'). \tag{10.46}$$

In three dimensions, such a relation is expressed as

$$\sum_n \Psi_n^*(\mathbf{r}')\Psi_n(\mathbf{r}) = \delta(\mathbf{r} - \mathbf{r}'). \tag{10.47}$$

This equation is called a ***closure*** relation. Some authors call it ***completeness*** relation. Here we use Cartesian coordinates. But such relations can be expressed in other coordinate systems also.

10.9 SOLVED PROBLEMS

PROBLEM 1 Prove that $\delta(x^2 - a^2) = \frac{1}{2|a|}\left[\delta(x-a) + \delta(x+a)\right]$ for $a \neq 0$.

Proof: From Eq. (10.23), we know that

$$\delta\big((x-a)(x-b)\big) = \frac{1}{|a-b|}\left[\delta(x-a) + \delta(x-b)\right].$$

Putting $b = -a$, we write

$$\delta\big((x-a)(x+a)\big) = \frac{1}{|2a|}\left[\delta(x-a) + \delta(x+a)\right]$$

or

$$\delta(x^2 - a^2) = \frac{1}{2|a|}\left[\delta(x-a) + \delta(x+a)\right].$$

PROBLEM 2 Prove that $\delta\left(\sqrt{x} - \sqrt{a}\right) = 2\sqrt{a}\,\delta(x-a)$ for $x > 0$.

Proof: Let us put

$$f(x) = \sqrt{x} - \sqrt{a} \tag{i}$$

We see that the function $f(x)$ has only one zero at $x = a$. Then, we have

$$f'(a) = \left[\frac{d}{dx} f(x)\right]_{x=a} = \left[\frac{1}{2} x^{-\frac{1}{2}}\right]_{x=a} = \frac{1}{2\sqrt{a}}. \tag{ii}$$

From Eq. (10.22), we know that

$$\delta\big(f(x)\big) = \sum_i \frac{1}{|f'(x_i)|}\delta(x - x_i) \tag{iii}$$

where x_i's are zeroes of $f(x)$.

Substituting Eqs. (i) and (ii) into Eq. (iii), we obtain

$$\delta\left(\sqrt{x}-\sqrt{a}\right)=\frac{1}{\left|\frac{1}{2\sqrt{a}}\right|}\delta(x-a)=2\sqrt{a}\,\delta(x-a).$$

PROBLEM 3 Prove that $\int_{-\infty}^{+\infty} f(x)\delta'(x)dx=-f'(0)$ where the primes denote derivatives.

Proof: Integrating the left-hand side by parts, we have

$$\int_{-\infty}^{+\infty} f(x)\delta'(x)dx=\left[f(x)\delta(x)\right]_{-\infty}^{+\infty}-\int_{-\infty}^{+\infty} f'(x)\delta(x)dx$$

$$=0-\int_{-\infty}^{+\infty} f'(x)\delta(x)dx, \text{ [as } \delta(x) \text{ vanishes at } x=\pm\infty]$$

$$=-f'(0) \text{ [by Eq. (10.10) or (10.12).}$$

PROBLEM 4 Prove that $\delta\left(x^2-7x+10\right)=\frac{1}{3}\left[\delta(x-2)+\delta(x-5)\right]$.

Proof: We know that $f(x)=x^2-7x+10$ has the zeroes at $x=2, 5$.

Further, we know that

$$f'(x)=2x-7.$$

Hence, we have

$$f'(2)=-3 \text{ and } f'(5)=3.$$

Using the formula,

$$\delta(f(x))=\frac{1}{|f'(x_i)|}\sum_i \delta(x-x_i)$$

We write

$$\delta\left(x^2-7x+10\right)=\frac{1}{|-3|}\delta(x-2)+\frac{1}{|3|}\delta(x-5)$$

$$=\frac{1}{3}\left[\delta(x-2)+\delta(x-5)\right].$$

PROBLEM 5 Evaluate

(i) $\int_{-\infty}^{+\infty} \sin(2x)\delta\left(x-\frac{\pi}{4}\right)dx$,

(ii) $\int_{-\infty}^{+\infty} e^{-3x}\delta'(x-2)dx$.

Solution:

(i) Using the formula $\int_{-\infty}^{+\infty} f(x)\delta(x-a)dx = f(a)$,

we have
$$\int_{-\infty}^{+\infty} \sin(2x)\delta\left(x-\frac{\pi}{4}\right)dx = \sin\frac{\pi}{2} = 1.$$

(ii) Integrating by parts, the given integral can be written as

$$\int_{-\infty}^{+\infty} e^{-3x}\delta'(x-2)dx = \left[e^{-3x}\delta(x-2)\right]_{-\infty}^{+\infty} - \int_{-\infty}^{+\infty} -3e^{-3x}\delta(x-2)dx$$

$$= 0 + 3\int_{-\infty}^{+\infty} e^{-3x}\delta(x-2)dx = 3e^{-6}.$$

PROBLEM 6 For a function $f(x) = \begin{cases} 0 & \text{for } x < a \\ ge^{-g(x-a)} & \text{for } x \geq a \end{cases}$

where $g > 0$, show that

(i) $\int_{-\infty}^{+\infty} f(x)dx = 1$ for all values of x,

(ii) $f(x) \to \delta(x-a)$ as $g \to \infty$.

Solution:

(i) We have

$$\int_{-\infty}^{+\infty} f(x)dx = \int_{-\infty}^{a} f(x)dx + \int_{a}^{\infty} f(x)dx$$

$$= \int_{-\infty}^{a} 0.dx + \int_{a}^{\infty} ge^{-g(x-a)}dx$$

$$= \int_a^{\infty} g e^{-g(x-a)} dx .$$

Putting $u = x - a$ on the right-hand side, we write

$$\int_{-\infty}^{+\infty} f(x)dx = \int_0^{\infty} g e^{-gu} du = \left[\frac{g e^{-gu}}{-g} \right]_{u=0}^{\infty} = 1 .$$

(ii) When $g \to \infty$, we have

for $x < a,\ f(x) = 0$,

for $x = a,\ f(x) = \lim_{g\to\infty} g e^{-g(x-a)} = \lim_{g\to\infty} \frac{g}{e^{g(x-a)}} = \infty$,

for $x > a,\ f(x) = \lim_{g\to\infty} \frac{g}{e^{g(x-a)}}$

$$= \lim_{g\to\infty} \frac{1}{(x-a)e^{g(x-a)}} \quad \text{[by L' Hospital's rule]}$$

$$= 0 .$$

From (i), we recall that $\int_{-\infty}^{+\infty} f(x)dx = 1$.

Thus, in the limit $g \to \infty$, we write

$$f(x) = \begin{cases} 0 & \text{for } x - a < 0 \\ \infty & \text{for } x - a = 0 \\ 0 & \text{for } x - a > 0 \end{cases}$$

$$= \begin{cases} \lim\limits_{\varepsilon\to 0} \frac{1}{\varepsilon} & \text{for } -\frac{\varepsilon}{2} \le x - a \le \frac{\varepsilon}{2} \\ 0 & \text{for } |x-a| > \frac{\varepsilon}{2} \end{cases}$$

$$= \delta(x - a) .$$

PROBLEM 7 Prove that

$$\delta(x-a) = \frac{d}{dx} H(x-a) .$$

Solution: We know that

$$H(x-a) = \begin{cases} 0 & \text{for } x < a , \\ 1 & \text{for } x > a . \end{cases}$$

For an arbitrary function $f(x)$, let us consider the integral

$$\int_{-\infty}^{+\infty} f(x)\frac{d}{dx}H(x-a)\,dx = \lim_{\substack{\nu\to\infty\\ \lambda\to\infty}} \int_{a-\lambda}^{a+\nu} f(x)\frac{d}{dx}H(x-a)\,dx$$

$$= \lim_{\substack{\nu\to\infty\\ \lambda\to\infty}} \left\{\left[f(x)H(x-a)\right]_{a-\lambda}^{a+\nu} - \int_{a-\lambda}^{a+\nu}\left[\frac{d}{dx}f(x)\right]H(x-a)\,dx\right\}$$

$$= \lim_{\substack{\lambda\to\infty\\ \nu\to\infty}} \left\{f(a+\nu)\cdot 1 - f(a-\lambda)\cdot 0 - \int_{a-\lambda}^{a+\nu} f'(x)H(x-a)\,dx\right\}$$

$$= \lim_{\nu\to\infty} \left\{f(a+\nu) - \int_{a}^{a+\nu} f'(x)\cdot 1\,dx\right\}$$

$$= \lim_{\nu\to\infty} \left\{f(a+\nu) - \left[f(x)\right]_a^{a+\nu}\right\}$$

$$= \lim_{\nu\to\infty} \left\{f(a+\nu) - f(a+\nu) + f(a)\right\}$$

$$= f(a) = \int_{-\infty}^{+\infty} f(x)\delta(x-a)\,dx.$$

Hence, we write

$$\frac{d}{dx}H(x-a) = \delta(x-a).$$

PROBLEM 8 For a function $g(x) = \frac{\sigma}{2}\operatorname{sech}^2\{\sigma(x-a)\}$ with $\sigma > 0$, prove that

$\int_{-\infty}^{+\infty} g(x)dx = 1$ and hence $g(x)$ tends to $\delta(x-a)$ as $\sigma \to \infty$.

Solution: We have

$$\int_{-\infty}^{+\infty} g(x)\,dx = \frac{\sigma}{2}\int_{-\infty}^{+\infty}\operatorname{sech}^2\{\sigma(x-a)\}\,dx = \frac{\sigma}{2}\int_{-\infty}^{+\infty}\operatorname{sech}^2 \sigma u\,du \qquad \text{where } u = x-a$$

$$= \lim_{k\to\infty}\frac{\sigma}{2}\int_{-k}^{+k}\operatorname{sech}^2 \sigma u\,du$$

$$= \lim_{k\to\infty} \sigma \int_0^k \operatorname{sech}^2 \sigma u \, du$$

$$= \lim_{k\to\infty} \sigma \left[\frac{\tanh \sigma u}{\sigma}\right]_0^k$$

$$= 1. \tag{i}$$

Moreover, we have

$$\lim_{\sigma\to\infty} g(x) = \lim_{\sigma\to\infty} \frac{\sigma}{2} \operatorname{sech}^2\{\sigma(x-a)\}$$

$$= \lim_{\sigma\to\infty} \frac{\sigma}{2} \left\{\frac{2}{e^{\sigma(x-a)} + e^{-\sigma(x-a)}}\right\}^2$$

$$= 2 \lim_{\sigma\to\infty} \frac{\sigma}{e^{2\sigma(x-a)} + e^{-2\sigma(x-a)} + 2}$$

$$= 2 \lim_{\sigma\to\infty} \frac{1}{2(x-a)\left[e^{2\sigma(x-a)} - e^{-2\sigma(x-a)}\right]} \quad \text{[by L' Hospital's rule]}$$

$$= \begin{cases} 0 & \text{for } x \neq a, \\ \infty & \text{for } x = a. \end{cases} \tag{ii}$$

Comparing Eqs. (i) and (ii) with the definition of delta function, viz.

$$\delta(x-a) = \begin{cases} 0 & \text{for } x \neq a, \\ \infty & \text{for } x = a, \end{cases}$$

together with $\int_{-\infty}^{+\infty} \delta(x-a)\,dx = 1$,

we conclude that $\lim_{\sigma\to\infty} g(x) = \delta(x-a)$.

PROBLEM 9 Obtain the completeness relation in the case of the Hermite-Gauss function

$$\Psi_n(x) = \left[2^n n! \sqrt{\pi}\right]^{-\frac{1}{2}} e^{-\frac{x^2}{2}} H_n(x); \; (n = 0, 1, 2, 3 \ldots)$$

where $H_n(x)$ are Hermite polynomials.

Solution: A function $f(x)$ can be expanded as

$$f(x)=\sum_{n=0}^{\infty} c_n \Psi_n(x). \quad \text{(i)}$$

Multiplying both sides by $\Psi_m^*(x)$ and integrating the results with respect to x within the interval from $x=-\infty$ to $x=\infty$, we have

$$\int_{-\infty}^{\infty} f(x)\Psi_m^*(x)dx=\sum_{n=0}^{\infty} c_n \int_{-\infty}^{\infty} \Psi_m^*(x)\Psi_n(x)dx$$

$$=\sum_{n=0}^{\infty} c_n \delta_{mn}=c_m.$$

Renaming the subscript, we write

$$c_n=\int_{-\infty}^{\infty} f(x)\Psi_n^*(x)dx. \quad \text{(ii)}$$

Substituting Eq. (ii) into Eq. (i), we obtain

$$f(x)=\sum_{n=0}^{\infty}\left[\int_{-\infty}^{\infty} f(x')\Psi_n^*(x')dx'\right]\Psi_n(x)$$

$$=\int_{-\infty}^{\infty}\left[\sum_{n=0}^{\infty}\Psi_n^*(x')\Psi_n(x)\right]f(x')dx'.$$

Here, we put a prime on the integration variable within the square bracket in order that it will not be confused with the variable of $\Psi_n(x)$.

Comparing the above equation with the equation

$$f(x)=\int_{-\infty}^{\infty}\delta(x-x')f(x')dx'$$

we write

$$\sum_{n=0}^{\infty}\Psi_n^*(x')\Psi_n(x)=\delta(x-x')$$

i.e.,
$$\sum_{n=0}^{\infty}\frac{1}{2^n n!\sqrt{\pi}}e^{-\left(x^2+x'^2\right)/2}H_n(x)H_n(x')=\delta(x-x').$$

PROBLEM 10 A piecewise continuous function is defined by

$$f(x)=\begin{cases}\cos x & \text{for } x<0,\\ 1+e^{-x} & \text{for } 0<x<2,\\ x & \text{for } x>2.\end{cases}$$

Find its generalized derivative.

Solution: At the discontinuity at $x=0$, amount of jump is $1+e^{0}-\cos 0=1$ and the delta functon is $\delta(x)$. At the discontinuity at $x = 2$, amount of jump is $2-(1+e^{-2})=1-e^{-2}$ and the delta function is $\delta(x-2)$.

Hence, by definition, the generalized derivative is

$$f'(x)=\delta(x)+\left(1-e^{-2}\right)\delta(x-2)+\begin{cases}-\sin x & \text{for } x<0,\\ -e^{-x} & \text{for } 0<x<2,\\ 1 & \text{for } x>2.\end{cases}$$

PROBLEM 11 A piecewise continuous function is defined by

$$f(x)=\begin{cases}-2 & \text{for } x<-2,\\ x+2 & \text{for } -2<x<-1,\\ -x & \text{for } -1<x<0,\\ \dfrac{x^2}{2}-2 & \text{for } 0<x<2,\\ 4-3(x-2)^2 & \text{for } x>2.\end{cases}$$

Find its generalized derivative.

Solution: At the discontinuity at $x=-2$, amount of jump is $-2+2-(-2)=2$ and the delta function is $\delta(x+2)$. At the discontinuity at $x=-1$, amount of jump is $-(-1)-(-1+2)=0$ and the delta function is $\delta(x+1)$. At the discontinuity at $x=0$, amount of jump is $0-2-0=-2$ and the delta function is $\delta(x)$.

At the discontinuity at $x=2$, amount of jump is $4-0-\left(\dfrac{2^2}{2}-2\right)=4$ and the delta function is $\delta(x-2)$.

Hence, by definition, the generalized derivative is

$$f'(x) = 2\delta(x+2) + 0\cdot\delta(x+1) - 2\delta(x) + 4\delta(x-2) + \begin{cases} 0 & \text{for } x < -2, \\ 1 & \text{for } -2 < x < -1, \\ -1 & \text{for } -1 < x < 0, \\ x & \text{for } 0 < x < 2, \\ -6(x-2) & \text{for } x > 2. \end{cases}$$

PROBLEM 12 For a function $f(x) = |x|$ show that $f''(x) = 2\delta(x)$.

Solution: Here, $f(x) = |x|$ is equivalent to $f(x) = \begin{cases} -x & \text{for } x < 0, \\ x & \text{for } x > 0. \end{cases}$

Therefore, we have $f'(x) = \begin{cases} -1 \text{ for } x < 0, \\ 1 \text{ for } x > 0. \end{cases}$

At $x = 0$ $f'(x)$ has an amount of jump of $[1-(-1) = 2]$ and delta function is $\delta(x-0) = \delta(x)$.

Hence, we have $f''(x) = 2\delta(x) + 0 = 2\delta(x)$.

PROBLEM 13 Prove that in the spherical polar coordinates $(r, \cos\theta, \phi)$

$$\delta(\mathbf{r}-\mathbf{r}') = \frac{\delta(r-r')\delta(\cos\theta - \cos\theta')\delta(\phi-\phi')}{r^2}.$$

Solution: In Eq. (10.42), it was established that in spherical polar coordinates (r, θ, ϕ) the delta function is given by

$$\delta(\mathbf{r}-\mathbf{r}') = \frac{\delta(r-r')\delta(\theta-\theta')\delta(\phi-\phi')}{r^2 \sin\theta}.$$

We know that the volume element is

$$dx\,dy\,dz = J\,dr\,d\theta\,d\phi \text{ (where } J \text{ is the Jacobian of transformation)}$$
$$= r^2 \sin\theta\,dr\,d\theta\,d\phi = -r^2 dr\,d(\cos\theta)d\phi.$$

Thus, the Jacobian of the new transformation is given by

$$J_{\text{New}} = -r^2.$$

Therefore, we write

$$\delta(\mathbf{r}-\mathbf{r}') = \frac{\delta(r-r')\delta(\cos\theta - \cos\theta')\delta(\phi-\phi')}{|J_{\text{New}}|}$$

$$= \frac{\delta(r-r')\delta(\cos\theta - \cos\theta')\delta(\phi-\phi')}{r^2}.$$

PROBLEM 14 Derive the spherical harmonic closure relation

$$\sum_{l=0}^{\infty}\sum_{m=-l}^{+l} Y_{lm}^*(\theta',\phi')Y_{lm}(\theta,\phi) = \frac{\delta(\theta-\theta')\delta(\phi-\phi')}{\sin\theta}$$

$$= \delta(\cos\theta - \cos\theta')\delta(\phi-\phi').$$

Derivation: Since the spherical harmonics form a complete set, a function $f(\theta,\phi)$ can be expanded as

$$f(\theta,\phi) = \sum_{l=0}^{\infty}\sum_{m=-l}^{+l} c_{lm}Y_{lm}(\theta,\phi). \tag{i}$$

Multiplying both sides by $Y_{l'm'}^*(\theta,\phi)$ and integrating the results over the entire solid angle, we write

$$\int f(\theta,\phi)Y_{l'm'}^*(\theta,\phi)d\Omega = \sum_{l=0}^{\infty}\sum_{m=-l}^{+l} c_{lm}\int_{\theta=0}^{\pi}\int_{\phi=0}^{2\pi} Y_{l'm'}^*(\theta,\phi)Y_{lm}(\theta,\phi)\sin\theta\, d\theta\, d\phi$$

$$= \sum_{l=0}^{\infty}\sum_{m=-l}^{+l} c_{lm}\delta_{ll'}\delta_{mm'} = c_{l'm'}. \tag{ii}$$

Here, $d\Omega = \sin\theta\, d\theta\, d\phi$ is an element of solid angle.

Renaming the subscript, we write

$$c_{lm} = \int f(\theta,\phi)\, Y_{lm}^*(\theta,\phi)d\Omega. \tag{iii}$$

Substituting Eq. (iii) in Eq. (i), we obtain

$$f(\theta,\phi) = \sum_{l=0}^{\infty}\sum_{m=-l}^{+l}\left[\int f(\theta',\phi')Y_{lm}^*(\theta',\phi')d\Omega'\right]Y_{lm}(\theta,\phi).$$

Here, we put a prime on the integration variable within the square bracket in order that it will not be confused with the variable of $Y_{lm}(\theta,\phi)$.

Interchanging the summation and the integration, we write

$$f(\theta,\phi) = \int\left[\sum_{l=0}^{\infty}\sum_{m=-l}^{+l} Y_{lm}^*(\theta',\phi')Y_{lm}(\theta,\phi)\right]f(\theta',\phi')d\Omega'. \tag{iv}$$

Comparing Eq. (iv) with the equation

$$f(\theta,\phi) = \int \delta(\Omega-\Omega')f(\theta',\phi')d\Omega',$$

we write

$$\sum_{l=0}^{\infty}\sum_{m=-l}^{+l} Y_{lm}^{*}(\theta',\phi')Y_{lm}(\theta,\phi)=\delta(\Omega-\Omega') \tag{v}$$

where Ω and Ω' are solid angles specified by the angles (θ,ϕ) and (θ',ϕ'), respectively.

In the three-dimensional Dirac-Delta function in spherical polar coordinates, the factors independent of the radial coordinate 'r' is the delta function $\delta(\Omega-\Omega')$.

Since the three-dimensional delta function in spherical polar coordinates is given by

$$\delta(\mathbf{r}-\mathbf{r}')=\frac{\delta(r-r')\delta(\theta-\theta')\delta(\phi-\phi')}{r^2\sin\theta}.$$

We have

$$\delta(\Omega-\Omega')=\frac{\delta(\theta-\theta')\delta(\phi-\phi')}{\sin\theta}.$$

Hence, Eq. (v) can be written as

$$\sum_{l=0}^{\infty}\sum_{m=-l}^{+l} Y_{lm}^{*}(\theta',\phi')Y_{lm}(\theta,\phi)=\delta(\Omega-\Omega')=\frac{\delta(\theta-\theta')\delta(\phi-\phi')}{\sin\theta}.$$

PROBLEM 15 Prove that

$$\nabla^2\left(\frac{1}{r}\right)=-4\pi\delta(\mathbf{r})=-4\pi\delta(x)\delta(y)\delta(z).$$

Solution: We have

$$\nabla^2\left(\frac{1}{r}\right)=\boldsymbol{\nabla}\cdot\left[\boldsymbol{\nabla}\left(\frac{1}{r}\right)\right]=\boldsymbol{\nabla}\cdot\left(-\frac{\mathbf{r}}{r^3}\right)=-\left[\boldsymbol{\nabla}\left(\frac{1}{r^3}\right)\right]\cdot\mathbf{r}-\frac{1}{r^3}(\boldsymbol{\nabla}\cdot\mathbf{r})$$

$$=3\left(\frac{\mathbf{r}}{r^5}\right)\cdot\mathbf{r}-\frac{3}{r^3}=\frac{3}{r^3}-\frac{3}{r^3}=0 \text{ for } r>0. \tag{i}$$

Also, we have

$$\iiint_V \nabla^2\left(\frac{1}{r}\right)dV=\iiint_V \boldsymbol{\nabla}\cdot\left[\boldsymbol{\nabla}\left(\frac{1}{r}\right)\right]dV \tag{ii}$$

where V is the volume of a sphere.

Using Gauss' divergence theorem, we write Eq. (ii) as

$$\iiint_V \nabla^2\left(\frac{1}{r}\right)dV=\iint_S \boldsymbol{\nabla}\left(\frac{1}{r}\right)\cdot\hat{r}\,dS$$

where S is the surface enclosing the volume V and $\hat{r}=\frac{\mathbf{r}}{r}$ is a unit vector in the direction of $\mathbf{r}$.

Thus, we have

$$\iiint_V \nabla^2\left(\frac{1}{r}\right)dV=\iint_S -\frac{\mathbf{r}}{r^3}\cdot\frac{\mathbf{r}}{r}dS$$

$$=-\iint_S \frac{dS}{r^2}=-\int_{\theta=0}^{\pi}\sin\theta\, d\theta\int_{\phi=0}^{2\pi}d\phi=-4\pi.$$

Thus, we have

$$\iiint_V \nabla^2\left(\frac{1}{r}\right)dV=-4\pi \text{ for } r=0.$$

Using the property of delta function, viz.

$$\iiint_V \delta(\mathbf{r})dV=1.$$

We write

$$\iiint_V \nabla^2\left(\frac{1}{r}\right)dV=-4\pi\iiint_V \delta(\mathbf{r})dV$$

$$=\iiint_V -4\pi\delta(\mathbf{r})dV.$$

Therefore, we have

$$\nabla^2\left(\frac{1}{r}\right)=-4\pi\delta(\mathbf{r})=-4\pi\delta(x)\delta(y)\delta(z)$$

which includes both

$$\nabla^2\left(\frac{1}{r}\right)=0 \text{ for } r>0 \text{ and } \nabla^2\left(\frac{1}{r}\right)=-4\pi \text{ for } r=0.$$

EXERCISE

1. Evaluate the following integrals:

(i) $\int_{-1}^{3} e^{x^2}\delta(x)dx,$ (ii) $\int_{1}^{3} e^{x}\delta(x)dx,$

(iii) $\int_{3}^{5} x^2\delta(x-2)\,dx,$ (iv) $\int_{-\infty}^{+\infty} \sin 2t\delta\left(t-\frac{\pi}{4}\right)dt,$

(v) $\int_{-1}^{4} \left(x^2-3x+2\right)\delta(x+2)\,dx,$ (vi) $\int_{-2}^{3} \left(x^2+5x+6\right)\delta(x-1)\,dx,$

(vii) $\int_{-2}^{+2} (2x+1)\delta(3x)\,dx.$

Ans. (i) 1, (ii) 0, (iii) 0, (iv) 1, (v) 0, (vi) 12, (vii) $\frac{1}{3}$.

2. Prove that $\nabla^2\left(\frac{1}{|\mathbf{r}-\mathbf{r}'|}\right) = -4\pi\delta(\mathbf{r}-\mathbf{r}')$

$$= -4\pi\delta(x-x')\delta(y-y')\delta(z-z').$$

3. Prove that the functions

$$\Psi_l(x) = \left(\frac{2l+1}{2}\right)^{\frac{1}{2}} P_l(x),\ l = 0, 1, 2, 3 \ldots$$

where $P_l(x)$ are Legendre polynomials, satisfy the closure relation

$$\sum_{l=0}^{\infty} \Psi_l^*(x')\Psi_l(x) = \delta(x-x')$$

or $\frac{1}{2}\sum_{l=0}^{\infty}(2l+1)P_l(x)P_l(x') = \delta(x-x')$ in the interval $-1 \le x \le 1$.

4. Prove that the Laguerre polynomials $L_n(x)$, $n = 0, 1, 2, 3 \ldots$ satisfy the closure relation

$$\sum_{n=0}^{\infty} e^{-\frac{(x+x')}{2}} L_n(x)L_n(x') = \delta(x-x').$$

5. Prove that the associated Laguerre polynomials $L_n^{\alpha}(x)$, $(n = 0, 1, 2, 3 \ldots)$ defined by

$$L_n^{\alpha}(x) = \sum_{r=0}^{n} \frac{(-1)^r\,\Gamma(n+\alpha+1)}{(n-r)!\,r!\,\Gamma(r+\alpha+1)} x^r$$, satisfy the closure relation

$$\sum_{n=0}^{\infty} (xx')^{\frac{\alpha}{2}} e^{-\frac{(x+x')}{2}} L_n^{\alpha}(x)L_n^{\alpha}(x') = \delta(x-x').$$

6. A piecewise continuous function is defined by

$$f(x) = \begin{cases} \sin x & \text{for } x < 0, \\ 1 + e^{-x} & \text{for } 0 < x < 3, \\ 2x + 1 & \text{for } x > 3. \end{cases}$$

Find its generalized derivative. Sketch profiles of $f(x)$ and its generalized derivative.

Ans. $f'(x) = 2\delta(x) + \left(6 - e^{-3}\right)\delta(x-3) + \begin{cases} \cos x \text{ for } x < 0, \\ -e^{-x} \text{ for } 0 < x < 3, \\ 2 \quad \text{for } x > 3. \end{cases}$

11
CHAPTER

Fourier Series and Fourier Transforms

11.1 ORTHOGONALITY PROPERTY OF SINE AND COSINE FUNCTIONS

The idea of orthogonality of functions was introduced in Chapter 2. Here, we study the orthogonality of sine and cosine functions. This orthogonality property is very important in the study of ***Fourier*** series.

The orthogonality property of sine and cosine functions in the interval $[-\pi, \pi]$ is specified by the following equations:

(i) $$\int_{-\pi}^{+\pi} \sin mx \sin nx \, dx = \pi\delta_{mn},$$

(ii) $$\int_{-\pi}^{+\pi} \cos mx \cos nx \, dx = \pi\, \delta_{mn},$$

(iii) $$\int_{-\pi}^{+\pi} \sin mx \cos nx \, dx = 0$$

where m and n are non-zero integers.

Proofs:

(i) For $m \neq n$, we have

$$\int_{-\pi}^{+\pi} \sin mx \sin nx \, dx = \int_0^{\pi} 2 \sin mx \sin nx \, dx$$

$$= \int_0^{\pi} \cos\{(m-n)x\} \, dx - \int_0^{\pi} \cos\{(m+n)x\} \, dx$$

$$= \frac{1}{m-n}\left[\sin\{(m-n)x\}\right]_0^{\pi} - \frac{1}{m+n}\left[\sin\{(m+n)x\}\right]_0^{\pi}$$

$$= 0.$$

For $m = n$, we have

$$\int_{-\pi}^{+\pi} \sin^2 nx\, dx = \int_0^{\pi} 2\sin^2 nx\, dx$$

$$= \int_0^{\pi} \left[1 - \cos(2nx)\right] dx = \pi .$$

Therefore, we write

$$\int_{-\pi}^{+\pi} \sin mx \sin nx\, dx = \pi\delta_{mn}. \tag{11.1}$$

(ii) For $m \neq n$, we have

$$\int_{-\pi}^{+\pi} \cos mx \cos nx\, dx = \int_0^{\pi} 2\cos mx \cos nx\, dx$$

$$= \int_0^{\pi} \cos\{(m+n)x\}dx + \int_0^{\pi} \cos\{(m-n)x\}dx$$

$$= \frac{1}{m+n}\left[\sin\{m+n\}x\right]_0^{\pi} + \frac{1}{m-n}\left[\sin\{(m-n)x\}\right]_0^{\pi} = 0.$$

For $m = n$, we have

$$\int_{-\pi}^{+\pi} \cos^2 nx\, dx = \int_0^{\pi} 2\cos^2 nx\, dx$$

$$= \int_0^{\pi} \{1 + \cos(2nx)\} dx = \pi .$$

Therefore, we write

$$\int_{-\pi}^{+\pi} \cos mx \cos nx\, dx = \pi\delta_{mn}. \tag{11.2}$$

(iii) For $m \neq n$, we have

$$\int_{-\pi}^{+\pi} \sin mx \cos nx\, dx = \frac{1}{2}\int_{-\pi}^{+\pi} \left[\sin\{(m+n)x\} + \sin\{(m-n)x\}\right] dx$$

$$= -\frac{1}{2} \cdot \frac{1}{m+n} \left[\cos\{(m+n)x\} \right]_{-\pi}^{+\pi} - \frac{1}{2} \cdot \frac{1}{m-n} \left[\cos\{(m-n)x\} \right]_{-\pi}^{+\pi} = 0.$$

For $m = n$, we have

$$\int_{-\pi}^{+\pi} \sin nx \cos nx \, dx = \frac{1}{2} \int_{-\pi}^{+\pi} \sin(2nx) dx$$

$$= -\frac{1}{2} \cdot \frac{1}{2n} \left[\cos(2nx) \right]_{-\pi}^{+\pi} = 0. \quad (11.3)$$

Therefore, we write

$$\int_{-\pi}^{+\pi} \sin mx \cos nx \, dx = 0.$$

Thus, the above three equations are proved (or established). Similarly, one can prove the following three equations

$$\int_{0}^{2\pi} \sin mx \sin nx \, dx = \pi \, \delta_{mn}, \quad (11.4)$$

$$\int_{0}^{2\pi} \cos mx \cos nx \, dx = \pi \, \delta_{mn}, \quad (11.5)$$

$$\int_{0}^{2\pi} \sin mx \cos nx \, dx = 0. \quad (11.6)$$

11.2 PERIODIC FUNCTION AND FOURIER SERIES IN TRIGONOMETRIC FORM

If a function $f(x)$ is such that $f(x) = f(P + x)$ for any positive constant P, then the function $f(x)$ is said to be a periodic function and the positive constant P is said to be the period of the function $f(x)$. All the fundamental trigonometric functions such as $\sin x$, $\cos x$, $\tan x$, etc. are periodic functions having the same period 2π.

Around the year 1822, the French physicist-cum-mathematician, Jean-Baptiste Joseph Fourier (21 March, 1768 - 16 May, 1830), while studying the theory of heat conduction, introduced the technique of representing a periodic function by a series containing sine and cosine terms. Later on, such a series was named after him and it became to be known as Fourier series. Thus, a ***Fourier series*** is defined as a representation of a periodic function by a series containing sine and cosine terms.

The Fourier series of a periodic function $f(x)$ having a period 2π can be represented by a series of the form

$$f(x)=\frac{a_0}{2}+\sum_{n=1}^{\infty}a_n\cos nx+\sum_{n=1}^{\infty}b_n\sin nx \tag{11.7}$$

where the coefficients (a_n and b_n) are known as Fourier coefficients and are to be determined through the orthogonality conditions of sine and cosine functions.

To determine the Fourier coefficients:

(i) To find a_0:

If the interval is $[-\pi, \pi]$:

Integrating both sides of Eq. (11.7) with respect to x in the interval from $x = -\pi$ to $x = +\pi$, we have

$$\int_{-\pi}^{+\pi} f(x)dx$$

$$=\int_{-\pi}^{+\pi}\frac{a_0}{2}dx+\sum_{n=1}^{\infty}a_n\int_{-\pi}^{+\pi}\cos nx\,dx+\sum_{n=1}^{\infty}b_n\int_{-\pi}^{+\pi}\sin nx\,dx$$

$$=\pi\, a_0\,.$$

Therefore, we have

$$a_0=\frac{1}{\pi}\int_{-\pi}^{+\pi}f(x)dx. \tag{11.8}$$

If the interval is $[0, 2\pi]$:

Integrating both sides of Eq. (11.7) with respect to x in the interval from $x = 0$ to $x = 2\pi$, we obtain

$$a_0=\frac{1}{\pi}\int_{0}^{2\pi}f(x)dx. \tag{11.8a}$$

(ii) To find $a_n\,(n\geq 1)$

If the interval is $\left[-\pi,\pi\right]$:

Multiplying both sides of Eq. (11.7) by $\cos mx$ and then integrating the results with respect to x in the interval from $x=-\pi$ to $x=+\pi$, we obtain

$$\int_{-\pi}^{+\pi} f(x)\cos mx\, dx = \frac{a_0}{2}\int_{-\pi}^{+\pi} \cos mx\, dx + \sum_{n=1}^{\infty} a_n \int_{\pi}^{+\pi} \cos mx \cos nx dx$$

$$+\sum_{n=1}^{\infty} b_n \int_{-\pi}^{+\pi} \cos mx \sin nx\, dx$$

$$= 0 + \pi\, a_m + 0.$$

Therefore, we have

$$a_m = \frac{1}{\pi}\int_{-\pi}^{+\pi} f(x)\cos mx\, dx.$$

Renaming the subscript, we write

$$a_n = \frac{1}{\pi}\int_{-\pi}^{+\pi} f(x)\cos nx\, dx. \tag{11.9}$$

If the interval is $[0, 2\pi]$:

Multiplying both sides of Eq. (11.7) by cos mx and then integrating the results with respect to x in the interval from $x = 0$ to $x = 2\pi$, we obtain

$$a_m = \frac{1}{\pi}\int_{0}^{2\pi} f(x)\cos mx\, dx.$$

Renaming the subscript, we write

$$a_n = \frac{1}{\pi}\int_{0}^{2\pi} f(x)\cos nx\, dx. \tag{11.9a}$$

(iii) To find $b_n\,(n \geq 1)$:

If the interval is $[-\pi, \pi]$:

Multiplying both sides of Eq. (11.7) by sin mx and then integrating the results with respect to x in the interval from $x = -\pi$ to $x = +\pi$, we obtain

$$\int_{-\pi}^{+\pi} f(x)\sin mx\, dx = \frac{a_0}{2}\int_{\pi}^{+\pi} \sin mx\, dx + \sum_{n=1}^{\infty} a_n \int_{-\pi}^{+\pi} \sin mx \cos nx\, dx$$

$$+\sum_{n=1}^{\infty} b_n \int_{-\pi}^{+\pi} \sin mx \sin nx\, dx$$

$$= 0 + 0 + \pi\, b_m.$$

Therefore, we have

$$b_m = \frac{1}{\pi}\int_{-\pi}^{+\pi} f(x)\sin mx\, dx.$$

Renaming the subscript, we write

$$b_n = \frac{1}{\pi}\int_{-\pi}^{+\pi} f(x)\sin nx\, dx. \tag{11.10}$$

If the interval is $[0, 2\pi]$:

Multiplying both sides of Eq. (11.7) by sin mx and then integrating the results with respect to x in the interval from $x = 0$ to $x = 2\pi$, we obtain

$$b_m = \frac{1}{\pi}\int_{0}^{2\pi} f(x)\sin mx\, dx.$$

Renaming the subscript, we write

$$b_n = \frac{1}{\pi}\int_{0}^{2\pi} f(x)\sin nx\, dx. \tag{11.10a}$$

11.3 FOURIER SERIES IN GENERALIZED INTERVAL

In many physical problems, the periods of relevant functions are not always 2π. It is, therefore, necessary to develop a Fourier series that will be valid over a wide range of periods and intervals. Here, let us obtain Fourier series of a periodic function $f(x)$ having a period $2L$ in an interval $[-L, L]$.

Here, we see that $2L$ is the interval of the variable x.

Therefore, 1 is the interval of the variable $\frac{x}{2L}$, and 2π is the interval of the variable $\frac{2\pi x}{2L} = \frac{\pi x}{L} = z$ (say).

Now, we have $x = \frac{Lz}{\pi}$.

Thus, the given function $f(x)$ of period $2L$ becomes the function

$$f(x) = f\left(\frac{Lz}{\pi}\right) = F(z) \text{ (say) of period } 2\pi.$$

Now, the function $F(z)$ can be represented by the Fourier series

$$F(z) = \frac{a_0}{2} + \sum_{n=1}^{\infty} a_n \cos nz + \sum_{n=1}^{\infty} b_n \sin nz \tag{11.11}$$

where

$$a_0 = \frac{1}{\pi}\int_{-\pi}^{+\pi} F(z)\, dz, \tag{11.12}$$

$$a_n = \frac{1}{\pi}\int_{-\pi}^{+\pi} F(z)\cos nz\,dz \tag{11.13}$$

and
$$b_n = \frac{1}{\pi}\int_{-\pi}^{+\pi} F(z)\sin nz\,dz. \tag{11.14}$$

Recalling that

$$F(z) = f(x),\ z = \frac{\pi x}{L} \text{ and } dz = \frac{\pi}{L}dx,$$

equations (11.11) to (11.14) yield

$$f(x) = \frac{a_0}{2} + \sum_{n=1}^{\infty} a_n \cos\left(\frac{n\pi x}{L}\right) + \sum_{n=1}^{\infty} b_n \sin\left(\frac{n\pi x}{L}\right) \tag{11.15}$$

where
$$a_0 = \frac{1}{L}\int_{-L}^{+L} f(x)dx, \tag{11.16}$$

$$a_n = \frac{1}{L}\int_{-L}^{+L} f(x)\cos\left(\frac{n\pi x}{L}\right)dx \tag{11.17}$$

and
$$b_n = \frac{1}{L}\int_{-L}^{+L} f(x)\sin\left(\frac{n\pi x}{L}\right)dx. \tag{11.18}$$

If the interval is from $x = 0$ to $x = 2L$, the period being $2L$ as above, the form of the Fourier series is still represented by Eq. (11.15) but the Fourier coefficients in this case become

$$a_0 = \frac{1}{L}\int_{0}^{2L} f(x)dx, \tag{11.19}$$

$$a_n = \frac{1}{L}\int_{0}^{2L} f(x)\cos\left(\frac{n\pi x}{L}\right)dx \tag{11.20}$$

and
$$b_n = \frac{1}{L}\int_{0}^{2L} f(x)\sin\left(\frac{n\pi x}{L}\right)dx. \tag{11.21}$$

It is seen that in finding the Fourier coefficients, the limits of integration are the end points of the specified interval and the multiplying factor in front of the integral is the reciprocal of half the period.

11.4 FOURIER SINE AND COSINE SERIES

Before we study the Fourier sine and cosine series, let us recapitulate the ideas of even and odd functions. A function $f(x)$ is said to be an even function if $f(-x) = f(x)$. The graph of an even function is symmetric. A function $f(x)$ is said to be an odd function if $f(-x) = -f(x)$.

If a function $f(x)$ is an even function, it has the property that

$$\int_{-a}^{+a} f(x)\,dx = 2\int_{0}^{a} f(x)\,dx.$$

If a function $f(x)$ is an odd function, it has the property that

$$\int_{-a}^{+a} f(x)\,dx = 0.$$

It can be easily seen that the product of two even functions is again an even function, the product of two odd functions is an even function, but the product of an even function and an odd function is an odd function. There are also functions which are neither even nor odd. A simple example is the exponential function $f(x) = e^x$.

Now, let us come to the main topic. Let us recall that the Fourier series of a function $f(x)$ having a period $2L$ in the interval $[-L, +L]$ is represented by

$$f(x) = \frac{a_0}{2} + \sum_{n=1}^{\infty} a_n \cos\left(\frac{n\pi x}{L}\right) + \sum_{n=1}^{\infty} b_n \sin\left(\frac{n\pi x}{L}\right)$$

where $$a_0 = \frac{1}{L}\int_{-L}^{+L} f(x)\,dx,$$

$$a_n = \frac{1}{L}\int_{-L}^{+L} f(x)\cos\left(\frac{n\pi x}{L}\right)dx$$

and $$b_n = \frac{1}{L}\int_{-L}^{+L} f(x)\sin\left(\frac{n\pi x}{L}\right)dx.$$

(i) If $f(x)$ is an odd function, we will have

$$a_0 = 0,\ a_n = 0.$$

Thus, the above Fourier series will contain only sine terms and it will be

$$f(x) = \sum_{n=1}^{\infty} b_n \sin\left(\frac{n\pi x}{L}\right). \tag{11.22}$$

The Fourier series of an odd function is called a ***Fourier sine series*** as it contains only sine terms.

(ii) If $f(x)$ is an even function, we will have

$$b_n = 0.$$

Thus, the above Fourier series will contain only the constant term and the cosine terms and it will be

$$f(x) = \frac{a_0}{2} + \sum_{n=1}^{\infty} a_n \cos\left(\frac{n\pi x}{L}\right). \tag{11.23}$$

The Fourier series of an even function is called a ***Fourier cosine series*** as it contains only cosine terms and a constant.

11.5 HALF RANGE FOURIER SERIES

In many physical and engineering problems, there is a practical need of applying Fourier series to functions which are defined merely on some finite interval. It is often required to obtain Fourier series of a function $f(x)$ over an interval $0 < x < L$, which is half the period. Whatever may be the nature of the function outside the interval $0 < x < L$, we may redefine the function to cover a range $-L < x < L$ so that the newly defined function is periodic (having a period $2L$) and is either even or odd. If the redefined periodic function is an even function, it is said to be even periodic extension of $f(x)$ and it will consist of only cosine terms (and a constant term). If the redefined periodic function is an odd function, it is said to be an odd periodic extension of $f(x)$ and it will consist of only sine terms.

In the interval $0 < x < L$, the newly defined function is identical to the original function.

In the case of **even periodic extension** of a function $f(x)$, $0 < x < L$, the redefined periodic function in the extended interval $[-L, L]$ is generally taken as

$$g(x) = \begin{cases} f(-x), & \text{for } -L < x < 0, \\ f(x) & \text{for } 0 < x < L. \end{cases} \tag{11.24}$$

In this case, the Fourier coefficients are given by

$$a_0 = \frac{1}{L}\int_{-L}^{+L} g(x)dx = \frac{2}{L}\int_0^L g(x)dx = \frac{2}{L}\int_0^L f(x)dx \tag{11.25}$$

as $g(x)$ is an even function and $g(x) = f(x)$ in the interval $0 < x < L$;

$$a_n = \frac{1}{L}\int_{-L}^{+L} g(x)\cos\left(\frac{n\pi x}{L}\right)dx = \frac{2}{L}\int_0^L g(x)\cos\left(\frac{n\pi x}{L}\right)dx$$

$$= \frac{2}{L}\int_0^L f(x)\cos\left(\frac{n\pi x}{L}\right)dx, \tag{11.26}$$

$$b_n = \frac{1}{L}\int_{-L}^{+L} g(x)\sin\left(\frac{n\pi x}{L}\right)dx = 0. \tag{11.27}$$

Thus, the Fourier series of the even periodic extension of $f(x)$ is written as

$$g(x) = \frac{a_0}{2} + \sum_{n=1}^{\infty} a_n \cos\left(\frac{n\pi x}{L}\right) \tag{11.28}$$

where a_0 and a_n are given by Eqs. (11.25) and (11.26).

In the case of **odd periodic extension** of a function $f(x)$, $0 < x < L$, the redefined periodic function in the extended interval $[-L, L]$ is generally taken as

$$g(x) = \begin{cases} -f(x) & \text{for } -L < x < 0, \\ f(x) & \text{for } 0 < x < L. \end{cases} \tag{11.29}$$

In the case of odd periodic extension, the Fourier coefficients are given by

$$a_0 = \frac{1}{L}\int_{-L}^{+L} g(x)dx = 0 \quad (\text{since } g(x) \text{ is odd function}), \tag{11.30}$$

$$a_n = \frac{1}{L}\int_{-L}^{+L} g(x)\cos\left(\frac{n\pi x}{L}\right)dx = 0, \tag{11.31}$$

$$b_n = \frac{1}{L}\int_{-L}^{+L} g(x)\sin\left(\frac{n\pi x}{L}\right)dx = \frac{2}{L}\int_{0}^{L} g(x)\sin\left(\frac{n\pi x}{L}\right)dx$$

$$= \frac{2}{L}\int_{0}^{L} f(x)\sin\left(\frac{n\pi x}{L}\right)dx. \tag{11.32}$$

Thus, the Fourier series of the odd periodic extension of $f(x)$ is written as

$$g(x) - \sum_{n=1}^{\infty} b_n \sin\left(\frac{n\pi x}{L}\right) \tag{11.33}$$

where b_n are given by Eq. (11.32).

The Fourier series representing such periodic extension of a function is known as half range Fourier series.

11.6 DIRICHLET'S CONDITIONS AND DIRICHLET'S THEOREM

For a function to have a valid Fourier series in a given interval, the German mathematician Peter Gustav Lejeune Dirichlet (February 13, 1805 – May 5, 1859), introduced the following conditions.

(i) the function must be periodic outside the given interval,

(ii) it must be bounded,

(iii) the function must be single-valued,

(iv) it may have only a finite number of finite disconuities, piecewise continuous, and

(v) the function has at most a finite number of maxima and minima.

The above conditions are known as ***Dirichlet's conditions***.

A function $f(x)$ is said to be bounded if the inequaliy $|f(x)| \leq M$ holds for some constant M for all values of x. A function $f(x)$ is said to be a single-valued function if it has only one value for a given value of x. A function $f(x)$ is said to be continuous at a point $x = a$ if $f(a)$ do exist and $\lim_{x \to a} f(x) = f(a)$.

Alternatively, a function $f(x)$ is said to be continuous at a point $x = a$ if the left and the right limits of the function at that point are aqual. A function $f(x)$ is said to be piecewise continuous in a finite interval $a \leq x \leq b$ if $f(x)$ is defined in that interval and is such that the interval can be subdivided into finitely many sub-intervals, in each of which $f(x)$ is continuous having finite limits as x approaches either end point of the sub-interval. A function $f(x)$ is said to be discontinuous at a point $x = a$ if the left hand limit of $f(x)$ as $x \to a$ is not equal to the right hand limit there.

Dirichlet's Theorem

In an interval $[-\pi, \pi]$ if a function $f(x)$ satisfies the above Dirichlet's conditions, then the partial sum

$$S_p(x) = \frac{a_0}{2} + \sum_{n=1}^{p} [a_n \cos nx + b_n \sin nx] \tag{11.34}$$

converges

(i) to $f(x)$ as $p \to \infty$ at values of x for which the function $f(x)$ is continuous

(ii) to $\frac{1}{2}[f(x+0) + f(x-0)]$ at points of discontinuity.

Here, the quantities $f(x + 0)$ and $f(x - 0)$ refer to limits from the right and from the left, respectively.

11.7 CONVERGENCE OF FOURIER SERIES

The Fourier series of a function $f(x)$ in an interval $[0, 2\pi]$ is given by

$$f(x) = \frac{a_0}{2} + \sum_{n=1}^{\infty} (a_n \cos nx + b_n \sin nx)$$

where the coefficients are given by

$$a_0 = \frac{1}{\pi} \int_0^{2\pi} f(x) dx,$$

$$a_n = \frac{1}{\pi}\int_0^{2\pi} f(x)\cos nx\,dx,$$

$$b_n = \frac{1}{\pi}\int_0^{2\pi} f(x)\sin nx\,dx.$$

Let us consider the partial sum

$$S_p(x) = \frac{a_0}{2} + \sum_{n=1}^{p}(a_n\cos nx + b_n\sin nx).$$

Substitution of the above coefficients yields

$$S_p(x) = \frac{1}{2\pi}\int_0^{2\pi} f(u)\,du + \sum_{n=1}^{p}\frac{1}{\pi}\int_0^{2\pi}\{f(u)\cos nu\,du\}\cos nx + \frac{1}{\pi}\int_0^{2\pi}\{f(u)\sin nu\,du\}\sin nx\Bigg]$$

where the integration variables are re-named. Thus, we write

$$S_p(x) = \frac{1}{\pi}\int_0^{2\pi}\left[\frac{1}{2} + \sum_{n=1}^{p}\{\cos nx\cos nu + \sin nx\sin nu\}\right] f(u)\,du$$

$$= \frac{1}{\pi}\int_0^{2\pi}\left[\frac{1}{2} + \sum_{n=1}^{p}\cos\{n(x-u)\}\right] f(u)\,du. \qquad (11.35)$$

Using the formula (to be proved next) that

$$\frac{1}{2} + \sum_{n=1}^{p}\cos nt = \frac{\sin\left\{\left(p+\frac{1}{2}\right)t\right\}}{2\sin\frac{t}{2}},$$

we write Eq. (11.35) as

$$S_p(x) = \frac{1}{2\pi}\int_0^{2\pi}\frac{\sin\left\{\left(p+\frac{1}{2}\right)(x-u)\right\}}{\sin\left(\frac{x-u}{2}\right)} f(u)\,du.$$

As the integrand is periodic here, we can write the above equation as

$$S_p(x) = \frac{1}{2\pi}\int_{-\pi+x}^{\pi+x}\frac{\sin\left\{\left(p+\frac{1}{2}\right)(x-u)\right\}}{\sin\left(\frac{x-u}{2}\right)} f(u)\,du. \qquad (11.36)$$

Now, let us introduce a new variable

$$\theta = -\frac{(x-u)}{2}.$$

Therefore, we have $u = 2\theta + x$ and $du = 2d\theta$.
Then Eq. (11.36) can be written as

$$S_p(x) = \frac{1}{2\pi}\int_{-\frac{\pi}{2}}^{\frac{\pi}{2}} \frac{\sin\left\{\left(p+\frac{1}{2}\right)(-2\theta)\right\}}{\sin(-\theta)} f(x+2\theta)2d\theta$$

$$= \frac{1}{\pi}\int_{-\frac{\pi}{2}}^{\frac{\pi}{2}} \frac{\sin\{(2p+1)\theta\}}{\sin\theta} f(x+2\theta)d\theta$$

$$= \frac{1}{\pi}\left[\int_{-\frac{\pi}{2}}^{0} \frac{\sin\{(2p+1)\theta\}}{\sin\theta} f(x+2\theta)d\theta + \int_{0}^{\frac{\pi}{2}} \frac{\sin\{(2p+1)\theta\}}{\sin\theta} f(x+2\theta)d\theta\right].$$

Replacing θ by $(-\theta)$ in the first integral, we write

$$S_p(x) = \frac{1}{\pi}\left[-\int_{\frac{\pi}{2}}^{0} \frac{\sin(2p+1)\theta}{\sin\theta} f(x-2\theta)d\theta + \int_{0}^{\frac{\pi}{2}} \frac{\sin\{(2p+1)\theta\}}{\sin\theta} f(x+2\theta)d\theta\right]$$

$$= \int_{0}^{\frac{\pi}{2}} \{f(x-2\theta)+f(x+2\theta)\}\frac{\sin\{(2p+1)\theta\}}{\pi\sin\theta}d\theta.$$

In the limit $p \to \infty$, we have

$$\lim_{p\to\infty} S_p(x) = \int_{0}^{\frac{\pi}{2}} \{f(x+2\theta)+f(x-2\theta)\}\left[\lim_{p\to\infty}\frac{\sin\{(2p+1)\theta\}}{\pi\sin\theta}\right]d\theta$$

$$= \int_{0}^{\frac{\pi}{2}} \{f(x+2\theta)+f(x-2\theta)\}\delta(\theta)d\theta$$

where the quantity within the square brackets is substituted by the Dirac-Delta function $\delta(\theta)$.

As $\delta(\theta)$ is an even function, the above integrand is an even function.

Hence, we have

$$\lim_{p\to\infty} S_p(x) = \frac{1}{2}\int_{-\frac{\pi}{2}}^{\frac{\pi}{2}} \{f(x+2\theta)+f(x-2\theta)\}\delta(\theta)\,d\theta$$

$$= \frac{1}{2}\int_{-\infty}^{+\infty} \{f(x+2\theta)+f(x-2\theta)\}\delta(\theta)\,d\theta$$

$$= \frac{1}{2}\{f(x+0)+f(x-0)\}. \tag{11.37}$$

Here, $f(x + 0)$ and $f(x - 0)$ are the limits of the function $f(x)$ from the right and from the left of the point x. If the function $f(x)$ is continuous at the point x, then both the above limits become equal to $f(x)$

Hence, for such a case, Eq. (11.37) yields

$$\lim_{p\to\infty} S_p(x) = f(x).$$

Note: To prove that $\frac{1}{2}+\sum_{n=1}^{p}\cos nt = \frac{\sin\left\{\left(p+\frac{1}{2}\right)t\right\}}{2\sin\frac{t}{2}}$:

We know that

$$\frac{1}{2}+\sum_{n=1}^{p}\cos nt = \frac{1}{2}+\frac{1}{2}\sum_{n=1}^{p}\left[e^{int}+e^{-int}\right]$$

$$= \frac{1}{2}+\frac{1}{2}\left[e^{it}+e^{2it}+e^{3it}+\cdots+e^{ipt}\right]+\frac{1}{2}\left[e^{-it}+e^{-2it}+e^{-3it}+\cdots+e^{-ipt}\right]$$

$$= \frac{1}{2}+\frac{1}{2}\cdot\frac{e^{it}-e^{i(p+1)t}}{1-e^{it}}+\frac{1}{2}\cdot\frac{e^{-it}-e^{-i(p+1)t}}{1-e^{-it}}$$

$$= \frac{1}{2}+\frac{1}{2}\left[\frac{e^{it}-1-e^{i(p+1)t}+e^{ipt}+e^{-it}-1-e^{-i(p+1)t}+e^{-ipt}}{2-\left(e^{it}+e^{-it}\right)}\right]$$

$$= \frac{1}{2}+\frac{\left[\frac{\left(e^{it}+e^{-it}\right)}{2}+\frac{\left(e^{ipt}+e^{-ipt}\right)}{2}-\frac{\left\{e^{i(p+1)t}+e^{-i(p+1)t}\right\}}{2}-1\right]}{2-\left(e^{it}+e^{-it}\right)}$$

$$=\frac{1}{2}+\frac{\cos t+\cos pt-\cos\{(p+1)t\}-1}{2-2\cos t}$$

$$=\frac{\cos pt-\cos\{(p+1)t\}}{2(1-\cos t)}$$

$$=\frac{\cos pt-\cos pt\cos t+\sin pt\sin t}{2(1-\cos t)}$$

$$=\frac{(1-\cos t)\cos pt+\sin pt\sin t}{2(1-\cos t)}$$

$$=\frac{2\sin^2\left(\frac{t}{2}\right)\cos pt+2\sin\frac{t}{2}\cos\frac{t}{2}\sin pt}{4\sin^2\left(\frac{t}{2}\right)}$$

$$=\frac{\sin\frac{t}{2}\cos pt+\cos\frac{t}{2}\sin pt}{2\sin\frac{t}{2}}=\frac{\sin\left\{\left(p+\frac{1}{2}\right)t\right\}}{2\sin\frac{t}{2}}.$$

Hence, proved.

11.8 PARSEVAL'S FORMULA IN TRIGONOMETRIC FORM

The French mathematician, Marc-Antoine Parseval des Chenes (April 27, 1755 – August 16, 1836), proposed a formula connecting the average of the square of a function to the Fourier coefficients of its Fourier series representation. This formula is known as ***Parseval's*** formula. Let us consider the Fourier series of a function $f(x)$ given by

$$f(x)=\frac{a_0}{2}+\sum_{n=1}^{\infty}\left[a_n\cos\left(\frac{n\pi x}{L}\right)+b_n\sin\left(\frac{n\pi x}{L}\right)\right]$$

in the interval $[-L, L]$.

Here, a_0, a_n, b_n are the Fourier coeffcients which can be determined by the use of orthonormality of sine and cosine functions.

Now, the average of the square of the function $f(x)$ over the interval $[-L, L]$ is

$$\frac{1}{2L}\int_{-L}^{L}[f(x)]^2dx=\frac{1}{2L}\int_{-L}^{L}\frac{a_0^2}{4}dx+\frac{1}{2L}\sum_{n=1}^{\infty}a_n^2\int_{-L}^{L}\cos^2\left(\frac{n\pi x}{L}\right)dx+\frac{1}{2L}\sum_{n=1}^{\infty}b_n^2\int_{-L}^{L}\sin^2\left(\frac{n\pi x}{L}\right)dx$$

(other terms vanish by orthogonality conditions).

Therefore, we write

$$\frac{1}{2L}\int_{-L}^{L}[f(x)]^2 dx = \frac{a_0^2}{4} + \sum_{n=1}^{\infty}\frac{a_n^2}{2} + \sum_{n=1}^{\infty}\frac{b_n^2}{2}$$

or

$$\frac{1}{L}\int_{-L}^{L}[f(x)]^2 dx = \frac{a_0^2}{2} + \sum_{n=1}^{\infty}(a_n^2 + b_n^2). \tag{11.38}$$

Equation (11.38) is known as Parseval's formula.

11.9 FOURIER INTEGRAL IN TRIGONOMETRIC FORM

The Fourier series of a periodic function $f(x)$ of period $2L$ in an interval $[-L, L]$ is given by

$$f(x) = \frac{a_0}{2} + \sum_{n=1}^{\infty}\left[a_n \cos\left(\frac{n\pi x}{L}\right) + b_n \sin\left(\frac{n\pi x}{L}\right)\right]$$

where

$$a_0 = \frac{1}{L}\int_{-L}^{L} f(x)dx,$$

$$a_n = \frac{1}{L}\int_{-L}^{L} f(x)\cos\left(\frac{n\pi x}{L}\right)dx,$$

$$b_n = \frac{1}{L}\int_{-L}^{L} f(x)\sin\left(\frac{n\pi x}{L}\right)dx.$$

Substituting these Fourier coefficients into the Fourier series and renaming the integration variables in the integrals of the Fourier coefficients in order not to confuse with the original variables, we write

$$f(x) = \frac{1}{2L}\int_{L}^{L} f(t)dt + \frac{1}{L}\sum_{n=1}^{\infty}\left[\int_{-L}^{L} f(t)\cos\left(\frac{n\pi t}{L}\right)\cos\left(\frac{n\pi x}{L}\right)dt + \int_{-L}^{L} f(t)\sin\left(\frac{n\pi t}{L}\right)\sin\left(\frac{n\pi x}{L}\right)dt\right]$$

$$= \frac{1}{2L}\int_{-L}^{L} f(t)dt + \frac{1}{L}\sum_{n=1}^{\infty}\int_{-L}^{L}\left[\cos\left(\frac{n\pi x}{L}\right)\cos\left(\frac{n\pi t}{L}\right) + \sin\left(\frac{n\pi x}{L}\right)\sin\left(\frac{n\pi t}{L}\right)\right] f(t)dt$$

$$= \frac{1}{2L}\int_{-L}^{L} f(t)\,dt + \frac{1}{L}\sum_{n=1}^{\infty}\int_{-L}^{L} \cos\left\{\frac{n\pi}{L}(t-x)\right\} f(t)\,dt$$

$$= \frac{1}{2\pi}\cdot\frac{\pi}{L}\int_{-L}^{L} f(t)\,dt + \frac{1}{\pi}\sum_{n=1}^{\infty}\frac{\pi}{L}\int_{-L}^{L} \cos\left\{\frac{n\pi}{L}(t-x)\right\} f(t)\,dt. \tag{11.39}$$

If we put $u = \frac{n\pi}{L}$, then we have $\Delta u = \frac{\pi}{L}\Delta n = \frac{\pi}{L}$.

Then, Eq. (11.39) can be written as

$$f(x) = \frac{1}{2\pi}\Delta u\int_{-L}^{L} f(t)\,dt + \frac{1}{\pi}\sum_{u=\pi/L}^{\infty}\Delta u\int_{-L}^{L}\cos\{u(t-x)\}\, f(t)\,dt. \tag{11.40}$$

Now, if $L \to \infty$ we have $\Delta u \to 0$.

In such case, the first term of Eq. (11.40) becomes

$$\lim_{\substack{L\to\infty \\ \Delta u\to 0}}\left[\frac{1}{2\pi}\Delta u\int_{-L}^{L} f(t)\,dt\right] = 0.$$

Also, we have

$$\lim_{L\to\infty}\sum_{u=\pi/L}^{\infty}\Delta u = \int_{0}^{\infty} du.$$

Hence, in the limit $L \to \infty$ we rewrite Eq. (11.40) as

$$f(x) = \frac{1}{\pi}\int_{u=0}^{\infty}\int_{t=-\infty}^{\infty} f(t)\cos\{u(t-x)\}\,dt\,du. \tag{11.41}$$

Equation (11.41) is known as ***Fourier integral*** in trigonometric form.

11.10 FOURIER SINE AND COSINE INTEGRALS

Equation (11.41) can be written as

$$f(x) = \frac{1}{\pi}\int_{u=0}^{\infty}\int_{t=-\infty}^{\infty} f(t)(\cos ut\cos ux + \sin ut\sin ux)\,dt\,du$$

$$= \frac{1}{\pi}\int_{u=0}^{\infty}\int_{t=-\infty}^{\infty} f(t)\cos ut\cos ux\,dt\,du$$

$$+ \frac{1}{\pi}\int_{u=0}^{\infty}\int_{t=-\infty}^{\infty} f(t)\sin ut\sin ux\,dt\,du \tag{11.42}$$

CASE I: If $f(t)$ is an odd function

In this case, the integrand in the first term of Eq. (11.42) is an odd function of t and hence the integral vanishes.

Therefore, we have

$$f(x)=\frac{1}{\pi}\int_{u=0}^{\infty}\int_{t=-\infty}^{\infty} f(t)\sin ut \sin ux \, dt \, du.$$

Here, the integrand is an even function of t and we write

$$f(x)=\frac{2}{\pi}\int_{0}^{\infty}\int_{0}^{\infty} f(t)\sin ut \sin ux \, dt \, du. \tag{11.43}$$

Relation (11.43) is called ***Fourier sine integral.***

CASE II: If $f(t)$ is an even function:

In this case, the integrand in the second term of Eq. (11.42) is an odd function of t and hence the integral vanishes.

Therefore, we have

$$f(x)=\frac{1}{\pi}\int_{u=0}^{\infty}\int_{t=-\infty}^{\infty} f(t)\cos ut \cos ux \, dt \, du$$

$$=\frac{2}{\pi}\int_{0}^{\infty}\int_{0}^{\infty} f(t)\cos ut \cos ux \, dt \, du \tag{11.44}$$

as the integrand is an even function of t.

Equation (11.44) is called ***Fourier cosine integral***.

11.11 GIBBS PHENOMENON

In mathematics, the ***Gibbs phenomenon*** refers to a manner in which the Fourier series of a periodic function overshoots or undershoots the value of the function near a discontinuity. This phenomenon was first reported by the English mathematician, Henry Wilbraham (July 25, 1825 – February 13, 1883). It was also investigated by the German mathematician, Paul David Gustav du Bois Reymond (December 2, 1831 – April 7, 1889). But their works remained unnoticed by majority of the people working in physics and mathematics. It was rediscovered by the American physicist, chemist-cum-mathematician Josiah Willard Gibbs (February 11, 1839 – April 28, 1903). To illustrate the penomenon, let us consider the behaviour of the Fourier series that represents a peiodic square wave.

The relevant function, known as the periodic step function, is defined by

$$f(x)=\begin{cases}-A & \text{for } -\dfrac{L}{2}<x<0, \\ A & \text{for } \ 0<x<\dfrac{L}{2}\end{cases} \tag{11.45}$$

with $f(x) = f(x + L)$.

As this function is an odd function, its Fourier series representation will be a sine series represented by

$$f(x)=\sum_{n=1}^{\infty} b_n \sin\left(\frac{2n\pi x}{L}\right) \tag{11.46}$$

where
$$b_n = \frac{2}{L}\int_{-\frac{L}{2}}^{+\frac{L}{2}} f(x)\sin\left(\frac{2n\pi x}{L}\right)dx$$

$$=\frac{2}{L}\left\{\int_{-\frac{L}{2}}^{0} -A\sin\left(\frac{2n\pi x}{L}\right)dx+\int_{0}^{\frac{L}{2}} A\sin\left(\frac{2n\pi x}{L}\right)dx\right\}$$

$$=\frac{2A}{n\pi}\left\{1-(-1)^n\right\}=\begin{cases}0 & \text{for } n=\text{even integer}, \\ \dfrac{4A}{n\pi} & \text{for } n=\text{odd integer}.\end{cases}$$

Substituting these values of b_n into Eq. (11.46), we write

$$f(x)=\frac{4A}{\pi}\left[\sin\left(\frac{2\pi x}{L}\right)+\frac{1}{3}\sin\left(\frac{6\pi x}{L}\right)+\frac{1}{5}\sin\left(\frac{10\pi x}{L}\right)+\ldots\right.$$
$$\left.+\frac{1}{2n-1}\sin\left\{\frac{2(2n-1)\pi x}{L}\right\}+\ldots\infty\right]. \tag{11.47}$$

From Eq. (11.47), the partial sum for n terms is given by

$$s_n(x)=\frac{4A}{\pi}\left[\sin\left(\frac{2\pi x}{L}\right)+\frac{1}{3}\sin\left(\frac{6\pi x}{L}\right)+\frac{1}{5}\sin\left(\frac{10\pi x}{L}\right)+\ldots\right.$$
$$\left.+\frac{1}{2n-1}\sin\left\{\frac{2(2n-1)\pi x}{L}\right\}\right]. \tag{11.48}$$

To find the locations of the maxima and the minima, we are to take the derivative of $s_n(x)$ and equate the result to zero.

Now, we have

$$s_n'(x)=\frac{ds_n(x)}{dx}=\frac{8A}{L}\left[\cos\left(\frac{2\pi x}{L}\right)+\cos\left(\frac{6\pi x}{L}\right)+\cos\left(\frac{10\pi x}{L}\right)+\ldots+\cos\left\{\frac{2(2n-1)\pi x}{L}\right\}\right]$$

$$=\frac{8A}{L}\mathrm{Re}\left[e^{\frac{2\pi xi}{L}}+e^{\frac{6\pi xi}{L}}+e^{\frac{10\pi xi}{L}}+\ldots+e^{\frac{2(2n-1)\pi xi}{L}}\right].$$

Here, Re $f(z)$ stands for the real part of the function $f(z)$.
We write,

$$s_n'(x)=\frac{8A}{L}\mathrm{Re}\left[e^{\frac{2\pi xi}{L}}\left\{1+e^{\frac{4\pi xi}{L}}+e^{\frac{8\pi xi}{L}}+\ldots+e^{\frac{4(n-1)\pi xi}{L}}\right\}\right]$$

$$=\frac{8A}{L}\mathrm{Re}\left[e^{\frac{2\pi xi}{L}}\left\{\frac{1-e^{\frac{4n\pi xi}{L}}}{1-e^{\frac{4\pi xi}{L}}}\right\}\right]$$

$$=\frac{8A}{L}\mathrm{Re}\left[e^{\frac{2\pi xi}{L}}\left\{\frac{e^{\frac{2n\pi xi}{L}}\left(e^{-\frac{2n\pi xi}{L}}-e^{\frac{2n\pi xi}{L}}\right)}{e^{\frac{2\pi xi}{L}}\left(e^{-\frac{2\pi xi}{L}}-e^{\frac{2\pi xi}{L}}\right)}\right\}\right]$$

$$=\frac{8A}{L}\mathrm{Re}\left[e^{\frac{2n\pi xi}{L}}\frac{\sin\left(\frac{2n\pi x}{L}\right)}{\sin\left(\frac{2\pi x}{L}\right)}\right]$$

$$=\frac{8A}{L}\frac{\cos\left(\frac{2n\pi x}{L}\right)\sin\left(\frac{2n\pi x}{L}\right)}{\sin\left(\frac{2\pi x}{L}\right)}$$

$$=\frac{4A}{L}\frac{\sin\left(\frac{4n\pi x}{L}\right)}{\sin\left(\frac{2\pi x}{L}\right)}. \tag{11.49}$$

For maxima and minima, we must have

$$s_n'(x)=0$$

or

$$\sin\left(\frac{4n\pi x}{L}\right)=0$$

or

$$\frac{4n\pi x}{L}=\pi, 2\pi, 3\pi, 4\pi, \ldots (2n-1)\pi.$$

The root $\frac{4n\pi x}{L}=0$ is discarded because, in that case, x must be zero (as n is not equal to zero) and then $\dfrac{\sin\left(\frac{4n\pi x}{L}\right)}{\sin\left(\frac{2\pi x}{L}\right)}$ will become indeterminate.

The maxima and the minima of $s_n(x)$ are identified in accordance with the conditions that its second order derivative $s_n''(x)$ has negative or positive values, respectively at the concerned points.

Differentiating Eq. (11.49), we obtain

$$s_n''(x)=\frac{4A}{L}\cdot\frac{\left[\frac{4n\pi}{L}\cos\left(\frac{4n\pi x}{L}\right)\sin\left(\frac{2\pi x}{L}\right)-\frac{2\pi}{L}\sin\left(\frac{4n\pi x}{L}\right)\cos\left(\frac{2\pi x}{L}\right)\right]}{\sin^2\left(\frac{2\pi x}{L}\right)}. \quad (11.50)$$

$s_n''(x)$ will have negative values at

$$\frac{4n\pi x}{L}=\pi, 3\pi, 5\pi, 7\pi, \ldots (2n-1)\pi.$$

Therefore, the partial sum $s_n(x)$ has maxima at

$$x=\frac{L}{4n}, \frac{3L}{4n}, \frac{5L}{4n}, \ldots \frac{(2n-1)L}{4n}. \quad (11.51)$$

$s_n''(x)$ will have positive values at

$$\frac{4n\pi x}{L}=2\pi, 4\pi, 6\pi, 8\pi, \ldots (2n-2)\pi.$$

Thus, the partial sum $s_n(x)$ has minima at

$$x=\frac{L}{2n}, \frac{2L}{2n}, \frac{3L}{2n}, \ldots \frac{(n-1)L}{2n}. \quad (11.52)$$

From Eq. (11.51), we see that the first maximum of $s_n(x)$ occurs at $x = \frac{L}{4n}$. At this point, Eq. (11.48) gives the maximum value of the partial sum $s_n(x)$ as

$$s_n\left(\frac{L}{4n}\right) = \frac{4A}{\pi}\left[\sin\left(\frac{\pi}{2n}\right) + \frac{1}{3}\sin\left(\frac{3\pi}{2n}\right) + \frac{1}{5}\sin\left(\frac{5\pi}{2n}\right) + \ldots + \frac{1}{2n-1}\sin\left\{\frac{(2n-1)\pi}{2n}\right\}\right]$$

$$= \frac{4A}{\pi}\sum_{r=1}^{n}\frac{1}{2r-1}\sin\left\{\frac{(2r-1)\pi}{2n}\right\}$$

$$= \frac{4A}{\pi}\sum_{r=1}^{n}\frac{\pi}{2n}\frac{2n}{(2r-1)\pi}\sin\left\{\frac{(2r-1)\pi}{2n}\right\}. \tag{11.53}$$

Putting $\xi_r = \dfrac{(2r-1)\pi}{2n}$, (11.54)

we have $\xi_{r-1} = \dfrac{(2r-3)\pi}{2n}$, $\Delta\xi_r = \xi_r - \xi_{r-1} = \dfrac{\pi}{n}$. (11.55)

Substituting Eqs. (11.54) and (11.55) into Eq. (11.53), we obtain

$$s_n\left(\frac{L}{4n}\right) = \frac{2A}{\pi}\sum_{r=1}^{n}\frac{\sin\xi_r}{\xi_r}\Delta\xi_r. \tag{11.56}$$

From Eq. (11.54), we see that in the limit $n \to \infty$,

(i) when $r = 1$, we have $\xi_r = 0$,

(ii) when $r = n$, we have $\xi_r = \pi$.

In the limit $n \to \infty, \Delta\xi_r \to 0$ and, therefore, Eq. (11.56) can be written as

$$s_n\left(\frac{L}{4n}\right) = \frac{2A}{\pi}\int_0^{\pi}\frac{\sin\xi}{\xi}\,d\xi \tag{11.57}$$

where we drop the suffix.

At the point of discontinuity, i.e., at $x = 0$, the magnitude of jump is

$$\left|f(0^+) - f(0^-)\right| = |A - (-A)| = 2A.$$

The fractional overshoot at the first maximum is

$$\frac{1}{2A}\left[s_n\left(\frac{L}{4n}\right)-A\right]=\frac{1}{\pi}\int_0^{\pi}\frac{\sin\xi}{\xi}d\xi-\frac{1}{2}$$

$$=\frac{1}{\pi}\int_0^{\pi}\frac{1}{\xi}\left(\xi-\frac{\xi^3}{3!}+\frac{\xi^5}{5!}-\frac{\xi^7}{7!}+\frac{\xi^9}{9!}-\ldots\infty\right)-\frac{1}{2}$$

$$=\frac{1}{\pi}\int_0^{\pi}\left(1-\frac{\xi^2}{3!}+\frac{\xi^4}{5!}-\frac{\xi^6}{7!}+\frac{\xi^8}{9!}-\ldots\infty\right)d\xi-\frac{1}{2}$$

$$=\frac{1}{\pi}\left[\pi-\frac{\pi^3}{3(3!)}+\frac{\pi^5}{5(5!)}-\frac{\pi^7}{7(7!)}+\frac{\pi^9}{9(9!)}-\ldots\infty\right]-\frac{1}{2}$$

$$=\left[1-\frac{\pi^2}{18}+\frac{\pi^4}{600}-\frac{\pi^6}{35280}+\frac{\pi^8}{3265920}-\ldots\infty\right]-\frac{1}{2}$$

$$\approx 0.0897\approx 9\%. \qquad (11.58)$$

Thus, it turns out that the Fourier series exceeds the height A of the square wave by an amount of about 9 per cent of the jump. More generally, at any jump-point of a piecewise smooth function with a jump of magnitude h, the nth partial sum of the Fourier series (for n very large) will overshoot this jump by approximately ($h \times 0.0897$ at one end and undershoot it by the same amount at the other end; thus the "jump" in the partial Fourier series will be about 18% larger than the original function. The quantity

$$G=\int_0^{\pi}\frac{\sin\theta}{\theta}d\theta=1.8519 \qquad (11.59)$$

is sometimes known as the Wilbraham–Gibbs constant.

11.12 FOURIER SERIES IN COMPLEX FORM

We know that the Fourier series of a periodic function $f(x)$ having a period '$2L$' in the interval $[-L, L]$ is expressed in trigonometric form as

$$f(x)=\frac{a_0}{2}+\sum_{n=1}^{\infty}a_n\cos\left(\frac{n\pi x}{L}\right)+\sum_{n=1}^{\infty}b_n\sin\left(\frac{n\pi x}{L}\right)$$

where $$a_0=\frac{1}{L}\int_{-L}^{+L}f(x)dx,$$

$$a_n = \frac{1}{L}\int_{-L}^{+L} f(x)\cos\left(\frac{n\pi x}{L}\right)dx,$$

$$b_n = \frac{1}{L}\int_{-L}^{+L} f(x)\sin\left(\frac{n\pi x}{L}\right)dx.$$

As trigonometric functions are related to exponential functions through Euler's formulae, the above Fourier series can be expressed in terms of exponential functions. We have

$$\cos\left(\frac{n\pi x}{L}\right) = \frac{1}{2}\left(e^{\frac{in\pi x}{L}} + e^{-\frac{in\pi x}{L}}\right),$$

$$\sin\left(\frac{n\pi x}{L}\right) = \frac{1}{2i}\left(e^{\frac{in\pi x}{L}} - e^{-\frac{in\pi x}{L}}\right).$$

Using these Euler's formulae, we can write the above Fourier series as

$$f(x) = \frac{a_0}{2} + \sum_{n=1}^{\infty}\frac{1}{2}a_n\left(e^{\frac{in\pi x}{L}} + e^{-\frac{in\pi x}{L}}\right) + \sum_{n=1}^{\infty}\frac{1}{2i}b_n\left(e^{\frac{in\pi x}{L}} - e^{-\frac{in\pi x}{L}}\right)$$

$$= \frac{a_0}{2} + \sum_{n=1}^{\infty}\frac{1}{2}(a_n - ib_n)e^{\frac{in\pi x}{L}} + \sum_{n=1}^{\infty}\frac{1}{2}(a_n + ib_n)e^{-\frac{in\pi x}{L}}$$

$$= c_0 + \sum_{n=1}^{\infty} c_n\, e^{\frac{in\pi x}{L}} + \sum_{n=1}^{\infty} c_{-n}\, e^{-\frac{in\pi x}{L}} \tag{11.60}$$

where $$c_0 = \frac{a_0}{2}, \tag{11.61}$$

$$c_n = \frac{1}{2}(a_n - ib_n), \tag{11.62}$$

$$c_{-n} = \frac{1}{2}(a_n + ib_n). \tag{11.63}$$

Equation (11.60) can be written as

$$f(x) = c_0\, e^0 + \sum_{n=1}^{\infty} c_n\, e^{\frac{in\pi x}{L}} + \sum_{n=-\infty}^{-1} c_n\, e^{\frac{in\pi x}{L}}$$

$$= \sum_{n=-\infty}^{+\infty} c_n\, e^{\frac{in\pi x}{L}}. \tag{11.64}$$

Equation (11.64) represents the Fourier series of $f(x)$ in complex form where the coefficients c_n are called the Fourier coefficients.

To find the Fourier Coefficients

Multiplying both sides of Eq. (11.64) by $e^{-im\pi x/L}$ and integrating the result with respect to x within the limit from $x = -L$ to $x = +L$, we obtain

$$\int_{-L}^{L} f(x)e^{-\frac{im\pi x}{L}}\,dx = \sum_{n=-\infty}^{+\infty} c_n \int_{-L}^{L} e^{\frac{i(n-m)\pi x}{L}}\,dx. \tag{11.65}$$

For $n = m$, the right-hand side of Eq. (11.65) becomes

$$\sum_{n=-\infty}^{+\infty} c_n \int_{-L}^{L} e^{\frac{i(n-m)\pi x}{L}}\,dx = c_m \int_{-L}^{L} dx = 2Lc_m.$$

For $n \neq m$, the right-hand side of Eq. (11.65) becomes

$$\sum_{n=-\infty}^{+\infty} c_n \int_{-L}^{L} e^{\frac{i(n-m)\pi x}{L}}\,dx$$

$$= \sum_{n=-\infty}^{+\infty} c_n \int_{-L}^{L} \left[\cos\left\{\frac{(n-m)\pi x}{L}\right\} + i\sin\left\{\frac{(n-m)\pi x}{L}\right\}\right] dx = 0.$$

In general, Eq. (11.65) can be written as

$$\int_{-L}^{L} f(x)e^{-\frac{im\pi x}{L}}\,dx = \sum_{n=-\infty}^{+\infty} 2L\,c_n\,\delta_{mn} = 2Lc_m.$$

Therefore, we obtain

$$c_m = \frac{1}{2L}\int_{-L}^{L} f(x)e^{-\frac{im\pi x}{L}}\,dx.$$

Renaming the subscript, we write

$$c_n = \frac{1}{2L}\int_{-L}^{L} f(x)e^{-\frac{in\pi x}{L}}\,dx \;(n = 0, \pm 1, \pm 2, \pm 3, \ldots \pm\infty). \tag{11.66}$$

Alternatively, the Fourier coefficients (c_n) in complex forms can also be obtained from the coefficients in trigonometric forms as in the following.

Using Eq. (11.61), we have

$$c_0 = \frac{1}{2}a_0 = \frac{1}{2L}\int_{-L}^{L} f(x)dx.$$

Using Eq. (11.62), we have, for $n = 1,2,3,\ldots\infty$,

$$c_n = \frac{1}{2}\left[\frac{1}{L}\int_{-L}^{L} f(x)\cos\left(\frac{n\pi x}{L}\right)dx - i\frac{1}{L}\int_{-L}^{L} f(x)\sin\left(\frac{n\pi x}{L}\right)dx\right]$$

$$= \frac{1}{2L}\int_{-L}^{L} f(x)\left[\cos\left(\frac{n\pi x}{L}\right) - i\sin\left(\frac{n\pi x}{L}\right)\right]dx$$

$$= \frac{1}{2L}\int_{-L}^{L} f(x)e^{-\frac{in\pi x}{L}}\,dx.$$

Similarly, using Eq. (11.63), we have, for n=1,2,3, . . . ∞,

$$c_{-n} = \frac{1}{2L}\int_{-L}^{L} f(x)\,e^{\frac{in\pi x}{L}}\,dx.$$

In general, we write

$$c_n = \frac{1}{2L}\int_{-L}^{L} f(x)e^{-\frac{in\pi x}{L}}\,dx \quad (n = 0, \pm 1, \pm 2, \pm 3, \ldots \pm\infty).$$

11.13 PARSEVAL'S FORMULA IN COMPLEX FORM

We know that in complex form, the Fourier series of a function $f(x)$ is given by

$$f(x) = \sum_{n=-\infty}^{+\infty} c_n\,e^{\frac{in\pi x}{L}}.$$

The square of the modulus of $f(x)$ is given by

$$|f(x)|^2 = f^*(x)f(x) = \left[\sum_{m=-\infty}^{+\infty} c_m^*\,e^{-\frac{im\pi x}{L}}\right]\left[\sum_{n=-\infty}^{+\infty} c_n\,e^{\frac{in\pi x}{L}}\right]$$

$$= \sum_{m=-\infty}^{+\infty}\sum_{n=-\infty}^{+\infty} c_m^*\,c_n\,e^{\frac{i(n-m)\pi x}{L}}.$$

Therefore, we have

$$\frac{1}{2L}\int_{-L}^{L}|f(x)|^2\,dx = \frac{1}{2L}\sum_{m=-\infty}^{+\infty}\sum_{n=-\infty}^{+\infty} c_m^*\,c_n\int_{-L}^{L} e^{\frac{i(n-m)\pi x}{L}}\,dx$$

$$= \sum_{m=-\infty}^{+\infty} \sum_{n=-\infty}^{+\infty} c_m^* c_n \delta_{mn} = \sum_{n=-\infty}^{+\infty} c_n^* c_n$$

$$= \sum_{n=-\infty}^{+\infty} |c_n|^2. \quad (11.67)$$

11.14 FOURIER INTEGRAL IN COMPLEX FORM

The Fourier integral in trigonometric form is

$$f(x) = \frac{1}{\pi} \int_{u=0}^{\infty} \int_{t=-\infty}^{\infty} f(t) \cos\{u(t-x)\} dt\, du$$

$$= \frac{1}{2\pi} \int_{-\infty}^{+\infty} \int_{-\infty}^{+\infty} f(t) \cos\{u(t-x)\} dt\, du$$

$$= \frac{1}{2\pi} \int_{-\infty}^{\infty} \int_{-\infty}^{\infty} f(t) \cos\{u(t-x)\} dt\, du$$

$$+ \frac{i}{2\pi} \int_{-\infty}^{\infty} \left[\int_{-\infty}^{\infty} \sin\{u(t-x)\} du \right] f(t) dt$$

where the last integral is equal to zero as $\sin\{u(t-x)\}$ is an odd function of u.

Thus, we write

$$f(x) = \frac{1}{2\pi} \int_{-\infty}^{\infty} \int_{-\infty}^{\infty} f(t) \left[\cos\{u(t-x)\} + i \sin\{u(t-x)\} \right] du\, dt$$

$$= \frac{1}{2\pi} \int_{-\infty}^{\infty} \int_{-\infty}^{\infty} f(t) e^{iu(t-x)} dt\, du. \quad (11.68)$$

Relation (11.68) is known as the ***Fourier Integral*** in complex form.

11.15 FOURIER TRANSFORMS

In an interval $[-L, L]$ the complex Fourier series of a periodic function $f(x)$ is given by

$$f(x) = \sum_{n=-\infty}^{\infty} c_n e^{\frac{in\pi x}{L}} \quad (11.69)$$

where
$$c_n = \frac{1}{2L} \int_{-L}^{L} f(x) e^{-in\pi x/L} dx. \quad (11.70)$$

Let us introduce a new variable that is defined by

$$k = \frac{n\pi}{L} \text{ yielding } \Delta k = \frac{\pi}{L} \text{ and } \left(\frac{\Delta k}{\pi}\right) L = 1 \text{ as } \Delta n = 1.$$

Now, Eq. (11.69) can be written as

$$f(x) = \sum_{n=\frac{kL}{\pi}=-\infty}^{\infty} \left(\frac{L}{\pi} c_n\right) e^{ikx} \Delta k = \sum_{n=\frac{kL}{\pi}=-\infty}^{\infty} C(k) e^{ikx} \Delta k \tag{11.71}$$

where $$C(k) = \frac{L}{\pi} c_n = \frac{1}{2\pi} \int_{-L}^{L} f(x) e^{-ikx} dx. \tag{11.72}$$

If we let $L \to \infty$ then $\Delta k \to 0$ and Eq. (11.71) can be written as

$$f(x) = \int_{-\infty}^{\infty} C(k) e^{ikx} dk. \tag{11.73}$$

If we define $F(k) = \sqrt{2\pi}\, C(k)$, in the limit $L \to \infty$ Eq. (11.72) yields

$$F(k) = \frac{1}{\sqrt{2\pi}} \int_{-\infty}^{\infty} f(x) e^{-ikx} dx. \tag{11.74}$$

Equation (11.73) yields

$$f(x) = \frac{1}{\sqrt{2\pi}} \int_{-\infty}^{\infty} F(k) e^{ikx} dk. \tag{11.75}$$

Here, $F(k)$ is called the ***Fourier transform*** of $f(x)$ and $f(x)$ is called the ***Fourier inverse (or reverse) transform*** of $F(k)$ Fourier transforms are useful in solving differential equations. They are also extensively used in Quantum Mechanics.

For simplicity, only Fourier transforms in one dimension are discussed in this text even though three-dimensional transforms are extensively used in quantum mechanics. Generalization to two- or three-dimensional case is easy if one is familiar with the one-dimensional case.

11.16 FOURIER SINE AND COSINE TRANSFORMS

Sine Transform

From Eq. (11.74), we know that for an odd function $f_s(x) = -f_s(-x)$, the Fourier transform is given by

$$F(k) = \frac{1}{\sqrt{2\pi}} \int_{-\infty}^{\infty} f_s(x) e^{-ikx} dx$$

$$= \frac{1}{\sqrt{2\pi}} \int_{-\infty}^{\infty} f_s(x)\cos kx\, dx - \frac{i}{\sqrt{2\pi}} \int_{-\infty}^{\infty} f_s(x)\sin kx\, dx. \tag{11.76}$$

Since $f_s(x)$ is an odd function, the integrand in the first term is also an odd function and hence the corresponding integral is zero. Thus, we write

$$F(k) = \frac{-i}{\sqrt{2\pi}} \int_{-\infty}^{\infty} f_s(x)\sin kx\, dx = \frac{-2i}{\sqrt{2\pi}} \int_{0}^{\infty} f_s(x)\sin kx\, dx$$

since the integrand here is an even function.

Defining $F_s(k) = i\,F(k)$, we write

$$F_s(k) = \sqrt{\frac{2}{\pi}} \int_{0}^{\infty} f_s(x)\sin kx\, dx. \tag{11.77}$$

Equation (11.77) is known as ***Fourier sine transform***.
The inverse transform is written as

$$f_s(x) = \sqrt{\frac{2}{\pi}} \int_{0}^{\infty} F_s(k)\sin kx\, dk. \tag{11.78}$$

Fourier Cosine Transform

For an even function $f_c(x) = f_c(-x)$, the Fourier transform is

$$F(k) = F_c(k) = \frac{1}{\sqrt{2\pi}} \int_{-\infty}^{\infty} f_c(x) e^{-ikx}\, dx$$

$$= \frac{1}{\sqrt{2\pi}} \int_{-\infty}^{\infty} f_c(x)\cos kx\, dx - \frac{i}{\sqrt{2\pi}} \int_{-\infty}^{\infty} f_c(x)\sin kx\, dx$$

Since $f_c(x)$ is an even function, the integrand of the second integral is an odd function and hence the corresponding integral is zero. Thus, we write

$$F_c(k) = \frac{1}{\sqrt{2\pi}} \int_{-\infty}^{\infty} f_c(x)\cos kx\, dx$$

$$= \sqrt{\frac{2}{\pi}} \int_{0}^{\infty} f_c(x)\cos kx\, dx. \tag{11.79}$$

Equation (11.79) is known as ***Fourier cosine transform***.

The inverse transform is written as

$$f_c(x)=\sqrt{\frac{2}{\pi}}\int_0^\infty F_c(k)\cos kx\,dk. \tag{11.80}$$

11.17 CONVOLUTION THEOREM FOR FOURIER TRANSFORMS

The *convolution* (or *Faltung*) of two functions $f(x)$ and $g(x)$ is defined as the integral

$$f(x)\star g(x)=\frac{1}{\sqrt{2\pi}}\int_{-\infty}^{\infty} f(x-u)g(u)\,du. \tag{11.81}$$

Here, Faltung is a German word having the meaning of 'folding'.

The ***convolution theorem*** for Fourier transforms states that the Fourier transform of the convolution of two functions $f(x)$ and $g(x)$ is the prodct of the individual Fourier transforms of $f(x)$ and $g(x)$.

Proof: If $FT[\Phi(x)]$ denotes the Fourier transform of the function $\Phi(x)$ we are to prove that

$$FT[f(x)\star g(x)]=\{FT[f(x)]\}\ \{FT[g(x)]\}=F(k)G(k).$$

We know that

$$f(x)\star g(x)=\frac{1}{\sqrt{2\pi}}\int_{-\infty}^{\infty} f(x-u)g(u)\,du.$$

Taking Fourier transform of both sides, we have

$$FT[f(x)\star g(x)]=FT\left[\frac{1}{\sqrt{2\pi}}\int_{-\infty}^{\infty} f(x-u)g(u)\,du\right]$$

$$=\frac{1}{\sqrt{2\pi}}\int_{-\infty}^{\infty}\left\{\frac{1}{\sqrt{2\pi}}\int_{-\infty}^{\infty} f(x-u)g(u)\,du\right\}e^{-ikx}dx$$

$$=\frac{1}{\sqrt{2\pi}}\int_{-\infty}^{\infty}\left\{\frac{1}{\sqrt{2\pi}}\int_{-\infty}^{\infty} f(x-u)e^{-ik(x-u)}dx\right\}e^{-iku}g(u)\,du$$

$$=\frac{1}{\sqrt{2\pi}}\int_{-\infty}^{\infty} g(u)e^{-iku}du\,\frac{1}{\sqrt{2\pi}}\int_{-\infty}^{\infty} f(t)e^{-ikt}dt$$

[where we put $(x-u=t)$]

$$= G(k)F(k) = F(k)G(k). \tag{11.82}$$

Hence proved.

11.18 PARSEVAL'S RELATION FOR FOURIER TRANSFORMS

If the Fourier transforms of two functions $f(x)$ and $g(x)$ are $F(k)$ and $G(k)$ respectively, then in the interval $[-\infty, +\infty]$, the integral of the product of $f(x)$ and the complex conjugate of $g(x)$ is equal to the integral of the product of $F(k)$ and the complex conjugate of $G(k)$. Mathematically, we write

$$\int_{-\infty}^{\infty} f(x)g^*(x)dx = \int_{-\infty}^{\infty} F(k)G^*(k)dk$$

where $g^*(x)$ denotes complex conjugate of $g(x)$.

This relation is known as the general form of Parseval's relation.

Proof: Using direct and inverse Fourier transforms, we have

$$\int_{-\infty}^{\infty} f(x)g^*(x)dx$$

$$= \int_{-\infty}^{\infty}\left[\frac{1}{\sqrt{2\pi}}\int_{-\infty}^{\infty} F(k)e^{ikx}dk\right]\left[\frac{1}{\sqrt{2\pi}}\int_{-\infty}^{\infty} G^*(k')e^{-ik'x}dk'\right]dx$$

where the variable in the second integral has been renamed.

$$= \int_{-\infty}^{\infty} F(k)dk\int_{-\infty}^{\infty} G^*(k')dk'\frac{1}{2\pi}\int_{-\infty}^{\infty} e^{-i(k'-k)x}dx$$

$$= \int_{-\infty}^{\infty} F(k)dk\int_{-\infty}^{\infty} G^*(k')dk'\delta(k'-k)$$

$$= \int_{-\infty}^{\infty} F(k)dk\, G^*(k) = \int_{-\infty}^{\infty} F(k)G^*(k)dk. \tag{11.83}$$

Hence, the general form of Parseval's relation has been proved.

Special Case: In a special case, if $g(x) = f(x)$, we obtain

$$\int_{-\infty}^{\infty} |f(x)|^2\, dx = \int_{-\infty}^{\infty} |F(k)|^2\, dk. \tag{11.84}$$

11.19 FOURIER TRANSFORMS OF DERIVATIVES

Using the exponential form, the Fourier transform of a function $f(x)$ is

$$F(k)=\frac{1}{\sqrt{2\pi}}\int_{-\infty}^{\infty} f(x)e^{-ikx}dx.$$

Therefore, the Fourier transform of the first order derivative of $f(x)$ is given by

$$F^{(1)}(k)=\frac{1}{\sqrt{2\pi}}\int_{-\infty}^{\infty}\frac{df(x)}{dx}e^{-ikx}dx. \tag{11.85}$$

Integrating the right-hand side of Eq. (11.85) by parts, we obtain

$$F^{(1)}(k)=\left[\frac{e^{-ikx}}{\sqrt{2\pi}}f(x)\right]_{-\infty}^{\infty}+ik\left[\frac{1}{\sqrt{2\pi}}\int_{-\infty}^{\infty}f(x)e^{-ikx}dx\right].$$

If $f(x)$ vanishes as $x\to\pm\infty$, we have

$$F^{(1)}(k)=ik\,F(k). \tag{11.86}$$

Generalizing to the nth order derivative, the Fourier transform of the nth order derivative of $f(x)$ is given by

$$F^{(n)}(k)=(ik)^n\,F(k) \tag{11.87}$$

provided all the integrated out terms vanish as $x\to\pm\infty$.

Thus, the Fourier transform of the second order derivative of $f(x)$ is given by

$$F^{(2)}(k)=-k^2F(k) \tag{11.88}$$

provided the integrated out terms vanish as $x\to\pm\infty$.

Cosine and Sine Transforms of the First and the Second Order Derivatives

The cosine transform of the first order derivative of a function $f(x)$ is

$$F_c^{(1)}(k)=\sqrt{\frac{2}{\pi}}\int_0^{\infty}\frac{df(x)}{dx}\cos kx\,dx$$

$$=\sqrt{\frac{2}{\pi}}\left\{\left[f(x)\cos kx\right]_0^{\infty}+k\int_0^{\infty}f(x)\sin kx\,dx\right\}$$

$$=-\sqrt{\frac{2}{\pi}}\,f(0)+k\sqrt{\frac{2}{\pi}}\int_0^{\infty}f(x)\sin kx\,dx$$

$$= kF_s(k) - \sqrt{\frac{2}{\pi}}\, f(0) \tag{11.89}$$

provided $f(x)$ vanishes as $x \to \pm\infty$.

The sine transform of the first order derivative of a function $f(x)$ is

$$F_s^{(1)}(k) = \sqrt{\frac{2}{\pi}} \int_0^{\infty} \frac{df(x)}{dx} \sin kx\, dx$$

$$= \sqrt{\frac{2}{\pi}} \left\{ \left[f(x)\sin kx \right]_0^{\infty} - k\int_0^{\infty} f(x)\cos kx\, dx \right\}$$

$$= -k\, F_c(k) \tag{11.90}$$

provided $f(x)$) and $\frac{df(x)}{dx}$ vanish as $x \to \pm\infty$.

The cosine transform of the second order derivative of a function $f(x)$ is

$$F_c^{(2)}(k) = \sqrt{\frac{2}{\pi}} \int_0^{\infty} \frac{d^2 f(x)}{dx^2} \cos kx\, dx$$

$$= \sqrt{\frac{2}{\pi}} \left\{ \left[\frac{df(x)}{dx} \cos kx \right]_0^{\infty} + k\int_0^{\infty} \frac{df(x)}{dx} \sin kx\, dx \right\}$$

$$= -\sqrt{\frac{2}{\pi}}\, f'(0) - k^2 F_c(k) \tag{11.91}$$

provided $f(x)$ and $\frac{df(x)}{dx}$ vanish as $x \to \pm\infty$.

Similarly, the sine transform of the second order derivative of a function $f(x)$ is

$$F_s^{(2)}(k) = \sqrt{\frac{2}{\pi}} \int_0^{\infty} \frac{d^2 f(x)}{dx^2} \sin kx\, dx$$

$$= \sqrt{\frac{2}{\pi}} \left\{ \left[\frac{df(x)}{dx} \sin kx \right]_0^{\infty} - k\int_0^{\infty} \frac{df(x)}{dx} \cos kx\, dx \right\}$$

$$= k\sqrt{\frac{2}{\pi}}\, f(0) - k^2 F_s(k) \tag{11.92}$$

provided $f(x)$ and $\frac{df(x)}{dx}$ vanish as $x \to \pm\infty$.

11.20 APPLICATIONS OF FOURIER SERIES AND TRANSFORMS

Fourier series and transforms have a wide range of applications. They have extensive applications in electrical engineering, vibration analysis, acoustics, interference and diffraction, signal processing, quantum mechanics, econometrics, thin-walled shell theory, designing and using antennas, image processing and filters, data processing and analysis, high-pass, low-pass and band-pass filters, signal and noise estimation—encoding time series and many more. The simplest application of Fourier transforms is in solving differential equations (ODE as well as PDE). The vast area of such applications is not studied in this text except the last cited one.

11.21 SOLVED PROBLEMS

PROBLEM 1 Obtain the Fourier series of the exponential function $f(x)=e^{kx}$ (k = a constant) in the interval $[-1,1]$ assuming that it is a periodic function of period 2.

Solution:

(A) Trigonometric Fourier Series

The trigonometric formula for Fourier series in an interval $[-L, L]$ is

$$f(x)=\frac{a_0}{2}+\sum_{n=1}^{\infty}a_n\cos\left(\frac{n\pi x}{L}\right)+\sum_{n=1}^{\infty}b_n\sin\left(\frac{n\pi x}{L}\right)$$

where $$a_0=\frac{1}{L}\int_{-L}^{L}f(x)dx,$$

$$a_n=\frac{1}{L}\int_{-L}^{L}f(x)\cos\left(\frac{n\pi x}{L}\right)dx,$$

$$b_n=\frac{1}{L}\int_{-L}^{L}f(x)\sin\left(\frac{n\pi x}{L}\right)dx.$$

Here, $L = 1$ and, therefore, we have

$$f(x)=e^{kx}=\frac{a_0}{2}+\sum_{n=1}^{\infty}a_n\cos(n\pi x)+\sum_{n=1}^{\infty}b_n\sin(n\pi x) \qquad \text{...(i)}$$

where $$a_0=\int_{-1}^{+1}f(x)dx, \qquad \text{...(ii)}$$

$$a_n = \int_{-1}^{+1} f(x)\cos(n\pi x)dx, \quad \text{...(iii)}$$

$$b_n = \int_{-1}^{+1} f(x)\sin(n\pi x)dx. \quad \text{...(iv)}$$

From Eq. (ii), we have

$$a_0 = \int_{-1}^{+1} e^{kx}dx = \frac{1}{k}(e^k - e^{-k}) = \frac{2}{k}\sinh k.$$

From Eq. (iii), we have

$$a_n = \int_{-1}^{+1} \cos(n\pi x)e^{kx}dx$$

$$= \left[\cos(n\pi x)\frac{e^{kx}}{k}\right]_{-1}^{+1} + \frac{n\pi}{k}\int_{-1}^{+1}\left\{\sin(n\pi x)e^{kx}dx\right\}$$

$$= \frac{1}{k}\left\{e^k \cos n\pi - e^{-k}\cos(-n\pi)\right\}$$

$$+ \frac{n\pi}{k}\left[\sin(n\pi x)\frac{e^{kx}}{k}\right]_{-1}^{+1} - \frac{n^2\pi^2}{k^2}\int_{-1}^{+1}\cos(n\pi x)e^{kx}dx$$

$$= \frac{(-1)^n}{k}(e^k - e^{-k}) + 0 - \frac{n^2\pi^2}{k^2}a_n.$$

Therefore, we have

$$\left(1 + \frac{n^2\pi^2}{k^2}\right)a_n = \frac{2(-1)^n}{k}\sinh k$$

yielding $$a_n = \frac{2k(-1)^n}{k^2 + n^2\pi^2}\sinh k.$$

From Eq. (iv), we have

$$b_n = \int_{-1}^{+1} \sin(n\pi x)e^{kx}dx$$

$$= \left[\sin(n\pi x)\frac{e^{kx}}{k}\right]_{-1}^{+1} - \frac{n\pi}{k}\int_{-1}^{+1}\cos(n\pi x)e^{kx}\,dx$$

$$= 0 - \frac{n\pi}{k}\left[\cos(n\pi x)\frac{e^{kx}}{k}\right]_{-1}^{+1} - \frac{n^2\pi^2}{k^2}\int_{-1}^{+1}\sin(n\pi x)e^{kx}\,dx$$

$$= -\frac{n\pi(-1)^n}{k^2}(e^k - e^{-k}) - \frac{n^2\pi^2}{k^2}b_n$$

or $\quad \left(1+\frac{n^2\pi^2}{k^2}\right)b_n = -\frac{2n\pi(-1)^n}{k^2}\sinh k$

yielding $\quad b_n = -\frac{2n\pi(-1)^n}{k^2+n^2\pi^2}\sinh k.$

Substituting the values of these coefficients into Eq. (i), we obtain

$$e^{kx} = 2\sinh k\left[\frac{1}{2k} + \sum_{n=1}^{\infty}\frac{(-1)^n}{k^2+n^2\pi^2}\{k\cos(n\pi x) - n\pi\sin n\pi x\}\right].$$

This equation represents the trigonometric Fourier series of the given function in the given interval.

(B) Complex Fourier Series

In complex form, Fourier series of a function $f(x)$ in an interval $[-L, L]$ is given by

$$f(x) = \sum_{n=-\infty}^{\infty} c_n e^{\frac{in\pi x}{L}}$$

where $\quad c_n = \frac{1}{2L}\int_{-L}^{L} f(x)e^{-\frac{in\pi x}{L}}\,dx.$

In the present problem, $f(x) = e^{kx}$ and $L = 1$.
Therefore, we have

$$f(x) = e^{kx} = \sum_{n=-\infty}^{\infty} c_n e^{in\pi x} \qquad \text{...(i)}$$

where $\quad c_n = \frac{1}{2}\int_{-1}^{+1} e^{kx}e^{-in\pi x}\,dx$

$$= \frac{1}{2}\int_{-1}^{+1} e^{(k-in\pi)x}\,dx$$

$$= \frac{1}{2}\frac{1}{(k-in\pi)}\left[e^{(k-in\pi)x}\right]_{-1}^{+1}$$

$$= \frac{1}{2}\frac{(k+in\pi)}{\left(k^2+n^2\pi^2\right)}\left[e^{k-in\pi} - e^{-k+in\pi}\right]$$

$$= \frac{(-1)^n (k+in\pi)}{\left(k^2+n^2\pi^2\right)}\frac{\left[e^k - e^{-k}\right]}{2} = (-1)^n \frac{(k+in\pi)}{\left(k^2+n^2\pi^2\right)}\sinh k.$$

Substituting this value of c_n into Eq. (i), we obtain

$$e^{kx} = \sinh k \sum_{n=-\infty}^{\infty} (-1)^n \frac{(k+in\pi)}{\left(k^2+n^2\pi^2\right)} e^{in\pi x}.$$

This equation represents the complex form of Fourier series of the given function in the given interval.

PROBLEM 2 Obtain a Fourier series representing the function $f(x)=e^x$ within the interval $-\pi < x < \pi$ and hence deduce a series representing $\dfrac{\pi}{\sinh \pi}$. Assume that $f(x)$ is a periodic function of period 2π.

Solution: For a periodic function $f(x)$ defined in an interval $-L < x < L$ the general formula for trigonometric Fourier series is

$$f(x) = \frac{a_0}{2} + \sum_{n=1}^{\infty} a_n \cos\left(\frac{n\pi x}{L}\right) + \sum_{n=1}^{\infty} b_n \sin\left(\frac{n\pi x}{L}\right)$$

where $$a_0 = \frac{1}{L}\int_{-L}^{+L} f(x)dx,$$

$$a_n = \frac{1}{L}\int_{-L}^{+L} f(x)\cos\left(\frac{n\pi x}{L}\right)dx,$$

and $$b_n = \frac{1}{L}\int_{-L}^{+L} f(x)\sin\left(\frac{n\pi x}{L}\right)dx.$$

In this problem, we have $L = \pi$.

Therefore, we write

$$f(x) = e^x = \frac{a_0}{2} + \sum_{n=1}^{\infty} a_n \cos nx + \sum_{n=1}^{\infty} b_n \sin nx \qquad \ldots\text{(i)}$$

where

$$a_0 = \frac{1}{\pi}\int_{-\pi}^{+\pi} e^x\,dx, \qquad \ldots\text{(ii)}$$

$$a_n = \frac{1}{\pi}\int_{-\pi}^{+\pi} e^x \cos nx\,dx, \qquad \ldots\text{(iii)}$$

and

$$b_n = \frac{1}{\pi}\int_{-\pi}^{+\pi} e^x \sin nx\,dx. \qquad \ldots\text{(iv)}$$

Now, from Eq. (ii), we have

$$a_0 = \frac{1}{\pi}(e^{\pi} - e^{-\pi}) = \frac{2\sinh\pi}{\pi}.$$

From Eq. (iii), we have

$$a_n = \frac{1}{\pi}\int_{-\pi}^{+\pi} \cos nx\, e^x dx$$

$$= \frac{1}{\pi}\left\{\left[\cos nx\, e^x\right]_{-\pi}^{+\pi} + n\int_{-\pi}^{+\pi} \sin nx\, e^x dx\right\}$$

$$= \frac{\cos n\pi}{\pi}(e^{\pi} - e^{-\pi}) + \frac{n}{\pi}\left[\sin nx\, e^x\right]_{-\pi}^{+\pi} - \frac{n^2}{\pi}\int_{-\pi}^{+\pi} \cos nx\, e^x dx$$

$$= \frac{2\sinh\pi}{\pi}\cos n\pi + 0 - n^2 a_n$$

or $\quad (1+n^2)a_n = \dfrac{2\sinh\pi}{\pi}\cos n\pi$

yielding $\quad a_n = \dfrac{2\sinh\pi}{\pi}\dfrac{(-1)^n}{1+n^2}.$

From Eq. (iv), we have

$$b_n = \frac{1}{\pi}\int_{-\pi}^{+\pi} \sin nx\, e^x dx$$

$$= \frac{1}{\pi}\left\{\left[\sin nx\, e^{x}\right]_{-\pi}^{+\pi} - n\int_{-\pi}^{+\pi} \cos nx\, e^{x}\,dx\right\}$$

$$= 0 - \frac{n}{\pi}\left\{\left[\cos nx\, e^{x}\right]_{-\pi}^{+\pi} + \int_{-\pi}^{+\pi} n \sin nx\, e^{x}\,dx\right\}$$

$$= -\frac{n\cos n\pi}{\pi}(e^{\pi} - e^{-\pi}) - n^2 b_n$$

or $$(1+n^2)b_n = -\frac{2\sinh\pi}{\pi} n\cos n\pi$$

giving $$b_n = -(-1)^n \frac{2\sinh\pi}{\pi}\frac{n}{1+n^2}.$$

Substituting the values of these Fourier coefficients into Eq. (i), we obtain

$$f(x) = e^{x} = \frac{\sinh\pi}{\pi} + \frac{2\sinh\pi}{\pi}\sum_{n=1}^{\infty}\frac{(-1)^n}{1+n^2}\cos nx - \frac{2\sinh\pi}{\pi}\sum_{n=1}^{\infty}\frac{n(-1)^n}{1+n^2}\sin nx$$

$$= \frac{2\sinh\pi}{\pi}\left[\left\{\frac{1}{2} - \frac{\cos x}{1+1^2} + \frac{\cos 2x}{1+2^2} - \frac{\cos 3x}{1+3^2} + \frac{\cos 4x}{1+4^2} - \ldots\infty\right\}\right.$$

$$\left. + \left\{\frac{\sin x}{1+1^2} - \frac{2\sin 2x}{1+2^2} + \frac{3\sin 3x}{1+3^2} - \frac{4\sin 4x}{1+4^2} + \ldots\infty\right\}\right].$$

Deduction: Putting $x = 0$, we obtain

$$1 = \frac{2\sinh\pi}{\pi}\left[\frac{1}{2} - \frac{1}{2} + \frac{1}{5} - \frac{1}{10} + \frac{1}{17} - \ldots\infty\right]$$

yielding $$\frac{\pi}{\sinh\pi} = 2\left[\frac{1}{5} - \frac{1}{10} + \frac{1}{17} - \frac{1}{26} + \ldots\infty\right].$$

PROBLEM 3 Obtain the Fourier series representing the periodic function $f(x) = x^2 + x$ in the interval $-L < x < L$. Prove that

$$\frac{1}{1^2} - \frac{1}{2^2} + \frac{1}{3^2} - \frac{1}{4^2} + \ldots\infty = \frac{\pi^2}{12}.$$

Solution: Let us assume the Fourier series of the given function as

$$f(x) = x^2 + x$$

$$= \frac{a_0}{2} + \sum_{n=1}^{\infty} a_n \cos\left(\frac{n\pi x}{L}\right) + \sum_{n=1}^{\infty} b_n \sin\left(\frac{n\pi x}{L}\right) \qquad \ldots\text{(i)}$$

where
$$a_0 = \frac{1}{L}\int_{-L}^{+L} f(x)dx, \qquad \ldots\text{(ii)}$$

$$a_n = \frac{1}{L}\int_{-L}^{+L} f(x)\cos\left(\frac{n\pi x}{L}\right)dx, \qquad \ldots\text{(iii)}$$

and
$$b_n = \frac{1}{L}\int_{-L}^{+L} f(x)\sin\left(\frac{n\pi x}{L}\right)dx. \qquad \ldots\text{(iv)}$$

From Eq. (ii), we obtain

$$a_0 = \frac{1}{L}\int_{-L}^{+L} (x^2 + x)dx = \frac{2L^2}{3}.$$

From Eq. (iii), we obtain

$$a_n = \frac{1}{L}\int_{-L}^{+L} (x^2 + x)\cos\left(\frac{n\pi x}{L}\right)dx$$

$$= \frac{2}{L}\int_{0}^{L} x^2 \cos\left(\frac{n\pi x}{L}\right)dx + \frac{1}{L}\int_{-L}^{+L} x\cos\left(\frac{n\pi x}{L}\right)dx$$

$$= \frac{2}{L}\left\{\frac{L}{n\pi}\left[x^2 \sin\left(\frac{n\pi x}{L}\right)\right]_0^L - \frac{2L}{n\pi}\int_0^L x\sin\left(\frac{n\pi x}{L}\right)dx + 0\right\}$$

$$= 0 + \frac{4}{n\pi}\left\{\frac{L}{n\pi}\left[x\cos\left(\frac{n\pi x}{L}\right)\right]_0^L - \frac{L}{n\pi}\int_0^L \cos\left(\frac{n\pi x}{L}\right)dx\right\}$$

$$= (-1)^n \frac{4L^2}{n^2\pi^2}.$$

From Eq. (iv), we obtain

$$b_n = \frac{1}{L}\int_{-L}^{+L} (x^2 + x)\sin\left(\frac{n\pi x}{L}\right)dx$$

$$=\frac{1}{L}\int_{-L}^{+L} x^2 \sin\left(\frac{n\pi x}{L}\right)dx+\frac{2}{L}\int_{0}^{L} x\sin\left(\frac{n\pi x}{L}\right)dx$$

$$=0+\frac{2}{L}\left\{-\frac{L}{n\pi}\left[x\cos\left(\frac{n\pi x}{L}\right)\right]_0^L+\frac{L}{n\pi}\int_0^L \cos\left(\frac{n\pi x}{L}\right)dx\right\}$$

$$=\frac{(-1)^{n+1}\,2L}{n\pi}.$$

Substituting the values of these Fourier coefficients into Eq. (i), we have

$$x^2+x=\frac{L^2}{3}+\sum_{n=1}^{\infty}(-1)^n\frac{4L^2}{n^2\pi^2}\cos\left(\frac{n\pi x}{L}\right)+\sum_{n=1}^{\infty}\frac{(-1)^{n+1}\,2L}{n\pi}\sin\left(\frac{n\pi x}{L}\right)$$

$$=\frac{L^2}{3}-\frac{4L^2}{\pi^2}\left[\frac{1}{1^2}\cos\left(\frac{\pi x}{L}\right)-\frac{1}{2^2}\cos\left(\frac{2\pi x}{L}\right)+\frac{1}{3^2}\cos\left(\frac{3\pi x}{L}\right)-\frac{1}{4^2}\cos\left(\frac{4\pi x}{L}\right)+\ldots\infty\right]$$

$$+\frac{2L}{\pi}\left[\sin\left(\frac{\pi x}{L}\right)-\frac{1}{2}\sin\left(\frac{2\pi x}{L}\right)+\frac{1}{3}\sin\left(\frac{3\pi x}{L}\right)-\frac{1}{4}\sin\left(\frac{4\pi x}{L}\right)+\ldots\infty\right].$$

This equation is the Fourier series representing the function $f(x)=x^2+x$ in the interval $[-L, L]$.

Putting $x = 0$, we obtain

$$0=\frac{L^2}{3}-\frac{4L^2}{\pi^2}\left(\frac{1}{1^2}-\frac{1}{2^2}+\frac{1}{3^2}-\frac{1}{4^2}+\ldots\infty\right)$$

or $$1-\frac{1}{2^2}+\frac{1}{3^2}-\frac{1}{4^2}+\ldots\infty=\frac{\pi^2}{12}.$$

PROBLEM 4 A periodic function of period '2' is defined by

$$f(x)=\begin{cases}-1 & \text{for } -1<x<0,\\ 2x & \text{for } 0<x<1\end{cases}$$

and $$f(x)=f(x+2).$$

Sketch a profile of the function and its periodic extension. Also obtain the Fourier series representation of $f(x)$ and deduce that

$$1-\frac{1}{3}+\frac{1}{5}-\frac{1}{7}+\ldots\infty=\frac{\pi}{4}.$$

Solution: The profile of the function and its periodic extension is shown in the figure below.

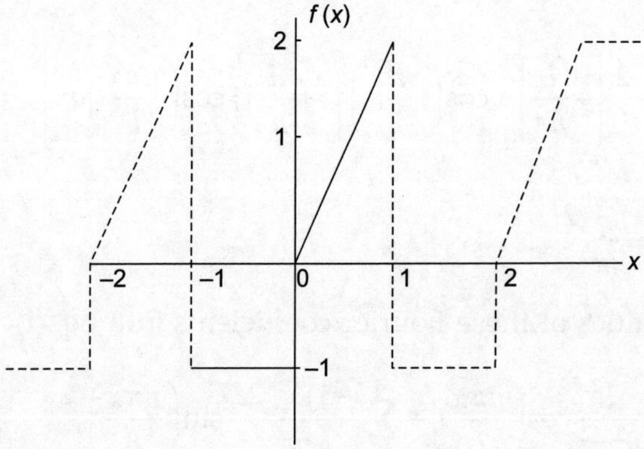

Fig. 11.1 Profile of the function $f(x)$ and its periodic extension.

The Fourier series of a function $f(x)$ in an interval $[-L, L]$ is expressed as

$$f(x)=\frac{a_0}{2}+\sum_{n=1}^{\infty}a_n\cos\left(\frac{n\pi x}{L}\right)+\sum_{n=1}^{\infty}b_n\sin\left(\frac{n\pi x}{L}\right)$$

where $$a_0=\frac{1}{L}\int_{-L}^{+L}f(x)dx,$$

$$a_n=\frac{1}{L}\int_{-L}^{+L}f(x)\cos\left(\frac{n\pi x}{L}\right)dx$$

and $$b_n=\frac{1}{L}\int_{-L}^{+L}f(x)\sin\left(\frac{n\pi x}{L}\right)dx.$$

Here, $L=1$ and hence we have

$$f(x)=\frac{a_0}{2}+\sum_{n=1}^{\infty}a_n\cos(n\pi x)+\sum_{n=1}^{\infty}b_n\sin(n\pi x) \quad \text{...(i)}$$

where $$a_0=\int_{-1}^{+1}f(x)dx, \quad \text{...(ii)}$$

$$a_n=\int_{-1}^{+1}f(x)\cos(n\pi x)dx \quad \text{...(iii)}$$

and
$$b_n = \int_{-1}^{+1} f(x)\sin(n\pi x)\,dx. \qquad \text{...(iv)}$$

From Eq. (ii), we obtain

$$a_0 = \int_{-1}^{0} f(x)dx + \int_{0}^{1} f(x)dx$$

$$= \int_{-1}^{0} (-1)\,dx + \int_{0}^{1} 2x\,dx = -1 + 1 = 0.$$

From Eq. (iii), we obtain

$$a_n = \int_{-1}^{0} f(x)\cos(n\pi x)\,dx + \int_{0}^{1} f(x)\cos(n\pi x)\,dx$$

$$= \int_{-1}^{0} (-1)\cos(n\pi x)\,dx + \int_{0}^{1} 2x\cos(n\pi x)\,dx$$

$$= \int_{0}^{-1} \cos(n\pi x)\,dx + \left[\frac{2x}{n\pi}\sin(n\pi x)\right]_0^1 - \frac{2}{n\pi}\int_{0}^{1}\sin(n\pi x)\,dx$$

$$= \left[\frac{1}{n\pi}\sin(n\pi x)\right]_0^{-1} + 0 + \left[\frac{2}{n^2\pi^2}\cos(n\pi x)\right]_0^1$$

$$= 0 + 0 + \frac{2}{n^2\pi^2}[\cos n\pi - \cos 0]$$

$$= \frac{2}{n^2\pi^2}\left[(-1)^n - 1\right]$$

$$= \begin{cases} 0 & \text{for } n = \text{even integer,} \\ -\dfrac{4}{n^2\pi^2} & \text{for } n = \text{odd integer.} \end{cases}$$

From Eq. (iv), we have

$$b_n = \int_{-1}^{0} (-1)\sin(n\pi x)\,dx + \int_{0}^{1} 2x\sin(n\pi x)\,dx$$

$$= \int_0^{-1} \sin(n\pi x)\,dx + \left[-\frac{2x}{n\pi}\cos(n\pi x)\right]_0^1 + \int_0^1 \frac{2}{n\pi}\cos(n\pi x)\,dx$$

$$= \left[-\frac{1}{n\pi}\cos(n\pi x)\right]_0^{-1} - \frac{2}{n\pi}[\cos n\pi - 0] + \left[\frac{2}{n^2\pi^2}\sin(n\pi x)\right]_0^1$$

$$= -\frac{1}{n\pi}\left[\cos(-n\pi) - \cos 0\right] - \frac{2}{n\pi}\cos n\pi + 0$$

$$= \frac{1}{n\pi}\left[1 - 3(-1)^n\right].$$

Substituting the values of the Fourier coefficients into Eq. (i), we obtain the Fourier series as

$$f(x) = -\frac{4}{\pi^2}\sum_{n=\text{odd}} \frac{1}{n^2}\cos(n\pi x) + \frac{1}{\pi}\sum_{n=1}^{\infty}\frac{1}{n}\left\{1 - 3(-1)^n\right\}\sin(n\pi x)$$

$$= -\frac{4}{\pi^2}\left[\cos(\pi x) + \frac{1}{3^2}\cos(3\pi x) + \frac{1}{5^2}\cos(5\pi x) + \ldots\infty\right]$$

$$+\frac{2}{\pi}\left[2\sin(\pi x) - \frac{1}{2}\sin(2\pi x) + \frac{2}{3}\sin(3\pi x)\right.$$

$$\left. - \frac{1}{4}\sin(4\pi x) + \frac{2}{5}\sin(5\pi x) + \ldots\infty\right].$$

Deduction: Putting $x = -\frac{1}{2}$, we obtain

$$-1 = -\frac{4}{\pi^2}\left[\cos\left(-\frac{\pi}{2}\right) + \frac{1}{3^2}\cos\left(-\frac{3\pi}{2}\right) + \frac{1}{5^2}\cos\left(-\frac{5\pi}{2}\right) + \ldots\infty\right]$$

$$+\frac{2}{\pi}\left[2\sin\left(-\frac{\pi}{2}\right) - \frac{1}{2}\sin(-\pi) + \frac{2}{3}\sin\left(-\frac{3\pi}{2}\right) - \frac{1}{4}\sin(-2\pi) + \frac{2}{5}\sin\left(-\frac{5\pi}{2}\right) + \ldots\infty\right]$$

or $$-1 = 0 + \frac{2}{\pi}\left[-2 - 0 + \frac{2}{3} - 0 - \frac{2}{5} - 0 + \frac{2}{7} - 0 + \ldots\infty\right]$$

$$= -\frac{4}{\pi}\left[1 - \frac{1}{3} + \frac{1}{5} - \frac{1}{7} + \ldots\infty\right].$$

Hence, we obtain

$$1-\frac{1}{3}+\frac{1}{5}-\frac{1}{7}+\ldots\infty=\frac{\pi}{4}.$$

PROBLEM 5 A function is defined by

$$f(x)=\begin{cases}0 & \text{for } -1<x<0, \\ x & \text{for } 0<x<1\end{cases}$$

and $$f(x)=f(x+2).$$

Find a Fourier series of the function and deduce that

$$1-\frac{1}{3}+\frac{1}{5}-\frac{1}{7}+\ldots\infty=\frac{\pi}{4}.$$

Solution: In the interval $-L<x<L$ the Fourier series of a function $f(x)$ is expressed as

$$f(x)=\frac{a_0}{2}+\sum_{n=1}^{\infty}a_n\cos\left(\frac{n\pi x}{L}\right)+\sum_{n=1}^{\infty}b_n\sin\left(\frac{n\pi x}{L}\right)$$

where $$a_0=\frac{1}{L}\int_{-L}^{+L}f(x)dx,$$

$$a_n=\frac{1}{L}\int_{-L}^{+L}f(x)\cos\left(\frac{n\pi x}{L}\right)dx$$

and $$b_n=\frac{1}{L}\int_{-L}^{+L}f(x)\sin\left(\frac{n\pi x}{L}\right)dx.$$

In the present problem, L = 1. Therefore, we have

$$f(x)=\frac{a_0}{2}+\sum_{n=1}^{\infty}a_n\cos(n\pi x)+\sum_{n=1}^{\infty}b_n\sin(n\pi x) \quad \ldots\text{(i)}$$

where $$a_0=\int_{-1}^{+1}f(x)dx, \quad \ldots\text{(ii)}$$

$$a_n=\int_{-1}^{+1}f(x)\cos(n\pi x)dx \quad \ldots\text{(iii)}$$

and $$b_n = \int_{-1}^{+1} f(x)\sin(n\pi x)\,dx. \qquad \text{...(iv)}$$

From Eq. (ii), we have

$$a_0 = \int_{-1}^{0} f(x)dx + \int_{0}^{1} f(x)dx$$

$$= \int_{-1}^{0} 0\cdot dx + \int_{0}^{1} x dx = \frac{1}{2}.$$

From Eq. (iii), we have

$$a_n = \int_{-1}^{0} f(x)\cos(n\pi x)dx + \int_{0}^{1} f(x)\cos(n\pi x)dx$$

$$= \int_{-1}^{0} 0\cdot\cos(n\pi x)dx + \int_{0}^{1} x\cos(n\pi x)dx$$

$$= 0 + \left[\frac{1}{n\pi}x\sin(n\pi x)\right]_0^1 - \frac{1}{n\pi}\int_{0}^{1}\sin(n\pi x)dx$$

$$= 0 + \frac{1}{n^2\pi^2}\left[\cos(n\pi x)\right]_0^1$$

$$= \frac{1}{n^2\pi^2}\left[(-1)^n - 1\right] = \begin{cases} 0 & \text{if } n \text{ is even integer,} \\ -\dfrac{2}{n^2\pi^2} & \text{if } n \text{ is odd integer.} \end{cases}$$

From Eq. (iv), we have

$$b_n = \int_{-1}^{0} 0\cdot\sin(n\pi x)dx + \int_{0}^{1} x\sin(n\pi x)dx$$

$$= 0 + \left[-\frac{1}{n\pi}x\cos(n\pi x)\right]_0^1 + \int_{0}^{1}\frac{1}{n\pi}\cos(n\pi x)dx$$

$$= -\frac{1}{n\pi}\left[\cos n\pi - 0\right] + 0 = \frac{(-1)^{n+1}}{n\pi}.$$

Substituting the values of these Fourier coefficients into Eq. (i), we obtain

$$f(x)=\frac{1}{4}-\frac{2}{\pi^2}\sum_{n=\text{odd}}\frac{1}{n^2}\cos(n\pi x)+\frac{1}{\pi}\sum_{n=1}^{\infty}\frac{(-1)^{n+1}}{n}\sin(n\pi x)$$

$$=\frac{1}{4}-\frac{2}{\pi^2}\left[\cos(\pi x)+\frac{\cos(3\pi x)}{3^2}+\frac{\cos(5\pi x)}{5^2}+\ldots\infty\right]$$

$$+\frac{1}{\pi}\left[\sin(\pi x)-\frac{\sin(2\pi x)}{2}+\frac{\sin(3\pi x)}{3}-\frac{\sin(4\pi x)}{4}+\frac{\sin(5\pi x)}{5}-\ldots\infty\right].$$

Deduction: Putting $x=\frac{1}{2}$, we obtain

$$\frac{1}{2}=\frac{1}{4}-0+\frac{1}{\pi}\left[1-0-\frac{1}{3}-0+\frac{1}{5}-0-\frac{1}{7}-0+\ldots\infty\right]$$

yielding
$$1-\frac{1}{3}+\frac{1}{5}-\frac{1}{7}+\ldots\infty=\frac{\pi}{4}.$$

PROBLEM 6 Obtain the Fourier series of a function defined by

$$f(x)=\begin{cases}-x-\pi & \text{for } -\pi<x<0,\\ x+\pi & \text{for } 0<x<\pi\end{cases}$$

and
$$f(x)=f(x+2\pi).$$

Find a Fourier series of the function and deduce that

$$1-\frac{1}{3}+\frac{1}{5}-\frac{1}{7}+\ldots\infty=\frac{\pi}{4}.$$

Solution: In the interval $-L<x<L$, the Fourier series of a function $f(x)$ is expressed as

$$f(x)=\frac{a_0}{2}+\sum_{n=1}^{\infty}a_n\cos\left(\frac{n\pi x}{L}\right)+\sum_{n=1}^{\infty}b_n\sin\left(\frac{n\pi x}{L}\right)$$

where
$$a_0=\frac{1}{L}\int_{-L}^{+L}f(x)dx,$$

$$a_n=\frac{1}{L}\int_{-L}^{+L}f(x)\cos\left(\frac{n\pi x}{L}\right)dx$$

and $$b_n = \frac{1}{L}\int_{-L}^{+L} f(x)\sin\left(\frac{n\pi x}{L}\right)dx.$$

In the present problem, $L = \pi$ and, therefore, we have

$$f(x) = \frac{a_0}{2} + \sum_{n=1}^{\infty} a_n \cos nx + \sum_{n=1}^{\infty} b_n \sin nx \qquad \ldots\text{(i)}$$

where $$a_0 = \frac{1}{\pi}\int_{-\pi}^{+\pi} f(x)dx, \qquad \ldots\text{(ii)}$$

$$a_n = \frac{1}{\pi}\int_{-\pi}^{+\pi} f(x)\cos nx\,dx \qquad \ldots\text{(iii)}$$

and $$b_n = \frac{1}{\pi}\int_{-\pi}^{+\pi} f(x)\sin nx\,dx. \qquad \ldots\text{(iv)}$$

Equation (ii) yields

$$a_0 = \frac{1}{\pi}\left\{\int_{-\pi}^{0} f(x)dx + \int_{0}^{\pi} f(x)dx\right\}$$

$$= \frac{1}{\pi}\left\{\int_{-\pi}^{0} (-x-\pi)dx + \int_{0}^{\pi}(x+\pi)dx\right\}$$

$$= \frac{1}{\pi}\left\{\int_{0}^{-\pi} (x+\pi)dx + \int_{0}^{\pi}(x+\pi)dx\right\}$$

$$= \frac{1}{\pi}\left\{\frac{\pi^2}{2} - \pi^2 + \frac{\pi^2}{2} + \pi^2\right\} = \pi.$$

Equation (iii) yields

$$a_n = \frac{1}{\pi}\left\{\int_{-\pi}^{0} f(x)\cos nx\,dx + \int_{0}^{\pi} f(x)\cos nx\,dx\right\}$$

$$= \frac{1}{\pi}\left\{\int_{-\pi}^{0} -(x+\pi)\cos nx\,dx + \int_{0}^{\pi}(x+\pi)\cos nx\,dx\right\}$$

$$= \frac{1}{\pi}\left\{\int_{0}^{-\pi}(x+\pi)\cos nx\,dx + \int_{0}^{\pi}(x+\pi)\cos nx\,dx\right\}$$

$$= \frac{1}{\pi}\left\{\int_{0}^{\pi}(-x+\pi)\cos(-nx)(-dx) + \int_{0}^{\pi}(x+\pi)\cos nx\,dx\right\}$$

$$= \frac{1}{\pi}\int_{0}^{\pi}(x-\pi+x+\pi)\cos nx\,dx = \frac{2}{\pi}\int_{0}^{\pi}x\cos nx\,dx$$

$$= \frac{2}{\pi}\left\{\left[\frac{1}{n}x\sin nx\right]_{0}^{\pi} - \int_{0}^{\pi}\frac{1}{n}\sin nx\,dx\right\}$$

$$= 0 + \frac{2}{\pi}\left[\frac{1}{n^2}\cos nx\right]_{0}^{\pi} = \frac{2}{\pi n^2}[\cos n\pi - \cos 0]$$

$$= \frac{2}{\pi n^2}\left\{(-1)^n - 1\right\} = \begin{cases} 0 & \text{if } n \text{ is even integer,} \\ -\dfrac{4}{\pi n^2} & \text{if } n \text{ is odd integer.} \end{cases}$$

Equation (iv) yields

$$b_n = \frac{1}{\pi}\left\{\int_{-\pi}^{0} f(x)\sin(nx)\,dx + \int_{0}^{\pi} f(x)\sin nx\,dx\right\}$$

$$= \frac{1}{\pi}\left\{\int_{-\pi}^{0}(-x-\pi)\sin nx\,dx + \int_{0}^{\pi}(x+\pi)\sin nx\,dx\right\}$$

$$= \frac{1}{\pi}\left\{\int_{\pi}^{0}(x-\pi)\sin(-nx)(-dx) + \int_{0}^{\pi}(x+\pi)\sin nx\,dx\right\}$$

$$= \frac{1}{\pi}\int_{0}^{\pi}(-x+\pi+x+\pi)\sin nx\,dx$$

$$= 2\int_{0}^{\pi}\sin nx\,dx = -2\left[\frac{1}{n}\cos nx\right]_{0}^{\pi}$$

$$= -\frac{2}{n}\left\{(-1)^n - 1\right\} = \begin{cases} 0 & \text{if } n \text{ is even integer,} \\ \dfrac{4}{n} & \text{if } n \text{ is odd integer.} \end{cases}$$

Now, substitution of the values of these Fourier coefficients into Eq. (i) yields

$$f(x)=\frac{\pi}{2}-\frac{4}{\pi}\sum_{n=\text{odd}}\frac{\cos nx}{n^2}+4\sum_{n=\text{odd}}\frac{\sin nx}{n}$$

$$=\frac{\pi}{2}-\frac{4}{\pi}\left\{\frac{\cos x}{1^2}+\frac{\cos 3x}{3^2}+\frac{\cos 5x}{5^2}+\ldots\infty\right\}+4\left\{\frac{\sin x}{1}+\frac{\sin 3x}{3}+\frac{\sin 5x}{5}+\ldots\infty\right\}.$$

This series represents the Fourier series of the given function.

Deduction: Putting $x=\frac{\pi}{2}$, we obtain

$$\frac{\pi}{2}+\pi=\frac{\pi}{2}-0+4\left\{\sin\frac{\pi}{2}+\frac{1}{3}\sin\frac{3\pi}{2}+\frac{1}{5}\sin\frac{5\pi}{2}+\frac{1}{7}\sin\frac{7\pi}{2}+\ldots\infty\right\}$$

or $$\pi=4\left(1-\frac{1}{3}+\frac{1}{5}-\frac{1}{7}+\ldots\infty\right)$$

or $$1-\frac{1}{3}+\frac{1}{5}-\frac{1}{7}+\ldots\infty=\frac{\pi}{4}.$$

PROBLEM 7 Obtain a Fourier series to represent the periodic function

$$f(x)=\begin{cases}0 & \text{for } -\pi<x\le 0,\\ \dfrac{\pi x}{4} & \text{for } 0<x<\pi\end{cases}$$

and deduce that

(i) $1-\frac{1}{3}+\frac{1}{5}-\frac{1}{7}+\ldots\infty=\frac{\pi}{4}$,

(ii) $1+\frac{1}{3^2}+\frac{1}{5^2}+\frac{1}{7^2}+\ldots\infty=\frac{\pi^2}{8}$.

Solution: In the interval $[-\pi,\pi]$ we assume the Fourier series as

$$f(x)=\frac{a_0}{2}+\sum_{n=1}^{\infty}a_n\cos nx+\sum_{n=1}^{\infty}b_n\sin nx \qquad \ldots\text{(i)}$$

where $$a_0=\frac{1}{\pi}\int_{-\pi}^{\pi}f(x)dx, \qquad \ldots\text{(ii)}$$

$$a_n=\frac{1}{\pi}\int_{-\pi}^{+\pi}f(x)\cos nx\,dx \qquad \ldots\text{(iii)}$$

and $$b_n = \frac{1}{\pi}\int_{-\pi}^{+\pi} f(x)\sin nx\,dx. \qquad \ldots\text{(iv)}$$

From Eq. (ii), we have

$$a_0 = \frac{1}{\pi}\left\{\int_{-\pi}^{0} f(x)dx + \int_{0}^{\pi} f(x)dx\right\}$$

$$= \frac{1}{\pi}\left\{\int_{-\pi}^{0} 0\cdot dx + \int_{0}^{\pi}\frac{\pi x}{4}dx\right\} = \frac{1}{4}\left[\frac{x^2}{2}\right]_0^{\pi} = \frac{\pi^2}{8}.$$

From Eq. (iii), we have

$$a_n = \frac{1}{\pi}\left\{\int_{-\pi}^{0} f(x)\cos nx\,dx + \int_{0}^{\pi} f(x)\cos nx\,dx\right\}$$

$$= \frac{1}{\pi}\left\{\int_{-\pi}^{0} 0\cdot\cos nx\,dx + \int_{0}^{\pi}\frac{\pi x}{4}\cos nx\,dx\right\}$$

$$= \frac{1}{4}\int_{0}^{\pi} x\cos nx\,dx$$

$$= \frac{1}{4}\left\{\left[\frac{x}{n}\sin nx\right]_0^{\pi} - \int_{0}^{\pi}\frac{1}{n}\sin nx\,dx\right\}$$

$$= \frac{1}{4}\left\{0 + \left[\frac{1}{n^2}\cos nx\right]_0^{\pi}\right\} = \frac{(-1)^n - 1}{4n^2} = \begin{cases} 0 & \text{if } n \text{ is even integer,} \\ -\dfrac{1}{2n^2} & \text{if } n \text{ is odd nteger.} \end{cases}$$

From Eq. (iv), we have

$$b_n = \frac{1}{\pi}\left\{\int_{-\pi}^{0} f(x)\sin nx\,dx + \int_{0}^{\pi} f(x)\sin nx\,dx\right\}$$

$$= \frac{1}{\pi}\left\{\int_{-\pi}^{0} 0\cdot\sin nx\,dx + \int_{0}^{\pi}\frac{\pi x}{4}\sin nx\,dx\right\}$$

$$= \frac{1}{4}\int_{0}^{\pi} x\sin nx\,dx$$

$$= \frac{1}{4}\left\{\left[-\frac{x}{n}\cos nx\right]_0^{\pi} + \int_0^{\pi}\frac{1}{n}\cos nx\,dx\right\}$$

$$= \frac{1}{4}\left\{-\frac{(-1)^n}{n}\pi + \left[\frac{1}{n^2}\sin nx\right]_0^{\pi}\right\} = \frac{(-1)^{n+1}}{4n}\pi.$$

Now, substituting the values of these Fourier coefficients into Eq. (i), we obtain

$$f(x) = \frac{\pi^2}{16} - \frac{1}{2}\sum_{n=\text{odd}}^{\infty}\frac{1}{n^2}\cos nx + \frac{\pi}{4}\sum_{n=1}^{\infty}\frac{(-1)^{n+1}}{n}\sin nx$$

$$= \frac{\pi^2}{16} - \frac{1}{2}\left[\cos x + \frac{\cos 3x}{3^2} + \frac{\cos 5x}{5^2} + \frac{\cos 7x}{7^2} + \ldots\infty\right]$$

$$+ \frac{\pi}{4}\left[\sin x - \frac{\sin 2x}{2} + \frac{\sin 3x}{3} - \frac{\sin 4x}{4} + \frac{\sin 5x}{5} - \ldots\infty\right].$$

Deduction:

(i) Putting $x = \frac{\pi}{2}$ in the above Fourier series of $f(x)$, we have

$$\frac{\pi^2}{8} = \frac{\pi^2}{16} - 0 + \frac{\pi}{4}\left[1 - 0 - \frac{1}{3} - 0 + \frac{1}{5} - \ldots\infty\right]$$

or
$$\frac{\pi^2}{16} = \frac{\pi}{4}\left[1 - \frac{1}{3} + \frac{1}{5} - \frac{1}{7} + \ldots\infty\right]$$

or $1 - \frac{1}{3} + \frac{1}{5} - \frac{1}{7} + \ldots\infty = \frac{\pi}{4}$.

(ii) Putting $x = 0$ in the above Fourier series of $f(x)$, we have

$$0 = \frac{\pi^2}{16} - \frac{1}{2}\left[1 + \frac{1}{3^2} + \frac{1}{5^2} + \frac{1}{7^2} + \ldots\infty\right]$$

or
$$1 + \frac{1}{3^2} + \frac{1}{5^2} + \frac{1}{7^2} + \ldots\infty = \frac{\pi^2}{8}.$$

PROBLEM 8 Find the Fourier series of the periodic function

$$f(x)=\begin{cases}0 & \text{for } -\pi \le x < 0,\\ 1 & \text{for } 0 \le x \le \dfrac{\pi}{2},\\ 0 & \text{for } \dfrac{\pi}{2} < x \le \pi.\end{cases}$$

Solution: The required Fourier series can be assumed as

$$f(x)=\frac{a_0}{2}+\sum_{n=1}^{\infty} a_n \cos nx+\sum_{n=1}^{\infty} b_n \sin nx \qquad \text{...(i)}$$

where
$$a_0=\frac{1}{\pi}\int_{-\pi}^{+\pi} f(x)dx, \qquad \text{...(ii)}$$

$$a_n=\frac{1}{\pi}\int_{-\pi}^{+\pi} f(x)\cos nx\,dx \qquad \text{...(iii)}$$

and
$$b_n=\frac{1}{\pi}\int_{-\pi}^{+\pi} f(x)\sin nx\,dx. \qquad \text{...(iv)}$$

From Eq. (ii), we have

$$a_0=\frac{1}{\pi}\left\{\int_{-\pi}^{0} f(x)dx+\int_{0}^{\frac{\pi}{2}} f(x)dx+\int_{\frac{\pi}{2}}^{\pi} f(x)dx\right\}$$

$$=\frac{1}{\pi}\left\{\int_{-\pi}^{0} 0\cdot dx+\int_{0}^{\frac{\pi}{2}} 1\cdot dx+\int_{\frac{\pi}{2}}^{\pi} 0\cdot dx\right\}=\frac{1}{2}.$$

From Eq. (iii), we have

$$a_n=\frac{1}{\pi}\left\{\int_{-\pi}^{0} f(x)\cos nx\,dx+\int_{0}^{\frac{\pi}{2}} f(x)\cos nx\,dx+\int_{\frac{\pi}{2}}^{\pi} f(x)\cos nx\,dx\right\}$$

$$= \frac{1}{\pi}\left\{0 + \int_0^{\frac{\pi}{2}} \cos nx \, dx + 0\right\}$$

$$= \frac{1}{\pi}\left[\frac{\sin nx}{n}\right]_0^{\frac{\pi}{2}} = \frac{1}{n\pi}\sin\frac{n\pi}{2}.$$

From Eq. (iv), we have

$$b_n = \frac{1}{\pi}\left\{\int_{-\pi}^{0} f(x)\sin nx \, dx + \int_0^{\frac{\pi}{2}} f(x)\sin nx \, dx + \int_{\frac{\pi}{2}}^{\pi} f(x)\sin nx \, dx\right\}$$

$$= \frac{1}{\pi}\left\{0 + \int_0^{\frac{\pi}{2}} \sin nx \, dx + 0\right\}$$

$$= \frac{1}{\pi}\left[-\frac{\cos nx}{n}\right]_0^{\frac{\pi}{2}} = \frac{1-\cos\frac{n\pi}{2}}{n\pi}.$$

Substituting the values of these Fourier coefficients into Eq. (i), we obtain

$$f(x) = \frac{1}{4} + \frac{1}{\pi}\sum_{n=1}^{\infty}\frac{1}{n}\sin\frac{n\pi}{2}\cos nx + \frac{1}{\pi}\sum_{n=1}^{\infty}\frac{1}{n}\left(1-\cos\frac{n\pi}{2}\right)\sin nx$$

$$= \frac{1}{4} + \frac{1}{\pi}\left[\cos x + \frac{1}{2}\sin\pi\cos 2x + \frac{1}{3}\sin\frac{3\pi}{2}\cos 3x + \frac{1}{4}\sin 2\pi\cos 4x\right.$$

$$\left. + \frac{1}{5}\sin\frac{5\pi}{2}\cos 5x + \frac{1}{6}\sin 3\pi\cos 6x + \ldots\infty\right]$$

$$+ \frac{1}{\pi}\left[\frac{\sin x}{1} + \frac{2\sin 2x}{2} + \frac{\sin 3x}{3} + \frac{\sin 5x}{5} + \frac{2\sin 6x}{6} + \ldots\infty\right]$$

$$= \frac{1}{4} + \frac{1}{\pi}\left[\cos x - \frac{\cos 3x}{3} + \frac{\cos 5x}{5} - \ldots\infty\right]$$

$$+ \frac{1}{\pi}\left[\frac{\sin x}{1} + \frac{2\sin 2x}{2} + \frac{\sin 3x}{3} + \frac{\sin 5x}{5} + \frac{2\sin 6x}{6} + \ldots\infty\right].$$

The above series is the required Fourier series of the given function.

PROBLEM 9 Find the Fourier series of the periodic function

$$f(x)=\begin{cases}2 & \text{in} -2<x<0,\\ x & \text{in } 0<x<2.\end{cases}$$

Sketch a profile of the function and also deduce that

$$1-\frac{1}{3}+\frac{1}{5}-\frac{1}{7}+\ldots\infty=\frac{\pi}{4}.$$

Solution: The Fourier series of a function $f(x)$ in an interval $[-L, L]$ has the form

$$f(x)=\frac{a_0}{2}+\sum_{n=1}^{\infty}a_n\cos\left(\frac{n\pi x}{L}\right)+\sum_{n=1}^{\infty}b_n\sin\left(\frac{n\pi x}{L}\right)$$

where $$a_0=\frac{1}{L}\int_{-L}^{+L}f(x)dx,$$

$$a_n=\frac{1}{L}\int_{-L}^{+L}f(x)\cos\left(\frac{n\pi x}{L}\right)dx$$

and $$b_n=\frac{1}{L}\int_{-L}^{+L}f(x)\sin\left(\frac{n\pi x}{L}\right)dx.$$

In this problem, $L = 2$ and hence we have

$$f(x)=\frac{a_0}{2}+\sum_{n=1}^{\infty}a_n\cos\left(\frac{n\pi x}{2}\right)+\sum_{n=1}^{\infty}b_n\sin\left(\frac{n\pi x}{2}\right) \quad \ldots\text{(i)}$$

with $$a_0=\frac{1}{2}\int_{-2}^{+2}f(x)dx, \quad \ldots\text{(ii)}$$

$$a_n=\frac{1}{2}\int_{-2}^{+2}f(x)\cos\left(\frac{n\pi x}{2}\right)dx \quad \ldots\text{(iii)}$$

and $$b_n=\frac{1}{2}\int_{-2}^{+2}f(x)\sin\left(\frac{n\pi x}{2}\right)dx. \quad \ldots\text{(iv)}$$

From Eq. (ii), we have

$$a_0 = \frac{1}{2}\left\{\int_{-2}^{0} f(x)dx + \int_{0}^{2} f(x)dx\right\}$$

$$= \frac{1}{2}\left\{\int_{-2}^{0} 2dx + \int_{0}^{2} x\,dx\right\}$$

$$= \frac{1}{2}\left\{[2x]_{-2}^{0} + \left[\frac{x^2}{2}\right]_{0}^{2}\right\} = \frac{1}{2}\{4+2\} = 3.$$

From Eq. (iii), we have

$$a_n = \frac{1}{2}\left\{\int_{-2}^{0} f(x)\cos\left(\frac{n\pi x}{2}\right)dx + \int_{0}^{2} f(x)\cos\left(\frac{n\pi x}{2}\right)dx\right\}$$

$$= \frac{1}{2}\left\{\int_{-2}^{0} 2\cos\left(\frac{n\pi x}{2}\right)dx + \int_{0}^{2} x\cos\left(\frac{n\pi x}{2}\right)dx\right\}$$

$$= \left[\frac{2}{n\pi}\sin\left(\frac{n\pi x}{2}\right)\right]_{-2}^{0} + \left[\frac{x}{n\pi}\sin\left(\frac{n\pi x}{2}\right)\right]_{0}^{2} - \int_{0}^{2}\frac{1}{n\pi}\sin\left(\frac{n\pi x}{2}\right)dx$$

$$= 0 + 0 + \left[\frac{2}{n^2\pi^2}\cos\left(\frac{n\pi x}{2}\right)\right]_{0}^{2} = \frac{2}{n^2\pi^2}\left[(-1)^n - 1\right]$$

$$= \begin{cases} 0 \text{ if } n \text{ is even integer,} \\ -\dfrac{4}{n^2\pi^2} \text{ if } n \text{ is odd integer.} \end{cases}$$

From Eq. (iv), we have

$$b_n = \frac{1}{2}\left\{\int_{-2}^{0} f(x)\sin\left(\frac{n\pi x}{2}\right)dx + \int_{0}^{2} f(x)\sin\left(\frac{n\pi x}{2}\right)dx\right\}$$

$$= \frac{1}{2}\left\{\int_{-2}^{0} 2\sin\left(\frac{n\pi x}{2}\right)dx + \int_{0}^{2} x\sin\left(\frac{n\pi x}{2}\right)dx\right\}$$

$$= -\left[\frac{2}{n\pi}\cos\left(\frac{n\pi x}{2}\right)\right]_{-2}^{0} - \left[\frac{x}{n\pi}\cos\left(\frac{n\pi x}{2}\right)\right]_{0}^{2} + \int_{0}^{2}\frac{1}{n\pi}\cos\left(\frac{n\pi x}{2}\right)dx$$

$$= -\frac{2}{n\pi}\left[1-(-1)^n\right] - \frac{2}{n\pi}(-1)^n + \left[\frac{2}{n^2\pi^2}\sin\left(\frac{n\pi x}{2}\right)\right]_0^2$$

$$= -\frac{2}{n\pi}.$$

Substituting the values of these Fourier coefficients into Eq. (i), we obtain

$$f(x) = \frac{3}{2} - \frac{4}{\pi^2}\sum_{n=\text{odd}}^{\infty}\frac{1}{n^2}\cos\left(\frac{n\pi x}{2}\right) - \frac{2}{\pi}\sum_{n=1}^{\infty}\frac{1}{n}\sin\left(\frac{n\pi x}{2}\right)$$

$$= \frac{3}{2} - \frac{4}{\pi^2}\left[\frac{1}{1^2}\cos\left(\frac{\pi x}{2}\right) + \frac{1}{3^2}\cos\left(\frac{3\pi x}{2}\right) + \frac{1}{5^2}\cos\left(\frac{5\pi x}{2}\right) + \ldots\infty\right]$$

$$-\frac{2}{\pi}\left[\sin\left(\frac{\pi x}{2}\right) + \frac{1}{2}\sin\left(\frac{2\pi x}{2}\right) + \frac{1}{3}\sin\left(\frac{3\pi x}{2}\right) + \frac{1}{4}\sin\left(\frac{4\pi x}{2}\right) + \frac{1}{5}\sin\left(\frac{5\pi x}{2}\right) + \ldots\infty\right].$$

This represents the required Fourier series.

Profile:

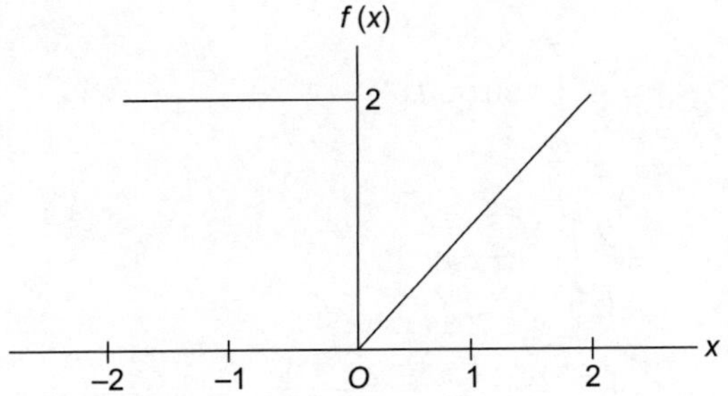

Fig. 11.2 Profile of the given function.

Deductions: Putting $x = 1$, we have

$$1 = \frac{3}{2} - 0 - \frac{2}{\pi}\left[1 + 0 - \frac{1}{3} + 0 + \frac{1}{5} + 0 - \frac{1}{7} + \ldots\infty\right]$$

or

$$1 - \frac{1}{3} + \frac{1}{5} - \frac{1}{7} + \ldots\infty = \frac{\pi}{4}.$$

PROBLEM 10 Find the Fourier series of the function defined by $f(x) = x\sin x$ and $f(x) = f(x+2\pi)$ in an interval $-\pi \le x \le \pi$.

Deduce that $\frac{1}{1\cdot 3}-\frac{1}{3\cdot 5}+\frac{1}{5\cdot 7}-\frac{1}{7\cdot 9}+\ldots\infty=\frac{\pi}{4}-\frac{1}{2}.$

Solution: Here, the function $f(x)=x\sin x$ is an even function and hence its Fourier series representation will not contain sine terms (i.e., b_n must be zero). If one does not have the idea of even or odd function and calculates the values of b_n its values will necessarily be zero.

Thus, the Fourier series can be expressed as

$$f(x)=\frac{a_0}{2}+\sum_{n=1}^{\infty}a_n\cos nx \qquad \ldots\text{(i)}$$

where $$a_0=\frac{1}{\pi}\int_{-\pi}^{+\pi}f(x)dx \qquad \ldots\text{(ii)}$$

and $$a_n=\frac{1}{\pi}\int_{-\pi}^{+\pi}f(x)\cos nx\,dx. \qquad \ldots\text{(iii)}$$

From Eq. (ii), we have

$$a_0=\frac{1}{\pi}\int_{-\pi}^{+\pi}x\sin x\,dx$$

$$=\frac{2}{\pi}\int_0^{\pi}x\sin x\,dx$$

$$=\frac{2}{\pi}\left\{\left[-x\cos x\right]_0^{\pi}+\int_0^{\pi}\cos x\,dx\right\}$$

$$=\frac{2}{\pi}\left\{-\pi\cos\pi+\left[\sin x\right]_0^{\pi}\right\}=2.$$

From Eq. (iii), we have

$$a_n=\frac{1}{\pi}\int_{-\pi}^{+\pi}x\sin x\cos nx\,dx$$

$$=\frac{2}{\pi}\int_0^{\pi}x\sin x\cos nx\,dx$$

$$=\frac{1}{\pi}\int_0^{\pi} x\left[\sin(n+1)x-\sin(n-1)x\right]dx$$

$$=\frac{1}{\pi}\int_0^{\pi} x\sin(n+1)x\,dx-\frac{1}{\pi}\int_0^{\pi} x\sin(n-1)x\,dx$$

$$=\frac{1}{\pi}\left\{\left[-\frac{x}{n+1}\cos(n+1)x\right]_0^{\pi}+\int_0^{\pi}\frac{1}{n+1}\cos(n+1)x\,dx\right\}$$

$$+\frac{1}{\pi}\left\{\left[\frac{x}{n-1}\cos(n-1)x\right]_0^{\pi}-\int_0^{\pi}\frac{1}{n-1}\cos(n-1)x\,dx\right\}$$

$$=-\frac{(-1)^{n+1}}{n+1}+0+\frac{(-1)^{n-1}}{n-1}-0=-\frac{2(-1)^n}{(n-1)(n+1)}\quad(n\neq 1).$$

We see that for $n = 1$ the value of a_1 cannot be determined as above. In this case, Eq. (iii) yields

$$a_1=\frac{1}{\pi}\int_{-\pi}^{+\pi} x\sin x\cos x\,dx=\frac{2}{\pi}\int_0^{\pi} x\sin x\cos x\,dx$$

$$=\frac{1}{\pi}\int_0^{\pi} x\sin 2x\,dx=\frac{1}{\pi}\left\{\left[-\frac{x}{2}\cos 2x\right]_0^{\pi}+\int_0^{\pi}\frac{1}{2}\cos 2x\,dx\right\}$$

$$=-\frac{1}{2}\cos 2\pi+0=-\frac{1}{2}.$$

Now, substituting the values of these Fourier coefficients into Eq. (i), we obtain

$$f(x)=x\sin x$$

$$=1-\frac{1}{2}\cos x-2\sum_{n=2}^{\infty}\frac{(-1)^n}{(n-1)(n+1)}\cos nx$$

$$=1-\frac{\cos x}{2}-2\left[\frac{\cos 2x}{1\cdot 3}-\frac{\cos 3x}{2\cdot 4}+\frac{\cos 4x}{3\cdot 5}-\frac{\cos 5x}{4\cdot 6}+\frac{\cos 6x}{5\cdot 7}-\ldots\infty\right].$$

Deduction: Putting $x=\frac{\pi}{2}$, we have

$$\frac{\pi}{2}=1-0-2\left[\frac{\cos\pi}{1\cdot 3}-0+\frac{\cos 2\pi}{3\cdot 5}-0+\frac{\cos 3\pi}{5\cdot 7}-0+\frac{\cos 4\pi}{7\cdot 9}-0+\ldots\infty\right]$$

or $$\frac{\pi}{2}-1=-2\left[-\frac{1}{1\cdot 3}+\frac{1}{3\cdot 5}-\frac{1}{5\cdot 7}+\frac{1}{7\cdot 9}-\ldots\infty\right]$$

or $$\frac{\pi}{4}-\frac{1}{2}=\frac{1}{1\cdot 3}-\frac{1}{3\cdot 5}+\frac{1}{5\cdot 7}-\frac{1}{7\cdot 9}+\ldots\infty.$$

PROBLEM 11 A periodic function of period $2L$ is defined by

$$f(x)=x^2,\ -L\le x\le L.$$

Sketch a profile of $f(x)$ and its periodic extension.
Find its Fourier series expansion and deduce that

(i) $1-\frac{1}{2^2}+\frac{1}{3^2}-\frac{1}{4^2}+\ldots\infty=\frac{\pi^2}{12}$,

(ii) $\frac{1}{1^2}+\frac{1}{2^2}+\frac{1}{3^2}+\frac{1}{4^2}+\ldots\infty=\frac{\pi^2}{6}$,

(iii) $\sum_{n=1}^{\infty}\frac{1}{n^4}=\frac{1}{1^4}+\frac{1}{2^4}+\frac{1}{3^4}+\frac{1}{4^4}+\ldots\infty=\frac{\pi^4}{90}$.

Solution:

Profile:

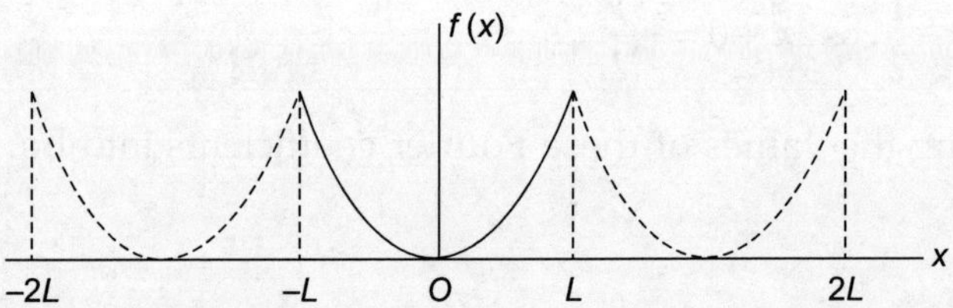

Fig. 11.3 Profile of the function $f(x)=x^2$ and its priodic extension.

Here, $f(x)$ is an even function and, therefore, its Fourier series will not contain sine terms, i.e., we will have $b_n = 0$.

Thus, the Fourier series can be expressed as

$$f(x)=\frac{a_0}{2}+\sum_{n=1}^{\infty}a_n\cos\left(\frac{n\pi x}{L}\right) \qquad \ldots\text{(i)}$$

where $$a_0 = \frac{1}{L}\int_{-L}^{+L} f(x)dx \qquad \text{...(ii)}$$

and $$a_n = \frac{1}{L}\int_{-L}^{+L} f(x)\cos\left(\frac{n\pi x}{L}\right)dx. \qquad \text{...(iii)}$$

From Eq. (ii), we have

$$a_0 = \frac{1}{L}\int_{-L}^{+L} x^2 dx = \frac{2}{L}\int_0^L x^2 dx = \frac{2}{L}\left[\frac{x^3}{3}\right]_0^L = \frac{2L^2}{3}.$$

From Eq. (iii), we have

$$a_n = \frac{1}{L}\int_{-L}^{+L} x^2 \cos\left(\frac{n\pi x}{L}\right)dx = \frac{2}{L}\int_0^L x^2 \cos\left(\frac{n\pi x}{L}\right)dx$$

$$= \frac{2}{L}\left\{\left[\frac{Lx^2}{n\pi}\sin\left(\frac{n\pi x}{L}\right)\right]_0^L - \int_0^L \frac{2Lx}{n\pi}\sin\left(\frac{n\pi x}{L}\right)dx\right\}$$

$$= 0 + \frac{4}{n\pi}\left\{\left[\frac{Lx}{n\pi}\cos\left(\frac{n\pi x}{L}\right)\right]_0^L - \int_0^L \frac{L}{n\pi}\cos\left(\frac{n\pi x}{L}\right)dx\right\}$$

$$= \frac{4L^2\cos n\pi}{n^2\pi^2} - 0 = \frac{4L^2(-1)^n}{n^2\pi^2}.$$

Now, substituting the values of these Fourier coefficients into Eq. (i), we obtain

$$f(x) = x^2 = \frac{L^2}{3} + \frac{4L^2}{\pi^2}\sum_{n=1}^{\infty}\frac{(-1)^n}{n^2}\cos\left(\frac{n\pi x}{L}\right)$$

$$= \frac{L^2}{3} - \frac{4L^2}{\pi^2}\left[\frac{1}{1^2}\cos\left(\frac{\pi x}{L}\right) - \frac{1}{2^2}\cos\left(\frac{2\pi x}{L}\right) + \frac{1}{3^2}\cos\left(\frac{3\pi x}{L}\right) - \frac{1}{4^2}\cos\left(\frac{4\pi x}{L}\right) + \ldots\infty\right].$$

Deduction:

(i) Putting $x = 0$ in the above Fourier series, we have

$$0 = \frac{L^2}{3} - \frac{4L^2}{\pi^2}\left[\frac{1}{1^2} - \frac{1}{2^2} + \frac{1}{3^2} - \frac{1}{4^2} + \ldots\infty\right]$$

or $$\frac{1}{1^2} - \frac{1}{2^2} + \frac{1}{3^2} - \frac{1}{4^2} + \ldots\infty = \frac{\pi^2}{12}.$$

(ii) Putting $x = L$ in the above Fourier series, we have

$$L^2 = \frac{L^2}{3} - \frac{4L^2}{\pi^2}\left[-\frac{1}{1^2} - \frac{1}{2^2} - \frac{1}{3^2} - \frac{1}{4^2} - \ldots \infty\right]$$

or $$\frac{2}{3} = \frac{4}{\pi^2}\left[\frac{1}{1^2} + \frac{1}{2^2} + \frac{1}{3^2} + \frac{1}{4^2} + \ldots \infty\right]$$

or $$\frac{\pi^2}{6} = \frac{1}{1^2} + \frac{1}{2^2} + \frac{1}{3^2} + \frac{1}{4^2} + \ldots \infty = \sum_{n=1}^{\infty}\frac{1}{n^2}.$$

Note: Here, the right-hand side is a Riemann zeta function $\zeta(2)$.

(iii) By Parseval's formula of Fourier series, we have

$$\frac{1}{2L}\int_{-L}^{+L}[f(x)]^2\,dx = \frac{a_0^2}{4} + \frac{1}{2}\sum_{n=1}^{\infty}\left(a_n^2 + b_n^2\right)$$

or $$\frac{1}{2L}\int_{-L}^{+L} x^4\,dx = \frac{1}{4}\left(\frac{2L^2}{3}\right)^2 + \frac{1}{2}\sum_{n=1}^{\infty}\left\{\left[\frac{4L^2(-1)^n}{n^2\pi^2}\right]^2 + 0\right\}$$

or $$\frac{1}{L}\int_0^L x^4\,dx = \frac{L^4}{5} = \frac{L^4}{9} + \frac{8L^4}{\pi^4}\sum_{n=1}^{\infty}\frac{1}{n^4}$$

or $$\sum_{n=1}^{\infty}\frac{1}{n^4} = \frac{\pi^4}{8}\left(\frac{1}{5} - \frac{1}{9}\right) = \frac{\pi^4}{90}$$

or $$\frac{1}{1^4} + \frac{1}{2^4} + \frac{1}{3^4} + \frac{1}{4^4} + \ldots = \frac{\pi^4}{90}.$$

PROBLEM 12 Represent the periodic function defined by $f(x) = L^2 - x^2\ (-L < x < L)$, $f(x) = f(x + 2L)$ in a Fourier series and deduce that

$$1 - \frac{1}{2^2} + \frac{1}{3^2} - \frac{1}{4^2} + \ldots \infty = \frac{\pi^2}{12}.$$

Solution: The function $f(x) = L^2 - x^2$ is an even function and, therefore, its Fourier series will not have sine-terms (i.e., $b_n = 0$).

We write the Fourier series as

$$f(x) = L^2 - x^2 = \frac{a_0}{2} + \sum_{n=1}^{\infty} a_n \cos\left(\frac{n\pi x}{L}\right) \qquad \text{...(i)}$$

where
$$a_0 = \frac{1}{L}\int_{-L}^{+L} f(x)dx \qquad \text{...(ii)}$$

and
$$a_n = \frac{1}{L}\int_{-L}^{+L} f(x)\cos\left(\frac{n\pi x}{L}\right)dx. \qquad \text{...(iii)}$$

From Eq. (ii), we have

$$a_0 = \frac{1}{L}\int_{-L}^{+L}\left(L^2 - x^2\right)dx$$

$$= \frac{2}{L}\int_0^L \left(L^2 - x^2\right)dx = \frac{2}{L}\left(L^3 - \frac{L^3}{3}\right) = \frac{4L^2}{3}.$$

From Eq. (iii), we have

$$a_n = \frac{1}{L}\int_{-L}^{+L}\left(L^2 - x^2\right)\cos\left(\frac{n\pi x}{L}\right)dx$$

$$= \frac{2}{L}\int_0^L \left(L^2 - x^2\right)\cos\left(\frac{n\pi x}{L}\right)dx$$

$$= 2L\int_0^L \cos\left(\frac{n\pi x}{L}\right)dx - \frac{2}{L}\int_0^L x^2 \cos\left(\frac{n\pi x}{L}\right)dx$$

$$= 0 - \frac{2}{L}\left\{\left[\frac{Lx^2}{n\pi}\sin\left(\frac{n\pi x}{L}\right)\right]_0^L - \int_0^L \frac{2Lx}{n\pi}\sin\left(\frac{n\pi x}{L}\right)dx\right\}$$

$$= \frac{4}{n\pi}\int_0^L x\sin\left(\frac{n\pi x}{L}\right)dx$$

$$= \frac{4}{n\pi}\left\{\left[-\frac{Lx}{n\pi}\cos\left(\frac{n\pi x}{L}\right)\right]_0^L + \int_0^L \frac{L}{n\pi}\cos\left(\frac{n\pi x}{L}\right)dx\right\}$$

$$= \frac{4}{n\pi}\left\{-\frac{L^2}{n\pi}\cos n\pi + 0\right\} = -\frac{4L^2(-1)^n}{n^2\pi^2}.$$

Substituting the values of these Fourier coefficients into Eq. (i), we obtain

$$f(x)=L^2-x^2=\frac{2L^2}{3}-\frac{4L^2}{\pi^2}\sum_{n=1}^{\infty}\frac{(-1)^n}{n^2}\cos\left(\frac{n\pi x}{L}\right)$$

$$=\frac{2L^2}{3}+\frac{4L^2}{\pi^2}\left[\frac{1}{1^2}\cos\left(\frac{\pi x}{L}\right)-\frac{1}{2^2}\cos\left(\frac{2\pi x}{L}\right)+\frac{1}{3^2}\cos\left(\frac{3\pi x}{L}\right)-\frac{1}{4^2}\cos\left(\frac{4\pi x}{L}\right)+\ldots\infty\right].$$

Deduction: Putting $x = 0$ in the above Fourier series, we have

$$L^2=\frac{2L^2}{3}+\frac{4L^2}{\pi^2}\left(\frac{1}{1^2}-\frac{1}{2^2}+\frac{1}{3^2}-\frac{1}{4^2}+\ldots\infty\right)$$

or $$\frac{1}{1^2}-\frac{1}{2^2}+\frac{1}{3^2}-\frac{1}{4^2}+\ldots\infty=\frac{\pi^2}{12}.$$

PROBLEM 13 Find the Fourier series of the function defined by $f(x) = x \cos x$, $f(x) = f(x + 2\pi)$ in the interval $-\pi < x < \pi$.

Solution: Here, the function is an odd function and its Fourier series will have only sine terms (i.e., $a_0 = a_n = 0$). Thus, the Fourier series will be

$$f(x)=x\cos x=\sum_{n=1}^{\infty}b_n\sin nx \qquad \ldots\text{(i)}$$

where $$b_n=\frac{1}{\pi}\int_{-\pi}^{+\pi}f(x)\sin nx\,dx$$

$$=\frac{1}{\pi}\int_{-\pi}^{+\pi}x\cos x\sin nx\,dx$$

$$=\frac{2}{\pi}\int_{0}^{\pi}x\sin nx\cos x\,dx$$

$$=\frac{1}{\pi}\int_{0}^{\pi}x\left[\sin\{(n+1)x\}+\sin\{(n-1)x\}\right]dx$$

$$=\frac{1}{\pi}\int_{0}^{\pi}x\sin\{(n+1)x\}\,dx+\frac{1}{\pi}\int_{0}^{\pi}x\sin\{(n-1)x\}\,dx$$

$$= \frac{1}{\pi}\left[\left[-x\frac{\cos\{(n+1)x\}}{n+1}\right]_0^{\pi} + \frac{1}{n+1}\int_0^{\pi}\cos\{(n+1)x\}\,dx\right]$$

$$+\frac{1}{\pi}\left[\left[-x\frac{\cos\{(n-1)x\}}{n-1}\right]_0^{\pi} + \frac{1}{n-1}\int_0^{\pi}\cos\{(n-1)x\}\,dx\right]$$

$$= -\frac{\cos\{(n+1)\pi\}}{n+1} + 0 - \frac{\cos\{(n-1)\pi\}}{n-1} + 0$$

$$= -\frac{(-1)^{n+1}}{n+1} - \frac{(-1)^{n-1}}{n-1}$$

$$= (-1)^n\left[\frac{1}{n+1} + \frac{1}{n-1}\right] = \frac{2n(-1)^n}{(n-1)(n+1)} \quad \text{for } n \neq 1.$$

For $n = 1$, we have

$$b_1 = \frac{1}{\pi}\int_{-\pi}^{+\pi} x\cos x \sin x\,dx$$

$$= \frac{2}{\pi}\int_0^{\pi} x\sin x\cos x\,dx$$

$$= \frac{1}{\pi}\int_0^{\pi} x\sin 2x\,dx$$

$$= \frac{1}{\pi}\left\{\left[-x\frac{\cos 2x}{2}\right]_0^{\pi} + \frac{1}{2}\int_0^{\pi}\cos 2x\,dx\right\}$$

$$= -\frac{\cos 2\pi}{2} + 0 = -\frac{1}{2}.$$

Now, substituting the values of these Fourier coefficients into Eq. (i), we obtain

$$f(x) = x\cos x$$

$$= -\frac{1}{2}\sin x + \sum_{n=2}^{\infty}\frac{2n(-1)^n}{(n-1)(n+1)}\sin nx$$

$$= -\frac{1}{2}\sin x + \frac{4}{1\cdot 3}\sin 2x - \frac{6}{2\cdot 4}\sin 3x + \frac{8}{3\cdot 5}\sin 4x - \ldots\infty.$$

PROBLEM 14 Obtain a Fourier series expansion of the periodic function $f(x)=x^3$ in the interval $-L<x<L$.

Solution: Here, the function $f(x)=x^3$ is an odd function and, therefore, its Fourier series will contain only sine terms (i.e., $a_0 = a_n = 0$).

Let us write the Fourier series as

$$f(x)=x^3=\sum_{n=1}^{\infty} b_n \sin\left(\frac{n\pi x}{L}\right) \qquad \ldots\text{(i)}$$

where
$$b_n=\frac{1}{L}\int_{-L}^{+L} f(x)\sin\left(\frac{n\pi x}{L}\right)dx$$

$$=\frac{1}{L}\int_{-L}^{+L} x^3 \sin\left(\frac{n\pi x}{L}\right)dx$$

$$=\frac{2}{L}\int_{0}^{L} x^3 \sin\left(\frac{n\pi x}{L}\right)dx$$

$$=\frac{2}{L}\left\{\left[-\frac{Lx^3}{n\pi}\cos\left(\frac{n\pi x}{L}\right)\right]_0^L+\int_0^L \frac{3Lx^2}{n\pi}\cos\left(\frac{n\pi x}{L}\right)dx\right\}$$

$$=-\frac{2L^3(-1)^n}{n\pi}+\frac{6}{n\pi}\int_0^L x^2\cos\left(\frac{n\pi x}{L}\right)dx$$

$$=-\frac{2L^3(-1)^n}{n\pi}+\frac{6}{n\pi}\left\{\left[\frac{Lx^2}{n\pi}\sin\left(\frac{n\pi x}{L}\right)\right]_0^L-\int_0^L \frac{2Lx}{n\pi}\sin\left(\frac{n\pi x}{L}\right)dx\right\}$$

$$=-\frac{2L^3(-1)^n}{n\pi}+\frac{6}{n\pi}\left\{0-\frac{2L}{n\pi}\int_0^L x\sin\left(\frac{n\pi x}{L}\right)dx\right\}$$

$$=-\frac{2L^3(-1)^n}{n\pi}+\frac{12L}{n^2\pi^2}\left\{\left[\frac{Lx}{n\pi}\cos\left(\frac{n\pi x}{L}\right)\right]_0^L-\int_0^L \frac{L}{n\pi}\cos\left(\frac{n\pi x}{L}\right)dx\right\}$$

$$=-\frac{2L^3(-1)^n}{n\pi}+\frac{12L}{n^2\pi^2}\left\{\frac{L^2(-1)^n}{n\pi}-0\right\}$$

$$= \frac{2L^3(-1)^n}{\pi}\left(-\frac{1}{n}+\frac{6}{n^3\pi^2}\right).$$

Substituting the values of these Fourier coefficients into Eq. (i), we obtain

$$f(x) = x^3 = \sum_{n=1}^{\infty}\frac{2L^3}{\pi}(-1)^n\left(-\frac{1}{n}+\frac{6}{n^3\pi^2}\right)\sin\left(\frac{n\pi x}{L}\right)$$

$$= \frac{2L^3}{\pi}\left[\left(1-\frac{6}{\pi^2}\right)\sin\left(\frac{\pi x}{L}\right)-\left(\frac{1}{2}-\frac{6}{2^3\pi^2}\right)\sin\left(\frac{2\pi x}{L}\right)+\left(\frac{1}{3}-\frac{6}{3^3\pi^2}\right)\sin\left(\frac{3\pi x}{L}\right)+\ldots\infty\right].$$

PROBLEM 15 Make even and odd periodic extensions of the function

$$f(x) = x(\pi - x) \quad \text{for} \quad 0 < x < \pi,$$

obtain the half-range Fourier series and deduce that

(a) $\displaystyle\sum_{n=1}^{\infty}\frac{(-1)^{n-1}}{n^2} = \frac{\pi^2}{12}$,

(b) $\displaystyle\frac{\pi^3}{32} = \sum_{n=0}^{\infty}\frac{(-1)^n}{(2n+1)^3}$,

(c) $\displaystyle\frac{1}{2^4}+\frac{1}{4^4}+\frac{1}{6^4}+\ldots\infty = \frac{\pi^4}{1440}$ and $\displaystyle\frac{1}{1^4}+\frac{1}{2^4}+\frac{1}{3^4}+\frac{1}{4^4}+\ldots\infty = \frac{\pi^4}{90}$,

(d) $\displaystyle\sum_{n=1}^{\infty}\frac{1}{n^6} = \frac{\pi^6}{945}$.

Even Periodic Extension

In this case, we assume the Fourier series as

$$f(x) = \frac{a_0}{2} + \sum_{n=1}^{\infty} a_n \cos nx\,dx \qquad \ldots\text{(i)}$$

where

$$a_0 = \frac{2}{\pi}\int_0^{\pi} f(x)dx \qquad \ldots\text{(ii)}$$

and $$a_n = \frac{2}{\pi}\int_0^{\pi} f(x)\cos nx\,dx. \qquad \text{...(iii)}$$

From Eq. (ii), we have

$$a_0 = \frac{2}{\pi}\int_0^{\pi}\left(\pi x - x^2\right)dx = \frac{2}{\pi}\left[\frac{\pi x^2}{2} - \frac{x^3}{3}\right]_0^{\pi} = \frac{\pi^2}{3}.$$

From Eq. (iii), we have

$$a_n = \frac{2}{\pi}\int_0^{\pi}\left(\pi x - x^2\right)\cos nx\,dx$$

$$= \frac{2}{\pi}\left\{\left[\frac{\left(\pi x - x^2\right)}{n}\sin nx\right]_0^{\pi} - \int_0^{\pi}\frac{(\pi - 2x)}{n}\sin nx\,dx\right\}$$

$$= \frac{2}{\pi}\left\{0 + \left[\frac{(\pi - 2x)}{n^2}\cos nx\right]_0^{\pi} + \int_0^{\pi}\frac{2}{n^2}\cos nx\,dx\right\}$$

$$= \frac{2}{\pi}\left\{-\frac{\pi(-1)^n}{n^2} - \frac{\pi}{n^2} + \left[\frac{2\sin nx}{n^3}\right]_0^{\pi}\right\}$$

$$= \frac{2}{\pi}\left\{-\frac{\pi(-1)^n}{n^2} - \frac{\pi}{n^2} + 0\right\}$$

$$= \begin{cases} -\dfrac{4}{n^2} & \text{when } n \text{ is even integer}, \\ 0 & \text{when } n \text{ is odd integer}. \end{cases}$$

Substituting the values of these Fourier coefficients into Eq. (i), we obtain

$$f(x) = x(\pi - x)$$

$$= \frac{\pi^2}{6} - 4\sum_{n=\text{even}}^{\infty}\frac{1}{n^2}\cos nx$$

$$= \frac{\pi^2}{6} - \left[\frac{\cos 2x}{1^2} + \frac{\cos 4x}{2^2} + \frac{\cos 6x}{3^2} + \frac{\cos 8x}{4^2} + \ldots\infty\right]. \qquad \text{...(iv)}$$

Odd Periodic Extension

In this case, we assume the Fourier series as

$$f(x) = \sum_{n=1}^{\infty} b_n \sin nx \qquad \ldots(v)$$

where
$$b_n = \frac{2}{\pi}\int_0^{\pi} f(x)\sin nx\,dx$$

$$= \frac{2}{\pi}\int_0^{\pi}(\pi x - x^2)\sin nx\,dx$$

$$= \frac{2}{\pi}\left\{\left[-(\pi x - x^2)\frac{\cos nx}{n}\right]_0^{\pi} + \int_0^{\pi}(\pi - 2x)\frac{\cos nx}{n}dx\right\}$$

$$= \frac{2}{\pi}\left\{0 + \left[\frac{(\pi - 2x)}{n^2}\sin nx\right]_0^{\pi} + \int_0^{\pi}\frac{2}{n^2}\sin nx\,dx\right\}$$

$$= \frac{2}{\pi}\left\{0 + \left[-\frac{2}{n^3}\cos nx\right]_0^{\pi}\right\}$$

$$= -\frac{4}{\pi n^3}\left\{(-1)^n - 1\right\}$$

$$= \begin{cases} 0 & \text{when } n \text{ is even integer,} \\ \dfrac{8}{\pi n^3} & \text{when } n \text{ is odd integer.} \end{cases}$$

Substituting the values of these Fourier coefficients into Eq. (v), we obtain

$$f(x) = x(\pi - x) = \frac{8}{\pi}\sum_{n=\text{odd}}^{\infty}\frac{\sin nx}{n^3}$$

$$= \frac{8}{\pi}\left[\frac{\sin x}{1^3} + \frac{\sin 3x}{3^3} + \frac{\sin 5x}{5^3} + \frac{\sin 7x}{7^3} + \ldots\infty\right]. \qquad \ldots(vi)$$

Deductions:

(a) Putting $x = \frac{\pi}{2}$ in Eq. (iv), we have:

$$\frac{\pi^2}{4}=\frac{\pi^2}{6}-\left[\frac{\cos\pi}{1^2}+\frac{\cos 2\pi}{2^2}+\frac{\cos 3\pi}{3^2}+\frac{\cos 4\pi}{4^2}+\ldots\infty\right]$$

$$=\frac{\pi^2}{6}+\left[\frac{1}{1^2}-\frac{1}{2^2}+\frac{1}{3^2}-\frac{1}{4^2}+\ldots\infty\right]$$

or $$\frac{\pi^2}{12}=\frac{1}{1^2}-\frac{1}{2^2}+\frac{1}{3^2}-\frac{1}{4^2}+\ldots\infty=\sum_{n=1}^{\infty}\frac{(-1)^{n-1}}{n^2}.$$

(b) Putting $x=\frac{\pi}{2}$ in Eq. (vi), we have

$$\frac{\pi^2}{4}=\frac{8}{\pi}\left[\frac{1}{1^3}-\frac{1}{3^3}+\frac{1}{5^3}-\frac{1}{7^3}+\ldots\infty\right]$$

or $$\frac{\pi^3}{32}=\frac{1}{1^3}-\frac{1}{3^3}+\frac{1}{5^3}-\frac{1}{7^3}+\ldots\infty$$

$$=\sum_{n=0}^{\infty}\frac{(-1)^n}{(2n+1)^3}.$$

(c) From the Fourier series of even periodic extension, Parseval's formula yields

$$\frac{1}{\pi}\int_0^{\pi}x^2(\pi-x)^2\,dx=\frac{a_0^2}{4}+\frac{1}{2}\sum a_n^2$$

or $$\frac{1}{\pi}\int_0^{\pi}\left(\pi^2x^2-2\pi x^3+x^4\right)dx=\frac{1}{4}\frac{\pi^4}{9}+\frac{1}{2}\sum_{n=\text{even}}\left(-\frac{4}{n^2}\right)^2$$

or $$\frac{1}{\pi}\left[\frac{\pi^2x^3}{3}-\frac{2\pi x^4}{4}+\frac{x^5}{5}\right]_0^{\pi}=\frac{\pi^4}{36}+8\sum_{n=\text{even}}\frac{1}{n^4}$$

or $$\frac{\pi^4}{30}=\frac{\pi^4}{36}+8\left(\frac{1}{2^4}+\frac{1}{4^4}+\frac{1}{6^4}+\frac{1}{8^4}+\ldots\infty\right)$$

or $$\frac{1}{2^4}+\frac{1}{4^4}+\frac{1}{6^4}+\frac{1}{8^4}+\ldots\infty=\frac{\pi^4}{1440}$$

or $$\frac{1}{2^4}\left(\frac{1}{1^4}+\frac{1}{2^4}+\frac{1}{3^4}+\frac{1}{4^4}+\ldots\infty\right)=\frac{\pi^4}{1440}$$

or
$$\frac{1}{1^4}+\frac{1}{2^4}+\frac{1}{3^4}+\frac{1}{4^4}+\ldots\infty=\frac{\pi^4}{1440}\times 16=\frac{\pi^4}{90}.$$

(d) From the Fourier series of odd periodic extension, Parseval's formula yields

$$\frac{1}{\pi}\int_0^{\pi} x^2(\pi-x)^2\,dx=\frac{1}{2}\sum b_n^2$$

or
$$\frac{\pi^4}{30}=\frac{1}{2}\sum_{n=\text{odd}}\left(\frac{8}{\pi n^3}\right)^2=\frac{32}{\pi^2}\sum_{n=\text{odd}}\left(\frac{1}{n^6}\right)$$

or
$$\frac{\pi^6}{960}=\sum_{n=\text{odd}}\left(\frac{1}{n^6}\right)=\left(\frac{1}{1^6}+\frac{1}{3^6}+\frac{1}{5^6}+\frac{1}{7^6}+\ldots\infty\right)$$

$$=\left(\frac{1}{1^6}+\frac{1}{2^6}+\frac{1}{3^6}+\frac{1}{4^6}+\frac{1}{5^6}+\frac{1}{6^6}+\ldots\infty\right)-\left(\frac{1}{2^6}+\frac{1}{4^6}+\frac{1}{6^6}+\ldots\infty\right)$$

$$=\sum_{n=1}^{\infty}\frac{1}{n^6}-\frac{1}{2^6}\sum_{n=1}^{\infty}\frac{1}{n^6}=\frac{63}{64}\sum_{n=1}^{\infty}\frac{1}{n^6}$$

or
$$\sum_{n=1}^{\infty}\frac{1}{n^6}=\frac{\pi^6}{960}\times\frac{64}{63}=\frac{\pi^6}{945}.$$

Problems on Triangular Waves

A triangular wave is a non-sinusoidal waveform named for its triangular shape. A function representing such a wave is a periodic, piecewise linear, continuous real function. The triangular wave contains only odd harmonics. A triangular or tent function is a function having a profile in the form of a triangle.

PROBLEM 16 A periodic triangular wave is represented by the function

$$f(x)=|x|,\ -L<x<L$$

or
$$f(x)=\begin{cases}-x & \text{for } -L<x<0,\\ x & \text{for } 0<x<L.\end{cases}$$

Sketch a profile of the function and its periodic extension.

Represent $f(x)$ by a Fourier series and deduce that

(i) $\frac{1}{1^2}+\frac{1}{3^2}+\frac{1}{5^2}+\ldots\infty=\frac{\pi^2}{8}$,

(ii) $\frac{1}{1^4}+\frac{1}{3^4}+\frac{1}{5^4}+\frac{1}{7^4}+\ldots\infty=\frac{\pi^4}{96}$.

Solution: The profile is shown in Fig. 11.4.

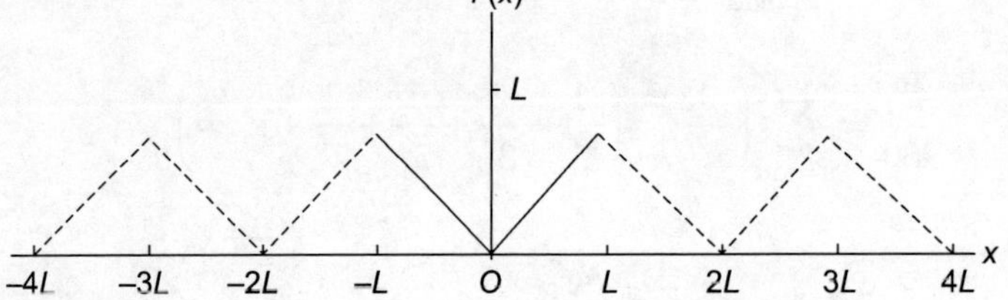

Fig. 11.4 Profile of $f(x)=|x|$ and its periodic extension.

Here, the function is an even function and its Fourier series will not have sine terms (i.e., $b_n = 0$). Thus, the Fourier series will be

$$f(x)=\frac{a_0}{2}+\sum_{n=1}^{\infty}a_n\cos\left(\frac{n\pi x}{L}\right) \qquad \ldots\text{(i)}$$

where
$$a_0=\frac{1}{L}\int_{-L}^{+L}f(x)dx \qquad \ldots\text{(ii)}$$

and
$$a_n=\frac{1}{L}\int_{-L}^{+L}f(x)\cos\left(\frac{n\pi x}{L}\right)dx\,. \qquad \ldots\text{(iii)}$$

From Eq. (ii), we have

$$a_0=\frac{1}{L}\left\{\int_{-L}^{0}f(x)dx+\int_{0}^{L}f(x)dx\right\}$$

$$=\frac{1}{L}\left\{\int_{-L}^{0}-x\,dx+\int_{0}^{L}xdx\right\}$$

$$=\frac{1}{L}\left\{\int_{0}^{-L}x\,dx+\int_{0}^{L}x\,dx\right\}$$

$$= \frac{1}{L}\left\{\frac{L^2}{2}+\frac{L^2}{2}\right\} = L.$$

From Eq. (iii), we have

$$a_n = \frac{1}{L}\left\{\int_{-L}^{0} f(x)\cos\left(\frac{n\pi x}{L}\right)dx + \int_{0}^{L} f(x)\cos\left(\frac{n\pi x}{L}\right)dx\right\}$$

$$= \frac{1}{L}\left\{\int_{-L}^{0} -x\cos\left(\frac{n\pi x}{L}\right)dx + \int_{0}^{L} x\cos\left(\frac{n\pi x}{L}\right)dx\right\}$$

$$= \frac{1}{L}\int_{0}^{-L} x\cos\left(\frac{n\pi x}{L}\right)dx + \frac{1}{L}\int_{0}^{L} x\cos\left(\frac{n\pi x}{L}\right)dx$$

$$= \frac{1}{L}\int_{0}^{L} (-x)\cos\left(-\frac{n\pi x}{L}\right)(-dx) + \frac{1}{L}\int_{0}^{L} x\cos\left(\frac{n\pi x}{L}\right)dx$$

$$= \frac{2}{L}\int_{0}^{L} x\cos\left(\frac{n\pi x}{L}\right)dx$$

$$= \frac{2}{L}\left\{\left[\frac{Lx}{n\pi}\sin\left(\frac{n\pi x}{L}\right)\right]_0^L - \frac{L}{n\pi}\int_{0}^{L}\sin\left(\frac{n\pi x}{L}\right)dx\right\}$$

$$= \frac{2}{L}\left\{0 + \frac{L^2}{n^2\pi^2}\left[\cos\left(\frac{n\pi x}{L}\right)\right]_0^L\right\}$$

$$= \frac{2L}{n^2\pi^2}\left\{(-1)^n - 1\right\} = \begin{cases} 0 & \text{if } n \text{ is even integer}, \\ -\dfrac{4L}{n^2\pi^2} & \text{if } n \text{ is odd integer}. \end{cases}$$

Now, substituting the values of these Fourier coefficients into Eq. (i), we obtain

$$f(x) = \frac{L}{2} + \sum_{n=\text{odd}}\left(-\frac{4L}{n^2\pi^2}\right)\cos\left(\frac{n\pi x}{L}\right)$$

$$= \frac{L}{2} - \frac{4L}{\pi^2}\left\{\frac{1}{1^2}\cos\left(\frac{\pi x}{L}\right) + \frac{1}{3^2}\cos\left(\frac{3\pi x}{L}\right) + \frac{1}{5^2}\cos\left(\frac{5\pi x}{L}\right) + \ldots\infty\right\}.$$

Deductions:

(i) Putting $x=0$ in the above Fourier series, we obtain

$$0=\frac{L}{2}-\frac{4L}{\pi^2}\left(\frac{1}{1^2}+\frac{1}{3^2}+\frac{1}{5^2}+\ldots\infty\right)$$

or $$\frac{1}{1^2}+\frac{1}{3^2}+\frac{1}{5^2}+\ldots\infty=\frac{\pi^2}{8}.$$

(ii) Using Parseval's formula in the above Fourier series, we have

$$\frac{1}{2L}\int_{-L}^{+L}\left[f(x)\right]^2dx=\frac{a_0^2}{4}+\frac{1}{2}\sum_{n=1}^{\infty}\left[a_n^2+b_n^2\right]$$

or $$\frac{1}{2L}\int_{-L}^{+L}x^2dx=\frac{L^2}{4}+\frac{1}{2}\sum_{n=\text{odd}}\left[\left(-\frac{4L}{\pi^2n^2}\right)^2+0\right]$$

or $$\frac{1}{L}\int_0^L x^2dx=\frac{L^2}{4}+\frac{8L^2}{\pi^4}\left[\frac{1}{1^4}+\frac{1}{3^4}+\frac{1}{5^4}+\ldots\infty\right]$$

or $$\frac{L^2}{3}=\frac{L^2}{4}+\frac{8L^2}{\pi^4}\left[\frac{1}{1^4}+\frac{1}{3^4}+\frac{1}{5^4}+\ldots\infty\right]$$

or $$\frac{1}{1^4}+\frac{1}{3^4}+\frac{1}{5^4}+\ldots\infty=\frac{\pi^4}{96}.$$

PROBLEM 17 A triangular wave function of period $2L$ is defined by

$$f(x)=\begin{cases}x & \text{for } 0<x<L\\ 2L-x & \text{for } L<x<2L\end{cases}$$

and $$f(x)=f(x+2L).$$

Using even periodic extension, sketch a profile of the function and its periodic extension. Also obtain a Fourier series representing the function $f(x)$.

Also deduce that $$\frac{1}{1^2}+\frac{1}{3^2}+\frac{1}{5^2}+\frac{1}{7^2}+\ldots\infty=\frac{\pi^2}{8}$$

and $$\frac{1}{1^4}+\frac{1}{3^4}+\frac{1}{5^4}+\ldots\infty=\frac{\pi^4}{96}.$$

Solution: Beyond the domain $0 < x < 2L$ we can extend the function as an even periodic function so that $f(-x) = f(x)$.

The profile of the given function and its even periodic extension is shown in Fig. 11.5.

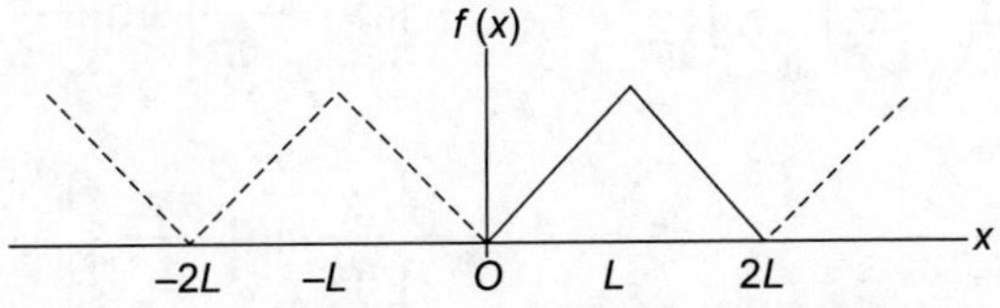

Fig. 11.5 Profile of the given function with even periodic extension.

In this case, we assume the Fourier series of the function as

$$f(x) = \frac{a_0}{2} + \sum_{n=1}^{\infty} a_n \cos\left(\frac{n\pi x}{L}\right) \quad \text{...(i)}$$

where

$$a_0 = \frac{1}{L}\int_0^{2L} f(x)dx \quad \text{...(ii)}$$

and

$$a_n = \frac{1}{L}\int_0^{2L} f(x)\cos\left(\frac{n\pi x}{L}\right)dx. \quad \text{...(iii)}$$

From Eq. (ii), we have

$$a_0 = \frac{1}{L}\left\{\int_0^{L} f(x)dx + \int_L^{2L} f(x)dx\right\}$$

$$= \frac{1}{L}\left\{\int_0^{L} x\,dx + \int_L^{2L} (2L - x)dx\right\}$$

$$= \frac{1}{L}\left\{\frac{L^2}{2} + 2L^2 - \frac{3L^2}{2}\right\} = L.$$

From Eq. (iii), we have

$$a_n = \frac{1}{L}\left\{\int_0^{L} f(x)\cos\left(\frac{n\pi x}{L}\right)dx + \int_L^{2L} f(x)\cos\left(\frac{n\pi x}{L}\right)dx\right\}$$

$$= \frac{1}{L}\left\{\int_0^{L} x\cos\left(\frac{n\pi x}{L}\right)dx + \int_L^{2L} (2L - x)\cos\left(\frac{n\pi x}{L}\right)dx\right\}$$

$$= \frac{1}{L}\int_0^L x\cos\left(\frac{n\pi x}{L}\right)dx + 2\int_L^{2L}\cos\left(\frac{n\pi x}{L}\right)dx - \frac{1}{L}\int_L^{2L}x\cos\left(\frac{n\pi x}{L}\right)dx$$

$$= \frac{1}{L}\left\{\left[\frac{Lx}{n\pi}\sin\left(\frac{n\pi x}{L}\right)\right]_0^L - \int_0^L \frac{L}{n\pi}\sin\left(\frac{n\pi x}{L}\right)dx\right\} + \frac{2L}{n\pi}\left[\sin\left(\frac{n\pi x}{L}\right)\right]_L^{2L}$$

$$-\left[\frac{x}{n\pi}\sin\left(\frac{n\pi x}{L}\right)\right]_L^{2L} + \int_L^{2L}\frac{1}{n\pi}\sin\left(\frac{n\pi x}{L}\right)dx$$

$$= 0 + \frac{L}{n^2\pi^2}\left[\cos\left(\frac{n\pi x}{L}\right)\right]_0^L + 0 - 0 - \frac{L}{n^2\pi^2}\left[\cos\left(\frac{n\pi x}{L}\right)\right]_L^{2L}$$

$$= \frac{L}{n^2\pi^2}\left\{(-1)^n - 1\right\} - \frac{L}{n^2\pi^2}\left\{1 - (-1)^n\right\}$$

$$= \frac{2L}{n^2\pi^2}\left\{(-1)^n - 1\right\} = \begin{cases} 0 & \text{if } n \text{ is even integer,} \\ -\dfrac{4L}{n^2\pi^2} & \text{if } n \text{ is odd integer.} \end{cases}$$

Substituting the values of these Fourier coefficients into Eq. (i), we obtain

$$f(x) = \frac{L}{2} - \frac{4L}{\pi^2}\sum_{n=\text{odd}}\frac{1}{n^2}\cos\left(\frac{n\pi x}{L}\right)$$

$$= \frac{L}{2} - \frac{4L}{\pi^2}\left[\frac{1}{1^2}\cos\left(\frac{\pi x}{L}\right) + \frac{1}{3^2}\cos\left(\frac{3\pi x}{L}\right) + \frac{1}{5^2}\cos\left(\frac{5\pi x}{L}\right) + \ldots\infty\right]. \qquad \ldots\text{(iv)}$$

Deductions: Putting $x = 0$ in Eq. (iv), we obtain

$$0 = \frac{L}{2} - \frac{4L}{\pi^2}\left[\frac{1}{1^2} + \frac{1}{3^2} + \frac{1}{5^2} + \frac{1}{7^2} + \ldots\infty\right]$$

or
$$\frac{1}{1^2} + \frac{1}{3^2} + \frac{1}{5^2} + \frac{1}{7^2} + \ldots\infty = \frac{\pi^2}{8}.$$

Further, Parseval's formula yields

$$\frac{a_0^2}{4} + \frac{1}{2}\sum_{n=1}^{\infty}\left(a_n^2 + b_n^2\right) = \frac{1}{2L}\int_0^{2L}\left[f(x)\right]^2 dx \text{ (Here, } b_n = 0)$$

or
$$\frac{L^2}{4} + \frac{1}{2}\sum_{n=\text{odd}}\left(-\frac{4L}{n^2\pi^2}\right)^2 \qquad [\text{as } b_n = 0]$$

$$= \frac{L^2}{4} + \frac{8L^2}{\pi^4}\left(\frac{1}{1^4} + \frac{1}{3^4} + \frac{1}{5^4} + \frac{1}{7^4} + \ldots \infty\right)$$

$$= \frac{1}{2L}\left\{\int_0^L [f(x)]^2 dx + \int_L^{2L} [f(x)]^2 dx\right\}$$

$$= \frac{1}{2L}\left\{\int_0^L x^2 dx + \int_L^{2L} \left(4L^2 - 4Lx + x^2\right) dx\right\}$$

$$= \frac{1}{2L}\left\{\left[\frac{x^3}{3}\right]_0^L + \left[4L^2x - 2Lx^2 + \frac{x^3}{3}\right]_L^{2L}\right\} = \frac{L^2}{3}.$$

Therefore, we have

$$\frac{1}{1^4} + \frac{1}{3^4} + \frac{1}{5^4} + \ldots \infty = \left(\frac{L^2}{3} - \frac{L^2}{4}\right)\frac{\pi^4}{8L^2} = \frac{\pi^4}{96}.$$

PROBLEM 18 A triangular wave is defined by a tent function

$$f(x) = \begin{cases} \dfrac{2k}{L}x & \text{for } 0 < x < \dfrac{L}{2}, \\ \dfrac{2k}{L}(L - x) & \text{for } \dfrac{L}{2} < x < L. \end{cases}$$

Sketch the profiles of the function with even and odd extensions respectively, and hence find the half range Fourier series using even and odd periodic extensions.

Also deduce that $\frac{1}{1^2} + \frac{1}{3^2} + \frac{1}{5^2} + \frac{1}{7^2} + \ldots \infty = \frac{\pi^2}{8}$

and $$\frac{1}{1^4} + \frac{1}{3^4} + \frac{1}{5^4} + \frac{1}{7^4} + \ldots \infty = \frac{\pi^4}{96}.$$

Solution: The above function may be extended perodically in two ways as detailed below.

I Even Periodic Extension

Beyond the domain $0 < x < L$ we can extend the function as an even periodic function so that $f(-x) = f(x)$ and $f(x) = f(x + L)$.

The profile of the given function and its even periodic extension is shown in Fig. 11.6.

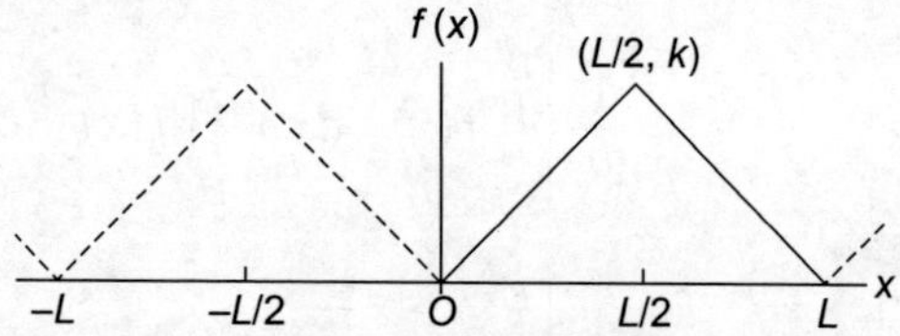

Fig. 11.6 Profile of the given function with even periodic extension.

Let us write the half-range cosine series as

$$f(x)=\frac{a_0}{2}+\sum_{n=1}^{\infty}a_n\cos\left(\frac{n\pi x}{L}\right) \quad \ldots\text{(i)}$$

where
$$a_0=\frac{2}{L}\int_0^L f(x)dx \quad \ldots\text{(ii)}$$

and
$$a_n=\frac{2}{L}\int_0^L f(x)\cos\left(\frac{n\pi x}{L}\right)dx. \quad \ldots\text{(iii)}$$

From Eq. (ii), we have

$$a_0=\frac{2}{L}\left\{\int_0^{\frac{L}{2}} f(x)dx+\int_{\frac{L}{2}}^{L} f(x)dx\right\}$$

$$=\frac{2}{L}\left\{\int_0^{\frac{L}{2}}\frac{2k}{L}x\,dx+\int_{\frac{L}{2}}^{L}\frac{2k}{L}(L-x)dx\right\}$$

$$=\frac{4k}{L^2}\left\{\left[\frac{x^2}{2}\right]_0^{\frac{L}{2}}+\left[Lx-\frac{x^2}{2}\right]_{\frac{L}{2}}^{L}\right\}=k.$$

From Eq. (iii), we have

$$a_n=\frac{2}{L}\left\{\int_0^{\frac{L}{2}} f(x)\cos\left(\frac{n\pi x}{L}\right)dx+\int_{\frac{L}{2}}^{L} f(x)\cos\left(\frac{n\pi x}{L}\right)dx\right\}$$

$$= \frac{2}{L}\left\{\int_0^{\frac{L}{2}} \frac{2k}{L} x\cos\left(\frac{n\pi x}{L}\right)dx + \int_{\frac{L}{2}}^{L} \frac{2k}{L}(L-x)\cos\left(\frac{n\pi x}{L}\right)dx\right\}$$

$$= \frac{4k}{L^2}\left\{\left[\frac{Lx}{n\pi}\sin\left(\frac{n\pi x}{L}\right)\right]_0^{\frac{L}{2}} - \int_0^{\frac{L}{2}} \frac{L}{n\pi}\sin\left(\frac{n\pi x}{L}\right)dx\right\}$$

$$+ \frac{4k}{L^2}\left\{\left[\frac{L(L-x)}{n\pi}\sin\left(\frac{n\pi x}{L}\right)\right]_{\frac{L}{2}}^{L} + \int_{\frac{L}{2}}^{L} \frac{L}{n\pi}\sin\left(\frac{n\pi x}{L}\right)dx\right\}$$

$$= \frac{4k}{L^2}\left\{\frac{L^2}{2n\pi}\sin\left(\frac{n\pi}{2}\right) + \left[\frac{L^2}{n^2\pi^2}\cos\left(\frac{n\pi x}{L}\right)\right]_0^{\frac{L}{2}}\right\}$$

$$+ \frac{4k}{L^2}\left\{0 - \frac{L^2}{2n\pi}\sin\left(\frac{n\pi}{2}\right) - \left[\frac{L^2}{n^2\pi^2}\cos\left(\frac{n\pi x}{L}\right)\right]_{\frac{L}{2}}^{L}\right\}$$

$$= \frac{4k}{n^2\pi^2}\left\{2\cos\left(\frac{n\pi}{2}\right) - 1 - (-1)^n\right\}.$$

Substituting the values of these Fourier coefficients into Eq. (i), we obtain

$$f(x) = \frac{k}{2} + \sum_{n=1}^{\infty} \frac{4k\left\{2\cos\left(\frac{n\pi}{2}\right) - (-1)^n - 1\right\}}{n^2\pi^2}\cos\left(\frac{n\pi x}{L}\right)$$

$$= \frac{k}{2} + \frac{4k}{\pi^2}\left\{-\frac{4}{2^2}\cos\left(\frac{2\pi x}{L}\right) - \frac{4}{6^2}\cos\left(\frac{6\pi x}{L}\right) - \frac{4}{10^2}\cos\left(\frac{10\pi x}{L}\right) - \infty\right\}$$

$$= \frac{k}{2} - \frac{4k}{\pi^2}\left\{\frac{1}{1^2}\cos\left(\frac{2\pi x}{L}\right) + \frac{1}{3^2}\cos\left(\frac{6\pi x}{L}\right) + \frac{1}{5^2}\cos\left(\frac{10\pi x}{L}\right) + \ldots\infty\right\}$$

$$= \frac{k}{2} - \frac{4k}{\pi^2}\sum_{n=0}^{\infty} \frac{1}{(2n+1)^2}\cos\left\{\frac{2(2n+1)\pi x}{L}\right\}. \qquad \ldots\text{(iv)}$$

Deductions: Putting $x = 0$, we obtain

$$0 = \frac{k}{2} - \frac{4k}{\pi^2}\left(\frac{1}{1^2} + \frac{1}{3^2} + \frac{1}{5^2} + \frac{1}{7^2} + \ldots \infty\right)$$

or
$$\frac{1}{1^2} + \frac{1}{3^2} + \frac{1}{5^2} + \frac{1}{7^2} + \ldots \infty = \frac{\pi^2}{8}.$$

Further, by Parseval's formula, we have

$$\frac{1}{L}\int_0^L [f(x)]^2 dx = \frac{a_0^2}{4} + \frac{1}{2}\sum\left(a_n^2 + b_n^2\right) \quad (\text{Here, } b_n = 0)$$

or
$$\frac{1}{L}\left\{\int_0^{\frac{L}{2}} [f(x)]^2 dx + \int_{\frac{L}{2}}^{L} [f(x)]^2 dx\right\}$$

$$= \frac{k^2}{4} + \frac{1}{2}\sum_{n=0}^{\infty}\left[\frac{4k}{n^2\pi^2}\left\{2\cos\left(\frac{n\pi}{2}\right) - 1 - (-1)^n\right\}\right]^2$$

or
$$\frac{1}{L}\left\{\int_0^{\frac{L}{2}} \frac{4k^2}{L^2} x^2 dx + \int_{\frac{L}{2}}^{L} \frac{4k^2}{L^2}\left(L^2 - 2Lx + x^2\right) dx\right\}$$

$$= \frac{k^2}{4} + \frac{8k^2}{\pi^4}\left(\frac{1}{1^4} + \frac{1}{3^4} + \frac{1}{5^4} + \frac{1}{7^4} + \ldots \infty\right)$$

or
$$\frac{4k^2}{L^3}\left\{\left[\frac{x^3}{3}\right]_0^{\frac{L}{2}} + \left[L^2 x - Lx^2 + \frac{x^3}{3}\right]_{\frac{L}{2}}^{L}\right\} = \frac{k^2}{3}$$

$$= \frac{k^2}{4} + \frac{8k^2}{\pi^4}\left(\frac{1}{1^4} + \frac{1}{3^4} + \frac{1}{5^4} + \frac{1}{7^4} + \ldots \infty\right).$$

Thus, we have

$$\frac{1}{1^4} + \frac{1}{3^4} + \frac{1}{5^4} + \frac{1}{7^4} + \ldots \infty = \left(\frac{k^2}{3} - \frac{k^2}{4}\right)\frac{\pi^4}{8k^2} = \frac{\pi^4}{96}.$$

II Odd Periodic Extension

Also we see that beyond the domain $0 < x < L$, we can extend the function as an odd periodic function so that $f(-x) = -f(x)$.

The profile of the given function and its odd periodic extension is shown in Fig. 11.7.

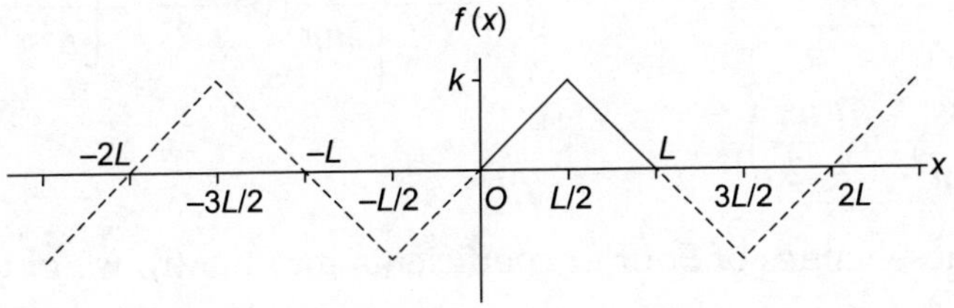

Fig. 11.7 Profile of the given function with odd periodic extension.

Let us write the half-range sine series as

$$f(x) = \sum_{n=1}^{\infty} b_n \sin\left(\frac{n\pi x}{L}\right) \qquad \ldots\text{(i)}$$

where
$$b_n = \frac{2}{L}\int_0^L f(x)\sin\left(\frac{n\pi x}{L}\right)dx$$

$$= \frac{2}{L}\left\{\int_0^{\frac{L}{2}} f(x)\sin\left(\frac{n\pi x}{L}\right)dx + \int_{\frac{L}{2}}^{L} f(x)\sin\left(\frac{n\pi x}{L}\right)dx\right\}$$

$$= \frac{2}{L}\left\{\int_0^{\frac{L}{2}} \frac{2k}{L}x\sin\left(\frac{n\pi x}{L}\right)dx + \int_{\frac{L}{2}}^{L} \frac{2k}{L}(L-x)\sin\left(\frac{n\pi x}{L}\right)dx\right\}$$

$$= \frac{4k}{L^2}\left\{\left[-\frac{Lx}{n\pi}\cos\left(\frac{n\pi x}{L}\right)\right]_0^{\frac{L}{2}} + \int_0^{\frac{L}{2}} \frac{L}{n\pi}\cos\left(\frac{n\pi x}{L}\right)dx\right\}$$

$$+ \frac{4k}{L^2}\left\{-\left[\frac{L(L-x)}{n\pi}\cos\left(\frac{n\pi x}{L}\right)\right]_{\frac{L}{2}}^{L} - \int_{\frac{L}{2}}^{L} \frac{L}{n\pi}\cos\left(\frac{n\pi x}{L}\right)dx\right\}$$

$$= \frac{4k}{L^2}\left\{-\frac{L^2}{2n\pi}\cos\left(\frac{n\pi}{2}\right)+\left[\frac{L^2}{n^2\pi^2}\sin\left(\frac{n\pi x}{L}\right)\right]_0^{\frac{L}{2}}\right\}$$

$$+\frac{4k}{L^2}\left\{\frac{L^2}{2n\pi}\cos\left(\frac{n\pi}{2}\right)-\left[\frac{L^2}{n^2\pi^2}\sin\left(\frac{n\pi x}{L}\right)\right]_{\frac{L}{2}}^{L}\right\}$$

$$= \frac{8k}{n^2\pi^2}\sin\left(\frac{n\pi}{2}\right).$$

Substituting these values of Fourier coefficients into Eq. (i), we obtain

$$f(x) = \sum_{n=1}^{\infty}\frac{8k}{n^2\pi^2}\sin\left(\frac{n\pi}{2}\right)\sin\left(\frac{n\pi x}{L}\right)$$

$$= \frac{8k}{\pi^2}\left[\frac{1}{1^2}\sin\left(\frac{\pi x}{L}\right)-\frac{1}{3^2}\sin\left(\frac{3\pi x}{L}\right)+\frac{1}{5^2}\sin\left(\frac{5\pi x}{L}\right)-\ldots\infty\right].$$

Deductions: At $x=\frac{L}{2}$, we have

$$f\left(\frac{L}{2}\right)=k=\frac{8k}{\pi^2}\left[\frac{1}{1^2}+\frac{1}{3^2}+\frac{1}{5^2}+\ldots\infty\right]$$

or $$\frac{1}{1^2}+\frac{1}{3^2}+\frac{1}{5^2}+\ldots\infty=\frac{\pi^2}{8}.$$

Further, as in the case of the even periodic extension, we have, by Parseval's formula,

$$\frac{a_0^2}{4}+\frac{1}{2}\sum_{n=1}^{\infty}\left(a_n^2+b_n^2\right)=\frac{1}{L}\int_0^L[f(x)]^2dx \quad \text{(Here, } a_0,\ a_n = 0)$$

or $$\frac{1}{2}\sum_{n=1}^{\infty}\left(\frac{8k}{n^2\pi^2}\right)^2\left(\sin\frac{n\pi}{2}\right)^2=\frac{1}{L}\left\{\int_0^{\frac{L}{2}}\frac{4k^2}{L^2}x^2dx+\int_{\frac{L}{2}}^{L}\frac{4k^2}{L^2}\left(L^2-2Lx+x^2\right)dx\right\}$$

or $$\frac{32k^2}{\pi^4}\left(\frac{1}{1^4}+\frac{1}{3^4}+\frac{1}{5^4}+\ldots\infty\right)=\frac{k^2}{3}$$

or $$\frac{1}{1^4}+\frac{1}{3^4}+\frac{1}{5^4}+\ldots\infty=\frac{\pi^4}{96}.$$

PROBLEM 19 Obtain the Fourier series of a function defined by

$$f(x)=\begin{cases}\frac{L}{2}+x & \text{for } -\frac{L}{2}<x<0,\\ \frac{L}{2}-x & \text{for } 0<x<\frac{L}{2}\end{cases}$$

and $$f(x)=f(x+L).$$

Deduce that $$\frac{\pi^2}{8}=\frac{1}{1^2}+\frac{1}{3^2}+\frac{1}{5^2}+\frac{1}{7^2}+\ldots\infty.$$

Also sketch a profile of the function and its periodic extension.

Solution: Since the function is an even function, its Fourier series will have the form

$$f(x)=\frac{a_0}{2}+\sum_{n=1}^{\infty}a_n\cos\left(\frac{2n\pi x}{L}\right) \qquad \ldots\text{(i)}$$

where $$a_0=\frac{2}{L}\int_{-\frac{L}{2}}^{+\frac{L}{2}}f(x)dx \qquad \ldots\text{(ii)}$$

and $$a_n=\frac{2}{L}\int_{-\frac{L}{2}}^{+\frac{L}{2}}f(x)\cos\left(\frac{2n\pi x}{L}\right)dx. \qquad \ldots\text{(iii)}$$

From Eq. (ii), we have

$$a_0=\frac{2}{L}\left\{\int_{-\frac{L}{2}}^{0}f(x)dx+\int_{0}^{\frac{L}{2}}f(x)dx\right\}$$

$$=\frac{2}{L}\left\{\int_{-\frac{L}{2}}^{0}\left(\frac{L}{2}+x\right)dx+\int_{0}^{\frac{L}{2}}\left(\frac{L}{2}-x\right)dx\right\}$$

$$=\frac{2}{L}\left\{\left[\frac{Lx}{2}+\frac{x^2}{2}\right]_{-\frac{L}{2}}^{0}+\left[\frac{Lx}{2}-\frac{x^2}{2}\right]_{0}^{\frac{L}{2}}\right\}$$

$$=\frac{L}{2}.$$

From Eq. (iii), we have

$$a_n = \frac{2}{L}\left\{\int_{-\frac{L}{2}}^{0} f(x)\cos\left(\frac{2n\pi x}{L}\right)dx + \int_{0}^{\frac{L}{2}} f(x)\cos\left(\frac{2n\pi x}{L}\right)dx\right\}$$

$$= \frac{2}{L}\int_{-\frac{L}{2}}^{0}\left(\frac{L}{2}+x\right)\cos\left(\frac{2n\pi x}{L}\right)dx + \frac{2}{L}\int_{0}^{\frac{L}{2}}\left(\frac{L}{2}-x\right)\cos\left(\frac{2n\pi x}{L}\right)dx$$

$$= \frac{2}{L}\int_{\frac{L}{2}}^{0}\left(\frac{L}{2}-x\right)\cos\left(-\frac{2n\pi x}{L}\right)(-dx) + \frac{2}{L}\int_{0}^{\frac{L}{2}}\left(\frac{L}{2}-x\right)\cos\left(\frac{2n\pi x}{L}\right)dx$$

$$= \frac{2}{L}\int_{0}^{\frac{L}{2}}\left(\frac{L}{2}-x\right)\cos\left(\frac{2n\pi x}{L}\right)dx + \frac{2}{L}\int_{0}^{\frac{L}{2}}\left(\frac{L}{2}-x\right)\cos\left(\frac{2n\pi x}{L}\right)dx$$

$$= \frac{4}{L}\int_{0}^{\frac{L}{2}}\left(\frac{L}{2}-x\right)\cos\left(\frac{2n\pi x}{L}\right)dx$$

$$= \frac{4}{L}\left\{\left[\left(\frac{L}{2}-x\right)\frac{L}{2n\pi}\sin\left(\frac{2n\pi x}{L}\right)\right]_0^{\frac{L}{2}} + \int_{0}^{\frac{L}{2}}\frac{L}{2n\pi}\sin\left(\frac{2n\pi x}{L}\right)dx\right\}$$

$$= \frac{4}{L}\left\{0 - \left[\frac{L^2}{4n^2\pi^2}\cos\left(\frac{2n\pi x}{L}\right)\right]_0^{\frac{L}{2}}\right\}$$

$$= -\frac{L}{n^2\pi^2}\left\{(-1)^n - 1\right\} = \begin{cases} 0 & \text{if } n \text{ is even integer,} \\ \dfrac{2L}{n^2\pi^2} & \text{if } n \text{ is odd integer.} \end{cases}$$

Substituting the values of these Fourier coefficients into Eq. (i), we obtain

$$f(x) = \frac{L}{4} + \frac{2L}{\pi^2}\sum_{n=\text{odd}}\left(\frac{1}{n^2}\right)\cos\left(\frac{2n\pi x}{L}\right)$$

$$= \frac{L}{4} + \frac{2L}{\pi^2}\left[\frac{1}{1^2}\cos\left(\frac{2\pi x}{L}\right) + \frac{1}{3^2}\cos\left(\frac{6\pi x}{L}\right) + \frac{1}{5^2}\cos\left(\frac{10\pi x}{L}\right) + \ldots\infty\right].$$

Deduction: At $x = 0$, we have

$$\frac{L}{2} = \frac{L}{4} + \frac{2L}{\pi^2}\left(\frac{1}{1^2} + \frac{1}{3^2} + \frac{1}{5^2} + \frac{1}{7^2} + \ldots\infty\right)$$

or

$$\frac{\pi^2}{8} = \frac{1}{1^2} + \frac{1}{3^2} + \frac{1}{5^2} + \frac{1}{7^2} + \ldots\infty.$$

A sketch of the profile of the function and its periodic extension is shown below (Fig. 11.8).

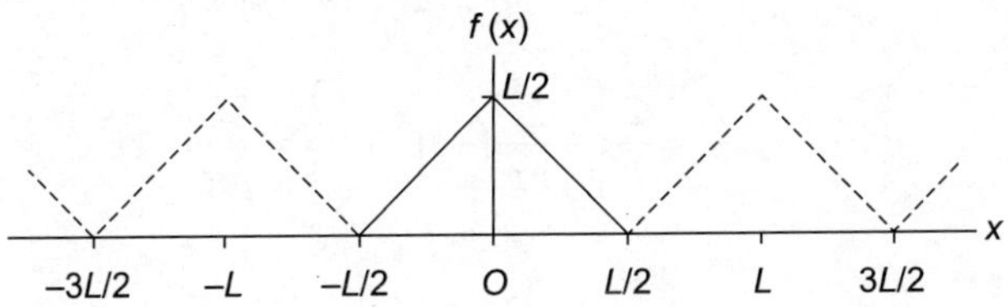

Fig. 11.8 Profile of the given function and its periodic extension.

Problems on Sawtooth Waves

A **sawtooth** wave is a wave whose profile has a waveform that increases linearly for a fixed interval, returns abruptly to a minimum level and repeats the pattern periodically, producing a shape resembling the teeth of a saw.

A **reverse sawtooth** wave is a wave whose profile has a waveform that decreases linearly for a fixed interval, returns abruptly to a maximum level and repeats the pattern periodically, producing a shape resembling the teeth of a reversed saw.

PROBLEM 20 Expand the function $f(x) = f(x + 2L) = x$ for $-L \le x \le L$ in a Fourier series and deduce the identities

(a) $\frac{\pi}{4} = 1 - \frac{1}{3} + \frac{1}{5} - \frac{1}{7} + \ldots\infty$

and (b) $\frac{\pi^2}{6} = \sum_{n=1}^{\infty}\frac{1}{n^2}$.

Sketch a profile of the function and its periodic extension.

Solution: Since $f(x) = x$ is an odd function, its Fourier series is given by

$$f(x) = \sum_{n=1}^{\infty} b_n \sin\left(\frac{n\pi x}{L}\right)$$

where
$$b_n = \frac{1}{L}\int_{-L}^{+L} f(x)\sin\left(\frac{n\pi x}{L}\right)dx = \frac{1}{L}\int_{-L}^{+L} x\sin\left(\frac{n\pi x}{L}\right)dx$$

$$= \frac{2}{L}\int_{0}^{L} x\sin\left(\frac{n\pi x}{L}\right)dx$$

$$= \frac{2}{L}\left\{\left[-\frac{xL}{n\pi}\cos\left(\frac{n\pi x}{L}\right)\right]_0^L + \int_0^L \frac{L}{n\pi}\cos\left(\frac{n\pi x}{L}\right)dx\right\}$$

$$= \frac{2}{L}\left\{-\frac{L^2}{n\pi}(-1)^n + 0\right\} = \frac{2L}{n\pi}(-1)^{n+1}.$$

Thus, we obtain

$$f(x) = x = \sum_{n=1}^{\infty} \frac{2L}{n\pi}(-1)^{n+1} \sin\left(\frac{n\pi x}{L}\right) \qquad \ldots(i)$$

Deductions:

(a) At $x = \frac{L}{2}$, we have

$$\frac{L}{2} = \sum_{n=1}^{\infty} \frac{2L}{n\pi}(-1)^{n+1} \sin\left(\frac{n\pi}{2}\right)$$

$$= \frac{2L}{\pi}\left[\sin\frac{\pi}{2} - \frac{1}{2}\sin\pi + \frac{1}{3}\sin\frac{3\pi}{2} - \frac{1}{4}\sin 2\pi + \frac{1}{5}\sin\frac{5\pi}{2} - \frac{1}{6}\sin 3\pi + \ldots\infty\right]$$

$$= \frac{2L}{\pi}\left[1 - 0 - \frac{1}{3} - 0 + \frac{1}{5} - 0 - \frac{1}{7} + \ldots\infty\right]$$

or $\frac{\pi}{4} = 1 - \frac{1}{3} + \frac{1}{5} - \frac{1}{7} + \ldots\infty.$

(b) Now, Parseval's formula yields

$$\frac{1}{2L}\int_{-L}^{+L} x^2 dx = \frac{1}{2}\sum_{n=1}^{\infty} b_n^2$$

$$\frac{1}{L}\int_0^L x^2 dx = \frac{1}{2}\sum_{n=1}^{\infty}\frac{4L^2}{n^2\pi^2}(-1)^{2n+2}$$

$$\frac{L^2}{3} = \frac{2L^2}{\pi^2}\sum_{n=1}^{\infty}\frac{1}{n^2}$$

$$\frac{\pi^2}{6} = \sum_{n=1}^{\infty}\frac{1}{n^2} = \frac{1}{1^2}+\frac{1}{2^2}+\frac{1}{3^2}+\ldots\infty.$$

Sketch of the profile of $f(x)$ and its periodic extension is shown in Fig. 11.9.

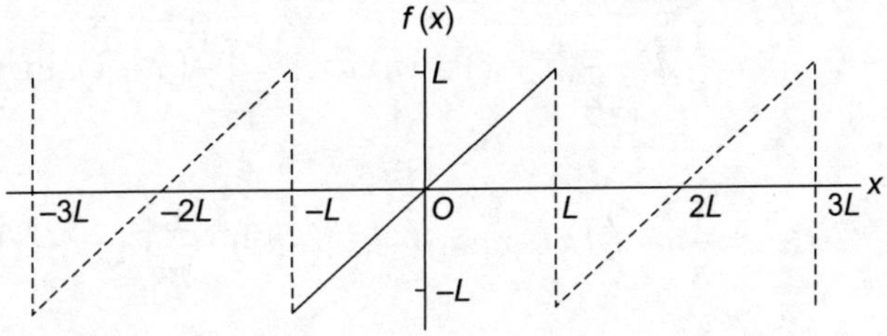

Fig. 11.9 Profile of $f(x)$ and its periodic extension

PROBLEM 21 A sawtooth wave is described by the function

$$f(x) = \begin{cases} -\frac{1}{2}(\pi + x) & \text{for} -\pi \le x < 0, \\ +\frac{1}{2}(\pi - x) & \text{for } 0 < x \le \pi. \end{cases}$$

Show that $f(x) = \sum_{n=1}^{\infty}\frac{\sin nx}{n}$ and deduce that

$$\frac{\pi}{4} = 1 - \frac{1}{3} + \frac{1}{5} - \frac{1}{7} + \ldots\infty.$$

Solution: A sketch of the profile of $f(x)$ and its periodic extension is shown in Fig. 11.10.

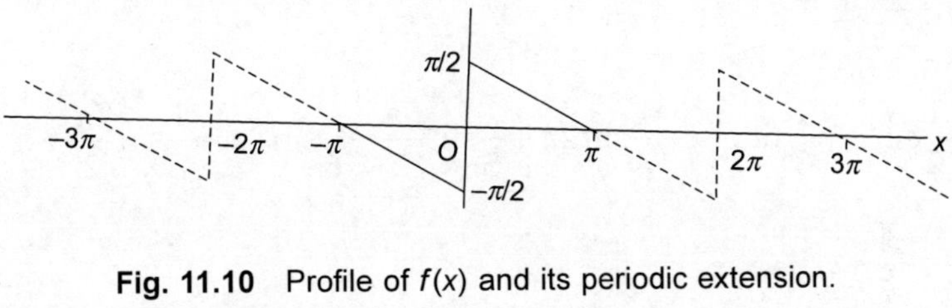

Fig. 11.10 Profile of $f(x)$ and its periodic extension.

As the function $f(x)$ is an odd function, we take its Fourier series as

$$f(x)=\sum_{n=1}^{\infty} b_n \sin nx$$

where
$$b_n=\frac{1}{\pi}\int_{-\pi}^{+\pi} f(x)\sin nx\,dx$$

$$=\frac{1}{\pi}\int_{-\pi}^{0} f(x)\sin nx\,dx+\frac{1}{\pi}\int_{0}^{\pi} f(x)\sin nx\,dx$$

$$=\frac{1}{\pi}\int_{-\pi}^{0} -\frac{1}{2}(\pi+x)\sin nx\,dx+\frac{1}{\pi}\int_{0}^{\pi}\frac{1}{2}(\pi-x)\sin nx\,dx$$

$$=\frac{1}{\pi}\int_{\pi}^{0} -\frac{1}{2}(\pi-x)\sin(-nx)(-dx)+\frac{1}{\pi}\int_{0}^{\pi}\frac{1}{2}(\pi-x)\sin nx\,dx$$

$$=\frac{1}{\pi}\int_{0}^{\pi}(\pi-x)\sin nx\,dx$$

$$=\frac{1}{\pi}\left\{\left[-\frac{(\pi-x)}{n}\cos nx\right]_0^{\pi}-\int_0^{\pi}\frac{1}{n}\cos nx\,dx\right\}$$

$$=\frac{1}{\pi}\left\{\frac{\pi}{n}-0\right\}=\frac{1}{n}.$$

Thus, we obtain the relevant Fourier series as

$$f(x)=\sum_{n=1}^{\infty}\frac{\sin nx}{n}=\sin x+\frac{\sin 2x}{2}+\frac{\sin 3x}{3}+\ldots\infty.$$

At $x=\frac{\pi}{2}$, we have

$$\frac{\pi}{4}=\sum_{n=1}^{\infty}\frac{1}{n}\sin\frac{n\pi}{2}$$

$$=\sin\frac{\pi}{2}+\frac{1}{2}\sin\pi+\frac{1}{3}\sin\frac{3\pi}{2}+\frac{1}{4}\sin 2\pi+\frac{1}{5}\sin\frac{5\pi}{2}+\frac{1}{6}\sin 3\pi+\frac{1}{7}\sin\frac{7\pi}{2}+\ldots\infty$$

$$=1-\frac{1}{3}+\frac{1}{5}-\frac{1}{7}+\ldots\infty.$$

PROBLEM 22 A function is defined by $f(x) = f(x+2L) = \frac{k}{2L}x, (0 \le x \le 2L)$ where k is a constant. Sketch a profile of the function and its periodic extension. Obtain a Fourier series representing the function.

Also deduce that

(a) $\frac{\pi}{4} = 1 - \frac{1}{3} + \frac{1}{5} - \frac{1}{7} + \ldots \infty$

and (b) $\frac{\pi^2}{6} = 1 + \frac{1}{2^2} + \frac{1}{3^2} + \frac{1}{4^2} + \ldots \infty$.

Solution: A sketch of the profile of $f(x)$ is shown in Fig. 11.11.

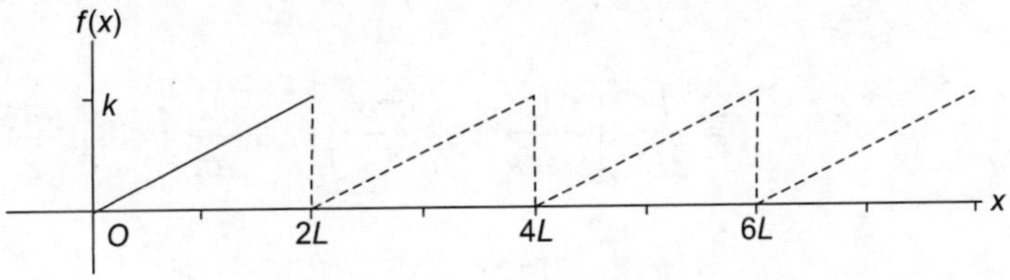

Fig. 11.11 Profile of $f(x)$ and its periodic extension.

Let the Fourier series of the function be

$$f(x) = \frac{a_0}{2} + \sum_{n=1}^{\infty} a_n \cos\left(\frac{n\pi x}{L}\right) + \sum_{n=1}^{\infty} b_n \sin\left(\frac{n\pi x}{L}\right) \quad \ldots\text{(i)}$$

where
$$a_0 = \frac{1}{L}\int_0^{2L} f(x)dx, \quad \ldots\text{(ii)}$$

$$a_n = \frac{1}{L}\int_0^{2L} f(x)\cos\left(\frac{n\pi x}{L}\right)dx \quad \ldots\text{(iii)}$$

and
$$b_n = \frac{1}{L}\int_0^{2L} f(x)\sin\left(\frac{n\pi x}{L}\right)dx. \quad \ldots\text{(iv)}$$

From Eq. (ii), we have

$$a_0 = \frac{1}{L}\int_0^{2L} \frac{kx}{2L}dx = \frac{k}{2L^2}\left[\frac{x^2}{2}\right]_0^{2L} = k.$$

From Eq. (iii), we have

$$a_n = \frac{1}{L}\int_0^{2L} \frac{kx}{2L}\cos\left(\frac{n\pi x}{L}\right)dx$$

$$= \frac{k}{2L^2}\left\{\left[\frac{Lx}{n\pi}\sin\left(\frac{n\pi x}{L}\right)\right]_0^{2L} - \int_0^{2L} \frac{L}{n\pi}\sin\left(\frac{n\pi x}{L}\right)dx\right\}$$

$$= \frac{k}{2L^2}\left\{0+\left[\frac{L^2}{n^2\pi^2}\cos\left(\frac{n\pi x}{L}\right)\right]_0^{2L}\right\} = 0.$$

From Eq. (iv), we have

$$b_n = \frac{1}{L}\int_0^{2L} \frac{kx}{2L}\sin\left(\frac{n\pi x}{L}\right)dx$$

$$= \frac{k}{2L^2}\left\{\left[-\frac{Lx}{n\pi}\cos\left(\frac{n\pi x}{L}\right)\right]_0^{2L} + \frac{L}{n\pi}\int_0^{2L} \cos\left(\frac{n\pi x}{L}\right)dx\right\}$$

$$= \frac{k}{2L^2}\left\{-\frac{2L^2}{n\pi}\cos(2n\pi)+0\right\} = -\frac{k}{n\pi}.$$

Substituting the values of these Fourier coefficients into Eq. (i), we obtain

$$f(x) = \frac{kx}{2L} = \frac{k}{2} - \frac{k}{\pi}\sum_{n=1}^{\infty}\frac{1}{n}\sin\left(\frac{n\pi x}{L}\right).$$

Deductions:

(a) At $x = \frac{L}{2}$, we have

$$\frac{k}{4} = \frac{k}{2} - \frac{k}{\pi}\sum_{n=1}^{\infty}\frac{1}{n}\sin\left(\frac{n\pi}{2}\right)$$

$$= \frac{k}{2} - \frac{k}{\pi}\left[\sin\frac{\pi}{2} + \frac{1}{2}\sin\pi + \frac{1}{3}\sin\frac{3\pi}{2} + \frac{1}{4}\sin 2\pi + \frac{1}{5}\sin\frac{5\pi}{2} + \frac{1}{6}\sin 3\pi + \ldots\infty\right]$$

$$= \frac{k}{2} - \frac{k}{\pi}\left[1 - \frac{1}{3} + \frac{1}{5} - \frac{1}{7} + \ldots\infty\right]$$

or $\quad 1 - \frac{1}{3} + \frac{1}{5} - \frac{1}{7} + \ldots\infty = \frac{\pi}{4}.$

(b) Parseval's theorem yields

$$\frac{1}{2L}\int_0^{2L}\left[\frac{kx}{2L}\right]^2 dx = \frac{a_0^2}{4}+\frac{1}{2}\sum_{n=1}^{\infty}\left(a_n^2+b_n^2\right)$$

or
$$\frac{k^2}{8L^3}\left[\frac{x^3}{3}\right]_0^{2L} = \frac{k^2}{4}+\frac{1}{2}\sum_{n=1}^{\infty}\left(0+\frac{k^2}{n^2\pi^2}\right)$$

or
$$\frac{k^2}{3} = \frac{k^2}{4}+\frac{k^2}{2\pi^2}\left(1+\frac{1}{2^2}+\frac{1}{3^2}+\frac{1}{4^2}+\ldots\infty\right)$$

or
$$\frac{\pi^2}{6} = 1+\frac{1}{2^2}+\frac{1}{3^2}+\frac{1}{4^2}+\ldots\infty.$$

PROBLEM 23 A function is defined by $f(x) = f(x + 2L) = L + x$ for $-L \le x \le L$.

Sketch its profile and periodic extension. Find its Fourier series and deduce the identities (a) and (b) in Problem 22.

Solution: A sketch of the profile is shown in the following figure (11.12).

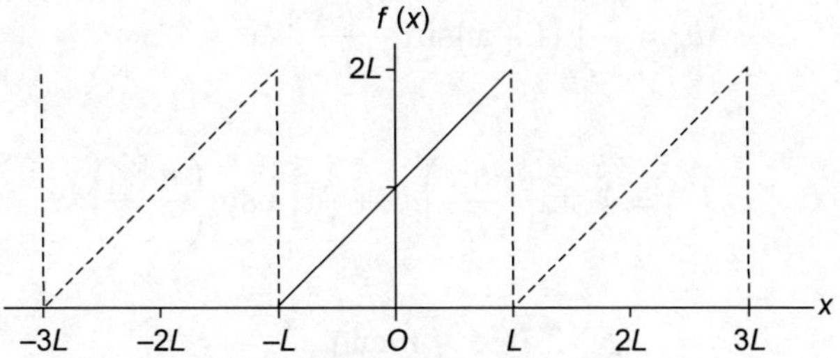

Fig. 11.12 Profile of $f(x)$ and its periodic extension.

Let the Fourier series of the given function be

$$f(x) = \frac{a_0}{2}+\sum_{n=1}^{\infty} a_n \cos\left(\frac{n\pi x}{L}\right)+\sum_{n=1}^{\infty} b_n \sin\left(\frac{n\pi x}{L}\right) \qquad \ldots\text{(i)}$$

where
$$a_0 = \frac{1}{L}\int_{-L}^{+L} f(x)dx, \qquad \ldots\text{(ii)}$$

$$a_n = \frac{1}{L}\int_{-L}^{+L} f(x)\cos\left(\frac{n\pi x}{L}\right)dx \qquad \ldots\text{(iii)}$$

and $$b_n = \frac{1}{L}\int_{-L}^{+L} f(x)\sin\left(\frac{n\pi x}{L}\right)dx. \qquad \text{...(iv)}$$

From Eq. (ii), we have

$$a_0 = \frac{1}{L}\int_{-L}^{+L} (L+x)\,dx = \int_{-L}^{+L} dx + \frac{1}{L}\int_{-L}^{+L} x\,dx = 2L.$$

From Eq. (iii), we have

$$a_n = \frac{1}{L}\int_{-L}^{+L} (L+x)\cos\left(\frac{n\pi x}{L}\right)dx$$

$$= \int_{-L}^{+L} \cos\left(\frac{n\pi x}{L}\right)dx + \frac{1}{L}\int_{L}^{+L} x\cos\left(\frac{n\pi x}{L}\right)dx$$

$$= 2\int_{0}^{L} \cos\left(\frac{n\pi x}{L}\right)dx + 0 = 0.$$

From Eq. (iv), we have

$$b_n = \frac{1}{L}\int_{-L}^{+L} (L+x)\sin\left(\frac{n\pi x}{L}\right)dx$$

$$= \int_{-L}^{+L} \sin\left(\frac{n\pi x}{L}\right)dx + \frac{1}{L}\int_{-L}^{+L} x\sin\left(\frac{n\pi x}{L}\right)dx$$

$$= 0 + \frac{2}{L}\int_{0}^{L} x\sin\left(\frac{n\pi x}{L}\right)dx$$

$$= \frac{2}{L}\left\{\left[-\frac{Lx}{n\pi}\cos\left(\frac{n\pi x}{L}\right)\right]_0^L + \int_0^L \frac{L}{n\pi}\cos\left(\frac{n\pi x}{L}\right)dx\right\}$$

$$= \frac{2}{L}\left\{-\frac{L^2}{n\pi}(-1)^n + 0\right\}$$

$$= \frac{2L}{n\pi}(-1)^{n+1}.$$

Substituting the values of these Fourier coefficients into Eq. (i), we have

$$f(x) = L + x$$

$$= L + \sum_{n=1}^{\infty} \frac{2L}{n\pi}(-1)^{n+1} \sin\left(\frac{n\pi x}{L}\right)$$

$$= L + \frac{2L}{\pi}\left[\sin\left(\frac{\pi x}{L}\right) - \frac{1}{2}\sin\left(\frac{2\pi x}{L}\right) + \frac{1}{3}\sin\left(\frac{3\pi x}{L}\right) - \frac{1}{4}\sin\left(\frac{4\pi x}{L}\right) + \ldots\infty\right].$$

Deductions:

(a) With $x = \frac{L}{2}$, we have

$$L + \frac{L}{2} = L + \frac{2L}{\pi}\left[1 - 0 - \frac{1}{3} - 0 + \frac{1}{5} - 0 - \ldots\infty\right]$$

or $$\frac{\pi}{4} = 1 - \frac{1}{3} + \frac{1}{5} - \frac{1}{7} + \ldots\infty.$$

(b) Parseval's formula yields

$$\frac{1}{2L}\int_{-L}^{+L}\left[f(x)\right]^2 dx = \frac{a_0^2}{4} + \frac{1}{2}\sum_{n=1}^{\infty}\left(a_n^2 + b_n^2\right)$$

or $$\frac{1}{2L}\int_{-L}^{+L}\left(L^2 + x^2 + 2Lx\right)dx = \frac{(2L)^2}{4} + \frac{1}{2}\sum_{n=1}^{\infty}\left(0 + \frac{4L^2}{n^2\pi^2}\right)$$

or $$L^2 + \frac{L^2}{3} + 0 = L^2 + \frac{2L^2}{\pi^2}\sum_{n=1}^{\infty}\frac{1}{n^2}$$

or $$\sum_{n=1}^{\infty}\frac{1}{n^2} = \frac{\pi^2}{6}.$$

PROBLEM 24 Expand the function $f(x) = x$ as a Fourier sine series in the half interval $0 \le x \le L$. Sketch a profile of the function with periodic extension.

Deduce that

(a) $\frac{\pi}{4} = 1 - \frac{1}{3} + \frac{1}{5} - \frac{1}{7} + \ldots\infty$

and (b) $\frac{\pi^2}{6} = \frac{1}{1^2} + \frac{1}{2^2} + \frac{1}{3^2} + \frac{1}{4^2} + \ldots\infty.$

Solution: Let the Fourier sine series be

$$f(x) = x = \sum_{n=1}^{\infty} b_n \sin\left(\frac{n\pi x}{L}\right)$$

where
$$b_n = \frac{2}{L}\int_0^L f(x)\sin\left(\frac{n\pi x}{L}\right)dx$$

$$= \frac{2}{L}\int_0^L x\sin\left(\frac{n\pi x}{L}\right)dx$$

$$= \frac{2}{L}\left\{\left[-\frac{Lx}{n\pi}\cos\left(\frac{n\pi x}{L}\right)\right]_0^L + \int_0^L \frac{L}{n\pi}\cos\left(\frac{n\pi x}{L}\right)dx\right\}$$

$$= \frac{2}{L}\left\{-\frac{L^2}{n\pi}\cos n\pi + 0\right\} + 0 = \frac{2L}{n\pi}(-1)^{n+1}.$$

Therefore, the Fourier sine series is given by

$$f(x) = x = \sum_{n=1}^{\infty} \frac{2L}{n\pi}(-1)^{n+1}\sin\left(\frac{n\pi x}{L}\right)$$

$$= \frac{2L}{\pi}\left[\sin\left(\frac{\pi x}{L}\right) - \frac{1}{2}\sin\left(\frac{2\pi x}{L}\right) + \frac{1}{3}\sin\left(\frac{3\pi x}{L}\right) - \frac{1}{4}\sin\left(\frac{4\pi x}{L}\right) + \frac{1}{5}\sin\left(\frac{5\pi x}{L}\right) - \ldots\infty\right].$$

Sketch of the function and its odd periodic extension is shown in the figure below (11.13).

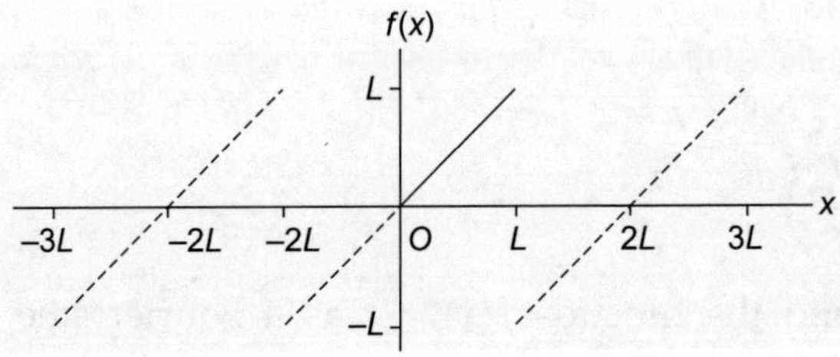

Fig. 11.13 Profile of $f(x)$ and its periodic extension.

Deductions:

(a) When $x = \frac{L}{2}$, we have

$$\frac{L}{2}=\frac{2L}{\pi}\left[\sin\frac{\pi}{2}-\frac{1}{2}\sin\pi+\frac{1}{3}\sin\frac{3\pi}{2}-\frac{1}{4}\sin 2\pi+\frac{1}{5}\sin\frac{5\pi}{2}-\ldots\infty\right]$$

or $$\frac{\pi}{4}=1-\frac{1}{3}+\frac{1}{5}-\frac{1}{7}+\ldots\infty.$$

(b) Parseval'sformula yields

$$\frac{1}{L}\int_0^L x^2dx=\frac{1}{2}\sum_{n=1}^{\infty}b_n^2$$

or $$\frac{L^2}{3}=\frac{1}{2}\sum_{n=1}^{\infty}\frac{4L^2}{n^2\pi^2}$$

or $$\frac{\pi^2}{6}=\sum_{n=1}^{\infty}\frac{1}{n^2}=\frac{1}{1^2}+\frac{1}{2^2}+\frac{1}{3^2}+\frac{1}{4^2}+\infty.$$

Problems on Square Waves

A square/rectangular wave is a wave whose graphical representation consists of square/rectangular loops. Such waves generally describe high frequency pulses.

PROBLEM 25 A periodic function of period $2L$ is defined by

$$f(x)=\begin{cases}-k & \text{for } -L<x<0,\\ k & \text{for } 0<x<L.\end{cases}$$

Sketch a profile of the function and its periodic extension. Obtain a Fourier series representing the function and deduce the identities

(a) $$\frac{\pi}{4}=1-\frac{1}{3}+\frac{1}{5}-\frac{1}{7}+\ldots\infty$$

and (b) $$\frac{\pi^2}{8}=1+\frac{1}{3^2}+\frac{1}{5^2}+\frac{1}{7^2}+\ldots\infty.$$

Solution: A sketch of the profile of the given function and its periodic extension is shown in the figure below (11.14).

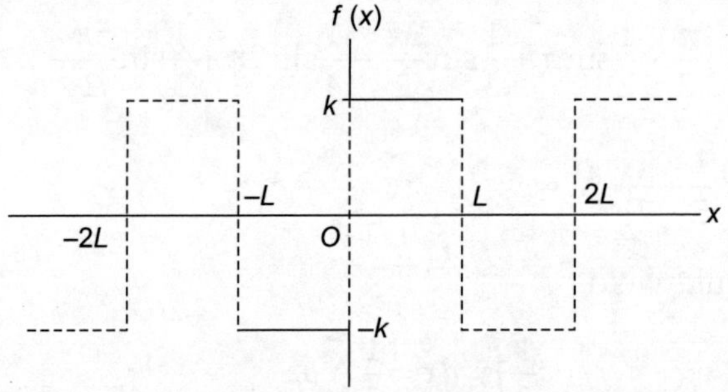

Fig. 11.14 Profile of $f(x)$ and its periodic extension.

Fourier Series

As the given function is an odd function, its Fourier series representation will contain only sine terms. Let the Fourier series be

$$f(x) = \sum_{n=1}^{\infty} b_n \sin\left(\frac{n\pi x}{L}\right) \qquad \ldots\text{(i)}$$

where
$$b_n = \frac{1}{L}\int_{-L}^{+L} f(x)\sin\left(\frac{n\pi x}{L}\right)dx$$

$$= \frac{1}{L}\int_{-L}^{0} f(x)\sin\left(\frac{n\pi x}{L}\right)dx + \frac{1}{L}\int_{0}^{L} f(x)\sin\left(\frac{n\pi x}{L}\right)dx$$

$$= \frac{1}{L}\int_{-L}^{0} -k\sin\left(\frac{n\pi x}{L}\right)dx + \frac{1}{L}\int_{0}^{L} k\sin\left(\frac{n\pi x}{L}\right)dx$$

$$= \frac{1}{L}\int_{L}^{0} -k\sin\left(-\frac{n\pi x}{L}\right)(-dx) + \frac{1}{L}\int_{0}^{L} k\sin\left(\frac{n\pi x}{L}\right)dx$$

$$= \frac{2}{L}\int_{0}^{L} k\sin\left(\frac{n\pi x}{L}\right)dx$$

$$= -\frac{2}{L}\cdot\frac{kL}{n\pi}\left[\cos\left(\frac{n\pi x}{L}\right)\right]_0^L$$

$$= \frac{2k}{n\pi}\left[1-(-1)^n\right] = \begin{cases} 0 & \text{if } n \text{ is even integer,} \\ \dfrac{4k}{n\pi} & \text{if } n \text{ is odd integer.} \end{cases}$$

Substituting the values of b_n into Eq. (i), we obtain

$$f(x)=\frac{4k}{\pi}\sum_{n=\text{odd}}\left(\frac{1}{n}\right)\sin\left(\frac{n\pi x}{L}\right)$$

$$=\frac{4k}{\pi}\left[\sin\left(\frac{\pi x}{L}\right)+\frac{1}{3}\sin\left(\frac{3\pi x}{L}\right)+\frac{1}{5}\sin\left(\frac{5\pi x}{L}\right)+\frac{1}{7}\sin\left(\frac{7\pi x}{L}\right)+\ldots\infty\right].$$

Deductions:

(a) When $x=\frac{L}{2}$, we have

$$k=\frac{4k}{\pi}\left[1-\frac{1}{3}+\frac{1}{5}-\frac{1}{7}+\ldots\infty\right]$$

or

$$\frac{\pi}{4}=1-\frac{1}{3}+\frac{1}{5}-\frac{1}{7}+\ldots\infty.$$

(b) Parseval's formula yields

$$\frac{1}{2L}\int_{-L}^{+L}[f(x)]^2dx=\frac{a_0^2}{4}+\frac{1}{2}\sum_{n=1}^{\infty}\left[a_n^2+b_n^2\right]$$

or

$$\frac{1}{2L}\int_{-L}^{+L}k^2dx=0+\frac{1}{2}\sum_{n=\text{odd}}\left[0+\frac{16k^2}{n^2\pi^2}\right]$$

or

$$k^2=\frac{8k^2}{\pi^2}\left[\frac{1}{1^2}+\frac{1}{3^2}+\frac{1}{5^2}+\frac{1}{7^2}+\ldots\infty\right]$$

or

$$\frac{\pi^2}{8}=\frac{1}{1^2}+\frac{1}{3^2}+\frac{1}{5^2}+\frac{1}{7^2}+\ldots\infty.$$

PROBLEM 26 A periodic function of period $2L$ is defined by

$$f(x)=\begin{cases}0 & \text{for}-L<x<0,\\ k & \text{for } 0<x<L.\end{cases}$$

Sketch a profile of the function and its periodic extension. Obtain a Fourier series representing the function and deduce the identities

(a) $\frac{\pi}{4}=1-\frac{1}{3}+\frac{1}{5}-\frac{1}{7}+\ldots\infty$

and (b) $\frac{\pi^2}{8} = 1 + \frac{1}{3^2} + \frac{1}{5^2} + \frac{1}{7^2} + \ldots \infty.$

Solution: A sketch of the profile of the given function and its periodic extension is shown in the figure below (11.15).

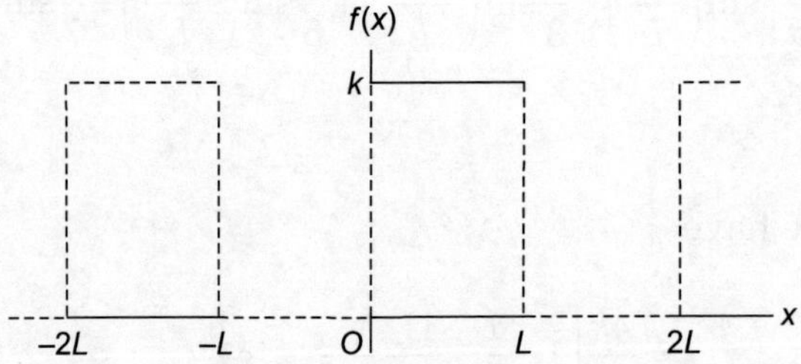

Fig. 11.15 Profile of $f(x)$ and its periodic extension.

Fourier Series

Let the Fourier series be

$$f(x) = \frac{a_0}{2} + \sum_{n=1}^{\infty} a_n \cos\left(\frac{n\pi x}{L}\right) + \sum_{n=1}^{\infty} b_n \sin\left(\frac{n\pi x}{L}\right) \quad \ldots\text{(i)}$$

where $$a_0 = \frac{1}{L}\int_{-L}^{+L} f(x)dx, \quad \ldots\text{(ii)}$$

$$a_n = \frac{1}{L}\int_{-L}^{+L} f(x)\cos\left(\frac{n\pi x}{L}\right)dx \quad \ldots\text{(iii)}$$

and $$b_n = \frac{1}{L}\int_{-L}^{+L} f(x)\sin\left(\frac{n\pi x}{L}\right)dx. \quad \ldots\text{(iv)}$$

From Eq. (ii), we have

$$a_0 = \frac{1}{L}\left\{\int_{-L}^{0} f(x)dx + \int_{0}^{L} f(x)dx\right\}$$

$$= \frac{1}{L}\left\{\int_{-L}^{0} 0 \cdot dx + \int_{0}^{L} k\,dx\right\} = k.$$

From Eq. (iii), we have

$$a_n = \frac{1}{L}\left\{\int_{-L}^{0} f(x)\cos\left(\frac{n\pi x}{L}\right)dx + \frac{1}{L}\int_{0}^{L} f(x)\cos\left(\frac{n\pi x}{L}\right)dx\right\}$$

$$= \frac{1}{L}\left\{\int_{-L}^{0} 0\cdot\cos\left(\frac{n\pi x}{L}\right)dx + \int_{0}^{L} k\cos\left(\frac{n\pi x}{L}\right)dx\right\}$$

$$= \frac{1}{L}\left\{0 + k\left[\frac{L}{n\pi}\sin\left(\frac{n\pi x}{L}\right)\right]_0^L\right\} = 0.$$

From Eq. (iv), we have

$$b_n = \frac{1}{L}\left\{\int_{-L}^{0} f(x)\sin\left(\frac{n\pi x}{L}\right)dx + \int_{0}^{L} f(x)\sin\left(\frac{n\pi x}{L}\right)dx\right\}$$

$$= \frac{1}{L}\left\{0 + k\int_{0}^{L}\sin\left(\frac{n\pi x}{L}\right)dx\right\}$$

$$= -\frac{k}{L}\cdot\frac{L}{n\pi}\left[\cos\left(\frac{n\pi x}{L}\right)\right]_0^L = \frac{k}{n\pi}\left\{1-(-1)^n\right\}$$

$$= \begin{cases} 0 & \text{if } n \text{ is even integer,} \\ \dfrac{2k}{n\pi} & \text{if } n \text{ is odd integer.} \end{cases}$$

Substituting the values of these Fourier coefficients into Eq. (i), we have

$$f(x) = \frac{k}{2} + \sum_{n=\text{odd}} \frac{2k}{n\pi}\sin\left(\frac{n\pi x}{L}\right)$$

$$= \frac{k}{2} + \frac{2k}{\pi}\left\{\sin\left(\frac{\pi x}{L}\right) + \frac{1}{3}\sin\left(\frac{3\pi x}{L}\right) + \frac{1}{5}\sin\left(\frac{5\pi x}{L}\right) + \frac{1}{7}\sin\left(\frac{7\pi x}{L}\right) + \infty\right\}.$$

Deductions:

(a) When $x = \frac{L}{2}$, we have

$$k = \frac{k}{2} + \frac{2k}{\pi}\left\{1 - \frac{1}{3} + \frac{1}{5} - \frac{1}{7} + \ldots\infty\right\}$$

or $$\frac{\pi}{4}=1-\frac{1}{3}+\frac{1}{5}-\frac{1}{7}+\ldots\infty.$$

(b) Parseval's theorem yields

$$\frac{1}{2L}\int_{-L}^{+L}[f(x)]^2\,dx=\frac{a_0^2}{4}+\frac{1}{2}\sum_{n=1}^{\infty}\left(a_n^2+b_n^2\right)$$

or $$\frac{1}{2L}\left\{\int_{-L}^{0}[f(x)]^2\,dx+\int_{0}^{L}[f(x)]^2\,dx\right\}=\frac{k^2}{4}+\frac{1}{2}\sum_{n=\text{odd}}\frac{4k^2}{n^2\pi^2}$$

or $$\frac{1}{2L}\left\{0+k^2L\right\}=\frac{k^2}{4}+\frac{2k^2}{\pi^2}\sum_{n=\text{odd}}\frac{1}{n^2}$$

or $$\frac{\pi^2}{8}=\sum_{n=\text{odd}}\frac{1}{n^2}=\frac{1}{1^2}+\frac{1}{3^2}+\frac{1}{5^2}+\frac{1}{7^2}+\infty.$$

PROBLEM 27 A periodic function of period 4 is defined by

$$f(x)=\begin{cases}0 & \text{for } -2<x<-1,\\ k & \text{for } -1<x<1,\\ 0 & \text{for } 1<x<2.\end{cases}$$

Sketch a profile of the function and its periodic extension. Obtain a Fourier series representing the function and deduce the identities

(a) $$\frac{\pi}{4}=1-\frac{1}{3}+\frac{1}{5}-\frac{1}{7}+\ldots\infty$$

and (b) $$\frac{\pi^2}{8}=1+\frac{1}{3^2}+\frac{1}{5^2}+\frac{1}{7^2}+\ldots\infty.$$

Solution: A sketch of the profile of the given function and its periodic extension is shown in the figure below (11.16).

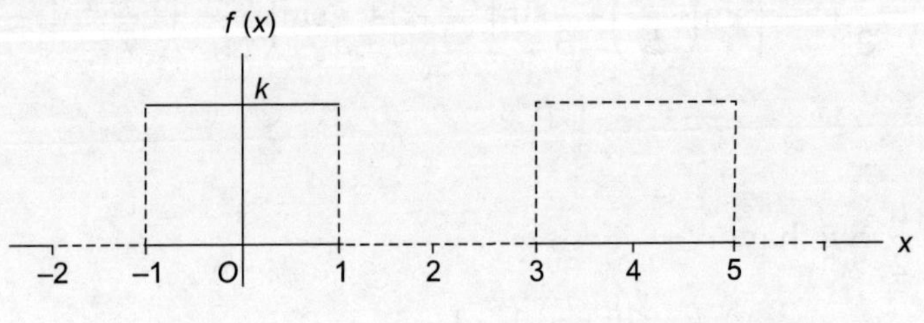

Fig. 11.16 Profile of $f(x)$ and its periodic extension.

Fourier Series

As $f(x)$ is an even function, its Fourier series can be written as

$$f(x)=\frac{a_0}{2}+\sum_{n=1}^{\infty}a_n\cos\left(\frac{n\pi x}{2}\right) \qquad \text{...(i)}$$

where
$$a_0=\frac{1}{2}\int_{-2}^{2}f(x)dx \qquad \text{...(ii)}$$

and
$$a_n=\frac{1}{2}\int_{-2}^{2}f(x)\cos\left(\frac{n\pi x}{2}\right)dx. \qquad \text{...(iii)}$$

From Eq. (ii), we have

$$a_0=\frac{1}{2}\left\{\int_{-2}^{-1}f(x)dx+\int_{-1}^{1}f(x)dx+\int_{1}^{2}f(x)dx\right\}$$

$$=\frac{1}{2}\left\{0+\int_{-1}^{+1}k\,dx+0\right\}=k.$$

From Eq. (iii), we have

$$a_n=\frac{1}{2}\left\{\int_{-2}^{-1}f(x)\cos\left(\frac{n\pi x}{2}\right)dx+\int_{-1}^{1}f(x)\cos\left(\frac{n\pi x}{2}\right)dx+\int_{1}^{2}f(x)\cos\left(\frac{n\pi x}{2}\right)dx\right\}$$

$$=\frac{1}{2}\left\{0+\int_{-1}^{1}k\cos\left(\frac{n\pi x}{2}\right)dx+0\right\}$$

$$=k\int_{0}^{1}\cos\left(\frac{n\pi x}{2}\right)dx=\frac{2k}{n\pi}\left[\sin\left(\frac{n\pi x}{2}\right)\right]_0^1$$

$$=\frac{2k}{n\pi}\sin\left(\frac{n\pi}{2}\right).$$

Substituting the values of the Fourier coefficients into Eq. (i), we obtain

$$f(x)=\frac{k}{2}+\sum_{n=1}^{\infty}\frac{2k}{n\pi}\sin\left(\frac{n\pi}{2}\right)\cos\left(\frac{n\pi x}{2}\right)$$

$$=\frac{k}{2}+\frac{2k}{\pi}\left\{\cos\left(\frac{\pi x}{2}\right)-\frac{1}{3}\cos\left(\frac{3\pi x}{2}\right)+\frac{1}{5}\cos\left(\frac{5\pi x}{2}\right)-\frac{1}{7}\cos\left(\frac{7\pi x}{2}\right)+\ldots\infty\right\}.$$

Deductions:

(a) At $x = 0$, we have

$$k = \frac{k}{2} + \frac{2k}{\pi}\left(1 - \frac{1}{3} + \frac{1}{5} - \frac{1}{7} + \dots \infty\right)$$

or
$$\frac{\pi}{4} = 1 - \frac{1}{3} + \frac{1}{5} - \frac{1}{7} + \dots \infty.$$

(b) Parseval's theorem yields

$$\frac{1}{4}\int_{-2}^{2}[f(x)]^2 dx = \frac{a_0^2}{4} + \frac{1}{2}\sum_{n=1}^{\infty} a_n^2$$

or
$$\frac{1}{4}\int_{-1}^{1} k^2 dx = \frac{k^2}{4} + \frac{1}{2}\sum_{n=1}^{\infty}\frac{4k^2}{n^2\pi^2}\sin^2\left(\frac{n\pi}{2}\right)$$

or
$$\frac{k^2}{2} = \frac{k^2}{4} + \frac{2k^2}{\pi^2}\left[\frac{1}{1^2} + \frac{1}{3^2} + \frac{1}{5^2} + \frac{1}{7^2} + \dots \infty\right]$$

or
$$\frac{\pi^2}{8} = \frac{1}{1^2} + \frac{1}{3^2} + \frac{1}{5^2} + \frac{1}{7^2} \dots \infty.$$

PROBLEM 28 A periodic function of period $2L$ is defined by

$$f(x) = \begin{cases} -k & \text{for } -L < x < -\frac{L}{2}, \\ 0 & \text{for } -\frac{L}{2} < x < \frac{L}{2}, \\ +k & \text{for } \frac{L}{2} < x < L. \end{cases}$$

Sketch a profile of the function and its periodic extension. Obtain a Fourier series representing the function .

Solution: A sketch of the profile of the given function and its periodic extension is shown in the figure below (11.17).

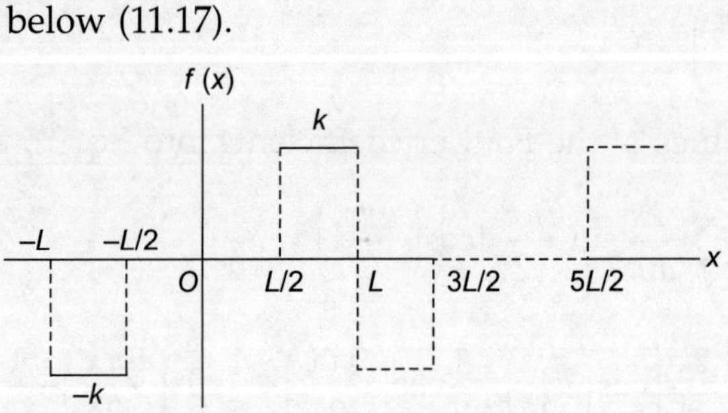

Fig. 11.17 Profile of $f(x)$ and its periodic extension.

As $f(x)$ is an odd function of period $2L$, its Fourier series can be expressed as

$$f(x) = \sum_{n=1}^{\infty} b_n \sin\left(\frac{n\pi x}{L}\right)$$

where $$b_n = \frac{1}{L}\int_{-L}^{+L} f(x)\sin\left(\frac{n\pi x}{L}\right)dx$$

$$= \frac{1}{L}\left\{\int_{-L}^{-\frac{L}{2}} f(x)\sin\left(\frac{n\pi x}{L}\right)dx + \int_{-\frac{L}{2}}^{\frac{L}{2}} f(x)\sin\left(\frac{n\pi x}{L}\right)dx + \int_{\frac{L}{2}}^{L} f(x)\sin\left(\frac{n\pi x}{L}\right)dx\right\}$$

$$= \frac{1}{L}\left\{\int_{-L}^{-\frac{L}{2}} -k\sin\left(\frac{n\pi x}{L}\right)dx + 0 + \int_{\frac{L}{2}}^{L} k\sin\left(\frac{n\pi x}{L}\right)dx\right\}$$

$$= \frac{1}{L}\left\{\int_{L}^{\frac{L}{2}} -k\sin\left(-\frac{n\pi x}{L}\right)(-dx) + 0 + \int_{\frac{L}{2}}^{L} k\sin\left(\frac{n\pi x}{L}\right)dx\right\}$$

$$= \frac{2k}{L}\int_{\frac{L}{2}}^{L} \sin\left(\frac{n\pi x}{L}\right)dx$$

$$= \frac{2k}{n\pi}\left\{\left[-\cos\left(\frac{n\pi x}{L}\right)\right]_{\frac{L}{2}}^{L}\right\}$$

$$= \frac{2k}{n\pi}\left\{\cos\left(\frac{n\pi}{2}\right) - \cos n\pi\right\}.$$

Thus, we obtain the Fourier series as

$$f(x) = \sum_{n=1}^{\infty} \frac{2k}{n\pi}\left\{\cos\left(\frac{n\pi}{2}\right) - \cos n\pi\right\}\sin\left(\frac{n\pi x}{L}\right)$$

$$= \frac{2k}{\pi}\left[\sin\left(\frac{\pi x}{L}\right) - \sin\left(\frac{2\pi x}{L}\right) + \frac{1}{3}\sin\left(\frac{3\pi x}{L}\right) + \frac{1}{5}\sin\left(\frac{5\pi x}{L}\right) + \ldots\infty\right].$$

Fouries Series for Rectifier Output

PROBLEM 29 An alternating current after passing through a half-wave rectifier has an output current described by

$$I(t)=\begin{cases} I_0 \sin \omega t \text{ for } 0<t<\dfrac{T}{2}, \\ 0 \qquad\quad \text{for } \dfrac{T}{2}<t<T \end{cases}$$

where I_0 is the peak value of current, ω is the circular frequency, t is the instantaneous time and T is the time period.

Sketch a profile of the function and its periodic extension. Obtain a Fourier series representation of the output current I and also deduce that

(a) $\dfrac{1}{1\cdot 3}+\dfrac{1}{3\cdot 5}+\dfrac{1}{5\cdot 7}+\ldots\infty=\dfrac{1}{2},$

(b) $\dfrac{1}{1\cdot 3}-\dfrac{1}{3\cdot 5}+\dfrac{1}{5\cdot 7}-\ldots\infty=\dfrac{\pi}{4}-\dfrac{1}{2}$

and (c) $\dfrac{1}{1^2\cdot 3^2}+\dfrac{1}{3^2\cdot 5^2}+\dfrac{1}{5^2\cdot 7^2}+\ldots\infty=\dfrac{\pi^2}{16}-\dfrac{1}{2}.$

Solution: A sketch of the profile of the given function and its periodic extension is shown in the figure below (11.18).

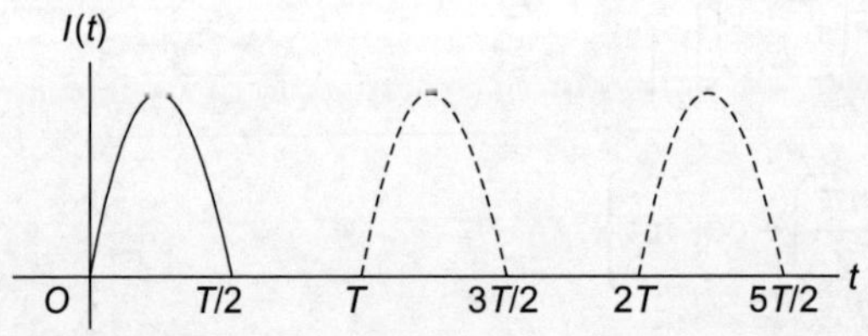

Fig. 11.18 Profile of output current $I(t)$ and its periodic extension.

Fourier Series

The Fourier series representation of the output current I can be written as

$$I(t)=\frac{a_0}{2}+\sum_{n=1}^{\infty} a_n \cos\left(\frac{2n\pi t}{T}\right)+\sum_{n=1}^{\infty} b_n \sin\left(\frac{2n\pi t}{T}\right).$$

Recalling the relation $\omega = \frac{2\pi}{T}$, we write

$$I(t) = \frac{a_0}{2} + \sum_{n=1}^{\infty} a_n \cos n\omega t + \sum_{n=1}^{\infty} b_n \sin n\omega t \quad \text{...(i)}$$

where

$$a_0 = \frac{2}{T}\int_0^T I(t)dt, \quad \text{...(ii)}$$

$$a_n = \frac{2}{T}\int_0^T I(t)\cos n\omega t\, dt \quad \text{...(iii)}$$

and

$$b_n = \frac{2}{T}\int_0^T I(t)\sin n\omega t\, dt. \quad \text{...(iv)}$$

To find the Fourier Coefficients

From Eq. (ii), we have

$$a_0 = \frac{2}{T}\left\{\int_0^{\frac{T}{2}} I_0 \sin \omega t\, dt + \int_{\frac{T}{2}}^{T} 0 \cdot dt\right\}$$

$$= \frac{2I_0}{T}\left[-\frac{\cos \omega t}{\omega}\right]_0^{\frac{T}{2}}$$

$$= \frac{2I_0}{2\pi}\{-\cos\pi + \cos 0\} = \frac{2I_0}{\pi}.$$

From Eq. (iii), we have

$$a_n = \frac{2}{T}\left\{\int_0^{\frac{T}{2}} I_0 \sin \omega t \cos n\omega t\, dt + \int_{\frac{T}{2}}^{T} 0 \cdot dt\right\}$$

$$= \frac{I_0}{T}\int_0^{\frac{T}{2}}\left[\sin\{(n+1)\omega t\} - \sin\{(n-1)\omega t\}\right]dt$$

$$= \frac{I_0}{T}\left[-\frac{\cos\{(n+1)\omega t\}}{\omega(n+1)} + \frac{\cos\{(n-1)\omega t\}}{\omega(n-1)}\right]_0^{\frac{T}{2}}$$

$$= \frac{I_0}{2\pi}\left[\frac{-\cos(n+1)\pi + \cos 0}{n+1} + \frac{\cos(n-1)\pi - \cos 0}{n-1}\right]$$

$$= \frac{I_0}{2\pi}\left[\frac{1+(-1)^n}{n+1} - \frac{(-1)^n + 1}{n-1}\right]$$

$$= -\frac{I_0}{\pi}\frac{1+(-1)^n}{(n-1)(n+1)} \quad (n \neq 1)$$

$$= \begin{cases} -\dfrac{2I_0}{\pi(n-1)(n+1)} & \text{if } n \text{ is even integer} \\ 0 & \text{if } n \text{ is odd integer}. \end{cases} \quad (\text{with } n \neq 1, \omega T = 2\pi),$$

When $n = 1$, Eq. (iii) yields

$$a_1 = \frac{2}{T}\int_0^T I(t)\cos\omega t \, dt$$

$$= \frac{2}{T}\left\{\int_0^{\frac{T}{2}} I_0 \sin\omega t \cos\omega t \, dt + \int_{\frac{T}{2}}^{T} 0 \cdot dt\right\}$$

$$= \frac{I_0}{T}\int_0^{\frac{T}{2}} \sin 2\omega t \, dt = \frac{I_0}{2\omega T}[-\cos 2\omega t]_0^{\frac{T}{2}}$$

$$= \frac{I_0}{2\omega T}[-\cos 2\pi + \cos 0] = 0.$$

From Eq. (iv), we have

$$b_n = \frac{2}{T}\int_0^T I(t)\sin n\omega t \, dt$$

$$= \frac{2}{T}\left\{\int_0^{\frac{T}{2}} I_0 \sin\omega t \sin n\omega t \, dt + \int_{\frac{T}{2}}^{T} 0 \cdot dt\right\}$$

$$= \frac{I_0}{T}\int_0^{\frac{T}{2}} \{\cos(n-1)\omega t - \cos(n+1)\omega t\} dt$$

$$= \frac{I_0}{T}\left[\frac{\sin(n-1)\omega t}{\omega(n-1)} - \frac{\sin(n+1)\omega t}{\omega(n+1)}\right]_0^{\frac{T}{2}}$$

$$= \frac{I_0}{\omega T}\left[\frac{\sin(n-1)\pi}{n-1} - \frac{\sin(n+1)\pi}{n+1}\right] \text{ (since, } \omega T = 2\pi)$$

$$= 0 \text{ for } n \neq 1.$$

For $n = 1$, Eq. (iv) yields

$$b_1 = \frac{2}{T}\int_0^T I(t)\sin\omega t\, dt$$

$$= \frac{2}{T}\left\{\int_0^{T/2} I_0 \sin^2(\omega t)\,dt + \int_{T/2}^{T} 0\cdot dt\right\}$$

$$= \frac{I_0}{T}\int_0^{\frac{T}{2}}(1-\cos 2\omega t)\,dt = \frac{I_0}{2}.$$

Substituting the values of Fourier coefficients into Eq. (i), we obtain the Fourier series as

$$I(t) = \frac{I_0}{\pi} + 0 + \sum_{n=\text{even}} -\frac{2I_0}{\pi(n-1)(n+1)}\cos n\omega t + \frac{I_0}{2}\sin\omega t + 0$$

$$= \frac{I_0}{\pi} - \frac{2I_0}{\pi}\sum_{n=\text{even}}\frac{\cos n\omega t}{(n-1)(n+1)} + \frac{I_0}{2}\sin\omega t$$

$$= \frac{I_0}{\pi} - \frac{2I_0}{\pi}\left[\frac{\cos 2\omega t}{1\cdot 3} + \frac{\cos 4\omega t}{3\cdot 5} + \frac{\cos 6\omega t}{5\cdot 7} + \ldots\infty\right] + \frac{I_0}{2}\sin\omega t.$$

In another form, restoring $\omega = \frac{2\pi}{T}$, we write

$$I(t) = \frac{I_0}{\pi} + \frac{I_0}{2}\sin\left(\frac{2\pi t}{T}\right) - \frac{2I_0}{\pi}\left[\frac{1}{1\cdot 3}\cos\left(\frac{4\pi t}{T}\right) + \frac{1}{3\cdot 5}\cos\left(\frac{8\pi t}{T}\right) + \frac{1}{5\cdot 7}\cos\left(\frac{12\pi t}{T}\right) + \ldots\infty\right].$$

Deduction:

(a) At $t = 0$, we have

$$I(0) = 0 = \frac{I_0}{\pi} - \frac{2I_0}{\pi}\left[\frac{1}{1\cdot 3} + \frac{1}{3\cdot 5} + \frac{1}{5\cdot 7} + \ldots \infty\right]$$

or $\dfrac{1}{1\cdot 3} + \dfrac{1}{3\cdot 5} + \dfrac{1}{5\cdot 7} + \ldots \infty = \dfrac{1}{2}.$

(b) At $t = \dfrac{T}{4}$ or $\omega t = \dfrac{\pi}{2}$, we have

$$I_0 \sin\frac{\pi}{2} = \frac{I_0}{\pi} + \frac{I_0}{2}\sin\frac{\pi}{2} - \frac{2I_0}{\pi}\left[\frac{1}{1\cdot 3}\cos\pi + \frac{1}{3\cdot 5}\cos 2\pi + \frac{1}{5\cdot 7}\cos 3\pi + \ldots \infty\right]$$

or $$I_0 = \frac{I_0}{\pi} + \frac{I_0}{2} + \frac{2I_0}{\pi}\left[\frac{1}{1\cdot 3} - \frac{1}{3\cdot 5} + \frac{1}{5\cdot 7} - \ldots \infty\right]$$

or $$\frac{\pi}{4} - \frac{1}{2} = \frac{1}{1\cdot 3} - \frac{1}{3\cdot 5} + \frac{1}{5\cdot 7} - \ldots \infty.$$

(c) Parseval's formula yields

$$\frac{1}{T}\int_0^T I^2 dt = \frac{a_0^2}{4} + \frac{1}{2}\sum_{n=1}^{\infty}\left(a_n^2 + b_n^2\right)$$

or $$\frac{1}{T}\left\{\int_0^{\frac{T}{2}} I_0^2 \sin^2 \omega t\, dt + \int_{\frac{T}{2}}^{T} 0\cdot dt\right\}$$

$$= \frac{I_0^2}{\pi^2} + \frac{1}{2}\sum_{n=\text{even}}\left\{-\frac{2I_0}{\pi(n-1)(n+1)}\right\}^2 + \frac{1}{2}\left(\frac{I_0^2}{4} + 0\right)$$

$$= \frac{I_0^2}{\pi^2} + \frac{2I_0^2}{\pi^2}\left(\frac{1}{1^2\cdot 3^2} + \frac{1}{3^2\cdot 5^2} + \frac{1}{5^2\cdot 7^2} + \ldots \infty\right) + \frac{I_0^2}{8}$$

or $$\frac{I_0^2}{2T}\int_0^{\frac{T}{2}}(1 - \cos 2\omega t)\,dt = \frac{I_0^2}{\pi^2} + \frac{I_0^2}{8} + \frac{2I_0^2}{\pi^2}\left(\frac{1}{1^2\cdot 3^2} + \frac{1}{3^2\cdot 5^2} + \frac{1}{5^2\cdot 7^2} + \ldots \infty\right)$$

or $$\frac{I_0^2}{4} = \frac{I_0^2}{\pi^2} + \frac{I_0^2}{8} + \frac{2I_0^2}{\pi^2}\left(\frac{1}{1^2\cdot 3^2} + \frac{1}{3^2\cdot 5^2} + \frac{1}{5^2\cdot 7^2} + \ldots \infty\right)$$

$$\text{or} \quad \frac{1}{1^2 \cdot 3^2}+\frac{1}{3^2 \cdot 5^2}+\frac{1}{5^2 \cdot 7^2}+\ldots\infty=\frac{\pi^2}{16}-\frac{1}{2}.$$

Note: If the output is given in terms of voltage, I_0 is to be replaced by V_0.

Deductions:

(a) At $t = 0$, we have

$$0=\frac{V_0}{\pi}-\frac{2V_0}{\pi}\left[\frac{1}{1\cdot 3}+\frac{1}{3\cdot 5}+\frac{1}{5\cdot 7}+\ldots\infty\right]$$

$$\text{or} \quad \frac{1}{1\cdot 3}+\frac{1}{3\cdot 5}+\frac{1}{5\cdot 7}+\ldots\infty=\frac{1}{2}.$$

(b) At $\omega t=\frac{\pi}{2}$ or $t=\frac{T}{4}$, we have

$$V_0\sin\frac{\pi}{2}=\frac{V_0}{\pi}-\frac{2V_0}{\pi}\left[\frac{1}{1\cdot 3}\cos\pi+\frac{1}{3\cdot 5}\cos 2\pi+\frac{1}{5\cdot 7}\cos 3\pi+\ldots\infty\right]+\frac{V_0}{2}\sin\frac{\pi}{2}$$

$$\text{or} \quad \frac{1}{2}=\frac{1}{\pi}+\frac{2}{\pi}\left[\frac{1}{1\cdot 3}-\frac{1}{3\cdot 5}+\frac{1}{5\cdot 7}-\ldots\infty\right]$$

$$\text{or} \quad \frac{\pi}{4}-\frac{1}{2}=\frac{1}{1\cdot 3}-\frac{1}{3\cdot 5}+\frac{1}{5\cdot 7}-\ldots\infty.$$

(c) Parseval's formula yields

$$\frac{1}{T}\int_0^T V^2 dt=\frac{a_0^2}{4}+\frac{1}{2}\sum_{n=1}^{\infty}\left(a_n^2+b_n^2\right)$$

$$\text{or} \quad \frac{V_0^2}{T}\int_0^{\frac{T}{2}}\sin^2\omega t\,dt=\frac{V_0^2}{\pi^2}+\frac{2V_0^2}{\pi^2}\left[\frac{1}{1^2\cdot 3^2}+\frac{1}{3^2\cdot 5^2}+\frac{1}{5^2\cdot 7^2}+\ldots\infty\right]+\frac{V_0^2}{8}$$

$$\text{or} \quad \frac{1}{2T}\int_0^{\frac{T}{2}}(1-\cos 2\omega t)dt=\frac{1}{\pi^2}+\frac{2}{\pi^2}\left[\frac{1}{1^2\cdot 3^2}+\frac{1}{3^2\cdot 5^2}+\frac{1}{5^2\cdot 7^2}+\ldots\infty\right]+\frac{1}{8}$$

$$\text{or} \quad \frac{1}{2T}\left[t-\frac{\sin 2\omega t}{2\omega}\right]_0^{\frac{T}{2}}=\frac{1}{4}=\frac{1}{\pi^2}+\frac{2}{\pi^2}\left[\frac{1}{1^2\cdot 3^2}+\frac{1}{3^2\cdot 5^2}+\frac{1}{5^2\cdot 7^2}+\ldots\infty\right]+\frac{1}{8}$$

or $$\frac{\pi^2}{16}-\frac{1}{2}=\frac{1}{1^2\cdot 3^2}+\frac{1}{3^2\cdot 5^2}+\frac{1}{5^2\cdot 7^2}+\ldots\infty.$$

PROBLEM 30 The output voltage of a full wave rectifier is given by

$$V(t)=V_0\sin\omega t,\, 0<t<\frac{T}{2}$$

where V_0 is the peak voltage, $\omega=\frac{2\pi}{T}$ is the circular frequency, t is instantaneous time and T is the time period.

Sketch a profile of the function and its periodic extension.

Find its Fourier series representation and deduce the identities (a), (b) and (c) of the previous problem.

Solution: A sketch of the profile of the given function and its periodic extension is shown in the figure below (11.19).

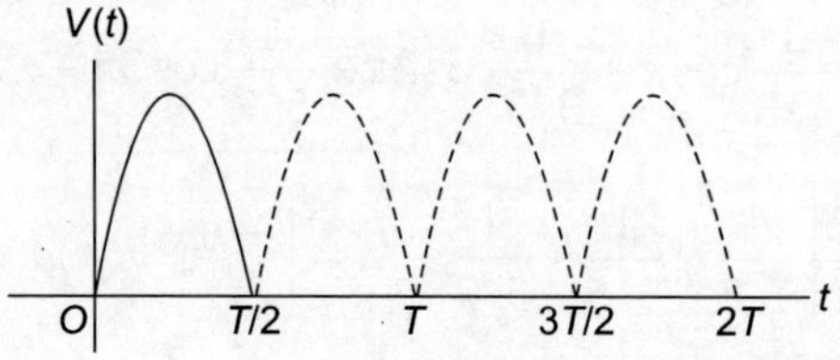

Fig. 11.19 Profile of output voltage $V(t)$ and its periodic extension.

Since the function is defined in half range, we use an even periodic extension to find the Fourier series.

The required Fourier series can be written as

$$V(t)=\frac{a_0}{2}+\sum_{n=1}^{\infty}a_n\cos n\omega t \qquad \ldots\text{(i)}$$

where $$a_0=\frac{2}{T/2}\int_0^{T/2}V(t)\,dt \qquad \ldots\text{(ii)}$$

and $$a_n=\frac{2}{T/2}\int_0^{T/2}V(t)\cos n\omega t\,dt\,. \qquad \ldots\text{(iii)}$$

To find the Fourier Constants

From Eq. (ii), we have

$$a_0 = \frac{4}{T}\int_0^{\frac{T}{2}} V_0 \sin\omega t\, dt$$

$$= -\frac{4V_0}{T}\left[\frac{\cos\omega t}{\omega}\right]_0^{\frac{T}{2}} = -\frac{4V_0}{\omega T}\left[\cos\frac{\omega T}{2} - \cos 0\right]$$

$$= -\frac{4V_0}{2\pi}[\cos\pi - \cos 0] = \frac{4V_0}{\pi} \quad (\text{as } \omega T = 2\pi).$$

From Eq. (iii), we have

$$a_n = \frac{4}{T}\int_0^{\frac{T}{2}} V_0 \sin\omega t \cos n\omega t\, dt$$

$$= \frac{2V_0}{T}\int_0^{\frac{T}{2}} \{\sin(n+1)\omega t - \sin(n-1)\omega t\}\, dt$$

$$= \frac{2V_0}{T}\left[\frac{-\cos(n+1)\omega t}{\omega(n+1)} + \frac{\cos(n-1)\omega t}{\omega(n-1)}\right]_0^{\frac{T}{2}}$$

$$= \frac{2V_0}{\omega T}\left[\frac{-\cos\left\{(n+1)\frac{\omega T}{2}\right\} + \cos 0}{n+1} + \frac{\cos\left\{(n-1)\frac{\omega T}{2}\right\} - \cos 0}{n-1}\right]$$

$$= \frac{2V_0}{\omega T}\left[\frac{-\cos\{(n+1)\pi\} + \cos 0}{n+1} + \frac{\cos\{(n-1)\pi\} - \cos 0}{n-1}\right]$$

$$= \frac{V_0}{\pi}\left[\frac{1+(-1)^n}{n+1} - \frac{1+(-1)^n}{n-1}\right]$$

$$= -\frac{2V_0\left[1+(-1)^n\right]}{\pi(n-1)(n+1)}$$

$$= \begin{cases} -\dfrac{4V_0}{\pi(n-1)(n+1)} & \text{with } n = \text{even integer}, \\ 0 & \text{with } n = \text{odd integer} \neq 1. \end{cases}$$

For $n = 1$, we have

$$a_1 = \frac{4V_0}{T}\int_0^{\frac{T}{2}} \sin\omega t \cos\omega t\, dt$$

$$= \frac{2V_0}{T}\int_0^{\frac{T}{2}} \sin 2\omega t\, dt = -\frac{2V_0}{T}\left[\frac{\cos 2\omega t}{2\omega}\right]_0^{\frac{T}{2}}$$

$$= -\frac{2V_0}{T}\left[\frac{\cos 2\pi - \cos 0}{2\omega}\right] = 0 \text{ (since } \omega T = 2\pi).$$

Substituting the values of the Fourier coefficients into Eq. (i), we obtain the Fourier series as

$$V(t) = \frac{2V_0}{\pi} - \frac{4V_0}{\pi}\sum_{n=\text{even}} \frac{\cos n\omega t}{(n-1)(n+1)}$$

$$= \frac{2V_0}{\pi} - \frac{4V_0}{\pi}\left[\frac{\cos 2\omega t}{1\cdot 3} + \frac{\cos 4\omega t}{3\cdot 5} + \frac{\cos 6\omega t}{5\cdot 7} + \ldots\infty\right].$$

Deductions:

(a) At $t = 0$, we have

$$0 = \frac{2V_0}{\pi} - \frac{4V_0}{\pi}\left[\frac{1}{1\cdot 3} + \frac{1}{3\cdot 5} + \frac{1}{5\cdot 7} + \ldots\infty\right]$$

or $$\frac{1}{1\cdot 3} + \frac{1}{3\cdot 5} + \frac{1}{5\cdot 7} + \ldots\infty = \frac{1}{2}.$$

(b) At $\omega t = \dfrac{\pi}{2}$ or $t = \dfrac{T}{4}$ we have

$$V_0 \sin\frac{\pi}{2} = \frac{2V_0}{\pi} - \frac{4V_0}{\pi}\left[\frac{1}{1\cdot 3}\cos\pi + \frac{1}{3\cdot 5}\cos 2\pi + \frac{1}{5\cdot 7}\cos 3\pi + \ldots\infty\right]$$

or $$1 = \frac{2}{\pi} + \frac{4}{\pi}\left[\frac{1}{1\cdot 3} - \frac{1}{3\cdot 5} + \frac{1}{5\cdot 7} - \ldots\infty\right]$$

Dividing both sides by $\frac{4}{\pi}$ and then transposing, we have

$$\frac{\pi}{4}-\frac{1}{2}=\frac{1}{1\cdot3}-\frac{1}{3\cdot5}+\frac{1}{5\cdot7}-\ldots\infty.$$

(c) Parseval's formula yields

$$\frac{2}{T}\int_0^{\frac{T}{2}}V^2dt=\frac{a_0^2}{4}+\frac{1}{2}\sum_{n=1}^{\infty}\left(a_n^2+b_n^2\right)$$

or $$\frac{2V_0^2}{T}\int_0^{\frac{T}{2}}\sin^2\omega t\,dt=\frac{4V_0^2}{\pi^2}+\frac{1}{2}\sum_{n=\text{even}}\left\{-\frac{4V_0}{\pi(n-1)(n+1)}\right\}^2 \quad (\text{as } b_n=0)$$

$$=\frac{4V_0^2}{\pi^2}+\frac{8V_0^2}{\pi^2}\left[\frac{1}{1^2\cdot3^2}+\frac{1}{3^2\cdot5^2}+\frac{1}{5^2\cdot7^2}+\ldots\infty\right]$$

or $$\frac{1}{T}\int_0^{\frac{T}{2}}(1-\cos2\omega t)\,dt=\frac{4}{\pi^2}+\frac{8}{\pi^2}\left[\frac{1}{1^2\cdot3^2}+\frac{1}{3^2\cdot5^2}+\frac{1}{5^2\cdot7^2}+\ldots\infty\right]$$

or $$\frac{1}{2}=\frac{4}{\pi^2}+\frac{8}{\pi^2}\left[\frac{1}{1^2\cdot3^2}+\frac{1}{3^2\cdot5^2}+\frac{1}{5^2\cdot7^2}+\ldots\infty\right]$$

or $$\frac{\pi^2}{16}-\frac{1}{2}=\frac{1}{1^2\cdot3^2}+\frac{1}{3^2\cdot5^2}+\frac{1}{5^2\cdot7^2}+\ldots\infty.$$

PROBLEM 31(a) The output voltage of a full-wave rectifier is given by

$$V(\theta)=\begin{cases}-V_0\sin\theta & \text{for } -\pi<\theta<0,\\ V_0\sin\theta & \text{for } 0<\theta<\pi\end{cases}$$

where V_0 is the peak voltage, $\theta=\omega t$ is the phase angle, ω is the circular frequency given by $\omega=\frac{2\pi}{T}$, t is instantaneous time and T is the time period.

Sketch a profile of the function and its periodic extension.

Find its Fourier series representation and deduce the identities (a), (b) and (c) of Problem 29.

Solution: A sketch of the profile of the given function and its periodic extension is shown in the figure below (11.20a).

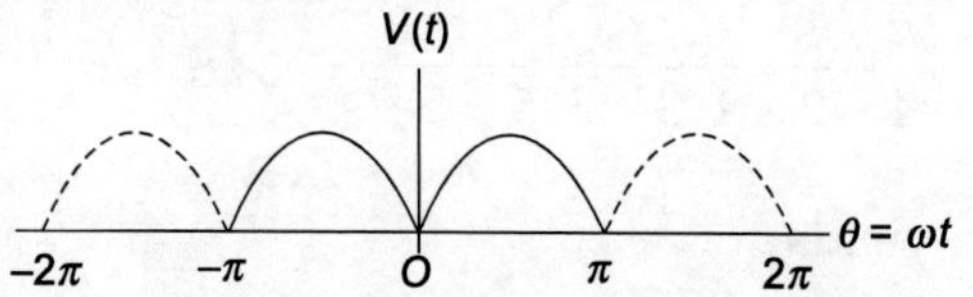

Fig. 11.20(a) Profile of output voltage $V(\theta)$ and its periodic extension.

As the function is an even function, the Fourier series can be written as

$$V(\theta)=\frac{a_0}{2}+\sum_{n=1}^{\infty}a_n\cos(n\theta) \qquad \ldots\text{(i)}$$

where
$$a_0=\frac{1}{\pi}\int_{-\pi}^{\pi}V(\theta)\,d\theta$$

$$=\frac{1}{\pi}\left\{\int_{-\pi}^{0}-V_0\sin\theta\,d\theta+\int_{0}^{\pi}V_0\sin\theta\,d\theta\right\}$$

$$=\frac{V_0}{\pi}\left\{\int_{\pi}^{0}-\sin(-\theta)(-d\theta)+\int_{0}^{\pi}\sin\theta\,d\theta\right\}$$

$$=\frac{2V_0}{\pi}\int_{0}^{\pi}\sin\theta\,d\theta$$

$$=\frac{2V_0}{\pi}[-\cos\theta]_0^{\pi}=\frac{2V_0}{\pi}(-\cos\pi+\cos 0)=\frac{4V_0}{\pi}$$

and
$$a_n=\frac{1}{\pi}\int_{-\pi}^{+\pi}V(\theta)\cos n\theta\,d\theta$$

$$=\frac{1}{\pi}\left\{\int_{-\pi}^{0}-V_0\sin\theta\cos n\theta\,d\theta+\int_{0}^{\pi}V_0\sin\theta\cos n\theta\,d\theta\right\}$$

$$=\frac{V_0}{\pi}\left\{\int_{\pi}^{0}-\sin(-\theta)\cos(-n\theta)(-d\theta)+\int_{0}^{\pi}\sin\theta\cos n\theta\,d\theta\right\}$$

$$=\frac{2V_0}{\pi}\int_{0}^{\pi}\sin\theta\cos n\theta\,d\theta$$

$$= \frac{V_0}{\pi}\int_0^{\pi}\{\sin(n+1)\theta - \sin(n-1)\theta\}\,d\theta$$

$$= \frac{V_0}{\pi}\left[\frac{-\cos(n+1)\theta}{(n+1)} + \frac{\cos(n-1)\theta}{(n-1)}\right]_0^{\pi}$$

$$= \frac{V_0}{\pi}\left[\frac{-(-1)^{n+1}+1}{n+1} + \frac{(-1)^{n-1}-1}{n-1}\right]$$

$$= \frac{V_0}{\pi}\left[\frac{1+(-1)^n}{n+1} - \frac{1+(-1)^n}{n-1}\right]$$

$$= -\frac{2V_0[1+(-1)^n]}{\pi(n-1)(n+1)} = \begin{cases} -\dfrac{4V_0}{\pi(n-1)(n+1)} & \text{if } n \text{ is an even integer,} \\ 0 & \text{if } n \text{ is an odd integer} \neq 1. \end{cases}$$

For $n = 1$, we have

$$a_1 = \frac{1}{\pi}\int_{-\pi}^{+\pi} V(\theta)\cos\theta\,d\theta$$

$$= \frac{1}{\pi}\left\{\int_{-\pi}^{0} -V_0\sin\theta\cos\theta\,d\theta + \int_0^{\pi} V_0\sin\theta\cos\theta\,d\theta\right\}$$

$$= \frac{V_0}{\pi}\left\{\int_{\pi}^{0} -\sin(-\theta)\cos(-\theta)(-d\theta) + \int_0^{\pi}\sin\theta\cos\theta\,d\theta\right\}$$

$$= \frac{2V_0}{\pi}\int_0^{\pi}\sin\theta\cos\theta\,d\theta$$

$$= \frac{V_0}{\pi}\int_0^{\pi}\sin 2\theta\,d\theta = \frac{V_0}{\pi}\left[-\frac{1}{2}\cos 2\theta\right]_0^{\pi} = 0.$$

Substituting the values of the Fourier coefficients into Eq. (i), we obtain

$$V(\theta) = \frac{2V_0}{\pi} - \frac{4V_0}{\pi}\sum_{n=\text{even}}\frac{\cos n\theta}{(n-1)(n+1)}$$

$$= \frac{2V_0}{\pi} - \frac{4V_0}{\pi}\left[\frac{\cos 2\theta}{1\cdot 3} + \frac{\cos 4\theta}{3\cdot 5} + \frac{\cos 6\theta}{5\cdot 7} + \ldots\infty\right].$$

Deductions: As in the previous problem, relation (a) can be deduced by putting $\theta = \omega t = 0$.

Relation (b) can be deduced by putting $\theta = \omega t = \frac{\pi}{2}$ and relation (c) can be deduced using Parseval's formula.

(b) The output voltage of a full-wave rectifier is given by

$$V(t) = \begin{cases} V_0 \sin \omega t & \text{for } 0 \le t \le \frac{T}{2}, \\ -V_0 \sin \omega t & \text{for } \frac{T}{2} \le t \le T \end{cases}$$

where V_0 is the peak voltage, T is the time period, ω is the circular frequency given by $\omega = \frac{2\pi}{T}$ and t is instantaneous time.

Sketch a profile of the function and its periodic extension.

Find its Fourier series representation and deduce the identities (a), (b) and (c) of Problem 29.

Solution:

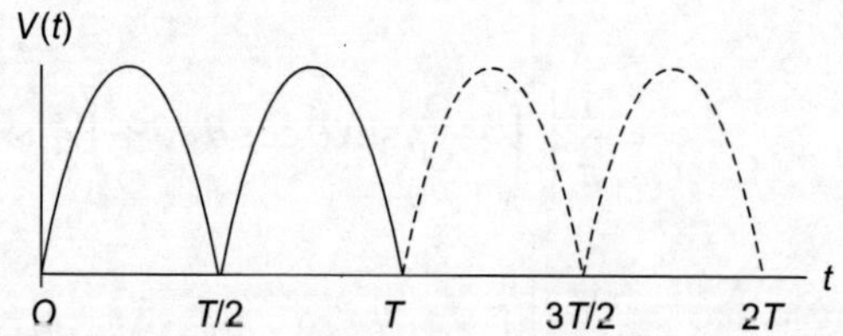

Fig. 11.20(b) Profile of output voltage $V(t)$ and its periodic extension.

The Fourier series can be written as

$$V(t) = \frac{a_0}{2} + \sum_{n=1}^{\infty} a_n \cos\left(\frac{2n\pi t}{T}\right) + \sum_{n=1}^{\infty} b_n \sin\left(\frac{2n\pi t}{T}\right) \quad \text{...(i)}$$

where $$a_0 = \frac{2}{T}\int_0^T V(t)dt, \quad \text{...(ii)}$$

$$a_n = \frac{2}{T}\int_0^T V(t)\cos\left(\frac{2n\pi t}{T}\right)dt \quad \text{...(iii)}$$

and $$b_n = \frac{2}{T}\int_0^T V(t)\sin\left(\frac{2n\pi t}{T}\right)dt. \quad \text{...(iv)}$$

From Eq. (ii), we have

$$a_0 = \frac{2}{T}\int_0^T V(t)\,dt = \frac{2}{T}\left\{\int_0^{\frac{T}{2}} V(t)\,dt + \int_{\frac{T}{2}}^{T} V(t)\,dt\right\}$$

$$= \frac{2}{T}\left\{\int_0^{\frac{T}{2}} V_0 \sin(\omega t)\,dt + \int_{\frac{T}{2}}^{T} -V_0 \sin(\omega t)\,dt\right\}$$

$$= \frac{2V_0}{T}\left\{\int_0^{\frac{T}{2}} \sin\left(\frac{2\pi t}{T}\right)dt - \int_{\frac{T}{2}}^{T} \sin\left(\frac{2\pi t}{T}\right)dt\right\}$$

$$= \frac{2V_0}{T}\left\{\frac{T}{2\pi}\left[-\cos\left(\frac{2\pi t}{T}\right)\right]_0^{\frac{T}{2}} + \frac{T}{2\pi}\left[\cos\left(\frac{2\pi t}{T}\right)\right]_{\frac{T}{2}}^{T}\right\}$$

$$= \frac{V_0}{\pi}[-\cos\pi + \cos 0 + \cos 2\pi - \cos\pi] = \frac{4V_0}{\pi}.$$

From Eq. (iii), we have

$$a_n = \frac{2}{T}\int_0^T V(t)\cos\left(\frac{2n\pi t}{T}\right)dt$$

$$= \frac{2}{T}\left\{\int_0^{\frac{T}{2}} V_0 \sin\omega t \cos\left(\frac{2n\pi t}{T}\right)dt - \int_{\frac{T}{2}}^{T} V_0 \sin\omega t \cos\left(\frac{2n\pi t}{T}\right)dt\right\}$$

$$= \frac{V_0}{T}\int_0^{\frac{T}{2}} 2\sin\left(\frac{2\pi t}{T}\right)\cos\left(\frac{2n\pi t}{T}\right)dt - \frac{V_0}{T}\int_{\frac{T}{2}}^{T} 2\sin\left(\frac{2\pi t}{T}\right)\cos\left(\frac{2n\pi t}{T}\right)dt$$

$$= \frac{V_0}{T}\int_0^{\frac{T}{2}}\left[\sin\left\{\frac{2(n+1)\pi t}{T}\right\} - \sin\left\{\frac{2(n-1)\pi t}{T}\right\}\right]dt$$

$$-\frac{V_0}{T}\int_{\frac{T}{2}}^{T}\left[\sin\left\{\frac{2(n+1)\pi t}{T}\right\}-\sin\left\{\frac{2(n-1)\pi t}{T}\right\}\right]dt$$

$$=\frac{V_0}{T}\cdot\frac{T}{2(n+1)\pi}\left[-\cos\left\{\frac{2(n+1)\pi t}{T}\right\}\right]_0^{\frac{T}{2}}+\frac{V_0}{T}\cdot\frac{T}{2(n-1)\pi}\left[\cos\left\{\frac{2(n-1)\pi t}{T}\right\}\right]_0^{\frac{T}{2}}$$

$$+\frac{V_0}{T}\cdot\frac{T}{2(n+1)\pi}\left[\cos\left\{\frac{2(n+1)\pi t}{T}\right\}\right]_{\frac{T}{2}}^{T}-\frac{V_0}{T}\cdot\frac{T}{2(n-1)\pi}\left[\cos\left\{\frac{2(n-1)\pi t}{T}\right\}\right]_{\frac{T}{2}}^{T}$$

$$=\frac{V_0}{2(n+1)\pi}\left[-\cos\{(n+1)\pi\}+\cos 0\right]+\frac{V_0}{2(n-1)\pi}\left[\cos\{(n-1)\pi\}-\cos 0\right]$$

$$+\frac{V_0}{2(n+1)\pi}\left[\cos\{2(n+1)\pi\}-\cos\{(n+1)\pi\}\right]$$

$$-\frac{V_0}{2(n-1)\pi}\left[\cos\{2(n-1)\pi\}-\cos\{(n-1)\pi\}\right]$$

$$=\frac{V_0}{2(n+1)\pi}\left[1-(-1)^{n+1}+1-(-1)^{n+1}\right]+\frac{V_0}{2(n-1)\pi}\left[(-1)^{n-1}-1-1+(-1)^{n-1}\right]$$

$$=\frac{V_0}{2(n+1)\pi}\left[2+2(-1)^n\right]-\frac{V_0}{2(n-1)\pi}\left[2+2(-1)^n\right]$$

$$=\frac{\left[1+(-1)^n\right]V_0}{\pi}\left[\frac{1}{n+1}-\frac{1}{n-1}\right]$$

$$=-\frac{2V_0}{\pi}\frac{[1+(-1)^n]}{(n-1)(n+1)}=\begin{cases}-\dfrac{4V_0}{\pi(n-1)(n+1)} & \text{if } n \text{ is an even integer,}\\ 0 & \text{if } n \text{ is an odd integer not equal to one.}\end{cases}$$

When $n = 1$, we have

$$a_1=\frac{2}{T}\int_0^T V(t)\cos\left(\frac{2\pi t}{T}\right)dt$$

$$=\frac{2}{T}\left\{\int_0^{\frac{T}{2}} V_0 \sin\omega t \cos\left(\frac{2\pi t}{T}\right)dt+\int_{\frac{T}{2}}^{T} -V_0 \sin\omega t \cos\left(\frac{2\pi t}{T}\right)dt\right\}$$

$$=\frac{V_0}{T}\int_0^{\frac{T}{2}} 2\sin\left(\frac{2\pi t}{T}\right)\cos\left(\frac{2\pi t}{T}\right)dt-\frac{V_0}{T}\int_{\left(\frac{T}{2}\right)}^{T} 2\sin\left(\frac{2\pi t}{T}\right)\cos\left(\frac{2\pi t}{T}\right)dt$$

$$=\frac{V_0}{T}\int_0^{\frac{T}{2}} \sin\left(\frac{4\pi t}{T}\right)dt-\frac{V_0}{T}\int_{\left(\frac{T}{2}\right)}^{T} \sin\left(\frac{4\pi t}{T}\right)dt=0.$$

From Eq. (iv), we have

$$b_n=\frac{2}{T}\left\{\int_0^{\frac{T}{2}} V_0 \sin\omega t \sin\left(\frac{2n\pi t}{T}\right)dt-\int_{\left(\frac{T}{2}\right)}^{T} V_0 \sin\omega t \sin\left(\frac{2n\pi t}{T}\right)dt\right\}$$

$$=\frac{2}{T}\left\{\int_0^{\frac{T}{2}} V_0 \sin\left(\frac{2\pi t}{T}\right)\sin\left(\frac{2n\pi t}{T}\right)dt-\int_{\left(\frac{T}{2}\right)}^{T} V_0 \sin\left(\frac{2\pi t}{T}\right)\sin\left(\frac{2n\pi t}{T}\right)dt\right\}$$

$$=\frac{V_0}{T}\int_0^{\frac{T}{2}} 2\sin\left(\frac{2\pi t}{T}\right)\sin\left(\frac{2n\pi t}{T}\right)dt-\frac{V_0}{T}\int_{\left(\frac{T}{2}\right)}^{T} 2\sin\left(\frac{2\pi t}{T}\right)\sin\left(\frac{2n\pi t}{T}\right)dt$$

$$=\frac{V_0}{T}\int_0^{\frac{T}{2}}\left[\cos\left\{(n-1)\frac{2\pi t}{T}\right\}-\cos\left\{(n+1)\frac{2\pi t}{T}\right\}\right]dt$$

$$-\frac{V_0}{T}\int_{\frac{T}{2}}^{T}\left[\cos\left\{(n-1)\frac{2\pi t}{T}\right\}-\cos\left\{(n+1)\frac{2\pi t}{T}\right\}\right]dt=0$$

Now, substituting the values of a_0, a_n and b_n into Eq. (i), we obtain the Fourier series as

$$V(t) = \frac{2V_0}{\pi} - \sum_{n=\text{even}} \frac{4V_0}{\pi(n-1)(n+1)} \cos\left(\frac{2n\pi t}{T}\right)$$

$$= \frac{2V_0}{\pi} - \frac{4V_0}{\pi}\left[\frac{1}{1\cdot 3}\cos\left(\frac{4\pi t}{T}\right) + \frac{1}{3\cdot 5}\cos\left(\frac{8\pi t}{T}\right) + \frac{1}{5\cdot 7}\cos\left(\frac{12\pi t}{T}\right) + \dots\infty\right].$$

Deductions:

(a) Putting $t = 0$, we obtain

$$V_0 \sin 0 = 0 = \frac{2V_0}{\pi} - \frac{4V_0}{\pi}\left[\frac{1}{1\cdot 3} + \frac{1}{3\cdot 5} + \frac{1}{5\cdot 7} + \dots\infty\right]$$

or
$$\frac{1}{2} = \frac{1}{1\cdot 3} + \frac{1}{3\cdot 5} + \frac{1}{5\cdot 7} + \dots\infty.$$

(b) Putting $t = \frac{T}{4}$ in the above Fourier series, we have

$$V_0 \sin\frac{\omega T}{4} = V_0 \sin\frac{\pi}{2} = V_0 = \frac{2V_0}{\pi} - \frac{4V_0}{\pi}\left[-\frac{1}{1\cdot 3} + \frac{1}{3\cdot 5} - \frac{1}{5\cdot 7} + \dots\infty\right]$$

or
$$\frac{\pi}{4} - \frac{1}{2} = \frac{1}{1\cdot 3} - \frac{1}{3\cdot 5} + \frac{1}{5\cdot 7} - \dots\infty \quad (\text{Here, } \omega T = 2\pi).$$

(c) Using Parseval's formula, we write

$$\frac{1}{T}\int_0^T [V(t)]^2\, dt = \frac{a_0^2}{4} + \frac{1}{2}\sum\left(a_n^2 + b_n^2\right)$$

or
$$\frac{1}{T}\left\{\int_0^{\frac{T}{2}} V_0^2 \sin^2\left(\frac{2\pi t}{T}\right) dt + \int_{\frac{T}{2}}^{T} V_0^2 \sin^2\left(\frac{2\pi t}{T}\right) dt\right\}$$

$$= \frac{1}{4}\left(\frac{4V_0}{\pi}\right)^2 + \frac{1}{2}\sum_{n=\text{even}}\left[-\frac{4V_0}{\pi(n-1)(n+1)}\right]^2$$

or
$$\frac{V_0^2}{T}\int_0^T \sin^2\left(\frac{2\pi t}{T}\right) dt = \frac{4V_0^2}{\pi^2} + \frac{8V_0^2}{\pi^2}\left[\frac{1}{1^2\cdot 3^2} + \frac{1}{3^2\cdot 5^2} + \frac{1}{5^2\cdot 7^2} + \dots\infty\right]$$

or $$\frac{V_0^2}{2T}\int_0^T\left\{1-\cos\left(\frac{4\pi t}{T}\right)\right\}dt=\frac{V_0^2}{2}=\frac{4V_0^2}{\pi^2}+\frac{8V_0^2}{\pi^2}\left[\frac{1}{1^2\cdot 3^2}+\frac{1}{3^2\cdot 5^2}+\frac{1}{5^2\cdot 7^2}+\ldots\infty\right]$$

or $$\frac{\pi^2}{16}-\frac{1}{2}=\frac{1}{1^2\cdot 3^2}+\frac{1}{3^2\cdot 5^2}+\frac{1}{5^2\cdot 7^2}+\ldots\infty.$$

Problems on Fourier Integrals and Fourier Transforms

PROBLEM 32 Find the Fourier integral of the function

$$f(x)=\begin{cases}1 & \text{for } |x|\le 1,\\ 0 & \text{for } |x|>1.\end{cases}$$

Hence evaluate $\int_0^\infty \frac{\sin u}{u}du$.

Solution: The required Fourier integral is

$$f(x)=\frac{1}{\pi}\int_0^\infty\int_{-\infty}^\infty f(t)\cos\{u(t-x)\}dt\,du$$

$$=\frac{1}{\pi}\int_0^\infty\int_{-1}^1\left[1\cdot\cos\{u(t-x)\}dt\right]du$$

$$=\frac{1}{\pi}\int_0^\infty\left[\frac{\sin\{u(t-x)\}}{u}\right]_{t=-1}^1 du$$

$$=\frac{1}{\pi}\int_0^\infty\frac{\sin\{u(1-x)\}-\sin\{u(-1-x)\}}{u}du$$

$$=\frac{1}{\pi}\int_0^\infty\frac{\sin\{u(1-x)\}+\sin\{u(1+x)\}}{u}du$$

$$=\frac{2}{\pi}\int_0^\infty\frac{\sin u\cos ux}{u}du.$$

Thus, we have

$$\int_0^\infty \frac{\sin u \cos ux}{u} du = \frac{\pi}{2} f(x) = \begin{cases} \dfrac{\pi}{2} & \text{for } |x| \le 1, \\ 0 & \text{for } |x| > 1. \end{cases}$$

For $x = 0$, we have

$$\int_0^\infty \frac{\sin u}{u} du = \frac{\pi}{2}.$$

PROBLEM 33 Find the Fourier integral of the function

$$f(x) = \begin{cases} 0 & \text{for } x < -1, \\ -1 & \text{for } -1 < x < 0, \\ 2 & \text{for } 0 < x < 1, \\ 0 & \text{for } x > 1. \end{cases}$$

Solution: The Fourier integral of the given function is

$$f(x) = \frac{1}{\pi} \int_0^\infty \int_{-\infty}^\infty f(t) \cos\{u(t-x)\} dt\, du$$

$$= \frac{1}{\pi} \int_0^\infty \left[\int_{-1}^0 (-1) \cos\{u(t-x)\} dt + \int_0^1 2\cos\{u(t-x)\} dt \right] du$$

$$= \frac{1}{\pi} \int_0^\infty \left\{ \left[\frac{-\sin\{u(t-x)\}}{u} \right]_{t=-1}^0 + \left[\frac{2\sin\{u(t-x)\}}{u} \right]_{t=0}^1 \right\} du$$

$$= \frac{1}{\pi} \int_0^\infty \left\{ \frac{-\sin(-ux) + \sin\{u(-1-x)\}}{u} + \frac{2\sin\{u(1-x)\} - 2\sin(-ux)}{u} \right\} du$$

$$= \frac{1}{\pi} \int_0^\infty \frac{3\sin ux - \sin\{u(1+x)\} + 2\sin\{u(1-x)\}}{u} du$$

$$= \frac{1}{\pi} \int_0^\infty \frac{3\sin ux - \sin u \cos ux - \cos u \sin ux + 2\sin u \cos ux - 2\cos u \sin ux}{u} du$$

$$= \frac{1}{\pi} \int_0^\infty \frac{3(1-\cos u)\sin ux + \sin u \cos ux}{u} du.$$

PROBLEM 34 Obtain the Fourier cosine integral of the function

$$f(x)=\begin{cases} x^2 & \text{for } 0<x<1, \\ 0 & \text{for } x>1. \end{cases}$$

Solution: The Fourier cosine integral is given by

$$f(x)=\frac{2}{\pi}\int_0^\infty\int_0^\infty f(t)\cos ut\cos ux\,dt\,du$$

$$=\frac{2}{\pi}\int_0^\infty\left\{\int_0^\infty f(t)\cos ut\,dt\right\}\cos ux\,du$$

$$=\frac{2}{\pi}\int_0^\infty\left\{\int_0^1 t^2\cos ut\,dt\right\}\cos ux\,du$$

$$=\frac{2}{\pi}\int_0^\infty\left\{\left[\frac{t^2\sin ut}{u}\right]_{t=0}^1-\int_0^1\frac{2t\sin ut}{u}dt\right\}\cos ux\,du$$

$$=\frac{2}{\pi}\int_0^\infty\left\{\frac{\sin u}{u}+\left[\frac{2t\cos ut}{u^2}\right]_{t=0}^1-\int_0^1\frac{2\cos ut}{u^2}dt\right\}\cos ux\,du$$

$$=\frac{2}{\pi}\int_0^\infty\left\{\frac{\sin u}{u}+\frac{2\cos u}{u^2}-\left[\frac{2\sin ut}{u^3}\right]_{t=0}^1\right\}\cos ux\,du$$

$$=\frac{2}{\pi}\int_0^\infty\left\{\frac{\sin u}{u}+\frac{2\cos u}{u^2}-\frac{2\sin u}{u^3}\right\}\cos ux\,du.$$

PROBLEM 35 A function is defined by

$$f(x)=\begin{cases} -2-x & \text{for } -2<x<-1, \\ x & \text{for } -1<x<1, \\ 2-x & \text{for } 1<x<2, \\ 0 & \text{otherwise.} \end{cases}$$

Sketch a graphical represetation of the function and obtain its Fourier integral.

Solution: A graphical representation of the given function is shown below (11.21).

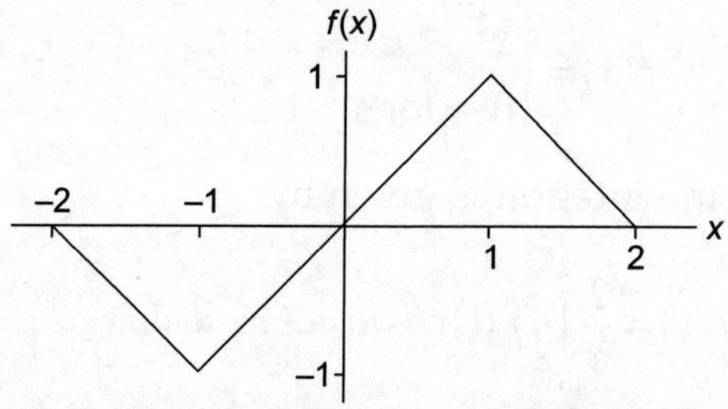

Fig. 11.21 Graphical representation of $f(x)$.

We see that the given function $f(x)$ is an odd function. Therefore, its Fourier integral will be Fourier sine integral given by

$$f(x)=\frac{2}{\pi}\int_0^{\infty}\int_0^{\infty} f(t)\sin ut \sin ux\, dt\, du$$

$$=\frac{2}{\pi}\int_0^{\infty}\left\{\int_0^{2} f(t)\sin ut\, dt\right\}\sin ux\, du$$

$$=\frac{2}{\pi}\int_0^{\infty}\left\{\int_0^{1} f(t)\sin ut\, dt+\int_1^{2} f(t)\sin ut\, dt\right\}\sin ux\, du$$

$$=\frac{2}{\pi}\int_0^{\infty}\left\{\int_0^{1} t\sin ut\, dt+\int_1^{2}(2-t)\sin ut\, dt\right\}\sin ux\, du$$

$$=\frac{2}{\pi}\int_0^{\infty}\left\{\left[\frac{-t\cos ut}{u}\right]_{t=0}^{1}+\int_0^{1}\frac{\cos ut}{u}dt+\left[\frac{(t-2)\cos ut}{u}\right]_{t=1}^{2}-\int_1^{2}\frac{\cos ut}{u}dt\right\}\sin ux\, du$$

$$=\frac{2}{\pi}\int_0^{\infty}\left\{\frac{-\cos u}{u}+\left[\frac{\sin ut}{u^2}\right]_{t=0}^{1}+\frac{\cos u}{u}-\left[\frac{\sin ut}{u^2}\right]_{t=1}^{2}\right\}\sin ux\, du$$

$$=\frac{2}{\pi}\int_0^{\infty}\left(\frac{2\sin u-\sin 2u}{u^2}\right)\sin ux\, du.$$

PROBLEM 36 Obtain the Fourier integral of the function

$$f(x)=\begin{cases} e^{kx} & \text{for } x<0, \\ e^{-kx} & \text{for } x>0. \end{cases}$$

Deduce that $\int_0^{\infty} \frac{\cos ux}{k^2+u^2}\,du = \frac{\pi}{2k} e^{-kx}$.

Solution: Here $f(x)$ is an even function. Therefore, its Fourier integral will be the Fourier cosine integral. The Fourier cosine integral is given by

$$f(x) = \frac{2}{\pi}\int_0^{\infty}\int_0^{\infty} f(t)\cos ut \cos ux \, dt\, du$$

$$= \frac{2}{\pi}\int_0^{\infty}\left\{\int_0^{\infty} e^{-kt} \cos ut \, dt\right\}\cos ux \, du$$

$$= \frac{1}{\pi}\int_0^{\infty}\left\{\int_0^{\infty} e^{-kt}\left[e^{iut} + e^{-iut}\right]dt\right\}\cos ux \, du$$

$$= \frac{1}{\pi}\int_0^{\infty}\left\{\int_0^{\infty}\left[e^{-(k-iu)t} + e^{-(k+iu)t}\right]dt\right\}\cos ux \, du$$

$$= \frac{1}{\pi}\int_0^{\infty}\left\{\left[-\frac{e^{-(k-iu)t}}{k-iu} - \frac{e^{-(k+iu)t}}{k+iu}\right]_{t=0}^{\infty}\right\}\cos ux \, du$$

$$= \frac{1}{\pi}\int_0^{\infty}\left\{\frac{1}{k-iu} + \frac{1}{k+iu}\right\}\cos ux \, du$$

$$= \frac{2k}{\pi}\int_0^{\infty}\frac{\cos ux}{k^2+u^2}\,du.$$

Deduction: For $k > 0, x > 0$, we have

$$e^{-kx} = \frac{2k}{\pi}\int_0^{\infty}\frac{\cos ux}{k^2+u^2}\,du$$

or
$$\int_0^{\infty}\frac{\cos ux}{k^2+u^2}\,du = \frac{\pi}{2k}e^{-kx}.$$

PROBLEM 37 Obtain Fourier integral of the functions

(a) $f(x)=\begin{cases}0 & \text{for } x<-1,\\ -5 & \text{for } -1<x<0,\\ 5 & \text{for } 0<x<1,\\ 0 & \text{for } x>1.\end{cases}$

(b) $f(x)=\begin{cases}|x| & \text{for } |x|\le\pi,\\ 0 & \text{for } |x|>\pi.\end{cases}$

Solution:

(a) Since the function $f(x)$ is an odd function, its Fourier integral will be the Fourier sine integral given by

$$f(x)=\frac{2}{\pi}\int_0^\infty\int_0^\infty f(t)\sin ut\,\sin ux\,dt\,du$$

$$=\frac{2}{\pi}\int_0^\infty\left\{\int_0^1 5\sin ut\,dt\right\}\sin ux\,du$$

$$=\frac{10}{\pi}\int_0^\infty\left[-\frac{\cos ut}{u}\right]_{t=0}^1\sin ux\,du$$

$$=\frac{10}{\pi}\int_0^\infty\frac{(1-\cos u)}{u}\sin ux\,du.$$

(b) In this problem, the function $f(x)$ is an even function. Therefore, its Fourier series will be the Fourier cosine integral given by

$$f(x)=\frac{2}{\pi}\int_0^\infty\int_0^\infty f(t)\cos ut\cos ux\,dt\,du$$

$$=\frac{2}{\pi}\int_0^\infty\left\{\int_0^\pi |t|\cos ut\,dt\right\}\cos ux\,du$$

$$=\frac{2}{\pi}\int_0^\infty\left\{\left[|t|\frac{\sin ut}{u}\right]_{t=0}^\pi-\int_0^\pi\frac{\sin ut}{u}dt\right\}\cos ux\,du$$

$$= \frac{2}{\pi}\int_0^{\infty}\left\{\frac{\pi \sin \pi u}{u} + \left[\frac{\cos ut}{u^2}\right]_{t=0}^{\pi}\right\}\cos ux\, du$$

$$= \frac{2}{\pi}\int_0^{\infty}\left\{\frac{\pi u \sin \pi u + \cos \pi u - 1}{u^2}\right\}\cos ux\, du.$$

PROBLEM 38 Find both the Fourier sine and cosine integrals of the function $f(x) = e^{-x}$ with $x \geq 0$.

Sketch the profiles of the given function and its odd and even periodic extensions.

Solution:

Fourier Sine Integral

The sine integral is given by

$$f(x) = e^{-x} = \frac{2}{\pi}\int_0^{\infty}\int_0^{\infty} f(t) \sin ut \sin ux\, dt\, du$$

$$= \frac{1}{\pi i}\int_0^{\infty}\left[\int_0^{\infty} e^{-t}\, 2i \sin ut\, dt\right]\sin ux\, du$$

$$= \frac{1}{\pi i}\int_0^{\infty}\left[\int_0^{\infty} e^{-t}(e^{iut} - e^{-iut})dt\right]\sin ux\, du$$

$$= \frac{1}{\pi i}\int_0^{\infty}\left[\int_0^{\infty}\left\{e^{-(1-iu)t} - e^{-(1+iu)t}\right\}dt\right]\sin ux\, du$$

$$= \frac{1}{\pi i}\int_0^{\infty}\left[-\frac{e^{-(1-iu)t}}{1-iu} + \frac{e^{-(1+iu)t}}{1+iu}\right]_{t=0}^{\infty}\sin ux\, du$$

$$= \frac{1}{\pi i}\int_0^{\infty}\left(\frac{1}{1-iu} - \frac{1}{1+iu}\right)\sin ux\, du$$

$$= \frac{2}{\pi}\int_0^{\infty}\frac{u \sin ux}{1+u^2}\, du.$$

Fourier Cosine Integral

The Fourier cosine integral is given by

$$f(x) = \frac{2}{\pi}\int_0^{\infty}\int_0^{\infty} f(t) \cos ut \cos ux\, dt\, du$$

$$= \frac{1}{\pi}\int_0^\infty \left[\int_0^\infty e^{-t}\, 2\cos ut\, dt\right]\cos ux\, du$$

$$= \frac{1}{\pi}\int_0^\infty \left[\int_0^\infty e^{-t}(e^{iut} + e^{-iut})dt\right]\cos ux\, du$$

$$= \frac{1}{\pi}\int_0^\infty \left[\int_0^\infty \left\{e^{-(1-iu)t} + e^{-(1+iu)t}\right\}dt\right]\cos ux\, du$$

$$= \frac{1}{\pi}\int_0^\infty \left[-\frac{e^{-(1-iu)t}}{1-iu} - \frac{e^{-(1+iu)t}}{1+iu}\right]_{t=0}^{\infty}\cos ux\, du$$

$$= \frac{1}{\pi}\int_0^\infty \left[\frac{1}{1-iu} + \frac{1}{1+iu}\right]\cos ux\, du$$

$$= \frac{2}{\pi}\int_0^\infty \frac{\cos ux}{1+u^2}du.$$

Profiles:

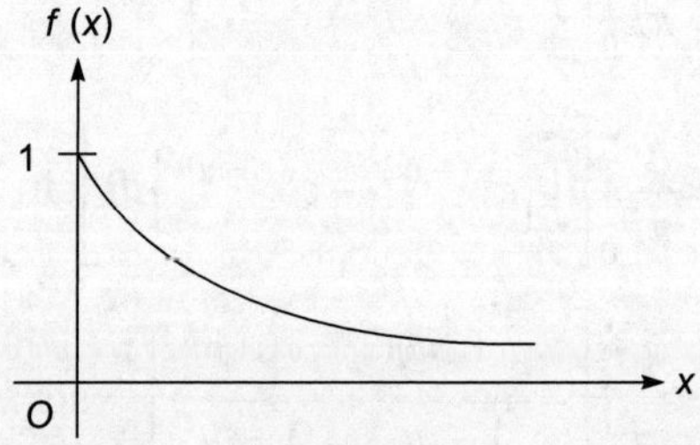

Fig. 11.22(a) Profile of the given function $f(x)$.

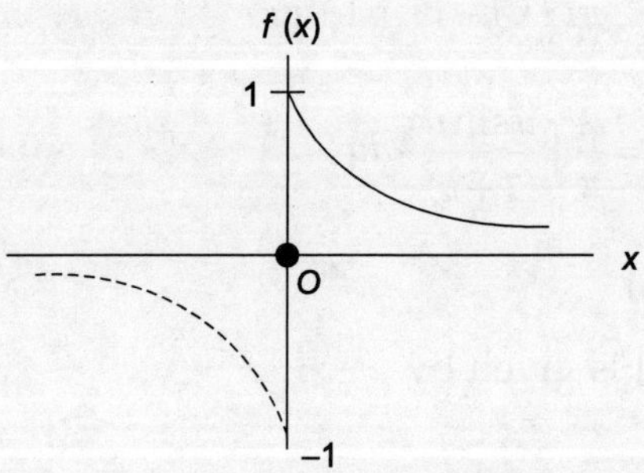

Fig. 11.22(b) Profile of the given function $f(x)$ with odd periodic extension.

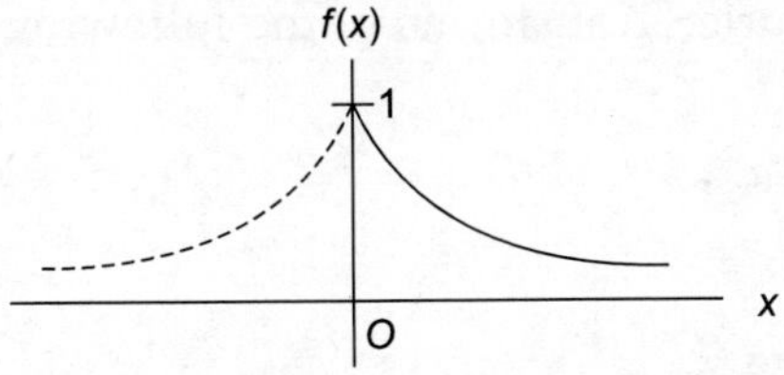

Fig. 11.22(c) Profile of the given function $f(x)$ with even periodic extension.

PROBLEM 39 Obtain the complex form of the Fourier integral of the box function

$$f(x) = \begin{cases} 1 & \text{for } -a < x < a, \\ 0 & \text{otherwise.} \end{cases}$$

Also, deduce that $\int_0^\infty \frac{\sin au}{u} du = \frac{\pi}{2}$ with $a > 0$.

Solution:

$$f(x) = \frac{1}{2\pi} \int_{-\infty}^{\infty} \int_{-\infty}^{\infty} f(t) e^{iu(t-x)} dt\, du$$

$$= \frac{1}{2\pi} \int_{-\infty}^{\infty} \left\{ \int_{-a}^{a} (1) e^{iut} dt \right\} e^{-iux} du$$

$$= \frac{1}{2\pi} \int_{-\infty}^{\infty} \left\{ \left[\frac{e^{iut}}{iu} \right]_{t=-a}^{a} \right\} e^{-iux} du$$

$$= \frac{1}{2\pi} \int_{-\infty}^{\infty} \left\{ \frac{e^{iau} - e^{-iau}}{iu} \right\} e^{-iux} du$$

$$= \int_{-\infty}^{\infty} \frac{\sin au}{\pi u} e^{-iux} du.$$

Deduction: At $x = 0$, we have

$$f(0) = 1 = \int_{-\infty}^{\infty} \frac{\sin au}{\pi u} du = \frac{2}{\pi} \int_0^\infty \frac{\sin au}{u} du$$

or

$$\int_0^\infty \frac{\sin au}{u} du = \frac{\pi}{2} \text{ with } a > 0.$$

PROBLEM 40 Obtain Fourier Transforms of the following functions:

(a) $f(x)=\begin{cases}1 & \text{for } -a\le x\le a,\\ 0 & \text{for } |x|>a.\end{cases}$

(b) $f(x)=\begin{cases}e^{-ax} & \text{for } a>0,\ x\ge 0,\\ 0 & \text{for } x<0.\end{cases}$

(c) $f(x)=e^{-a|x|}$ for $a > 0$.

Solution:

(a) The Fourier transform of the given function $f(x)$ is given by

$$FT[f(x)] = F(k) = \frac{1}{\sqrt{2\pi}}\int_{-\infty}^{\infty} f(x)e^{-ikx}\,dx$$

$$= \frac{1}{\sqrt{2\pi}}\int_{-a}^{a}(1)e^{-ikx}\,dx$$

$$= \frac{1}{\sqrt{2\pi}}\left[-\frac{e^{-ikx}}{ik}\right]_{x=-a}^{a}$$

$$= \frac{1}{\sqrt{2\pi}}\left[\frac{e^{ika}-e^{-ika}}{ik}\right] = \sqrt{\frac{2}{\pi}}\,\frac{\sin ak}{k}.$$

(b) The Fourier transform of the given function $f(x)$ is given by

$$F(k) = \frac{1}{\sqrt{2\pi}}\int_{-\infty}^{\infty} f(x)e^{-ikx}\,dx$$

$$= \frac{1}{\sqrt{2\pi}}\int_{0}^{\infty} e^{-ax}\cdot e^{-ikx}\,dx$$

$$= \frac{1}{\sqrt{2\pi}}\int_{0}^{\infty} e^{-(a+ik)x}\,dx$$

$$= \frac{1}{\sqrt{2\pi}}\left[-\frac{e^{-(a+ik)x}}{a+ik}\right]_{x=0}^{\infty}$$

$$= \frac{1}{\sqrt{2\pi}}\cdot\frac{1}{a+ik}.$$

(c) The Fourier transform of the given function $f(x)$ is given by

$$F(k)=\frac{1}{\sqrt{2\pi}}\int_{-\infty}^{\infty} f(x)e^{-ikx}dx$$

$$=\frac{1}{\sqrt{2\pi}}\left\{\int_{-\infty}^{0} f(x)e^{-ikx}dx+\int_{0}^{\infty} f(x)e^{-ikx}dx\right\}$$

$$=\frac{1}{\sqrt{2\pi}}\left\{\int_{-\infty}^{0} e^{ax}\cdot e^{-ikx}dx+\int_{0}^{\infty} e^{-ax}\cdot e^{-ikx}dx\right\}$$

$$=\frac{1}{\sqrt{2\pi}}\left\{\int_{-\infty}^{0} e^{(a-ik)x}dx+\int_{0}^{\infty} e^{-(a+ik)x}dx\right\}$$

$$=\frac{1}{\sqrt{2\pi}}\left\{\left[\frac{e^{(a-ik)x}}{a-ik}\right]_{x=-\infty}^{0}+\left[-\frac{e^{-(a+ik)x}}{a+ik}\right]_{x=0}^{\infty}\right\}$$

$$=\frac{1}{\sqrt{2\pi}}\left\{\frac{1}{a-ik}+\frac{1}{a+ik}\right\}=\sqrt{\frac{2}{\pi}}\frac{a}{a^2+k^2}.$$

PROBLEM 41 Find the Fourier transform of the Gaussian function $f(x)=Ae^{-ax^2}$ where A and a are positive constants.

Solution: The Fourier transform of $f(x)$ is given by

$$FT[f(x)]=F(k)=\frac{1}{\sqrt{2\pi}}\int_{-\infty}^{\infty} f(x)e^{-ikx}dx$$

$$=\frac{1}{\sqrt{2\pi}}\int_{-\infty}^{\infty} Ae^{-ax^2}\cdot e^{-ikx}dx$$

$$=\frac{A}{\sqrt{2\pi}}\int_{-\infty}^{\infty} e^{-a\left(x^2+i\frac{k}{a}x\right)}dx$$

$$=\frac{Ae^{-\frac{k^2}{4a}}}{\sqrt{2\pi}}\int_{-\infty}^{\infty} e^{-a\left(x+\frac{ik}{2a}\right)^2}dx$$

$$=\frac{Ae^{-\frac{k^2}{4a}}}{\sqrt{2\pi a}}\int_{-\infty}^{\infty} e^{-y^2}dy \text{ where we put } y=\sqrt{a}\left(x+\frac{ik}{2a}\right).$$

Recalling gamma function integral, $\int_{-\infty}^{\infty} e^{-y^2} dy = \Gamma\left(\frac{1}{2}\right) = \sqrt{\pi}$, we write

$$F(k) = \frac{A}{\sqrt{2a}} e^{-\frac{k^2}{4a}}.$$

PROBLEM 42 Obtain the Fourier transform of the function $f(x) = e^{-a|x|+ibx}$ and deduce that

$$\int_{-\infty}^{\infty} \frac{\cos kx}{a^2 + (b-k)^2} dk = \frac{\pi}{a} e^{-a|x|} \cos bx$$

and
$$\int_{-\infty}^{\infty} \frac{\sin kx}{a^2 + (b-k)^2} dk = \frac{\pi}{a} e^{-a|x|} \sin bx.$$

Solution: The Fourier transform of $f(x)$ is given by

$$F(k) = \frac{1}{\sqrt{2\pi}} \int_{-\infty}^{\infty} f(x) e^{-ikx} dx$$

$$= \frac{1}{\sqrt{2\pi}} \left[\int_{-\infty}^{0} f(x) e^{-ikx} dx + \int_{0}^{\infty} f(x) e^{-ikx} dx \right]$$

$$= \frac{1}{\sqrt{2\pi}} \left[\int_{-\infty}^{0} e^{ax+ibx} \cdot e^{-ikx} dx + \int_{0}^{\infty} e^{-ax+ibx} \cdot e^{-ikx} dx \right]$$

$$= \frac{1}{\sqrt{2\pi}} \left[\int_{-\infty}^{0} e^{\{a+i(b-k)\}x} dx + \int_{0}^{\infty} e^{-\{a-i(b-k)\}x} dx \right]$$

$$= \frac{1}{\sqrt{2\pi}} \left\{ \left[\frac{e^{\{a+i(b-k)\}x}}{a+i(b-k)} \right]_{-\infty}^{0} + \left[-\frac{e^{-\{a-i(b-k)\}x}}{a-i(b-k)} \right]_{0}^{\infty} \right\}$$

$$= \frac{1}{\sqrt{2\pi}} \left\{ \frac{1}{a+i(b-k)} + \frac{1}{a-i(b-k)} \right\}$$

$$= \sqrt{\frac{2}{\pi}} \frac{a}{a^2 + (b-k)^2}.$$

Deductions: The inverse Fourier transform of $F(k)$ is given by

$$f(x)=\frac{1}{\sqrt{2\pi}}\int_{-\infty}^{\infty}F(k)e^{ikx}dk$$

or
$$e^{-a|x|+ibx}=\frac{1}{\sqrt{2\pi}}\int_{-\infty}^{\infty}\sqrt{\frac{2}{\pi}}\frac{ae^{ikx}}{a^2+(b-k)^2}dk$$

or
$$e^{-a|x|}(\cos bx+i\sin bx)=\frac{a}{\pi}\int_{-\infty}^{\infty}\frac{\cos kx+i\sin kx}{a^2+(b-k)^2}dk$$

or
$$e^{-a|x|}\cos bx+ie^{-a|x|}\sin bx=\frac{a}{\pi}\int_{-\infty}^{\infty}\frac{\cos kx}{a^2+(b-k)^2}dk+i\frac{a}{\pi}\int_{-\infty}^{\infty}\frac{\sin kx}{a^2+(b-k)^2}dk.$$

Equating real parts from both sides, we obtain

$$\int_{-\infty}^{\infty}\frac{\cos kx}{a^2+(b-k)^2}dk=\frac{\pi}{a}e^{-a|x|}\cos bx.$$

Again, equating imaginary parts from both sides, we obtain

$$\int_{-\infty}^{\infty}\frac{\sin kx}{a^2+(b-k)^2}dk=\frac{\pi}{a}e^{-a|x|}\sin bx.$$

PROBLEM 43 Find the Fourier transform of the Dirac-Delta function, $\delta(x)$.

Solution: The Fourier transform of Dirac–Delta function, $\delta(x)$, is given by

$$\delta(k)=\frac{1}{\sqrt{2\pi}}\int_{-\infty}^{\infty}\delta(x)e^{-ikx}dx=\frac{1}{\sqrt{2\pi}}e^0=\frac{1}{\sqrt{2\pi}}$$

where use has been made of the formula

$$\int_{-\infty}^{\infty}\delta(x)f(x)dx=f(0).$$

PROBLEM 44 Find the Fourier transform of the function

$$f(x)=\begin{cases}x^2 & \text{for } |x|\le a,\\ 0 & \text{for } |x|>a.\end{cases}$$

Solution: The Fourier transform is given by

$$F(k)=\frac{1}{\sqrt{2\pi}}\int_{-\infty}^{\infty} f(x)e^{-ikx}\,dx$$

$$=\frac{1}{\sqrt{2\pi}}\int_{-a}^{a} x^2 e^{-ikx}\,dx$$

$$=\frac{1}{\sqrt{2\pi}}\left\{\left[-\frac{x^2 e^{-ikx}}{ik}\right]_{x=-a}^{a}+\int_{-a}^{a}\frac{2xe^{-ikx}}{ik}\,dx\right\}$$

$$=\frac{1}{\sqrt{2\pi}}\left\{\frac{a^2(e^{ika}-e^{-ika})}{ik}+\left[-\frac{2x\,e^{-ikx}}{(ik)^2}\right]_{x=-a}^{a}+2\int_{-a}^{a}\frac{e^{-ikx}}{(ik)^2}\,dx\right\}$$

$$=\frac{1}{\sqrt{2\pi}}\left\{\frac{2a^2}{k}\sin ka+\frac{2a(e^{-ika}+e^{ika})}{k^2}+2\left[-\frac{e^{-ikx}}{(ik)^3}\right]_{-a}^{a}\right\}$$

$$=\frac{1}{\sqrt{2\pi}}\left\{\frac{2a^2}{k}\sin ka+\frac{4a}{k^2}\cos ka+2\frac{e^{-ika}-e^{ika}}{ik^3}\right\}$$

$$=\frac{1}{\sqrt{2\pi}}\left\{\left(\frac{2a^2}{k}-\frac{4}{k^3}\right)\sin ka+\frac{4a}{k^2}\cos ka\right\}.$$

PROBLEM 45 Find the Fourier transform of the function

$$f(x)=\begin{cases} a-|x| & \text{for } |x|\le a\ge 0,\\ 0 & \text{for } |x|>a\end{cases}$$

and hence prove that

(i) $\displaystyle\int_0^{\infty}\frac{\sin^2 t}{t^2}\,dt=\frac{\pi}{2}$, (ii) $\displaystyle\int_0^{\infty}\frac{\sin^4 t}{t^4}\,dt=\frac{\pi}{3}$.

Solution: The Fourier transform of the function $f(x)$ is given by

$$F(k)=\frac{1}{\sqrt{2\pi}}\int_{-\infty}^{\infty} f(x)e^{-ikx}\,dx$$

$$= \frac{1}{\sqrt{2\pi}}\left\{\int_{-\infty}^{0} f(x)e^{-ikx}dx + \int_{0}^{\infty} f(x)e^{-ikx}dx\right\}$$

$$= \frac{1}{\sqrt{2\pi}}\left\{\int_{-a}^{0} (a+x)e^{-ikx}dx + \int_{0}^{a} (a-x)e^{-ikx}dx\right\}$$

$$= \frac{1}{\sqrt{2\pi}}\left\{\left[\frac{(a+x)e^{-ikx}}{-ik}\right]_{x=-a}^{0} - \int_{-a}^{0} \frac{e^{-ikx}}{-ik}dx\right\}$$

$$+ \frac{1}{\sqrt{2\pi}}\left\{\left[\frac{(a-x)e^{-ikx}}{-ik}\right]_{x=0}^{a} + \int_{0}^{a} \frac{e^{-ikx}}{-ik}dx\right\}$$

$$= \frac{1}{\sqrt{2\pi}}\left\{-\frac{a}{ik} - \left[\frac{e^{-ikx}}{(-ik)^2}\right]_{x=-a}^{0}\right\} + \frac{1}{\sqrt{2\pi}}\left\{\frac{a}{ik} + \left[\frac{e^{-ikx}}{(-ik)^2}\right]_{x=0}^{a}\right\}$$

$$= \frac{1}{\sqrt{2\pi}}\left\{\frac{(1-e^{ika})}{k^2} - \frac{(e^{-ika}-1)}{k^2}\right\} = \sqrt{\frac{2}{\pi}}\frac{(1-\cos ka)}{k^2}.$$

(i) Now, inverse Fourier transform yields

$$f(x) = \frac{1}{\sqrt{2\pi}}\int_{-\infty}^{\infty} F(k)e^{ikx}dk$$

$$= \frac{1}{\sqrt{2\pi}}\int_{-\infty}^{\infty} \sqrt{\frac{2}{\pi}}\frac{(1-\cos ka)}{k^2}e^{ikx}dk$$

$$= \frac{1}{\pi}\int_{-\infty}^{\infty} \frac{(1-\cos ka)}{k^2}(\cos kx + i\sin kx)dk$$

$$= \frac{1}{\pi}\int_{-\infty}^{\infty} \frac{(1-\cos ka)\cos kx}{k^2}dk + i\frac{1}{\pi}\int_{-\infty}^{\infty} \frac{(1-\cos ka)\sin kx}{k^2}dk$$

$$= \frac{2}{\pi}\int_{0}^{\infty} \frac{(1-\cos ka)\cos kx}{k^2}dk + 0$$

At $x = 0$, we have

$$f(0) = a = \frac{2}{\pi}\int_0^\infty \frac{1-\cos ka}{k^2}\,dk = \frac{2}{\pi}\int_0^\infty \frac{2\sin^2\left(\frac{ka}{2}\right)}{k^2}\,dk$$

Putting $\frac{ka}{2} = t$ or $k = \frac{2t}{a}$ we have

$$a = \frac{4}{\pi}\int_0^\infty \frac{a^2\sin^2 t}{4t^2}\cdot\frac{2}{a}\,dt$$

or $$\int_0^\infty \frac{\sin^2 t}{t^2}\,dt = \frac{\pi}{2}.$$

(ii) Parseval's formula for Fourier transforms yields

$$\int_{-\infty}^{\infty} |f(x)|^2\,dx = \int_{-\infty}^{\infty} |F(k)|^2\,dk$$

or $$\int_{-a}^{a} \left|(a-|x|)\right|^2 dx = \frac{2}{\pi}\int_{-\infty}^{\infty} \frac{(1-\cos ka)^2}{k^4}\,dk$$

or $$2\int_0^a (a-x)^2\,dx = \frac{4}{\pi}\int_0^\infty \frac{\left\{2\sin^2\left(\frac{ka}{2}\right)\right\}^2}{k^4}\,dk$$

or $$\int_0^a (a-x)^2\,dx = \frac{8}{\pi}\int_0^\infty \frac{\sin^4\left(\frac{ka}{2}\right)}{k^4}\,dk$$

Putting $\frac{ka}{2} = t$ or $k = \frac{2t}{a}$, we have

$$\int_0^a (a^2 - 2ax + x^2)\,dx = \frac{8}{\pi}\int_0^\infty \frac{\sin^4 t}{t^4}\,\frac{a^4}{2^4}\,\frac{2}{a}\,dt$$

or $$\left[a^2x - ax^2 + \frac{x^3}{3}\right]_0^a = \frac{a^3}{\pi}\int_0^\infty \frac{\sin^4 t}{t^4}\,dt$$

or
$$\frac{a^3}{3} = \frac{a^3}{\pi}\int_0^\infty \frac{\sin^4 t}{t^4}dt$$

or
$$\int_0^\infty \frac{\sin^4 t}{t^4}dt = \frac{\pi}{3}.$$

PROBLEM 46 Find the Fourier sine and cosine transforms of the function

$$f(x) = e^{-ax} \text{ (a > 0)}.$$

Deduce that $\displaystyle\int_0^\infty \frac{dt}{(a^2+t^2)(b^2+t^2)} = \frac{\pi}{2ab(a+b)}.$

Solution: The Fourier sine transform is given by

$$F_s(k) = \sqrt{\frac{2}{\pi}}\int_0^\infty f(x)\sin kx\,dx$$

$$= \sqrt{\frac{2}{\pi}}\int_0^\infty e^{-ax}\sin kx\,dx$$

$$= \sqrt{\frac{2}{\pi}}\int_0^\infty e^{-ax}\frac{(e^{ikx}-e^{-ikx})}{2i}dx$$

$$= \sqrt{\frac{2}{\pi}}\frac{1}{2i}\left\{\int_0^\infty e^{-(a-ik)x}dx - \int_0^\infty e^{-(a+ik)x}dx\right\}$$

$$= \sqrt{\frac{2}{\pi}}\frac{1}{2i}\left\{\left[\frac{e^{-(a-ik)x}}{-(a-ik)}\right]_0^\infty - \left[\frac{e^{-(a+ik)x}}{-(a+ik)}\right]_0^\infty\right\}$$

$$= \sqrt{\frac{2}{\pi}}\frac{1}{2i}\left\{\frac{1}{a-ik} - \frac{1}{a+ik}\right\} = \sqrt{\frac{2}{\pi}}\left(\frac{k}{a^2+k^2}\right).$$

The Fourier cosine transform is given by

$$F_c(k) = \sqrt{\frac{2}{\pi}}\int_0^\infty f(x)\cos kx\,dx$$

$$= \sqrt{\frac{2}{\pi}}\int_0^\infty e^{-ax}\frac{(e^{ikx}+e^{-ikx})}{2}dx$$

$$= \sqrt{\frac{2}{\pi}} \frac{1}{2} \left\{ \int_0^\infty e^{-(a-ik)x} dx + \int_0^\infty e^{-(a+ik)x} dx \right\}$$

$$= \sqrt{\frac{2}{\pi}} \frac{1}{2} \left\{ \left[\frac{e^{-(a-ik)x}}{-(a-ik)} \right]_0^\infty + \left[\frac{e^{-(a+ik)x}}{-(a+ik)} \right]_0^\infty \right\}$$

$$= \sqrt{\frac{2}{\pi}} \left(\frac{a}{a^2+k^2} \right).$$

Deduction: The Fourier cosine transform of the function $f(x) = e^{-ax}$ is given by

$$F_c(k) = \sqrt{\frac{2}{\pi}} \left(\frac{a}{a^2+k^2} \right)$$

The Fourier cosine transform of another function $g(x) = e^{-bx}$ is given by

$$G_c(k) = \sqrt{\frac{2}{\pi}} \left(\frac{b}{b^2+k^2} \right)$$

Parseval's theorem of Fourier transforms yields

$$\int_0^\infty F_c(k) G_c^*(k) dk = \int_0^\infty f(x) g^*(x) dx$$

or
$$\frac{2}{\pi} \int_0^\infty \frac{ab\, dk}{(a^2+k^2)(b^2+k^2)} = \int_0^\infty e^{-ax} \cdot e^{-bx} dx$$

$$= \int_0^\infty e^{-(a+b)x} dx = \left[\frac{e^{-(a+b)x}}{-(a+b)} \right]_0^\infty = \frac{1}{a+b}$$

or
$$\int_0^\infty \frac{dk}{(a^2+k^2)(b^2+k^2)} = \frac{\pi}{2ab(a+b)}.$$

Renaming the variable, we write

$$\int_0^\infty \frac{dt}{(a^2+t^2)(b^2+t^2)} = \frac{\pi}{2ab(a+b)}.$$

PROBLEM 47 Find the Fourier transform of the function defined by

$$f(x)=\begin{cases} e^{-ax} & \text{for } x>0, \\ -e^{ax} & \text{for } x<0 \text{ with } a>0. \end{cases}$$

Deduce that for $x > 0$,

(i) $\displaystyle\int_{-\infty}^{\infty} \frac{ke^{ikx}}{a^2+k^2}\,dk = i\pi\, e^{-ax}$

and (ii) $\displaystyle\int_{0}^{\infty} \frac{k\sin kx}{a^2+k^2}\,dk = \frac{\pi}{2}e^{-ax}.$

Solution: The Fourier transform is given by

$$FT[f(x)]=\frac{1}{\sqrt{2\pi}}\int_{-\infty}^{\infty} f(x)e^{-ikx}\,dx$$

or

$$F(k)=\frac{1}{\sqrt{2\pi}}\left\{\int_{-\infty}^{0} f(x)e^{-ikx}\,dx+\int_{0}^{\infty} f(x)e^{-ikx}\,dx\right\}$$

$$=\frac{1}{\sqrt{2\pi}}\left\{\int_{-\infty}^{0} -e^{ax}e^{-ikx}\,dx+\int_{0}^{\infty} e^{-ax}e^{-ikx}\,dx\right\}$$

$$=\frac{1}{\sqrt{2\pi}}\left\{-\int_{-\infty}^{0} e^{(a-ik)x}\,dx+\int_{0}^{\infty} e^{-(a+ik)x}\,dx\right\}$$

$$=\frac{1}{\sqrt{2\pi}}\left\{-\left[\frac{e^{(a-ik)x}}{a-ik}\right]_{-\infty}^{0}+\left[\frac{e^{-(a+ik)x}}{-(a+ik)}\right]_{0}^{\infty}\right\}$$

$$=\frac{1}{\sqrt{2\pi}}\left\{-\frac{1}{a-ik}+\frac{1}{a+ik}\right\}=\sqrt{\frac{2}{\pi}}\frac{(-ik)}{a^2+k^2}.$$

Deductions:

(i) Now, for $x > 0$ Fourier inverse transform yields

$$f(x)=FIT[F(k)]=\frac{1}{\sqrt{2\pi}}\int_{-\infty}^{\infty} F(k)e^{ikx}\,dk$$

or
$$e^{-ax} = \frac{1}{\sqrt{2\pi}} \int_{-\infty}^{\infty} \sqrt{\frac{2}{\pi}} \frac{(-ik)}{a^2+k^2} e^{ikx} dk$$

or
$$\int_{-\infty}^{\infty} \frac{ke^{ikx}}{a^2+k^2} dk = i\pi e^{-ax}.$$

(ii) The above equation can also be written as

$$\int_{-\infty}^{\infty} \frac{k(\cos kx + i \sin kx)}{a^2+k^2} dk = i\pi e^{-ax}$$

or
$$\int_{-\infty}^{\infty} \frac{k \cos kx}{a^2+k^2} dk + i \int_{-\infty}^{\infty} \frac{k \sin kx}{a^2+k^2} dk = i\pi e^{-ax}.$$

The first integral on the left-hand side vanishes as the integrand is an odd function of k.

Hence, we write

$$\int_{-\infty}^{\infty} \frac{k \sin kx}{a^2+k^2} dk = \pi e^{-ax}.$$

As the integrand is an even function of k, we write

$$2\int_{0}^{\infty} \frac{k \sin kx}{a^2+k^2} dk = \pi e^{-ax}$$

or
$$\int_{0}^{\infty} \frac{k \sin kx}{a^2+k^2} dk = \frac{\pi}{2} e^{-ax}.$$

PROBLEM 48 Show that $\frac{d^m F(k)}{dk^m} = (-i)^m F\left(x^m f(x)\right)$ and hence

$$F\left(x^m e^{-\frac{x^2}{2}}\right) = i^m \frac{d^m}{dk^m}\left(e^{-\frac{k^2}{2}}\right)$$

where $F(k)$ denotes the Fourier transform of $f(x)$.

Solution: The Fourier transform of $f(x)$ is given by

$$F(k) = \frac{1}{\sqrt{2\pi}} \int_{-\infty}^{\infty} f(x) e^{-ikx} dx$$

Differentiating both sides m times with respect to k, we have

$$\frac{d^m F(k)}{dk^m} = \frac{1}{\sqrt{2\pi}} \int_{-\infty}^{\infty} f(x)\left[\frac{d^m}{dk^m} e^{-ikx}\right] dx$$

$$= (-i)^m \frac{1}{\sqrt{2\pi}} \int_{-\infty}^{\infty} \left[x^m f(x)\right] e^{-ikx} dx$$

or

$$\frac{d^m F(k)}{dk^m} = (-i)^m F\left(x^m f(x)\right).$$

Therefore, we have

$$F\left(x^m f(x)\right) = i^m \frac{d^m F(k)}{dk^m}$$

Putting $f(x) = e^{-\frac{x^2}{2}}$, we have $F(k) = e^{-\frac{k^2}{2}}$

and

$$F\left(x^m e^{-\frac{x^2}{2}}\right) = i^m \frac{d^m}{dk^m}\left(e^{-\frac{k^2}{2}}\right).$$

PROBLEM 49 Prove that the convolution of two Gaussian functions is yet another Gaussian function.

Solution: Let us consider two Gaussian functions $f(x) = e^{-a^2x^2}$ and $g(x) = e^{-b^2x^2}$.
Then, their convolution is given by

$$f(x) \star g(x) = \frac{1}{\sqrt{2\pi}} \int_{-\infty}^{\infty} f(x-u) g(u) du$$

$$= \frac{1}{\sqrt{2\pi}} \int_{-\infty}^{\infty} e^{-a^2(x-u)^2} e^{-b^2u^2} du$$

$$= \frac{e^{-a^2x^2}}{\sqrt{2\pi}} \int_{-\infty}^{\infty} e^{-\left(a^2+b^2\right)u^2 + 2a^2xu} du.$$

Recalling the integral $\int_{-\infty}^{\infty} e^{-\alpha x^2 + \beta x} dx = \sqrt{\frac{\pi}{\alpha}}\, e^{\frac{\beta^2}{4\alpha}}$ the above convolution becomes

$$f(x) \star g(x) = \frac{e^{-a^2x^2}}{\sqrt{2\pi}} \sqrt{\frac{\pi}{a^2+b^2}}\, e^{\frac{4a^4x^2}{4(a^2+b^2)}}$$

$$= \sqrt{\frac{1}{2(a^2+b^2)}}\, e^{-\left(\frac{a^2b^2}{a^2+b^2}\right)x^2},$$

which is also a Gaussian function.

PROBLEM 50 Prove that the convolution of a delta function, $\delta(x-x_0)$, with an arbitrary function, $f(x)$, yields the function f with a shift of origin and multiplied by $(2\pi)^{-1/2}$, that is,

$$\delta(x-x_0) \star f(x) = \frac{1}{\sqrt{2\pi}} f(x-x_0).$$

Solution: We write the convolution in the form

$$\delta(x-x_0) \star f(x) = \frac{1}{\sqrt{2\pi}} \int_{-\infty}^{\infty} \delta(x-x_0-u) f(u)\,du.$$

Recalling the symmetry property of delta function, viz., $\delta(x-a) = \delta(a-x)$ and also the delta function integral, viz.

$$\int_{-\infty}^{\infty} \delta(x-a) f(x) dx = f(a)$$

We write the above equation as

$$\delta(x-x_0) \star f(x) = \frac{1}{\sqrt{2\pi}} \int_{-\infty}^{+\infty} \delta(u-(x-x_0)) f(u)\,du$$

$$= \frac{1}{\sqrt{2\pi}} f(x-x_0).$$

Fourier Transforms in Solving Differential Equations

Application of Fourier transform to a given differential equation transforms the equation into a simpler one that can be solved easily. Then the inversrse Fourier transform of the solution of the transformed equation gives the required solution of the given differential equation.

PROBLEM 51 Using Fourier transforms, solve the Airy equation

$$\frac{d^2u}{dx^2} - xu = 0,\ \lim_{|x|\to\infty} u(x) = 0.$$

Solution: We know that $FT\left[\frac{du}{dx}\right] = ik\,FT[u(x)] = ik\,U(k)$, ...(i)

$$FT\left[\frac{d^2u}{dx^2}\right] = (ik)^2\,FT[u(x)] = -k^2U(k) \qquad \text{...(ii)}$$

where
$$FT[u(x)] = U(k) = \frac{1}{\sqrt{2\pi}}\int_{-\infty}^{\infty} u(x)e^{-ikx}dx \qquad \text{...(iii)}$$

is the Fourier transform of $u(x)$.

From Eq. (iii), we have

$$\frac{dU(k)}{dk} = \frac{1}{\sqrt{2\pi}}\int_{-\infty}^{\infty} -ix\,u(x)e^{-ikx}dx$$

$$= i\left[\frac{1}{\sqrt{2\pi}}\int_{-\infty}^{\infty} -xu(x)e^{-ikx}dx\right]$$

or
$$-\frac{i\,dU(k)}{dk} = \frac{1}{\sqrt{2\pi}}\int_{-\infty}^{\infty} -xu(x)e^{-ikx}dx = FT[-xu(x)]. \qquad \text{...(iv)}$$

Applying Fourier transform to the given differential equation and using Eqs. (ii) and (iv), we obtain

$$-k^2U(k) - \frac{i\,dU(k)}{dk} = 0$$

or
$$\frac{dU}{U} = ik^2dk.$$

Solving this first order ODE, we obtain

$$U(k) = Ce^{ik^3/3} \qquad \text{...(v)}$$

where C is a constant.

Now, the solution of the given differential equation will be the Fourier inverse transform of Eq. (v).

Thus, the required solution of the given differential equation is

$$u(x) = FIT\left[U(k)\right] = \frac{1}{\sqrt{2\pi}}\int_{-\infty}^{\infty} U(k)e^{ikx}dk$$

$$= \frac{1}{\sqrt{2\pi}}\int_{-\infty}^{\infty} C\,e^{ik^3/3}\cdot e^{ikx}dk = \frac{C}{\sqrt{2\pi}}\int_{-\infty}^{\infty} e^{i\left(kx+k^3/3\right)}dk.$$

If we choose $C = \frac{1}{\sqrt{2\pi}}$ the above solution is the so-called *Airy* function.

PROBLEM 52 The one-dimensional classical wave equation is expressed as

$$\frac{\partial^2 u(x,t)}{\partial x^2} = \frac{1}{v^2}\frac{\partial^2 u(x,t)}{\partial t^2}, \quad (-\infty < x < \infty,\ t \geq 0)$$

with the boundary conditions $u(x,t) \to 0, \frac{\partial u}{\partial x} \to 0$ as $x \to \pm\infty$ and the initial conditions $u(x,0) = f(x), \left(\frac{\partial u}{\partial t}\right)_{t=0} = u_t(x,0) = 0$ (v is the constant speed of the wave). Solve the equation.

Solution: Performing Fourier transform of the given equation, we have

$$\frac{1}{\sqrt{2\pi}}\int_{-\infty}^{\infty}\frac{\partial^2 u(x,t)}{\partial x^2}e^{-ikx}dx = \frac{1}{\sqrt{2\pi}}\int_{-\infty}^{\infty}\left[\frac{1}{v^2}\frac{\partial^2 u(x,t)}{\partial t^2}\right]e^{-ikx}dx$$

$$= \frac{1}{v^2}\frac{\partial^2}{\partial t^2}\left[\frac{1}{\sqrt{2\pi}}\int_{-\infty}^{\infty}u(x,t)e^{-ikx}dx\right]$$

or

$$(ik)^2 U(k,t) = \frac{1}{v^2}\frac{\partial^2 U(k,t)}{\partial t^2}$$

where $U(k,t) = \frac{1}{\sqrt{2\pi}}\int_{-\infty}^{\infty}u(x,t)e^{-ikx}dx$ is the Fourier transform of $u(x,t)$.

Thus, we write $\frac{\partial^2 U(k,t)}{\partial t^2} + v^2k^2U(k,t) = 0$ whose solution is

$$U(k,t) = C_1 e^{ivkt} + C_2 e^{-ivkt} \quad \text{...(i)}$$

where C_1 and C_2 are arbitrary constants.

From the first initial condition, viz., $u(x,0)=f(x)$ Fourier transformation yields

$$U(k,0)=F(k)=C_1+C_2. \quad \text{...(ii)}$$

The second initial condition , on Fourier transformation, yields

$$\left(\frac{\partial U}{\partial t}\right)_{t=0}=ikvC_1-ikvC_2=0. \quad \text{...(iii)}$$

From Eqs. (ii) and (iii), we obtain $C_1=C_2=\dfrac{F(k)}{2}$.

Hence, Eq. (i) becomes

$$U(k,t)=\frac{F(k)}{2}e^{ivkt}+\frac{F(k)}{2}e^{-ivkt}. \quad \text{...(iv)}$$

The solution of the given PDE is the Fourier inverse transform of Eq. (iv) and it is given by

$$u(x,t)=\frac{1}{\sqrt{2\pi}}\int_{-\infty}^{\infty}U(k,t)e^{ikx}dk$$

$$=\frac{1}{\sqrt{2\pi}}\int_{-\infty}^{\infty}\left[\frac{F(k)}{2}e^{ivkt}+\frac{F(k)}{2}e^{-ivkt}\right]e^{ikx}dk$$

$$=\frac{1}{2}\frac{1}{\sqrt{2\pi}}\int_{-\infty}^{\infty}F(k)e^{ik(x-vt)}dk+\frac{1}{2}\frac{1}{\sqrt{2\pi}}\int_{-\infty}^{\infty}F(k)e^{ik(x+vt)}dk$$

$$=\frac{1}{2}\phi(x-vt)+\frac{1}{2}\phi(x+vt) \text{ (say).}$$

PROBLEM 53 The temperature distribution in a very long heat conducting uniform rod is described by the one-dimensional heat equation,

$$\frac{\partial^2 T(x,t)}{\partial x^2}=\frac{1}{h}\frac{\partial T}{\partial t}$$

where $T(x,t)$ is the temperature at position x at time t and h is the thermal diffusivity of the medium.

Given boundary condition is that $T(x,t)$ is defined on $-\infty<x<+\infty$ and given initial condition is that $T(x,0)=f(x)$.

Using Fourier transform method, solve the above PDE.

Solution: The Fourier transform of $T(x,t)$ with respect to the variable x is given by

$$F(k,t)=\frac{1}{\sqrt{2\pi}}\int_{-\infty}^{\infty}T(x,t)e^{-ikx}dx. \qquad \ldots\text{(i)}$$

Taking the Fourier transform of the given heat equation, we obtain

$$\frac{1}{\sqrt{2\pi}}\int_{-\infty}^{\infty}\frac{\partial^2 T(x,t)}{\partial x^2}e^{-ikx}dx=\frac{1}{h}\frac{1}{\sqrt{2\pi}}\int_{-\infty}^{\infty}\frac{\partial T(x,t)}{\partial t}e^{-ikx}dx$$

or
$$(ik)^2F(k,t)=\frac{1}{h}\frac{\partial}{\partial t}\left[\frac{1}{\sqrt{2\pi}}\int_{-\infty}^{\infty}T(x,t)e^{-ikx}dx\right]$$

or
$$-k^2F(k,t)=\frac{1}{h}\frac{\partial F(k,t)}{\partial t}$$

or
$$\frac{\partial F(k,t)}{F}=-hk^2\partial t.$$

The above equation has the solution

$$F(k,t)=\phi(k)e^{-hk^2t} \qquad \ldots\text{(ii)}$$

where $\phi(k)$ is a function of k.

Application of initial condition to Eq. (i) yields

$$F(k,0)=\frac{1}{\sqrt{2\pi}}\int_{-\infty}^{\infty}T(x,0)e^{-ikx}dx$$

$$=\frac{1}{\sqrt{2\pi}}\int_{-\infty}^{\infty}f(x)e^{-ikx}dx.$$

Putting $t = 0$, Eq. (ii) yields)

$$F(k0)=\phi(k).$$

Therefore, we have

$$\phi(k)=\frac{1}{\sqrt{2\pi}}\int_{-\infty}^{\infty}f(x)e^{-ikx}dx.$$

Substituting this expression for $\phi(k)$ in Eq. (ii), we obtain

$$F(k,t)=\frac{1}{\sqrt{2\pi}}\int_{-\infty}^{\infty}f(x)e^{-hk^2t-ikx}dx.$$

Inverting the Fourier transform, we obtain

$$T(x,t)=\frac{1}{\sqrt{2\pi}}\int_{-\infty}^{\infty}F(k,t)e^{ikx}dk$$

$$=\frac{1}{\sqrt{2\pi}}\int_{-\infty}^{\infty}\left[\frac{1}{\sqrt{2\pi}}\int_{-\infty}^{\infty}f(\xi)e^{-hk^2t-ik\xi}d\xi\right]e^{ikx}dk$$

where the integration variable in the expression for $F(k,t)$ has been renamed.

Hence, we have

$$T(x,t)=\frac{1}{2\pi}\int_{-\infty}^{\infty}\left[\int_{-\infty}^{\infty}e^{-hk^2t+ik(x-\xi)}dk\right]f(\xi)d\xi$$

$$=\frac{1}{2\pi}\int_{-\infty}^{\infty}\left[\sqrt{\frac{\pi}{ht}}e^{\frac{i^2(x-\xi)^2}{4ht}}\right]f(\xi)d\xi$$

$$=\frac{1}{2\sqrt{\pi ht}}\int_{-\infty}^{\infty}f(\xi)e^{-\frac{(x-\xi)^2}{4ht}}d\xi\,.$$

EXERCISE

1. Prove that in the interval $-\frac{T}{2}<t<\frac{T}{2}$ the trigonmetric form of the Fourier series of the function $f(t) = e^{kt}$ (where k is a constant) is given by

$$e^{kt}=2\sinh\left(\frac{kT}{2}\right)\left[\frac{1}{kT}+\sum_{n=1}^{\infty}\frac{4(-1)^n}{k^2T^2+4n^2\pi^2}\left\{\frac{kT}{2}\cos\left(\frac{2n\pi t}{T}\right)-n\pi\sin\left(\frac{2n\pi t}{T}\right)\right\}\right].$$

2. Prove that in the interval $-\frac{\pi}{\omega}<t<\frac{\pi}{\omega}$ the complex form of the Fourier series of the function $f(t) = e^{kt}$ is given by

$$e^{kt}=\frac{1}{\pi}\sinh\left(\frac{\pi k}{\omega}\right)\sum_{n=-\infty}^{\infty}(-1)^n\frac{k\omega+in\omega^2}{k^2+n^2\omega^2}e^{in\omega t}.$$

3. Prove that the Fourier series of $f(x) = x^2 - x$ in the interval $-L < x < L$ is

$$x^2-x=\frac{L^2}{3}+\frac{4L^2}{\pi^2}\sum_{n=1}^{\infty}\frac{(-1)^n}{n^2}\cos\left(\frac{n\pi x}{L}\right)+\frac{2L}{\pi}\sum_{n=1}^{\infty}\frac{(-1)^n}{n}\sin\left(\frac{n\pi x}{L}\right).$$

Deduce that $1-\frac{1}{2^2}+\frac{1}{3^2}-\frac{1}{4^2}+\ldots\infty=\frac{\pi^2}{12}$.

4. Obtain a Fourier series representation of a function $f(x)=\frac{x^2}{4}+x$ in the interval $-\pi < x < \pi$ and hence deduce that

$$1-\frac{1}{2^2}+\frac{1}{3^2}-\frac{1}{4^2}+\ldots\infty=\frac{\pi^2}{12}.$$

Ans: $f(x)=\frac{\pi^2}{12}+\sum_{n=1}^{\infty}\frac{(-1)^n}{n^2}\cos nx-2\sum_{n=1}^{\infty}\frac{(-1)^n}{n}\sin nx.$

5. Find the Fourier series representation of the periodic function of period $2L$ defined by

$$f(x) = x \sin x, \quad -L < x < L$$

[**Hint:** In Solved Problem 10, replace π by L.]

6. Find the Fourier series expansion of the periodic function defined by

$$f(x) = x^2, \quad -\pi < x < \pi.$$

Deduce that $\frac{1}{1^2}-\frac{1}{2^2}+\frac{1}{3^2}-\frac{1}{4^2}+\ldots\infty=\frac{\pi^1}{12}$.

[**Hint:** In Solved Problem 11, replace L by π]

7. A periodic function of periode 4 is defined by

$$f(x) = x^2, \quad -2 < x < 2.$$

Find its Fourier series representation and deduce that

$$\frac{1}{1^2}-\frac{1}{2^2}+\frac{1}{3^2}-\frac{1}{4^2}+\ldots\infty=\frac{\pi^2}{12}.$$

[**Hint:** In Solved Problem 11, replace L by 2.]

8. Obtain a Fourier series expansion of the periodic function $f(x) = x^3$ in the interval $-\pi < x < \pi$.

[**Hint:** In Solved Problem 13, replace L by π.]

9. Obtain a Fourier series expansion of the periodic function

$$f(x) = L^2 + x^2, \ (-L < x < L)$$

and deduce that

$$\frac{1}{1^2}-\frac{1}{2^2}+\frac{1}{3^2}-\frac{1}{4^2}+\ldots\infty=\frac{\pi^2}{12}.$$

Ans: $f(x)=\frac{4L^2}{3}+\frac{4L^2}{\pi^2}\sum_{n=1}^{\infty}\frac{(-1)^n}{n^2}\cos\left(\frac{n\pi x}{L}\right)$

10. Obtain a Fourier series expansion of the periodic function

$$f(x) = L^2 + x^2, \quad (-\pi < x < \pi).$$

11. Obtain a Fourier series expansion of the periodic function

$$f(x) = L^2 - x^2, \quad (-\pi < x < \pi).$$

12. A triangular wave is represented by the function

$$f(x) = |x|, \quad -\pi < x < \pi$$

or
$$f(x)=\begin{cases}-x & \text{for } -\pi<x<0,\\ x & \text{for } 0<x<\pi.\end{cases}$$

Sketch a profile of the function and its periodic extension.
Represent $f(x)$ by a Fourier series and deduce that

(i) $\frac{1}{1^2}+\frac{1}{3^2}+\frac{1}{5^2}+\ldots\infty=\frac{\pi^2}{8}$

and (ii) $\frac{1}{1^4}+\frac{1}{3^4}+\frac{1}{5^4}+\ldots\infty=\frac{\pi^4}{96}$.

13. A triangular wave is represented by the function

$$f(x)=\begin{cases}x & \text{for } -L<x<0,\\ -x & \text{for } 0<x<L.\end{cases}$$

Sketch a profile of the function and its periodic extension.
Represent $f(x)$ by a Fourier series and deduce relations (i) and (ii) of Problem 12.

Ans: $f(x)=-\frac{L}{2}+\frac{2L}{\pi^2}\sum_{n=1}^{\infty}\frac{\{1-(-1)^n\}}{n^2}\cos\left(\frac{n\pi x}{L}\right)$

14. Solve Problem 13 with $L = \pi$.

15. We are given a tent function defined by

$$f(x)=\begin{cases} x & \text{for } 0<x<\pi, \\ 2\pi - x & \text{for } \pi<x<2\pi \end{cases}$$

and $$f(x) = f(x + 2\pi).$$

Find the Fourier series of the function and deduce the identities (i) and (ii) of Problem 12.

[**Hint:** In Solved Problem 17, put $L = \pi$]

16. A tent function defined by

$$f(x)=\begin{cases} x & \text{for } 0<x<1, \\ 2-x & \text{for } 1<x<2 \end{cases}$$

and $$f(x) = f(x + 2).$$

Find the Fourier series of the function and deduce the dentities (i) and (ii) of Problem 12.

[**Hint:** In Solved Problem 17, put $L = 1$]

17. A triangular wave is described by the tent function

$$f(x)=\begin{cases} x & \text{for } 0<x<\frac{L}{2}, \\ L-x & \text{for } \frac{L}{2}<x<L. \end{cases}$$

Sketch the profiles of the function with even and odd extensions respectively, and find the half-range Fourier series using even and odd periodic extensions. Find the Fourier series of the function and deduce the dentities (i) and (ii) in Problem 12.

[**Hint:** In Solved Problem 18, put $\frac{2k}{L} = 1$]

18. If a function is defined by

$$f(x)=\begin{cases} wx & \text{for } 0<x<\frac{L}{2}, \\ w(L-x) & \text{for } \frac{L}{2}<x<L. \end{cases}$$

Sketch the profiles of the function with even and odd extensions respectively, and find the half-range Fourier series using even and odd periodic extensions.

Deduce the identities (i) and (ii) in Problem 12.

[**Hint:** In Solved Problem 18, put $2k/L = w$]

19. If a function is defined by

$$f(x)=\begin{cases} x & \text{for } 0<x<\dfrac{\pi}{2}, \\ \pi - x & \text{for } \dfrac{\pi}{2}<x<\pi, \end{cases}$$

sketch the profiles of the function with even and odd extensions respectively, and find the half-range Fourier series using even and odd periodic extensions. Also deduce the identities (i) and (ii) in Problem 12.

[**Hint:** In Solved Problem 18, put $L=\pi, k=\dfrac{\pi}{2}$ such that $\dfrac{2k}{L}=1$]

20. If a function is defined by

$$f(x)=\begin{cases} 2x & \text{for } 0<x<1, \\ 2(2-x) & \text{for } 1<x<2, \end{cases}$$

sketch the profiles of the function with even and odd extensions respectively, and find the half-range Fourier series using even and odd periodic extensions. Deduce the identities (i) and (ii) in Problem 12.

[**Hint:** In Solved Problem 18, put $k = L = 2$]

21. If we define $f(x)=\begin{cases} \pi x & \text{for } 0<x<1, \\ \pi(2-x) & \text{for } 1<x<2, \end{cases}$

obtain a half-range Fourier series and deduce that

$$\frac{1}{1^4}+\frac{1}{3^4}+\frac{1}{5^4}+\frac{1}{7^4}+\ldots\infty=\frac{\pi^4}{96}.$$

[**Hint:** Same as Problem 18 if $w = \pi$, $L = 2$]

22. We are given a function $f(x)=\begin{cases} \dfrac{1}{2}+x & \text{for } -\dfrac{1}{2}<x<0, \\ \dfrac{1}{2}-x & \text{for } 0<x<\dfrac{1}{2}. \end{cases}$

Sketch a profile of the function and its periodic extension. Also deduce the identity (i) in Problem 12.

[**Hint:** Put $L = 1$ in Solved Problem 19.]

23. Expand a function $f(x) = f(x + 2\pi) = x$ for $-\pi \le x \le \pi$ in a Fourier series. Sketch a profile of the function and its periodic extension. Also deduce the identities

$$\text{(a)}\quad \frac{\pi}{4}=1-\frac{1}{3}+\frac{1}{5}-\frac{1}{7}+\ldots\infty$$

$$\text{and (b)}\quad \frac{\pi^2}{6}=\sum_{n=1}^{\infty}\frac{1}{n^2}.$$

[**Hint:** In Solved Problem 20, put $L = \pi$]

24. Prove that the Fourier series of the odd periodic extension of the function

$$f(x)=\begin{cases} x(h-1) & \text{for } 0<x<h, \\ h(x-1) & \text{for } h<x<1, \end{cases}$$

is given by $f(x)=-\dfrac{2}{\pi^2}\displaystyle\sum_{n=1}^{\infty}\frac{\sin(n\pi h)}{n^2}\sin(n\pi x)$.

25. A sawtooth wave is described by the function

$$f(x)=\begin{cases} -\dfrac{1}{2}(L+x) \text{ for } -L\le x<0, \\ +\dfrac{1}{2}(L-x) \text{ for } 0<x\le L. \end{cases}$$

Sketch a profile of $f(x)$ and its periodic extension.
Obtain a Fourier series representing the above function and deduce that

$$\frac{\pi}{4}=1-\frac{1}{3}+\frac{1}{5}-\frac{1}{7}+\ldots\infty.$$

26. A function is defined by $f(x)=f(x+2L)=\dfrac{k}{L}x, -L\le x\le L$ where k is a constant. Sketch a profile of $f(x)$ and its periodic extension. Also obtain a Fourier series representing the function.

[**Hint:** Same as Solved Problem 20 if we put $k = L$]

27. A function is defined by $f(x) = f(x + 2L) = x$, $(0 \le x \le 2L)$. Sketch a profile of the function and its periodic extension. Obtain the Fourier series representing the function.

Also deduce that

$$\text{(a)}\quad \frac{\pi}{4}=1-\frac{1}{3}+\frac{1}{5}-\frac{1}{7}+\ldots\infty$$

and
$$\text{(b)}\quad \frac{\pi^2}{6}=1+½^2+\frac{1}{3^2}+¼^2+\ldots\infty.$$

[**Hint:** In Solved Problem 22, put $k = 2L$]

28. Find the Fourier series representing the function

$$f(x) = x \ (0 \le x \le 2\pi)$$

and sketch its profile from $x = -4\pi$ to $x = 4\pi$.

Deduce the identities (a) and (b) of Problem 27.

29. Obtain the Fourier series representing the function $f(x)=f(x+L)=\frac{kx}{L}$ $(0 \le x \le L)$ where k is a constant and deduce the identities (a) and (b) of Problem 27.

30. A function is defined by

$$f(x) = f(x + 2\pi) = \pi + x \text{ for } -\pi \le x \le \pi.$$

Sketch a profile of it and its periodic extension. Find a Fourier series of it and deduce

$$\text{(a)}\quad \frac{\pi}{4}=1-\frac{1}{3}+\frac{1}{5}-\frac{1}{7}+\ldots\infty$$

and
$$\text{(b)}\quad \frac{\pi^2}{6}=\sum_{n=1}^{\infty}\frac{1}{n^2}.$$

31. Expand the function $f(x) = x$ as a Fourier sine series in the half interval $0 \le x \le \pi$.

Sketch a profile of the function with periodic extension.

Deduce that

$$\text{(a)}\quad \frac{\pi}{4}=1-\frac{1}{3}+\frac{1}{5}-\frac{1}{7}+\ldots\infty$$

and
$$\text{(b)}\quad \frac{\pi^2}{6}=\frac{1}{1^2}+½^2+\frac{1}{3^2}+¼^2+\ldots\infty.$$

32. Solve Problem 31 when the interval is $0 \le x \le 1$.

33. A periodic function of period 2π is defined by

$$f(x)=\begin{cases}-k & \text{for } -\pi < x < 0, \\ k & \text{for } 0 < x < \pi.\end{cases}$$

Sketch a profile of the function and its periodic extension. Obtain the Fourier series representing the function and deduce the identities

(a) $\frac{\pi}{4}=1-\frac{1}{3}+\frac{1}{5}-\frac{1}{7}+\ldots\infty$

and (b) $\frac{\pi^2}{8}=1+\frac{1}{3^2}+\frac{1}{5^2}+\frac{1}{7^2}+\ldots\infty.$

34. A periodic function of period $2L$ is defined by

$$f(x)=\begin{cases}k & \text{for } -L<x<0\\ -k & \text{for } 0<x<L\end{cases}$$

Sketch a profile of the function and its periodic extension. Obtain the Fourier series representing the function and deduce the identities

(a) $\frac{\pi}{4}=1-\frac{1}{3}+\frac{1}{5}-\frac{1}{7}+\ldots\infty$

and (b) $\frac{\pi^2}{8}=1+\frac{1}{3^2}+\frac{1}{5^2}+\frac{1}{7^2}+\ldots\infty.$

35. A periodic function of period $2L$ is defined by

$$f(x)=\begin{cases}k & \text{for } -L<x<0\\ 0 & \text{for } 0<x<L\end{cases}$$

Sketch a profile of the function and its periodic extension. Obtain the Fourier series representing the function and deduce the identities

(a) $\frac{\pi}{4}=1-\frac{1}{3}+\frac{1}{5}-\frac{1}{7}+\ldots\infty$

and (b) $\frac{\pi^2}{8}=1+\frac{1}{3^2}+\frac{1}{5^2}+\frac{1}{7^2}+\ldots\infty.$

36. A periodic function of period 2π is defined by

$$f(x)=\begin{cases}+1 & \text{for } -\pi<x<-\frac{\pi}{2},\\ 0 & \text{for } -\frac{\pi}{2}<x<\frac{\pi}{2},\\ -1 & \text{for } \frac{\pi}{2}<x<\pi.\end{cases}$$

Sketch a profile of the function and its periodic extension. Obtain the Fourier series representing the function .

37. A periodic function of period $2L$ is defined by

$$f(x)=\begin{cases}0 & \text{for } 0<x<L,\\ k & \text{for } L<x<2L.\end{cases}$$

Sketch a profile of the function and its periodic extension. Obtain the Fourier series representing the function and deduce the identities

(a) $\frac{\pi}{4}=1-\frac{1}{3}+\frac{1}{5}-\frac{1}{7}+\ldots\infty$

and (b) $\frac{\pi^2}{8}=1+\frac{1}{3^2}+\frac{1}{5^2}+\frac{1}{7^2}+\ldots\infty.$

Ans: $f(x)=\frac{k}{2}-\frac{2k}{\pi}\left[\frac{1}{1}\sin\left(\frac{\pi x}{L}\right)+\frac{1}{3}\sin\left(\frac{3\pi x}{L}\right)+\frac{1}{5}\sin\left(\frac{5\pi x}{L}\right)+\ldots\infty\right]$

38. The output voltage of a half-wave rectifier is given by

$$V(t)=\begin{cases}V_0\sin\omega t & \text{for } 0<\omega t<\pi,\\ 0 & \text{for } \pi<\omega t<2\pi,\end{cases}$$

where V_0 is the peak value of voltage, ω is the circular frequency, and t is the instantaneous time.

Sketch a profile of the function and its periodic extension. Obtain the Fourier series representation of the output voltage V and also deduce that

(a) $\frac{1}{1\cdot 3}+\frac{1}{3\cdot 5}+\frac{1}{5\cdot 7}+\ldots\infty=\frac{1}{2},$

(b) $\frac{1}{1\cdot 3}-\frac{1}{3\cdot 5}+\frac{1}{5\cdot 7}-\ldots\infty=\frac{\pi}{4}-\frac{1}{2}$

and (c) $\frac{1}{1^2\cdot 3^2}+\frac{1}{3^2\cdot 5^2}+\frac{1}{5^2\cdot 7^2}+\ldots\infty=\frac{\pi^2}{16}-\frac{1}{2}.$

39. The output voltage of a half-wave rectifier is give by

$$V(t)=\begin{cases}0 & \text{for } -\pi<\omega t<0,\\ V_0\sin\omega t & \text{for } 0<\omega t<\pi,\end{cases}$$

where V_0 is the peak value of voltage, ω is the circular frequency, and t is the instantaneous time.

Sketch a profile of the function and its periodic extension. Obtain the Fourier series representation of the output voltage V and also deduce that

(a) $\frac{1}{1\cdot 3}+\frac{1}{3\cdot 5}+\frac{1}{5\cdot 7}+\ldots\infty=\frac{1}{2}$,

(b) $\frac{1}{1\cdot 3}-\frac{1}{3\cdot 5}+\frac{1}{5\cdot 7}-\ldots\infty=\frac{\pi}{4}-\frac{1}{2}$

and (c) $\frac{1}{1^2\cdot 3^2}+\frac{1}{3^2\cdot 5^2}+\frac{1}{5^2\cdot 7^2}+\ldots\infty=\frac{\pi^2}{16}-\frac{1}{2}$.

40. The output current of a full-wave rectifier is given by

$$I(t)=\begin{cases} I_0 \sin\omega t & \text{for } 0<t<\frac{T}{2}, \\ -I_0 \sin\omega t & \text{for } \frac{T}{2}<t<T, \end{cases}$$

where I_0 is the peak value of current, T is the time period, ω is the circular frequency given by $\omega=\frac{2\pi}{T}$, and t is instantaneous time.

Find its Fourier series representation and deduce the identities (a), (b) and (c) of the previous Problem.

41. The output current of a full-wave rectifier is given by

$$I(t)=\begin{cases} \sin\omega t & \text{for } -\pi<\omega t<0, \\ -\sin\omega t & \text{for } 0<\omega t<\pi, \end{cases}$$

where T is the time period, ω is the circular frequency given by $\omega=\frac{2\pi}{T}$, and t is instantaneous time.

Find its Fourier series representation and deduce the identities (a), (b) and (c) of the previous Problem.

42. Find the Fourier series of the following periodic functions

(a) $f(x)=\begin{cases} 1 & \text{for } -1<x<0, \\ x & \text{for } 0<x<1. \end{cases}$

Ans. $\frac{3}{4}+\sum_{n=1}^{\infty}\left\{\frac{(-1)^n-1}{n^2\pi^2}\cos(n\pi x)-\frac{\sin(n\pi x)}{n\pi}\right\}$

(b) $f(x)=\begin{cases}0 & \text{for } -\pi<x<0,\\ \sin x & \text{for } 0\le x<\pi.\end{cases}$

Ans. $\dfrac{1}{\pi}+\dfrac{\sin x}{2}+\dfrac{1}{\pi}\sum_{n=2}^{\infty}\dfrac{(-1)^n+1}{1-n^2}\cos nx$

(c) $f(x)=\begin{cases}1 & \text{for } -5<x<0,\\ 1+x & \text{for } 0\le x<5.\end{cases}$

Ans. $\dfrac{9}{4}+5\sum_{n=1}^{\infty}\left\{\dfrac{(-1)^n-1}{n^2\pi^2}\cos\left(\dfrac{n\pi x}{5}\right)+\dfrac{(-1)^{n+1}}{n\pi}\sin\left(\dfrac{n\pi x}{5}\right)\right\}$

43. Find the Fourier integral of the function

$$f(x)=\begin{cases}0 & \text{for } x<0,\\ x & \text{for } 0<x<3,\\ 0 & \text{for } x>3.\end{cases}$$

Ans: $f(x)=\dfrac{1}{\pi}\int_0^{\infty}\left\{\left(\dfrac{3\sin 3u}{u}+\dfrac{\cos 3u-1}{u^2}\right)\cos ux-\left(\dfrac{3\cos 3u}{u}-\dfrac{\sin 3u}{u^2}\right)\sin ux\right\}du$

44. Find the Fourier cosine integral of the functions

(a) $f(x)=\begin{cases}x & \text{for } 0<x<a,\\ 0 & \text{otherwise.}\end{cases}$

Ans: $\dfrac{2}{\pi}\int_0^{\infty}\left[\dfrac{a\sin au}{u}+\dfrac{\cos au-1}{u^2}\right]\cos ux\,du$

(b) $f(x)=\begin{cases}x & \text{for } 0<x<1,\\ 2-x & \text{for } 1<x<2,\\ 0 & \text{for } x>2.\end{cases}$

Ans: $\dfrac{2\cos\{u(1-x)\}-\cos ux-\cos\{u(2-x)\}}{u^2}$

(c) $f(x)=\begin{cases}1-\cos x & \text{for } -\dfrac{\pi}{2}<x<\dfrac{\pi}{2},\\ 0 & \text{otherwise.}\end{cases}$

$$\textbf{Ans:}\quad f(x)=\frac{2}{\pi}\int_0^\infty\left\{\frac{\sin(u\pi/2)}{u}-\frac{\cos(u\pi/2)}{1-u^2}\right\}\cos ux\,du$$

45. Find the Fourier sine integrals of

(a) $f(x)=\begin{cases}1 & \text{for } 0\le x\le\pi,\\ 0 & \text{otherwise.}\end{cases}$

$$\textbf{Ans:}\quad \frac{2}{\pi}\int_0^\infty\frac{1-\cos\pi u}{u}\sin ux\,du$$

(b) $f(x) = e^{-ax}$ $(a > 0)$.

$$\textbf{Ans:}\quad \frac{2}{\pi}\int_0^\infty\frac{u\sin ux}{a^2+u^2}du$$

46. Obtain Fourier integrals of the functions

(a) $f(x)=\begin{cases}0 & \text{for } x<0,\\ e^{-x} & \text{for } x>0.\end{cases}$

$$\textbf{Ans:}\quad \frac{1}{\pi}\int_0^\infty\frac{\cos ux+u\sin ux}{1+u^2}du$$

(b) $f(x) = e^{-|x|}\sin x$.

$$\textbf{Ans:}\quad \frac{4}{\pi}\int_0^\infty\frac{u\sin ux}{4+u^4}du$$

47. Obtain both Fourier sine and cosine integrals in respect of the following functions

(a) $f(x) = e^{-kx}$ for $k > 0$, $x > 0$,

$$\textbf{Ans:}\quad \frac{2}{\pi}\int_0^\infty\frac{u\sin ux}{k^2+u^2}du,\ \frac{2k}{\pi}\int_0^\infty\frac{\cos ux}{k^2+u^2}du$$

(b) $f(x) = xe^{-2x}$ for $x > 0$.

$$\textbf{Ans:}\quad \frac{8}{\pi}\int_0^\infty\frac{u\sin ux}{\left(4+u^2\right)^2}du,\ \frac{2}{\pi}\int_0^\infty\frac{\left(4-u^2\right)\cos ux}{\left(4+u^2\right)^2}du$$

48. Find the complex form of the Fourier integral of the function

$$f(x)=\begin{cases} e^{-kx} & \text{for } k>0, x>0, \\ 0, & \text{otherwise.} \end{cases}$$

Ans: $\frac{1}{2\pi}\int_{-\infty}^{\infty}\frac{e^{-iux}du}{k-iu}$

49. Obtain Fourier transforms of the following functions:

(a) $f(x)=\begin{cases} x & \text{for } |x|<a, \\ 0 & \text{for } |x|>a. \end{cases}$

Ans: (a) $\frac{1}{\sqrt{2\pi}}\left[\frac{2i(ka\cos ka-\sin ka)}{k^2}\right]$

(b) $f(x)=e^{-\frac{x^2}{2}}$.

Ans: (b) $e^{-\frac{k^2}{2}}$

(c) $f(x)=\begin{cases} \frac{1}{a}\sqrt{\frac{\pi}{2}} & \text{for } |x|<a, \\ 0 & \text{for } |x|>0. \end{cases}$

Ans: (c) $\frac{\sin ka}{ka}$

50. The displacement of a damped linear harmonic oscillator is defined as a function of time (t) by

$$f(t)=\begin{cases} 0 & \text{for } t<0, \\ e^{-\gamma t}\sin\omega_0 t & \text{for } t\geq 0, \end{cases}$$

where γ is a constant known as damping coefficient and ω_0 is the circular frequency of the undamped linear harmonic oscillator. If ω is the circular frequency of the damped linear harmonic oscillator, show that the Fourier transform of $f(t)$ is

$$F(\omega)=\frac{1}{2}\left[\frac{1}{\omega+\omega_0-i\gamma}-\frac{1}{\omega-\omega_0-i\gamma}\right].$$

51. Obtain the Fourier sine and cosine transforms of the function

$$f(x) = e^{-2x}$$

and also find the values of the integrals

(i) $\int_0^{\infty} \frac{dx}{(x^2+4)^2}$,

(ii) $\int_0^{\infty} \frac{x^2 dx}{(x^2+4)^2}$.

52. Verify the convolution theorem of Fourier transforms using an arbitrary function and a Dirac-Delta function.

53. Considering two Gaussian functions defined by $f(x) = \exp\left[-x^2/\alpha^2\right]$ and $g(x) = \exp\left[-x^2/\beta^2\right]$ verify the convolution theorem of Fourier transforms.

54. The temperature distribution in a very long heat conducting uniform rod is described by the one-dimensional heat equation,

$$\frac{\partial^2 T(x, t)}{\partial x^2} = \frac{1}{h}\frac{\partial T}{\partial t}$$

where $T(x, t)$ is the temperature at position x at time t and h is the thermal diffusivity of the medium.

Given boundary condition, $T(x, t)$, is defined on $-\infty < x < +\infty$ and the given initial condition is $T(x, 0) = \delta(x)$.

Using Fourier transform method, solve the above PDE.

Ans: $T(x,t) = \frac{1}{2\sqrt{\pi h t}} e^{-\frac{x^2}{4ht}}$

55. Prove that the Fourier transform of the function $f(x) = e^{-\frac{x^2}{2}}$ is $F(k) = e^{-\frac{k^2}{2}}$.

56. Using another approach simpler than that given earlier in the text, prove that

$$\frac{1}{2} + \sum_{n=1}^{p} \cos nt = \frac{\sin\left\{\left(p+\frac{1}{2}\right)t\right\}}{2\sin\frac{t}{2}}.$$

12 CHAPTER

Partial Differential Equations and Boundary Value Problems

12.1 INTRODUCTION

A partial differential equation (PDE) is a differential equation having partial derivatives of a dependent variable (function). The ***order*** of a PDE is the order of the highest derivative that appears in that equation. A partial differential equation is said to be **linear if**, in each term, the power of the dependent variable or its derivatives is unity and these are not multiplied together, their coefficients being constants or are functions of the independent variables.

A general form of a first order linear partial differential equation in two independent variables x and y is often expressed as

$$p(x,y)\frac{\partial u(x,y)}{\partial x}+q(x,y)\frac{\partial u(x,y)}{\partial y}+r(x,y)\,u(x,y)=s(x,y)$$

where p, q, r, s are constants or known functions of x and y, and u is a function of x and y.

A general form of a second order linear partial differential equation in two independent variables, x and y, is often expressed as

$$A(x,y)\frac{\partial^2 u(x,y)}{\partial x^2}+B(x,y)\frac{\partial^2 u(x,y)}{\partial x\,\partial y}+C(x,y)\frac{\partial^2 u(x,y)}{\partial y^2}$$
$$+L(x,y)\frac{\partial u(x,y)}{\partial x}+M(x,y)\frac{\partial u(x,y)}{\partial y}+N(x,y)\,u(x,y)=S(x,y) \quad (12.1)$$

where u is a function of x and y, and A, B, C, L, M, N, S are constants or functions of x and y.

12.2 CLASSIFICATION OF SECOND ORDER LINEAR PARTIAL DIFFERENTIAL EQUATION IN TWO INDEPENDENT VARIABLES

Second order linear partial differential equations in two independent variables are classified as stated below.

(i) Equation (12.1) is said to be ***parabolic*** if $B^2 = 4AC$,
(ii) Equation (12.1) is said to be ***elliptic*** if $B^2 < 4AC$,
(iii) Equation (12.1) is said to be ***hyperbolic*** if $B^2 > 4AC$.

EXAMPLES

1. Let us consider the one-dimensional heat conduction equation, viz.

$$\frac{\partial^2 T(x,t)}{\partial x^2} - \frac{1}{h}\frac{\partial T(x,t)}{\partial t} = 0.$$

Here, $A = 1, B = 0, C = 0.$

Then, $B^2 = 0 = 4\,AC.$

Hence, the above differential equation is a ***parabolic*** one.

2. Next, let us consider two-dimensional Laplace's equation, viz.

$$\frac{\partial^2 u(x,y)}{\partial x^2} + \frac{\partial^2 u(x,y)}{\partial y^2} = 0.$$

Here, $A = 1, B = 0, C = 1.$

Then, $B^2 = 0, 4AC = 4$, that is, $B^2 < 4\,AC.$

Hence, the above differential equation is an ***elliptic*** one.

3. As a last example, let us consider the one-dimensional wave equation, viz.,

$$\frac{\partial^2 u(x,t)}{\partial x^2} - \frac{1}{v^2}\frac{\partial^2 u(x,t)}{\partial t^2} = 0.$$

Here, $A = 1, B = 0, C = -\frac{1}{v^2}.$

Then, $B^2 = 0, 4AC = -\frac{4}{v^2}.$

Thus, $B^2 > 4AC.$

Hence, the above differential equation is a ***hyperbolic*** one.

12.3 BOUNDARY CONDITIONS, INITIAL CONDITIONS AND BOUNDARY VALUE PROBLEMS

Boundary Conditions

A physical problem can often be modelled by a differential equation together with certain conditions which are to be satisfied at certain fixed points of an interval. Such conditions are generally referred to as ***boundary conditions*** (since such fixed points are generally chosen at the boundaries or endpoints of the interval).

Classification of Boundary Conditions

(A) Based on Homogeneity Condition, boundary conditions to be satisfied in a physical problem are classified into the following two classes.

(i) Homogeneous boundary condition is the boundary condition that is satisfied at a point 'a' of an interval by a function $f(x)$ if

$$c_1 f(a) + c_2 f'(a) = 0$$

where c_1 and c_2 are constants.

(ii) Nonhomogeneous boundary condition is the boundary condition that is satisfied at a point 'a' of an interval by a function $f(x)$ if

$$c_1 f(a) + c_2 f'(a) = B \ (\neq 0)$$

where B, c_1 and c_2 are constants.

(B) Based on the Nature of the Function(s) to be Specified at the Boundary, Boundary conditions are catagorized into 4 types as detailed below.

(i) Dirichlet boundary condition is a condition in which a function $u(x)$ is specified at a boundary point. It was named after the German mathematician, Peter Gustav Lejeune Dirichlet (13 February, 1805 - 5 May, 1859).

(ii) Neumann boundary condition is a condition in which the normal component of the gradient of a function $u(x)$, i.e., $(\nabla u)_n$ is specified at a boundary point. It was named after the German mathematician, Carl Gottfried Neumann (7 May, 1832 - 27 March, 1925).

(iii) Robin boundary condition is a condition in which for a function $u(x)$, the value of $(\nabla u)_n + hu$ (where h is a constant) is specified at a boundary point. This condition

is just the weighted combination of Dirichlet boundary condition and the Neumann boundary condition. It was named after the French mathematician, Victor Gustave Robin (1855 - 1897).

(iv) Mixed boundary conditions are boundary conditions of different types which are specified at different boundary points.

Initial Condition: *An initial condition*, also called a seed value in some contexts, is a value of an evolving variable at some instant of time designated as the initial time (typically taken as $t = 0$).

Boundary Value Problem: *A boundary value problem (BVP)* is a problem of solving a differential equation subject to certain boundary conditions. Boundary value problems are generally solved by certain methods, namely: (i) method of separation of variables, (ii) Green's function method, (iii) method of integral transforms like (a) Fourier transform, (b) Laplace's transform, (c) Hankel transform, (d) Mellin's transform, (e) Spectral transform, etc. Of these various methods, only the method of separation of variables is discussed in this text.

Initial Value Problem: An *initial value problem (IVP)* is a problem of solving a differential equation subject to certain initial conditions.

Cauchy Problem: An initial value-boundary value problem is often called a Cauchy problem. It was named after the French mathematician-cum-physicist, ***Baron Augustin Louis Cauchy*** (21 August, 1789 – 23 May, 1857).

12.4 METHOD OF SEPARATION OF VARIABLES

In this method, the solution of a given partial differential equation (PDE) is assumed as the product of two or more functions and then the product function is substituted in the given PDE. A further simplication yields a set of ordinary differential equations which can be easily solved. Substitution of such individual functions in the assumed solution (product function) gives the solution of the given PDE. Applications of this method in solving certain partial differential equations are discussed in various sections.

12.5 TO DERIVE EQUATION OF TRANSVERSE (FREE) VIBRATIONS OF A STRETCHED STRING OR ONE-DIMENSIONAL CLASSICAL WAVE EQUATION

Let us consider a uniform string of length (L) stretched taut along the x-axis between its two endpoints at $x = 0$ and $x = L$. Let ρ be the mass per unit length of the string.

Let us assume that when the string is set into transverse vibrations in a vertical plane, $u(x, t)$ is the deflection of a point distant (x) from $x = 0$ at time t and the deflection is in the xu-plane. For simplicity, we are to make the following assumptions:

1. The string is homogeneous, i.e., its mass per unit length is constant.
2. The string is perfectly elastic and offers no resistance to bending.
3. The tension along the string is sufficiently large so that gravitational effect on it can be neglected.
4. The transverse vibrations of the string are of small amplitudes so that the deflection and the slope at every point of the string always remain small.
5. The tension along the string is uniform throughout and no other external forces act on the string.

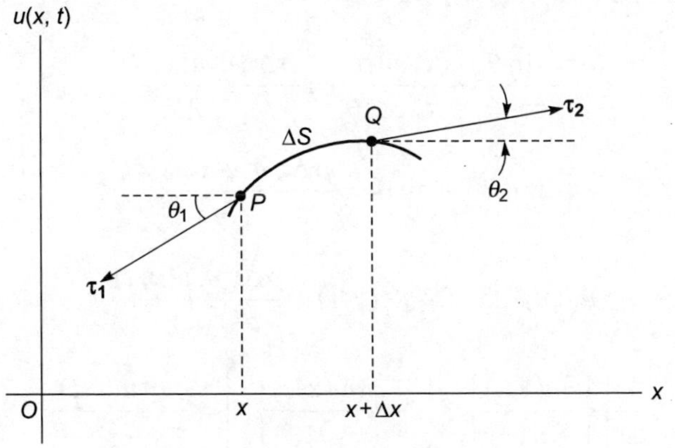

Fig. 12.1 A section of vibrating string

Let us consider the force acting on a small section (Δs) of the string vibrating in a vertical plane. Let $\boldsymbol{\tau}_1$ and $\boldsymbol{\tau}_2$ be the tensions at the endpoints P and Q of the section. Let $\boldsymbol{\tau}_1$ and $\boldsymbol{\tau}_2$ make angles θ_1 and θ_2 with the undisturbed (horizontal) length of the string. Since there is no motion in the horizontal plane along the string, the horizontal components of the tension should cancel out.

Therefore, we have

$$\tau_2 \cos\theta_2 - \tau_1 \cos\theta_1 = 0$$

or $$\tau_2 \cos\theta_2 = \tau_1 \cos\theta_1 = \tau \text{ (constant)} \tag{12.2}$$

where $$\tau_1 = |\boldsymbol{\tau}_1| \text{ etc.}$$

The magnitude of net vertical force acting on the corresponding element (Δs) of the string is then

$$\tau_2 \sin\theta_2 - \tau_1 \sin\theta_1.$$

Since $\Delta s = \Delta x\left[1+\left(\dfrac{\Delta u}{\Delta x}\right)^2\right]^{1/2} \cong \Delta x$, mass of the section ($\Delta s$) of the string between x and $(x+\Delta x)$ is $\rho\,\Delta s \cong \rho\,\Delta x$.

By Newton's second law of motion, the force acting on this section is

$$(\rho\,\Delta x)\frac{\partial^2 u(x,t)}{\partial t^2}.$$

Thus, we write

$$\tau_2 \sin\theta_2 - \tau_1 \sin\theta_1 = (\rho\,\Delta x)\frac{\partial^2 u(x,t)}{\partial t^2}. \tag{12.3}$$

Dividing Eq. (12.3) by Eq. (12.2), we obtain

$$\frac{\tau_2 \sin\theta_2}{\tau_2 \cos\theta_2} - \frac{\tau_1 \sin\theta_1}{\tau_1 \cos\theta_1} = \frac{\rho\,\Delta x}{\tau}\frac{\partial^2 u(x,t)}{\partial t^2}$$

or

$$\tan\theta_2 - \tan\theta_1 = \frac{\rho\,\Delta x}{\tau}\frac{\partial^2 u(x,t)}{\partial t^2}$$

or

$$u_x(x+\Delta x,t) - u_x(x,t) = \frac{\rho\,\Delta x}{\tau}\frac{\partial^2 u(x,t)}{\partial t^2}$$

or

$$\frac{1}{\Delta x}\left[\left\{\frac{\partial u(x,t)}{\partial x}\right\}_{x+\Delta x} - \left\{\frac{\partial u(x,t)}{\partial x}\right\}_{x}\right] = \frac{\rho}{\tau}\frac{\partial^2 u(x,t)}{\partial t^2}$$

In the limit $\Delta x \to 0$, we have

$$\lim_{\Delta x\to 0}\frac{1}{\Delta x}\left[\left\{\frac{\partial u(x,t)}{\partial x}\right\}_{x+\Delta x} - \left\{\frac{\partial u(x,t)}{\partial x}\right\}_{x}\right] = \frac{\rho}{\tau}\frac{\partial^2 u(x,t)}{\partial t^2}$$

or

$$\frac{\partial^2 u(x,t)}{\partial x^2} = \frac{1}{c^2}\frac{\partial^2 u(x,t)}{\partial t^2} \tag{12.4}$$

where $c = \sqrt{\dfrac{\tau}{\rho}}$ represents the constant speed of the propagation of the wave.

Here, c is the speed with which a disturbance is propagated along the string, that is, it is the speed of the transverse wave produced. Equation (12.4) represents the equation for the transverse vibrations of a stretched string. This equation is called one-dimensional classical wave equation.

Confirmation of the Nature of *c*

Dimensionally, we have

dimension of
$$c=\sqrt{\frac{MLT^{-2}}{ML^{-1}}}=LT^{-1}$$

which is the dimension of velocity or speed.

12.6 SOLUTION OF ONE-DIMENSIONAL WAVE EQUATION (EQUATION FOR TRANSVERSE VIBRATIONS OF A STRETCHED STRING) BY THE SEPARATION OF VARIABLES METHOD

The vertical deflection, $u(x, t)$, of a vibrating string of length (L), fixed at its two ends at $x = 0$ and $x = L$ is given by the one-dimensional classical wave equation

$$\frac{\partial^2 u(x,t)}{\partial x^2}=\frac{1}{c^2}\frac{\partial^2 u(x,t)}{\partial t^2}, 0 \le x \le L, t \ge 0 \tag{12.5}$$

with the boundary conditions

$$u(0,t)=0, u(L,t)=0$$

and initial conditions

$$u(x, 0) = f(x), \left(\frac{\partial u}{\partial t}\right)_{t=0} = g(x).$$

Here, $f(x)$ is the initial deflection at a point distant (x) from the end at $x = 0$ and $g(x)$ is the initial speed at that point.

We are to solve this boundary value problem (BVP) by the method of separation of variables. In accordance with this method, we assume a solution in the form

$$u(x,t)=X(x)T(t) \ne 0 \tag{12.6}$$

where X is a function of coordinate x and T is a function of time t which are to be determined later.

Substituting Eq. (12.6) into Eq. (12.5), we have

$$T\frac{d^2X}{dx^2}=\frac{X}{c^2}\frac{d^2T}{dt^2}.$$

Dividing both sides of the equation by the product XT, we obtain

$$\frac{1}{X}\frac{d^2X}{dx^2}=\frac{1}{c^2 T}\frac{d^2T}{dt^2}.$$

In this equation, the left-hand side is a function of x only whereas the right-hand side is a function of t only and their equality is possible only when both of them are equal to a single constant. Such a constant is known as the ***separation constant***.

Thus, we write

$$\frac{1}{X}\frac{d^2X}{dx^2} = \frac{1}{c^2 T}\frac{d^2T}{dt^2} = \lambda \tag{12.7}$$

where λ is a separation constant.

Now, Eq. (12.7) yields the following two equations

$$\frac{d^2X}{dx^2} - \lambda X = 0 \tag{12.8}$$

and

$$\frac{d^2T}{dt^2} - \lambda c^2 T = 0. \tag{12.9}$$

Then, we are to solve Eqs. (12.8) and (12.9) with the following transformed boundary conditions.

To find the transformed boundary conditions:

The first boundary condition is transformed into the relation

$$u(0, t) = X(0)T(t) = 0.$$

Since $T(t) \neq 0$ for all values of t, we have

$$X(0) = 0. \tag{12.10}$$

In a similar manner, the second boundary condition implies

$$X(L) = 0. \tag{12.11}$$

Equations (12.10) and (12.11) are the transformed boundary conditions.

Appropriate Choice of the Separation Constant (λ)

We know that the solution of the given one-dimensional wave equation can be obtained if we know the solutions of Eqs. (12.8) and (12.9).

To solve these two equations, we are to make an appropriate choice of λ which is to be determined later. We consider three possible cases as in the following.

CASE 1: $\lambda > 0$

In this case, let us take $\lambda = p^2$ = a positive number. Then, the general solution of Eq. (12.8) is

$$X(x) = A_1 e^{-px} + A_2 e^{px}$$

where A_1 and A_2 are arbitrary constants.

Using the transformed boundary conditions given in Eqs. (12.10) and (12.11), the above equation yields two simultaneous equations, viz.,

$$\begin{cases} X(0) = A_1 + A_2 = 0, \\ X(L) = A_1 e^{-pL} + A_2 e^{pL} = 0. \end{cases}$$

Solving these two simultaneous equations, we obtain

$$A_1 = A_2 = 0\,.$$

Thus, we obtain $X(x) = 0$ and hence $u(x, t) = 0$. The solution is trivial and hence, it is to be neglected.

CASE 2: $\lambda = 0$

In this case, the general solution of Eq. (12.8) is

$$X(x) = B_1 x + B_2$$

where B_1 and B_2 are arbitrary constants.

Using the transformed boundary conditions given in Eqs. (12.10) and (12.11), the above equation yields

$$X(0) = B_2 = 0$$

and

$$X(L) = B_1 L = 0\,.$$

Then, we have

$$B_1 = B_2 = 0.$$

Thus, we have $X(x) = 0$ giving $u(x, t) = 0$.

In this case also, the solution is trivial and hence, it is to be neglected.

CASE 3: $\lambda < 0$

In this case we take $\lambda = -k^2$ where k^2 is a positive constant.

Then, Eqs. (12.8) and (12.9) become respectively

$$\frac{d^2X}{dx^2} + k^2 X = 0 \tag{12.12}$$

and

$$\frac{d^2T}{dt^2} + k^2 c^2 T = 0. \tag{12.13}$$

Now, the general solution of Eq. (12.12) is

$$X(x) = C_1 \cos kx + C_2 \sin kx \tag{12.14}$$

where C_1 and C_2 are arbitrary constants.

Using the transformed boundary conditions given in Eqs. (12.10) and (12.11), we have

$$X(0) = C_1 \cos 0 + C_2 \sin 0 = 0 \quad \text{yielding } C_1 = 0$$

and

$$X(L) = C_1 \cos kL + C_2 \sin kL = 0 \text{ yielding } C_2 \sin kL = 0.$$

Since $C_2 = 0$ leads to a trivial solution, it is not to be considered. We consider only

$$\sin kL = 0 \text{ or, } kL = n\pi \text{ where } n = 1, 2, 3, \ldots, \text{etc.}$$

Here, the case of $n=0$ is not considered as it leads to a trivial solution.

The cases of $n = -1, -2, -3 \ldots$ etc., are also not considered as they impose only a minus sign to those solutions with $n = 1, 2, 3 \ldots$ etc., and these are not new solutions.

Thus, the values of the constant (k) are given by

$$k_n = \frac{n\pi}{L} \quad (n = 1, 2, 3 \ldots). \tag{12.15}$$

Now, as $C_1 = 0$, Eq. (12.14) has the solution

$$X_n(x) = C_2 \sin\frac{n\pi x}{L} \quad (n = 1, 2, 3 \ldots) \tag{12.16}$$

with the values of k given in Eq. (12.15). We rewrite Eq. (12.13) as

$$\frac{d^2T}{dt^2} + \left(\frac{n\pi c}{L}\right)^2 T = 0.$$

With the values of n cited above, this equation has the general solution

$$T_n = D_n \cos\frac{n\pi ct}{L} + E_n \sin\frac{n\pi ct}{L} \tag{12.17}$$

where D_n and E_n are arbitrary constants.

Recalling Eq. (12.6), we write the solutions of Eq. (12.5) as

$$u_n(x,t) = \left[C_2 D_n \cos\frac{n\pi ct}{L} + C_2 E_n \sin\frac{n\pi ct}{L}\right]\sin\frac{n\pi x}{L}$$

$$= \left[\alpha_n \cos\frac{n\pi ct}{L} + \beta_n \sin\frac{n\pi ct}{L}\right]\sin\frac{n\pi x}{L} \tag{12.18}$$

where $\alpha_n = C_2 D_n$, $\beta_n = C_2 E_n$, $n = 1, 2, 3 \ldots$

Superposition Principle

Let us recall the ***superposition principle*** which states that if $u_1, u_2, u_3, \ldots, u_n$ are solutions of a homogeneous linear partial differential equation, then the linear combination

$$u = c_1u_1 + c_2u_2 + c_3u_3 + \ldots + c_nu_n$$

where c_j ($j = 1, 2, 3 \ldots n$) are constants, or simply the sum

$$u = u_1 + u_2 + u_3 + \ldots + u_n$$

is also a solution of the partial differential equation.

By the superposition principle, the general solution of the above boundary value problem (BVP) is given by

$$u(x,t)=\sum_{n=1}^{\infty}u_n(x,t)$$

$$=\sum_{n=1}^{\infty}\left[\alpha_n\cos\frac{n\pi ct}{L}+\beta_n\sin\frac{n\pi ct}{L}\right]\sin\frac{n\pi x}{L}. \tag{12.19}$$

Putting $\alpha_n = C_n \cos\varepsilon_n$, $\beta_n = C_n \sin\varepsilon_n$

such that $$C_n=\sqrt{\alpha_n^2+\beta_n^2} \text{ and } \varepsilon_n=\text{arc tan}\left(\frac{\beta_n}{\alpha_n}\right),$$

we write Eqs. (12.18) and (12.19) respectively, as

$$u_n(x,t)=C_n\sin\frac{n\pi x}{L}\cos\left(\frac{n\pi ct}{L}-\varepsilon_n\right) \tag{12.20}$$

and $$u(x,t)=\sum_{n=1}^{\infty}C_n\sin\frac{n\pi x}{L}\cos\left(\frac{n\pi ct}{L}-\varepsilon_n\right). \tag{12.21}$$

To find the coefficients α_n and β_n

These two coefficients are determined by applying the initial conditions cited before to Eq. (12.19).

Applying the first initial condition, viz., $u(x, 0) = f(x)$ to Eq. (12.19), we have

$$u(x,0)=f(x)=\sum_{n=1}^{\infty}\alpha_n\sin\frac{n\pi x}{L}.$$

Multiplying both sides by $\sin\frac{m\pi x}{L}$ and integrating the results with respect to x within the limit from $x = 0$ to $x = L$, we obtain

$$\int_0^L f(x)\sin\frac{m\pi x}{L}dx=\sum_{n=1}^{\infty}\alpha_n\int_0^L\sin\frac{m\pi x}{L}\sin\frac{n\pi x}{L}dx$$

$$=\alpha_m\int_0^L\sin^2\left(\frac{m\pi x}{L}\right)dx=\frac{L}{2}\alpha_m$$

[since the integrals vanish for $m \neq n$].

Thus, we have

$$\alpha_m = \frac{2}{L}\int_0^L f(x)\sin\frac{m\pi x}{L}\,dx.$$

Renaming the subscript, we write

$$\alpha_n = \frac{2}{L}\int_0^L f(x)\sin\frac{n\pi x}{L}\,dx. \tag{12.22}$$

Applying the second initial condition, viz., $\left(\frac{\partial u}{\partial t}\right)_{t=0} = g(x)$ to Eq. (12.19), we obtain

$$g(x) = \left(\frac{\partial u}{\partial t}\right)_{t=0}$$

$$= \sum_{n=1}^{\infty}\left[\frac{n\pi c}{L}\left\{-\alpha_n \sin\frac{n\pi ct}{L} + \beta_n \cos\frac{n\pi ct}{L}\right\}\sin\frac{n\pi x}{L}\right]_{t-0}$$

$$= \sum_{n=1}^{\infty}\frac{n\pi c}{L}\beta_n \sin\frac{n\pi x}{L}.$$

Multiplying both sides by $\sin\frac{m\pi x}{L}$ and then integrating both sides with respect to x within the limit from $x = 0$ to $x = L$, we obtain

$$\int_0^L g(x)\sin\frac{m\pi x}{L}\,dx = \sum_{n=1}^{\infty}\frac{n\pi c}{L}\beta_n\int_0^L \sin\frac{m\pi x}{L}\sin\frac{n\pi x}{L}\,dx$$

$$= \frac{m\pi c}{L}\beta_m\int_0^L \sin^2\left(\frac{m\pi x}{L}\right)dx = \frac{m\pi c}{L}\beta_m\frac{L}{2}$$

or

$$\beta_m = \frac{2}{m\pi c}\int_0^L g(x)\sin\frac{m\pi x}{L}\,dx.$$

Renaming the subscript, we write

$$\beta_n = \frac{2}{n\pi c}\int_0^L g(x)\sin\frac{n\pi x}{L}\,dx. \tag{12.23}$$

The solution of the vibrating string problem is, therefore, given by the series (12.19) where the coefficients α_n and β_n are determined by the formulae (12.22) and (12.23).

Another form of the solution is given in Eq. (12.21).

Now, we examine physical significance of the problem. The functions, $u_n(x, t)$, given in Eq. (12.18) are called ***eigenfunctions*** or *characteristic functions* and the quantities $\left(\omega_n = \dfrac{n\pi c}{L}\right)$ are called ***eigenfrequencies***. The quantities ω_n $(n = 1, 2, 3 \ldots)$ form a discrete spectrum of ***circular or radian frequencies*** and the quantities (ν_n) (pronounced as $(nu)_n$) $= \dfrac{\omega_n}{2\pi} = \dfrac{nc}{2L}$ are called ordinary frequencies.

From Eq. (12.19), we see that

$$u(x, t) = u_1(x, t) + u_2(x, t) + u_3(x, t) + \ldots .$$

A particular eigenfunction, $u_n(x, t)$, is said to correspond to the nth ***normal mode*** of vibration or the nth ***harmonic*** vibration. [A normal mode of vibration is one in which all points of the string vibrate with the same period and pass through their equilibrium positions at the same time.] The first normal mode or the first harmonic corresponding to $n = 1$ is called the ***fundamental mode*** and all other harmonics for $n > 1$ are called ***overtones.*** To find the frequency of fundamental mode, we have

$$\nu_1 = \frac{\omega_1}{2\pi} = \frac{1}{2\pi}\left(\frac{\pi c}{L}\right) = \frac{c}{2L} = \frac{1}{2L}\sqrt{\frac{\tau}{\rho}} \tag{12.24}$$

where τ is the magnitude of tension on the string and ρ is the mass per unit length of the string.

The time period for fundamental mode is

$$T_1 = \frac{1}{\nu_1} = \frac{2\pi}{\omega_1} = \frac{2L}{c} = 2L\sqrt{\frac{\rho}{\tau}}. \tag{12.25}$$

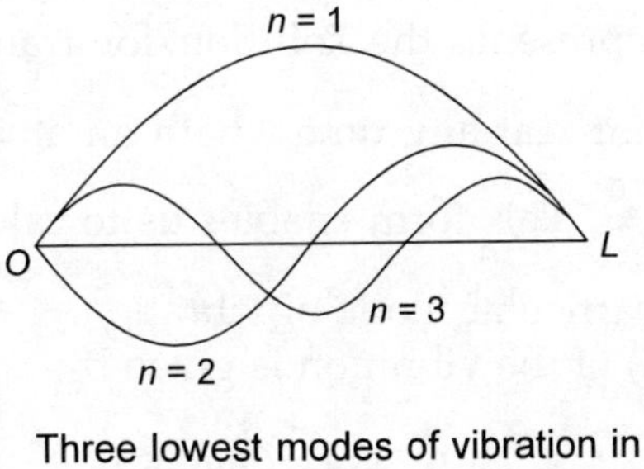

Fig. 12.2 Three lowest modes of vibration in a string

From the solution for the nth mode of vibration, i.e., from the expression for $u_n(x, t)$, we see that the solution vanishes at $x = 0, \dfrac{L}{n}, \dfrac{2L}{n}, \dfrac{3L}{n} \ldots \dfrac{(n-1)L}{n}, L$. These points are points with zero displacement. In the transverse vibrations of a stretched string fixed at the ends, excluding the two endpoints, the remaining points with zero displacement are known as ***nodes***. A superposition of forward waves with waves reflected from a fixed end produces a pattern of several visible loops on the string. For such a pattern, the waves are called ***stationary waves***. The points with maximum displacement are called ***antinodes***. Such patterns are illustrated in Fig. 12.3 (a, b, c) shown below.

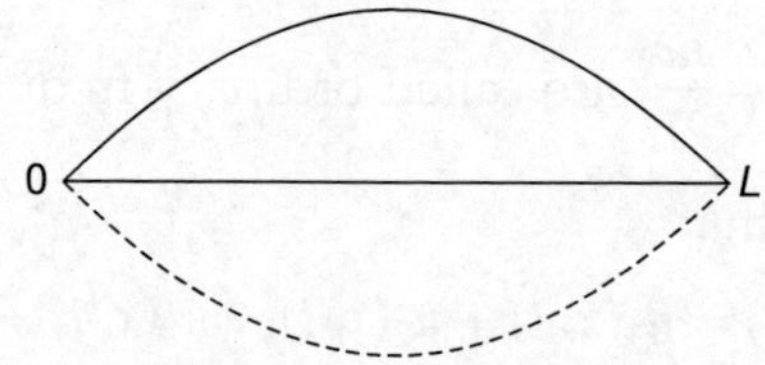

Fig. 12.3(a) Stationary wave of the first harmonic vibration.

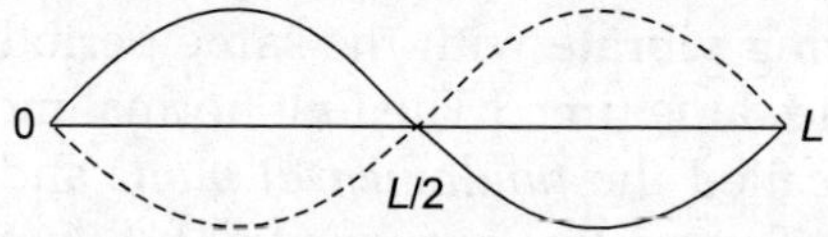

Fig. 12.3(b) Stationary wave of the second harmonic vibration.

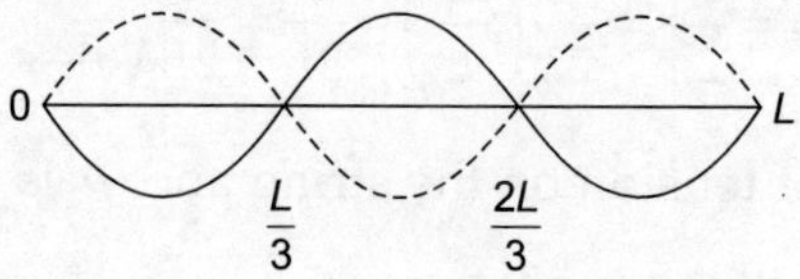

Fig. 12.3(c) Stationary wave of the third harmonic vibration.

Expressions for the Kinetic and the Potential Energies for the Vibration in the *n*th Mode

We see that Eq. (12.20) represents the solution for transverse vibrations of the stretched string at any point x at any time t with an amplitude $\left[C_n \sin\frac{n\pi x}{L}\right]$ and circular frequency, $\omega_n = \frac{n\pi c}{L}$. This form enables us to calculate the kinetic and the potential energies of that particular mode of vibration.

The kinetic energy (K.E.) of the vibration is given by

$$\text{K.E.} = K_n = \int_0^L \frac{1}{2}\rho\left(\frac{\partial u_n}{\partial t}\right)^2 dx.$$

Using Eq. (12.20), we have

$$K_n = \left[\frac{1}{2}\rho C_n^2\left(\frac{n\pi c}{L}\right)^2 \sin^2\left(\frac{n\pi ct}{L} - \varepsilon_n\right)\right]\int_0^L \sin^2\left(\frac{n\pi x}{L}\right)dx$$

$$=\left[\frac{1}{2}\rho C_n^2\omega_n^2\sin^2(\omega_n t-\varepsilon_n)\right]\frac{1}{2}\int_0^L\left\{1-\cos\left(\frac{2n\pi x}{L}\right)\right\}dx$$

$$=\frac{1}{4}\rho LC_n^2\omega_n^2\sin^2(\omega_n t-\varepsilon_n). \tag{12.26}$$

An elementary derivation of the expression for the potential energy is stated in the following. Let dx be an element of the unstretched string (in equilibrium), and the same element when displaced from equilibrium, has a length, ds.

Therefore, the increase in the length of the string is given by

$$ds-dx=\sqrt{dx^2+du_n^2}-dx=dx\left[\left\{1+\left(\frac{\partial u_n}{\partial x}\right)^2\right\}^{\frac{1}{2}}-1\right]\approx\frac{1}{2}\left(\frac{\partial u_n}{\partial x}\right)^2dx.$$

The potential energy of the element is given by $dV_n=\tau(ds-dx)=\frac{1}{2}\tau\left(\frac{\partial u_n}{\partial x}\right)^2dx$ where τ is the tension on the string.

Integrating both sides with respect to x within the limit from x = 0 to $x = L$, we obtain the total potential energy of the vibrating string in that particular mode as

$$V_n=\int_0^L\frac{1}{2}\tau\left(\frac{\partial u_n}{\partial x}\right)^2dx. \tag{12.27}$$

Using Eq. (12.20), we have

$$V_n=\frac{1}{2}\tau C_n^2\left(\frac{n\pi}{L}\right)^2\cos^2\left(\frac{n\pi ct}{L}-\varepsilon_n\right)\int_0^L\cos^2\left(\frac{n\pi x}{L}\right)dx.$$

Using $\tau=\rho c^2$ and $\omega_n=\frac{n\pi c}{L}$, we obtain

$$V_n=\frac{1}{4}\rho LC_n^2\,\omega_n^2\cos^2\left(\frac{n\pi ct}{L}-\varepsilon_n\right). \tag{12.28}$$

Using Eqs. (12.26) and (12.28), the total energy of the string vibrating in the nth normal mode is given by

$$E_n=K_n+V_n=\frac{1}{4}\rho LC_n^2\,\omega_n^2=\text{constant}. \tag{12.29}$$

Finally, the total energy of the system for all modes of vibration is given by

$$E=\sum_{n=1}^{\infty}E_n=\frac{1}{4}\rho L\sum_{n=1}^{\infty}(C_n\omega_n)^2=\text{another constant}. \tag{12.30}$$

A Special Case

In a special case, if the initial velocity is zero (e.g., the case of plucked string), we have

$$\left(\frac{\partial u}{\partial t}\right)_{t=0} = g(x) = 0.$$

Then, we have $\beta_n = 0$ and Eq. (12.19) becomes

$$\begin{aligned} u(x, t) &= \sum_{n=1}^{\infty} \alpha_n \sin\left(\frac{n\pi x}{L}\right)\cos\left(\frac{n\pi ct}{L}\right) \\ &= \sum_{n=1}^{\infty} \frac{1}{2}\alpha_n \left[\sin\left\{\frac{n\pi}{L}(x-ct)\right\} + \sin\left\{\frac{n\pi}{L}(x+ct)\right\}\right] \\ &= \frac{1}{2}F(x-ct) + \frac{1}{2}F(x+ct) \end{aligned} \tag{12.31}$$

where $\quad F(x-ct) = \alpha_n \sin\left\{\frac{n\pi}{L}(x-ct)\right\}$

and $\quad F(x+vt) = \alpha_n \sin\left\{\frac{n\pi}{L}(x+ct)\right\}.$

Here, $F(x-ct)$ represents wave travelling towards the right (positive side of x-axis) and $F(x+ct)$ represents wave travelling towards the left (negative side of x-axis).

12.7 D'ALEMBERT'S SOLUTION OF ONE-DIMENSIONAL WAVE EQUATION

This solution was proposed by the French mathematician, ***Jean-Baptiste Le Rond d'Alembert*** (17 November 1717 - 29 October 1783). The one-dimensional wave equation for vibrations of a stretched string is given by

$$\frac{\partial^2 u}{\partial x^2} = \frac{1}{c^2}\frac{\partial^2 u}{\partial t^2} \tag{12.32}$$

where $u = u(x, t)$ is the deflection at the point x at time t.

Let us introduce the new variables $\xi = x - ct$ and $\eta = x + ct$.

Now, we have

$$\frac{\partial \xi}{\partial x} = \frac{\partial \eta}{\partial x} = 1$$

Then, $$\frac{\partial u}{\partial x} = \frac{\partial u}{\partial \xi}\frac{\partial \xi}{\partial x} + \frac{\partial u}{\partial \eta}\frac{\partial \eta}{\partial x} = \left(\frac{\partial}{\partial \xi} + \frac{\partial}{\partial \eta}\right)u.$$

Therefore, we have

$$\frac{\partial^2 u}{\partial x^2}=\frac{\partial}{\partial x}\left(\frac{\partial u}{\partial x}\right)=\left(\frac{\partial}{\partial \xi}+\frac{\partial}{\partial \eta}\right)\left(\frac{\partial u}{\partial \xi}+\frac{\partial u}{\partial \eta}\right)=\frac{\partial^2 u}{\partial \xi^2}+2\frac{\partial^2 u}{\partial \xi \partial \eta}+\frac{\partial^2 u}{\partial \eta^2}. \tag{12.33}$$

Also, we have

$$\frac{\partial \xi}{\partial t}=-c \text{ and } \frac{\partial \eta}{\partial t}=c.$$

Then, we have

$$\frac{\partial u}{\partial t}=\frac{\partial u}{\partial \xi}\frac{\partial \xi}{\partial t}+\frac{\partial u}{\partial \eta}\frac{\partial \eta}{\partial t}=c\left(\frac{\partial}{\partial \eta}-\frac{\partial}{\partial \xi}\right)u.$$

Therefore, we have

$$\begin{aligned}\frac{\partial^2 u}{\partial t^2}&=c\left(\frac{\partial}{\partial \eta}-\frac{\partial}{\partial \xi}\right)\left\{c\left(\frac{\partial u}{\partial \eta}-\frac{\partial u}{\partial \xi}\right)\right\}\\&=c^2\left[\frac{\partial^2 u}{\partial \eta^2}-2\frac{\partial^2 u}{\partial \xi \partial \eta}+\frac{\partial^2 u}{\partial \xi^2}\right].\end{aligned} \tag{12.34}$$

Substituting Eqs. (12.33) and (12.34) into Eq. (12.32), we obtain

$$2\frac{\partial^2 u}{\partial \xi\, \partial \eta}=-2\frac{\partial^2 u}{\partial \xi\, \partial \eta}$$

or

$$4\frac{\partial^2 u}{\partial \xi\, \partial \eta}=0$$

or

$$\frac{\partial}{\partial \xi}\left(\frac{\partial u}{\partial \eta}\right)=0$$

or

$$\frac{\partial u}{\partial \eta}=\phi(\eta) \text{ where } \phi(\eta) \text{ is a function of } \eta \text{ only.}$$

Integrating both sides with respect to η, we have

$$u=\int \phi(\eta)\partial \eta+F(\xi)$$

where $F(\xi)$ is a function of ξ only.

Putting $G(\eta)=\int \phi(\eta)\partial \eta$, we write

$$u=u(\xi,\eta)=F(\xi)+G(\eta).$$

In terms of the original variables (x,t), we write a solution of the given one-dimensional wave equation as

$$u(x,t)=F(x-ct)+G(x+ct). \tag{12.35}$$

This solution is known as d'Alembert's solution.

Let the initial conditions be

$$u(x,0)=h(x)$$

or

$$F(x)+G(x)=h(x) \tag{12.36}$$

and

$$\left[\frac{\partial u(x,t)}{\partial t}\right]_{t=0}=0.$$

The last equation can be written as

$$\left[\frac{\partial F(\xi)}{\partial \xi}\frac{\partial \xi}{\partial t}+\frac{\partial G(\eta)}{\partial \eta}\frac{\partial \eta}{\partial t}\right]_{t=0}=0$$

or

$$-c\frac{dF(x)}{dx}+c\frac{dG(x)}{dx}=0$$

or

$$\frac{d}{dx}[G(x)-F(x)]=0.$$

Therefore, we have

$$G(x)-F(x)=a \tag{12.37}$$

where a is a constant.

Solving Eqs. (12.36) and (12.37), we obtain

$$F(x)=\frac{1}{2}[h(x)-a]$$

and

$$G(x)=\frac{1}{2}[h(x)+a].$$

Therefore, we have

$$F(x-vt)=\frac{1}{2}[h(x-vt)-a]$$

and

$$G(x+vt)=\frac{1}{2}[h(x+vt)+a].$$

Substituting these two functions into Eq. (12.35), we write the d'Alembert's solution as

$$u(x,t)=\frac{1}{2}h(x-vt)+\frac{1}{2}h(x+vt). \tag{12.38}$$

This equation is similar to Eq. (12.31).

12.8 VIBRATIONS OF PLUCKED STRING

Let a stretched string of length L, fixed at its ends, be set into vibrations by plucking at some point and then releasing freely. Playing a guitar is a simple example. The relevant wave equation is

$$\frac{\partial^2 u}{\partial x^2} = \frac{1}{c^2}\frac{\partial^2 u}{\partial t^2} \tag{12.39}$$

where $u = u(x, t)$ is the deflection at a point distant x from the end $x = 0$ at a time t and c is the constant speed of wave propagation.

The general solution of the above equation is

$$u(x, t) = \sum_{n=1}^{\infty} u_n(x, t)$$

$$= \sum_{n=1}^{\infty}\left[\alpha_n \cos\frac{n\pi ct}{L} + \beta_n \sin\frac{n\pi ct}{L}\right]\sin\frac{n\pi x}{L} \tag{12.40}$$

where the coefficients α_n and β_n are to be determined.

Let the string be plucked at a point distant a from the end $x = 0$. Let h be the initial deflection at the plucking point. In the case of a plucked string, the initial deflection of a point is non-zero whereas the initial velocity is zero.

Let us find the initial deflection, $f(x)$, at an arbitrary point x on the string.

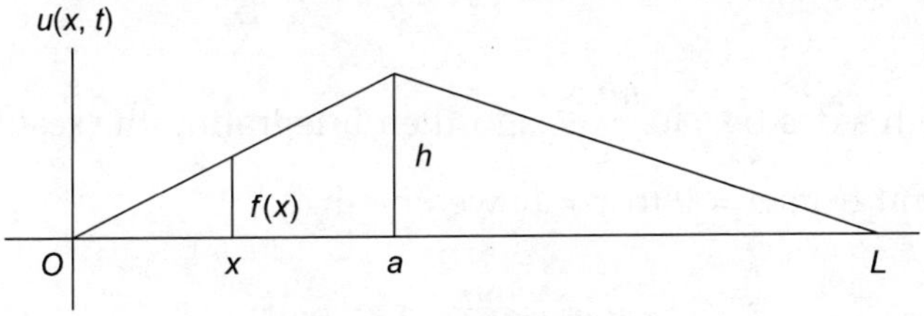

Fig. 12.4(a) Plucked string

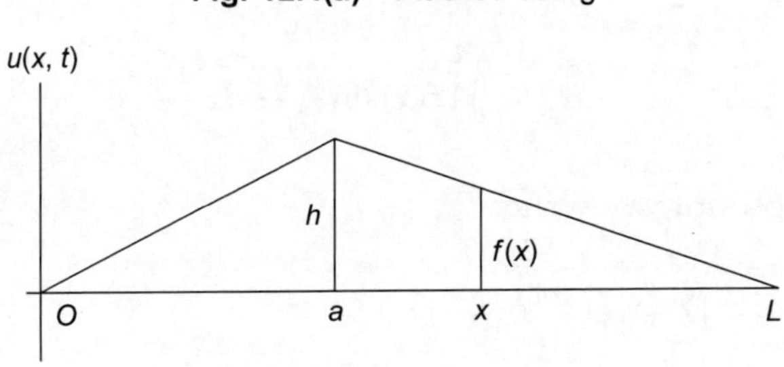

Fig. 12.4(b) Plucked string

Initial Conditions

In the vibrations of a plucked string, there is an initial deflection but no initial velocity.

From Figure 12.4 (a), for $0<x<a,$ we have

$$\frac{f(x)}{x}=\frac{h}{a} \quad \text{or, } f(x)=\frac{h}{a}x. \tag{12.41}$$

Also, from Figure 12.4 (b), for $a<x<L$, we have

$$\frac{f(x)}{L-x}=\frac{h}{L-a} \qquad \text{or, } f(x)=\frac{h(L-x)}{L-a}. \tag{12.42}$$

In general, we write the Initial deflection of a point at a distance x at time $t = 0$ as

$$u(x,0)=f(x)=\begin{cases} \dfrac{h}{a}x & \text{for } 0<x<a, \\ \dfrac{h(L-x)}{L-a} & \text{for } a<x<L. \end{cases} \tag{12.43}$$

The initial speed is

$$\left(\frac{\partial u}{\partial t}\right)_{t=0}=g(x)=0. \tag{12.44}$$

Initially, at $t = 0$, Eq. (12.40) yields

$$u(x,0)=f(x)=\sum_{n=1}^{\infty}\alpha_n \sin\frac{n\pi x}{L}.$$

Multiplying both sides by $\sin\dfrac{m\pi x}{L}$ and then integrating the results with respect to x within the limit from $x = 0$ to $x = L$, we obtain

$$\int_0^L f(x)\sin\frac{m\pi x}{L}dx=\alpha_m\frac{L}{2}$$

or

$$\alpha_m=\frac{2}{L}\int_0^L f(x)\sin\frac{m\pi x}{L}dx.$$

Renaming the subscript, we write

$$\alpha_n=\frac{2}{L}\int_0^L f(x)\sin\frac{n\pi x}{L}dx$$

$$=\frac{2}{L}\left\{\int_0^a \frac{h}{a}x\sin\frac{n\pi x}{L}dx+\int_a^L \frac{h(L-x)}{L-a}\sin\frac{n\pi x}{L}dx\right\}$$

$$= \frac{2}{L}\left\{-\left[\frac{hL}{n\pi a}x\cos\frac{n\pi x}{L}\right]_0^a + \frac{hL}{n\pi a}\int_0^a \cos\frac{n\pi x}{L}dx\right\}$$

$$+\frac{2}{L}\left\{\left[-\frac{hL(L-x)}{n\pi(L-a)}\cos\frac{n\pi x}{L}\right]_a^L - \frac{hL}{n\pi(L-a)}\int_a^L \cos\frac{n\pi x}{L}dx\right\}$$

$$= \frac{2}{L}\left\{-\frac{hL}{n\pi}\cos\frac{n\pi a}{L} + \left[\frac{hL^2}{n^2\pi^2 a}\sin\frac{n\pi x}{L}\right]_0^a\right\}$$

$$+\frac{2}{L}\left\{0+\frac{hL}{n\pi}\cos\frac{n\pi a}{L} - \left[\frac{hL^2}{n^2\pi^2(L-a)}\sin\frac{n\pi x}{L}\right]_a^L\right\}$$

$$= \frac{2}{L}\left\{\frac{hL^2}{n^2\pi^2 a}\sin\frac{n\pi a}{L} - 0 + \frac{hL^2}{n^2\pi^2(L-a)}\sin\frac{n\pi a}{L}\right\}$$

$$= \frac{2hL^2}{n^2\pi^2 a(L-a)}\sin\frac{n\pi a}{L}. \qquad (12.45)$$

Since the initial speed in this problem is zero, we have

$$\left(\frac{\partial u}{\partial t}\right)_{t=0} = g(x) = 0.$$

Using Eq. (12.40), we obtain

$$\frac{\partial}{\partial t}\left[\sum_{n=1}^{\infty}\left(\alpha_n\cos\frac{n\pi ct}{L} + \beta_n\sin\frac{n\pi ct}{L}\right)\sin\frac{n\pi x}{L}\right]_{t=0} = 0$$

or
$$\sum_{n=1}^{\infty}\frac{n\pi c}{L}\beta_n\sin\frac{n\pi x}{L} = 0.$$

Multiplying both sides by $\sin\frac{m\pi x}{L}$ and integrating the results with respect to x within the limit from $x = 0$ to $x = L$, we obtain

$$\sum_{n=1}^{\infty}\frac{n\pi c}{L}\beta_n\int_0^L \sin\frac{m\pi x}{L}\sin\frac{n\pi x}{L}dx = 0$$

or
$$\frac{m\pi c}{L}\beta_m\int_0^L \sin^2\left(\frac{m\pi x}{L}\right)dx = 0$$

or
$$\frac{m\pi c}{2L}\beta_m\int_0^L\left(1-\cos\frac{2m\pi x}{L}\right)dx=0$$

or
$$\frac{m\pi c}{2L}L\beta_m=0$$

or
$$\beta_m=0.$$

Renaming the subscript, we write

$$\beta_n=0. \tag{12.46}$$

Substituting the values of α_n and β_n from Eqs. (12.45) and (12.46) into Eq. (12.40), the deflection of the plucked string at a point x at time t is given by

$$u(x,t)=\sum_{n=1}^{\infty}u_n(x,t)$$
$$=\sum_{n=1}^{\infty}\left[\frac{2hL^2}{n^2\pi^2a(L-a)}\sin\frac{n\pi a}{L}\sin\frac{n\pi x}{L}\right]\cos\frac{n\pi ct}{L}. \tag{12.47}$$

In Eq. (12.47), the quantities within the square brackets represent the amplitude of vibration. As this amplitude is inversely proportional to n^2, we infer that the amplitudes of higher harmonics fall off rapidly.

For fundamental mode of vibration, we have

$$u_1(x,t)=\left[\frac{2hL^2}{a(L-a)\pi^2}\sin\frac{\pi a}{L}\sin\frac{\pi x}{L}\right]\cos\frac{\pi ct}{L}. \tag{12.48}$$

In this mode, the fundamental circular frequency is given by

$$\omega_1=\frac{\pi c}{L} \tag{12.49}$$

and the fundamental ordinary frequency is given by

$$\nu_1=\frac{\omega_1}{2\pi}=\frac{c}{2L}=\frac{1}{2L}\sqrt{\frac{\tau}{\rho}}$$

where τ is the tension on the string and ρ is the mass per unit length of the string.

For the second harmonic, we have

$$u_2(x,t)=\left[\frac{hL^2}{2a(L-a)\pi^2}\sin\frac{2\pi a}{L}\sin\frac{2\pi x}{L}\right]\cos\frac{2\pi ct}{L}. \tag{12.50}$$

In this case, $\omega_2=\dfrac{2\pi c}{L}$ and $\nu_2=\dfrac{\omega_2}{2\pi}=\dfrac{c}{L}=\dfrac{1}{L}\sqrt{\dfrac{\tau}{\rho}}$ and so on.

Special Cases

CASE I: If the string is plucked at its middle point:

In this case $a = \frac{L}{2}$ and, therefore, $\sin\frac{n\pi a}{L} = \sin\frac{n\pi}{2} = 0$ for $n = 2, 4, 6 \ldots$

Therefore, we see that

$$u_2(x, t) = 0 = u_4(x, t) = u_6(x, t) = u_8 = \ldots$$

We infer that if the string is plucked at its midpoint, all the even harmonic vibrations are absent.

CASE II: If the string is plucked at a point distant one-third of its length from the end $x = 0$:

In this case, $a = \frac{L}{3}$ Then, $\sin\frac{n\pi a}{L} = \sin\frac{n\pi}{3} = 0$ for $n = 3, 6, 9, 12 \ldots$

and
$$u_3(x, t) = 0 = u_6(x, t) = u_9(x, t) = u_{12} = \ldots$$

We infer that if the string is plucked at a point distant one-third of its length from the end at the origin, all harmonic vibrations having orders of integral multiples of 3 are absent.

In similar manner, other cases with different plucking points can be studied. In general, we say that if a string is set into vibrations by plucking at a point distant $\frac{L}{n}$ from the end at the origin and then released freely, all harmonics of integral multiples of n will be absent. This is what we call ***Young's theorem***.

In the transverse vibrations of a plucked string, all harmonics which have a node at the plucked point are always absent. In such vibrations, there is an initial deflection but no initial velocity.

Effect of Touching the String

After setting into vibrations, if the string is touched at some point, the vibrations stop, because all harmonics which have a node at the plucked point are absent from the beginning and those harmonics which do not have a node at the plucked point stop by damping.

12.9 VIBRATIONS OF STRUCK STRING

A stretched string can be set into vibrations by striking it at some point. Playing a ***piano*** or a ***santoor*** is a suitable example. The time of contact between the hammer and the string is assumed to be extremely small and thus the external force applied to the striking point is of the nature of an impulse. Due to this impulse, there is an initial velocity of the striking point. Therefore, in this problem, the initial conditions differ from those of a plucked string. We may say that the initial conditions of a plucked string are ***static*** whereas the initial conditions of a struck string are ***dynamic***. It is assumed that the motion is free with a given initial velocity imparted to an extremely short length of the string. The waves then move in opposite directions from the striking point and get reflected from the fixed ends.

Let a stretched string of length L be fixed at its endpoints at $x = 0$ and $x = L$. Let the string be struck at $x = a$ imparting an initial speed, v.

Initial Conditions

In the vibrations of a struck string, there is no initial deflection but an initial velocity.

Mathematically, we write

$$u(x, 0) = f(x) = 0. \tag{12.51}$$

Let us find the initial speed $g(x)$ at any point x of the string.

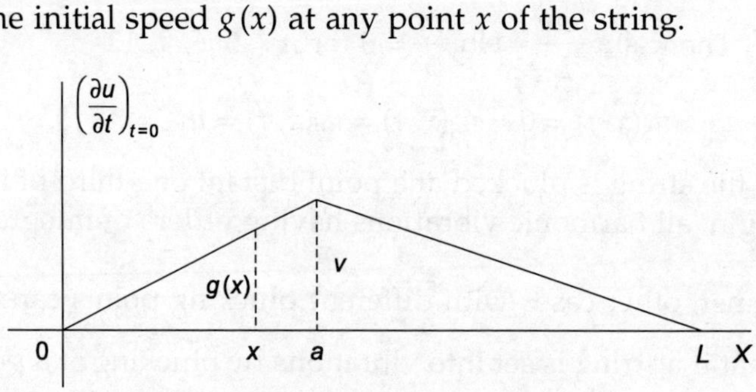

Fig. 12.5(a) Struck string

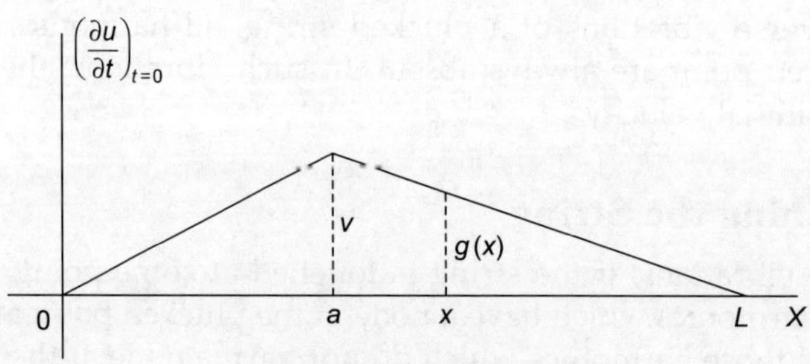

Fig. 12.5(b) Struck string

From Figure 12.5 (a), for $0 < x < a$, we have

$$\frac{g(x)}{x} = \frac{v}{a} \quad \text{or,} \quad g(x) = \frac{v}{a}x. \tag{12.52}$$

Also, from Figure 12.5 (b), for $a < x < L$, we have

$$\frac{g(x)}{L-x} = \frac{v}{L-a} \quad \text{or,} \quad g(x) = \frac{v(L-x)}{L-a}. \tag{12.53}$$

In general, the initial speed is written as

$$\left(\frac{\partial u}{\partial t}\right)_{t=0} = g(x) = \begin{cases} \frac{v}{a}x & \text{for } 0 \le x \le a, \\ \frac{v(L-x)}{L-a} & \text{for } a \le x \le L. \end{cases} \tag{12.54}$$

We know that the deflection at a point x of the vibrating string at time t is given by

$$u(x,t) = \sum_{n=1}^{\infty} u_n(x,t)$$

$$= \sum_{n=1}^{\infty}\left[\alpha_n \cos\frac{n\pi ct}{L} + \beta_n \sin\frac{n\pi ct}{L}\right]\sin\frac{n\pi x}{L} \tag{12.55}$$

where α_n and β_n are to be determined from the initial conditions.

Using Eq. (12.51) in Eq. (12.55), we write

$$u(x,0) = 0 = \sum_{n=1}^{\infty} \alpha_n \sin\frac{n\pi x}{L}.$$

Multiplying both sides by $\sin\frac{m\pi x}{L}$ and then integrating the results with respect to x within the limit from $x = 0$ to $x = L$, we have

$$0 = \alpha_m \int_0^L \sin^2\left(\frac{m\pi x}{L}\right)dx = \alpha_m \frac{L}{2} \text{ yielding } \alpha_m = 0.$$

Renaming the subscript, we write $\alpha_n = 0$.

Then, Eq. (12.55) becomes

$$u(x,t) = \sum_{n=1}^{\infty} u_n(x,t)$$

$$= \sum_{n=1}^{\infty} \beta_n \sin\frac{n\pi ct}{L}\sin\frac{n\pi x}{L}. \tag{12.56}$$

Using Eq. (12.54) in Eq. (12.56), we have

$$\sum_{n=1}^{\infty} \beta_n \frac{\partial}{\partial t}\left[\sin\frac{n\pi ct}{L}\sin\frac{n\pi x}{L}\right]_{t=0} = g(x)$$

or

$$\sum_{n=1}^{\infty} \frac{n\pi c}{L}\beta_n \sin\frac{n\pi x}{L} = g(x)$$

Multiplying both sides by $\sin\frac{m\pi x}{L}$ and then integrating the results with respect to x within the limit from $x = 0$ to $x = L$, we have

$$\sum_{n=1}^{\infty}\frac{n\pi c}{L}\beta_n\int_0^L \sin\frac{m\pi x}{L}\sin\frac{n\pi x}{L}dx = \int_0^L g(x)\sin\frac{m\pi x}{L}dx$$

or
$$\frac{m\pi c}{L}\beta_m\int_0^L \sin^2\left(\frac{m\pi x}{L}\right)dx = \int_0^a \frac{v}{a}x\sin\frac{m\pi x}{L}dx + \int_a^L \frac{v(L-x)}{L-a}\sin\frac{m\pi x}{L}dx$$

or
$$\frac{m\pi c}{L}\beta_m\frac{L}{2} = \left[-\frac{Lv}{am\pi}x\cos\frac{m\pi x}{L}\right]_0^a + \int_0^a \frac{Lv}{am\pi}\cos\frac{m\pi x}{L}dx$$
$$+\left[-\frac{Lv(L-x)}{m\pi(L-a)}\cos\frac{m\pi x}{L}\right]_a^L - \int_a^L \frac{Lv}{m\pi(L-a)}\cos\frac{m\pi x}{L}dx$$

or
$$\frac{m\pi c}{2}\beta_m = -\frac{Lv}{m\pi}\cos\frac{m\pi a}{L} + \left[\frac{vL^2}{am^2\pi^2}\sin\frac{m\pi x}{L}\right]_0^a$$
$$+\frac{Lv}{m\pi}\cos\frac{m\pi a}{L} - \left[\frac{vL^2}{m^2\pi^2(L-a)}\sin\frac{m\pi x}{L}\right]_a^L$$
$$= \frac{vL^2}{am^2\pi^2}\sin\frac{m\pi a}{L} + \frac{vL^2}{m^2\pi^2(L-a)}\sin\frac{m\pi a}{L}$$
$$= \frac{vL^3}{m^2\pi^2 a(L-a)}\sin\frac{m\pi a}{L}$$

or
$$\beta_m = \frac{2vL^3}{m^3\pi^3 ca(L-a)}\sin\frac{m\pi a}{L}.$$

Renaming the subscript, we write

$$\beta_n = \frac{2vL^3}{n^3\pi^3 ca(L-a)}\sin\frac{n\pi a}{L}. \tag{12.57}$$

Substituting Eq. (12.57) into Eq. (12.56), we write the displacement at a point x at time t as

$$u(x,t) = \sum_{n=1}^{\infty} u_n(x,t)$$

$$= \sum_{n=1}^{\infty} \left[\frac{2vL^3}{n^3\pi^3 ca(L-a)} \sin\frac{n\pi a}{L} \sin\frac{n\pi x}{L} \right] \sin\frac{n\pi ct}{L}. \tag{12.58}$$

It is seen that the amplitude of the nth mode of normal vibration of the struck string is inversely proportional to n^3 whereas the amplitude in the case of the plucked string is inversely proportional to n^2. Therefore, the convergence of the infinite series representing the deflection at x at time t is more rapid in the case of the struck string.

For fundamental mode, $n = 1, \omega_1 = \frac{\pi c}{L}, f_1 = \frac{\omega_1}{2\pi} = \frac{c}{2L} = \frac{1}{2L}\sqrt{\frac{\tau}{\rho}}$.

If the string is struck at the middle point, then $a = \frac{L}{2}$, then, $\sin\frac{n\pi a}{L} = \sin\frac{n\pi}{2}$ which is zero for n = 2, 4, 6, 8 ...

In such cases, the amplitude in Eq. (12.58) vanishes. Thus, if the string is struck at the middle, all even harmonics are absent. If it is struck at $a = \frac{L}{3}$ all harmonics which are multiples of 3 are absent. Similarly, we can discuss other higher harmonics.

Effect of Touching the String

After setting into vibrations, if the string is touched at some point, the vibrations stop, because all harmonics which have a node at the struck point are absent from the beginning and those harmonics which do not have a node at the struck point stop by damping.

12.10 FORCED VIBRATIONS OF A STRETCHED STRING

In this problem, there is an external periodic force applied to the string whereas external forces are totally absent in the case of free vibration of the string.

Let the two ends of a stretched string of length L be fixed at the points $x = 0$ and $x = L$. Let ρ be the mass per unit length of the string. We make the same assumptions as in the case of free vibrations of the strethed string. Let the vibrations occur in a vertical plane.

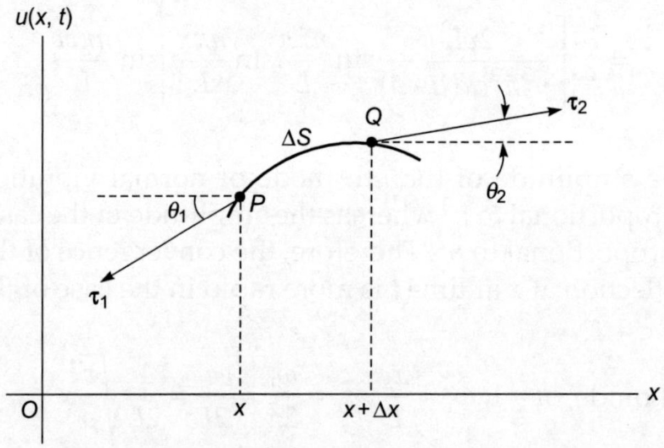

Fig. 12.6 Vibrations of a stretched string

Let us consider the force acting on a small section (Δs) of the string vibrating in the vertical plane. Let $\boldsymbol{\tau}_1$ and $\boldsymbol{\tau}_2$ be the tensions at the endpoints, P and Q, of the section. Let $\boldsymbol{\tau}_1$ and $\boldsymbol{\tau}_2$ make angles θ_1 and θ_2 with the undisturbed (horizontal) length of the string. Since there is no motion in the horizontal plane along the string, the horizontal components of the tension should cancel out.

Therefore, we have

$$\tau_2 \cos \theta_2 - \tau_1 \cos \theta_1 = 0$$

or

$$\tau_2 \cos \theta_2 = \tau_1 \cos \theta_1 = \tau \text{ (constant)} \tag{12.59}$$

where $\tau_1 = |\boldsymbol{\tau}_1|$ etc.

If $q(x, t)$ is the force per unit length applied externally to the section, then the magnitude of the net vertical force acting on the element (Δs) of the string is

$$\tau_2 \sin \theta_2 - \tau_1 \sin \theta_1 + q(x, t)\, \Delta s.$$

Since $\Delta s = \Delta x \left[1+\left(\frac{\Delta u}{\Delta x}\right)^2\right]^{\frac{1}{2}} \cong \Delta x$ the above vertical force acting on the element (Δs) becomes equal to

$$\tau_2 \sin \theta_2 - \tau_1 \sin \theta_1 + q(x, t)\, \Delta x.$$

Also, the mass of the section (Δs) of the string between x and $(x + \Delta s)$ is $\rho\Delta s \cong \rho\, \Delta x$. By Newton's second law of motion, the force acting on this section is

$$(\rho\, \Delta x)\frac{\partial^2 u(x,t)}{\partial t^2}.$$

Thus, we write

$$\tau_2 \sin\theta_2 - \tau_1 \sin\theta_1 + q(x,t)\Delta x = (\rho\,\Delta x)\frac{\partial^2 u(x,t)}{\partial t^2}. \tag{12.60}$$

Dividing Eq. (12.60) by Eq. (12.59), we obtain

$$\frac{\tau_2 \sin\theta_2}{\tau_2 \cos\theta_2} - \frac{\tau_1 \sin\theta_1}{\tau_1 \cos\theta_1} + \frac{q}{\tau}\Delta x = \frac{\rho\,\Delta x}{\tau}\frac{\partial^2 u(x,t)}{\partial t^2}$$

or
$$\tan\theta_2 - \tan\theta_1 + \frac{q}{\tau}\Delta x = \frac{\rho\,\Delta x}{\tau}\frac{\partial^2 u(x,t)}{\partial t^2}$$

or
$$u_x(x+\Delta x, t) - u_x(x,t) + \frac{q}{\tau}\Delta x = \frac{\rho\,\Delta x}{\tau}\frac{\partial^2 u(x,t)}{\partial t^2}$$

or
$$\frac{1}{\Delta x}\left[\left\{\frac{\partial u(x,t)}{\partial x}\right\}_{x+\Delta x} - \left\{\frac{\partial u(x,t)}{\partial x}\right\}_{x}\right] + \frac{q}{\tau} = \frac{\rho}{\tau}\frac{\partial^2 u(x,t)}{\partial t^2}$$

In the limit $\Delta x \to 0$, we have

$$\lim_{\Delta x\to 0}\frac{1}{\Delta x}\left[\left\{\frac{\partial u(x,t)}{\partial x}\right\}_{x+\Delta x} - \left\{\frac{\partial u(x,t)}{\partial x}\right\}_{x}\right] + \frac{q}{\tau} = \frac{\rho}{\tau}\frac{\partial^2 u(x,t)}{\partial t^2}$$

or
$$\frac{\partial^2 u(x,t)}{\partial x^2} + \frac{q}{\rho c^2} = \frac{1}{c^2}\frac{\partial^2 u(x,t)}{\partial t^2} \tag{12.61}$$

where $c = \sqrt{\dfrac{\tau}{\rho}}$ represents the constant speed of the propagation of the wave.

Here, c is the speed with which a disturbance is propagated along the string, that is, it is the speed of the transverse wave produced. Equation (12.61) represents the equation for the forced transverse vibrations of a stretched string. The analysis of the forced vibration problem is somewhat simpler when $q(x,t)$ is either a function of x only or a function of t only.

12.11 WAVE EQUATION FOR FREE VIBRATIONS OF A STRETCHED RECTANGULAR MEMBRANE (TWO-DIMENSIONAL CLASSICAL WAVE EQUATION IN CARTESIAN COORDINATES)

To derive the two-dimensional wave equation that can describe the free transverse vibrations of a streched rectangular elastic membrane, we make the following assumptions.

1. The mass per unit area of the membrane is constant, i.e., the membrane is homogeneous.
2. The membrane is perfectly flexible and so thin that it does not offer any resistance to bending.

3. The magnitude of tension per unit length τ caused by stretching the membrane is the same at all points and in all directions and does not change during motion.
4. In the rest position, the membrane is plane.
5. During vibrational motion, the deflection, $u(x,y,t)$, of the membrane is small compared to the size of the membrane, and all angles of inclination are small.

Let a thin rectangular membrane be stretched and then fixed along its entire boundary in the xy-plane. Let a free vibration of the membrane occur in the vertical direction. We are to find a differential equation for the deflection $u(x, y, t)$ at any point (x, y) at any time t. Let us consider an infinitesimal rectangular section of the membrane with sides Δx and Δy If τ is the tensile force per unit length, then the forces acting on the sides of the element are $\tau\Delta x$ and $\tau\Delta y$ as shown in Figure 12.7. Here, $\tau = |\boldsymbol{\tau}|$. Since the membrane is assumed to be perfectly elastic and flexible, these forces are tangent to the moving membrane at every instant.

Horizontal Components of the Forces

The horizontal components are obtained by multiplying the magnitudes of the forces by the cosines of the angles of inclination. Since these angles are small, their cosines are close to unity and hence the horizontal components of the forces at opposite sides are approximately equal. Therefore, the motion of the particles of the membrane in a horizontal direction will be negligibly small and hence it is neglected. From this we conclude that we can regard the motion of the membrane as purely transversal; that is, all particles move in vertical planes.

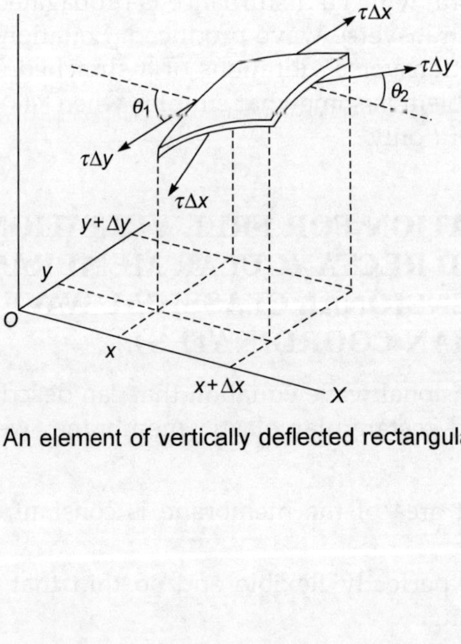

Fig. 12.7(a) An element of vertically deflected rectangular membrane

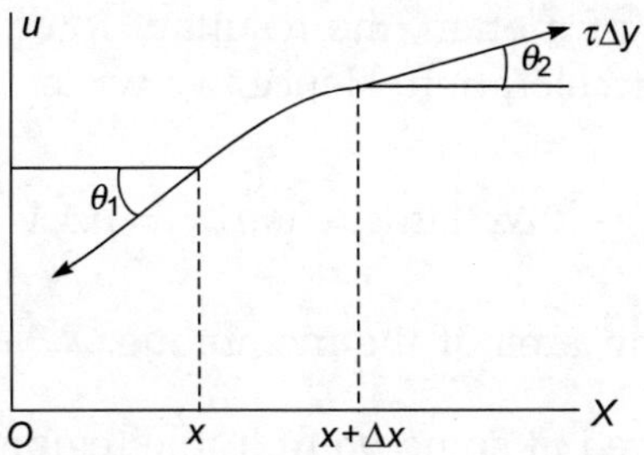

Fig. 12.7(b) Forces acting on the element in the *xu*-plane

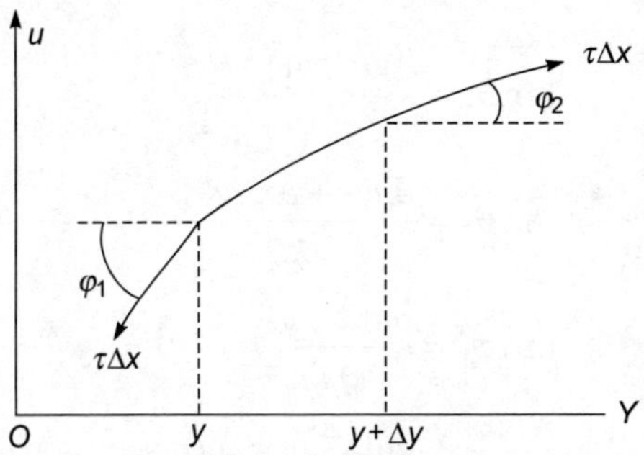

Fig. 12.7(c) Forces acting on the element in the *yu*-plane

Vertical Components of the Forces

The vertical components along the left edge and the right edge of the element are respectively,

$$-\tau\Delta y \sin\theta_1 \text{ and } \tau\Delta y \sin\theta_2$$

where θ_1 and θ_2 are the angles of inclination in the middle of the left and the right edges of the element.

Similarly, the vertical components along the back edge and the front edge of the element are respectively,

$$-\tau\Delta x \sin\varphi_1 \text{ and } \tau\Delta x \sin\varphi_2$$

where φ_1 and φ_2 are the angles of inclination in the middle of the back and the front edges of the element.

Thus, the resultant of the vertical components of forces acting on the rectangular element becomes

$$\tau\,\Delta y \sin\theta_2 - \tau\,\Delta y \sin\theta_1 + \tau\,\Delta x \sin\varphi_2 - \tau\,\Delta x \sin\varphi_1.$$

Since the slopes are small, the sines of the angles are approximately equal to their tangents and, therefore, the resultant of the vertical components of the forces acting on the element becomes

$$\tau\,\Delta y\left(\tan\theta_2 - \tan\theta_1\right) + \tau\,\Delta x\left(\tan\varphi_2 - \tan\varphi_1\right).$$

By Newton's second law of motion, the resultant force acting on the element is equal to its mass times the acceleration. Hence, we write

$$\tau\,\Delta y(\tan\theta_2 - \tan\theta_1) + \tau\,\Delta x(\tan\varphi_2 - \tan\varphi_1) = \rho\Delta A \frac{\partial^2 u(x,y,t)}{\partial t^2} \tag{12.62}$$

where ρ is the mass per unit area of the membrane, $\Delta A \cong \Delta x \Delta y$ is the area of the element, and $\dfrac{\partial^2 u}{\partial t^2}$ is computed at some point in the region under consideration.

From differential calculus, we know that

$$\tan\theta_1 = \frac{\partial u(x, y_1)}{\partial x}$$

$$\tan\theta_2 = \frac{\partial u(x+\Delta x, y_2)}{\partial x}$$

$$\tan\varphi_1 = \frac{\partial u(x_1, y)}{\partial y}$$

and

$$\tan\varphi_2 = \frac{\partial u(x_2, y+\Delta y)}{\partial y}$$

where x_1 and x_2 are values of x between x and $x+\Delta x$ and y_1 and y_2 are the values of y between y and $y+\Delta y$.

Substituting these relations in Eq. (12.62), we obtain

$$\tau\,\Delta y\left[\frac{\partial}{\partial x}u(x+\Delta x, y_2) - \frac{\partial}{\partial x}u(x, y_1)\right]$$

$$+\tau\,\Delta x\left[\frac{\partial}{\partial y}u(x_2, y+\Delta y) - \frac{\partial}{\partial y}u(x_1, y)\right] = \rho\,\Delta x\,\Delta y\frac{\partial^2 u(x,y,t)}{\partial t^2}.$$

Dividing both sides by $\rho\,\Delta x\,\Delta y$, we obtain

$$\frac{\tau}{\rho}\frac{1}{\Delta x}\left[\frac{\partial}{\partial x}u(x+\Delta x, y_2) - \frac{\partial}{\partial x}u(x, y_1)\right]$$

$$+\frac{\tau}{\rho}\frac{1}{\Delta y}\left[\frac{\partial}{\partial y}u(x_2, y+\Delta y) - \frac{\partial}{\partial y}u(x_1, y)\right] = \frac{\partial^2 u(x,y,t)}{\partial t^2}$$

If we let Δx and Δy approach zero, we obtain

$$\frac{\tau}{\rho}\left(\frac{\partial^2 u}{\partial x^2} + \frac{\partial^2 u}{\partial y^2}\right) = \frac{\partial^2 u}{\partial t^2}$$

or
$$\left(\frac{\partial^2 u}{\partial x^2}+\frac{\partial^2 u}{\partial y^2}\right)=\frac{1}{c^2}\frac{\partial^2 u}{\partial t^2} \tag{12.63}$$

where $c=\sqrt{\frac{\tau}{\rho}}$ is the speed of the transverse wave on the surface of the membrane. Equation (12.63) gives the deflection at a point (x, y) at time t in the free vibrations of a retangular membrane and this equation is called the two-dimensional wave equation. If there is an external force (q) per unit area acting on the membrane, the above equation takes the form

$$\left(\frac{\partial^2 u}{\partial x^2}+\frac{\partial^2 u}{\partial y^2}\right)+\frac{q}{\rho c^2}=\frac{1}{c^2}\frac{\partial^2 u}{\partial t^2}. \tag{12.64}$$

12.12 GENERAL SOLUTION OF WAVE EQUATION FOR FREE VIBRATIONS OF A STRETCHED RECTANGULAR MEMBRANE (TWO-DIMENSIONAL CLASSICAL WAVE EQUATION IN CARTESIAN COORDINATES)

Let us determine a solution of the PDE describing free transverse vibrations of a rectangular membrane. Let a rectangular membrane of length a and width b be stretched by a tension of magnitude τ and let it be kept fixed along its boundary. Let the membrane undergo free transverse vibrations with small amplitude. The deflection $u(x, y, t)$ of a point (x, y) of the membrane at time t satisfies the PDE given by

$$\frac{\partial^2 u}{\partial x^2}+\frac{\partial^2 u}{\partial y^2}=\frac{1}{c^2}\frac{\partial^2 u}{\partial t^2}\ (0 \le x \le a,\ 0 \le y \le b,\ t \ge 0) \tag{12.65}$$

where c is the speed of wave propagation.

Boundary Condition

The boundary condition of this problem is $u=0$ on the boundary, that is,

$$u(0,y,t)=u(a,y,t)=0, \tag{12.66}$$

$$u(x,0,t)=u(x,b,t)=0. \tag{12.67}$$

Initial Conditions

The initial conditions of the problem are

(i) Initial deflection of point (x, y) is $u(x, y, 0) = f(x, y)$ (12.68) and (ii) Initial speed of point (x, y) is $\left(\frac{\partial u}{\partial t}\right)_{t=0} = u_t(x, y, 0) = g(x, y)$. (12.69)

Seeking a General Solution of the Problem

Using the method of separation of variables, we assume the solution of Eq. (12.65) as

$$u(x, y, t) = S(x, y)\, T(t) \tag{12.70}$$

where $S(x, y)$ is a spatial function of x and y and $T(t)$ is a temporal function of t only.

Substituting Eq. (12.70) into Eq. (12.65), we obtain

$$T\left(\frac{\partial^2 S}{\partial x^2} + \frac{\partial^2 S}{\partial y^2}\right) = \frac{S}{c^2}\frac{d^2 T}{dt^2}.$$

Dividing both sides by the product ST, we obtain

$$\frac{1}{S}\left(\frac{\partial^2 S}{\partial x^2} + \frac{\partial^2 S}{\partial y^2}\right) = \frac{1}{c^2 T}\frac{d^2 T}{dt^2}$$

In the above equation, the expression on the left-hand side is a function of x and y whereas the expression on the right-hand side is a function of t only. Therefore, their equality demands that the expression on each side is equal to the same constant. Hence, we write

$$\frac{1}{S}\left(\frac{\partial^2 S}{\partial x^2} + \frac{\partial^2 S}{\partial y^2}\right) = \frac{1}{c^2 T}\frac{d^2 T}{dt^2} = -\lambda^2 \text{ (a constant)}$$

which yields the two equations

$$\frac{\partial^2 S}{\partial x^2} + \frac{\partial^2 S}{\partial y^2} + \lambda^2 S = 0 \tag{12.71}$$

and

$$\frac{d^2 T}{dt^2} + \omega^2 T = 0 \tag{12.72}$$

where

$$\omega = \lambda c. \tag{12.73}$$

Equation (12.71) is the two-dimensional form of the famous ***Helmholtz equation***

$$\nabla^2 S + \lambda^2 S = 0 \text{ (where } \lambda^2 \text{ is a constant)}$$

This equation is named after the German physicist, ***Hermann Ludwig Ferdinand von Helmholtz*** (31 August 1821 - 8 September 1894)].

As S is a function of x and y, Eq. (12.71) needs further separation of variables. Therefore, let us put

$$S(x, y) = X(x)\, Y(y). \tag{12.74}$$

Using this, we write Eq. (12.71) as

$$Y\frac{d^2X}{dx^2}+X\frac{d^2Y}{dy^2}+\lambda^2XY=0.$$

Dividing both sides by XY and then transposing, we have

$$\frac{1}{X}\frac{d^2X}{dx^2}=-\left(\frac{1}{Y}\frac{d^2Y}{dy^2}+\lambda^2\right).$$

In the above equation, the left-hand side is function of x only whereas the right-hand side is of y only and hence each of them must be equal to a constant say, $-k^2$

That is, $$\frac{1}{X}\frac{d^2X}{dx^2}=-\left(\frac{1}{Y}\frac{d^2Y}{dy^2}+\lambda^2\right)=-k^2$$

Thus, we obtain the following two equations.

$$\frac{d^2X}{dx^2}+k^2X=0, \tag{12.75}$$

and $$\frac{d^2Y}{dy^2}+p^2Y=0 \tag{12.76}$$

where we put $$p^2=\lambda^2-k^2. \tag{12.77}$$

General solutions of Eqs. (12.75) and (12.76) are respectively

$$X(x) = A\cos kx + B\sin kx \tag{12.78}$$

and $$Y(y) = C\cos py + D\sin py \tag{12.79}$$

where A, B, C, D are arbitrary constants.

Application of Boundary Conditions

On the boundaries, $x = 0, x = a, y = 0, y = b$; we know that $u = ST$ and hence $S = XY$ is always zero at the boundaries for all t. Therefore, we write

$$X(0) = 0,\ X(a) = 0,\ Y(0) = 0,\ Y(b) = 0.$$

Thus, we obtain

$$X(0) = A = 0 \text{ and } X(a) = B\sin ka = 0.$$

As B is an arbitrary constant, we must have

$$ka = m\pi \text{ or}, k=\frac{m\pi}{a} \text{ where } m = 1, 2, 3, 4 \ldots \tag{12.80}$$

Here, $m = 0$ is excluded because it gives a trivial solution.
Also, $m = -1, -2, -3 \ldots$ are neglected as they do not yield new solutions.
In a similar fashion, we obtain

$$Y(0) = C = 0 \text{ and } Y(b) = D \sin pb = 0$$

yielding
$$p = \frac{n\pi}{b} \text{ where } n = 1, 2, 3, 4 \ldots \tag{12.81}$$

In this way, application of boundary conditions to Eqs. (12.78) and (12.79) and introduction of the subscripts (m and n) to B, D, X and Y respectively results in the solutions

$$X_m(x) = B_m \sin\frac{m\pi x}{a} \quad m = 1, 2, 3, 4 \ldots \tag{12.82}$$

and
$$Y_n(y) = D_n \sin\frac{p\pi y}{b} \quad n = 1, 2, 3, 4 \ldots \tag{12.83}$$

Using Eq. (12.74), we write the solution of the spatial Eq. (12.71) as

$$S_{mn}(x, y) = A_{mn} \sin\frac{m\pi x}{a} \sin\frac{n\pi y}{b} \quad (m = 1, 2, 3 \ldots, n = 1, 2, 3 \ldots) \tag{12.84}$$

where $A_{mn} = B_m D_n$.

Now, from Eq. (12.77), we have

$$\lambda^2 = k^2 + p^2.$$

For particular m and n, using Eqs. (12.80) and (12.81), we have

$$\lambda_{mn} = \sqrt{\left(\frac{m\pi}{a}\right)^2 + \left(\frac{n\pi}{b}\right)^2} = \pi\sqrt{\frac{m^2}{a^2} + \frac{n^2}{b^2}}\,. \tag{12.85}$$

Also, for particular m and n, using Eqs. (12.73) and (12.85), we have

$$\omega_{mn} = \pi c\sqrt{\frac{m^2}{a^2} + \frac{n^2}{b^2}}\,. \tag{12.86}$$

We know that the temporal equation, Eq. (12.72), has the general solution

$$T_{mn}(t) = M_{mn} \cos \omega_{mn}t + N_{mn} \sin \omega_{mn}t \tag{12.87}$$

where M_{mn} and N_{mn} are arbitrary constants and ω_{mn} are given in Eq. (12.86).

Using Eqs. (12.70), (12.84) and (12.87), we write the solutions of Eq. (12.65) as

$$u_{mn}(x, y, t) = (M_{mn} \cos\omega_{mn}t + N_{mn} \sin\omega_{mn}t) A_{mn} \sin\frac{m\pi x}{a} \sin\frac{n\pi y}{b}$$

$$= (\alpha_{mn} \cos\omega_{mn}t + \beta_{mn} \sin\omega_{mn}t) \sin\frac{m\pi x}{a} \sin\frac{n\pi y}{b} \tag{12.88}$$

where $\quad \alpha_{mn} = M_{mn}A_{mn}, \quad \beta_{mn} = N_{mn}A_{mn},$
$\quad m = 1, 2, 3 \ldots, \quad n = 1, 2, 3 \ldots$

Here, u_{mn} are called the *eigenfunctions* or *characteristic functions* and ω_{mn} are called the *eigenvalues* or *characteristic values* of the vibrating membrane. In this problem, ω_{mn} cannot be called circular frequencies but the quantity $\frac{\omega_{mn}}{2\pi}$ represents ordinary frequencies.

By the superposition principle, we write the general solution of Eq. (12.65) as

$$u(x, y, t) = \sum_{m=1}^{\infty}\sum_{n=1}^{\infty} u_{mn}(x, y, t)$$

$$= \sum_{m=1}^{\infty}\sum_{n=1}^{\infty} (\alpha_{mn}\cos\omega_{mn}t + \beta_{mn}\sin\omega_{mn}t)\sin\frac{m\pi x}{a}\sin\frac{n\pi y}{b} \tag{12.89}$$

where α_{mn} and β_{mn} are to be determined from the initial conditions of the problem.

To Determine the Coefficients α_{mn} and β_{mn}

Using the first initial condition shown in Eq. (12.68), Eq. (12.89) yields

$$u(x, y, 0) = f(x, y) = \sum_{m=1}^{\infty}\sum_{n=1}^{\infty} \alpha_{mn}\sin\frac{m\pi x}{a}\sin\frac{n\pi y}{b}.$$

Multiplying both sides by $\sin\frac{r\pi x}{a}$ and then integrating the results with respect to x within the limit from $x = 0$ to $x = a$, we obtain

$$\int_0^a f(x, y)\sin\frac{r\pi x}{a}dx = \sum_{n=1}^{\infty}\alpha_{rn}\sin\frac{n\pi y}{b}\int_0^a \sin^2\left(\frac{r\pi x}{a}\right)dx$$

$$= \frac{a}{2}\sum_{n=1}^{\infty}\alpha_{rn}\sin\frac{n\pi y}{b}.$$

Again, multiplying both sides by $\sin\frac{s\pi y}{b}$ and then integrating the results with respect to y within the limit from $y = 0$ to $y = b$, we obtain

$$\int_{x=0}^{a}\int_{y=0}^{b} f(x, y)\sin\frac{r\pi x}{a}\sin\frac{s\pi y}{b}dx\,dy = \frac{a}{2}\alpha_{rs}\int_0^b \sin^2\left(\frac{s\pi y}{b}\right)dy$$

$$= \alpha_{rs}\frac{a}{2}\frac{b}{2}$$

or
$$\alpha_{rs} = \frac{4}{ab}\int_{x=0}^{a}\int_{y=0}^{b} f(x,y)\sin\frac{r\pi x}{a}\sin\frac{s\pi y}{b}\,dx\,dy.$$

Renaming the subscripts, we write

$$\alpha_{mn} = \frac{4}{ab}\int_{x=0}^{a}\int_{y=0}^{b} f(x,y)\sin\frac{m\pi x}{a}\sin\frac{n\pi y}{b}\,dx\,dy. \tag{12.90}$$

Using the second initial condition shown in Eq. (12.69), Eq. (12.89) yields

$$g(x,y) = \left(\frac{\partial u}{\partial t}\right)_{t=0} = \sum_{m=1}^{\infty}\sum_{n=1}^{\infty}\omega_{mn}\beta_{mn}\sin\frac{m\pi x}{a}\sin\frac{n\pi y}{b}.$$

Multiplying both sides by $\sin\frac{r\pi x}{a}$ and then integrating the results with respect to x within the limit from $x = 0$ to $x = a$, we obtain

$$\int_0^a g(x,y)\sin\frac{r\pi x}{a}\,dx = \sum_{n=1}^{\infty}\omega_{rn}\beta_{rn}\sin\frac{n\pi y}{b}\int_0^a \sin^2\left(\frac{r\pi x}{a}\right)dx$$

$$= \frac{a}{2}\sum_{n=1}^{\infty}\omega_{rn}\beta_{rn}\sin\frac{n\pi y}{b}.$$

Again, multiplying both sides by $\sin\frac{s\pi y}{b}$ and then integrating the results with respect to y within the limit from $y = 0$ to $y = b$, we obtain

$$\int_{x=0}^{a}\int_{y=0}^{b} g(x,y)\sin\frac{r\pi x}{a}\sin\frac{s\pi y}{b}\,dx\,dy = \frac{a}{2}\omega_{rs}\beta_{rs}\int_0^b \sin^2\left(\frac{s\pi y}{b}\right)dy$$

$$= \frac{ab}{4}\omega_{rs}\beta_{rs}$$

or
$$\beta_{rs} = \frac{4}{ab\,\omega_{rs}}\int_{x=0}^{a}\int_{y=0}^{b} g(x,y)\sin\frac{r\pi x}{a}\sin\frac{s\pi y}{b}\,dx\,dy.$$

Renaming the subscripts, we write

$$\beta_{mn} = \frac{4}{ab\,\omega_{mn}}\int_{x=0}^{a}\int_{y=0}^{b} g(x,y)\sin\frac{m\pi x}{a}\sin\frac{n\pi y}{b}\,dx\,dy. \tag{12.91}$$

Thus, Eq. (12.89) together with Eqs. (12.90) and (12.91) gives the required solution of the problem. It is noteworthy that the loci of points which do not move during the vibrations of the membrane are called ***nodal lines***.

Nodal lines of some solutions are illustrated in Figure 12.8.

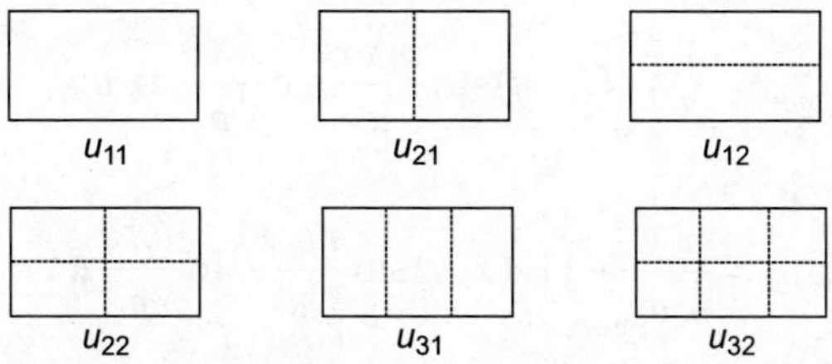

Fig. 12.8 Nodal lines for some modes of vibrations of rectangular membrane

12.13 VIBRATING SQUARE MEMBRANE

Let us consider the vibrations of a very thin square membrane of sides a fixed along the boundary.

The differential equation of the problem is

$$\frac{\partial^2 u}{\partial x^2}+\frac{\partial^2 u}{\partial y^2}=\frac{1}{c^2}\frac{\partial^2 u}{\partial t^2} \quad (0 \le x \le a,\ 0 \le y \le a,\ t \ge 0)$$

where c is the constant speed of wave propagation on the membrane.

This problem is similar to that of the rectangular membrane where we are to put $a = b$.

Boundary Condition

The boundary condition of this problem is $u = 0$ on the boundary, that is

$$u(0, y, t) = u(a, y, t) = 0,$$
$$u(x, 0, t) = u(x, a, t) = 0.$$

Initial Conditions

The initial conditions of the problem are:

(i) Initial displacement of point (x, y) is $u(x, y, 0) = f(x, y)$

and (ii) Initial speed of point (x, y) is $\left(\frac{\partial u}{\partial t}\right)_{t=0} = u_t(x,y,0) = g(x, y)$.

General Solution of the Problem

Following the same procedure described in the case of rectangular membrane, we obtain the solutions

$$u_{mn}(x,y,t)=(\alpha_{mn}\cos\omega_{mn}t+\beta_{mn}\sin\omega_{mn}t)\sin\frac{m\pi x}{a}\sin\frac{n\pi y}{a}$$

with $m = 1, 2, 3 \ldots, n = 1, 2, 3 \ldots;$

$$\alpha_{mn} = \frac{4}{a^2}\int_0^a\int_0^a f(x,y)\sin\frac{m\pi x}{a}\sin\frac{n\pi y}{a}\,dx\,dy,$$

$$\beta_{mn} = \frac{4}{a^2\omega_{mn}}\int_0^a\int_0^a g(x,y)\sin\frac{m\pi x}{a}\sin\frac{n\pi y}{a}\,dx\,dy$$

and $$\omega_{mn} = \frac{\pi c}{a}\sqrt{m^2+n^2}.$$

Here, we see that $\omega_{mn} = \omega_{nm}$.

However, the corresponding eigenfunctions (u_{mn} and u_{nm}) are not equal for $m \neq n$ since

$$\sin\frac{m\pi x}{a}\sin\frac{n\pi y}{a} \neq \sin\frac{n\pi x}{a}\sin\frac{m\pi y}{a}.$$

For $\omega_{11} = \frac{\pi c}{a}\sqrt{2}$, we find a solution

$$u_{11} = \left[\alpha_{11}\cos\left(\frac{\pi ct}{a}\sqrt{2}\right) + \beta_{11}\sin\left(\frac{\pi ct}{a}\sqrt{2}\right)\right]\sin\frac{\pi x}{a}\sin\frac{\pi y}{a}. \tag{12.92}$$

This solution yields no nodal line.

For $\omega_{12} = \omega_{21} = \frac{\pi c}{a}\sqrt{5}$, the solutions

$$u_{12} = \left[\alpha_{12}\cos\left(\frac{\pi ct}{a}\sqrt{5}\right) + \beta_{12}\sin\left(\frac{\pi ct}{a}\sqrt{5}\right)\right]\sin\frac{\pi x}{a}\sin\frac{2\pi y}{a} \tag{12.93}$$

and $$u_{21} = \left[\alpha_{21}\cos\left(\frac{\pi ct}{a}\sqrt{5}\right) + \beta_{21}\sin\left(\frac{\pi ct}{a}\sqrt{5}\right)\right]\sin\frac{2\pi x}{a}\sin\frac{\pi y}{a} \tag{12.94}$$

are two different solutions and their sum is yet another different solution.

Here, u_{12} shows a nodal line at $y = \frac{a}{2}$ and u_{21} shows a nodal line at $x = \frac{a}{2}$.

For $\omega_{22} = \frac{2\pi c}{a}\sqrt{2}$, u_{22} is a solution showing nodal lines at $x = \frac{a}{2}$ and $y = \frac{a}{2}$.

For $\omega_{13} = \omega_{31} = \frac{\pi c}{a}\sqrt{10}$, we have three solutions, namely, u_{13}, u_{31} and their sum.

The solution u_{13} has nodal lines at $y = \frac{a}{3}, \frac{2a}{3}$ and the solution u_{31} has nodal lines at $x = \frac{a}{3}, \frac{2a}{3}$.

We see that even more than two functions may correspond to the same numerical value of ω_{mn}. For example, the four functions u_{18}, u_{81}, u_{47}, u_{74} correspond to the value

$$\omega_{18} = \omega_{81} = \omega_{47} = \omega_{74} = \frac{\pi v}{a}\sqrt{65}$$

since $$1^2 + 8^2 = 8^2 + 1^2 = 4^2 + 7^2 = 7^2 + 4^2 = 65.$$

12.14 WAVE EQUATION FOR TRANSVERSE VIBRATIONS OF A CIRCULAR MEMBRANE (TWO-DIMENSIONAL CLASSICAL WAVE EQUATION IN PLANE POLAR CO-ORDINATES)

We know that in two-dimensional Cartesian (rectangular) coordinates, the wave equation is written as

$$\nabla^2 u = \frac{\partial^2 u}{\partial x^2} + \frac{\partial^2 u}{\partial y^2} = \frac{1}{c^2}\frac{\partial^2 u}{\partial t^2} \tag{12.95}$$

where $u = u(x, y, t)$ is the deflection at a point (x, y) at time t and c is the speed of wave propagation.

Let us introduce plane polar coordinates (r, θ) defined by $x = r\cos\theta$, $y = r\sin\theta$ such that

$$r = \sqrt{x^2 + y^2} \text{ and } \theta = \text{arc tan}\frac{y}{x}. \tag{12.96}$$

We know that $\dfrac{\partial u}{\partial x} = \dfrac{\partial u}{\partial r}\dfrac{\partial r}{\partial x} + \dfrac{\partial u}{\partial \theta}\dfrac{\partial \theta}{\partial x}$.

Also, we have $\dfrac{\partial^2 u}{\partial x^2} = \dfrac{\partial}{\partial x}\left(\dfrac{\partial u}{\partial r}\dfrac{\partial r}{\partial x}\right) + \dfrac{\partial}{\partial x}\left(\dfrac{\partial u}{\partial \theta}\dfrac{\partial \theta}{\partial x}\right)$

$$= \left[\frac{\partial}{\partial x}\left(\frac{\partial u}{\partial r}\right)\right]\frac{\partial r}{\partial x} + \frac{\partial u}{\partial r}\frac{\partial^2 r}{\partial x^2} + \left[\frac{\partial}{\partial x}\left(\frac{\partial u}{\partial \theta}\right)\right]\frac{\partial \theta}{\partial x} + \frac{\partial u}{\partial \theta}\frac{\partial^2 \theta}{\partial x^2}$$

$$= \left[\left\{\frac{\partial}{\partial r}\left(\frac{\partial u}{\partial r}\right)\right\}\frac{\partial r}{\partial x} + \left\{\frac{\partial}{\partial \theta}\left(\frac{\partial u}{\partial r}\right)\right\}\frac{\partial \theta}{\partial x}\right]\frac{\partial r}{\partial x} + \frac{\partial u}{\partial r}\frac{\partial^2 r}{\partial x^2}$$

$$+ \left[\left\{\frac{\partial}{\partial r}\left(\frac{\partial u}{\partial \theta}\right)\right\}\frac{\partial r}{\partial x} + \left\{\frac{\partial}{\partial \theta}\left(\frac{\partial u}{\partial \theta}\right)\right\}\frac{\partial \theta}{\partial x}\right]\frac{\partial \theta}{\partial x} + \frac{\partial u}{\partial \theta}\frac{\partial^2 \theta}{\partial x^2}$$

$$= \left[\left(\frac{\partial^2 u}{\partial r^2}\right)\frac{\partial r}{\partial x} + \left(\frac{\partial^2 u}{\partial r \partial \theta}\right)\frac{\partial \theta}{\partial x}\right]\frac{\partial r}{\partial x} + \frac{\partial u}{\partial r}\frac{\partial^2 r}{\partial x^2}$$

$$+ \left[\left(\frac{\partial^2 u}{\partial r \partial \theta}\right)\frac{\partial r}{\partial x} + \frac{\partial^2 u}{\partial \theta^2}\frac{\partial \theta}{\partial x}\right]\frac{\partial \theta}{\partial x} + \frac{\partial u}{\partial \theta}\frac{\partial^2 \theta}{\partial x^2}. \quad (12.97)$$

From Eq. (12.96), we have

$$\frac{\partial r}{\partial x} = \frac{1}{2}(x^2 + y^2)^{-\frac{1}{2}}(2x) = \frac{x}{r},$$

$$\frac{\partial \theta}{\partial x} = \frac{1}{1 + \left(\frac{y}{x}\right)^2}\left(-\frac{y}{x^2}\right) = -\frac{y}{r^2}.$$

Then, we have

$$\frac{\partial^2 r}{\partial x^2} = \frac{r - x\frac{\partial r}{\partial x}}{r^2} = \frac{r - x\frac{x}{r}}{r^2} = \frac{r^2 - x^2}{r^3} = \frac{y^2}{r^3},$$

$$\frac{\partial^2 \theta}{\partial x^2} = \frac{\partial}{\partial x}\left(-\frac{y}{r^2}\right) = -y(-2r^{-3})\frac{\partial r}{\partial x} = \frac{2xy}{r^4}.$$

Substituting these expressions of the derivatives into Eq. (12.97), we have

$$\frac{\partial^2 u}{\partial x^2} = \frac{x^2}{r^2}\frac{\partial^2 u}{\partial r^2} - 2\frac{xy}{r^3}\frac{\partial^2 u}{\partial r \partial \theta} + \frac{y^2}{r^3}\frac{\partial u}{\partial r} + \frac{y^2}{r^4}\frac{\partial^2 u}{\partial \theta^2} + \frac{2xy}{r^4}\frac{\partial u}{\partial \theta}. \quad (12.98)$$

In a similar fashion, we have

$$\frac{\partial^2 u}{\partial y^2} = \frac{y^2}{r^2}\frac{\partial^2 u}{\partial r^2} + 2\frac{xy}{r^3}\frac{\partial^2 u}{\partial r \partial \theta} + \frac{x^2}{r^3}\frac{\partial u}{\partial r} + \frac{x^2}{r^4}\frac{\partial^2 u}{\partial \theta^2} - \frac{2xy}{r^4}\frac{\partial u}{\partial \theta}. \quad (12.99)$$

Adding Eqs. (12.98) and (12.99), we obtain

$$\nabla^2 u = \frac{\partial^2 u}{\partial x^2} + \frac{\partial^2 u}{\partial y^2} = \frac{\partial^2 u}{\partial r^2} + \frac{1}{r}\frac{\partial u}{\partial r} + \frac{1}{r^2}\frac{\partial^2 u}{\partial \theta^2}.$$

Substituting this equation into Eq. (12.95), we obtain

$$\frac{\partial^2 u}{\partial r^2} + \frac{1}{r}\frac{\partial u}{\partial r} + \frac{1}{r^2}\frac{\partial^2 u}{\partial \theta^2} = \frac{1}{c^2}\frac{\partial^2 u}{\partial t^2}. \quad (12.100)$$

Equation (12.100) represents two-dimensional classical wave equation in plane polar coordinates.

12.15 TO FIND THE DEFLECTION OF A POINT IN A VIBRATING FINITE CIRCULAR MEMBRANE FIXED ALONG THE BOUNDARY

CASE I: Symmetric Case

Let us consider a uniform circular membrane of radius a fixed along the entire boundary, i.e., along its circumference. Let this circular membrane undergo transverse vibrations having a radial symmetry about its centre. Choosing the centre as the origin of plane polar coordinates, the deflection $u(r, t)$ at a point distant r at time t is given by the partial differential equation

$$\frac{\partial^2 u}{\partial r^2}+\frac{1}{r}\frac{\partial u}{\partial r}=\frac{1}{c^2}\frac{\partial^2 u}{\partial t^2} \tag{12.101}$$

where c is the constant speed of propagation of the wave produced.

In the above equation, there should be no term depending on the angular co-ordinate θ as the vibrations have radial symmetry.

Boundary Condition

Boundary condition of the problem is $u(a, t) = 0$ for all $t \geq 0$. (12.102)

Initial Conditions

The initial conditions of the problem are:

(i) $u(r, 0) = f(r)$ (initial deflection),

(ii) $u_t(r, 0) = \left(\frac{\partial u}{\partial t}\right)_{t=0} = g(r)$ (initial speed). (12.103)

Seeking the Solution of the Problem

Using the method of separation of variables, let us assume the solution as

$$u(r, t) = R(r)\,T(t) \tag{12.104}$$

where R is function of r only and T is function of t only.

Substituting Eq. (12.104) into Eq. (12.101), we obtain

$$T\frac{d^2R}{dr^2}+\frac{T}{r}\frac{dR}{dr}=\frac{R}{c^2}\frac{d^2T}{dt^2}.$$

Dividing both sides by $R(r)\,T(t)$, we obtain

$$\frac{1}{R}\frac{d^2R}{dr^2}+\frac{1}{rR}\frac{dR}{dr}=\frac{1}{c^2 T}\frac{d^2T}{dt^2}=-k^2 \text{ (a constant).}$$

This leads to the two linear homogeneous ODEs

$$\frac{d^2R}{dr^2}+\frac{1}{r}\frac{dR}{dr}+k^2R=0 \tag{12.105}$$

and

$$\frac{d^2T}{dt^2}+\omega^2T=0 \tag{12.106}$$

where

$$\omega=ck. \tag{12.107}$$

If we multiply both sides of Eq. (12.105) by r^2, we obtain

$$r^2\frac{d^2R}{dr^2}+r\frac{dR}{dr}+k^2r^2R=0$$

or

$$\xi^2\frac{d^2R}{d\xi^2}+\xi\frac{dR}{d\xi}+\xi^2R=0 \text{ (where } \xi = kr). \tag{12.108}$$

Equation (12.108) is a Bessel equation equation having a general solution

$$R(\xi) = AJ_0\,(\xi) + B\,N_0\,(\xi)$$

or

$$R(r) = AJ_0\,(kr) + B\,N_0(kr)$$

where A and B are arbitrary constants and J_0 and N_0 are Bessel functions of the first and second kind with zero order respectively.

Since $N_0(0)$ is infinite, finiteness of deflection of the membrane at the centre or at any point requires that $B = 0$. Therefore, we write

$$R(r) = AJ_0\,(kr). \tag{12.109}$$

Also, the general solution of Eq. (12.106) is

$$T(t) = C\cos\omega t + D\sin\omega t \tag{12.110}$$

where C and D are arbitrary constants.

Application of Boundary Condition

Through Eqs. (12.104), (12.109) and (12.110); the boundary condition $u(a, t) = 0$ yields $AJ_0\,(ka)\,(C\cos\omega t + D\sin\omega t) = 0$ for all t.

Thus, we must have, as A and t are arbitrary,

$$J_0(ka) = 0$$

i.e.,

$$ka = \alpha_m\ (m = 1, 2, 3\ \ldots)$$

where α_m are zeroes of J_0 such that $\alpha_1 = 2.4048$, $\alpha_2 = 5.5201$, $\alpha_3 = 8.6537$, etc.

Therefore, for a particular value of m, we have

$$k=k_m=\frac{\alpha_m}{a},\ m = 1, 2, 3\ \ldots \tag{12.111}$$

Thus, Eqs. (12.109) and (12.110) yield

$$R_m(r) = AJ_0(k_m r) = AJ_0\left(\frac{\alpha_m}{a} r\right), \; m = 1, 2, 3 \ldots \tag{12.112}$$

and $$T_m(t) = C_m \cos \omega_m t + D_m \sin \omega_m t \tag{12.113}$$

Using Eqs. (12.104), (12.112) and (12.113), we write the solutions of Eq. (12.101) as

$$u_m(r,t) = AJ_0\left(\frac{\alpha_m}{a} r\right)\left[C_m \cos \omega_m t + D_m \sin \omega_m t\right]$$

$$= \left[A_m \cos \omega_m t + B_m \sin \omega_m t\right] J_0\left(\frac{\alpha_m}{a} r\right) \text{ (with } m = 1, 2, 3 \ldots) \tag{12.114}$$

where $A_m = A\,C_m$ and $B_m = A\,D_m$ are arbitrary constants.

Here, u_m are called eigenfunctions and ω_m (circular) eigenfrequencies. The vibration of the membrane corresponding to the eigenfunction (u_m) is called the mth normal mode. This mode of vibration has an ordinary frequency of $\frac{\omega_m}{2\pi}$ cycles per second.

For the normal mode with $m = 1$ all the points of the membrane move up (or down) at the same time and hence there is no nodal line between the centre and the circumference. For the normal mode with $m = 2$ there is a circular nodal line at some value of r given by $\frac{\alpha_2}{a} r = \alpha_1$ or, $r = \frac{\alpha_1}{\alpha_2}\, a = 0.4356a$. Thus, this mode of vibration of the circular membrane of radius a has a nodal line in the form of a concentric circle of radius $0.4356a$. In general, the normal mode of vibration corresponding to the eigenfunction u_m has $m - 1$ concentric nodal lines. The following two figures show the graph of the Bessel function of the first order and the nodal lines of the vibrations of the circular membrane.

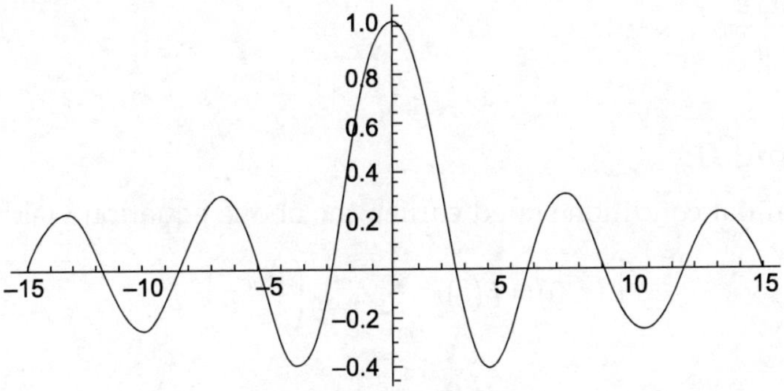

Fig. 12.9 Graph of $J_0(x)$

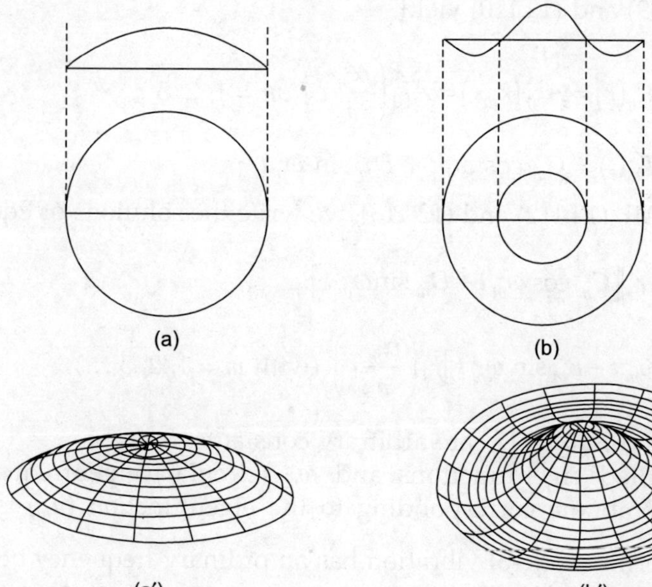

Fig. 12.10 Vibrations of different modes and nodal lines: Figures (a) and (a′) show fundamental vibration whereas Figures (b) and (b′) show 2nd harmonic vibration.

By the superposition principle, the general solution of Eq. (12.101) is

$$u(r,t)=\sum_{m=1}^{\infty}u_m(r,t)$$

$$=\sum_{m=1}^{\infty}\left[A_m\cos\omega_m t+B_m\sin\omega_m t\right]J_0\left(\frac{\alpha_m}{a}r\right)\quad m=1,2,3\ldots \tag{12.115}$$

where $\omega_m=\dfrac{\alpha_m c}{a}$.

To Find A_m and B_m

Using the first initial condition stated earlier, the above equation yields

$$u(r,0)=f(r)=\sum_{m=1}^{\infty}A_mJ_0\left(\frac{\alpha_m}{a}r\right).$$

Multiplying this equation by $rJ_0\left(\dfrac{\alpha_l}{a}r\right)$ and then integrating with respect to r within the limit from $r=0$ to $r=a$, we obtain

$$\int_0^a rf(r)J_0\left(\frac{\alpha_l}{a}r\right)dr = \sum_{m=1}^{\infty} A_m \int_0^a J_0\left(\frac{\alpha_l}{a}r\right)J_0\left(\frac{\alpha_m}{a}r\right)rdr$$

$$= A_l \frac{a^2}{2} J_1^2(\alpha_l) \text{ (by orthonormality of Bessel functions).}$$

Then we have

$$A_l = \frac{2}{a^2 J_1^2(\alpha_l)} \int_0^a f(r) J_0\left(\frac{\alpha_l}{a}r\right) r\,dr.$$

Renaming the subscript, we write

$$A_m = \frac{2}{a^2 J_1^2(\alpha_m)} \int_0^a f(r) J_0\left(\frac{\alpha_m}{a}r\right) r\,dr. \tag{12.116}$$

Using the second initial condition, Eq. (12.115) yields

$$g(r) = \left(\frac{\partial u}{\partial t}\right)_{t=0} = \left[\sum_{m=1}^{\infty} \omega_m \{-A_m \sin \omega_m t + B_m \cos \omega_m t\} J_0\left(\frac{\alpha_m}{a}r\right)\right]_{t=0}$$

$$= \sum_{m=1}^{\infty} \omega_m B_m J_0\left(\frac{\alpha_m}{a}r\right).$$

Multiplying both sides by $rJ_0\left(\frac{\alpha_l}{a}r\right)$ and integrating the results with respect to r within the limit from $r = 0$ to $r = a$, we obtain

$$\int_0^a g(r) J_0\left(\frac{\alpha_l}{a}r\right) r\,dr = \sum_{m=1}^{\infty} \omega_m B_m \int_0^a J_0\left(\frac{\alpha_l}{a}r\right) J_0\left(\frac{\alpha_m}{a}r\right) r\,dr$$

$$= \omega_l B_l \frac{a^2}{2} J_1^2(\alpha_l)$$

or

$$B_l = \frac{2}{\omega_l a^2 J_1^2(\alpha_l)} \int_0^a g(r) J_0\left(\frac{\alpha_l}{a}r\right) r\,dr.$$

Renaming the subscript, we write

$$B_m = \frac{2}{\omega_m a^2 J_1^2(\alpha_m)} \int_0^a g(r) J_0\left(\frac{\alpha_m}{a}r\right) r\,dr$$

$$= \frac{2}{ac\alpha_m J_1^2(\alpha_m)} \int_0^a g(r) J_0\left(\frac{\alpha_m}{a}r\right) r\,dr. \tag{12.117}$$

Equation (12.115) together with Eqs. (12.116) and (12.117) gives the solution of Eq. (12.101).

Note: If the initial speed is $g(r) = 0$ then $B_m = 0$.

CASE II: Unsymmetric Case

In this case, the vibrations of the finite circular membrane of radius a is unsymmetric. Let $u(r, \theta, t)$ be the deflection of a vibrating point located at (r, θ) at time t. The relevant wave equation is

$$\frac{\partial^2 u}{\partial r^2} + \frac{1}{r}\frac{\partial u}{\partial r} + \frac{1}{r^2}\frac{\partial^2 u}{\partial \theta^2} = \frac{1}{c^2}\frac{\partial^2 u}{\partial t^2} \tag{12.118}$$

where c is the constant speed of the propagation of the wave.

Boundary Condition

Boundary conditions of the problem are

$$\begin{aligned} &u(a, \theta, t) = 0, && 0 \le \theta \le 2\pi,\ t \ge 0; \\ &|u(0, \theta, t)| < \infty, && 0 \le \theta \le 2\pi,\ t \ge 0; \\ &u(r, \theta + 2\pi, t) = u(r, \theta, t), && 0 \le r \le a,\ 0 \le \theta \le 2\pi,\ t \ge 0. \end{aligned} \tag{12.119}$$

Initial Conditions

The initial conditions of the problem are:

(i) Initial deflection, $u(r, \theta, 0) = f(r, \theta),\ 0 \le r < a,\ 0 < \theta \le 2\pi$;

(ii) Initial speed, $u_t(r, \theta, 0) = \left(\dfrac{\partial u}{\partial t}\right)_{t=0} = g(r, \theta),\ 0 \le r < a,\ 0 \le \theta \le 2\pi.$ (12.120)

Seeking the Solution of the Problem

Using the method of separation of variables, let us assume the solution as

$$u(r, \theta, t) = S(r, \theta)\, T(t) \tag{12.121}$$

where S is function of r and θ and T is function of t only.

Substituting Eq. (12.121) into Eq. (12.118), we obtain

$$T\left(\frac{\partial^2 S}{\partial r^2} + \frac{1}{r}\frac{\partial S}{\partial r} + \frac{1}{r^2}\frac{\partial^2 S}{\partial \theta^2}\right) = \frac{S}{c^2}\frac{d^2 T}{dt^2}.$$

Dividing both sides by $S(r, \theta)\, T(t)$, we obtain

$$\frac{1}{S}\left(\frac{\partial^2 S}{\partial r^2} + \frac{1}{r}\frac{\partial S}{\partial r} + \frac{1}{r^2}\frac{\partial^2 S}{\partial \theta^2}\right) = \frac{1}{c^2 T}\frac{d^2 T}{dt^2}.$$

In the above equation, the left-hand side is function of r and θ whereas the right-hand side is function of t only and hence each of them must be equal to some constant say $-k^2$. That is,

$$\frac{1}{S}\left(\frac{\partial^2 S}{\partial r^2}+\frac{1}{r}\frac{\partial S}{\partial r}+\frac{1}{r^2}\frac{\partial^2 S}{\partial \theta^2}\right)=\frac{1}{c^2 T}\frac{d^2 T}{dt^2}=-k^2.$$

Thus, we obtain two equations

$$\frac{\partial^2 S}{\partial r^2}+\frac{1}{r}\frac{\partial S}{\partial r}+\frac{1}{r^2}\frac{\partial^2 S}{\partial \theta^2}+k^2 S=0 \tag{12.122}$$

and

$$\frac{d^2 T}{dt^2}+\omega^2 T=0 \tag{12.123}$$

where

$$\omega = ck. \tag{12.124}$$

Let's assume the solution of Eq. (12.122) as

$$S(r,\theta)=R(r)\Theta(\theta). \tag{12.125}$$

Substituting this equation into Eq. (12.122), we obtain

$$\Theta\frac{d^2 R}{dr^2}+\frac{\Theta}{r}\frac{dR}{dr}+\frac{R}{r^2}\frac{d^2\Theta}{d\theta^2}+k^2 R\Theta=0.$$

Multiplying both sides by $\frac{r^2}{R\Theta}$ and then transposing, we obtain

$$\frac{1}{\Theta}\frac{d^2\Theta}{d\theta^2}=-\left[\frac{r^2}{R}\frac{d^2 R}{dr^2}+\frac{r}{R}\frac{dR}{dr}+k^2 r^2\right]=-m^2 \text{ (another constant)}.$$

Thus, we obtain two more equations

$$r^2\frac{d^2 R}{dr^2}+r\frac{dR}{dr}+(k^2 r^2-m^2)R=0 \tag{12.126}$$

and

$$\frac{d^2\Theta}{d\theta^2}+m^2\,\Theta=0. \tag{12.127}$$

Equation (12.126) is a Bessel equation having the solution

$$R(r) = A\,J_m(kr) + B\,N_m\,(kr)\ (m = 0, 1, 2, 3\ \ldots)$$

where A, B are arbitrary constants and J_m and N_m are mth order Bessel functions of the first and the second kinds. Finiteness of the deflection at the centre of the circular membrane needs the constant B to be zero and hence we have

$$R(r) = A\,J_m\,(kr). \tag{12.128}$$

Now, Eqs. (12.123) and (12.127) have the respective solutions

$$T(t) = C\cos\omega t + D\sin\omega t \tag{12.129}$$

and

$$\Theta(\theta) = M\cos m\theta + N\sin m\theta \tag{12.130}$$

where C, D, M, N are arbitrary constants.

Through Eq. (12.121) and (12.125), the above equations yield the solutions

$$u_{mn}(r,\theta,t) = [M\cos m\theta + N\sin m\theta][C\cos\omega t + D\sin\omega t]A J_m(kr). \tag{12.131}$$

Application of Boundary Conditions

Applying the first boundary condition given in Eq. (12.119) to Eq. (12.131), we have

$$J_m(ka) = 0, \text{ i.e., } ka = \alpha_{mn} \ (n = 1, 2, 3 \ldots),$$

where α_{mn} is the nth positive zero of the Bessel function $J_m(x)$.

Thus, for particular values of m and n, the value of k is given by

$$k_{mn} = \frac{\alpha_{mn}}{a} \tag{12.132}$$

and using Eq. (12.124), the corresponding value of ω is given by

$$\omega_{mn} = \frac{c\alpha_{mn}}{a}. \tag{12.133}$$

Thus, for particular values of m and n, the solutions of the problem are

$$u_{mn}(r,\theta,t) = [A_m\cos m\theta + B_m\sin m\theta] \times\left[C_{mn}\cos\left(\frac{\alpha_{mn}}{a}ct\right) + D_{mn}\sin\left(\frac{\alpha_{mn}}{a}ct\right)\right]J_m\left(\frac{\alpha_{mn}}{a}r\right) \tag{12.134}$$

where A_m, B_m, C_{mn}, D_{mn} are arbitrary constants for particular values of m and n.

By the superposition principle, the general solution of the problem is obtained as

$$\begin{aligned} u(r,\theta,t) &= \sum_{m=0}^{\infty}\sum_{n=1}^{\infty} u_{mn}(r,\theta,t) \\ &= \sum_{m=0}^{\infty}\sum_{n=1}^{\infty}[A_m\cos m\theta + B_m\sin m\theta] \\ &\quad \times\left[C_{mn}\cos\left(\frac{\alpha_{mn}}{a}ct\right) + D_{mn}\sin\left(\frac{\alpha_{mn}}{a}ct\right)\right]J_m\left(\frac{\alpha_{mn}}{a}r\right). \end{aligned} \tag{12.135}$$

The values of the arbitrary constants A_m, B_m, C_{mn}, D_{mn} are to be determined from the initial conditions of a particular problem.

12.16 DERIVATION OF THE HEAT CONDUCTION EQUATION IN ONE DIMENSION(EQUATION OF HEAT CONDUCTION IN A ROD)

Let us consider the problem of heat conduction along a straight rod of uniform cross section having a cross-sectional area A. Let us choose the x-axis along the axis of the rod such that the ends of the rod are at $x = 0$ and $x = L$. Let us assume that:

(i) The rod is homogeneous, that is, its mass per unit volume is constant.

(ii) The specific heat and the thermal conductivity of the material of the rod are constants.

(iii) The flow of heat within the rod takes place only in the direction of increasing x.

(iv) The lateral surface of the rod is insulated so that no heat escapes from this surface.

(v) No heat is generated inside the rod.

Let ρ be the mass per unit volume of the rod and let s and K be respectively the specific heat and the coefficient of thermal conductivity of the material of the rod.

Now, to derive the partial differential equation satisfied by the temperature $u(x, t)$ at point x at time t, we need two empirical laws of heat conduction:

(a) The quantity (Q) of heat required to raise the temperature of a body of mass (m) and specific heat (s) by an amount u is $Q = msu$. Therefore, the rate of heat flow is given by

$$\frac{\partial Q}{\partial t} = ms\frac{\partial u(x,t)}{\partial t}. \tag{12.136}$$

(b) The rate of heat flow $\frac{\partial Q}{\partial t}$ across a surface is proportional to the face area A and the temperature gradient $\frac{\partial u}{\partial x}$ at that surface. The constant of proportionality is called the coefficient of thermal conductivity of the material of the medium.

Therefore, the rate of heat flow across the cross-sectional area A is given by

$$Q_t = \frac{\partial Q}{\partial t} = -KA\frac{\partial u}{\partial x} \tag{12.137}$$

where the minus sign is due to the fact that heat flows in the direction of decreasing temperature.

Let us consider the following Figure (12.11). In the figure, if the slice between x and $x + \Delta x$ is very thin, then $u(x, t)$ can be taken as the temperature of every point in that slice. The mass of the slice is then $m = \rho A\Delta x$.

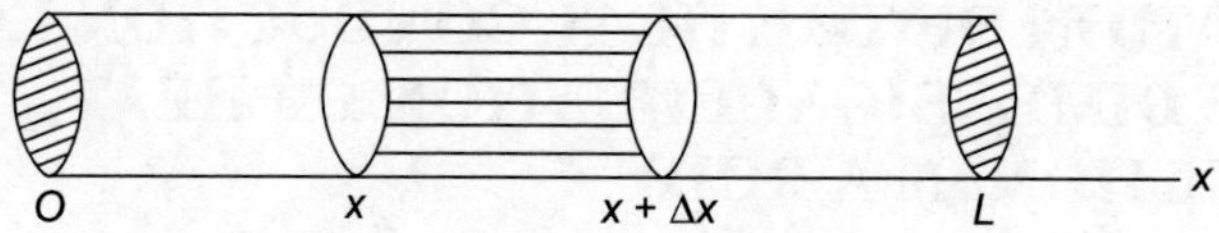

Fig. 12.11 One-dimensional heat flow along a rod

Substituting the above mass m into Eq. (12.136), we have

$$\frac{\partial Q}{\partial t} = \rho s\, A\, \Delta x \frac{\partial u}{\partial t}. \tag{12.138}$$

When heat flows in the positive x-direction, we find from Eq. (12.137) that heat is acquired in the slice at the net rate

$$\frac{\partial Q}{\partial t} = -KA\frac{\partial u(x,t)}{\partial x} - \left[-KA\frac{\partial u(x+\Delta x,t)}{\partial x}\right]$$

$$= KA\left[\frac{\partial u(x+\Delta x,t)}{\partial x} - \frac{\partial u(x,t)}{\partial x}\right]. \tag{12.139}$$

Equating Eqs. (12.138) and (12.139), we obtain

$$K\left[\frac{\partial u(x+\Delta x,t)}{\partial x} - \frac{\partial u(x,t)}{\partial x}\right] = \rho s\, \Delta x \frac{\partial u}{\partial t}.$$

In the limit $\Delta x \to 0$, we write

$$\lim_{\Delta x \to 0} \frac{1}{\Delta x}\left[\frac{\partial u(x+\Delta x,t)}{\partial x} - \frac{\partial u(x,t)}{\partial x}\right] = \frac{\rho s}{K}\frac{\partial u}{\partial t}$$

or

$$\frac{\partial^2 u}{\partial x^2} = \frac{1}{D}\frac{\partial u}{\partial t} \tag{12.140}$$

where $D = \dfrac{K}{\rho s}$ is a constant known as the ***thermal diffusivity*** of the rod. The thermal diffusivity of a material is the ratio of its coefficient of thermal conductivity to the product of its density and specific heat. In SI units, the thermal diffusivity is expressed in metre square per second. Equation (12.140) is known as *diffusion equation*.

In a special case, let there be a heat generation inside the rod. If the rate of generation per unit volume is g, $\left[g = \dfrac{\text{Heat energy}}{\text{time} \times \text{volume}}\right]$, the rate at which heat is generated in the slice is $Ag\Delta x$ In this case, the right-hand side of Eq. (12.139) will have an additional term, $Ag\Delta x$. Therefore, Eq. (12.140) will become

$$\frac{\partial^2 u}{\partial x^2} + \frac{g}{\rho s D} = \frac{1}{D}\frac{\partial u}{\partial t}. \qquad (12.141)$$

12.17 GENERAL SOLUTION OF ONE-DIMENSIONAL HEAT FLOW EQUATION IN THE CASE OF A FINITE ROD WHEN BOTH ENDS ARE KEPT AT ZERO DEGREE FOR ALL TIME

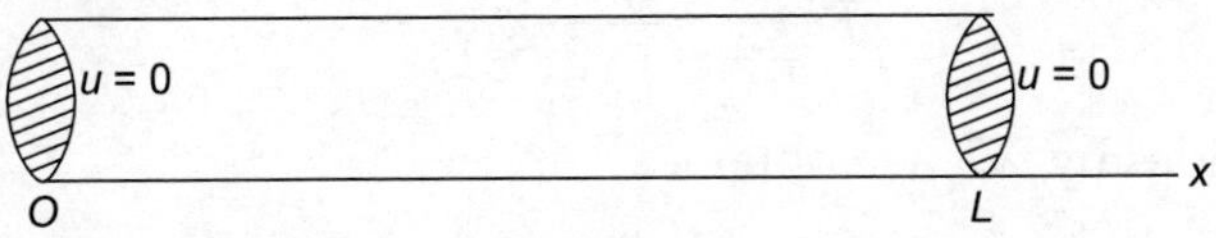

Fig. 12.12 Temperature distribution in a rod of length L

Let us consider a thin uniform rod of length L with an initial temperature $f(x)$ throughout and whose ends are held at temperature zero for all time $t \geq 0$. If the assumptions made in section 12.16 in the derivation of one-dimensional heat equation are all satisfied, then the temperature distribution $u(x, t)$ along the rod is determined by the boundary value problem

$$\frac{\partial^2 u}{\partial x^2} = \frac{1}{D}\frac{\partial u}{\partial t}, \; (0 < x < L, t > 0) \qquad (12.142)$$

where D is the diffusivity of the material of the rod. Here, we have

$$u(0, t) = 0 = u(L, t) \text{ for } t \geq 0, \text{ (boundary conditions)}, \qquad (12.143)$$

$$u(x, 0) = f(x), 0 < x < L, \text{ (initial condition)}. \qquad (12.144)$$

A model of this boundary value problem is shown in Fig. 12.13.

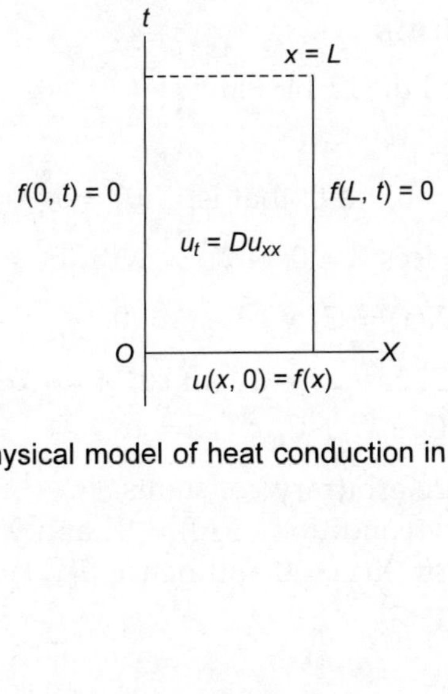

Fig. 12.13 Physical model of heat conduction in one dimension

Seeking the Solution

Let us assume the solution as

$$u(x, t) = X(x)\, T(t) \tag{12.145}$$

where X is function of x only, T is function of t only and these are to be determined later.

Substituting Eq. (12.145) into Eq. (12.142), we have

$$T\frac{d^2X}{dx^2} = \frac{X}{D}\frac{dT}{dt}.$$

Dividing both sides by XT, we obtain

$$\frac{1}{X}\frac{d^2X}{dx^2} = \frac{1}{DT}\frac{dT}{dt}.$$

In the above equation, the left-hand side is function of x only whereas the right-hand side is function of t only and hence their equality exists only when both of them are equal to the same constant, say λ. Here, λ is called separation constant. Therefore, we write

$$\frac{1}{X}\frac{d^2X}{dx^2} = \frac{1}{DT}\frac{dT}{dt} = \lambda.$$

Thus, we obtain two ODEs as given below.

$$\frac{d^2X}{dx^2} - \lambda X = 0 \tag{12.146}$$

and
$$\frac{dT}{dt} - D\lambda T = 0. \tag{12.147}$$

New Boundary Conditions

The boundary conditions for Eq. (12.143) are:

(i) $u(0, t) = X(0)\, T(t) = 0$

and (ii) $u(L, t) = X(L)\, T(t) = 0,\ t \geq 0$; that is, $X(0) = 0,\ X(L) = 0$ for all t. (12.148)

Making three different choices $\lambda = 0, -\alpha^2, \alpha^2$ with $\alpha^2 > 0$, Eq. (12.146) yields

$$X(x) = C_1 x + C_2,\ \lambda = 0, \tag{12.149}$$

$$X(x) = C_3 \cos \alpha x + C_4 \sin \alpha x,\ \lambda = -\alpha^2 \tag{12.150}$$

and
$$X(x) = C_5 e^{-\alpha x} + C_6 e^{\alpha x},\ \lambda = \alpha^2 \tag{12.151}$$

where $C_1, C_2, C_3, C_4, C_5, C_6$ are arbitrary constants.

On applying the boundary conditions $X(0) = 0$ and $X(L) = 0$, Eqs. (12.149) and (12.151) yield only the solution $X(x) = 0$ and hence $u(x, t) = X(x)\, T(t) = 0$.

Such trivial solutions are neglected and, therefore, the only admissible solution of Eq. (12.146) is the solution given in Eq. (12.150).

When the boundary condition $X(0) = 0$ is applied to Eq. (12.150), we obtain

$$C_3 = 0.$$

When the boundary condition $X(L) = 0$ is applied to Eq. (12.150), we obtain

$$0 = 0 + C_4 \sin \alpha L.$$

Since C_4 is an arbitrary constant, we have

$$\sin \alpha L = 0, \text{ that is, } \alpha L = n\pi$$

or
$$\alpha = \frac{n\pi}{L} \quad (n = 1, 2, 3 \ldots).$$

For a particular value of n, we write

$$\alpha_n = \frac{n\pi}{L}. \tag{12.152}$$

Here, $n = 0$ is not considered as it gives a trivial solution and also $n = -1, -2, -3$ ... are not considered as they do not lead to new solutions.

For particular values of n, substituting the values of C_3 and α_n into Eq. (12.150), we obtain the solutions of Eq. (12.146) as

$$X_n(x) = C_{4n} \sin \frac{n\pi x}{L}. \tag{12.153}$$

Now, Eq. (12.147) can be written as

$$\frac{dT}{T} = D\lambda\, dt = -D\alpha^2 dt \quad \text{as} \quad \lambda = -\alpha^2.$$

Integrating both sides, we obtain,

$$T(t) = Ae^{-D\alpha^2 t} \quad \text{where } A \text{ is another constant.}$$

For particular values of n, we write

$$T_n(t) = A_n e^{-D\alpha_n^2 t} = A_n e^{-D\left(\frac{n^2\pi^2}{L^2}\right)t}. \tag{12.154}$$

Using Eqs. (12.145), (12.153) and (12.154), we write the solutions of the BVP as

$$u_n(x, t) = b_n e^{-Dn^2\pi^2 t/L^2} \sin \frac{n\pi x}{L} \tag{12.155}$$

where $b_n = A_n C_{4n}$ are constants to be determined from the given initial conditions.

By the superposition principle, the general solution of the given BVP is given by

$$u(x, t) = \sum_{n=1}^{\infty} u_n(x, t)$$

$$= \sum_{n=1}^{\infty} b_n e^{-Dn^2\pi^2 t/L^2} \sin\frac{n\pi x}{L}. \tag{12.156}$$

To Find the Coefficients b_n

Using the initial condition given in Eq. (12.144), we have

$$f(x) = u(x,0) = \sum_{n=1}^{\infty} b_n \sin\frac{n\pi x}{L}.$$

Multiplying both sides by $\sin\frac{m\pi x}{L}$ and integrating the results with respect to x within the limit from $x = 0$ to $x = L$, we have

$$\int_0^L f(x)\sin\frac{m\pi x}{L}dx = \sum_{n=1}^{\infty} b_n \int_0^L \sin\frac{m\pi x}{L}\sin\frac{n\pi x}{L}dx = b_m\frac{L}{2}$$

or

$$b_m = \frac{2}{L}\int_0^L f(x)\sin\frac{m\pi x}{L}dx.$$

Renaming the subscript, we write

$$b_n = \frac{2}{L}\int_0^L f(x)\sin\frac{n\pi x}{L}dx. \tag{12.157}$$

Using Eqs. (12.156) and (12.157), we write the general solution of the given BVP as

$$u(x,t) = \frac{2}{L}\sum_{n=1}^{\infty}\left(\int_0^L f(w)\sin\frac{n\pi w}{L}dw\right)e^{-\frac{Dn^2\pi^2 t}{L^2}}\sin\frac{n\pi x}{L}. \tag{12.158}$$

Here, the integration variables are renamed.

12.18 HEAT CONDUCTION IN A FINITE ROD WITH NONHOMOGENEOUS BOUNDARY CONDITIONS (i.e., WHEN THE ENDS ARE KEPT AT DIFFERENT FIXED TEMPERATURES)

Let us consider the problem of heat conduction in a finite uniform rod of length L whose ends are kept at different fixed temperatures. Let us suppose that the end $x = 0$ of the bar is held at a constant temperature T_1 and the other end at $x = L$ is maintained at another constant temperature T_2.

The governing PDE is

$$\frac{\partial^2 u}{\partial x^2} = \frac{1}{D}\frac{\partial u}{\partial t} \tag{12.159}$$

where $u(x, t)$ is the temperature at a point x at time t and D is the diffusivity. Boundary conditions of the problem are

$$u(0, t) = T_1 \text{ and } u(L, t) = T_2,\ t \geq 0. \tag{12.160a}$$

The initial condition is

$$u(x, 0) = f(x). \tag{12.160b}$$

We know that after a long time, that is, when $t \to \infty$, there will be a steady temperature distribution $u_{\text{steady}}(x) = v(x)$ which is independent of time t and the initial conditions.

From Eq. (12.159), we see that under the steady condition, $v(x)$ satisfies the equation

$$\frac{d^2 v(x)}{dx^2} = 0. \tag{12.161}$$

Here, $v(x)$ must also satisfy the boundary conditions

$$v(0) = T_1 \text{ and } v(L) = T_2. \tag{12.162}$$

Equation (12.161) has the solution

$$v(x) = C_2 x + C_1$$

where C_1 and C_2 are arbitrary constants.

Thus, the steady-state temperature distribution is a linear function of x.

Applying the boundary condition given in Eq. (12.162), we obtain

$$v(0) = C_1 = T_1$$

and

$$v(L) = C_2 L + T_1 = T_2 \text{ or, } C_2 = \frac{T_2 - T_1}{L}.$$

Using these values of C_1 and C_2, we write

$$v(x) = \left(\frac{T_2 - T_1}{L}\right)x + T_1. \tag{12.163}$$

If there exists a solution $w(x, t)$ as the transient temperature distribution such that

$$u(x, t) = v(x) + \omega(x, t) \tag{12.164}$$

then the problem will be solved since $v(x)$ is already found.

Substituting Eq. (12.164) into Eq. (12.159), we obtain

$$\frac{d^2 v}{dx^2} + \frac{\partial^2 w}{\partial x^2} = \frac{1}{D}\left(\frac{\partial v}{\partial t} + \frac{\partial w}{\partial t}\right)$$

Since $\frac{d^2v}{dx^2}=0$ and $\frac{\partial v}{\partial t}=0$,

we write
$$\frac{\partial^2 w}{\partial x^2}=\frac{1}{D}\frac{\partial w}{\partial t} \tag{12.165}$$

with the newly transformed boundary conditions

$$\left.\begin{aligned} w(0,t)&=u(0,t)-v(0)=T_1-T_1=0\\ w(L,t)&=u(L,t)-v(L)=T_2-T_2=0.\end{aligned}\right] \tag{12.166}$$

Also, the newly transformed initial condition is

$$w(x,0)=u(x,0)-v(x)=f(x)-v(x)=\phi(x) \text{ (say).} \tag{12.167}$$

Now, Eq. (12.165), under the conditions given by Eqs. (12.166) and (12.167), can be solved by the procedure stated in the previous section.

Thus, we obtain the solutions for n = 1, 2, 3 … as

$$w_n(x,t)=b_n e^{-Dn^2\pi^2\frac{t}{L^2}}\sin\frac{n\pi x}{L} \tag{12.168}$$

where
$$\begin{aligned} b_n&=\frac{2}{L}\int_0^L \phi(x)\sin\frac{n\pi x}{L}dx\\ &=\frac{2}{L}\int_0^L [f(x)-v(x)]\sin\frac{n\pi x}{L}dx\\ &=\frac{2}{L}\int_0^L\left[f(x)-\left\{\frac{(T_2-T_1)x}{L}+T_1\right\}\right]\sin\frac{n\pi x}{L}dx\\ &=\frac{2}{L}\int_0^L\left[f(x)+\frac{(T_1-T_2)x}{L}-T_1\right]\sin\frac{n\pi x}{L}dx. \end{aligned} \tag{12.169}$$

Therefore, we write

$$\begin{aligned} u_n(x,t)&=v(x)+w_n(x)\\ &=\frac{(T_2-T_1)x}{L}+T_1+w_n(x,t)\\ &=T_1+\frac{(T_2-T_1)x}{L}+b_n e^{-Dn^2\pi^2\frac{t}{L^2}}\sin\frac{n\pi x}{L}. \end{aligned} \tag{12.170}$$

By the superposition principle, the general solution of the given problem is given by

$$u(x,t)=\sum_{n=1}^{\infty}u_n(x,t)$$

$$=T_1+\frac{(T_2-T_1)x}{L}+\sum_{n=1}^{\infty}b_n e^{-Dn^2\pi^2\frac{t}{L^2}}\sin\frac{n\pi x}{L} \tag{12.171}$$

where b_n is given by Eq. (12.169).

12.19 HEAT CONDUCTION IN A FINITE ROD INSULATED AT BOTH ENDS

Let us consider the problem of heat conduction in a uniform rod of finite length L. The ends of the bar at $x = 0$ and $x = L$ are insulated instead of being held at constant temperatures. The boundary value–initial value problem that describes the temperature in this rod is modelled by the equations

$$\frac{\partial^2 u}{\partial x^2}=\frac{1}{D}\frac{\partial u}{\partial t},\quad 0\le x\le L,\, t\ge 0, \tag{12.172}$$

$$u(x,0)=f(x),\ 0\le x\le L;\ \text{(initial condition)}, \tag{12.173}$$

$$\left[\frac{\partial u(x,t)}{\partial x}\right]_{x=0}=0,\left[\frac{\partial u(x,t)}{\partial x}\right]_{x=L}=0,\, t\ge 0\,,\ \text{(boundary condition)} \tag{12.174}$$

where $f(x)$ is supposed to be a given function.

To solve the problem by the method of separation of variables, we let

$$u(x,\,t) = X(x)\,T(t). \tag{12.175}$$

Substituting Eq. (12.175) into Eq. (12.172), we obtain

$$T\frac{d^2X}{dx^2}=\frac{X}{D}\frac{dT}{dt}.$$

Dividing both sides by $X(x)\,T(t)$, we obtain

$$\frac{1}{X}\frac{d^2X}{dx^2}=\frac{1}{DT}\frac{dT}{dt}=\lambda \tag{12.176}$$

where λ is a separation constant.

Thus, we arrive at the two equations,

$$\frac{d^2X}{dx^2}-\lambda X=0 \tag{12.177}$$

and

$$\frac{dT}{dt}-D\lambda T=0. \tag{12.178}$$

Substituting Eq. (12.175) in the boundary conditions given in Eq. (12.174), we obtain the new boundary conditions

$$X'(0) = 0 \text{ and } X'(L) = 0. \tag{12.179}$$

Here, the prime denotes derivative.

(i) If we choose $\lambda = 0$, Eq. (12.177) becomes

$$\frac{d^2X}{dx^2} = 0$$

having the solution $X(x) = C_2 x + C_1$ where C_1 and C_2 are arbitrary constants. Boundary conditions shown in Eq. (12.179) require that $C_2 = 0$. But these boundary conditions cannot determine C_1 and we have $X(x) = C_1$. For $\lambda = 0$, Eq. (12.178) yields

$$T(t) = \text{constant}.$$

Therefore, the choice $\lambda = 0$ leads to a constant solution

$$u_0(x, t) = C_1 \times \text{a constant} = A_0 \text{ (constant)}. \tag{12.180}$$

(ii) If $\lambda > 0$, we write $\lambda = \alpha^2$ where α is a real positive number. Then, Eq. (12.177) becomes

$$\frac{d^2X}{dx^2} - \alpha^2 X = 0 \text{ yielding the solution}$$

$X(x) = C_3 \cosh \alpha x + C_4 \sinh \alpha x$ [where C_3 and C_4 are arbitrary constants.] In this case the boundary conditions can be satisfied if we choose $C_3 = C_4 = 0$. But, this choice is unacceptable as it leads to a trivial solution to the given problem.

(iii) If $\lambda < 0$, we let $\lambda = \alpha^2$ where α^2 is real and positive. Then, Eq. (12.177) becomes

$$\frac{d^2X}{dx^2} + \alpha^2 X = 0$$

yielding a general solution in the form

$$X(x) = C_5 \cos \alpha x + C_6 \sin \alpha x \tag{12.181}$$

where C_5 and C_6 are arbitrary constants.

Applying the first boundary condition shown in Eq. (12.179), we have

$$X'(0) = [-\alpha C_5 \sin \alpha x + \alpha C_6 \cos \alpha x]_{x=0} = 0$$

or

$$C_6 = 0 \text{ as } (\alpha \neq 0). \tag{12.182}$$

Therefore, Eq. (12.181) becomes

$$X(x) = C_5 \cos \alpha x. \tag{12.183}$$

Applying the second boundary condition shown in Eq. (12.179), we have

$$X'(L) = -\alpha C_5 \sin \alpha L = 0.$$

Since, C_5 is arbitrary and $\alpha \neq 0$, we must have

$$\sin \alpha L = 0 \text{ or, } \alpha = \frac{n\pi}{L} \text{ where } n = 0, 1, 2, 3, \ldots$$

For a particular n, we write

$$\alpha = \alpha_n = \frac{n\pi}{L}. \tag{12.184}$$

Thus, for a particular n, substituting Eq. (12.184) into Eq. (12.183), we write

$$X_n(x) = C_{5n} \cos \alpha_n x = C_{5n} \cos \frac{n\pi x}{L}. \tag{12.185}$$

With $\lambda = -\alpha^2$ Eq. (12.178) becomes

$$\frac{dT}{dt} = -D\alpha^2 T \text{ having the solution}$$

$$T(t) = C_7 e^{-D\alpha^2 t}.$$

For a particular n, we write

$$T_n(t) = C_{7n} e^{-D\alpha_n^2 t} = C_{7n} e^{-\frac{Dn^2\pi^2 t}{L^2}}. \tag{12.186}$$

Using Eqs. (12.175), (12.180), (12.185) and (12.186), we write the solutions of Eq. (12.172) as

$$u_0(x, t) = A_0,$$

$$u_n(x,t) = A_n \cos \frac{n\pi x}{L} e^{-\frac{Dn^2\pi^2 t}{L^2}}, \tag{12.187}$$

where $A_n = C_{5n} C_{7n}$.

By the principle of superposition, any linear combination of solutions is also a solution.

Therefore, we write the general solution of the given problem as

$$u(x, t) = A_0 + \sum_{n=1}^{\infty} A_n \cos \frac{n\pi x}{L} e^{-\frac{Dn^2\pi^2 t}{L^2}}$$

$$= \sum_{n=0}^{\infty} A_n \cos \frac{n\pi x}{L} e^{-\frac{Dn^2\pi^2 t}{L^2}} \tag{12.188}$$

where the constants A_n are to be determined from the initial conditions of the problem.

To Determine A_0 and A_n

Applying the initial condition stated in Eq. (12.173), we have

$$u(x,0) = A_0 + \sum_{n=1}^{\infty} A_n \cos\frac{n\pi x}{L} = f(x). \tag{12.189}$$

Integrating both sides of Eq. (12.189) with respect to x within the limit from $x = 0$ to $x = L$, we obtain

$$\int_0^L A_0 dx + \sum_{n=1}^{\infty} A_n \int_0^L \cos\frac{n\pi x}{L} dx = \int_0^L f(x) dx$$

or

$$A_0 L = \int_0^L f(x) dx$$

or

$$A_0 = \frac{1}{L}\int_0^L f(x) dx. \tag{12.190}$$

Multiplying both sides of Eq. (12.189) by $\cos\frac{m\pi x}{L}$ and integrating the results with respect to x within the limit from $x = 0$ to $x = L$, we obtain

$$A_0 \int_0^L \cos\frac{m\pi x}{L} dx + \sum_{n=1}^{\infty} A_n \int_0^L \cos\frac{m\pi x}{L} \cos\frac{n\pi x}{L} dx = \int_0^L f(x) \cos\frac{m\pi x}{L} dx$$

or

$$0 + A_m \cdot \frac{L}{2} = \int_0^L f(x) \cos\frac{m\pi x}{L} dx$$

or

$$A_m = \frac{2}{L}\int_0^L f(x) \cos\frac{m\pi x}{L} dx.$$

Re-naming the subscript, we write

$$A_n = \frac{2}{L}\int_0^L f(x) \cos\frac{n\pi x}{L} dx. \tag{12.191}$$

Equation (12.188), together with the Eqs. (12.190) and (12.191), gives the solution of the given problem.

When $t \to \infty$, that is, in the steady state, we have

$$\lim_{t\to\infty} u(x,t) = A_0 = \frac{1}{L}\int_0^L f(x)dx.$$

12.20 HEAT CONDUCTION IN A SEMI-INFINITE ROD

Let us consider the problem of heat conduction in a very long uniform rod. Let the rod be so long that we may assume it extending from $x = 0$ to $x = \infty$. We treat such a rod as semi-infinite. If there exists no heat generation inside the rod, the temperature $u(x, t)$at point x at time t is given by the PDE

$$\frac{\partial^2 u}{\partial x^2} = \frac{1}{D}\frac{\partial u}{\partial t}. \tag{12.192}$$

Let us suppose that the temperature at the end $x = 0$ of the rod is kept constant, say, $u(0, t) = 0$ in some temperature scale. Further, it is considered that the temperature u at the other end remains finite, i.e., $u(x, t) =$ finite as $x \to \infty$.

Thus, the above equation must satisfy the following conditions

$$u(0, t) = 0,\ t > 0 \quad \text{(boundary condition)}, \tag{12.193}$$

$$u(x, \infty) = \text{finite} \quad \text{(boundedness condition)}, \tag{12.194}$$

$$u(x, 0) = f(x) \quad \text{(initial condition)}. \tag{12.195}$$

Using the method of separation of variables, let us put

$$u(x, t) = X(x)\, T(t). \tag{12.196}$$

Substituting Eq. (12.196) in Eq. (12.192), we obtain

$$\frac{1}{X}\frac{d^2X}{dx^2} = \frac{1}{DT}\frac{dT}{dt} = \lambda,\ \text{(a constant)}. \tag{12.197}$$

As known before, we know that the choices $\lambda = 0$ and $\lambda =$ positive real constant are not admissible as these lead to trivial solution. Therefore, we choose λ to be a negative real constant, say, $\lambda = -\alpha^2$

Now, we write

$$\frac{1}{X}\frac{d^2X}{dx^2} = \frac{1}{DT}\frac{dT}{dt} = -\alpha^2.$$

Thus, we obtain the following two equations

$$\frac{d^2X}{dx^2} + \alpha^2 X = 0 \tag{12.198}$$

and

$$\frac{dT}{dt} = -D\alpha^2 T. \tag{12.199}$$

Using the boundary condition as well as the boundedness condition stated earlier, we obtain the new boundary condition and boundedness condition

$$X(0)=0 \text{ for } t\geq 0 \text{ and } \lim_{x\to\infty} X(x)=\text{finite}. \tag{12.200}$$

Equation (12.198) has the general solution

$$X(x) = C_1 \cos \alpha x + C_2 \sin \alpha x$$

where C_1, C_2 are arbitrary constants.

Applying the new boundary condition given in Eq. (12.200), we obtain

$$X(0) = C_1 = 0.$$

Therefore, we have

$$X(x) = C_2 \sin \alpha x. \tag{12.201}$$

Equation (12.199) has the solution

$$T(t)=C_3\, e^{-D\alpha^2 t} \tag{12.202}$$

where C_3 is another arbitrary constant.

Since the parameter α may take on any value, we must use an integral (the continuous analogue of a sum or series) to include all possibilities. Thus, u should have the form

$$u(x,t)=\int_0^\infty C(\alpha)\sin\alpha x\, e^{-D\alpha^2 t}\,d\alpha. \tag{12.203}$$

As $C = C_2C_3$ is arbitrary, we consider it as function of α, that is, we replace $C=C_2C_3$ by $C(\alpha)$. The initial condition demands that we must choose $C(\alpha)$ such that

$$u(x,0)=\int_0^\infty C(\alpha)\sin\alpha x\,d\alpha = f(x).$$

Considering this integral as a Fourier integral, we obtain

$$C(\alpha)=\frac{2}{\pi}\int_0^\infty f(x)\sin\alpha x\,dx. \tag{12.204}$$

Substituting Eq. (12.204) in Eq. (12.203), we write the solution of the given problem as

$$u(x,t)=\int_0^\infty\left[\frac{2}{\pi}\int_0^\infty f(p)\sin\alpha p\,dp\right]\sin\alpha x\, e^{-D\alpha^2 t}\,d\alpha$$

$$=\frac{2}{\pi}\int_0^\infty\int_0^\infty f(p)\sin\alpha p\sin\alpha x\, e^{-D\alpha^2 t}\,dp\,d\alpha$$

$$= \frac{1}{\pi}\int_0^\infty\int_0^\infty f(p)\left[\cos(p-x)\alpha - \cos(p+x)\alpha\right]e^{-D\alpha^2 t}\,dp\,d\alpha. \quad (12.205)$$

12.21 HEAT CONDUCTION IN AN INFINITE ROD

If we are to study the heat conduction near the centre of a very long rod, we may assume that the rod extends from $-\infty$ to ∞. Then, there are no boundary conditions and the problem to be solved is modelled as in the following.

The temperature $u(x, t)$ at the point x at time t is described by the PDE

$$\frac{\partial^2 u}{\partial x^2} = \frac{1}{D}\frac{\partial u}{\partial t} \quad \text{where } D \text{ is the diffusivity.} \quad (12:206)$$

There is an initial condition,

$$u(x, 0) = f(x), -\infty \le x \le \infty. \quad (12.207)$$

Further, there is a boundedness condition,

$$|u(x, t)| = \text{finite for all } t. \quad (12.208)$$

Using the method of separation of variables, let us put

$$u(x, t) = X(x)T(t). \quad (12.209)$$

Substituting Eq. (12.209) in Eq. (12.206), we obtain

$$\frac{1}{X}\frac{d^2X}{dx^2} = \frac{1}{DT}\frac{dT}{dt} = -\alpha^2, \text{ (a constant).}$$

Thus, we obtain the following two equations

$$\frac{d^2X}{dx^2} + \alpha^2 X = 0 \quad (12.210)$$

and

$$\frac{dT}{dt} = -D\alpha^2 T. \quad (12.211)$$

Respective solutions of the above two equations are

$$X(x) = C_1 \cos\alpha x + C_2 \sin\alpha x$$

and

$$T(t) = C_3\, e^{-D\alpha^2 t}$$

where C_1, C_2 and C_3 are arbitrary constants.

As in the previous section, we write the solution of the problem as

$$u(x, t) = \int_0^\infty \left[A(\alpha)\cos\alpha x + B(\alpha)\sin\alpha x\right]e^{-D\alpha^2 t}\,d\alpha \quad (12.212)$$

where $A = C_1C_3$ and $B = C_2C_3$ are replaced by $A(\alpha)$ and $B(\alpha)$ respectively.

At $t=0$, using the initial condition, we have

$$u(x,0)=\int_0^{\infty}[A(\alpha)\cos\alpha x+B(\alpha)\sin\alpha x]d\alpha=f(x).$$

As this is clearly a Fourier integral problem, we must have

$$A(\alpha)=\frac{1}{\pi}\int_{-\infty}^{+\infty}f(x)\cos\alpha x\,dx \tag{12.213}$$

and

$$B(\alpha)=\frac{1}{\pi}\int_{-\infty}^{+\infty}f(x)\sin\alpha x\,dx. \tag{12.214}$$

Equation (12.212), together with Eqs. (12.213) and (12.214), gives the general solution of the problem.

12.22 HEAT CONDUCTION EQUATION OR DIFFUSION EQUATION IN HIGHER DIMENSIONS

Let us consider the problem of heat conduction through a body. We know that in a body, heat flows in the direction of decreasing temperature. In deriving an equation for heat flow in a body, we make the following assumptions.

(i) The temperature $u(x,y,z,t)$ at a point (x,y,z) at time t is a finite and continuous point function.
(ii) Within the body, there is an isothermal surface.
(iii) The temperature gradient (∇u) and the direction of heat flow is normal to an isothermal surface at a point under consideration.
(iv) The temperature gradient (∇u) is finite and continuous and in the direction of increasing temperature, u.
(v) The body is isotropic.
(vi) The rate (or velocity) of heat flow per unit area in the isotropic medium is given by

$$\mathbf{v}=-K\nabla u$$

where K is the coefficient of thermal conductivity of the the body.

Let us consider an arbitrary region of a heat conducting solid of volume V and surface area S. The total heat flux across the surface S or the quantity of heat leaving the surface per unit time (i.e., the rate of decrease of heat in the region) is

$$\iint_S \mathbf{v}\cdot\hat{n}\,dS=-\iint_S K\nabla u\cdot\hat{n}\,dS$$

Using divergence theorem, we write

$$\iint_S \mathbf{v}\cdot\hat{n}\,dS = -\iiint_V K\,\nabla\cdot(\nabla u)\,dV = -\iiint_V K\,\nabla^2 u\,dV. \tag{12.215}$$

The amount of heat in the region of volume V and surface S is

$$\iiint_V \rho s\,u\,dV$$

where ρ and s are the density and and the specific heat of the material of the body, respectively.

The rate of decrease of heat in the region is

$$-\frac{\partial}{\partial t}\iiint_V \rho s u\,dV = -\iiint_V \rho s\frac{\partial u}{\partial t}\,dV. \tag{12.216}$$

Equating Eqs. (12.215) and (12.216), we have

$$K\nabla^2 u = \rho s\frac{\partial u}{\partial t}$$

or

$$\nabla^2 u = \frac{1}{D}\frac{\partial u}{\partial t} \tag{12.217}$$

where $D = \dfrac{K}{\rho s}$ is the thermal diffusivity of the material of the body.

12.23 HEAT FLOW IN A RECTANGULAR PLATE

Let us consider the problem of heat conduction in a thin rectangular plate in which the temperature u is a function of position (x, y) and time t. Let us consider a thin rectangular plate bounded by $x = 0$, $x = b$, $y = 0$, $y = c$.

CASE I: Boundaries held at zero degree of some temperature scale and the initial temperature is $f(x, y)$ throughout:

The relevant PDE is

$$\frac{\partial^2 u}{\partial x^2} + \frac{\partial^2 u}{\partial y^2} = \frac{1}{D}\frac{\partial u}{\partial t},\ 0 \le x \le b, 0 \le y \le c, t \ge 0 \tag{12.218}$$

subject to the boundary conditions

$$\begin{aligned}
&u(0, y, t) = 0, 0 \le y \le c, t \ge 0,\\
&u(b, y, t) = 0, 0 \le y \le c, t \ge 0,\\
&u(x, 0, t) = 0, 0 \le x \le b, t \ge 0,\\
&u(x, c, t) = 0, 0 \le x \le b, t \ge 0
\end{aligned}$$

and the initial condition

$$u(x, y, 0) = f(x, y), 0 \le x \le b, 0 \le y \le c.$$

To solve the problem, let us put

$$u(x, y, t) = S(x, y)T(t) \tag{12.219}$$

where S is function of x and y and T is function of t only.

Substituting Eq. (12.219) into Eq. (12.218), we obtain

$$T\left(\frac{\partial^2 S}{\partial x^2} + \frac{\partial^2 S}{\partial y^2}\right) = \frac{S}{D}\frac{dT}{dt}.$$

Dividing both sides by $S(x, y)T(t)$, we have

$$\frac{1}{S}\left(\frac{\partial^2 S}{\partial x^2} + \frac{\partial^2 S}{\partial y^2}\right) = \frac{1}{DT}\frac{dT}{dt} = -\lambda$$ where λ is a positive real constant.

Thus, we arrive at the equations

$$\frac{\partial^2 S}{\partial x^2} + \frac{\partial^2 S}{\partial y^2} + \lambda S = 0 \tag{12.220}$$

and

$$\frac{dT}{dt} = -\lambda DT. \tag{12.221}$$

Further, let us put

$$S(x, y) = X(x)Y(y). \tag{12.222}$$

Substituting Eq. (12.222) into Eq. (12.220), we obtain

$$Y\frac{d^2X}{dx^2} + X\frac{d^2Y}{dy^2} + \lambda XY = 0.$$

Dividing both sides by $X(x)Y(y)$ and transposing, we have

$$\frac{1}{X}\frac{d^2X}{dx^2} = -\frac{1}{Y}\frac{d^2Y}{dy^2} - \lambda = -\alpha^2$$

where α^2 is a positive real constant.

Thus, we arrive at two equations, viz.

$$\frac{d^2X}{dx^2} + \alpha^2 X = 0 \tag{12.223}$$

and

$$\frac{d^2Y}{dy^2} + \beta^2 Y = 0 \tag{12.224}$$

where $$\beta^2 = \lambda - \alpha^2 \text{ or, } \lambda = \alpha^2 + \beta^2. \tag{12.225}$$

General solutions of Eqs. (12.223) and (12.224) are respectively

$$X(x) = A_1 \cos\alpha x + A_2 \sin\alpha x \tag{12.226}$$

and $$Y(y) = B_1 \cos\beta y + B_2 \sin\beta y \tag{12.227}$$

where A_1, A_2, B_1, B_2 are arbitrary constants.

Also, Eq. (12.221) has the solution

$$T(t) = C\,e^{-\lambda Dt} = C\,e^{-D(\alpha^2+\beta^2)t}. \tag{12.228}$$

Using the boundary conditions stated earlier, we arrive at the new boundary conditions

$$X(0) = 0,\ X(b) = 0,\ Y(0) = 0,\ Y(c) = 0.$$

Applying the first boundary conditions to Eq. (12.226), we obtain

$$X(0) = A_1 = 0.$$

Therefore, Eqs. (12.226) becomes

$$X(x) = A_2 \sin \alpha x. \tag{12.229}$$

Applying the second boundary condition, we obtain

$$X(b) = A_2 \sin \alpha b = 0 \text{ yielding } \sin \alpha b = 0 \text{ as } A_2 \text{ is arbitrary.}$$

Thus, we have

$$\alpha b = m\pi,\ m = 1, 2, 3 \ldots$$

or $$\alpha = \frac{m\pi}{b}$$

For various values of m, we write

$$\alpha_m = \frac{m\pi}{b} \tag{12.230}$$

and $$X_m(x) = A_m \sin\alpha_m x = A_m \sin\frac{m\pi x}{b},\ m = 1, 2, 3 \ldots \tag{12.231}$$

Applying the third and the fourth boundary conditions to Eq. (12.227), we obtain for various values of n,

$$Y_n(y) = B_n \sin\beta_n y = B_n \sin\frac{n\pi y}{c},\ n = 1, 2, 3\ldots \tag{12.232}$$

where $$\beta_n = \frac{n\pi}{c}.$$

Now, with various values of m and n, Eq.(12:228) yields

$$T_{mn}(t) = C_{mn} e^{-D\left(\frac{m^2\pi^2}{b^2}+\frac{n^2\pi^2}{c^2}\right)t}$$

Thus, for various values of m and n, we write various solutions of Eq. (12.218) as

$$u_{mn}(x,y,t) = X_m(x) Y_n(y) T_{mn}(t)$$

$$= N_{mn} \sin\frac{m\pi x}{b} \sin\frac{n\pi y}{c} e^{-D\left(\frac{m^2\pi^2}{b^2}+\frac{n^2\pi^2}{c^2}\right)t} \tag{12.233}$$

where $N_{mn} = A_m B_n C_{mn}$ is a constant.

By the superposition principle, the general solution of the given problem is written as

$$u(x, y, t) = \sum_{m=1}^{\infty}\sum_{n=1}^{\infty} u_{mn}(x, y, t)$$

$$= \sum_{m=1}^{\infty}\sum_{n=1}^{\infty} N_{mn} \sin\frac{m\pi x}{b} \sin\frac{n\pi y}{c} e^{-D\left(\frac{m^2\pi^2}{b^2}+\frac{n^2\pi^2}{c^2}\right)t}. \tag{12.234}$$

Here, the coefficient N_{mn} is to be determined from the initial condition of the problem.

To Determine N_{mn}

Using the initial condition, we have

$$u(x, y, 0) = \sum_{m=1}^{\infty}\sum_{n=1}^{\infty} N_{mn} \sin\frac{m\pi x}{b} \sin\frac{n\pi y}{c} = f(x,y). \tag{12.235}$$

Multiplying both sides of Eq. (12.235) by $\sin\frac{p\pi x}{b}$ and integrating the results with respect to x within the limit from $x = 0$ to $x = b$, we obtain

$$\sum_{m=1}^{\infty}\sum_{n=1}^{\infty} N_{mn} \sin\frac{n\pi y}{c} \int_0^b \sin\frac{p\pi x}{b} \sin\frac{m\pi x}{b} dx = \int_0^b f(x,y) \sin\frac{p\pi x}{b} dx$$

or

$$\sum_{n=1}^{\infty} N_{pn} \sin\frac{n\pi y}{c} \cdot \frac{b}{2} = \int_0^b f(x, y) \sin\frac{p\pi x}{b} dx. \tag{12.236}$$

Multiplying both sides of Eq. (12.236) by $\sin\frac{q\pi y}{c}$ and integrating the results with respect to y within the limit from $y = 0$ to $y = c$, we obtain

$$\frac{b}{2}\sum_{n=1}^{\infty} N_{pn} \int_0^c \sin\frac{q\pi y}{c} \sin\frac{n\pi y}{c} dy = \int_{y=0}^{c}\int_{x=0}^{b} f(x, y) \sin\frac{p\pi x}{b} \sin\frac{q\pi y}{c} dx\, dy$$

or
$$\frac{b}{2}\cdot N_{pq}\cdot\frac{c}{2}=\int_{y=0}^{c}\int_{x=0}^{b} f(x,y)\sin\frac{p\pi x}{b}\sin\frac{q\pi y}{c}\,dx\,dy$$

or
$$N_{pq}=\frac{4}{bc}\int_{y=0}^{c}\int_{x=0}^{b} f(x,y)\sin\frac{p\pi x}{b}\sin\frac{q\pi y}{c}\,dx\,dy\,.$$

Renaming the subscripts, we write

$$N_{mn}=\frac{4}{bc}\int_{y=0}^{c}\int_{x=0}^{b} f(x,y)\sin\frac{m\pi x}{b}\sin\frac{n\pi y}{c}\,dx\,dy\,. \tag{12.237}$$

Equation (12.234), together with Eq. (12.237), gives the general solution of the problem.

CASE II: Two opposite sides kept perfectly insulated and the other two opposite sides at zero degree:

Let the plate of length b and width c be perfectly insulated at the faces $x = 0$ and $x = b$. Let the other two sides $y = 0$ and $y = c$ be maintained at zero temperature. Let the initial temperature distribution be $f(x, y)$ Then, we seek the solution of the two-dimensional heat conduction equation

$$\frac{\partial^2 u}{\partial x^2}+\frac{\partial^2 u}{\partial y^2}=\frac{1}{D}\frac{\partial u}{\partial t} \tag{12.238}$$

where D is the thermal diffusivity of the plate.

The boundary conditions are

$$\left[\frac{\partial u}{\partial x}\right]_{x=0}=0,\left[\frac{\partial u}{\partial x}\right]_{x=b}=0,\ u(x,0,t)=0,\ u(x,c,t)=0. \tag{12.239}$$

The initial condition is

$$u(x, y, 0) = f(x, y). \tag{12.240}$$

As in Case I, putting

$$u(x, y, t) = S(x, y)\,T(t) \text{ and } S(x, y) = X(x)\,Y(y),$$

we obtain

$$X(x) = A_1 \cos \alpha x + A_2 \sin \alpha x, \tag{12.241}$$

$$Y(y) = B_1 \cos \beta y + B_2 \sin \beta y \tag{12.242}$$

and
$$T(t) = C\,e^{-\lambda D t} \tag{12.243}$$

where α, β, $\lambda(= \alpha^2 + \beta^2)$, A_1, A_2, B_1, B_2 and C are arbitrary constants.

The new boundary conditions are

$$\left[\frac{dX}{dx}\right]_{x=0}=0,\left[\frac{dX}{dx}\right]_{x=b}=0,\ Y(0)=0,\ Y(c)=0.$$

Applying the first boundary condition to Eq. (12.241), we have

$$-\alpha A_1 \sin 0 + \alpha A_2 \cos 0 = 0 \text{ or, } A_2 = 0 \text{ as } \alpha \neq 0.$$

Using this, Eq. (12.241) yields

$$X(x) = A_1 \cos \alpha x.$$

Applying the second boundary condition to this equation, we obtain

$$\alpha A_1 \sin \alpha b = 0, \alpha \neq 0, A_1 \neq 0 \quad \text{or} \quad \alpha b = m\pi, (m = 0, 1, 2, 3 \ldots).$$

Therefore, for particular values of m, we have

$$\alpha = \alpha_m = \frac{m\pi}{b}, (m = 0, 1, 2, 3 \ldots)$$

and consequently, $X_m(x) = A_m \cos \dfrac{m\pi x}{b} \quad (m = 0, 1, 2, 3 \ldots)$ (12.244)

where A_m is arbitrary constant. Here, $m = 0$ is also included as it yields a solution.

Applying the third boundary condition to Eq. (12.242), we have

$$B_1 \cos 0 + B_2 \sin 0 = 0 \text{ giving } B_1 = 0.$$

Thus, Eq. (12.242) yields

$$Y(y) = B_2 \sin \beta y.$$

Applying the fourth boundary condition to this equation, we obtain for particular values of n,

$$\beta = \beta_n = \frac{n\pi}{c}, (n = 1, 2, 3 \ldots)$$

and

$$Y_n(y) = B_n \sin \frac{n\pi y}{c}, (n = 1, 2, 3 \ldots) \tag{12.245}$$

where B_n is an arbitrary constant.

Also, Eq. (12.243) yields

$$T_{mn}(t) = C_{mn} e^{-D\left(\frac{m^2\pi^2}{b^2} + \frac{n^2\pi^2}{c^2}\right)t} \quad [\text{where } C_{mm} = \text{constant}]. \tag{12.246}$$

Recalling that $u(x, y, t) = S(x, y)\, T(t) = X(x)\, Y(y)\, T(t)$, we write for particular values of m and n,

$$u_{mn}(x,y,t) = X_m(x) Y_n(y) T_{mn}(t)$$

$$= N_{mn} \cos \frac{m\pi x}{b} \sin \frac{n\pi y}{c} e^{-D\left(\frac{m^2\pi^2}{b^2} + \frac{n^2\pi^2}{c^2}\right)t} \tag{12.247}$$

where $N_{mn} = A_m\, B_n\, C_{mm}$ is an arbitrary constant.

These are the various solutions of Eq. (12.238).

By the superposition principle, the general solution of the problem is

$$u(x, y, t) = \sum_{m=0}^{\infty}\sum_{n=1}^{\infty} u_{mn}(x, y, t)$$

$$= \sum_{m=0}^{\infty}\sum_{n=1}^{\infty} N_{mn} \cos\frac{m\pi x}{b} \sin\frac{n\pi y}{c} e^{-D\left(\frac{m^2\pi^2}{b^2}+\frac{n^2\pi^2}{c^2}\right)t} \tag{12.248}$$

where the coefficient (N_{mn}) is to be determined from the initial condition.

To Determine N_{mn}

Applying the initial condition, we have

$$u(x,y,0) = \sum_{m=0}^{\infty}\sum_{n=1}^{\infty} N_{mn} \cos\frac{m\pi x}{b} \sin\frac{n\pi y}{c}$$

$$= \sum_{n=1}^{\infty} N_{0n} \sin\frac{n\pi y}{c} + \sum_{m=1}^{\infty}\sum_{n=1}^{\infty} N_{mn} \cos\frac{m\pi x}{b} \sin\frac{n\pi y}{c} = f(x, y). \tag{12.249}$$

Multiplying both sides by $\sin\frac{q\pi y}{c}$ and then integrating the results with respect to y within the limit from $y = 0$ to $y = c$, we have

$$N_{0q}\int_0^c \sin^2\left(\frac{q\pi y}{c}\right) dy + \sum_{m=1}^{\infty} N_{mq} \cos\frac{m\pi x}{b}\int_0^c \sin^2\left(\frac{q\pi y}{c}\right) dy = \int_0^c f(x, y)\sin\frac{q\pi y}{c} dy$$

or

$$\frac{c}{2}N_{0q} + \sum_{m=1}^{\infty} \frac{c}{2} N_{mq} \cos\frac{m\pi x}{b} = \int_0^c f(x, y)\sin\frac{q\pi y}{c} dy.$$

Integrating both sides with respect to x within the limit from $x = 0$ to $x = b$, we have

$$\frac{c}{2}N_{0q}\int_0^b dx + \sum_{m=1}^{\infty} \frac{c}{2} N_{mq}\int_0^b \cos\frac{m\pi x}{b} dx = \int_{y=0}^{c}\int_{x=0}^{b} f(x, y)\sin\frac{q\pi y}{c} dx\, dy$$

or

$$\frac{bc}{2}N_{0q} + 0 = \int_{y=0}^{c}\int_{x=0}^{b} f(x, y)\sin\frac{q\pi y}{c} dx\, dy$$

or

$$N_{0q} = \frac{2}{bc}\int_{y=0}^{c}\int_{x=0}^{b} f(x, y)\sin\frac{q\pi y}{c} dx\, dy.$$

Renaming the subscript, we write

$$N_{0n} = \frac{2}{bc} \int_{y=0}^{c} \int_{x=0}^{b} f(x, y) \sin \frac{n\pi y}{c} \, dx \, dy. \tag{12.250}$$

Multiplying both sides of Eq. (12.249) by $\cos \frac{p\pi x}{b}$, integrating the results with respect to x within the limit from $x = 0$ to $x = b$ and further multiplying both sides by $\sin \frac{q\pi y}{c}$ and integrating the results with respect to y within the limit from $y = 0$ to $y = c$, we obtain

$$N_{pq} = \frac{4}{bc} \int_{y=0}^{c} \int_{x=0}^{b} f(x, y) \cos \frac{p\pi x}{b} \sin \frac{q\pi y}{c} \, dx \, dy, \; p \geq 1.$$

Renaming the subscripts, we write

$$N_{mn} = \frac{4}{bc} \int_{y=0}^{c} \int_{x=0}^{b} f(x, y) \cos \frac{m\pi x}{b} \sin \frac{n\pi y}{c} \, dx \, dy, \; m \geq 1. \tag{12.251}$$

Equation (12.248), together with Eqs. (12.250) and (12.251) gives the solution of the problem in this case.

CASE III: All the four sides are perfectly insulated:
Let the plate of length b and width c be perfectly insulated at the faces $x = 0$, $x = b$, $y = 0$, $y = c$. Let the initial temperature distribution be $f(x, y)$. We seek the solution of the two-dimensional heat conduction equation

$$\frac{\partial^2 u}{\partial x^2} + \frac{\partial^2 u}{\partial y^2} = \frac{1}{D} \frac{\partial u}{\partial t} \tag{12.252}$$

where D is the thermal diffusivity of the plate.

The boundary conditions are

$$\left[\frac{\partial u}{\partial x}\right]_{x=0} = 0, \quad \left[\frac{\partial u}{\partial x}\right]_{x=b} = 0, \quad \left[\frac{\partial u}{\partial y}\right]_{y=0} = 0, \quad \left[\frac{\partial u}{\partial y}\right]_{y=c} = 0. \tag{12.253}$$

The initial condition is

$$u(x, y, 0) = f(x, y). \tag{12.254}$$

As in Case I, putting

$$u(x, y, t) = S(x, y)\, T(t) \text{ and } S(x, y) = X(x)\, Y(y),$$

we obtain

$$X(x) = A_1 \cos \alpha x + A_2 \sin \alpha x, \tag{12.255}$$

$$Y(y) = B_1 \cos \beta y + B_2 \sin \beta y \tag{12.256}$$

and
$$T(t) = C\, e^{-\lambda D t} \tag{12.257}$$

where α, β, $\lambda (= \alpha^2 + \beta^2)$, A_1, A_2, B_1, B_2 and C are arbitrary constants.

The new boundary conditions are

$$\left[\frac{dX}{dx}\right]_{x=0} = 0, \left[\frac{dX}{dx}\right]_{x=b} = 0, \left[\frac{dY}{dy}\right]_{y=0} = 0, \left[\frac{dY}{dy}\right]_{y=c} = 0.$$

Applying the first boundary condition to Eq. (12.255), we have

$$-\alpha A_1 \sin 0 + \alpha A_2 \cos 0 = 0 \text{ or } A_2 = 0 \text{ as } \alpha \neq 0.$$

Using this value of A_2, Eq. (12.255) yields

$$X(x) = A_1 \cos \alpha x.$$

Applying the second boundary condition to this equation, we obtain

$$\alpha A_1 \sin \alpha b = 0, \alpha \neq 0, A_1 \neq 0 \quad \text{or} \quad \alpha b = m\pi, (m = 0, 1, 2, 3 \ldots).$$

Therefore, for particular values of m, we have

$$\alpha = \alpha_m = \frac{m\pi}{b}, (m = 0, 1, 2, 3 \ldots)$$

and consequently, $X_m(x) = A_m \cos \dfrac{m\pi x}{b}, (m = 0, 1, 2, 3 \ldots)$ (12.258)

where A_m is arbitrary constant. Here, $m = 0$ is also included as it yields a solution.

Similarly, applying the third and the fourth boundary conditions to Eq. (12.256), we obtain for particular values of n,

$$\beta = \beta_n = \frac{n\pi}{c}, (n = 0, 1, 2, 3 \ldots)$$

and
$$Y_n(y) = B_n \cos \frac{n\pi y}{c}, (n = 0, 1, 2, 3 \ldots) \tag{12.259}$$

where B_n is an arbitrary constant. Here, $n = 0$ is also included as it yields a solution.

Thus, Eq. (12.257) yields

$$T_{mn}(t) = C_{mn} e^{-D\left(\frac{m^2\pi^2}{b^2} + \frac{n^2\pi^2}{c^2}\right)t}. \tag{12.260}$$

Recalling that $u(x, y, t) = S(x, y)T(t) = X(x)Y(y)T(t)$, we write for particular values of m and n,

$$u_{mn}(x, y, t) = X_m(x)\, Y_n(y)\, T_{mn}(t)$$

$$= N_{mn} \cos\frac{m\pi x}{b} \cos\frac{n\pi y}{c} e^{-D\left(\frac{m^2\pi^2}{b^2}+\frac{n^2\pi^2}{c^2}\right)t} \tag{12.261}$$

where $N_{mn} = A_m B_n C_{mn}$ is arbitrary constant.

These are the various solutions of Eq. (12.252).

By superposition principle, the general solution of the problem is written as

$$u(x, y, t) = \sum_{m=0}^{\infty}\sum_{n=0}^{\infty} u_{mn}(x,y,t) = \sum_{m=0}^{\infty}\sum_{n=0}^{\infty} N_{mn} \cos\frac{m\pi x}{b} \cos\frac{n\pi y}{c} e^{-D\left(\frac{m^2\pi^2}{b^2}+\frac{n^2\pi^2}{c^2}\right)t} \tag{12.262}$$

where the coefficient (N_{mn}) is to be determined from the initial condition.

To Determine N_{mn}

Applying the initial condition, we have

$$u(x,y,0) = f(x, y) = \sum_{m=0}^{\infty}\sum_{n=0}^{\infty} N_{mn} \cos\frac{m\pi x}{b} \cos\frac{n\pi y}{c}$$

$$= N_{00} + \sum_{m=1}^{\infty} N_{m0} \cos\frac{m\pi x}{b} + \sum_{n=1}^{\infty} N_{0n} \cos\frac{n\pi y}{c}$$

$$+ \sum_{m=1}^{\infty}\sum_{n=1}^{\infty} N_{mn} \cos\frac{m\pi x}{b} \cos\frac{n\pi y}{c}. \tag{12.263}$$

Integrating both sides with respect to x and y, respectively within the limits from $x = 0$, $y = 0$ to $x = b$, $y = c$, we obtain

$$\int_{y=0}^{c}\int_{x=0}^{b} f(x, y)dx\, dy = N_{00} \int_{y=0}^{c}\int_{x=0}^{b} dx\, dy + 0 = bc\, N_{00}$$

or
$$N_{00} = \frac{1}{bc}\int_{y=0}^{c}\int_{x=0}^{b} f(x, y)dx\, dy. \tag{12.264}$$

Multiplying both sides of Eq. (12.263) by $\cos\frac{p\pi x}{b}$ $(p \geq 1)$ and then integrating the results with respect to x within the limit from $x = 0$ to $x = b$, we have

$$\int_0^b f(x, y)\cos\frac{p\pi x}{b} dx = 0 + N_{p0}\int_0^b \cos^2\left(\frac{p\pi x}{b}\right)dx + 0 + \sum_{n=1}^{\infty} N_{pn} \cos\frac{n\pi y}{c}\int_0^b \cos^2\left(\frac{p\pi x}{b}\right)dx$$

$$= \frac{b}{2}N_{p0} + \sum_{n=1}^{\infty}\frac{b}{2}N_{pn} \cos\frac{n\pi y}{c}.$$

Integrating both sides with respect to y within the limit from $y = 0$ to $y = c$, we have

$$\int_{y=0}^{c}\int_{x=0}^{b} f(x, y)\cos\frac{p\pi x}{b}\,dx\,dy = \frac{b}{2}N_{p0}\int_{0}^{c} dy + 0 = \frac{bc}{2}N_{p0}$$

or
$$N_{p0} = \frac{2}{bc}\int_{y=0}^{c}\int_{x=0}^{b} f(x, y)\cos\frac{p\pi x}{b}\,dx\,dy.$$

Renaming the subscript, we write

$$N_{m0} = \frac{2}{bc}\int_{y=0}^{c}\int_{x=0}^{b} f(x, y)\cos\frac{m\pi x}{b}\,dx\,dy\,(m \geq 1). \tag{12.265}$$

Similarly, multiplying both sides of Eq. (12.263) by $\cos\frac{q\pi y}{c}(q \geq 1)$ and then integrating the results with respect to y within the limit from $y = 0$ to $y = c$ and further integrating the results with respect to x within the limit from $x = 0$ to $x = b$ we have

$$N_{0q} = \frac{2}{bc}\int_{y=0}^{c}\int_{x=0}^{b} f(x, y)\cos\frac{q\pi y}{c}\,dx\,dy.$$

Renaming the subscripts, we write

$$N_{0n} = \frac{2}{bc}\int_{y=0}^{c}\int_{x=0}^{b} f(x, y)\cos\frac{n\pi y}{c}\,dx\,dy\,(n \geq 1). \tag{12.266}$$

Multiplying both sides of Eq. (12.263) by $\cos\frac{p\pi x}{b}$, integrating the result with respect to x within the limit from $x = 0$ to $x = b$, further multiplying both sides by $\cos\frac{q\pi y}{c}$ and integrating the result with respect to y within the limit from $y = 0$ to $y = c$, we obtain

$$N_{pq} = \frac{4}{bc}\int_{y=0}^{c}\int_{x=0}^{b} f(x, y)\cos\frac{p\pi x}{b}\cos\frac{q\pi y}{c}\,dxdy,\ p \geq 1, q \geq 1.$$

Renaming the subscripts, we write

$$N_{mn} = \frac{4}{bc}\int_{y=0}^{c}\int_{x=0}^{b} f(x, y)\cos\frac{m\pi x}{b}\cos\frac{n\pi y}{c}\,dx\,dy,\ m \geq 1, n \geq 1. \tag{12.267}$$

Equation (12.262), together with Eqs. (12.264), (12.265), (12.266) and (12.267) gives the solution of the problem in this case.

12.24 HEAT CONDUCTION IN A CIRCULAR PLATE

Let us find the temperature distribution in a circular plate of radius a when the initial temperature is $f(r)$ and the rim is kept at zero temperature of some scale. It is assumed that the faces are perfectly insulated and the temperature distribution is radially symmetric.

The general form of equation for heat conduction is

$$\nabla^2 u = \frac{1}{D}\frac{\partial u}{\partial t}$$

where $u(r, t)$ is the temperature at a point distant r from the centre at time t and D is the thermal diffusivity of the plate. In plane polar coordinates, the equation is

$$\frac{\partial^2 u}{\partial r^2} + \frac{1}{r}\frac{\partial u}{\partial r} = \frac{1}{D}\frac{\partial u}{\partial t}. \tag{12.268}$$

The boundary condition is

$$u(a, t) = 0 \tag{12.269}$$

and the initial condition is

$$u(r, 0) = f(r). \tag{12.270}$$

Further, there exists a finiteness condition

$$u(0, t) < \infty. \tag{12.271}$$

Using the method of separation of variables, let us put

$$u(r, t) = R(r)\, T(t). \tag{12.272}$$

Substituting Eq. (12.272) into Eq. (12.268), we obtain

$$T\left[\frac{d^2R}{dr^2} + \frac{1}{r}\frac{dR}{dr}\right] = \frac{R}{D}\frac{dT}{dt}.$$

Dividing both sides by $R(r)T(t)$, we obtain

$$\frac{1}{R}\left[\frac{d^2R}{dr^2} + \frac{1}{r}\frac{dR}{dr}\right] = \frac{1}{DT}\frac{dT}{dt} = \lambda, \text{ (a constant)}.$$

Thus, we obtain two equations,

$$\frac{d^2R}{dr^2} + \frac{1}{r}\frac{dR}{dr} - \lambda R = 0 \tag{12.273}$$

and $$\frac{dT}{dt} = \lambda DT. \tag{12.274}$$

Choice of the constant λ:

Choice I: $\lambda = 0$:
In this case, Eq. (12.273) becomes

$$\frac{d^2R}{dr^2} + \frac{1}{r}\frac{dR}{dr} = 0.$$

Putting $r = e^z$ or $z = \ln r$, we arrive at

$$\frac{d^2R}{dz^2} = 0 \text{ yielding } R(z) = Az + B$$

or $$R(r) = A \ln r + B$$

where A and B are arbitrary constants.

Also, Eq. (12.274) becomes $\frac{dT}{dt} = 0$ yielding $T(t) = C$, a constant.

Thus, we obtain

$$u(r, t) = R(r)\, T(t) = C[A \ln r + B] = A_1 \ln r + A_2$$

where $A_1 = CA$, $A_2 = CB$.

The finiteness condition $u(0, t) < \infty$ demands that $A_1 = 0$. Further, the boundary condition $u(a, t) = 0$ requires that $A_2 = 0$. Thus, we have $u(r, t) = 0$ which is not admissible. Therefore, the choice $\lambda = 0$ is not admissible.

Choice II: $\lambda > 0$:
In this case, let us put $\lambda = \alpha^2$ (positive real number). Then Eqs. (12.273) and (12.274) become respectively

$$r^2\frac{d^2R}{dr^2} + r\frac{dR}{dr} - \alpha^2 r^2 R = 0 \text{ and } \frac{dT}{dt} = \alpha^2 DT.$$

The first equation has the general solution

$$R(r) = C_1 I_0(\alpha r) + C_2 K_0(\alpha r)$$

where C_1 and C_2 are arbitrary constants and I_0 and K_0 are modified Bessel functions of the first and the second kinds of order zero.

The other equation has the solution

$$T(t) = Be^{\alpha^2 Dt}$$

where B is arbitrary constant.

Thus, we obtain

$$u(r,t) = R(r)T(t) = \left[B_1 I_0(\alpha r) + B_2 K_0(\alpha r)\right] e^{\alpha^2 Dt}$$

where $$B_1 = BC_1 \text{ and } B_2 = BC_2.$$

Finiteness of temperature at the centre for all time t discards K_0 to be a solution. Hence, we have

$$u(r,t) = B_1 I_0(\alpha r) e^{\alpha^2 Dt}.$$

This solution is inadmissible as it yields $u(a,t) \to \infty$ as $t \to \infty$ which is absurd. Hence, the choice $\lambda > 0$ is not admissible.

Choice III: $\lambda < 0$:

In this case, let us put $\lambda = -\mu^2$ (negative real number). Then Eqs. (12.273) and (12.274) become respectively

$$r^2 \frac{d^2R}{dr^2} + r\frac{dR}{dr} + \mu^2 r^2 R = 0 \text{ and } \frac{dT}{dt} = -\mu^2 DT$$

The first equation has the general solution

$$R(r) = C_3 J_0(\mu r) + C_4 N_0(\mu r)$$

where C_3 and C_4 are arbitrary constants and J_0 and N_0 are Bessel functions of the first and the second kinds of order zero.

The other equation has the solution

$$T(t) = Be^{-\mu^2 Dt}$$

where B is arbitrary constant.

Thus, we obtain

$$u(r,t) = R(r)\,T(t) = [M J_0(\mu r) + P N_0(\mu r)]e^{-\mu^2 Dt}$$

where $M = BC_3$ and $P = BC_4$ are arbitrary constants.

Finiteness of temperature at the centre for all time t discards N_0 to be a solution. Hence, we have

$$u(r,t) = M J_0(\mu r) e^{-\mu^2 Dt}. \tag{12.275}$$

Applying the boundary condition $u(a, t) = 0$, we have

$$M J_0(\mu a) e^{-\mu^2 Dt} = 0 \text{ for all } t.$$

Since M is an arbitrary constant, we have

$$J_0(\mu a) = 0 \text{ or, } \mu a = k_n$$

where k_n ($n = 1, 2, 3 \ldots$) are zeroes of the Bessel function of the first kind having the order zero.

Therefore, for a particular n, we write

$$\mu = \mu_n = \frac{k_n}{a}.$$

For various values of n, using Eq. (12.275), the solutions of heat conduction equation (12.268) are

$$u_n(r,t) = M_n J_0\left(\frac{k_n}{a} r\right) e^{-\frac{k_n^2}{a^2}Dt}. \tag{12.276}$$

By the superposition principle, the general solution of the problem is

$$u(r,t) = \sum_{n=1}^{\infty} u_n(r,t) = \sum_{n=1}^{\infty} M_n J_0\left(\frac{k_n}{a} r\right) e^{-\frac{k_n^2}{a^2}Dt} \tag{12.277}$$

where the coefficients (M_n) are to be determined from the initial condition of the problem.

To Determine M_n:

Applying the initial condition, Eq. (12.277) yields

$$u(r,0) = f(r) = \sum_{n=1}^{\infty} M_n J_0\left(\frac{k_n}{a} r\right)$$

or

$$u(\rho,0) = f(\rho) = \sum_{n=1}^{\infty} M_n J_0(k_n \rho) \tag{12.278}$$

where we put $\rho = \frac{r}{a}$.

Multiplying both sides of Eq. (12.278) by $\rho J_0(k_m \rho)$ and integrating the results with respect to ρ within the limit from $\rho = 0$ to $\rho = 1$, we obtain

$$\int_0^1 f(\rho) J_0(k_m\rho)\rho d\rho = \sum_{n=1}^{\infty} M_n \int_0^1 J_0(k_m\rho) J_0(k_n\rho)\rho d\rho$$

$$= M_m \frac{1}{2} J_1^2(k_m) \text{ [using orthonormality of Bessel functions].}$$

Therefore, we have

$$M_m = \frac{2}{J_1^2(k_m)} \int_0^1 f(\rho) J_0(k_m\rho)\rho d\rho.$$

Renaming the subscript, we write

$$M_n = \frac{2}{J_1^2(k_n)} \int_0^1 f(\rho) J_0(k_n\rho)\rho d\rho$$

$$= \frac{2}{a^2 J_1^2(k_n)} \int_0^a f\left(\frac{r}{a}\right) J_0\left(\frac{k_n}{a} r\right) r dr \tag{12.279}$$

Equation (12.277), together with Eq. (12.279), gives the general solution of the problem.

12.25 HEAT CONDUCTION IN A RECTANGULAR PARALLELEPIPED OR CUBOID

Let us consider the problem of heat conduction in a rectangular parallelepiped or cuboid which is placed in the first octant. Let a, b and c be the lengths of its edges along the axes of x, y and z, respectively. Let the faces be maintained at zero degree temperature of some scale. The temperature $u(x, y, z, t)$ at the point (x, y, z) at time t is given by the equation

$$\nabla^2 u = \frac{1}{D} \frac{du}{dt}$$

or

$$\frac{\partial^2 u}{\partial x^2} + \frac{\partial^2 u}{\partial y^2} + \frac{\partial^2 u}{\partial z^2} = \frac{1}{D} \frac{\partial u}{\partial t} \tag{12.280}$$

where D is the thermal diffusivity of the parallelepiped or cuboid.

The boundary conditions of the problem are

$$u(0, y, z, t) = 0, u(a, y, z, t) = 0;$$
$$u(x, 0, z, t) = 0, u(x, b, z, t) = 0;$$
$$u(x, y, 0, t) = 0, u(x, y, c, t) = 0.$$

The initial condition of the problem is

$$u(x, y, z, 0) = f(x, y, z).$$

Let us assume the solution of the problem as

$$u(x, y, z, t) = W(x, y, z)\, T(t). \tag{12.281}$$

Substituting Eq. (12.281) into Eq. (12.280), we obtain

$$T\nabla^2 W = \frac{W}{D} \frac{dT}{dt}.$$

Dividing both sides by $W(x, y, z)\, T(t)$, we have

$$\frac{1}{W} \nabla^2 W = \frac{1}{DT} \frac{dT}{dt} = -\lambda^2 \text{ (a constant).}$$

Thus, we obtain two equations

$$\nabla^2 W + \lambda^2 W = 0 \text{ (Helmholtz equation)} \tag{12.282}$$

and
$$\frac{dT}{dt} = -\lambda^2 DT. \tag{12.283}$$

Equation (12.283) has the solution

$$T(t) = E\,e^{-\lambda^2 Dt} \tag{12.284}$$

where E is arbitrary constant.

Further, let us put

$$W(x, y, z) = S(x, y)\, Z(z). \tag{12.285}$$

Substitution of Eq. (12.285) into Eq. (12.282) (which is known as Helmholtz equation), yields the equations

$$Z\left(\frac{\partial^2 S}{\partial x^2} + \frac{\partial^2 S}{\partial y^2}\right) + S\frac{d^2 Z}{dz^2} + \lambda^2 SZ = 0.$$

Dividing both sides by $S(x, y)\, Z(z)$ and then transposing, we obtain

$$\frac{1}{Z}\frac{d^2 Z}{dz^2} = -\frac{1}{S}\left(\frac{\partial^2 S}{\partial x^2} + \frac{\partial^2 S}{\partial y^2}\right) - \lambda^2 = -\alpha_n^2 \text{ (another constant).}$$

Thus, we obtain two equations

$$\frac{d^2 Z}{dz^2} + \alpha_n^2 Z = 0 \tag{12.286}$$

and
$$\frac{\partial^2 S}{\partial x^2} + \frac{\partial^2 S}{\partial y^2} + (\lambda^2 - \alpha_n^2)S = 0. \tag{12.287}$$

Let us further put

$$S(x, y) = X(x)\, Y(y). \tag{12.288}$$

Substitution of Eq. (12.288) into Eq. (12.287) results in

$$Y\frac{d^2 X}{dx^2} + X\frac{d^2 Y}{dy^2} + (\lambda^2 - \alpha_n^2)XY = 0.$$

Dividing both sides by $X(x)\, Y(y)$ and transposing, we obtain

$$\frac{1}{Y}\frac{d^2 Y}{dy^2} = -\frac{1}{X}\frac{d^2 X}{dx^2} - (\lambda^2 - \alpha_n^2) = -\alpha_m^2 \text{ (another constant).}$$

We obtain two equations

$$\frac{d^2 Y}{dy^2} + \alpha_m^2 Y = 0 \tag{12.289}$$

and
$$\frac{d^2X}{dx^2}+\alpha_l^2X=0 \tag{12.290}$$

where we put

$$\alpha_l^2=\lambda^2-\alpha_m^2-\alpha_n^2 \text{ or, } \lambda^2=\alpha_l^2+\alpha_m^2+\alpha_n^2. \tag{12.291}$$

Now, for particular values of l, m, n; Eqs. (12.286), (12.289) and (12.290) have the general solutions

$$Z_n(z)=C_1\cos(\alpha_n z)+C_2\sin(\alpha_n z), \tag{12.292}$$
$$Y_m(y)=B_1\cos(\alpha_m y)+B_2\sin(\alpha_m y), \tag{12.293}$$
$$X_l(x)=A_1\cos(\alpha_l x)+A_2\sin(\alpha_l x) \tag{12.294}$$

where $A_j, B_j, C_j\ (j=1,2)$ are arbitrary constants.

Thus, for particular values of l, m, n; we obtain the solutions of Eq. (12.280) as

$$u_{lmn}(x,y,z,t)=\left[A_1\cos(\alpha_l x)+A_2\sin(\alpha_l x)\right]\left[B_1\cos(\alpha_m y)+B_2\sin(\alpha_m y)\right]$$
$$\times\left[C_1\cos(\alpha_n z)+C_2\sin(\alpha_n z)\right]E\,e^{-\lambda^2 DT}. \tag{12.295}$$

Applying the boundary condition, $u(0,y,z,t)=0$, we have $A_1=0$.
Next, applying the boundary condition, $u(a, y, z, t) = 0$ in Eq. (12.295), we have

$$A_2\sin a\alpha_l=0 \text{ yielding } \alpha_l=\frac{l\pi}{a}, A_2\neq 0, l=1, 2, 3 \ldots$$

Similarly, applying the remaining boundary conditions succssively to Eq. (12.295), we obtain

$$B_1=0, \alpha_m=\frac{m\pi}{b}, B_2\neq 0, m=1,2,3 \ldots$$

and
$$C_1=0, \alpha_n=\frac{n\pi}{c}, C_2\neq 0, n=1,2,3 \ldots$$

Therefore, Eq. (12.295) yields the solutions

$$u_{lmn}(x,y,z,t)=A_{2l}\,B_{2m}\,C_{2n}\sin\left(\frac{l\pi x}{a}\right)\sin\left(\frac{m\pi y}{b}\right)\sin\left(\frac{n\pi z}{c}\right)\times E_{lmn}e^{-\left(\frac{l^2}{a^2}+\frac{m^2}{b^2}+\frac{n^2}{c^2}\right)\pi^2 Dt}$$

$$=R_{lmn}\sin\left(\frac{l\pi x}{a}\right)\sin\left(\frac{m\pi y}{b}\right)\sin\left(\frac{n\pi z}{c}\right)e^{-\left(\frac{l^2}{a^2}+\frac{m^2}{b^2}+\frac{n^2}{c^2}\right)\pi^2 Dt} \tag{12.296}$$

where use has been made of Eq. (12.291), $R_{lmn} = A_{2l}\, B_{2m}\, C_{2n} E_{lmn}$ and $l = 1, 2, 3 \ldots$; $m = 1, 2, 3 \ldots$; $n = 1, 2, 3 \ldots$

Using the superposition principle, we write the general solution of the given heat equation as

$$u(x, y, z, t) = \sum_{l=1}^{\infty}\sum_{m=1}^{\infty}\sum_{n=1}^{\infty} u_{lmn}(x, y, z, t)$$

$$= \sum_{l=1}^{\infty}\sum_{m=1}^{\infty}\sum_{n=1}^{\infty} R_{lmn} \sin\left(\frac{l\pi x}{a}\right)\sin\left(\frac{m\pi y}{b}\right)\sin\left(\frac{n\pi z}{c}\right) e^{-\left(\frac{l^2}{a^2}+\frac{m^2}{b^2}+\frac{n^2}{c^2}\right)\pi^2 Dt}. \quad (12.297)$$

To Find the Coefficient R_{lmn}:

Application of the initial condition to Eq. (12.297) yields

$$R_{lmn} = \frac{8}{abc}\int_0^c\int_0^b\int_0^a f(x,y,z)\sin\left(\frac{l\pi x}{a}\right)\sin\left(\frac{m\pi y}{b}\right)\sin\left(\frac{n\pi z}{c}\right) dx\, dy\, dz. \quad (12.298)$$

Equation (12.297), together with Eq. (12.298), gives the general solution of the given heat conduction problem.

12.26 CLASSICAL WAVE EQUATIONS

The general form of classical wave equation is

$$\nabla^2 u = \frac{1}{c^2}\frac{\partial^2 u}{\partial t^2} \quad (12.299)$$

where ∇^2 is the Laplacian operator, u is the displacement of a point at time t and c is the constant speed of wave propagation.

12.26A ONE-DIMENSIONAL WAVE EQUATION

The one-dimensional wave equation is

$$\frac{\partial^2 u}{\partial x^2} = \frac{1}{c^2}\frac{\partial^2 u}{\partial t^2} \quad (12.300)$$

where $u(x, t)$ is the deflection of the point x at time t and c is the constant speed of wave propagation.

This equation was solved in the case of transverse vibrations of a stretched string in Sections 12.6 and 12.7.

12.26B TWO-DIMENSIONAL WAVE EQUATION

I. In Cartesian Coordinates:

The two-dimensional wave equation is

$$\frac{\partial^2 u}{\partial x^2}+\frac{\partial^2 u}{\partial y^2}=\frac{1}{c^2}\frac{\partial^2 u}{\partial t^2} \tag{12.301}$$

where $u(x, y, t)$ is the deflection of the point (x, y) at time t and c is the constant speed of wave propagation.

This equation was solved in the case of transverse vibrations of a stretched rectangular membrane and of a square membrane in Sections 12.12 and 12.13.

II. In Plane Polar Coordinates:

The two-dimensional wave equation is written as

$$\frac{\partial^2 u}{\partial r^2}+\frac{1}{r}\frac{\partial u}{\partial r}+\frac{1}{r^2}\frac{\partial^2 u}{\partial \theta^2}=\frac{1}{c^2}\frac{\partial^2 u}{\partial t^2} \tag{12.302}$$

where $u(r, \theta, t)$ is the deflection of the point (r, θ) at time t and c is the constant speed of wave propagation.

This equation was solved in the case of transverse vibrations of a stretched circular membrane in Section 12.15.

12.26C THREE-DIMENSIONAL WAVE EQUATION

I. In Cartesian Coordinates:

The three-dimensional wave equation is

$$\nabla^2 u=\frac{\partial^2 u}{\partial x^2}+\frac{\partial^2 u}{\partial y^2}+\frac{\partial^2 u}{\partial z^2}=\frac{1}{v^2}\frac{\partial^2 u}{\partial t^2} \tag{12.303}$$

where $u(x, y, z, t)$ is the displacement of the point (x, y, z) at time t and v is the constant speed of wave propagation.

This equation describes the propagation of waves due to an initial disturbance in a rectangular parallelepiped of sides a, b and c, respectively.

Let there be the boundary conditions

$$u(0, y, z, t) = 0,\ u(a, y, z, t) = 0, \tag{12.304}$$

$$u(x, 0, z, t) = 0,\ u(x, b, z, t) = 0, \tag{12.305}$$

$$u(x, y, 0, t) = 0,\ u(x, y, c, t) = 0. \tag{12.306}$$

Further, let there be the initial conditions

$$u(x, y, z, 0) = f(x, y, z) \text{ for } 0 \le x \le a,\ 0 \le y \le b,\ 0 \le z \le c \tag{12.307}$$

and

$$\left[\frac{\partial u(x, y, z, t)}{\partial t}\right]_{t=0} = g(x, y, z) \text{ for } 0 \le x \le a,\ 0 \le y \le b,\ 0 \le z \le c. \tag{12.308}$$

Let us assume a nontrivial solution in the form

$$u(x, y, z, t) = R(x, y, z)\ T(t). \tag{12.309}$$

Then, we obtain two separated equations as

$$\nabla^2 R + \lambda^2 R = 0 \text{ (Helmholtz equation)} \tag{12.310}$$

and

$$\frac{d^2T}{dt^2} + \lambda^2 v^2 T = 0 \tag{12.311}$$

where $-\lambda^2$ is a separation constant.

Further, let us write a nontrivial solution of Eq. (12.310) as

$$R(x, y, z) = X(x)\ Y(y)\ Z(z). \tag{12.312}$$

Substitution of this into Eq. (12.310) yields

$$\frac{d^2X}{dx^2} - \mu X = 0 \tag{12.313}$$

$$\frac{d^2Y}{dy^2} - \eta Y = 0 \tag{12.314}$$

and

$$\frac{d^2Z}{dz^2} + \left(\lambda^2 + \mu + \eta\right) Z = 0 \tag{12.315}$$

where μ and η are new separation constants. It can be seen that $\mu \ge 0$ and $\eta \ge 0$ cannot yield nontrivial solutions and hence we take $\mu = -\alpha^2$ and $\eta = -\beta^2$ and

$$\lambda^2 + \mu + \eta = \lambda^2 - \alpha^2 - \beta^2 = \gamma^2$$

or

$$\lambda^2 = \alpha^2 + \beta^2 + \gamma^2.$$

Then, Eqs. (12.313), (12.314) and (12.315) are rewritten as

$$\frac{d^2X}{dx^2} + \alpha^2 X = 0 \tag{12.316}$$

$$\frac{d^2Y}{dy^2} + \beta^2 Y = 0 \tag{12.317}$$

and

$$\frac{d^2Z}{dz^2} + \gamma^2 Z = 0. \tag{12.318}$$

Equation (12.316) has the general solution

$$X(x) = A \sin \alpha x + B \cos \alpha x$$

where A and B are arbitrary constants.

Applying the boundary conditions given in Eq. (12.304), we require that $B = 0$ and with $l = 1, 2, 3 \ldots$ we must have $\alpha_l = \frac{l\pi}{a}$.

Thus, we write the solutions of Eq. (12.316) as

$$X_l(x) = A_l \sin\frac{l\pi x}{a}. \tag{12.319}$$

Similarly, Eq. (12.317) has the general solution

$$Y(y) = C \sin \beta y + D \cos \beta y$$

where C and D are arbitrary constants.

Applying the boundary conditions given in Eq. (12.305), we require that $D = 0$ and with $m = 1, 2, 3 \ldots$ we must have $\beta_m = \frac{m\pi}{b}$.

Thus, we write the solutions of Eq. (12.317) as

$$Y_m(y) = C_m \sin\frac{m\pi y}{b}. \tag{12.320}$$

Further, Eq. (12.318) has the general solution

$$Z(z) = E \sin \gamma z + K \cos \gamma z$$

where E and K arbitrary constants.

Applying the boundary conditions given in Eq. (12.306), we require that $K = 0$ and with $n = 1, 2, 3 \ldots$ we must have $\gamma_n = \frac{n\pi}{c}$.

Thus, we write the solutions of Eq. (12.318) as

$$Z_n(x) = E_n \sin\frac{n\pi z}{c}. \tag{12.321}$$

Also, the general solution of Eq. (12.311) can be written as

$$T(t) = G \cos \lambda v t + H \sin \lambda v t \tag{12.322}$$

where G and H are arbitrary constants and $\lambda = \sqrt{\alpha_l^2 + \beta_m^2 + \gamma_n^2}$

Collecting Eqs. (12.319) to (12.322), we write the solutions of Eq. (12.303) as

$$u_{lmn}(x, y, z, t) = \{M_{lmn} \cos \lambda v t + N_{lmn} \sin \lambda v t\} \sin\frac{l\pi x}{a} \sin\frac{m\pi y}{b} \sin\frac{n\pi z}{c} \tag{12.323}$$

where $M_{lmn} = GA_l C_m E_n$ and $N_{lmn} = HA_l C_m E_n$ are constants, and $l, m, n = 1, 2, 3 \ldots$.

By the superposition principle, the general solution of the three-dimensional wave equation in Cartesian coordinate system is written as

$$u(x,y,z,t)=\sum_{l=1}^{\infty}\sum_{m=1}^{\infty}\sum_{n=1}^{\infty}u_{lmn}(x,y,z,t)$$

$$=\sum_{l=1}^{\infty}\sum_{m=1}^{\infty}\sum_{n=1}^{\infty}\{M_{lmn}\cos\lambda vt+N_{lmn}\sin\lambda vt\}\sin\frac{l\pi x}{a}\sin\frac{m\pi y}{b}\sin\frac{n\pi z}{c}. \quad (12.324)$$

The coefficients M_{lmn} are determined from the initial condition $u(x,y,z,0)=f(x,y,z)$ and they are found to be

$$M_{lmn}=\frac{8}{abc}\int_{x=0}^{a}\int_{y=0}^{b}\int_{z=0}^{c}f(x,y,z)\sin\frac{l\pi x}{a}\sin\frac{m\pi y}{b}\sin\frac{n\pi z}{c}\,dx\,dy\,dz. \quad (12.325)$$

The coefficients N_{lmn} are determined from the initial condition

$$\left[\frac{\partial u(x,y,z,t)}{\partial t}\right]_{t=0}=g(x,y,z)$$

and they are found to be

$$N_{lmn}=\frac{8}{\lambda vabc}\int_{x=0}^{a}\int_{y=0}^{b}\int_{z=0}^{c}g(x,y,z)\sin\frac{l\pi x}{a}\sin\frac{m\pi y}{b}\sin\frac{n\pi z}{c}\,dx\,dy\,dz \quad (12.326)$$

where $\lambda=\pi\sqrt{\frac{l^2}{a^2}+\frac{m^2}{b^2}+\frac{n^2}{c^2}}.$

Here, $\omega=\lambda v$ is the angular frequency of the wave.

II. In Circular Cylindrical Coordinates

In circular cylindrical coordinates, the three-dimensional wave equation is written as

$$\nabla^2u=\frac{\partial^2u}{\partial r^2}+\frac{1}{r}\frac{\partial u}{\partial r}+\frac{1}{r^2}\frac{\partial^2u}{\partial\theta^2}+\frac{\partial^2u}{\partial z^2}=\frac{1}{v^2}\frac{\partial^2u}{\partial t^2} \quad (12.327)$$

[with $0\le r\le\infty, 0\le\theta\le 2\pi, 0\le z\le\infty$]

where $u(r,\theta,z,t)$ is the displacement of the point (r,θ,z) at time t and v is the constant speed of wave propagation.

Using the method of separation of variables, let us assume the solution of Eq. (12.327) as

$$u(r,\theta,z,t)=S(r,\theta,z)T(t). \quad (12.328)$$

Substituting Eq. (12.328) into Eq. (12.327), we obtain

$$T\left[\frac{\partial^2S}{\partial r^2}+\frac{1}{r}\frac{\partial S}{\partial r}+\frac{1}{r^2}\frac{\partial^2S}{\partial\theta^2}+\frac{\partial^2S}{\partial z^2}\right]=\frac{S}{v^2}\frac{d^2T}{dt^2}.$$

Dividing both sides by ST, we have

$$\frac{1}{S}\left[\frac{\partial^2 S}{\partial r^2}+\frac{1}{r}\frac{\partial S}{\partial r}+\frac{1}{r^2}\frac{\partial^2 S}{\partial \theta^2}+\frac{\partial^2 S}{\partial z^2}\right]=\frac{1}{v^2 T}\frac{d^2 T}{dt^2}=-k^2 \tag{12.329}$$

where $-k^2$ is a separation constant.

Thus, we obtain two equations

$$\frac{d^2 T}{dt^2}+\omega^2 T=0 \text{ (where } \omega = kv) \tag{12.330}$$

and
$$\frac{\partial^2 S}{\partial r^2}+\frac{1}{r}\frac{\partial S}{\partial r}+\frac{1}{r^2}\frac{\partial^2 S}{\partial \theta^2}+\frac{\partial^2 S}{\partial z^2}+k^2 S=0. \tag{12.331}$$

Here, ω denotes the angular or circular frequency of the propagating wave. Let us assume the solution of Eq. (12.331) as

$$S(r,\theta,z)=R(r)\,\Theta(\theta)Z(z) \tag{12.332}$$

Substituting Eq. (12.332) into Eq. (12.331), we have

$$\Theta Z\frac{d^2 R}{dr^2}+\frac{\Theta Z}{r}\frac{dR}{dr}+\frac{RZ}{r^2}\frac{d^2\Theta}{d\theta^2}+R\,\Theta\frac{d^2 Z}{dz^2}+k^2 R\,\Theta\, Z=0.$$

Dividing both sides by $R\,\Theta\, Z$ and then transposing, we have

$$\frac{1}{Z}\frac{d^2 Z}{dz^2}=-\left[\frac{1}{R}\frac{d^2 R}{dr^2}+\frac{1}{rR}\frac{dR}{dr}+\frac{1}{r^2\Theta}\frac{d^2\Theta}{d\theta^2}+k^2\right]=-k_z^2$$

where $-k_z^2$ (with k_z = integer) is another separation constant.

Thus, we obtain two equations

$$\frac{d^2 Z}{dz^2}+k_z^2 Z=0 \tag{12.333}$$

and
$$\frac{1}{R}\frac{d^2 R}{dr^2}+\frac{1}{rR}\frac{dR}{dr}+\frac{1}{r^2\,\Theta}\frac{d^2\Theta}{d\theta^2}+k^2=k_z^2. \tag{12.334}$$

Multiplying both sides of Eq. (12.334) by r^2 and then transposing, we write

$$\frac{1}{\Theta}\frac{d^2\Theta}{d\theta^2}=-\left[\frac{r^2}{R}\frac{d^2 R}{dr^2}+\frac{r}{R}\frac{dR}{dr}+\left(k^2-k_z^2\right)r^2\right]=-m^2$$

where $-m^2$ (with m = integer) is another separation constant.

Thus, we obtain two equations, viz.,

$$\frac{d^2\Theta}{d\theta^2}+m^2\Theta=0 \tag{12.335}$$

and
$$r^2\frac{d^2R}{dr^2}+r\frac{dR}{dr}+(k_r^2r^2-m^2)R=0 \tag{12.336}$$

where
$$k_r^2=k^2-k_z^2.$$

Equation (12.330) has the solutions

$$T_k(t)=a_k\cos(kvt)+b_k\sin(kvt),\ (\omega=kv) \tag{12.337}$$

where a_k, b_k are arbitrary constants.

For a particular pair of $k_r=\sqrt{k^2-k_z^2}$ and m, Eq. (12.336) has a general solution in either of the forms

$$\begin{aligned}R_{kz}(r)&=B_{kz}J_m(k_r\,r)+C_{kz}N_m(k_r\,r)\\&=B_{kz}J_m\left(\sqrt{k^2-k_z^2}\,r\right)+C_{kz}N_m\left(\sqrt{k^2-k_z^2}\,r\right)\end{aligned} \tag{12.338a}$$

or
$$R_{kz}(r)=D_{kz}H_m^{(1)}\left(\sqrt{k^2-k_z^2}\,r\right)+E_{kz}H_m^{(2)}\left(\sqrt{k^2-k_z^2}\,r\right) \tag{12.338b}$$

where $B_{kz}, C_{kz}, D_{kz}, E_{kz}$ are arbitrary constants; $J_m\left(\sqrt{k^2-k_z^2}\,r\right)$ is a Bessel function of the first kind, $N_m\left(\sqrt{k^2-k_z^2}\,r\right)$ is a Neumann function, $H_m^{(1)}\left(\sqrt{k^2-k_z^2}\,r\right)$ is a Hankel function of the first kind and $H_m^{(2)}\left(\sqrt{k^2-k_z^2}\,r\right)$ is a Hankel function of the second kind.

For a particular m, Eq. (12.335) has the general solution

$$\Theta_m(\theta)=L_m\cos m\theta+M_m\sin m\theta \tag{12.339}$$

where L_m and M_m are arbitrary constants.

For a particular value of k_z, Eq. (12.333) has the general solution

$$Z_{k_z}(z)=P_{k_z}\cos(k_zz)+Q_{k_z}\sin(k_zz) \tag{12.340}$$

where P_{k_z} and Q_{k_z} are arbitrary constants.

Now, collecting Eqs. (12.337) to (12.340) and using Eq. (12.328), we write for particular values of k, k_z and m, the solutions of the wave equation in cylindrical co-ordinates as

$$\begin{aligned}u_{k,k_z,m}(r,\theta,z,t)&=R_m(r)[L_m\cos m\theta+M_m\sin m\theta]\left[P_{k_z}\cos(k_zz)+Q_{k_z}\sin(k_zz)\right]\\&\quad\times\left[a_k\cos(kvt)+b_k\sin(kvt)\right]\end{aligned} \tag{12.341}$$

where $R_m(r)$ is given by either Eq. (12.338a) or (12.338b).

By the superposition theorem, the general solution may be obtained as the sum of the individual solutions for all possible values of m and k_z.

Thus, we write

$$u(r,\theta,z,t)=\sum_k\sum_{k_z}\sum_m u_{k,k_z,m}(r,\theta,z,t)$$

$$=\sum_k\sum_{k_z}\sum_m\{R_m(r)[L_m\cos m\theta+M_m\sin m\theta][P_{k_z}\cos(k_z z)+Q_{k_z}\sin(k_z z)]$$

$$\times[a_k\cos(kvt)+b_k\sin(kvt)]\}. \quad (12.342)$$

Here, the arbitrary constants are to be determined from boundary conditions and the initial conditions of the relevant problem. The solution with $H_m^{(1)}\left(\sqrt{k^2-k_z^2}\,r\right)$ corresponds to a diverging outgoing wave and the solution with $H_m^{(2)}\left(\sqrt{k^2-k_z^2}\,r\right)$ corresponds to a converging incoming wave. If the region under consideration includes the origin, the Neumann function solution should be discarded.

III. In Spherical Polar Coordinates:

In spherical polar coordinates, the three-dimensional wave equation is written as

$$\nabla^2 u=\frac{1}{r^2}\frac{\partial}{\partial r}\left(r^2\frac{\partial u}{\partial r}\right)+\frac{1}{r^2\sin\theta}\frac{\partial}{\partial\theta}\left(\sin\theta\frac{\partial u}{\partial\theta}\right)+\frac{1}{r^2\sin^2\theta}\frac{\partial^2 u}{\partial\phi^2}$$

$$=\frac{1}{v^2}\frac{\partial^2 u}{\partial t^2},(0\le r\le\infty,0\le\theta\le\pi,0\le\phi\le 2\pi) \quad (12.343)$$

where $u(r,\theta,\phi,t)$ is the deflection of the point (r,θ,ϕ) at time t and v is the constant speed of wave propagation.

We assume the solution of the above equation as

$$u(r,\theta,\phi,t)=\psi(r,\theta,\phi)T(t). \quad (12.344)$$

Substituting Eq. (12.344) into Eq. (12.343), we obtain

$$\frac{T}{r^2}\frac{\partial}{\partial r}\left(r^2\frac{\partial\psi}{\partial r}\right)+\frac{T}{r^2\sin\theta}\frac{\partial}{\partial\theta}\left(\sin\theta\frac{\partial\psi}{\partial\theta}\right)+\frac{T}{r^2\sin^2\theta}\frac{\partial^2\psi}{\partial\phi^2}=\frac{\psi}{v^2}\frac{d^2T}{dt^2}.$$

Dividing both sides by ψT, we obtain

$$\frac{1}{r^2\psi}\left[\frac{\partial}{\partial r}\left(r^2\frac{\partial\psi}{\partial r}\right)+\frac{1}{\sin\theta}\frac{\partial}{\partial\theta}\left(\sin\theta\frac{\partial\psi}{\partial\theta}\right)+\frac{1}{\sin^2\theta}\frac{\partial^2\psi}{\partial\phi^2}\right]=\frac{1}{v^2T}\frac{d^2T}{dt^2}=-k^2$$

where $-k^2$ is a separation constant.

Thus, we obtain two equations, viz.,

$$\frac{d^2T}{dt^2}+(kv)^2T=0 \tag{12.345}$$

and

$$\frac{1}{r^2}\frac{\partial}{\partial r}\left(r^2\frac{\partial\psi}{\partial r}\right)+\frac{1}{r^2\sin\theta}\frac{\partial}{\partial\theta}\left(\sin\theta\frac{\partial\psi}{\partial\theta}\right)+\frac{1}{r^2\sin^2\theta}\frac{\partial^2\psi}{\partial\phi^2}+k^2\psi=0. \tag{12.346}$$

Further, let us put

$$\psi(r,\theta,\phi)=R(r)S(\theta,\phi). \tag{12.347}$$

Then, Eq. (12.346) can be written as

$$\frac{S}{r^2}\frac{d}{dr}\left(r^2\frac{dR}{dr}\right)+\frac{R}{r^2\sin\theta}\frac{\partial}{\partial\theta}\left(\sin\theta\frac{\partial S}{\partial\theta}\right)+\frac{R}{r^2\sin^2\theta}\frac{\partial^2 S}{\partial\phi^2}+k^2RS=0.$$

Multiplying both sides by $\frac{r^2}{RS}$ and then transposing, we obtain

$$\frac{1}{S\sin\theta}\frac{\partial}{\partial\theta}\left(\sin\theta\frac{\partial S}{\partial\theta}\right)+\frac{1}{S\sin^2\theta}\frac{\partial^2 S}{\partial\phi^2}=-\left[\frac{1}{R}\frac{d}{dr}\left(r^2\frac{dR}{dr}\right)+k^2r^2\right]=-l(l+1)$$

where $-l(l+1)$ is a separation constant with l = 0, 1, 2, 3 ….

The above equation leads to two equations, viz.,

$$\frac{1}{\sin\theta}\frac{\partial}{\partial\theta}\left(\sin\theta\frac{\partial S}{\partial\theta}\right)+\frac{1}{\sin^2\theta}\frac{\partial^2 S}{\partial\phi^2}+l(l+1)S=0 \tag{12.348}$$

and

$$r^2\frac{d^2R}{dr^2}+2r\frac{dR}{dr}+\left\{k^2r^2-l(l+1)\right\}R=0. \tag{12.349}$$

Now, we see that Eq. (12.349) is a differential equation satisfied by spherical Bessel functions. It has a general solution of the form

$$R_k(r)=A_k\,j_l(kr)+B_k n_l(kr) \tag{12.350}$$

where A_k, B_k are arbitrary constants and $j_l(kr)$ and $n_l(kr)$ are spherical Bessel function and spherical Neumann function respectively defined by

$$j_l(kr)=\sqrt{\frac{\pi}{2kr}}\,J_{l+\frac{1}{2}}(kr) \tag{12.351}$$

and

$$n_l(kr)=\sqrt{\frac{\pi}{2kr}}\,N_{l+\frac{1}{2}}(kr). \tag{12.352}$$

Here, $J_{l+\frac{1}{2}}(kr)$ is Bessel function of the first kind of order $l+\frac{1}{2}$ and $N_{l+\frac{1}{2}}(kr)$ is Neumann function or Bessel function of the second kind of order $l+\frac{1}{2}$.

Also Eq. (12.348) has the general solution

$$S_{l,m}(\theta,\phi)=Y_{l,m}(\theta,\phi)=(-1)^m\sqrt{\frac{(2l+1)}{4\pi}\frac{(l-m)!}{(l+m)!}}\,e^{im\phi}P_l^m(\cos\theta) \tag{12.353}$$

where $P_l^m(\cos\theta)$ is associated Legendre polynomial and $Y_{l,m}(\theta,\phi)$ is a spherical harmonic.

Further, Eq. (12.345) has the general solution

$$T_k(t)=C_k\cos(kvt)+D_k\sin(kvt) \tag{12.354}$$

where C_k and D_k are arbitrary constants.

Now collecting Eqs. (12.344), (12.347), (12.350), (12.353) and (12.354), we write solutions of the wave equation in spherical polar co-ordinates as

$$\begin{aligned}u_{k,l,m}(r,\theta,\phi,t)&=R_k(r)S_{l,m}(\theta,\phi)T_k(t)\\&=\{A_k j_l(kr)+B_k n_l(kr)\}Y_{l,m}(\theta,\phi)\{C_k\cos(kvt)+D_k\sin(kvt)\}.\end{aligned} \tag{12.355}$$

Using superposition theorem, sum of all solutions with different values of k, l, and m is the general solution of the wave equation in spherical polar coordinates.

Thus, the general solution is

$$\begin{aligned}u(r,\theta,\phi,t)&=\sum_k\sum_l\sum_m u_{k,l,m}(r,\theta,\phi,t)\\&=\sum_k\sum_l\sum_m\{A_k j_l(kr)+B_k n_l(kr)\}Y_{l,m}(\theta,\phi)\{C_k\cos(kvt)+D_k\sin(kvt)\}.\end{aligned} \tag{12.356}$$

If the region under consideration includes the origin, the solution with spherical Neumann function should be discarded.

12.27 LAPLACE'S EQUATION

Laplace's equation is a second order PDE (partial differential equation) having the form

$$\nabla^2 u = 0 \tag{12.357}$$

where

$$\nabla^2\equiv\frac{\partial^2}{\partial x^2}+\frac{\partial^2}{\partial y^2}+\frac{\partial^2}{\partial z^2} \tag{12.358}$$

is an operator known as Laplacian operator.

This formula was first developed by the French mathematician, Pierre-Simon Laplace (23 March 1749 - 5 March 1827).

12.27A LAPLACE'S EQUATION IN ONE DIMENSION

In one dimension, Laplace's equation is written as

$$\frac{d^2u(x)}{dx^2} = 0.$$

Solution: Integrating both sides of the above equation once with respect to x, we obtain

$$\frac{du}{dx} = c_1$$

where c_1 is an integration constant.

Integrating both sides once again, we obtain

$$u(x) = c_1 x + c_2$$

where c_2 is another integration constant.

12.27B LAPLACE'S EQUATION IN TWO DIMENSIONS

I: In Cartesian Coordinates:

In two dimensions, Laplace's equation in Cartesian coordinates is written as

$$\frac{\partial^2 u}{\partial x^2} + \frac{\partial^2 u}{\partial y^2} = 0 \text{ where } u = u(x, y). \tag{12.359}$$

Solution: Using the method of separation of variables, let us put

$$u(x, y) = X(x)Y(y) \tag{12.360}$$

where $X(x)$ is function of x and $Y(y)$ is function of y.

Substituting Eq. (12.360) into Eq. (12.359), we obtain

$$Y\frac{d^2X}{dx^2} + X\frac{d^2Y}{dy^2} = 0.$$

Dividing both sides by XY and then transposing, we obtain

$$\frac{1}{X}\frac{d^2X}{dx^2} = -\frac{1}{Y}\frac{d^2Y}{dy^2} = -k^2$$

where $(-k^2)$ is a separation constant.

Thus, we obtain two equations, viz.

$$\frac{d^2X}{dx^2} + k^2X = 0 \text{ and } \frac{d^2Y}{dy^2} - k^2Y = 0 \tag{12.361}$$

whose solutions are

$$X(x) = A_1 \cos(kx) + A_2 \sin(kx) \tag{12.362}$$

and

$$Y(y) = B_1 e^{ky} + B_2 e^{-ky} = C_1 \cosh(ky) + C_2 \sinh(ky) \tag{12.363}$$

where A_1, A_2, B_1, B_2, C_1, C_2 are arbitrary constants.

Thus, the solution of two-dimensional Laplace's equation in Cartesian coordinates is written as

$$\begin{aligned} u(x, y) &= [A_1 \cos(kx) + A_2 \sin(kx)][B_1 e^{ky} + B_2 e^{-ky}] \\ &= [A_1 \cos(kx) + A_2 \sin(kx)][C_1 \cosh(ky) + C_2 \sinh(ky)]. \end{aligned} \tag{12.364}$$

Here, the arbitrary constants are to be determined from boundary conditions.

II. In Plane Polar Coordinates:

In plane polar coordinates, the Laplacian operator is expressed as

$$\nabla^2 \equiv \frac{\partial^2}{\partial r^2} + \frac{1}{r}\frac{\partial}{\partial r} + \frac{1}{r^2}\frac{\partial^2}{\partial \theta^2} \tag{12.365}$$

where (r, θ) are plane polar coordinates.

Therefore, Laplace's equation in plane polar coordinates is written as

$$\nabla^2 u = \frac{\partial^2 u}{\partial r^2} + \frac{1}{r}\frac{\partial u}{\partial r} + \frac{1}{r^2}\frac{\partial^2 u}{\partial \theta^2} = 0 \tag{12.366}$$

where $u = u(r, \theta)$.

Solution: Using the method of separation of variables, let us put

$$u(r, \theta) = R(r)\,\Theta(\theta) \tag{12.367}$$

where $R(r)$ is function of r and $\Theta(\theta)$ is function of θ.

Substituting Eq. (12.367) into Eq. (12.366), we obtain

$$\Theta \frac{d^2 R}{dr^2} + \frac{\Theta}{r}\frac{dR}{dr} + \frac{R}{r^2}\frac{d^2\Theta}{d\theta^2} = 0.$$

Multiplying both sides by $\dfrac{r^2}{R\Theta}$ and then transposing, we obtain

$$\frac{1}{\Theta}\frac{d^2\Theta}{d\theta^2} = -\left[\frac{r^2}{R}\frac{d^2R}{dr^2} + \frac{r}{R}\frac{dR}{dr}\right] = -l^2, \text{ where } (-l^2) \text{ is a separation constant.}$$

From this equation, we obtain two equations, viz.

$$\frac{d^2\Theta}{d\theta^2}+l^2\Theta=0 \tag{12.368}$$

and

$$r^2\frac{d^2R}{dr^2}+r\frac{dR}{dr}-l^2R=0. \tag{12.369}$$

Equation (12.368) has a general solution

$$\Theta(\theta)=C_l\cos(l\theta)+D_l\sin(l\theta) \tag{12.370}$$

where C_l and D_l are arbitrary constants.

To solve Eq. (12.369), let us put $r = e^s$ or, $s = \ln r$.

Then, we have

$$\frac{dR}{dr}=\frac{dR}{ds}\frac{ds}{dr}=\frac{1}{r}\frac{dR}{ds}=e^{-s}\frac{dR}{ds} \text{ and}$$

$$\frac{d^2R}{dr^2}=e^{-s}\frac{d}{ds}\left(e^{-s}\frac{dR}{ds}\right)=e^{-2s}\frac{d^2R}{ds^2}-e^{-2s}\frac{dR}{ds}.$$

Using these derivatives, Eq. (12.369) yields

$$\frac{d^2R}{ds^2}-l^2R=0.$$

The general solution of this equation is expressed as

$$R(r)=A_l e^{ls}+B_l e^{-ls}=A_l r^l+B_l r^{-l} \tag{12.371}$$

where A_l and B_l are also arbitrary constants.

Recalling Eqs. (12.367), (12.370) and (12.371); we write the solutions of Eq. (12.369) as

$$u_l(r,\theta)=\left(A_l r^l+B_l r^{-l}\right)\left\{C_l\cos(l\theta)+D_l\sin(l\theta)\right\}. \tag{12.372}$$

Using superposition theorem, we write a general solution in the form

$$u(r,\theta)=\sum_l u_l(r,\theta)=\sum_l\left(A_l r^l+B_l r^{-l}\right)\left\{C_l\cos(l\theta)+D_l\sin(l\theta)\right\}. \tag{12.373}$$

Here, the arbitrary constants are to be determined from boundary conditions.

12.27C LAPLACE'S EQUATION IN THREE DIMENSIONS

In three dimensions, Laplace's equation is studied here in three coordinate systems, namely, cartesian, circular cylindrical and spherical polar coordinate systems.

I. In Three-Dimensional Cartesian Coordinates:

In three-dimensional Cartesian coordinates, Laplace's equation is written as

$$\nabla^2 u = \frac{\partial^2 u}{\partial x^2} + \frac{\partial^2 u}{\partial y^2} + \frac{\partial^2 u}{\partial z^2} = 0 \tag{12.374}$$

where $u = u(x,y,z)$ is function of x, y, z.

Solution: Using the method of separation of variables, let us put

$$u(x,y,z) = X(x)Y(y)Z(z). \tag{12.375}$$

Substituting Eq. (12.375) into Eq. (12.374), we obtain

$$Y(y)Z(z)\frac{d^2X}{dx^2} + Z(z)X(x)\frac{d^2Y}{dy^2} + X(x)Y(y)\frac{d^2Z}{dz^2} = 0.$$

Dividing both sides by XYZ and then transposing, we obtain

$$\frac{1}{X}\frac{d^2X}{dx^2} = -\left[\frac{1}{Y}\frac{d^2Y}{dy^2} + \frac{1}{Z}\frac{d^2Z}{dz^2}\right] = -k_1^2$$

where $(-k_1^2)$ is a separation constant.

Thus, we obtain two equations, viz.

$$\frac{d^2X}{dx^2} + k_1^2 X = 0 \tag{12.376}$$

and

$$\frac{1}{Y}\frac{d^2Y}{dy^2} + \frac{1}{Z}\frac{d^2Z}{dz^2} = k_1^2$$

or

$$\frac{1}{Y}\frac{d^2Y}{dy^2} = -\frac{1}{Z}\frac{d^2Z}{dz^2} + k_1^2 = -k_2^2$$

where $(-k_2^2)$ is another separation constant.

Thus, we obtain two more equations, viz.

$$\frac{d^2Y}{dy^2} + k_2^2\, Y = 0 \tag{12.377}$$

and

$$\frac{d^2Z}{dz^2} - \left(k_1^2 + k_2^2\right)Z = 0. \tag{12.378}$$

Now, Eqs. (12.376), (12.377) and (12.378) have the respective general solutions

$$X_{k_1}(x) = A_{k_1}\cos(k_1 x) + B_{k_1}\sin(k_1 x) \tag{12.379}$$

$$Y_{k_2}(y) = C_{k_2}\cos(k_2 y) + D_{k_2}\sin(k_2 y) \tag{12.380}$$

and $$Z_{k_1 k_2}(z) = E_{k_1k_2} e^{\left(\sqrt{k_1^2+k_2^2}\right)z} + L_{k_1k_2} e^{-\left(\sqrt{k_1^2+k_2^2}\right)z}$$

$$= M_{k_1k_2} \cosh\left\{\left(\sqrt{k_1^2+k_2^2}\right)z\right\} + N_{k_1k_2} \sinh\left\{\left(\sqrt{k_1^2+k_2^2}\right)z\right\} \tag{12.381}$$

where $A_{k_1}, B_{k_1}, C_{k_2}, D_{k_2}, E_{k_1k_2}, L_{k_1k_2}, M_{k_1k_2}, N_{k_1k_2}$ are arbitrary constants.

Recalling Eqs. (12.375), (12.379), (12.380) and (12.381); we write the solutions of Eq. (12.374) as

$$u_{k_1k_2}(x, y, z) = X_{k_1} Y_{k_2} Z_{k_1k_2}$$

$$= \left[A_{k_1} \cos(k_1 x) + B_{k_1} \sin(k_1 x)\right]\left[C_{k_2} \cos(k_2 y) + D_{k_2} \sin(k_2 y)\right]$$

$$\times\left[M_{k_1k_2} \cosh\left\{\left(\sqrt{k_1^2+k_2^2}\right)z\right\} + N_{k_1k_2} \sinh\left\{\left(\sqrt{k_1^2+k_2^2}\right)z\right\}\right]. \tag{12.382}$$

Using superposition theorem, we write a general solution in the form

$$u(x, y, z) = \sum_{k_1}\sum_{k_2} u_{k_1k_2}(x, y, z) = \sum_{k_1}\sum_{k_2} X_{k_1} Y_{k_2} Z_{k_1k_2}. \tag{12.383}$$

Here, the arbitrary constants are to be determined from boundary conditions.

II. In Circular Cylindrical Coordinates:

In circular cylindrical coordinates, the Laplacian operator is expressed as

$$\nabla^2 \equiv \frac{\partial^2}{\partial r^2} + \frac{1}{r}\frac{\partial}{\partial r} + \frac{1}{r^2}\frac{\partial^2}{\partial \theta^2} + \frac{\partial^2}{\partial z^2}.$$

Therefore, the three-dimensional Laplace's equation in circular cylindrical coordinates is written as

$$\nabla^2 u = \frac{\partial^2 u}{\partial r^2} + \frac{1}{r}\frac{\partial u}{\partial r} + \frac{1}{r^2}\frac{\partial^2 u}{\partial \theta^2} + \frac{\partial^2 u}{\partial z^2} = 0 \tag{12.384}$$

where $u = u(r, \theta, z)$.

Solution: Using the method of separation of variables, let us put

$$u(r, \theta, z) = R(r)\,\Theta(\theta) Z(z). \tag{12.385}$$

Substituting Eq. (12.385) into Eq. (12.384), we obtain

$$\Theta Z \frac{d^2 R}{dr^2} + \frac{\Theta Z}{r}\frac{dR}{dr} + \frac{RZ}{r^2}\frac{d^2\Theta}{d\theta^2} + R\Theta \frac{d^2 Z}{dz^2} = 0.$$

Multiplying both sides by $\dfrac{r^2}{R\Theta Z}$ and then transposing, we obtain

$$\frac{1}{\Theta}\frac{d^2\Theta}{d\theta^2}=-\left[\frac{r^2}{R}\frac{d^2R}{dr^2}+\frac{r}{R}\frac{dR}{dr}+\frac{r^2}{Z}\frac{d^2Z}{dz^2}\right]=-n^2$$

where $(-n^2)$ is a separation constant.

Thus, we obtain two equations, viz.

$$\frac{d^2\Theta}{d\theta^2}+n^2\Theta=0 \tag{12.386}$$

and

$$\frac{r^2}{R}\frac{d^2R}{dr^2}+\frac{r}{R}\frac{dR}{dr}+\frac{r^2}{Z}\frac{d^2Z}{dz^2}=n^2.$$

Dividing both sides of the last equation by r^2 and then transposing, we obtain

$$\frac{1}{R}\frac{d^2R}{dr^2}+\frac{1}{rR}\frac{dR}{dr}-\frac{n^2}{r^2}=-\frac{1}{Z}\frac{d^2Z}{dz^2}=-k^2$$

where $(-k^2)$ is another separation constant.

Thus, we obtain two more equations, viz.

$$\frac{d^2z}{dz^2}-k^2Z=0 \tag{12.387}$$

and

$$r^2\frac{d^2R}{dr^2}+r\frac{dR}{dr}+(k^2r^2-n^2)R=0. \tag{12.388}$$

Here, we see that Eq. (12.388) is a Bessel equation having a general solution given by

$$R_n(r)=A_nJ_n(kr)+B_nN_n(kr) \tag{12.389}$$

where A_n and B_n are arbitrary constants and $J_n(kr)$ and $N_n(kr)$ are Bessel functions of the first and the second kinds. $N_n(kr)$ is also known as Neumann function.

Further, Eqs. (12.386) and (12.387) have the respective general solutions

$$\Theta_n(\theta)=C_n\cos(n\theta)+D_n\sin(n\theta) \tag{12.390}$$

and

$$Z_k(z)=L_k\cosh(kz)+M_k\sinh(kz) \tag{12.391}$$

where C_n, D_n, L_k, M_k are all arbitrary constants.

Using Eqs. (12.389), (12.390), (12.391) and recalling Eq. (12.385), we write the solutions of Eq. (12.384) as

$$\begin{aligned}u_{kn}(r,\theta,z)&=R_n(r)\,\Theta_n(\theta)Z_k(z)\\&=\left[A_nJ_n(kr)+B_nN_n(kr)\right]\left[C_n\cos(n\theta)+D_n\sin(n\theta)\right]\\&\qquad\times\left[L_k\cosh(kz)+M_k\sinh(kz)\right].\end{aligned} \tag{12.392}$$

Using the superposition theorem, we write a general solution of Eq. (12.384) in the form

$$u(r,\theta,z)=\sum_k\sum_n u_{kn}(r,\theta,z)=\sum_k\sum_n R_n(r)\Theta_n(\theta)Z_k(z). \tag{12.393}$$

Note: If the region under consideration includes the origin of the coordinate system, the Neumann function or Bessel function of the second kind should be dropped out from the R_n solution. Further, if we choose k^2 as separation constant in place of $(-k^2)$ the solution R_n will be obtained in terms of modified Bessel functions of the first and second kinds.

III. In Spherical Polar Coordinates:

In spherical polar coordinates, the Laplacian operator is expressed as

$$\nabla^2 \equiv \frac{1}{r^2}\frac{\partial}{\partial r}\left(r^2\frac{\partial}{\partial r}\right)+\frac{1}{r^2\sin\theta}\frac{\partial}{\partial\theta}\left(\sin\theta\frac{\partial}{\partial\theta}\right)+\frac{1}{r^2\sin^2\theta}\frac{\partial^2}{\partial\phi^2}. \tag{12.394}$$

Therefore, the three-dimensional Laplace's equation in circular cylindrical coordinates is written as

$$\nabla^2 u=\frac{1}{r^2}\frac{\partial}{\partial r}\left(r^2\frac{\partial u}{\partial r}\right)+\frac{1}{r^2\sin\theta}\frac{\partial}{\partial\theta}\left(\sin\theta\frac{\partial u}{\partial\theta}\right)+\frac{1}{r^2\sin^2\theta}\frac{\partial^2 u}{\partial\phi^2}=0 \tag{12.395}$$

where $u=u(r,\theta,\phi)$.

Solution: Using the method of separation of variables, let us put

$$u(r,\theta,\phi)=R(r)\Theta(\theta)\Phi(\phi). \tag{12.396}$$

Substituting Eq. (12.396) into Eq. (12.395), we obtain

$$\frac{\Theta\Phi}{r^2}\frac{d}{dr}\left(r^2\frac{dR}{dr}\right)+\frac{R\Phi}{r^2\sin\theta}\frac{d}{d\theta}\left(\sin\theta\frac{d\Theta}{d\theta}\right)+\frac{R\Theta}{r^2\sin^2\theta}\frac{d^2\Phi}{d\phi^2}=0.$$

Multiplying both sides by $\dfrac{r^2\sin^2\theta}{R\Theta\Phi}$, we obtain

$$\frac{\sin^2\theta}{R}\frac{d}{dr}\left(r^2\frac{dR}{dr}\right)+\frac{\sin\theta}{\Theta}\frac{d}{d\theta}\left(\sin\theta\frac{d\Theta}{d\theta}\right)+\frac{1}{\Phi}\frac{d^2\Phi}{d\phi^2}=0$$

or

$$\frac{1}{\Phi}\frac{d^2\Phi}{d\phi^2}=-\left[\frac{\sin^2\theta}{R}\frac{d}{dr}\left(r^2\frac{dR}{dr}\right)+\frac{\sin\theta}{\Theta}\frac{d}{d\theta}\left(\sin\theta\frac{d\Theta}{d\theta}\right)\right]=-m^2$$

where $-m^2$ is a separation constant with $m = 0, \pm1, \pm2, \pm3 \ldots$

Thus, we obtain two equations, viz.,

$$\frac{d^2\Phi}{d\phi^2}+m^2\Phi=0 \tag{12.397}$$

and $$\frac{\sin^2\theta}{R}\frac{d}{dr}\left(r^2\frac{dR}{dr}\right)+\frac{\sin\theta}{\Theta}\frac{d}{d\theta}\left(\sin\theta\frac{d\Theta}{d\theta}\right)-m^2=0.$$

Dividing both sides of the last equation by $\sin^2\theta$ and then transposing, we have

$$\frac{1}{\Theta\sin\theta}\frac{d}{d\theta}\left(\sin\theta\frac{d\Theta}{d\theta}\right)-\frac{m^2}{\sin^2\theta}=-\frac{1}{R}\frac{d}{dr}\left(r^2\frac{dR}{dr}\right)=-l(l+1)$$

where $-l(l+1)$ is another separation constant with $l=0,1,2,3,\ \ldots$

The above equation leads to two equations as shown below

$$\frac{d}{dr}\left(r^2\frac{dR}{dr}\right)-l(l+1)R=0$$

or $$r^2\frac{d^2R}{dr^2}+2r\frac{dR}{dr}-l(l+1)R=0 \tag{12.398}$$

and $$\frac{1}{\sin\theta}\frac{d}{d\theta}\left(\sin\theta\frac{d\Theta}{d\theta}\right)+\left\{l(l+1)-\frac{m^2}{\sin^2\theta}\right\}\Theta=0$$

or $$\frac{d^2\Theta}{d\theta^2}+\frac{\cos\theta}{\sin\theta}\frac{d\Theta}{d\theta}+\left\{l(l+1)-\frac{m^2}{\sin^2\theta}\right\}\Theta=0. \tag{12.399}$$

If we put $\Theta(\theta)=w(x)$ and $x=\cos\theta$ then $\sin\theta=\sqrt{1-x^2}$ and $\sin^2\theta=1-x^2$.

Also, we have

$$\frac{d\Theta}{d\theta}=\frac{d\Theta}{dx}\frac{dx}{d\theta}=-\sin\theta\frac{dw}{dx}=-\sqrt{1-x^2}\frac{dw}{dx}$$

and $$\frac{d^2\Theta}{d\theta^2}=\frac{d}{d\theta}\frac{d\Theta}{d\theta}=-\sqrt{1-x^2}\frac{d}{dx}\left(-\sqrt{1-x^2}\frac{dw}{dx}\right)=(1-x^2)\frac{d^2w}{dx^2}-x\frac{dw}{dx}.$$

Using the above transformations, Eq. (12.399) can be written as

$$(1-x^2)\frac{d^2w}{dx^2}-2x\frac{dw}{dx}+\left\{l(l+1)-\frac{m^2}{1-x^2}\right\}w=0. \tag{12.400}$$

Now, we see that Eq. (12.397) can have a solution in the form

$$\Phi_m(\phi) = A_m e^{im\phi} \tag{12.401}$$

where $i=\sqrt{-1}$, m = 0, ±1, ±2, ±3 … and A_m is a constant such that

$$\int_0^{2\pi} \Phi_m(\phi)\Phi_m^*(\phi)\,d\phi = 1.$$

Here, the asterik (*) symbol denotes complex conjugation.

Thus, we obtain

$$A_m = \frac{1}{\sqrt{2\pi}}$$

and hence
$$\Phi_m(\phi) = \frac{1}{\sqrt{2\pi}} e^{im\phi}. \tag{12.402}$$

To solve Eq. (12.398), let us put $r = e^s$ or, $s = \ln r$

Then, we have

$$\frac{dR}{dr} = \frac{dR}{ds}\frac{ds}{dr} = \frac{1}{r}\frac{dR}{ds} = e^{-s}\frac{dR}{ds}$$

and
$$\frac{d^2R}{dr^2} = e^{-s}\frac{d}{ds}\left(e^{-s}\frac{dR}{ds}\right) = e^{-2s}\frac{d^2R}{ds^2} - e^{-2s}\frac{dR}{ds}.$$

Using these derivatives in Eq. (12.398), we write

$$\frac{d^2R}{ds^2} + \frac{dR}{ds} - l(l+1)R = 0. \tag{12.403}$$

Equation (12.403) has the solution

$$R_l(s) = A_l e^{ls} + B_l e^{-(l+1)s}$$

where A_l and B_l are arbitrary constants.

Rcalling the transformation $r = e^s$ we write the solution of Eq. (12.398) as

$$R_l(r) = A_l r^l + B_l r^{-(l+1)}. \tag{12.404}$$

Further, we see that Eq. (12.400) is associated Legendre equation and, therefore, its solution can be written as

$$w_{lm}(x) = \sqrt{\frac{(2l+1)}{2}\frac{(l-m)!}{(l+m)!}}P_l^m(x). \tag{12.405}$$

Further, recalling that $\Theta(\theta) = w(x)$ and $x = \cos\theta$ we write the solution of Eq. (12.399) as

$$\Theta_{lm}(\theta) = \sqrt{\frac{(2l+1)}{2}\frac{(l-m)!}{(l+m)!}} P_l^m(\cos\theta). \tag{12.406}$$

Recalling Eq. (12.396) and using Eqs. (12.402), (12.404) and (12.406), we write the solutions of Eq. (12.395) in the form

$$u_{lm}(r,\theta,\phi) = R_l(r)\,\Theta_{lm}(\theta)\Phi_m(\phi)$$

$$= \left\{A_l r^l + B_l r^{-(l+1)}\right\}\left[\sqrt{\frac{(2l+1)}{2}\frac{(l-m)!}{(l+m)!}} P_l^m(\cos\theta)\frac{1}{\sqrt{2\pi}}e^{im\phi}\right]. \tag{12.407}$$

Introducing a conventional factor of $(-1)^m$ in the factors inside the square bracket in the right-hand side expression, we define a function known as **'spherical harmonic'** by

$$Y_{lm}(\theta,\phi) = (-1)^m\sqrt{\frac{(2l+1)}{4\pi}\frac{(l-m)!}{(l+m)!}} P_l^m(\cos\theta)e^{im\phi} \tag{12.408}$$

[Some authors use the notation $Y_l^m(\theta,\phi)$].

It is to be noted that the functions (spherical harmonics) are the solutions of the angular part of Laplace's equation in spherical polar coordinates.

Now, using this newly defined spherical harmonic, we write the solutions of Eq. (12.395) (i.e., Laplace's equation in three-dimensional spherical polar coordinated) as

$$u_{lm}(r,\theta,\phi) = \left\{A_l r^l + B_l r^{\ (l+1)}\right\} Y_{lm}(\theta,\phi). \tag{12.409}$$

Using the superposition theorem, the general solution of Eq. (12.395) is written as

$$u(r,\theta,\phi) = \sum_l\sum_m u_{lm}(r,\theta,\phi) = \sum_l\sum_m \left\{A_l r^l + B_l r^{-(l+1)}\right\} Y_{lm}(\theta,\phi). \tag{12.410}$$

The arbitrary constants are to be determined from boundary conditions.

12.28 SOLVED PROBLEMS

PROBLEM 1 A taut string of length L, fastened at both ends at $x = 0$ and $x = L$ is set into vibrations by plucking initially at its midpoint, producing an initial deflection (h) at that point. Taking the initial velocity at every point of the string as zero, find the deflection $u(x,t)$ of the string at a point x at time t.

Solution: Referring to Section 12.8, for a stretched string of length (L) plucked at a point distant (a) from the end at $x = 0$ and released to undergo transverse vibrations, the deflection at any pont (x) at time (t) is given by

$$u(x \cdot t) = \sum_{n=1}^{\infty} \left[\frac{2hL^2}{a(L-a)n^2\pi^2} \sin\frac{n\pi a}{L} \sin\frac{n\pi x}{L} \right] \cos\frac{n\pi ct}{L} \quad [c = \text{speed of wave}].$$

In this problem, we have, $a = \frac{L}{2}$.

Referring to Fig. 12.4, at any point x, the initial displacement $f(x)$ is given by

$$f(x) = \begin{cases} \dfrac{2h}{L}x & \text{for } 0 \le x \le \dfrac{L}{2}, \\ \dfrac{2h(L-x)}{L} & \text{for } \dfrac{L}{2} \le x \le L. \end{cases}$$

It is also given that the initial speed, $g(x)$, is zero anywhere on the string. The boundary conditions are

$$u(0,t) = 0 \text{ and } u(L,t) = 0$$

and initial conditions are $u\left(\frac{L}{2}, 0\right) = f\left(\frac{L}{2}\right) = h$ and $\left.\frac{\partial u}{\partial t}\right|_{t=0} = g(x) = 0$.

Since the string is set into vibrations after plucking at its midpoint, even harmonic vibrations will be absent. Thus, we have

$$u(x \cdot t) = \sum_{m=0}^{\infty} \frac{8h}{(2m+1)^2\pi^2} \sin\left\{\frac{(2m+1)\pi}{2}\right\} \sin\left\{\frac{(2m+1)\pi x}{L}\right\} \cos\left\{\frac{(2m+1)\pi ct}{L}\right\}$$

$$= \frac{8h}{\pi^2}\left[\sin\left(\frac{\pi x}{L}\right)\cos\left(\frac{\pi ct}{L}\right) - \frac{1}{3^2}\sin\left(\frac{3\pi x}{L}\right)\cos\left(\frac{3\pi ct}{L}\right) + \frac{1}{5^2}\sin\left(\frac{5\pi x}{L}\right)\cos\left(\frac{5\pi ct}{L}\right) - \ldots \infty\right].$$

PROBLEM 2 A uniform string of length $2L$ is stretched and then fastened at both ends. The midpoint of the string is then plucked making an initial deflection (h) perpendicular to the equilibrium position of the string and then released from rest. Find the deflection of a point of the string lying at a distance (x) from the midpoint of the string at time (t).

Solution: As in Problem 1, for a stretched string of length L plucked at a point distant a from the end at $x = 0$ and released to undergo transverse vibrations, the deflection at any point x at time t is given by

$$u(x \cdot t) = \sum_{n=1}^{\infty} \left[\frac{2hL^2}{a(L-a)n^2\pi^2} \sin\frac{n\pi a}{L} \sin\frac{n\pi x}{L} \right] \cos\frac{n\pi ct}{L}.$$

In the present problem, L is to be replaced by $2L$, a by L and x by $L \pm x$.

The boundary conditions are

$$u(0,t)=0 \text{ and } u(2L,t)=0$$

and initial conditions are

$$u(L,0)=f(L)=h \text{ and } \left.\frac{\partial u}{\partial t}\right|_{t=0}=g(x)=0.$$

Since the string is set into vibrations after plucking at its midpoint, even harmonic vibrations will be absent.

Thus, we have

$$u(x\cdot t)=\sum_{m=0}^{\infty}\frac{8h}{(2m+1)^2\pi^2}\sin\left\{\frac{(2m+1)\pi}{2}\right\}\sin\left\{\frac{(2m+1)\pi(L\pm x)}{2L}\right\}\cos\left\{\frac{(2m+1)\pi ct}{2L}\right\}$$

$$=\frac{8h}{\pi^2}\left[\sin\left\{\frac{\pi(L\pm x)}{2L}\right\}\cos\left(\frac{\pi ct}{2L}\right)-\frac{1}{3^2}\sin\left\{\frac{3\pi(L\pm x)}{2L}\right\}\cos\left(\frac{3\pi ct}{2L}\right)\right.$$

$$\left.+\frac{1}{5^2}\sin\left\{\frac{5\pi(L\pm x)}{2L}\right\}\cos\left(\frac{5\pi ct}{2L}\right)-\ldots\infty\right].$$

PROBLEM 3 Suppose that a string of length 2 metres is plucked in the middle such that the initial deflection at the plucking point is 0.1 metre. It is then released to set into transverse vibrations. Taking the initial speed everywhere as zero and the speed of wave propagation as 1 metre per second, find the deflection at any point x at time t.

Solution: This problem is exactly problem 1 with L = 2 metres, h = 0.1 metre, c = 1 met/sec.

Substituting these values in the solution of Problem 1, we write the deflection at x at time t as

$$u(x,t)=\frac{0\cdot 8}{\pi^2}\left[\sin\left(\frac{\pi x}{2}\right)\cos\left(\frac{\pi t}{2}\right)-\frac{1}{3^2}\sin\left(\frac{3\pi x}{2}\right)\cos\left(\frac{3\pi t}{2}\right)+\frac{1}{5^2}\sin\left(\frac{5\pi x}{2}\right)\cos\left(\frac{5\pi t}{2}\right)-\ldots\infty\right].$$

PROBLEM 4 How far apart are the nodes on a string 80 cm long vibrating with a frequency of 1600 Hz assuming the speed of wave propagation on the string as 320 m/s?

Solution: Here, the wavelength is given by

$$\lambda=\frac{\text{speed}}{\text{frequency}}=\frac{320}{1600}=0.2\text{ metre}=20\text{ cm}$$

As the string is 80 cm long, it will fit 4 wavelengths of 20 cm each.

We know that nodes must be spaced every half wavelength. Therefore, we should have $4 \times 2 - 1 = 7$ nodes spaced by every 10 cm.

PROBLEM 5 A light string of length L is stretched and then fastened at the two ends. Initially, the string is set into vibrations by deflecting any point at distance x from the end $x = 0$ by an amount $u(x, 0) = f(x) = (Lx - x^2)$ and then releasing it at time $t = 0$. Find the deflection at any point of the vibrating string at a distance x from the end $x = 0$ at time t.

Solution: The equation for transverse vibrations of the string is given by

$$\frac{\partial^2 u}{\partial x^2} = \frac{1}{c^2}\frac{\partial^2 u}{\partial t^2} \tag{i}$$

where c is the speed of propagation of the wave along the string and $u(x, t)$ is the deflection of a point distant x from the origin at one end at time t.

The boundary conditions are

$$u(0, t) = 0 \text{ and } u(L, t) = 0.$$

The initial conditions are

$$u(x, 0) = f(x) = (Lx - x^2) \text{ and } \left.\frac{\partial u}{\partial t}\right|_{t=0} = g(x) = 0.$$

Putting $\quad u(x, t) = X(x)T(t),$ (ii)

Equation (i) can be written as

$$T\frac{d^2X}{dx^2} = \frac{X}{c^2}\frac{d^2T}{dt^2}.$$

Dividing both sides by XT, we write

$$\frac{1}{X}\frac{d^2X}{dx^2} = \frac{1}{c^2T}\frac{d^2T}{dt^2} = -k^2$$

where $(-k^2)$ is a separation constant.

Now, the transformed boundary conditions are

$$X(0)T(t) = 0 \text{ and } X(L)T(t) = 0 \text{ for all '}t\text{', i.e., } X(0) = X(L) = 0 \text{ for all } t.$$

Thus, we obtain the following two equations

$$\frac{d^2X}{dx^2} + k^2X = 0 \tag{iii}$$

and

$$\frac{d^2T}{dt^2} + \omega^2T = 0 \text{ with } \omega = kc. \tag{iv}$$

Equation (iii) has the general solution

$$X(x) = A\cos(kx) + B\sin(kx) \tag{v}$$

and Eq. (iv) has the general solution

$$T(t) = C\cos(\omega t) + D\sin(\omega t) \tag{vi}$$

where A, B, C and D are all arbitrary constants.

Now, applying the first transformed boundary condition to Eq. (v), we write

$$X(0) = A + 0 = A = 0.$$

Substituting this value of A in Eq. (v), we obtain

$$X(x) = B\sin(kx). \tag{vii}$$

Next, applying the second transformed boundary condition to Eq. (vii), we write

$$X(L) = B\sin(kL) = 0.$$

For arbitrary values of B, we have $\sin(kl) = 0$

giving $$kl = n\pi \text{ or, } k = \frac{n\pi}{L} \tag{viii}$$

where n = 1, 2, 3 … .

Here, n = 0 is excluded as it leads to a trivial solution. Also, n = –1, –2, –3 … are also neglected as they do not lead to independent solutions.

Substituting this value of k into Eq. (vii), we write for a particular value of n,

$$X_n(x) = B_n \sin\left(\frac{n\pi x}{L}\right) \tag{ix}$$

where B_n are arbtrary constants for n = 1, 2, 3 … .

Thus, for these different values of n, Eq. (vi) yields

$$T_n(x) = C_n\cos(\omega_n t) + D_n\sin(\omega_n t) = C_n\cos\left(\frac{n\pi ct}{L}\right) + D_n\sin\left(\frac{n\pi ct}{L}\right) \tag{x}$$

where C_n and D_n are arbitrary constants.

Recalling Eqs. (ix) and (x) and using Eq. (ii), we write the solutions of Eq. (i) as

$$u_n(x,t) = X_n(x)T_n(t) = \left[\alpha_n\cos\left(\frac{n\pi ct}{L}\right) + \beta_n\sin\left(\frac{n\pi ct}{L}\right)\right]\sin\left(\frac{n\pi x}{L}\right) \tag{xi}$$

where $\alpha_n = B_nC_n$ and $\beta_n = B_nD_n$ are arbitrary constants.

Applying superposition theorem, we write the general solution of Eq. (i) as

$$u(x,t) = \sum_{n=1}^{\infty} u_n(x,t) = \sum_{n=1}^{\infty}\left[\alpha_n\cos\left(\frac{n\pi ct}{L}\right) + \beta_n\sin\left(\frac{n\pi ct}{L}\right)\right]\sin\left(\frac{n\pi x}{L}\right). \tag{xii}$$

Applying the first initial condition, Eq. (xii) yields the initial displacement

$$u(x,0)=\sum_{n=1}^{\infty}\alpha_n \sin\left(\frac{n\pi x}{L}\right)=f(x)=(Lx-x^2).$$

Multiplying both sides of the above equation by $\sin\left(\frac{m\pi x}{L}\right)$ and integrating the results with respect to x within the limit from $x = 0$ to $x = L$, we write

$$\sum_{n=1}^{\infty}\alpha_n\int_0^L \sin\left(\frac{m\pi x}{L}\right)\sin\left(\frac{n\pi x}{L}\right)dx=\int_0^L (Lx-x^2)\sin\left(\frac{m\pi x}{L}\right)dx.$$

Using orthogonality of sine functions, we have

$$\alpha_m\int_0^L \sin^2\left(\frac{m\pi x}{L}\right)dx = L\int_0^L x\sin\left(\frac{m\pi x}{L}\right)dx-\int_0^L x^2\sin\left(\frac{m\pi x}{L}\right)dx$$

or

$$\alpha_m\cdot\frac{L}{2}=L\left\{\left[-\frac{Lx}{m\pi}\cos\left(\frac{m\pi x}{L}\right)\right]_{x=0}^{L}+\int_0^L\frac{L}{m\pi}\cos\left(\frac{m\pi x}{L}\right)dx\right\}$$

$$+\left[\frac{L}{m\pi}x^2\cos\left(\frac{m\pi x}{L}\right)\right]_{x=0}^{L}-\int_0^L\frac{2Lx}{m\pi}\cos\left(\frac{m\pi x}{L}\right)dx$$

$$=-\frac{L^3}{m\pi}(-1)^m+\frac{L^3}{m^2\pi^2}\left[\sin\left(\frac{m\pi x}{L}\right)\right]_0^L+\frac{L^3}{m\pi}(-1)^m$$

$$-\frac{2L^2}{m^2\pi^2}\left[x\sin\left(\frac{m\pi x}{L}\right)\right]_0^L+\frac{2L^2}{m^2\pi^2}\int_0^L\sin\left(\frac{m\pi x}{L}\right)dx$$

$$=-\frac{2L^3}{m^3\pi^3}\left[\cos\left(\frac{m\pi x}{L}\right)\right]_0^L=-\frac{2L^3}{m^3\pi^3}\left\{(-1)^m-1\right\}.$$

Therefore, we have

$$\alpha_m=\frac{4L^2}{m^3\pi^3}\left\{1-(-1)^m\right\}=\begin{cases}0 & \text{for even } m,\\ \dfrac{8L^2}{m^3\pi^3} & \text{for odd } m.\end{cases}$$

Replacing m by n, we write

$$\alpha_n=\frac{4L^2}{n^3\pi^3}\left\{1-(-1)^n\right\}=\begin{cases}0 & \text{for even } n,\\ \dfrac{8L^2}{n^3\pi^3} & \text{for odd } n.\end{cases}\qquad\text{(xiii)}$$

Further, applying the second initial condition, Eq. (xii) yields the initial speed

$$g(x)=\left.\frac{\partial u}{\partial t}\right|_{t=0}=\sum_{n=1}^{\infty}\frac{n\pi c}{L}\beta_n\sin\left(\frac{n\pi x}{L}\right)=0.$$

Multiplying both sides of the above equation by $\sin\left(\frac{m\pi x}{L}\right)$ and integrating the results with respect to x within the limit from $x = 0$ to $x = L$, we obtain $\beta_n = 0$.

Substituting the values of α_n and β_n in Eq. (xii), we obtain the general solution of Eq. (i) as

$$u(x,t)=\frac{8L^2}{\pi^3}\sum_{n=0}^{\infty}\frac{1}{(2n+1)^3}\sin\left\{\frac{(2n+1)\pi x}{L}\right\}\cos\left\{\frac{(2n+1)\pi ct}{L}\right\}$$

$$=\frac{8L^2}{\pi^3}\left[\sin\left(\frac{\pi x}{L}\right)\cos\left(\frac{\pi ct}{L}\right)+\frac{1}{3^3}\sin\left(\frac{3\pi x}{L}\right)\cos\left(\frac{3\pi ct}{L}\right)+\frac{1}{5^3}\sin\left(\frac{5\pi x}{L}\right)\cos\left(\frac{5\pi ct}{L}\right)+\ldots\infty\right].$$

PROBLEM 6 A light string is stretched and fastened at two points L apart. It is then set into transverse vibrations by plucking it at some point and then releasing it. If $u(x,0)=f(x)=A\sin\left(\frac{\pi x}{L}\right)$ is the initial deflection at a point distant x from one end at time $t = 0$ prove that the deflection at x at time t is given by

$$u(x,t)=A\sin\left(\frac{\pi x}{L}\right)\cos\left(\frac{\pi ct}{L}\right)\quad [A = \text{constant}]$$

where c is the constant speed of propagation of the transverse waves along the string.

Solution: The relevant wave equation is written as

$$\frac{\partial^2 u}{\partial x^2}=\frac{1}{c^2}\frac{\partial^2 u}{\partial t^2} \tag{i}$$

where c is the constant speed of wave propagation and $u(x,t)$ is the deflection at x at time t.

The boundary conditions are the same as in Problem 5.

To solve the above equation by the method of separation of variables under the given boundary conditions, follow the steps in Problem 5 up to Eq. (xii).

Thus, the general solution of Eq. (i) is written as

$$u(x,t)=\sum_{n=1}^{\infty}u_n(x,t)$$

$$=\sum_{n=1}^{\infty}\left\{\alpha_n\cos\left(\frac{n\pi ct}{L}\right)+\beta_n\sin\left(\frac{n\pi ct}{L}\right)\right\}\sin\left(\frac{n\pi x}{L}\right)$$

$$= \sum_{n=1}^{\infty} \alpha_n \cos\left(\frac{n\pi ct}{L}\right)\sin\left(\frac{n\pi x}{L}\right) + \sum_{n=1}^{\infty} \beta_n \sin\left(\frac{n\pi ct}{L}\right)\sin\left(\frac{n\pi x}{L}\right). \quad \text{(ii)}$$

In this case of plucked string, the initial conditions are

$$u(x,0) = f(x) = A\sin\left(\frac{\pi x}{L}\right) \text{ and } \left(\frac{\partial u}{\partial t}\right)\bigg|_{t=0} = g(x) = 0.$$

Using the first initial condition, Eq. (ii) yields

$$\sum_{n=1}^{\infty} \alpha_n \sin\left(\frac{n\pi x}{L}\right) = A\sin\left(\frac{\pi x}{L}\right).$$

Multiplying both sides by $\sin\left(\frac{\pi x}{L}\right)$ and integrating the results with respect to x within the limit from $x = 0$ to $x = L$, we obtain

$$\sum_{n=1}^{\infty} \alpha_n \int_0^L \sin\left(\frac{n\pi x}{L}\right)\sin\left(\frac{\pi x}{L}\right)dx = A\int_0^L \sin^2\left(\frac{\pi x}{L}\right)dx$$

or

$$\alpha_1 \int_0^L \sin^2\left(\frac{\pi x}{L}\right)dx = A\int_0^L \sin^2\left(\frac{\pi x}{L}\right)dx.$$

Here, we obtain $\alpha_1 = A$ and all other alphas, say α_2, α_3, α_4, ... are all zero. Hence, Eq. (ii) becomes

$$u(x,t) = \alpha_1 \cos\left(\frac{\pi ct}{L}\right)\sin\left(\frac{\pi x}{L}\right) + \sum_{n=1}^{\infty} \beta_n \sin\left(\frac{n\pi ct}{L}\right)\sin\left(\frac{n\pi x}{L}\right). \quad \text{(iii)}$$

Applying the second initial condition of the problem in Eq. (iii), we obtain

$$\left(\frac{\partial u}{\partial t}\right)\bigg|_{t=0} = \sum_{n=1}^{\infty} \frac{n\pi c}{L} \beta_n \sin\left(\frac{n\pi x}{L}\right) = 0.$$

Multiplying both sides by $\sin\left(\frac{m\pi x}{L}\right)$ and integrating the results with respect to x within the limit from $x = 0$ to $x = L$, we obtain $\beta_m = 0$.

Replacing the subscript m by n, we have $\beta_n = 0$.

Substituting the values of β_n in Eq. (iii), we write

$$u(x,t) = A\cos\left(\frac{\pi ct}{L}\right)\sin\left(\frac{\pi x}{L}\right) \text{ as } \alpha_1 = A \text{ and } \alpha_2 = \alpha_3 = \ldots = 0.$$

PROBLEM 7 A stretched string of length L is fastened at both ends. As a result of being struck somewhere, the string is imparted an initial velocity of magnitude $g(x)$ at a point distant x from the end $x=0$ at time $t=0$. Show that the deflection of the string at a point x at time t is given by

$$u(x,t)=\sum_{n=1}^{\infty}\left[\frac{2}{n\pi c}\int_0^L g(s)\sin\left(\frac{n\pi s}{L}\right)ds\right]\sin\left(\frac{n\pi x}{L}\right)\sin\left(\frac{n\pi ct}{L}\right)$$

where c is the constant speed of the propagation of transverse waves along the string.

Solution: The relevant wave equation is written as

$$\frac{\partial^2 u}{\partial x^2}=\frac{1}{c^2}\frac{\partial^2 u}{\partial t^2} \tag{i}$$

where $u(x,t)$ is the deflection at x at time t and c is the constant speed of the propagation of transverse waves along the string.

To solve the above equation by the method of separation of variables under the given boundary conditions, follow the steps in Problem 5 up to Eq. (xii).

Thus, we write the general solution of Eq. (i) as

$$u(x,t)=\sum_{n=1}^{\infty}u_n(x,t)$$

$$=\sum_{n=1}^{\infty}\left\{\alpha_n\cos\left(\frac{n\pi ct}{L}\right)+\beta_n\sin\left(\frac{n\pi ct}{L}\right)\right\}\sin\left(\frac{n\pi x}{L}\right). \tag{ii}$$

In this case of struck string, the initial conditions are

$$u(x,0)=f(x)=0 \text{ and } \left(\frac{\partial u}{\partial t}\right)_{t=0}=g(x).$$

Using the first initial condition, Eq. (ii) yields

$$\sum_{n=1}^{\infty}\alpha_n\sin\left(\frac{n\pi x}{L}\right)=0$$

Multiplying both sides of the above equation by $\sin\left(\frac{m\pi x}{L}\right)$ and integrating the results with respect to x within the limit from $x=0$ to $x=L$, we obtain $\alpha_n=0$

Hence, Eq. (ii) becomes

$$u(x,t)=\sum_{n=1}^{\infty}\beta_n\sin\left(\frac{n\pi ct}{L}\right)\sin\left(\frac{n\pi x}{L}\right). \tag{iii}$$

Further, using the second initial condition, Eq. (iii) yields

$$\sum_{n=1}^{\infty}\frac{n\pi c}{L}\beta_n \sin\left(\frac{n\pi x}{L}\right)=g(x).$$

Multiplying both sides of the above equation by $\sin\left(\frac{m\pi x}{L}\right)$ and integrating the results with respect to x within the limit from $x = 0$ to $x = L$, we obtain

$$\sum_{n=1}^{\infty}\frac{n\pi c}{L}\beta_n\int_0^L \sin\left(\frac{m\pi x}{L}\right)\sin\left(\frac{n\pi x}{L}\right)dx=\int_0^L g(x)\sin\left(\frac{m\pi x}{L}\right)dx$$

or

$$\frac{m\pi c}{L}\beta_m\int_0^L \sin^2\left(\frac{m\pi x}{L}\right)dx=\int_0^L g(x)\sin\left(\frac{m\pi x}{L}\right)dx$$

or

$$\frac{m\pi c}{L}\beta_m\frac{L}{2}=\int_0^L g(x)\sin\left(\frac{m\pi x}{L}\right)dx$$

yielding

$$\beta_m=\frac{2}{m\pi c}\int_0^L g(x)\sin\left(\frac{m\pi x}{L}\right)dx.$$

Replacing the subscript m by n, we write

$$\beta_n=\frac{2}{n\pi c}\int_0^L g(x)\sin\left(\frac{n\pi x}{L}\right)dx. \tag{iv}$$

Finally, substituting Eq. (iv) into Eq. (iii), we obtain

$$u(x,t)=\sum_{n=1}^{\infty}\left\{\frac{2}{n\pi c}\int_0^L g(s)\sin\left(\frac{n\pi s}{L}\right)ds\right\}\sin\left(\frac{n\pi ct}{L}\right)\sin\left(\frac{n\pi x}{L}\right).$$

Here, the integration variables are renamed.

PROBLEM 8 A light stretched string fixed at the two ends $x=0$ and $x=\pi$ is initially at rest in its equilibrium position. It is then set into transverse vibrations by imparting an initial velocity of magnitude $g(x)=0.2\sin x$ at a point x. Find the deflection at any time t.

Solution: This is similar to Problem 7 with $L=\pi$ and $g(x)=0.2\sin x$

The deflection at a point x is given by

$$u(x,t)=\sum_{n=1}^{\infty}\left\{\frac{2}{n\pi c}\int_0^{\pi} 0.2\sin s\,\sin(ns)\,ds\right\}\sin(nct)\sin(nx)$$

$$= \left\{ \frac{0.4}{\pi c} \int_0^{\pi} \sin^2 s \, ds \right\} \sin(ct) \sin x = \frac{0.2}{c} \sin(ct) \sin x.$$

PROBLEM 9 A rectangular membrane described by $0 \le x \le a, 0 \le y \le b$ is fixed along its entire boundary. If the membrane is set into transverse vibrations starting from rest, find the deflection $u(x, y, t)$ of the membrane at any point (x, y) at time t if the initial deflection is $f(x, y) = xy(a - x)(b - y)$.

Solution: The relevant wave equation is

$$\frac{\partial^2 u}{\partial x^2} + \frac{\partial^2 u}{\partial y^2} = \frac{1}{c^2} \frac{\partial^2 u}{\partial t^2} \tag{i}$$

where $u = u(x, y, t)$ is the deflection of a point (x, y) at time t and c is the constant speed of wave propagation along the surface of the membrane.

The boundary conditions are

$$u(0, y, t) = 0,\ u(a, y, t) = 0,\ u(x, 0, t) = 0,\ u(x, b, t) = 0. \tag{ii}$$

The initial conditions are

$$u(x, y, 0) = 0 = f(x, y) = xy\,(a - x)\,(b - y) \text{ and } \left(\frac{\partial u}{\partial t}\right)\bigg|_{t=0} = g(x, y) = 0. \tag{iii}$$

Solving Eq. (i) with the given boundary conditions as in Article 12.12, we arrive at

$$u(x, y, t) = \sum_{m=1}^{\infty}\sum_{n=1}^{\infty} \{\alpha_{mn} \cos(\omega_{mn} t) + \beta_{mn} \sin(\omega_{mn} t)\} \sin\frac{m\pi x}{a} \sin\frac{n\pi y}{b} \tag{iv}$$

where α_{mn} and β_{mn} are to be determined from the initial conditions and

$$\omega_{mn} = \pi c \sqrt{\frac{m^2}{a^2} + \frac{n^2}{b^2}}. \tag{v}$$

To Find α_n and β_n:

Applying the second initial condition given in Eq. (iii) to Eq. (iv), we obtain

$$\sum_{m=1}^{\infty}\sum_{n=1}^{\infty} \omega_{mn} \sin\frac{m\pi x}{a} \sin\frac{n\pi y}{b} \{-\alpha_{mn} \sin(\omega_{mn} t) + \beta_{mn} \cos(\omega_{mn} t)\}\bigg|_{t=0} = 0$$

or

$$\sum_{m=1}^{\infty}\sum_{n=1}^{\infty} \beta_{mn}\, \omega_{mn} \sin\frac{m\pi x}{a} \sin\frac{n\pi y}{b} = 0.$$

Multiplying both sides by $\sin\frac{r\pi x}{a}$ and integrating the results with respect to x within the limit from $x = 0$ to $x = a$, we obtain

$$\sum_{n=1}^{\infty}\beta_{rn}\,\omega_{rn}\sin\frac{n\pi y}{b}\int_{x=0}^{a}\sin^2\left(\frac{r\pi x}{a}\right)dx=0$$

or
$$\sum_{n=1}^{\infty}\frac{a}{2}\beta_{rn}\,\omega_{rn}\sin\frac{n\pi y}{b}=0.$$

Again, multiplying both sides by $\sin\frac{s\pi y}{b}$ and integrating the results with respect to y within the limit from $y = 0$ to $y = b$, we obtain

$$\frac{ab}{4}\beta_{rs}\,\omega_{rs}=0 \text{ yielding } \beta_{rs}=0.$$

Thus, $\beta_{mn}=0$.

Substituting β_{mn} in Eq. (iv), we have

$$u(x,y,t)=\sum_{m=1}^{\infty}\sum_{n=1}^{\infty}\alpha_{mn}\cos(\omega_{mn}t)\sin\frac{m\pi x}{a}\sin\frac{n\pi y}{b}. \quad \text{(vi)}$$

Applying the first initial condition given in Eq. (iii) to Eq. (vi), we have

$$\sum_{m=1}^{\infty}\sum_{n=1}^{\infty}\alpha_{mn}\sin\frac{m\pi x}{a}\sin\frac{n\pi y}{b}=xy(a-x)(b-y).$$

Multiplying both sides by $\sin\frac{r\pi x}{a}$ and integrating the results with respect to x within the limit from $x = 0$ to $x = a$, we obtain

$$\sum_{n=1}^{\infty}\frac{a}{2}\alpha_{rn}\sin\frac{n\pi y}{b}=\int_{x=0}^{a}xy(a-x)(b-y)\sin\frac{r\pi x}{a}dx.$$

Again, multiplying both sides by $\sin\frac{s\pi y}{b}$ and integrating the results with respect to y within the limit from $y = 0$ to $y = b$, we obtain

$$\frac{ab}{4}\alpha_{rs}=\int_{x=0}^{a}\int_{y=0}^{b}xy(a-x)(b-y)\sin\frac{r\pi x}{a}\sin\frac{s\pi y}{b}dx\,dy$$

$$=\int_{0}^{a}(ax-x^2)\sin\frac{r\pi x}{a}dx\int_{0}^{b}(by-y^2)\sin\frac{s\pi y}{b}dy. \quad \text{(vii)}$$

Here, $I_1=\int_{0}^{a}(ax-x^2)\sin\frac{r\pi x}{a}dx$

$$=\left[-\frac{a}{r\pi}(ax-x^2)\cos\frac{r\pi x}{a}\right]_{x=0}^{a}+\frac{a}{r\pi}\int_{0}^{a}(a-2x)\cos\frac{r\pi x}{a}dx$$

$$=0+\frac{a^2}{r^2\pi^2}\left[(a-2x)\sin\frac{r\pi x}{a}\right]_0^a+\frac{2a^2}{r^2\pi^2}\int_0^a \sin\frac{r\pi x}{a}dx$$

$$=0+0+\frac{2a^3}{r^3\pi^3}\left[-\cos\frac{r\pi x}{a}\right]_0^a=\frac{2a^3}{r^3\pi^3}\left[1-(-1)^r\right]. \qquad \text{(viii)}$$

Similarly, we have

$$I_2=\int_0^b (by-y^2)\sin\frac{s\pi y}{b}dy=\frac{2b^3}{s^3\pi^3}\left[1-(-1)^s\right]. \qquad \text{(ix)}$$

Substituting Eqs. (viii) and (ix) into Eq. (vii), we write

$$\alpha_{rs}=\frac{16a^2b^2}{r^3s^3\pi^6}\left[1-(-1)^r\right]\left[1-(-1)^s\right]$$

$$=\begin{cases}\dfrac{64\,a^2b^2}{r^3s^3\pi^3} & \text{when } r \text{ and } s \text{ are both odd integers,}\\ 0, & \text{otherwise.}\end{cases}$$

Hence, we write,

$$\alpha_{mn}=\frac{64\,a^2b^2}{(2m+1)^3\,(2n+1)^3\,\pi^6} \text{ with } m=0,1,2,3\ldots \text{ and } n=0,1,2,3\ldots \qquad \text{(x)}$$

Substituting the values of α_{mn} and β_{mn} into Eq. (vi), we write

$$u(x,y,t)=\frac{64\,a^2b^2}{\pi^6}\sum_{m=0}^{\infty}\sum_{n=0}^{\infty}\frac{1}{(2m+1)^3\,(2n+1)^3}\cos(\omega_{mn}t)\sin\left\{\frac{(2m+1)\pi x}{a}\right\}\sin\left\{\frac{(2n+1)\pi y}{b}\right\}$$

with $\omega_{mn}=\pi c\sqrt{\dfrac{(2m+1)^2}{a^2}+\dfrac{(2n+1)^2}{b^2}}$.

PROBLEM 10 A rectangular membrane bounded by $0\le x\le a$, $0\le y\le b$ is fixed along its boundary. It is struck at its central point and any point $P(x,y)$ is imparted an initial speed $g(x,y)$ while the initial deflection there is zero. If the membrane undergoes transverse vibrations, find the deflection of the point $P(x,y)$ at time t.

Solution: The relevant wave equation is

$$\frac{\partial^2 u}{\partial x^2}+\frac{\partial^2 u}{\partial y^2}=\frac{1}{c^2}\frac{\partial^2 u}{\partial t^2} \qquad \text{(i)}$$

where $u = u(x, y, t)$ is the deflection of a point $P(x, y)$ at time t and c is the constant speed of wave propagation along the surface of the membrane.

The boundary conditions are

$$u(0, y, t) = 0, u(a, y, t) = 0, u(x, 0, t) = 0, u(x, b, t) = 0. \tag{ii}$$

The initial conditions are

$$u(x, y, 0) = f(x, y) = 0 \text{ and } \left(\frac{\partial u}{\partial t}\right)\bigg|_{t=0} = g(x, y). \tag{iii}$$

Solving Eq. (i) with the given boundary conditions as in the previous problem, we arrive at

$$u(x, y, t) = \sum_{m=1}^{\infty}\sum_{n=1}^{\infty}\{\alpha_{mn}\cos(\omega_{mn}t) + \beta_{mn}\sin(\omega_{mn}t)\}\sin\frac{m\pi x}{a}\sin\frac{n\pi y}{b} \tag{iv}$$

where α_{mn} and β_{mn} are to be determined from the initial conditions and

$$\omega_{mn} = \pi c\sqrt{\frac{m^2}{a^2} + \frac{n^2}{b^2}}. \tag{v}$$

To Find α_n and β_n:

Applying the first initial condition given in Eq. (iii) to Eq. (iv), we have

$$\sum_{m=1}^{\infty}\sum_{n=1}^{\infty}\alpha_{mn}\sin\frac{m\pi x}{a}\sin\frac{n\pi y}{b} = 0.$$

Multiplying both sides by $\sin\frac{r\pi x}{a}$ and integrating the results with respect to x within the limit from $x = 0$ to $x = a$, we obtain

$$\sum_{n=1}^{\infty}\frac{a}{2}\alpha_{rn}\sin\frac{n\pi y}{b} = 0.$$

Again, multiplying both sides by $\sin\frac{s\pi y}{b}$ and integrating the results with respect to y within the limit from $y = 0$ to $y = b$, we obtain

$$\frac{ab}{4}\alpha_{rs} = 0.$$

Therefore, we have $\alpha_{rs} = 0$ Hence, $\alpha_{mn} = 0$.

Substituting α_{mn} into Eq. (iv), we obtain

$$u(x, y, t) = \sum_{m=1}^{\infty}\sum_{n=1}^{\infty}\beta_{mn}\sin(\omega_{mn}t)\sin\frac{m\pi x}{a}\sin\frac{n\pi y}{b}. \tag{vi}$$

Applying the second initial condition given in Eq. (iii) to Eq. (vi), we obtain

$$\sum_{m=1}^{\infty}\sum_{n=1}^{\infty}\beta_{mn}\,\omega_{mn}\sin\frac{m\pi x}{a}\sin\frac{n\pi y}{b}=g(x,y).$$

Multiplying both sides by $\sin\frac{r\pi x}{a}$ and integrating the results with respect to x within the limit from $x = 0$ to $x = a$, we obtain

$$\sum_{n=1}^{\infty}\beta_{rn}\,\omega_{rn}\sin\frac{n\pi y}{b}\int_{x=0}^{a}\sin^2\left(\frac{r\pi x}{a}\right)dx=\int_{x=0}^{a}g(x,y)\sin\frac{r\pi x}{a}dx$$

or

$$\frac{a}{2}\beta_{rn}\omega_{rn}\sin\frac{n\pi y}{b}=\int_{x=0}^{a}g(x,y)\sin\frac{r\pi x}{a}dx.$$

Again, multiplying both sides by $\sin\frac{s\pi y}{b}$ and integrating the results with respect to y within the limit from $y = 0$ to $y = b$, we obtain

$$\frac{ab}{4}\beta_{rs}\,\omega_{rs}=\int_{x=0}^{a}\int_{y=0}^{b}g(x,y)\sin\frac{r\pi x}{a}\sin\frac{s\pi y}{b}dx\,dy$$

yielding

$$\beta_{rs}=\frac{4}{ab\omega_{rs}}\int_{x=0}^{a}\int_{y=0}^{b}g(x,y)\sin\frac{r\pi x}{a}\sin\frac{s\pi y}{b}dx\,dy.$$

Hence, we write

$$\beta_{mn}=\frac{4}{ab\,\omega_{mn}}\int_{x=0}^{a}\int_{y=0}^{b}g(x,y)\sin\frac{m\pi x}{a}\sin\frac{n\pi y}{b}dx\,dy.$$

Substituting β_{mn} into Eq. (vi), we obtain

$$u(x,y,t)$$
$$=\sum_{m=1}^{\infty}\sum_{n=1}^{\infty}\frac{4}{ab\,\omega_{mn}}\left[\int_{x=0}^{a}\int_{y=0}^{b}g(p,q)\sin\frac{m\pi p}{a}\sin\frac{n\pi q}{b}dp\,dq\right]\sin(\omega_{mn}t)\sin\frac{m\pi x}{a}\sin\frac{n\pi y}{b}.$$

Here, the integration variables are re-named.

PROBLEM 11 Find the deflection $u(r,t)$ of a circular membrane of unit radius and unit velocity of wave propagation on its surface if the initial deflection is $f(r) = K(1 - r^2)$ where K is a constant and the initial speed is zero. (Assume a radial symmetry.)

Solution: Recalling Eq. (12.115), we write the deflection at a point distant r from the centre at time t as

$$u(r,t)=\sum_{m=1}^{\infty}u_m(r,t)$$

$$= \sum_{m=1}^{\infty}\left[A_m \cos \omega_m t + B_m \sin \omega_m t\right] J_0\left(\frac{\alpha_m}{a} r\right) m = 1, 2, 3 \ldots$$

where α_m is the mth zero of the Bessel function J_0, $\omega_m = \frac{\alpha_m c}{a} = \alpha_m$ (as the wave speed c and radius a are equal to unity), A_m and B_m are constants to be determined.

Thus, we have

$$u(r,t) = \sum_{m=1}^{\infty}\left[A_m \cos \alpha_m t + B_m \sin \alpha_m t\right] J_0(\alpha_m r). \tag{i}$$

The boundary condition used for obtaining the above equation is

$$u(a, t) = u(1, t) = 0 \text{ for all } t.$$

The initial conditions of the problem are

$$u(r, 0) = f(r) = K(1 - r^2)$$

and

$$\left.\frac{\partial u}{\partial t}\right|_{t=0} = g(r) = 0.$$

Applying the first initial condition to Eq. (i), we write

$$\sum_{m=1}^{\infty} A_m J_0(\alpha_m r) = K(1 - r^2). \tag{ii}$$

Multiplying both sides of Eq. (ii) by $rJ_0(\alpha_l r)$ and integrating the results with respect to r within the limits from $r = 0$ to $r = 1$, we have

$$\sum_{m=1}^{\infty} A_m \int_0^1 J_0(\alpha_l r) J_0(\alpha_m r) r\, dr = K\int_0^1 (r - r^3) J_0(\alpha_l r)\, dr.$$

Using orthogonality condition of Bessel functions, we obtain

$$A_l \int_0^1 J_0(\alpha_l r) J_0(\alpha_l r) r\, dr = K\int_0^1 r J_0(\alpha_l r)\, dr - K\int_0^1 r^3 J_0(\alpha_l r)\, dr$$

or

$$A_l \frac{1}{2}\left[J_1(\alpha_l)\right]^2 = \frac{K}{\alpha_l^2}\int_0^{\alpha_l} pJ_0(p)\, dp - \frac{K}{\alpha_l^4}\int_0^{\alpha_l} p^3 J_0(p)\, dp$$

$$= \frac{K}{\alpha_l^2}\int_0^{\alpha_l} p J_0(p)\, dp - \frac{K}{\alpha_l^4}\int_0^{\alpha_l} p^2 \{p J_0(p)\}\, dp \tag{iii}$$

Here, we put $p = \alpha_l r$

Using the recurrence relation of Bessel functions, viz.

$$\frac{d}{dx}\left[x^n J_n(x)\right] = x^n J_{n-1}(x), \text{ or, } \int x^n J_{n-1}(x)\, dx = x^n J_n(x)$$

we write

$$\int_0^{\alpha_l} p\,J_0(p)\,dp = \alpha_l\,J_1(\alpha_l) \tag{iv}$$

and
$$\int_0^{\alpha_l} p^2\{p\,J_0(p)\}dp = \left[p^2\int pJ_0(p)\,dp\right]_0^{\alpha_l} - \int_0^{\alpha_l} 2p\left\{\int pJ_0(p)\,dp\right\}dp$$

$$= \left[p^2\{p\,J_1(p)\}\right]_0^{\alpha_l} - \int_0^{\alpha_l} 2p\{p\,J_1(p)\}\,dp$$

$$= \left[p^3 J_1(p)\right]_0^{\alpha_l} - 2\int_0^{\alpha_l} p^2 J_1(p)\,dp$$

$$= \left[p^3 J_1(p)\right]_0^{\alpha_l} - \left[2p^2 J_2(p)\right]_0^{\alpha_l}$$

$$= \alpha_l^3 J_1(\alpha_l) - 2\alpha_l^2 J_2(\alpha_l). \tag{v}$$

Using Eqs. (iv) and (v), we write Eq. (iii) as

$$A_l \frac{1}{2}\left[J_1(\alpha_l)\right]^2 = \frac{K}{\alpha_l^2}\cdot \alpha_l\,J_1(\alpha_l) - \frac{K}{\alpha_l^4}\left\{\alpha_l^3 J_1(\alpha_l) - 2\alpha_l^2 J_2(\alpha_l)\right\}$$

$$= \frac{2K}{\alpha_l^2} J_2(\alpha_l).$$

Therefore, we have

$$A_l = \frac{4K}{\left[\alpha_l\,J_1(\alpha_l)\right]^2} J_2(\alpha_l) \text{ giving } A_m = \frac{4K}{\left[\alpha_m J_1(\alpha_m)\right]^2} J_2(\alpha_m). \tag{vi}$$

Applying the second initial condition to Eq. (i), we can obtain

$$B_m = 0. \tag{vii}$$

Substituting the values of A_m and B_m into Eq. (i), we have

$$u(r,t) = \sum_{m=1}^{\infty} \frac{4K}{\left[\alpha_m\,J_1(\alpha_m)\right]^2} J_2(\alpha_m)\cos(\alpha_m t)\,J_0(\alpha_m r).$$

PROBLEM 12 A thin uniform rod of length L with lateral insulation is initially at a uniform temperature θ. Its ends are suddenly cooled to 0° C and are kept constant at that temperature. Find the temperature distribution $u(x, t)$.

Solution: Let $u(x, t)$ be the temperature of the rod at a point distant x from the end $x = 0$ at time t.

The equation for one-dimensional heat flow is

$$\frac{\partial^2 u}{\partial x^2}=\frac{1}{D}\frac{\partial u}{\partial t} \tag{i}$$

where D is the diffusivity of the medium of the rod.

Using the method of separation of variables, let us put

$$u(x, t) = X(x)\, T(t). \tag{ii}$$

Now, Eq. (i) yields

$$T\frac{d^2X}{dx^2}=\frac{X}{D}\frac{dT}{dt}.$$

Dividing both sides by XT, we have

$$\frac{1}{X}\frac{d^2X}{dx^2}=\frac{1}{DT}\frac{dT}{dt}=-\alpha^2 \text{ (a constant)}.$$

Thus, we obtain two equations

$$\frac{d^2X}{dx^2}+\alpha^2X=0 \text{ and } \frac{dT}{T}=-D\alpha^2dt$$

whose general solutions are respectively

$$X(x)=C_1\cos\alpha x+C_2\sin\alpha x \tag{iii}$$

and

$$T(t)=C_3\, e^{-D\alpha^2t} \tag{iv}$$

where C_1, C_2, C_3 are arbitrary constants.

Thus, we have

$$u(x,t)=X(x)T(t)=\{C_1\cos\alpha x+C_2\sin\alpha x\}C_3\, e^{-D\alpha^2t}. \tag{v}$$

Boundary conditions of the problem are

1. $u(0, t) = X(0)\, T(t) = 0$ yielding $X(0) = 0$

and 2. $u(L, t) = X(L)\, T(t) = 0$ yielding $X(L) = 0$.

Applying the first boundary condition to Eq. (v), we have $C_1 = 0$

Therefore, Eq. (v) becomes

$$u(x,t)=C\, e^{-D\alpha^2t}\sin\alpha x \tag{vi}$$

where $C = C_2C_3$ = constant.

Applying the second boundary condition to Eq. (vi), we obtain

$$\sin\alpha L = 0 \text{ giving } \alpha=\frac{n\pi}{L} \text{ with } n = 1, 2, 3 \ldots$$

Here, $n = 0$ is excluded as it leads to a trivial solution and negative values of n are also neglected as these don't provide new independent solutions.

Thus, for a particular n, we write

$$\alpha = \alpha_n = \frac{n\pi}{L} \text{ and } C = C_n.$$

Now, for a patcular value of n, Eq. (vi) yields

$$u_n(x,t) = C_n\, e^{-D\left(\frac{n\pi}{L}\right)^2 t} \sin\left(\frac{n\pi x}{L}\right). \tag{vii}$$

By the superposition principle, we write the general solution of the problem as

$$u(x,t) = \sum_{n=1}^{\infty} u_n(x,t) = \sum_{n=1}^{\infty} C_n\, e^{-D\left(\frac{n\pi}{L}\right)^2 t} \sin\left(\frac{n\pi x}{L}\right). \tag{viii}$$

The initial condition of the problem is

$$u(x,0) = \sum_{n=1}^{\infty} C_n \sin\left(\frac{n\pi x}{L}\right) = \theta.$$

Multiplying both sides by $\sin\left(\frac{m\pi x}{L}\right)$ and integrating the results with respect to x within the limit from $x = 0$ to $x = L$, we have

$$\sum_{n=1}^{\infty} C_n \int_0^L \sin\left(\frac{m\pi x}{L}\right) \sin\left(\frac{n\pi x}{L}\right) dx = \theta \int_0^L \sin\left(\frac{m\pi x}{L}\right) dx$$

or
$$C_m \int_0^L \sin^2\left(\frac{m\pi x}{L}\right) dx = -\frac{\theta L}{m\pi}\left[\cos\left(\frac{m\pi x}{L}\right)\right]_0^L$$

or
$$\frac{L}{2} C_m = -\frac{\theta L}{m\pi}[\cos m\pi - \cos 0]$$
$$= \frac{\theta L}{m\pi}\left[1-(-1)^m\right].$$

Therefore, we have

$$C_m = \frac{2\theta}{m\pi}\left[1-(-1)^m\right] = \begin{cases} 0 & \text{if } m \text{ is an even integer,} \\ \dfrac{4\theta}{m\pi} & \text{if } m \text{ is an odd integer.} \end{cases}$$

Thus, we write

$$C_n = \begin{cases} 0 & \text{if } n \text{ is an even integer,} \\ \dfrac{4\theta}{n\pi} & \text{if } n \text{ is an odd integer.} \end{cases}$$

With this value of C_n, we finally write Eq. (viii) as

$$u(x,t)=\sum_{n=\text{odd}}^{\infty}\frac{4\theta}{n\pi}e^{-D\left(\frac{n\pi}{L}\right)^2 t}\sin\left(\frac{n\pi x}{L}\right)$$

$$=\frac{4\theta}{\pi}\sum_{n=0}^{\infty}\frac{1}{(2n+1)}e^{-\frac{D(2n+1)^2\pi^2 t}{L^2}}\sin\left\{\frac{(2n+1)\pi x}{L}\right\}.$$

PROBLEM 13 The heat flow along a uniform bar of length 10 cm of a homogeneous material is governed by the partial differential equation $u_t = Du_{xx}$ where D is the diffusivity of the medium of the bar. The ends of the bar are kept constant at 0°C. The initial temperature of the laterally insulated bar is $f(x)=x(10-x)$. Find the temperature of the bar at a distance x from one end at an instant, t.

Solution: Let $u(x,t)$ be the temperature of the rod at a point distant x from the end $x=0$ at time t.

The equation for one-dimensional heat flow is

$$\frac{\partial^2 u}{\partial x^2}=\frac{1}{D}\frac{\partial u}{\partial t}. \tag{i}$$

As in Problem 12, using the method of separation of variables under similar boundary conditions, we can obtain the solution of the above PDE as

$$u(x,t)=\sum_{n=1}^{\infty}u_n(x,t)=\sum_{n=1}^{\infty}C_n e^{-D\left(\frac{n\pi}{10}\right)^2 t}\sin\left(\frac{n\pi x}{10}\right) \tag{ii}$$

where C_n is an arbitrary constant.

The initial condition of the problem is

$$u(x\ 0) = f(x) = x(10 - x)$$

or

$$\sum_{n=1}^{\infty}C_n\sin\left(\frac{n\pi x}{10}\right)=10x-x^2.$$

Multiplying both sides by $\sin\left(\frac{m\pi x}{10}\right)$ and integrating the results with respect to x within the limit from $x = 0$ to $x = 10$, we have

$$\sum_{n=1}^{\infty}C_n\int_0^{10}\sin\left(\frac{m\pi x}{10}\right)\sin\left(\frac{n\pi x}{10}\right)dx=\int_0^{10}(10x-x^2)\sin\left(\frac{m\pi x}{10}\right)dx$$

or

$$C_m\int_0^{10}\sin^2\left(\frac{m\pi x}{10}\right)dx=\int_0^{10}10x\sin\left(\frac{m\pi x}{10}\right)dx-\int_0^{10}x^2\sin\left(\frac{m\pi x}{10}\right)dx$$

or $$5\,C_m = -\frac{100}{m\pi}\left[x\cos\left(\frac{m\pi x}{10}\right)\right]_0^{10} + \frac{100}{m\pi}\int_0^{10}\cos\left(\frac{m\pi x}{10}\right)dx$$

$$+\frac{10}{m\pi}\left[x^2\cos\left(\frac{m\pi x}{10}\right)\right]_0^{10} - \frac{20}{m\pi}\int_0^{10}x\cos\left(\frac{m\pi x}{10}\right)dx$$

$$= -\frac{100}{m\pi}[10\cos m\pi - 0] + 0 + \frac{10}{m\pi}[100\cos m\pi - 0]$$

$$-\frac{200}{m^2\pi^2}\left[x\sin\left(\frac{m\pi x}{10}\right)\right]_0^{10} + \frac{200}{m^2\pi^2}\int_0^{10}\sin\left(\frac{m\pi x}{10}\right)dx$$

$$= -\frac{2000}{m^3\pi^3}\left[\cos\left(\frac{m\pi x}{10}\right)\right]_0^{10} = -\frac{2000}{m^3\pi^3}[\cos m\pi - \cos 0]$$

$$= \frac{2000}{m^3\pi^3}\left[1-(-1)^m\right].$$

Therefore, we have

$$C_m = \frac{400}{m^3\pi^3}\left[1-(-1)^m\right] = \begin{cases} 0 & \text{if } m \text{ is an even integer,} \\ \dfrac{800}{m^3\pi^3} & \text{if } m \text{ is an odd integer.} \end{cases}$$

That is, $$C_n = \begin{cases} 0 & \text{if } n \text{ is an even integer,} \\ \dfrac{800}{n^3\pi^3} & \text{if } n \text{ is an odd integer.} \end{cases}$$

Substituting this value of C_n into Eq. (ii), we obtain

$$u(x,t) = \frac{800}{\pi^3}\sum_{n=\text{odd}}^{\infty}\frac{1}{n^3}e^{-D\left(\frac{n\pi}{10}\right)^2 t}\sin\left(\frac{n\pi x}{10}\right)$$

$$= \frac{800}{\pi^3}\sum_{n=0}^{\infty}\frac{1}{(2n+1)^3}e^{-D(2n+1)^2\pi^2 t/100}\sin\left\{\frac{(2n+1)\pi x}{10}\right\}.$$

PROBLEM 14 A uniform rod of length L has its ends A and B maintained at a uniform temperature of 100°C. If the temperature at the end B is suddenly reduced to 0°C and maintained at that temperature, find the temperature $u(x, t)$ of the rod at a distance x from the end A at time t.

Solution: Recalling Eq. (12.171), we have

$$u(x,t) = T_1 + \frac{(T_2 - T_1)x}{L} + \sum_{n=1}^{\infty} b_n e^{-Dn^2\pi^2 \frac{t}{L^2}} \sin\frac{n\pi x}{L} \quad \text{(i)}$$

where D is the thermal diffusivity of the material of the rod.

Here, $T_1 = 100°C, T_2 = 0°C$

Now, we have

$$u(x,t) = 100 + \frac{(0-100)x}{L} + \sum_{n=1}^{\infty} b_n e^{-Dn^2\pi^2 \frac{t}{L^2}} \sin\frac{n\pi x}{L}. \quad \text{(ii)}$$

The initial condition is $u(x, 0) = f(x) = 100$.

Therefore, we have

$$100 - \frac{100\,x}{L} + \sum_{n=1}^{\infty} b_n \sin\frac{n\pi x}{L} = 100$$

or

$$\sum_{n=1}^{\infty} b_n \sin\frac{n\pi x}{L} = \frac{100\,x}{L}.$$

Multiplying both sides by $\sin\left(\frac{m\pi x}{L}\right)$ and integrating the results with respect to x within the interval from $x = 0$ to $x = L$, we obtain

$$\sum_{n=1}^{\infty} b_n \int_0^L \sin\left(\frac{m\pi x}{L}\right)\sin\left(\frac{n\pi x}{L}\right)dx = \int_0^L \frac{100\,x}{L}\sin\left(\frac{m\pi x}{L}\right)dx$$

or

$$b_m \int_0^L \sin^2\left(\frac{m\pi x}{L}\right)dx = \left[-\frac{100\,x}{m\pi}\cos\left(\frac{m\pi x}{L}\right)\right]_0^L + \int_0^L \frac{100}{m\pi}\cos\left(\frac{m\pi x}{L}\right)dx$$

or

$$\frac{L}{2}b_m = -\frac{100\,L}{m\pi}\cos m\pi + 0 = -\frac{100\,L}{m\pi}(-1)^m.$$

Therefore, we obtain

$$b_m = -\frac{200}{m\pi}(-1)^m$$

i.e.,

$$b_n = -\frac{200}{n\pi}(-1)^n.$$

Substituting the values of b_n into Eq. (ii), we write

$$u(x,t) = \frac{100(L-x)}{L} - \sum_{n=1}^{\infty} (-1)^n \frac{200}{n\pi} e^{-Dn^2\pi^2 \frac{t}{L^2}} \sin\frac{n\pi x}{L}.$$

PROBLEM 15 A silver rod 10 cm long is heated to a uniform temperature of 100° C and its ends are suddenly cooled to 0°C. Find an expression for the temperature at any point at any time $t>0$. Find how long will it take for the centre to cool to 5°C. (Given, thermal diffussivity of silver, $D=1.71\ \text{cm}^2/\text{sec}$.)

Solution: Recalling Eq. (12.171), the temperature distribution is given by

$$u(x,t)=T_1+\frac{(T_2-T_1)x}{L}+\sum_{n=1}^{\infty}b_n e^{-Dn^2\pi^2\frac{t}{L^2}}\sin\frac{n\pi x}{L}. \qquad \text{(i)}$$

Here, L = 10 cm, $u(x, 0) = f(x)$ = 100°C, $T_1 = T_2$ = 0°C.
Therefore, we have

$$u(x,t)=\sum_{n=1}^{\infty}b_n e^{-1.71\times 10^{-2}\ n^2\pi^2 t}\sin\frac{n\pi x}{10}. \qquad \text{(ii)}$$

Applying the initial condition, viz., $u(x, 0) = f(x)$ = 100°C, Eq. (ii) yields

$$\sum_{n=1}^{\infty}b_n\sin\frac{n\pi x}{10}=100.$$

Multiplying both sides by $\sin\frac{m\pi x}{10}$ and integrating the results with respect to x within the limit from x = 0 to x = 10, we obtain

$$\sum_{n=1}^{\infty}b_n\int_0^{10}\sin\frac{m\pi x}{10}\sin\frac{n\pi x}{10}dx=100\int_0^{10}\sin\frac{m\pi x}{10}dx$$

or
$$5\,b_m=-\frac{1000}{m\pi}\left[\cos\frac{m\pi x}{10}\right]_0^{10}=\frac{1000}{m\pi}\left[1-(-1)^m\right]$$

or
$$b_m=\frac{200}{m\pi}\left[1-(-1)^m\right].$$

Thus, we write

$$b_n=\begin{cases}0 & \text{if } n \text{ is an even integer,}\\ \dfrac{400}{n\pi} & \text{if } n \text{ is an odd integer.}\end{cases}$$

Substituting these values of b_n into Eq. (ii), we have

$$u(x,t)=\frac{400}{\pi}\sum_{n=\text{odd}}^{\infty}\frac{1}{n}e^{-1.71\times 10^{-2}\,n^2\pi^2 t}\sin\frac{n\pi x}{10}$$

$$= \frac{400}{\pi}\sum_{n=0}^{\infty}\left[\frac{1}{(2n+1)}e^{-1.71\times 10^{-2}(2n+1)^2\pi^2 t}\right]\sin\left\{\frac{(2n+1)\pi x}{10}\right\}.$$

At the middle of the rod, $x = 5$ cm. When the temperature of the midpoint is 5°C, we write

$$u(5,t) = \frac{400}{\pi}\sum_{n=0}^{\infty}\frac{1}{(2n+1)}e^{-1.71\times 10^{-2}(2n+1)^2\pi^2 t}\sin\left\{\frac{(2n+1)\pi}{2}\right\} = 5$$

or

$$\sum_{n=0}^{\infty}\frac{1}{(2n+1)}e^{-1.71\times 10^{-2}(2n+1)^2\pi^2 t}\sin\left\{\frac{(2n+1)\pi}{2}\right\} = \frac{\pi}{80}$$

or

$$e^{-1.71\times 10^{-2}\pi^2 t} \cong \frac{\pi}{80}$$

or

$$e^{1.71\times 10^{-2}\pi^2 t} \cong \frac{80}{\pi}$$

or

$$1.71\times 10^{-2}\pi^2 t \cong \ln\left(\frac{80}{\pi}\right)$$

Therefore,

$$t = \frac{10^2\times\ln(80/\pi)}{1.71\times\pi^2} \cong 19.18 \text{ sec}.$$

PROBLEM 16 The temperature distribution in a homogeneous uniform bar of length π which is perfectly insulated at both ends, is governed by the partial differential equation $u_t = u_{xx}$. Asuming the uniform initial temperature as $u(x,0) = f(x) = \cos 2x$ find the time dependent temperature distribution at any instant of time.

Solution: Solving the PDE by the method of separation of variables as in Problem 12, we write

$$u(x,t) = X(x)T(t) = \{C_1\cos\alpha x + C_2\sin\alpha x\}C_3\, e^{-\alpha^2 t}$$

$$= \{B\cos\alpha x + C\sin\alpha x\}e^{-\alpha^2 t} \quad \text{(i)}$$

where $B = C_1C_3$ and $C = C_2C_3$ are arbitrary constants and α is a constant to be determined later.

The boundary conditions of the problem are

$$\left.\frac{\partial u}{\partial x}\right|_{x=0} = 0 \text{ and } \left.\frac{\partial u}{\partial x}\right|_{x=\pi} = 0 \text{ for all } t.$$

The initial condition is $u(x, 0) = f(x) = \cos 2x$.

Application of the first boundary condition to Eq. (i) yields

$$\left.\frac{\partial u}{\partial x}\right|_{x=0} = 0 \text{ for all } t.$$

or $$\alpha\{-B\sin\alpha x + C\cos\alpha x\}\big|_{x=0} = 0$$

or $$C = 0.$$

With this value of C, Eq. (i) is written as

$$u(x,t) = B\,e^{-\alpha^2 t}\cos\alpha x. \tag{ii}$$

Applying the second boundary condition to Eq. (ii), we obtain

$$\left.\frac{\partial u}{\partial x}\right|_{x=\pi} = 0 \text{ for all } t$$

or $-\alpha B\sin\alpha\pi = 0$ for arbitrary α and B.

Therefore, we must have $\alpha = 0, 1, 2, 3 \ldots$.

For a particular value of α say, $\alpha = n$, we write Eq. (ii) as

$$u_n(x,t) = B_n\,e^{-n^2 t}\cos nx \tag{iii}$$

Using the superposition principle, the general solution of this problem is written as

$$u(x,t) = \sum_{n=0}^{\infty} u_n(x,t) = \sum_{n=0}^{\infty} B_n e^{-n^2 t}\cos nx. \tag{iv}$$

To Find the Value of B_n

Applying the initial condition, viz., $u(x, 0) = f(x) = \cos 2x$ to Eq. (iv), we have

$$\sum_{n=0}^{\infty} B_n \cos nx = \cos 2x$$

Multiplying both sides by $\cos 2x$ and integrating the results with respect to x within the limit from $x = 0$ to $x = \pi$, we obtain

$$\sum_{n=0}^{\infty} B_n \int_0^{\pi} \cos 2x \cos nx\, dx = \int_0^{\pi} \cos^2(2x)\,dx$$

or $$B_2 \int_0^{\pi} \cos^2(2x)\,dx = \int_0^{\pi} \cos^2(2x)\,dx.$$

Therefore, $B_2 = 1$ and all other B's are zero.

Hence, Eq. (iv) becomes

$$u(x,t) = B_2\,e^{-2^2 t}\cos 2x = e^{-4t}\cos 2x.$$

PROBLEM 17 A uniform rod of length L, whose lateral surface is thermally insulated, is initially at a temperature θ_0. At time $t = 0$, one end is suddenly cooled to zero and subsequently maintained at this temperature while the other end is kept thermally insulated. Find the time dependent temperature distribution in the rod.

Solution: As in Problem 12, we obtain the temperature distribution as

$$u(x,t) = \{B\cos\alpha x + C\sin\alpha x\}e^{-D\alpha^2 t} \tag{i}$$

where B, C, α are arbitrary constants and D is the diffusivity.

The boundary conditions of the problem are

$$\left.\frac{\partial u}{\partial x}\right|_{x=0} = 0 \text{ and } u(L,t) = 0 \text{ for all } t.$$

The initial condition of the problem is

$$u(x, 0) = f(x) = \theta_0.$$

Applying the first boundary condition to Eq. (i), we obtain

$$\left.\frac{\partial u}{\partial x}\right|_{x=0} = [-\alpha B\sin\alpha x + \alpha C\cos\alpha x]_{x=0}\, e^{-D\alpha^2 t} = 0 \text{ for all } t,$$

or $$\alpha C = 0 \text{ or, } C = 0.$$

Therefore, Eq. (i) becomes

$$u(x,t) = B\cos\alpha x\, e^{-D\alpha^2 t}. \tag{ii}$$

Applying the second boundary condition to Eq. (ii), we obtain

$$u(L,t) = B\cos\alpha L\, e^{-D\alpha^2 t} = 0 \text{ for all } t.$$

As B is arbitrary, we have $\cos \alpha L = 0$

giving $$\alpha L = \frac{(2n+1)\pi}{2} \text{ or, } \alpha = \frac{(2n+1)\pi}{2L} \text{ with } n = 0, 1, 2, 3 \ldots$$

For a particular n, we obtain $\alpha_n = \dfrac{(2n+1)\pi}{2L}$.

Therefore, for a particular n, we write Eq. (ii) as

$$u_n(x,t) = B_n\, e^{-\frac{(2n+1)^2\pi^2 Dt}{4L^2}} \cos\left\{\frac{(2n+1)\pi x}{2L}\right\} \tag{iii}$$

where B_n are constants to be determined.

By superposition principle, the general solution of the problem is written as

$$u(x,t)=\sum_{n=0}^{\infty}u_n(x,t)=\sum_{n=0}^{\infty}B_n\, e^{-\frac{(2n+1)^2\pi^2Dt}{4L^2}}\cos\left\{\frac{(2n+1)\pi x}{2L}\right\}. \tag{iv}$$

Applying the initial condition of the problem to Eq. (iv), we have

$$u(x,0)=\sum_{n=0}^{\infty}B_n\cos\left\{\frac{(2n+1)\pi x}{2L}\right\}=\theta_0.$$

Multiplying both sides by $\cos\left\{\dfrac{(2m+1)\pi x}{2L}\right\}$ and integrating the results with respect to x within the interval form $x = 0$ to $x = L$, we have

$$\sum_{n=0}^{\infty}B_n\int_0^L\cos\left\{\frac{(2m+1)\pi x}{2L}\right\}\cos\left\{\frac{(2n+1)\pi x}{2L}\right\}dx=\theta_0\int_0^L\cos\left\{\frac{(2m+1)\pi x}{2L}\right\}dx$$

or
$$B_m\int_0^L\cos^2\left\{\frac{(2m+1)\pi x}{2L}\right\}dx=\frac{2L\theta_0}{(2m+1)\pi}\left[\sin\left\{\frac{(2m+1)\pi x}{2L}\right\}\right]_0^L$$

or
$$B_m\frac{L}{2}=\frac{2L\theta_0}{(2m+1)\pi}\sin\left\{\frac{(2m+1)\pi}{2}\right\}$$

or
$$B_m=\frac{4\theta_0}{(2m+1)\pi}\sin\left\{\frac{(2m+1)\pi}{2}\right\}.$$

This means that $B_n=\dfrac{4\theta_0}{(2n+1)\pi}\sin\left\{\dfrac{(2n+1)\pi}{2}\right\}$.

Substitution of B_n into Eq. (iv) results in

$$u(x,t)=\frac{4\theta_0}{\pi}\sum_{n=0}^{\infty}\frac{1}{(2n+1)}\sin\left\{\frac{(2n+1)\pi}{2}\right\}e^{-\frac{(2n+1)^2\pi^2Dt}{4L^2}}\cos\left\{\frac{(2n+1)\pi x}{2L}\right\}$$

$$=\frac{4\theta_0}{\pi}\sum_{n=0}^{\infty}\frac{(-1)^n}{(2n+1)}e^{-\frac{(2n+1)^2\pi^2Dt}{4L^2}}\cos\left\{\frac{(2n+1)\pi x}{2L}\right\}.$$

PROBLEM 18 Find the time dependent temperature distribution within a solid cube of sides L initially maintained at a temperature $u = 0$ which at time $t = 0$, is immersed in a heat bath at a temperature u_0.

Solution: Let the temperature distribution be given by $u(x,y,z,t)$. The governing PDE is

$$\frac{\partial^2 u}{\partial x^2}+\frac{\partial^2 u}{\partial y^2}+\frac{\partial^2 u}{\partial z^2}=\frac{1}{D}\frac{\partial u}{\partial t} \quad \text{where } D \text{ is the diffusivity.}$$

The boundary conditions of the problem are

$$u = u_0 \text{ at } x = 0, L;\ u = u_0 \text{ at } y = 0, L;\ u = u_0 \text{ at } z = 0, L \text{ for all } t$$

As in Article 12.25, if we take the solution of this problem as

$$u(x, y, z, t)=\sum_{l=1}^{\infty}\sum_{m=1}^{\infty}\sum_{n=1}^{\infty}R_{lmn}\, e^{-D\pi^2(l^2+m^2+n^2)t/L^2}\sin\frac{l\pi x}{L}\sin\frac{m\pi y}{L}\sin\frac{n\pi z}{L} \tag{i}$$

where R_{lmn} is a constant to be determined, the given boundary conditions are not satisfied as the right hand side of Eq. (i) vanishes at the boundaries. Boundary conditions are satisfied if we add u_0 to the right-hand side of Eq. (i), i.e., if we take

$$u(x, y, z, t)=u_0+\sum_{l=1}^{\infty}\sum_{m=1}^{\infty}\sum_{n=1}^{\infty}R_{lmn}\, e^{-D\pi^2\frac{(l^2+m^2+n^2)t}{L^2}}\sin\frac{l\pi x}{L}\sin\frac{m\pi y}{L}\sin\frac{n\pi z}{L}. \tag{ii}$$

Initial condition of the problem is

$$u(x, y, z, 0) = f(x, y, z) = 0.$$

Applying this initial condition to Eq. (ii), we write

$$u_0+\sum_{l=1}^{\infty}\sum_{m=1}^{\infty}\sum_{n=1}^{\infty}R_{lmn}\sin\frac{l\pi x}{L}\sin\frac{m\pi y}{L}\sin\frac{n\pi z}{L}=0$$

or

$$\sum_{l=1}^{\infty}\sum_{m=1}^{\infty}\sum_{n=1}^{\infty}R_{lmn}\sin\frac{l\pi x}{L}\sin\frac{m\pi y}{L}\sin\frac{n\pi z}{L}=-u_0.$$

This equation yields

$$R_{lmn}=-\frac{8}{L^3}\int_{x=0}^{L}\int_{y=0}^{L}\int_{z=0}^{L}u_0\sin\frac{l\pi x}{L}\sin\frac{m\pi y}{L}\sin\frac{n\pi z}{L}\,dx\,dy\,dz$$

$$=-\frac{8u_0}{L^3}\int_0^L\sin\frac{l\pi x}{L}dx\int_0^L\sin\frac{m\pi y}{L}dy\int_0^L\sin\frac{n\pi z}{L}dz$$

$$=-\frac{8u_0}{L^3}\left[-\frac{L}{l\pi}\cos\frac{l\pi x}{L}\right]_0^L\left[-\frac{L}{m\pi}\cos\frac{m\pi y}{L}\right]_0^L\left[-\frac{L}{n\pi}\cos\frac{n\pi z}{L}\right]_0^L$$

$$=\frac{8u_0}{\pi^3 lmn}[\cos l\pi-\cos 0][\cos m\pi-\cos 0][\cos n\pi-\cos 0]$$

$$= \frac{8u_0}{\pi^3 lmn}\left[(-1)^l - 1\right]\left[(-1)^m - 1\right]\left[(-1)^n - 1\right]$$

$$= \begin{cases} -\dfrac{64\,u_0}{\pi^3 lmn} & \text{when } l, m, n \text{ are all odd,} \\ 0 & \text{otherwise.} \end{cases}$$

Substituting this value into Eq. (ii), we obtain

$$u(x, y, z, t) = u_0\left[1 - \frac{64}{\pi^3}\sum_{l=0}^{\infty}\sum_{m=0}^{\infty}\sum_{n=0}^{\infty}\frac{\exp\{-D\pi^2\{(2l+1)^2+(2m+1)^2+(2n+1)^2\}t\}/L^2}{(2l+1)(2m+1)(2n+1)}\right.$$

$$\left.\times \sin\left\{\frac{(2l+1)\pi x}{L}\right\}\sin\left\{\frac{(2m+1)\pi y}{L}\right\}\sin\left\{\frac{(2n+1)\pi z}{L}\right\}\right].$$

PROBLEM 19 Show that a function $\Psi(\vec{r}, t) = Ae^{i(\vec{k}\cdot\vec{r} - \omega t)}$ where A is a constant, $\vec{K}$ is a constant vector called the propagation vector of the wave, and ω (omega) is the angular frequency of the wave, satisfies the three-dimensional classical wave equation with the propagation speed given by $c = \omega/k$ where $k = |\vec{k}|$

Solution: We have

$$\Psi(\vec{r}, t) = Ae^{i(k_1 x + k_2 y + k_3 z - \omega t)} \tag{i}$$

where k_1, k_2, k_3 are the components of the vector $\vec{k}$ in Cartesian coordinates along the axes of x, y and z, respectively.

From Eq. (i), we have

$$\frac{\partial^2\Psi}{\partial x^2} = (ik_1)^2\,\Psi = -k_1^2\Psi$$

$$\frac{\partial^2\Psi}{\partial y^2} = (ik_2)^2\,\Psi = -k_2^2\Psi$$

and
$$\frac{\partial^2\Psi}{\partial z^2} = (ik_3)^2\,\Psi = -k_3^2\Psi.$$

Adding together the above three equations, we obtain

$$\nabla^2\Psi = -\left(k_1^2 + k_2^2 + k_3^2\right)\Psi = -k^2\Psi$$

or
$$-\frac{1}{k^2}\nabla^2\Psi = \Psi. \tag{ii}$$

Again, from Eq. (i), we obtain

$$\frac{\partial^2 \Psi}{\partial t^2} = (-i\omega)^2 \Psi = -\omega^2 \Psi$$

or
$$-\frac{1}{\omega^2}\frac{\partial^2 \Psi}{\partial t^2} = \Psi. \tag{iii}$$

From Eqs. (ii) and (iii), we have

$$-\frac{1}{k^2}\nabla^2 \Psi = -\frac{1}{\omega^2}\frac{\partial^2 \Psi}{\partial t^2}$$

or
$$\nabla^2 \Psi = \frac{k^2}{\omega^2}\frac{\partial^2 \Psi}{\partial t^2}$$

or
$$\nabla^2 \Psi = \frac{1}{c^2}\frac{\partial^2 \Psi}{\partial t^2} \quad \text{where } c = \frac{\omega}{k}.$$

This equation is the classical wave equation in three-dimensional Cartesian coordinates.

PROBLEM 20 Write down the three-dimensional equation of heat flow through a medium of thermal conductivity, K. Show that in the steady state condition, the equation reduces to Laplace's equation.

If heat flows out radially from a point source at the centre of a sphere, using spherical polar coordinates, show that the steady state solution has the form

$$u(r) = \frac{Q}{4\pi K r} + C$$

where Q is the rate at which heat flows out isothermally from the point source at $r = 0$ and C is a constant.

Further, if the temperatures of two concentric spheres of radii r_1 and r_2 with the point source at the centre are u_1 and u_2 respectively, show that

$$Q = \frac{4\pi K(u_1 - u_2) r_1 r_2}{r_2 - r_1}.$$

Solution: The three-dimensional equation of heat flow through the medium is written as

$$\nabla^2 u = \frac{1}{D}\frac{\partial u}{\partial t} \tag{i}$$

where $D = \dfrac{K}{\rho s}$ is the diffussivity of the medium of density ρ and specific heat, s.

In the steady state condition, we have $\dfrac{\partial u}{\partial t} = 0$.

Therefore, Eq. (i) reduces to the Laplace's equation

$$\nabla^2 u = 0. \tag{ii}$$

Using spherical polar co-ordinates, the above Laplace's equation can be written as

$$\frac{1}{r^2}\frac{\partial}{\partial r}\left(r^2\frac{\partial u}{\partial r}\right) + \frac{1}{r^2 \sin\theta}\frac{\partial}{\partial \theta}\left(\sin\theta\frac{\partial u}{\partial \theta}\right) + \frac{1}{r^2 \sin^2\theta}\frac{\partial^2 u}{\partial \phi^2} = 0.$$

As heat flows out radially, the above Laplace's equation must be independent of the angles θ and ϕ.

Thus, we have

$$\frac{1}{r^2}\frac{d}{dr}\left(r^2\frac{du}{dr}\right) = 0$$

or

$$r^2\frac{du}{dr} = A \text{ (a constant)}$$

or

$$du = A\frac{dr}{r^2}.$$

Integrating both sides, we write

$$u = -\frac{A}{r} + C \tag{iii}$$

where C is another constant.

Considering a sphere of radius r, it is known that

$$Q = -K \cdot 4\pi r^2 \cdot \frac{du}{dr} = -K \cdot 4\pi r^2 \cdot \frac{A}{r^2} = -4\pi K A.$$

Therefore, we have

$$A = -\frac{Q}{4\pi K}.$$

Substituting this value of A into Eq. (iii), we obtain

$$u = \frac{Q}{4\pi K r} + C. \tag{iv}$$

Considering the sphere of radius r_1 Eq. (iv) yields

$$u_1 = \frac{Q}{4\pi K r_1} + C. \tag{v}$$

Considering the sphere of radius r_2 Eq. (iv) yields

$$u_2 = \frac{Q}{4\pi K r_2} + C. \quad \text{(vi)}$$

Subtraction of Eq. (vi) from Eq. (v) results in

$$u_1 - u_2 = \frac{Q}{4\pi K}\left(\frac{1}{r_1} - \frac{1}{r_2}\right)$$

or $$Q = \frac{4\pi K(u_1 - u_2) r_1 r_2}{r_2 - r_1}. \text{ Hence shown.}$$

EXERCISE

1. A horizontally stretched string of length 150 cm is fixed at its ends. It is plucked at a point distant 50 cm from the end $x = 0$ by a vertical height of 2 cm and then released from rest. Show that the deflection of the string at any point x at time t is given by

$$u(x, t) = \frac{18}{\pi^2} \sum_{n=1}^{\infty} \frac{1}{n^2} \sin\frac{n\pi}{3} \sin\frac{n\pi x}{150} \cos\frac{n\pi ct}{150}$$

where c is the constant speed of wave propagation.

2. A horizontally stretched string of length L is fixed at its ends. It is plucked at a point distant $\frac{L}{4}$ from the end $x = 0$ by a vertical height h and then released from rest. Show that the deflection of the string at any point x at time t is given by

$$u(x, t) = \frac{32\, h}{3\pi^2} \sum_{n=1}^{\infty} \frac{1}{n^2} \sin\frac{n\pi}{4} \sin\frac{n\pi x}{L} \cos\frac{n\pi ct}{L}$$

where c is the constant speed of wave propagation.

3. A rectangular membrane described by $0 \le x \le a, 0 \le y \le b$ is fixed along its entire boundary. If the membrane is set into transverse vibrations starting from rest, find the deflection $u(x, y, t)$ of the membrane at any point (x, y) at time t if the initial deflection is $f(x, y) = xy$.

Ans. $$u(x, y, t) = \frac{4ab}{\pi^2} \sum_{m=1}^{\infty}\sum_{n=1}^{\infty} \frac{(-1)^{m+n}}{mn} \sin\frac{m\pi x}{a} \sin\frac{n\pi y}{b} \cos\omega_{mn} t$$

where $\omega_{mn} = \pi c\sqrt{\frac{m^2}{a^2} + \frac{n^2}{b^2}}$, c being the constant speed of wave propagation.

4. Suppose that a thin uniform rod of length L is first immersed in boiling water so that its temperature is uniformly 100°C. Then imagine that it is removed from the water at time $t = 0$ with its ends immediately put into ice so that these ends are kept at temperature 0°C. Find the temperature, $u(x, t)$, under these circumstances.

$$\textbf{Ans. } u(x,t) = \frac{400}{\pi}\sum_{n=0}^{\infty}\frac{1}{(2n+1)}e^{-D(2n+1)^2\pi^2 t/L^2}\sin\left\{\frac{(2n+1)\pi x}{L}\right\}$$

5. Let an aluminum rod of length 20 cm be initially at the uniform temperature of 25°C. Suppose that at time $t = 0$ the end $x = 0$ is heated to 60°C, while the end $x = 20$ cm is cooled to 0°C and both are thereafter maintained at those temperatures. Find the time-dependent temperature distribution in the rod at any time t. (For aluminium, $D = 0.86$ cm^2/sec.)

$$\textbf{Ans. } u(x,t) = 60 - 3x - \sum_{n=1}^{\infty}\frac{10}{n\pi}\left\{7 + 5(-1)^n\right\}e^{-0.86\,n^2\pi^2\,t/20^2}\sin\frac{n\pi x}{20}$$

6. A homogeneous uniform rod of length L has its lateral surface perfectly insulated and is so thin that heat flow in it can be regarded as one dimensional. Initially, the rod is at a uniform temperature of 100°C throughout. The end at $x = L$ is suddenly cooled to 50°C and maintained thereafter at that value while the temperature at the end at $x = 0$ is maintained at the temperature of 100°C. Find the temperature at any point of the rod at any subsequent time.

$$\textbf{Ans. } u(x,t) = 100 - \frac{50x}{L} - \frac{100}{\pi}\sum_{n=1}^{\infty}\frac{(-1)^n}{n}e^{-Dn^2\pi^2 t/L^2}\sin\left(\frac{n\pi x}{L}\right)$$

13

CHAPTER

Beta and Gamma Functions

13.1 BETA FUNCTIONS

(i) Most generally, beta function is defined in the form of a definite integral given by

$$B(m,n)=\int_0^1 x^{m-1}(1-x)^{n-1}\,dx \tag{13.1}$$

where m and n are both real positive, or complex numbers whose real parts are positive numbers. If m and n are negative real or complex numbers whose real parts are negative numbers, the above definition is not valid.

(ii) In the first definition, if we put $x=\dfrac{y}{1+y}$ or $y=\dfrac{x}{1-x}$, we have

$$dx=\frac{(1+y)dy-ydy}{(1+y)^2}=\frac{dy}{(1+y)^2}.$$

Since $y = 0$ when $x = 0$ and $y = \infty$ when $x = 1$, we write Eq. (13.1) as

$$\begin{aligned}B(m,n)&=\int_0^1 x^{m-1}(1-x)^{n-1}\,dx\\&=\int_0^\infty \frac{y^{m-1}}{(1+y)^{m-1}}\cdot\left(1-\frac{y}{1+y}\right)^{n-1}\frac{dy}{(1+y)^2}\\&=\int_0^\infty \frac{y^{m-1}}{(1+y)^{m+n}}\,dy.\end{aligned}$$

We write $B(m,n)=\displaystyle\int_0^\infty \frac{x^{m-1}}{(1+x)^{m+n}}\,dx.$ (13.2)

(iii) From the second definition, we have

$$B(m,n)=\int_0^\infty \frac{x^{m-1}}{(1+x)^{m+n}}dx.$$

Putting $x=\frac{a}{b}u$, we have $dx=\frac{a}{b}du$

Then, we write

$$B(m,n)=\int_0^\infty \frac{a^{m-1}}{b^{m-1}}u^{m-1}\frac{1}{\left(1+\frac{a}{b}u\right)^{m+n}}\cdot\frac{a}{b}du$$

$$=a^m b^n\int_0^\infty \frac{u^{m-1}}{(au+b)^{m+n}}du=a^m b^n\int_0^\infty \frac{x^{m-1}}{(ax+b)^{m+n}}dx. \quad (13.3)$$

(iv) From the second definition, we have

$$B(m,n)=\int_0^\infty \frac{x^{m-1}}{(1+x)^{m+n}}dx$$

$$=\int_0^1 \frac{x^{m-1}}{(1+x)^{m+n}}dx+\int_1^\infty \frac{x^{m-1}}{(1+x)^{m+n}}dx. \quad (13.4)$$

In the second integral on the right-hand side of the above equation, let us put $x=\frac{1}{u}$.

Then, we have $dx=-\frac{du}{u^2}$.

Thus, the second integral becomes equal to

$$\int_1^0 \frac{1}{u^{m-1}}\cdot\frac{1}{\left(1+\frac{1}{u}\right)^{m+n}}\cdot\left(-\frac{du}{u^2}\right)=\int_0^1 \frac{u^{n-1}}{(1+u)^{m+n}}du$$

$$=\int_0^1 \frac{x^{n-1}}{(1+x)^{m+n}}dx.$$

Thus, Eq. (13.4) is written as

$$B(m,n)=\int_0^1 \frac{x^{m-1}}{(1+x)^{m+n}}dx+\int_0^1 \frac{x^{n-1}}{(1+x)^{m+n}}dx$$

$$= \int_0^1 \frac{(x^{m-1} + x^{n-1})}{(1+x)^{m+n}} dx. \tag{13.5}$$

(v) We know that

$$B(m, n) = \int_0^1 x^{m-1} (1-x)^{n-1} dx.$$

Putting $x = \frac{1}{a} u$, we have $dx = \frac{1}{a} du$.

Now, we write

$$B(m, n) = \int_0^a \frac{1}{a^{m-1}} u^{m-1} \left(1 - \frac{u}{a}\right)^{n-1} \frac{du}{a}$$

$$= \frac{1}{a^{m+n-1}} \int_0^a u^{m-1} (a-u)^{n-1} du.$$

We can write

$$B(m, n) = \frac{1}{a^{m+n-1}} \int_0^a x^{m-1} (a-x)^{n-1} dx. \tag{13.6}$$

(vi) We know that

$$B(m, n) = \int_0^1 x^{m-1} (1-x)^{n-1} dx$$

Putting $x = \sin^2 \theta$, we have $dx = 2 \sin \theta \cos \theta \, d\theta$.

Now, we write

$$B(m, n) = \int_0^{\frac{\pi}{2}} (\sin^2 \theta)^{m-1} (\cos^2 \theta)^{n-1} (2 \sin \theta \cos \theta \, d\theta)$$

$$= 2 \int_0^{\pi/2} \sin^{2m-1} \theta \cos^{2n-1} \theta \, d\theta.$$

We can arrive at the same result by putting $x = \cos^2 \theta$.

(vii) We know that

$$B(m, n) = \int_0^1 x^{m-1} (1-x)^{n-1} dx.$$

Putting $x=\dfrac{u-a}{b-a}$ (where a and b are constants), we write

$$B(m,n)=\int_a^b\left(\frac{u-a}{b-a}\right)^{m-1}\left(1-\frac{u-a}{b-a}\right)^{n-1}\frac{du}{b-a}$$

$$=\frac{1}{(b-a)^{m+n-1}}\int_a^b(u-a)^{m-1}(b-u)^{n-1}\,du. \tag{13.7}$$

13.2 SYMMETRY PROPERTY OF BETA FUNCTIONS

Beta functions are symmetrical in m and n, i.e., $B(m,n)=B(n,m)$.

Proof: We have $B(m,n)=\int_0^1 x^{m-1}(1-x)^{n-1}\,dx.$

Putting $x=1-u$, or, $u=1-x$, we write

$$B(m,n)=\int_1^0(1-u)^{m-1}u^{n-1}(-du)$$

$$=\int_0^1 u^{n-1}(1-u)^{m-1}\,du=B(n,m).$$

Hence proved.

13.3 GAMMA FUNCTION

Gamma function is also defined in a number of ways.

(i) Gamma function is generally defined as an improper integral of the form

$$\Gamma(n)=\int_0^\infty x^{n-1}e^{-x}dx \tag{13.8}$$

where n is a real positive number, or a complex number whose real part is a positive number. If n is a negative real number or a complex number whose real part is a negative number, then the above definition is not valid.

(ii) In the above definition, viz., $\Gamma(n)=\int_0^\infty x^{n-1}e^{-x}dx$, if we put $x=au$, we will have

$$\Gamma(n)=\int_0^\infty (au)^{n-1}e^{-au}(adu)$$

$$= a^n \int_0^\infty u^{n-1} e^{-au} du$$

$$= a^n \int_0^\infty x^{n-1} e^{-ax} dx. \tag{13.9}$$

(iii) We know that

$$\Gamma(n) = \int_0^\infty x^{n-1} e^{-x} dx$$

Putting $x = u^{\frac{1}{n}}$, we have $dx = \frac{1}{n} u^{\frac{1}{n}-1} du$.

Then, we write

$$\Gamma(n) = \int_0^\infty \left(u^{\frac{1}{n}}\right)^{n-1} \left(e^{-u^{\frac{1}{n}}}\right)\left(\frac{1}{n} u^{\frac{1}{n}-1} du\right)$$

$$= \frac{1}{n}\int_0^\infty e^{-u^{1/n}} du = \frac{1}{n}\int_0^\infty e^{-x^{1/n}} dx. \tag{13.10}$$

(iv) We know that

$$\Gamma(n) = \int_0^\infty x^{n-1} e^{-x} dx$$

Putting $x = u^2$, we have $dx = 2u\,du$.

Therefore, we write

$$\Gamma(n) = \int_0^\infty (u^2)^{n-1} e^{-u^2} (2u\,du)$$

$$= 2\int_0^\infty u^{2n-1} e^{-u^2} du = 2\int_0^\infty x^{2n-1} e^{-x^2} dx. \tag{13.11}$$

(v) We know that

$$\Gamma(n) = \int_0^\infty x^{n-1} e^{-x} dx$$

Putting $e^{-x} = u$ or $x = \ln\left(\frac{1}{u}\right)$, we have $dx = -\frac{du}{u}$.

Thus, we write

$$\Gamma(n)=\int_{1}^{0}\left[\ln\left(\frac{1}{u}\right)\right]^{n-1}u\left(-\frac{du}{u}\right)=\int_{0}^{1}\left[\ln\left(\frac{1}{u}\right)\right]^{n-1}du$$

$$=\int_{0}^{1}\left[\ln\left(\frac{1}{x}\right)\right]^{n-1}dx. \tag{13.12}$$

(vi) We know that

$$\Gamma(n)=\int_{0}^{\infty}x^{n-1}e^{-x}dx$$

$$=\int_{0}^{\infty}x^{n-1}\left[\lim_{m\to\infty}\left(1-\frac{x}{m}\right)^{m}\right]dx$$

$$=\lim_{m\to\infty}\int_{0}^{m}x^{n-1}\left(1-\frac{x}{m}\right)^{m}dx.$$

Putting $\frac{x}{m}=u$ or, $x=mu$, we have $dx = m\,du$.

Then, we write

$$\Gamma(n)=\lim_{m\to\infty}\int_{0}^{1}(mu)^{n-1}(1-u)^{m}\,mdu$$

$$=\lim_{m\to\infty}m^{n}\int_{0}^{1}(1-u)^{m}u^{n-1}du.$$

Performing integration by parts on the right-hand side, we have

$$\Gamma(n)=\lim_{m\to\infty}m^{n}\left\{\left[(1-u)^{m}\frac{u^{n}}{n}\right]_{0}^{1}+\int_{0}^{1}m(1-u)^{m-1}\frac{u^{n}}{n}du\right\}$$

$$=\lim_{m\to\infty}m^{n}\frac{m}{n}\int_{0}^{1}(1-u)^{m-1}u^{n}du$$

$$=\lim_{m\to\infty}m^{n}\frac{m}{n}\left\{\left[(1-u)^{m-1}\frac{u^{n+1}}{(n+1)}\right]_{0}^{1}+\int_{0}^{1}(m-1)(1-u)^{m-2}\frac{u^{n+1}}{(n+1)}du\right\}$$

$$=\lim_{m\to\infty}m^{n}\frac{m(m-1)}{n(n+1)}\int_{0}^{1}(1-u)^{m-2}u^{n+1}du$$

Repeating the integration by parts up to m times, we obtain

$$\Gamma(n)=\lim_{m\to\infty} m^n \frac{m(m-1)(m-2)(m-3)\dots 3.2.1}{n(n+1)(n+2)(n+3)\dots(n+m-1)}\int_0^1 u^{n+m-1}du$$

$$=\lim_{m\to\infty} m^n \frac{m(m-1)(m-2)(m-3)\dots 3.2.1}{n(n+1)(n+2)(n+3)\dots(m+n-1)}\left[\frac{u^{m+n}}{(m+n)}\right]_0^1$$

$$=\lim_{m\to\infty} m^n \frac{m(m-1)(m-2)(m-3)\dots 3.2.1}{n(n+1)(n+2)(n+3)\dots(m+n-1)(m+n)}. \qquad (13.13)$$

13.4 PROPERTIES OF GAMMA FUNCTIONS

(i) $\Gamma(1) = 1$.

Proof: We have $\Gamma(n)=\int_0^\infty x^{n-1}e^{-x}dx$

Putting $n = 1$, we write

$$\Gamma(1)=\int_0^\infty e^{-x}dx=[-e^{-x}]_0^\infty=-e^{-\infty}+e^0=1. \qquad (13.14)$$

(ii) $\Gamma(n + 1) = n\Gamma(n)$.

Proof: In the formula $\Gamma(n)=\int_0^\infty x^{n-1}e^{-x}dx$,

replacing n by $(n + 1)$, we write

$$\Gamma(n+1)=\int_0^\infty x^n e^{-x}dx$$

$$=\left[x^n\left(-e^{-x}\right)\right]_0^\infty-\int_0^\infty nx^{n-1}\left(-e^{-x}\right)dx$$

$$=0+n\int_0^\infty x^{n-1}e^{-x}dx=n\Gamma(n). \qquad (13.15)$$

(iii) $\Gamma(n + 1) = n!$ when n is a non-negative integer.

Proof: We know that $\Gamma(n + 1) = n\,\Gamma(n)$.

Repeated application of the formula itself leads to

$$\Gamma(n+1) = n(n-1)(n-2)(n-3)\ldots 3.2.1.\Gamma(1)$$
$$= n! \text{ as } [\Gamma(1) = 1]. \tag{13.16}$$

(iv) $\Gamma(0) = \infty$.

Proof: From the formula $\Gamma(n+1) = n\Gamma(n)$, we have

$$\Gamma(n) = \frac{\Gamma(n+1)}{n}.$$

Putting $n = 0$, we write

$$\Gamma(0) = \frac{\Gamma(1)}{0} = \frac{1}{0} = \infty. \tag{13.17}$$

(v) $\Gamma(-n) = \infty$ for $n = 0, 1, 2, 3, \ldots$

Proof: From the formula

$$\Gamma(n) = \lim_{m\to\infty} m^n \frac{1.2.3\ldots(m-1)m}{n(n+1)(n+2)(n+3)\ldots(n+m-1)(n+m)},$$

we see that it becomes infinite when n = 0, –1, –2, –3, ... That is, $\Gamma(0) = \infty$, $\Gamma(-1) = \infty$, $\Gamma(-2) = \infty$, $\Gamma(-3) = \infty$, ... Therefore, we write, in general, $\Gamma(-n) = \infty$ if n = 0, 1, 2, 3, ...

(vi) $\Gamma\left(\frac{1}{2}\right) = \sqrt{\pi}$.

Proof: We know that

$$\Gamma(n) = 2\int_0^\infty x^{2n-1}e^{-x^2}\,dx.$$

Putting $n = \frac{1}{2}$, we write

$$\Gamma\left(\frac{1}{2}\right) = 2\int_0^\infty e^{-x^2}\,dx = \int_{-\infty}^\infty e^{-x^2}\,dx.$$

Also, we write

$$\Gamma\left(\frac{1}{2}\right) = 2\int_0^\infty e^{-y^2}\,dy \cdot = \int_{-\infty}^\infty e^{-y^2}\,dy.$$

Thus, we write

$$\left[\Gamma\left(\frac{1}{2}\right)\right]^2 = \int_{-\infty}^{\infty} e^{-x^2}dx \int_{-\infty}^{\infty} e^{-y^2}dy$$

$$= \int_{-\infty}^{\infty}\int_{-\infty}^{\infty} e^{-(x^2+y^2)}dxdy.$$

Transforming the two-dimensional Cartesian coordinate system into plane polar coordinates, we write

$$\left[\Gamma\left(\frac{1}{2}\right)\right]^2 = \int_{r=0}^{\infty}\int_{\phi=0}^{2\pi}\left[e^{-r^2}\left(r\,dr\,d\phi\right)\right]$$

$$= \left[\int_0^{\infty} e^{-r^2} r\,dr\right]\left[\int_0^{2\pi} d\phi\right]$$

$$= \pi\cdot 2\int_0^{\infty} e^{-r^2} r^{2\times 1-1}dr = \pi\Gamma(1) = \pi.$$

Therefore, we obtain

$$\Gamma\left(\frac{1}{2}\right) = \sqrt{\pi}. \tag{13.18}$$

13.5 RELATION BETWEEN BETA FUNCTION AND GAMMA FUNCTION

We know that

$$\Gamma(m) = 2\int_0^{\infty} x^{2m-1}e^{-x^2}dx$$

and

$$\Gamma(n) = 2\int_0^{\infty} y^{2n-1}e^{-y^2}dy.$$

Therefore, we write

$$\Gamma(m)\Gamma(n) = \left[2\int_0^{\infty} x^{2m-1}e^{-x^2}dx\right]\left[2\int_0^{\infty} y^{2n-1}e^{-y^2}dy\right]$$

$$= 4\int_0^{\infty}\int_0^{\infty}\left[x^{2m-1}y^{2n-1}e^{-(x^2+y^2)}\right]dxdy.$$

Since both of x and y vary from zero to infinity, the above integral is to be evaluated within the first quadrant of two-dimensional Cartesian coordinate system.

Now, changing the above coordinate system into plane polar coordinate system, we have

$$x = r\cos\phi,\ y = r\sin\phi,\ x^2 + y^2 = r^2,\ dxdy = r\,dr\,d\phi.$$

Therefore, we write

$$\Gamma(m)\Gamma(n) = 4\int_{r=0}^{\infty}\int_{\phi=0}^{\frac{\pi}{2}} (r\cos\phi)^{2m-1}(r\sin\phi)^{2n-1} e^{-r^2}\, r\,dr\,d\phi$$

$$= \left[2\int_0^{\infty} r^{2(m+n)-1} e^{-r^2}\,dr\right]\left[2\int_0^{\frac{\pi}{2}} \cos^{2m-1}\phi\,\sin^{2n-1}\phi\,d\phi\right]$$

$$= \Gamma(m+n)B(m,n).$$

Hence, we obtain

$$B(m,n) = \frac{\Gamma(m)\Gamma(n)}{\Gamma(m+n)}. \tag{13.19}$$

13.6 REFLECTION FORMULA OF GAMMA FUNCTIONS

Let us first derive a relevant formula, i.e., sine formula.

We know that

$$\sin x = x - \frac{x^3}{3!} + \frac{x^5}{5!} - \frac{x^7}{7!} + \ldots\infty. \tag{13.20}$$

We have

$$\lim_{x\to 0}\frac{\sin x}{x} = \lim_{x\to 0}\left[1 - \frac{x^2}{3!} + \frac{x^4}{5!} - \frac{x^6}{7!} + \ldots\infty\right] = 1. \tag{13.21}$$

Since the function $\sin x$ has roots given by $x = 0, \pm\pi, \pm 2\pi, \pm 3\pi, \ldots\infty$; we write

$$\sin x = A(x-0)(x-\pi)(x+\pi)(x-2\pi)(x+2\pi)(x-3\pi)(x+3\pi)\ldots\infty$$

$$= Ax\left(x^2 - \pi^2\right)\left(x^2 - 2^2\pi^2\right)\left(x^2 - 3^2\pi^2\right)\ldots\infty. \tag{13.22}$$

In the limit $x \to 0$, we have

$$\lim_{x\to 0}\frac{\sin x}{x} = \lim_{x\to 0} A\left(x^2 - \pi^2\right)\left(x^2 - 2^2\pi^2\right)\left(x^2 - 3^2\pi^2\right)\ldots\infty$$

or
$$1 = A\left(-\pi^2\right)\left(-2^2\pi^2\right)\left(-3^2\pi^2\right)\ldots\infty.$$

Therefore, we have

$$A = \frac{1}{\left(-\pi^2\right)\left(-2^2\pi^2\right)\left(-3^2\pi^2\right)\ldots\infty}. \tag{13.23}$$

Substituting this value of A into Eq. (13.22) , we have

$$\sin x = x\left(1-\frac{x^2}{\pi^2}\right)\left(1-\frac{x^2}{2^2\pi^2}\right)\left(1-\frac{x^2}{3^2\pi^2}\right)\ldots\infty$$
$$= x\prod_{m=1}^{\infty}\left(1-\frac{x^2}{m^2\pi^2}\right)$$

or
$$\frac{1}{\prod_{m=1}^{\infty}\left(1-\frac{x^2}{m^2\pi^2}\right)} = \frac{x}{\sin x}. \tag{13.24}$$

Now, to derive the reflection formula of gamma function, let us proceed as in the following.

We know that

$$\Gamma(n)\Gamma(1-n) = \Gamma(n)\Gamma(-n+1) = -n\Gamma(n)\Gamma(-n).$$

Now, we write

$$\Gamma(n)\Gamma(1-n)$$
$$= -n\left[\lim_{m\to\infty} m^n \frac{(1.2.3\ldots m)}{n(n+1)(n+2)(n+3)\ldots(n+m)}\right]$$
$$\times\left[\lim_{m\to\infty} m^{-n}\frac{(1.2.3\ldots m)}{(-n)(-n+1)(-n+2)(-n+3)\ldots(-n+m)}\right]$$

$$= \frac{1}{n}\lim_{m\to\infty}\frac{1^2.2^2.3^2\ldots m^2}{\left(1^2-n^2\right)\left(2^2-n^2\right)\left(3^2-n^2\right)\ldots\left(m^2-n^2\right)}$$

$$= \frac{1}{n}\lim_{m\to\infty}\frac{1}{\left(1^2-\frac{n^2}{1^2}\right)\left(1-\frac{n^2}{2^2}\right)\left(1-\frac{n^2}{3^2}\right)\ldots\left(1-\frac{n^2}{m^2}\right)}$$

$$= \frac{1}{n}\left[\frac{1}{\prod_{m=1}^{\infty}\left(1-\frac{n^2}{m^2}\right)}\right] = \frac{1}{n}\left[\frac{1}{\prod_{m=1}^{\infty}\left\{1-\frac{(n\pi)^2}{m^2\pi^2}\right\}}\right].$$

Using Eq. (13.24), we have

$$\frac{1}{\prod_{m=1}^{\infty}\left(1-\frac{(n\pi)^2}{m^2\pi^2}\right)}=\frac{n\pi}{\sin n\pi}$$

Hence, we write

$$\Gamma(n)\Gamma(1-n)=\frac{1}{n}\left[\frac{n\pi}{\sin n\pi}\right]=\frac{\pi}{\sin n\pi} \tag{13.25}$$

which is valid for $0<n<1$.

This formula is known as reflection formula of gamma functions.

Alternative Method: We know that

$$\begin{aligned}\Gamma(n)\Gamma(1-n)&=\frac{\Gamma(n)\Gamma(1-n)}{\Gamma(n+1-n)}\\&=B(n,1-n)\\&=\int_0^{\infty}\frac{u^{n-1}}{(1+u)^{n+1-n}}du=\int_0^{\infty}\frac{u^{n-1}}{1+u}du.\end{aligned}$$

Let us introduce a new variable x such that $u=e^x$.

Then, we have $du=e^x dx$.

Also, we know that u = 0 corresponds to $x=-\infty$ and $u=\infty$ corresponds to $x=\infty$

Therefore, we write

$$\Gamma(n)\Gamma(1-n)=\int_{-\infty}^{+\infty}\frac{(e^x)^{n-1}}{1+e^x}e^x dx=\int_{-\infty}^{+\infty}\frac{e^{nx}}{1+e^x}dx.$$

The above integral can be evaluated by the use of contour integration with calculus of residues. The evaluation of the integral is left as an exercise. The value of this integral is $\frac{\pi}{\sin n\pi}$.

Hence, we write

$$\Gamma(n)\Gamma(1-n)=\frac{\pi}{\sin n\pi}.$$

13.7 CONTOUR INTEGRAL FOR RECIPROCAL OF GAMMA FUNCTION

A Hankel contour is a contour C in the complex plane that begins at $-\infty$, continues towards the origin along a line parallel to the real axis just below it up to a point

$-\sigma$, turns along a circle $\{|\xi|=\sigma\}$ enclosing the origin in the counter clockwise sense and then returns to $-\infty$ along a line just above the real axis.

Let us consider the integral

$$I=\int_C e^{\xi}\xi^{-z}d\xi. \tag{13.26}$$

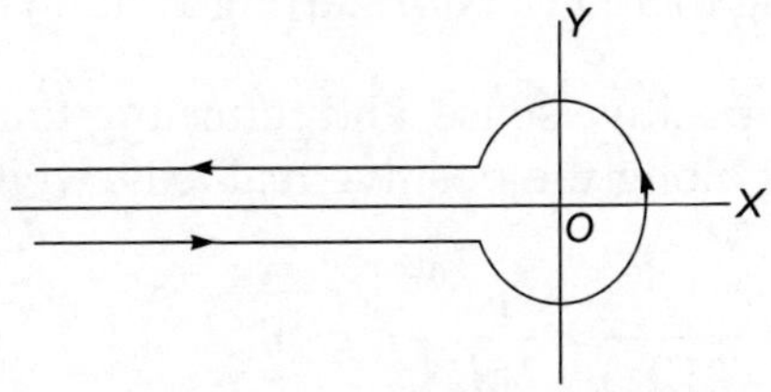

Fig. 13.1 Hankel contour

We know that below the real axis, $\arg(\xi)=-\pi$ and above the real axis, $\arg(\xi)=\pi$ Using the Hankel contour stated above, we can write

$$I=\int_{-\infty}^{-\sigma} e^{\xi}\left(-\xi e^{-i\pi}\right)^{-z}d\xi+\int_{-\pi}^{+\pi} e^{\sigma e^{i\theta}}\left(\sigma e^{i\theta}\right)^{-z}i\sigma e^{i\theta}d\theta+\int_{-\sigma}^{-\infty} e^{\xi}\left(-\xi e^{i\pi}\right)^{-z}d\xi$$

where $\xi=\sigma e^{i\theta}$ and $\theta=\arg(\xi)$ on the circumference of the small circle of radius σ with centre at the origin.

If we set $s=-\xi$ and make $\sigma\to 0$, we have

$$I=\int_{\infty}^{0} e^{-s}\left(se^{-i\pi}\right)^{-z}(-ds)+\int_{0}^{\infty} e^{-s}\left(se^{i\pi}\right)^{-z}(-ds)$$

$$=\int_{0}^{\infty} e^{-s}\left(se^{-i\pi}\right)^{-z}ds-\int_{0}^{\infty} e^{-s}\left(se^{i\pi}\right)^{-z}ds$$

$$=\left(e^{i\pi z}-e^{-i\pi z}\right)\int_{0}^{\infty} e^{-s}s^{-z}ds$$

$$=2i\sin\pi z\Gamma(1-z)=2\pi i\frac{\sin\pi z}{\pi}\Gamma(1-z)$$

$$=2\pi i\left[\frac{1}{\Gamma(z)\Gamma(1-z)}\right]\Gamma(1-z)=2\pi i\frac{1}{\Gamma(z)}. \tag{13.27}$$

Here, the following formula has been used.

$$\Gamma(z)\Gamma(1-z)=\frac{\pi}{\sin\pi z}. \tag{13.28}$$

From Eqs. (13.26) and (13.27), we write

$$\frac{1}{\Gamma(z)} = \frac{1}{2\pi i} I = \frac{1}{2\pi i} \int_C e^{\xi} \xi^{-z} d\xi. \tag{13.29}$$

Writing $\int_{-\infty}^{(0+)}$ for $\int_C$ meaning thereby a path starting at minus infinity on the real axis, encircling the origin O in a positive sense, and returning to minus infinity along the real axis, respecting the cut along the positive real axis, we write

$$\frac{1}{\Gamma(z)} = \frac{1}{2\pi i} \int_{-\infty}^{(0+)} e^{\xi} \xi^{-z} d\xi. \tag{13.29a}$$

Alternatively, a Hankel contour is also a contour C in the complex plane that begins at $+\infty$, continues towards the origin along a line parallel to the real axis just above it up to a point σ, turns along a circle $\{|\xi| = \sigma\}$ enclosing the origin in the counterclockwise sense and then returns to $+\infty$ along a line just below the real axis. With such a contour, the reciprocal of gamma function is given by

$$\frac{1}{\Gamma(z)} = \frac{i}{2\pi} \int_C e^{-\xi} (-\xi)^{-z} d\xi \tag{13.30}$$

or,

$$\frac{1}{\Gamma(z)} = \frac{i}{2\pi} \int_{\infty}^{(0+)} e^{-\xi} (-\xi)^{-z} d\xi. \tag{13.31}$$

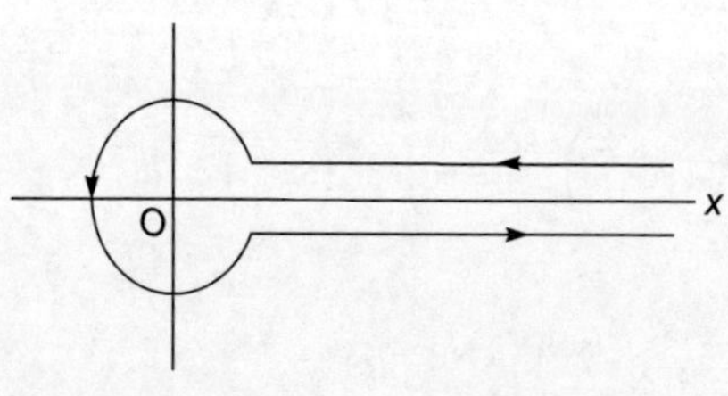

Fig. 13.2 An alternative of Hankel contour

13.8 LEGENDRE DUPLICATION FORMULA FOR GAMMA FUNCTION

We know that

$$\frac{\Gamma(m)\Gamma(n)}{\Gamma(m+n)} = B(m, n). \tag{13.32}$$

With $m=n$, we write

$$\frac{\Gamma(n)\Gamma(n)}{\Gamma(2n)} = B(n,n) = 2\int_0^{\frac{\pi}{2}} \cos^{2n-1}\theta \sin^{2n-1}\theta \, d\theta$$

$$= 2^{2-2n}\int_0^{\frac{\pi}{2}} (2\sin\theta\cos\theta)^{2n-1} d\theta$$

$$= \left(2^{2-2n}\right)\int_0^{\pi/2} \sin^{2n-1}(2\theta)\, d\theta .$$

Putting $t = 2\theta$, we have $d\theta = \frac{dt}{2}$.
Then, we write

$$\frac{\Gamma(n)\Gamma(n)}{\Gamma(2n)} = (2^{1-2n})\int_0^{\pi} \sin^{2n-1} t \, dt$$

$$= \left(2^{1-2n}\right)2\int_0^{\frac{\pi}{2}} \sin^{2n-1} t \, dt$$

$$= \left(2^{1-2n}\right)2\int_0^{\frac{\pi}{2}} \cos^{2\left(\frac{1}{2}\right)-1} t \sin^{2n-1} t \, dt$$

$$= (2^{1-2n})B\left(\frac{1}{2}, n\right) = (2^{1-2n})\frac{\Gamma\left(\frac{1}{2}\right)\Gamma(n)}{\Gamma\left(n+\frac{1}{2}\right)}$$

or
$$\frac{\Gamma(n)}{\Gamma(2n)} = (2^{1-2n})\frac{\sqrt{\pi}}{\Gamma\left(n+\frac{1}{2}\right)}$$

or
$$\frac{2^{2n-1}}{\sqrt{\pi}}\Gamma(n)\Gamma\left(n+\frac{1}{2}\right) = \Gamma(2n). \tag{13.33}$$

This formula is known Legendre duplication formula.

13.9 SOME FUNCTIONS RELATED TO GAMMA FUNCTION

(a) The Psi or Digamma Function and Polygamma Functions

The psi function or digamma function is the logarithmic derivative of the gamma function and it is denoted by $\psi(z)$.

Thus, we have

$$\psi(z)=\frac{d}{dz}\ln[\Gamma(z)]=\frac{\Gamma'(z)}{\Gamma(z)}. \tag{13.34}$$

Most of the properties of this function can be obtained directly from the corresponding properties of the gamma function.

For example, we have

$$\Gamma(z+1)=z\Gamma(z).$$

Therefore, we have

$$\ln[\Gamma(z+1)]=\ln z+\ln[\Gamma(z)].$$

Differentiating both sides with respect to z, we write

$$\frac{\Gamma'(z+1)}{\Gamma(z+1)}=\frac{1}{z}+\frac{\Gamma'(z)}{\Gamma(z)}$$

or

$$\psi(z+1)=\frac{1}{z}+\psi(z). \tag{13.35}$$

The derivatives of psi function are known as polygamma functions.

The first order derivative of psi function, i.e., $\psi'(z)$ is known as trigamma function, the second order derivative of psi function, i.e., $\psi''(z)$ is known as tetragamma function and so on.

(b) Incomplete Gamma Function and Complementary Incomplete Gamma Function

The *incomplete gamma function*, $\gamma(\alpha,z)$, is defined by

$$\gamma(\alpha,z)=\int_0^z t^{\alpha-1}e^{-t}dt, \text{Re}\,\alpha>0. \tag{13.36}$$

Using power series expansion of e^{-t}, a term-by-term integration will lead to

$$\gamma(\alpha,z)=z^{\alpha}\sum_{n=0}^{\infty}\frac{(-z)^n}{n!(n+\alpha)}, \text{Re}\,\alpha>0. \tag{13.37}$$

The *complementary incomplete gamma function* is defined by

$$\Gamma(\alpha,z)=\int_z^{\infty} t^{\alpha-1}e^{-t}dt. \tag{13.38}$$

Since $\int_0^z t^{\alpha-1}e^{-t}dt + \int_z^\infty t^{\alpha-1}e^{-t}dt = \int_0^\infty t^{\alpha-1}e^{-t}dt,$

we write $\gamma(\alpha, z) + \Gamma(\alpha, z) = \Gamma(\alpha).$ (13.39)

(c) The Error Function and Complementary Error Function

The *error function* and *complementary error function* are defined respectively by

$$\operatorname{erf}(z) = \frac{2}{\sqrt{\pi}}\int_0^z e^{-t^2}\,dt \tag{13.40}$$

and

$$\operatorname{erf}c(z) = \frac{2}{\sqrt{\pi}}\int_z^\infty e^{-t^2}\,dt. \tag{13.41}$$

We see that $\operatorname{erf}(z) + \operatorname{erf}c(z) = 1.$ (13.42)

13.10 EVALUATION OF INTEGRALS OF THE TYPE $\int_0^{\pi/2} \sin^p\theta\cos^q\theta\,d\theta$

To evaluate such an integral, let us put

$$p = 2m-1,\; q = 2n-1. \tag{13.43}$$

Then, we write

$$\int_0^{\frac{\pi}{2}} \sin^p\theta\cos^q\theta\,d\theta = \frac{1}{2}\cdot 2\int_0^{\frac{\pi}{2}} \sin^{2m-1}\theta\cos^{2n-1}\theta\,d\theta$$

$$= \frac{1}{2}B(m, n) = \frac{1}{2}\frac{\Gamma(m)\Gamma(n)}{\Gamma(m+n)} \tag{13.44}$$

where m and n are given by Eq. (13.43) as $m = \frac{p+1}{2}, n = \frac{q+1}{2}$

Using these values of m and n, Eq. (13.44) is written as

$$\int_0^{\frac{\pi}{2}} \sin^p\theta\cos^q\theta\,d\theta = \frac{1}{2}B\left(\frac{p+1}{2}, \frac{q+1}{2}\right) = \frac{\Gamma\left(\frac{p+1}{2}\right)\Gamma\left(\frac{q+1}{2}\right)}{2\,\Gamma\left(\frac{p+q}{2}+1\right)}. \tag{13.45}$$

13.11 STIRLING APPROXIMATION

We know that

$$\Gamma(n+1)=\int_0^\infty e^{-x}x^n dx.$$

Introducing a transformation $x=n(1+u)$, we write

$$\Gamma(n+1)=\int_{-1}^{\infty} e^{-n(1+u)}\{n(1+u)\}^n n du$$

$$=n^{n+1}e^{-n}\int_{-1}^{\infty} e^{-nu}(1+u)^n du$$

$$=n^{n+1}e^{-n}\int_{-1}^{\infty} e^{-nu}e^{\ln(1+u)^n} du$$

$$=n^{n+1}e^{-n}\int_{-1}^{\infty} e^{-nu+n\ln(1+u)} du$$

$$=n^{n+1}e^{-n}\int_{-1}^{\infty} e^{n\{-u+\ln(1+u)\}} du. \tag{13.46}$$

Here, the function $f(u)=-u+\ln(1+u)$ is zero at $u=0$ and for other values of u, the function $f(u)$ is less than zero.

Thus, the integrand of the above integral becomes unity with $u=0$ for any value of n whereas this integrand is very small for other values of u when n is very large.

We know that

$$f(u)=-u+\ln(1+u)=-u+u-\frac{u^2}{2}+\frac{u^3}{3}-\ldots\infty$$

For small u, we have the approximation

$$f(u)=-u+\ln(1+u)\cong-\frac{u^2}{2}$$

Therefore, for small u and very large n, we have the approximation

$$\int_{-1}^{\infty} e^{n\{-u+\ln(1+u)\}} du \cong \int_{-\infty}^{\infty} e^{-\frac{nu^2}{2}} du.$$

Now, Eq. (13.46) yields

$$\Gamma(n+1)\cong n^{n+1}e^{-n}\int_{-\infty}^{\infty} e^{-\frac{nu^2}{2}} du.$$

Putting $t = u\sqrt{\frac{n}{2}}$ or, $u = t\sqrt{\frac{2}{n}}$, we have $du = \sqrt{\frac{2}{n}}\,dt$.

Now, we write

$$\Gamma(n+1) \cong n^{n+1}e^{-n}\sqrt{\frac{2}{n}}\int_{-\infty}^{+\infty} e^{-t^2}\,dt$$

$$= n^{n+\frac{1}{2}}e^{-n}\sqrt{2\pi} \qquad \left[\text{since } \int_{-\infty}^{+\infty} e^{-t^2}\,dt = \Gamma\left(\frac{1}{2}\right) = \sqrt{\pi}\right].$$

If n is a very large positive integer, we write

$$n! \cong n^{n+\frac{1}{2}}e^{-n}\sqrt{2\pi}.$$

Taking natural logarithm on both sides, we write

$$\ln(n!) \cong \left(n+\frac{1}{2}\right)\ln n - n + \frac{1}{2}\ln(2\pi)$$

$$= n\ln n - n + \frac{1}{2}\left[\ln n + \ln(2\pi)\right]$$

$$\cong n\ln n - n \text{ when } n \text{ is very large.} \qquad (13.47)$$

Equation (13.47) is a famous formula known as ***Stirling approximation*** or ***Stirling formula.*** This formula is named after the famous Scottish mathematician, ***James Stirling*** (May, 1692 - 5 December, 1770). It is frequently used in the study of Statistical Physics.

13.12 ELLIPTIC INTEGRALS

For an ellipse, the parametric equations for an arc are given by

$$x = a\cos\theta, \qquad y = b\sin\theta \qquad 0 \le \theta \le \phi \qquad (13.48)$$

where a is the semi-major axis, b is the semi-minor axis of the ellipse, θ is the angular position of an arbitrary point within the arc and ϕ is that of the endpoint of the arc.

Using the formula for arc length from differential calculus, we find the length of the elliptic arc as

$$l = \int_0^{\phi}\sqrt{a^2\cos^2\theta + b^2\sin^2\theta}\,d\theta$$

$$= \int_0^{\phi}\sqrt{a^2 - \left(a^2 - b^2\right)\sin^2\theta}\,d\theta. \qquad (13.49)$$

Since the eccentricity of the ellipse is defined by

$$\varepsilon = \frac{1}{a}\sqrt{a^2 - b^2}, \tag{13.50}$$

we write Eq. (13.49) as

$$l = a\int_0^{\phi}\sqrt{1-\varepsilon^2 \sin^2\theta}\, d\theta. \tag{13.51}$$

Because of its origin, the integral in Eq. (13.51) is called an ***elliptic integral***. Since such integrals were extensively studied by the French mathematician, Adrien Marie Legendre (Sep. 18, 1752 to Jan. 10, 1833), such integrals are also known as Legendre elliptic integrals. There are two classes of elliptic integrals, namely: (1) incomplete elliptic integrals and (2) complete elliptic integrals. Each class has three kinds of integrals.

The three kinds of incomplete elliptic integrals are (i) incomplete elliptic integral of the first kind, (ii) incomplete elliptic integral of the second kind, and (iii) incomplete elliptic integral of the third kind.

The incomplete elliptic integral of the first kind is defined by the integral

$$F(\phi, m) = \int_0^{\phi}\frac{d\theta}{\sqrt{1-m^2\sin^2\theta}} \quad (0 < m < 1). \tag{13.52}$$

The incomplete elliptic integral of the second kind is defined by the integral

$$E(\phi, m) = \int_0^{\phi}\sqrt{1-m^2\sin^2\theta}\, d\theta \quad (0 < m < 1). \tag{13.53}$$

The incomplete elliptic integral of the third kind is defined by the integral

$$\Pi(\phi, a^2, m) = \int_0^{\phi}\frac{d\phi}{\left(1-a^2\sin^2\theta\right)\sqrt{\left(1-m^2\sin^2\theta\right)}} \left(0 < m < 1, a \neq 0, m\right). \tag{13.54}$$

Here, ϕ is known as the amplitude and m is known as the modulus.

In a similar manner, the three kinds of complete elliptic integrals are (i) complete elliptic integral of the first kind, (ii) complete elliptic integral of the second kind, and (iii) complete elliptic integral of the third kind.

The complete elliptic integral of the first kind is defined by the integral

$$K(m) = \int_0^{\frac{\pi}{2}}\frac{d\theta}{\sqrt{1-m^2\sin^2\theta}} \quad (0 < m < 1). \tag{13.55}$$

The complete elliptic integral of the second kind is defined by the integral

$$E(m)=\int_0^{\frac{\pi}{2}}\sqrt{1-m^2\sin^2\theta}\,d\theta \quad (0<m<1). \tag{13.56}$$

The complete elliptic integral of the third kind is defined by the integral

$$\Pi\left(a^2,m\right)=\int_0^{\frac{\pi}{2}}\frac{d\phi}{\left(1-a^2\sin^2\theta\right)\sqrt{\left(1-m^2\sin^2\theta\right)}}\left(0<m<1,a\neq 0,m\right). \tag{13.57}$$

Elliptic functions: Based on these elliptic integrals, some functions which are the inverse functions of the above integrals are defined. These newly defined functions are known as elliptic functions. Two important kinds of elliptic functions are (i) Jacobi elliptic function and (ii) Weierstrass elliptic function. Details of such elliptic functions will not be discussed in this text. Those who have interest on these functions are suggested to consult with advanced texts.

13.13 EVALUATION OF THREE SPECIAL TYPES OF INTEGRAL

(i) $\int_0^{\infty}e^{-ax}x^{n-1}\cos bx\,dx,$

(ii) $\int_0^{\infty}e^{-ax}x^{n-1}\sin bx\,dx,$

(iii) $\int_{-\infty}^{+\infty}e^{-ax^2+bx}dx.$

Solution: (i) and (ii)

Let us consider an integral

$$I=\int_0^{\infty}e^{-(a+ib)x}x^{n-1}dx \tag{13.58}$$

where a and b are constants and $i=\sqrt{-1}$.

Putting $(a+ib)x=u$, or, $x=\frac{u}{a+ib}$, we have $dx=\frac{du}{a+ib}$.

Now, Eq. (13.58) is written as

$$I=\int_0^{\infty}e^{-(a+ib)x}x^{n-1}dx=\int_0^{\infty}e^{-u}\left(\frac{u}{a+ib}\right)^{n-1}\frac{du}{a+ib}$$

or
$$\int_0^\infty e^{-ax}(\cos bx - i\sin bx)x^{n-1}dx = (a+ib)^{-n}\int_0^\infty e^{-u}u^{n-1}du$$

$$= (a+ib)^{-n}\,\Gamma(n). \tag{13.59}$$

In the expression on right-hand side, let us put

$$a = r\cos\theta \quad \text{and} \quad b = r\sin\theta$$

yielding
$$r = \sqrt{a^2+b^2} \quad \text{and} \quad \theta = \tan^{-1}\left(\frac{b}{a}\right). \tag{13.60}$$

With these new substitutions, Eq. (13.59) can be written as

$$\int_0^\infty e^{-ax}(\cos bx - i\sin bx)x^{n-1}dx = \{r(\cos\theta + i\sin\theta)\}^{-n}\,\Gamma(n)$$

$$= r^{-n}(\cos n\theta - i\sin n\theta)\Gamma(n)$$

or
$$\int_0^\infty e^{-ax}x^{n-1}\cos bx dx - i\int_0^\infty e^{-ax}x^{n-1}\sin bx\,dx$$

$$= \frac{\Gamma(n)\cos n\theta}{r^n} - i\frac{\Gamma(n)\sin n\theta}{r^n}. \tag{13.61}$$

Equating real parts from both sides of Eq. (13.61), we obtain

$$\int_0^\infty e^{-ax}x^{n-1}\cos bx\,dx = \frac{\Gamma(n)\cos n\theta}{r^n} = \frac{\Gamma(n)\cos n\theta}{(a^2+b^2)^{\frac{n}{2}}}. \tag{13.62}$$

Further, equating imaginary parts from both sides of Eq. (13.61), we obtain

$$\int_0^\infty e^{-ax}x^{n-1}\sin bx\,dx = \frac{\Gamma(n)\sin n\theta}{r^n} = \frac{\Gamma(n)\sin n\theta}{(a^2+b^2)^{\frac{n}{2}}}. \tag{13.63}$$

(iii) We have

$$\int_{-\infty}^{+\infty} e^{-ax^2+bx}dx = \int_{-\infty}^{+\infty} e^{-a\left\{x^2-\frac{b}{a}x\right\}}dx$$

$$= \int_{-\infty}^{+\infty} e^{-a\left\{x^2-2x\frac{b}{2a}+\left(\frac{b}{2a}\right)^2-\left(\frac{b}{2a}\right)^2\right\}}dx$$

$$= e^{\frac{b^2}{4a}}\int_{-\infty}^{+\infty} e^{-a\left(x-\frac{b}{2a}\right)^2}dx.$$

Putting $\left(x-\frac{b}{2a}\right)\sqrt{a}=t$, we have $dx=\frac{dt}{\sqrt{a}}$.

Then, the above integral can be written as

$$\int_{-\infty}^{+\infty} e^{-ax^2+bx}dx=\frac{1}{\sqrt{a}}e^{\frac{b^2}{4a}}\int_{-\infty}^{+\infty} e^{-t^2}dt$$

$$=\frac{1}{\sqrt{a}}e^{\frac{b^2}{4a}}\Gamma\left(\frac{1}{2}\right)=\sqrt{\frac{\pi}{a}}e^{\frac{b^2}{4a}}. \tag{13.64}$$

13.13.1 Deductions

In Eqs. (13.62) and (13.63), if we put $a=0,\ \theta=\tan^{-1}\left(\frac{b}{0}\right)=\frac{\pi}{2}$, we have

$$\int_0^{\infty} x^{n-1}\cos bx\,dx=\frac{\Gamma(n)}{b^n}\cos\frac{n\pi}{2}, \tag{13.65}$$

$$\int_0^{\infty} x^{n-1}\sin bx\,dx=\frac{\Gamma(n)}{b^n}\sin\frac{n\pi}{2}. \tag{13.66}$$

13.14 SOLVED PROBLEMS

PROBLEM 1 Obtain the values of $\Gamma\left(-\frac{1}{2}\right),\Gamma\left(-\frac{3}{2}\right)$ and $\Gamma\left(-\frac{5}{2}\right)$.

Solution: Using the formula $\Gamma(n+1)=n\Gamma(n)$, we write

$$\Gamma(n)=\frac{1}{n}\Gamma(n+1).$$

(i) Putting $n=-\frac{1}{2}$, we write

$$\Gamma\left(-\frac{1}{2}\right)=-2\Gamma\left(\frac{1}{2}\right)=-2\sqrt{\pi}.$$

(ii) Putting $n=-\frac{3}{2}$, we write

$$\Gamma\left(-\frac{3}{2}\right)=-\frac{2}{3}\Gamma\left(-\frac{3}{2}+1\right)=-\frac{2}{3}\Gamma\left(-\frac{1}{2}\right)=-\frac{2}{3}\left(-2\sqrt{\pi}\right)=\frac{4\sqrt{\pi}}{3}.$$

(iii) Putting $n=-\frac{5}{2}$, we write

$$\Gamma\left(-\frac{5}{2}\right)=-\frac{2}{5}\Gamma\left(-\frac{5}{2}+1\right)=-\frac{2}{5}\Gamma\left(-\frac{3}{2}\right)=-\frac{2}{5}\cdot\frac{4\sqrt{\pi}}{3}=-\frac{8\sqrt{\pi}}{15}.$$

PROBLEM 2 Evaluate $\int_0^\infty \sqrt{x}\,e^{-x}dx$ and $\int_0^1 \left[\ln\left(\frac{1}{x}\right)\right]^4 dx$.

Solution:

(i) We have

$$\int_0^\infty \sqrt{x}\,e^{-x}\,dx=\int_0^\infty x^{\frac{3}{2}-1}\,e^{-x}dx=\Gamma\left(\frac{3}{2}\right)=\frac{\sqrt{\pi}}{2}.$$

(ii) We have

$$\int_0^1 \left[\ln\left(\frac{1}{x}\right)\right]^4 dx=\int_0^1 \left[\ln\left(\frac{1}{x}\right)\right]^{5-1} dx=\Gamma(5)=4!.$$

PROBLEM 3 Evaluate

(i) $\int_0^1 \left(1-x^n\right)^{\frac{1}{n}} dx,$

(ii) $\int_0^1 x^{m-1}\left[\ln\left(\frac{1}{x}\right)\right]^{n-1} dx,$

(iii) $\int_0^1 \frac{dx}{\sqrt{x\ln\left(\frac{1}{x}\right)}}.$

Solution:

(i) Putting $t=x^n$ or, $x=t^{\frac{1}{n}}$, we have $dx=\frac{1}{n}t^{\frac{1}{n}-1}dt$.

Then, we write

$$\int_0^1 \left(1-x^n\right)^{\frac{1}{n}} dx=\int_0^1 (1-t)^{\frac{1}{n}}\cdot\frac{1}{n}t^{\frac{1}{n}-1}dt=\frac{1}{n}\int_0^1 t^{\frac{1}{n}-1}(1-t)^{\frac{1}{n}+1-1}dt$$

$$=\frac{1}{n}B\left(\frac{1}{n},\frac{1}{n}+1\right)=\frac{1}{n}\frac{\Gamma\left(\frac{1}{n}\right)\Gamma\left(\frac{1}{n}+1\right)}{\Gamma\left(\frac{2}{n}+1\right)}=\frac{1}{n}\cdot\frac{\frac{1}{n}\Gamma\left(\frac{1}{n}\right)\Gamma\left(\frac{1}{n}\right)}{\frac{2}{n}\Gamma\left(\frac{2}{n}\right)}=\frac{1}{2n}\frac{\Gamma\left(\frac{1}{n}\right)\Gamma\left(\frac{1}{n}\right)}{\Gamma\left(\frac{2}{n}\right)}.$$

(ii) Putting $\ln\left(\frac{1}{x}\right)=t$ or, $\frac{1}{x}=e^t$, we have

$$x=e^{-t} \quad \text{and} \quad dx=-e^{-t}dt.$$

Thus, we write

$$\int_0^1 x^{m-1}\left[\ln\left(\frac{1}{x}\right)\right]^{n-1}dx=\int_\infty^0 \left(e^{-t}\right)^{m-1}t^{n-1}\left(-e^{-t}dt\right)$$

$$=\int_0^\infty e^{-mt}t^{n-1}dt.$$

Again, putting $mt=u$ or, $t=\frac{u}{m}$, we have $dt=\frac{du}{m}$.
Then, we write

$$\int_0^1 x^{m-1}\left[\ln\left(\frac{1}{x}\right)\right]^{n-1}dx=\int_0^\infty e^{-u}\left(\frac{u}{m}\right)^{n-1}\frac{du}{m}$$

$$=\frac{1}{m^n}\int_0^\infty e^{-u}u^{n-1}du=\frac{\Gamma(n)}{m^n},\ (m,n>0).$$

(iii) We have $\displaystyle\int_0^1 \frac{dx}{\sqrt{x\ln\left(\frac{1}{x}\right)}}=\int_0^1 x^{-\frac{1}{2}}\left[\ln\left(\frac{1}{x}\right)\right]^{-\frac{1}{2}}dx$

$$=\int_0^1 x^{\frac{1}{2}-1}\left[\ln\left(\frac{1}{x}\right)\right]^{\frac{1}{2}-1}dx=\frac{\Gamma\left(\frac{1}{2}\right)}{\left(\frac{1}{2}\right)^{1/2}}=\sqrt{2\pi}.$$

Here, we use the result of (ii)

PROBLEM 4 Evaluate

(i) $\displaystyle\int_0^\infty x^4e^{-x^4}dx,$

(ii) $\displaystyle\int_0^\infty \frac{x^c}{c^x}dx\ (c>1).$

Solution:

(i) Putting $u = x^4$ or, $x = u^{\frac{1}{4}}$, we have $dx = \frac{1}{4}u^{-\frac{3}{4}}du$.

Then, we write $\int_0^\infty x^4 e^{-x^4} dx = \int_0^\infty u e^{-u} \frac{1}{4} u^{-\frac{3}{4}} du$

$$= \frac{1}{4}\int_0^\infty u^{\frac{1}{4}} e^{-u} du = \frac{1}{4}\int_0^\infty u^{\frac{5}{4}-1} e^{-u} du = \frac{1}{4}\Gamma\left(\frac{5}{4}\right).$$

(ii) Putting $c^x = e^t$, we have $x \ln c = t$.

Therefore, we have $x = \frac{t}{\ln c}$ and $dx = \frac{dt}{\ln c}$.

Hence, the given integral can be written as

$$\int_0^\infty \frac{x^c}{c^x} dx = \int_0^\infty \frac{t^c}{(\ln c)^c} \frac{1}{e^t} \frac{dt}{\ln c}$$

$$= \frac{1}{(\ln c)^{c+1}} \int_0^\infty t^c e^{-t} dt = \frac{1}{(\ln c)^{c+1}} \Gamma(c+1).$$

PROBLEM 5 Evaluate

(i) $\int_0^1 \frac{x^2 dx}{\sqrt{1-x^3}}$,

(ii) $\int_0^3 \frac{dx}{\sqrt{3x-x^2}}$.

Solution:

(i) Putting $x^3 = u$ or, $x = u^{\frac{1}{3}}$, we have $dx = \frac{1}{3}u^{-\frac{2}{3}}du$.

Now, the given integral can be written as

$$\int_0^1 \frac{x^2 dx}{\sqrt{1-x^3}} = \int_0^1 \frac{u^{\frac{2}{3}}}{\sqrt{1-u}} \frac{1}{3} u^{-\frac{2}{3}} du$$

$$= \frac{1}{3}\int_0^1 u^0 (1-u)^{-\frac{1}{2}} du = \frac{1}{3} B\left(1, \frac{1}{2}\right) = \frac{1}{3} \cdot \frac{\Gamma(1)\Gamma\left(\frac{1}{2}\right)}{\Gamma\left(\frac{3}{2}\right)}$$

$$= \frac{1}{3} \cdot \frac{\Gamma\left(\frac{1}{2}\right)}{\frac{1}{2}\Gamma\left(\frac{1}{2}\right)} = \frac{2}{3}.$$

(ii) Putting $t = \frac{x}{3}$, or, $x = 3t$, we have $dx = 3dt$.

Now, the given integral can be written as

$$\int_0^3 \frac{dx}{\sqrt{3x - x^2}} = \int_0^1 \frac{3dt}{\sqrt{9t - 9t^2}}$$

$$= \int_0^1 \frac{dt}{\sqrt{t(1-t)}} = \int_0^1 t^{-\frac{1}{2}} (1-t)^{-\frac{1}{2}}\, dt$$

$$= \int_0^1 t^{\frac{1}{2}-1} (1-t)^{\frac{1}{2}-1}\, dt = B\left(\frac{1}{2}, \frac{1}{2}\right) = \frac{\Gamma\left(\frac{1}{2}\right)\Gamma\left(\frac{1}{2}\right)}{\Gamma(1)} = \pi.$$

PROBLEM 6 If n is a positive integer, prove that

$$1.3.5.7 \ldots (2n-3)(2n-1) = \frac{2^n \Gamma\left(n + \frac{1}{2}\right)}{\sqrt{\pi}}.$$

Solution: We have

$$1.3.5.7 \ldots (2n-1) = 2^n \left[\frac{1}{2} \cdot \frac{3}{2} \cdot \frac{5}{2} \cdot \frac{7}{2} \cdots \frac{(2n-3)}{2} \cdot \frac{(2n-1)}{2}\right]$$

$$= \frac{2^n \Gamma\left(\frac{1}{2}\right)\left[\frac{1}{2} \cdot \frac{3}{2} \cdot \frac{5}{2} \cdot \frac{7}{2} \cdots \frac{(2n-3)}{2} \cdot \frac{(2n-1)}{2}\right]}{\Gamma\left(\frac{1}{2}\right)}$$

$$= \frac{2^n \Gamma\left(n + \frac{1}{2}\right)}{\sqrt{\pi}}$$ Hence proved.

PROBLEM 7 Express the product $2.5.8 \ldots (3n-1)$ in terms of gamma functions when n is a positive integer.

Solution: We have

$$2.5.8\ldots(3n-1)=3^n\left[\frac{2}{3}\cdot\frac{5}{3}\cdot\frac{8}{3}\cdots\left(n-\frac{1}{3}\right)\right]$$

$$=\frac{3^n\Gamma\left(\frac{2}{3}\right)\left[\frac{2}{3}\cdot\frac{5}{3}\cdot\frac{8}{3}\cdots\left(n-\frac{1}{3}\right)\right]}{\Gamma\left(\frac{2}{3}\right)}=\frac{3^n\Gamma\left(n+\frac{2}{3}\right)}{\Gamma\left(\frac{2}{3}\right)}.$$

PROBLEM 8 Evaluate

(i) $\int_0^\infty e^{-x^4}\,dx$,

(ii) $\int_0^{\frac{\pi}{2}} \frac{\sin^4 x}{\sqrt{\cos x}}\,dx$.

Solution:

(i) Putting $x^2=u$ or, $x=u^{\frac{1}{2}}$, we have $dx=\frac{1}{2}u^{-\frac{1}{2}}du$.

Now, the given integral can be written as

$$\int_0^\infty e^{-x^4}\,dx=\int_0^\infty e^{-u^2}\frac{1}{2}u^{-\frac{1}{2}}du$$

$$=\frac{1}{4}\cdot 2\int_0^\infty u^{2\left(\frac{1}{4}\right)-1}e^{-u^2}\,du$$

$$=\frac{1}{4}\Gamma\left(\frac{1}{4}\right)=\Gamma\left(\frac{5}{4}\right).$$

(ii) The given integral is written as

$$\int_0^{\frac{\pi}{2}} \frac{\sin^4 x}{\sqrt{\cos x}}\,dx=\int_0^{\frac{\pi}{2}} \sin^4 x(\cos x)^{-\frac{1}{2}}\,dx$$

$$=\frac{1}{2}\cdot\frac{\Gamma\left(\frac{4+1}{2}\right)\Gamma\left(\frac{-\frac{1}{2}+1}{2}\right)}{\Gamma\left(\frac{4+1}{2}+\frac{1-\frac{1}{2}}{2}\right)}=\frac{1}{2}\cdot\frac{\Gamma\left(\frac{5}{2}\right)\Gamma\left(\frac{1}{4}\right)}{\Gamma\left(\frac{11}{4}\right)}.$$

PROBLEM 9 Evaluate

(i) $\int_0^1 x^m (\ln x)^n \, dx$,

(ii) $\int_0^1 (x \ln x)^3 \, dx$,

(iii) $\int_0^1 \frac{dx}{\sqrt{-\ln x}}$.

Solution:

(i) Let us put $\ln x = -t$.

Then, $x = e^{-t}$ and $dx = -e^{-t} dt$.

Now, the given integral is written as

$$\int_0^1 x^m (\ln x)^n \, dx = \int_\infty^0 e^{-mt} (-t)^n \left(-e^{-t} dt\right)$$

$$= (-1)^n \int_0^\infty t^n e^{-(m+1)t} dt.$$

Further, let us put $(m+1)t = u$ or, $t = \frac{u}{m+1}$.

Then, we have $dt = \frac{du}{m+1}$.

Thus, the given integral is written as

$$\int_0^1 x^m (\ln x)^n \, dx = (-1)^n \int_0^\infty \left(\frac{u}{m+1}\right)^n e^{-u} \frac{du}{m+1}$$

$$= \frac{(-1)^n}{(m+1)^{n+1}} \int_0^\infty u^n e^{-u} du = \frac{(-1)^n}{(m+1)^{n+1}} \Gamma(n+1).$$

(ii) This is just Problem 9 (i) with $m = n = 3$

Therefore, we have

$$\int_0^1 (x \ln x)^3 \, dx = \frac{(-1)^3}{4^4} \Gamma(4) = -\frac{3}{128}.$$

(iii) Let us put $\ln x = -t$ or, $x = e^{-t}$.

Then, we have $dx = -e^{-t}dt$.

Now, the given integral is written as

$$\int_0^1 \frac{dx}{\sqrt{-\ln x}} = \int_\infty^0 t^{-\frac{1}{2}}\left(-e^{-t}\right)dt = \int_0^\infty t^{\frac{1}{2}-1}e^{-t}dt = \Gamma\left(\frac{1}{2}\right) = \sqrt{\pi}.$$

PROBLEM 10 Evaluate

(i) $\int_0^\infty e^{-x^{1/n}}dx$,

(ii) $\int_0^2 \frac{x^2dx}{\sqrt{2-x}}$.

Solution:

(i) Let us put $x^{\frac{1}{n}} = u$ or, $x = u^n$.

Then, we have $dx = nu^{n-1}du$.

Now, the given integral is written as

$$\int_0^\infty e^{-x^{\frac{1}{n}}}dx = \int_0^\infty e^{-u}nu^{n-1}du$$

$$= n\int_0^\infty e^{-u}u^{n-1}du = n\Gamma(n) = \Gamma(n+1).$$

(ii) Let us put $x = 2u$ or, $u = \frac{x}{2}$. Then, we have $dx = 2du$.

Now, the given integral is written as

$$\int_0^2 \frac{x^2dx}{\sqrt{2-x}} = \int_0^1 \frac{(2u)^2}{\sqrt{2-2u}}2du = \frac{8}{\sqrt{2}}\int_0^1 u^2(1-u)^{-\frac{1}{2}}du$$

$$= \frac{8}{\sqrt{2}}\int_0^1 u^{3-1}(1-u)^{\frac{1}{2}-1}du = 4\sqrt{2}\,B\left(3,\frac{1}{2}\right)$$

$$= 4\sqrt{2}\frac{\Gamma(3)\Gamma\left(\frac{1}{2}\right)}{\Gamma\left(\frac{7}{2}\right)} = 8\sqrt{2}\frac{\Gamma\left(\frac{1}{2}\right)}{\frac{5}{2}\cdot\frac{3}{2}\cdot\frac{1}{2}\cdot\Gamma\left(\frac{1}{2}\right)} = \frac{64\sqrt{2}}{15}.$$

PROBLEM 11 Evaluate

(i) $\int_0^\infty x^{2s+1}e^{-ax^2}\,dx,$

(ii) $\int_0^\infty x^{2s}e^{-ax^2}\,dx.$

Solution:

Putting $ax^2 = u$ or, $x = \frac{1}{\sqrt{a}}u^{1/2},$

we have $dx = \frac{1}{2\sqrt{a}}u^{-1/2}du.$

(i) Now, the given integral is written as

$$\int_0^\infty x^{2s+1}e^{-ax^2}dx = \int_0^\infty \left(\frac{1}{\sqrt{a}}u^{\frac{1}{2}}\right)^{2s+1} e^{-u}\left(\frac{1}{2\sqrt{a}}\right)u^{-1/2}du$$

$$= \frac{1}{2a^{s+1}}\int_0^\infty u^s e^{-u}du = \frac{\Gamma(s+1)}{2a^{s+1}}.$$

(ii) The given integral is written as

$$\int_0^\infty x^{2s}e^{-ax^2}dx = \int_0^\infty \left(\frac{1}{\sqrt{a}}u^{\frac{1}{2}}\right)^{2s} e^{-u}\left(\frac{1}{2\sqrt{a}}\right)u^{-1/2}du$$

$$= \frac{1}{2a^{s+\frac{1}{2}}}\int_0^\infty u^{s-\frac{1}{2}}e^{-u}du$$

$$= \frac{1}{2a^{s+\frac{1}{2}}}\int_0^\infty u^{s+\frac{1}{2}-1}e^{-u}du$$

$$= \frac{1}{2a^{s+\frac{1}{2}}}\Gamma\left(s+\frac{1}{2}\right) = \frac{1}{2a^{s+\frac{1}{2}}}\left(s-\frac{1}{2}\right)\left(s-\frac{3}{2}\right)\left(s-\frac{5}{2}\right)\dots\frac{5}{2}\cdot\frac{3}{2}\cdot\frac{1}{2}\cdot\Gamma\left(\frac{1}{2}\right)$$

$$= \frac{1}{2a^s\sqrt{a}}\cdot\frac{(2s-1)(2s-3)(2s-5)\dots 5.3.1\cdot\sqrt{\pi}}{2^s} = \frac{(2s-1)!!}{2^{s+1}a^s}\sqrt{\frac{\pi}{a}}.$$

Here, $(2s-1)!! = 1.3.5\dots(2s-5)(2s-3)(2s-1)$ is called a double factorial.

PROBLEM 12 Prove that

(i) $\Gamma(n)=\int_0^1\left[\ln\left(\frac{1}{t}\right)\right]^{n-1}dt,$

(ii) $\int_1^\infty\frac{(x-1)^p}{x^2}dx=\frac{p\pi}{\sin p\pi}.$

Solution:

(i) Putting $\ln\left(\frac{1}{t}\right)=u$ or, $t=e^{-u}$ we have $dt=-e^{-u}du$.

Now, the given integral is written as

$$\int_0^1\left[\ln\left(\frac{1}{t}\right)\right]^{n-1}dt=\int_\infty^0 u^{n-1}\left(-e^{-u}du\right)=\int_0^\infty u^{n-1}e^{-u}du=\Gamma(n).$$

(ii) Putting $x=\frac{1}{u}$ or, $u=\frac{1}{x}$, we have $dx=-u^{-2}du$.

Now, the given integral is written as

$$\int_1^\infty\frac{(x-1)^p}{x^2}dx=\int_1^0\frac{\left(\frac{1}{u}-1\right)^p}{u^{-2}}\left(-u^{-2}du\right)$$

$$=\int_0^1 u^{-p}(1-u)^p\,du=\int_0^1 u^{(1-p)-1}(1-p)^{(1+p)-1}\,du$$

$$=B(1-p,1+p)=\frac{\Gamma(1-p)\Gamma(1+p)}{\Gamma(2)}$$

$$=p\Gamma(p)\Gamma(1-p)=\frac{p\pi}{\sin p\pi}.$$

Here, reflection formula of gamma functions has been used.

PROBLEM 13 Evaluate

(i) $\int_0^\infty x^m e^{-ax^n}dx,$

(ii) $\int_0^1 x^2\left(1-x^2\right)^{-\frac{1}{2}}dx.$

Solution:

(i) Let us put $u = ax^n$ or, $x = a^{-1/n}u^{1/n}$.

Then, we have $dx = \frac{1}{n}a^{-1/n}u^{\frac{1}{n}-1}du$.

Now, the given integral is written as

$$\int_0^\infty x^m e^{-ax^n}dx = \int_0^\infty a^{-\frac{m}{n}}u^{\frac{m}{n}}e^{-u}\left(\frac{1}{n}a^{-\frac{1}{n}}u^{\frac{1}{n}-1}du\right)$$

$$= \frac{1}{n}a^{-\left(\frac{m+1}{n}\right)}\int_0^\infty u^{\frac{m+1}{n}-1}e^{-u}du$$

$$= \frac{1}{n}a^{-\left(\frac{m+1}{n}\right)}\Gamma\left(\frac{m+1}{n}\right).$$

(ii) Let us put $x^2 = u$. Then, we have $x = u^{\frac{1}{2}}$ and $dx = \frac{1}{2}u^{-\frac{1}{2}}du$.

Now, the given integral is written as

$$\int_0^1 x^2\left(1-x^2\right)^{-\frac{1}{2}}dx = \int_0^1 u(1-u)^{-\frac{1}{2}}\cdot\frac{1}{2}u^{-\frac{1}{2}}du$$

$$= \frac{1}{2}\int_0^1 u^{\frac{1}{2}}(1-u)^{-\frac{1}{2}}du = \frac{1}{2}\int_0^1 u^{\frac{3}{2}-1}(1-u)^{\frac{1}{2}-1}du$$

$$= \frac{1}{2}B\left(\frac{3}{2},\frac{1}{2}\right) = \frac{1}{2}\frac{\Gamma\left(\frac{3}{2}\right)\Gamma\left(\frac{1}{2}\right)}{\Gamma(2)} = \frac{\pi}{4}.$$

PROBLEM 14 Prove that

(i) $\int_0^3 \left(9-x^2\right)^{\frac{3}{2}}dx = \frac{243}{16}\pi$,

(ii) $\int_0^{\pi/2}\sqrt{\tan\theta}\,d\theta = \int_0^{\pi/2}\sqrt{\cot\theta}\,d\theta = \frac{\pi}{\sqrt{2}}$.

Solution:

(i) Let us put $x^2 = 9u$ or, $x = 3u^{\frac{1}{2}}$. Then, $dx = \frac{3}{2}u^{-\frac{1}{2}}du$.

Now, the given integral is written as

$$\int_0^3 \left(9-x^2\right)^{\frac{3}{2}} dx = \int_0^1 \{9(1-u)\}^{\frac{3}{2}} \cdot \frac{3}{2} u^{-\frac{1}{2}} du$$

$$= \frac{81}{2}\int_0^1 u^{\frac{1}{2}-1}(1-u)^{\frac{5}{2}-1} du = \frac{81}{2} B\left(\frac{1}{2}, \frac{5}{2}\right)$$

$$= \frac{81}{2}\frac{\Gamma\left(\frac{1}{2}\right)\Gamma\left(\frac{5}{2}\right)}{\Gamma(3)} = \frac{81}{2}\frac{\sqrt{\pi}\,\frac{3}{2}\cdot\frac{1}{2}\sqrt{\pi}}{2} = \frac{243\pi}{16}.$$

(ii) We have

$$\int_0^{\frac{\pi}{2}} \sqrt{\tan\theta}\, d\theta = \int_0^{\frac{\pi}{2}} \sin^{\frac{1}{2}}\theta \cos^{-\frac{1}{2}}\theta d\theta$$

$$= \frac{1}{2}\cdot B\left(\frac{\frac{1}{2}+1}{2}, \frac{-\frac{1}{2}+1}{2}\right)$$

$$= \frac{1}{2}B\left(\frac{3}{4}, \frac{1}{4}\right) = \frac{1}{2}B\left(\frac{1}{4}, \frac{3}{4}\right)$$

$$= \frac{1}{2}\cdot 2\int_0^{\frac{\pi}{2}} \sin^{2\left(\frac{1}{4}\right)-1}\theta \cos^{2\left(\frac{3}{4}\right)-1}\theta d\theta$$

$$= \int_0^{\frac{\pi}{2}} \sin^{-\frac{1}{2}}\theta \cos^{\frac{1}{2}}\theta d\theta = \int_0^{\frac{\pi}{2}} \sqrt{\cot\theta}\, d\theta$$

$$= \frac{1}{2}\frac{\Gamma\left(\frac{1}{4}\right)\Gamma\left(\frac{3}{4}\right)}{\Gamma(1)} = \frac{1}{2}\Gamma\left(\frac{1}{4}\right)\Gamma\left(1-\frac{1}{4}\right) = \frac{1}{2}\frac{\pi}{\sin\frac{\pi}{4}} = \frac{\pi}{\sqrt{2}}.$$

PROBLEM 15 Show that $B(m, n) = a^m b^n \int_0^{\infty} \frac{x^{m-1} dx}{(ax+b)^{m+n}}$ and deduce that

$$\int_0^{\frac{\pi}{2}} \frac{\sin^{2m-1}\theta \cos^{2n-1}\theta}{\left(a\sin^2\theta + b\cos^2\theta\right)^{m+n}} d\theta = \frac{\Gamma(m)\Gamma(n)}{2a^m b^n \Gamma(m+n)}.$$

Solution:

First Part: The first part is derived in Eq. (13.3) of Article 13.1 (iii).

Second Part: Putting $x = \tan^2\theta$, we have $dx = 2\tan\theta\sec^2\theta\, d\theta$

Then, we write the first relation as

$$\frac{1}{a^m b^n} B(m,n) = \int_0^{\frac{\pi}{2}} \frac{\tan^{2m-2}\theta}{\left(a\tan^2\theta + b\right)^{m+n}} 2\tan\theta\sec^2\theta\, d\theta$$

or

$$\frac{\Gamma(m)\Gamma(n)}{2a^m b^n \Gamma(m+n)} = \int_0^{\frac{\pi}{2}} \frac{\sin^{2m-1}\theta\cos^{2m+2n-2}\theta\, d\theta}{\cos^{2m-1}\theta\left(a\sin^2\theta + b\cos^2\theta\right)^{m+n}}$$

$$= \int_0^{\frac{\pi}{2}} \frac{\sin^{2m-1}\theta\cos^{2n-1}\theta\, d\theta}{\left(a\sin^2\theta + b\cos^2\theta\right)^{m+n}}.$$

PROBLEM 16 Prove that $\displaystyle\int_0^1 \frac{x^{m-1}(1-x)^{n-1}}{(x+a)^{m+n}} dx = \frac{\Gamma(m)\Gamma(n)}{a^n(1+a)^m\Gamma(m+n)}$.

Solution: Let us put $\dfrac{x}{x+a} = \dfrac{u}{1+a}$. Then, we have $u = \dfrac{(1+a)x}{x+a}$ or, $x = \dfrac{au}{1+a-u}$.

Also, we have $dx = \dfrac{a(1+a)du}{(1+a-u)^2}$.

Now, we have

$$1 - x = \frac{(1+a)(1-u)}{1+a-u} \quad \text{and} \quad x + a = \frac{a(1+a)}{1+a-u}.$$

The given integral can be written as

$$\int_0^1 \frac{x^{m-1}(1-x)^{n-1}}{(x+a)^{m+n}} dx$$

$$= \int_0^1 \left(\frac{au}{1+a-u}\right)^{m-1} \left\{\frac{(1+a)(1-u)}{1+a-u}\right\}^{n-1} \left\{\frac{1+a-u}{a(1+a)}\right\}^{m+n} \left\{\frac{a(1+a)du}{(1+a-u)^2}\right\}$$

$$= \frac{1}{a^n(1+a)^m} \int_0^1 u^{m-1}(1-u)^{n-1} du$$

$$= \frac{1}{a^n(1+a)^m} B(m,n) = \frac{\Gamma(m)\Gamma(n)}{a^n(1+a)^m \Gamma(m+n)}.$$

PROBLEM 17 Evaluate (i) $\int_0^{\frac{\pi}{2}} \sin^2\theta \cos^4\theta \, d\theta$ and (ii) $\int_0^{2a} x\sqrt{2ax - x^2}\, dx$.

Solution:

(i) Using the formula

$$\int_0^{\frac{\pi}{2}} \sin^p\theta \cos^q\theta d\theta = \frac{1}{2} B\left(\frac{p+1}{2}, \frac{q+1}{2}\right) = \frac{1}{2} \frac{\Gamma\left(\frac{p+1}{2}\right)\Gamma\left(\frac{q+1}{2}\right)}{\Gamma\left(\frac{p+q+2}{2}\right)},$$

we write

$$\int_0^{\frac{\pi}{2}} \sin^2\theta \cos^4\theta \, d\theta = \frac{1}{2} \frac{\Gamma\left(\frac{3}{2}\right)\Gamma\left(\frac{5}{2}\right)}{\Gamma(4)} = \frac{1}{2} \cdot \frac{1}{3!} \cdot \frac{1}{2} \cdot \Gamma\left(\frac{1}{2}\right) \cdot \frac{3}{2} \cdot \frac{1}{2} \cdot \Gamma\left(\frac{1}{2}\right) = \frac{\pi}{32}.$$

(ii) The given integral is

$$\int_0^{2a} x\sqrt{2ax - x^2}\, dx = \int_0^{2a} x^{\frac{3}{2}} \sqrt{2a - x}\, dx.$$

Putting $x = 2at$ or, $t = \frac{x}{2a}$, we have $dx = 2a\,dt$.

Now, the given integral can be written as

$$\int_0^{2a} x\sqrt{2ax - x^2}\, dx = \int_0^1 (2at)^{\frac{3}{2}} \{2a(1-t)\}^{\frac{1}{2}}\, 2adt$$

$$= (2a)^3 \int_0^1 t^{\frac{3}{2}} (1-t)^{\frac{1}{2}}\, dt$$

$$= 8a^3 B\left(\frac{5}{2}, \frac{3}{2}\right) = 8a^3 \frac{\Gamma\left(\frac{5}{2}\right)\Gamma\left(\frac{3}{2}\right)}{\Gamma(4)}$$

$$= \frac{8a^3}{3!} \cdot \frac{3}{2} \cdot \frac{1}{2} \cdot \Gamma\left(\frac{1}{2}\right) \cdot \frac{1}{2} \cdot \Gamma\left(\frac{1}{2}\right) = \frac{\pi a^3}{2}.$$

PROBLEM 18 Evaluate

(i) $\int_0^1 x^m \left(1-x^n\right)^p dx$,

(ii) $\int_0^2 x\left(8-x^3\right)^{\frac{1}{3}} dx$.

Solution:

(i) Putting $x^n = t$ or, $x = t^{\frac{1}{n}}$, we have $dx = \frac{1}{n} t^{\frac{1}{n}-1} dt$.

Then, the given integral can be written as

$$\int_0^1 x^m \left(1-x^n\right)^p dx = \int_0^1 t^{\frac{m}{n}} (1-t)^p \frac{1}{n} t^{\frac{1}{n}-1} dt$$

$$= \frac{1}{n}\int_0^1 t^{\frac{m+1}{n}-1} (1-t)^p \, dt = \frac{1}{n} B\left(\frac{m+1}{n}, p+1\right)$$

$$= \frac{1}{n} \frac{\Gamma\left(\frac{m+1}{n}\right)\Gamma(p+1)}{\Gamma\left(\frac{m+1}{n}+p+1\right)}.$$

(ii) Putting $x^3 = 8t$, or, $x = 2t^{\frac{1}{3}}$ we have $dx = \frac{2}{3} t^{\frac{1}{3}-1} dt$.

Then, the given integral can be written as

$$\int_0^2 x\left(8-x^3\right)^{\frac{1}{3}} dx = \int_0^1 2t^{\frac{1}{3}} (8-8t)^{\frac{1}{3}} \frac{2}{3} t^{\frac{1}{3}-1} dt$$

$$= \frac{8}{3}\int_0^1 t^{\frac{2}{3}-1} (1-t)^{\frac{1}{3}} dt = \frac{8}{3} B\left(\frac{2}{3}, \frac{4}{3}\right)$$

$$= \frac{8}{3} \frac{\Gamma\left(\frac{2}{3}\right)\Gamma\left(\frac{4}{3}\right)}{\Gamma(2)} = \frac{8}{3}\Gamma\left(\frac{2}{3}\right)\frac{1}{3}\Gamma\left(\frac{1}{3}\right)$$

$$= \frac{8}{9}\Gamma\left(\frac{1}{3}\right)\Gamma\left(1-\frac{1}{3}\right) = \frac{8}{9}\frac{\pi}{\sin\frac{\pi}{3}} = \frac{16\pi}{9\sqrt{3}}.$$

Here, we use reflection formula.

PROBLEM 19 If m is a positive integer, prove that

$$B(m,n)=\frac{(m-1)!}{n(n+1)(n+2)(n+3)\dots(m+n-1)}.$$

Solution: We know that if m is a positive integer

$$B(m,n)=\frac{\Gamma(m)\Gamma(n)}{\Gamma(m+n)}=\frac{(m-1)!\,\Gamma(n)}{(m+n-1)(m+n-2)(m+n-3)\dots(n+2)(n+1)(n)\Gamma(n)}$$

$$=\frac{(m-1)!}{n(n+1)(n+2)(n+3)\dots(m+n-1)}.$$

PROBLEM 20 Evaluate

(i) $\int_0^1 \frac{dx}{\sqrt{1-x^n}}$,

(ii) $\int_0^1 \frac{dx}{\left(1-x^n\right)^{\frac{1}{n}}}$,

(iii) $\int_0^\infty \frac{x^8\left(1-x^6\right)}{(1+x)^{24}}dx$.

Solution:

(i) Putting $x^n = u$ or, $x=u^{\frac{1}{n}}$, we have $dx=\frac{1}{n}u^{\frac{1}{n}-1}du$.

Then, the given integral can be written as

$$\int_0^1 \frac{dx}{\sqrt{1-x^n}}=\int_0^1 \frac{1}{\sqrt{1-u}}\frac{1}{n}u^{\frac{1}{n}-1}du$$

$$=\frac{1}{n}\int_0^1 u^{\frac{1}{n}-1}(1-u)^{\frac{1}{2}-1}du=\frac{1}{n}B\left(\frac{1}{n},\frac{1}{2}\right)$$

$$=\frac{1}{n}\frac{\Gamma\left(\frac{1}{n}\right)\Gamma\left(\frac{1}{2}\right)}{\Gamma\left(\frac{1}{n}+\frac{1}{2}\right)}=\frac{\sqrt{\pi}}{n}\frac{\Gamma\left(\frac{1}{n}\right)}{\Gamma\left(\frac{n+2}{2n}\right)}.$$

(ii) Putting $x^n = u$ or, $x = u^{\frac{1}{n}}$, we have $dx = \frac{1}{n}u^{\frac{1}{n}-1}du$.
Then, the given integral can be written as

$$\int_0^1 \frac{dx}{\left(1-x^n\right)^{\frac{1}{n}}} = \int_0^1 (1-u)^{-\frac{1}{n}} \frac{1}{n} u^{\frac{1}{n}-1} du$$

$$= \frac{1}{n}\int_0^1 u^{\frac{1}{n}-1}(1-u)^{1-\frac{1}{n}-1} du$$

$$= \frac{1}{n}B\left(\frac{1}{n}, 1-\frac{1}{n}\right) = \frac{1}{n}\frac{\Gamma\left(\frac{1}{n}\right)\Gamma\left(1-\frac{1}{n}\right)}{\Gamma(1)} = \frac{1}{n}\cdot\frac{\pi}{\sin\frac{\pi}{n}}.$$

$$\left[\text{using } \Gamma(m)\Gamma(1-m) = \frac{\pi}{\sin m\pi}\right]$$

(iii) We have

$$\int_0^\infty \frac{x^8\left(1-x^6\right)}{(1+x)^{24}}dx = \int_0^\infty \frac{x^8}{(1+x)^{24}}dx - \int_0^\infty \frac{x^{14}}{(1+x)^{24}}dx$$

$$= \int_0^\infty \frac{x^{9-1}}{(1+x)^{9+15}}dx - \int_0^\infty \frac{x^{15-1}}{(1+x)^{15+9}}dx$$

$$= B(9,15) - B(15,9) = 0.$$

PROBLEM 21 Prove that

$$\int_0^1 \frac{x^n dx}{\sqrt{1-x^2}} = \begin{cases} \dfrac{1.3.5\ldots(n-1)}{2.4.6\ldots n}\cdot\dfrac{\pi}{2} & \text{for } n = \text{positive even integer,} \\[2ex] \dfrac{2.4.6\ldots(n-1)}{1.3.5\ldots n} & \text{for } n = \text{positive odd integer.} \end{cases}$$

Solution: Putting $x^2 = u$ or, $x = u^{\frac{1}{2}}$ we have $dx = \frac{1}{2}u^{-\frac{1}{2}}du$.

Thus, the given integral can be written as

$$\int_0^1 \frac{x^n dx}{\sqrt{1-x^2}} = \int_0^1 u^{\frac{n}{2}}(1-u)^{-\frac{1}{2}}\left(\frac{1}{2}\right)u^{-\frac{1}{2}}du$$

$$= \frac{1}{2}\int_0^1 u^{\frac{n+1}{2}-1}(1-u)^{\frac{1}{2}-1}\,du$$

$$= \frac{1}{2}B\left(\frac{n+1}{2},\frac{1}{2}\right) = \frac{1}{2}\frac{\Gamma\left(\frac{n+1}{2}\right)\Gamma\left(\frac{1}{2}\right)}{\Gamma\left(\frac{n}{2}+1\right)}$$

$$= \begin{cases} \dfrac{1}{2}\cdot\dfrac{\left[\left(\frac{n-1}{2}\right)\left(\frac{n-3}{2}\right)\left(\frac{n-5}{2}\right)\dots\frac{5}{2}\cdot\frac{3}{2}\cdot\frac{1}{2}\Gamma\left(\frac{1}{2}\right)\right]\Gamma\left(\frac{1}{2}\right)}{\frac{n}{2}\left(\frac{n-2}{2}\right)\left(\frac{n-4}{2}\right)\left(\frac{n-6}{2}\right)\dots 3.2.1} & \text{for } n = \text{positive even integer,} \\[3ex] \dfrac{1}{2}\cdot\dfrac{\left[\left(\frac{n-1}{2}\right)\left(\frac{n-3}{2}\right)\left(\frac{n-5}{2}\right)\dots 3.2.1\right]\Gamma\left(\frac{1}{2}\right)}{\frac{n}{2}\left(\frac{n-2}{2}\right)\left(\frac{n-4}{2}\right)\left(\frac{n-6}{2}\right)\dots\frac{5}{2}\cdot\frac{3}{2}\cdot\frac{1}{2}\Gamma\left(\frac{1}{2}\right)} & \text{for } n = \text{positive odd integer,} \end{cases}$$

$$= \begin{cases} \dfrac{1}{2}\cdot\dfrac{[(n-1)(n-3)(n-5)\dots 5.3.1]\pi}{n(n-2)(n-4)(n-6)\dots 6.4.2} & \text{for } n = \text{positive even integer,} \\[3ex] \dfrac{(n-1)(n-3)(n-5)\dots 6.4.2}{n(n-2)(n-4)(n-6)\dots 5.3.1} & \text{for } n = \text{positive odd integer,} \end{cases}$$

$$= \begin{cases} \dfrac{1.3.5\dots(n-1)}{2.4.6\dots n}\cdot\dfrac{\pi}{2} & \text{for } n = \text{positive even integer,} \\[3ex] \dfrac{2.4.6\dots(n-1)}{1.3.5\dots n} & \text{for } n = \text{positive odd integer.} \end{cases}$$

PROBLEM 22 Evaluate

(i) $\displaystyle\int_0^\infty \frac{x^4 dx}{(1+x)^7}$,

(ii) $\int_0^\infty \frac{y\,dy}{\left(1+y^3\right)^2}$.

Solution:

(i) We have

$$\int_0^\infty \frac{x^4\,dx}{(1+x)^7} = \int_0^\infty \frac{x^{5-1}dx}{(1+x)^{5+2}} = B(5,2) = \frac{\Gamma(5)\Gamma(2)}{\Gamma(7)} = \frac{1}{30}.$$

(ii) Putting $y^3 = u$ or, $y = u^{\frac{1}{3}}$ we have $dy = \frac{1}{3}u^{-\frac{2}{3}}du$.

Then, we write the given integral as

$$\int_0^\infty \frac{y\,dy}{\left(1+y^3\right)^2} = \int_0^\infty \frac{u^{\frac{1}{3}}}{(1+u)^2}\left(\frac{1}{3}\right)u^{-\frac{2}{3}}du$$

$$= \frac{1}{3}\int_0^\infty \frac{u^{-\frac{1}{3}}}{(1+u)^2}du = \frac{1}{3}\int_0^\infty \frac{u^{\frac{2}{3}-1}}{(1+u)^{\frac{2}{3}+\frac{4}{3}}}du$$

$$= \frac{1}{3}B\left(\frac{2}{3},\frac{4}{3}\right) = \frac{1}{3}\frac{\Gamma\left(\frac{2}{3}\right)\Gamma\left(\frac{4}{3}\right)}{\Gamma(2)}$$

$$= \frac{1}{3}\Gamma\left(\frac{2}{3}\right)\frac{1}{3}\Gamma\left(\frac{1}{3}\right) = \frac{1}{9}\Gamma\left(\frac{1}{3}\right)\Gamma\left(1-\frac{1}{3}\right)$$

$$= \frac{1}{9}\frac{\pi}{\sin\frac{\pi}{3}} = \frac{2\pi}{9\sqrt{3}}.$$

PROBLEM 23 Prove that $\int_a^b \frac{dx}{(b-x)^{1-n}(x-a)^n} = \frac{\pi}{\sin n\pi}$ $(0<n<1)$.

Solution: Let us introduce a new variable θ such that

$$b-x = \cos^2\theta \quad \text{and} \quad x-a = \sin^2\theta.$$

Then, from both of them, we have $dx = 2\sin\theta\cos\theta\,d\theta$.

Also, when $x=a$, we have $\theta = 0$ and when $x=b$, we have $\theta = \frac{\pi}{2}$.

Then, the given integral can be written as

$$\int_a^b \frac{dx}{(b-x)^{1-n}(x-a)^n} = \int_a^b (b-x)^{n-1}(x-a)^{-n}\,dx$$

$$= \int_0^{\frac{\pi}{2}} \left(\cos^2\theta\right)^{n-1}\left(\sin^2\theta\right)^{-n} 2\sin\theta\cos\theta\,d\theta$$

$$= 2\int_0^{\frac{\pi}{2}} \sin^{1-2n}\theta\cos^{2n-1}\theta\,d\theta$$

$$= 2\int_0^{\frac{\pi}{2}} \sin^{2(1-n)-1}\theta\cos^{2n-1}\theta\,d\theta$$

$$= B(1-n,n) = \frac{\Gamma(1-n)\Gamma(n)}{\Gamma(1)}$$

$$= \Gamma(n)\Gamma(1-n) = \frac{\pi}{\sin n\pi}.$$

PROBLEM 24 Evaluate

(i) $\int_{-1}^{1} (1+x)^a (1-x)^b\,dx$,

(ii) $\int_{-1}^{1} \sqrt{\frac{1+x}{1-x}}\,dx$.

Solution:

(i) Putting $x = 2u-1$ or, $u = \frac{x+1}{2}$, we have $dx = 2du$

Now, the given integral can be written as

$$\int_{-1}^{1} (1+x)^a (1-x)^b\,dx = \int_0^1 (2u)^a \{2(1-u)\}^b\,2du$$

$$= 2^{a+b+1}\int_0^1 u^a (1-u)^b\,du$$

$$= 2^{a+b+1} B(a+1, b+1)$$

$$= 2^{a+b+1} \frac{\Gamma(a+1)\Gamma(b+1)}{\Gamma(a+b+2)}.$$

Alternatively, putting $x = \cos 2\theta$ we have

$$dx = -2 \sin 2\theta \, d\theta = -4 \sin\theta \cos\theta \, d\theta.$$

Then, the given integral can be written as

$$\int_{-1}^{1} (1+x)^a (1-x)^b \, dx$$

$$= \int_{\frac{\pi}{2}}^{0} (1+\cos 2\theta)^a (1-\cos 2\theta)^b (-4\sin\theta \cos\theta \, d\theta)$$

$$= 4\int_{0}^{\frac{\pi}{2}} \left(2\cos^2\theta\right)^a \left(2\sin^2\theta\right)^b \sin\theta \cos\theta \, d\theta$$

$$= 2^{a+b+1} \times 2\int_{0}^{\frac{\pi}{2}} \cos^{2(a+1)-1}\theta \sin^{2(b+1)-1}\theta \, d\theta$$

$$= 2^{a+b+1} B(a+1, b+1) \text{ etc.}$$

(ii) This problem is a special case of problem (i) with $a = \frac{1}{2}, b = -\frac{1}{2}$.

Therefore, proceed as in problem (ii) to give the result

$$\int_{-1}^{1} \sqrt{\frac{1+x}{1-x}} \, dx = \pi.$$

PROBLEM 25 Prove that $\int_{0}^{\pi/2} \cos^{2n+1}\theta \, d\theta = \frac{2^{2n}(n!)^2}{(2n+1)!}$ when n is a positive integer.

Solution: Using the formula

$$\int_{0}^{\pi/2} \sin^p\theta \cos^q\theta \, d\theta = \frac{1}{2} \frac{\Gamma\left(\frac{p+1}{2}\right)\Gamma\left(\frac{q+1}{2}\right)}{\Gamma\left(\frac{p+q+2}{2}\right)},$$

we write

$$\int_0^{\frac{\pi}{2}} \cos^{2n+1}\theta\, d\theta = \frac{1}{2}\frac{\Gamma\left(\frac{0+1}{2}\right)\Gamma\left(\frac{2n+2}{2}\right)}{\Gamma\left(\frac{2n+3}{2}\right)}$$

$$= \frac{1}{2}\frac{\Gamma\left(\frac{1}{2}\right)\Gamma(n+1)}{\Gamma\left(\frac{2n+3}{2}\right)} = \frac{1}{2}\frac{n!\Gamma\left(\frac{1}{2}\right)}{\left(\frac{2n+1}{2}\right)\left(\frac{2n-1}{2}\right)\left(\frac{2n-3}{2}\right)\ldots\frac{5}{2}\cdot\frac{3}{2}\cdot\frac{1}{2}\Gamma\left(\frac{1}{2}\right)}$$

$$= \frac{1}{2}\cdot\frac{2^{n+1}n!}{(2n+1)(2n-1)(2n-3)\ldots 5.3.1}$$

$$= \frac{2^n n!\left[(2n)(2n-2)(2n-4)\ldots 6.4.2\right]}{(2n+1)(2n)(2n-1)(2n-2)(2n-3)(2n-4)\ldots 6.5.4.3.2.1}$$

$$= \frac{2^{2n}(n!)^2}{(2n+1)!}.$$

PROBLEM 26 Evaluate $\int_a^b \sqrt[n]{(x-a)(b-x)}dx$.

Solution: Let us put $t = \frac{x-a}{b-a}$ or, $x = (b-a)t + a$.

Then, we have $dx = (b-a)dt,\ x-a = (b-a)t,\ b-x = (b-a)(1-t)$.
Thus, the given integral can be written as

$$\int_a^b \sqrt[n]{(x-a)(b-x)}\,dx = \int_0^1 \sqrt[n]{(b-a)^2 t(1-t)}\,(b-a)dt$$

$$= (b-a)^{\frac{2}{n}+1}\int_0^1 t^{\frac{1}{n}}(1-t)^{\frac{1}{n}}\,dt = (b-a)^{\frac{2}{n}+1} B\left(\frac{1}{n}+1, \frac{1}{n}+1\right)$$

$$= (b-a)^{\frac{2}{n}+1}\frac{\Gamma\left(\frac{1}{n}+1\right)\Gamma\left(\frac{1}{n}+1\right)}{\Gamma\left(\frac{2}{n}+2\right)}.$$

PROBLEM 27 Evaluate the complete elliptic integral of the first kind

$$K\left(\frac{1}{\sqrt{2}}\right)=\int_0^{\pi/2}\frac{d\theta}{\sqrt{1-\frac{1}{2}\sin^2\theta}}.$$

Solution: The given integral is

$$\int_0^{\frac{\pi}{2}}\frac{d\theta}{\sqrt{1-\frac{1}{2}\sin^2\theta}}=\sqrt{2}\int_0^{\frac{\pi}{2}}\frac{d\theta}{\sqrt{2-\sin^2\theta}}=\frac{\sqrt{2}}{2}\int_0^{\pi}\frac{d\theta}{\sqrt{2-\sin^2\theta}}$$

$$=\frac{1}{\sqrt{2}}\int_0^{\pi}\frac{d\theta}{\sqrt{2-\sin^2\theta}}=\frac{1}{\sqrt{2}}\int_0^{\pi}\frac{d\theta}{\sqrt{2-4\sin^2\frac{\theta}{2}\cos^2\frac{\theta}{2}}}$$

$$=\frac{1}{\sqrt{2}}\int_0^{\pi}\frac{d\theta}{\sqrt{2-\dfrac{4\tan^2\frac{\theta}{2}}{\sec^4\frac{\theta}{2}}}}=\frac{1}{\sqrt{2}}\int_0^{\pi}\frac{d\theta}{\sqrt{2-\dfrac{4\tan^2\frac{\theta}{2}}{\left\{1+\tan^2\frac{\theta}{2}\right\}^2}}}. \quad \ldots\text{(i)}$$

Let us introduce a variable

$$t=\tan\frac{\theta}{2}. \quad \ldots\text{(ii)}$$

Then,

$$dt=\frac{1}{2}\sec^2\left(\frac{\theta}{2}\right)d\theta=\frac{1}{2}\left\{1+\tan^2\left(\frac{\theta}{2}\right)\right\}d\theta=\frac{1}{2}\left(1+t^2\right)d\theta.$$

Therefore, we have

$$d\theta=\frac{2\,dt}{1+t^2}. \quad \ldots\text{(iii)}$$

Using Eqs. (ii) and (iii) in Eq. (i), we write

$$\int_0^{\frac{\pi}{2}}\frac{d\theta}{\sqrt{1-\frac{1}{2}\sin^2\theta}}=\frac{1}{\sqrt{2}}\int_0^{\infty}\frac{1}{\sqrt{2-\dfrac{4t^2}{\left(1+t^2\right)^2}}}\frac{2dt}{\left(1+t^2\right)}$$

$$= \frac{1}{\sqrt{2}} \int_0^{\infty} \frac{1}{\sqrt{\frac{2\left(1+t^4\right)}{\left(1+t^2\right)^2}}} \frac{2dt}{\left(1+t^2\right)} = \int_0^{\infty} \frac{dt}{\sqrt{1+t^4}}. \qquad \ldots\text{(iv)}$$

Further, let us put $t^2 = \tan x$.

Then, we have $2t\,dt = \sec^2 x dx$ giving $dt = \dfrac{\sec^2 x}{2\sqrt{\tan x}} dx$.

Now, Eq. (iv) can be written as

$$\int_0^{\frac{\pi}{2}} \frac{d\theta}{\sqrt{1-\frac{1}{2}\sin^2\theta}} = \int_0^{\frac{\pi}{2}} \frac{1}{\sqrt{1+\tan^2 x}} \frac{\sec^2 x}{2\sqrt{\tan x}} dx$$

$$= \frac{1}{2}\int_0^{\frac{\pi}{2}} \frac{1}{\sec x} \frac{\sec^2 x}{\sqrt{\tan x}} dx = \frac{1}{2}\int_0^{\frac{\pi}{2}} (\sin x)^{-\frac{1}{2}} (\cos x)^{-\frac{1}{2}} dx$$

$$= \frac{1}{4} \cdot 2\int_0^{\frac{\pi}{2}} (\sin x)^{2\left(\frac{1}{4}\right)-1} (\cos x)^{2\left(\frac{1}{4}\right)-1} dx$$

$$= \frac{1}{4} \cdot \frac{\Gamma\left(\frac{1}{4}\right)\Gamma\left(\frac{1}{4}\right)}{\Gamma\left(\frac{1}{2}\right)} = \frac{\Gamma\left(\frac{1}{4}\right)\Gamma\left(\frac{1}{4}\right)}{4\sqrt{\pi}}.$$

PROBLEM 28 Evaluate $\int_{-\infty}^{+\infty} e^{-3x^2+4x} dx$.

Solution: Using Eq. (13.64), we have

$$\int_{-\infty}^{+\infty} e^{-3x^2+4x} dx = \sqrt{\frac{\pi}{3}} e^{\frac{4^2}{4\times 3}} = \sqrt{\frac{\pi}{3}} e^{\frac{4}{3}}.$$

PROBLEM 29 Evaluate $\int_0^{\infty} \cos\left(bu^{\frac{1}{n}}\right) du$ and obtain the values of

(i) $\int_0^{\infty} \cos(u^2) du$,

(ii) $\int_{-\infty}^{\infty} \cos\left(\frac{\pi u^2}{2}\right) du.$

Solution: From Eq. (13.62), we know that

$$\int_0^{\infty} e^{-ax} x^{n-1} \cos bx dx = \frac{\Gamma(n)\cos n\theta}{\left(a^2+b^2\right)^{n/2}} \quad \text{where} \quad \theta = \text{arc}\tan\left(\frac{b}{a}\right).$$

When $a = 0$, we have $\theta = \frac{\pi}{2}$.

Therefore, in this case, we write

$$\int_0^{\infty} x^{n-1} \cos bx dx = \frac{\Gamma(n)}{b^n} \cos\left(\frac{n\pi}{2}\right).$$

Putting $x^n = u$ or, $x = u^{\frac{1}{n}}$, we have $nx^{n-1}dx = du$.

Therefore, we have

$$x^{n-1}dx = \frac{du}{n}.$$

Now, we write

$$\int_0^{\infty} x^{n-1} \cos bx\, dx = \int_0^{\infty} \cos\left(bu^{\frac{1}{n}}\right) \frac{du}{n} = \frac{\Gamma(n)}{b^n} \cos\left(\frac{n\pi}{2}\right).$$

Therefore, we have

$$\int_0^{\infty} \cos\left(bu^{\frac{1}{n}}\right) du = \frac{n\Gamma(n)}{b^n} \cos\left(\frac{n\pi}{2}\right) = \frac{\Gamma(n+1)}{b^n} \cos\left(\frac{n\pi}{2}\right). \qquad \ldots\text{(i)}$$

First deduction: In Eq. (i), putting $b = 1$ and $n = \frac{1}{2}$, we write

$$\int_0^{\infty} \cos\left(u^2\right) du = \frac{1}{2}\Gamma\left(\frac{1}{2}\right)\cos\left(\frac{\pi}{4}\right) = \frac{\sqrt{\pi}}{2\sqrt{2}}.$$

Second deduction: We know that

$$\int_{-\infty}^{\infty} \cos\left(\frac{\pi u^2}{2}\right) du = 2\int_0^{\infty} \cos\left(\frac{\pi u^2}{2}\right) du.$$

From Eq. (i) putting $b=\frac{\pi}{2}$ and $n=\frac{1}{2}$, we write

$$\int_0^\infty \cos\left(\frac{\pi}{2}u^2\right)du = \frac{\Gamma\left(\frac{1}{2}+1\right)}{\left(\frac{\pi}{2}\right)^{1/2}}\cos\left(\frac{\pi}{4}\right)$$

or
$$\frac{1}{2}\int_{-\infty}^{\infty}\cos\left(\frac{\pi u^2}{2}\right)du = \sqrt{\frac{2}{\pi}}\,\frac{1}{2}\sqrt{\pi}\,\frac{1}{\sqrt{2}} = \frac{1}{2}$$

or
$$\int_{-\infty}^{\infty}\cos\left(\frac{\pi u^2}{2}\right)du = 1.$$

PROBLEM 30 Evaluate $\int_0^\infty e^{-ax^2}\cos bx\,dx$.

Solution: Let us consider the integral

$$\int_{-\infty}^{\infty} e^{-ax^2}e^{ibx}dx = \int_{-\infty}^{\infty} e^{-a\left(x^2-\frac{ib}{a}x\right)}dx$$

$$= \int_{-\infty}^{\infty} e^{-a\left\{x^2-2x\frac{ib}{2a}+\left(\frac{ib}{2a}\right)^2-\left(\frac{ib}{2a}\right)^2\right\}}dx$$

$$= e^{-\frac{b^2}{4a}}\int_{-\infty}^{\infty} e^{-a\left(x-\frac{ib}{2a}\right)^2}dx.$$

Putting $\sqrt{a}\left(x-\frac{ib}{2a}\right)=u$, we have $dx=\frac{du}{\sqrt{a}}$.

Then the above integral can be written as

$$\int_{-\infty}^{\infty} e^{-ax^2}(\cos bx + i\sin bx)dx = e^{-\frac{b^2}{4a}}\int_{-\infty}^{\infty} e^{-u^2}\frac{du}{\sqrt{a}}$$

or
$$\int_{-\infty}^{\infty} e^{-ax^2}\cos bx\,dx + i\int_{-\infty}^{\infty} e^{-ax^2}\sin bx\,dx = \frac{e^{-\frac{b^2}{4a}}}{\sqrt{a}}2\int_0^\infty e^{-u^2}du$$

$$= \frac{e^{-\frac{b^2}{4a}}}{\sqrt{a}} \Gamma\left(\frac{1}{2}\right) = \sqrt{\frac{\pi}{a}}\, e^{-\frac{b^2}{4a}}.$$

Equating real parts from both sides of the equation, we obtain

$$\int_{-\infty}^{\infty} e^{-ax^2} \cos bx\, dx = \sqrt{\frac{\pi}{a}}\, e^{-\frac{b^2}{4a}}.$$

Hence, we have

$$\int_{0}^{\infty} e^{-ax^2} \cos bx\, dx = \frac{1}{2}\sqrt{\frac{\pi}{a}}\, e^{-\frac{b^2}{4a}}.$$

PROBLEM 31 Prove that $\int_0^\infty \frac{\sinh^p x}{\cosh^q x} dx = \frac{1}{2} B\left(\frac{p+1}{2}, \frac{q-p}{2}\right)$.

Solution: Putting $\cosh x = \frac{e^x + e^{-x}}{2} = t$, we have

$$\sinh x\, dx = dt \quad \text{or,} \quad dx = \frac{dt}{\sinh x} = \frac{dt}{\sqrt{\cosh^2 x - 1}} = \frac{dt}{\sqrt{t^2 - 1}}.$$

Now, the given integral can be written as

$$\int_0^\infty \frac{\sinh^p x}{\cosh^q x} dx = \int_0^\infty \frac{\left(\sqrt{\cosh^2 x - 1}\right)^p}{(\cosh x)^q} dx$$

$$= \int_1^\infty \frac{\left(t^2 - 1\right)^{p/2}}{t^q} \frac{dt}{\sqrt{t^2 - 1}} = \int_1^\infty \left(t^2 - 1\right)^{\frac{p-1}{2}} t^{-q} dt. \qquad \ldots\text{(i)}$$

Further, putting $t^2 = \frac{1}{w}$, or, $t = w^{-\frac{1}{2}}$, we have $dt = -\frac{1}{2} w^{-\frac{3}{2}} dw$.

Then, Eq. (i) can be written as

$$\int_0^\infty \frac{\sinh^p x}{\cosh^q x} dx = \int_1^0 \left(\frac{1}{w} - 1\right)^{\frac{p-1}{2}} w^{\frac{q}{2}} \left(-\frac{1}{2} w^{-\frac{3}{2}} dw\right)$$

$$= \frac{1}{2}\int_0^1 w^{\frac{q-p}{2}-1} (1-w)^{\frac{p+1}{2}-1} dw$$

$$= \frac{1}{2} B\left(\frac{q-p}{2}, \frac{p+1}{2}\right) = \frac{1}{2} B\left(\frac{p+1}{2}, \frac{q-p}{2}\right).$$

PROBLEM 32 If n is a positive integer, directly prove the duplication formula

$$\Gamma(2n)=\frac{2^{2n-1}}{\sqrt{\pi}}\Gamma(n)\Gamma\left(n+\frac{1}{2}\right).$$

Solution: We write

$$\begin{aligned}\Gamma(2n)&=(2n-1)!\\&=1.2.3.4.5.6\ldots(2n-4)(2n-3)(2n-2)(2n-1)\\&=\left[2.4.6\ldots(2n-4)(2n-2)\right]\left[1.3.5.7\ldots(2n-3)(2n-1)\right]\\&=2^{n-1}(n-1)!\frac{2^n\cdot\Gamma\left(\frac{1}{2}\right)\left[\frac{1}{2}\cdot\frac{3}{2}\cdot\frac{5}{2}\cdot\frac{7}{2}\cdots\left(\frac{2n-3}{2}\right)\left(\frac{2n-1}{2}\right)\right]}{\Gamma\left(\frac{1}{2}\right)}\\&=\frac{2^{2n-1}}{\sqrt{\pi}}\Gamma(n)\Gamma\left(n+\frac{1}{2}\right).\end{aligned}$$

EXERCISE

1. Evaluate

(i) $\int_0^\infty \sqrt{x}\,e^{-\sqrt[3]{x}}\,dx,$ **Ans.** (i) $\frac{315}{16}\sqrt{\pi}$

(ii) $\int_0^\infty \sqrt[4]{x}\,e^{\sqrt{x}}\,dx.$ (ii) $\frac{3\sqrt{\pi}}{2}$

2. Prove that

(i) $\dfrac{\Gamma\left(-\frac{1}{2}\right)}{\Gamma\left(\frac{1}{2}\right)}=-2,$

(ii) $\int_0^1 (x\ln x)^4\,dx=\frac{4!}{5^5}.$

3. Evaluate

(i) $\int_0^\infty x^4 e^{-x^2}\,dx$, **Ans.** (i) $\frac{3\sqrt{\pi}}{8}$

(ii) $\int_0^\infty x^{3/2} e^{-3x}\,dx$. (ii) $\frac{\sqrt{3\pi}}{36}$

4. Evaluate

(i) $\int_0^\infty \frac{e^{-st}}{\sqrt{t}}\,dt$, **Ans.** (i) $\sqrt{\frac{\pi}{s}}$

(ii) $\int_0^1 \left[\ln\left(\frac{1}{t}\right)\right]^{-\frac{1}{2}} dt$. (ii) $\sqrt{\pi}$

5. Evaluate

(i) $\int_0^4 \frac{dt}{\sqrt{4t-t^2}}$, **Ans.** (i) π

(ii) $\int_0^{\pi/2} \sin^6\theta\cos^4\theta\,d\theta$. (ii) $\frac{3\pi}{512}$

6. Evaluate

(i) $\int_0^{\frac{\pi}{2}} \sin^6\theta\,d\theta$, **Ans.** (i) $\frac{5\pi}{32}$

(ii) $\int_0^1 \sqrt{1-x^n}\,dx$. (ii) $\frac{\sqrt{\pi}}{(n+2)} \frac{\Gamma\left(\frac{1}{n}\right)}{\Gamma\left(\frac{n+2}{2n}\right)}$

7. Evaluate

(i) $\int_0^{\frac{\pi}{2}} \sin^3\theta\cos^{\frac{5}{2}}\theta\,d\theta$, **Ans.** (i) $\frac{8}{77}$

(ii) $\int_0^1 x\left(1-x^3\right)^{\frac{1}{3}} dx$. (ii) $\frac{2\pi}{9\sqrt{3}}$

8. Show that $\int_0^{\pi/2} \tan^n \theta \, d\theta = \frac{\pi}{2} \sec \frac{n\pi}{2}$.

9. Show that $\Gamma(m)\Gamma(1-m) = \frac{\pi}{\sin m\pi}$ for $0 \le x \le 1$ and deduce that

   (i) $\Gamma(1-m)\Gamma(1+m) = \frac{m\pi}{\sin m\pi}$,

   (ii) $\Gamma\left(\frac{1}{2}\right) = \sqrt{\pi}$.

10. Evaluate $\int_0^{\pi/2} \sin^{m-1}\theta \cos^{n-1}\theta \, d\theta$. **Ans.** $\frac{1}{2} \frac{\Gamma\left(\frac{m}{2}\right)\Gamma\left(\frac{n}{2}\right)}{\Gamma\left(\frac{m+n}{2}\right)}$

11. Show that $\Gamma\left(k + \frac{1}{2}\right) = \frac{1.3.5.7 \ldots (2k-1)}{2^k}$.

12. Evaluate

    (i) $\int_0^1 \frac{x^4 dx}{\sqrt{1-x^2}}$, **Ans.** (i) $\frac{3\pi}{16}$

    (ii) $\int_0^\infty \frac{x^3 dx}{(1+x)^5}$, (ii) $\frac{1}{4}$

    (iii) $\int_0^1 (x \ln x)^4 \, dx$. (iii) $\frac{4!}{5^5}$

13. Evaluate

    (i) $\int_0^1 (x^2)^p (1-x^2)^q \, dx$, **Ans.** (i) $\frac{1}{2} \frac{\Gamma\left(p+\frac{1}{2}\right)\Gamma(q+1)}{\Gamma\left(p+q+\frac{3}{2}\right)}$

    (ii) $\int_0^1 x^k \ln x \, dx$. (ii) $-\frac{1}{(k+1)^2}$

14. Prove that $\int_0^\infty \frac{x^5}{5^x}dx = \frac{5!}{(\ln 5)^6}$.

15. Evaluate

(i) $\int_2^4 \sqrt{(x-2)(4-x)}\,dx$, **Ans.** (i) $\frac{\pi}{2}$,

(ii) $\int_3^7 \sqrt[4]{(x-3)(7-x)}\,dx$, (ii) $\frac{2\left[\Gamma\left(\frac{1}{4}\right)\right]^2}{3\sqrt{\pi}}$

(iii) $\int_{-1}^1 (1+x)^{p-1}(1-x)^{q-1}\,dx$. (iii) $2^{p+q-1}\frac{\Gamma(p)\Gamma(q)}{\Gamma(p+q)}$

16. Evaluate

(i) $\int_{-\infty}^\infty e^{-4x^2+3x}dx$, **Ans.** (i) $\sqrt{\frac{\pi}{4}}e^{\frac{9}{16}}$

(ii) $\int_{-\infty}^\infty e^{-ax^2-bx}dx$, (ii) $\sqrt{\frac{\pi}{a}}e^{\frac{b^2}{4a}}$

(iii) $\int_0^\infty e^{-3x^2}\cos 2x\,dx$. (iii) $\frac{1}{2}\sqrt{\frac{\pi}{3}}e^{-\frac{1}{3}}$

14

CHAPTER

Theory of Errors and Fitting of Curves

14.1 OBSERVATIONS AND MEASUREMENTS

Observation and measurement play very significant roles in the development of physics and other physical sciences. An *observation* is the action or process of closely observing or monitoring something or someone and a *measurement* is the determination of the size or magnitude of something by comparing that unknown quantity with some standard quantity of equal nature, known as *measurement unit*.

14.2 ORDER OF MAGNITUDE, OBSERVED VALUE, ACCURACY AND PRECISION

The *order of magnitude* of a number N refers to the smallest power of ten needed to represent that number. The order of magnitude b of a number N is given by

$$N = a \times 10^b \quad \text{where} \quad 0.5 < a \leq 5.$$

EXAMPLES

(1) The order of magnitude of the number 0.5 is –1 since we can write $0.5 = 5 \times 10^{-1}$.

(2) The order of magnitude of the number 5 is zero since we can write $5 = 5 \times 10^0$.

(3) The order of magnitude of the number 25 is 1 since we can write $25 = 2.5 \times 10^1$.

If a person performs a measurement of a physical quantity, the result obtained by him is called an *observed value* or the *measured value*.

In a measurement of a physical quantity, *accuracy* refers to the closeness of an observed value to a standard or known value of that quantity. If a student is asked to measure the mass of a specific gravity bottle whose actual mass is 30 gm (not disclosed) and if he finds the observed value of mass as 20 gm, then his measurement is not accurate. If another student performs the measurement and finds the mass of the specific gravity bottle as 29.98 gm, then his result has a high accuracy. Closer the observed value to the standard value, higher is the degree of accuracy. On the other hand, in a set of measurements of a physical quantity, *precision* refers to the closeness of the observed values to one another. In measuring the mass of the above specific gravity bottle, if a student measures five times and in every measurement he obtains the same mass of 20 gm, then his experiment is said to have a high precision but no accuracy. A good experiment is one that has results of both high accuracy and high precision.

14.3 MISTAKES AND ERRORS

Any faulty measurement that is due to carelessness of an experimenter is referred to as a *mistake*. It can be avoided if utmost care is taken on the part of the experimenter. Mistakes cannot be meaningfully classified.

On the other hand, an *error* is defined as the difference between the observed (measured) value and the actual (standard) value.

Error and its Types

Basically, an *error* is defined as the difference between the actual or true value and the measured or observed value of a physical quantity,

Errors can be classified into three types, namely: (i) *personal errors,* (ii) *systematic errors,* and (iii) *random errors.* As personal errors are not important, many authors consider only the last two types.

(i) ***Personal errors*** are those errors which are associated with special traits, habits or whims of an experimenter. Such errors can also be treated as mistakes.

(ii) ***Systematic errors*** are those errors which are governed by some definite rule. Once a knowledge of the rule is available, we can minimize or even eliminate such errors. Faulty graduation of a thermometer, instrumental error or zero error of a slide calipers and index error in optical bench experiments are examples of systematic errors.

(iii) ***Random errors*** are those errors which are not associated with any constant or systematic cause, i.e., their law of action is unknown. Random errors are the most important errors discussed in physics.

14.4 PROPAGATION OF ERRORS

Propagation of errors is the effect(s) of the variables' random errors (or uncertainties) on the error of a function of those variables.

(i) Additive Formulae (Sum and Difference)

Sum

If a result R is calculated from two individual measurements x and y with errors or uncertainties Δx and Δy, respectively such that

$$R = x + y,$$

we write,

measured value of $\quad x = x_{\text{best}} \pm \Delta x,$

measured value of $\quad y = y_{\text{best}} \pm \Delta y,$

probable value of $\quad R = R_{\text{best}} \pm \Delta R = x_{\text{best}} + y_{\text{best}} \pm \Delta x \pm \Delta y.$

Here, x_{best} means the best estimate value of x and similarly, for others.

The highest probable value of $R = x_{\text{best}} + y_{\text{best}} + \Delta x + \Delta y$

and the lowest probable value of $R = x_{\text{best}} + y_{\text{best}} - \Delta x - \Delta y.$

We can say that the value of R lies between the above to numbers and the error in R is $\Delta R \approx \Delta x + \Delta y$.

Independent errors may cancel each other with some probability since if x is measured somewhat too large and y somewhat to small, the error in R may be very small and it is negligible. Such partial and statistical cancellation is readily accounted for if the errors are *added in quadrature.* If we square two or more individual numbers, add the squared numbers and take the square root of the sum, then the numbers are said to be *added in quadrature*.

If the (random) errors in x and y are independent, then the error in the result R will be given by

$$\Delta R = \sqrt{(\Delta x)^2 + (\Delta y)^2}. \tag{14.1}$$

Using the triangle law from the following triangle, we see that

$$\sqrt{(\Delta x)^2 + (\Delta y)^2} < \Delta x + \Delta y. \tag{14.2}$$

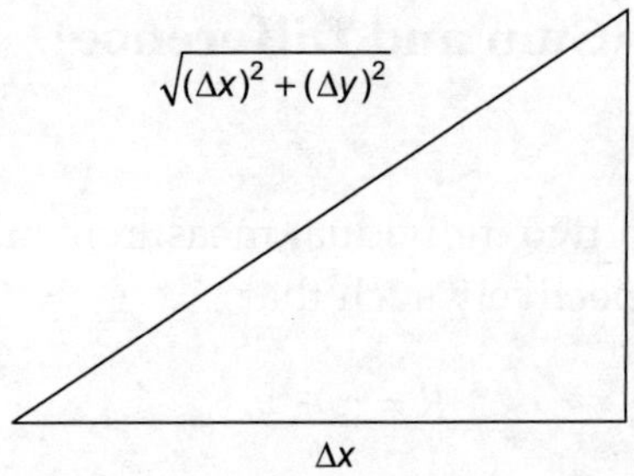

Fig. 14.1

Difference

Proceeding as in the case for sum, we arrive at

the highest probable value of $R = x_{\text{best}} - y_{\text{best}} + \Delta x + \Delta y$

and the lowest probable value of $R = x_{\text{best}} - y_{\text{best}} - \Delta x - \Delta y$.

We can say that the value of R lies between the above two numbers and the error in R is $\Delta R \approx \Delta x + \Delta y$.

If the errors are independent, we have

$$\Delta R = \sqrt{(\Delta x)^2 + (\Delta y)^2}. \tag{14.1}$$

In a generalization of Eq. (14.1), if several quantities $u, v, x, \ldots z$ are measured with errors or uncertainties $\Delta u, \Delta v, \Delta x, \ldots \Delta z$, then the error in the measurement of the quantity $R = u + v + x + \ldots + z$ will be

$$\Delta R = \sqrt{(\Delta u)^2 + (\Delta v)^2 + \ldots + (\Delta z)^2}. \tag{14.3}$$

(ii) Product and Quotient

Product

If a result $R = xy$ is to be calculated from two individual measurements x and y with errors or uncertainties Δx and Δy, respectively, then we write

measured value of $x = x_{\text{best}} \pm \Delta x$,

measured value of $y = y_{\text{best}} \pm \Delta y$,

probable value of $R = R_{\text{best}} \pm \Delta R = \dfrac{x_{\text{best}}}{y_{\text{best}}} \pm \Delta R$.

Now, the highest probable value of $R = xy$ is

$$(x_{\text{best}} + \Delta x)(y_{\text{best}} + \Delta y) = x_{\text{best}}\, y_{\text{best}} + \Delta x\, y_{\text{best}} + x_{\text{best}}\Delta y + \Delta x \Delta y$$

$$= x_{\text{best}}\, y_{\text{best}}\left(1 + \frac{\Delta x}{x_{\text{best}}} + \frac{\Delta y}{y_{\text{best}}}\right)$$

where $\Delta x \Delta y \to 0$ assuming that Δx and Δy are small.

The lowest probable value of R is

$$(x_{\text{best}} - \Delta x)(y_{\text{best}} - \Delta y) = x_{\text{best}}\, y_{\text{best}}\left(1 - \frac{\Delta x}{x_{\text{best}}} - \frac{\Delta y}{y_{\text{best}}}\right).$$

We can say that the value of R lies between the above two numbers and the error in R is

$$\Delta R \approx x_{\text{best}}\, y_{\text{best}}\left(\frac{\Delta x}{x_{\text{best}}} + \frac{\Delta y}{y_{\text{best}}}\right).$$

Dividing both sides by $R_{\text{best}} = x_{\text{best}}\, y_{\text{best}}$, we obtain

$$\frac{\Delta R}{R_{\text{best}}} = \frac{\Delta x}{x_{\text{best}}} + \frac{\Delta y}{y_{\text{best}}}. \qquad (14.4)$$

If the errors are random and independent, we write

$$\frac{\Delta R}{R_{\text{best}}} = \sqrt{\left(\frac{\Delta x}{x_{\text{best}}}\right)^2 + \left(\frac{\Delta y}{y_{\text{best}}}\right)^2}. \qquad (14.5)$$

Quotient

Let us consider $R = \dfrac{x}{y}$, $R_{\text{best}} = \dfrac{x_{\text{best}}}{y_{\text{best}}}$,

measured value of $x = x_{\text{best}} \pm \Delta x$,

measured value of $y = y_{\text{best}} \pm \Delta y$,

probable value of $R = R_{\text{best}} \pm \Delta R = \dfrac{x_{\text{best}}}{y_{\text{best}}} \pm \Delta R$.

Now, the highest probable value of $R = \frac{x}{y}$ is

$$\frac{(x_{\text{best}} + \Delta x)}{(y_{\text{best}} - \Delta y)} = \frac{(x_{\text{best}} + \Delta x)}{y_{\text{best}}}\left(1 - \frac{\Delta y}{y_{\text{best}}}\right)^{-1} = \frac{(x_{\text{best}} + \Delta x)}{y_{\text{best}}}\left(1 + \frac{\Delta y}{y_{\text{best}}}\right)$$

$$= \frac{(x_{\text{best}} + \Delta x)(y_{\text{best}} + \Delta y)}{(y_{\text{best}})^2}$$

$$= \frac{x_{\text{best}}\, y_{\text{best}} + \Delta x\, y_{\text{best}} + x_{\text{best}}\, \Delta y + (\Delta x)(\Delta y)}{(y_{\text{best}})^2}$$

$$= \frac{x_{\text{best}}}{y_{\text{best}}}\left(1 + \frac{\Delta x}{x_{\text{best}}} + \frac{\Delta y}{y_{\text{best}}}\right).$$

The lowest probable value of $R = \frac{x}{y}$ is

$$\frac{(x_{\text{best}} - \Delta x)}{(y_{\text{best}} + \Delta y)} = \frac{(x_{\text{best}} - \Delta x)}{y_{\text{best}}}\left(1 + \frac{\Delta y}{y_{\text{best}}}\right)^{-1} = \frac{(x_{\text{best}} - \Delta x)}{y_{\text{best}}}\left(1 - \frac{\Delta y}{y_{\text{best}}}\right)$$

$$= \frac{(x_{\text{best}} - \Delta x)(y_{\text{best}} - \Delta y)}{(y_{\text{best}})^2}$$

$$= \frac{x_{\text{best}}\, y_{\text{best}} - \Delta x\, y_{\text{best}} - x_{\text{best}}\, \Delta y + (\Delta x)(\Delta y)}{(y_{\text{best}})^2}$$

$$= \frac{x_{\text{best}}}{y_{\text{best}}}\left(1 - \frac{\Delta x}{x_{\text{best}}} - \frac{\Delta y}{y_{\text{best}}}\right).$$

We can say that the value of R lies between the above two numbers and the error in R is

$$\Delta R \approx \frac{x_{\text{best}}}{y_{\text{best}}}\left(\frac{\Delta x}{x_{\text{best}}} + \frac{\Delta y}{y_{\text{best}}}\right).$$

Dividing both sides by $R_{\text{best}} = \frac{x_{\text{best}}}{y_{\text{best}}}$, we obtain

$$\frac{\Delta R}{R_{\text{best}}} = \frac{\Delta x}{x_{\text{best}}} + \frac{\Delta y}{y_{\text{best}}}. \tag{14.4}$$

If the errors are random and independent, we write

$$\frac{\Delta R}{R_{\text{best}}}=\sqrt{\left(\frac{\Delta x}{x_{\text{best}}}\right)^2+\left(\frac{\Delta y}{y_{\text{best}}}\right)^2}. \tag{14.5}$$

We simply write

$$\frac{\Delta R}{R}=\sqrt{\left(\frac{\Delta x}{x}\right)^2+\left(\frac{\Delta y}{y}\right)^2}. \tag{14.6}$$

Here, the quantity $\frac{\Delta R}{R}$ is called *fractional* or *proportional error* in *R*.

In a generalization of Eq. (14.6), if several quantities $a, b, c, \ldots m$ and $u, v, x, \ldots z$ are measured with errors or uncertainties $(\Delta a, \Delta b, \ldots \Delta m)$ and $(\Delta u, \Delta v, \ldots \Delta z)$; then in the measurement of the quantity

$$R=\frac{a\times b\times\ldots\times m}{u\times v\times\ldots\times z},$$

the computed value of fractional error is

$$\frac{\Delta R}{R}=\sqrt{\left(\frac{\Delta a}{a}\right)^2+\left(\frac{\Delta b}{b}\right)^2+\ldots+\left(\frac{\Delta y}{y}\right)^2+\left(\frac{\Delta z}{z}\right)^2}. \tag{14.7}$$

(iii) A Quantity Raised to a Power

If a measurement of a quantity *x* has an error or uncertainty Δx, let us calculate the uncertainty in quantity $Q = x^n$. For simplicity when *n* is a positive integer, we have

$$Q = x\times x\times x\times\ldots \text{ up to } n \text{ factors.}$$

Then, by propagation of errors for multiplicative case, we write the fractional uncertainty in *Q* as

$$\frac{\Delta Q}{Q_{\text{best}}}=\frac{\Delta x}{x_{\text{best}}}+\frac{\Delta x}{x_{\text{best}}}+\ldots \text{ up to } n \text{ terms} = n\frac{\Delta x}{x_{\text{best}}}$$

or simply

$$\frac{\Delta Q}{Q}=n\frac{\Delta x}{x}. \tag{14.8}$$

If $n=-s$ (*s* being positive integer) is a negative integer, we have

$$Q=x^{-s}=\frac{1}{s}\times\frac{1}{s}\times\ldots \text{ up to } s \text{ factors,}$$

$$\frac{\Delta Q}{Q_{\text{best}}}=\frac{\Delta x}{x_{\text{best}}}+\frac{\Delta x}{x_{\text{best}}}+\ldots \text{ up to } s \text{ terms}$$

or $$\frac{\Delta Q}{Q}=\frac{\Delta x}{x}+\frac{\Delta x}{x}+\ldots \text{ up to } s \text{ terms} = s\frac{\Delta x}{x}.$$

In general, we write

$$\frac{\Delta Q}{Q_{\text{best}}}=|n|\frac{\Delta x}{x_{\text{best}}} \tag{14.9}$$

or $$\frac{\Delta Q}{Q}=|n|\frac{\Delta x}{x}. \tag{14.10}$$

(iv) Error Propagation for a Function of a Single Variable

Let $q(x)$ be a function of a single variable x. Let the best estimate of x be x_{best}. Then, the best estimate of $q(x)$ is $q_{\text{best}}=q(x_{\text{best}})$. If Δx is the error in x, then the measured value of x is $x_{\text{best}} \pm \Delta x$.

Let the maximum value of $q(x)$ be $q_{\max}=q(x_{\text{best}}+\Delta x)$ i.e., $q(x)$ is maximum when x is $(x_{\text{best}}+\Delta x)$.

Now, uncertainty in q is

$$\Delta q = q_{\max} - q_{\text{best}} = q(x_{\text{best}}+\Delta x)-q(x_{\text{best}})$$

or simply $$\Delta q = \Delta x\left[\frac{q(x+\Delta x)-q(x)}{\Delta x}\right]=\frac{dq}{dx}\Delta x. \tag{14.11}$$

Here, the expression within the square brackets is equal to $\frac{dq}{dx}$ when the error Δx is small enough .

In general, if $v, w\ldots z$ are measured with uncertainties $\Delta v, \Delta w, \ldots \Delta z$ and if we are to compute a quantity $q(v, w, x, \ldots z)$ where q is a function of the variables $v, w \ldots z$; then the error or uncertainty in q will be

$$\Delta q = \frac{\partial q}{\partial v}\Delta v + \frac{\partial q}{\partial w}\Delta w + \ldots + \frac{\partial q}{\partial z}\Delta z. \tag{14.12}$$

If the errors are random and independent, we have

$$\Delta q = \sqrt{\left(\frac{\partial q}{\partial v}\Delta v\right)^2 + \left(\frac{\partial q}{\partial w}\Delta w\right)^2 + \ldots + \left(\frac{\partial q}{\partial z}\Delta z\right)^2}. \tag{14.13}$$

14.5 DEVIATION, STANDARD DEVIATION, VARIANCE, STANDARD ERROR AND SOME OTHER TERMS

In a measurement, *deviation* means the observed or measured value minus the actual value. In a set of a large number of measurements, *individual deviation* means an individual observed value minus the (arithmetic) mean value. In a set of N observations or measurements, if $x_i\,(i=1,2,3,\ldots N)$ are the individual observed values, then the *average* value or (arithmetic) *mean* value is $\overline{x}=\frac{\sum_{i=1}^{N} x_i}{N}$ and individual deviations are $d_i = x_i - \overline{x}\,(i=1,2,3,\ldots N)$. Mean value is equivalently known as *expectation value* and it is also represented by $< x >$. A data set containing a number of elements or items or measurement is called a sample. The total number of observations or measurements in a set of measurements is called *sample size.*

In a sufficiently large finite number of measurements N, having observed values $x_i\,(i=1,2,3,\ldots N)$, if $\overline{x}$ is the mean value, then the *standard deviation* is defined by

$$\sigma = \sqrt{\frac{1}{N}\sum_{i=1}^{N} d_i^2} = \sqrt{\frac{1}{N}\sum_{i=1}^{N}\left(x_i-\overline{x}\right)^2}\,. \tag{14.14}$$

The square of standard deviation $((\sigma^2)$ is known as *variance.* Thus, we write variance

$$=\frac{1}{N}\sum_{i=1}^{N}\left(x_i-\overline{x}\right)^2. \tag{14.15}$$

From a large number N of independent observations of x having the values $x_i\,(i=1,2,3,\ldots N)$, we are provided originally with N independent equations. After finding the mean value $\overline{x}$, there exists the constraint condition that $\sum_{i=1}^{N}\left(x_i-\overline{x}\right)=0$. The number of independent equations, i.e., the degree of freedom N is thereby reduced to $(N-1)$.

Hence, in many practical purposes, the standard deviation for a large number of observations or measurements having N observed values given by $x_i\,(i=1,2,3,\ldots N)$ and a mean value $\overline{x}$ is often expressed as

$$\sigma = \sqrt{\frac{1}{(N-1)}\sum_{i=1}^{N}\left(x_i-\overline{x}\right)^2}\,. \tag{14.16}$$

From Eqs. (14.14) and (14.15), we see that

$$\text{Variance} = \sigma^2. \tag{14.17}$$

If one repeats the set of N observations up to K times, i.e., if there are K sets of observations having mean values, $\bar{x}_1, \bar{x}_2, \bar{x}_3, \ldots \bar{x}_K$ each based on N observations, we can find a grand mean

$$\bar{\bar{x}} = \bar{m} = \frac{\bar{x}_1 + \bar{x}_2 + \bar{x}_3 + \ldots \bar{x}_N}{K}.$$

Then, we are to prove that the standard deviation of the mean values is given by

$$\sigma_{\bar{x}} = \frac{\sigma}{\sqrt{N}}. \tag{14.18}$$

Proof of the above Formula

Let there be N independent random variables $x = x_1, x_2, x_3, \ldots x_N$; each of them identically distributed with variance σ^2. We are first to prove that variance of a sum is just the sum of the variances, i.e.,

$$\operatorname{var}\left(\sum_{i=1}^{N} x_i\right) = \sum_{i=1}^{N} \operatorname{var}(x_i). \tag{14.19}$$

For simplicity, let us tackle the variance of $(x+y)$.
Then,

$$\begin{aligned}
\operatorname{var}(x+y) &= \text{mean of}\left[(x+y) - \text{mean}(x+y)\right]^2 \\
&= E\left[(x+y) - E(x+y)\right]^2 \\
&= E\left[\{x - E(x)\} + \{y - E(y)\}\right]^2 \\
&= E\left[\{x - E(x)\}^2 + \{y - E(y)\}^2 + 2\{x - E(x)\}\{y - E(y)\}\right] \\
&= E\{x - E(x)\}^2 + E\{y - E(y)\}^2 + 2E\left[\{x - E(x)\}\{y - E(y)\}\right].
\end{aligned} \tag{14.20}$$

Here, $E(\psi) = E\{\psi\} = E[\psi]$ denote mathematical expectation or mean value of ψ.
We know that for a random variable x having a finite number of finite outcomes $x_1, x_2, x_3, \ldots x_n$ occurring with probabilities $p_1, p_1, p_3, \ldots p_n$ respectively, the expectation of x is defined by

$$\begin{aligned}
E(x) = E\{x\} = E[x] &= p_1 x_1 + p_2 x_2 + p_3 x_3 + \ldots + p_n x_n \\
&= \sum_{i=1}^{n} p_i x_i.
\end{aligned}$$

Thus, we have

$$E(xy)=\sum_{i=1}^{n}\sum_{j=i}^{n}p_ix_i\,p_jy_j=\sum_{i=1}^{n}p_ix_i\sum_{j=1}^{n}p_jy_j=E(x)E(y). \tag{14.21}$$

Using the property in Eq. (14.21) in the last term of Eq. (14.20), we write

$$\begin{aligned}\text{var}(x+y)&=E[x-E(x)]^2+E[y-E(y)]^2+2E[x-E(x)]E[y-E(y)]\\&=E[x-E(x)]^2+E[y-E(y)]^2+2[E(x)-E(E(x))][E(y)-E(E(y))]\\&=E[x-E(x)]^2+E[y-E(y)]^2+2[E(x)-E(x)][E(y)-E(y)]\\&=E[x-E(x)]^2+E[y-E(y)]^2\\&=\text{var}(x)+\text{var}(y).\end{aligned} \tag{14.22}$$

Generalizing Eq. (14.22), we obtain Eq. (14.19).

Now, using the formula $\text{var}(kx)=k^2\ \text{var}(x)$ we write the variance of mean as

$$\text{var}\left(\frac{1}{N}\sum_{i=1}^{N}x_i\right)=\frac{1}{N^2}\text{var}\left(\sum_{i=1}^{N}x_i\right)=\frac{1}{N^2}N\sigma^2=\frac{\sigma^2}{N}.$$

Taking square root on both sides, standard deviation of the mean values is given by

$$\sigma_{\overline{x}}=\sqrt{\text{var}\left(\frac{1}{N}\sum_{i=1}^{N}x_i\right)}=\frac{\sigma}{\sqrt{N}}. \tag{14.23}$$

Standard deviation of the mean values is also known as *standard error*.

Probable error: The *probable error* or the median deviation is the median of the deviations or differences of individual scores or measurements from their average. It may be computed by determining the median of deviations or differences. It is also defined as the standard error multiplied by 0.6745. Here, we have to recapitulate the terms mean (or expected value or expectation value), median and mode. Mean is the average value of all the numbers in a data set.

Arithmetic mean: For a non-group data set of N variables $x=x_1, x_2, x_3, \ldots x_N$; the *arithmetic mean* is given by

$$E(x)=\overline{x}=m=\frac{\sum_{i=1}^{N}x_i}{N}. \tag{14.24}$$

If x_1 occurs f_1 times, x_2 occurs f_2 times, …x_N occurs f_N times, that is, if $f_1, f_2, f_3, \ldots f_N$ are the frequencies of the variables $x_1, x_2, x_3, \ldots x_N$ respectively; the arithmetic mean may also be written as

$$E(x) = \bar{x} = m = \frac{\sum_{i=1}^{N} f_1 x_1}{\sum_{i=1}^{N} f_i} = \frac{\sum_{i=1}^{N} f_i x_i}{N} = \sum_{i=1}^{N} p_i x_i \tag{14.25}$$

where $p_i = \dfrac{f_i}{N}$ is the probability that $x = x_i$.

The mean deviation is defined by

$$\text{M.D.} = \sum_i \frac{f_i |x_i - \bar{x}|}{\sum_i f_i} = \frac{1}{N} \sum_i f_i |x_i - \bar{x}|. \tag{14.25a}$$

Median: A *median* is the measure of the central item of a data set when the values of items are arranged in ascending or descending order of magnitude.

For a non-group data set of N numbers or items, the *median* is the measure of (i) the $\dfrac{(N+1)}{2}$th item if N is odd, and (ii) the $\dfrac{(N+2)}{2}$th item if N is even.

Mode: *Mode* is the measure or size of the variable that occurs most frequently in a data set.

For a grouped data set, there should be only *estimated mean*, *estimated median* and *estimated mode.*

For a grouped data set having N class intervals or ranges,

$$\text{Estimated mean} = \frac{\sum_i f_i x_i}{\sum_i f_i} = \frac{\sum_i f_i x_i}{N} = \sum_{i=1}^{N} p_i x_i$$

where f_i is the frequency of the ith class interval and x_i is the midpoint of the ith class interval and $p_i = f_i/N$ is the probability that $x = x_i$.

For a grouped data set having N class intervals or ranges, the *median* is given by

$$\text{Estimated median} = L + \left(\frac{\frac{N}{2} - C}{f}\right) w \tag{14.26}$$

where L is the lower limit of the median class interval, f is the frequency of the median class interval, C is the cumulative frequency of the class interval preceding the median class interval and w is the size or width of the median class interval. Here, *frequency* is the number of times a particular class interval occurs. *Cumulative frequency* is defined as the running total of frequencies. It is the sum of all the previous frequencies up to the current class interval under consideration.

$$\text{Estimated mode} = L + \frac{f_m - f_{m-1}}{(f_m - f_{m-1}) + (f_m - f_{m+1})} \times w \tag{14.27}$$

where

L is the lower limit of the modal class interval,

f_m is the frequency of the modal class interval,

f_{m-1} is the frequency of the class interval preceding the modal class interval,

f_{m+1} is the frequency of the class interval succeeding the modal class interval, and

w is the width of the modal class interval.

Other Terms used in Error Analysis of Physical Experiments

Most probable value: When a sufficiently large number of readings are taken, then the mean value is known as the *most probable value*.

Absolute error: The magnitude of the difference between the experimentally measured value and the known standard value is called absolute error. Mathematically,

$$\text{Absolute error} = x_i - \bar{x}.$$

Relative error: It is the ratio of absolute error to the known standard value or true value.

$$\text{Relative error} = \frac{\text{Measured or observed value} - \text{True value}}{\text{True value}}.$$

Percentage error: It is the relative error multiplied by 100.

14.6 FREQUENCY DISTRIBUTION AND ITS TYPES

Frequency distribution is a data set organized to show the frequency of occurrence of each possible outcome of a repeatable event or item observed a number of times. *Probability* is a measure of how likely the outcome of an event is. *A probability distribution* is a mathematical function that defines the relationship between the outcomes and their likelihood of an event. It is a function which assigns to each possible outcome of a random variable x, a number between 0 and 1. That number is the probability associated with that outcome, and it describes the likelihood of occurrence of the outcome. Frequency distribution and probability distribution are similar if the sample size is large. A *frequency distribution curve* is a curve obtained by plotting the frequency of occurrence of values in a data set against the values themselves for a series of identical measurements of a statistical process.

A *random variable* is a variable whose possible values are the numerical outcomes of a random event. Its value varies with every trial of the experiment.

There are two types of random variables—*discrete* and *continuous*. If a random variable can take only a finite number of discrete values, then it is said to be a *discrete random variable*. For example, in tossing a coin, if the outcome with head "up" is assigned the value 1 and the outcome with head "down" is assigned the value 0, then the outcome denoted by x is a discrete random variable having the values 0, 1. If a random variable can take up an infinite number of possible values in a given range, it is said to be a *continuous random variable*. For example, if a person has a lifespan of 100 years, his life that may be represented by x, is a continuous random variable as it can have continuous values starting from zero and ending at 100 years since he may die just at birth ($x = 0$) or at any time before 100 years or just after he attains $x = 100$ years.

Out of the various types of discrete frequency distributions studied in statistics, here we discuss only three types related to physics. They are: (i) *binomial distribution*, (ii) *normal distribution*, and (iii) *Poisson distribution*.

(i) Binomial Distribution

This type of distribution is an extension of a distribution known as **Bernoulli distribution** that was proposed by the Swiss Mathematician, James Bernoulli, also known as Jacob (or Jacques) Bernoulli, (27 December 1654-16 August 1705).

A **Bernoulli distribution** is a distribution in which a random variable (x) takes any one of only two possible values in a single event. A Bernoulli event is an event having only two possible outcomes. A coin toss having two outcomes of "head" or "tail, answering a question with "yes" or "no"; flight of an aircraft with "safe" or "crash"; my fight with John Cena with my "win" or "loss" are suitable examples of Bernoulli event. In the coin toss, the probability of the outcome with "head" up is ½ and that with the " tail" up is also ½ . In answering a question with yes or no, the probability of answering "yes" is ½ and that with "no" is also ½. Thus, in the first two examples, the probabilities of the outcome in each event are equal, each being ½. In each of the last two examples, the probabilities are not equal, one being comparatively very small. In the "flight of aircraft", the probability that it crashes is extremely small as compared to the probability that it is safe. In "my fight with John Cena", the probability of my win is extremely small as compared with my loss. We see that in Bernoulli distribution, the two probabilities of the outcome may or may not be equal.

The *Bernoulli distribution* is a discrete probability distribution of a random variable that can take a binary, Boolean output: 1 with probability p, and 0 with probability $(1 - p)$. It is the outcome of a single event. If you appear an examination, the outcome (result) must be either a pass (success) or a failure. If your success is labelled with 1 having a probability p, while your failure is labelled with 0 having a probability $(1 - p)$, the discrete probability distribution function is a Bernoulli distribution. In a Bernoulli distribution, the discrete probability distribution function is given by

$$P_x = \begin{cases} p^x(1-p)^{1-x} & \text{if } x = 0,1, \\ 0 & \text{otherwise} \end{cases} = \begin{cases} p & \text{if } x = 1, \\ 1-p & \text{if } x = 0. \end{cases} \tag{14.28}$$

If a Bernoulli event is repeated a large number of times, say N, then the distribution becomes a binomial distribution. The *binomial distribution* is a frequency distribution where only two mutually exclusive outcomes are possible in a large number of repeated events or trials. Here, mutually exclusive outcomes mean that the occurrence of one discards the occurrence of the other.

The binomial distribution is used to obtain the probability of observing x successes in N trials. If $P(x; N, p)$ is the probability that N Bernoulli trials with probabilities p for success and $q = 1-p$ for failure result in x successes and $N - x$ failures, then

$$P(x;N,p) = \binom{N}{x} p^x q^{N-x} = \frac{N!}{x!(N-x)!} p^x q^{N-x}. \tag{14.29}$$

Here,

$$N_{C_x} = \binom{N}{x} = \frac{N!}{x!(N-x)!}$$

is the number of ways of placing the x successes among the N trials.

Since this probability is a term of the binomial expansion of $(p+q)^N$ this type of distribution is called binomial distribution.

We know that

$$(p+q)^N = p^N + Np^{N-1}q + \frac{N!}{2!(N-2)!} p^{N-2} q^2 + \frac{N!}{3!(N-3)!} p^{N-3} q^3 + \ldots$$

$$+ \frac{N!}{x!(N-x)!} p^x q^{N-x} + \frac{N!}{(x+1)!(N-x-1)!} p^{x+1} q^{N-x-1} + \ldots + q^N. \tag{14.30}$$

From Eqs. (14.29) and (14.30), we see that

$$(p+q)^N = P_N + P_{N-1} + P_{N-2} + \ldots + P_x + \ldots + P_1 + P_0 = 1. \tag{14.31}$$

Equation (14.29) can be written as

$$P(x;N,p) = \frac{N!}{x!(N-x)!} p^x q^{N-x} = \frac{N!}{x!(N-x)!} p^x (1-p)^{N-x}. \tag{14.32}$$

We see that binomial distribution contains two independent parameters, N and p, which are both integers.

In the case of binomial distribution having the parameters N, p the mean, the standard deviation, and the variance are respectively given by

$$m = Np, \sigma = \sqrt{Np(1-p)}, \text{var} = \sigma^2 = Np(1-p). \tag{14.33}$$

Proof: We have

$$m = E(x) = \sum_{x=0}^{N} xP(x;N,p) = \sum_{x=0}^{N} x\frac{N!}{x!(N-x)!}p^x(1-p)^{N-x}$$

$$= \sum_{x=1}^{N} \frac{N!}{(x-1)!(N-x)!}p^x(1-p)^{N-x}.$$

Putting $x-1=y$ in the expression on the right-hand side, we write

$$m = E(x) = \sum_{y=0}^{N-1} \frac{N!}{y!(N-y-1)!}p^{y+1}(1-p)^{N-y-1}$$

$$= Np\sum_{y=0}^{N-1} \frac{(N-1)!}{y!(N-1-y)!}p^y(1-p)^{N-1-y}$$

$$= Np\{p+(1-p)\}^{N-1} = Np.$$

Variance is given by

$$\text{var}(x) = E\{x-E(x)\}^2 = E(x^2) - \{E(x)\}^2$$

$$= \sum_{x=0}^{N} x^2 P(x;N,p) - (Np)^2 = \sum_{x=0}^{N}\{x(x-1)+x\}P(x;N,p) - (Np)^2$$

$$= \sum_{x=0}^{N} x(x-1)P(x;N,p) + \sum_{x=0}^{N} xP(x;N,p) - (Np)^2$$

$$= \sum_{x=0}^{N} x(x-1)\frac{N!}{x!(N-x)!}p^x(1-p)^{N-x} + Np - (Np)^2$$

$$= \sum_{x=2}^{N} \frac{N!}{(x-2)!(N-x)!}p^x(1-p)^{N-x} + Np - (Np)^2$$

$$= \sum_{y=0}^{N-2} \frac{N!}{y!(N-y-2)!}p^{y+2}(1-p)^{N-y-2} + Np - (Np)^2 \left[\text{where } y = x-2\right]$$

$$= N(N-1)p^2\sum_{y=0}^{N-2} \frac{(N-2)!}{y!(N-2-y)!}p^y(1-p)^{N-2-y} + Np - (Np)^2$$

$$= N(N-1)p^2\{p+(1-p)\}^{N-2} + Np - (Np)^2$$

$$= N(N-1)p^2 + Np - (Np)^2 = Np(1-p).$$

Standard deviation is given by

$$\sigma = \sqrt{\text{Variance}} = \sqrt{Np(1-p)}.$$

Properties of a Binomial Distribution

(a) Trials are all independent.

(b) There are only two possible outcomes in a trial.

(c) The total number of identical trials is N.

(d) The probability of each outcome is the same for all trials. (Trials are identical.)

(e) If the probabilities p and q are equal, the distribution is symmetric and if p and q are not equal, the distribution is asymmetric.

(f) As the value of N becomes larger and larger, binomial distribution approaches more and more to normal distribution.

As an example of binomial distribution, let us consider tossing of four fair coins, then the number of heads in an outcome is represented by a discrete random variable x having values x = 0 head, 1 head, 2 heads, 3 heads, and 4 heads. Also, we have

Number of trials is N = 4, probability of occurring head in a single toss is $p = 1/2$.

Probability of occurrence of 0 head is $P\left(0;4,\frac{1}{2}\right)$

$$= {}^4C_0\left(\frac{1}{2}\right)^0\left(1-\frac{1}{2}\right)^4 = \frac{4!}{0!(4-0)!}\left(\frac{1}{2}\right)^0\left(1-\frac{1}{2}\right)^4 = \frac{1}{16} = 0.0625.$$

Probability of occurrence of 1 head is

$$P\left(1;\,4,\frac{1}{2}\right) = \frac{4!}{1!(4-1)!}\left(\frac{1}{2}\right)^1\left(1-\frac{1}{2}\right)^{4-1} = 0.25.$$

Probability of occurrence of 2 heads is

$$P\left(2;\,4,\frac{1}{2}\right) = \frac{4!}{2!\,2!}\left(\frac{1}{2}\right)^2\left(\frac{1}{2}\right)^2 = 0.375.$$

Probability of occurrence of 3 heads is

$$P\left(3;\,4,\frac{1}{2}\right) = \frac{4!}{3!\,1!}\left(\frac{1}{2}\right)^3\left(\frac{1}{2}\right) = 0.25.$$

Probability of occurrence of 4 heads is

$$P\left(4;\,4,\frac{1}{2}\right) = 0.0625.$$

$$\text{Mean, } m = \frac{0+1+2+3+4}{5} = Np = 2.$$

$$\text{Variance} = Np(1-p) = 1, \text{ standard deviation } = \sigma = \sqrt{1} = 1.$$

A bar graph for binomial distribution with $N = 4, p = \frac{1}{2}$ is shown in Figure 14.2.

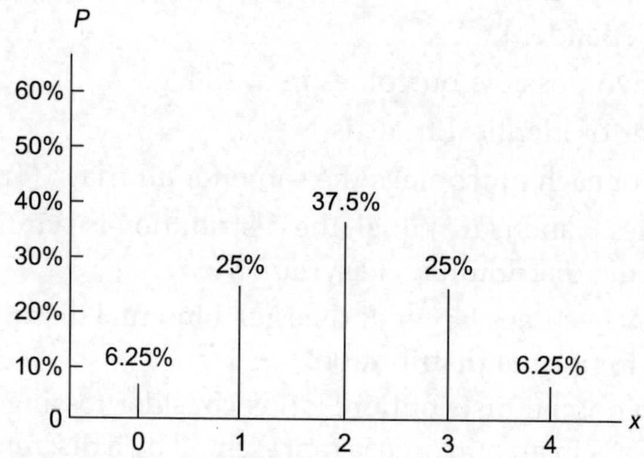

Fig. 14.2 Binomial distribution with $N = 4, p = \frac{1}{2}$.

(ii) Normal Distribution

The normal or Guassian distribution is an analytical approximation to a binomial distribution when the number of events (*N*) is very large. It is a distribution which is symmetric about the mean indicating that the data near the mean are more frequent in occurrence than those away from the mean. Since the distribution curve is a bell curve, it is also known as bell-shaped distribution.

The French mathematician Abraham de-Moivre (26 May 1667 – 27 November 1754) tried to generalize binomial distribution in the year 1733 to give a new distribution law that was later known as normal distribution. Since this type of distribution was first used in astronomical observations by the German mathematician-cum-physicist, Johann Carl Friedich Gauss (30 April 1777– 23 February 1855), it is also known as *Gaussian* distribution. In binomial distribution, when the number of trials (*N*) is very large, the distribution can be treated as a continuous distribution. Hence, normal or Gaussian distribution is a continuous distribution.

A normal distribution in a variable x with a mean m and a variance σ^2 is a statistical distribution with probability density function given by

$$P(x) = \frac{1}{\sigma\sqrt{2\pi}} e^{-\frac{(x-m)^2}{2\sigma^2}} \tag{14.34}$$

where e = 2.7183 is the base of natural logarithm and σ is the standard deviation.

A normal distribution curve also known as Gaussian curve is illustrated in Figure 14.3.

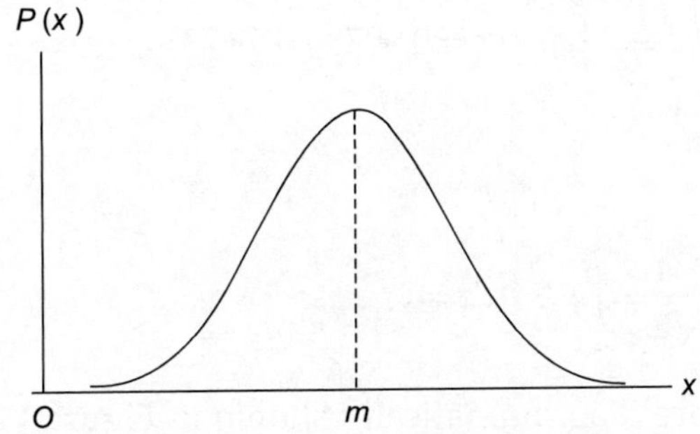

Fig. 14.3 A normal distribution curve.

Areas Under the Normal Distribution Curve

The probability of a continuous normal variable (x) to be found in a particular interval ($a \le x \le b$) is the area under the curve bounded by $x = a$ and $x = b$ and is given by

$$A = \int_a^b P(x)\,dx = \int_a^b \frac{1}{\sigma\sqrt{2\pi}} e^{-\frac{(x-m)^2}{2\sigma^2}}\,dx. \tag{14.35}$$

Putting $z = \dfrac{x-m}{\sigma}$, we have $dx = \sigma dz$. Then, we write

$$A = \frac{1}{\sqrt{2\pi}} \int_{\frac{a-m}{\sigma}}^{\frac{b-m}{\sigma}} e^{-z^2/2}\,dz. \tag{14.36}$$

1. Area between $m-\sigma$ and $m+\sigma$:

   In this case, $A_1 = \dfrac{1}{\sqrt{2\pi}} \displaystyle\int_{-1}^{+1} e^{-z^2/2}\,dz = 0.68269 = 68.27\%$.

2. Area between $m-2\sigma$ and $m+2\sigma$:

   In this case, $A_2 = \dfrac{1}{\sqrt{2\pi}} \displaystyle\int_{-2}^{+2} e^{-\frac{z^2}{2}}\,dz = 0.95450 = 95.45\%$.

3. Area between $m-3\sigma$ and $m+3\sigma$:

In this case, $A_3 = \frac{1}{\sqrt{2\pi}} \int_{-3}^{+3} e^{-\frac{z^2}{2}} dz = 0.99730 = 99.73\%$.

4. Whole area:

In this case, $A = \frac{1}{\sqrt{2\pi}} \int_{-\infty}^{+\infty} e^{-z^2/2} dz = 1$.

These area portions are diagrammatically shown in Figure 14.4.

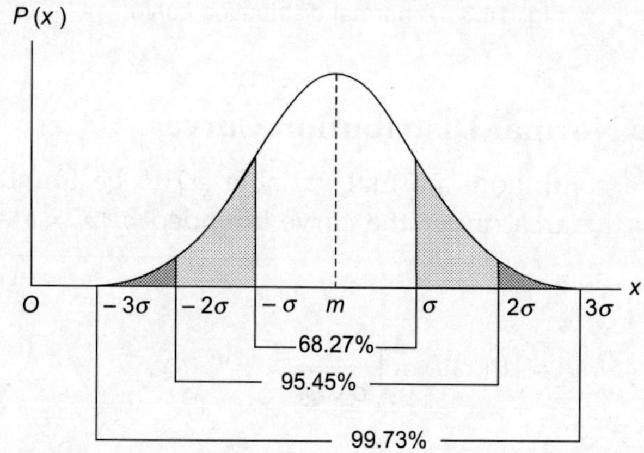

Fig. 14.4 Diagram showing areas of different portions.

To find the mean: We have

$$\text{Mean, } E(x) = \int_{-\infty}^{+\infty} xP(x)dx = \int_{-\infty}^{+\infty} x\left[\frac{1}{\sigma\sqrt{2\pi}} e^{-\frac{(x-m)^2}{2\sigma^2}} dx\right].$$

If we put $z = \frac{x-m}{\sigma}$, then $dx = \sigma\, dz$ and we write

$$E(x) = \frac{1}{\sqrt{2\pi}} \int_{-\infty}^{+\infty} (\sigma z + m) e^{-\frac{z^2}{2}} dz$$

$$= \frac{\sigma}{\sqrt{2\pi}} \int_{-\infty}^{+\infty} z e^{-z^2/2} dz + \frac{m}{\sqrt{2\pi}} \int_{-\infty}^{+\infty} e^{-z^2/2} dz.$$

Here, the integral in the first term is zero as the integrand is an odd function and the integral in the second term is $\sqrt{2\pi}$.

Hence, we obtain

$$\text{Mean, } E(x) = m. \tag{14.37}$$

To find the variance: The variance is given by

$$\operatorname{var}(x) = E\left[(x-m)^2\right] = \int_{-\infty}^{+\infty} (x-m)^2 P(x)\,dx$$

$$= \frac{1}{\sigma\sqrt{2\pi}} \int_{-\infty}^{+\infty} (x-m)^2 e^{-\frac{(x-m)^2}{2\sigma^2}}\,dx.$$

Putting $z = \frac{x-m}{\sigma}$, we have $x - m = \sigma z$ and $dx = \sigma\, dz$.

Hence, we write

$$\operatorname{var}(x) = \frac{\sigma^2}{\sqrt{2\pi}} \int_{-\infty}^{+\infty} z^2 e^{-z^2/2} dz = \frac{2\sigma^2}{\sqrt{2\pi}} \int_0^{\infty} z^2 e^{-z^2/2} dz.$$

Further, if we put $\frac{z}{\sqrt{2}} = u$, we have $z = \sqrt{2}\,u$ and $dz = \sqrt{2}\,du$.
Then, we write

$$\operatorname{var}(x) = \frac{2\sigma^2}{\sqrt{2\pi}} \int_0^{\infty} 2u^2 e^{-u^2} \sqrt{2}\,du = \frac{2\sigma^2}{\sqrt{\pi}} \Gamma\left(\frac{3}{2}\right) = \sigma^2. \tag{14.38}$$

A normal distribution is called a *standard normal distribution* if the mean is zero and the standard deviation is one (i.e., if $m = 0, \sigma = 1$).

Properties of Normal Distribution

Normal distribution has the following properties.

1. It requires only two parameters, namely, the mean and the variance (or standard deviation.
2. The total area under the normal distribution curve is unity.
3. The mean is at the middle dividing the area into two equal halves and hence the distribution curve is symmetric.
4. The mean, the median and the mode are equal.

(iii) Poisson Distribution

Poisson distribution is a discrete distribution that describes all random processes whose probability of occurrence is small but constant so that the product of sample size and the probability of occurrence is also constant. The Poisson distribution can also be considered a limiting case of binomial distribution for those random processes in which the probability of occurrence of an outcome is very small $(\text{i.e., } p \ll 1)$ while the number of trials (N) is very large and the mean value ($m = Np$) remains constant. This type of distribution was first proposed by the French mathematician Simeon – Denis Poisson (21 June 1781 – 25 April 1840) in the year 1837.

The probability distribution formula can be derived from that of binomial distribution as a limiting case. From binomial distribution, we have

$$P(x,Np) = \frac{N!}{x!(N-x)!}p^x(1-p)^{N-p}$$

$$= \frac{N(N-1)(N-2)(N-3)\ldots(N-x+1)}{x!}p^x(1-p)^{N-x}.$$

Using $p = \frac{m}{N}$, we write

$$P(x,N,p) = \frac{N(N-1)(N-2)(N-3)\ldots(N-x+1)}{x!}\left(\frac{m}{N}\right)^x\left(1-\frac{m}{N}\right)^{N-x}$$

$$= 1\left(1-\frac{1}{N}\right)\left(1-\frac{2}{N}\right)\left(1-\frac{3}{N}\right)\ldots\left(1-\frac{x-1}{N}\right)\left(1-\frac{m}{N}\right)^N\frac{m^x}{x!}\left(1-\frac{m}{N}\right)^{-x}.$$

In the limit $N \to \infty$ while m and p remain constant, we have

$$\lim_{N\to\infty} 1\left(1-\frac{1}{N}\right)\left(1-\frac{2}{N}\right)\left(1-\frac{3}{N}\right)\ldots\left(1-\frac{x-1}{N}\right) = 1$$

and

$$\lim_{N\to\infty}\left(1-\frac{m}{N}\right)^{-x} = 1.$$

Hence, we can write

$$P(x) = \frac{m^x}{x!}\left[\lim_{N\to\infty}\left(1-\frac{m}{N}\right)^N\right] = \frac{m^x}{x!}e^{-m}. \qquad (14.39)$$

To find mean and variance:

$$\text{Mean, } E(x) = \sum_{x=0}^{\infty} xP(x) = \sum_{x=0}^{\infty} x\frac{m^x}{x!}e^{-m}$$

$$= m\sum_{x=1}^{\infty}\frac{m^{x-1}}{(x-1)!}e^{-m} = m\sum_{y=0}^{\infty}\frac{m^y}{y!}e^{-m} \quad \text{where } y = x-1$$

$$= m, \text{ since } \sum_{y=0}^{\infty}\frac{m^y}{y!}e^{-m} = \text{sum of probabilities} = 1.$$

$$\text{var}(x) = \sum_{x=0}^{\infty}(x-m)^2 P(x) = \sum_{x=0}^{\infty}\left(x^2 - 2mx + m^2\right)\frac{m^x}{x!}e^{-m}$$

$$= \sum_{x=0}^{\infty} x^2 \frac{m^x}{x!}e^{-m} - 2m\sum_{x=0}^{\infty} x\frac{m^x}{x!}e^{-m} + m^2\sum_{x=0}^{\infty}\frac{m^x}{x!}e^{-m}$$

$$= \sum_{x=0}^{\infty}\left\{\left(x^2 - x\right) + x\right\}\frac{m^x}{x!}e^{-m} - (2m)m + m^2 \text{ (using the previous result)}$$

$$= \sum_{x=0}^{\infty} x(x-1)\frac{m^x}{x!}e^{-m} + \sum_{x=0}^{\infty} x\frac{m^x}{x!}e^{-m} - m^2$$

$$= m^2\sum_{x=2}^{\infty}\frac{m^{x-2}}{(x-2)!}e^{-m} + m - m^2$$

$$= m^2\sum_{y=0}^{\infty}\frac{m^y}{y!}e^{-m} + m - m^2$$

$$= m^2 + m - m^2 = m.$$

Standard deviation $= \sigma = \sqrt{\text{Variance}} = \sqrt{m}$.

Properties of Poisson Distribution

1. The mean and the variance are equal.
2. There is a single parameter in this distribution.
3. It is also known as *uni-parametric distribution*.

14.7 CURVE FITTING

Curve fitting is a problem that arises frequently in science and engineering. *Curve fitting* is a process of constructing a curve, or a mathematical relationship between the dependent variables and the independent variables that has the best fit to a series of data points. The objective of curve fitting is to theoretically describe experimental data with a model (function or equation) and to find the parameters associated with this model. It is used to find the "best fit" line or curve for a given series of data points. Broadly speaking, curve fitting is of two types, namely: (i) interpolation, and (ii) smoothing.

(i) Interpolation is the process of estimating other values of a function $f(x)$ from certain known values of the function, the estimated value being within the range of the known values. If the estimated value goes beyond the range of the known values, it is not interpolation but *extrapolation*.

If $x_0 < x_1 < \ldots < x_n$ and $y_0 = f(x_0), y_1 = f(x_1), \ldots y_n = f(x_n)$ are known values and if $x_0 < x < x_n$ the estimated value of $f(x)$ is said to be an interpolation whereas if $x < x_0$ or $x > x_n$ the estimated value of $f(x)$ is said to be an extrapolation. The various methods of interpolation will not be discussed in this context. As interpolation has limited applications in physical sciences, it will not be considered.

(ii) Smoothing refers to estimating a smooth trend, usually by means of weighted averages of observations. A particular simple class of curve fitting by smoothing is the *least squares curve fitting*. Of the different types of curve fitting, the method of least squares curve fitting is the simplest and it is the most suitable for applications to physical experiments. In drawing fitting curves, one should have a clear idea of error bars. In drawing experimental graphs, an *error bar* is the extent of uncertainty in the abscissa or ordinate value of a plotted point. If the coordinate values of a point are $x \pm \Delta x$ and $y \pm \Delta y$ then the line drawn from $x - \Delta x$ to $x + \Delta x$ is called the error bar in the direction of abscissa and the line drawn from $y - \Delta y$ to $y + \Delta y$ is called error bar in the direction of ordinate. The true point will then lie within the rectangle or square (or box) perpendicularly bisected by these two bars.

The *principle of least squares* is a principle that the most probable value of an observed quantity available from a set of observations is the one for which the sum of the squares of the residuals or errors is minimum. When a quantity is deduced from a series of observations, the *residuals* or *errors* will be the differences between the adopted value and each of the several observed values. The *least squares fitting* is a statistical procedure of finding the best fitting line or curve for a set of data points by minimizing the sum of the squares of the residuals or errors of points from the plotted curve. The *best fitting line* or *curve* is the line or curve that minimizes the sum of the squares of the errors or residuals.

General Procedure of Least Squares Fitting

For a set of N data points (x_i, y_i) with $i = 1, 2, 3 \ldots N$ plotted in xy-plane, let us consider a function $f(\alpha_1, \alpha_2, \alpha_3 \ldots \alpha_N, x)$ with N parameters $\alpha_1, \alpha_2, \alpha_3 \ldots \alpha_N$ such that the various y_i's will be well approximated by the function $f(\alpha_1, \alpha_2, \alpha_3 \ldots \alpha_N, x_i)$ Then, the residuals or errors are given by $r_i = y_i - f(\alpha_1, \alpha_2, \alpha_3 \ldots \alpha_N, x_i)$. The least squares fitting has the condition that the sum of the squares of the various residuals will be minimum, i.e.,

$$R^2 = \sum_{i=1}^{N} r_i^2 = \sum_{i=1}^{N} \left[y_i - f(\alpha_1, \alpha_2, \alpha_3 \ldots \alpha_N, x_i)\right]^2 \text{ is minimum.} \tag{14.40}$$

This means that the partial derivatives of the quantity R^2 defined above with respect to the parameters α_j will be zero, i.e.,

$$\frac{\partial(R^2)}{\partial\alpha_j} = -2\sum_{i=1}^{N}(y_i - f)\frac{\partial f}{\partial\alpha_j} = 0 \quad (j = 1, 2, 3 \ldots N). \tag{14.41}$$

Thus, we will obtain a system of N algebraic equations in the parameters, α_j.

Solving the parameters (α_j) from the system of equations and substituting them into the assumed function $f(\alpha_1, \alpha_2, \alpha_3 \ldots \alpha_N, x) = F(x)$ the best fitting curve or line is the graph of the function $F(x)$. If the assumed function (f) and hence $F(x)$ is a linear function represented by a straight line, then the curve fitting is said to be a *linear fitting* and if the assumed function is a nonlinear function represented by a curve, the curve fitting is said to be a *nonlinear fitting*.

Linear fitting: If the fitting line is a straight line, the fitting is said to be a linear fitting. In this case, the assumed function is the equation of a straight line say $f(\alpha_1, \alpha_2, x) = \alpha_1 x + \alpha_2$. Then, Eq. (14.40) yields

$$R^2 = \sum_{i=1}^{N}\left[y_i - (\alpha_1 x_i + \alpha_2)\right]^2 = \sum_{i=1}^{N}(y_i - \alpha_1 x_i - \alpha_2)^2 \tag{14.42}$$

where α_1 and α_2 are two parameters.

Using Eqs. (14.41) and (14.42), we have

$$\frac{\partial(R^2)}{\partial\alpha_1} = -2\sum_{i=1}^{N}(y_i - \alpha_1 x_i - \alpha_2)x_i = 0, \tag{14.43}$$

$$\frac{\partial(R^2)}{\partial\alpha_2} = -2\sum_{i=1}^{N}(y_i - \alpha_1 x_i - \alpha_2) = 0. \tag{14.44}$$

Equations (14.43) and (14.44) can be re written as

$$\left(\sum_{i=1}^{N} x_i^2\right)\alpha_1 + \left(\sum_{i=1}^{N} x_i\right)\alpha_2 = \sum_{i=1}^{N} x_i\, y_i \tag{14.45}$$

and

$$\left(\sum_{i=1}^{N} x_i\right)\alpha_1 + \left(\sum_{i=1}^{N} 1\right)\alpha_2 = \sum_{i=1}^{N} y_i. \tag{14.46}$$

In matrix form, we can write the above system of equations as

$$\begin{pmatrix} \sum_{i=1}^{N} x_i^2 & \sum_{i=1}^{N} x_i \\ \sum_{i=1}^{N} x_i & \sum_{i=1}^{N} 1 \end{pmatrix} \begin{pmatrix} \alpha_1 \\ \alpha_2 \end{pmatrix} = \begin{pmatrix} \sum_{i=1}^{N} x_i y_i \\ \sum_{i=1}^{N} y_i \end{pmatrix}. \tag{14.47}$$

Solving the above system of equations, we obtain

$$\alpha_1 = \frac{N\sum_{i=1}^{N} x_i y_i - \sum_{i=1}^{N} x_i \sum_{i=1}^{N} y_i}{N\sum_{i=1}^{N} x_i^2 - \left(\sum_{i=1}^{N} x_i\right)^2} = \frac{\sum_{i=1}^{N} x_i y_i - N\overline{xy}}{\sum_{i=1}^{N} x_i^2 - N(\overline{x})^2} = m = \text{slope} \tag{14.48}$$

where $\overline{x} = \frac{1}{N}\sum_{i=1}^{N} x_i$

and
$$\alpha_2 = \frac{\left(\sum_{i=1}^{N} y_i\right)\left(\sum_{i=1}^{N} x_i^2\right) - \left(\sum_{i=1}^{N} x_i\right)\left(\sum_{i=1}^{N} x_i y_i\right)}{N\sum_{i=1}^{N} x_i^2 - \left(\sum_{i=1}^{N} x_i\right)^2}$$

$$= \frac{\overline{y}\left(\sum_{i=1}^{N} x_i^2\right) - \overline{x}\left(\sum_{i=1}^{N} x_i y_i\right)}{\sum_{i=1}^{N} x_i^2 - N(\overline{x})^2} = c = \text{intercept on } y\text{-axis.} \tag{14.49}$$

Now, the best fit line is a straight line whose slope (m) is given by Eq. (14.48) and whose intercept on y-axis is given by Eq. (14.49). The best fit line will then be

$$y = mx + c = \alpha_1 x + \alpha_2. \tag{14.50}$$

Nonlinear fitting: In this case, the assumed function (f) is a nonlinear function but the procedure remains the same. Some cases of nonlinear curve fitting are discussed below.

(a) Polynomial Curve Fitting

In this case, let us take

$$f(\alpha_0, \alpha_1, \alpha_2 \ldots \alpha_N, x) = \alpha_0 + \alpha_1 x + \alpha_2 x^2 + \alpha_3 x^3 + \ldots + \alpha_N x^N$$

$$= \alpha_0 + \sum_{j=1}^{N} \alpha_j x^j. \tag{14.51}$$

As before, we must have

$$R^2 = \sum_{i=1}^{N} r_i^2 = \sum_{i=1}^{N}\left[y_i - f(\alpha_0, \alpha_1, \alpha_2 \ldots \alpha_N, x_i)\right]^2$$

$$= \sum_{i=1}^{N}\left[y_i - \left(\alpha_0 + \alpha_1 x_i + \alpha_2 x_i^2 + \alpha_3 x_i^3 + \ldots + \alpha_N x_i^N\right)\right]^2$$

$$= \sum_{i=1}^{N}\left[y_i - \left(\alpha_0 + \sum_{j=1}^{N}\alpha_j x_i^j\right)\right]^2. \tag{14.52}$$

Minimizing Eq. (14.52), we are to calculate the values of α_0, α_j. We must have

$$\frac{\partial(R^2)}{\partial\alpha_0} = -2\sum_{i=1}^{N}\left[y_i - \left(\alpha_0 + \sum_{j=1}^{N}\alpha_j x_i^j\right)\right] = 0,$$

$$\frac{\partial(R^2)}{\partial\alpha_1} = -2\sum_{i=1}^{N}\left[y_i - \left(\alpha_0 + \sum_{j=1}^{N}\alpha_j x_i^j\right)\right]x_i = 0,$$

$$\frac{\partial(R^2)}{\partial\alpha_2} = -2\sum_{i=1}^{N}\left[y_i - \left(\alpha_0 + \sum_{j=1}^{N}\alpha_j x_i^j\right)\right]x_i^2 = 0,$$

$$\vdots \quad = \quad \vdots \qquad \vdots \qquad = \vdots$$

$$\frac{\partial(R^2)}{\partial\alpha_N} = -2\sum_{i=1}^{N}\left[y_i - \left(\alpha_0 + \sum_{j=1}^{N}\alpha_j x_i^j\right)\right]x_i^N = 0.$$

From the above system of equations, we write

$$\left(\sum_{i=1}^{N}1\right)\alpha_0 + \left(\sum_{i=1}^{N}x_i\right)\alpha_1 + \left(\sum_{i=1}^{N}x_i^2\right)\alpha_2 \ldots + \left(\sum_{i=1}^{N}x_i^N\right)\alpha_N = \sum_{i=1}^{N}y_i,$$

$$\left(\sum_{i=1}^{N}x_i\right)\alpha_0 + \left(\sum_{i=1}^{N}x_i^2\right)\alpha_1 + \left(\sum_{i=1}^{N}x_i^3\right)\alpha_2 \ldots + \left(\sum_{i=1}^{N}x_i^{N+1}\right)\alpha_N = \sum_{i=1}^{N}x_i y_i,$$

$$\left(\sum_{i=1}^{N}x_i^2\right)\alpha_0 + \left(\sum_{i=1}^{N}x_i^3\right)\alpha_1 + \left(\sum_{i=1}^{N}x_i^4\right)\alpha_2 \ldots + \left(\sum_{i=1}^{N}x_i^{N+2}\right)\alpha_N = \sum_{i=1}^{N}x_i^2 y_i,$$

$$\vdots \qquad \vdots \qquad \vdots \qquad \vdots = \qquad \vdots$$

$$\left(\sum_{i=1}^{N}x_i^N\right)\alpha_0 + \left(\sum_{i=1}^{N}x_i^{N+1}\right)\alpha_1 + \left(\sum_{i=1}^{N}x_i^{N+2}\right)\alpha_2 + \left(\sum_{i=1}^{N}x_i^{N+N}\right)\alpha_N = \sum_{i=1}^{N}x_i^N y_i.$$

This system of equations can be written in matrix form as:

$$\begin{pmatrix} N & \sum_{i=1}^{N} x_i & \sum_{i=1}^{N} x_i^2 & \dots & \sum_{i=1}^{N} x_i^N \\ \sum_{i=1}^{N} x_i & \sum_{i=1}^{N} x_i^2 & \sum_{i=1}^{N} x_i^3 & \dots & \sum_{i=1}^{N} x_i^{N+1} \\ \sum_{i=1}^{N} x_i^2 & \sum_{i=1}^{N} x_i^3 & \sum_{i=1}^{N} x_i^4 & \dots & \sum_{i=1}^{N} x_i^{N+2} \\ \vdots & \vdots & \vdots & \vdots & \vdots \\ \sum_{i=1}^{N} x_i^N & \sum_{i=1}^{N} x_i^{N+1} & \sum_{i=1}^{N} x_i^{N+2} & \dots & \sum_{i=1}^{N} x_i^{2N} \end{pmatrix} \begin{pmatrix} \alpha_0 \\ \alpha_1 \\ \alpha_2 \\ \vdots \\ \alpha_N \end{pmatrix} = \begin{pmatrix} \sum_{i=1}^{N} y_i \\ \sum_{i=1}^{N} x_i y_i \\ \sum_{i=1}^{N} x_i^2 y_i \\ \vdots \\ \sum_{i=1}^{N} x_i^N y_i \end{pmatrix}.$$

Now, we write

$$\begin{pmatrix} \alpha_0 \\ \alpha_1 \\ \alpha_2 \\ \vdots \\ \alpha_N \end{pmatrix} = Inv \left[\begin{pmatrix} N & \sum_{i=1}^{N} x_i & \sum_{i=1}^{N} x_i^2 & \dots & \sum_{i=1}^{N} x_i^N \\ \sum_{i=1}^{N} x_i & \sum_{i=1}^{N} x_i^2 & \sum_{i=1}^{N} x_i^3 & \dots & \sum_{i=1}^{N} x_i^{N+1} \\ \sum_{i=1}^{N} x_i^2 & \sum_{i=1}^{N} x_i^3 & \sum_{i=1}^{N} x_i^4 & \dots & \sum_{i=1}^{N} x_i^{N+2} \\ \vdots & \vdots & \vdots & \dots & \vdots \\ x_i^N & x_i^{N+1} & \sum_{i=1}^{N} x_i^{N+2} & \dots & \sum_{i=1}^{N} x_i^{2N} \end{pmatrix} \right] \begin{pmatrix} \sum_{i=1}^{N} y_i \\ \sum_{i=1}^{N} x_i y_i \\ \sum_{i=1}^{N} x_i^2 y_i \\ \vdots \\ \sum_{i=1}^{N} x_i^N y_i \end{pmatrix}. \qquad (14{:}53)$$

Then, we can find the values of $\alpha_0, \alpha_1, \alpha_2 \dots \alpha_N$.
In this way, we can find the fitting curve.

(b) Line Fitting of Certain Nonlinear Functions by Linearization

(i) Exponential:

In such a case, without linearization, we can perform the curve fitting if we take

$$f(k_1, k_2, x) = k_2\, e^{k_1 x}.$$

But, it will be simpler if we linearize by taking logarithm on both sides to give

$$\ln f = \ln k_2 + k_1 x.$$

The fitting curve will be given by

$$\ln y = \alpha_1 x + \alpha_2 \text{ with } \alpha_1 = k_1, \alpha_2 = \ln k_2$$

where α_1 and α_2 are given by Eqs. (14.48) and (14.49).

Thus, we have

$$\alpha_1 = k_1 = \frac{N\sum_{i=1}^{N} x_i \ln y_i - \sum_{i=1}^{N} x_i \sum_{i=1}^{N} \ln y_i}{N\sum_{i=1}^{N} x_i^2 - \left(\sum_{i=1}^{N} x_i\right)^2} \tag{14.54}$$

and
$$\alpha_2 = \ln k_2 = \frac{\left(\sum_{i=1}^{N} \ln y_i\right)\left(\sum_{i=1}^{N} x_i^2\right) - \left(\sum_{i=1}^{N} x_i\right)\left(\sum_{i=1}^{N} x_i \ln y_i\right)}{N\sum_{i=1}^{N} x_i^2 - \left(\sum_{i=1}^{N} x_i\right)^2}. \tag{14.55}$$

Now, the fitting curve will be the graph of the equation

$$\ln y = \alpha_1 x + \alpha_2. \tag{14.56}$$

This equation will yield a straight line if $(\ln y)$ is plotted along y-axis and, thus, we get a linearized fitting.

(ii) Logarithmic:

If the fitting curve is the graph given by the equation $e^{y-a} = x^b$ or, $y = a + b\ln x$, we have

$$\alpha_1 = b = \frac{N\sum_{i=1}^{N} y_i \ln x_1 - \sum_{i=1}^{N} \ln x_i \sum_{i=1}^{N} y_i}{N\sum_{i=1}^{N} (\ln x_i)^2 - \left(\sum_{i=1}^{N} \ln x_2\right)^2}, \tag{14.57}$$

$$\alpha_2 = a = \frac{\left(\sum_{i=1}^{N} y_i\right)\left\{\sum_{i=1}^{N} (\ln x_i)^2\right\} - \left(\sum_{i=1}^{N} \ln x_i\right)\cdot\left(\sum_{i=1}^{N} y_i \ln x_i\right)}{N\sum_{i=1}^{N} (\ln x_i)^2 - \left(\sum_{i=1}^{N} \ln x_i\right)^2}. \tag{14.58}$$

If $\ln x$ is plotted along x-axis, the fitting curve will become a straight line and, thus, the fitting is said to be linearized.

(iii) Power law:

If the fitting curve is the graph of the equation $y = Bx^A$,

we have $\ln y = A\ln x + \ln B$.

Here, A and B can be calculated as usual.

If ln x is plotted along the abscissa and ln y along the ordinate, we will have a straight line graph and we have a linearized fitting.

14.8 ERRORS IN THE SLOPE AND INTERCEPT OF A FITTED LINE (OF LINEAR FITTING)

After drawing a best fit line on the graph, in order to determine the error in the slope and the error in the intercept, one is required to draw the maximum and the minimum line of the given data.

The *maximum line* is the straight line graph having the greatest slope and passing through the right bottom of the error box of the first point in the graph and the left top of the error box of the last point in the graph.

The *minimum line* is the straight line graph having the smallest slope and passing through the left top of the error box of the first point in the graph and the right bottom of the error box of the last point in the graph.

The slopes of the maximum and the minimum lines are given by

$$\text{Max. slope} = \left(\frac{\Delta y}{\Delta x}\right)_{\text{max}},$$

$$\text{Min. slope} = \left(\frac{\Delta y}{\Delta x}\right)_{\text{min}}.$$

The half of the difference between the maximum and the minimum slopes is called the error in the slope. An illustration is presented in Fig. 14.5.

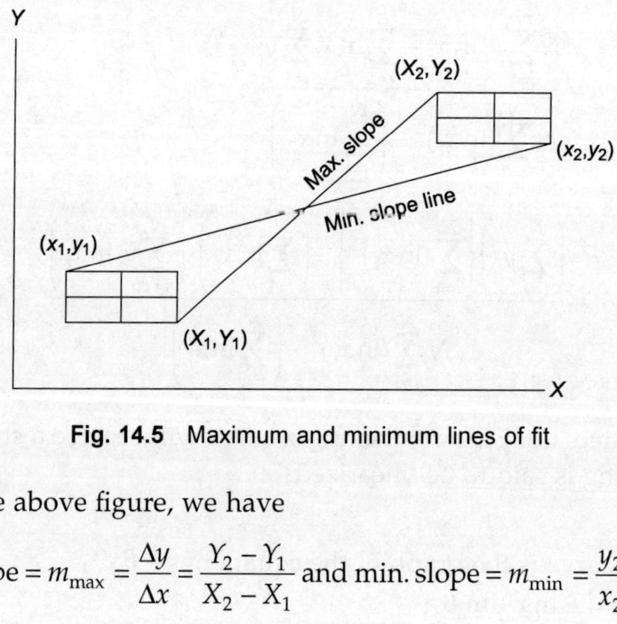

Fig. 14.5 Maximum and minimum lines of fit

Now, from the above figure, we have

$$\text{Max. slope} = m_{\text{max}} = \frac{\Delta y}{\Delta x} = \frac{Y_2 - Y_1}{X_2 - X_1} \text{ and min. slope} = m_{\text{min}} = \frac{y_2 - y_1}{x_2 - x_1}. \qquad (14.59)$$

Then, error in slope, $\Delta m = \pm\frac{1}{2}(m_{max} - m_{min})$

$$= \pm\frac{1}{2}\left(\frac{Y_2 - Y_1}{X_2 - X_1} - \frac{y_2 - y_1}{x_2 - x_1}\right). \tag{14.60}$$

The intercepts on the maximum line and the minimum line can be read from the graphs and let c_{max} and c_{min}, be the respective values. Alternatively, these intercepts can also be found from the equation of straight line by writing in the form

$$\text{Intercept} = c = y - mx.$$

Here, c may be c_{max} or c_{min}, m may be m_{max} or m_{min}, the values of x and y are to be taken from the points of the line in use.

Then, the error in intercept is given by

$$\Delta c = \pm\frac{1}{2}(c_{max} - c_{min}). \tag{14.61}$$

14.9 GOODNESS OF FIT/CHI-SQUARED TEST

The *goodness of fit* of a statistical model is a measure of how well it fits a set of observations to the theoretically expected values. It measures the discrepancy between observed values and the values expected theoretically in the model under consideration. A chi-squared (χ^2) test is a test that measures how expectations compare with actual observations. It was first introduced in the year 1900 by the English mathematician, Karl Pearson (27 March 1857-27 April 1936). The chi-squared test allows evaluation of the probability that a given data set follows Poisson distribution. It is a test for goodness of fit. Chi-square is a quantity defined by

$$\chi^2 = \sum_{i=1}^{N}\frac{(O_i - E_i)^2}{E_i} = \sum_{i=1}^{N}\frac{(x_i - \bar{x})^2}{\bar{x}} = \frac{\sum_{i=1}^{N}(x_i - \bar{x})^2}{\bar{x}} \tag{14.62}$$

where O_i is an observed frequency, E_i is an expected (theoretical) frequency, and N is the number of observations (or measurements). The *null hypothesis is* a hypothesis which states that here is no relationship between the variables being studied (one variable does not affect the other). The results are due to chance and are not significant in terms of supporting the idea being investigated. It is denoted by H_0. An *alternative hypothesis* is one which opposes the null hypothesis. It is denoted by H_a. A low probability value demands rejection of null hypothesis. The probability value below which the null hypothesis is rejected is called α (alpha) level. For a particular degree of freedom, the curve obtained by plotting probability values against the chi-square values is known as a chi-square curve.

Using the formula for chi-square given above, the value of chi-square is calculated from the experimental data. Then, from a probability chi-square table, the probability p corresponding to the calculated chi-square value is found out. If p lies within the range $0.1 \le p \le 0.9$ the null hypothesis is to be accepted. If the value p lies outside this range, the null hypothesis is to be rejected.

The steps in performing a Pearson's chi-square test to counting data are stated as in the following.

1. The given data is to be divided into at least five class intervals, each interval containing at least five numbers (counts).
2. The mean (m) or average $(\bar{x})$ is to be calculated.
3. The value of χ^2 (chi-square) is to be calculated using the formula for it.
4. The degree of freedom is to be found out (usually $F = N - 1$).
5. The calculated value of chi-square is to be compared with those in a chi-square table. If this value of chi-square corresponds to the probability range $0.1 \le P \le 0.9$ the null hypothesis is very likely true and if the value corresponds to probability outside the above range, the null hypothesis is very unlikely.

A chi-square table is shown below. Here, d.f. stands for degree of freedom, the numbers in the row between two lines on the top of table shows a range of probabilities and the numbers inside the table are the values of chi-square in the probability range shown above.

Chi-Square Table

df \ p	**0.010**	**0.025**	**0.050**	**0.100**	**0.200**	**0.300**	**0.500**	**0.700**	**0.800**	**0.900**	**0.950**	**0.975**	**0.990**
1	6.63	5.02	3.84	2.71	1.64	1.07	0.45	0.15	0.06	0.02	0.00	0.00	0.00
2	9.21	7.38	5.99	4.61	3.22	2.41	1.39	0.71	0.45	0.21	0.10	0.05	0.02
3	11.34	9.35	7.81	6.25	4.64	3.66	2.37	1.42	1.01	0.58	0.35	0.22	0.11
4	13.28	11.14	9.49	7.78	5.99	4.88	3.36	2.19	1.65	1.06	0.71	0.48	0.30
5	15.09	12.83	11.07	9.24	7.29	6.06	4.35	3.00	2.34	1.61	1.15	0.83	0.55
6	16.81	14.45	12.59	10.64	8.56	7.23	5.35	3.83	3.07	2.20	1.64	1.24	0.87
7	18.48	16.01	14.07	12.02	9.80	8.38	6.35	4.67	3.82	2.83	2.17	1.69	1.24
8	20.09	17.53	15.51	13.36	11.03	9.52	7.34	5.53	4.59	3.49	2.73	2.18	1.65
9	21.67	19.02	16.92	14.68	12.24	10.66	8.34	6.39	5.38	4.17	3.33	2.70	2.09
10	23.21	20.48	18.31	15.99	13.44	11.78	9.34	7.27	6.18	4.87	3.94	3.25	2.56
16	32.00	28.85	26.30	23.54	20.47	18.42	15.34	12.62	11.15	9.31	7.96	6.91	5.81
17	33.41	30.19	27.59	24.77	21.61	19.51	16.34	13.53	12.00	10.09	8.67	7.56	6.41

18	34.81	31.53	28.87	25.99	22.76	20.60	17.34	14.44	12.86	10.86	9.39	8.23	7.01
19	36.19	32.85	30.14	27.20	23.90	21.69	18.34	15.35	13.72	11.65	10.12	8.91	7.63
20	37.57	34.17	31.41	28.41	25.04	22.77	19.34	16.27	14.58	12.44	10.85	9.59	8.26
21	38.93	35.48	32.67	29.62	26.17	23.86	20.34	17.18	15.44	13.24	11.59	10.28	8.90
22	40.29	36.78	33.92	30.81	27.30	24.94	21.34	18.10	16.31	14.04	12.34	10.98	9.54
23	41.64	38.08	35.17	32.01	28.43	26.02	22.34	19.02	17.19	14.85	13.09	11.69	10.20
24	42.98	39.36	36.42	33.20	29.55	27.10	23.34	19.94	18.06	15.66	13.85	12.40	10.86
25	44.31	40.65	37.65	34.38	30.68	28.17	24.34	20.87	18.94	16.47	14.61	13.12	11.52
30	50.89	46.98	43.77	40.26	36.25	33.53	29.34	25.51	23.36	20.60	18.49	16.79	14.95
31	52.19	48.23	44.99	41.42	37.36	34.60	30.34	26.44	24.26	21.43	19.28	17.54	15.66
32	53.49	49.48	46.19	42.58	38.47	35.66	31.34	27.37	25.15	22.27	20.07	18.29	16.36
33	54.78	50.73	47.40	43.75	39.57	36.73	32.34	28.31	26.04	23.11	20.87	19.05	17.07
34	56.06	51.97	48.60	44.90	40.68	37.80	33.34	29.24	26.94	23.95	21.66	19.81	17.79
35	57.34	53.20	49.80	46.06	41.78	38.86	34.34	30.18	27.84	24.80	22.47	20.57	18.51
40	63.69	59.34	55.76	51.81	47.27	44.16	39.34	34.87	32.34	29.05	26.51	24.43	22.16
41	64.95	60.56	56.94	52.95	48.36	45.22	40.34	35.81	33.25	29.91	27.33	25.21	22.91
42	66.21	61.78	58.12	54.09	49.46	46.28	41.34	36.75	34.16	30.77	28.14	26.00	23.65
43	67.46	62.99	59.30	55.23	50.55	47.34	42.34	37.70	35.07	31.63	28.96	26.79	24.40
44	68.71	64.20	60.48	56.37	51.64	48.40	43.34	38.64	35.97	32.49	29.79	27.57	25.15
45	69.96	65.41	61.66	57.51	52.73	49.45	44.34	39.58	36.88	33.35	30.61	28.37	25.90
46	71.20	66.62	62.83	58.64	53.82	50.51	45.34	40.53	37.80	34.22	31.44	29.16	26.66
47	72.44	67.82	64.00	59.77	54.91	51.56	46.34	41.47	38.71	35.08	32.27	29.96	27.42
48	73.68	69.02	65.17	60.91	55.99	52.62	47.34	42.42	39.62	35.95	33.10	30.75	28.18
49	74.92	70.22	66.34	62.04	57.08	53.67	48.33	43.37	40.53	36.82	33.93	31.55	28.94

14.10 SOLVED PROBLEMS

PROBLEM 1 Directly measuring two quantities x and y, one is to find a quantity $Q = 5xy$. Find the error in Q.

Solution: Error in Q is

$$\Delta Q = 5xy\left(\frac{\Delta x}{x}+\frac{\Delta y}{y}\right) = Q\left(\frac{\Delta x}{x}+\frac{\Delta y}{y}\right).$$

PROBLEM 2 In measuring the side of a cube of sides 3 cm, there exists an error ± 0.02 cm. Find the error in finding the volume of the cube and hence find the volume of the cube with the error.

Solution: We denote a side by L. Then, the volume is given by $V = L^3$.

The error in finding V is $\Delta V = 3L^2 \Delta L = 3 \times 3^2 \times 0.02 \text{ cm}^3 = 0.54 \text{ cm}^3$.

The volume of the cube with error is $27 \pm 0.54 \text{ cm}^3$.

PROBLEM 3 In an examination, 20 students of physics secure marks as detailed below.

53, 55, 56, 56, 58, 58, 59, 60, 61, 61, 62, 62, 62, 64, 65, 65, 68, 68, 69, 70.

(a) Working with ungrouped data, find the mean, median, mode, variance, mean deviation and standard deviation.

(b) Working with grouped data, find the quantities mentioned above.

Solution: Let us rewrite the data as

53, 55, 56, 56, 58, 58, 59, 60, 61, 61, 62, 62, 62, 64, 65, 65, 68, 68, 69,70.

We denote the individual marks by x_i the frequency by f_i and the number of individual marks by N.

(a) From the above data, we make an ungrouped table as the following.

Value x_i	Frequency f_i	$f_i x_i$	Mean $\overline{x}$	$\lvert x_i - \overline{x}\rvert$	$f_i \lvert x_i - \overline{x}\rvert$	$f_i (x_i - \overline{x})^2$
53	1	53		8.6	8.6	73.96
55	1	55		6.6	6.6	43.56
56	2	112		5.6	11.2	62.72
58	2	116		3.6	7.2	25.92
59	1	59		2.6	2.6	6.76
60	1	60		1.6	1.6	2.56
61	2	122	61.6	0.6	1.2	0.72
62	3	186		0.4	1.2	0.48
64	1	64		2.4	2.4	5.76
65	2	130		3.4	6.8	23.12
68	2	136		6.4	12.8	81.92
69	1	69		7.4	7.4	54.76
70	1	70		8.4	8.4	70.56
	$\Sigma f_i = N = 20$	$\Sigma f_i x_i = 1232$	$\overline{x} = \sum_i \frac{f_i x_i}{N}$		$\sum f_i \lvert x_i - \overline{x}\rvert = 78$	$\sum f_i (x_i - \overline{x})^2 = 452.8$

Now, we have

$$\text{Mean } = \overline{x} = \frac{1232}{20} = 61.6.$$

Here, $$N = 20(\text{even}).$$

$$\text{Median } = \left(\frac{N+2}{2}\right)\text{th number in increasing order} = 11\text{th number} = 62.$$

Mode = Number with highest frequency = 62.

$$\text{Mean Dev. } = \frac{\sum_i f_i |x_i - \overline{x}|}{N} = \frac{78}{20} = 3.9.$$

$$\text{Variance} = \sum_i \frac{f_i (x_i - \overline{x})^2}{N} = \frac{452.8}{20} = 22.64.$$

$$\text{The standard deviation } = \sigma = \sqrt{\text{Variance}} = \sqrt{22.64} = 4.76.$$

(b) Let us make grouped data table as the following.

Class interval	**Mid-point** x_i	**Frequency** f_i	$f_i x_i$	**Estimated mean** $\overline{x}$	$\lvert x_i - \overline{x}\rvert$	$f_i \lvert x_i - \overline{x}\rvert$	$f_i (x_i - \overline{x})^2$
51 – 55	53	2	106		8.5	17	144.5
56 – 60	58	6	348	61.5	3.5	21	73.5
61 – 65	63	8	504		1.5	12	18.0
66 – 70	68	4	272		6.5	26	169.0
		$N = \Sigma f_i = 20$	$\sum_i f_i x_i = 1230$	$\overline{x} = \frac{\sum_i f_i x_i}{N} = 61.5$		$\Sigma f_i \lvert x_i - \overline{x}\rvert = 76$	$\sum_i f_i (x_i - \overline{x})^2 = 405$

$$\text{Estimated mean, } \overline{x} = \frac{\sum_i f_i x_i}{N} = \frac{1230}{20} = 61.5.$$

$$\text{Estimated mean dev. } \sum_i \frac{f_i |x_i - \overline{x}|}{N} = \frac{76}{20} = 3.8.$$

$$\text{Estimated variance} = \frac{\sum_i f_i (x_i - \bar{x})^2}{N} = \frac{405}{20} = 20.25.$$

$$\text{Estimated standard deviation, } \sigma = \sqrt{\text{Variance}} = \sqrt{20.25} = 4.5.$$

Here, median class is the $\left(\frac{4+2}{2}\right)$th class, i.e., $61-65$.

$$\text{Estimated median, } L + \left(\frac{\frac{N}{2} - C}{f}\right) w = 61 + \frac{\left\{\frac{20}{2} - (6+2)\right\} 5}{8} = 62.25.$$

$$\text{Estimated mode, } L + \frac{f_m - f_{m-1}}{(f_m - f_{m-1}) + (f_m - f_{m+1})} \times w$$

$$= 61 + \frac{8-6}{8-6+8-4} \times 5 = 62.67.$$

PROBLEM 4 For a specific purpose, a committee having 25 members is formed with persons having the respective ages of 61, 63, 63, 63, 64, 64, 66, 66, 66, 66, 69, 70, 70 ,70, 70, 70, 70, 70, 72, 72, 74, 74, 75, 75, 75 years.

(a) Working with ungrouped data, find the mean, median, mode, variance, and standard deviation.

(b) Working with grouped data, find the quantities mentioned above.

(a) **Ungrouped data:**

Let us rewrite the data as 61, 63, 63, 63, 64, 64, 66, 66, 66, 66, 69, 70, 70 ,70, 70, 70, 70, 70, 72, 72, 74, 74, 75, 75, 75.

Here, $N = 25$ (odd).

Median $= \left(\frac{N+1}{2}\right)$th number in increasing order = 13th number = 70.
Mode = 70.

Let us make an ungrouped data table as the following.

x_i	f_i	$f_i x_i$	$\bar{x}$	$\lvert x_i - \bar{x}\rvert$	$f_i \lvert x_i - \bar{x}\rvert$	$f_i (x_i - \bar{x})^2$
61	1	61		7.72	7.72	59.5984
63	3	189		5.72	17.16	98.1552
64	2	128		4.72	9.44	44.5568
66	4	264		2.72	10.88	29.5936
69	1	69		0.28	0.28	0.0784
70	7	490	68.72	1.28	8.96	11.4688
72	2	144		3.28	6.56	21.5168
74	2	148		5.28	10.56	55.7568
75	3	225		6.28	18.84	118.3152
	$N = \Sigma f_i = 25$	$\Sigma f_i x_i = 1718$	$\frac{1718}{25} = 68.72$		$\Sigma f_i \lvert x_i - \bar{x}\rvert = 90.4$	$\Sigma f_i (x_i - \bar{x})^2 = 439.04$

Now, we have

$$\text{Mean, } \bar{x} = \frac{\sum_i f_i x_i}{N} = \frac{1718}{25} = 68.72.$$

Mode, Number with highest frequency = 70.

$$\text{Mean Dev. } = \frac{\sum_i f_i \lvert x_i - \bar{x}\rvert}{N} = \frac{90.4}{25} = 3.616.$$

$$\text{Variance} = \sum_i \frac{f_i (x_i - \bar{x})^2}{N} = \frac{439.04}{25} = 17.5616.$$

$$\text{Standard deviation, } \sigma = \sqrt{\text{Variance}} = \sqrt{17.5616} = 4.19.$$

(b) Grouped data:

Let us make a grouped data table as the following.

Class interval	Mid pt. x_i	Freq. f_i	$f_i x_i$	Estimated mean $\overline{x}$	$\lvert x_i - \overline{x}\rvert$	$f_i \lvert x_i - \overline{x}\rvert$	$f_i (x_i - \overline{x})^2$
61 – 65	63	6	378		5.2	31.2	162.24
66 – 70	68	12	816	68.2	0.2	2.4	0.48
71 – 75	73	7	511		4.8	33.6	161.28
		$N = \Sigma f_i = 25$	$\Sigma f_i x_i = 1705$	$\overline{x} = \dfrac{\sum f_i x_i}{N} = 68.2$		$\Sigma f_i \lvert x_i - \overline{x}\rvert = 67.2$	$\Sigma f_i (x_i - \overline{x})^2 = 324$

$$\text{Estimated mean, } \quad \frac{\sum f_i x_i}{N} = \frac{1705}{25} = 68.2.$$

$$\text{Estimated median, } L + \left(\frac{\frac{N}{2} - C}{f}\right) w = 66 + \frac{\frac{25}{2} - 6}{12} \times 5 = 68.71.$$

$$\text{Estimated mode} = L + \frac{f_m - f_{m-1}}{(f_m - f_{m-1}) + (f_m - f_{m+1})} \times w$$

$$= 66 + \frac{12 - 6}{12 - 6 + 12 - 7} \times 5 = 68.73.$$

$$\text{Estimated mean dev.} = \frac{\sum_i f_i \lvert x_i - \overline{x}\rvert}{N} = \frac{67.2}{25} = 2.688.$$

$$\text{Estimated variance} = \sum_i \frac{f_i (x_i - \overline{x})^2}{N} = \frac{324}{25} = 12.96.$$

$$\text{Estimated standard deviation, } \sigma = \sqrt{\text{Variance}} = \sqrt{12.96} = 3.6.$$

PROBLEM 5 In binomial distribution, prove that the most probable number of successes of N independent trials is $[(N+1)p]$, where the symbol $[k]$ denotes the greatest integral value less than k.

Solution: The chances of x successes to be most probable, it will not be less than that of $(x-1)$ or $(x+1)$ successes. Thus, we have

$$\frac{N!}{(x-1)!(N-x+1)!}p^{x-1}q^{N-x+1} \le \frac{N!}{x!\,(N-x)!}p^{x}q^{N-x} \ge \frac{N!}{(x+1)!(N-x-1)!}p^{x+1}q^{N-x-1}$$

or

$$\frac{q}{(N-x)(N-x+1)p} \le \frac{1}{x(N-x)} \ge \frac{p}{x(x+1)q}.$$

Thus, we have

$$qx \le (N-x+1)p \text{ or, } (p+q)x \le (N+1)p$$

and

$$q(x+1) \ge (N-x)p \text{ or, } (p+q)x \ge Np-q.$$

As $p+q=1$, we write

$$x \le (N+1)p,$$

$$x \ge Np-(1-p), \text{ i.e., } x \ge (N+1)p-1.$$

We can write as

$$(N+1)p-1 \le x \le (N+1)p.$$

Hence, the most probable number of successes is the greatest integral value less than $(N+1)p$.

PROBLEM 6 In a bike race of 20 bike riders, the probability that a bike will face accident is 0.05. Find the probability that one bike rider faces accident.

Solution: Here, number of trials, $N = 20$. Probability of failure in a single trial, $q = 0.05$

Probability of success in a single trial, $p = 1 - q = 0.95$.

One accident means one failure and (20 – 1) successes.

Therefore, probability of one failure = Probability of 19 successes.

Using binomial distribution, the required probability is given by

$$P(19) = {}^{20}C_{19}\,p^{19}q^{1} = \frac{20!}{19!\,1!}(0.95)^{19}(0.05)^{1} = 0.3774.$$

Thus, probability that a bike rider will face accident is 0.3774.

PROBLEM 7 An electrician is performing electric fitting of a building. He purchases a packet of three dozens of electric switches and fits them. If the defective rate of the switches is 10%, find the probability that (a) all the switches work, (b) three of them do not work, and (c) at most two of them do not work.

Solution: We are to use binomial distribution.

Here, $q = 10\% = 0.1, p = 1 - q = 0.9.$

(a) The probability that all the 36 switches work is given by

$$P(36) = {}^{36}C_{36}\, p^{36} q^0 = \frac{36!}{36!\,0!}(0.9)^{36}(0.1)^0 = 0.0225.$$

(b) The probability that three of the switches do not work is given by

$$P(33) = {}^{36}C_{33}\, p^{33} q^3 = \frac{36!}{33!\,3!}(0.9)^{33}(0.1)^3 = 0.2206.$$

(c) In this case, 34 or 35 or 36 switches work.

The required probability is given by

$$P(x \geq 34) = P(34) + P(35) + P(36)$$

$$= {}^{36}C_{34}\, p^{34} q^2 + {}^{36}C_{35}\, p^{35} q^1 + {}^{36}C_{36}\, p^{36} q^0$$

$$= \frac{36!}{34!\,2!}(0.9)^{34}(0.1)^2 + \frac{36!}{35!\,1!}(0.9)^{35}(0.1)^1 + \frac{36!}{36!\,0!}(0.9)^{36}(0.1)^0$$

$$= 0.1752 + 0.0901 + 0.0225 = 0.2878.$$

PROBLEM 8 In a city, 95 per cent of the height of the people ranges from 1.1 metres to 1.9 metres. Assuming a normal distribution, calculate the mean height of the people and also find the standard deviation.

Solution: Here, mean height $= \frac{1.1\text{ m} + 1.9\text{ m}}{2} = 1.5\text{ m}$.

In the normal distribution curve, 95% of area is 2 standard deviations (S.D.) away on either side of the mean.

Therefore, we have, $4\text{ S.D.} = 1.9\text{ m} - 1.1\text{ m} = 0.8\text{ m}$

or $$1\text{ S.D.} = \frac{0.8\text{ m}}{4} = 0.2\text{ m}.$$

PROBLEM 9 If x and y are independent random variables obeying Poisson distributions with parameters (means) m and μ respectively, prove that the sum $(x + y)$ has also a Poisson distribution with parameter $(m + \mu)$.

Solution: By the law of total probability, we have

$$P(x+y=N)=\sum_{k=0}^{N}P(x=k,y=N-k)$$

$$=\sum_{k=0}^{N}P(x=k)P(y=N-k)$$

$$=\sum_{k=0}^{N}\left[\frac{m^x}{x!}\times e^{-m}\right]\left[\frac{\mu^y}{y!}\times e^{-\mu}\right]$$

$$=\sum_{k=0}^{N}\left[\frac{m^k}{k!}\times e^{-m}\right]\left[\frac{\mu^{N-k}}{(N-k)!}\times e^{-\mu}\right]$$

$$=\frac{e^{-(m+\mu)}}{N!}\sum_{k=0}^{N}\frac{N!}{k!(N-k)!}m^k\mu^{N-k}$$

$$=\frac{e^{-(m+\mu)}}{N!}(m+\mu)^N=P(x+y=N).$$ Hence, proved.

PROBLEM 10 A gamma ray source emits on an average 4 gamma photons in 5 seconds. Find the probability that exactly 3 gamma photons are emitted during that interval of time.

Solution: Using Poisson distribution, the probability is given by

$$P(x)=\frac{m^x}{x!}e^{-m}$$

where m denotes mean or average.

Therefore, $$P(x=3)=\frac{4^3}{3!}\times e^{-4}=0.1954.$$

PROBLEM 11 In type-writing a document of several pages, mistakes occur in accordance with Poisson distribution at an average rate of 10 per cent per page. Find the probability that the first two pages contain no mistakes.

Solution: Here, mean or average per page is 10% = 0.1.

Mean or average for two pages, $m=0.1\times2=0.2$.
Using Poisson distribution, the required probability is given by

$$P(x=0)=\frac{m^x}{x!}e^{-m}=\frac{(0.2)^0}{0!}\times e^{-0.2}=0.8187.$$

PROBLEM 12 With the help of a G.M. counter, a set of 2 minutes counts of a standard radioactive source is recorded in a certain laboratory. Using 10 observations, apply chi-squared test to determine whether the recorded data reflect proper working of the G.M. counter system. The data of counts with frequencies are: 240, 241, 243, 243, 247, 262, 263, 263, 263, 267.

Solution: Let us make a table as the following.

Counts x_i	**Freq.** f_i	$f_i x_i$	$\bar{x} = \frac{\sum f_i x_i}{\sum f_i}$	$f_i (x_i - \bar{x})^2$
240	1	240		174.24
241	1	241		148.84
243	2	486		208.08
247	1	247	253.2	38.44
262	1	262		77.44
263	3	789		288.12
267	1	267		190.44
	$N = \Sigma f_i = 10$	$\Sigma f_i x_i = 2532$		$\Sigma f_i (x_i - \bar{x})^2 = 1125.6$

Degree of freedom, $F = N - 1 = 9$.

We have $\chi^2 = \dfrac{1125.6}{253.2} = 4.45$ corresponding to probability value, $P = 0.84$.

Thus, we say that the above data reflect proper working of the counter system.

EXERCISE

1. For 19 consecutive years, a certain nuclear reactor is producing electric power (in megawatt) at the rates detailed below.

   81, 84, 85, 86, 87, 89, 91, 94, 95, 95, 95, 95, 95, 99, 99, 99, 100, 100, 100.

   Working with ungrouped data, find the mean, median, mode, mean deviation and standard deviation.

   Again, working with group data, find the estimated mean, median, mode, mean deviation and standard deviation. Do the respective values fairly agree?
2. In an examination, the marks obtained by 40 students are shown by the following group data.

Marks obtained:	11 – 20,	21 – 30,	31 – 40,	41 – 50,	51 – 60,	61 – 70,	71 – 80.
No. of students:	2,	3,	8,	14,	8,	3,	2.

Find the estimated mean, median, mode, mean deviation, variance, and standard deduction.

3. Out of 30 circular rings rolling on the ground, if 10% of the rings are to fall on the ground, find the probability that at most two rings fall on the ground. (**Ans.** 0.4114.)
4. A driver starts driving a car in the streets of a city for a certain destination. Up to his destination, there are 15 traffic crossings. If the probability that the driver must be stopped by traffic signal at any crossing is 0.2, find the probability that he will be stopped at 1 or more of the 15 crossings. (**Ans.** 0.9648.)
5. 68 per cent of the ages of teachers in the government colleges of a state are in the range of 45 to 55 years. Assuming a normal distribution, calculate the mean age of the teachers and also find the standard deviation. (**Ans.** 50 years, 5 years)
6. A telecommunication company is using underground fibre-optic cables for signal transmission. In these cables, if defects occur in accordance with Poisson distribution at an average rate of 1.65 per km. Find the probability that the first three kilometres of the cables contain no mistakes. (**Ans.** 0.0071.)
7. In a physics laboratory, a student records beta particle counts from a standard beta source with the help of a Geiger Muller counter for a fixed period of four minutes eight times. The recorded data are 210, 218, 248, 234, 225, 224, 235, 236. Determine whether the counter system works satisfactorily.

References

1. Agarwal, Ravi P. O'Regan, Donal, *Ordinary and Partial Differential Equations*, Springer Science+Business Media, LLC, 233 Spring Street, New York, USA, 2009.
2. Andrews, George E., Askey, Richard and Roy, Ranjan, *Special Functions* (1st Pb. Ed.), Cambridge University Press, New York, USA , 2000.
3. Andrews, Larry C., *Special Functions of Mathematics for Engineers* (2nd Ed.), SPIE PRESS and Oxford University Press, Bellingham, USA, 1998.
4. Arfken, George B., Weber, Hans J., *Mathematical Methods for Physicists* (6th Ed.), Elsevier Academic Press, Burlington, USA, 2005.
5. Beals, Richard and Wong, Roderick, *Special Functions : A Graduate Text*, Cambridge University Press, New York, USA, 2010.
6. Bell, W. W., *Special Functions for Scientists and Engineers*, D. Van Nostrand Co. Ltd., London, UK, 1968.
7. Bleistein, Norman and Handelsman, Richard A., *Asymptotic Expansions of Integrals*, Dover Publications, Inc., New York, USA, 1986.
8. Boas, Mary L., *Mathematical Methods in the Physical Sciences* (3rd Ed.), John Wiley ans Sons, Inc., New Jersey, USA, 2006.
9. Boyce, William E, DiPrima, Richard C, *Elementary Differential Equations and Boundary Value Problems* (7th Ed.), John Wiley & Sons, Inc., New York, USA, 2001.
10. Butkov, Eugene, *Mathematical Physics*, Addison-Wesley Publishing Co., Massachusetts, USA, 1973.
11. Carlson, B. C., *Special Functions of Applied Mathematics*, Academic Press, Inc., London, UK, 1977.
12. Chow, Tall L., *Mathematical Methods for Physicists: A Concise Introduction*, Cambridge University Press, New York, USA, 2000.

13. Copson, E. T., *Asymptotic Expansions*, Cambridge University Press, Cambridge, U.K., 1965.

14. Dym, H and McKean, H. P., *Fourier Series and Integrals*, Academic Press, New York, USA, 1973.

15. Erdelyi, A. (editor), *Higher Transcendental Functions* (Vol. I & II), McGraw Hill Book Company, Inc., New York, USA, 1981.

16. Erdelyi, A., *Asymptotic Expansions*, Dover Publications, Inc., New York, USA, 1956.

17. Fock, Vladimir A., *Fundamentals of Quantum Mechanics* (3rd Printing), MIR Publishers, Moscow, USSR, 1986.

18. Folland, Gerald B., *Fourier Analysis and its Applications*, Brooks/Cole Publishing Co. California, USA, 1992.

19. Frank W. J. Olver, NIST Handbook of Mathematical Functions, Cambridge University Press, NY, USA, 2010.

20. Ghatak, Ajoy K., Goyal, I. C. and Chua, S. J., *Mathematical Physics*, Macmillan India Ltd. , New Delhi, India, 1995.

21. Harper, Charlie, *Introduction to Mathematical Physics*, Prentice Hall India Ltd., New Delhi, India, 2012.

22. Hochstadt, Harry, *The Functions of Mathematical Physics*, Dover Publications, Inc. New York, USA, 1986.

23. Irving, J. and Mullineux, N., *Mathematics in Physics and Engineering*, Academic Press, New York, USA, 1959.

24. Jeffreys, Harold and Jeffreys, B. Swirles, Methods of Mathematical Physics (3rd Ed), Cambridge University Press, Cambridge, UK, 1956.

25. Korenev, B. G., *Bessel Functions and Their Applications*, Taylor and Francis, London, UK, 2002.

26. Kraut, Edgar A., *Fundamentals of Mathematical Physics*, Dover Publications, Inc., New York, USA, 2007.

27. Kreyszig, Erwin, *Advanced Engineering Mathematics* (5th Ed.), New Age International (P) Ltd., New Delhi, India, 1997.

28. Lebedev, N . N., *Special Functions and Their Applications*, Prentice Hall, Inc., New Jersey, USA, 1965.

29. Luke, Yudell L., *The Special Functions and Their Approximations Vol.* 1, Academic Press, New York, USA, 1969.

30. Mathews, John and Walker, R. L., *Mathematical Methods of Physics* (2nd Ed.), W. A. Benjamin, Inc., Massachusetts, USA, 1970.

31. Mathai, A. M. and Haubold, H. J., *Special Functions for Applied Scientists*, Springer Science+Business Media, LLC, USA, 2008.
32. Murray, J. D., *Asymptotic Analysis*, Springer Verlag, New York, Inc., USA, 1984.
33. Myint-U, Tyn and Debnath, Lokenath, *Linear Partial Differential Equations for Scientists and Engineers* (4th Ed.), Birkhauser, Boston, USA, 2007.
34. Nikiforov, Arnold F. and Uvarov, Vasilli B., *Special Functions of Mathematical Physics*, Birkhauser Verlag Basel, Switzerland, 1988.
35. Olver, Frank W. J., *Asymptotics and Special Functions*, A. K. Peters Ltd., Wellesley, Massachusetts, USA, 1997.
36. O'Neil, Peter V., *Advanced Engineering Mathematics* (7th Ed.), Cengage Learning, Stamford, USA, 2012.
37. Pain, H. J., *The Physics of Vibrations and Waves* (6th Ed.), John Wiley and Sons Ltd., West Sussex, UK, 2005.
38. Powers, David L., *Boundary Value Problems and Partial Differential Equations* (5th Ed.), Elsevier Academic Press, Burlington, USA, 2006.
39. Riley, K. F., Hobson, M. P. and Bence, S. J., *Mathematical Methods for Physics and Engineering* (3rd Ed.), Cambridge University Press, New York, USA, 2006.
40. Selvadurai, A.P. S., *Partial Differential Equations in Mechanics* 1 (1st Ed.), Springer-Verlag Berdin Heidelberg New York, 2000.
41. Simmons, George F. and Krantz, Steven G., *Differential Equations: Theory, Technique and Practice*, McGraw-Hill Companies, Inc., New York, USA, 2007.
42. Sneddon, Ian N., *Special Functions of Mathematical Physics and Chemistry* (2nd Ed.), Oliver and Boyd Ltd., Edinburgh, UK, 1961.
43. Sommerfeld Arnold, *Partial Differential Equations in Physics*, Academic Press Inc., New York, USA, 1953.
44. Stuart, R. D., *An Introduction to Fourier Analysis*, Methuen and Co. Ltd., London, UK, 1966.
45. Szego, Gabor, *Orthogonal Polynomials* (4th Ed.), American Mathematical Society, Providence, Rhode Island, USA, 1975.
46. Temme, Nico M., *Special Functions: An Introduction to the Classical Functions of Mathematical Physics*, John Wiley and Sons, Inc., New York, USA, 1996.
47. Tolstov, Georgi P., *Fourier Series*, Dover Publications Inc, New York, USA, 1976.
48. Vallee, Olivier and Soares, Manuel., *Airy Functions and Applications to Physics*, Imperial College Press, London, UK, 2004.
49. Wang, Z. X. and Guo, D. R., *Special Functions*, World Scientific Publishing Co. Pte. Ltd., Singapore, 1989.

50. Watson, G. N., *A Treatise on the Theory of Bessel Functions* (2nd Ed.), University Press, Cambridge, UK, 1966.
51. Wong, Chun Wa, *Introduction to Mathematical Physics: Methods and Concepts*, Oxford University Press, Inc., New York, USA, 1991.
52. Wyld, H. W., *Mathematical Methods for Physics*, Perseus Books Publishing, Massachusetts, USA, 1999.

Index

C

D

E

F

G

H